What Is "An Integrated Approach"?

One of the key themes in this book is that the body is an integrated set of systems. One of your tasks as you study will be to construct for yourself this global view of the body, its systems, and the many processes that keep the systems working. You will find many features here that are designed to help you develop this integrated view by learning to analyze, synthesize, and evaluate information.

Free Student Aid.

Log on.

Tune in.

Succeed.

To help you succeed in Human Physiology, your professor has arranged for you to enjoy access to great media resources, including an interactive CD-ROM and The Physiology Place™. You'll find that these resources that accompany your textbook will enhance your course materials.

What your system needs to use these media resources:

WINDOWS™
250 MHz Intel Pentium processor or greater
Windows 98, NT, 2000, XP
32 MB RAM installed, 64 preferred
800 X 600 screen resolution, thousands of colors
Browser: Internet Explorer 5.0 or Netscape
Communicator 4.7, 7.0
Plug-Ins: Flash player, QuickTime
NOTE: Use of Netscape 6.0 and 6.1 are not recommended due to a known compatibility issue between Netscape 6.0 and 6.1 and the Flash and Shockwave plug-ins.

MACINTOSH™
233 MHz PowerPC
OS 9.2 or higher
32 MB RAM minimum
800 x 600 screen resolution, thousands of colors
Browser: Internet Explorer 5.0 or Netscape
Communicator 4.7, 7.0
Plug-Ins: Flash player, QuickTime
NOTE: Use of Netscape 6.0 and 6.1 are not recommended due to a known compatibility issue between Netscape 6.0 and 6.1 and the Flash and Shockwave plug-ins.

Got technical questions?

For technical support, please visit www.aw.com/techsupport and complete the appropriate online form. Technical support is available Monday-Friday, 9 a.m. to 6 p.m. Eastern (US and Canada).

Here's your personal ticket to success:

How to log in to www.physiologyplace.com:

1. Go to www.physiologyplace.com.
2. Click *Human Physiology: An Integrated Approach,* Third Edition by Dee Silverthorn.
3. Click "Register."
4. Scratch off the silver foil coating below to reveal your pre-assigned access code.
5. Enter your pre-assigned access code exactly as it appears below.
6. Complete the online registration form to create your own personal user Login Name and Password.
7. Once your personal Login Name and Password are confirmed by email, go back to www.physiologyplace.com, type in your new Login Name and Password, and click "Log In."

Your Access Code is:

If there is no silver foil covering the access code above, the code may no longer be valid. In that case, you need to either:
- Purchase a new student access kit at your campus bookstore.
- Purchase access online using a major credit card. Go to www.physiologyplace.com, click *Human Physiology*, and click **Buy Now**.

Important: Please read the License Agreement, located on the launch screen before using The Physiology Place or InterActive Physiology®. By using the web site, you indicate that you have read, understood, and accepted the terms of this agreement.

HUMAN PHYSIOLOGY

AN INTEGRATED APPROACH

THIRD EDITION

Dee Unglaub Silverthorn, Ph.D.

University of Texas – Austin

WITH

William C. Ober, M.D.
Illustration Coordinator

Claire W. Garrison, R.N.
Illustrator

Andrew C. Silverthorn, M.D.
Clinical Consultant

WITH CONTRIBUTIONS BY

Bruce R. Johnson, Ph.D.
Cornell University

San Francisco Boston New York
Cape Town Hong Kong London Madrid Mexico City
Montreal Munich Paris Singapore Sydney Tokyo Toronto

Publisher: Daryl Fox
Executive Editor: Leslie Berriman
Development Manager: Claire Brassert
Development Editor: Anne A. Reid
Associate Project Editor: Ziki Dekel
Editorial Assistant: Michael Roney
Executive Producer: Lauren Fogel
Media Project Editor: Kim Neumann
Art and Design Coordinator: Melissa Morgan
Photo Researcher: Travis Amos
Text and Cover Design: Jim Gibson
Production Coordination: Elm Street Publishing Services, Inc.
Compositor: Preparé/Emilcomp
Manufacturing Buyer: Stacey Weinberger
Executive Marketing Manager: Lauren Harp

Cover Photo: Henry S. Lucid, M.D.

Credits can be found on page 878.

ISBN 0-13-102015-3

Library of Congress Cataloging-in-Publication Data

Silverthorn, Dee Unglaub, 1948–
    Human Physiology: an integrated approach / Dee Unglaub Silverthorn; with William C.
  Ober, illustration coordinator, Claire W. Garrison, illustrator, Andrew C. Silverthorn,
  clinical consultant; with contributions by Bruce R. Johnson.
      p. cm
    Includes index
    ISBN 0-13-10215-3 (alk. paper)
    1. Human physiology. 1. Title
    QP34.5.S55 2004
    612—dc21   2003052278

1 2 3 4 5 6 7—VHC—07 06 05 04 03

PEARSON
Benjamin
Cummings

www.aw-bc.com

Contents in Brief

About the Author

DEE UNGLAUB SILVERTHORN studied biology as an undergraduate at Tulane University and went on to earn a Ph.D. in marine science at the University of South Carolina. Her research interest is epithelial transport, and work in her laboratory currently focuses on transport properties of the chick allantoic membrane. She began her teaching career in the Physiology Department at the Medical University of South Carolina but over the years has taught a wide range of students, from medical and college students to those still preparing for higher education. At the University of Texas she lectures in physiology, coordinates student laboratories in physiology, and instructs graduate students in a course on developing teaching skills in the life sciences. She has substantial experience with active learning in the classroom and has given workshops on this subject at regional, national, and international conferences. Dee is currently editor-in-chief of *Advances in Physiology Education* and past-chair of the Teaching Section of the American Physiological Society. She works with members of the International Union of Physiological Sciences to improve physiology education in developing countries. She is also a member of the Human Anatomy and Physiology Society, the Society for Comparative and Integrative Biology, the Association for Biology Laboratory Education, and the Society for College Science Teaching. Her free time is spent creating multimedia fiber art and enjoying the Texas hill country with her husband, Andy, and their three dogs.

About the Illustrators

Art Coordinator and Illustrator **DR. WILLIAM C. OBER** studied as an undergraduate at Washington and Lee University and obtained an M.D. from the University of Virginia in Charlottesville. In addition to his medical school education, he studied in the Department of Art as Applied to Medicine at Johns Hopkins University. After graduation, Dr. Ober completed training as a resident in family practice. He is now an instructor in the Division of Sports Medicine at the University of Virginia. During the summer, he teaches biological illustration at Shoals Marine Laboratory, where he is part of the core faculty. Dr. Ober's professional life now focuses on medical and scientific illustration.

Illustrator **CLAIRE W. GARRISON** chose to make medical illustration her career after almost two decades as a pediatric and obstetric nurse. After receiving a degree in art from Mary Baldwin College, she became Dr. Ober's associate in 1986. Ms. Garrison is also a core faculty member at Shoals. The texts illustrated by Dr. Ober and Ms. Garrison have won numerous awards from groups including the following: Association of Medical Illustrators (Award of Excellence); Chicago Book Clinic (Award for Art and Design); Printing Industries of America (Award of Excellence); the Society of Academic Authors (Talbot Prize) and Bookbinders West. They have also received the Art Director's Award.

About the Clinical Consultant

ANDREW C. SILVERTHORN, M.D. is a physician in private practice in Austin, Texas. He is a graduate of the United States Military Academy (West Point), served in the infantry in Vietnam, and upon his return entered medical school at the Medical University of South Carolina in Charleston. He completed his family practice residency at the University of Texas Medical Branch, Galveston, where he was chief resident, and he is board-certified by the American Board of Family Practice. When Andy is not busy seeing patients he may be found on the golf course, playing golf or running with his chocolate lab, Lady Godiva.

Owner's Manual: How to Use This Book

Welcome to Human Physiology!

As you begin your study of the human body, you should be prepared to make maximum use of the resources available to you, including your instructor, the library, the Internet, and your

textbook. One of my goals in this book is to provide you not only with information about how the human body functions but also with tips for studying and problem solving. Many of these study aids have been developed with the input of my students, so I think you may find them particularly helpful.

On the following pages, I have put together a brief tour of the special features of the book, especially those that you may not have encountered previously in textbooks. Please take a few minutes to read this section so that you can make optimum use of the text as you study. If you take advantage of features such as the Concept Checks, the Figure and Graph Questions, and Running Problems, you will find that you begin to think about physiology in a different way.

One of your tasks as you study will be to construct for yourself a global view of the body, its systems, and the many processes that keep the systems working. This "big picture" is what physiologists call the integration of systems, and it is a key theme in the book. In order to integrate information, however, you must do more than simply memorize it. You need to truly understand it and be able to use it to solve problems that you have never encountered before. If you are headed for a career in the health professions, you will do this in the clinics. If you are headed for a career in biology, you will solve problems in the laboratory, field, or classroom. Analyzing, synthesizing, and evaluating information are skills that you need to develop while you are in school, and I hope that the features of this book will help you with this goal.

Good luck with your studies!

Warmest regards, Dr. Dee (as my students call me)
silverthorn@mail.utexas.edu

CHAPTER OUTLINE

Each chapter begins with an outline listing the major headings for that chapter. You will find a page reference next to each heading. Use this outline to preview the chapter before you start reading. Do you see any connections between these topics and concepts you have learned in previous chapters or other courses?

BACKGROUND BASICS

This handy list at the beginning of each chapter shows you which important concepts you should have mastered prior to beginning the chapter. If any of these topics seem unfamiliar, use the page references to revisit the subjects you need to review.

CONCEPT LINKS

 These encircled "chain-links" with page references are an extension of the Background Basics, linking you to concepts discussed earlier in the text. They help you find material that you might have forgotten or that might be helpful for understanding new concepts.

FIGURE REFERENCE LOCATORS

The colored circles found next to the first reference to each figure in the text function as place markers, making it easier for you to toggle between an illustration and the running narrative associated with that illustration.

CONCEPT CHECKS

These breaks in the text, placed at intervals throughout each chapter, act as stopping points where you can check your understanding of what you have just read. Some of these questions are factual, while others ask you to think about or apply what you have learned. Answers to the Concept Check questions appear at the end of each chapter.

REFLEX PATHWAY AND CONCEPT MAPS

The reflex pathways and concept maps organize material into a logical, visual format that links both physiological themes and their supporting details into a unified reference. Like a schematic or a blueprint, concept maps provide detailed visual accounts of systems and processes and guide you to a better understanding of coordinated physiological function. Colors and shapes are used to key different processes and properties inside the maps. For example, the stimulus or starting point is always a yellow oval, while the integrating center is always a red diamond. You will learn to create your own maps by completing the mapping questions found in the end-of-chapter exercises.

ANATOMY SUMMARIES

To understand physiology, you must also have a good understanding of basic anatomy. This book includes several Anatomy Summary figures that help you visualize the anatomy of a physiological system, from a macro to micro perspective. As you move across the figure, components of a structure are enlarged, all the way down to the tissue level. These summaries allow you to see all of the essential features of each system in a single figure, either as a review or for the first time.

FOCUS ON ORGAN FIGURES

 There are several large focus features highlighting the anatomy and physiology of organs that are often overlooked in physiology texts, including skin, the liver, and the pineal gland.

FIGURE AND GRAPH QUESTIONS

 In an effort to help you interpret quantitative information and apply concepts, a number of figures and graphs throughout the book include questions that ask you to consider what you really know about the information you have just encountered. Answers to Figure and Graph Questions appear at the end of every chapter.

FOCUS BOXES

You will find three kinds of small boxes in this book. All of these boxes were developed with the goal of helping you understand the role of physiology in science and medicine today:

BIOTECHNOLOGY boxes discuss physiology-related applications and laboratory techniques from the fast-moving world of biotechnology.

CLINICAL boxes focus on clinical applications and pathologies. In addition to being interesting, these boxes will help you understand homeostasis and normal function.

EMERGING CONCEPTS boxes give you a glimpse into new and exciting developments in physiological research. Some topics, such as stem cells, that were highlighted in Emerging Concepts boxes in earlier editions have become fundamental concepts in physiology and are now part of the main text.

An additional feature to all three of these focus boxes is a DIABETES THEME, which is used to illustrate the complexity of human physiology. A diabetes theme box bears the icon of a Biotechnology box, a Clinical box, or an Emerging Concepts box, depending on the information presented. Because diabetes has such a widespread effect on the body, it makes a perfect example of integrated physiology. Some boxes explain the effects of diabetes and then show how the body must compensate to maintain homeostasis. Others describe biotechnological advances or breakthroughs in the treatment of diabetes.

RUNNING PROBLEMS

Each chapter includes a relevant, real-life problem that appears in segments throughout the chapter. After the opening scenario is established, segments on subsequent pages present additional information and ask questions that prompt you to utilize information you have learned. Use the numbers at the bottom of each box to navigate to the next segment of the problem.

Problem Conclusion: At the end of the chapter, you can check your problem-solving strategy against the Problem Conclusion, which lists the questions from each segment, the relevant facts associated with each question, and the "integration and analysis" steps that were required to successfully analyze the running problem.

Question: To find out more about how to answer a question (or to check your solution), examine the Facts and Integration and Analysis columns to the right of each question.

Facts: As you study the chapter, identify relevant facts from this and from earlier chapters. Use the facts listed here to check your assessment.

Integration and Analysis: As you work through the chapter, try to analyze the information you've learned to arrive at an answer. Using the explanation under Integration and Analysis, check to see if your reasoning is sound.

CHAPTER SUMMARY

The end-of-chapter summary begins with a narrative paragraph that highlights the physiological themes encountered in the chapter. This is followed by enumerated points and page references, providing you with a quick review of every major topic in the chapter.

END-OF-CHAPTER QUESTIONS

The end-of-chapter questions have been organized into four learning levels to allow you to develop your problem-solving skills through a logical progression of exercises—from factual to conceptual to real-world to quantitative. Answers to Level One, Two, and Three questions can be found on the companion website. Answers to Level Four questions appear at the end of the chapter.

Level One questions review basic facts and terms. Successfully answering these questions gives you a foundation for more conceptual questions.

Level Two questions test your ability to understand the key concepts.

Level Three exercises are real-world scenarios designed to develop your problem-solving skills. You will be more successful dealing with Level Three questions if you have successfully worked through the first two levels. To check your answers or for additional help, you can go to the book's website.

Level Four questions are quantitative and require graphing or some mathematical manipulation. To successfully answer Level Four questions, you must first understand the facts and the concepts and then be able to manipulate relevant data to develop a quantitative picture of the physiology at hand.

EXPLORATIONS

Found at the end of each chapter following the chapter exercises, *Explorations* are interactive activities that utilize media sources, ranging from print and online journals to magazines to Internet sites. Each *Exploration* examines topics relevant to the chapter material. *Explorations* can also be found at *www.physiologyplace.com*.

COMBINED GLOSSARY AND INDEX

Students like the accessibility of definitions in a glossary, but sometimes you want additional information from the text itself. Usually this means flipping from the Glossary to the Index. Our combined Glossary-Index eliminates the need to flip back and forth between both.

Preface to the Third Edition

IN THE THREE YEARS SINCE THE SECOND EDITION was published, the human genome has been sequenced. Biomedical scientists are now turning their attention from genes to proteins and to the role of those proteins in the integrated functioning of the intact organism. The study of physiology, which for a short period in the late twentieth century seemed to be in a decline, has experienced a rebirth. Research with genetically engineered animals has shown us that simply knowing the sequence of a protein does not tell us what that protein does. In more than one instance, researchers have deleted a protein in genetically engineered animals, only to find that the protein was not as essential to the animal's function as they believed. Now more than ever, scientists must turn to whole animal and tissue studies to understand the intricate interworkings of the body.

There is so much new and exciting information being reported daily that one of the challenges of writing an undergraduate textbook is deciding what to include and how to include it. The goal of this text is to present basic themes and concepts of physiology so that teachers and students can establish a mental model of how the human body functions.

KEY THEMES OF THIS BOOK

In the third edition we kept the four approaches to learning physiology that have proved so popular since this book was first published in 1998.

1. A Focus on Problem Solving

One of the most valuable skills we can teach our students is the ability to think critically and use the information they learn to solve new problems. A number of features in the book are designed to help students practice these skills as they read, including Concept Checks, Running Problems, Explorations, and the Figure and Graph Questions, many of which are new to this edition.

A number of features in this book, many of them new to this edition, are designed to help students practice critical thinking and the use of information to solve new problems.

2. An Emphasis on Integration

The three chapters on Integrative Physiology (Chapters 13, 20, and 25) ask students to step back and look at the big picture to discover how the separate physiological processes work together in an integrated system. Special attention is given to the cooperative function of multiple systems in the human body when homeostasis is disturbed.

3. Cellular and Molecular Physiology

Most physiology research today is being done at the cellular and molecular level, and this edition continues to integrate physiology with cellular and molecular biology. There have been many exciting developments in molecular medicine and physiology since the first edition.

For example, electrical signaling is not just for neurons and muscles any more, as you will learn in Chapter 5. and scientists have learned enough about membrane receptors to classify them into four major categories (Chapter 6). The information on membrane receptors is integrated with discussions of enzymes and membrane transporters to emphasize the commonalities of protein-protein binding: saturation, specificity, competition, and isoforms.

4. Physiology as a Dynamic Field

I have tried to present physiology as a dynamic discipline with numerous unanswered questions that merit further investigation and research. It is essential that students appreciate that many of the "facts" they are learning are really only our current theories, especially as new information emerges from the union of the human genome project and basic research. The demand for systems physiologists is increasing as research efforts focus on applying information uncovered by cellular and molecular research to clinical observations.

The physiology students of today are the next generation of scientists and health care providers—the people who will try to answer the unanswered questions. It is my hope that this book will provide an integrated view of physiology so that tomorrow's scientists and health care providers will enter their chosen professions with respect for the complexity of the human body and a clear vision of the potential of physiological and biomedical research.

NEW TO THIS EDITION

The third edition of *Human Physiology: An Integrated Approach* builds upon the thorough coverage of integrative and molecular physiology topics that have been the foundation of this book since its first publication. The text has been revised with shorter chapters, clearer figures and diagrams, and extensive content updates.

Chapter-by-Chapter Content Updates

The latest findings in research and medicine are integrated throughout the third edition.

Chapter 1
- Updated information on the Human Genome project and introduced Proteomics
- Introduced the concept of emergent properties of a system
- Revised list of concepts in physiology
- Revised and expanded section on experimental design to include prospective and retrospective studies, nocebo effect, and meta-analysis
- Revised and expanded mapping and graphing features; added histograms to graph types

Explorations
1. Placebo surgery: analysis of a clinical trial
2. Variability in a human population: collecting pulse data to show normal variability

Chapter 2
- New box on antioxidants and free radicals
- Updated the periodic table

Explorations
1. Evaluating research and web reports on the efficacy of chromium picolinate
2. How to evaluate websites: the TILT tutorial (*http://tilt.lib.utsystem.edu*)
3. Scandal and the periodic table: discusses ethics in science
4. *Life* magazine issue on imaging

Chapter 3
- Revised and numbered list of cytoskeleton functions; new table of cytoskeleton proteins
- Introduced term *motor proteins* with examples
- Moved 2e box on matrix, metalloproteinases and metastasis into text
- Revised organization of the cell junction section and updated terminology
  - Concept of less permanent adhesions vs. more permanent junctions
  - New classifications: occluding and anchoring junctions
  - New map of cell junctions (Figure 3-15)
  - New table on major cell adhesion molecules (CAMs)
- Revised information on stem cells

Explorations
1. Cell junctions: metastasis and pemphigus
2. The Pap smear: 2001 Bethesda System and testing for human papilloma virus (HPV)
3. Apoptosis: web search
4. Matrix metalloproteinases: reading on their role in cell migration

Chapter 4
- Updated information about the ATP yield in glucose oxidation
- New Navigator Figure for metabolism
- Updated terminology
- Introduced the Human Proteomics Initiative
- Revised section on protein sorting introduces targeting sequence

- New section on post-translational modification includes protein folding, molecular chaperones, methylation, and the role of ubiquitin in protein destruction

Explorations
1. Animations of metabolism: web search
2. Biomedical engineering
3. Readings on metabolism

Chapter 5
- Updated information on caveolae
- Updated terminology
- New Figure 5-7 compares channels and carriers in one figure
- Revised mechanism for Na^+-K^+-ATPase (Figure 5-24)
- Introduced endosome, phagosome, and constitutive processes
- Define *absorption* and *secretion*
- Two new graphs, one of body fluid volumes and one of ion concentrations (Figures 5-32 and 5-33), both with graph questions
- Clarified explanation of osmolarity versus tonicity
- New Level Four question with a graph to interpret

Explorations
1. Caveolae and caveolin: discovered in 1950s, rediscovered in 1990s
2. Cystic fibrosis: research and CF mouse model
3. A Nobel prize for the Na^+-K^+-ATPase

Chapter 6
- Contact-dependent signals added as a 4th method of cell-to-cell communication
- Updated gap junction information
- Added lipid paracrines and COX2 inhibitors in the signal pathways
- Added carbon monoxide as a signal molecule
- Added the concept of regulated variable to homeostasis discussion
- New Navigator Figure for cell communication (Figure 6-3)
- New Figure 6-22 shows steps of a reflex and introduces new conventions
- New Figure 6-29 compares endocrine, neuroendocrine, and neural pathways
- Revised Figure 6-30 compares multiple reflex patterns

Explorations
1. The importance of proteins: pharmacology and nomenclature
2. Communication and signal transduction: *Science's STKE* maps of pathways
3. Walter Cannon: link to biography
4. Biological rhythms and chronotherapy

Chapter 7
- Running problem on Graves' disease is now based on golfer Ben Crenshaw
- The Anatomy Summary (Figure 7-2) now includes hormone abbreviations
- Added diabetes box on the discovery of insulin
- The *What Makes a Chemical a Hormone?* section shows how the classic definition is being challenged as we learn more about chemical signaling
- Added *Diffuse Endocrine System*

- Updated discussion of pheromones, introduced term *ectohormone*, and revised pheromone box
- New section on termination of hormone action
- The *Control of Hormone Release* section is reorganized to be consistent with terminology and concepts from Chapter 6
- The pituitary section is reorganized for clarity
- Pathologies: the try-it problem was moved from a table into Figure 7-21
- *Hormone Evolution* is expanded and discusses the importance of animal research
- Pineal Focus box has a new graph showing melatonin secretion over a day
- Chapter questions added two more Level Four questions, one with graph to interpret and one that asks students to create a graph

Explorations
1. Endocrinology past and present: will endocrinology survive as a discipline?
2. Pheromones in humans: data analysis

Chapter 8
- New box and information about neural stem cells
- Added kiss-and-run pathway and fusion pores; new box: *Synaptic Vesicles Kiss and Run*
- Goldman equation moved from an appendix into text and a quantitative question added at the end of the chapter
- Added anaxonic neurons
- Updated information in dendritic spines
- New photo of neuronal growth cone

Explorations
1. Dynamic action potentials: finding animations on the web
2. Sequencing the mouse genome: a gateway to understanding human disease
3. Defining search terms: shows students the difference between index terms in a database and using search terms in Google

Chapter 9
- New Running Problem: infantile spasms (a form of epilepsy)
- Emphasis on emergent properties of neural networks
- Expanded section on evolution of nervous systems (new Figure 9-1)
- Expanded discussion and new Figures 9-2 and 9-3 on nervous system development
- Updated information on source of neural stem cells
- New Figure 9-6 for blood-brain barrier
- New Figure 9-8 showing spinal reflexes
- Anatomy summary of brain (Figure 9-9) now includes bones of the skull
- New Figure 9-11 of the basal nuclei
- New model of brain function (Figure 9-14)
- Expanded discussion and new Figure 9-19 of diffuse modulatory systems
- New box: adenosine and that "java jolt" discusses adenosine as a neuromodulator
- Box on depression moved into text discussion of mood disorders
- New box on Alzheimer's disease

Explorations
1. Does hormone replacement therapy decrease Alzheimer's risk? Evaluation of experimental design
2. Spinal cord injuries
3. Sleep, dreams, and naps: readings and web searches

Chapter 10
- Rearranged general properties section to go from receptors to CNS integration
- Deleted nociceptors as type of receptor because they respond to multiple stimuli
- Moved discussion of primary and secondary sensory neurons to somatic senses
- New Figure 10-9 showing somatic sensory pathways in the CNS
- Expanded discussion of pain and itch
- Expanded discussion of endogenous opiates
- Updated acupuncture information
- Revised model of olfactory processing.
- Updated taste model to reflect latest research.
- Added presbycusis, age-related hearing loss
- Reorganized vision section to put anatomy first
- New Figure 10-27 showing external anatomy of the eye, lacrimal gland, orbit, and nasolacrimal duct
- New box on melanopsin, a new visual pigment
- New box on glaucoma
- New section and Figure 10-29 on neural pathways for vision and pupillary reflexes

Explorations
1. Pheromones and the human VNO: evaluating research
2. Keeping up with changing models

Chapter 11
- Added chromaffin cells to description of the adrenal medulla
- Added Level Four question on graph interpretation

Explorations
1. Nicotine addiction
2. *Clostridium botulinum* – Friend or foe?

Chapter 12
- Updated model of myosin action during contraction (Figure 12-9)
- New Navigator Figure for skeletal muscle contraction (Figure 12-7)
- New theory that myosin binding state influences Ca^{2+}-troponin binding
- Updated theory of muscle fatigue and added Figure 12-14
- Completely revised smooth muscle section to reflect the latest research findings:
  - Smooth muscle types can be classified into 6 major groups
  - New overview of smooth muscle regulation (Figure 12-30)
  - New: store-operated Ca^{2+} channels, IP_3 receptor-channels of sarcoplasmic reticulum

Explorations
1. Muscle fatigue: data and graph analysis
2. Visualizing muscle contraction: finding animations on the web
3. Neuromuscular disorders: other diseases of interest

Chapter 13

- Updated muscle spindle discussion
- New Figure 13-12 of corticospinal pathways
- Revised Figures 13-9 and 13-11 for central control of movement
- Moved information on Parkinson's disease from box to text and updated it

Explorations

1. Retraining the spinal cord (after spinal cord injuries)
2. Parkinson's disease (and Michael J. Fox)

Chapter 14

- Updated control of myocardial contraction via summing of Ca^{2+} sparks
- Updated Figure 14-11 to show Ca^{2+} sparks
- Introduced *hyperpolarization-activated cyclic nucleotide-gated channels* (HCN channels)
- Revised Figure 14-13 on length-tension relationships
- Updated Running Problem for ER treatment of heart attack

Explorations

1. The electrocardiogram: history, readings, and web searches for tutorials
2. Cardiac pathologies: arrhythmias and long QT syndrome caused by a channelopathy

Chapter 15

- Revised information on pericytes
- Combined Tables 15-1 and 15-2, and moved table to section on chemical control of resistance
- Updated hypertension to include inflammatory hypothesis
- Added photo of malnutrition and ascites
- Two new Level Three questions, one with graph and one with figure
- Added new box on nitric oxide: from dynamite to vasodilation

Explorations

1. Cardiovascular disease—not just for the elderly: congenital heart defects
2. Interpreting data: cause or association? C-reactive protein and homocysteine

Chapter 16

- Moved information on stem cells from umbilical cord blood to an Exploration
- Revised Figure 16-8 on iron metabolism

Explorations

1. Stem cells from umbilical cord blood
2. EPO and the Olympics: blood doping and sports
3. Clinical applications—the CBC and Chem tests

Chapter 17

- New shorter chapter covers the mechanics of breathing
- Modified discussion of Law of LaPlace to correct misconception about how the law applies to the alveoli
- New Level Four question on using helium dilution to measure lung volume

Explorations

1. Pneumothorax: true story of a dramatic almost-fatal pneumothorax

2. Congestive heart failure: interaction of cardiovascular and pulmonary function

Chapter 18

- New shorter chapter focuses on gas exchange, transport, and the regulation of ventilation
- New Running Problem on high altitude acclimatization
- Simplified Figures 18-3 and 18-14 summarizing gas transport
- New Figures 18-6 and 18-13 for oxygen and carbon dioxide transport
- Oxygen sensing by glomus cells moved from box into text with new Figure 18-17
- New Level Three and Level Four questions

Explorations

1. Training at high altitude: is it different than taking EPO?
2. An interactive lung model from Johns Hopkins: predictions and questions to answer

Chapter 19

- Updated information on tubuloglomerular feedback

Explorations

1. Clearance—Napoleon's hair and garlic breath: clearance isn't just for kidneys
2. Artificial kidneys: dialysis

Chapter 20

- Re-designed Figure 19-17, the complex dehydration responses
- Updated information on renin secretion

Exploration

1. The countercurrent multiplier: web simulations

Chapter 21

- Updated and reorganized *Regulation of GI Function*
- Discussion of digestion divided into cephalic, gastric, and intestinal phases
- Text and numbered steps in the art make processes easier to follow
- New Level Four question with graph of transport in "everted sac" preparation

Explorations

1. Heartburn and parietal cells: how drugs for heartburn work and web search for parietal cell animations
2. Dry mouth = Guilty; information and application questions on salivary secretion

Chapter 22

- Shortened chapter focuses on metabolism and its control
- Updated information on control of food intake moved to start of chapter
- Control of body temperature moved to end of chapter
- Updated information on insulin action
- New Figure 22-10 for insulin signal transduction
- Updated information on diabetes and its treatment
- New Running Problem on anorexia and eating disorders
- New Focus boxes on ketogenic diets and discovering peptides by "reverse engineering" from orphan receptors

- New Level Three questions ask students to work with graphs and apply information

Explorations
1. Diabetes and the discovery of insulin: link to article by Minkowski describing his work
2. Beta cell transplants

Chapter 23
- Shortened chapter focuses on endocrinology of growth and long-term control of metabolism
- New information on pro-opiomelanocortin POMC (new Figure 23-6)
- New information on melanocortins and melanocortin receptors (MC-R)
- New: sodium-iodide cotransporter
- New photograph for acromegaly
- New Figure 23-2 showing synthesis pathways of steroid hormones
- New Figure 23-10 showing steps of thyroid hormone synthesis
- New box on revised children's growth charts
- New box on diabetes, melanocortins, and the *agouti* mouse
- New Level Three and Level Four questions ask students to apply information

Explorations
1. Androstenedione and DHEA: Mark McGwire and over-the-counter prohormones
2. Growth hormone and diabetic nephropathy: evidence for an association
3. Learning more about endocrinology: links to the Endocrine Society and other sites

Chapter 24
- Information updated to reflect the latest findings in immunology

Explorations
1. Mind, body, and health: humor therapy and the effect of laughter on the immune system
2. HIV vaccine: the latest in the search

Chapter 25
- Following the lead of the American College of Sports Medicine, removed the definition of exercise
- Updated information on sympathetic cholinergic vasodilator system

Explorations
1. Exercise for your health: readings and website for calculating exercise heart rate range
2. Lactate threshold: test used in athletic training

Chapter 26
- Updated information on CRH as a signal that induces labor
- Updated terminology
- Updated information on hPL and gestational diabetes
- Expanded discussion of gestational diabetes mellitus
- Updated information on hormone replacement therapy
- New Level Four question asks students to interpret graph

Explorations
1. The hunt for a female Viagra: analysis of experimental design
2. Determining gender: pathologies of sex determination and development

An Enhanced Art Program
Every piece of art has been revised so that figures and diagrams are clearer, making it easier to understand complex physiological processes. Key changes include:

- New **Navigator Figures** break physiological processes into clear, easy-to-follow steps. Similar to a road map with a highlighted path, the *Navigator Figures* begin with a "big picture" overview and are repeated at appropriate intervals in the discussion to help students keep sight of the larger view without getting lost in the details.
- **Text boxes** are embedded with key figures, providing a "running commentary" that helps students interpret illustrations.
- **Numbered steps** have been added to several reflex maps, process maps, and other figures to follow the sequence of events for response and feedback loops.
- Improved **color consistency** in diagrams creates visual categories: for example, the stimulus or starting point is always a yellow oval, while the integrating center is always a red diamond.
- Redesigned **anatomy summaries** are organized from left to right, making it easier to follow the progression of structures from a macro level to a micro level.

Shorter Chapters
The two longest chapters from the second edition have been split into smaller, more manageable pieces. **Respiration** is now covered by Chapter 17: Mechanics of Breathing, and Chapter 18: Gas Exchange and Transport. **Metabolism** is divided into Chapter 22: Metabolism and Energy Balance, and Chapter 23: Endocrine Control of Growth and Metabolism.

InterActive Physiology® CD-ROM

This award-winning tutorial series now comes packaged with every new copy of the text. The program uses detailed animations and engaging quizzes to help students advance beyond memorization to a genuine understanding of the most difficult topics in physiology. InterActive Physiology® is referenced in the chapter summaries throughout the text.

Explorations
Found at the end of each chapter, these enrichment activities encourage independent learning through readings in print and online journals and magazines or through web searches. *Explorations* ask students to read and evaluate experiments on topics of interest. They also teach students how to search for information in databases and indexes, and how to evaluate the information they find through web searches. These *Explorations* replace the *Media Labs* from the previous edition.

SUPPLEMENTS

IN THE TWENTY-FIRST CENTURY, education has become a multidimensional endeavor. The textbook remains the basic teaching tool in a physiology course, but students with different learning preferences also use computer animations and simulations, Web resources, videos, and printed supplements.

How we help our students learn is almost as important as what we teach. They must understand that memorization is not sufficient, that the "facts" they memorize today may be "wrong" tomorrow. We can help them recognize patterns and themes, create mental models, and bridge the gap between organ system physiology in the textbook and physiology as it is practiced in the clinics and the research laboratory.

It is time-consuming, however, for instructors to be creative and inspiring teachers while dealing with the administrative demands of a course. Who has time to surf for animations or write problems for small group work? To meet the need for this type of curriculum material, we have assembled an array of innovative supplements for instructors and students.

Instructor Supplements

INSTRUCTOR'S RESOURCE CD-ROM: IMAGES, ANIMATIONS, AND LECTURE PRESENTATION (0-13-140182-3). This cross-platform CD-ROM includes all of the art from the textbook presented in labeled, unlabeled, and leadered-only formats, available in both jpeg and PowerPoint® files. The PowerPoint® lecture series integrates both textbook figure art and selected animations from the InterActive Physiology® CD-ROM with key chapter content in editable bullet-point outlines. All figure art, animations, and tables are also accessible from within each chapter's index page to provide a comprehensive and customizable lecture tool for use in the classroom.

INSTRUCTOR'S RESOURCE MANUAL (0-13-140180-7). Co-authored by Dee Silverthorn, this complete resource shows instructors how to integrate active learning into their classes. It includes activities for classroom use, in-depth commentary on selected "hot topics," chapter summaries and outlines, key word lists, reading and website lists, lists of applicable media for each chapter, and extra quantitative (Level Four) problems.

TRANSPARENCY ACETATES (0-13-140181-5). Features approximately 400 key illustrations from the text.

PRINTED TEST BANK (0-13-140188-2) and **COMPUTERIZED TEST BANK (0-13-112537-0).** Based on the textbook's four-level learning system, the test bank contains more than 3,000 questions, updated for this edition. They include multiple choice, matching, completion, short answer, essay, quantitative analysis, and graph questions. The computerized version comes on a cross-platform CD-ROM and utilizes TestGen® test software for easy test preparation.

COURSE MANAGEMENT—WEBCT *(http://cms.aw.com/webct)*, **BLACKBOARD** *(http://cms.aw.com/blackboard)*, and **COURSECOMPASS™** *(http://cms.aw.com/coursecompass)*. Assets from The Physiology Place website and the Silverthorn Computerized Test Bank are available as pre-built, easy-to-use courses in any of these three course management systems. Professors can use these powerful tools to create sophisticated web-based educational programs.

Student Supplements

INTERACTIVE PHYSIOLOGY® CD-ROM. Integrated with every new copy of the text, this award-winning tutorial series uses detailed animations and engaging quizzes to help students advance beyond memorization to a genuine understanding of the most difficult topics in A&P. Co-developed by Benjamin Cummings and adam.com, InterActive Physiology® is referenced throughout the third edition in chapter summaries.

THE PHYSIOLOGY PLACE WEBSITE *(http://www.physiology-place.com)*. Features include three levels of quizzing, case studies, *Explorations*, web links, muscle review, bone review, histology, glossary, flashcards, and Ask an Expert. Instructors will have access to a syllabus manager and instructor resources with PowerPoint® art files from the textbook in labeled, unlabeled, and leadered formats.

STUDENT WORKBOOK (0-13140176-9). Designed to work with the textbook and to teach students to learn independently, this workbook came out of the guide Dee Silverthorn developed for her own students. It features helpful study aides such as guided note-taking, try-it-yourself activities, review questions, chapter summaries, lab exercises, and reading lists.

PHYSIOEX™ V4.0 FOR HUMAN PHYSIOLOGY CD-ROM (0-8053-6153-7). This program now features a dramatically improved histology tutorial and ten laboratory simulations that allow students to perform physiological experiments that may be too costly, difficult, or time-consuming to conduct in a traditional wet lab setting. Students can adjust the variables of each lab, perform experiments repeatedly, and even design their own labs. Available on both CD-ROM and the web, each experiment includes step-by-step instructions and review materials in the accompanying PhysioEx™ V4.0 for Human Physiology laboratory manual.

ACKNOWLEDGMENTS

WRITING A TEXTBOOK OF THIS SCOPE is a group project that requires the talent and expertise of many people. I would particularly like to acknowledge the contribution of Bruce Johnson, Cornell University, Department of Neurobiology and Behavior, whose expertise and guidance shaped the first edition chapters on the central nervous system and motor control. Bruce is a superb neurobiologist and educator, and in the third edition he worked closely with me to update the chapters on neurobiology and ensure that they reflect the latest developments in that rapidly changing field.

Many other people devoted time and energy to making this book a reality, and I would like to make an attempt at thanking them all. I apologize in advance to anyone whose name I've left out, and I take full responsibility for any errors and omissions.

Reviewers

I am particularly grateful to the instructors who reviewed one or more chapters of the third edition. They include:

- Cynthia Beck, George Mason University
- David Bernard, University of Texas–Arlington
- Sunny Boyd, University of Notre Dame
- John Capeheart, University of Houston–Downtown
- Carlos Crocker, San Francisco State University
- Arianne Dantas, University of Melbourne
- Andrew Goliszek, North Carolina A&T State University
- David Hammerman, Long Island University
- John Lepri, University of North Carolina–Greensboro
- Jennifer Lundmark, California State University–Sacramento
- Kerry McDonald, University of Missouri
- Stacia Moffett, Washington State University
- David Muehleisen, University of Utah
- Tim Mullican, Dakota Wesleyan University
- Scott Nunes, University of San Francisco
- Linda Ogren, University of California–Santa Cruz
- Amanda Starnes, Emory University
- Benjamin Walcott, SUNY Stonybrook
- Burl Williams, Jr., East Tennessee State University

Many other instructors and students took time to write or email queries or suggestions for clarification. I am always delighted to have input, and I apologize if I have omitted any names from this list.

- Mark Bracken, Utah Valley State College
- Jorge Golowasch, Rutgers University
- Chester Harbut, Cerritos Community College
- Karen Hart, Peninsula College
- Ivan Ho, National University of Singapore
- Rob Kemm, University of Melbourne
- Stanley Kosasih, Victoria University
- Denis Loiselle, University of Auckland
- Tim Mulligan, Dakota Wesleyan University
- Bruce O'Gara, Humboldt State University
- Richard Quinn, University of Texas–Austin
- Christine Wade, University of Wyoming
- Kira Wennstrom, University of Washington
- Theresa Woodruff, Northwestern University
- Yuri Zagvazdin, Nova Southeastern University

I would also like to thank Patti Thorn and the faculty members who participated in our NSF-supported project nicknamed "Integrative Themes in Physiology" for their comments and insight on teaching physiology.

Specialty Reviews

No one can be an expert in every area of physiology, and I am deeply thankful for my friends and colleagues who did everything from answer frantic emails with questions to review entire chapters. Even with their help, there may be errors, for which I take full responsibility.

- Beverly Bishop, SUNY Buffalo
- Arthur Dalley, Vanderbilt University
- Michael Joyner, Mayo Clinic
- Henry S. Lucid, M.D., Austin, TX
- Jeffrey Pommerville, Maricopa Community Colleges
- Elizabeth Weiss, University of Texas–Austin
- Withrow Gil Wier, University of Maryland, Baltimore School of Medicine
- Richard Wilcox, University of Texas–Austin
- Jack Wilmore, Texas A&M University

Reviewers for Previous Editions

I would like to acknowledge again the contribution of the faculty who reviewed the first and second editions.

- Brenda Alston-Mills, North Carolina State University
- Roland M. Bagby, University of Tennessee
- Patricia J. Berger, University of Utah
- Ronald Beumer, Armstrong State College
- Sunny K. Boyd, University of Notre Dame
- Timothy Bradley, University of California–Irvine
- Alan P. Brockway, University of Colorado–Denver
- David Byman, Pennsylvania State University Worthington–Scranton Campus
- Robert G. Carroll, East Carolina University School of Medicine
- Annabelle Cohen, CUNY The College of Staten Island
- Kim Cooper, Arizona State University
- Charles L. Costa, Eastern Illinois University
- Catherine Dallemagne, Queensland University
- Douglas Eaton, Emory University
- Gordon Garrett, Utah Valley State College
- Lori K. Garrett, Danville Area Community College
- Barbara Goodman, University of South Dakota
- Margaret M. Gould, Georgia State University
- Tammy Greco, Eastern Michigan University
- Dan Hardy, M.D., Austin, Texas
- Janet L. Haynes, Long Island University–Brooklyn
- Richard W. Heninger, Brigham Young University
- Tracy M. Hodgson, University of Pittsburgh
- Katja Hoehn, Mount Royal College
- Carol Hoffman, San Jose City College
- Bruce Johnson, Cornell University
- William C. Kleinelp, Middlesex Community College
- James Larimer, University of Texas
- Daniel L. Mark, Penn Valley Community College

- Paul Matsudaira, Massachusetts Institute of Technology
- Roberta Meehan, Denver, Colorado
- Wayne Meyer, Austin College
- Joel Michael, Rush Medical College
- Alice C. Mills, Middle Tennessee State University
- Ron Mobley, Wake Technical Community College
- Harold Modell, National Resource for Computers Life Science Education
- John G. Moner, University of Massachusetts–Amherst
- Patricia L. Munn, Longview Community College
- Joanne Nosky, Okanagan Valley College of Massage Therapy
- William Percy, University of South Dakota
- Dell Redding, Evergreen Valley College
- Carol Rivard, Fanshawe College
- Mary Anne Rokitka, SUNY Buffalo
- Allen Rovick, Rush Medical College
- Jeffrey Sands, Emory University
- Evelyn Schlenker, University of South Dakota
- Virginia Scofield, University of Southern California
- Norm Scott, University of Waterloo
- Brian R. Shmaefsky, Kingwood College
- Sabyasachi Sircar, Delhi, India
- James Stockand, University of Texas Health Science Center–San Antonio
- Judy Sullivan, Antelope Valley College
- Roger Tanner-Theis, University of Oklahoma (Retired)
- Deborah Taylor, Kansas City Kansas Community College
- David Thomas, Fanshawe College
- Richard L. Walker, University of Calgary
- Curt Walter, Dixie College
- Donald Whitmore, University of Texas–Arlington
- Ronald L. Wiley, Miami University of Ohio
- Jack Wilmore, Texas A&M University
- David A. Woodman, University of Nebraska–Lincoln
- Leona Young, Emory University

Photographs

I would like to thank the following colleagues who generously provided micrographs from their research:

- Daniel Casellas, Groupe Rein & Hypertension IURC, Montpellier, France
- Flora M. Love, University of Texas
- Jane Lubisher, University of Texas
- Carole L. Moncman, University of Texas
- L. Gabriel Navar, Tulane University
- Young-Jin Son, University of Texas

Supplements

Pat Munn at Longview Community College in Kansas helped write the end-of-chapter questions. Pat and Alice Mills from University of Tennessee–Martin wrote the Test Bank questions. Damian Hill worked with me to revise the Instructor's Resource Guide and Student Workbook. My colleagues, Cynthia Gill and Jan Melton Machart, worked on the end-of-chapter *Explorations* with me and provided moral support when deadlines loomed.

The Publisher

Making a book is a complex process that requires the energy and talents of many people. The development of this edition was complicated by Pearson Education's decision to transfer the anatomy and physiology lists from Prentice Hall, where this book was born, to its sister company, Benjamin Cummings. The teams at both companies worked hard to achieve a seamless transition, and that made the move much easier.

At Prentice Hall my energetic editor, Halee Dinsey, directed the revision plans with vision and enthusiasm. Her tireless assistant, Susan Ziegler, managed the minutiae and coordinated the requests for reviews. I will miss working with them and everyone in the Prentice Hall organization.

During the transition period, Anne A. Reid, my development editor, and Bill Ober and Claire Garrison, my art co-authors, provided support and encouragement. Annie's calm and gentle advice and eagle eye have guided me through two revisions now, and I could not have done them without her. Bill and Claire used their wisdom and talents to make the art in this edition even more clear and beautiful. They worked closely with Jim Gibson, the talented designer who designed the cover and interior pages, to create a coordinated art program that makes the art and layout of this edition appealing and user-friendly.

The team at Benjamin Cummings did a wonderful job of taking a project in progress and seeing it to completion. Daryl Fox and Linda Davis made me feel immediately at home. Leslie Berriman, my Benjamin Cummings editor, came onboard and jumped right into the complexities of book, supplements, and associated projects. Ziki Dekel at Benjamin Cummings and Becky Dodson at Elm Street Publishing Services ably coordinated the production process. Travis Amos tracked down new photographs for the book. Kim Neumann and Michael Roney masterfully coordinated the wonderful group of colleagues who contributed to the media and print supplements, including Kelly Johnson, University of Kansas–Lawrence; John Lepri, University of North Carolina–Greensboro; and Howard Booth, Eastern Michigan University.

Special thanks go to Lauren Harp and the sales teams at Benjamin Cummings and Pearson International who welcomed me at the national sales meeting. They provide an invaluable link between the author and faculty around the world who teach physiology, and I look forward to working with them more in the future.

Special Thanks

As always, I would like to thank my students who looked for errors and areas that needed improvement. In this edition, Susan Johnson wrote the excellent running problem that appears in Chapter 9. Theresa Marie Long, Karen Lasky, and Francisco Serna assisted with research on special topics.

My graduate teaching assistants have played a huge role in my teaching ever since I arrived at the University of Texas, and their input has helped shape how I teach. Many of them are now faculty members themselves. I would particularly like to thank:

- Patti Thorn, Ph.D.
- Karina Loyo-Garcia
- Carol C. Linder, Ph.D.
- Peter English, Ph.D.
- Jan Melton-Machart, Ph.D.
- Ari Berman
- Lawrence Brewer, Ph.D.
- Kurt Venator, Ph.D.

Finally, special thanks go to a group of my colleagues for their continuing support: Ruth Buskirk, Judy Edmiston, Cindy Gill, Jeanne Lagowski, and Marilla Svinicki (University of Texas), Penelope Hansen (Memorial University, St. John's), Mary Anne Rokitka (SUNY Buffalo), Rob Carroll (East Carolina University School of Medicine), and Joel Michael (Rush Medical College).

I also must thank my family, friends, and colleagues for their continued understanding and support during the chaos that seems to accompany my book revisions. A special thank you to my friends, David and Kristy Brake and Damian Hill, who have supported this project since its inception. And, as always, the biggest thank you to my husband Andy, without whose support nothing would ever get written.

A Work in Progress

One of the most rewarding aspects of writing a textbook is the opportunity it has given me to meet or communicate with other instructors and students. In the years since the first edition was published, I have heard from people around the world, and have had the pleasure of hearing how the book has been incorporated into their teaching and learning.

Because science textbooks are revised every three or four years, they are always works in progress. I invite you to contact me or my publisher with any suggestions, corrections, or comments about this third edition. I am most reachable through e-mail at *silverthorn@mail.utexas.edu*. You can reach my editor at the following address:

Applied Sciences
Benjamin Cummings
1301 Sansome Street
San Francisco, CA 94111

Dee U. Silverthorn
University of Texas
Austin, Texas

Contents

4 *Cellular Metabolism*   84

5 *Membrane Dynamics*   124

10 *Sensory Physiology*   321

11 *Efferent Division: Autonomic and Somatic Motor Control*   369

UNIT 3

Integration of Function

UNIT 4

Metabolism, Growth, and Aging

For Andy, with whom I have shared the journey

Introduction to Physiology

1

Physiology is not a science or a profession but a point of view.
—Ralph W. Gerard

For most of recorded history, humans have been interested in how their bodies work. Early Egyptian, Indian, and Chinese writings describe attempts by physicians to treat various diseases and to restore health. Although some ancient remedies such as camel dung and powdered sheep horn may seem bizarre, we are still using others, such as blood-sucking leeches and chemicals derived from medicinal plants. The way we use these treatments has changed as we have learned more about the human body. To appropriately treat disease and injury, we must first understand the human body in its healthy state.

Physiology is the study of the functions of a living organism and its component parts, including all chemical and physical processes. The term *physiology* literally means "knowledge of nature." Aristotle (384–322 B.C.) used the word in this broad sense to describe the functioning of all living organisms, not just of the human body. However, Hippocrates (ca. 460–377 B.C.), considered the father of medicine, used the word *physiology* to mean "the healing power of nature," and the field became closely associated with medicine. By the sixteenth century in Europe, physiology had been formalized as the study of the vital functions of the human body, although today the term is again used to refer to the study of the functions of all animals and plants.

In contrast, **anatomy** is the study of structure, with much less emphasis on function. Despite this distinction, anatomy and physiology cannot truly be separated. The function of a tissue or organ is closely tied to its structure, and the structure of an organism presumably evolved to provide an efficient physical base for its function.

There has never been a more exciting time to study human physiology. Today we benefit from centuries of work by physiologists who constructed a foundation of knowledge about how the human body functions. Since the 1970s, rapid advances in the fields of cellular and molecular biology have supplemented this work. A few decades ago we thought that we would find the key to the secret of life by sequencing the human genome. However, this deconstructionist view of biology has proved to have its limitations because living organisms are much more than the simple sum of their parts.

Sequencing genes and proteins in the molecular biology laboratory does not tell us the significance of these substances in the intact animal. Molecular biologists are turning to physiologists to help them understand the functions of the intact organism. Phys-

iologists are trained to think about the **integration** of function across all levels, from molecules to the living body. (To integrate means to bring varied elements together to create a unified whole.) Today, as we try to reconcile molecular genetics with what we know about how our bodies work, the skills of physiologists are in more demand than ever.

In this chapter, we examine the fundamental principles of physiology and explore how we know what we know about the way the human body functions.

▶ LEVELS OF ORGANIZATION

To understand how the human body functions as a coordinated entity, we must first examine its component parts. Figure 1-1 ●

● **Figure 1-1** *Levels of organization and the related fields of study*

Table 1-1: Organ Systems of the Human Body

SYSTEM NAME	ORGANS OR TISSUES	REPRESENTATIVE FUNCTIONS
Circulatory	Heart, blood vessels, blood	Transport of materials throughout the body
Digestive	Stomach, intestines, liver, pancreas	Conversion of food into particles that can be taken into the body; elimination of some wastes
Endocrine	Thyroid gland, adrenal gland	Coordination of body function through synthesis and release of regulatory molecules
Immune	Thymus, spleen, lymph nodes	Defense against foreign invaders
Integumentary	Skin	Protection from external environment
Musculoskeletal	Skeletal muscles, bones	Support and movement
Nervous	Brain, spinal cord	Coordination of body function through electrical signals and release of regulatory molecules
Reproductive	Ovaries and uterus, testes	Perpetuation of the species
Respiratory	Lungs, airways	Exchange of oxygen and carbon dioxide between the internal and external environments
Urinary	Kidneys, bladder	Maintenance of water and solutes in the internal environment; waste removal

shows the different **levels of organization** of living organisms, ranging from the parts of a single organism to groups of the same species (*populations*) and populations of different species living together in *ecosystems* and the *biosphere*. The different disciplines of chemistry and biology related to the study of each level are shown along the top of the figure. There is considerable overlap between the different fields, and these artificial divisions will vary according to who is defining them. One distinguishing feature of physiology is that it encompasses many levels of organization.

At a fundamental level, atoms of elements link together to form molecules. One of the great, and possibly unsolvable, mysteries of science is how interactions between certain groups of molecules result in the unique properties of a living organism. The smallest unit of structure capable of carrying out all life processes is the **cell**. Cells are collections of molecules separated from the external environment by a barrier called the cell membrane.

Simple organisms are composed of only one cell, but complex organisms have many cells with different structural and functional specializations. Collections of cells that carry out related functions are known as **tissues** [*texere*, to weave]. Tissues form structural and functional units known as **organs** [*organon*, tool], and groups of organs integrate their functions to create **organ systems.** The human body has 10 physiological organ systems (Table 1-1).

Figure 1-2 ● is a schematic diagram showing the interrelationships of some of the systems in the human body. The **integumentary system**, composed of the skin, forms a protective

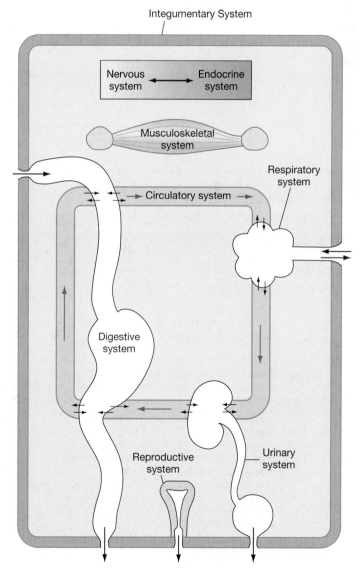

● **Figure 1-2** *The integration between systems of the body* This schematic figure shows relationships between some of the systems of the human body. The lumens of some hollow organs (shown in white) open to the external environment.

boundary that separates the body's internal environment from the external environment of the outside world. The **musculoskeletal system** provides support and body movement.

Four systems exchange material between the internal and external environments. The **respiratory system** exchanges gases, the **digestive system** takes up nutrients and water and eliminates wastes, the **urinary system** removes excess water and waste material, and the **reproductive system** produces eggs or sperm. The major organs of these four systems are hollow, shown by the white space in Figure 1-2. These interior spaces are essentially extensions of the external environment, and material that enters them is not truly part of the **internal environment** until it crosses the tissue wall of the organ. The interior of a hollow organ or tube is called the **lumen** [*lumin*, window].

An interesting illustration of the difference between the internal and external environment involves the bacterium *Escherichia coli*. This organism normally lives and reproduces in the lumen of the large intestine, an interior cavity that is continuous with the external environment. In this location, the organism does not harm the host. However, if the intestinal wall is punctured by disease or accident and *E. coli* enters

the internal environment of the body, a serious infection can result.

The **immune system** of the body protects the internal environment from foreign invaders such as the *E. coli* bacterium. To carry out that function, the cells of the immune system are positioned to intercept material that may enter through the exchange surfaces or through a break in the skin. In addition, immune tissues are closely associated with the **circulatory system**, which distributes material throughout the body.

The **nervous** and **endocrine systems** coordinate body function. Note that the figure shows them as a continuum rather than as two distinct systems. Why? Because as we have learned more about the integrative nature of physiological function, the lines between these systems have blurred. How do we classify nerve cells that secrete hormones as endocrine cells do? How should we classify responses that begin in the nervous system and end with the endocrine system? One of the current challenges in physiology is integrating information from the different body systems into a cohesive picture of the living human body. One way that physiologists do this is by using maps or flow charts, schematic representations of structure and function (see *Focus on Mapping*, p. 5).

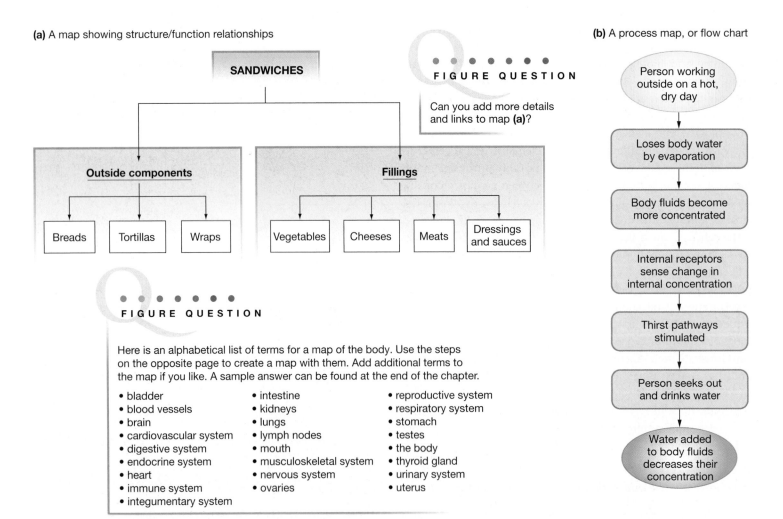

(a) A map showing structure/function relationships

SANDWICHES

Outside components

Breads | Tortillas | Wraps

Fillings

Vegetables | Cheeses | Meats | Dressings and sauces

FIGURE QUESTION

Can you add more details and links to map **(a)**?

FIGURE QUESTION

Here is an alphabetical list of terms for a map of the body. Use the steps on the opposite page to create a map with them. Add additional terms to the map if you like. A sample answer can be found at the end of the chapter.

- bladder
- blood vessels
- brain
- cardiovascular system
- digestive system
- endocrine system
- heart
- immune system
- integumentary system
- intestine
- kidneys
- lungs
- lymph nodes
- mouth
- musculoskeletal system
- nervous system
- ovaries
- reproductive system
- respiratory system
- stomach
- testes
- the body
- thyroid gland
- urinary system
- uterus

(b) A process map, or flow chart

Person working outside on a hot, dry day

↓

Loses body water by evaporation

↓

Body fluids become more concentrated

↓

Internal receptors sense change in internal concentration

↓

Thirst pathways stimulated

↓

Person seeks out and drinks water

↓

Water added to body fluids decreases their concentration

● **Figure 1-3** *Maps for physiology*

Mapping

Mapping is a nonlinear way of organizing material that has been shown to improve understanding and retention. It is a useful study tool because creating a map requires thinking about the importance and relationship of information. Studies have shown that when people interact with information by organizing it in their own way *before* they load it into memory, improved long-term learning results.

Mapping is not just a study technique. Experts in a field make maps when they are trying to integrate newly acquired information into their knowledge base, and they may create two or three versions of a map before they are satisfied that it represents their understanding. Scientists map out the steps in their experiments. Health care professionals create maps to guide them through patient diagnosis and treatment.

A map can take a variety of forms but usually consists of terms linked by arrows. A map may also include graphs or pictures. In general, there are three types of associations between the terms in a map: part/type, description/characteristic, and cause/effect. The linking arrows of a map may be labeled with explanatory phrases or may be drawn in different colors to represent different types of links. Maps in physiology usually focus on the relationships between anatomical structures and physiological processes (*structure/function maps*), or normal homeostatic control pathways and responses to abnormal (pathophysiological) events (*process maps or flow charts*). Figures 1-3a ●, 5-11, and 5-12 are examples of structure/function maps. Figures 1-3b and 7-10 are examples of flow charts.

Many maps appear in this textbook, and they may serve as the starting point for maps of your own. However, the real benefit from mapping comes from preparing maps yourself. By mapping the material on your own, you think about the relationships between terms, organize concepts into a hierarchical structure, and look for similarities and differences between items. Such interaction with the material helps you process it into long-term memory instead of simply memorizing it and forgetting it.

Getting Started

1. First, select the terms or concepts to map. (In this text, the end-of-chapter questions include at least one list of terms to map.) Sometimes it's helpful to write out the terms onto individual slips of paper or onto 1-by-2-inch sticky notes so that you can rearrange the map more easily.

2. Usually the most difficult part of mapping is deciding where to begin. Start by grouping terms in an organized fashion. You may find that you want to put some terms into more than one group. Make a note of these terms, as they will probably have several arrows pointing to them or leading away from them.

3. Now try to create some hierarchy with your terms. You may arrange the terms on a piece of paper or on a table or on the floor. In a structure/function map, start at the top with the most general, important, or overriding concept from which all the others stem. In a process map or flow chart, start with the first event to occur. Next, break down the key idea into progressively more specific parts using the other concepts, or follow the event through its time course. Use arrows to point the direction of linkages and include horizontal links to tie concepts together. The downward development of the map will generally mean the passage of time or an increase in complexity.

   You may find that some of your arrows cross each other. Sometimes this can be avoided by rearranging the items on the map. Labeling the linking arrows with explanatory words may be useful. For example, *channel proteins* —— *form* ——→ *open channels*.

   Color can be very effective on maps. You can use colors for different types of links or for different categories of terms. You may also add figures, diagrams, and graphs that are associated with specific terms in your map.

4. Once you have created your map, sit back and think about it. Are all the items in the right place? You may want to move them around once you see the big picture. Revise your map to fill in the picture with new concepts or to correct wrong linkages. Review by recalling the main concept and then moving to the more specific details. Ask yourself questions like, "What is the cause? effect? parts involved? main characteristics?" to jog your memory.

5. A useful way to study with a map is to trade maps with a classmate and try to understand each other's maps. Your maps will almost certainly NOT look the same! That's okay. Remember that your map reflects the way you think about the subject, which may be different than the way someone else thinks about it. Did one of you put in something the other forgot? Did one of you have an incorrect relationship between two items?

6. Practice making maps. The study questions in each chapter will give you some ideas of what you should be mapping. Your instructor can help if you don't know how to get started.

PHYSIOLOGY IS AN INTEGRATIVE SCIENCE

Physiology cuts across many levels of organization. At the systems level, there are relatively few unanswered questions about how the human body works. Most of these questions involve the nervous control of function, such as the nervous control mechanisms that govern breathing. The bulk of research in physiology today is focused on the cellular and molecular level because many questions about how the human body works at that level remain unanswered. Nevertheless, explaining what happens in test tubes or isolated cells can only partially answer questions about how the intact body works. We must also figure out how events in a single cell influence neighboring cells, tissues, organs, and systems throughout the body. In this respect, physiologists are like children throwing rocks into a pond. They want to see what kind of splash the rocks make where they enter the water, but they also want to know where the ripples go and what happens when they encounter other ripples.

The integration of function among systems is a special focus of physiology. Traditionally, students have studied the cardiovascular system and the regulation of blood pressure in one unit, then studied the kidneys and the control of body fluid volume in a separate unit. But body fluid volume plays a key role in determining blood pressure, so how can the two units be completely separated? Creating an integrated view that links together the different systems is one of the most intriguing and challenging aspects of physiology.

Another challenge of physiological research is unraveling the emergent properties of complex systems. An **emergent property** is some property of a system that is greater than the simple sum of the system's individual parts. An emergent property is not a property of any single component of the system and it cannot be predicted to exist based on what we know about the individual components. Some of the most complex emergent properties are emotion, intelligence, and other aspects of brain function that cannot be predicted from knowing the individual properties of nerve cells.

FUNCTION AND PROCESS

Function and process are two related concepts in physiology. The **function** of a physiological system or event is the "why" of the system or event: why does the system exist? This way of thinking about a subject is called the **teleological approach** to science. For example, the teleological answer to the question of why red blood cells transport oxygen is "because red blood cells bring oxygen to the cells that need it." This answer explains the reason red blood cells transport oxygen but says nothing about *how* the cells transport oxygen.

In contrast, physiological **processes**, or **mechanisms**, are the "how" of a system. The **mechanistic approach** to physiology examines process. The mechanistic answer to the question of why red blood cells transport oxygen is "because they contain hemoglobin molecules that combine reversibly with oxygen mole-

cules." This very concrete answer explains exactly how oxygen transport occurs but says nothing about the significance of oxygen transport to the intact animal.

Students often confuse the two approaches to thinking about physiology, and studies have shown that even medical students tend to answer questions with teleological rather than mechanistic explanations. Sometimes the fault lies with the instructor, who asks "why" a physiological event occurs when he or she wants the mechanistic answer. Staying aware of the two approaches will help prevent confusion.

Although function and process seem to be two sides of the same coin, it is possible to study processes, particularly at the cellular and subcellular level, without understanding their function in the life of the organism. As biological knowledge becomes more complex, and as research focuses on the subcellular and molecular levels, scientists sometimes become so involved in the processes they study that they fail to step back and look at what those processes mean to cells, organ systems, or the intact animal. One role of physiology is to integrate function and process into a cohesive picture.

THE EVOLUTION OF PHYSIOLOGICAL SYSTEMS

We human beings are large, mobile, terrestrial animals whose bodies are about 60% water. Our anatomy and physiology are designed to allow us to survive in a dry, highly variable external environment. This fact becomes more obvious when we examine the adaptations of other animals to their habitats.

Life probably originated in tropical seas, a stable environment where salinity, oxygen content, and pH vary little and where light and temperature cycle in predictable ways. The earliest living organisms had an internal environment almost identical to that of seawater, just as today's marine invertebrates do. If external conditions changed, conditions inside the primitive organisms changed as well. Even today, marine invertebrates cannot tolerate significant changes in salinity and pH, as you know if you have ever maintained a saltwater aquarium. In both ancient and modern times, most marine organisms have relied on the constancy of their external environment to keep their internal environment in balance.

As animals migrated from the ancient seas into estuaries, where fresh and sea water mix, then into freshwater environments and onto the land, their external environment was no longer constant. Salinity varies when rains send fresh water into the estuaries or as the sun evaporates tidal pools among the rocks. Temperature fluctuates widely from day to night and from season to season. Animals that live in fresh water must cope with the influx of water into their body fluids, while animals that live on land constantly lose internal water to the dry air around them.

Coping with the demands of variable external environments required the evolution of a variety of physiological, structural, and behavioral mechanisms. For terrestrial animals, these mechanisms include anatomical, metabolic, and behavioral adaptations to reduce water loss (impermeable body surfaces, internalized surfaces for gas exchange, and highly efficient kidneys that conserve water). Fertilization and fetal development, two processes that require aqueous environments, often take place internally. In the absence of support from the buoyancy of water, terrestrial animals require rigid skeletons to support their weight. The result of all these adaptations is an efficient organism designed for survival in a hostile terrestrial world.

▶ HOMEOSTASIS

Although the human body as a whole is adapted to cope with a variable external environment, its individual cells are much less tolerant of change. Only a small minority of cells in a multicellular organism are actually in direct contact with the external environment (Fig. 1-4 ●). The vast majority of the cells are sheltered from the outside world by the buffer zone of the **extracellular fluid** [*extra-*, outside of], the body fluid that surrounds the cells and makes up one-third of the body's total volume. The bulk of the body's water is found in the **intracellular fluid** inside cells [*intra-*, within].

Extracellular fluid is the body's internal environment and serves as the interface between the external environment and the cells. When conditions outside the body change, the changes are reflected in the composition of the extracellular fluid, which in turn affects the cells. But our cells are not very tolerant of changes in their surroundings. As a result, a variety of mechanisms have evolved that maintain the composition of the extracellular fluid within a narrow range of values. The body's ability to maintain internal stability is known as **homeostasis** [*homeo-*, similar + -*stasis,* condition]. Homeostasis and the regulation of the internal environment are central precepts of physiology and create an underlying theme of each chapter in this book.

Failure to maintain homeostasis disrupts normal function and may result in a disease state or **pathological** condition [*pathos,* suffering]. Diseases generally can be divided into two groups according to their origin: those in which the problem arises from internal failure of some normal physiological process and those that originate from some outside source. Internal causes of disease include the abnormal growth of cells, which may cause cancer or benign tumors, the production of antibodies by the body against its own tissues (autoimmune diseases), and the premature death of cells or the failure of cell processes. Inherited disorders are also considered to have internal causes. External causes of disease include toxic chemicals, physical trauma, and foreign invaders such as viruses and bacteria. In both internally and externally caused diseases, when homeostasis is disturbed, the body attempts to compensate (Fig. 1-5 ●). If the compensation is successful, homeostasis is restored. If compensation fails, illness or disease often results. The study of body functions in a disease state is known as **pathophysiology.**

You will encounter many examples of pathophysiology as we study the various systems of the body. One very common

● **Figure 1-4** *The internal and external environments* Few cells in the body are able to exchange material with the external environment. Most cells are in contact with the internal environment, composed of the extracellular fluid.

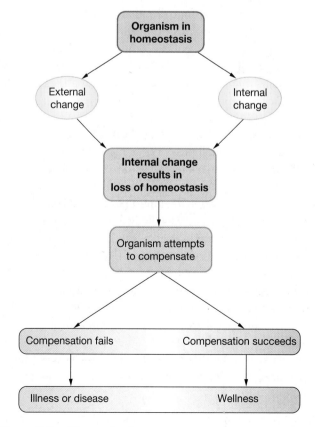

● **Figure 1-5** *Homeostasis*

pathological condition in the United States is **diabetes mellitus**, a metabolic disorder characterized by abnormally high blood glucose concentrations. Although we speak of diabetes as if it were a single disease, it is actually a whole family of diseases with varying causes and manifestations. You will learn more about diabetes in the focus boxes scattered throughout the chapters of this book. The influence of this one disease on many systems of the body makes it an excellent example of the integrative nature of physiology.

▶ THEMES IN PHYSIOLOGY

Recall the quotation that opens this chapter: "Physiology is not a science or a profession but a point of view."* Physiologists pride themselves on relating the processes they study to the functioning of the animal as a whole. This integration of body system function is one of the more difficult aspects of learning physiology because it requires remembering the processes of several systems simultaneously.

The amount of information to be learned in a physiology course may seem overwhelming at times, but there are some strategies you can employ to help simplify the process. One is the use of flow charts and maps (see the *Focus on Mapping*, 🐾 p. 5). Another technique is learning to recognize repeating patterns and themes.

Learning scientists have found that to develop expertise in a field, a novice must do more than simply acquire knowledge. Memorizing facts and learning new terminology are important and necessary first steps, but the ability to solve problems requires a conceptual framework or "big picture" of the field. In this book we will help you build a big picture by explicitly describing seven basic physiological concepts or themes (Table 1-2). As you work through the text, you will find that these concepts, with variations, occur repeatedly in the different organ systems.

Homeostasis and Control Systems As discussed in the previous section, a central precept of human physiology is homeostasis, the maintenance of a relatively stable internal environment. The body has certain key functions, such as blood pressure and blood glucose concentration, that must be regulated, or kept within a particular operating range, if the body is to remain healthy. These key functions are **regulated variables** that are monitored and regulated by a **physiological control system** (Fig. 1-6●).

If a variable moves out of its desirable range, specialized **sensors** are activated and send a signal to an **integrating center** [*integrare*, to restore]. The integrating center is often a nerve cell or endocrine cell that evaluates information coming from the sensor and initiates a response that is designed to bring the regulated variable back into the desired range. The muscles and other tissues controlled by integrating centers are known as **effectors**

---

*Ralph W. Gerard, *Mirror to Physiology: A Self Survey of Physiological Science* (Washington, D.C.: American Physiology Society, 1958).

Table 1-2: Key Concepts in Physiology

1. Homeostasis and control systems
2. Structure/function relationships
   Molecular interactions
   Mechanical properties of cells, tissues, and organs
   Compartmentation
3. Communication
4. Movement of substances across membranes
5. Biological energy
6. Mass balance
7. Mass flow

[*efficere*, to carry out] because they effect a change. The different types of control systems will be discussed in Chapter 6.

Structure/Function Relationships The integration of structure and function is another physiological theme. Structural influences on function range from molecular interactions to the division of the body into discrete intracellular and extracellular compartments.

Molecular interactions The ability of individual molecules to bind to or react with other molecules is essential for biological function. Examples of molecular interactions include enzymes that speed up chemical reactions and signal molecules that bind to membrane proteins.

Mechanical properties of cells, tissues, and organs The properties of physiological systems are often a direct result of their anatomy. Some of the mechanical properties of cells and tissues that influence function include *compliance* (ability to stretch), *elastance* (ability to return to the unstretched state), strength, flexibility, and fluidity. In addition, specialized structures in the body are designed to function as biological pumps, filters, or motors.

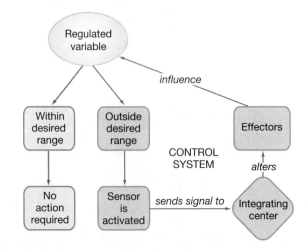

● **Figure 1-6** *Physiological control systems keep regulated variables within a desired range during homeostasis*

Compartmentation of the body and of cells Cell **membranes** separate cells from each other and from the extracellular fluid. Similarly, within a cell, membranes divide the cell into discrete units or organelles. This **compartmentation**, or the presence of separate compartments within the body, allows different areas to specialize their functions.

Communication Integration and homeostasis both require that the cells of the body communicate with one another rapidly and efficiently. Most cell-to-cell communication uses chemical signals, with the nervous system adding speed by sending electrical signals. Information may pass from one cell to its neighbors (local communication) or from one part of the body to another (long distance communication). Movement of blood in the circulatory system aids long distance chemical communication.

Communication between the internal environment of the cells and the extracellular fluid requires information transfer across the cell membrane. Some chemical messengers are able to enter the cell but others are unable to cross the barrier created by the cell membrane. In these instances, the message must be transferred across the cell membrane and translated into an intracellular signal, a process known as **signal transduction.** Chemical communication and signal transduction are discussed in more detail in Chapter 6.

Movement of Substances across Membranes

Movement of water, gases, and other molecules between the intracellular and extracellular compartments requires crossing the barrier of the cell membrane. Specialized processes have evolved that allow cells to be somewhat selective about what materials they allow in or out. The details of movement across biological membranes will be discussed in Chapter 5.

Biological Energy Living processes require the continuous input of energy. Where does this energy come from, and how is it stored? As we answer those questions, you will learn how energy in the body is used for synthesis and breakdown of molecules, to transport molecules across cell membranes, and to create movement.

Mass Balance Another important concept in physiology is the **law of mass balance.** This law says that if the amount of a substance in the body is to remain constant, any gain must be offset by an equal loss (Fig. 1-7 ●). For example, to maintain constant body temperature, heat gain from the external environment and from metabolism must be offset by heat loss back to the external environment.

Most substances enter the body from the outside environment, although some, like heat, can be produced internally. The major routes for loss from the body are the urinary and digestive systems (in the urine and feces), the respiratory system, and the integumentary system (skin). Sensors that detect changes in the internal and external environments monitor some aspects of mass balance. When mass balance is disturbed, physiological and

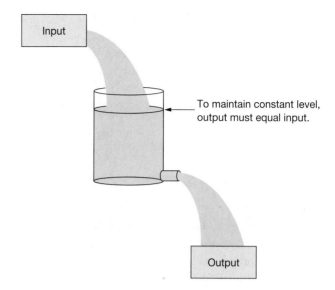

● **Figure 1-7** *Mass balance in the body*

behavioral responses bring the body back into balance. The law of mass balance is summarized by the following equation:

Total amount (or **load**) of substance x in the body

$$= \text{intake} + \text{production} - \text{output}$$

Substances whose concentrations are maintained through mass balance include oxygen and carbon dioxide, water, salts, and hydrogen ions. In addition, our bodies have regulatory mechanisms to keep temperature and energy stores within an acceptable range.

Mass Flow Mass balance in the body is usually discussed as a function of time. We talk about sodium intake per day or carbon dioxide output per minute. Thus, the rate of intake, production, or output is an important consideration. This is expressed as **mass flow,** where

$$\frac{\text{Mass flow}}{\text{(amount/min)}} = \frac{\text{concentration}}{\text{(amount/vol)}} \times \frac{\text{volume flow}}{\text{(vol/min)}}$$

For example, a person is given an intravenous infusion of glucose containing 50 grams of glucose per liter. If the infusion is given at a rate of 2 milliliters per minute, the mass flow for glucose would be

50 g glucose/1000 mL solution $\times$ 2 mL solution/min

$$= 0.1 \text{ g glucose/min}$$

In physiology we are also concerned with the movement of substances within and between compartments of the body. Blood flows in the circulatory system; air flows into and out of the lungs; gases, nutrients, and wastes move into and out of cells. In each instance, movement or flow of the substance is created by a driving force such as a pressure or concentration difference (a **gradient**). In many situations, flow is opposed by friction or forces that create **resistance** to flow. The concepts of gradients, flow, and resistance will be reintroduced in Chapter 5.

The seven themes just described will appear over and over again in the subsequent chapters of the book. Look for them in the summary material at the end of the chapters and in the end-of-chapter questions.

CONCEPT CHECK | Answers: p. 18

1. If a person eats 12 milligrams of salt in a day and excretes 11 milligrams of it in the urine, what happened to the remaining 1 milligram?

▶ THE SCIENCE OF PHYSIOLOGY

How do we know what we know about the physiology of the human body? Observation and experimentation are the key elements of **scientific inquiry.** An investigator observes an event or phenomenon and, using prior knowledge, generates a **hypothesis** [*hypotithenai,* to assume], or logical guess, about how the event takes place. The next step is to test the hypothesis by designing an experiment in which the investigator manipulates some aspect of the phenomenon.

Good Scientific Experiments Must Be Carefully Designed

A common type of biological experiment removes or alters some element, or *variable,* that the investigator thinks is an essential part of the observed phenomenon. That altered element is the **independent variable.** For example, a biologist notices that birds at a feeder seem to eat more in the winter than in the summer. She generates a hypothesis that cold temperatures cause birds to increase their food intake. To test her hypothesis, she designs an experiment in which she will keep birds at different temperatures and monitor how much food they eat. In her experiment, temperature, the manipulated element, is the independent variable. Food intake, which is hypothesized to be dependent on temperature, becomes the **dependent variable.**

CONCEPT CHECK | Answers: p. 18

2. Students in the laboratory execute an experiment in which they drink different volumes of water and measure their urine output in the hour following drinking. What are the independent and dependent variables in the experiment?

An essential feature of any experiment is a **control.** A control group is usually a duplicate of the experimental group in every respect except that the manipulated variable is not changed from its normal value. For example, in the bird-feeding experiment, the control group would be a set of birds maintained at warm temperatures but otherwise treated exactly like the birds held at cold temperatures. The purpose of the control is to ensure that any observed changes are due to the experimental manipulation and not to changes in some other variable. For example, if the in-

vestigator in the bird experiment changed to a different food, and food intake increased, she could not determine if the increased food intake was due to temperature or to more palatable food unless she had a control group that was also fed the different food.

During an experiment, the investigator carefully collects information, or **data** [plural; singular *datum,* a thing given], about the effect that the manipulated (independent) variable has on the observed (dependent) variable. Once the investigator feels that she has sufficient information to draw a conclusion, she begins to analyze the data. Analysis can take many forms and usually includes statistical analysis to determine if apparent differences are statistically significant. A common format for presenting data is a graph (see Fig. 1-8 ●, *Focus on Graphs,* p. 12).

If one experiment supports the hypothesis that cold causes birds to eat more, then the experiment should be repeated to ensure that the results were not an unusual one-time event. This step is called **replication.** When data support a hypothesis in multiple experiments, the hypothesis may become a **scientific theory.**

Most information presented in textbooks like this one is based on theoretical **models** that scientists have developed from the best available experimental evidence. On occasion, new experimental evidence that does not support a current model is published. In that case, the model must be revised to fit the available evidence. Thus, although you may learn a physiological "fact" while using this textbook, in 10 years that fact may be inaccurate because of what scientists have discovered in the interval. For example, in 1970 students learned that the cell membrane was a "butter sandwich," composed of two layers of proteins with a layer of fats sandwiched between. But in 1972 scientists presented a very different model of the membrane, one in which globules of proteins float within a double layer of fats. As a result, textbook writers had to revise their descriptions of cell membranes, and students who had learned the butter sandwich model had to revise their mental model of the membrane.

Where do our scientific models for human physiology come from? We have learned much of what we know from experiments on animals ranging from squid to rats. In many instances, the physiological processes are identical or similar enough that we can extrapolate from the animal model to humans. It is important to use nonhuman models because experiments using human subjects can be difficult to perform.

However, not all studies done on animals can be applied to humans. For example, an antidepressant that had been used safely by Europeans for years was undergoing stringent testing required by the U.S. Food and Drug Administration before it could be sold in this country. When beagle dogs were given the drug for a period of months, the dogs started dying from heart problems. Scientists were alarmed until further research showed that beagles have a unique genetic makeup that causes them to break down the drug into a more toxic substance. The drug was perfectly safe in other kinds of dogs and in humans, and it was subsequently approved for human use.

The Results of Human Experiments Can Be Difficult to Interpret

There are many reasons it is difficult to carry out physiological experiments in humans, including variability, psychological factors, and ethical considerations.

Variability Human populations have tremendous genetic and environmental **variability**. Although physiology books usually present *average* values for various functions such as blood pressure, these average values simply represent a number that falls somewhere near the middle of a wide range of values. Thus, to show significant differences between experimental and control groups in a human experiment, an investigator would, ideally, have to include large numbers of identical subjects.

However, getting two groups of people who are *identical* in every respect is impossible. The researcher then must attempt to recruit subjects who are *similar* in as many aspects as possible. You may have seen advertisements in newspapers requesting volunteers: "Healthy males between 18 and 25, nonsmokers, within 10% of ideal body weight, to participate in a study ...." The variability inherent in even a select group of humans must be taken into account when doing experiments with human subjects, because variability may affect the researcher's ability to accurately interpret the significance of data collected on that group.

One way to reduce variability within a test population, whether human or animal, is to do a **crossover study.** In a crossover study, each subject acts both as experimental animal and control. Thus, the individual's response to the treatment can be compared with the individual's own control value. This method is particularly effective when there is wide variability within a population. For example, in a test of blood pressure medication, subjects might be divided into two groups. Group A takes an inactive substance (**placebo,** from the Latin for "I shall be pleasing") for the first half of the experiment, then changes to the experimental drug for the second half. Group B starts with the experimental drug, then changes to the placebo. This scheme allows the researcher to assess the effect of the drug on each individual. Statistically, the resulting data can be analyzed using different methods that look at the changes in each individual rather than at changes in the collective group data.

Placebo and Nocebo Effects Another significant variable in human studies is the psychological aspect of administering a treatment. If you give someone a pill and tell the person that it will help them, there is a strong possibility that it will have that effect, even if the pill contains only sugar or an inert substance. This is a well-documented phenomenon called the **placebo effect.** Similarly, if you tell people that a drug they are taking may have specific adverse side effects, those people will report a higher incidence of the side effects than a similar group of people who were not warned. This phenomenon is called a **nocebo** effect, from the Latin word *nocere,* to do harm. The placebo and nocebo effects show the ability of our minds to alter the physiological functioning of our bodies.

In setting up an experiment with human subjects, we must try to control the placebo effect. The simplest way to do this is with a **blind study,** in which the subject does not know whether he or she is receiving the drug or the placebo. Even this can fail, however, if the researcher who is assessing the subject knows which type of treatment the subject is receiving. The researchers' expectations of what the treatment will or will not do may color their measurements or interpretations. To avoid this outcome, researchers often use **double-blind studies,** in which a third party, not involved in the experiment, is the only one who knows which group is receiving the experimental treatment and which group is receiving the control treatment. The most sophisticated experimental design for minimizing psychological effects is the **double-blind crossover study,** in which the control group in the first half of the experiment becomes the experimental group in the second half, and vice versa.

Ethical Considerations Ethical questions arise when humans are used as experimental subjects, particularly when the subjects are people suffering from a disease or other illness. Is it ethical to withhold a new and promising treatment from the control group? A noteworthy example occurred a few years ago when researchers were testing the efficacy of a treatment for dissolving blood clots in heart attack victims. The survival rate among the treated patients was so much higher that testing was halted so that control groups could also be given the experimental drug.

In contrast, tests on some anticancer agents have shown that the experimental treatments were less effective in stopping the spread of cancer than were the standard treatments used by the controls. Was it ethical to undertreat the experimental patients by depriving them of the more-effective current medical practice? Most studies now are evaluated as they progress to minimize the possibility that participants will be harmed by their participation.

A recent trial on hormone replacement therapy in postmenopausal women was halted early when investigators realized that women taking a pill with two different hormones were developing cardiovascular disease and breast cancer at a higher rate than women on placebo pills. On the other hand, the women receiving hormones also had a *lower* rate of colon cancer and bone fractures. The investigators decided that the risks associated with taking the hormones exceeded the potential benefits, and the study was stopped. To learn more about this clinical trial and the pros and cons of hormone replacement therapy, go to *www.nlm.nih.gov/medlineplus/hormonereplacementtherapy.html*, the website of the U.S. National Library of Medicine.

Human Studies Can Be Designed in Many Formats

Almost daily the newspapers carry articles about clinical trials studying the efficacy of drugs or other medical treatments. There are many different aspects of experimental design that can affect the validity and applicability of the results of these trials. For example, some trials are carried out for only a limited period of time

Graphs

Graphs are pictorial representations of the relationship between two (or more) different variables, plotted in a rectangular region (Fig. 1-8a ●). We use graphs to present a large amount of numerical data in a small space, emphasize comparisons between variables, or show trends over time. A viewer can extract information much more rapidly from a graph than from a table of numbers or from a written description. A well-constructed graph should contain (in very abbreviated form) everything the reader needs to know about the data, including the purpose of the experiment, how the experiment was conducted, and the results.

All scientific graphs have common features. The independent variable that was manipulated by the experimenter is graphed on the horizontal *x*-axis. The dependent variable that was measured by the experimenter is plotted on the vertical *y*-axis. If the experimental design is valid and the hypothesis is correct, changes in the independent variable (*x*-axis) will cause changes in the dependent variable (*y*-axis). In other words, *y* is a function of *x*. Mathematically this relationship can be expressed as $y = f(x)$.

CONCEPT CHECK   Answers: p. 18

3. In the following experiment, what were the independent and dependent variables? Draw a graph and put labels for the correct variable on each axis. Draw trend lines or bars that might approximate the data collected.

   (a) Students in a physiology laboratory collected heart rate data at rest and when they exercised. Heart rate increased with exercise.

   (b) In the same experiment the female students had higher resting heart rates than the male students.

Each axis of a graph is divided into units represented by evenly spaced tick marks on the axis. A label tells what variable the axis represents (time, temperature, amount of food consumed) and in what units it is marked (days, degrees Celsius, grams per day). The intersection of the two axes is called the *origin*. The origin usually, but not always, has a value of zero for both axes. A graph should have a title or legend that describes what the graph represents. If multiple groups are shown on one graph, the lines or bars representing the groups may have labels, or a key may show what group each symbol or color represents.

Most graphs that you will encounter in physiology display data as bars (bar graphs or histograms), lines (line graphs), or dots (scatter plots). Four typical types of graphs are shown in Figure 1-8b–e. **Bar graphs** (Fig. 1-8b ●) are used when the independent variables are distinct entities. A **histogram** (Fig. 1-8c ●) is a specialized bar graph that shows the distribution of one variable over a range. The *x*-axis is divided into units (called "bins" in some computer graphing programs) and the *y*-axis indicates how many pieces of data are associated with each bin.

● Figure 1-8

(a) The standard features of a graph include units and labels on the axes, a key, and a figure legend.

Line graphs (Fig. 1-8d ●) are commonly used when the independent variable on the *x*-axis is a continuous phenomenon, such as time, temperature, or weight. Each point on the graph may represent the average of a set of observations. Because the independent variable is a continuous function, the points on the graph can be connected with a line (point-to-point connections or a mathematically calculated "best fit" line or curve). Connecting the points allows the reader to **interpolate,** or estimate values between the measured values.

Scatter plots (Fig. 1-8e ●) show the relationship between two variables, such as time spent studying for an exam and performance on that exam. Usually each point on the plot represents one member of a test population. Individual points on a scatter plot are never connected by a line, but a "best fit" line or curve may indicate a trend in the data.

Here are some questions to ask when you are trying to extract information from a graph:

1. What variable does each axis represent?

2. What is the relationship between the variables represented by the axes? This relationship can usually be expressed by substituting the labels on the axes into the following statement: *y* varies with *x*. For example, in graph (b), the canaries' daily food intake varied with the type of diet.

3. Are any trends shown in the graph? For line graphs and scatter plots, is the line horizontal (no change in the dependent variable when the independent variable changes), or does it have a slope? Is the line a straight line or is it a curve? For bar graphs, are the bars the same or different heights? Is there a trend in the direction of change if they are different heights?

(b) Bar graph. Each bar shows a distinct variable. The bars are lined up side by side along one axis so that they can be easily compared with one another. Scientific bar graphs traditionally have the bars running vertically.

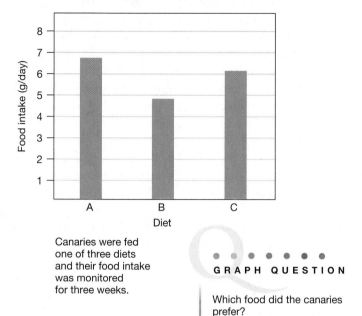

Canaries were fed one of three diets and their food intake was monitored for three weeks.

GRAPH QUESTION

Which food did the canaries prefer?

(c) Histogram. A histogram quantifies the distribution of one variable over a range of values.

The distribution of student scores on a 10-point quiz is plotted on a histogram.

GRAPH QUESTION

How many students took the quiz?

(d) Line graph. The x-axis frequently represents time; the points represent individual observations. The points may be connected by lines, in which case the slope of the line between two points shows the rate at which the variable changed.

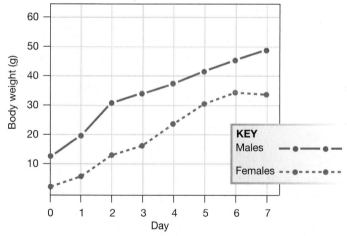

KEY
Males ●——●——●
Females ●·····●·····●

Male and female mice were fed a standard diet and weighed daily.

GRAPH QUESTION

When did male mice increase their body weight the fastest?

(e) Scatter plot. Each point represents one member of a test population. The individual points of a scatter plot are never connected by lines, but a best fit line may be estimated to show a trend in the data, or better yet, the line may be calculated by a mathematical equation.

Student scores were directly related to the amount of time they spent studying.

GRAPH QUESTIONS

For graphs (d) and (e), answer the following:
• What was the investigator trying to determine?

• What are the independent and dependent variables?
• What are the results or trends indicated by the data?

on a limited number of people, such as studies being conducted for the U.S. Food and Drug Administration's drug approval process. There have been several instances in the past few years where drugs approved as a result of such studies have later been withdrawn from the market because longer use of the drug by more people uncovered adverse side effects, including deaths.

Other studies are designed to be **longitudinal studies** that are carried out for a long period of time. One of the most famous longitudinal studies that is still ongoing is the Framingham Heart Study (*www.framingham.com/heart*), started in 1948. Framingham is a **prospective study** [*prospectus,* outlook, looking forward] that recruited healthy people and followed them for years to determine what factors contribute to the development of cardiovascular disease. This study has already made important contributions to health care and is continuing with the adult children of the original participants.

Additional study designs that you may encounter in the literature include cross-sectional and retrospective studies. **Cross-sectional studies** survey a population for the prevalence of a disease or condition. Data from cross-sectional studies determine trends to be investigated further, such as whether age group or socioeconomic status is associated with a higher risk of developing the condition being surveyed. **Retrospective studies** [*retro,* backwards + *spectare,* to look] match groups of people with a disease to a similar but healthy control group to see if development of the disease can be associated with a particular variable.

Often studies that are published have results that do not agree with the results of other published studies. In some cases, this is a result of limitations of the experimental design, such as a small number of subjects that may not be representative of larger populations. In other cases, it may be due to small but potentially significant differences in the experimental designs of the different studies. One way that scientists attempt to resolve contradictory results is to perform a **meta-analysis** of the data [*meta-,* at a higher level]. A meta-analysis combines the data from a group of similar studies and uses sophisticated statistical techniques to extract significant trends or findings from the group as a whole. For example, a number of studies were done to assess whether glucosamine and chondroitin, two dietary supplements, can improve degenerative joint disease. However, the individual

studies had small numbers (<50) of subjects and used different dosing regimens, making comparison of their results difficult. A group of researchers then performed a meta-analysis using statistical methods and concluded that in most of the studies, glucosamine did have a beneficial effect on joint disease.*

The difficulty of using human subjects in experiments is one reason that scientists use animals to develop many of our scientific models. Since the 1970s, physiological research has increasingly enhanced animal experimentation with techniques developed by cellular biologists and molecular geneticists. As we have come to understand the fundamentals of how chemical signals in the body are received and interpreted by the cells, we have unlocked the mysteries of many processes. In doing so, we also have come closer to being able to treat many diseases by correcting their cause rather than simply alleviating their symptoms.

More and more, medicine is turning to therapies based on interventions at the molecular level. A classic example is the treatment of cystic fibrosis, an inherited disease in which the mucus of the lungs and digestive tracts is unusually thick. For many years, patients with this condition had few treatment options, and most died at a young age. However, basic research into the mechanisms by which cells move salt and water across their membranes provided clues to the underlying cause of cystic fibrosis: a defective protein in the membrane of certain cells. Once molecular geneticists found the gene that coded for that protein, the door was opened to the possibility of replacing the defective gene in cystic fibrosis patients with the gene for the normal protein. But without the basic research into how cells and tissues carry out their normal function, this treatment would never have been developed.

As you read this book and learn what we know currently about how the human body works, keep in mind that many of the facts presented in it are just theories. There still are many questions in physiology waiting for investigators to find the answers.

---

*"Glucosamine and chondroitin for treatment of osteoarthritis: A systematic quality assessment and meta-analysis," JAMA 283(11):1469–75, 2000, March 15. (*www.jama.ama-assn.org*)

CHAPTER SUMMARY

1. **Physiology** is the study of the normal functioning of a living organism and its component parts. (p. 2)

Levels of Organization

2. Physiologists study many **levels of organization** of living organisms, from molecules to populations. (p. 3; Fig. 1-1)

3. The **cell** is the smallest unit of structure capable of carrying out all life processes. (p. 3)

4. Collections of cells that carry out related functions group into **tissues** and **organs**. (p. 3)

5. The human body has 10 physiological organ systems: **integumentary, musculoskeletal, respiratory, digestive, urinary, immune, circulatory, nervous, endocrine,** and **reproductive.** (pp. 3–4; Fig. 1-2)

6. The **lumens** of many hollow organs open to the external environment. Material in the lumens is not part of the **internal environment** until it crosses the wall of the organ. (p. 4)

Physiology Is an Integrative Science

7. Physiology is an integrative science that studies how the different systems of the body coordinate in the intact animal. Many complex functions in the body are **emergent proper-** ties that cannot be predicted from the properties of the individual component parts. (p. 6)

Function and Process

8. The **function** of a physiological system or event is the "why" of the system; the **process** by which events occur is the "how" of a system. A **teleological approach** to physiology explains why events happen; the **mechanistic approach** explains how they happen. (p. 6)

The Evolution of Physiological Systems

9. The human body as a whole is adapted to cope with a variable external environment, but most individual cells of the body are much less tolerant of change. For this reason, the body attempts to keep the internal environment relatively constant. (p. 7)

Homeostasis

10. **Homeostasis** is the body's coordinated response to change in order to maintain internal stability. (p. 7)

11. Failure to maintain homeostasis may result in illness or disease. (p. 7; Fig. 1-5)

Themes in Physiology

 IP Fluids and Electrolytes: Water Homeostasis

12. Key themes in physiology include homeostasis and control systems, structure-function relationships such as **compartmentation,** communication, movement of substances across **membranes,** biological energy, **mass balance,** and **mass flow.** (pp. 8–9)

The Science of Physiology

13. Observation and experimentation are the key elements of **scientific inquiry.** A **hypothesis** is a logical guess about how an event takes place. (p. 10)

14. In scientific experimentation, the factor manipulated by the investigator is the **independent variable.** The observed factor is the **dependent variable.** All well-designed experiments have a **control** to ensure that observed changes are due to the experimental manipulation and not to an outside factor. (p. 10)

15. **Data,** the information collected during the experiment, are analyzed and presented, often as a graph. (p. 10)

16. A **scientific theory** is a hypothesis that has been supported by data on multiple occasions. Sometimes new experimental evidence is presented that does not support a theory or **model,** and the model must be revised. (p. 10)

17. Animal experimentation is an important part of learning about human physiology because there is a tremendous amount of **variability** within human populations and because it is difficult to control human experiments. In addition, ethical questions arise when using humans as experimental animals. (p. 11)

18. In a **crossover study,** each subject acts both as experimental subject and control. For half the experiment, the subject takes an inactive substance known as a **placebo.** One difficulty in human experiments is the **placebo and nocebo effects,** in which changes take place even if the treatment is inactive. (p. 11)

19. In a **blind study,** the subjects do not know whether they are receiving the experimental treatment or a placebo. In a **double-blind study,** a third party removed from the experiment is the only one who knows which group is the experimental and which is the control. In a **double-blind crossover study,** the control group in the first half of the experiment becomes the experimental group in the second half, and vice versa. (p. 11)

20. **Meta-analysis** of data combines data from many studies to look for trends. (p. 14)

QUESTIONS

LEVEL ONE *Reviewing Facts and Terms*

1. Define physiology. Describe the relationship between physiology and anatomy.

2. Name the different levels of organization in the biosphere and the disciplines that study different levels.

3. Name the 10 systems of the body and give their major function(s).

4. What is meant by the phrase "Physiology is an integrative science"?

5. Define homeostasis. What aspects of human function are maintained homeostatically?

6. Name six major themes in physiology.

LEVEL TWO *Reviewing Concepts*

7. **Mapping exercise:**
   (a) Make a map of the law of mass balance using the following phrases. You may use terms more than once.

body contents	external environment	lungs
body synthesis	input	output
digestive system	kidneys	skin

   (b) Make a large map showing the organization of the human body. Use all levels of organization in the body (see Fig. 1-1) and all 10 organ systems. Try to include functions of all components on the map and remember that some structures may share functions. (Hint: Start with the human body as the most important term. You may also draw the outline of a body and make your map using it as the basis.)

8. Distinguish between the following groups of terms:
   (a) tissues and organs
   (b) x-axis and y-axis on a graph
   (c) dependent and independent variables
   (d) teleological and mechanistic approaches
   (e) the internal and external environments for a human
   (f) blind, double-blind, and crossover studies

9. Name as many organs or body structures that have a lumen as you can. Do all of these lumens connect directly with the external environment? Which ones do not?

10. Which systems are responsible for coordinating body function? For protecting the body from outside invaders? Which systems exchange material with the external environment and what do they exchange?

LEVEL THREE *Problem Solving*

11. A group of biology majors went to a mall and asked passersby, "Why does blood flow?" These are some of the answers they received. Which answers are teleological and which are mechanistic? (Not all answers are correct, but they can still be classified.)
    (a) Because of gravity
    (b) To bring oxygen and food to the cells
    (c) Because if it didn't flow, we would die
    (d) Because of the pumping action of the heart

12. Although dehydration is one of the most serious physiological obstacles that animals on land must overcome, there are others. Think of as many as you can, and think of different ways that different terrestrial animals have to overcome these obstacles. (Hint: Think of humans, insects, and amphibians; also think of as many different terrestrial habitats as you can.)

LEVEL FOUR *Quantitative Problems*

13. A group of students wanted to see what effect a diet deficient in vitamin D would have on the growth of baby guppies. They fed the guppies a special diet with low vitamin D and measured their length every third day for three weeks. The data they collected are in the table below:

Day	0	3	6	9	12	15	18	21
Average length (mm)	6	7	9	12	14	16	18	21

   (a) What is the dependent variable and what is the independent variable in this experiment?
   (b) What was the control in this experiment?
   (c) Make a fully labeled graph with a legend, using the data in the table.
   (d) During what period of time was growth slowest? Most rapid? (Use your graph to answer this question.)

14. You performed an experiment in which you measured the volumes of slices of potato, then soaked the slices in solutions of varying salinities for 30 minutes. At the end of 30 minutes, you again measured the volumes of the slices. The results are shown at the top of the next page.
   (a) What is the independent variable in this experiment? What is the dependent variable?

% change in volume after 30 minutes

SOLUTION	SAMPLE 1	SAMPLE 2	SAMPLE 3
Distilled water	+10%	+8%	+11%
1% salt (NaCl)	0%	+0.5%	−1%
9% salt (NaCl)	−8%	−12%	−11%

(b) Was there a control in this experiment or can you tell from the information given? If there was a control, what was it?

(c) Graph the results of the experiment using the most appropriate type of graph.

15. If a woman weighs 121 pounds and 51% of her total body weight is water, what is her total body water volume in liters? Assume that 2.2 pounds = 1 kilogram, and that 1 kilogram of water = 1 liter. Use the correct abbreviations for these units in your answer.

16. At the end of the semester an intermediate-level class of 25 male weight lifters was measured for aerobic fitness and for their midarm muscle circumference. The relationship between those two variables is graphed in Figure 1-9 ●.

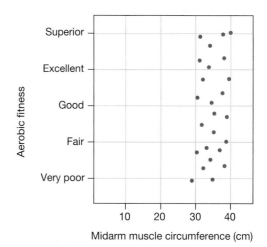

● **Figure 1-9**

(a) What kind of graph is this?

(b) What question were the investigators asking?

(c) In one sentence, summarize the relationship between the two variables plotted on the graph.

Explorations

Explorations are also available at *www.physiologyplace.com*

Introduction

This is the first in a series of Explorations that you will find at the end of each chapter. These Explorations will challenge you to enhance, expand, and assess your understanding of the material presented in the chapter.

Exploration 1: Placebo Surgery

"A controlled trial of arthroscopic surgery for osteoarthritis of the knee," *New England Journal of Medicine* 347(2):81–88, 2002, July 11.*

A recent study was performed on human volunteers to see whether two procedures performed during arthroscopic surgery [*arthro-*, joint + *scopium*, to look at] are effective in relieving knee pain associated with osteoarthritis, or degenerative joint disease [*osteon*, bone + *arthro-*, joint + *-itis*, inflammation]. The volunteers were up to 75 years old and were recruited from a Veterans Affairs Medical Center. They were 93% male and 60% white. One-third of the subjects in the study had placebo oper-

---

*This is one standard format for the citation of scientific articles published in journals. Et al. is an abbreviation for *et alia*, meaning "and others." The journal name is spelled out here but is often shortened using standard abbreviations: *N Engl J Med*. The number 347 refers to the volume number and (2) means it is the second issue within that volume. The numbers after the full colon are the inclusive page numbers for the article.

ations, that is, they were given anesthesia and their knees were cut open but the remainder of the treatment procedure was not done. The other two-thirds of the subjects had one of the two treatment procedures performed. Subjects were followed for two years. They answered questions about their knee pain and function and were given an objective walking and stair-climbing test. At the end of the study the results showed no significant difference in knee function or perception of pain between subjects getting one of the standard treatments and those getting the placebo operation.

1. Do you think that it is ethical to perform placebo surgeries on humans who are suffering from a painful condition, even if the subjects are informed that they might receive the placebo operation and not the standard treatment?

2. Give two possible explanations for the decreased pain in the placebo operation subjects.

3. Read the article online at *www.nejm.org* and answer the following questions:
   (a) Analyze and critique the experimental design of this study. Are the results of this study applicable to everyone with knee pain?
   (b) Was this study a blind, double-blind, or double-blind crossover design?
   (c) Why did the investigators feel it was necessary to include a placebo operation in this study?

For peer critiques of this study, read the editorials published in the same issue: *New England Journal of Medicine* 347(2):132–33 and 137–39, 2002, July 11.

Exploration 2: Variability in a Human Population

Although we talk in physiology about the *average* values for various functions, these average values simply represent a number in the middle of a wide range of normal values. To show how a simple physiological function can have a wide range of "normal" values, record resting heart rates of a group of your classmates. Have each subject sit quietly and take his/her pulse for 30 seconds. Convert to beats/minute and record. Repeat and calculate an average pulse for each subject. Record the subject's gender and average resting pulse on a data sheet.

1. What question(s) can you ask with this data?
2. Create one or more graphs with the data to answer your questions.

ANSWERS

Answers to Concept Check Questions

Page 10

1. The remaining 1 milligram of salt remains in the body.

Page 10

2. The independent variable is the amount of water the students drink. The dependent variable is their urine output.

Page 12

3. (a) Exercise intensity is the independent variable (*x*-axis) and heart rate is the dependent variable (*y*-axis). The trend line should slope upward to the right, showing that as exercise intensity increases, heart rate increases. A line graph would be appropriate for this data but a bar graph could also be used if the intensity of exercise was considered to be the same in all subjects. (b) To show the difference between males and females, the graphs from (a) would either have separate lines or separate bars for males and females.

Answers to Figure Questions

Page 4

Fig. 1-3: (a) You might include different kinds of breads, meats, etc., or add a category of sandwich characteristics, such as temperature (hot, cold) or layers (single, club).

Page 13

Fig. 1-8b: The birds preferred diet A.

Fig. 1-8c: Twenty-four students took the quiz.

Fig. 1-8d: Male mice (blue line) increased their weight most between day 1 and day 2.

Fig. 1-8e: (a) In graph 1-8d the investigator was looking at changes of body weight over time in male and female mice. In graph 1-8e the investigator was trying to determine if there was a relationship between the amount of time spent studying for an examination and the student's score on that examination. (b) In graph 1-8d the independent variable was time and the dependent variable was body weight. In graph 1-8e the independent variable was number of hours spent studying and the dependent variable was the student score. (c) Graph 1-8d shows that male mice weigh more than female mice from the start of the experiment and that body weight increases with time. The rate of increase is about the same in males and females, indicated by the nearly parallel lines. Graph 1-8e shows that more hours spent studying resulted in higher exam scores.

Page 4

Fig. 1-10 ●. This sample map shows one possible way to map the terms. Notice how some items are linked to more than one other item. The map does not show many terms that could be included.

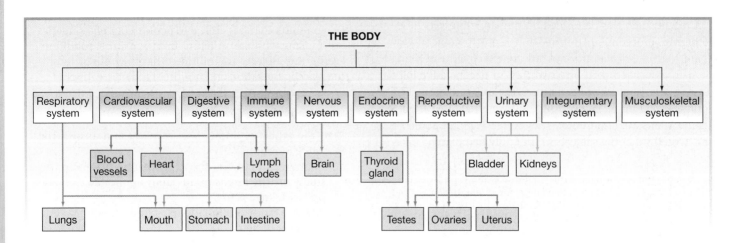

● **Figure 1-10** *Sample answer for graph question on p. 4* Notice that this map uses color to distinguish different elements in the map. This type of coding is optional when you make your own maps.

Page 16

13. (a) The independent variable was time (x-axis). The dependent variable was the length (y-axis). (b) There was no control in this experiment. A proper control would have been a similar group of fish treated identically except for being fed a diet with normal levels of vitamin D. (c) A graph should have time in days on the x-axis and length on the y-axis. A line graph is most appropriate for these data. (d) Growth was slowest from Day 0 to Day 3. Growth was most rapid between Days 6 and 9 and between Days 18 and 21.

14. (a) The independent variable was the concentration of the soaking solution. The dependent variable was volume change of the potato slices. (b) The control was the volume measurements before soaking. (c) A bar graph or a line graph could be used to graph these data. A line graph would allow you to estimate volume change at intermediate salt concentrations such as 5%.

Page 17

15. Her weight is 55 kg. Her total body water is 28.05 L (121 lb/2.2 lb/kg × 51%).

16. (a) This graph is a scatter plot. (b) The investigators were asking if there is a relationship between midarm muscle circumference and aerobic fitness. (c) There appears to be no relationship between midarm muscle circumference and aerobic fitness.

Atoms, Ions, and Molecules

2

Protein motion measured using a fluorescent probe

RUNNING PROBLEM

Chromium Supplements

"Lose weight while gaining muscle," the ads promise. "Prevent heart disease." "Stabilize blood sugar." What is this miracle substance? It's chromium picolinate, a nutritional supplement being marketed to consumers looking for a quick fix. But does it work and is it safe? Some athletes, like Stan—the star halfback on the college football team—swear by it. Stan takes 500 micrograms of chromium picolinate daily. Many researchers, however, are skeptical and feel that the necessity for and safety of chromium supplements have not been established.

One of the things that distinguishes ours from all earlier generations is this, that we have seen our atoms.
—**Karl Kelchner Darrow**

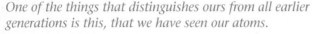

Chromium, zinc, copper, and iron... The list reads like the inventory of a metal company but it actually comes from the label on a bottle of multivitamin-multimineral pills. Why are we supposed to ingest these metals and what do they do in our bodies? Look at magazine articles touting vitamins C and E as antioxidants—what is an oxidant and why do we need antioxidants? To answer these questions we must start at the lowest level of organization in the human body and look at its smallest particles, the atom and its three fundamental subunits: the proton, the neutron, and the electron. We then examine how atoms link together to form molecules. Because the human body is 60% water and most of its molecules are dissolved, we will study how to make solutions and express the concentration of molecules dissolved in them. Finally, we will introduce the biologically important molecules or biomolecules: carbohydrates, lipids, proteins, and nucleotides.

▶ ELEMENTS AND ATOMS

The simplest kinds of matter are called **elements**. Some are substances that you encounter in everyday life: the oxygen that you breathe, the carbon in the soot of a barbecue grill. Elements are the building blocks of all matter, including the human body. Three of the elements, oxygen, carbon, and hydrogen, make up more than 90% of the body's mass. These elements and others bind together into more complex substances called molecules. To understand what distinguishes one element from another and how they combine into molecules, we will first look at atoms.

Atoms Are Composed of Protons, Neutrons, and Electrons

All elements are composed of small particles called **atoms**, once thought to be the smallest particles of matter [*atomos*, indivisible]. We now know, however, that an atom itself is composed of three types of even smaller particles: protons, neutrons, and electrons. These subatomic particles are distinguished from each other by their different electrical charges and masses. A **proton** has one positive charge (written +), an **electron** has one negative charge (written −), and a **neutron** has no charge, or zero charge—that is, it is neutral. Like the poles of a magnet, oppositely charged atomic particles (+ and −) attract, while those with like charge (+ and +, or − and −) repel. An atom usually has an equal number of protons and electrons, giving it an overall electrical charge of zero.

Protons and neutrons have nearly equal masses of about 1 **atomic mass unit** or **amu** (1 amu = 1.6605×10^{-27} kg). Electrons are much smaller: it takes the mass of 1836 electrons to equal the mass of one neutron. The arrangement of protons, neutrons, and electrons in an atom is always the same. The protons and neutrons of the atom, which account for almost all of its mass, are clustered into a dense body called the **nucleus** [*nuculeus*, little nut; plural, nuclei]. The space around the nucleus, almost all of the atom's volume, contains the rapidly moving, lightweight, negatively charged electrons, held in their orbit by the positively charged protons (Fig. 2-1 ●).

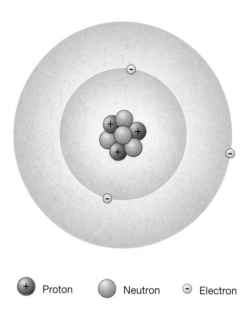

+ Proton   ⬤ Neutron   ⊖ Electron

● **Figure 2-1** *Atomic structure* An atom has a central nucleus of protons and neutrons, surrounded by rapidly moving, lightweight electrons, one for each proton in the nucleus.

Atoms are very tiny, having diameters in the range of 1 angstrom to 5 angstroms (1 angstrom = 1 Å = 10^{-10} m). Most of their volume is empty space. To get a feel for how little of an atom's volume is composed of the subatomic particles, imagine that the atom has a diameter equal to the length of a football field. The nucleus would then be the size of a large apple in the center of the field, and tiny electrons, smaller than a pea, would move through the remainder of the space.

If all atoms have protons, neutrons, and electrons, what makes an atom of one element different from that of another? For example, why is carbon so unlike oxygen? The next three sections will answer those questions by showing what happens to an atom when it gains or loses one of its subatomic particles.

An Element Is Distinguished by the Unique Number of Protons in Its Nucleus

There are more than 100 different elements in the world, about 90 of which occur naturally. The remainder are manufactured. All of the elements that we know of are listed in a **periodic table of the elements**, as shown in Figure 2-2 ●. Each box in the table contains the name of an element and its one- to three-letter symbol. Some of the symbols come from the original Latin names for the elements and thus do not match our modern names. For example, the element sodium is abbreviated Na, from its Latin name *natrium*.

CONCEPT CHECK   Answers: p. 48

1. Look at the periodic table in Fig. 2-2 and find nine other elements whose symbols are not derived from their modern names.

Periodic Table of the Elements

Atomic number = number of protons

6	— Name
Carbon	
C	— Symbol
12.0	— Atomic mass

☐ Major essential elements
☐ Minor essential elements
☐ Not believed essential for life

Transitional metals

Note: Numbers in parentheses are mass numbers (the total number of protons and neutrons in the nucleus) of the most stable or best-known isotope of radioactive elements.

Modern name	Latin name	Symbol
Copper	Cuprium	Cu
Iron	Ferrum	Fe
Potassium	Kalium	K
Sodium	Natrium	Na

Group 1	2	3	4	5	6	7	8	9	10	11	12	13	14	15	16	17	18
Hydrogen **H** 1.0																	Helium **He** 4.0
Lithium **Li** 6.9	Beryllium **Be** 9.0											Boron **B** 10.8	Carbon **C** 12.0	Nitrogen **N** 14.0	Oxygen **O** 16.0	Fluorine **F** 19.0	Neon **Ne** 20.2
Sodium **Na** 23.0	Magnesium **Mg** 24.3											Aluminum **Al** 27.0	Silicon **Si** 28.1	Phosphorus **P** 31.0	Sulfur **S** 32.1	Chlorine **Cl** 35.5	Argon **Ar** 39.9
Potassium **K** 39.1	Calcium **Ca** 40.1	Scandium **Sc** 45.0	Titanium **Ti** 47.9	Vanadium **V** 50.9	Chromium **Cr** 52.0	Manganese **Mn** 54.9	Iron **Fe** 55.8	Cobalt **Co** 58.9	Nickel **Ni** 58.7	Copper **Cu** 63.5	Zinc **Zn** 65.4	Gallium **Ga** 69.7	Germanium **Ge** 72.6	Arsenic **As** 74.9	Selenium **Se** 79.0	Bromine **Br** 79.9	Krypton **Kr** 83.8
Rubidium **Rb** 85.5	Strontium **Sr** 87.6	Yttrium **Y** 88.9	Zirconium **Zr** 91.2	Niobium **Nb** 92.9	Molybdenum **Mo** 95.9	Technetium **Tc** (98)	Ruthenium **Ru** 101.1	Rhodium **Rh** 102.9	Palladium **Pd** 106.4	Silver **Ag** 107.9	Cadmium **Cd** 112.4	Indium **In** 114.8	Tin **Sn** 118.7	Antimony **Sb** 121.8	Tellurium **Te** 127.6	Iodine **I** 126.9	Xenon **Xe** 131.3
Cesium **Cs** 132.9	Barium **Ba** 137.3	Lanthanum 57* **La** 138.9	Hafnium **Hf** 178.5	Tantalum **Ta** 181.0	Tungsten **W** 183.9	Rhenium **Re** 186.2	Osmium **Os** 190.2	Iridium **Ir** 192.2	Platinum **Pt** 195.1	Gold **Au** 197.0	Mercury **Hg** 200.6	Thallium **Tl** 204.4	Lead **Pb** 207.2	Bismuth **Bi** 209.0	Polonium **Po** (209)	Astatine **At** (210)	Radon **Rn** (222)
Francium **Fr** (223)	Radium **Ra** 226.0	Actinium 89** **Ac** (227)	Rutherfordium **Rf** (261)	Dubnium **Db** (262)	Seaborgium **Sg** (266)	Bohrium **Bh** (264)	Hassium **Hs** (269)	Meitnerium **Mt** (268)	Ununnillium **Uun** (271)	Unununium **Uuu** (272)	Ununbium **Uub** (277)		Ununquadium **Uuq** (289)				

* Lanthanides:

| 58 Cerium **Ce** 140.1 | 59 Praseodymium **Pr** 140.9 | 60 Neodymium **Nd** 144.2 | 61 Promethium **Pm** (145) | 62 Samarium **Sm** 150.4 | 63 Europium **Eu** 152.0 | 64 Gadolinium **Gd** 157.3 | 65 Terbium **Tb** 158.9 | 66 Dysprosium **Dy** 162.5 | 67 Holmium **Ho** 164.9 | 68 Erbium **Er** 167.3 | 69 Thulium **Tm** 168.9 | 70 Ytterbium **Yb** 173.0 | 71 Lutetium **Lu** 175.0 |

** Actinides:

| 90 Thorium **Th** 232.0 | 91 Protactinium **Pa** 231.0 | 92 Uranium **U** 238.0 | 93 Neptunium **Np** (237) | 94 Plutonium **Pu** (244) | 95 Americium **Am** (243) | 96 Curium **Cm** (247) | 97 Berkelium **Bk** (247) | 98 Californium **Cf** (251) | 99 Einsteinium **Es** (252) | 100 Fermium **Fm** (257) | 101 Mendelevium **Md** (258) | 102 Nobelium **No** (259) | 103 Lawrencium **Lr** (262) |

● **Figure 2-2** *Periodic table of the elements*

Only about one-fourth of the elements are commonly encountered in biological systems. Those that are necessary for life are called **essential elements**. Eleven **major essential elements** are found in the human body in fairly large amounts. Carbon (C), hydrogen (H), oxygen (O), and nitrogen (N) are by far the most common essential elements because they make up most biological molecules. The major essential elements are shown in pink squares in the periodic table in Figure 2-2.

The essential elements that are found in very low concentrations in living tissues are called **trace elements**, or minor essential elements (blue squares in the periodic table in Fig. 2-2). All of the trace elements are essential in some way to cell function, and the list of such elements continues to change as scientists discover more about how cells work. Many modern vitamin pills contain minor essential elements such as selenium, chromium, manganese, and molybdenum.

The need for trace elements is evident only when they are absent from a person's diet or present in insufficient quantities. For example, a lack of iodine in the diet causes low thyroid hormone production, which is then reflected by development of physical signs and symptoms such as fatigue and inability to tolerate cold. One trace element that has received a lot of attention in recent years is chromium, usually marketed to the American public as chromium picolinate. A deficiency of chromium in the body has been postulated (but not proven) to be associated with the metabolic disease diabetes mellitus ("sugar diabetes") in some people.

The atoms of different elements are distinguished from one another by the fact that each element has a unique number of protons in its nucleus. For example, the lightest element, hydrogen, has only one proton in its nucleus. The next element, helium (He), has two protons, followed by lithium (Li) with three protons, and so on. The number of protons in one atom of an element is called the **atomic number** of the element. The periodic table is arranged by atomic number, starting at hydrogen, with an atomic number of 1. The end point of the periodic table is constantly shifting as scientists find evidence for additional very short-lived elements (see Explorations at the end of the chapter for the story of a scandal associated with this research).

The average **atomic mass** of an element is shown below the symbol of the element. Atomic mass is calculated by adding the mass of the protons and neutrons in one atom of the element. For example, one atom of carbon has six protons and six neutrons. Because the mass of each proton or neutron is one atomic mass unit, carbon's atomic mass is 12 amu. However, in the periodic table, not all atomic masses are whole numbers. For example, the box for chlorine (Cl), with 17 protons in its nucleus, shows an atomic mass of 35.5. This is because chlorine comes in two forms, one with 18 neutrons and one with 20. The atomic mass in the periodic table is a weighted average of the different forms of the element. In the next section we will discuss these different forms, which are known as isotopes.

CONCEPT CHECK   Answers: p. 48

Put the correct answers in the right column with the subatomic particles in the left column. Answers may be used more than once.

2. electron   (a) has atomic mass of 1 amu

3. neutron   (b) found in the nucleus

4. proton   (c) negatively charged

                     (d) changing the number of these in an atom creates a new element

Use the periodic table of the elements in Figure 2-2 to answer the following questions:

5. Which element has 30 protons?

6. How many electrons does one atom of sodium (Na) have?

7. How many electrons does one atom of calcium (Ca) have?

8. What is the average atomic mass of nitrogen?

9. What element is illustrated by the atom in Figure 2-1?

RUNNING PROBLEM

What is chromium picolinate? Chromium (Cr) is an essential element that is linked to normal glucose metabolism. In the diet, chromium is found in brewer's yeast, broccoli, mushrooms, and apples. Chromium in food and in chromium chloride supplements is poorly absorbed from the digestive tract, so a scientist developed and patented the compound chromium picolinate. Picolinate, derived from amino acids, enhances chromium uptake at the intestine. The recommended daily allowance (RDA) of chromium is 50 micrograms. As we've seen, Stan takes several times this amount.

Question 1: Locate chromium on the periodic table of the elements. What is chromium's atomic number? Atomic mass? How many electrons does one atom of chromium have? Which elements close to chromium are also essential elements?

Isotopes of an Element Are Atoms with Different Numbers of Neutrons

The number of protons in one atom of any given element is constant and distinguishes that element from all others. However, the number of neutrons in the atoms of an element is variable (Fig. 2-3 ●). Atoms of an element that have different numbers of neutrons are known as **isotopes** of the element [*iso-*, same + *topos*, place (that is, the same place in the periodic table)]. The symbol for an isotope shows the atomic mass number written to the upper left of the chemical letter symbol.

For example, ^1H, hydrogen-1, is the most common naturally occurring isotope of hydrogen. It has a single proton in its nucleus (Fig. 2-4 ●). Hydrogen also has two other isotopes, hydrogen-2, or *deuterium* (^2H), and hydrogen-3, or *tritium* (^3H), which are formed by the addition of one or two neutrons to the nucleus

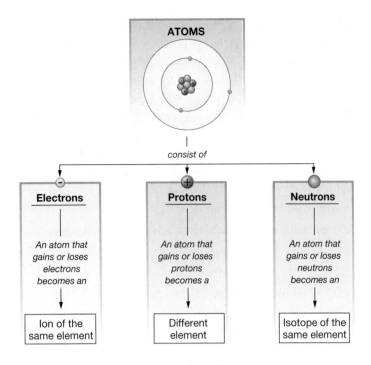

ATOMS

consist of

Electrons	Protons	Neutrons
An atom that gains or loses electrons becomes an	An atom that gains or loses protons becomes a	An atom that gains or loses neutrons becomes an
Ion of the same element	Different element	Isotope of the same element

● **Figure 2-3** *A map showing the relationship between elements, ions, isotopes, and atoms*

A typical hydrogen atom has one proton in the nucleus and one orbiting electron, but no neutrons.

(a) Hydrogen-1, H

A neutron has been added to the nucleus, creating the isotope ^2H.

(b) Hydrogen-2, deuterium, ^2H

Two neutrons have been added to the nucleus, creating ^3H.

(c) Hydrogen-3, tritium, ^3H

● **Figure 2-4** *The three isotopes of hydrogen*

(Fig. 2-4b,c). Adding one neutron to hydrogen's one proton results in an atomic mass of 2 (^2H); adding two neutrons yields an atomic mass of 3 (^3H). All isotopes of a single element have the same chemical properties because these properties are affected by the electron configuration and not by the number of neutrons.

Isotopes are important in biology and medicine, because some are unstable and emit energy. These unstable isotopes are called **radioisotopes**, and the energy they emit is called **radiation**. Every element has at least one radioisotope, most of which are produced artificially by using high-energy particle accelerators to "shoot" extra neutrons into a stable nucleus. A small number of elements, such as carbon and uranium, have naturally occurring radioisotopes.

Both artificial and natural radioisotopes have medical uses. For example, radioisotopes such as thallium-201 (^{201}Tl) produce images of the inside of the body for *diagnostic procedures* (those that identify disease). Radioisotopes are also used to treat cancer, an example of *therapeutic use*. The use of radioisotopes in the diagnosis and treatment of disease is a specialty known as *nuclear medicine*.

BIOTECHNOLOGY

Radioisotopes and Nuclear Medicine

Of the 2000 known isotopes of the elements, unstable isotopes that emit radiation outnumber the stable isotopes four to one. Some of these isotopes occur naturally, but many have been made artificially, including all those with atomic numbers of 84 or higher. Radioisotopes emit three types of radiation: alpha (α), beta (β), and gamma (γ). Alpha and beta radiation are composed of fast-moving particles (protons + neutrons, or electrons) ejected from an unstable atom. Gamma radiation is composed of high-energy waves rather than particles. Gamma radiation is able to penetrate matter much more deeply than α or β particles. X-rays are similar to gamma radiation and are used to create images of the body's interior.

Radioisotopes are widely used in medicine for diagnosis and treatment. In some instances, the radioisotope is used by itself. For example, ^{131}I, a radioactive isotope of ^{127}I, is useful for diagnosis and treatment of thyroid gland disorders. ^{131}I emits both α and β radiation. In small doses it does little damage, but larger doses can be administered to selectively destroy thyroid tissue when the gland is overactive. Another widely used radioisotope for diagnostic procedures is technetium-99, an element that can be bound to carrier molecules that target it to specific tissues.

CONCEPT CHECK   Answers: p. 48

10. What are the atomic masses (in amu) of the three hydrogen isotopes shown in Figure 2-4?

Electrons Form Bonds Between Atoms and Capture Energy

Electrons, the third subatomic particle, play a significant role in physiology. The arrangement of electrons around the nucleus determines the ability of an atom to bind with other atoms and form molecules. The gain or loss of electrons changes the electrical charge on an atom and plays an essential role in communication throughout the body. Finally, electrons on certain atoms can capture energy from their environment and transfer it to other molecules, so that the energy can be used for movement and other life processes.

The electrons in an atom do not move around the nucleus at random. Instead, they are arranged in a series of **energy levels**, or **shells**. The lowest energy level, designated as level 1, is closest to the nucleus. As distance from the nucleus increases, energy levels increase. Each energy level is divided into sublevels called, for historical reasons, *s*, *p*, *d*, and *f*. Each of these sublevels may be further subdivided into a certain number of **orbitals**, each of which holds a maximum of two electrons (Fig. 2-5 ●).

For example, the first energy level or shell, 1, has one sublevel, 1*s*, with one orbital. Thus, the first energy level can hold a total of two electrons. The second shell, 2, has two sublevels (2*s* and 2*p*) with a total of four orbitals, so it can hold a total of eight electrons. If an energy sublevel has more than one orbital, each orbital is assigned one electron before any orbital gets two (Fig. 2-5b). The nitrogen atom is a good example of this principle: its highest energy level, 2*p*, has one electron in each of its three orbitals. The most stable electron configuration for an element is one in which all orbitals in the outer shell are filled with electrons. Neon, a completely unreactive element, is an example of an element whose outer electron shell is totally filled (Fig. 2-5b).

Electrons are able to absorb energy from their environment and "jump" up one or more energy levels to a higher orbital. **High-energy electrons** are not stable and they eventually fall back to their stable orbital position, emitting energy as they do so. The released energy can be captured by other atoms or may be emitted as radiation. The bioluminescence of fireflies is visible light emitted by high-energy electrons returning to their normal low-energy state. When biologists speak of "high-energy electrons" that capture and transfer energy in the cell, they usually mean the electron in a hydrogen atom that jumps out of its 1*s* orbital into an unoccupied higher orbital.

When an orbital is filled with two electrons, the electrons are said to be paired. Paired electrons are stable. An atom having an unpaired electron has a higher probability of losing the unpaired electron to another atom or gaining an unpaired electron from another atom. In some cases, two atoms with unpaired electrons share the electrons. When electrons are transferred or shared in that manner, each participating atom ends up with a stable orbital of paired electrons. The transfer and sharing of electrons brings us to our next topic: the bonds formed between atoms.

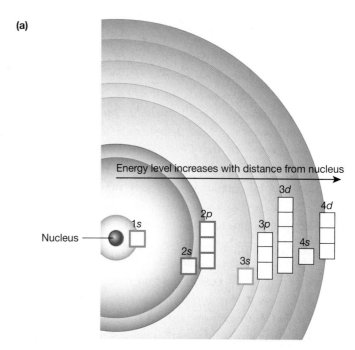

(a)

The first eight energy levels of electrons in an atom. The lowest energy level, 1*s*, is closest to the nucleus. Within an energy level, the electrons move in orbitals, represented by boxes.

(b)

Element	Total electrons	Orbital diagram

	Element	Total electrons	1*s*	2*s*	2*p*	3*s*
H	Hydrogen	1	↿			
Li	Lithium	3	⇅	↿		
Be	Beryllium	4	⇅	⇅		
B	Boron	5	⇅	⇅	↿	
C	Carbon	6	⇅	⇅	↿ ↿	
N	Nitrogen	7	⇅	⇅	↿ ↿ ↿	
O	Oxygen	8	⇅	⇅	⇅ ↿ ↿	
Ne	Neon	10	⇅	⇅	⇅ ⇅ ⇅	
Na	Sodium	11	⇅	⇅	⇅ ⇅ ⇅	↿

Each arrow represents one electron and each box represents an orbital. Within an energy level (e.g., 2*p*), each orbital gets one electron before any orbital gets two. In the most stable configuration, all occupied orbitals have two electrons.

● **Figure 2-5 *Energy levels of electrons in an atom***
(a) Energy levels of electrons in an atom. (b) The electron configurations of some of the lighter elements.

CONCEPT CHECK   Answers: p. 48

11. Which is more stable: a paired or an unpaired electron?

12. Which has more energy: an electron in its lowest possible orbit or an electron that has jumped up one or more orbits?

Imaging in Medicine

When X-rays were discovered in 1895 by Wilhelm Roentgen, they created such a stir that they influenced all aspects of society, not just medicine. Marcel Duchamp's famous painting of *Nude Descending a Staircase* may have been inspired by the skeletal interior revealed by these high-energy emissions. X-rays were our only means of visualizing the interior structures of the body until the 1950s, when ultrasound was introduced. Ultrasound is a technique similar to sonar in the ocean, in which an image is created from sound waves that are reflected as they bounce off tissues and fluid. In the 1970s, computerized tomography (CT scans), positron emission tomography (PET scans [*tomos*, a section]), and magnetic resonance imaging (MRI) came into use. CT scans are series of X-rays that create images of very thin slices of the body. CT images are assembled by computer rather than projected onto a photographic plate as X-rays are, so CT can be used to visualize the three-dimensional structure of an organ. PET scans are made using radioisotopes that emit *positrons*: tiny positively charged particles released when a proton converts into a neutron. Because the body does not normally contain positron-emitting chemicals, these radioisotopes must be administered to a subject before a PET scan can be taken. For MRI scans, the subject is placed inside a strong magnetic field that increases the energy level of the hydrogen nucleus. When the nucleus "jumps" back to its lower energy level, it emits radiation that is captured to create an image of the internal structure.

MOLECULES AND BONDS

The transfer or sharing of electrons is a critical part of forming **bonds,** or links between atoms. When two or more atoms link by sharing electrons, they make units known as **molecules.** Some molecules, such as those of oxygen (O_2) and nitrogen (N_2) gas, are made of two atoms of the same element. But most molecules contain more than one element and are known as **compounds.** For example, oxygen gas (O_2) is a molecule but not a compound because it has only oxygen atoms. In contrast, glucose, a molecule composed of carbon, hydrogen, and oxygen, is also a compound.

What Does the Chemical Formula of a Molecule Tell Us?

The **chemical formula** of a molecule is a form of notation that describes the type and number of atoms in the molecule. Each element in the molecule is represented by its chemical symbol fol-

lowed by a numeric subscript showing the number of atoms of the element. Thus water, which has two hydrogen atoms and one oxygen atom, is written as H_2O. Glucose is written $C_6H_{12}O_6$, meaning it has six atoms each of carbon and oxygen, and twelve atoms of hydrogen.

From the chemical formula of a molecule, we can calculate its **molecular weight**. This is the weight of one molecule, expressed in atomic mass units or, more often, in *daltons* (1 amu = 1 Da).* To calculate molecular weight, multiply the atomic mass of each element in the molecule by the number of atoms of the element, then add the results. For example, the molecular weight of glucose, $C_6H_{12}O_6$, would be calculated:

ELEMENT	NUMBER OF ATOMS IN THE MOLECULE	×	ATOMIC MASS OF THE ELEMENT	
Carbon	6		12.0 amu =	72.0
Hydrogen	12		1.0 amu =	12.0
Oxygen	6		16.0 amu =	96.0
			SUM	180.0 amu
			or	180 Da

13. Calculate the molecular weight of water, H_2O.

Although the chemical formula gives the number of atoms in a molecule, it doesn't tell you which atom is linked to which. To understand that aspect of molecular structure, you need to know the different types of bonds that atoms can form with each other. There are four common bond types, two strong and two weak.

1. **Covalent bonds** are strong bonds that result when two atoms share a pair of electrons, one from each atom. Covalent bonds usually require the input of energy to break them apart.
2. **Ionic bonds** result when an atom has such a strong attraction for electrons that it pulls one or more electrons completely away from another atom.
3. **Hydrogen bonds** are weak attractive forces between hydrogen atoms and certain other atoms.
4. **Van der Waals forces** are another type of weak attractive force that acts between all molecules.

We will discuss each type of bond further in the sections below.

Covalent Bonds Are Formed When Adjacent Atoms Share Electrons

Covalent bonds are formed when atoms share one or more pairs of electrons. Each atom contributes one electron to each pair.

---

*In recognition of John Dalton, who proposed the modern theory of the atom around 1803.

Most of the bonds that link atoms together into molecules are co-valent bonds. It is possible to predict how many covalent bonds an atom will form by knowing how many unpaired electrons the atom has in its outer shell. Figure 2-6● shows the electron con-figurations of the three most plentiful elements in the body: hy-drogen, carbon, and oxygen.

One atom of hydrogen, the lightest element, has one proton in its nucleus and one unpaired electron in its outer shell (Fig. 2-6a). Because it has only one electron to share, it always forms a single covalent bond. Chemists represent the single elec-tron by a single dot in the *electron-dot shorthand.*

An atom of carbon, with an atomic number of 6, has six pro-tons in the nucleus and therefore six electrons surrounding the nucleus. Two electrons fill the first electron shell, while the outer-most second shell has an additional four electrons (Fig. 2-6b). There are four unoccupied places in the second shell, which has

(a) Hydrogen has one proton in the nucleus and one electron in the lowest energy level.

(b) Carbon has six electrons, two in the first shell and four in the 8-place second shell.

(c) Oxygen has eight electrons, two in the first shell and six in the second. A shell is most stable when filled with electrons.

● ● ● ● ● ●
FIGURE QUESTION

How many electrons do these elements need to fill their outer shells?

● **Figure 2-6** *Electron configuration of the three most com-mon elements in the body* An electron-dot model shows the number of electrons found in the outer shell of an atom. Each electron is represented by a single dot.

space for eight electrons. An atom is most stable when its outer electron shell is filled. This means that carbon, with four of eight possible outer electrons, binds to other atoms to find the four electrons it needs to complete its outer shell. When carbon is sharing all four of its outermost electrons with other elements, it forms four covalent bonds.

Each atom of oxygen has eight electrons: two electrons com-pletely fill the first shell and six are found in the eight-place sec-ond shell (Fig. 2-6c). Because oxygen needs only two additional electrons to complete its outer shell, it can get them by sharing two of its outer electrons in covalent bonds. A water molecule is a simple example of covalent bonds formed by oxygen. In each molecule of water, two hydrogen atoms share their electrons with two electrons in the outer shell of the oxygen atom.

We can use electron-dot shorthand to show how atoms form covalent bonds in molecules. In the next example, one carbon combines with four hydrogens to form a molecule of methane, an odorless gas that is a principal constituent of natural gas. Each hydrogen fills its two-electron outer shell by sharing its one elec-tron. At the same time, the carbon atom fills its eight-place outer shell with the four shared hydrogen electrons. When the outer electron shells are filled, even if it is with shared electrons, the atoms are more stable than if the shells are unfilled.

Methane CH_4

Another form of chemical shorthand shows the paired elec-trons of a covalent bond as a single line drawn between the chemical symbols for the elements of the molecule. Sometimes adjacent atoms share two pairs of electrons rather than just one. In this case the bond is called a **double bond** and is represented by a double line ($=$). The following example shows two carbon atoms sharing two pairs of electrons:

Ethylene C_2H_4

Because two of the four carbon electrons are shared in the carbon-carbon bond, this leaves only two electrons per carbon to bond with hydrogen. But if you count all the bonds made by the carbons, once again each carbon forms four bonds and each hy-drogen forms one. The arrangement of bonds is different but the number of electrons being shared has not changed. The chemical formula for this molecule, called ethylene, is C_2H_4.

CONCEPT CHECK   Answers: p. 48

14. How many covalent bonds can nitrogen form? (Hint: see Fig. 2-5b)

15. How do you calculate the atomic number of an element?

16. How do atoms form covalent bonds?

17. The number of electrons in the outer shell determines how many covalent bonds an atom can form. Use this fact to decide whether a hydrogen atom can form a double bond. Why or why not?

18. Why does carbon always form four bonds to other molecules?

19. Write the chemical formula for this molecule:

Citric acid

20. What is the molecular weight of the citric acid molecule above?

Functional Groups There are several combinations of atoms, or **functional groups**, that occur repeatedly in biological molecules. The atoms in functional groups tend to move from molecule to molecule as a single unit. For example, hydroxyl groups, —OH, common on many biological molecules, are added and removed as a group rather than as single hydrogen or oxygen atoms. Functional groups usually attach to molecules by single covalent bonds, represented as a single line such as the one on the left of the hydroxyl group above. The most common functional groups are listed in Table 2-1.

Table 2-1: Common Functional Groups

Notice that oxygen, with two electrons to share, sometimes forms a double bond with another atom.

	SHORTHAND	BOND STRUCTURE
Carboxyl (acid)	—COOH	—C(=O)OH
Hydroxyl	—OH	—O—H
Amino	—NH₂	—N(H)(H)
Phosphate	—H₂PO₄	—O—P(=O)(OH)OH

Molecular Shape Molecules can have many shapes. Molecular shape is important because it is closely related to molecular function. Probably the best-known molecular shape is the double-helix of DNA, the molecule that stores genetic information. Figure 2-7● shows how atoms are linked in two biological molecules. Notice the carbons linked in a straight chain in the leucine

Antioxidants and Free Radicals

One of the buzz words for marketing vitamins and other dietary supplements is "Antioxidant!" **Antioxidants** are molecules that prevent damage to our cells by scavenging free radicals. But what are free radicals and why are they harmful? To answer that question, we must turn to chemistry and the structure of atoms.

Our bodies are constantly exposed to radiation from natural sources, such as the sun, and from manufactured sources, such as microwave ovens and television sets. Some types of radiation alter the distribution of electrons in molecules, creating unstable molecules known as **free radicals**. A free radical is a molecule with at least one unpaired electron. Because electrons are most stable when they occur in pairs, free radicals try to "steal" an electron from another molecule. When this happens, the donor molecule is left with an unpaired electron, making it a free radical. The end result of this "electron theft" is a chain reaction of free radical production within the body that can disrupt normal cell function. Antioxidant molecules stop the chain reaction by giving up electrons without themselves becoming free radicals.

One common free radical is the electrically neutral hydroxy free radical, OH, formed when a hydroxyl ion, OH⁻, loses an electron. Another common free radical is superoxide or •O₂⁻. This free radical is constantly manufactured by the body during normal metabolism when a neutral oxygen molecule (O₂) gains an extra electron (•). Free radicals are thought to contribute to aging and the development of certain diseases such as some cancers. Many antioxidants occur naturally in fruits and vegetables. The most common antioxidants promoted in vitamin supplements are vitamins C and E.

molecule, represented in Figure 2-7a. In the glucose molecule (Fig. 2-7b), the atoms join together to form a closed circle, or **ring**.

The three-dimensional structure of molecules is difficult to show on paper. The angles of the bonds between atoms may cause molecules to take on characteristic shapes. Two well-known examples are the double helix of the DNA molecule and the alpha helix of many proteins. These molecules are shown in Figures 2-16 and 2-18 (pp. 39 and 41).

Polar and Nonpolar Molecules Electron pairs in covalent bonds are not always evenly shared between the linked atoms. In some molecules, the nucleus of one atom has a stronger

(a)

$$H-\overset{\overset{\displaystyle H}{|}}{\underset{\underset{\displaystyle H}{|}}{C}}-\overset{\overset{\displaystyle H}{|}}{\underset{\underset{\displaystyle CH_3}{|}}{C}}-\overset{\overset{\displaystyle H}{|}}{\underset{\underset{\displaystyle H}{|}}{C}}-\overset{\overset{\displaystyle H}{|}}{\underset{\underset{\displaystyle NH_2}{|}}{C}}-C\overset{\displaystyle O}{\underset{\displaystyle OH}{<}}$$

Leucine, an amino acid, is a straight chain of carbon atoms with hydrogens and other functional groups attached.

(b) Glucose, $C_6H_{12}O_6$, has a ring of five carbon atoms and one oxygen atom.

is the same as:

This glucose molecule is drawn with all of its carbons and hydrogens shown.

In the shorthand version, the carbons at each corner of the ring are omitted, as are the hydrogens attached to the carbons.

FIGURE QUESTIONS

• Write the chemical formula of leucine.
• What functional groups do leucine and glucose have?

● **Figure 2-7** *Chemical structures and formulas of some biological molecules*

attraction for the shared electrons than the other nucleus does. The electron pair spends more time near the nucleus with the stronger attraction, giving that atom a slightly more negative charge. Meanwhile the atom that is farther from the electrons develops a slightly more positive charge. For both atoms, the change in charge is only a fraction of the charge carried by a single electron or proton. Molecules that develop regions of partial positive and negative charge are called **polar molecules** because they can be said to have positive and negative poles. Certain elements, particularly nitrogen and oxygen, have a strong attraction for electrons and are often found in polar molecules.

A good example of a polar molecule is water (H_2O). When the hydrogen atoms form covalent bonds with oxygen, their electrons are pulled away from the hydrogen nuclei toward the larger and stronger nucleus of the oxygen atom. This leaves the two hydrogen atoms of the molecule with a partial positive charge, and the single oxygen atom with a partial negative charge from the unevenly shared electrons (see Fig. 2-8 ●). These partial charges are represented by the lowercase Greek letter delta: δ^+ indicates a positive pole of the molecule, and δ^- indicates a negative pole. Note that the net charge for the entire water molecule is zero.

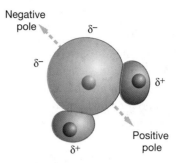

Water molecule

● **Figure 2-8** *Water is a polar molecule* A water molecule develops separate regions, or poles, of partial positive (δ^+) and partial negative (δ^-) charge when the oxygen nucleus pulls the hydrogen electrons toward itself.

A **nonpolar molecule** is one whose shared electrons are distributed so evenly that there are no regions of partial positive or negative charge. For example, molecules composed mostly of carbon and hydrogen tend to form nonpolar bonds. This is because carbon does not attract electrons as strongly as oxygen does.

The polarity of a molecule is important in determining whether it will dissolve in water. Polar molecules generally dissolve easily. Nonpolar molecules do not.

CONCEPT CHECK   Answers: p. 48

21. Based on what you know from experience about the tendency of the following substances to dissolve, predict whether they are polar or nonpolar molecules: table sugar, vegetable oil.

RUNNING PROBLEM

One advertising claim for chromium is that it improves transfer of glucose—the simple sugar that cells use to fuel all their activities—from the bloodstream into cells. In the disease diabetes mellitus, cells are unable to take up glucose from the blood efficiently. It seemed logical, therefore, to test whether addition of chromium to the diet would enhance glucose uptake in people with diabetes. In one Chinese study, diabetic patients receiving 500 micrograms of chromium picolinate twice a day showed significant improvement in their diabetes but patients receiving 100 micrograms or a placebo did not.

Question 2: If people have a chromium deficiency, would you predict that their blood glucose level would be lower or higher than normal? From the results of the Chinese study, can you conclude that all people with diabetes suffer from a chromium deficiency?

Ionic Bonds Form When Atoms Gain or Lose Electrons

When an atom or molecule completely gains or loses one or more electrons, it becomes an **ion**. Generally, the electrons that are gained have been released by another atom or molecule. Table 2-2 lists the most common ions that you will encounter in physiology. Notice that some of the ions are single atoms, and others are functional groups.

A molecule that gains electrons acquires one negative charge (−1) for each electron added. Such negatively charged ions are called **anions**. By convention, ions are indicated by a superscript notation following the symbol of the element, with the number of charges first (if greater than one) and the sign of the charge second. For example, a chloride ion has one extra electron (Cl^-), and a phosphate ion has two (HPO_4^{2-}).

Any molecule that loses electrons has one positive charge (+1) for each electron lost. This is due to the protons that are left behind without matching electrons. Positively charged ions are called **cations**. Hydrogen loses its only electron and ends up as a single proton, H^+. The ion formed by a calcium atom that loses two electrons is abbreviated as Ca^{2+}.

A chemical bond between a cation and an anion is called an **ionic bond**. Figure 2-9● shows how sodium and chlorine atoms become ions and how those ions form ionic bonds. A sodium (Na) atom has only one electron in its eight-place outer electron shell, while a chlorine (Cl) atom has seven of the eight electrons needed to fill its outer shell. The less stable sodium atom gives up its electron to chlorine, resulting in a positively charged sodium

ion (Na^+) and a negatively charged chloride ion (Cl^-). In table salt, the solid form of NaCl, the opposite charges attract, creating a neatly ordered crystal of alternating Na^+ and Cl^- ions held together by ionic bonds.

(a)

Step 1: Sodium gives up its one weakly held electron to chlorin creating sodium and chloride ions, Na+ and Cl⁻.

Sodium atom   Chlorine atom

Step 2: The sodium and chloride ions both have stable outer shells that are filled with electrons.

Sodium ion (Na⁺)   Chloride ion (Cl⁻)

Step 3: The Na+ and Cl⁻ ions are attracted to each other because of their opposite charges.

Sodium chloride (NaCl) molecule

Table 2-2: Important Ions of the Body

CATIONS		ANIONS	
Na^+	Sodium	Cl^-	Chloride
K^+	Potassium	HCO_3^-	Bicarbonate
Ca^{2+}	Calcium	HPO_4^{2-}	Phosphate
H^+	Hydrogen	SO_4^{2-}	Sulfate
Mg^{2+}	Magnesium		

(b)

Sodium ions (Na^+)   Chloride ions (Cl^-)

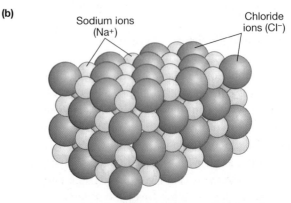

In the solid state, ionic bonds between Na+ and Cl⁻ create a **sodium chloride** (NaCl) crystal.

● **Figure 2-9** *Ions and ionic bonds*

RUNNING PROBLEM

Chromium is found in several ionic forms. The chromium usually found in biological systems and in dietary chromium supplements is the cation Cr^{3+}. This ion is called trivalent because it has a net charge of +3. The hexavalent cation, Cr^{6+}, with a charge of +6, is used in industries such as in the manufacturing of stainless steel and the chrome plating of metal parts.

Question 3: How many electrons have been lost from the hexavalent ion of chromium? From the trivalent ion?

20   23   29   R   33   35   43

Match the descriptions on the right with the correct bond type on the left. Items may be used more than once.

22. ionic bond

23. polar bond

(a) atoms gain or lose electrons

(b) electrons in the bond are shared equally between atoms

(c) electrons in the bond are attracted more strongly to one atom

24. Can a bond be both nonpolar and covalent? Explain.

25. A negatively charged ion is called a/an _____ and a positively charged ion is called a/an _____.

26. Name two elements whose presence contributes to a molecule becoming a polar molecule.

27. A magnesium ion Mg^{2+} has *gained* / *lost* how many (choose one) *protons* / *neutrons* / *electrons*?

28. The H^+ ion is also called a proton. Why is it given that name?

Hydrogen Bonds Are Weak Bonds Between Molecules or Regions of a Molecule

A **hydrogen bond** is the weak attraction of a hydrogen atom to a nearby oxygen, nitrogen, or fluorine atom. Hydrogen bonds may occur between atoms in neighboring molecules or between atoms in different parts of the same molecule. For example, hydrogen bonds cause large biological molecules such as proteins to fold back on themselves, creating a three-dimensional shape that is essential for the function of the protein.

The hydrogen bonds of water are a good example of bonds between adjacent molecules. The positive hydrogen pole of one water molecule is attracted to the negative oxygen poles of as many as four other water molecules. As a result, the molecules line up with their neighbors in a somewhat ordered fashion (Fig. 2-10 ●). The weak interaction between the positive hydrogen regions and the negative oxygen regions is the hydrogen bond.

Hydrogen bonding in water is responsible for the **surface tension** of water, the attractive force between water molecules that causes water to form spherical droplets when falling or to bead up when spilled onto a nonabsorbent surface (Fig. 2-10b). These weak forces of attraction also make it difficult to separate water molecules, as you may have noticed in trying to pick up a wet glass that has become "stuck" to a slick table top by a thin film of water.

Van der Waals Forces Are Weak Attractions Between All Molecules

Van der Waals forces are even weaker bonds than hydrogen bonds. They are nonspecific attractions between the nucleus of any atom and the electrons of nearby atoms. Two atoms that are weakly attracted to each other by van der Waals forces move together until they are so close that their electrons begin to repulse

(a) The polar regions of adjacent water molecules allow them to form hydrogen bonds with one another.

(b) Surface tension created by hydrogen bonds holds a drop of water in a hemispheric shape.

● **Figure 2-10**   *Hydrogen bonds of water*

each other. Consequently, van der Waals forces allow atoms to pack closely together and occupy a minimum amount of space. A single van der Waals attraction between atoms is very weak, but multiple van der Waals forces supplement the hydrogen bonds that hold proteins in their three-dimensional shapes.

29. What do dishwashing detergents do that keeps water from beading up on glasses?

▶ SOLUTIONS AND SOLUTES

Substances that dissolve in liquid are collectively known as **solutes**. The liquids into which they dissolve are known as **solvents**. The combination of solutes dissolved in solvent is called a **solution**. In biological solutions, water is the universal solvent. The human body is about 60% water. This section examines some of the properties of biological solutions.

Not All Molecules Dissolve in Aqueous Solutions

Life as we know it is established on water-based, or *aqueous*, solutions that resemble dilute seawater in their ionic composition. Na^+, K^+, and Cl^- are the main ions; other ions make up a lesser proportion. All molecules and cell components are either dissolved or suspended in these saline solutions. It is useful, therefore, to classify molecules on the basis of their ability to dissolve in water.

The degree to which a molecule is able to dissolve in a solvent is its **solubility**: the more easily it dissolves, the higher its solubility. Molecules that dissolve readily in water are **hydrophilic** [*hydro-*, water + *-philic*, loving]. Most are polar or ionic molecules whose positive and negative regions can form hydrogen bonds with water molecules (Fig. 2-11 ●). Molecules that do not dissolve readily in water are **hydrophobic** [*-phobic*, hating]. They tend to be nonpolar molecules that are unable to make hydrogen bonds with water molecules.

The lipids (fats and oils) are the most hydrophobic group of biological molecules. They cannot interact with water molecules when placed in an aqueous solution, so they tend to form separate layers, like oil floating on vinegar in a bottle of salad dressing. For hydrophobic molecules such as lipids to dissolve in body fluids, they must be combined with a hydrophilic molecule that can effectively carry them into solution. For example, cholesterol, a common animal fat, is a hydrophobic molecule. Fat from a piece of meat dropped into a glass of warm water will float to the top, undissolved. Similarly, cholesterol in the blood will not dissolve unless it is combined with special water-soluble carrier molecules.

Sodium chloride in solution

● **Figure 2-11** *Sodium chloride dissolves in water*
Interaction with polar water molecules breaks ionic bonds that held the NaCl crystal together.

In addition to its role as medium for dissolved substances, water has other properties that affect the way our bodies work. The surface tension of water is important in lung function, as you will see in Chapter 17. Water also has the ability to capture and store large amounts of heat. The heat-retaining property of water is significant in heat loss from the body through sweating. Temperature control in the body is discussed in Chapter 22.

30. Why does table salt (NaCl) dissolve in water?

There Are Several Ways to Express the Concentration of a Solution

The **concentration** of a solution is the amount of solute per unit volume of solution. There are many different ways to express concentration in physiology. The amount of solute can be described as the weight of the solute before it dissolves or as the number of molecules or ions of solute in the solution. Weight is usually given in grams (g) or milligrams (mg). Numbers of solute molecules are given in moles (mol), and numbers of solute ions are given in equivalents (eq). Volume is usually expressed as liters (L) or milliliters (mL).

31. What are the two components of a solution?
32. Do polar or nonpolar molecules dissolve more easily in water?
33. A molecule that dissolves easily is said to be hydro_____ic.
34. The concentration of a solution is expressed as:
    (a) amount of solvent/volume of solute
    (b) amount of solute/amount of solvent
    (c) amount of solute/volume of solute
    (d) amount of solute/volume of solvent

Moles and Molarity   A mole [*moles*, a mass] is defined as 6.02×10^{23} atoms, ions, or molecules of a substance. Thus, one mole of a substance has the same number of particles as one mole of any other substance, just as a dozen eggs has the same number of items as a dozen roses. However, the weight of a mole depends on the weight of its particles.

The weight of one mole of a substance is equal to the atomic mass or molecular weight of the substance, expressed in grams. This value is called the **gram molecular weight**. For example, the molecular weight of glucose is 180 daltons. Therefore, one mole of glucose weighs 180 grams. One mole of NaCl, with the same number of molecules, weighs only 58.5 grams.

Solution concentrations in biology are frequently given as **molarity**, the number of moles of solute in a liter of solution. Molarity is abbreviated as mol/L or, simply, M. A one molar solution of glucose (1 mol/L, 1 M) contains 6.02×10^{23} molecules of

The hexavalent form of chromium used in industry is known to be toxic to humans. In 1992, officials at California's Hazard Evaluation System and Information Service warned that inhaling chromium dust, mist, or fumes placed chrome and stainless steel workers at increased risk for lung cancer. Officials found no risk to the public from normal contact with chrome surfaces or stainless steel. The first link between the biological trivalent form of chromium (Cr^{3+}) and cancer came from *in vitro* studies [*vitrum*, glass] in which isolated cells were kept alive in tissue culture. In these experiments, cells exposed to fairly high levels of chromium picolinate developed "potentially cancerous changes."

Question 4: From this information, can you conclude that hexavalent and trivalent chromium are equally toxic?

glucose per liter. It is made by dissolving 180 grams (one mole) of glucose in enough water to make one liter of solution.

Typical biological solutions are so dilute that solute concentrations are usually expressed as **millimoles** per liter (mmol/L or mM). A millimole is 1/1000th of a mole [*milli-*, 1/1000].

Answers: p. 49

35. How much does a mole of potassium chloride weigh?

Equivalents Concentrations of ions are sometimes expressed in equivalents per liter instead of moles per liter. An **equivalent** (eq) is equal to the molarity of an ion times the number of charges the ion carries. A sodium ion with a charge of +1 has one equivalent per mole. A calcium ion, Ca^{2+}, has two equivalents per mole. The same rule is true for negatively charged ions. One mole of chloride ions (Cl^-) has one equivalent per mole, one mole of phosphate ions (HPO_4^{2-}) has two equivalents. A **milliequivalent** (meq) is 1/1000th of an equivalent. Milliequivalents are useful for expressing the concentration of dilute biological solutions. Many clinical measurements, such as the concentration of sodium ions in the blood, are reported in milliequivalents per liter (meq/L).

Weight/Volume and Percent Solutions In the laboratory or pharmacy, scientists cannot measure out solutes by the mole. Instead, they use the more conventional measurement of weight. The solute concentration of these solutions may then be expressed as a percentage of the total solution, or **percent solution**. A 10% solution means that there are 10 parts of solute per 100 parts *of total solution*.

If the solute is normally a solid at room temperature, the percent solution is expressed as *weight/volume*. A 5% glucose solution would be made by weighing out 5 grams of glucose and adding

water to make a final volume of 100 mL. If the solute is a liquid, such as concentrated hydrochloric acid, the solution is made using *volume/volume* measurements. To make 0.1% HCl, add 0.1 mL of concentrated acid to enough water to give a final volume of 100 mL.

Another convention is common in medicine. The concentrations of drugs and other chemicals in the body are often expressed as milligrams per deciliter (mg/dL). A **deciliter** (dL) is 1/10 of a liter, or 100 mL. A second way of expressing this, no longer commonly used, is mg%, where % means per 100 parts or 100 mL. A solution of 20 mg/dL could also be expressed as 20 mg%.

Answers: p. 49

36. Which solution is more concentrated: a 100 mM solution of glucose or a 0.1 M solution of glucose?

37. In making a 5% solution of glucose, why wouldn't you weigh out 5 grams of glucose and add it to 100 mL of water?

The Concentration of Hydrogen Ions in the Body Is Expressed in pH Units

One of the important ionic solutes in the body is the hydrogen ion, H^+. The concentration of H^+ in body fluids determines the body's *acidity*, a physiological parameter that is closely regulated because H^+ in solution can interfere with hydrogen bonding and van der Waals forces. These bonds are responsible for the shapes of many important molecules, and when the bonds are disrupted, the shape of a molecule can change, thereby destroying the molecule's ability to function.

Where do hydrogen ions in body fluids come from? Some of them come from the separation of water (H_2O) into H^+ and OH^- ions. Others come from ionized molecules that release H^+ ions when they dissolve in water. A molecule that contributes H^+ to a solution is called an **acid**. For example, carbonic acid is an acid made in the body from CO_2 (carbon dioxide) and water. In solution, carbonic acid separates into a bicarbonate ion and a hydrogen ion:

$$CO_2 + H_2O \rightleftharpoons H_2CO_3 \text{ (carbonic acid)} \rightleftharpoons H^+ + HCO_3^-$$

Note that when the H^+ is part of the intact carbonic acid molecule, it does not contribute to acidity. *It is only free H^+ that affects H^+ concentration.*

Some molecules decrease the H^+ concentration of a solution by combining with free H^+. These molecules are called **bases**. Molecules that produce hydroxide ions, OH^-, in solution are bases because the hydroxide combines with H^+ to form water:

$$H^+ + OH^- \rightleftharpoons H_2O$$

Another molecule that acts as a base is ammonia, NH_3. It reacts with free H^+ to form an ammonium ion:

$$NH_3 + H^+ \rightleftharpoons NH_4^+$$

The concentration of H^+ in body fluids is measured in terms of **pH**. The expression pH stands for "power of hydrogen" and is calculated from the negative log of the hydrogen ion concentration: $pH = -\log[H^+]$ [● Appendix A for logarithms]. The square brackets around the symbol for a solute are the shorthand notation for "concentration": $[H^+]$ = hydrogen ion concentration.

Another way to write the pH equation is $pH = \log(1/[H^+])$. This equation states that pH is inversely related to the H^+ concentration. In other words, as the H^+ concentration goes up, the pH goes down. The pH of a solution is measured on a numeric scale between 0 and 14.

Pure water has a pH value of 7.0, meaning its H^+ concentration is 1×10^{-7} M. Water is considered neutral. Solutions that have gained H^+ from an acid have higher H^+ concentrations. Because $[H^+]$ is inversely proportional to pH, **acidic** solutions have a pH less than 7. Solutions that have lost H^+ to a base have a lower H^+ concentration than pure water; they are called **alkaline**. They have a pH value greater than 7. Remember, because pH is proportional to the reciprocal of the $[H^+]$, as the hydrogen ion concentration *increases*, the pH number *decreases*.

The pH scale is logarithmic, meaning that a change in pH value of 1 unit indicates a 10-fold increase or decrease in the hydrogen ion concentration. For example, if the pH of a solution is altered from a pH of 8 to a pH of 6, there has been a 100-fold (10^2 or 10×10) increase in the H^+ concentration.

Figure 2-12● shows a pH scale and the pH values of various solutions. The normal pH of blood in the human body is 7.40, slightly alkaline. Regulation of the body's pH within a narrow range is critical because a blood pH more acidic than 7.00 (pH < 7.00) or more alkaline than 7.70 (pH > 7.70) is incompatible with life.

Buffers are a key factor in the body's ability to maintain normal pH. A **buffer** is any molecule that moderates changes in pH. Many buffers are anions that have a strong affinity for H^+ molecules. When free H^+ is added to a solution containing buffers, the buffers bond to the H^+, thereby diminishing any change in pH.

The bicarbonate ion, HCO_3^-, is an important buffer in the human body. Sodium bicarbonate (baking soda) is a buffer used to adjust the pH of swimming pools and spas. The following equation shows how sodium bicarbonate buffers a solution to which hydrochloric acid (HCl) has been added. When placed in solution, HCl dissociates into H^+ and Cl^-, creating a high H^+ concentration and low pH. If sodium bicarbonate is present, some of the bicarbonate ions combine with some of the H^+ to form undissociated carbonic acid. The removal of free H^+ lowers the H^+ concentration and raises the pH:

$$H^+ + Cl^- + HCO_3^- + Na^+ \rightleftharpoons H_2CO_3 + Cl^- + Na^+$$

Hydrochloric acid	Sodium bicarbonate	Carbonic acid	Sodium chloride (table salt)

CONCEPT CHECK   Answers: p. 49

38. To be classified as an acid, a molecule must do what when dissolved in water?

39. pH is an expression of the concentration of what in a solution?

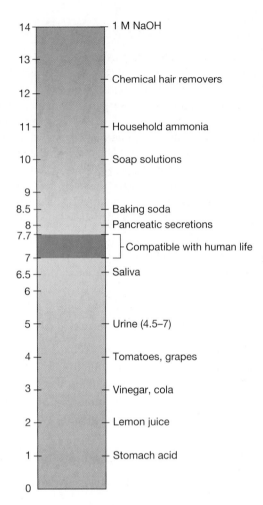

● **Figure 2-12** *pH scale* This logarithmic scale is a measure of the hydrogen ion concentration. For example, at a pH of 14, the hydrogen ion concentration is 10^{-14} moles/liter.

40. When the body becomes more acidic, does the pH increase or decrease?

41. How can urine, stomach acid, and saliva have pH values outside the pH range that is compatible with life and yet be part of the living body?

▶ BIOMOLECULES

Most of the molecules of concern in human physiology have three elements in common: carbon (C), hydrogen (H), and oxygen (O). Molecules that contain carbon are known as **organic molecules**, because it was once thought that they existed in or were derived from only plants and animals. Organic molecules associated with living organisms are also called **biomolecules**.

There are four kinds of biomolecules: carbohydrates, lipids, proteins, and nucleotides. The first three groups are used by the body for energy and as the building blocks for cellular components. The fourth group, the nucleotides, includes DNA and RNA, the structural components of genetic material. Compounds that carry energy, such as adenosine triphosphate (ATP), or regulate

RUNNING PROBLEM

Stan has been taking chromium picolinate because he heard that it would increase his strength and muscle mass. Then a friend told him that the Federal Trade Commission (FTC) said that there was no evidence to show that chromium would help build muscle. In one study, a group of researchers gave high daily doses of chromium picolinate to football players during a two-month training period. By the end of the study the players who took chromium supplements had not increased muscle mass or strength any more than players who did not take the supplement.

Question 5: Based on the FTC announcement and this study, which did not support enhanced muscle development from chromium supplements, and on the studies that suggest that chromium picolinate might cause cancer, do you think that Stan should continue taking chromium picolinate?

metabolism, such as cyclic adenosine monophosphate (cAMP), are also nucleotides. Each group of biomolecules has a characteristic composition and molecular structure, which we describe briefly in the following sections.

Carbohydrates Are the Most Abundant Type of Biomolecule

Carbohydrates get their name from their structure: literally, carbons with water. The basic formula for a carbohydrate is $(CH_2O)_n$, showing that for each carbon there are two hydrogens and one oxygen, the same H : O ratio found in water. The *n* after the parentheses represents the number of repetitions of the CH_2O unit: $C_nH_{2n}O_n$. For example, glucose, $C_6H_{12}O_6$, has an *n* of 6. Glucose is an example of a simple sugar, the smallest type of carbohydrate.

The names of all the simple sugars end with the suffix *-ose*. There are two kinds of simple sugars: the **monosaccharides** [*mono-*, one + *sakcharon*, sugar] and the **disaccharides** [*di-*, two]. The monosaccharides are the building blocks of complex carbohydrates and have either five carbons, like **ribose**, or six carbons, like **glucose (dextrose)**, **fructose**, and **galactose**.

When two monosaccharides bond to each other, they form a disaccharide molecule. Typical disaccharides are **maltose**, **lactose** (milk sugar), and **sucrose** (table sugar). Figure 2-13 ● shows the composition of most of these simple sugars.

When many glucose molecules join together, they form very large molecules. These complex carbohydrate molecules are known as **polysaccharides** [*poly-*, many]. A large molecule made up of repeating units is called a **polymer**. Thus, the complex carbohydrates are all glucose polymers (Fig. 2-13). Because these polymers have only one type of molecule, glucose, the differences in the polysaccharides come from the manner in which the glucose molecules are linked.

All living cells store glucose for energy in the form of a polysaccharide, and some cells produce polysaccharides for structural purposes as well. For example, yeasts and bacteria make a glucose storage polymer called *dextran*. Animal cells make a storage polysaccharide called **glycogen** that is found in all tissues of the body. Many invertebrate animals make a structural polysaccharide called *chitin*. Plants make two types of polysaccharides: the storage molecule **starch**, which humans can digest, and the structural molecule **cellulose**, which humans cannot digest. It is unfortunate that we are unable to digest cellulose and obtain its energy, because it is the most abundant organic molecule on earth.

CONCEPT CHECK   Answers: p. 49

42. True or false? All organic molecules are biomolecules.

43. What is the basic formula of a carbohydrate?

Lipids Are the Most Diverse Biomolecules

Lipids are biomolecules made up of carbon, hydrogen, and oxygen, like the carbohydrates, but as a rule they contain much less oxygen. An important characteristic of this group is that they are not very soluble in water, because of their nonpolar structure. Technically, lipids are called fats if they are solid at room temperature and oils if they are liquid at room temperature. Most lipids derived from animal sources, like lard and butter, are fats, whereas most plant lipids are oils.

Lipids are the most diverse group of biomolecules. In addition to the true lipids, this category includes three lipid-related substances: phospholipids, steroids, and eicosanoids (Fig. 2-14 ●).

The true lipids and phospholipids are most similar to each other in structure. Both groups contain a simple 3-carbon molecule known as **glycerol** and long carbon-chain molecules known as **fatty acids**. Fatty acids are long chains of carbon atoms bound to hydrogens, with a carboxyl (—COOH) group at one end of the chain.

Fatty acids are described as **saturated** if there are no double bonds between carbons; **monounsaturated** if there is one double bond in the molecule; or **polyunsaturated** if there are two or more double bonds in the molecule (Fig. 2-14). You have probably heard these terms in commercials for margarine and cooking oil: manufacturers like to boast that their products are "High in polyunsaturates!" Research studies are trying to relate the degree of fat saturation in our diets to the development of heart disease. Saturated fats, such as the fat that marbles a piece of choice beef, are generally considered to contribute to the development of artery-clogging fatty deposits. For a time, it was thought that the more unsaturated fats were the least likely to be harmful, but recent studies suggest that monounsaturated fats such as olive oil may be the most healthful.

Fatty acids in the body link to glycerol to form mono-, di-, or triglycerides. **Triglycerides** (more correctly known as **triacylglycerol**) are the most important form of lipid in the body; more

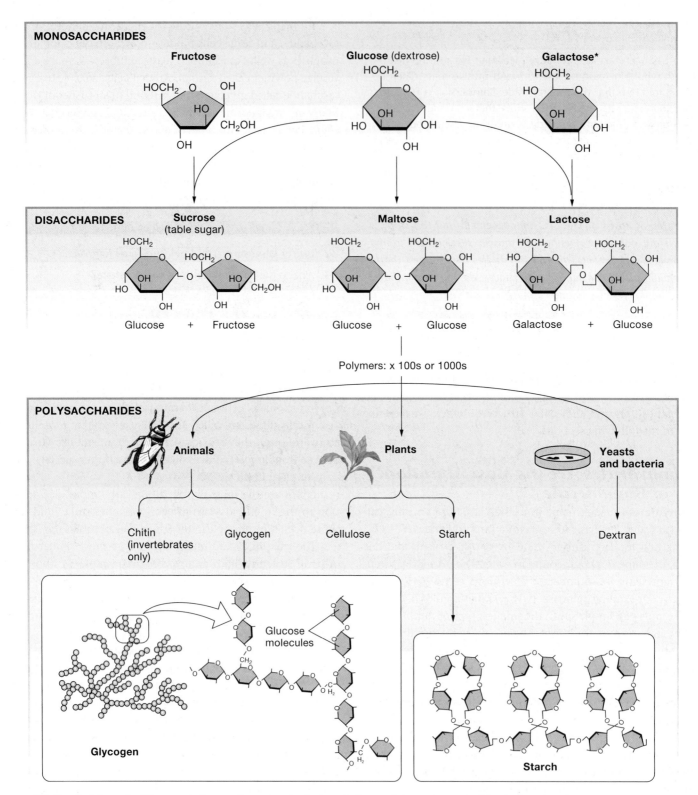

* Notice that the only difference between glucose and galactose is the spatial arrangement of the hydroxyl groups.

● **Figure 2-13** *Carbohydrates*

than 90% of our lipids are in this form. Triglyceride levels in the blood are also important as a predictor of artery disease; an elevated fasting triglyceride level is linked to a higher risk of disease.

The lipid-related molecules include phospholipids, steroids, and eicosanoids. Most **phospholipids** are diglycerides with a phosphate group attached to the single carbon that lacks a fatty acid. Phospholipids are important components of cell membranes.

Steroids are lipid-related molecules whose structure includes four linked carbon rings (Fig. 2-14). Cholesterol is the source of

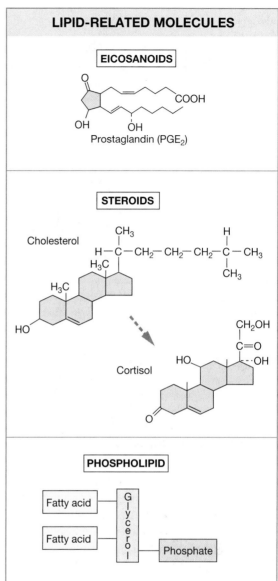

● **Figure 2-14** *Lipids and lipid-related molecules* Palmitic acid is shown with all bonds and atoms. Oleic acid, another fatty acid, is shown without symbols for the carbons and hydrogens in the middle of the molecule. Palmitic acid is a saturated fatty acid with no double bonds between carbons. Oleic acid is a monounsaturated fatty acid with one double bond.

steroids in the human body. It is also an important component of animal cell membranes.

The **eicosanoids** are modified 20-carbon fatty acids that are found in animals [*eikosi*, twenty]. These molecules all contain one open five- or six-carbon ring that makes the long carbon chain appear to double back on itself (Fig. 2-14). The main eicosanoids are thromboxanes, leukotrienes, and prostaglandins. As you will learn in later chapters, eicosanoids are regulators of various physiological functions.

Proteins Are the Most Versatile of the Biomolecules

Proteins are large molecules made from smaller building-block molecules called **amino acids** (Fig. 2-15). Twenty different amino acids commonly occur in natural proteins, and these building blocks may be assembled in an almost infinite number of combinations. All amino acids have a similar core structure: a central carbon atom is linked to a hydrogen atom, a nitrogen-containing amino group ($-NH_2$), a carboxyl group ($-COOH$), and a group of atoms designated "R" that is different in each of the amino acids (Fig. 2-15a ●). The R groups differ in their size, shape, and ability to form hydrogen bonds or ions. Because of the different R groups, each of the different amino acids reacts with other molecules in unique ways.

The human body can synthesize all but nine of the 20 protein-forming amino acids. Those nine, which must be obtained from dietary proteins, are called the **essential amino acids** (Table 2-3). Because of nitrogen in the amino group, proteins are the major dietary source of nitrogen.

(a) Model amino acid

All amino acids have a carboxyl group (-COOH), an amino group ($-NH_2$) and a hydrogen attached to the same carbon. The fourth bond of the carbon attaches to a variable "R" group.

(b) Glycine

(c) Isoleucine

(d) Methionine

Note the sulfur atom **-S-** in the "R" group.

(e) Tyrosine

● **Figure 2-15** *Amino acid structure*

Table 2-3: The Amino Acids of the Human Body

Each amino acid has both a three-letter abbreviation of its name and a single-letter abbreviation, used to spell out the sequence of amino acids in large proteins. The essential amino acids are marked with a star. These nine amino acids cannot be made by the body and must be obtained from the diet.

AMINO ACID	THREE-LETTER ABBREVIATION	ONE-LETTER SYMBOL
Alanine	Ala	A
Arginine	Arg	R
Asparagine	Asn	N
Asparagine or aspartic acid	Asx	B
Aspartic acid	Asp	D
Cysteine	Cys	C
Glutamic acid	Glu	E
Glutamine	Gln	Q
Glutamine or glutamic acid	Glx	Z
Glycine	Gly	G
*Histidine	His	H
*Isoleucine	Ile	I
*Leucine	Leu	L
*Lysine	Lys	K
*Methionine	Met	M
*Phenylalanine	Phe	F
Proline	Pro	P
Serine	Ser	S
*Threonine	Thr	T
*Tryptophan	Trp	W
Tyrosine	Tyr	Y
*Valine	Val	V

* Essential amino acids

There are also some amino acids that do not occur in proteins but that have important physiological functions. They include *homocysteine*, a sulfur-containing amino acid that occurs normally in the body but which in excess is associated with heart disease; γ-*amino butyric acid* (gamma-amino butyric acid or *GABA*), a chemical made by nerve cells; and *creatine*, a molecule that stores energy when it binds to a phosphate group.

When two amino acids link together, the amino group of one is joined to the carboxyl group of the other, forming a **peptide bond**. By linking many molecules in this way, long chains of up to hundreds of amino acids can be formed. If there are two to nine amino acids in the amino acid chain, the molecule is called a **peptide**. A chain of 10 to 100 amino acids is called a **polypeptide**. A chain of more than 100 amino acids is called a **protein**.

The sequence of amino acids in the chain is called the **primary structure** of the protein (Fig. 2-16●). For example, the primary structure of thyrotropin-releasing hormone, a peptide made in the brain, is glutamic acid-histidine-proline. The primary structure of a protein is genetically determined and is essential to proper function. For example, a substitution of only one amino acid causes sickle cell disease, a condition in which red blood cells cannot function normally.

As the amino acid chain forms, it takes on its **secondary structure**, which is the spatial arrangement of the amino acids in the chain. Some amino acid chains curl into a spiral, or **α-helix**, whereas others create a **β-pleated sheet**. The secondary structure is determined by the angle of the bond between adjacent amino acids and is stabilized by hydrogen bonding between different parts of the molecule. Proteins that are destined for structural uses may be composed almost entirely of sheets, because this configuration is very stable. Other proteins combine sheets with helices.

The three-dimensional shape of an amino acid chain is its **tertiary structure**. Some long protein chains fold up into a ball or other compact shape when the R groups of amino acids in the α-helix are attracted to each other. The attraction may be between adjacent amino acids or between amino acids that are at different ends of the molecule. In some proteins, the tertiary structure is

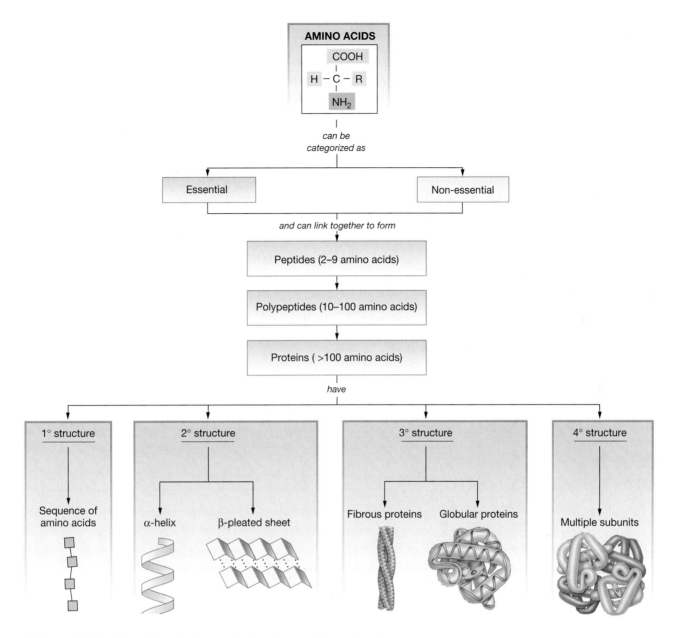

AMINO ACIDS

COOH
|
H — C — R
|
NH₂

can be categorized as

Essential   Non-essential

and can link together to form

Peptides (2–9 amino acids)

Polypeptides (10–100 amino acids)

Proteins (>100 amino acids)

have

1° structure — Sequence of amino acids

2° structure — α-helix / β-pleated sheet

3° structure — Fibrous proteins / Globular proteins

4° structure — Multiple subunits

● **Figure 2-16**   *Map of levels of organization in protein molecules*

held in place by hydrogen bonds. In other proteins, sulfur atoms in two different *cysteine* molecules bond covalently to each other in a *disulfide* (S—S) *bond*.

Proteins are often classified as either globular or fibrous proteins based on their shape. **Globular proteins** have amino acid chains folded into ball-like shapes. **Fibrous proteins** are found as pleated sheets or in long chains that wind around each other.

A protein with more than one polypeptide chain is said to have **quaternary structure** (Fig. 2-16). Hemoglobin, the oxygen-carrying pigment found in red blood cells, has a quaternary structure composed of four subunits: two polypeptide alpha chains of 141 amino acids each and two polypeptide beta chains of 146 amino acids each. The hemoglobin molecule is illustrated in Figure 2-16 under the label "Multiple subunits."

Proteins, peptides, and amino acids serve a number of important functions in the body. The fibrous proteins, which are insoluble in water, are important structural components of cells and tissues. Examples include *collagen*, a fibrous protein found in many types of connective tissue, and *keratin*, a fibrous protein found in hair and nails. The globular proteins are soluble in water. They act as *carriers* for insoluble lipids in the blood, binding to them and making them soluble. They may also serve as *enzymes* that increase the rate of chemical reactions. Soluble proteins also act as cell-to-cell messengers in the form of hormones and neurotransmitters and as defense molecules to help fight foreign invaders.

44. What is the chemical formula of an amino group? Of a carboxyl group?

45. Why must we include essential amino acids in our diet but not the other amino acids?

46. What aspect of protein structure allows proteins to have more versatility than lipids or carbohydrates?

Some Molecules Combine Carbohydrates, Proteins, and Lipids

Not all biomolecules are pure protein, pure carbohydrate, or pure lipid. **Conjugated proteins** are protein molecules combined with another kind of biomolecule. Proteins combine with lipids to form **lipoproteins**. Lipoproteins are found mostly in cell membranes. They also carry hydrophobic molecules such as cholesterol in the blood: the lipid section of the molecule interacts with insoluble cholesterol while the protein section makes the entire lipoprotein-cholesterol complex soluble in water.

Proteins combine with carbohydrates to form **glycoproteins**. A "glyco-" molecule is composed of a carbohydrate bonded to either a protein or a lipid (**glycolipid**). Glycoproteins and glycolipids, like lipoproteins, are important components of cell membranes (Chapter 5). The carbohydrate portions of these molecules are usually found on the external surface of the cell, where they can form hydrogen bonds with

● **Figure 2-17** *Map of nucleotide structure*

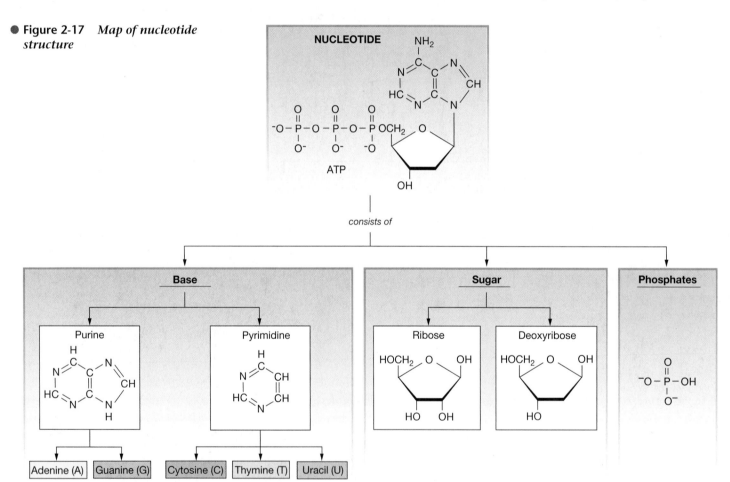

NUCLEOTIDES		Bases	Sugar	Phosphate groups	Other component
	ATP	= Adenine	+ Ribose	+ 3	—
	ADP	= Adenine	+ Ribose	+ 2	—
	cAMP	= Adenine	+ Ribose	+ 1	—
	NAD	= Adenine	+ 2 Ribose	+ 2	+ Nicotinamide
NUCLEIC ACIDS:					
	DNA	= A,G,C,T	+ Deoxyribose	+ 1 per nucleotide	—
	RNA	= A,G,C,U	+ Ribose	+ 1 per nucleotide	—

water and other molecules in body fluids. Glycolipids are believed to contribute to the stability of the cell membrane. Certain glycoproteins and glycolipids act as cell membrane receptors and cell markers. *Membrane receptors* bind to molecules outside the cell. The *cell markers* identify cells, like a label, so that the defense cells of the body can tell the difference between "self" and foreign invaders.

Nucleotides Transmit and Store Energy and Information

Nucleotides play an important role in many of the processes of the cell, including the storage and transmission of genetic information and the transfer of energy. A single **nucleotide** is a three-part molecule composed of (1) one or more phosphate groups, (2) a five-carbon sugar, and (3) a carbon-nitrogen ring structure called a **base** (Fig. 2-17 ●). Every nucleotide contains one of two possible sugars: **ribose** or **deoxyribose** [*de-*, without; *-oxy-*, oxygen]. There are also two types of bases: the purines and the pyrimidines. The **purines** have a double ring structure; the **pyrimidines** have a single ring. There are two different purines: **adenine** and **guanine**. There are three different pyrimidines: **cytosine**, **thymine**, and **uracil**.

Different nucleotides are made up of sugars, phosphate groups, and various combinations of bases. The smallest nucleotides include the energy-transferring compounds **ATP** (adenosine triphosphate) and **ADP** (adenosine diphosphate). These molecules contain the base adenine, the sugar ribose, and two or three phosphate groups (Fig. 2-17). A related nucleotide is **cyclic AMP** (cyclic adenosine monophosphate, or cAMP), a molecule important in the transfer of signals between the extracellular fluid and the cell. Other energy-transferring nucleotides include **NAD** (nicotinamide adenine dinucleotide) and **FAD** (flavin adenine dinucleotide). Molecules that take part in energy transfer and storage are discussed in more detail in Chapter 4.

DNA and RNA are nucleotide polymers, or **nucleic acids**. They store genetic information within the cell and transmit it to future generations of cells. One complete DNA molecule with its associated proteins forms a *chromosome.*

DNA stands for *deoxyribonucleic acid*; **RNA** is short for *ribonucleic acid*. In each case, the name describes the type of sugar found in the nucleotides of the polymer. DNA contains the bases adenine, guanine, cytosine, and thymine. RNA also contains adenine, guanine, and cytosine but has uracil instead of thymine. Both polymers are formed by linking nucleotides into long chains, similar to the long amino acid chains of proteins. The sugar of one nucleotide links to the phosphate of the next, creating a chain or backbone of alternating sugar-phosphate-sugar-phosphate groups. The bases extend freely to the side (Fig. 2-18 ●).

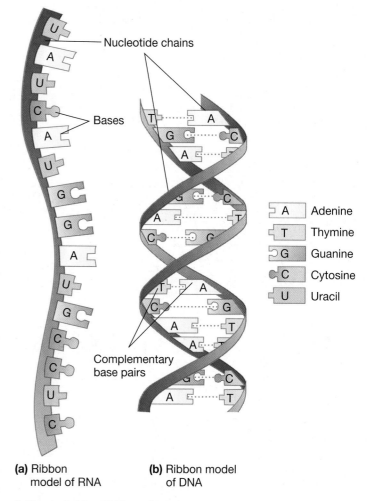

A	Adenine
T	Thymine
G	Guanine
C	Cytosine
U	Uracil

(a) Ribbon model of RNA   **(b)** Ribbon model of DNA

● **Figure 2-18** *RNA and DNA*

The DNA molecule is composed of two linked nucleotide chains that form a *double helix*, a three-dimensional structure that was first described by J. D. Watson and F. H. C. Crick, in 1953. The double helix is formed when two DNA chains link through complementary base pairing. Because of space limitations, the larger purines (adenine or guanine) must always link to a smaller pyrimidine (cytosine or thymine). Adenine (A) always pairs with thymine (T), and guanine (G) always pairs with cytosine (C). These base pairs (A—T and G—C) hold the double strand of DNA in its spiral shape. Base pairing and DNA replication are reviewed in Appendix B.

RNA usually takes the form of a single chain, although it is temporarily bound to DNA while it is being formed. RNA plays an important role in protein synthesis, discussed in Chapter 4.

This chapter looked at how atoms combine to form molecules, how molecules dissolve to form solutions, and how different molecular building blocks combine into biologically important compounds (see Figure 2-19 ● for a summary). The next chapter will consider the role of biomolecules in the structure of cells and tissues.

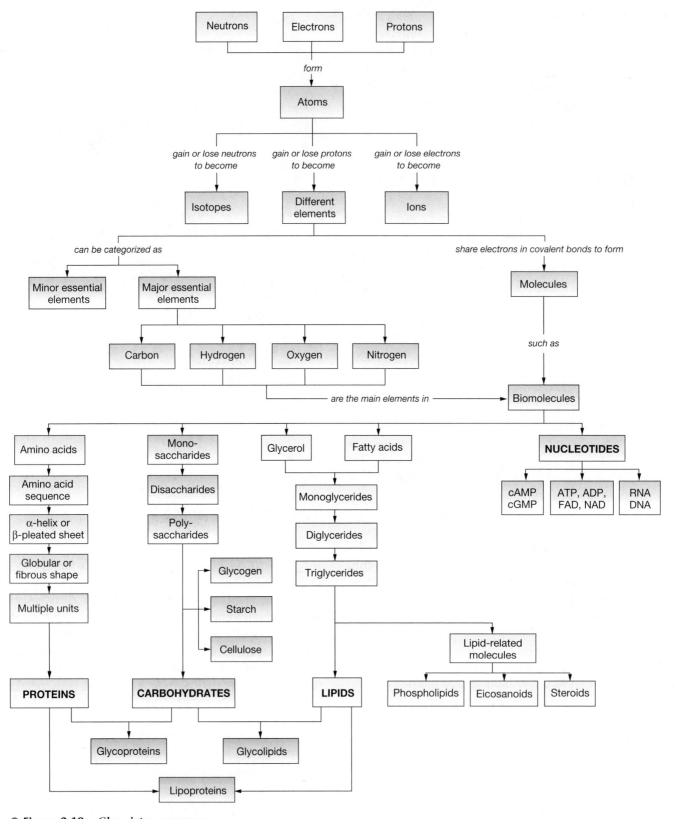

● **Figure 2-19** *Chemistry summary*

Chromium Supplements

In this running problem, you learned that the claims made for chromium picolinate have not been supported with evidence from controlled scientific experiments. You also learned that hexavalent chromium is toxic, and that evidence suggests that some forms of the biological trivalent form of chromium may be toxic.

Further check your understanding of this running problem by comparing your answers to those in the summary table.

	QUESTION	FACTS	INTEGRATION & ANALYSIS
1a	Locate chromium on the periodic table of the elements.	The periodic table organizes the elements according to their atomic number.	N/A*
1b	What is chromium's atomic number? Atomic mass?	Reading from the table, chromium (Cr) has an atomic number of 24 and an average atomic mass of 52.	N/A
1c	How many electrons does one atom of chromium have?	The atomic number of an element is equal to the number of protons in one atom. An atom has equal numbers of protons and electrons.	The atomic number of chromium is 24, therefore one atom of chromium has 24 protons and 24 electrons.
1d	Which elements close to chromium are also essential elements?	Reading from the table, the essential elements close to chromium include molybdenum, manganese, and iron.	N/A
2a	If people have a chromium deficiency, would you predict that their blood glucose level would be lower or higher than normal?	Chromium helps move glucose from the blood into cells.	If chromium is absent/lacking, less glucose will leave the blood and blood glucose would be elevated.
2b	From the results of the Chinese study, can you conclude that all people with diabetes suffer from chromium deficiency?	Higher doses of chromium supplements lowered elevated blood glucose levels but lower doses had no effect. This is only one study and no information is given about similar studies elsewhere.	We have insufficient evidence from the information presented to draw a conclusion about the role of chromium deficiency in diabetes.
3	How many electrons have been lost from the hexavalent ion of chromium? From the trivalent ion?	For each electron lost from an ion, a positively charged proton is left behind in the nucleus of the ion.	The hexavalent ion of chromium, Cr^{6+}, has six unmatched protons and has lost six electrons. The trivalent ion, Cr^{3+}, has lost three electrons.
4	From this information, can you conclude that hexavalent and trivalent chromium are equally toxic?	The hexavalent form of chromium is used in industry and when inhaled, has been linked to an increased risk of lung cancer. Enough studies have shown an association that California's Hazard Evaluation System and Information Service has issued warnings to chromium workers. Evidence to date for the toxicity of trivalent chromium in chromium picolinate comes from studies done on isolated cells in tissue culture.	Although the toxicity of Cr^{6+} is well established, the toxicity of Cr^{3+} has not been conclusively determined. Studies performed on cells *in vitro* may not be applicable to humans. Additional studies need to be performed in which animals are given reasonable doses of chromium picolinate for an extended time period.

* Not applicable

(continues on next page)

	QUESTION	FACTS	INTEGRATION & ANALYSIS
5	Based on the FTC announcement and this study, which did not support enhanced muscle development from chromium supplements, and on the studies that suggest that chromium picolinate might cause cancer, do you think that Stan should continue taking chromium picolinate?	No research evidence supports a role for chromium picolinate in increasing muscle mass or strength. Other research suggests that chromium picolinate may cause cancerous changes in isolated cells.	The evidence presented suggests that for Stan, there is no benefit from taking chromium picolinate and there may be risks. Using risk-benefit analysis, the evidence supports stopping the supplements. However, the decision is Stan's personal responsibility. He should remain informed of new developments that would change the risk-benefit analysis.

CHAPTER SUMMARY

This chapter introduces biomolecules and ions, whose chemical properties underlie many of the key themes in physiology. To understand molecular interactions, communication, energy storage and transfer, and the mechanical properties of cells and tissues, you need a good background in basic chemistry.

Elements and Atoms

1. **Atoms** are composed of protons, neutrons, and electrons. A **proton** has a mass of about 1 **atomic mass unit** (amu), or **dalton** (Da), and an electrical charge of +1. A **neutron** has a mass of about 1 amu and no electrical charge. An **electron** has negligible mass, compared with protons and neutrons, and a charge of −1. (p. 21; Fig. 2-1)

2. **Major essential elements** are necessary for life and are present in large quantities in the body. **Trace elements** are necessary elements found in tiny amounts. (p. 23)

3. **Elements** are distinguished from each other by their **atomic number**, which is the number of protons in the nucleus of one atom. (p. 23)

4. The **mass** of an atom is calculated by adding the number of protons and neutrons in its nucleus. (p. 23)

5. Atoms of one element that have different numbers of neutrons in their nuclei are known as **isotopes** of the element. Radioactive isotopes emit energy called **radiation**. (pp. 23–24; Fig. 2-4)

6. The electrons around an atom are arranged in **energy levels** centered on the nucleus. The energy levels are subdivided into **orbitals**. An atom is most stable when all its outermost orbitals are occupied by pairs of electrons. (p. 25)

7. Electrons form **bonds** between atoms and capture energy. (p. 26)

Molecules and Bonds

8. Two or more atoms linked together create a **molecule**. Molecules that include atoms of more than one element are called **compounds**. (p. 26)

9. The **molecular weight** of a molecule is determined by adding the atomic masses of all its atoms. (p. 26)

10. **Covalent bonds** are strong bonds formed when adjacent atoms share one or more pairs of electrons. Each pair of shared electrons forms a single covalent bond. A **double bond** forms when two atoms are sharing two pairs of electrons. (pp. 26–27)

11. Molecules whose atoms share electrons unevenly are called **polar molecules**. If electrons are shared evenly by the atoms in a molecule, the molecule is a **nonpolar molecule**. (p. 29)

12. An atom or molecule that gains or loses electrons acquires an electrical charge and is called an **ion**. (p. 30; Fig. 2-9)

13. Atoms form **cations** when they lose electrons and **anions** when they gain electrons. **Ionic bonds** are strong bonds formed when oppositely charged ions are attracted to each other. (p. 30)

14. **Hydrogen bonds** form when hydrogens in polar molecules are attracted to oxygen, nitrogen, or fluorine atoms. Hydrogen bonds may form within a molecule or between adjacent molecules. Hydrogen bonding among water molecules is responsible for the **surface tension** of water. (p. 31; Fig. 2-10)

15. **Van der Waals forces** are weak bonds that form when molecules are attracted to each other. (p. 31)

Solutions and Solutes

 IP Fluids and Electrolytes: Acid-Base Homeostasis

16. Molecules that dissolve in liquid are called **solutes**. The liquid in which they dissolve is called the **solvent**. The universal solvent for biological solutions is water. (p. 31)

17. The ease with which a molecule dissolves is called its **solubility**. **Hydrophilic molecules** dissolve easily in water, whereas **hydrophobic molecules** do not. (p. 32)

18. A **mole** is 6.02×10^{23} particles of an element or compound. The weight of a mole in grams is known as the **gram molecular weight** of the element or compound. (p. 32)

19. The **concentration** of a solution is the amount of solute per unit volume of solution. A one **molar** (1 M) solution has one mole of solute per liter of solution. The concentration of most biological solutions is expressed in **millimoles per liter** of solution (mM). (pp. 32–33)

20. Ion concentrations may be expressed in milliequivalents, where one **equivalent** equals the molarity of the ion times its electrical charge. (p. 33)

21. The **pH** of a solution is a measure of its hydrogen ion concentration. **Acids** are molecules that contribute H^+ to a solution. **Bases** are molecules that remove H^+ from solution. The more acidic the solution, the lower the pH. (p. 33; Fig. 2-12)

22. **Buffers** are molecules that moderate pH changes. (p. 34)

Biomolecules

23. There are four major groups of **biomolecules**, which all contain carbon, hydrogen, and oxygen: carbohydrates, lipids, proteins, and nucleotides. (p. 34)

24. **Carbohydrates** have the general formula $(CH_2O)_n$. **Monosaccharides** are the simplest building blocks of complex carbohydrates. Two monosaccharides join to form **disaccharides**. **Polysaccharides** are glucose **polymers** used for the storage of energy and as structural components of cells. (p. 35; Fig. 2-13)

25. **Lipids** contain carbon and hydrogen with little oxygen. Most are nonpolar. **Fatty acids** are long chains of carbon atoms bound to hydrogens. **Triglycerides** are composed of glycerol and three fatty acids. (p. 35; Fig. 2-14)

26. **Proteins** are the most versatile biomolecule because 20 different **amino acids** serve as their building blocks. (p. 37; Fig. 2-15)

27. The **primary structure** of a protein is the sequence of its amino acids. The **secondary structure** is its spatial arrangement into helixes or sheets. The **tertiary structure** is the three-dimensional shape of the folded amino acid chain, and the **quaternary structure** is the arrangement of multiple amino acid chains in a single protein. (pp. 38–39; Fig. 2-16)

28. **Globular proteins** are soluble. They act as carriers, messengers, defense molecules, and enzymes. **Fibrous proteins** are not very soluble. They serve as structural components of cells and tissues. (p. 39)

29. Proteins, lipids, and carbohydrates combine to form **glycoproteins**, **glycolipids**, or **lipoproteins**. (p. 40)

30. Nucleotides are important in the transmission and storage of information and the transfer of energy. (p. 41)

31. A **nucleotide** is composed of phosphate groups, a sugar, and either a purine or a pyrimidine base. Small nucleotides include ATP, ADP, and cyclic AMP. Nucleotide polymers are the **nucleic acids** DNA and RNA. (p. 41; Fig. 2-17)

QUESTIONS

LEVEL ONE *Reviewing Facts and Terms*

1. List three major essential elements found in the human body.

2. When atoms of elements are bound to one another into compounds, such as H_2O or CO_2, one unit is called a _____.

3. (a) The subatomic particle with a mass of about 1 amu and a charge of +1 is called the_____. (b) The subatomic particle with a mass of about 1 amu and a neutral charge is the_____. (c) The remaining subatomic particle is the electron, with a negligible mass but a charge of _____.

4. Which subatomic particle in an element determines its atomic number?

5. Name the element associated with each of these symbols: Ca, C, O, Na, N, K, H, and P.

6. Isotopes of an element have the same number of _____ and _____, but differ in their number of _____.

7. Unstable isotopes emit energy called _____. Using this energy for the diagnosis or treatment of disease is a specialty called _____.

8. The smallest subdivision of an electron shell is called a(n) _____, and each holds _____ electrons.

9. When electrons in an atom or molecule are stable, are they paired or unpaired?

10. When an atom of an element gains or loses one or more electrons, it is called a/an _____ of that element.

11. Match each type of bond with its description:
    (a) covalent bond
    (b) ionic bond
    (c) hydrogen bond
    (d) van der Waals forces

    1. weak attractive forces between hydrogen and oxygen or nitrogen
    2. shares a pair of electrons
    3. weak attractive forces between molecules
    4. atoms gain or lose electrons

12. In mixing compounds with water, the polarity of the compounds helps predict their solubility. When a compound's electrons are shared evenly, the compound is said to be _____; if shared unevenly, the compound is _____. Which group dissolves more readily in water?

13. The pH of a solution is an abbreviation conveying what information about that solution? A solution with a pH less than 7 is _____; a solution with a pH greater than 7 is _____.

14. A _____ is a molecule that moderates changes in the pH of a solution by binding to free H^+ or OH^- ions.

15. List the four kinds of biomolecules. Give an example of each kind that is relevant to physiology.

16. Match each carbohydrate with its description:
    (a) starch
    (b) chitin
    (c) glucose
    (d) cellulose
    (e) glycogen
    (f) lactose

    1. most abundant carbohydrate on earth
    2. disaccharide, found in milk
    3. storage form of glucose for animals
    4. storage form of glucose for plants
    5. structural polysaccharide of invertebrates
    6. monosaccharide, found in blood

17. Proteins may be combined with other types of molecules. They are called _____ when combined with fatty components. They are called _____ when combined with carbohydrates.

18. Match each lipid with its description:
    (a) triglycerides
    (b) eicosanoids
    (c) steroids
    (d) oils
    (e) phospholipids

    1. most common form of lipid in the body
    2. liquid at room temperature, usually from plants
    3. important component of cell membrane
    4. structure composed of carbon rings
    5. modified 20-carbon fatty acids

19. Match these terms pertaining to proteins and amino acids:
    (a) the building blocks
    (b) must be included in our diet
    (c) proteins that speed the rate of chemical reactions
    (d) sequence of amino acids in a protein
    (e) protein chains folded into a ball-shaped structure

    1. essential amino acids
    2. primary structure
    3. amino acids
    4. globular proteins
    5. enzymes
    6. tertiary structure
    7. fibrous proteins

20. List the components of a nucleotide.

LEVEL TWO *Reviewing Concepts*

21. **Mapping exercises:** Make each list of terms into a separate map.

    List 1: The atom
    anion
    atomic mass
    atomic number
    cation
    covalent bond
    electron
    electron shell
    element
    ion
    isotope
    neutron
    nucleus
    orbital
    proton

    List 2: Solutions
    concentration
    equivalent
    hydrogen bond
    hydrophilic
    hydrophobic
    molarity
    mole
    nonpolar molecule
    polar molecule
    solubility
    solute
    solvent
    water

22. Use the periodic table of the elements in Figure 2-2 to answer the following questions: An atom of sodium has 11 protons in its nucleus. (a) How many electrons does the atom have? (b) What is the electrical charge of the atom? (c) How many neutrons does the atom have? (d) If this atom loses one electron, it would be called a(n) _____? (e) What would be the electrical charge of the substance formed in (d)? (f) Write the correct chemical symbol for the ion referred to in (d). (g) What does the sodium atom become if it loses a proton from its nucleus? (h) Write the correct chemical symbol for the atom referred to in (g).

23. A solution with $[H^+]$ of 10^{-3} M is _____ (acidic/basic?), whereas a solution with an $[H^+]$ of 10^{-10} M is _____ (acidic/basic?). Give the pH for each of these solutions.

24. Name three nucleotides, and tell why each one is important.

25. Explain the progression of protein structure from primary to quaternary.

26. Compare the structure of DNA and RNA.

27. Distinguish between purines and pyrimidines.

LEVEL THREE *Problem Solving*

28. You have been summoned to assist with the autopsy of an alien being whose remains have been brought to your lab. The chemical analysis returns with 33% C, 40% O, 4% H, 14% N, and 9% P. From this information you conclude that the cells do contain nucleotides, possibly even DNA or RNA. Your assistant is demanding that you tell him how you knew this. What do you tell him?

29. A cell carrying out work normally produces CO_2. The harder a cell works, the more CO_2 it produces. CO_2 is carried in the blood according to the following equation: $CO_2 + H_2O \rightleftharpoons H_2CO_3 \rightleftharpoons H^+ + HCO_3^-$. What effect does hard work by your muscle cells have on the pH of the blood?

30. When the radioactive isotope of iodine, ^{131}I, emits beta radiation, it converts a neutron to a proton and electron. What is the atomic mass, atomic number, and name of the resulting atom?

LEVEL FOUR *Quantitative Problems*

31. Write the chemical formulas for each molecule depicted. Calculate the molecular weight of each molecule. Predict whether each molecule is polar or nonpolar.

(a)

(b) $O=C=O$

(c)

(d) $H_2N-\overset{\displaystyle COOH}{\underset{\displaystyle CH_3}{C}}-H$

(e)

32. Calculate the amount of NaCl you would weigh out to make one liter of 0.9% NaCl. Explain how you would make a liter of this solution.

33. A 1.0 M NaCl solution has 58.5 g of salt per liter. (a) How many molecules of NaCl are present in this solution? (b) How many millimoles of NaCl are present? (c) How many equivalents of Na^+ are present? (d) Convert this measurement to a percent solution.

34. How would you make 200 mL of a 5% glucose solution? Calculate the molarity of this solution. How many millimoles of glucose are present in 500 mL of this solution? (Hint: what is the molecular weight of glucose?)

Explorations

Explorations are also available at *www.physiologyplace.com*

Exploration 1: Chromium Picolinate

A number of well-designed controlled studies about the influence of chromium picolinate on muscle mass and body fat content have now been published. The subjects in these studies ranged from young athletes to older men and women. In none of the studies did chromium picolinate increase muscle mass or decrease body fat more than the exercise regimens followed by both the control and chromium supplementation groups. There is still intense interest in chromium, however, because of its possible role in reducing blood glucose levels in diabetic patients. A meta-analysis of human studies on that subject concluded that chromium had no effect on blood glucose in nondiabetic subjects and that the evidence is still inconclusive for an effect in diabetics. Research is also continuing on the potential harmful effects of chromium picolinate supplements.

You can use your understanding of chromium picolinate to analyze the validity of claims made about this and other nutritional supplements. To get some appreciation of the type of information offered to the public in general, go to *www.google.com*, a search engine developed by Stanford University. Type in the words *chromium picolinate*. Look at the first five search results and decide if the public is being given the full story on this nutritional supplement.

Here are some published papers on chromium and other dietary supplements:

- "Chromium as a supplement," *Annual Review of Nutrition* 19: 279–302, 1999.

- "Glucose and insulin responses to dietary chromium supplements: A meta-analysis," *American Journal of Clinical Nutrition* 72: 585–593, 2002.
- "The role of chromium in nutrition and therapeutics and as a potential toxin," *Nutrition Reviews* 57(11): 329–335, 1999.

Exploration 2: Web Searches

How do you search for information published on the World Wide Web? Should you believe everything you read? For an award-winning set of interactive tutorials on web searching, go to *http://tilt.lib.utsystem.edu/* (notice that there is no www in this address). These tutorials will help you learn how to search and evaluate websites.

One authoritative site for information on selected dietary supplements is the National Institues of Health nutritional supplement website (*http://www.cc.nih.gov/ccc/supplements*). Check out *http://www.quackwatch.com/03HealthPromotion/antioxidants.html* to learn more about antioxidants.

Exploration 3: Scandal and the Periodic Table

Science can seem pretty cut-and-dry sometimes. Look at the periodic table of the elements in Figure 2-2. Pretty boring, huh? But behind some of those squares are fascinating stories of science at work. For example, in recent years nuclear scientists have been working to discover how heavy an element could be. In 1999 a group at the Lawrence Berkeley National Laboratory in California announced the discovery of the world's heaviest element, atomic number 118. But in 2002 that same laboratory retracted its findings, saying that one of its scientists had faked the results! Subsequently, former colleagues of the same scientist revealed that they now believe data from the discovery of elements 110 and 112 may have been falsified. By the time you read this, those elements may also have been removed from the periodic table of the elements pending duplication of the research.

To see the current version of the periodic table or to print out your own copy, go to Web Elements at *www.webelements.com*. To read more about the controvesy surrounding element 118, go to Google (*www.google.com*) and type "*element 118*" in the SEARCH box (the quotes will create a search for the phrase).

The Web Elements site also contains detailed information about elements. Find chromium in the periodic table and click on its box to see what this website says about the element. Or choose another trace element, such as selenium (Se), and find out whether it, like chromium, occurs in both toxic and nontoxic compounds.

Exploration 4: Imaging

Life magazine devoted its February 1997 issue to "A fantastic voyage through the human body: A 21st century look at how doctors will see us—and heal us." The articles range from historical drawings of the human body as people understood it in past centuries to modern images made using CT, PET, and MRI techniques. *Life* stopped publication in 1999, so you will have to go to a library to read this issue.

ANSWERS

Answers to Concept Check Questions

Page 21

1. Nine other elements with symbols derived from their Latin names are copper (Cu), iron (Fe), potassium (K), silver (Ag), tin (Sn), antimony (Sb), gold (Au), mercury (Hg), and lead (Pb).

Page 23

2. (c)

3. (a), (b)

4. (a), (b), (d)

5. Zinc has 30 protons.

6. Sodium has 11 electrons.

7. One atom of calcium has 20 electrons.

8. The atomic mass of nitrogen is 14.0 amu.

9. The element illustrated is lithium (atomic number 3).

Page 24

10. The atomic masses are: (a) 1 amu, (b) 2 amu, (c) 3 amu.

Page 25

11. Paired electrons are more stable.

12. Electrons that jump to higher orbits have more energy.

Page 26

13. Molecular weight of water is 18 amu.

Pages 27–28

14. Nitrogen can form three covalent bonds.

15. Atomic number is the number of protons in the nucleus of one atom.

16. Atoms share electrons to form covalent bonds.

17. Hydrogen cannot form a double bond because it only has one electron to share.

18. One carbon atom has four of eight possible electrons in its outer shell, and it obtains the other four electrons needed to fill the shell through covalent bonds.

19. $C_6H_8O_7$

20. The molecular weight of the citric acid molecule is 204.

Page 29

21. Table sugar dissolves easily so it would be polar; vegetable oil does not dissolve so it would be nonpolar.

Page 31

22. (a)

23. (c)

24. Yes, a bond can be nonpolar and covalent. A covalent bond is any bond in which electrons are shared between two atoms. The

terms *nonpolar* and *polar* describe whether the electrons are equally shared or whether one atom attracts electrons more strongly than the other.

25. Negatively charged ions are anions; positively charged ions are cations.

26. Oxygen and nitrogen strongly attract electrons and tend to form polar bonds.

27. Mg^{2+} has lost two negatively charged electrons.

28. The hydrogen ion is called a proton because loss of its one electron leaves behind one proton.

Page 31

29. Dishwashing detergents are surfactants that disrupt the hydrogen bonds of water. This disrupts the surface tension of the water and prevents beading.

Page 32

30. Na^+ and Cl^- ions form hydrogen bonds with the polar water molecules. This disrupts the ionic bonds that held the NaCl crystal together.

Page 32

31. Solvent and solute.

32. Polar molecules dissolve more easily.

33. Hydrophilic.

34. (d)

Page 33

35. A mole of KCl weighs 74.6 grams.

Page 33

36. 1 M = 1000 mM, therefore 0.1 M = 100 mM.

37. The 5 g of glucose will add volume to the solution and you will no longer have 100 mL of solution.

Page 34

38. It must dissociate into one or more H^+ ions plus cations.

39. pH is the concentration of H^+ ions.

40. As H^+ concentration increases, pH decreases. Therefore, increasing acidity means decreasing pH.

41. Urine, stomach acid, and saliva are all contained in the lumens of hollow organs, where they are not part of the body's internal environment (see Fig. 1-2 on p. 3).

Page 35

42. False.

43. $C_nH_{2n}O_n$

Page 40

44. Amino is NH_2. Carboxyl is COOH.

45. Our bodies cannot make the essential amino acids. We must therefore obtain them through our diet. We can synthesize the other amino acids from nutrients.

46. Proteins are composed of 20 different amino acids that can be linked in an almost infinite number of sequences. These combinations allow proteins to take on different forms, such as the globular and fibrous proteins.

Answers to Figure Questions

Page 27

Fig. 2-6: Hydrogen needs one electron, carbon four electrons, and oxygen two electrons to fill their outer shells.

Page 29

Fig. 2-7: (a) $C_6H_{13}NO_2$ (b) Leucine has carboxyl and amino groups. Glucose has hydroxyl groups.

Answers to Level Four Quantitative Problems

Page 47

31. (a) $C_6H_{12}O_6$; m.w. 180; polar (b) CO_2; m.w. 44; polar (c) H_2O; m.w. 18; polar (d) $C_3H_7O_2N$; m.w. 89; polar (e) $C_5H_{10}O_5$; m.w. 150; polar

32. 0.9% solution = 0.9 g/100 mL. Weigh out 9 g NaCl, place into a 1 liter flask, and dissolve in enough water to yield 1 L of solution.

33. (a) 1 M NaCl contains 6.02×10^{23} molecules of NaCl. (b) 1 M solution contains 1000 millimoles. (c) This equals one equivalent of Na^+. (d) 58.5 g/liter = 5.85 g/100 mL = 5.85% solution.

34. 5% glucose = 5 g/100 mL or 10 g in 200 mL of solution. Molarity: 5 g/100 mL = 50 g/L × 1 mole/180 g = 0.278 moles/L or 278 millimoles/L (278 mM). 500 mL of 5% glucose would have 25 g glucose × 1 mole/180 g = 139 millimoles glucose in 500 mL.

Cells and Tissues

3

Developing cell in tissue culture showing microtubules (red), actin (green), and DNA (blue)

RUNNING PROBLEM

The Pap Smear

Dr. George Papanicolaou has saved the lives of millions of women by popularizing the Pap test, a screening method that detects the early signs of cervical cancer. In the past 50 years the Pap test has become widely used, and deaths from cervical cancer have dropped nearly 70%. If detected early, cervical cancer is one of the most treatable forms of cancer. Today, Jan Melton, who had an abnormal Pap test several months ago, returns to her family physician for a repeat test. The results may determine whether she should undergo treatment for cervical cancer.

53   63   70   76   79

BACKGROUND BASICS

Units of measure: **inside back cover** ǀ Compartmentation: **101** ǀ Extracellular fluid: **7** ǀ Hydrophobic molecules: **32** ǀ Proteins: **38** ǀ Compound molecules: **40** ǀ pH: **33** ǀ DNA and RNA: **41**

Cells are organisms, and entire animals and plants are aggregates of these organisms.

—Theodor Schwann, 1839

Imagine a world in which people with spinal cord injuries no longer suffer permanent loss of function, people with diabetes mellitus no longer require shots of insulin, or people with failing hearts no longer wait for someone to die so that they can have a heart transplant. These scenarios may seem like wishful thinking or science fiction, but with each passing day, they get closer to becoming reality. Scientists have spent years deconstructing the body into its tiniest components, and now they are hard at work figuring out how to integrate those components into functional cells, tissues, and organs.

A vast collection of molecules makes up the **cell**, the basic functional unit of most living organisms. Individual cells carry out the same life-sustaining functions as multicellular animals. They sense and respond to changes in their environment, and they communicate with neighboring cells by releasing chemicals or by generating electrical signals. Cells take in oxygen and nutrients, they extract energy from the nutrients for growth, repair, and reproduction, and they get rid of wastes such as carbon dioxide.

Much of what we know about cells comes from studies of simple organisms that consist of one cell. But humans are much more complex, with trillions of cells in their bodies. It has been estimated that there are more than 200 different types of cells within the human body, each cell with its own characteristic structure and function. In studying physiology, you must be aware both of how individual cells carry out their functions and of how they cooperate and communicate with each other in the human body.

Cytology [*kytos*, cell + *logia*, explaining] is the study of cells. In recent years, the word *cytology* has been used mostly to refer to the study of cell structure, whereas the more general term *cell biology* has referred mostly to the study of cell function. The development of new research techniques has greatly expanded our understanding of how cells do what they do, and so we are better able to apply that knowledge to the study of tissues, organs, and the body as a whole.

Cooperation between cells is a central theme of physiology. This chapter looks first at the structure and function of individual cells. It then examines how groups of cells with similar functions join together to form the tissues and organs of the body.

▶ STUDYING CELLS AND TISSUES

The first recorded use of magnifying lenses to study the composition of biological material took place in the last half of the 1600s. In the Netherlands, Antoine van Leeuwenhoek discovered that pond water was teeming with "wee beasties" too small to be seen with the naked eye, while the English scientist Robert Hooke around 1665 used a simple light microscope to investigate the structure of cork, the dead tissue layer under the bark of trees. It was Hooke who named the holes in cork "cells," from the Latin word *cella*, meaning a small room. But Hooke's cork cells, like the rooms of a building, were only walls and did not contain living material. Not until the early part of the 1800s, with improved microscopes, did we begin to understand that living cells were filled with a semifluid substance, and that the cell was the fundamental structural unit of life.

The light microscope makes it possible to examine both living cells and preserved cells. But even at the highest magnification and with the aid of dyes that stain different cellular components, it is possible to see only the general shape and size of cells or the larger internal structures such as the nucleus. Because most internal components are too small to be seen with the light microscope, the fine details of cell structure went undescribed until the electron microscope came into widespread use in the 1950s.

Figure 3-1 ● shows a cell as seen under a light microscope, by *transmission electron microscopy* [*transmittere*, to send across], and

(a) Light microscopy   **(b) Transmission electron microscopy**   **(c) Scanning electron microscopy**

Cilia

Nucleus

Cilia

● **Figure 3-1**   *One cell type viewed by three types of microscopes* by *scanning electron microscopy*. The scanning electron microscope has greatly added to the understanding of cells and tissues by making it possible to view their fine structure in a realistic three-dimensional fashion rather than as thin flat slices that cut through internal structures. The next section of this chapter examines the details of cellular structure as seen with the aid of these microscopic techniques.

▶ CELLULAR ANATOMY

Animals begin life as a single cell, the **zygote**, or fertilized ovum (egg). As the zygote divides into two cells, then four cells, then eight, each daughter cell looks identical. But mature cells are not all alike. By a process known as **differentiation**, developing cells take on more than 200 different shapes and functions.

How does differentiation occur? Each cell inherits the same genetic information in its DNA, but no one cell uses all of it. During differentiation, only selected genes are activated, transforming the cell into a specialized unit. In most cases, the final shape and size of a cell and its contents reflect its function. Figure 3-2 ● shows some of the different cells in the human body. Although these mature cells look very different from each other, they all started out alike and they retain many features in common.

The structural organization of a cell can be compared to that of a medieval walled city. The city is separated from the surrounding countryside by a high wall, with entry and exit strictly controlled through gates that can be opened and closed. The city inside the walls is divided into streets and a diverse collection of houses and shops with varied functions. Within the city, a ruler

in the castle oversees the everyday comings and goings of the city's inhabitants. Because the city depends on food and raw material from outside the walls, the ruler negotiates with the farmers in the countryside. Foreign invaders are always a threat, so the city ruler communicates and cooperates with the rulers of neighboring cities.

In the cell, the outer boundary is the cell membrane. Like the city wall, it controls the movement of material between the cell interior and the outside by opening and closing "gates," made of protein. The inside of the cell is divided into compartments rather than into shops and houses. Each of these compartments has a specific purpose that contributes to the function of the cell as a whole. In the cell, DNA in the nucleus is the "ruler in the castle," controlling both the internal workings of the cell and its interaction with other cells. Like the city, the cell depends on supplies from its external environment. It must also communicate and cooperate with other cells in order to keep the body functioning in a coordinated fashion.

Figure 3-3 ● is a map of cell structure. Recall from Chapter 1 [● p. 7] that the cells of the body are surrounded by the dilute salt solution known as **extracellular fluid** (ECF) [*extra-*, outside]. This is the external environment of the cell from which the cell gains nutrients and into which it puts wastes and the chemicals that it uses to communicate with other cells. The cell membrane separates the inside environment of the cell from the extracellular fluid.

Internally the cell is divided into the cytoplasm and the nucleus. The cytoplasm consists of a fluid portion, called the cytosol, and assorted structures, collectively known as organelles.

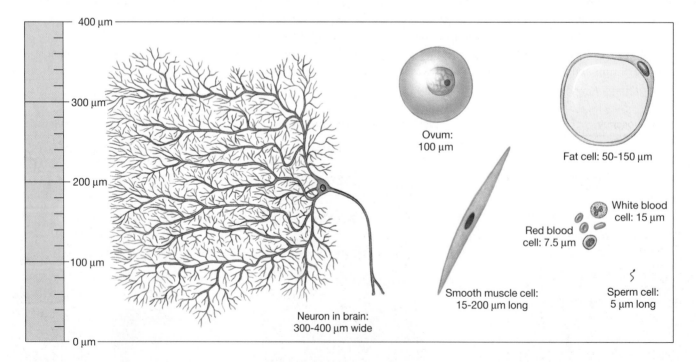

● **Figure 3-2** *Representative cell types in the human body* Each cell has the genetic potential to become any type of cell, but selective activation of genes during differentiation determines the ultimate size, shape, and function of the cell.

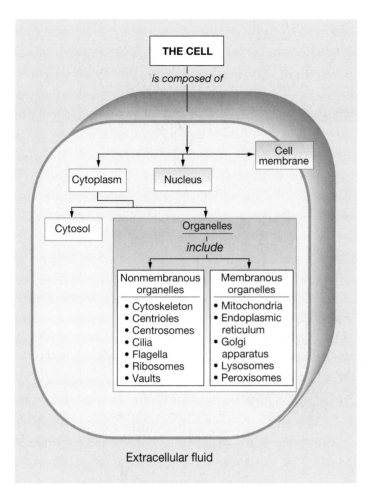

● **Figure 3-3**   *A map for the study of cell structure*

The organelles work in an integrated manner, each organelle taking on one or more of the cell's functions. A typical cell from the lining of the small intestine is shown in Figure 3-4d ●. It has most of the structures found in animal cells.

The Cell Membrane Separates the Cell from Its Environment

The **cell membrane** (also called the **plasma membrane** or *plasmalemma*) separates the inside of the cell from the surrounding extracellular fluid. Structurally, the membrane is a bilayer (double layer) of phospholipid molecules (Fig. 3-5 ●, p. 56). They are arranged so that the hydrophilic phosphate groups face the aqueous solutions inside and outside the cell, leaving the hydrophobic lipid portions hidden in the center of the membrane. The membrane is studded with protein molecules, like raisins in a slice of bread, and the extracellular surface has glycoproteins and glycolipids [● p. 40].

Functionally, the cell membrane serves as both a gateway and a barrier. All substances that enter or leave the cell must cross the cell membrane, which regulates exchange between the cell and the extracellular fluid. Because of the importance of membranes in physiology, we will cover the details of membrane structure and function separately in Chapter 5.

The Cytoplasm Includes Cytosol and Organelles

The **cytoplasm** includes all material inside the cell membrane, except for the nucleus. The cytoplasm has two components:

1. **Cytosol**, or intracellular fluid: The *cytosol* is a semi-gelatinous substance, separated from the extracellular fluid by the cell membrane. The cytosol contains dissolved nutrients and proteins, ions, waste products, and particles of insoluble materials known collectively as **inclusions**. Among the most common inclusions are stored nutrients such as glycogen granules and lipid droplets. Also suspended in the cytosol are the:

2. **Organelles**: These "little organs" are like the organs of the body in that each has a specific role to play in the overall function of the cell. For example, mitochondria generate most of the cell's ATP, and lysosomes filled with enzymes act as the digestive system of the cell. Organelles are often classified as *nonmembranous* or *membranous* according to whether they have a membrane around them.

Nonmembranous Organelles Are in Direct Contact with the Cytosol

Nonmembranous organelles do not have boundary membranes and so are in direct contact with the cytosol. Movement of material between these organelles and the cytosol does not require transport across a membrane. Nonmembranous organelles can be divided into two groups: those made from RNA and protein, and those made from insoluble protein fibers. The latter include the cytoskeleton, centrosomes and centrioles, cilia, and flagella.

Ribosomes Participate in Protein Synthesis   Ribosomes (Fig. 3-6 ●, p. 56) are small dense granules of RNA and protein that manufacture proteins under the direction of the cell's DNA (see Chapter 4). Most ribosomes are attached to the cytosolic side of the endoplasmic reticulum membrane and are called **fixed ribosomes**. Ribosomes also float freely in the cytosol, where some

● **Figure 3-4** *ANATOMY SUMMARY* *Levels of Organization—System to Cell*

SYSTEM

ORGAN

TISSUE

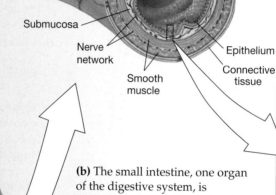

Mesentery

Nerve

Vein

Artery

(a) The function of the digestive system is to take in food and water from the external environment, process it, and bring it into the internal environment of the body.

Submucosa

Nerve network

Smooth muscle

Epithelium

Connective tissue

(c) A cross section of part of the small intestine, showing the different tissues that compose it. Nutrients and water are absorbed across the epithelium and enter either the blood vessels or the lymph vessel.

Lymph vessel

Lumen

Epithelium

Capillary network

Nerve

(b) The small intestine, one organ of the digestive system, is responsible for mechanical and chemical digestion of food so that it can be absorbed. The intestine secretes enzymes for chemical digestion. The products of digestion are absorbed across the wall of the intestine into the blood.

C E L L

(d) The microvilli on the side of the cell that faces the lumen of the intestine increase the surface area available for absorption. This epithelial cell also has most of the organelles of a typical cell.

N A V I G A T O R

Look for this navigator throughout the chapter to help you locate the placement of the various organelles.

(e)

Sub-cellular structure (organelles)	Characteristics	Function
Microvilli	Extensive folding of the cell membrane	Increase surface area for absorption
Cell membrane	Bilayer of phospholipid molecules with inserted proteins	Acts as a gateway for and barrier to movement of material between the interior of the cell and the extracellular fluid
Cytosol	Semi-gelatinous substance	Contains dissolved nutrients, ions, wastes, insoluble inclusions; suspends the organelles
Centrioles	Bundles of microtubules	Direct movement of DNA during cell division
Lysosomes and peroxisomes	Membrane-bound vesicles filled with enzymes	Digest bacteria, old organelles; metabolize fatty acids
Mitochondrion	Double wall with central matrix	Produces most of cell's ATP
Golgi apparatus	Hollow membranous sacs	Modify and package proteins
Nucleus	Central lumen with a two-membrane outer envelope with pores	Contains DNA to direct all functions of the cell
Nucleolus	Region of DNA, RNA, and protein	Contains the genes that direct synthesis of ribosomal RNA
Rough endoplasmic reticulum	Membrane tubules that are continuous with the outer nuclear membrane; edged with ribosomes	Site of protein synthesis
Smooth endoplasmic reticulum	Same as rough ER without ribosomes	Synthesis of fatty acids, steroids, and lipids
Ribosomes	Granules of RNA and protein	Assemble amino acids into proteins
Microtubules and microfilaments	Protein fibers	Provide strength and support; enable cell motility; transport

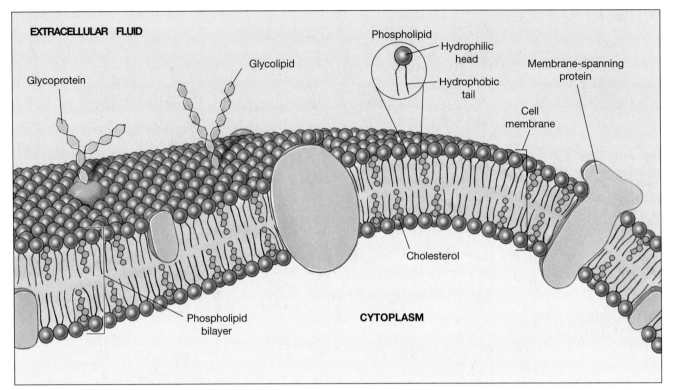

EXTRACELLULAR FLUID

Glycoprotein

Glycolipid

Phospholipid

Hydrophilic head

Hydrophobic tail

Membrane-spanning protein

Cell membrane

Cholesterol

Phospholipid bilayer

CYTOPLASM

● **Figure 3-5** *The cell membrane* The cell membrane is a double layer of phospholipid molecules. Proteins of various kinds are inserted into and through the phospholipid bilayer. Carbohydrates bind to proteins and lipids on the extracellular surface, creating glycoproteins and glycolipids.

● ● ● ● ● ● ●
N A V I G A T O R

The navigator shows the location of this organelle in the large cell shown in Figure 3-4d.

● **Figure 3-6** *Ribosomes are non-membranous organelles composed of RNA and protein*

Free ribosomes in cytosol

Fixed ribosomes are attached to the membranes of

Endoplasmic reticulum

Small subunit

Large subunit

Ribosome

Table 3-1: Diameter of Protein Fibers in the Cytoplasm

	DIAMETER	TYPE OF PROTEIN	FUNCTIONS
Microfilaments	5–9 nm	Actin (globular)	Cytoskeleton; associates with myosin for muscle contraction
Intermediate filaments	10 nm	Myosin, neurofilament protein, keratin (filaments)	Cytoskeleton, hair and nails, protective barrier of skin. Myosin associates into thick filaments for muscle contraction.
Microtubules	25 nm	Tubulin (globular)	Movement of cilia, flagella, and chromosomes. Intracellular transport of organelles. Cytoskeleton.

form groups of 10 to 20 known as **polyribosomes**. Ribosomes are most numerous in cells that synthesize proteins, such as enzymes, for export out of the cell.

Vaults Are Complexes of Proteins and RNA As scientists develop new techniques for looking at the fine structure of cells, they continue to find new organelles. One of the newest organelles, first described in 1986, is the **vault**, a particle composed of RNA and three different proteins. Vaults are found in all cells with nuclei (eukaryotic cells) and are often located close to protein fibers. One of the vault proteins is found in increased amounts in some cancer cells and is associated with resistance to certain drugs in those cells.

There Are Three Sizes of Protein Fibers in the Cytoplasm The protein fibers in the cytoplasm are classified by their diameter (Table 3-1). The thinnest, **microfilaments**, are made of the protein **actin**. Somewhat larger **intermediate filaments** are made of several different types of protein, including *myosin* in muscle, *keratin* in hair and skin, and *neurofilament* in nerve cells. In muscle, intermediate myosin filaments combine to form **thick filaments**. The largest protein fiber organelles are the **microtubules**, made of a protein called **tubulin**. Microtubules and other fibers link together to create the internal scaffolding of

the cell, the cytoskeleton. In addition, microtubules combine to form the more complex structures of centrioles, cilia, and flagella.

The Cytoskeleton Is a Changeable Organelle The **cytoskeleton** [Greek *kytos*, cell + skeleton] is a flexible, changeable three-dimensional scaffolding of actin microfilaments, intermediate filaments, and microtubules that extends throughout the cytoplasm. Some cytoskeleton fibers are permanent, but most are synthesized or disassembled according to the cell's needs. Because of the cytoskeleton's changeable nature, its organizational details are complex and will not be discussed in detail here.

The cytoskeleton has at least five important functions.

1. **Cell shape**. The protein scaffolding of the cytoskeleton provides mechanical strength to the cell, and in some cells, plays an important role in determining the shape of the cell. Figure 3-7a● shows how cytoskeletal fibers help support

(a)

(b)

Microvilli increase cell surface area. They are supported by microfilaments.

Microfilaments form a network just inside the cell membrane.

Microtubules are the largest cytoskeleton fiber.

Intermediate filaments include myosin and keratin.

● Figure 3-7
The cytoskeleton and cytoplasmic protein fibers

microvilli [*micros*, small + *villus*, tuft of hair], fingerlike extensions of the cell membrane that increase the surface area for absorption of material.

2. **Internal organization**. Cytoskeletal fibers stabilize the positions of organelles. Figure 3-7b is a diagram of a typical cell with organelles held in place by cytoskeletal fibers. Note, however, that this figure is only a snapshot of one moment in the cell's life. The interior arrangement and composition of a cell is dynamic, changing from minute to minute in response to the needs of the cell, just as the inside of the walled city is always in motion. One disadvantage of the static illustrations in textbooks is that they are unable to accurately represent movement and the dynamic nature of many physiological processes.

3. **Intracellular transport**. The cytoskeleton helps transport materials into the cell and within the cytoplasm, serving as an intracellular "railroad track" system for moving vesicles and organelles. This function is particularly important in cells of the nervous system, where material must be transported over intracellular distances as long as a meter.

4. **Assembly of cells into tissues**. Fibers of the cytoskeleton connect with protein fibers in the extracellular space, linking cells to each other and to support material outside the cells. In addition to providing mechanical strength to the tissue, these linkages allow the transfer of information from one cell to another.

5. **Movement**. Finally, the cytoskeleton enables cell movement. Mobile cells in the human body include white blood cells that squeeze out of blood vessels and travel to sites of infection, as well as growing nerve cells that send out long extensions as they elongate. Hairlike extensions of the cell membrane called cilia and flagella are able to move in the extracellular fluid because of their cytoskeleton.

Intracellular transport and movement are facilitated by special **motor proteins** that use energy from ATP to slide or step along cytoskeletal fibers. Some motor proteins move vesicles or organelles from one location within the cell to another. Others allow filaments to slide past each other, as occurs when muscles contract.

There are three groups of motor proteins: myosins, kinesins, and dyneins. **Myosin**, an intermediate filament, is best known for its role in muscle contraction. **Kinesins** and **dyneins** are associated with movement along microtubules.

Centrosomes and Centrioles Are Associated with Microtubules

Centrosomes are regions of darkly staining material that are often found close to the cell nucleus. Centrosomes usually act as the cell's **microtubule-organizing center**, where tubulin molecules are assembled into microtubules that lengthen outward through the cytoplasm. In most animal cells, the centrosome contains two **centrioles**, shown in the typical cell of Figure 3-4d. Each centriole is a cylindrical bundle of 27 microtubules, arranged in nine triplets (Fig. 3-8a ●). In cell division, the centrioles direct the movement of DNA strands (see Appendix B). Cells that have lost their ability to undergo cell division, such as mature nerve cells, lack centrioles.

Cilia and Flagella Are Movable Hairlike Structures

Cilia [singular, *cilium*, Latin for eyelash] are short, hairlike structures projecting from the cell surface like the bristles of a brush. The surface of a cilium is a continuation of the cell membrane, and its core contains nine pairs of microtubules surrounding a central pair (Fig. 3-8c). The microtubules terminate just inside the cell at the *basal body*. Cilia beat rhythmically back and forth when the microtubule pairs in their core slide past each other with the help of the motor protein dynein. Ciliary movement creates currents that sweep fluids or secretions across the cell surface.

Flagella [singular, *flagellum*, Latin for whip] have the same microtubule arrangement as cilia but are considerably longer. In addition, a flagellated animal cell usually only has one or two flagella, whereas ciliated cells may have one surface almost totally covered with cilia. Flagella are found on free-floating single cells; the only such cell in humans is the male sperm cell (see Fig. 3-2). The function of a flagellum is to push the cell through fluid with wavelike movements of the flagellum, just as undulating contractions of a snake's body push it headfirst through its environment. Flagella bend and move by the same basic mechanism as cilia.

CONCEPT CHECK   Answers: p. 83

1. Name the three sizes of cytoplasmic protein fibers.

2. What are thick filaments?

3. What is the diffrence between cytoplasm and cytosol?

4. What do motor proteins do?

5. What is the difference between a cilium and a flagellum?

6. How would the absence of a flagellum affect a sperm cell?

Membranous Organelles Create Compartments for Specialized Functions

Membranous organelles are separated from the cytosol by one or more phospholipid membranes similar in structure to the cell membrane. Many membranous organelles have hollow interiors. (The cavity of a hollow tube, organ, or organelle is known as its **lumen**, from the Latin word for light or window. Various anatomical structures have lumens.)

The internal membranes surrounding organelles create compartments within the cell that allow it to isolate substances and separate different functions. For example, the membrane barrier might allow an organelle to contain substances that could be harmful to the cell, such as digestive enzymes. The process of protein synthesis is another example of compartmentation and sepa-

(a) Centrioles direct DNA movement during cell division.

Microtubules form centrioles

(b) A pair of centrioles.

● **Figure 3-8** *Centrioles, cilia, and flagella*

(c) Cilia and flagella have 9 pairs of microtubules surrounding a central pair.

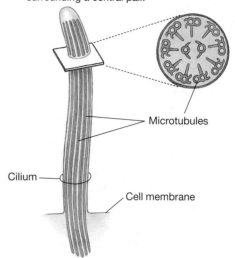

Microtubules

Cilium

Cell membrane

(d) The beating of cilia and flagella creates fluid movement.

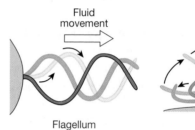

Fluid movement

Fluid movement

Flagellum

Cilia

ration of function. Proteins are synthesized in one organelle, modified in another, and stored in a third. Figure 3-4e shows five types of membranous organelles: mitochondria, the endoplasmic reticulum, the Golgi apparatus, lysosomes, and peroxisomes.

Mitochondria Are the Powerhouse of the Cell

Mitochondria [singular, *mitochondrion*; *mitos*, thread + *chondros*, granule] are small spherical to elliptical organelles with an unusual double wall (Fig. 3-9 ●). The outer membrane of the wall gives the mitochondrion its shape. The inner membrane is folded into leaflets or tubules called **cristae**. The double membrane creates two separate compartments within the mitochondrion. In the center, inside the inner membrane, is the compartment called the mitochondrial **matrix**, which contains enzymes, ribosomes, granules, and DNA. Between the outer and inner membranes is the **intermembrane space**, a compartment that plays an important role in the production of ATP by the mitochondria.

Mitochondria are where most ATP is generated, which is why they are nicknamed the cell's "powerhouse." The number of mitochondria in a particular cell depends on the cell's energy needs. Cells such as skeletal muscle cells, which use a lot of energy, have many more mitochondria than less active cells, such as adipose (fat) cells.

Mitochondria are unusual organelles in two ways. First of all, in their matrix they have their own unique DNA. This **mitochondrial DNA**, along with matrix ribosomes, means that mitochondria can manufacture some of their own proteins. Why is this true of mitochondria and not other organelles? This question has been the subject of intense scrutiny. According to the *prokaryotic endosymbiont theory*, mitochondria are the descendants of bacteria that invaded cells millions of years ago. The bacteria developed a mutually beneficial relationship with their hosts and soon became an integral part of the host cell. Supporting evidence for this theory is the fact that our mitochondria contain DNA, RNA, and enzymes similar to the type found in bacteria but unlike those found in our own nuclei.

The second unusual characteristic of mitochondria is their ability to replicate themselves even when the cell to which they belong is not undergoing cell division. This process is aided by the presence of mitochondrial DNA that allows the organelles to direct their own duplication. Mitochondrial replication takes place by budding, with small daughter mitochondria pinching off an enlarged parent. Cells such as exercising muscle cells, which are subjected to increased energy demands over a period of time, may meet the demand for more ATP by increasing the number of mitochondria in their cytoplasm.

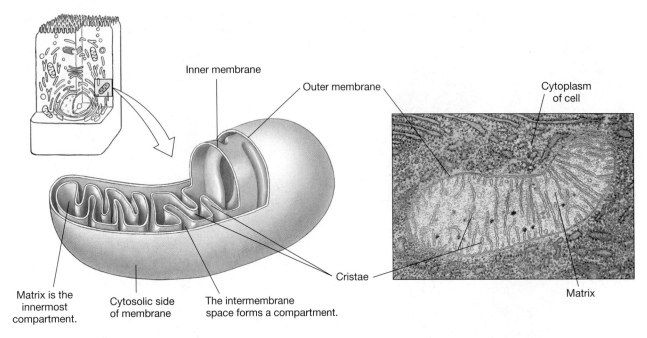

● **Figure 3-9** *Mitochondria* Mitochondria have a double wall structure with a heavily folded inner membrane.

The Endoplasmic Reticulum Is the Site of Protein and Lipid Synthesis The endoplasmic reticulum, or ER, is a network of interconnected membrane tubes (Figs. 3-6 and 3-10 ●) that are a continuation of the outer nuclear membrane. The name *reticulum* comes from the Latin word for *net* and refers to the netlike appearance of the tubules. Electron micrographs reveal that there are two forms of endoplasmic reticulum: rough endoplasmic reticulum and smooth endoplasmic reticulum. Rough endoplasmic reticulum has a granular or rough appearance from rows of ribosomes dotting its cytoplasmic surface. Smooth endoplasmic reticulum lacks the ribosomes and appears

as smooth membrane tubes. Both types of endoplasmic reticulum have the same three major functions: synthesis, storage, and transport of biomolecules.

The **smooth endoplasmic reticulum** (SER) is the main site for the synthesis of fatty acids, steroids, and lipids [● p. 35]. Phospholipids for the cell membrane are produced here, and cholesterol is modified into steroid hormones, such as the sex hormones estrogen and testosterone. The smooth endoplasmic reticulum of liver and kidney cells detoxifies or inactivates drugs. In skeletal muscle cells, a modified form of smooth endoplasmic reticulum stores calcium ions (Ca^{2+}) to be used in muscle contraction.

● **Figure 3-10** *The endoplasmic reticulum*

The **rough endoplasmic reticulum** (RER) is the main site for the synthesis of proteins. Proteins are assembled on ribosomes attached to the cytoplasmic surface of the rough endoplasmic reticulum, then inserted into the lumen, where they undergo chemical modification. Most of these proteins are packaged into spherical membrane-bounded **transport vesicles** [*vesicula*, bladder] that pinch off from the tips of the endoplasmic reticulum. The transport vesicles shuttle their contents across the cytosol to the Golgi apparatus.

The Golgi Apparatus Packages Proteins into Membrane-Bound Vesicles

The **Golgi apparatus** (Fig. 3-11 ●) was first described by Camillo Golgi in 1898. For years, some investigators thought that this organelle was just a result of the fixation process needed to prepare tissues for viewing under the light microscope. However, we now know from electron microscope studies that the Golgi apparatus is indeed another membranous organelle. Its function is to modify proteins made by the rough endoplasmic reticulum and package them into membrane-bound vesicles.

The Golgi apparatus consists of five or six hollow curved sacs stacked on top of each other like a series of hot water bottles and connected so that they share a single lumen. The convex side of the stack faces the rough endoplasmic reticulum and receives the transport vesicles from it. These transport vesicles fuse with the membranes of the Golgi sac and discharge their contents into its lumen.

As the proteins move through the sac, they may be modified by enzymes. For example, long proteins may be cleaved into smaller proteins or carbohydrates may be attached to proteins to make glycoproteins. Finally, the processed proteins are enclosed in membrane-bounded vesicles that pinch off from the concave face of the Golgi apparatus and move out into the cytosol.

Cytoplasmic vesicles are of two kinds: secretory vesicles and storage vesicles. **Secretory vesicles** contain proteins that will be exported to other parts of the body. (Secretion is the process by which a cell releases a substance into the extracellular space.) The contents of most **storage vesicles**, however, never leave the cytoplasm. Lysosomes are the major storage vesicles of the cell.

Lysosomes Are the Intracellular Digestive System

Lysosomes [*lysis*, dissolution + *soma*, body] are small, spherical storage vesicles that appear as membrane-bound granules in the cytoplasm (Fig. 3-12 ●). Lysosomes contain powerful enzymes and act as the digestive system of the cell. They take up bacteria or old organelles, such as mitochondria, and use the enzymes to break them down into their component molecules. Those molecules that can be reused are reabsorbed into the cytosol, while the rest are dumped out of the cell. As many as 50 types of enzymes have been identified from lysosomes of different cell types.

The digestive enzymes of lysosomes are not always kept isolated within the membranes of the organelle. Occasionally, lysosomes release their enzymes to dissolve extracellular support

Secretory vesicles leave filled with newly packaged proteins.

Secretory vesicle

Transport vesicles bring newly synthesized proteins from the endoplasmic reticulum to the Golgi apparatus.

● **Figure 3-11** *The Golgi apparatus*

Lysosomes and **peroxisomes** are vesicles filled with enzymes.

Peroxi-some

Peroxi-some   Lysosome

Peroxi-some   Lyso-some   Mito-chondrion

● **Figure 3-12** *Lysosomes and peroxisomes*

material, such as the hard calcium carbonate portion of bone. The inappropriate release of lysosomal enzymes has been implicated in certain disease states: for example, the inflammation and destruction of joint tissue in *rheumatoid arthritis*.

In other instances, cells allow the enzymes of their lysosomes to come in contact with the cytoplasm, leading to self-digestion of all or part of the cell. When muscles *atrophy* (shrink) from lack of use or the uterus diminishes in size after pregnancy, the loss of cell mass is due to the action of lysosomes.

Because lysosomal enzymes are so powerful, one of the questions that puzzled researchers was why these enzymes do not normally destroy the cell that contains them. What they discovered was that these enzymes are activated only by very acid conditions, 100 times more acid than the normal cytoplasm. When lysosomes are first pinched off from the Golgi apparatus, their interior pH is about the same as that of the cytosol, 7.0–7.3. The enzymes are inactive at this pH. Their inactivity serves as a form of insurance. If the lysosome breaks or accidentally releases the enzyme, it will not harm the cell. However, as the lysosome sits in the cytoplasm, it accumulates H^+ ions in a process that uses energy. The pH inside the vesicle drops to 4.8–5.0, and the enzymes are activated. Once this has occurred, the enzymes are capable of breaking down various biomolecules.

In inherited conditions known as *lysosomal storage diseases*, lysosomes are not effective because they lack specific enzymes.

One of the best-known lysosomal storage diseases is the fatal inherited condition known as *Tay-Sachs disease*. Infants with Tay-Sachs disease have defective lysosomes that fail to break down glycolipids in nerve cells. Accumulation of glycolipids in the cells causes symptoms of nervous system dysfunction, including blindness and loss of coordination. Most of the afflicted infants die in early childhood.

Peroxisomes Contain Enzymes That Neutralize Toxins Peroxisomes are storage vesicles that are even smaller than lysosomes. For years, they were thought to be a kind of lysosome, but we now know that they contain a different set of enzymes. Their main function appears to be to degrade long-chain fatty acids and potentially toxic foreign molecules.

Peroxisomes get their name from the fact that the reactions inside them generate hydrogen peroxide (H_2O_2), a toxic molecule. The peroxisomes rapidly convert this peroxide to oxygen and water using the enzyme *catalase*. Peroxisomal disorders disrupt the normal processing of lipids and can severely disrupt the normal function of the nervous system by altering the structure of nerve cell membranes.

CONCEPT CHECK   Answers: p. 83

7. What is the anatomical difference between rough endoplasmic reticulum and smooth endoplasmic reticulum? What is the functional difference?

8. How do lysosomes differ from peroxisomes?

9. How does the physiological theme of compartmentation apply to membranous organelles in general and to mitochondria in particular?

10. Microscopic examination of a cell reveals many mitochondria. What does this observation imply about the cell's energy requirements?

11. You are examining tissue from a previously unknown species of fish. You discover a tissue with large amounts of smooth endoplasmic reticulum in its cells. What is one possible function of these cells?

The Nucleus Is the Cell's Control Center

The **nucleus** of the cell contains DNA, the genetic material that ultimately controls all cell processes. How the nucleus receives information about conditions in the cell or elsewhere in the body and responds with fine-tuned control of the cell's protein synthesis is an active area of biological research. Figure 3-13 ● illustrates the structure of a typical nucleus. The boundary, or **nuclear envelope**, is a two-membrane structure that separates the nucleus from the cytoplasmic compartment. The outer membrane of the envelope is connected with the endoplasmic reticulum, and both membranes of the envelope are pierced here and there by round holes, or **pores**.

Communication between the nucleus and cytosol occurs through the **nuclear pore complexes**, large protein complexes

Nucleolus contains DNA that controls synthesis of ribosomal RNA.

Chromatin is DNA and protein.

Nuclear envelope is a double membrane that separates the nucleus from the cytoplasm.

Nuclear pores regulate movement of material into and out of the nucleus.

Inner membrane of nuclear envelope

Broken edge of outer membrane

Outer membrane of nuclear envelope

● **Figure 3-13** *The nucleus*

with a central channel. Ions and small molecules move freely through this channel when it is open, but proteins and RNA must be transported by a process that uses energy. This requirement allows the cell to restrict large molecules such as DNA to the nucleus and various enzymes to either the cytoplasm or the nucleus.

RUNNING PROBLEM

During a Pap test for cervical cancer, cell samples are swabbed from the cervix (neck) of the uterus. The cells are then smeared onto a glass slide and sent to a laboratory for examination by a trained cytologist. The cytologist looks for **dysplasia**, a change in the size and shape of the cells that is indicative of cancerous changes [*dys-*, abnormal + *-plasia*, growth or cell multiplication]. Cancer cells can usually be recognized by a large nucleus surrounded by a relatively small amount of cytoplasm. Jan's first Pap test showed all the hallmarks of dysplasia.

Question 2: What property of cancer cells explains the large size of their nucleus and the relatively small amount of cytoplasm?

In electron micrographs of cells that are not dividing, the nucleus appears filled with randomly scattered granular material, or **chromatin**, composed of DNA and associated proteins. A nucleus usually also contains from one to four larger dark-staining bodies of DNA, RNA, and protein called **nucleoli** [singular, *nucleolus*, little nucleus]. Nucleoli contain the genes and proteins that control the synthesis of RNA for ribosomes (see Appendix B).

▶ TISSUES OF THE BODY

Despite the amazing variety of intracellular structures, no single cell can carry out all the processes of the mature human body. Instead, cells assemble into larger units called **tissues**, collections of cells held together by specialized connections called **cell junctions** and by other support structures. Tissues range in complexity from simple tissues containing only one cell type, such as the lining of blood vessels, to complex tissues containing many cell types and extensive extracellular material, such as connective tissue. The cells of most tissues work together to achieve a common purpose.

The study of tissue structure and function is known as **histology** [*histos*, tissue]. Histologists describe tissues by their physical features: (1) the shape and size of the cells, (2) how the cells are arranged in the tissue (in layers, scattered, and so on), (3) how the cells are connected to each other, and (4) the amount of extracellular material that is present in the tissue. There are four **primary tissue types** in the human body: epithelial, connective, muscle, and *neural*, or nerve. Before we consider each tissue type specifically, let's examine how cells link together to form tissues.

Extracellular Matrix Has Many Functions

Matrix is extracellular material that is synthesized and secreted by the cells of a tissue. For years, scientists believed that matrix was an inert substance whose only function was to hold cells together. However, experimental evidence now shows that matrix plays a vital role in many physiological processes ranging from growth and development to cell death. A number of disease states are associated with overproduction or disruption of matrix, including chronic heart failure and the spread of cancerous cells throughout the body (*metastasis*).

The composition of matrix varies from tissue to tissue but all matrix has two basic components: proteoglycans and insoluble protein fibers. **Proteoglycans** are glycoproteins, proteins covalently bound to polysaccharide chains [p. 40]. Insoluble protein fibers such as collagen, *fibronectin*, and *laminin* provide strength and anchor cells to the matrix. Attachments between the matrix and cell membrane or cytoskeleton proteins are one means of communication between the cell and its external environment.

The amount of matrix in a tissue is highly variable. Nerve and muscle tissue have very little matrix, but the connective tissues, such as cartilage, bone, and blood, have extensive matrix that occupies as much volume as their cells. The consistency of matrix varies from watery (blood and lymph) to rigid (bone).

Cell Junctions Hold Cells Together to Form Tissues

Tissues are composed of cells connected to each other and to the extracellular matrix. During growth and development, cells form *cell-cell adhesions* that may be transient or that develop into more permanent **cell junctions**. Cell junctions can be grouped into three categories: gap junctions, tight junctions, and anchoring junctions. Gap junctions allow direct cell-to-cell communication. Tight junctions are *occluding* [*occludere*, to close up] junctions that block movement of material between cells. Anchoring junctions hold cells to each other and to the extracellular matrix.

1. **Gap junctions** are the simplest cell-cell junctions. They create cytoplasmic bridges between adjoining cells so that chemical and electrical signals pass rapidly from one cell to the next (Fig. 3-14c). Cylindrical proteins, called *connexins*, interlock to create passageways that look like hollow rivets with narrow channels through their centers. The channels are able to open and close, regulating the movement of small molecules and ions through them. Gap junctions were once thought to occur only in certain muscle and nerve cells, but we now know they are important in cell-to-cell communication in many tissues, including the liver, pancreas, ovary, and thyroid gland.

2. **Tight junctions** in humans and other vertebrates are designed to restrict the movement of material between the cells they link. In tight junctions the cell membranes of adjacent cells partly fuse together with the help of proteins called *claudins* and *occludins*, thereby making a barrier (Fig. 3-14a). Like many physiological processes, the barrier properties of tight junctions are dynamic and can be altered depending on the body's needs.

   In physiology, tight junctions in the intestinal tract and kidney prevent most substances from moving freely between the external and internal environments and thus allow cells to regulate what enters and leaves the body. Tight junctions also create the so-called *blood-brain barrier* that prevents potentially harmful substances in the blood from reaching the extracellular fluid of the brain.

3. **Anchoring junctions** attach cells to each other (cell-cell junctions) or to the extracellular matrix (cell-matrix junctions). Anchoring junctions have been compared to buttons or zippers that tie cells together and hold them in position within a tissue. The protein linkage of anchoring cell junctions is very strong, allowing sheets of tissue in skin and lining body cavities to resist damage from stretching and twisting. Even the tough protein fibers of anchoring junctions can be broken, however. If you have shoes that rub against your skin, the stress can shear the proteins connecting the different skin layers. When fluid accumulates in the resulting space and the layers separate, a *blister* results.

   Tissues held together with anchoring junctions are like a picket fence, where spaces between the connecting bars allow material to pass from one side of the fence to the other. In contrast, tissues held together with tight junctions are more like a solid brick wall: very little can pass from one side of the wall to the other between the bricks.

Cell Adhesion Molecules Cell adhesion molecules, or CAMs, are membrane-spanning proteins with loops and tails that extend into the cytoplasm and extracellular fluid (see Fig. 5.5, p. 129). CAMs are responsible both for cell junctions and for transient cell adhesions (Table 3-2). Cell-cell or cell-matrix adhesions mediated by CAMS are essential for normal growth and development. For example, growing nerve cells creep across matrix with the help of *nerve-cell adhesion molecules*, or *N-CAMs*. Cell adhesion helps white blood cells escape from the circulation and move into infected tissues, and it allows clumps of platelets to cling to damaged blood vessels. In these examples, cell adhesion is not a permanent event and the bonds between CAMs and matrix are weak.

Anchoring Junctions In vertebrates, cell-cell anchoring junctions are created by CAMs called **cadherins** that connect

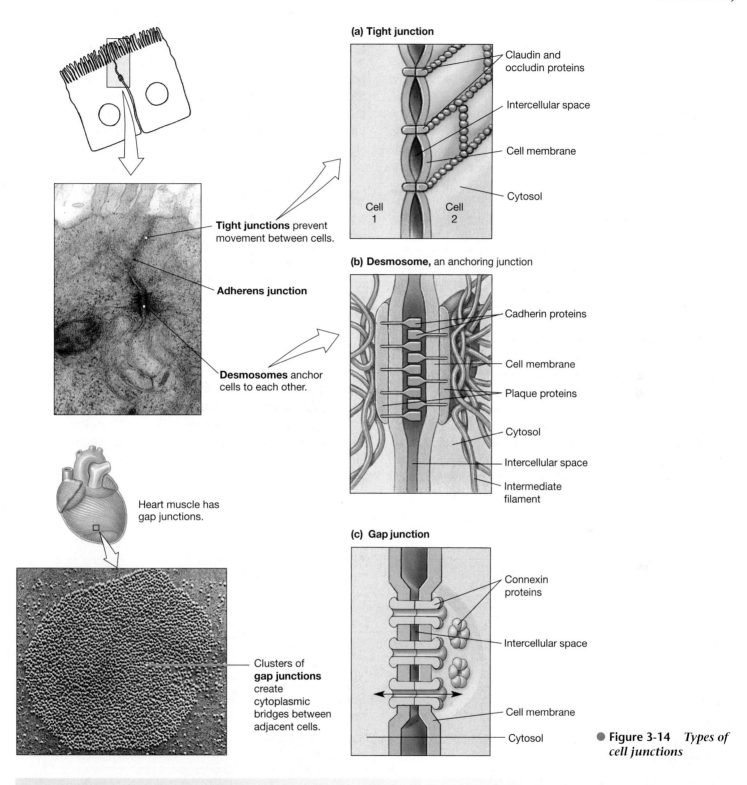

(a) **Tight junction**

Claudin and occludin proteins

Intercellular space

Cell membrane

Cytosol

Cell 1   Cell 2

Tight junctions prevent movement between cells.

Adherens junction

Desmosomes anchor cells to each other.

(b) **Desmosome,** an anchoring junction

Cadherin proteins

Cell membrane

Plaque proteins

Cytosol

Intercellular space

Intermediate filament

Heart muscle has gap junctions.

(c) **Gap junction**

Connexin proteins

Intercellular space

Cell membrane

Cytosol

Clusters of **gap junctions** create cytoplasmic bridges between adjacent cells.

● **Figure 3-14** *Types of cell junctions*

Table 3-2: Major Cell Adhesion Molecules

NAME	EXAMPLES
Cadherins	Cell-cell junctions such as adherens junctions and desmosomes
Integrins	Primarily found in cell-matrix junctions
N-CAMs (Nerve-cell adhesion molecules)	Nerve cell growth during development of the nervous system
Selectins	Cell-cell adhesions when blood cells bind to blood vessel walls

with each other across the intercellular space. Cell-cell junctions take the form of either adherens junctions or desmosomes. **Adherens junctions** (Fig. 3-14b) link actin microfilaments in adjacent cells together. **Desmosomes** [*desmos*, band + *soma*, body] attach to intermediate filaments of the cytoskeleton.

Desmosomes are the strongest cell-cell junctions. In electron micrographs they can be recognized by the dense glycoprotein bodies, or *plaques*, that lie just inside the cell membranes in the region where the two cells connect (Fig. 3-14b). Desmosomes may be small points of contact between two cells (spot desmosomes) or bands that encircle the entire cell (belt desmosomes).

The disappearance of anchoring cell junctions is a characteristic of cancer and probably contributes to the spread, or *metastasis*, of cancerous cells throughout the body. Cancer cells lose their anchoring junctions because they have fewer cadherin molecules and are not bound as tightly to neighboring cells. Once a cancer cell is released from its moorings, it secretes protein-digesting enzymes known as *proteases*. These enzymes, especially those called *matrix metalloproteinases* (MMPs), dissolve the extracellular matrix so that escaping cancer cells can invade adjacent tissues or enter the bloodstream. Researchers are investigating ways to block MMP enzymes to see if they can prevent metastasis.

Cell-matrix attachments use CAMs called **integrins**. Integrins are membrane proteins that can also bind to signal molecules in the cell's environment, transferring information carried by the signal across the cell membrane into the cytoplasm. As with cell-cell junctions, there are two types of cell-matrix connections. **Hemidesmosomes** [*hemi-*, half] are strong junctions that anchor intermediate fibers of the cytoskeleton to fibrous matrix proteins such as *laminin*. **Focal adhesions** tie intracellular actin fibers to different matrix proteins such as *fibronectin*. A summary map of cell junctions is shown in Figure 3-15●. Now that you understand how cells are held together into tissues, we will look at the four different tissue types in the body: (1) epithelial, (2) connective, (3) muscle, and (4) nervous tissues.

CONCEPT CHECK   Answers: p. 83

12. Name the three functional categories of cell junctions.

13. Which type of cell junction:
    (a) restricts movement of materials between cells?
    (b) allows direct movement of substances from the cytoplasm of one cell to the cytoplasm of an adjacent cell?
    (c) provides the strongest cell-cell junction?
    (d) anchors actin fibers in the cell to the extracellular matrix?

Epithelia Provide Protection and Regulate Exchange

The **epithelial tissues**, or **epithelia** (singular *epithelium*), protect the internal environment of the body and regulate the exchange of material between the internal and external environments. They cover exposed surfaces, such as the skin, and line

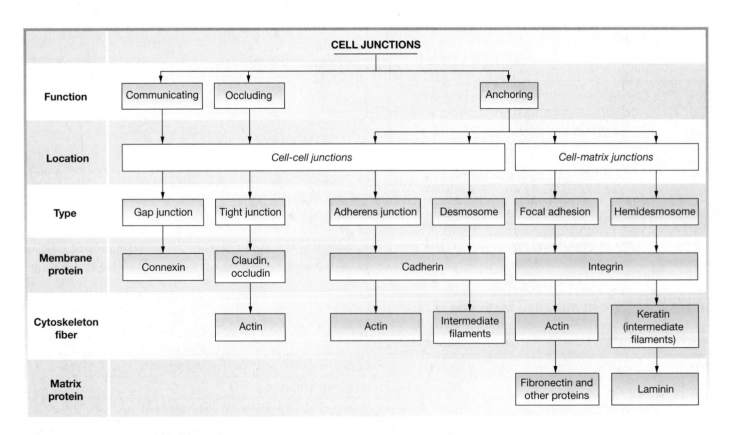

● **Figure 3-15**   *A map of cell junctions*

internal passageways, such as the digestive tract. *Any substance that enters or leaves the internal environment of the body must cross an epithelium.*

Some epithelia, such as those of the skin and mucous membranes of the mouth, act as a barrier to keep water in the body and invaders such as bacteria out. Other epithelia, such as those in the kidney and intestinal tract, control the movement of material between the external environment and the extracellular fluid of the body. Nutrients, gases, and wastes often must cross several different epithelia in their passage between cells and the outside world.

Another type of epithelium is specialized to manufacture and secrete chemicals into the blood or to the external environment. Sweat and saliva are examples of substances secreted by epithelia into the environment. Hormones, signal molecules used to maintain homeostasis, are secreted into the blood.

Structure of Epithelia   Epithelia typically consist of one or more layers of cells connected to each other, with a thin layer of extracellular matrix lying between the epithelial cells and their underlying tissues (Fig. 3-16 ●). This matrix layer, called the **basal lamina** [*bassus*, low; *lamina*, a thin plate], or **basement membrane**, is composed of a network of collagen and laminin filaments embedded in proteoglycans. The protein filaments hold the epithelial cells to the underlying cell layers, just as cell junctions hold the individual cells in the epithelium to each other.

The cell junctions in epithelia are variable. Physiologists classify epithelia either as "leaky" or "tight," depending on how easily substances pass between the cells. In a leaky epithelium, anchoring junctions leave gaps or pores that allow molecules to pass across the epithelium between the cells. A typical leaky epithelium is the wall of the capillary blood vessels, where all dissolved molecules except for large proteins can pass between the cells. In a tight epithelium, such as the kidney, adjacent cells are bound to each other by tight junctions and in-

folded membranes that create a barrier to movement between the cells. To cross a tight epithelium, most substances must enter the cells and go *through* them. The tightness of an epithelium is directly related to how selective it is about what can move across it. Some epithelia, such as those of the intestine, have the ability to alter the tightness of their junctions according to the body's needs.

Types of Epithelia   Epithelial tissues can be divided into two general types: (1) sheets of tissue that lie on the surface of the body or that line the inside of tubes and hollow organs and (2) secretory epithelia that synthesize and release substances into the extracellular space. Histologists classify sheet epithelia by the number of cell layers in the tissue and by the shape of the cells in the surface layer. This classification scheme recognizes two types of layering—**simple** (one cell thick) and **stratified** (multiple cell layers) [*stratum*, layer + *facere*, to make]—and three cell shapes—**squamous** [*squama*, flattened plate or scale], **cuboidal**, and **columnar**. However, physiologists are more concerned with the functions of these tissues, so we will divide epithelia into five groups according to their function.

There are five functional types of epithelia: exchange, transporting, ciliated, protective, and secretory. **Exchange epithelia** permit rapid exchange of material, in particular, gases. **Transporting epithelia** are selective about what can cross them and are found primarily in the intestinal tract and the kidney. **Ciliated epithelia** are located in the airways of the respiratory system and in the female reproductive tract. **Protective epithelia** are found on the surface of the body and just inside the openings of body cavities. The **secretory epithelia** synthesize and release secretory products into the external environment or into the blood.

Figure 3-17 ● shows the distribution of these types of epithelia in the different systems of the body. Notice that most epithelia face the external environment on one surface and the extracellular fluid on the other. The only exception to this rule is the endocrine glands. Table 3-3 summarizes the types of epithelia.

Exchange epithelia   The **exchange epithelia** are composed of very thin, flattened cells that allow gases (CO_2 and O_2) to pass rapidly through the cells. This type of epithelium lines the blood vessels and the lungs, the two major sites of gas exchange in the body. In the smallest blood vessels, called capillaries, gaps or pores in the epithelium also allow molecules smaller than proteins to pass *between* the cells, making this a leaky epithelium (Fig. 3-18a ●). Histologists classify thin exchange tissue as *simple squamous epithelium* because it has a single layer of thin, flattened cells. The simple squamous epithelium lining the heart and blood vessels is also called the **endothelium**.

Transporting epithelia   The **transporting epithelia** actively and selectively regulate the exchange of nongaseous material, such as ions and nutrients, between the internal and external

Epithelial cells attach to the basal lamina using cell adhesion molecules.

Basal lamina is an acellular matrix layer that is secreted by the epithelial cells.

● **Figure 3-16**   *Epithelial tissue*

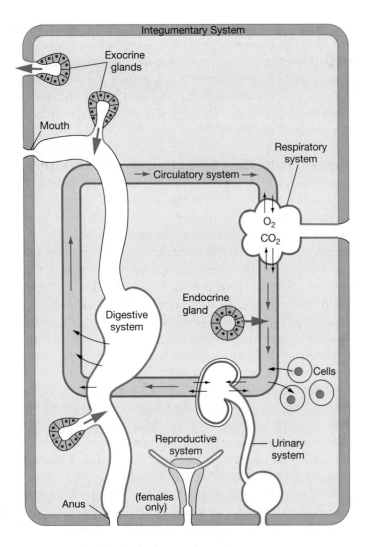

Integumentary System

Exocrine glands

Mouth

Respiratory system

→ Circulatory system →

O_2
CO_2

Digestive system

Endocrine gland

Cells

Reproductive system
(females only)

Urinary system

Anus

FIGURE QUESTIONS

• What systems have transporting epithelia? exchange epithelia? ciliated epithelia? protective epithelia?
• Where do secretions from endocrine glands go?
• Where do secretions from exocrine glands go?

KEY

⇄ exchange

secretion

secretory epithelium

——— exchange epithelium
▬▬▬ protective epithelium
——— ciliated epithelium
━━━ transporting epithelium

● Figure 3-17 *Distribution of epithelia in the body*

Table 3-3: Types of Epithelia

	NUMBER OF CELL LAYERS	CELL SHAPE	SPECIAL FEATURES	WHERE FOUND
Exchange	One	Flattened	Pores between cells to permit easy passage of molecules	Lungs, lining blood vessels
Transporting	One	Columnar or cuboidal	Tight junctions prevent movement between cells; surface area increased by folding of cell membrane into villi, or fingers	Intestine, kidney, some exocrine glands
Protective	Many	Flattened in surface layers; polygonal in deeper layers	Cells tightly connected by many desmosomes	Skin, lining cavities, such as the mouth, that open to the environment
Ciliated	One	Cuboidal to columnar	One side covered with cilia to move fluid across surface	Nose, trachea, and upper airways; female reproductive tract
Secretory	One to many	Columnar to polygonal	Protein-secreting cells filled with membrane-bound secretory granules and extensive RER. Steroid-secreting cells contain lipid droplets and extensive SER	Exocrine glands, including pancreas, sweat glands, salivary glands; endocrine glands, such as thyroid, gonads

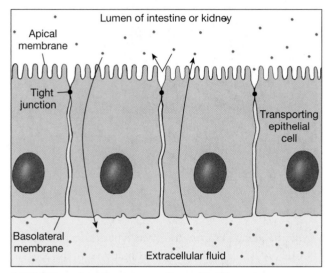

(a) Leaky exchange epithelium allows movement through gaps between the cells.

(b) Tight junctions in a transporting epithelium prevent movement between the cells. To cross the epithelium, substances must go through two cell membranes.

● **Figure 3-18** *Movement of substances across tight and leaky epithelia*

environments. These epithelia are found lining the hollow tubes of the digestive system and the kidney, where lumens open into the external environment [● p. 4].

There are several characteristics of transporting epithelia (Fig. 3-18b):

1. **Cell shape**. The cells of transporting epithelia are much thicker than those of exchange epithelia, so they act as a barrier as well as an entry point. The cell layer is only one cell thick, but the cells are cuboidal or columnar in shape.
2. **Membrane modifications**. The **apical membrane**, the surface of the cell that faces the lumen, has **microvilli**, finger-like projections that increase the surface area available for transport. A cell with microvilli has at least 20 times the surface area of a cell without them. In addition, the **basolateral membrane**, the side of the cell facing the extracellular fluid, may also have folds that increase the cell's surface area.
3. **Cell junctions**. The cells of transporting epithelia are firmly attached to adjacent cells by moderately tight to very tight junctions. This means that material must move into an epithelial cell on one side of the tissue and out of the cell on the other in order to cross the epithelium.

4. **Cell organelles**. Most cells that transport materials have numerous mitochondria to provide energy for transport processes (discussed further in Chapter 5).

The properties of transporting epithelia differ. For example, glucose can be moved from the lumen of the small intestine to the extracellular fluid, but cannot be moved across the epithelium of the large intestine. Furthermore, the transport properties of an epithelium can be regulated and modified in response to various stimuli. Hormones, for example, affect the transport of ions by the epithelium of the kidney. You will learn more about transporting epithelia when you study the kidney and digestive systems.

Ciliated epithelia Ciliated epithelia are nontransporting tissues found lining the respiratory system and parts of the female reproductive tract. The surface of the tissue facing the lumen is covered with cilia that beat in a coordinated, rhythmic fashion, moving fluid and particles across the surface of the tissue. Injury to the cilia or to their epithelial cells can stop ciliary movement. Paralysis of the ciliated epithelium lining the respiratory tract is one effect of smoking. It is considered to contribute to the higher incidence of respiratory infection in smokers, because the mucus that traps bacteria is no longer swept up out of the lungs by the cilia. Figure 3-19 ● shows the ciliated columnar epithelium of the trachea.

Protective epithelia The **protective epithelia** are designed to prevent any exchange between the internal and external environments. They are stratified tissues, composed of many stacked layers of cells. The epidermis [*epi*, upon + *derma*, skin] and linings of the mouth, pharynx, esophagus, urethra, and vagina are all protective epithelia. The multiple cell layers of this tissue type protect areas subject to mechanical or chemical stresses. Some of these tissues are toughened further by the secretion of *keratin* [*keras*, horn], the same insoluble protein abundant in hair and nails.

Because the protective epithelia are subjected to irritating chemicals, bacteria, and other destructive forces, the cells in them have a short life span. In the deeper layers, new cells are produced continually, displacing the older cells at the surface. Each time you wash your face, you scrub off dead cells on the surface layer. As the skin ages, the rate of cell turnover declines. Tretinoin (Retin-A®), a drug derived from vitamin A, speeds up cell division and surface shedding so that the treated skin develops a more youthful appearance.

Secretory epithelia Secretory epithelia are composed of cells that produce a substance inside the cell and release it into the extracellular space in a process known as **secretion**. Secretory cells may be scattered among other epithelial cells, or they may group together to form a multicellular **gland**. There are two types of secretory glands: exocrine glands and endocrine glands.

(a) Ciliated epithelium lining the airways.

(b) Scanning electron micrograph showing ciliated cells mixed with cells covered with shorter microvilli.

Cilia

Microvilli

Golgi apparatus

Nucleus

Basement membrane

Mitochondria

● **Figure 3-19** *Ciliated epithelia*

Many kinds of cancer develop in epithelial cells that are subject to damage or trauma. The cervix consists of two types of epithelia. Secretory epithelium with mucus-secreting glands lines the inside of the cervix, while a protective epithelium covers the outside of the cervix. At the opening of the cervix, these two types of epithelia come together. As Jan lies on the examining table, her physician uses a small spatula and a little brush to take samples of cells from both inside and outside her cervix. She then smears the cells on a slide and sprays them with fixative. The slides will now be sent to a lab for evaluation.

Question 3: What kind of damage or trauma are cervical epithelial cells normally subjected to? Which of its two types of epithelia is more likely to be affected by trauma?

50   53   63   🏃   76   79

Exocrine glands release their secretions into the external environment [*exo-*, outside + *krinein*, to secrete]. This may be onto the surface of the skin or onto an epithelium lining one of the internal passageways, such as the airways of the lung or the lumen of the intestine. In effect, an exocrine secretion leaves the body. This is important to know, if you consider that some exocrine secretions, like stomach acid, have a pH that is incompatible with life. Most exocrine glands release their products through open tubes known as **ducts**. Sweat glands, mammary glands found in the breast, salivary glands, the liver, and the pancreas are all exocrine glands.

Exocrine gland cells produce two types of secretions. **Serous secretions** are watery solutions, and many of them contain enzymes. Tears, sweat, and digestive enzyme solutions are all serous exocrine secretions. **Mucous** secretions (also called **mucus**) are sticky solutions containing glycoproteins and proteoglycans. **Goblet cells**, shown in Figure 3-20●, are single exocrine cells that produce mucus. Mucus acts as a lubricant for food to be swallowed, as a trap for foreign particles and microorganisms inhaled or ingested, and as a protective barrier between the epithelium and the environment. Some exocrine glands contain more than one type of secretory cell, and they produce both serous and mucous secretions. For example, the salivary glands release mixed secretions.

Endocrine secretions, or **hormones**, are released by ductless glands or single cells directly into the extracellular space. Hormones then enter the blood for distribution to other portions of the body, where they regulate or coordinate the activities of various tissues, organs, and organ systems. Some of the best-known endocrine glands are the pancreas, the thyroid gland, the gonads, and the pituitary gland. For years, it was thought that all hormones were produced by cells grouped together into endocrine glands. We now know that isolated endocrine cells occur scattered in the epithelial lining of the digestive tract, in the tubules of the kidney, and in the walls of the heart.

Figure 3-21● shows the epithelial origin of endocrine and exocrine glands. During development, epithelial cells grow downward into the supporting tissues. Exocrine glands retain a link to the parent epithelium in the form of a hollow passageway or duct that transports the secretion to its destination. Endocrine glands lose the connecting cells and secrete their hormones into the bloodstream.

Nucleus

Mucus

Golgi apparatus

Basal lamina

Goblet cells secrete mucus into the lumen of hollow organs such as the intestine.

● **Figure 3-20** *Goblet cells*

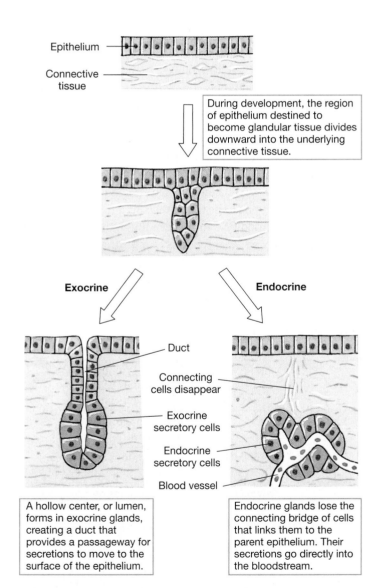

Epithelium

Connective tissue

During development, the region of epithelium destined to become glandular tissue divides downward into the underlying connective tissue.

Exocrine

Endocrine

Duct

Connecting cells disappear

Exocrine secretory cells

Endocrine secretory cells

Blood vessel

A hollow center, or lumen, forms in exocrine glands, creating a duct that provides a passageway for secretions to move to the surface of the epithelium.

Endocrine glands lose the connecting bridge of cells that links them to the parent epithelium. Their secretions go directly into the bloodstream.

● **Figure 3-21** *Development of endocrine and exocrine glands from epithelium*

Connective Tissues Provide Support and Barriers

Connective tissues, the second major tissue type, provide structural support and sometimes a physical barrier that, along with specialized cells, helps defend the body from foreign invaders such as bacteria. The distinguishing characteristic of connective tissues is the presence of extensive extracellular matrix containing widely scattered cells. The cells secrete and modify the matrix of the tissue. Connective tissues range from blood, to the support tissues for the skin and internal organs, to cartilage and bone.

Structure of Connective Tissue   The matrix of connective tissue is a **ground substance** of proteoglycans and water in which insoluble fibrous protein fibers are arranged, much like suspended pieces of fruit in a gelatin salad. The consistency of ground substance is highly variable, depending on the type of connective tissue. At one extreme is the watery matrix of blood, and at the other extreme is the hardened matrix of bone. In between are solutions of proteoglycans that vary in consistency from syrupy to gelatinous. The term *ground substance* is sometimes used interchangeably with *matrix*.

The actual cells of connective tissue lie embedded within the extracellular matrix. Connective tissue cells are described as *fixed* if they remain in one place, or *mobile* if they can move from place to place. **Fixed cells** are responsible for local maintenance, tissue repair, and energy storage. The **mobile cells** of connective tissue are responsible mainly for defense. The distinction between fixed and mobile cells is not absolute, because at least one cell type is found in both fixed and mobile forms.

Although matrix itself is nonliving, the connective tissue cells constantly modify it by adding, deleting, or rearranging molecules. The suffix *-blast* [*blastos*, sprout] on a connective tissue cell name often indicates a cell that is either growing or actively secreting extracellular matrix. **Fibroblasts**, for example, are connective tissue cells that secrete collagen-rich matrix. Cells that are actively breaking down matrix are identified by the suffix *-clast* [*klastos*, broken]. Cells that are neither growing, secreting, nor breaking down matrix may be given the suffix *-cyte*, meaning "cell." Remembering these suffixes should help you remember the functional difference between cells with similar names, such as the osteoblast, osteocyte, and osteoclast, three cell types found in bone.

In addition to secreting glycoprotein ground substance, connective tissue cells produce the fibers of the matrix. Four types of fiber proteins are found in the matrix of connective tissue, aggregated into insoluble fibers. **Collagen** [*kolla*, glue + *-genes*, produced] is the most abundant protein in the human body, almost one-third of its dry weight. It is also the most diverse of the four

protein types, with at least 12 different variations. Collagen is found almost everywhere connective tissue is found, from the skin to muscles and bones. Individual collagen molecules pack together to form collagen fibers, flexible but inelastic fibers whose strength per unit weight exceeds that of steel. The amount and arrangement of collagen fibers is one of the distinguishing characteristics of different types of connective tissue.

Three other protein fibers in connective tissue are elastin, fibrillin, and fibronectin. **Elastin** is a coiled, wavy protein that returns to its original length after being stretched. This property is known as *elastance*. Elastin combines with the very thin, straight fibers of **fibrillin** to form filaments and sheets of elastic fibers. These two fibers are important in elastic tissues such as the lungs, blood vessels, and skin. As mentioned earlier, **fibronectin** connects cells to extracellular matrix at focal adhesions. Fibronectins also play an important role in wound healing and in blood clotting.

Types of Connective Tissue   Table 3-4 describes in detail the different types of connective tissue. The most common types of connective tissue are loose and dense connective tissue, adipose tissue, cartilage, bone, and blood.

Loose connective tissues (Fig. 3-22 ●) are the elastic tissues that underlie skin and provide support for small glands. **Dense connective tissues** are tissues whose primary function is strength or flexibility. Important examples of this tissue type are tendons, ligaments, and the sheaths that surround muscles and nerves. In these tissues, collagen fibers are the dominant type.

Tendons (Fig. 3-23 ●) attach skeletal muscles to bones. **Ligaments** connect one bone to another. Because ligaments contain elastic fibers in addition to collagen fibers, they have a limited ability to stretch, whereas tendons cannot stretch.

Table 3-4: Types of Connective Tissue

TISSUE NAME	GROUND SUBSTANCE	FIBER TYPE AND ARRANGEMENT	MAIN CELL TYPES	WHERE FOUND
Loose connective tissue	Gel; more ground than fibers and cells	Collagen, elastic, reticular; random	Fibroblasts	Skin, around blood vessels and organs, under epithelia
Dense, irregular connective tissue	More fibers than ground	Mostly collagen; random	Fibroblasts	Muscle and nerve sheaths
Dense, regular connective tissue	More fibers than ground	Collagen, parallel	Fibroblasts	Tendons and ligaments
Adipose	Very little	None	Brown fat, white fat	Depends on age and sex
Blood	Aqueous	None	Blood cells	In blood and lymph vessels
Cartilage	Firm but flexible; hyaluronic acid	Collagen	Chondroblasts	Joint surfaces, spine, ear, nose, larynx
Bone	Rigid due to calcium salts	Collagen	Osteoblasts and osteoclasts	Bones

Reticular fibers
Melanocytes
Fixed macrophage
White blood cell
Capillary
Red blood cell
Fat cells (adipocytes)

Mast cell
Elastic fibers
Collagen fibers
Fibroblasts are cells that secrete matrix proteins.
Free macrophage
Ground substance is the matrix of loose connective tissue.

Loose connective tissue

Light micrograph of loose connective tissue

● **Figure 3-22** *Cells and fibers of connective tissue* Macrophages and mast cells are cells that defend against foreign invaders, such as bacteria.

● **Figure 3-23**
Tendons and ligaments

(a)

(b)

Muscle

Tendon

Ligaments attach bone to bone.

Tendons attach bone to muscle.

(c) Collagen fibers of tendon are densely packed into parallel bundles.

Adipose tissue is made up of **adipocytes**, or fat cells (Fig. 3-24 ●). An adipocyte of **white fat** typically contains a single enormous lipid droplet that occupies most of the volume of the cell. This is the most common form of adipose tissue in adults. **Brown fat** is composed of adipose cells that contain multiple lipid droplets rather than a single large droplet. This type of fat is almost completely lacking in adults, but it plays an important role in temperature regulation in infants.

Blood is an unusual connective tissue that is characterized by its watery matrix, consisting of a dilute solution of ions and dissolved organic molecules. The matrix lacks insoluble protein fibers but contains a large variety of soluble proteins. We will discuss these in Chapter 16.

Cartilage and bone together are considered supporting connective tissues. These tissues have a dense ground substance that contains closely packed fibers. **Cartilage** is found in structures such as the nose, ears, knee, and windpipe. It is solid, flexible, and notable for its lack of blood supply. Without a blood supply, nutrients and oxygen must reach the cells of cartilage by diffusion. This is a slow process, which means that damaged cartilage heals slowly.

Lipid droplets displace cytosol in adipose cells.

Adipocytes (fat cells)

Adipose tissue

● **Figure 3-24** *Adipose tissue*

The fibrous matrix of **bone** is said to be *calcified* because it contains mineral deposits, primarily calcium salts, such as calcium phosphate. These minerals give the bone strength and rigidity. We will examine the structure and formation of bone along with calcium metabolism in Chapter 23. Figure 3-25 ● is a study map showing the components of connective tissues.

BIOTECHNOLOGY

Growing New Cartilage

Someone bite your ear off? Replacing cartilage that is lost or damaged is moving from the realm of science fiction into reality. Researchers are mixing *chondrocytes*, the cells that synthesize the matrix of cartilage, with a biodegradable "glue" made from a protein that helps form blood clots. The mixture has been implanted in animals and found to produce normal cartilage when the chondrocytes secrete matrix. In one experiment, bovine (cow) chondrocytes were grown on a three-dimensional scaffold shaped like a human ear. Similar experiments are now under way with human cells. The day may come not too far in the future when we can grow new joints, noses, and tracheas to replace lost or damaged ones.

CONCEPT CHECK   Answers: p. 83

20. What is the distinguishing characteristic of connective tissues?

21. Name four types of protein fibers found in connective tissue matrix and give the characteristics of each.

22. Name six types of connective tissues.

23. Blood is a connective tissue with two components: plasma and cells. What is the matrix in this connective tissue?

24. Why does torn cartilage heal more slowly than a cut in the skin?

Muscle and Neural Tissues Are Excitable

Muscle and neural tissues are collectively called the *excitable tissues* because of their ability to generate electrical signals called *action potentials*. Both of these tissue types have minimal extracellular matrix, usually limited to a supportive basal lamina. Some types of muscle and nerve cells are also notable for their gap junctions that play an important role in the rapid conduction of electrical signals from cell to cell.

Muscle tissue has the ability to contract and produce force and movement. There are three types of muscle tissue in the body: cardiac muscle, in the heart; smooth muscle, which makes up most internal organs; and skeletal muscle. Most skeletal muscles are attached to bone and are responsible for gross movement of the body. We will discuss muscle tissue in more detail in Chapter 12.

Neural tissue has two types of cells. **Neurons**, or nerve cells, carry information in the form of chemical and electrical signals from one part of the body to another. They are concentrated in the brain and spinal cord but also include a network of cells that extends to virtually every part of the body. **Glial cells**, or neuroglia, are the support cells for neurons. We will discuss the anatomy of neural tissue in Chapter 8.

A summary of the characteristics of the four tissue types can be found in Table 3-5.

▶ TISSUE REMODELING

Growth is a process that most people associate with the period from birth to adulthood. But cell birth, growth, and death continue throughout a person's life. The tissues of the body are constantly changing as cells die and are replaced.

Apoptosis Is a Tidy Form of Cell Death

Cell death occurs two ways, one messy and one tidy. In **necrosis**, cells die from physical trauma, toxins, or lack of oxygen when their blood supply is cut off. Necrotic cells swell, their organelles deteriorate, and finally the cells rupture. The cell contents thus released include digestive enzymes that damage adjacent cells

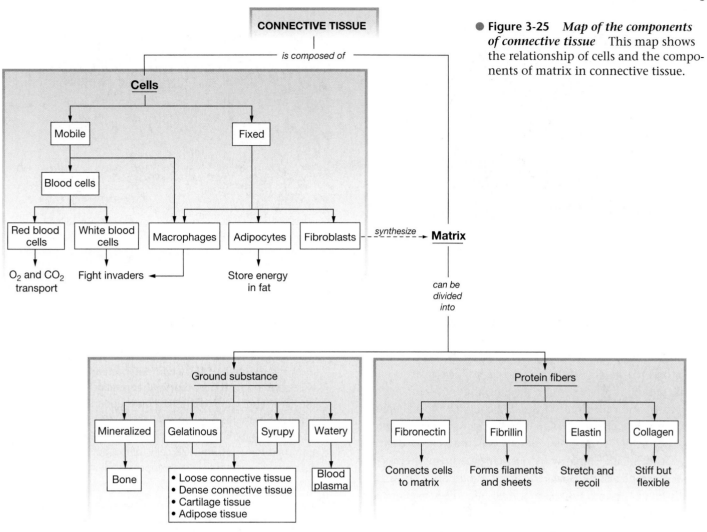

● **Figure 3-25** *Map of the components of connective tissue* This map shows the relationship of cells and the components of matrix in connective tissue.

Table 3-5: Characteristics of the Four Tissue Types

	EPITHELIAL	CONNECTIVE	MUSCLE	NERVE
Matrix amount	Minimal	Extensive	Absent	Absent
Matrix type	Basement membrane	Varied—protein fibers in ground substance that ranges from liquid to gelatinous to firm to calcified	NA	NA
Unique features	No direct blood supply	Cartilage has no blood supply	Able to generate electrical signals, force, and movement	Able to generate electrical signals
Surface features of cells	Microvilli, cilia	NA	NA	NA
Locations	Covers body surface; lines cavities and hollow organs and tubes. Secretory glands	Supports skin and other organs. Cartilage, bone, and blood	Makes up skeletal muscles, hollow organs, and tubes	Throughout body. Concentrated in brain and spinal cord
Cell arrangement and shapes	Variable number of layers, from one to many; cells flattened, cuboidal, or columnar	Cells not in layers; usually randomly scattered in matrix; cell shape irregular to round	Cells linked in sheets or elongated bundles; cells shaped in elongated, thin cylinders. Heart muscle cells may be branched	Cells isolated or networked; cell appendages highly branched and/or elongated

(a) Jan's abnormal Pap test.

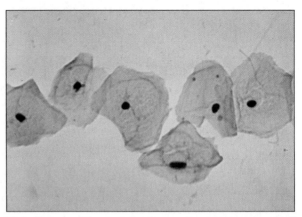

(b) Jan's second Pap test. Are these cells normal or abnormal?

● **Figure 3-26** *Pap smears of cervical cells*

and trigger an inflammatory response. You see necrosis when you have a red area of skin surrounding a scab.

In contrast, cells that undergo programmed cell death, or **apoptosis** [ap-oh-TOE-sis or ap-op-TOE-sis; *apo-* apart, away + *ptosis*, falling], do not disrupt their neighbors when they die. Apoptosis, also called cell suicide, is a complex process regulated by multiple chemical signals. Some signals keep apoptosis from occurring while other signals tell the cell to destruct. When the suicide signal wins out, chromatin in the nucleus condenses, the cell pulls away from its neighbors, shrinks, and finally breaks up into tidy membrane-bound *blebs* that are gobbled up by neighboring cells or by wandering cells of the immune system.

Apoptosis is a normal event in the life of an organism. During development, apoptosis removes unneeded cells, such as half the cells in the developing brain, and the webs of skin between fingers and toes. In adults, cells that are subject to wear and tear from exposure to the outside environment may live only a day or two before undergoing apoptosis. For example, it has been estimated that the intestinal epithelium is completely replaced with new cells every two to five days.

CONCEPT CHECK   Answers: p. 83

25. What are some features of apoptosis that distinguish it from cell death due to injury?

Stem Cells Can Create New Specialized Cells

If cells in the adult body are constantly dying, where do their replacements come from? This question is still being answered and is one of the hottest topics in biological research today. The paragraphs below describe what we currently know.

All cells in the body are derived from the single cell formed at conception. That cell and others that follow reproduce

themselves by undergoing the cell division process known as **mitosis** (see Appendix B). The very earliest cells in the life of a human being are said to be **totipotent** [*totus*, entire] and have the ability to develop into any and all types of specialized cells. Any totipotent cell has the potential to become a functioning organism.

After about day 4 of development, the totipotent cells of the embryo begin to specialize, or *differentiate*. As they do so, they narrow their potential fates and become **pluripotent** [*plures*, many]. Pluripotent cells can develop into many different cell types but not all cell types. An isolated pluripotent cell cannot develop into an organism.

As differentiation continues, pluripotent cells develop into the various tissues of the body. As the cells specialize and mature, many lose the ability to reproduce themselves. Others, called **stem cells**, retain the ability to divide. In adults, these stem cells usually can become only one particular cell type, such as a blood cell. Undifferentiated cells in a tissue that retain the ability to divide and develop into specialized cells of that tissue are said to be **multipotent** [*multi*, many]. Some of the most-studied multipotent adult stem cells are found in bone marrow and give rise to blood cells.

Biologists used to believe that nerve and muscle cells, which are highly specialized in their mature forms, could not be replaced when they died. But exciting new research indicates that stem cells for these tissues do exist in the body. However, they occur in very small numbers, are difficult to isolate, and do not thrive in the laboratory. Naturally occurring adult stem cells are unable to replace large masses of dead or dying tissue that occur as a result of diseases such as strokes or heart attacks. Consequently, one goal of stem cell research is to find a source of pluripotent or multipotent cells that could be grown in the laboratory and implanted to treat *degenerative* diseases, those in which cells degenerate and die.

One example of a degenerative disease is Parkinson's disease, in which certain types of nerve cells in the brain die. Embryos and fetal tissue are rich sources of stem cells but the use of these two tissues is controversial and accompanied by many legal and ethical questions. Some researchers hope that adult stem cells will show **plasticity**, the ability to specialize into a different cell of a type than that for which they were destined.

BIOTECHNOLOGY

The Bioartificial Pancreas

When organs such as the kidney, heart, or liver fail, the solution is often to transplant a replacement organ from a donor. But in one form of diabetes mellitus, in which the insulin-producing beta cells (β-cells) of the pancreas fail, the solution is not so easy. Most of the diabetic patient's pancreas is functioning normally, and only the beta cells need to be replaced. A variety of solutions have been proposed and are in the trial stage. For patients who have kidney failure due to their diabetes, combined human kidney-pancreas transplants have been performed. A second technique being investigated encloses isolated beta cells from humans or pigs in a semipermeable capsule. This bioartificial pancreas allows glucose and nutrients to enter and insulin secreted by the beta cells to leave. Both types of transplantation may be complicated by rejection reactions in which the body recognizes the transplants as foreign and tries to destroy them, so research is continuing.

There are still many challenges facing us before stem cell therapy becomes a reality. One is finding a good source of stem cells. A second major challenge is determining the chemical signals that tell stem cells when to differentiate and what cell type to become. And even once these two challenges are overcome and donor stem cells are implanted, the body may recognize that the new cells are foreign tissue and try to reject them.

Stem cell research is an excellent example of the dynamic nature of science. In the time lag between writing of this chapter (summer, 2002) and when you read it, there will almost certainly be new and exciting developments. There may also be legislation and laws regulating stem cell research and use. For the latest research findings you should check authoritative websites such as that sponsored by the U.S. National Institutes of Health (*www.nih.gov/news/stemcell*).

▶ ORGANS

Groups of tissues that carry out related functions may form structures known as **organs**. The organs of the body contain the four types of tissue in various combinations. The skin is an excellent example of an organ that incorporates all four types of tissue into an integrated whole. We think of skin as a thin layer that covers the external surfaces of the body, but in reality it is the heaviest single organ, at about 16% of an adult's total body weight! If it were flattened out, it would cover a surface of between 1.2 and 2.3 square meters, about the size of a couple of card-table tops. Its size and weight make skin one of the most important organs of the body. The structure of this organ is highlighted in *Focus on the Skin*.

There are other organs, like the skin, whose functions do not fit neatly into any one chapter of this book. We will highlight several of them in special *Organ Focus Features* throughout the book. These illustrated boxes discuss structure and functions of these versatile organs so that you can gain an appreciation for the way different tissues combine for a united purpose.

As we consider the systems of the body in the succeeding chapters, you will see how diverse cells, tissues, and organs carry out the processes of the living body. Although the cells have different structures and different functions, they have one need in common: a continuous supply of energy. Without energy, cells cannot survive, let alone carry on all the other processes of daily living. The next chapter looks at cellular metabolism: how cells capture and harness the energy released by chemical reactions.

(a) The layers of the skin

Hair follicles secrete the nonliving keratin shaft of hair.

Sebaceous glands are exocrine glands that secrete a lipid mixture.

Arrector pili muscles pull hair follicles into a vertical position when the muscle contracts, creating "goose bumps."

Sweat glands secrete a dilute salt fluid to cool the body.

Sensory receptors monitor external conditions.

Epidermis consists of multiple cell layers that create a protective barrier.

The **dermis** is loose connective tissue that contains exocrine glands, blood vessels, muscles, and nerve endings.

Hypodermis contains adipose tissue for insulation

Apocrine glands in the genitalia, anus, axillae (*axilla*, armpit), and eyelids release waxy or viscous milky secretions in response to fear or sexual excitement.

Sensory nerve

Artery   Vein

Blood vessels extend upward into the dermis.

(b) Epidermis. The skin surface is a mat of linked keratin fibers left behind when old epithelial cells die.

Phospholipid matrix acts as the skin's main waterproofing agent.

Surface keratinocytes produce keratin fibers.

Desmosomes anchor epithelial cells to each other.

Epidermal cell

Basal lamina

(c) Connection between epidermis and dermis

Hemidesmosomes tie epidermal cells to fibers of the basal lamina.

Basal lamina or basement membrane is an acellular layer between epidermis and dermis.

RUNNING PROBLEM CONCLUSION

The Pap Smear

In this running problem, you learned that the Pap test can detect the early changes that herald cervical cancer. The diagnosis is not always simple because the change in cell cytology from normal to cancerous is a continuum and can be subject to individual interpretation. In 2001 a group of clinicians, cytotechnologists, pathologists, and patient advocates representing 44 professional societies from over 20 countries met to review the evidence on which the guidelines for interpreting Pap smears are based. Their recommendations shaped the 2001 Bethesda System terminology guidelines (http://bethesda2001.cancer.gov).

Check your understanding of the running problem by comparing your answers to those in the summary table below.

	QUESTION	FACTS	INTEGRATION AND ANALYSIS
1	Why does the treatment of cancer focus on killing the cancerous cells?	Cancerous cells divide uncontrollably and fail to coordinate with other normal cells. Cancerous cells fail to differentiate into specialized cells.	Unless removed, cancerous cells will displace normal cells. This may cause destruction of normal tissues. In addition, because cancerous cells do not become specialized, they cannot carry out the same functions as the specialized cells that they displace.
2	What property of cancer cells explains the large size of their nucleus and the relatively small amount of cytoplasm?	Cancerous cells divide uncontrollably. Dividing cells must duplicate their DNA prior to cell division. [🔹 Appendix B]	Actively reproducing cells are likely to have more DNA in their nucleus as they prepare to divide, so their nuclei tend to be larger. Each cell division splits the cytoplasm between the two daughter cells. If division is occurring rapidly, the daughter cells may not have time to synthesize new cytoplasm, so the amount of cytoplasm is less than in a normal cell.
3a	What kind of damage or trauma are cervical epithelial cells normally subjected to?	The cervix is the neck of the uterus.	The cervix is subject to trauma or damage during sexual intercourse and childbirth.
3b	Which of its two types of epithelia is more likely to be affected by trauma?	The cervix consists of secretory epithelium with mucus-secreting glands lining the inside, and protective epithelium that covers the outside.	Protective epithelium is composed of multiple layers of cells and is designed to protect areas from mechanical and chemical stress [🔹 p. 69]. Therefore, the secretory epithelium with its single cell layer is more easily damaged.
4	Has Jan's dysplasia improved or worsened? What evidence do you have to support your answer?	The slide from Jan's first Pap smear shows abnormal cells with large nuclei and little cytoplasm. These abnormal cells do not appear in the second smear.	The disappearance of the abnormal cells indicates that Jan's dysplasia has resolved.

50 53 63 70 76 🏃

CHAPTER SUMMARY

Cell biology and histology illustrate one of the major themes in physiology: *compartmentation*. In this chapter you learned how a cell is subdivided into compartments: the nucleus and the membranous organelles. You also learned how cells form tissues that create larger compartments within the body. A second theme in this chapter is the *mechanical properties* of cells and tissues that arise from the protein fibers of the cytoskeleton, from cell junctions, and from the molecules that make up extracellular matrix, part of the "glue" that helps hold tissues together.

Cellular Anatomy

1. The **cell** is the basic functional unit of most living organisms. (p. 51)

2. The **cell membrane** is a phospholipid bilayer with embedded proteins that separates the inside of the cell from the **extracellular fluid**. (pp. 52–53; Fig. 3-5)

3. The cytoplasm consists of semigelatinous **cytosol** with dissolved nutrients, ions, wastes, insoluble **inclusions**, and **organelles** to carry out specific functions. (p. 53)

4. Important nonmembranous organelles include **ribosomes**, which take part in protein synthesis, and insoluble protein fibers in three sizes: **microfilaments, intermediate filaments**, and **microtubules**. (pp. 53,57; Fig. 3-6)

5. The **cytoskeleton** provides strength, support, and internal organization; aids transport of materials within the cell; links cells together; and enables motility in certain cells. (p. 57; Fig. 3-7)

6. **Centrosomes** are the microtubule-organizing center of the cell. **Centrioles** aid the movement of chromosomes during cell division. (p. 58; Fig. 3-8a)

7. **Cilia** move fluid or secretions across the cell surface. **Flagella** propel sperm through body fluids. (p. 58; Fig. 3-8c, d)

8. Membranes around membranous organelles divide the cytosol into compartments whose functions can be separated. (p. 58)

9. **Mitochondria** generate most ATP for the cell. (p. 59; Fig. 3-9)

10. The **smooth endoplasmic reticulum** is the primary site of lipid synthesis. The **rough endoplasmic reticulum** is the primary site of protein synthesis. (pp. 60–61; Fig. 3-10)

11. The **Golgi apparatus** packages proteins into vesicles. **Secretory vesicles** secrete their contents into the extracellular fluid. (p. 61; Fig. 3-11)

12. **Lysosomes** and **peroxisomes** are **storage vesicles** that contain digestive enzymes. (pp. 61–62; Fig. 3-12)

13. The **nucleus** contains DNA, the genetic material that ultimately controls all cell processes, in the form of **chromatin**. The **nuclear envelope** has **nuclear pore complexes** that allow controlled chemical communication between the nucleus and cytosol. **Nucleoli** are nuclear areas that control the synthesis of RNA for ribosomes. (pp. 62–63; Fig. 3-13)

Tissues of the Body

IP Muscular: Anatomy Review—Skeletal Muscle Tissue

14. There are four **primary tissue types** in the human body: epithelial, connective, muscle, and neural. (p. 64)

15. Extracellular **matrix** secreted by cells provides support and a means of cell-cell communication. It is composed of proteoglycans and insoluble protein fibers. (p. 64)

16. Cell junctions fall into three categories. **Gap junctions** allow chemical and electrical signals to pass directly from cell to cell. **Tight junctions** prevent the movement of material between cells. **Anchoring junctions** hold cells to each other or to the extracellular matrix. (p. 64; Fig. 3-14)

17. Membrane proteins called **cell adhesion molecules (CAMs)** are essential in cell adhesion and in anchoring junctions. (p. 64)

18. **Desmosomes** and **adherens junctions** anchor cells to each other. **Focal adhesions** and **hemidesmosomes** anchor cells to matrix. (p. 66)

19. **Epithelial tissues** protect the internal environment, regulate the exchange of material, or manufacture and secrete chemicals. (p. 66)

20. **Exchange epithelia** permit rapid exchange of material, particularly gases. **Transporting epithelia** actively regulate the selective exchange of nongaseous material between the internal and external environments. **Ciliated epithelia** move fluid and particles across the surface of the tissue. **Protective** epithelia help prevent exchange between the internal and external environments. The **secretory epithelia** release secretory products into the external environment or the blood. (pp. 67–69; Fig. 3-17)

21. **Exocrine glands** release their secretions into the external environment through **ducts**. **Endocrine glands** are ductless glands that release their secretions, called **hormones**, directly into the extracellular fluid. (p. 70; Fig. 3-21)

22. **Connective tissues** have extensive extracellular matrix that provides structural support and a physical barrier. (p. 71)

23. **Loose connective tissues** are the elastic tissues that underlie skin. **Dense connective tissues**, including **tendons and ligaments**, have strength or flexibility due to collagen. **Adipose tissue** stores fat. **Blood** is characterized by a watery matrix. **Cartilage** is solid and flexible and has no blood supply. The fibrous matrix of **bone** is hardened by deposits of calcium salts. (pp. 72–74)

24. Muscle and neural tissues are called excitable tissues because of their ability to generate electrical signals called action potentials. **Muscle tissue** has the ability to contract and produce force and movement. There are three types of muscle: cardiac, smooth, and skeletal. (p. 74)

25. **Neural tissue** includes **neurons** that use electrical and chemical signals to transmit information from one part of the body to another and support cells known as **glia**. (p. 74)

Tissue Remodeling

26. Cell death occurs by **necrosis**, which adversely affects neighboring cells, and by **apoptosis**, programmed cell death that does not disturb the tissue. (pp. 74, 76)

27. **Stem cells** are cells that are able to reproduce themselves and differentiate into specialized cells. Stem cells are most plentiful in embryos but are also found in the adult body. (p. 76)

Organs

28. **Organs** are formed by groups of tissues that carry out related functions. The organs of the body contain the four types of tissues in various ratios. (p. 77)

QUESTIONS

LEVEL ONE *Reviewing Facts and Terms*

1. What are the two primary types of biomolecules found in the cell membrane?

2. Define and distinguish between inclusions, membranous organelles, and nonmembranous organelles. Give an example of each.

3. Define cytoskeleton. List five functions of the cytoskeleton.

4. Match the terms with their descriptions:
   (a) cilia
   (b) centriole
   (c) flagellum
   (d) centrosome

   1. in human cells, appears as single, long, whiplike tail
   2. short, hairlike structures that beat to produce currents in fluids
   3. a bundle of microtubules that aids in mitosis
   4. the microtubule-organizing center

5. Match the organelle with its function:
   (a) endoplasmic reticulum
   (b) Golgi apparatus
   (c) lysosomes
   (d) mitochondria
   (e) peroxisomes

   1. "powerhouse" of the cell where most ATP is produced
   2. these degrade long-chain fatty acids and toxic foreign molecules
   3. network of membranous tubules that synthesize biomolecules
   4. digestive system of cell, degrading or recycling components
   5. modifies and packages proteins into vesicles

6. What process activates the enzymes inside lysosomes?

7. Exocrine glands produce secretions that are either watery, called _____, such as tears and sweat, or stickier solutions called _____.

8. _____ glands release hormones, which enter the blood and are carried throughout the body to regulate and coordinate the various activities of organs or systems.

9. List the four major tissue types. Give an example and location of each.

10. The largest and heaviest organ in the body is the _____.

11. Match the proteins to their functions. Functions may be used more than once.
    (a) cadherin
    (b) CAM
    (c) collagen
    (d) connexin
    (e) elastin
    (f) fibrillin
    (g) fibronectin
    (h) integrin
    (i) occludin

    1. membrane protein used to form cell junctions
    2. matrix glycoprotein used to anchor cells
    3. protein found in gap junctions
    4. matrix protein found in connective tissue

12. What types of glands can be found within the skin? Name the secretion of each type.

13. The term *matrix* is used in reference to an organelle and tissues. Compare the meanings of the term in these two contexts.

LEVEL TWO *Reviewing Concepts*

14. **Mapping exercise:** Compile this list of terms into a map of cell structure. Add functions where appropriate.

actin	microtubule
cell membrane	microtubule-organizing center
centriole	mitochondria
centrosome	myosin
cilia	neurofilament
cytoplasm	nonmembranous organelle
cytoskeleton	nucleus
cytosol	peroxisome
extracellular matrix	RER
flagella	ribosome
Golgi apparatus	secretory vesicle
intermediate filament	SER
keratin	storage vesicle
lysosome	thick filament
membranous organelle	tubulin
microfilament	vault

15. List, compare, and contrast the three types of intercellular junctions and their subtypes. Give an example of where each type can be found in the body and describe its function in that location.

16. Which would have more rough endoplasmic reticulum: pancreatic cells that manufacture the protein hormone insulin, or adrenal cortex cells that synthesize the steroid hormone cortisol?

17. A number of organelles can be considered vesicles. Define *vesicle* and describe at least three examples.

18. Explain why a stratified epithelium offers more protection than a simple epithelium.

19. Sketch a short series of columnar cells. Label the apical and basolateral borders of these typical epithelial cells. Briefly explain the different kinds of activities that may take place at each cell border, as well as how the structures and junctions found on each side of the cell support these functions.

20. Compare and contrast the structure, locations, and functions of bone and cartilage.

21. Differentiate between the terms in each set below:
   (a) lumen/wall   (b) cytoplasm/cytosol
   (c) histology/cytology   (d) myosin/keratin

22. When a tadpole turns into a frog, its tail shrinks and is reabsorbed. Is this an example of necrosis or apoptosis? Defend your answer.

23. Match the structures from the chapter to the basic physiological themes in the right column and give an example or explanantion for each match. A structure may match with more than one theme.

(a) cell junctions	1. communication
(b) cell membrane	2. molecular interactions
(c) cytoskeleton	3. compartmentation
(d) organelles	4. mechanical properties
(e) cilia	5. use of biological energy

24. Define the term *organ*. Explain the specific traits exhibited by the following body parts that allow them to qualify as organs: skin, stomach, uterus, heart. What similarities do they share?

25. In some instance, the extracellular matrix can be quite rigid. How then might developing and expanding tissues cope with a rigid matrix to make space for themselves?

LEVEL THREE *Problem Solving*

26. One result of cigarette smoking is paralysis of the cilia that line the respiratory passageways. What function do these cilia serve? Based on what you have read in this chapter, why is it harmful when they no longer beat? What health problems would you expect to arise? How does this explain the hacking cough common among smokers?

27. Cancer is abnormal, uncontrolled cell division. What property of epithelial tissues might (and does) make them more prone to develop this?

28. What might happen to normal physiological function if matrix metalloproteinases are inhibited by drugs?

Explorations

Explorations are also available at *www.physiologyplace.com*

This chapter introduced you to the structure and function of tissues, cells, and organelles. The following Explorations will help you gain an appreciation of how important understanding cell biology is in medicine and health care.

Exploration 1: Cell Junctions

All cell junctions are not alike. Tight junctions vary from tissue to tissue and may allow water, ions, and other solutes to pass between the cells. See:

"Molecular structure of tight junctions and their role in epithelial transport," *News in Physiological Sciences* 16: 126–29, June 2001. (free access at *www.nips.org*)

The loss of normal cell junctions plays a role in a number of diseases and in metastasis, the spread of cancer throughout the body. Diseases in which cell junctions are destroyed or fail to form can have disfiguring and painful symptoms, such as blistering skin. One such disease is *pemphigus*, a condition in which the body attacks some of its own cell junction proteins (*www.pemphigus.org*). Read more about cell junctions and disease in:

"Integrins and health," *Scientific American* 276(5): 68–75, May 1997.

"Molecular physiology and pathophysiology of tight junctions. I. Tight junction structure and function: Lessons from mutant animals and proteins," *American Journal of Physiology Gastrointestinal and Liver Physiology* 279: G250–254, 2000. (free access at *www.ajpgi.org*)

Exploration 2: The Pap Smear

As a result of the 2001 Bethesda System recommendations, abnormal pap smears may now be tested for the presence of DNA from the human papilloma virus (HPV). To learn more about the association between HPV and cervical cancer, go to the National Cancer Institute homepage (*www.nci.nih.gov*) and search for "HPV." This site also contains information about cervical cancer.

The 2001 Bethesda System recommendations and the rationale behind them are presented in an editorial and two articles published in the April 24, 2002, issue of *JAMA*, pp. 2140, 2114, and 2120 (*www.jama.ama-assn.org*).

Is the extra expense of the automated PAPNET test cost-effective? To see what is being said about this issue, do a Google search (*www.google.com*) and type in PAPNET. You will find published articles such as this one: *http://www.mja.com.au/public/issues/oct21/farnsworth/farnsworth.html*. This link should take you to an article in the *Medical Journal of Australia* in which Pap smears were evaluated both manually and using PAPNET.

Exploration 3: Apoptosis

Apoptosis is a fast-growing field of research. To see some photographs of cells undergoing apoptosis, go to *www.cellsalive.com/apop.htm*. What factors might influence whether a cell undergoes apoptosis instead of attempting to repair itself? How might understanding the mechanisms of apoptosis be of benefit in the treatment of cancer? To learn more about apoptosis, follow the links at the bottom of the cellsalive page.

Exploration 4: Matrix Metalloproteinases

Extracellular matrix (ECM) is a substance of varying composition that provides support to surrounding cells. A web article, written by Eugene Russo, introduces us to **matrix metalloproteases** (MMPs), enzymes that specifically degrade proteins commonly found in the ECM. Go to *www.the-scientist.com* and look up "MMPs May Provide Clues to Multiple Ailments" in *The Scientist* 13(14):6, 1999, July 5. (This site requires free registration to view articles.) Can you envision how understanding cell migration might aid in the treatment of cancer? How might MMPs be involved in diseases such as arthritis that affect the ECM? Can MMPs be involved in blood vessel migration?

ANSWERS

Answers to Concept Check Questions

Page 58

1. Microfilaments, intermediate filaments, microtubules.
2. Thick filaments are associations of the intermediate filament myosin.
3. Cytoplasm is everything inside the cell membrane except for the nulceus. Cytosol is the semi-gel substance in which organelles and inclusions are suspended.
4. Motor proteins use energy to create movement along cytoskeleton fibers.
5. Cilia are short, usually occur in multiples on a cell, and move fluid or substances across the cell surface. Flagella are longer, usually occur singly on human sperm, and are used to propel a cell through a fluid.
6. Without a flagellum a sperm would be unable to swim to find an egg to fertilize.

Page 62

7. Rough ER has ribosomes attached to the cytoplasmic side of its membrane that are not present on smooth ER. Rough ER synthesizes proteins; smooth ER synthesizes lipids.
8. Lysosomes contain enzymes to break down bacteria and old organelles. Peroxisomes contain enzymes to break down fatty acids and foreign molecules.

9. The membranes create compartments that are physically isolated from the cytosol. The double membrane of mitochondria creates two different compartments inside the organelle.
10. Large numbers of mitochondria suggest that the cell has a high energy requirement because mitochondria are the site of greatest energy production in the cell.
11. Large amounts of smooth endoplasmic reticulum suggest that the tissue synthesizes large amounts of lipids, fatty acids, or steroids or that it detoxifies foreign molecules.

Page 66

12. Cell junctions are occluding (tight), communicating (gap), or anchoring.
13. (a) occluding, (b) gap, (c) desmosomes, (d) focal adhesions.

Page 71

14. Protective, secretory, transporting, ciliated, and exchange.
15. Secretion is the process by which a cell releases a substance into the environment.
16. Endocrine glands do not have ducts and they secrete into the blood. Exocrine glands have ducts and secrete into the external environment.

17. Hemidesmosomes attach to laminin (Fig. 3-15).
18. No, skin would have many layers of cells in order to protect the internal environment. A simple squamous epithelium that is one cell thick with flattened cells would not be a protective epithelium.
19. The cell is an endocrine cell because it secretes its product into the extracellular space for distribution in the blood.

Page 74

20. Connective tissues have extensive matrix.
21. Collagen provides strength and flexibility; elastin and fibrillin provide elasticity; fibronectin helps anchor cells to matrix.
22. Bone, cartilage, blood, dense connective tissues (ligaments and tendons), loose connective tissue, adipose.
23. The plasma, or liquid portion of blood, could be considered the matrix.
24. Cartilage lacks a blood supply, so oxygen and nutrients for repair must reach the cells by diffusion, a slow process.

Page 76

25. Apoptosis is a tidy form of cell death that removes cells without disrupting their neighbors. Necrosis releases digestive enzymes that damage neighboring cells.

Answers to Figure Questions

Page 68

Fig. 3-17: Transporting epithelia (purple) are found primarily in the digestive tract and kidney, where they regulate the movement of substances into and out of the extracellular fluid. The lungs and the blood vessels are lined with exchange epithelium (red). Ciliated epithelia (blue) are found in the airways of the respiratory tract and in part of the female reproductive system. The outer surface of the body and the openings of the body cavities are covered with protective epithelium (brown). Endocrine glands (without ducts) secrete their hormones into the blood. Exocrine glands, with ducts, secrete their products outside the body—onto the surface of the skin or into the lumen of an organ that opens into the environment outside the body (Fig. 1-2, p. 3).

Cellular Metabolism

4

Transketolase, an enzyme for sugar metabolism, with cofactors (white)

There is no good evidence that ... life evades the second law of thermodynamics, but in the downward course of the energy-flow it interposes a barrier and dams up a reservoir which provides potential for its own remarkable activities.

—F. G. Hopkins, 1933

RUNNING PROBLEM

Tay-Sachs Disease

In many ultra-orthodox Jewish communities of the United States—in which arranged marriages are the norm—the rabbi is entrusted with an important, life-saving task. He keeps a confidential record of individuals known to carry the gene for Tay-Sachs disease, a fatal, inherited condition that strikes one in 3600 American Jews of Eastern European descent. Babies born with this disease rarely live beyond age four, and there is no cure. Based on the family trees he constructs, the rabbi can avoid pairing two individuals who carry the deadly gene.

Sarah and David, who met while working on their college newspaper, are not orthodox Jews. But both are aware that their Jewish ancestry might put any children they would have at risk for Tay-Sachs disease. Six months before their wedding, they decide to see a genetic counselor to determine whether they are carriers of the gene for Tay-Sachs disease.

91 98 100 108 117

BACKGROUND BASICS

Forms of energy and work (Appendix A): **835** ı DNA and RNA (Appendix B): **838** ı Emergent properties: **6** ı Hydrogen bonds: **31** ı Covalent bonds: **26** ı ATP: **41** ı Organelles: **9** ı Electrons: **21** ı Protein structure: **41** ı Carbohydrates: **35** ı Nucleotides: **44** ı Graphing: **12** ı Lipids: **35**

Christine Schmidt, Ph.D., and her graduate students seed isolated endothelial cells onto an engineered matrix and watch them grow. They know that if their work is successful, the tissue that results might someday help replace a blood vessel in the body. Just as a child playing with building blocks assembles them into a house, the bioengineer and her students create tissue from cells. In both examples, someone familiar with the starting components, building blocks or cells, could predict what the final product will be: blocks make buildings, cells make tissues.

Why then can't biologists, knowing the characteristics of nucleic acids, proteins, lipids, and carbohydrates, explain how combinations of these molecules acquire the remarkable attributes of a living cell? How can living cells carry out processes that far exceed what we would predict from understanding the properties of their components? The answer is *emergent properties* [p. 6], those distinctive traits that cannot be predicted from the simple sum of the component parts.

The history of the Human Genome Project (HGP; *www.ornl. gov/hgmis/*) illustrates how our thinking about emergent properties has evolved. In 1990, when the HGP started, scientists predicted that learning the sequence of DNA would unlock all the secrets of life. But as work progressed, they discovered that one nucleotide sequence in a strand of DNA could direct production of many different proteins. As a result, identifying genes became only the starting point of the project. The focus now has shifted to the protein equivalent of the genome, the **proteome**. The Human Genome Project has been complemented by the Human Proteomics Initiative, an ambitious undertaking to catalog the structure and function of all proteins in the body (*http://us.expasy. org/sprot/hpi/*).

Proteins—the most versatile biomolecule—have many shapes and diverse functions. They form spirals or sheets or fold back on themselves to create complex three-dimensional structures with pockets, channels, or protruding knobs. Proteins become insoluble structural fibers [p. 39] of the cytoskeleton and extracellular matrix [p. 64], form passageways through cell membranes, or dissolve in the cytoplasm and extracellular fluid. One of the most important functions of soluble proteins is the role they play in the extraction, transfer, and storage of biological energy, some of the chemical processes that are collectively called *metabolism*.

Energy is essential for processes we associate with living things: growth, movement, reproduction, and the maintenance of a relatively constant internal environment. Without these processes, a cell is like a ghost town, filled with buildings that are slowly crumbling into ruin. People must provide energy for growth and maintenance, for building new buildings and repairing old ones.

Similar energy-dependent processes take place within a cell. Cells need energy to import raw materials, make new molecules, and repair or recycle aging parts with the help of specialized proteins. The ability of cells to extract energy from the external environment and use that energy to maintain themselves as organized, functioning units is one of their most outstanding characteristics. In this chapter, we look at the cell processes through which the human body obtains energy and maintains its ordered systems.

▶ ENERGY IN BIOLOGICAL SYSTEMS

Energy cycling between the environment and living organisms is one of the fundamental concepts of biology. All cells use energy from their environment to grow, make new parts, and reproduce. Plants trap radiant energy from the sun and store it as chemical bond energy through photosynthesis (Fig. 4-1 ●). They extract carbon and oxygen from carbon dioxide, nitrogen from the soil, and hydrogen and oxygen from water to make biomolecules such as glucose and amino acids.

Animals, on the other hand, cannot trap energy from the sun or use carbon and nitrogen from the air and soil to synthesize biomolecules. They must import energy and biomolecules synthesized by plants or obtained by other animals from plants. Thus, energy from the sun trapped by photosynthesis is the ultimate energy source for all animals, including humans.

Animals extract energy from these biomolecules through the process of *respiration*, which consumes oxygen and produces carbon dioxide and water that recycle to be used again by plants. The process of energy extraction is never totally efficient. As a result, some chemical bond energy is returned into the environment as heat.

If animals ingest more energy than they need for immediate use, the excess energy is stored in chemical bonds, just as it is in plants. Glycogen (a glucose polymer) and lipid molecules are the main energy stores in animals [p. 35]. These storage molecules are available for use at times when an animal's energy needs exceed its food intake.

CONCEPT CHECK	Answers: p. 122

1. Which classes of biomolecules always include nitrogen in their chemical makeup?

Energy Is Used to Perform Work

All living organisms obtain, store, and use energy to fuel their activities. **Energy** can be defined as the capacity to do work. But what is work? We use the word in everyday life to mean various things, from hammering a nail to sitting at a desk writing a paper. In biological systems, work has three basic forms: chemical work, transport work, and mechanical work.

Chemical work enables cells and organisms to grow, maintain a suitable internal environment, and store information needed for reproduction and normal activities. Protein synthesis for wound repair is an example of chemical work.

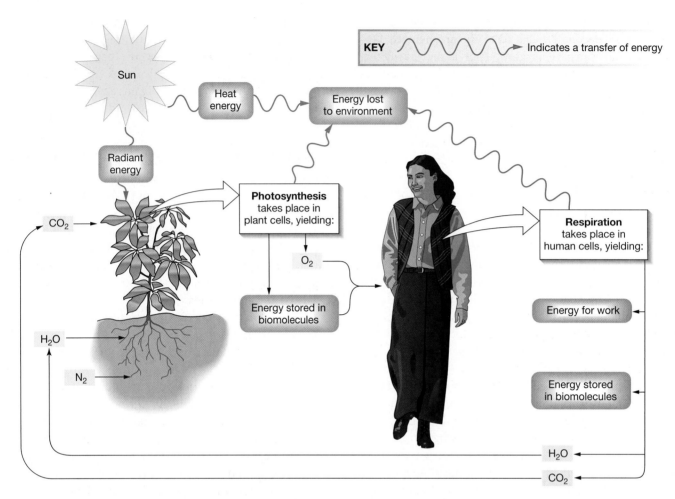

● **Figure 4-1** *Energy transfer in the environment* Plants trap radiant energy from the sun. Animals eat the plants and either use the energy or store it.

Transport work enables cells to move ions, molecules, and larger particles through the cell membrane and through membranes of organelles within the cell. Transport work is particularly useful for creating **concentration gradients**, distributions of molecules in which the concentration is higher on one side of a membrane than on the other. For example, certain types of endoplasmic reticulum [● p. 60] use energy to import calcium ions from the cytosol. This creates a high calcium concentration inside the organelle and a low concentration in the cytosol. If calcium is then released back into the cytosol, it creates a "calcium signal" that causes the cell to perform an action such as muscle contraction.

Mechanical work in animals is used for movement. At the cellular level, movement includes organelles that move around within the cell, changes in cell shape, and the beating of cilia and flagella [● p. 58]. In whole animals, movement usually involves muscle contraction. Most mechanical work, whether in cells or entire animals, is mediated by motor proteins that make up certain intracellular fibers and filaments of the cytoskeleton [● p. 57].

Energy Comes in Two Forms: Kinetic and Potential

Energy can be classified in various ways.* We often think of energy in terms that we deal with daily: thermal energy, electrical energy, mechanical energy. We speak of energy stored in chemical bonds. Each type of energy has its own characteristics. However, all share an ability to appear in two forms: as kinetic energy or as potential energy.

Kinetic energy is the energy of motion [*kinetikos*, motion]. A ball rolling down a hill, perfume spreading through the air, electric current flowing through power lines, heat warming a frying pan, and molecules moving across biological membranes are all examples of kinetic energy.

Potential energy is stored energy. Potential energy may be stored in the position of a molecule with respect to its concentration gradient. In chemical bonds, potential energy is stored in the

_____
*To review the common forms of energy and their standard units, see Appendix A.

position of the electrons that form the bond [● p. 26]. A key feature of all forms of energy is the ability of potential energy to become kinetic energy and vice versa.

Energy Can Be Converted from One Form to Another

Recall that a general definition of energy is the capacity to do work. Work always involves movement and therefore is associated with kinetic energy. Potential energy also can be used to perform work, but it must be converted to kinetic energy. The conversion from potential to kinetic energy is never 100% efficient, and a certain amount of energy is lost to the environment, usually as heat.

The amount of energy lost in the transformation depends on the *efficiency* of the process. Many physiological processes in the human body are not very efficient. For example, 70% of the energy used in physical exercise is lost as heat rather than transformed into the work of muscle contraction.

Figure 4-2 ● summarizes the relationship of kinetic energy and potential energy:

1. Kinetic energy of the moving ball is transformed into potential energy as the ball is pushed up the ramp (Fig. 4-2a).
2. Potential energy is stored in the stationary ball at the top of the ramp (Fig. 4-2b). No work is being performed, but the capacity to do work is stored in the position of the ball.
3. The potential energy of the ball becomes kinetic energy when the ball rolls down the ramp (Fig. 4-2c). Some kinetic energy is lost to the environment as heat due to friction between the ball and the air and ramp.

In biological systems, potential energy is stored in concentration gradients and chemical bonds. It is transformed into kinetic energy when needed to do chemical, transport, or mechanical work.

Thermodynamics Is the Study of Energy Use

Two basic rules govern the transfer of energy in biological systems and in the universe as a whole. The **first law of thermodynamics** states that the total amount of energy in the universe is constant. The universe is considered to be a "closed system"—nothing enters and nothing leaves. Energy can be converted from one form to another, but the total amount of energy in a closed system never changes.

The human body is not a closed system, however. As an open system, it exchanges material and energy with its surroundings. Our bodies cannot create energy, so they import it from outside in the form of food. By the same token, our bodies lose energy, especially in the form of heat, to the environment. Energy that stays within the body can be changed from one form to another or used to do work.

The **second law of thermodynamics** states that natural spontaneous processes move from a state of order (nonrandomness) to a condition of randomness or disorder, also known as **entropy**. To create and maintain order in an open system such as the body requires the input of energy. Disorder occurs when open systems lose energy to their surroundings without regaining it.

The analogy of the deserted town illustrates the second law. When people put all their energy into activities outside the town, the town slowly falls into disrepair and becomes less organized. Similarly, without constant input of energy, a cell is unable to maintain its ordered internal environment. As the cell loses organization, its ability to carry out normal functions disappears, and it dies.

In the remainder of this chapter you will learn how cells obtain energy from and store energy in the chemical bonds of biomolecules, a process called metabolism. Through chemical reactions, cells transfer the potential energy of chemical bonds into kinetic energy for growth, maintenance, reproduction, and movement.

(a)

(b)

(c)

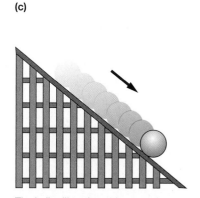

Work is used to push a ball up a ramp.

The ball sitting at the top of the ramp has potential energy, the potential to do work.

The ball rolling down the ramp is converting the potential energy to kinetic energy. However, the conversion is not totally efficient, and some energy is lost as heat due to friction between the ball, ramp, and air.

● **Figure 4-2** *Potential energy, kinetic energy, and work*

2. Name two ways animals store energy in their bodies.

3. What is the difference between potential energy and kinetic energy?

4. What is entropy?

▶ CHEMICAL REACTIONS

Living organisms are characterized by their ability to extract energy from the environment and use it to support life processes. The study of energy flow through biological systems is a field known as **bioenergetics** [*bios*, life + *en-*, in + *ergon*, work]. In a biological system, chemical reactions are a critical means of transferring energy from one part of the system to another. This section explains chemical reactions, discusses different types of reactions, and presents the general mechanisms for trapping, releasing, or transferring energy during reactions.

Energy Is Transferred Between Molecules During Reactions

In **chemical reactions**, a substance becomes a different substance, usually by breaking and/or making covalent bonds. A reaction begins with one or more molecules called **reactants** (or *substrates*) and ends with one or more **products** (Table 4-1). In this discussion, we will consider a reaction that begins with two reactants and ends with two products:

$$A + B \longrightarrow C + D$$

The speed with which a reaction takes place, the **reaction rate**, is the disappearance rate of the reactants (A and B) or the appearance rate of the products (C and D). Reaction rate is measured as change in concentration during a certain time period and is often expressed as molarity per second (M/sec).

Energy Transfer in Reactions The purpose of chemical reactions in cells is to transfer energy from one molecule to another or to use energy stored in the reactants to do work. The potential energy stored in the chemical bonds of a molecule is

Table 4-1: Types of Chemical Reactions

REACTION TYPE	REACTANTS (SUBSTRATES)		PRODUCTS
Combination	A + B	$\longrightarrow$	C
Decomposition	C	$\longrightarrow$	A + B
Single displacement*	L + MX	$\longrightarrow$	LX + M
Double displacement*	LX + MY	$\longrightarrow$	LY + MX

*X and Y represent atoms, ions, or chemical groups.

known as the **free energy** of the molecule. Generally, complex molecules have more chemical bonds and therefore higher free energies. For example, a large glycogen molecule has more free energy than a single glucose molecule, which in turn has more free energy than the carbon dioxide and water from which it was synthesized. The high free energy of complex molecules such as glycogen is the reason these molecules are used to store energy in cells.

To understand how chemical reactions transfer energy between molecules, we should answer two questions. First, how do reactions get started? The energy required to initiate a reaction is known as the *activation energy*. Second, what happens to the free energy of the products and reactants during a reaction? The difference in free energy between reactants and products is known as the *net free energy change of the reaction*.

Activation Energy Activation energy is the initial input of energy required to bring reactants into a position that allows them to react with each other. This "push" needed to start the reaction is shown in Figure 4-3 ● as the little hill up which the rock must be pushed before it can roll by itself down the slope. A reaction with low activation energy will proceed spontaneously when the reactants are brought together. You can demonstrate a spontaneous reaction by pouring a little vinegar onto some baking soda and watching the two react to form carbon dioxide. Reactions with high activation energies will not proceed spontaneously or will proceed too slowly to be useful. For example, if you pour vinegar over a pat of butter, no observable reaction will take place.

Endergonic and Exergonic Reactions One characteristic property of any chemical reaction is the free energy change that occurs as the reaction moves from reactants to products. If the free energy of the products is greater than that of the reactants,

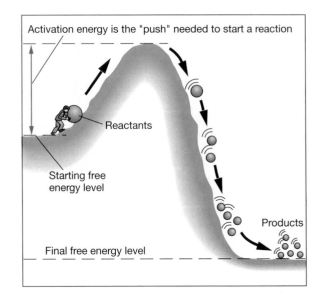

● **Figure 4-3** *Activation energy*

(a) **Exergonic reactions** release energy because the products have less energy than the reactants.

(b) **Endergonic reactions** trap some activation energy in the products, which then have more free energy than the reactants.

KEY —— Substrates —— Activation of reaction —— Reaction process —— Products

● **Figure 4-4** *Exergonic and endergonic reactions*

some of the activation energy was trapped in the chemical bonds of the products. A reaction that requires net input of energy is called an **endergonic reaction** [*end(o)*, within + *ergon*, work]. If the free energy of the products is lower than the free energy of the reactants, the reaction released energy. A reaction that releases energy is called an **exergonic reaction** [*ex-*, out + *ergon*, work].

A plot of an exergonic, or *energy-producing*, reaction is shown in Figure 4-4a ●. Note that the products C and D have a lower free energy than the reactants A and B. This indicates the release of energy during the reaction. The energy released can be used by other molecules to do work or may be given off as heat. In a few cases, released energy is stored as potential energy in a concentration gradient. An important biological example of an exergonic reaction is the combination of ATP and water to form ADP, inorganic phosphate (P_i), and H^+. Energy is released during this reaction when the high-energy phosphate bond of the ATP molecule is broken:

$$ATP + H_2O \longrightarrow ADP + P_i + H^+ + energy$$

During endergonic, or *energy-utilizing*, reactions, part of the activation energy added to the reactants is retained by the products. A plot of net free energy change for an endergonic reaction, $E + F \longrightarrow G + H$, is shown in Figure 4-4b. Compare this endergonic reaction with the energy-producing reaction of Figure 4-4a, in which all activation energy added to the reactants was released during the reaction.

Some of the energy added to an endergonic reaction remains trapped in the chemical bonds of the products. Thus, energy-utilizing reactions are often *synthesis* reactions in which complex molecules are made from smaller molecules. For example, many glucose molecules link together to create the glucose polymer glycogen. The complex glycogen molecule has higher free energy than the simple glucose molecules used to make it.

If a reaction traps energy in one direction, it will release energy in the reverse direction. For example, the energy trapped in the bonds of glycogen during its synthesis will be released when glycogen is broken back down into glucose.

Coupled Reactions Where does the activation energy for metabolic reactions come from? The simplest source of activation energy is to couple an exergonic reaction that produces energy to an endergonic reaction that requires energy. Some of the most familiar coupled reactions are those that use the energy released by breaking the high-energy bond of ATP to drive an endergonic reaction:

$$E + F \xrightarrow{\quad ATP \quad ADP + P_i \quad} G + H$$

In this type of coupled reaction, the two reactions take place simultaneously and in the same location, so that the energy from ATP can be used immediately to drive the endergonic reaction of E and F.

However, it is not always practical for reactions to be directly coupled like this. Consequently, living cells have developed ways to trap and save energy released by exergonic reactions. The most common method for trapping energy is in the form of high-energy electrons carried on nucleotides [● p. 41]. The nucleotide molecules NADH, $FADH_2$, and NADPH all capture energy in the electrons of their hydrogen atoms (Fig. 4-5 ●). NADH and $FADH_2$ usually transfer most of their energy to ATP, which can then be used to drive endergonic reactions.

Reversible and Irreversible Reactions A chemical reaction that can proceed in both directions is called a **reversible reaction**. In a reversible reaction, the forward reaction $A + B \longrightarrow C + D$ and its reverse reaction $C + D \longrightarrow A + B$

89

Exergonic reaction

● **Figure 4-5** *Energy transfer and storage in biological reactions* Energy released by exergonic reactions can be trapped in the high-energy electrons of NADH, FADH$_2$, or NADPH. Energy that is not trapped is given off as heat.

are both likely. If a reaction proceeds in one direction but not the other, it is an **irreversible reaction**.

The net free energy change of a reaction plays an important role in determining whether that reaction can be reversed, because the net free energy change of the forward reaction contributes to the activation energy of the reverse reaction. For example, look at the activation energy of the reaction C + D ⟶ A + B in Figure 4-6 ●. This reaction is the reverse of the reaction shown in Figure 4-4a. Because there was a large release of energy in the forward reaction A + B ⟶ C + D, the reverse reaction (Fig. 4-6) has a large activation energy. As you will recall, the larger the activation energy, the less likely it is that the reaction will proceed spontaneously. Theoretically all reactions could be reversed with enough energy input, but some reactions have such large activation energies that they are essentially irreversible.

In your study of physiology, you will encounter a few irreversible reactions. However, most biological reactions are reversible: if the reaction A + B ⟶ C + D is possible, then so is the reaction C + D ⟶ A + B. Reversible reactions are shown with arrows that point in both directions: A + B ⇌ C + D. One of the main reasons that many biological reactions are reversible is that they are aided by the specialized proteins known as enzymes.

CONCEPT CHECK   Answers: p. 122

5. What is the difference between endergonic and exergonic reactions?

6. If you mix baking soda and vinegar together in a bowl, the mixture reacts and foams up, releasing carbon dioxide gas. Name the reactant(s) and product(s) in this reaction.

7. Do you think this reaction is endergonic or exergonic? Do you think it is reversible? Defend your answers.

▶ ENZYMES

Enzymes are biological *catalysts*, proteins that speed up the rate of chemical reactions without themselves being changed. Without enzymes, most chemical reactions in the cell would go so slowly that the cell would be unable to live. Because an enzyme is not permanently changed or used up in the reaction that it catalyzes, we might write it in a reaction equation this way:

$$A + B + enzyme \longrightarrow C + D + enzyme$$

This way of writing the reaction shows that the enzyme participates with reactants A and B but is unchanged at the end of the reaction. A more common shorthand for enzymatic reactions shows the name of the enzyme above the reaction arrow, like this:

$$A + B \xrightarrow{enzyme} C + D$$

In enzymatically catalyzed reactions, the reactants are called **substrates**.

Enzymes Lower the Activation Energy of Reactions

How does an enzyme increase the rate of a reaction? In thermodynamic terms, it lowers the activation energy, making it more likely that the reaction will start (Fig. 4-7 ●). Enzymes accomplish

GRAPH QUESTION

Is this an endergonic or exergonic reaction?

Activation energy

Free energy of molecule

A+B

Net free energy change

C+D

Time ⟶

KEY

— Substrates
— Activation of reaction
— Reaction process
— Products

● **Figure 4-6** *Some reactions have large activation energies*

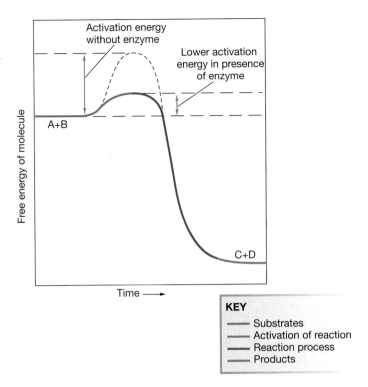

KEY
— Substrates
— Activation of reaction
— Reaction process
— Products

● **Figure 4-7** *Enzymes lower the activation energy of reactions*

of one million molecules of CO_2 and water to carbonic acid. In the absence of enzyme, it takes nearly 100 seconds for only one molecule of CO_2 and water to be converted to carbonic acid. Without enzymes in cells, biological reactions would go so slowly that the cell would be unable to obtain enough energy to live.

Enzymes Bind to Their Substrates

How does an enzyme bring the substrates into the best position for them to react? Most enzymes are large protein molecules with complex three-dimensional shapes [● p. 39]* On each enzyme molecule is a region known as the **binding site**, part of the protein molecule that actually binds to the substrates. When enzyme-substrate binding takes place, the substrate molecules are brought close to each other and to the enzyme's **active site**, the region that promotes reaction of the substrates with each other.

For many years, it was thought that the binding site was shaped to fit the substrates exactly, the way a key fits into a lock. A representation of this *lock-and-key model* of enzyme activity is shown in Figure 4-8a ●. In recent decades, however, scientists have discovered that the binding site and the substrates do not need to fit one another exactly.

The binding site needs only to attract the substrates to that region of the enzyme. Then, as the binding site and the substrates begin to interact, the binding site changes shape to fit more closely to the substrates. This **induced-fit model** of enzyme-substrate interaction is shown in Figure 4-8b. According to this model, the binding site has an intermediate shape that can change to fit either the substrate or the product molecules. This trait would enable an enzyme to bind to either reactants or products and thus be able to catalyze a single reaction in both directions. In fact, enzymes can do this. Induced fit explains reversibility better than the older lock-and-key model, in which the binding site was matched to either the substrates or the products, but not to both.

Enzyme Specificity Most enzymes react with only one set of substrates or with a group of similar substrates. The ability of an enzyme to catalyze a certain reaction or a group of closely related reactions is called **specificity**. Some enzymes are very specific about the reactions that they catalyze. For example, the sole function of the enzyme *glucokinase* is to attach a phosphate group onto glucose molecules when they enter a cell. Other enzymes act on whole groups of molecules. The enzymes known as *peptidases* act on peptide bonds of polypeptides without regard for which two amino acids are joined by those bonds. Peptidases therefore are not very specific in their action.

An Enzyme's Name Gives Information About Its Function Most enzyme names are important clues to their function in a cell. The names of most enzymes are instantly

this by binding to the reactant molecules and bringing them together into the best position for reacting with each other. Without enzymes, the reaction would depend on the random collision of the molecules to bring them into alignment. Notice in the figure that neither the starting nor ending free energy content of the substrates and products has changed.

The rate of a reaction catalyzed by an enzyme is much more rapid than the same reaction taking place in the absence of the enzyme. For example, consider the enzyme *carbonic anhydrase*, which converts CO_2 and water to carbonic acid. A single molecule of carbonic anhydrase takes one second to catalyze the conversion

RUNNING PROBLEM

Tay-Sachs disease is a particularly devastating condition. Normally, lysosomes within cells contain enzymes that digest old, worn-out parts of the cell. In Tay-Sachs, hexosaminidase A, an enzyme that digests specific glycolipids called gangliosides, is absent. As a result, gangliosides accumulate in nerve cells in the brain, causing them to swell and function abnormally. Infants with Tay-Sachs disease slowly lose muscle control and brain function. They usually die before age four.

Question 1: Hexosaminidase A is also required to remove gangliosides from the light-sensitive cells of the eye. Based on this information, what is another symptom of Tay-Sachs disease besides loss of muscle control and brain function?

---

*Recently, researchers discovered that RNA can sometimes act as a catalyst.

(a) Lock-and-key model
In this older model, the enzyme binding site was an exact match for the substrates.

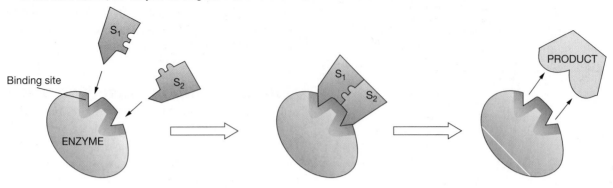

(b) Induced-fit model
In this model, the enzyme binds most tightly to an intermediate form of substrate or product. This model is more compatible with reversible reactions that are catalyzed by one enzyme.

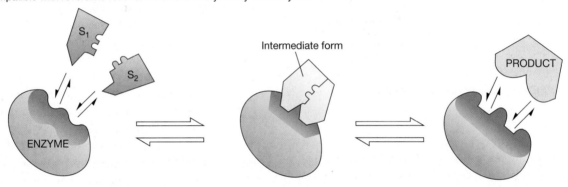

● **Figure 4-8** *Two models of enzyme binding sites*

recognizable by the suffix -ase. The first part of the enzyme's name (everything that precedes the suffix) usually refers to the type of reaction, the substrate upon which the enzyme acts, or both. For example, glucokinase, mentioned previously, has glucose as its substrate and as a kinase adds a phosphate group [● p. 28] to the substrate. Addition of a phosphate group is called **phosphorylation**. The peptidases mentioned above break up peptides into smaller peptides or amino acids by cleaving the peptide bond that links the amino acids.

Note that a few enzymes have two names. These enzymes were discovered before 1972, when the current standards for naming enzymes were first adopted. As a result, they have both a new name and a commonly used older name. Pepsin and trypsin, two different digestive enzymes, are examples of older enzyme names.

A few enzymes come in a variety of related forms; these are known as isozymes [*iso-*, equal]. **Isozymes** are enzymes that catalyze the same reaction but under different conditions or in different tissues. Their structures are slightly different, which causes the variability in their activity. Lactate dehydrogenase, for example, has several isozymes, including one found primarily in the heart and a second found in skeletal muscle and the liver.

Enzymes and their isozymes have an important role in the medical diagnosis of certain conditions. For example, in the hours following a heart attack, damaged heart muscle cells release enzymes into the blood. One way to determine whether a person's chest pain was indeed due to a heart attack is to look for elevated levels of heart isozymes in the blood. Some diagnostically important enzymes and the diseases of which they are suggestive are listed in Table 4-2.

CONCEPT CHECK   Answers: p. 122

8. Name the substrate for each of the following enzymes: lactase, peptidase, lipase, sucrase.

9. How does the induced-fit model of enzyme action differ from the lock-and-key model?

10. What would be a biological advantage of having isozymes of one enzyme?

Some Enzymes Must Be Activated

Some enzymes are not ready to catalyze reactions when they are first synthesized. These enzymes are produced instead as inactive molecules called *proenzymes* or *zymogens*. When the enzymes are

Table 4-2: Diagnostically Important Enzymes

Elevated blood levels of these enzymes are suggestive of the pathologies listed.

ENZYME	RELATED DISEASES
Acid phosphatase*	Cancer of the prostate
Alkaline phosphatase	Diseases of bone or liver
Amylase	Pancreatic disease
Creatine kinase (CPK)	Myocardial infarction (heart attack), muscle disease
Glutamate dehydrogenase (GDH)	Liver disease
Lactate dehydrogenase (LDH)	Myocardial infarction, liver disease, excess break-down of red blood cells

*A newer test for a molecule called prostate specific antigen (PSA) has replaced the test for acid phosphatase in the diagnosis of prostate cancer.

needed, one or more portions of the molecule are chopped off, and the inactive protein becomes an active enzyme (Fig. 4-9 ●). This process is known as *proteolytic activation* [*proteo-*, protein + *lysis*, to dissolve]. The enzymes involved in blood clotting and digestion are among those that are produced as proenzymes. Protein hormones as well as enzymes undergo proteolytic activation.

Separation of Proteins by Electrophoresis

Many enzymes that have isozymes are complex structures with multiple protein chains. For example, lactate dehydrogenase has two kinds of subunits, named H and M, that are assembled into *tetramers*—groups of four. The isozymes include H_4, H_2M_2, and M_4. One way to determine which isozymes are present in a tissue sample is to use a technique known as **electrophoresis**. In this technique, a solution of proteins is placed at one end of a container filled with a polyacrylamide polymer gel. An electric current is passed through the gel, causing the negatively charged proteins to move toward the positively charged end of the gel. The rate at which the proteins move depends upon their size, shape, and the electrical charge on the amino acids that make up the protein. The proteins move along the gel at different rates and become separated from each other, causing each of them to appear as an individual band of color when stained with a dye called Coomassie blue or with silver. Electrophoresis can be used to separate mixtures of charged macromolecules and is also used to identify DNA.

Chymotrypsinogen is the inactive form of the enzyme.

Activation by removing peptide fragments

Chymotrypsin, the active enzyme, has 3 chains held together by disulfide (S—S) bonds.

● **Figure 4-9** *Proteolytic activation of the enzyme chymotrypsinogen*

If an enzyme has both an active and an inactive form, the inactive enzyme is often named by adding the suffix *-ogen* to the name of the active enzyme. For example, *pepsinogen* is the inactive form of the digestive enzyme *pepsin*.

Some Enzymes Require Cofactors or Coenzymes

Sometimes the activation of an enzyme requires the presence of an additional molecule or ion called a **cofactor**. Cofactors may be either inorganic molecules or nonprotein organic molecules. Inorganic cofactors are ions, such as Ca^{2+} or Mg^{2+}. They must attach to the enzyme before the substrates can bind to the binding site. This process is shown in Figure 4-10 ●.

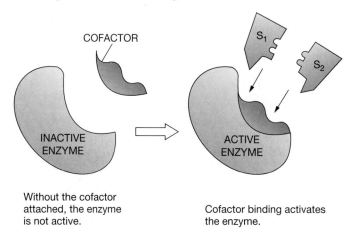

COFACTOR

S_1

S_2

INACTIVE ENZYME

ACTIVE ENZYME

Without the cofactor attached, the enzyme is not active.

Cofactor binding activates the enzyme.

● **Figure 4-10** *Cofactors bind to the binding site*

Organic cofactors are called **coenzymes**. Coenzymes do not alter the enzyme's binding site as inorganic cofactors do. Instead, coenzymes act as receptors and carriers for atoms or functional groups that are removed from the substrates during the reaction. Although coenzymes are needed for metabolic reactions to take place, they are not required in large amounts.

Many chemicals that we call **vitamins** are the precursors of coenzymes. The water-soluble vitamins, such as the B vitamins, vitamin C, folic acid, biotin, and pantothenic acid, become coenzymes required for various metabolic reactions. For example, vitamin C is needed for adequate collagen synthesis.

Modulators Alter Enzyme Activity

The ability of an enzyme to speed up a reaction can be altered by various factors, including temperature, pH, or molecules that interact with the enzyme. A factor that influences enzyme activity is known as a **modulator**. If a modulator *activates* an enzyme, the reaction rate catalyzed by the enzyme will increase. If a modulator *inactivates* the enzyme, the reaction rate will decrease and may even stop completely.

There are two basic mechanisms by which modulation takes place. The modulator either (1) changes the ability of the substrate to bind to the binding site or (2) changes the ability of the enzyme to alter the activation energy of the reaction. The different types of modulation are discussed in the following sections and are summarized in Table 4-3.

Acidity and Temperature
Both temperature and pH can have dramatic effects on protein structure and function. Changes in temperature and pH can disrupt the bonds that hold a protein in its tertiary configuration, causing the protein to lose its shape

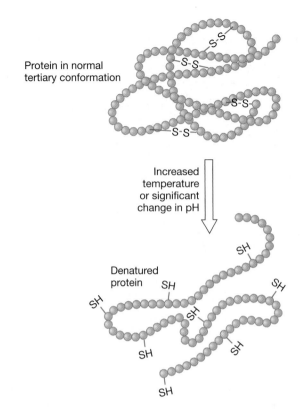

● **Figure 4-11** *Denaturation of proteins results when intramolecular bonds are broken* When an enzyme is denatured, it loses its activity along with its three-dimensional shape.

(Fig. 4-11 ●). If you have ever fried an egg, you have watched this transformation happen to the egg white protein albumin as it changes from a slithery clear state to a firm white state.

Hydrogen ions in high enough concentration to be called acids have a similar effect on proteins. During preparation of

Table 4-3: Factors That Affect Reaction Rate

Essential for enzymatic reaction*	
Cofactors	Required for substrate binding at active site.
Coenzymes	Act as receptors for atoms or groups.
Proteolytic activation	Converts inactive to active form by removal of part of molecule. Examples: digestive enzymes, blood-clotting enzymes.

Modulators that alter reaction rate	
1. Changes substrate binding at active site	
Competitive inhibitor	Competes directly with substrate by binding to active site.
Allosteric modulator	Binds to enzyme away from active site and changes active site. Reversible.
(activator or inhibitor)	Examples: Feedback inhibition, regulatory proteins.
pH and temperature	Alter three-dimensional shape of enzyme by disrupting hydrogen or S—S bonds. May be irreversible if protein denatures.
2. Changes ability of enzyme to decrease activation energy of the reaction	
Covalent modulation	Binds covalently to enzyme and changes its ability to alter activation energy. Example: addition or removal of phosphate groups.

*Not all enzymes require these, but those that do must have these factors in order to catalyze a reaction.

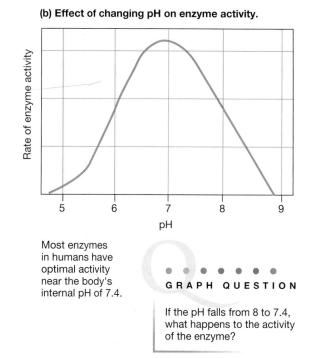

(a) Effect of changing temperature on enzyme activity.

Rate of enzyme activity

Temperature (°C)
20 30 40 50 60

This enzyme denatures around 50° C.

GRAPH QUESTION

Is the enzyme more active at 30° C or at 48° C?

(b) Effect of changing pH on enzyme activity.

Rate of enzyme activity

pH
5 6 7 8 9

Most enzymes in humans have optimal activity near the body's internal pH of 7.4.

GRAPH QUESTION

If the pH falls from 8 to 7.4, what happens to the activity of the enzyme?

● **Figure 4-12** *Effect of pH and temperature on enzyme activity*

ceviche, the national dish of Ecuador, raw fish is marinated in lime juice. The acidic lime juice contains hydrogen ions that disrupt hydrogen bonds in the muscle proteins of the fish, causing the proteins to lose their configuration. As a result, the meat becomes more firm and opaque, just as it would if it were cooked with heat.

Small changes in pH and temperature can increase or decrease the activity of enzymes (Fig. 4-12 ●). However, once the changes exceed some critical point, the structure of the enzyme is altered so much that its activity is destroyed. When this occurs, the enzyme is said to be *denatured*. In a few cases, activity can be restored if the modulator is removed. The protein then resumes its original shape as if nothing had happened. Usually, however, denaturation is a permanent loss of activity. There is certainly no way to unfry an egg or uncook a piece of fish. The potentially disastrous influence of temperature and pH on enzymes and other proteins is one reason that these factors are so closely regulated by the body.

Chemical Modulators Chemical modulators are molecules that bind to enzymes and alter their catalytic ability. There are three types of chemical modulators: competitive inhibitors, allosteric modulators, and covalent modulators.

Competitive inhibitors bind to the enzyme at its binding site, blocking the site and thus preventing the substrate from binding with the enzyme (Fig. 4-13 ●). Some competitive inhibitors are alternative substrates that can also be acted upon by the enzyme. Others are simply molecules that bind to the enzyme and block the binding site without being acted upon themselves. They are like the guy who slips into the front of the movie ticket line to chat with his girlfriend, the cashier. He has no interest in buying a ticket, but he prevents the people in line behind him from getting their tickets for the movie.

Allosteric modulators [*allo-*, other + *stereos*, solid (as a shape)] bind to the enzyme away from the binding site and change the shape of the binding site (Fig. 4-14 ●). This type of modulation can either (1) increase the probability of enzyme-substrate binding and enhance enzyme activity or (2) decrease the affinity of the binding site for the substrate and inhibit enzyme activity. Allosteric modulation also occurs in proteins other than enzymes. For example, in Chapter 18 you will learn how the oxygen-binding ability of the respiratory protein hemoglobin changes with allosteric modulation by carbon dioxide, H^+, and several other factors.

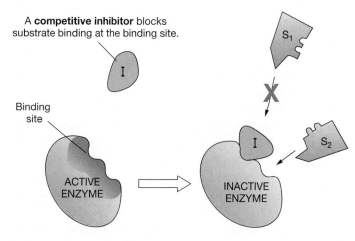

A **competitive inhibitor** blocks substrate binding at the binding site.

Binding site

ACTIVE ENZYME

INACTIVE ENZYME

● **Figure 4-13** *Competitive inhibition*

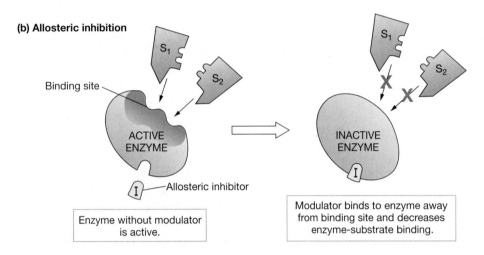

● **Figure 4-14** *Allosteric modulation*

Covalent modulators are atoms or functional groups that use covalent bonds to bind to enzymes and alter the enzymes' properties. Like allosteric modulators, covalent modulators may either increase or decrease the activity of the enzyme. One of the most common covalent modulators is the phosphate group. Many enzymes in the cell can be either activated or inactivated when a phosphate group forms a covalent bond with them. The ability to turn reactions on and off or up and down allows the cell to regulate the flow of biomolecules through synthetic and energy-producing pathways.

One of the best known chemical modulators is the antibiotic penicillin. Alexander Fleming discovered this compound in 1928, when he noticed that *Penicillium* mold inhibited bacterial growth. By 1938, researchers had extracted the active ingredient penicillin from the mold and used it to treat infections in humans. Yet it was not until 1965 that researchers figured out exactly how the antibiotic works. Penicillin is a competitive inhibitor that binds to a key bacterial enzyme by mimicking the normal substrate. In doing so, it forms bonds that are unbreakable and thereby irreversibly inhibits the enzyme. Without the enzyme, the bacterium is unable to make a rigid cell wall. Without the rigid wall, the bacterium swells, ruptures, and dies.

CONCEPT CHECK Answers: p. 122

11. Name two processes in enzymatic reactions that an enzyme modulator can affect.

12. Match the modulator to its action(s).

    (a) Allosteric
    (b) Competitive
    (c) Covalent

    1. Bind by covalent bonds
    2. Bind away from the binding site
    3. Bind to the binding site
    4. Inhibit activity only
    5. Inhibit or enhance activity

Enzyme and Substrate Concentration Affect Reaction Rate

We measure the rate of an enzymatic reaction by monitoring either how fast the products are synthesized or how fast the substrates are consumed. Reaction rate can be altered by a number of factors, including changes in temperature, the amount of enzyme present, or the substrate concentration. In mammals we consider temperature to be essentially constant. This leaves enzyme amount and substrate concentration as the two main variables that affect reaction rate.

Reaction Rate Is Directly Related to the Amount of Enzyme Present

Enzyme amount is a major determinant of the rate of an enzymatic reaction. If there is no enzyme, most metabolic reactions go very slowly. If enzyme is present, the rate of the reaction will be proportional to the amount of enzyme.

The graph in Figure 4-15 ● shows the results of a typical experiment in which the amount of substrate remains constant while the amount of enzyme is varied. As the graph shows, an increase in the amount of enzyme present causes an increase in the rate of the reaction.

As an analogy, think of the checkout lines in a supermarket. Imagine that each cashier is an enzyme, the waiting customers are substrates, and those leaving the store with their purchases are products. One hundred people will get checked out faster when there are 25 lines open than when there are only 10 lines. Likewise, in a reaction, the presence of more enzyme molecules means that more binding sites are available to interact with the substrate molecules. As a result, reactants are converted into products more rapidly.

Regulating enzyme concentration is an important strategy that cells use to control their physiological processes. Cells alter the amount of an enzyme by influencing both its synthesis and its breakdown. If enzyme synthesis exceeds breakdown, enzyme accumulates and the reaction catalyzed by the enzyme speeds up. If enzyme breakdown exceeds synthesis, the amount of enzyme decreases, as does the reaction rate. Even when the amount of enzyme is constant, there is still a steady turnover of enzyme molecules.

Reaction Rate Can Reach a Maximum

If the concentration of enzyme is constant, then the reaction rate will vary according to the concentration of substrate. Figure 4-16 ● shows the results of a typical experiment in which the enzyme concentration is constant, but the concentration of substrate varies. At low substrate concentrations, the reaction rate is directly propor-

In this experiment, the amount of enzyme was constant.
At the maximum rate the enzyme is said to be saturated.

● **Figure 4-16** *Effect of changing substrate concentration on rate*

tional to the substrate concentration. But once the concentration of substrate molecules increases beyond a certain level, the enzyme molecules have no more binding sites free to bind substrate molecules. The enzyme is catalyzing reactions as fast as it can, and the rate of reaction reaches a maximum value. This condition is known as **saturation**.

An analogy to saturation appeared in the early days of television on the *I Love Lucy* show. Lucille Ball was working on the conveyor belt of a candy factory, loading chocolates into the little paper cups of a candy box. Initially, the belt was moving slowly, and she had no difficulty removing the candy and putting it into the box. Gradually, the belt brought candy to her more rapidly, and she had to increase her packing speed to keep up. Finally, the belt was bringing candy to her so fast that she couldn't pack it all in the boxes because she was working at her maximum rate. That was Lucy's saturation point. (Her solution was to stuff the candy into her mouth as well as into the box!)

CONCEPT CHECK   Answers: p. 122

13. What happens to the rate of an enzymatic reaction as the amount of enzyme present decreases?

14. What happens to the rate of an enzymatic reaction when the enzyme has reached saturation?

Reversible Reactions Obey the Law of Mass Action

The study of reaction rates in cells is complicated by the fact that some reactions are reversible. This means that the forward reaction $A + B \longrightarrow C + D$ can reverse, so that $C + D \longrightarrow A + B$. In the reverse reaction, the products become the reactants and the reactants become the products.*

What then determines in which direction the reaction is to go? The answer is that reversible reactions go to a state of

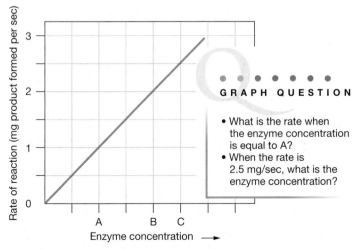

In this experiment, the substrate amount remains constant.

● **Figure 4-15** *Effect of changing enzyme concentration on rate*

---

*The naming of forward and reverse directions is arbitrary.

Law of mass action:
when a reaction is at equilibrium, the ratio of the products and substrates will remain constant.

(a) Reaction at equilibrium

$$A + B \underset{r_2}{\overset{r_1}{\rightleftharpoons}} C + D$$

Rate of reaction in forward direction (r_1) = rate of reaction in reverse direction (r_2)

(b) Equilibrium disturbed

Add A and B to system

$$A + B \underset{r_2}{\overset{r_1}{\rightleftharpoons}} C + D$$

Reaction rate r_1 increases: $r_1 > r_2$
(Some of the added substrate is converted into product.)

(c) Equilibrium restored

$$A + B \underset{r_2}{\overset{r_1}{\rightleftharpoons}} C + D$$

$r_1 = r_2$

The ratio of substrates to products is always the same at equilibrium.

● **Figure 4-17** *Law of mass action* Although this figure shows the amount of A + B equal to C + D at equilibrium, the actual ratio can vary, depending on the reaction.

equilibrium, where the rate of the reaction in the forward direction (A + B ⟶ C + D) is exactly equal to the rate of the reverse reaction (C + D ⟶ A + B), as shown in Figure 4-17a ●. At equilibrium, there is no net change in the amount of reactant or product. As fast as A and B convert to C and D, the reverse reaction takes place: C and D turn back into A and B at an equal rate.

But what happens if some factor disturbs this equilibrium by increasing the amount of A and B within the system? You learned in the previous section that reaction rate is proportional to substrate concentration. If the concentration of A and B in the system goes up, so does the rate of the forward reaction (Fig. 4-17b). More A and B are converted into products C and D. But because the reaction is reversible, as the product concentration increases, so does the rate of the reverse reaction. The system swings back and forth until finally it reaches equilibrium again (Fig. 4-17c).

Reversible reactions like this obey the **law of mass action**, a simple relationship that holds for chemical reactions whether in a test tube or in a cell. You may have learned this law in chemistry as *Le Châtelier's principle*. In very general terms, the law of mass action says that when a reaction is at equilibrium, the ratio of the substrates to the products is always the same.

If the concentration of a substrate (A or B) or product (C or D) changes, the equilibrium will be disturbed. The system then ad-

justs the substrate and product concentrations until the equilibrium ratio is restored. The law of mass action is important in physiology because as the concentration of substrates changes, the change is reflected in the concentrations of the products.

Enzymatic Reactions Can Be Categorized

Most reactions catalyzed by enzymes can be classified into one of four categories: oxidation-reduction, hydrolysis-dehydration, exchange-addition-subtraction, and ligation reactions. Table 4-4 summarizes these categories.

Oxidation-Reduction Reactions Oxidation-reduction reactions are the most important reactions in energy extraction and transfer within the cell. These reactions transfer electrons or protons (H^+) between atoms, ions, or molecules. A molecule that gains electrons or loses H^+ is said to be **reduced**. One way to think of this is to remember that adding negatively charged electrons *reduces* the electric charge on the molecule. Likewise, the loss of a positive proton leaves behind an electron with a negative charge. Conversely, molecules, atoms, or ions that lose electrons or gain H^+ are said to be **oxidized**.

Hydrolysis-Dehydration Reactions Hydrolysis and dehydration reactions are important in the breakdown and synthesis of large biomolecules. In **dehydration reactions**, a water molecule is one of the products. In many dehydration reactions, two molecules combine into one, losing water in the process. For example, the monosaccharides glucose and fructose join to make one sucrose molecule [● p. 35]. In the process, one molecule loses a hydroxyl group (—OH) and the other molecule loses a hydrogen to create water, H_2O. When a dehydration reaction results in the synthesis of a new molecule, the process is known as *dehydration synthesis*.

Table 4-4: Types of Chemical Reactions

REACTION TYPE	WHAT HAPPENS	REPRESENTATIVE ENZYMES
1. Oxidation-reduction	Add or subtract electrons or H^+	
(a) Oxidized	Transfer electrons from donor to oxygen	Oxidase
	Remove electrons and H^+	Dehydrogenase
(b) Reduced	Gain electrons	Reductase
2. Hydrolysis-dehydration	Add or subtract a water	Hydrolase
(a) Hydrolysis	Split large molecules by adding water	Protease, lipase
(b) Dehydration	Remove water. Used to make large molecules from several smaller ones	
3. Transfer chemical groups	Add, subtract, or exchange groups between molecules	
(a) Exchange reaction	Phosphate	Kinase
	Amino group	Transaminase
(b) Add a group	Phosphate	Phosphorylase
	Amino group	Aminase
(c) Subtract a group	Phosphate	Phosphatase
	Amino group	Deaminase
4. Ligation	Join two substrates using energy from ATP	Synthetase

In a **hydrolysis reaction** [*hydro*, water + *lysis*, to loosen or dissolve], a substrate changes into one or more products by adding water. In these reactions, the covalent bonds of the water molecule are broken so the water reacts as a hydroxyl group (—OH) and a hydrogen (—H). For example, an amino acid can be removed from the end of a peptide with a hydrolysis reaction (see Fig. 4-27a, p. 109).

When an enzyme name consists of the substrate name with the suffix *-ase*, the enzyme causes a hydrolysis reaction. One example is lipase, an enzyme that breaks up large lipids into smaller lipids by hydrolysis. A peptidase is an enzyme that removes an amino acid from a peptide.

Addition-Subtraction-Exchange Reactions

An **addition reaction** adds a functional group to one or more of the reactants. A **subtraction reaction** removes a functional group from one or more of the reactants. Functional groups are exchanged between or among reactants during **exchange reactions**.

For example, phosphate groups may be transferred between molecules during addition, subtraction, or exchange reactions. The transfer of phosphate groups is an important means of covalent modulation, turning reactions on, off, up, or down. Several types of enzymes catalyze reactions that transfer phosphate groups. The enzymes named *kinases* transfer a phosphate group from a substrate to an ADP molecule to create ATP, or from an ATP molecule to a substrate. For example, creatine kinase transfers a phosphate group from creatine phosphate to ADP, forming an ATP and leaving behind creatine.

The addition, subtraction, or exchange of amino groups [● p. 28] is also important in metabolism. Removal of an amino group from an amino acid or peptide is a **deamination** reaction, illustrated in Figure 4-27b, p. 109. Addition of an amino group is **amination**, and the transfer of an amino group from one molecule to another is called **transamination**.

Ligation Reactions

Ligation reactions join two molecules together using enzymes known as *synthetases* and energy from ATP. An example of a ligation reaction is the synthesis of acetyl coenzyme A from fatty acids and coenzyme A. Acetyl coenzyme A is an important molecule in metabolism, as you will learn in the next section.

CONCEPT CHECK   Answers: p. 122

15. Match the reaction type or enzyme in the left column to the group or particle that is involved.

  (a) Kinase   1. Amino group
  (b) Oxidation   2. Electrons
  (c) Hydrolysis   3. Phosphate group
  (d) Transaminase   4. Water

▶ METABOLISM

Metabolism refers to all chemical reactions that take place within an organism. These reactions extract energy from nutrient biomolecules (such as proteins, carbohydrates, and lipids) and synthesize or break down molecules. Metabolism is often divided into **catabolism**, reactions that produce energy through the breakdown of large biomolecules, and **anabolism**, energy-utilizing reactions that result in the synthesis of large biomolecules. Anabolic and catabolic reactions take place simultaneously in cells throughout the body, so that at any given moment, some biomolecules are being synthesized while others are being broken down.

The energy released from or stored in the chemical bonds of biomolecules during metabolism is commonly measured in kilocalories (kcal). A **kilocalorie** is the amount of energy needed to raise the temperature of 1 liter of water by 1 degree Celsius. One kilocalorie is the same as a Calorie, with a capital C, used for quantifying the energy content of food.

Much of the energy released during catabolism is trapped in the high-energy phosphate bonds of ATP or in the high-energy electrons [● p. 25] of NADH, FADH$_2$, or NADPH. From these temporary carriers, energy is transferred to ATP or the covalent bonds of biomolecules through anabolic reactions.

Metabolism is a highly coordinated process in which activities at any given moment are matched to the needs of the cell. The enzymatic reactions of metabolism form a network of interconnected chemical reactions, or **pathways**. Each step of a pathway is a different enzymatic reaction, and the reactions of a pathway proceed in sequence. Starting substrate A is changed into product B, which then becomes the substrate for the next reaction. In that reaction, B is changed into C, and so forth:

$$A \longrightarrow B \longrightarrow C \longrightarrow D$$

We call the molecules of the pathway **intermediates** because the products of one reaction become the substrates for the next. You will sometimes hear metabolic pathways called *intermediary metabolism*. Certain intermediates, called *key intermediates*, participate in more than one pathway and act as the branch points for channeling substrate in one direction or another. Glucose is a key intermediate in several metabolic pathways.

In many ways, a network of metabolic pathways is similar to a detailed road map (Fig. 4-18 ●). Just as a map shows a network of roads that connect various cities and towns, metabolism can be thought of as a network of chemical reactions connecting various intermediate products. Each city or town is a different chemical intermediate. One-way roads are irreversible reactions, and big cities with roads to several destinations are key interme-

diates. Just as there may be more than one way to get from one place to another, there can be several pathways between chemical intermediates.

Cells Regulate Their Metabolic Pathways

How do cells regulate the flow of molecules through their metabolic pathways? They do so in five basic ways:

1. By controlling the enzyme concentration
2. By producing allosteric and covalent modulators
3. By using two different enzymes to catalyze reversible reactions
4. By isolating enzymes within intracellular organelles
5. By maintaining an optimum ratio of ATP to ADP

(a)

Section of road map

(b)

Glycogen
Glucose
Glucose 6-phosphate
Fructose 6-phosphate
Fructose
Fructose 1-phosphate
Fructose 1,6-bisphosphate
Ribose 5-phosphate
Glycerol
DHAP
Glucose 3-phosphate

DHAP = dihydroxyacetone phosphate

Metabolic pathways drawn like a road map

● **Figure 4-18** *A group of metabolic pathways resembles a road map* Cities on the map are equivalent to intermediates in metabolism. In metabolism, there may be more than one way to go from one intermediate to another, just as on the map there may be many ways to get from one city to another.

- **Figure 4-19** *Feedback inhibition* The accumulation of product Z inhibits the first step of the pathway. As the cell consumes Z in another metabolic reaction, the inhibition is removed and the pathway resumes.

We discussed the effect of changing enzyme concentration in an earlier section of this chapter. The other four methods are described below.

Enzyme Modulation Enzyme modulation, described on pages 94–95 ✎, is frequently controlled by hormones and other signals coming from outside the cell. This type of outside regulation is a key element in the integrated control of the body's metabolism following a meal or during periods of fasting.

Metabolic pathways may also have their own built-in form of modulation, called **feedback inhibition**. In this form of modulation, the end product of a pathway, shown as Z in Figure 4-19 ●, acts as an inhibitory modulator of the pathway. As the pathway proceeds and Z accumulates, the enzyme catalyzing the conversion of A to B is inhibited. This slows down production of Z until the cell can use it up. Once the levels of Z fall, the inhibition is removed and the pathway starts to run again. Because Z is the end product of the pathway, this type of inhibition is sometimes called *end-product inhibition.*

Enzymes and Reversible Reactions Cells can use reversible reactions to regulate the rate and direction of metabolism. If a single enzyme can run the reaction in either direction (Fig. 4-20a ●), the reaction will go to a state of equilibrium as determined by the law of mass action. These reactions therefore cannot be closely regulated except by modulators and by control of the amount of enzyme. But if a reaction requires two different enzymes, one for the forward and one for the reverse reaction, the cell can regulate the reaction more closely (Fig. 4-20b). If there is no enzyme for the reverse reaction, the reaction is irreversible (Fig. 4-20c).

Isolation of Enzymes Within the Cell Many enzymes of metabolism are localized in specific parts of the cell. Some, like the enzymes of carbohydrate metabolism, are found dissolved in the cytosol, whereas others are isolated within specific organelles. Mitochondria, endoplasmic reticulum, Golgi apparatus, and lysosomes all contain enzymes that are not found in the cytosol. This separation of pathways allows the cell to control metabolism by regulating the movement of substrate between the different compartments. The isolation of enzymes within organelles is an example of *compartmentation* [✎ p. 9].

Ratio of ATP to ADP The energy status of the cell is one final mechanism that can influence metabolic pathways. Through complex regulation, the ratio of ATP to ADP in the cell determines whether pathways that result in ATP synthesis are turned on or off. When ATP levels are high, production of ATP drops. When ATP levels are low, the cell sends substrates through pathways that result in more ATP synthesis. In the next section, we look further into the role of ATP in cellular metabolism.

ATP Transfers Energy Between Reactions

The usefulness of metabolic pathways as suppliers of energy is often measured in terms of the net amount of adenosine triphosphate, or ATP, that the pathways can yield. ATP is a nucleotide molecule containing three phosphate groups [✎ p. 40]. The third phosphate group is attached by a covalent bond in an energy-requiring reaction. Energy is stored in this **high-energy phosphate bond** and then released when the bond is broken during removal of the phosphate group. This relationship is shown by the following reaction:

$$ADP + P_i + energy \longleftrightarrow ADP{\sim}P \ (= ATP)$$

The squiggle indicates a high-energy bond and P_i is the abbreviation for an inorganic phosphate group. Estimates of the amount of free energy released by breaking a high-energy phosphate bond range from 7 to 12 kcal per mole of ATP.

ATP is more important as a carrier of energy than as an energy storage molecule. For one thing, cells have a limited amount of ATP. A resting adult human would require 40 kg (88 pounds) of ATP to support one day's worth of metabolic activity, far more than our cells could store. Instead, the body acquires most of its daily energy requirement from the chemical bond energy stored in complex biomolecules. Metabolic reactions transfer that energy to the high-energy bonds of ATP, or in a few cases, into high-energy bonds of the related nucleotide, *guanosine triphosphate*, or **GTP**.

(a)

Reversible reaction using one enzyme for both directions

(b)

Reversible reaction requiring two enzymes

(c)

Glucose + PO₄

hexokinase

Glucose 6-phosphate

Irreversible reaction goes in only one direction

- **Figure 4-20** *Reversible and irreversible metabolic reactions* Most cells of the body can phosphorylate glucose to glucose 6-phosphate but cannot make glucose by dephosphorylation, because they lack the enzyme glucose 6-phosphatase (c). The liver and kidney are the primary sites for glucose synthesis (b).

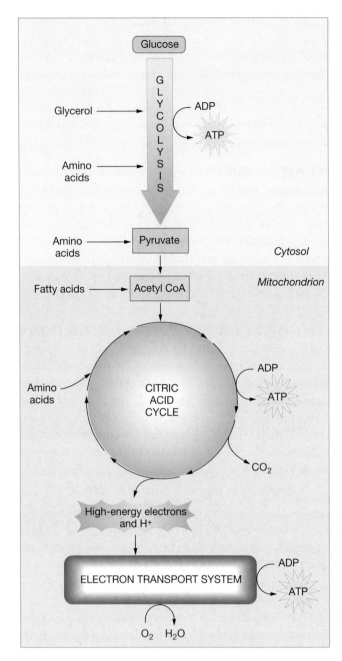

● **Figure 4-21** *Overview of aerobic pathways for ATP production* ATP can be made through aerobic pathways using the chemical bond energy from glucose, amino acids, or glycerol and fatty acids.

The metabolic pathways that yield the most ATP molecules are those that require oxygen, the **aerobic**, or *oxidative*, pathways. **Anaerobic** [*an-*, without + *aer*, air] pathways that can proceed without oxygen also produce ATP molecules, but in much smaller quantities. The lower ATP yield of anaerobic pathways means that most animals (including humans) are unable to survive for extended periods on anaerobic metabolism alone. In the next section we will consider how the three major groups of biomolecules are metabolized in order to transfer energy to ATP.

NAVIGATOR

This icon represents the different steps in the metabolic summary figure. Look for it in the figures that follow to help you navigate your way through metabolism.

CONCEPT CHECK   Answers: pp. 122–123

16. Name five ways that cells regulate the movement of substrates through metabolic pathways.

17. In what part of the molecule does ATP trap and store energy? In what part of the molecule does NADH store energy?

18. What is the difference between aerobic and anaerobic pathways?

▶ ATP PRODUCTION

The catabolic pathways that extract energy from biomolecules and transfer it to ATP are summarized in Figure 4-21●. Aerobic production of ATP follows two common pathways: glycolysis and the citric acid (tricarboxylic acid) cycle. Both pathways produce small amounts of ATP directly, but their most important contributions to ATP synthesis are high-energy electrons carried by NADH and $FADH_2$ to the electron transport system in the mitochondria. The electron transport system, in turn, transfers energy from the electrons to the high-energy phosphate bond of ATP. At various points, the process produces carbon dioxide and water. The water can be used by the cell, but carbon dioxide is a waste product and must be removed from the body.

All three types of biomolecules used for energy (carbohydrates, proteins, and lipids) use the enzymes of glycolysis, the citric acid cycle, and the electron transport system to produce ATP. Carbohydrates enter the pathways in the form of glucose. Lipids are broken down into fatty acids and glycerol [● p. 35], which enter the pathways at different points. Proteins are broken down into assorted amino acids, which also enter at various points. Because glucose is the only molecule that follows the entire pathway, we look first at glucose catabolism. We will then see how protein and lipid catabolism differ from glucose catabolism.

The pathways for ATP production are a good example of compartmentation. The enzymes of glycolysis are located in the cytosol while the enzymes of the citric acid cycle are in the mitochondria.

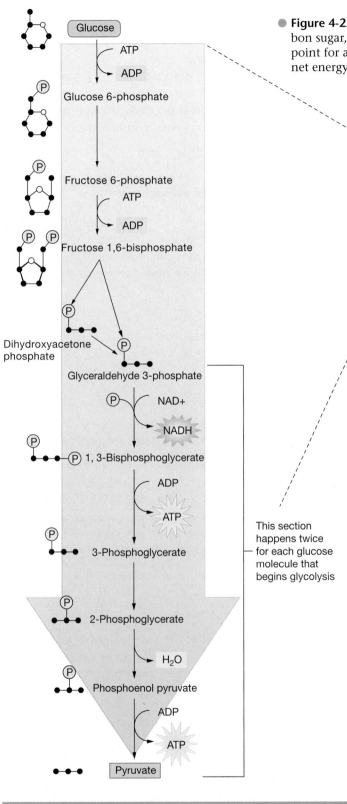

● **Figure 4-22** *Glycolysis* The glycolysis pathway converts glucose, a six-carbon sugar, into two three-carbon pyruvate molecules. Pyruvate is the branch point for aerobic and anaerobic metabolism of glucose. Aerobic glycolysis has a net energy yield of two ATP and two NADH.

KEY
● = Carbon
○ = Oxygen
Ⓟ = Phosphate group
(side groups not shown)

Glycolysis Converts Glucose and Glycogen into Pyruvate

The main catabolic pathway of the cytosol is the series of reactions known as **glycolysis** [*glyco-*, sweet + *lysis*, dissolve]. During glycolysis, a molecule of glucose is converted by a series of enzymatically catalyzed reactions into two pyruvate molecules, producing a net release of energy. Thus, glycolysis is considered to be an exergonic pathway even though some steps of the pathway require energy input. A portion of the energy released during glycolysis is used to phosphorylate ADP molecules, trapping the energy in ATP molecules. The events of glycolysis can be summarized as follows:

$$\text{Glucose} + 2\,\text{NAD}^+ + \text{ADP} + 2\,\text{P}_i \longrightarrow$$
$$2\,\text{Pyruvate} + 2\,\text{ATP} + 2\,\text{NADH} + 2\,\text{H}^+ + 2\,\text{H}_2\text{O}$$

Notice that glycolysis does not require oxygen. It therefore serves as a common pathway for aerobic and anaerobic catabolism of glucose.

Figure 4-22 ● shows the steps of glycolysis. In the first step, glucose is phosphorylated with the aid of ATP to glucose 6-phosphate.* Glucose 6-phosphate is a key intermediate that leads into glycolysis as well as to other pathways we will not consider. Note that by the time the six-carbon glucose molecule reaches the end of glycolysis, it has been split into two molecules of pyruvate, each containing three carbons.

Two steps of glycolysis are endergonic and require energy input from two ATP. Other steps are exergonic and produce four

This section happens twice for each glucose molecule that begins glycolysis

CONCEPT CHECK Answers: p. 123

19. Match the components to the molecules in which they are part:

(a) Amino acids
(b) Fatty acids
(c) Glycerol
(d) Glucose

1. Carbohydrates
2. Lipids
3. Polysaccharides
4. Proteins
5. Triglycerides

ATP and two NADH for each glucose that enters the pathway. The net energy yield from each glucose is therefore two ATP and two NADH. (See Table 4-5 for a summary of energy yield of various catabolic pathways.)

At normal cell conditions, many organic acids such as pyruvic acid, lactic acid, and glutamic acid are ionized and are therefore named as the corresponding anion: pyruvate, lactate, and glutamate. The suffix -ate identifies the anions of organic and some inorganic acids.

CONCEPT CHECK   Answers: p. 123

20. Do endergonic reactions release energy or trap it in the products?

Anaerobic Metabolism Converts Pyruvate into Lactate

Pyruvate is a branch point for metabolic pathways, like the hub cities on the road map. Depending on the cell's needs and condi-

tion, pyruvate can be shuttled off into one of two pathways (Fig. 4-23 ●). If the cell contains adequate oxygen, pyruvate continues into the citric acid cycle.

If the cell lacks sufficient oxygen for aerobic pathways (a condition that may be caused by many factors, including strenuous exercise), pyruvate is converted into lactate* with the assistance of the enzyme lactate dehydrogenase mentioned earlier:

$$\text{Pyruvate} \xrightarrow[\text{Lactate dehydrogenase}]{\text{NADH} \quad \text{NAD}^+} \text{Lactate}$$

The conversion of pyruvate to lactate changes the NADH produced earlier in glycolysis back to NAD^+ when a hydrogen atom is transferred to the lactate molecule. As a result, the net energy yield for **anaerobic metabolism** of one glucose molecule is two ATP and no NADH.

---

*Lactate is the anion of lactic acid.

Table 4-5: Summary of Energy Yield from Catabolic Pathways

Aerobic metabolism

PATHWAY	USES	MAKES	NET ENERGY YIELD
Glycolysis			
glucose $\longrightarrow$ 2 pyruvate	2 ATP	4 ATP	+2 ATP
		2 NADH	+2 NADH*
Aerobic Metabolism			
2 pyruvate $\longrightarrow$ 2 acetyl CoA	—	2 NADH	+2 NADH
		2 CO_2	
2 acetyl CoA through citric acid cycle		2 ATP	+2 ATP
		6 NADH	+6 NADH
		2 $FADH_2$	+2 $FADH_2$
Total yield for 1 glucose through aerobic metabolism:			+4 ATP
			+2 NADH cytoplasm*
			+8 NADH mitochondria
			+2 $FADH_2$

Maximal ATP yield after mitochondrial oxidative phosphorylation:

4 ATP	4 ATP
2 cytoplasmic NADH × 1.5–2.5 ATP/NADH	3–5 ATP
8 mitochondrial NADH × 2.5 ATP/NADH	20 ATP
2 $FADH_2$ × 1.5 ATP/$FADH_2$	3 ATP
	= 30–32 ATP

Anaerobic metabolism

PATHWAY	USES	MAKES	NET ENERGY YIELD
Glucose $\longrightarrow$ 2 lactate	2 ATP	4 ATP	+2 ATP
	2 NADH	2 NADH	0 NADH

*Cytoplasmic NADH molecules cannot enter mitochondria. Their ATP yield depends on whether their electrons go to NADH or to $FADH_2$ in the mitochondria.

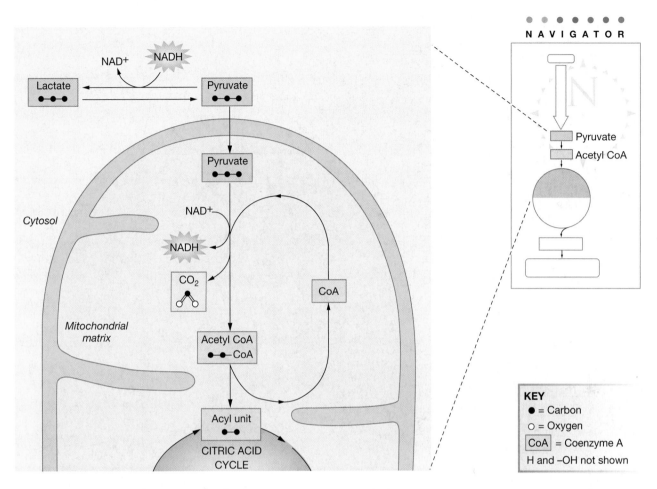

KEY
● = Carbon
○ = Oxygen
CoA = Coenzyme A
H and –OH not shown

● **Figure 4-23** *Pyruvate metabolism* Anaerobic metabolism of pyruvate creates lactate (lactic acid). Under aerobic conditions, pyruvate is transported into mitochondria and converted to acetyl CoA. The acyl unit of acetyl CoA feeds into the citric acid cycle.

CONCEPT CHECK Answers: p. 123

21. Using Figure 4-22, identify the steps of glycolysis that are:
    (a) obviously exergonic
    (b) obviously endergonic
    (c) catalyzed by kinases
    (d) catalyzed by dehydrogenases (*Hint*: see Table 4-4)
    (e) dehydration reactions

22. Lactate dehydrogenase acts on lactate by (adding or removing?) _____ and _____. This process is called (oxidation or reduction?).

Pyruvate Enters the Citric Acid Cycle in Aerobic Metabolism

If the cell has adequate oxygen for **aerobic metabolism**, then pyruvate molecules formed from glucose during glycolysis are transported into the mitochondria (Fig. 4-23). Once in the mitochondrial matrix, pyruvate is converted into another key intermediate called *acetyl coenzyme A*, or **acetyl CoA** for short. As its name suggests, this molecule has two parts: a two-carbon acyl unit, derived from pyruvate, and a coenzyme.

Coenzyme A is made from the vitamin pantothenic acid. Like enzymes, it is not consumed during metabolic reactions and can be reused. The synthesis of acetyl CoA from pyruvate and coenzyme A is an exergonic reaction that produces one NADH. During the reaction, one of pyruvate's three carbon atoms combines with oxygen and is released as carbon dioxide (CO_2), leaving acetyl CoA with only two carbons.

Acetyl CoA releases its two-carbon acyl unit into a cyclical pathway known as the **citric acid cycle**, or *tricarboxylic acid cycle* (Fig. 4-24 ●). This pathway was first described by Sir Hans A. Krebs, so it is also known as the *Krebs cycle*. Because Sir Hans Krebs described other metabolic cycles, we will avoid confusion by using the term citric acid cycle. The citric acid cycle makes a never-ending circle, adding carbons from an acetyl CoA with each turn of the cycle and producing ATP, high-energy electrons, and carbon dioxide.

The two-carbon acyl unit enters the cycle by combining with a four-carbon oxaloacetate molecule that is the last intermediate in the cycle. The resulting six-carbon citrate molecule then goes through a series of reactions until it completes the cycle as another oxaloacetate molecule. Most of the energy released by the

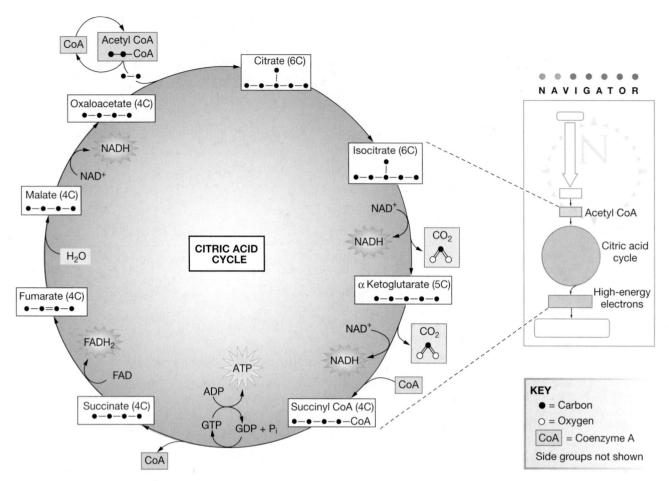

● **Figure 4-24**   *The citric acid cycle*   A two-carbon acyl unit from acetyl CoA combines with a four-carbon oxaloac-etate molecule to make citrate. Each full turn of the cycle releases two carbons as carbon dioxide and produces one ATP, three NADH, and one $FADH_2$.

reactions of the cycle is captured as **high-energy electrons** on three NADH and one $FADH_2$. However, some energy is used to make the high-energy phosphate bond of one ATP, and the re-mainder is given off as heat. In two of the reactions, carbon and oxygen are removed in the form of carbon dioxide. By the end of the cycle, the four-carbon oxaloacetate molecule that remains is ready to begin the cycle again.

In the next step of aerobic metabolism, NADH and $FADH_2$ transfer their high-energy electrons to the electron transport sys-tem and return to the citric acid cycle as NAD^+ and FAD^+.

The Electron Transport System Transfers Energy from NADH and FADH₂ to ATP

The final step in aerobic ATP production is the transfer of energy from the electrons of NADH and $FADH_2$ to ATP. This transfer is made possible by a group of mitochondrial proteins known as the **electron transport system**, located in the inner mitochondrial membrane [● p. 59]. The protein complexes of the electron transport system include enzymes and the iron-containing pro-teins known as **cytochromes**.

According to the **chemiosmotic theory**, as high-energy elec-trons pass through the electron transport system, their energy is used to move H^+ from the mitochondrial matrix into the inter-membrane space. The energy stored in this concentration gradi-ent is then transferred to the high-energy bond of ATP when the H^+ move back across the membrane. The synthesis of ATP using the electron transport system is called **oxidative phosphoryla-tion**, because the electron transport system requires oxygen to act as the final acceptor of electrons and H^+.

The process of oxidative phosphorylation is shown in detail in Figure 4-25 ●. Mitochondrial NADH and $FADH_2$ produced in the citric acid cycle give up high-energy electrons. Two at a time, the electrons pass through the protein complexes, losing energy with each transfer. During three of the transfers, enough energy is released to perform the work of transporting H^+ ions into the in-termembrane space. During the other transfers, the released ener-gy is given off as heat.

By the end of the electron transport chain, electrons have given up the usable portion of their stored energy. At that point, each two electrons combine with two H^+ from the pool of H^+ ions in the matrix. The resulting hydrogen atoms then combine with an oxygen atom, creating a molecule of water, H_2O.

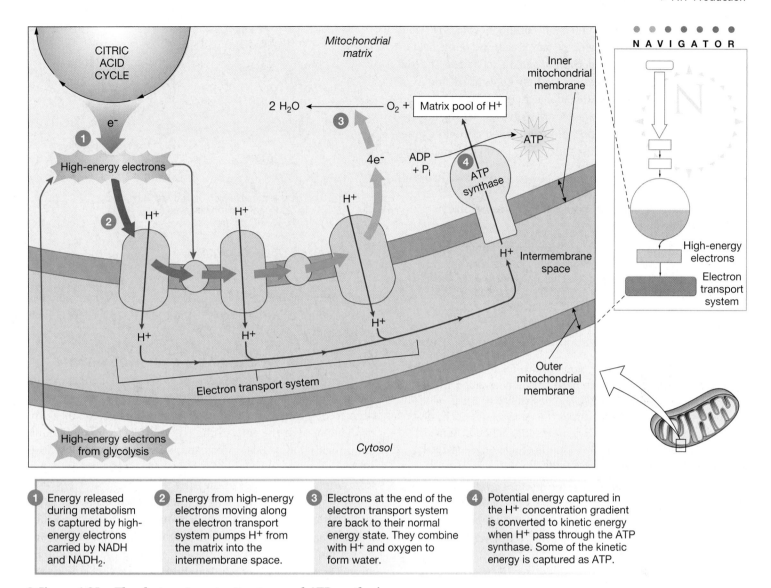

1. Energy released during metabolism is captured by high-energy electrons carried by NADH and NADH$_2$.

2. Energy from high-energy electrons moving along the electron transport system pumps H$^+$ from the matrix into the intermembrane space.

3. Electrons at the end of the electron transport system are back to their normal energy state. They combine with H$^+$ and oxygen to form water.

4. Potential energy captured in the H$^+$ concentration gradient is converted to kinetic energy when H$^+$ pass through the ATP synthase. Some of the kinetic energy is captured as ATP.

● **Figure 4-25**   *The electron transport system and ATP synthesis*

CONCEPT CHECK   Answers: p. 123

23. What is phosphorylation?

ATP Production Is Coupled to Movement of Hydrogen Ions Across the Inner Mitochondrial Membrane

As we have seen, the energy released by electrons moving through the electron transport system is stored as potential energy by H$^+$ ions concentrated in the intermembrane space. This stored energy is converted into chemical energy when the ions move back into the mitochondrial matrix through a protein known as the F_1F_0 *ATPase,* or **ATP synthase.** As H$^+$ ions move back into the matrix through a pore in the enzyme, the synthase transfers their kinetic energy to the high-energy phosphate bond of ATP. Because energy conversions are never completely efficient, a portion of the energy is released as heat. For each three H$^+$ that shuttle through the enzyme, a maximum of one ATP is formed.

The Maximum Energy Yield of One Glucose Molecule Is 30–32 ATP

If we tally up the maximum potential energy yield for the catabolism of one glucose molecule through aerobic pathways, the total comes to 30–32 ATP (Table 4-5). Note that we say potential yield. This is because often the mitochondria do not work up to capacity. There are various reasons for this, including the fact that there is a certain amount of leakage of H$^+$ ions back into the matrix.

A second source of variability in the ATP produced per glucose comes from the two cytosolic NADH molecules produced during glycolysis. These NADH molecules are unable to enter mitochondria and must transfer their electrons through membrane carriers. Inside the mitochondrion some of these electrons go to FADH$_2$, which has a potential yield of only 1.5 ATP compared to 2.5 ATP made by mitochondrial NADH. Thus, if cytosolic electrons go to NADH, an additional 2 ATP molecules can be produced.

Despite the variability, under all conditions the ATP yield from aerobic metabolism of glucose far exceeds the puny yield of two ATP from anaerobic conversion of glucose to lactate. The low efficiency of anaerobic metabolism severely limits its usefulness in most vertebrate cells, whose metabolic energy demand is greater than can be met in this way. Some cells, such as exercising muscle, can tolerate anaerobic metabolism for a limited period of time. Eventually, however, they must shift back to aerobic metabolism.

CONCEPT CHECK   Answers: p. 123

24. Name two differences between aerobic and anaerobic metabolism of glucose.

25. What is the role of oxygen in oxidative phosphorylation?

26. How is the separation of mitochondria into two compartments essential to the chemiosmotic theory of ATP synthesis?

RUNNING PROBLEM

David and Sarah had their blood drawn for the genetic test several weeks ago and have been anxiously awaiting the results. Today, they returned to the hospital to hear the news. The tests show that Sarah carries the gene for Tay-Sachs disease but David does not. This means that although some of their children may be carriers of the Tay-Sachs gene like Sarah, none of the children will develop the disease.

Question 4: The Tay-Sachs gene is a recessive gene (t). If Sarah is a carrier of the Tay-Sachs gene (Tt) but David is not (TT), what is the chance that any child of theirs will be a carrier? (Consult a general biology or genetics text if you need help solving this problem.)

Large Biomolecules Can Be Used to Make ATP

We have now examined how glucose is used for ATP production, but how are large biomolecules such as glycogen, protein, and fat used to make ATP? The answer is a series of pathways that converts these biomolecules into intermediates that enter glycolysis and the citric acid cycle (see overview in Figure 4-21, p. 102).

Glycogen Converts to Glucose   Glycogen, a glucose polymer, is the primary carbohydrate storage molecule in animals. Glycogen can be broken down through the pathways of **glycogenolysis** into glycolysis intermediates. About 10% of stored glycogen is hydrolyzed back to plain glucose molecules (Fig. 4-26●). These molecules must then be phosphorylated to glucose 6-phosphate with the help of an ATP.

The other 90% of glycogen molecules are converted directly to glucose 6-phosphate in a reaction that splits a glucose molecule from the polymer with the help of an inorganic phosphate from the cytosol. Because the direct phosphorylation of glycogen

The direct conversion of glycogen to glucose 6-phosphate saves the cell one ATP per glucose.

● **Figure 4-26**   *Glycogen catabolism*

to glucose 6-phosphate saves an ATP, this pathway has a net yield of one additional ATP molecule (maximum 31–33 ATP for aerobic metabolism).

Proteins Can Be Catabolized to Produce ATP   The first step in protein catabolism is the digestion of proteins into smaller polypeptides by enzymes called *proteases*. Once that has been accomplished, other enzymes known as *peptidases* break the peptide bonds at the ends of the polypeptide, freeing individual amino acids (Fig. 4-27●). Most amino acids used for ATP synthesis come not from breaking down protein in cells, however, but from surplus amino acids in the diet.

Amino acids can be converted into intermediates for either glycolysis or the citric acid cycle. The first step in this conversion is *deamination*, or removal of the amino group, creating an ammonia molecule and an organic acid. Some of the organic acids created by deamination include pyruvate, acetyl CoA, and several intermediates of the citric acid cycle. The former amino acids thus enter the pathways of aerobic metabolism and become part of the ATP production process previously described.

The amino groups of amino acids are removed as ammonia (NH_3). They then rapidly pick up hydrogen ions (H^+) to become ammonium ions (NH_4^+). Because both ammonia and ammonium are toxic, liver cells rapidly convert them into urea (CH_4N_2O). Urea is the main nitrogenous waste of the body and is excreted by the kidneys.

Lipids Yield More Energy per Unit Weight than Glucose or Proteins   Lipids are the primary fuel storage molecule of the body, because they have a higher energy content than proteins or carbohydrates. In the reactions known as **lipolysis**, lipids are broken down by lipases into glycerol and fatty acids. Glycerol feeds into glycolysis about halfway through the pathway and goes through the same reactions as glucose from that point on (Fig. 4-28●).

The long carbon chains of fatty acids are not as easy to turn into ATP as glucose or amino acids. A fatty acid must first be transported from the cytosol into the mitochondria. There it is

(a) Protein catabolism

Proteins are broken into amino acids by hydrolysis of their peptide bonds.

Protein or Peptide

Hydrolysis of peptide bond

H_2O

Peptide + Amino acid

(b) Deamination

Removal of the amino group from an amino acid creates ammonia and an organic acid.

$NAD + H_2O$ → $NADH + H^+$

Deamination

NH_3 + Ammonia

Amino acid

Organic acid

Glycolysis or citric acid cycle

(c) Ammonia is toxic and must be converted to urea.

NH_3 → NH_4^+ → Urea
Ammonia   Ammonium

H^+

FIGURE QUESTION

The use of water in reaction (b) makes the reaction:
(1) hydration
(2) hydrolysis
(3) dehydration

● **Figure 4-27** *Protein catabolism and deamination*

① Triglycerides break down into glycerol and 3 fatty acids.

② Glycerol becomes a glycolysis substrate.

③ β-oxidation chops 2-carbon acyl units off the fatty acids.

④ Acyl units become acetyl CoA and can be used in the citric acid cycle.

● **Figure 4-28** *Lipolysis*

slowly disassembled by chopping two-carbon units off the end of the chain, one unit at a time, in the process of **beta-oxidation**.

In most cells, the two-carbon units from fatty acids are converted into acetyl CoA, which feeds directly into the citric acid cycle. Because of the many acetyl CoA molecules that can be produced from a single fatty acid, lipids contain 9 kcal of stored energy per gram, compared with 4 kcal per gram for either proteins or carbohydrates.

The liver is one tissue that is unable to convert the two-carbon units into acetyl CoA and oxidize them to CO_2 and water. For this reason, liver cells must use alternative pathways to break down fatty acids. One result of these alternative pathways is the formation of **ketone bodies**, strong metabolic acids that can seriously disrupt the body's pH balance. We will cover the details of fatty acid metabolism in Chapter 22.

27. Is the digestion of proteins to polypeptides more likely to be a dehydration reaction or a hydrolysis reaction?

28. How do enzymes convert an amino acid to an organic acid?

▶ SYNTHETIC PATHWAYS

The metabolic reactions discussed to this point have been catabolic processes that break up large molecules into smaller ones to trap energy in the high-energy phosphate bond of ATP. We now examine anabolic reactions, those that synthesize polysaccharides, lipids, and proteins. To carry out synthetic reactions, the cell must have an adequate supply of organic nutrients, as well as energy in the form of ATP.

Glycogen Can Be Made from Glucose

Glycogen, the main storage form of glucose in the body, is found in all cells, but the liver and skeletal muscles contain especially high concentrations. Glycogen in skeletal muscles provides a ready energy source for muscle contraction. Glycogen in the liver acts as the main source of glucose for the body in periods between meals. It is estimated that the liver keeps about a four-hour supply of glucose stored as glycogen. Once that glycogen is exhausted by catabolism, the cells must turn to fats and proteins as their source of energy.

The synthesis of glycogen is essentially the reverse of glycogen breakdown (discussed in the previous section). Individual glucose molecules can be linked together into glycogen, or glucose 6-phosphate from glycolysis can be synthesized into the branching glucose chains of glycogen by removal of the phosphate group (see Fig. 4-26).

Glucose Can Be Made from Glycerol or Amino Acids

Glucose is one of the key metabolic intermediates in the body for several reasons. As you have just learned, the aerobic metabolism of glucose is the most efficient way to make ATP, and so glucose is the primary substrate for energy production in all cells. More importantly, it is normally the only substrate for ATP synthesis in nervous tissue. If the brain is deprived of glucose for any period of time, its cells begin to die. For this reason, the body has multiple metabolic pathways that it can use to make glucose, ensuring a continuous supply for the brain.

The simplest source of glucose is glycogen breakdown, or glycogenolysis, as previously discussed. But what happens when all the glycogen stores in the body are used up? In that case, cells can turn to fats and proteins for the intermediates needed to make glucose.

The production of glucose from nonglucose precursors such as proteins or glycerol is a process known as **gluconeogenesis**, or, literally, "birth of new glucose." The pathway is similar to glycolysis run in reverse, but uses different enzymes for some steps. Gluconeogenesis can start with glycerol, various amino acids, or lactate (Fig. 4-29 ●). In all cells, the process can go as far as glu-

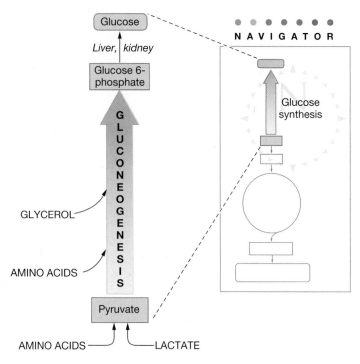

● **Figure 4-29   *Gluconeogenesis***   Glycerol, amino acids, and lactate can be used to make glucose by a series of reactions that are similar to the glycolysis pathway in reverse. Only the liver and kidneys are able to convert glucose 6-phosphate to glucose.

cose 6-phosphate (G-6-P). However, only liver and kidney cells have significant amounts of *glucose 6-phosphatase*, the enzyme that removes the phosphate group from G-6-P to make glucose. During periods of fasting, the liver is the primary source of glu-

Insulin and Metabolism

Why do people with diabetes have elevated levels of blood glucose ("blood sugar")? Part of the answer lies with the disruption of glucose metabolism when insulin is absent or when cells do not respond to it. In a normal person, insulin promotes cellular uptake and anabolic metabolism of glucose by stimulating the appropriate enzymes. Glucose absorbed from the digestive system goes through glycolysis to make ATP or is stored as glycogen. Fatty acids are stored as fats, and amino acids are used to make proteins. In the absence of insulin, the enzymes associated with catabolic pathways are more active. Fats undergo beta-oxidation, amino acids go through gluconeogenesis to become glucose, and glycogen stores are turned back into glucose. The production of glucose within cells coupled to the inability of many cells to absorb glucose creates the elevated "blood sugar" levels that are a hallmark of diabetes.

cose synthesis, producing about 90% of the glucose made from the breakdown of fats and proteins.

CONCEPT CHECK   Answers: p. 123

29. What is the difference between hexokinase and glucose 6-phosphatase? (*Hint*: See Table 4-4.)

Acetyl CoA Is an Important Precursor for Lipid Synthesis

Lipids are so diverse that generalizing about their synthesis is difficult. Generally, lipids are synthesized by enzymes in the smooth endoplasmic reticulum and in the cytosol. Fatty acids are made in the cytosol by an enzyme called *fatty acid synthetase*, which links the two-carbon acyl groups of acetyl CoA together (Fig. 4-30 ●). The process also uses hydrogens and high-energy electrons from NADPH.

Glycerol can be made from glucose or glycolysis intermediates. The combination of glycerol and fatty acids into triglycerides takes place in the smooth endoplasmic reticulum. The phosphorylation steps that turn triglycerides into phospholipids also take place in the endoplasmic reticulum.

Cholesterol [● Fig. 2-14, p. 37] is an important component of cell membranes, and it forms the skeleton of steroid hormones. Cholesterol is usually obtained from the diet, but it is such an important molecule for cells that the body will synthesize it if the diet is deficient. Even vegetarians who eat no animal products (vegans) have substantial amounts of cholesterol in their cells.

The entire structure of cholesterol can be made from acetyl CoA. From there, it is a simple matter to add or subtract functional groups to the carbon rings to change cholesterol into various hormones and other steroids. As you will learn later, all the steroid hormones are very similar in structure.

Proteins Are the Key to Cell Function

Proteins are the molecules that run the cell from day to day. As enzymes, they control the synthesis and breakdown of carbohydrates, lipids, structural proteins, and signal molecules. Proteins in the cell membrane and organelle membranes regulate the movement of molecules into and out of the cell, or within the cell. Other proteins form the structural skeleton of cells and tissues. Thus, protein synthesis is critical to cell function.

The power of proteins arises from their tremendous variability and specificity. Protein synthesis can be compared to creating a language with an alphabet of 20 letters (amino acids) and words up to hundreds of letters in length (peptides and proteins). The language is a rich one, with thousands of different words with different meanings. Like a language, a "spelling error" during the synthesis of a protein can change the function of the protein, just as changing the last letter turns the word "foot" into "food."

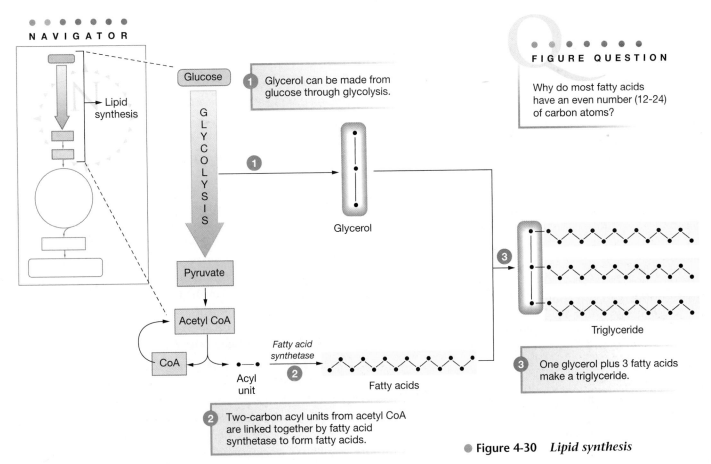

NAVIGATOR

→ Lipid synthesis

GLYCOLYSIS

Glucose

① Glycerol can be made from glucose through glycolysis.

Glycerol

Pyruvate

Acetyl CoA

CoA

Acyl unit

Fatty acid synthetase

② Two-carbon acyl units from acetyl CoA are linked together by fatty acid synthetase to form fatty acids.

Fatty acids

FIGURE QUESTION

Why do most fatty acids have an even number (12-24) of carbon atoms?

Triglyceride

③ One glycerol plus 3 fatty acids make a triglyceride.

● **Figure 4-30** *Lipid synthesis*

In many inherited diseases, the substitution of a single amino acid alters the structure of a protein dramatically. One classic example is sickle cell anemia. In this inherited disease, a change of one amino acid alters the shape of hemoglobin enough so that red blood cells take on a sickle shape, which causes them to get tangled up and block small blood vessels.

The Protein "Alphabet" Begins with DNA   One of the mysteries of biology until the 1960s was the question of how only four bases in the DNA molecule—adenine (A), guanine (G), cytosine (C), and thymine (T)—could code for more than 20 different amino acids. If each base controlled the synthesis of one amino acid, a cell could only make four different amino acids. If pairs of bases represented different amino acids, it could make 4^2, or 16, possible combinations. Because we have 20 amino acids, this is still not satisfactory. But if triplets of bases were the code for different molecules, DNA could create 4^3 or 64 different amino acids. These triplets, called **codons**, are indeed the way information is encoded in DNA (Fig. 4-31 ●).

Of the 64 possible triplet combinations, one DNA triplet (TAC) acts as the initiator or "start codon" that signifies the beginning of a coding sequence. Three codons are recognized as terminator or "stop codons" that show where the sequence ends. The remaining 60 triplets all code for amino acids. Methionine and tryptophan have only one codon each, but the other amino acids have between two and six different codons each. Thus, the information contained in the DNA sequence determines the amino acid sequence of proteins.

How is information translated from DNA into a chain of linked amino acids? First, the codon sequence of DNA is transferred to **messenger RNA (mRNA)** in the process known as

transcription. The mRNA leaves the nucleus and enters the cytosol, where it directs **translation**, the assembly of amino acids into proteins. Two other forms of RNA, *transfer RNA* and *ribosomal RNA*, are needed to help mRNA complete the process of amino acid assembly.

During Transcription, DNA Guides the Synthesis of a Complementary mRNA Molecule

Because DNA is a very large molecule, it cannot pass through the nuclear envelope to the cytosol where protein synthesis takes place. Instead, a smaller single chain of mRNA is made by using DNA as a template during transcription [*trans*, over + *scribe*, to write]. Messenger RNA is formed by **base pairing** in the same way that the double strand of DNA is formed (see Appendix B for a review of DNA synthesis).

During transcription, part of the DNA chain unwinds and the complementary RNA base pairs come in to make a new RNA strand (Fig. 4-32 ●). The main difference between DNA and RNA synthesis is that RNA uses the base uracil (U) in place of thymine (T). Each adenine on the DNA pairs with a uracil on the RNA; the DNA thymine continues to pair with adenine.

CONCEPT CHECK   Answers: p. 123

30. Use the genetic code in Fig. 4-31 to write the DNA triplets that correspond to the three stop codons.

The synthesis of mRNA from the DNA template requires an enzyme known as *RNA polymerase*, plus magnesium or manganese ions, and energy in the form of high-energy phosphate bonds:

DNA template + nucleotides A, U, C, G

↓ RNA polymerase, Mg^{2+} or Mn^{2+}, and energy

mRNA + DNA template

How does the cell know which of the thousands of base pairs of DNA to use to make a strand of mRNA? It turns out that the information a cell needs to make a protein is contained within a segment of DNA known as a gene. A **gene** is a region of DNA that contains all the information needed to make a functional piece of mRNA. We used to believe that a gene coded for only one protein (the one gene–one protein theory), but we now know that one gene can create many proteins.

Alternative Splicing Creates Multiple Proteins from One Sequence of DNA

The gene itself is a nucleotide string that includes the codon sequence for the polypeptide to be synthesized, as well as some noncoding segments. The designation of sequences of nucleotides as coding or noncoding is not fixed for a given gene. This allows a single piece of DNA to create more than one protein

● **Figure 4-31   *The genetic code***   This table shows the genetic code as it appears in the codons of mRNA. The three-letter abbreviations to the right of the brackets indicate the amino acids that these codons represent. The start and stop codons are also marked.

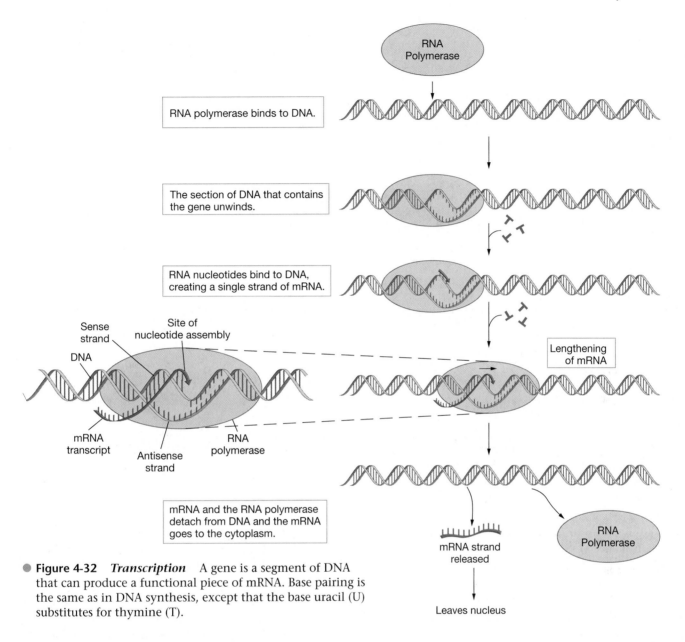

RNA Polymerase

RNA polymerase binds to DNA.

The section of DNA that contains the gene unwinds.

RNA nucleotides bind to DNA, creating a single strand of mRNA.

Sense strand
Site of nucleotide assembly
DNA
mRNA transcript
Antisense strand
RNA polymerase

Lengthening of mRNA

mRNA and the RNA polymerase detach from DNA and the mRNA goes to the cytoplasm.

mRNA strand released

RNA Polymerase

Leaves nucleus

● **Figure 4-32** *Transcription* A gene is a segment of DNA that can produce a functional piece of mRNA. Base pairing is the same as in DNA synthesis, except that the base uracil (U) substitutes for thymine (T).

(see below). Each gene is preceded by a **promoter** region that must be activated before transcription can begin. The promoter region is not transcribed into mRNA.

Transcription begins when the gene is activated. Activation requires the presence of protein **initiation factors** that bind to the promoter region. This, in turn, enables the DNA to bind to **RNA polymerase**. This enzyme moves along a segment of the double helix strand of the DNA molecule and "unwinds" it by breaking the hydrogen bonds between the paired nucleotide bases. One strand of DNA, called the *sense strand*, serves as the guide for mRNA synthesis, while the other *antisense strand* sits idly by.

With the aid of RNA polymerase, nucleotide bases in the sense strand of the DNA molecule pair with the corresponding RNA bases, which are then linked together. For example, a DNA segment with the bases AGTAC will pair with the complementary RNA bases UCAUG. During transcription of mRNA, nucleotides

are linked at an average rate of 40 per second. In humans, the largest mRNAs may contain as many as 5000 nucleotides, and their transcription may take over a minute—a long time for a cellular process. When RNA polymerase reaches the stop codon of the gene, it stops adding nucleotides and releases the mRNA.

The initial chain of newly synthesized mRNA has the complementary sequence to the sense strand of DNA. That means the mRNA contains both segments that encode proteins (**exons**) and noncoding segments called **introns**. The noncoding introns must be edited out of the nucleotide sequence before mRNA leaves the nucleus.

mRNA processing is accomplished by enzymes that clip segments of nucleotides out of the middle or off the ends of the mRNA strand. Other enzymes then splice the remaining pieces back together. The result is a smaller piece of mRNA than was initially made from the DNA template, a piece that now contains the coding sequence for a distinct protein (Fig. 4-33 ●).

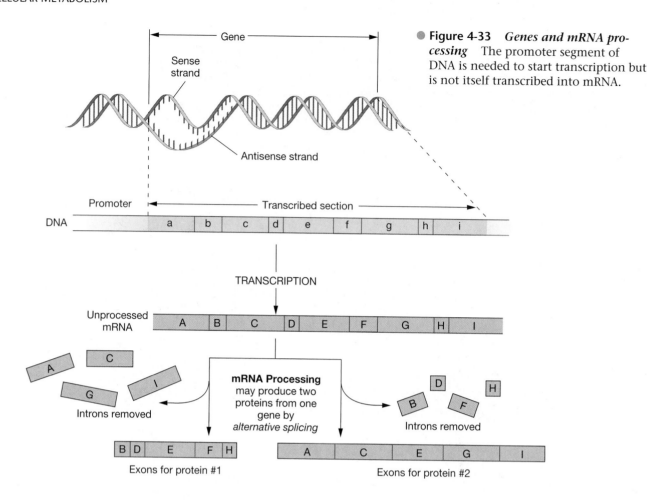

● **Figure 4-33** *Genes and mRNA processing* The promoter segment of DNA is needed to start transcription but is not itself transcribed into mRNA.

One advantage of mRNA processing is that it allows a single nucleotide sequence on the DNA to code for more than one protein. Segments of RNA that are removed one time can be left in the next time, producing a finished mRNA with a different sequence. This phenomenon is known as **alternative splicing**. The closely related forms of a single enzyme known as *isozymes* are probably made by alternative splicing of a single gene.

After mRNA has been processed, it exits the nucleus through nuclear pores and goes to ribosomes in the cytosol. There the mRNA directs construction of protein.

CONCEPT CHECK   Answers: p. 123

31. Explain in one or two sentences the relationship between mRNA, nucleotides, introns, exons, mRNA processing, and proteins.

Translation of mRNA Produces a String of Amino Acids

Protein synthesis is a coordinated effort that requires cooperation between mRNA, ribosomal RNA, and transfer RNA. Upon arrival in the cytosol, processed mRNA binds to ribosomes. Ribosomes, nonmembranous organelles, are two-part structures made of protein and several types of **ribosomal RNA (rRNA)**. Each ribosome has two subunits, one large and one small, that come together

when protein synthesis begins. The small ribosomal subunit binds the mRNA, then adds the large subunit so that the mRNA is sandwiched in the middle (Fig. 4-34 ●). Now the complex of ribosome and mRNA is ready to begin the process of translation.

In translation, the triplet code of the mRNA codons must be matched to the proper amino acid. This is done with the assistance of a third form of RNA, **transfer RNA (tRNA)** (Fig. 4-34). With the aid of a cytosolic enzyme and ATP, one portion of the tRNA binds to its designated amino acid. Another region of the tRNA molecule contains the **anticodon** for that amino acid. The anticodon is the tRNA base triplet that pairs with the mRNA codon for the amino acid. For example, an mRNA codon for the amino acid lysine is AAA. The corresponding anticodon is UUU, found on the tRNA carrying lysine.

As translation begins, tRNAs carrying amino acids bind to the corresponding anticondons of ribosomal mRNA. This puts newly arrived amino acids into the correct orientation to link to the growing peptide chain. Dehydration synthesis creates a *peptide bond* between the amino group (NH_2) of the newly arrived amino acid and the carboxyl end (COOH) of the peptide chain.

When the last amino acid is joined, the mRNA, the polypeptide, and the ribosomal subunits separate. The ribosomes are ready for a new round of protein synthesis, but the mRNA is broken down by enzymes known as **ribonucleases**. Some forms of

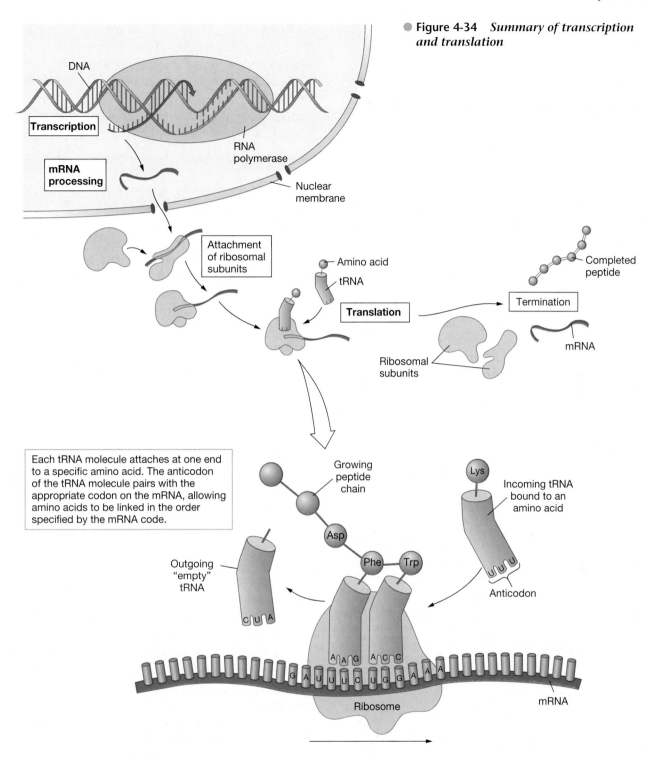

● **Figure 4-34**   *Summary of transcription and translation*

DNA

Transcription

RNA polymerase

mRNA processing

Nuclear membrane

Attachment of ribosomal subunits

Amino acid

tRNA

Translation

Termination

Completed peptide

mRNA

Ribosomal subunits

Each tRNA molecule attaches at one end to a specific amino acid. The anticodon of the tRNA molecule pairs with the appropriate codon on the mRNA, allowing amino acids to be linked in the order specified by the mRNA code.

Growing peptide chain

Lys

Incoming tRNA bound to an amino acid

Asp

Phe   Trp

Outgoing "empty" tRNA

Anticodon

Ribosome

mRNA

mRNA are broken down quite rapidly, while others may linger in the cytosol and be translated many times.

Protein Sorting Directs Proteins to Their Destination

One of the amazing aspects of protein synthesis is the way specific proteins go from the ribosomes directly to where they are needed within the cell. It is as if each protein has an address label on it that tells the cell where it should go. Those proteins destined for one of the membranous organelles are identified by a special segment of amino acids known as a **signal sequence** or **targeting sequence**. This tag directs the protein to the proper organelle and allows it to be transported through the organelle membrane.

Some proteins synthesized on cytoplasmic ribosomes and polyribosomes [● p. 57] do not have signal sequences and remain in the cytoplasm when released from the ribosome. Others are directed to the mitochondria or peroxisomes [● p. 62]. Many peptides are synthesized on ribosomes attached to the rough endoplasmic reticulum. As these proteins are being made, the signal sequence directs them through the membrane and into the lumen of the endoplasmic reticulum.

Final Protein Structure Is Determined by Post-Translational Modification

The polypeptide sequence that comes off a ribosome is the primary structure of a protein [p. 38]. However, the newly created protein is far from its final form. The amino acid chain can:

1. fold into various three-dimensional shapes
2. create cross-links between different regions of its peptide chain
3. be cleaved (split) into fragments
4. attach other molecules or groups
5. assemble with other peptide chains into a *polymeric* (many part) protein

Protein folding creates the tertiary structure of the proteins. Cleavage of the peptide chain, attachment of molecules or groups, and cross-linkages are three general types of **post-translation modification**, changes made to the protein molecule after translation. Over 100 different types of post-translational modification have been described thus far. Assembly of proteins into *polymers* creates the quaternary structure of the protein.

Protein Folding Polypeptide chains released from ribosomes are free to take on their final three-dimensional shape. The peptide chain first forms its secondary structure, which may be an α-helix or a β-pleated sheet [p. 38]. The molecule then folds into its final shape as hydrogen bonds, covalent bonds, and ionic bonds form between amino acids in the chain. Research studies show that some protein folding takes place spontaneously but that it is often facilitated by helper proteins called *molecular chaperones*.

The three-dimensional shape of proteins is often essential for proper function. Misfolded proteins, along with other proteins the cell wishes to destroy, are tagged with a protein called *ubiquitin* and sent to *proteasomes*, cylindrical cytoplasmic enzyme complexes whose function is to break down proteins.

Cross-linkage Some protein folding is held in place by relatively weak hydrogen bonds and ionic bonds. However, other proteins form strong covalent bonds between different parts of the chain. These bonds are often **disulfide bonds** (S—S) between two amino acids. For example, the three chains of the digestive enzyme chymotrypsin are held together by disulfide bonds (Fig. 4-9, p. 93).

Cleavage Some biologically active proteins, such as enzymes and hormones, are synthesized initially as inactive molecules that must have segments removed before they become active. Chymotrypsin, shown in Figure 4-9 (p. 93), must have two small peptide fragments removed before it can catalyze a reaction. Post-translation processing of peptide hormones is discussed in Chapter 7.

Addition of other molecules or groups Proteins can be modified by the addition of sugars, creating glycoproteins, or by combination with lipids to make lipoproteins [p. 40]. The two most common chemical groups added to proteins are phosphate groups, PO_4^{2-} and methyl groups, $—CH_3$. Addition of a phosphate group is called *phosphorylation* and addition of a methyl group is called *methylation*.

CONCEPT CHECK   Answers: p. 123

32. What would removal of a phosphate group be called?

33. Name three general types of post-translational modification of proteins.

Assembly into multimeric proteins Many complex proteins have quarternary structure with multiple subunits, where protein chains assemble into dimers, trimers, or tetramers. One example is the enzyme lactate dehydrogenase, described in the Biotechnology box on page 93. Another example is the hemoglobin molecule, with four protein chains, shown in Figure 2-16 (p. 39).

CONCEPT CHECK   Answers: p. 123

34. Would hemoglobin be a monomer, dimer, trimer, or tetramer?

Proteins Made on the Endoplasmic Reticulum Enter the Secretory Pathway

Many proteins undergo post-translational modification within the endoplasmic reticulum. Once the peptide chain enters the lumen, the signal sequence is enzymatically removed. The protein is then ready to be modified. Once segments have been cleaved off or molecules added, the proteins are packaged into small membrane-bounded *transport vesicles* that bud off the endoplasmic reticulum. These vesicles cross the cytosol to the adjacent Golgi apparatus where they fuse with the Golgi membranes and dump their contents into the lumen (Fig. 4-35).

Once inside the Golgi, the proteins may be modified further. Finally, when the protein reaches the end of the Golgi apparatus, it is again packaged into vesicles for delivery to its final destination. Vesicles that bud off the Golgi apparatus may become lysosomes, intracellular storage vesicles, or secretory vesicles, depending on their contents.

The many ways that proteins can be modified after synthesis add to the complexity of the human proteome. Not only must we know the sequence of a protein, we must know how it is processed, where the protein occurs in or outside the cell, and what it does. Scientists working on the Human Genome Project predicted that our DNA would code for about 30,000 proteins, but they were not taking into account alternative splicing or post-translational modifications. Scientists working on the Human Proteomics Initiative are predicting that we will find over one million different proteins. The magnitude of this project means that we will be working on it for many years into the future.

1 Newly transcribed mRNA goes to the cytoplasm.

2 mRNA on ribosomes is translated into a string of amino acids, which is directed into the lumen of the endoplasmic reticulum. Post-translation modification begins here.

3 Transport vesicles take proteins to the Golgi apparatus.

4 Post-translational modification continues in the Golgi.

5 Proteins that will leave the cell are packaged in secretory vesicles.

6 Proteins that remain in the cell are packaged in lysosomes or storage vesicles.

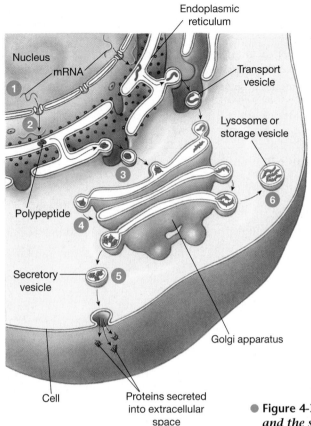

● **Figure 4-35** *Post-translational modification and the secretory pathway*

RUNNING PROBLEM CONCLUSION

Tay-Sachs Disease

In this running problem you learned that Tay-Sachs disease is an incurable, recessive genetic disorder in which the enzyme that breaks down gangliosides in cells is missing. One in 27 Americans of Eastern European Jewish descent carries the gene for this disorder. You have also learned that a blood test can detect the presence of the deadly gene. Check your understanding of this running problem by comparing your answers to those in the summary table.

To read more on Tay Sachs disease, go to the website of the National Tay Sachs & Allied Diseases Association (*www.ntsad.org*) or see "Tay-Sachs Disease Carrier screening, prenatal diagnosis, and the molecular era. An international perspective, 1970 to 1993" in *JAMA, Journal of the American Medical Association* 270 (19):2307–2315, Nov. 17, 1993.

	QUESTION	FACTS	INTEGRATION AND ANALYSIS
1	What is another symptom of Tay-Sachs disease besides loss of muscle control and brain function?	Hexosaminidase A breaks down gangliosides. In Tay-Sachs disease, the enzyme is absent and gangliosides accumulate in cells, including light-sensitive cells of the eye, and cause them to function abnormally.	Damage to light-sensitive cells of the eye could cause vision problems and even blindness.
2	How could you test whether Sarah and David are carriers of the Tay-Sachs gene?	Carriers of the gene have lower-than-normal levels of hexosaminidase A.	Run tests to determine the average enzyme levels in known carriers of the disease (i.e., people who have had children with Tay-Sachs disease) and in people who have little likelihood of being carriers. Compare the enzyme levels of suspected carriers to the averages for the known carriers and noncarriers.

(continues on next page)

	QUESTION	FACTS	INTEGRATION AND ANALYSIS
3	Why is the new test for the Tay-Sachs disease more accurate than the old test?	The new test detects the defective gene. The old test analyzed levels of the enzyme produced by the gene.	The new test is a direct way to test if a person is a carrier. The old test was an indirect way. It is possible for factors other than a defective gene to alter a person's enzyme level. Can you think of some? (See answers at end of chapter.)
4	The Tay-Sachs gene is a recessive gene (t). What is the chance that any child of a carrier and a noncarrier will be a carrier?	Tt × TT ↓ TT Tt TT Tt	Each child has a 50% chance of being a carrier (Tt).

84 91 98 100 108

CHAPTER SUMMARY

The major theme of this chapter is *biological energy* and how it is acquired, transferred, and used to do biological work. Energy is stored in large biomolecules such as fats and glycogen and is extracted from them through the processes of metabolism. Reactions and processes that require energy often extract it from the high-energy bond of ATP, a pattern you will see repeated as you learn more about the organ systems of the body.

Two other themes in the chapter involve different kinds of *structure-function relationships*: molecular interactions and compartmentation. *Molecular interactions* are important in proteins, where the function of a protein often depends on its folding into the proper shape, and in the interactions between enzymes and

their substrates, which also depend heavily on molecular shapes. For example, this chapter presented two models of enzyme-substrate interaction, the older lock-and-key model and the newer induced-fit model. (Note that as new models are presented, they may either displace an older model or become an alternative model.)

The final theme presented is *compartmentation* of enzymes, which is essential for organizing and separating metabolic processes. Glycolysis takes place in the cytosol of the cell, for example, as does the anaerobic metabolism of pyruvate to lactate. The citric acid cycle occurs in the mitochondria, and the electron transport chain is located on the inner mitochondrial membrane.

Energy in Biological Systems

1. **Energy** is the capacity to do work. **Chemical work** enables cells and organisms to grow, reproduce, and carry out normal activities. **Transport work** enables cells to move molecules to create concentration gradients. **Mechanical work** is used for movement. (pp. 85–86)

2. **Kinetic energy** is the energy of motion. **Potential energy** is stored energy. (p. 86; Fig. 4-2)

Chemical Reactions

3. A **chemical reaction** begins with one or more **reactants** and ends with one or more **products. Reaction rate** is measured as the change in concentration of products with time. (p. 88)

4. The energy stored in the chemical bonds of a molecule and available to perform work is the **free energy** of the molecule. (p. 88)

5. **Activation energy** is the initial input of energy required to begin a reaction. (p. 88; Fig. 4-3)

6. **Exergonic reactions** are energy producing. **Endergonic reactions** are energy utilizing. (p. 89; Fig. 4-4)

7. Metabolic pathways couple exergonic reactions to endergonic reactions. (p. 89)

8. Energy for driving endergonic reactions is stored in ATP. (p. 89)

9. A reaction that can proceed in both directions is called a **reversible reaction**. If a reaction can proceed in one direction but not the other, it is an **irreversible reaction**. The net free energy change of a reaction determines whether that reaction can be reversed. (pp. 89–90)

Enzymes

10. **Enzymes** are biological catalysts that speed up the rate of chemical reactions without themselves being changed. Enzymes bind to reactant molecules, bringing them together into the best position for reacting with each other. In reactions catalyzed by enzymes, the reactants are called **substrates**. (p. 90)

11. The part of the enzyme that binds to the substrates is called the **binding site**. The **induced-fit model** of enzyme-substrate interaction says that the binding site has an intermediate shape that can change to fit either the substrate or the product molecules. (p. 91; Fig. 4-8)

12. **Specificity** is the ability of an enzyme to catalyze a certain reaction or a group of closely related reactions. (p. 91)

13. Some enzymes are produced as inactive precursors and must be activated. This may require the presence of a **cofactor**. Organic cofactors are called **coenzymes**. (pp. 93–94; Fig. 4-10)

14. Enzyme activity is altered by temperature, pH, and modulator molecules. A **modulator** changes the ability of the enzyme to (1) bind the substrate or (2) alter activation energy of the reaction. (p. 94)

15. **Competitive inhibitors** block the binding site of the enzyme directly. **Allosteric modulators** bind to the enzyme away from the binding site and change the shape of the binding site. **Covalent modulators** affect the ability of an enzyme to decrease the activation energy of the reaction. (pp. 95–96; Figs. 4-13 and 4-14)

16. Factors that affect the rate of enzymatic reactions include modulation, the enzyme concentration, and the ratio of the concentrations of substrates and products. (p. 96)

17. When an enzyme is **saturated** with substrate, it works at its maximum possible rate. (p. 97)

18. Reversible reactions go to a state of **equilibrium**, where the rate of the reaction in the forward direction is exactly equal to the rate of the reverse reaction. Reversible reactions obey the **law of mass action**: when a reaction is at equilibrium, the ratio of substrates to products is always the same. If the concentration of a substrate or product changes, the equilibrium will be disturbed. (p. 98; Fig. 4-17)

19. Most reactions can be classified as **oxidation-reduction**, **hydrolysis-dehydration**, **addition-subtraction-exchange**, or **ligation** reactions. (pp. 98–99; Tbl. 4-4)

Metabolism

20. All the chemical reactions in the body are known collectively as **metabolism**. **Catabolic reactions** release energy and break down large biomolecules. **Anabolic reactions** require a net input of energy and synthesize large biomolecules. (p. 99)

21. Cells regulate the flow of molecules through their metabolic pathways by (1) controlling the enzyme concentration, (2) producing allosteric and covalent modulators, (3) using different enzymes to catalyze reversible reactions, (4) isolating enzymes within intracellular organelles, or (5) maintaining an optimum ratio of ATP to ADP. (p. 100)

22. **Aerobic pathways** require oxygen and yield the most ATP. **Anaerobic pathways** can proceed without oxygen, but produce ATP in much smaller quantities. (p. 102)

ATP Production

IP Muscular: Muscle Metabolism

23. Through **glycolysis**, one molecule of glucose is converted into two pyruvate molecules, two ATP, two NADH, and two H^+. Glycolysis does not require the presence of oxygen. (p. 103; Fig. 4-22)

24. In **anaerobic metabolism**, pyruvate is converted into lactate with a net yield of two ATP for each glucose molecule. (p. 104; Fig. 4-23)

25. **Aerobic metabolism** of pyruvate through the **citric acid cycle** yields ATP, carbon dioxide, water, and high-energy electrons captured by NADH and $FADH_2$. (p. 105; Fig. 4-24)

26. **High-energy electrons** from NADH and $FADH_2$ give up their energy as they pass through the **electron transport system**. Their energy is trapped in the high-energy bonds of ATP. (p. 106; Fig. 4-25)

27. Glycogen and lipids are the primary energy storage molecules in animals. Lipids are broken down for ATP production in the process of **beta-oxidation**. (p. 109; Fig. 4-28)

Synthetic Pathways

28. In **gluconeogenesis**, amino acids and glycerol can be converted to glucose. (p. 110; Fig. 4-29)

29. Fat and cholesterol synthesis requires acetyl CoA. (p. 111; Fig. 4-30)

30. Protein synthesis is controlled by nuclear **genes** made of DNA. The nucleotide code from a gene is transcribed into a matching code on **messenger RNA (mRNA)**. **Alternative splicing** of mRNA in the nucleus allows one gene to code for multiple proteins. (pp. 112–114; Fig. 4-33)

31. The mRNA goes to the cytoplasm and, with the assistance of **transfer RNA** and **ribosomal RNA**, assembles amino acids into the sequence designated by the gene. This process is called **translation**. (pp. 112–114; Fig. 4-34)

32. **Post-translational modification** and protein folding convert the newly synthesized protein to its finished form. (p. 116)

33. Proteins may be modified in the rough endoplasmic reticulum or the Golgi apparatus. In the Golgi, they are packaged into membrane-bound vesicles that become lysosomes, storage vesicles, or secretory vesicles. (p. 116; Fig. 4-35)

QUESTIONS

LEVEL ONE *Reviewing Facts and Terms*

1. List the three basic forms of work and give a physiological example of each.

2. Explain the difference between potential and kinetic energy.

3. State the two laws of thermodynamics in your own words.

4. The sum of all chemical processes through which cells obtain and store energy is called _____.

5. In the reaction $CO_2 + H_2O \longrightarrow H_2CO_3$, water and carbon dioxide are the _____, and H_2CO_3 is the product. Because this reaction is catalyzed by an enzyme, it is also appropriate to call water and carbon dioxide _____. The speed at which this reaction occurs is called the reaction _____, often expressed as molarity/second.

6. _____ are protein molecules that speed up chemical reactions by (increasing or decreasing?) the activation energy of the reaction.

7. Match these terms with their definitions (not all terms are used):
   - (a) a reaction that can run in either direction
   - (b) the part of a protein molecule that actually binds the substrates
   - (c) a reaction that releases energy
   - (d) the ability of an enzyme to catalyze one reaction but not another
   - (e) boost of energy needed to get a reaction started
   - (f) the ability of a protein to alter its shape to fit more closely with that of the substrate

   1. exergonic
   2. endergonic
   3. activation energy
   4. reversible
   5. irreversible
   6. induced fit
   7. binding site
   8. specificity
   9. free energy
   10. saturation

8. Since 1972, enzymes have been designated by adding the suffix _____ to their name.

9. An ion, such as Ca^{2+} or Mg^{2+}, that must be present in order for an enzyme to work is called a _____. Organic molecules that must be present for an enzyme to function are called _____. The precursors of these organic molecules come from _____ in our diet.

10. A protein whose structure is altered to the point that its activity is destroyed is said to be _____.

11. In an oxidation-reduction reaction, where electrons are moved between molecules, the molecule that gains an electron is said to be _____, and the one that loses an electron is _____.

12. The removal of H_2O from reacting molecules is called _____. Using H_2O to break down polymers, such as starch, is called _____.

13. The removal of an amino group (NH_2) from a molecule (such as an amino acid) is called _____. Transfer of an amino group from one molecule to the carbon skeleton of another molecule (to form a different amino acid) is called _____. What happens to the amino group removed from an amino acid?

14. In metabolism, _____ reactions release energy and result in the breakdown of large biomolecules, and _____ reactions require a net input of energy and result in the synthesis of large biomolecules. In what units do we measure the energy of metabolism?

15. Metabolic regulation where the last product of a metabolic pathway (the end product) accumulates and slows or stops reactions earlier in the pathway is called _____.

16. Explain how H^+ movement across the inner mitochondrial membrane results in ATP synthesis.

17. List the two carrier molecules that deliver high-energy electrons to the electron transport system.

18. The breakdown of lipids for energy is termed _____. The long chains of fatty acids are broken into two-carbon acetyl CoA molecules by a pathway called _____. Which pathway do these acetyl CoA molecules join in order to produce ATP?

19. How many kilocalories per gram are stored in carbohydrates, proteins, and fats, respectively?

20. Which cellular organelles are involved in lipid synthesis?

LEVEL TWO *Reviewing Concepts*

21. Create maps using the following terms.

Map 1: Metabolism

acetyl CoA
amino acids
ATP
citric acid cycle
CO_2
cytosol
electron transport system
$FADH_2$
fatty acids
gluconeogenesis
glucose
glycerol
glycogen
glycolysis
high-energy electrons
lactate
mitochondria
NADH
oxygen
pyruvate
water

Map 2: Protein synthesis

alternative splicing
base pairing
DNA
exon
gene
initiation factors
intron
mRNA processing
mRNA
nucleotides (A, C, G, T, A, U)
promoter
RNA polymerase
sense strand
start codon
stop codon
transcription

22. When bonds are broken during a chemical reaction, what are the three possible fates for the potential energy found in those bonds?

23. Match the following theme(s) with the appropriate metabolic processes.

    (a) Glycolysis takes place in the cytoplasm; oxidative phosphorylation takes place in the mitochondria.
    (b) The electron transport system traps energy in a hydro-

    1. Biological energy
    2. Compartmentation
    3. Molecular interactions

gen ion concentration gradient.
    (c) Proteins are modified in the endoplasmic reticulum.
    (d) Metabolic reactions are often coupled to the reaction $ATP \longrightarrow ADP + P_i$.
    (e) Some proteins have S—S bonds between nonadjacent amino acids.
    (f) Enzymes catalyze biological reactions.
    (g) Allosteric modulators change the binding site of enzymes.

24. Define, compare, and contrast the following terms: competitive inhibition, allosteric modulation, covalent modulation.

25. Explain why it is advantageous for a cell to store or secrete an enzyme in an inactive form.

26. Compare the energy yields from the aerobic breakdown of one glucose to CO_2 and H_2O to the yield from one glucose going through anaerobic glycolysis ending with lactate. What are the advantages each pathway offers?

27. What is the advantage of converting glycogen to glucose 6-phosphate rather than to glucose?

28. Briefly describe the processes of transcription and translation. Which organelles are involved in these processes?

29. On what molecule does the anticodon appear? Explain the role of this molecule in protein synthesis.

30. Is ATP an example of potential or kinetic energy?

31. If ATP releases energy to drive a chemical reaction, would you suspect the activation energy of that reaction to be large or small? Explain.

LEVEL THREE *Problem Solving*

32. Transporting epithelial cells in the intestine do not use glucose as their primary energy source. What molecules might they use instead?

33. Given the following strand of DNA, list the sequence of bases that would appear in the matching mRNA. For the under-

lined triplets, underline the corresponding mRNA codon and list the appropriate anticodon.

    DNA: CGCTACAAGTCAGGTACCGTAACG

    mRNA:

    Anticodons:

LEVEL FOUR *Quantitative Problems*

34. The graph in the right column shows the free energy change for the reaction $A + B \longrightarrow D$. Is this an endergonic or exergonic reaction?

35. If the protein-coding portion of a piece of edited mRNA is 450 nucleotides long, how many amino acids will be in the corresponding polypeptide? (*Hint*: The start codon is translated into an amino acid but the stop codon is not.)

Explorations

Explorations are also available at *www.physiologyplace.com*

Exploration 1: Animations of Metabolism

Dynamic processes such as enzyme action or movement of substrate through metabolic pathways are difficult to visualize from the static drawings in a textbook, even when they show the steps in sequence (for example, the drawings of transcription in Fig. 4-32). However, you can find many animations on the web, some of them created by students. An excellent animated primer on DNA and genetics that includes the details of DNA replication, transcription, and translation can be found at *www.dnaftb.org*.

To see some animations on enzymes, do a Google search (*www.google.com*) using the search terms *activation energy animation* or *enzyme animation*. One site that contains these animations and more is NetBioChem (*www.mcphu.edu/netbiochem/NetWelco.htm*). Under the TOPICS button, select GRAPHICS to look at the animations on enzymology and bioenergetics.

Exploration 2: Biomedical Engineering

Biomedical engineering (BME) is a rapidly growing field with researchers in almost every level of physiology, from tiny nanomachines to artificial hearts (nano = 1×10^{-9}). The home page for the Whitaker Foundation (*www.whitaker.org*) has more information about the field of biomedical engineering (see "BME @ A Glance") as well as a news archive that presents exciting developments. To learn more about Dr. Christine Schmidt and her tissue engineering research, go to *www.bme.utexas.edu/faculty/schmidt*.

Exploration 3: Readings on Metabolism

Although you can find information on the Internet, there is still no substitute for a well-written book or article. *Scientific American* is a periodical that publishes scientific review articles for the educated layperson. *News in Physiological Sciences* (*NIPS*), published by the American Physiological Society, publishes review articles. NIPS articles over a year old can be accessed free at *www.nips.org*.

Some articles related to topics in this chapter include:

"Death to sperm mitochondria—A ubiquitin clue to why mitochondrial DNA comes only from Mom," *Scientific American*, March 1999.

"The tall, the deep, the long/designer genomes—extreme engineering," *Scientific American*, Winter 1999.

"The cellular chamber of doom—A look at proteasomes," *Scientific American*, January 2001.

"A nonconventional role of molecular chaperones: Involvement in the cytoarchitecture," *News in Physiological Sciences* 16:123–126, June 2001.

"The human proteomics initiative (HPI)," *Trends in Biotechnology* 19(5):178–181, May 2001.

ANSWERS

Answers to Concept Check Questions

Page 85

1. Amino acids and nucleotides always contain nitrogen.

Page 88

2. Energy in the body is stored in chemical bonds and in concentration gradients.

3. Kinetic energy is energy in motion: something is happening. Potential energy is stored energy: something is waiting to happen.

4. Entropy is a state of randomness or disorder.

Page 90

5. Endergonic reactions use energy; exergonic reactions release energy.

6. The reactants are baking soda and vinegar; the product is carbon dioxide.

7. The foaming-up indicates that energy is being released, so this is an exergonic reaction. The relatively large amount of energy released indicates that the reaction is not readily reversible.

Page 92

8. The substrates are lactose (lactase), peptides (peptidase), lipids (lipase), and sucrose (sucrase).

9. In the induced-fit model, both reactants (substrates) and products can combine with the active site. In the lock-and-key model, the enzyme fits with the reactants but not the products.

10. By having isozymes, one reaction could be catalyzed under a variety of conditions.

Page 96

11. Modulators may alter the enzyme's ability to decrease activation energy or alter its ability to bind to substrate.

12. (a) 2, 5 (b) 3, 4 (c) 1, 5

Page 97

13. As the amount of enzyme present decreases, the reaction rate decreases.

14. If an enzyme has reached saturation, the rate is at its maximum.

Page 99

15. (a) 3; (b) 2; (c) 4; (d) 1.

Page 102

16. Cells regulate substrate movement by (1) controlling the amount of enzyme, (2) producing allosteric and covalent mod-

ulators, (3) using two different enzymes to catalyze reversible reactions, (4) isolating enzymes within intracellular organelles, and (5) altering the ADP:ATP ratio in the cell.

17. ATP traps energy in a phosphate bond. NADH stores energy in high-energy electrons.

18. Aerobic pathways require sufficient quantities of molecular oxygen (O_2) in the cell. Anaerobic pathways can proceed without oxygen.

Page 103

19. (a) 4; (b) 2, 5; (c) 2, 5; (d) 1, 3

Page 104

20. Endergonic reactions trap energy in the products.

Page 105

21. (a) Obviously exergonic reactions release energy that can be trapped by ATP or NADH (3 steps).
    (b) Obviously endergonic reactions require energy input from ATP (2 steps).
    (c) Kinases add a phosphate group to one of the reactants (4 steps).
    (d) Dehydrogenases remove an electron and H. In the step from glyceraldehyde 3-phosphate to 1,3-bisphosphoglycer-

ate, NAD^+ acquires an electron and an H, suggesting that this step is catalyzed by a dehydrogenase.
    (e) Dehydration reactions produce water as one of the products (1 step).

22. Lactate dehydrogenase removes an electron and a hydrogen atom. Loss of the negatively charged electron means the lactate has been oxidized.

Page 107

23. Phosphorylation is the addition of a phosphate group.

Page 108

24. Anaerobic metabolism of glucose can proceed in the absence of oxygen; aerobic metabolism requires oxygen. Anaerobic metabolism produces a maximum yield of two ATP per glucose; aerobic metabolism has a maximum yield of 30–32 ATP.

25. Oxygen acts as an acceptor of electrons and hydrogen ions at the termination point of the electron transport system.

26. According to the chemiosmotic theory, energy is trapped in a hydrogen ion concentration gradient and selectively released as the ions travel across the membrane separating the matrix and intermembrane space.

Page 110

27. The digestion of proteins to polypeptides is a hydrolysis reaction because large molecules are split into smaller molecules by adding water.

28. Enzymes convert amino acids to organic acids by removing the amino group.

Page 111

29. Hexokinase adds a phosphate group; a phosphatase removes a phosphate group.

Page 112

30. The DNA triplets are ATT, ATC, and ACT.

Page 114

31. During mRNA processing some nucleotide sequences called introns are cut out of the mRNA. The remaining segments, the exons, are spliced back together and provide the code for a protein.

Page 116

32. Removal of a phosphate group is dephosphorylation.

33. Three types of post-translational modification are cleavage, addition of groups, and cross-linking.

34. Hemoglobin with four subunits is a tetramer.

Answers to Figure Questions

Page 90

Fig. 4-6: The graph shows an endergonic reaction.

Page 95

Fig. 4-12: (a) The enzyme is more active at 30°C. (b) When pH goes from 8 to 7.4, the activity increases.

Page 97

Fig. 4-15: (a) At enzyme concentration A, the rate is 1 mg product formed per sec. (b) At a re-

action rate of 2.5 mg/sec, the enzyme concentration is C.

Page 97

Fig. 4-16: When substrate concentration is 200 mg/mL, the rate is 4 mg product formed per sec.

Page 109

Fig. 4-27: Reaction (b) is a hydrolysis reaction just like reaction (a).

Page 111

Fig. 4-30: Most fatty acids have an even number of carbons because they are made from two-carbon acyl units. Some fatty acids with odd numbers of carbons are found in marine organisms.

Answers to Level Four Quantitative Problems

Page 121

Exergonic

Page 121

The polypeptide will have 149 amino acids.

Answer to Running Problem Summary

Page 118

Question 3: What factors other than a defective gene could alter enzyme levels? Decreased protein synthesis or increased protein breakdown in the cell might decrease the enzyme level even though the gene was normal.

Membrane Dynamics

5

Acetylcholine receptor-channel

RUNNING PROBLEM

Cystic Fibrosis

Over 100 years ago, midwives performed an unusual test on the infants they delivered: the midwife would lick the infant's forehead. A salty taste meant that the child was destined to die of a mysterious disease that withered the flesh and robbed the breath. Today, a similar "sweat test" will be performed in a major hospital—this time with state-of-the-art techniques—on Daniel Biller, a 2-year-old with a history of weight loss and respiratory problems. The name of the mysterious disease? Cystic fibrosis.

BACKGROUND BASICS

Polar and nonpolar molecules: **28–29** I Protein and lipid structure: **35–38** I Cell junctions: **63** I Microscopy: **51–52** I Molarity: **32** I Solutions: **31** I Cytoskeleton: **57** I Types of epithelia: **67** I Enzymes **90**

Since the cell membrane is not visible, evidence for its existence is indirect, deriving from several types of investigation.

—Bradley T. Sheer, in his textbook *General Physiology*, 1953

U p to this point, we have been concerned with the structures and processes of individual cells. However, the human body is made of trillions of cells that must communicate and cooperate with each other while maintaining their separate identities. The cell membrane plays a key role in this process because it is the barrier that separates the intracellular compartment from signals outside the cell.

Cell membranes are increasingly the focus of physiological and medical research. Almost daily, you can read in the newspaper about new medical treatments, such as drugs for high blood pressure, depression, gout, and stomach ulcers, that target membrane processes. Membrane research is also helping us answer basic physiological questions that were mysterious "black boxes" for many years. A good example of a physiological mechanism that was not explained until recently is secretion of the hormone insulin.

In 1942 researchers accidentally discovered that chemicals called sulfonylureas lower blood glucose levels by stimulating insulin secretion. This property made them an ideal drug for the treatment of diabetes, and by the 1950s, sulfonylureas went into clinical use, even though no one understood the mechanism by which they act. It took nearly 40 more years before we learned the intracellular pathways that lead to insulin secretion. The mechanisms involved in insulin release demonstrate basic aspects of membrane biology and are described in the last section of this chapter.

In this chapter and the next, you will learn more about four of the fundamental themes that are the basis of physiology [◐ p. 8]. Chapter 5 discusses the division of the body into intracellular and extracellular compartments by the cell membrane. You will learn how molecules cross cell membranes and how the selectivity of the cell membrane allows the intracellular and extracellular compartments to differ in their chemical and electrical properties. In Chapter 6 we examine how signals outside the cell are communicated to the cytoplasm, and we consider the role of cell-cell communication in maintaining homeostasis.

▶ MEMBRANES IN THE BODY

The word *membrane* has come to have two different meanings in physiology. The simplest membrane, the **plasma membrane** or **cell membrane**, is the membrane that surrounds the cell. But confusion sometimes arises because the word *membrane* is also used to describe epithelial tissues that line a cavity or separate two compartments [◐ p. 67]. Anatomically, we speak of mucous membranes in the mouth and vagina, the peritoneal membrane that lines the inside of the abdomen, the pleural membrane that covers the surface of the lungs, and the pericardial membrane that surrounds the heart. These visible membranes are actually tissues: thin, translucent layers of cells. To confuse matters further, epithelial membranes are often depicted in book illustrations as a single line, leading students to think of them as if they were similar in structure to the cell membrane (Fig. 5-1 ◐).

Cell membranes, on the other hand, are a double layer (*bilayer*) of phospholipids [◐ p. 36] with protein molecules inserted in them. Material moving into and out of cells must cross the cell membrane, and anything that enters and leaves the body through an epithelial tissue passes through two cell membranes. If a membrane allows a substance to pass through it, it is said to be **permeable** to the substance [*permeare*, to pass through]. If a membrane does not allow a substance to pass, it is said to be **impermeable** [*im-*, not].

▶ CELL MEMBRANES

Phospholipid membranes surround the contents of the cytoplasm and divide the interior of the cell into compartments such as the nucleus, mitochondria, endoplasmic reticulum, and Golgi apparatus [◐ p. 58]. In this chapter we will be concerned with the external cell boundary called the cell membrane. (We will use the term *cell membrane* rather than *plasma membrane* to avoid confusion with blood plasma.) The general functions of the cell membrane include:

The **pericardial membrane** is a tissue that surrounds the heart.

Heart

Loose connective tissue

Seen magnified, the pericardial membrane is a layer of flattened epithelial cells supported by connective tissue.

Cell

Each cell of the pericardial membrane has a cell membrane surrounding it.

The **cell membrane** is a phospholipid bilayer.

● **Figure 5-1** *Membranes in the body*

1. **Physical isolation**: The cell membrane is a physical barrier that separates the inside of the cell (the cytosol or intracellular fluid) from the surrounding extracellular fluid.

2. **Regulation of exchange with the environment**: The cell membrane controls the entry of ions and nutrients, the elimination of wastes, and the release of secretory products.

3. **Communication between the cell and its environment**: The cell membrane is in direct contact with both the cytosol (the intracellular fluid, or **ICF**) and the extracellular fluid, or **ECF** [❧ p. 7]. It contains receptors that allow the cell to recognize and respond to molecules or changes in its external environment. Any alteration in the cell membrane may affect the cell's activities.

4. **Structural support**: Cell membrane proteins hold proteins of the cytoskeleton in place to maintain cell shape. They also create specialized junctions between adjacent cells or between cells and the extracellular matrix. These junctions stabilize the structure of tissues [❧ p. 64].

To understand how the cell membrane can carry out such diverse functions, we will first examine the current model of cell membrane structure.

Membranes Are Mostly Lipid and Protein

The structure of cell membranes has been a matter of great interest ever since scientists first observed cells with a microscope. By the 1890s, scientists had concluded that the outer membrane of cells was made of a thin layer of lipids that acted as a barrier between the aqueous interior and the watery environment outside the cell. However, this simple and uniform structure did not account for the highly variable properties of membranes found in different types of cells. How could water cross the cell membrane to enter a red blood cell but not be able to enter certain cells of the kidney tubule? The explanation had to lie in the structure of the cell membrane.

To isolate and analyze cell membranes, researchers in the first decades of the twentieth century ground up cells. They discovered that the cell membrane and all membranes within the cell consist of a combination of lipids and proteins with a small amount of carbohydrate. The ratio of protein to lipid varies widely, depending on the source of the membrane (Table 5-1). Generally, the more metabolically active a membrane is, the more proteins it contains. For example, the inner mitochondrial membrane, which contains enzymes for oxidative phosphorylation [❧ p. 106], is three-quarters protein.

The chemical analysis of membranes was useful but it still did not explain how lipids and proteins were structurally arranged in the membrane. Studies in the 1920s suggested that there was enough lipid in a given area of membrane to create a double layer. This model was further modified in the 1930s to account for the presence of proteins. With the introduction of electron microscopy, scientists saw the membrane for the first time. The 1960s model of the membrane was a "butter sandwich"—a double layer of lipids sandwiched between two layers of proteins.

By the early 1970s, freeze-fracture electron micrographs had revealed the actual three-dimensional arrangement of lipids and proteins within cell membranes. Because of what scientists learned from looking at freeze-fractured membranes, S. J. Singer and G. L. Nicolson in 1972 proposed the **fluid mosaic model** of the membrane. Figure 5-2● highlights the major features of this contemporary model of membrane structure. The phospholipids of the membrane are arranged in a bilayer, with a variety of proteins inserted partly or completely through the bilayer. Carbohydrates on the extracellular surface are attached to both the proteins and the lipids, forming a "sugar coating." All cell membranes have similar components and are of uniform thickness, about 8 nm.

Membrane Lipids Form a Barrier Between the Cytoplasm and Extracellular Fluid

Two main types of lipids make up the cell membrane: phospholipids and cholesterol. **Phospholipids** are made of a glycerol backbone (Fig. 5-3a●) with two fatty acid chains extending to one side and a phosphate group extending to the other [❧ p. 28]. The glycerol-phosphate "head" of the molecule is polar and thus hydrophilic. The fatty acid "tail" is nonpolar and thus hydrophobic (Fig. 5-3b).

When placed in an aqueous solution, phospholipids orient themselves so that the polar side of the molecule interacts with the water molecules while the nonpolar fatty acid tails hide by putting the polar heads between themselves and the water. This arrangement can be seen in three different structures: the micelle, the liposome, and the phospholipid bilayer (Fig. 5-3b). **Micelles** are small droplets of phospholipid, arranged so that the interior is filled with hydrophobic fatty acid tails. Micelles are important in the digestion and absorption of fats in the digestive tract.

Table 5-1: Composition of Selected Membranes

MEMBRANE	PROTEIN	LIPID	CARBOHYDRATE
Red blood cell membrane	49%	43%	8%
Myelin membrane around nerve cells	18%	79%	3%
Inner mitochondrial membrane	76%	24%	0%

Carbohydrate group of **glycoprotein**

Peripheral proteins are loosely attached to integral proteins.

Carbohydrate group of **glycolipid**

Membrane splits into layers in freeze-fracture electron microscopy

Extracellular surface of membrane

Membrane-spanning **integral proteins** are tightly bound to the membrane.

Intracellular surface of membrane

Cholesterol molecules insert themselves into the lipid layer.

Phospholipid heads face the aqueous intracellular and extracellular compartments.

● **Figure 5-2**   *The fluid mosaic model of the membrane*

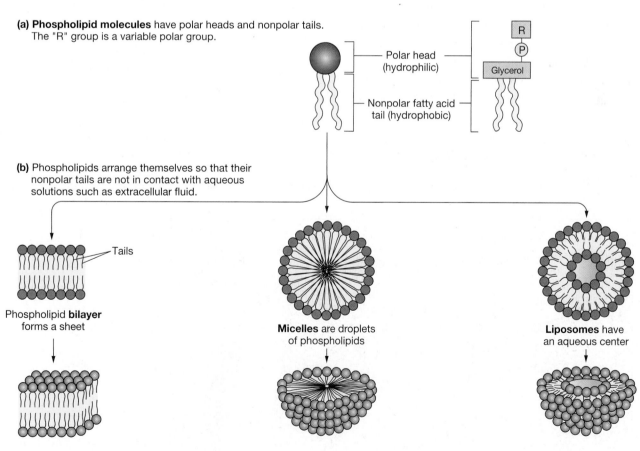

(a) Phospholipid molecules have polar heads and nonpolar tails. The "R" group is a variable polar group.

R
P
Glycerol

Polar head (hydrophilic)

Nonpolar fatty acid tail (hydrophobic)

(b) Phospholipids arrange themselves so that their nonpolar tails are not in contact with aqueous solutions such as extracellular fluid.

Tails

Phospholipid **bilayer** forms a sheet

Micelles are droplets of phospholipids

Liposomes have an aqueous center

● **Figure 5-3**   *Membrane phospholipids*

Liposomes are larger hollow spheres with phospholipid bilayer walls. This arrangement leaves a hollow center with an aqueous core that can be loaded with water-soluble molecules. Liposomes are being used as a medium for the delivery of drugs through the skin.

The **phospholipid bilayer** is a sheetlike structure with two layers of phospholipids arranged so that the hydrophobic tails are sandwiched between two layers of hydrophilic heads. The major role of the bilayer is structural: to form a barrier across which only lipid-soluble molecules can migrate and to provide a framework for the membrane proteins. A few membrane lipids participate in

Liposomes for Beauty and Health

Many people first hear the term *liposome* in connection with cosmetic skin creams that promise to deliver special ingredients to the cells that need them. But the cosmetic manufacturers have adopted a medical technique developed to enhance the delivery of drugs. In medicine, the centers of liposomes are filled with drugs or with fragments of DNA for gene therapy. Then the liposomes are applied to the skin or injected into the bloodstream. The newest generation of liposomes has attached proteins that bind to receptors on specific cells. In this way, toxic drugs designed for anticancer chemotherapy can be targeted specifically to diseased tissues, enhancing their effectiveness and reducing their side effects on normal tissues. In one study, liposomes loaded with antibodies against the potent anticancer drug doxorubicin (DXR) were applied to the scalp of mice receiving DXR treatments. The liposomes were absorbed into hair follicles, where the antibodies protected the rapidly dividing follicle cells from DXR. This prevented the drug-induced hair loss usually associated with chemotherapy.

communication between the extracellular environment and the cell.

Although phospholipids are the major lipid of membranes, cholesterol is also a significant part of many cell membranes. Cholesterol molecules, which are mostly hydrophobic, insert themselves between the hydrophilic heads of phospholipids and extend back into the central lipid portion of the bilayer (Fig. 5-2). Cholesterol helps make membranes impermeable to small water-soluble molecules and keeps the membrane flexible over a wide range of temperatures.

Membrane Proteins May Be Loosely or Tightly Bound to the Membrane

According to some estimates, membrane proteins may be nearly one-third of all the proteins coded in our DNA. Each cell has between 10 and 50 different types of proteins inserted into its phospholipid bilayer. Anatomically, membrane proteins are described as being either peripheral [*peripheria*, circumference] or integral. **Peripheral proteins**, also called associated proteins, attach themselves loosely to membrane-spanning proteins or to the polar regions of phospholipids. They can be removed without disrupting the integrity of the membrane itself. Peripheral proteins include enzymes and some structural binding proteins that anchor the cytoskeleton to the membrane (Fig. 5-4 ⬤).

Integral proteins are tightly bound into the phospholipid bilayer. The only way an integral protein can be removed is by disrupting the membrane structure with detergents or other harsh methods that destroy the membrane's integrity. Most integral proteins span the entire membrane, but a few extend only partway into the lipid core of the membrane.

Membrane-Spanning Proteins Membrane-spanning proteins are integral proteins that extend all the way across the membrane. They can be classified into families according to how many transmembrane segments they have. Many physiologically important membrane proteins have 7 transmembrane segments, while others may have as few as 1 or as many as 12 or more.

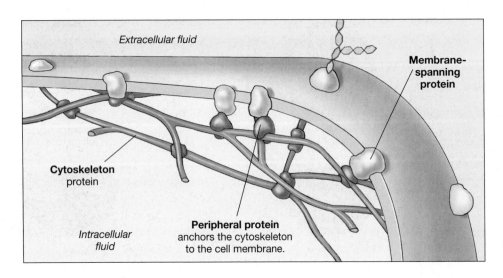

● Figure 5-4 *The cytoskeleton is anchored to the cell membrane*

Figure 5-5 *Membrane-spanning proteins* Membrane-spanning proteins have loops of peptide chains that extend into cytoplasm and extracellular fluid.

When proteins cross the membrane more than once, amino acid chains between the membrane-spanning sections protrude into the cytoplasm and the extracellular fluid (Fig. 5-5 ●). Carbohydrates may attach to the extracellular loops. Phosphate groups may attach to the intracellular loops. Phosphorylation of proteins is one regulatory method cells use to alter protein function.

According to the fluid mosaic model of the cell membrane, membrane proteins may move laterally from location to location, directed by fibers of the cytoskeleton that run just under the membrane surface. On the other hand, some membrane-spanning proteins are immobile, anchored to cytoskeleton proteins (Fig. 5-4). The ability of the cytoskeleton to restrict the movement of membrane proteins allows cells to develop *polarity*, in which different faces of the cell have different proteins and therefore different properties. This is particularly important in the cells of the transporting epithelia, as you will learn later in this chapter.

Membrane Proteins Function as Structural Proteins, Enzymes, Receptors, and Transporters

For physiologists, classifying membrane proteins by their function is more useful than classifying them by their structure. Cell membrane proteins function as (1) structural elements, (2) enzymes, (3) receptors, and (4) transporters that move molecules into and out of the cell. Many physiological processes you will be studying involve membrane enzymes, receptors, and transporters.

Structural Proteins The **structural proteins** have three major roles. The first is to connect the membrane and the cytoskeleton to maintain the shape of the cell. The microvilli of transporting epithelia [● p. 69] are one example of membrane shaping by the cytoskeleton. The flattened disk shape of the red blood cell is another example.

The second role of structural membrane proteins is to create cell junctions that hold tissues together, such as tight junctions and gap junctions [● p. 64]. Finally, the membrane-spanning proteins help attach cells to the extracellular matrix. They serve as anchors that link cytoskeleton fibers to extracellular collagen and other proteins fibers [● Fig. 3-14, p. 65]. In addition, the sugar chains of membrane glycoproteins mingle with extracellular proteoglycans, creating a "sugar coating" on the cell.

Enzymes **Membrane-associated enzymes** catalyze chemical reactions that take place on the cell's external surface or just inside the cytoplasm. For example, enzymes on the luminal surface of cells in the small intestine are responsible for digesting peptides and carbohydrates. Enzymes attached to the intracellular surface of many cell membranes play an important part in transferring signals from the extracellular environment to the cytoplasm of the cell.

Receptors **Receptor proteins** of the cell membrane are part of the body's chemical signaling system. Each receptor is *specific* for a certain molecule or family of related molecules. It will combine only with those molecules, just as enzymes recognize and bind only to specific substrates. Two major types of receptors are those that bind to chemical signal molecules such as hormones and those that act as cell recognition molecules (the *major histocompatibility complex* proteins, to be discussed in Chapter 24).

The molecule that binds to a receptor is called the **ligand** [*ligare*, to bind or tie]. The binding of receptor and ligand usually triggers another event at the membrane, such as activation of a

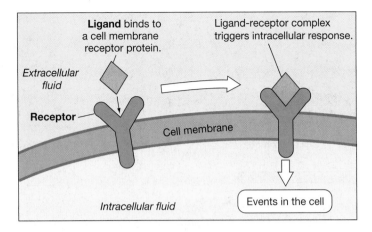

● Figure 5-6 *Cell membrane receptor* Cell membrane receptors combine with molecules known as ligands.

membrane enzyme (Fig. 5-6●). An example of a ligand is the hormone insulin, which combines with an insulin receptor on a cell membrane to exert its effects. You will learn more about receptors in Chapter 6.

Transporters The fourth major function of membrane proteins is their role as transporters. Many molecules enter or leave the cell using transport proteins. We will classify **transport proteins** into two categories: channels and carriers. **Channel proteins** create water-filled passageways that directly link the intracellular and extracellular compartments (Fig. 5-7●). **Carrier proteins** bind to the substrates that they carry but never form a direct connection between the intracellular fluid (ICF) and extracellular fluid (ECF). Carriers will be open to one side of the membrane or the other, but not to both at once, as channels are.

Why do cells need both channels and carriers? The answer lies in the different properties of the two transporters. Channel proteins allow more rapid transport across the membrane but are not as selective about what they transport. Carriers, while slower, are better at discriminating between closely related molecules. Carriers can also move larger molecules than channels can.

Channel proteins Channel proteins are made of membrane-spanning protein subunits that zigzag back and forth across the membrane, creating a cluster of cylinders that surround a narrow water-filled pore (Fig. 5-8●). Movement through channels is restricted primarily to water and ions. When water-filled channels

are open, tens of millions of ions per second can whisk through them unimpeded.

Channels are named according to the substance(s) they allow to pass. Most cells have water channels made from a protein called *aquaporin*. In addition, more than 100 different types of **ion channels** have been identified. Ion channels may be specific for one ion or may allow ions of similar size and charge to pass. There are Na^+ channels, K^+ channels, and nonspecific *monovalent* ("one charge") cation channels that transport Na^+, K^+, and lithium ions (Li^+).

The selectivity of a channel is determined by the diameter of its pore and by the electrical charge of the amino acids that line the channel. If the channel amino acids are positively charged, positive ions will be repelled and negative ions will pass through the channel. On the other hand, the nonspecific cation channel mentioned above must have a negative charge that attracts cations but prevents the passage of Cl^- or other anions.

Channel proteins are like narrow doorways into the cell. If the door is closed, nothing can go through. If the door is open, there is a continuous passage between the two rooms connected by the doorway. The open or closed state of a channel is determined by regions of the protein molecule that act like swinging "gates."

According to current models, many channel "gates" are part of the cytoplasmic side of the membrane protein (Fig. 5-9b●). Such a gate can be envisioned as a ball on a chain that swings up and blocks the mouth of the channel. A few channels have a gate in the middle of the protein, and one type of channel in nerve cells has two different gates.

Channels can be classified according to whether their gates are usually open or closed. **Open channels** spend most of their time in the open state, allowing ions to move back and forth across the membrane without regulation. Their gates may occasionally flicker closed, but for the most part, these channels behave as if they have no gates. Open channels are sometimes called *leak channels* or *pores*, as in *water pores*.

Gated channels spend most of their time in the closed state, which allows these channels to regulate the movement of mole-

(a) **Open channels** create a water-filled pore.

(b) **Carriers** never form an open channel between the two sides of the membrane.

ECF

Cell membrane

ICF

Carrier open to *ICF*

Same carrier open to *ECF*

● Figure 5-7 *Transport proteins of the cell membrane*

Many channels are made of multiple protein subunits that assemble in the membrane.

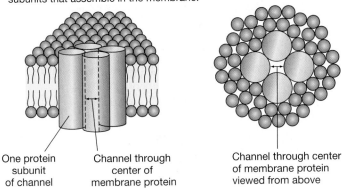

One protein subunit of channel | Channel through center of membrane protein | Channel through center of membrane protein viewed from above

● **Figure 5-8** *Structure of channel proteins* Hydrophilic amino acids facing the center of the channel create a water-filled passage that allows ions and very small molecules, such as water and urea, to pass through.

cules through them. If a gated channel is open, molecules move through the channel just as they move through open channels. If a gated channel is closed, which it may be much of the time, it allows no movement between the intracellular and extracellular fluid.

What controls the opening and closing of gated channels? The gating of these channels is controlled by intracellular messenger molecules or extracellular ligands (**chemically gated channels**), by the electrical state of the cell (**voltage-gated channels**), or by a physical change such as increased temperature or a blow that puts tension on the membrane and pops the channel open (**mechanically gated channels**). You will encounter many variations of these channel types as you study physiology.

RUNNING PROBLEM

Cystic fibrosis is a debilitating disease caused by a defect in a gated channel protein that normally transports chloride ions (Cl⁻). The channel protein—called the cystic fibrosis transmembrane regulator, or CFTR—is located in the cell membranes of epithelia lining the airways, sweat glands, and pancreas. The open state of the channel is regulated through the binding of nucleotides to specific regions of the channel. In people with cystic fibrosis, the CFTR channel is nonfunctional or absent. As a result, chloride transport (either secretion or reabsorption) across the epithelium is impaired.

Question 1: Is CFTR a chemically gated, voltage-gated, or mechanically gated channel protein?

124 🏃 148 154 156 163

Carrier proteins The second type of transport protein is the carrier protein. Carrier proteins bind with specific molecules, called *substrates*, and carry them across the membrane by changing conformation or shape. Small organic molecules, such as glucose and amino acids, cross membranes only on carriers. At the same time, ions such as Na⁺ and K⁺ may move by carriers as well as through channels.

(a) Open channels have gates (not shown) but spend most of their time in the open state.

(b) Gated channels are usually closed. They open in response to chemical, mechanical, or electrical signals.

● **Figure 5-9** *Gating of channel proteins*

Carriers are large, complex proteins with multiple subunits. The conformation change required of the carrier protein makes this mode of transport much slower than movement through open channels. Only 1000 to 1,000,000 molecules per second can be moved by each protein, compared to the tens of millions of ions per second that move through a channel.

Carrier proteins differ from channel proteins in another way: they never create a continuous passage between the inside and outside of the cell. If channels are like doorways, then carrier proteins are like revolving doors that allow movement between inside and outside without creating an open hole. Carrier proteins can transport molecules across a membrane in both directions, like a revolving door at a hotel, or they can restrict their transport to one direction, like the revolving doors at amusement parks that allow you out of the park but not back in.

Carriers can form passageways, but these openings always have two gates, one on the extracellular side and one on the intracellular side. One gate is always closed, preventing free exchange between inside and outside. In this respect, carrier proteins are like the Panama Canal (Fig. 5-10a ●). Picture the canal with only two gates, one on the Atlantic side and one on the Pacific side. Only one gate at a time is open. When the Atlantic gate is closed, the canal opens into the Pacific. A ship enters the canal from the Pacific and the gate closes behind it. Now the canal is isolated from both oceans with the ship trapped in the middle. Then the Atlantic gate opens, making the canal continuous with the Caribbean Sea. The ship sails out of the gate, having crossed the barrier of the land without the canal ever forming a direct connection between the two bodies of water.

Movement across the membrane through carrier proteins is similar (Fig. 5-9b). The substrate binds to the carrier on one side of the membrane. This binding changes the conformation of the protein so that the open gate closes. After a brief transition where both gates are closed, the opposite gate opens to the other side of the membrane. The carrier then releases the substrate into the opposite compartment, having brought it through the membrane without creating a direct connection between the extracellular and intracellular compartments.

131

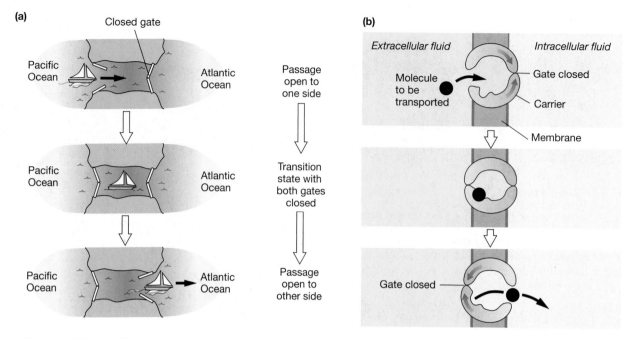

● **Figure 5-10** *Facilitated diffusion by means of a carrier protein* Carrier proteins, like the canal illustrated, never form a continuous passageway between the extracellular and intracellular fluids.

The mechanisms by which molecules cross membranes, either on transporters or by moving through the phospholipid bilayer, are the subject of a later section in this chapter, Movement Across Membranes.

5. Name four functions of membrane proteins.

6. What kinds of molecules pass through open channels?

7. Name three ways channels differ from carriers.

8. If a channel is lined with amino acids that have a net positive charge, which of the following ions is/are likely to move freely through the channel? Na^+, Cl^-, K^+, Ca^{2+}.

9. Why doesn't glucose cross the membrane through open channels?

Membrane Carbohydrates Attach to Both Lipids and Proteins

Most membrane carbohydrates are glucose polymers attached either to membrane proteins (*glycoproteins*) or to membrane lipids (*glycolipids*). They are found exclusively on the external surface of the cell, where they form a protective layer known as the **glycocalyx** [*glykys*, sweet + *kalyx*, husk or pod]. The glycoproteins on the surface of cells play a key role in the body's immune response.

Figure 5-11● is a summary map organizing the structure of the cell membrane. Figure 5-12● is a map showing the structure and function of membrane proteins. The next two sections of the chapter discuss how membranes divide the body into compartments, and how membrane proteins regulate the movement of molecules between these compartments.

● **Figure 5-11** *Map of cell membrane structure*

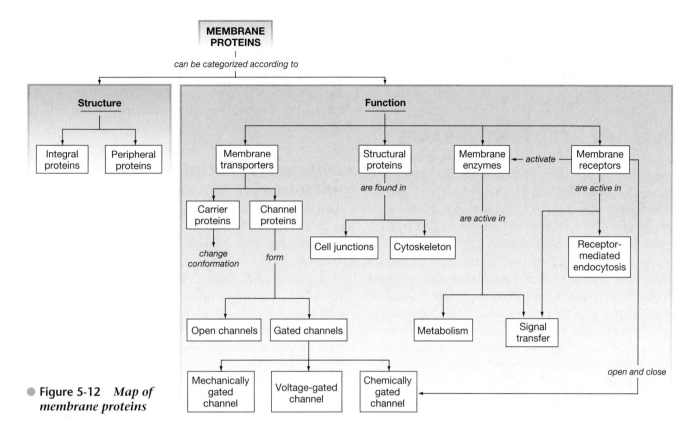

● **Figure 5-12** *Map of membrane proteins*

▶ BODY FLUID COMPARTMENTS

Cell membranes divide the inside of the body into compartments, just as the membranes of organelles create compartments inside the cell. Most cells of the body are not in direct contact with the outside world. Instead they are surrounded by extracellular fluid [● Fig. 1-4, p. 7]. If we think of all the cells of the body together as a single unit, we can divide the body into two main compartments: (1) the intracellular fluid (ICF) within the cells and (2) the extracellular fluid (ECF) outside the cells. These compartments are separated by the barrier of the cell membrane.

The extracellular fluid can be further subdivided. The dividing "wall" in this case is the tissue wall of the circulatory system. **Plasma**, the fluid portion of the blood, lies within the circulatory system and forms one extracellular compartment. The fluid that lies outside the circulatory system and directly bathes the cells is known as **interstitial fluid** [*inter-*, between + *stare*, to stand]. Figure 5-13 ● shows the relationships between these compartments.

In many parts of the circulatory system, no exchange between the plasma and the interstitial fluid is possible due to the thick walls of the blood vessels. However, in the smallest blood vessels, called *capillaries*, the walls are made of leaky exchange epithelium. The capillary walls act like a sieve, retaining blood cells and large proteins but allowing the remainder of the plasma to cross freely through spaces between the cells [● Fig. 3-18a, p. 69].

Although many materials move freely between the blood and the interstitial fluid, exchange between the intracellular and extracellular compartments is restricted by the cell membrane.

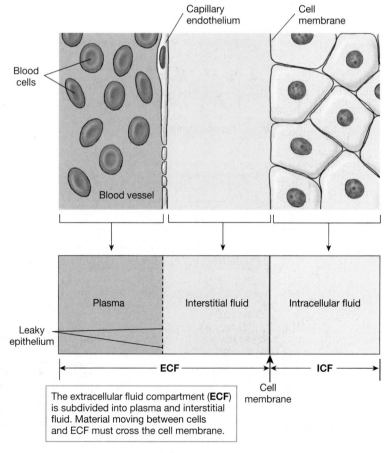

The extracellular fluid compartment (**ECF**) is subdivided into plasma and interstitial fluid. Material moving between cells and ECF must cross the cell membrane.

● **Figure 5-13** *Body fluid compartments* The box in the bottom half schematically illustrates the relationships between the compartments.

133

Whether or not a substance enters a cell depends on the properties of the cell membrane and of the substance itself. To understand how cells can be selective about the materials that enter them, we must look in detail at the mechanisms by which molecules cross membranes.

MOVEMENT ACROSS MEMBRANES

Some molecules, such as water, oxygen, carbon dioxide, and lipids, move easily across cell membranes. On the other hand, ions, larger polar molecules, and very large molecules, such as proteins, enter cells with more difficulty or may not enter at all. Membrane permeability is variable and can be changed by altering the proteins or lipids of the membrane.

Two properties of a molecule will influence its movement across cell membranes: the size of the molecule and its lipid-solubility or polarity [p. 28]. Very small or lipid-soluble substances can cross directly through the phospholipid bilayer. Larger or less lipid-soluble molecules are excluded from crossing the cell membrane unless a cell has a specific mechanism for transporting them across. Usually, membrane proteins are the mediators that cells use to transport molecules across their membranes.

There are two ways to categorize how molecules move across membranes. One scheme describes movement according to whether it takes place through the phospholipid bilayer or with the aid of a membrane protein (Fig. 5-14 ●). A different scheme classifies movement according to its energy requirements. **Passive transport** does not require input of energy. **Active transport** requires the input of energy from another source, such as the high-energy phosphate bond of ATP. In this section we examine all the different ways that substances move into and out of cells.

Diffusion Uses Only the Energy of Molecular Movement

All molecules are constantly in motion. Gas molecules and molecules in solution constantly move from one place to another, bouncing off other molecules or off the sides of a container. When molecules start out concentrated in one area of an enclosed space, their motion causes them to spread out gradually until they are evenly distributed throughout the available space. This process is known as diffusion.

Diffusion may be defined as the movement of molecules from an area of higher concentration of the molecule to an area of lower concentration of the molecule.* If you leave a bottle of

---

*Some texts use the term *diffusion* to mean any random movement of molecules. They call molecular movement along a concentration gradient *net diffusion*. To simplify matters, we will use the term *diffusion* to mean movement of molecules down a concentration gradient.

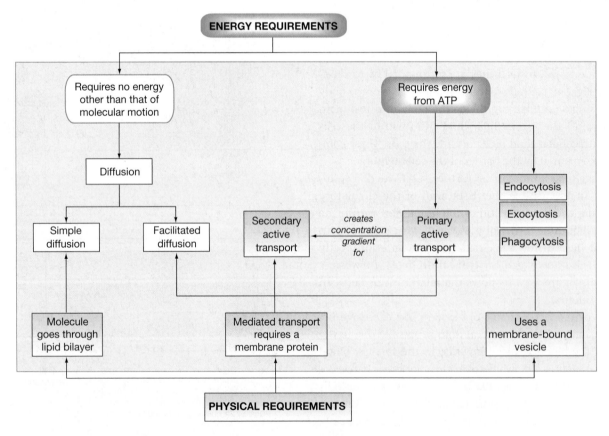

● **Figure 5-14** *Map of the ways molecules move across cell membranes* Movement of substances across cell membranes can be classified by the energy requirements of transport (top part of map) or according to whether transport uses diffusion, a membrane protein, or a vesicle (bottom part of map).

cologne open and later notice its fragrance across the room, it is because the aromatic molecules in the cologne have diffused from where they are more concentrated (in the bottle) to where they are less concentrated (across the room).

Diffusion has the following seven properties:

1. *Diffusion is a passive process.* Diffusion does not require the input of energy from another source, such as the energy stored in ATP. Diffusion uses only the kinetic energy possessed by all molecules.

2. *Molecules move from an area of higher concentration to an area of lower concentration.* A difference in the concentration of a substance between two places is called a **concentration gradient**. We say that molecules diffuse *down the gradient*, from higher concentration to lower concentration (Fig. 5-15 ●). The rate of diffusion depends on the magnitude of the concentration gradient. The larger the concentration difference, the faster diffusion takes place. For example, when you open a bottle of ammonia, the rate of diffusion is most rapid as the ammonia molecules first escape from the bottle into the air. Later, when the ammonia has spread evenly throughout the room, the rate of diffusion has dropped to zero because there is no longer a concentration gradient.

3. *There will be net movement of molecules until the concentration is equal everywhere.* Once the molecules have distributed themselves evenly through the system, **equilibrium** occurs [*aequus*, equal + *libra*, balance] and diffusion stops. Individual molecules are still moving at equilibrium, but for each molecule that exits one area, another one enters. The *dynamic* [*dynamis*, power] equilibrium state in diffusion means that the concentration has equalized throughout the system.

4. *Diffusion is rapid over short distances but much slower over long distances.* Albert Einstein studied the diffusion of molecules in solution and found that the time required for a molecule to diffuse from point A to point B is proportional to the square of the distance from A to B. In other words, if the distance doubles from 1 to 2, the time needed for diffusion increases from 1^2 to 2^2.

CONCEPT CHECK   Answers: p. 169

10. If the distance doubles from 1 to 2, diffusion takes how many times as long?

What does the slow rate of diffusion over long distances mean for biological systems? In humans, nutrients take five seconds to diffuse from the blood to a cell that is 100 μm from the nearest capillary. At that rate, it would take years for nutrients to diffuse from the small intestine to cells in the big toe, and the cells would starve to death. To overcome the limitations of diffusion over distance, organisms have developed various transport mechanisms that speed up the movement of molecules. Most multicellular organisms have some form of circulatory system to bring oxygen and nutrients rapidly from the point at which they enter the body to the cells.

5. *Diffusion is directly related to temperature.* At higher temperatures, molecules move faster. Because diffusion is a result of molecular movement, as temperature increases so does the rate of diffusion. Generally, changes in temperature do not significantly affect diffusion in humans because we maintain a relatively constant body temperature.

6. *Diffusion is inversely related to molecular size.* The larger a molecule is, the more slowly it will diffuse. Albert Einstein showed that friction between the surface of a particle and the medium through which it diffuses is a source of resistance to movement. He calculated that diffusion is inversely proportional to the radius of the molecule: the larger the molecule, the slower its diffusion through a given medium.

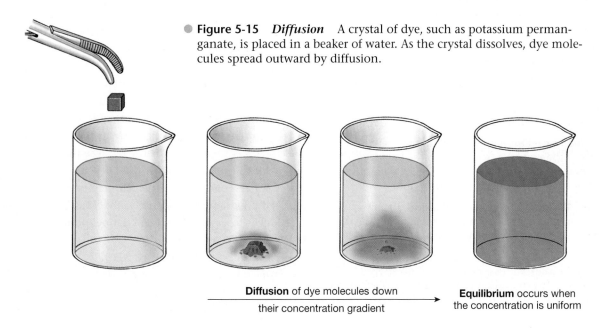

● **Figure 5-15 *Diffusion*** A crystal of dye, such as potassium permanganate, is placed in a beaker of water. As the crystal dissolves, dye molecules spread outward by diffusion.

Diffusion of dye molecules down their concentration gradient ⟶

Equilibrium occurs when the concentration is uniform

7. *Diffusion can take place in an open system or across a partition that separates two systems*. Diffusion of ammonia or cologne within a room is an example of diffusion taking place in an open system. There are no barriers to molecular movement, and the molecules spread out to fill the entire system. Diffusion can take place between two different systems, such as the intracellular and extracellular compartments, but only if the partition dividing the two compartments allows the diffusing molecules to cross.

For example, if you close the top of an open bottle of ammonia, diffusion cannot spread molecules into the room because the bottle is not permeable to the ammonia. However, if you cover the open mouth of the bottle with a "vegetable-keeper" plastic bag with tiny holes in it, you will begin to smell the ammonia in the room because the bag is permeable to the ammonia. Similarly, if a cell membrane is permeable to a molecule, that molecule can enter or leave the cell by diffusion. If the membrane is not permeable, the molecule cannot cross. Table 5-2 summarizes these points.

Table 5-2: Rules for Simple Diffusion

1. Diffusion is a passive process that does not require an outside energy source. It uses only the kinetic energy of molecular movement.

2. Molecules diffuse from an area of higher concentration to an area of lower concentration. The larger the concentration difference, the faster diffusion takes place.

3. There will be net movement of molecules until the concentration is equal. Molecular movement continues, however, so this is called a dynamic equilibrium.

4. Diffusion is rapid over short distances but much slower over long distances.

5. Diffusion is directly related to temperature. At higher temperatures, molecules move faster and diffusion is more rapid.

6. Diffusion is inversely related to molecular size. The larger a molecule is, the slower it will diffuse.

7. Diffusion can take place in an open system or across a partition that separates two systems.

Diffusion across a membrane

8. The rate of diffusion through a cell membrane depends on the ability of the molecule to dissolve in the lipid layer of the membrane. Only lipid-soluble (lipophilic) molecules can diffuse across the membrane by simple diffusion.

9. The rate of diffusion through a membrane is directly proportional to the surface area of the membrane. The larger the surface area, the more molecules can diffuse across per unit time.

10. The rate of diffusion across a membrane is inversely proportional to the thickness of the membrane. The thicker the membrane, the slower diffusion will take place.

In summary, diffusion is the passive movement of molecules down their concentration gradient due to random molecular movement. Diffusion is slower over long distances and slower for large molecules. When the concentration of the diffusing molecule is equal throughout the system, the system has come to equilibrium and diffusion stops, although the random movement of molecules continues.

Lipophilic Molecules Can Diffuse Through the Phospholipid Bilayer

The ability of cell membranes to act as barriers is a direct result of their hydrophobic lipid core. Water is the primary solvent of the body, and many vital nutrients, ions, and other molecules dissolve in it because of its polar nature. But substances that are hydrophilic and dissolve in water are lipophobic as a rule: they do not readily dissolve in lipids. For this reason, the lipid layer between the cytoplasm and the cell's exterior acts as a boundary that prevents hydrophilic molecules from crossing.

Substances that can pass through the lipid center of the cell membrane move by diffusion. Diffusion directly across the phospholipid bilayer of a membrane is called **simple diffusion** and has the following properties in addition to the properties of diffusion listed earlier.

1. *The rate of diffusion depends on the ability of the molecule to dissolve in the lipid layer of the membrane*. Most molecules in solution can mingle with the polar phosphate-glycerol heads of the membrane's phospholipid bilayer, but only nonpolar molecules that are lipid-soluble (lipophilic) can traverse the central lipid core of the membrane. As a rule, only lipids, steroids, and small lipophilic molecules can move across the membrane by simple diffusion.

Water, although a polar molecule, can diffuse across some phospholipid membranes. For years, it was thought that the polar nature of the water molecule prevented it from moving through the lipid center of the bilayer, but experiments done with artificial membranes have shown that the small size of the water molecule allows it to slip between the lipid tails in many membranes. How readily water passes through the lipid layer depends on the composition of the phospholipid bilayer. Membranes with high cholesterol content are less permeable to water than those with low cholesterol content, presumably because the cholesterol molecules fill the spaces between the fatty acid tails and thus exclude water. For example, the cell membranes of some sections of the kidney are essentially impermeable to water unless the cells insert special water channel proteins into the phospholipid bilayer.

2. *The rate of diffusion across a membrane is directly proportional to the surface area of the membrane*. In other words, the larger the surface area, the more molecules can diffuse across per unit time. This may seem obvious, but it has important implica-

tions in physiology. One striking example of how a change in surface area affects diffusion is the lung disease emphysema. As lung tissue is destroyed, the surface area available for diffusion of oxygen decreases. Consequently, less oxygen can move into the body. In severe cases, the oxygen that reaches the cells is not enough to sustain any muscular activity and the patient is confined to bed.

3. *The rate of diffusion across a membrane is inversely proportional to the thickness of the membrane.* The thicker the membrane, the slower diffusion will take place. This factor comes into play in certain lung conditions in which the exchange epithelium of the lung is thickened with scar tissue. This slows diffusion so that the oxygen entering the body is not adequate to meet metabolic needs.

The rules for simple diffusion across membranes are summarized in Table 5-2. They can be combined mathematically into an equation known as **Fick's law of diffusion**. In an abbreviated form, Fick's law says that:

Rate of diffusion is proportional to

$$\frac{\text{available surface area} \times \text{concentration gradient}}{\text{resistance of membrane} \times \text{membrane thickness}}$$

The resistance of the membrane is the most complex of the four terms because it has several factors that influence it: (1) the size of the diffusing molecule, (2) the lipid-solubility of the molecule, and (3) the composition of the lipid layer across which it is diffusing. As molecular size increases, resistance of the membrane to diffusion increases. As lipid solubility of the diffusing molecule

increases, resistance of the membrane to diffusion decreases. Changes in the lipid composition of the membrane may alter how easily diffusing molecules can slip between the individual phospholipids. For example, cholesterol molecules in membranes pack themselves into the spaces between the fatty acids tails and retard passage of molecules through those spaces (see Fig. 5-2). Figure 5-16 ● illustrates the principles of Fick's law.

Lipophobic molecules cannot enter a cell by simple diffusion, therefore cells must make special provisions for their transport. Smaller lipophobic molecules and ions cross cell membranes through protein carriers. Ions also cross membranes through open water-filled channels. However, the movement of ions through channels is influenced by electric gradients as well as concentration gradients. This means that ion movement is not simple diffusion and cannot be predicted by using Fick's law. We will discuss ion movement in the last sections of this chapter. Macromolecules and particles such as bacteria are too large to be transported by protein carriers. They must move into and out of cells by way of membrane vesicles.

CONCEPT CHECK   Answers: p. 169

11. Where does the energy for diffusion come from?

12. Which is more likely to cross a cell membrane by simple diffusion: a fatty acid or glucose?

13. What happens to the rate of diffusion in each of the following cases?
    (a) membrane thickness increases
    (b) concentration gradient increases
    (c) surface area decreases

    (continues on next page)

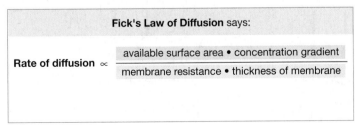

● **Figure 5-16** *Fick's law of diffusion*   Fick's law of diffusion mathematically relates factors that influence the rate of simple diffusion across a membrane.

14. Two compartments are separated by a membrane that is permeable only to water and yellow dye molecules. Compartment A is filled with a solution of yellow dye, compartment B is filled with a solution of equal concentration of blue dye. If the system is left undisturbed for a long period of time, what color will compartment A be: yellow, blue, or green? (Remember, yellow + blue makes green.)

15. What keeps atmospheric oxygen from diffusing into our bodies across our skin? (Hint: What kind of epithelium is skin?)

Carrier-Mediated Transport Exhibits Saturation, Specificity, and Competition

Movement of substances across cell membranes with the aid of a carrier protein is known as **mediated transport**. Some mediated transport carriers move only one kind of molecule, a process known as **uniport** (Fig. 5-17a ●). However, it is common to find carriers that move two or even three different kinds of molecules. A protein that moves more than one kind of molecule at one time is called a **cotransporter**. If the molecules are moving in the same direction, whether into or out of the cell, the process is called **symport** [*sym-*, together + *portare*, to carry]. If the molecules are being carried in different directions, the process is called **antiport** [*anti*, opposite + *portare*, to carry]. Symport and antiport are shown in Figure 5-17b, c.

If mediated transport is passive and moves molecules down their concentration gradient, and if net transport stops when concentrations are equal inside the cell and out, the process is known as **facilitated diffusion**. If protein-mediated transport requires energy from ATP or another outside source and moves a substance against its concentration gradient, the process is known as **active transport**.

Both passive and active forms of mediated transport demonstrate three properties: specificity, competition, and saturation. These concepts were introduced in the discussion on enzymes [● p. 97]. Mediated transport shares these properties with enzymes because both processes depend on the interaction of a substrate with a protein.

Specificity Specificity refers to the ability of a carrier to move only one molecule or a group of closely related molecules. One example of specificity is the family of transporters known as the GLUT transporters. These membrane proteins move six-carbon sugars (*hexoses*), such as glucose, mannose, galactose, and fructose [● p. 35], across cell membranes. But GLUT transporters will not transport the disaccharide maltose or a form of glucose that is not normally found in nature. Thus we can say that GLUT transporters are specific for the naturally occurring six-carbon monosaccharides.

Competition Competition is a property closely related to specificity. A transporter may move several members of a related group of substrates, but those substrates will compete with each other for the binding sites on the transport proteins. For example, the GLUT transporters move the family of hexose sugars, but each different GLUT transporter has a "preference" for one or more hexoses, based on its binding affinity.

The results of an experiment demonstrating competition are shown in Figure 5-18 ●. In the initial part of this graph, GLUT transporters are moving galactose into a cell at a constant rate. At the arrow, glucose is added to the extracellular fluid. Because the transporters have a higher binding affinity for glucose than for galactose, glucose replaces galactose on some of the binding sites. When galactose is displaced from some of the transporters, the rate of galactose transport into the cell decreases. In this example, glucose and galactose are competing for the transporter.

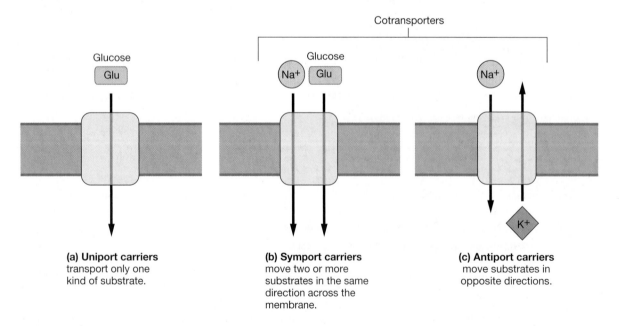

Cotransporters

Glucose

(a) Uniport carriers transport only one kind of substrate.

Glucose

(b) Symport carriers move two or more substrates in the same direction across the membrane.

(c) Antiport carriers move substrates in opposite directions.

● **Figure 5-17** *Types of carrier-mediated transport*

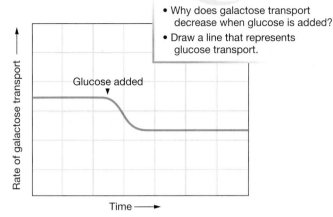

GRAPH QUESTIONS

• Why does galactose transport decrease when glucose is added?
• Draw a line that represents glucose transport.

Glucose and galactose use the same membrane transporter. The transporter has a higher affinity for glucose than for galactose.

● **Figure 5-18** *Graph of competition*

Sometimes the competing molecule is not transported but merely blocks the transport of another substrate. In this case, the competing molecule is a *competitive inhibitor* [● p. 95]. In the glucose transport system, the disaccharide maltose is an inhibitor (Fig. 5-19 ●). It competes with glucose for the binding site, but, once bound, is too large to be moved across the membrane.

Competition between transported substrates has been put to good use in medicine. For example, gout is a disease caused by elevated uric acid in the plasma. One method of decreasing plasma uric acid is to enhance its excretion in the urine. Normally, the kidney's organic acid transporter reclaims uric acid from the urine. However, if an organic acid called probenecid is administered, the carrier binds to probenecid instead of to uric acid, preventing the reabsorption of uric acid. As a result, more uric acid leaves the body in the urine, lowering the uric acid concentration in the plasma.

EMERGING CONCEPTS

The GLUT Transporter Family

Facilitated diffusion carriers for the hexose sugars are found in every cell of the body. For many years, scientists assumed that there were different variants of the carrier. In some cells, glucose transport was regulated by hormones, yet in other cells, it was independent of hormonal action. It was not until the 1980s that the first glucose transporter was isolated. To date, six glucose transporters have been identified. The first transporter, isolated from red blood cells, was named GLUT1. This protein is the facilitated diffusion carrier that moves glucose and other hexoses into most cells of the body. The GLUT2 transporter is found in the hepatocytes of the liver and on the basolateral membrane of transporting epithelial cells in the kidney and intestine. GLUT3 transports glucose and other hexoses into neurons. The GLUT4 transporter is found in adipose tissue and muscle. It is the glucose transporter regulated by the hormone insulin. GLUT5 is actually a fructose transporter and is the transporter present on the apical membrane of the intestinal epithelial cells. The protein named GLUT6 turned out not to be a transporter, but by the time that was discovered, GLUT7 had already been named. GLUT7 is an intracellular transporter found on the endoplasmic reticulum membrane. The restriction of different GLUT transporters to different tissues is an important feature in the metabolism and homeostasis of glucose.

Saturation The rate of substrate transport depends on both the substrate concentration and the number of carrier molecules, just as the rate of enzyme activity depends on both enzyme and substrate concentration [● p. 96]. For a fixed number of carriers, however, as substrate concentration increases, the rate increases

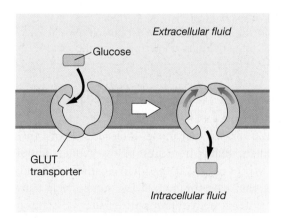

(a) The **GLUT transporter** brings glucose across cell membranes.

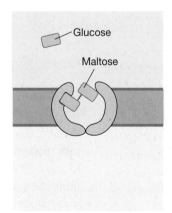

(b) Maltose is a competitive inhibitor that binds to the GLUT transporter but is not itself carried across the membrane.

● **Figure 5-19** *Competitive inhibition*

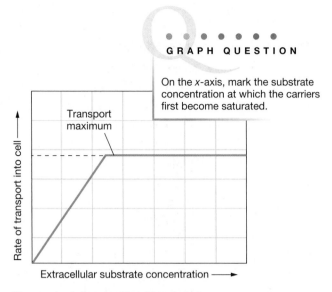

GRAPH QUESTION

On the *x*-axis, mark the substrate concentration at which the carriers first become saturated.

Transport rate is proportional to substrate concentration until the carriers are saturated.

● **Figure 5-20** *Graph showing saturation of carrier-mediated transport*

up to a maximum, the point at which all carriers are filled with substrate. At this point, the carriers are said to have reached saturation. At **saturation** the carriers are working at their maximum rate, and a further increase in substrate concentration will have no effect. Figure 5-20● shows saturation represented graphically.

For an analogy, think of the carriers as doors into a concert hall. Each door has a maximum number of people that it can allow to enter the hall in a given period of time. Suppose all the doors together can allow a maximum of 100 people per minute to enter the hall. This is the maximum transport rate, also called the **transport maximum**. When the concert hall is empty, three maintenance people enter the doors every hour. The transport rate is 3 people/60 minutes, or 0.05 people/minute, well under the maximum. For the local dance recital, about 50 people per minute go through the doors, still well under the maximum. But when the most popular rock group of the day appears in concert, thousands of people gather outside. When the doors open, thousands of people are clamoring to get in. But the doors will only allow 100 people/minute into the hall. They are working at the maximum rate, so it does not matter whether there are 1000 or 3000 people trying to get in. The transport rate is saturated at 100 people per minute.

How can cells avoid having transport reach saturation? One way is to increase the number of carriers in the membrane. This would be like opening more doors into the concert hall. Under some circumstances, cells are able to insert additional carriers into their membranes. Under other circumstances, a cell may withdraw carriers to decrease movement of a molecule into or out of the cell.

All forms of mediated transport show saturation, specificity, and competition. But they differ in other ways. Passive mediated transport, better known as facilitated diffusion, requires no input

of energy. Active transport requires energy input from ATP, either directly or indirectly.

CONCEPT CHECK   Answers: p. 169

16. What would you call a carrier that moves two substrates in opposite directions across the membrane?

17. Using the door analogy above, can you think of another way a cell might increase its maximum transport rate?

Facilitated Diffusion Is Diffusion That Uses Membrane Proteins

Some polar molecules appear to move into and out of cells by diffusion, even though we know from their chemical properties that they are unable to pass easily through the lipid core of the cell membrane. Most of these polar molecules travel by facilitated diffusion, in which carrier proteins transport them across the cell membrane. Glucose and amino acids are examples of molecules that enter or leave cells with the help of facilitated diffusion carriers.

Facilitated diffusion has the same properties as simple diffusion. The molecules move down their concentration gradient, the process requires no input of energy, and net movement stops at equilibrium, when concentration inside the cell equals concentration outside the cell (Fig. 5-21●):

$$[\text{glucose}]_{ECF} = [\text{glucose}]_{ICF}*$$

Cells can avoid reaching equilibrium by keeping the concentration of substrate in the cell low. With glucose, for example, this is accomplished by phosphorylating glucose (Fig. 5-22●). As soon as glucose enters the cell, it is phosphorylated to glucose 6-phosphate, the first step of glycolysis [● p. 103]. Addition of the phosphate group keeps the concentration of plain glucose inside the cell low and also prevents glucose from leaving the cell.

CONCEPT CHECK   Answers: p. 169

18. Liver cells are able to convert glycogen to glucose and make intracellular glucose concentrations higher than the extracellular glucose concentration. What do you think happens to facilitated diffusion of glucose when this occurs?

Active Transport Requires the Input of Energy from ATP

Active transport is a process that moves molecules *against* their concentration gradient, that is, from areas of lower concentration to areas of higher concentration. Rather than creating an equilibrium state, where the concentration of the molecule is equal throughout the system, active transport creates a state of

---

*In this book we will use the notation of brackets around a solute name to indicate concentration.

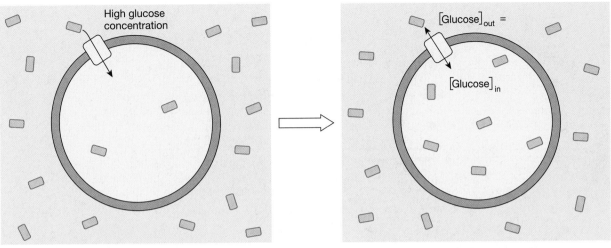

Facilitated diffusion brings glucose into the cell down its concentration gradient.

Diffusion stops at equilibrium, when the glucose concentration inside and outside the cell is equal.

● **Figure 5-21**
Diffusion stops at equilibrium

*dis*equilibrium by making concentration differences more pronounced. Moving molecules against their concentration gradient requires the input of outside energy, just as pushing a ball up a hill requires energy [● Fig. 4-2, p. 87]. The energy for active transport comes directly or indirectly from the high-energy phosphate bond of ATP.

Active transport can be divided into primary active transport and secondary active transport. In **primary (direct) active transport**, the energy for transport comes directly from the high-energy phosphate bond of ATP. In **secondary (indirect) active transport**, the energy to push molecules against their gradient comes from potential energy stored in a concentration gradient [● p. 87]. All secondary active transport ultimately depends on primary active transport because the concentration gradients that drive it are created using energy from ATP.

The mechanism for both types of active transport carriers appears to be similar to that for facilitated diffusion carriers. The substrate(s) binds to the protein carrier, which then changes conformation, releasing the substrate into the opposite compartment. Active transport differs from facilitated diffusion because these conformation changes require energy input.

Primary Active Transport Primary active transport uses ATP as its energy source. Consequently, many of the primary active transporters are known as **ATPases**. You may recall from Chapter 4 that the suffix *-ase* signifies an enzyme, and the stem (ATP) is the substrate upon which the enzyme is acting [● p. 91]. ATPases hydrolyze ATP to ADP and inorganic phosphate (P_i), releasing usable energy in the process. Most of the ATPases you will encounter in your study of physiology are listed in Table 5-3. ATPases are sometimes called *pumps*, as in the sodium-potassium pump.

The sodium-potassium pump, or **Na^+-K^+-ATPase**, is probably the single most important transport protein in animal cells, because it maintains the concentration gradients of Na^+ and K^+ across the cell membrane. The transporter is arranged in the cell membrane so that it pumps 3 Na^+ out of the cell and 2 K^+ into the cell for each ATP consumed (Fig. 5-23 ●). In some cells, the energy needed to move these ions uses 30% of all the ATP produced by the cell. The current model of how the Na^+-K^+-ATPase works is illustrated in Figure 5-24 ●.

● **Figure 5-22** *Diffusion of glucose into cells*

Table 5-3: Primary Active Transporters

NAMES	TYPE OF TRANSPORT
Na^+-K^+-ATPase or sodium-potassium pump	Antiport
Ca^{2+}-ATPase	Uniport
H^+-ATPase or proton pump	Uniport
H^+-K^+-ATPase	Antiport

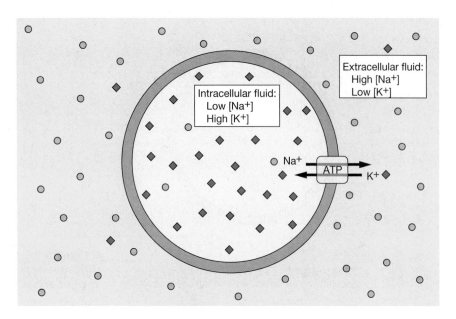

The Na$^+$-K$^+$-ATPase uses energy from ATP to pump Na$^+$ out of the cell and K$^+$ into the cell.

● **Figure 5-23** *The Na$^+$-K$^+$-ATPase*

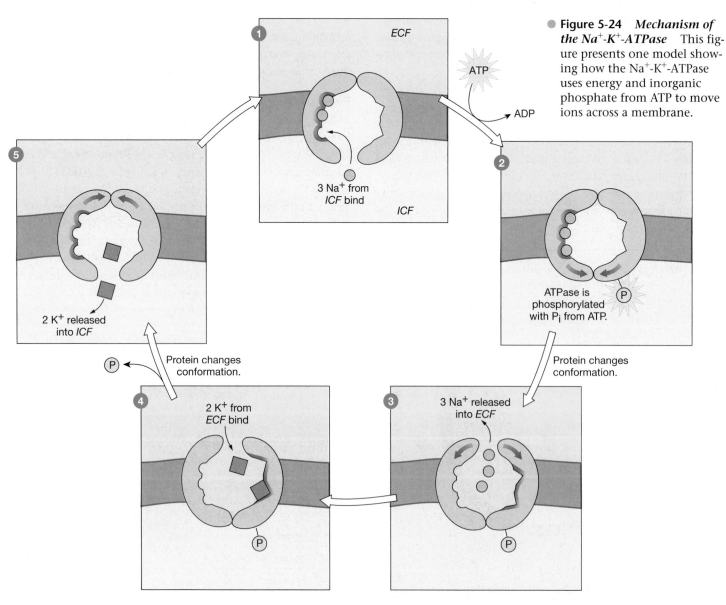

● **Figure 5-24** *Mechanism of the Na$^+$-K$^+$-ATPase* This figure presents one model showing how the Na$^+$-K$^+$-ATPase uses energy and inorganic phosphate from ATP to move ions across a membrane.

The sodium concentration gradient, with Na^+ concentrated in the extracellular fluid and low inside the cell, is a source of potential energy that the cell can harness for other functions. For example, nerve cells use the sodium gradient to transmit electrical signals. Epithelial cells use it to drive the uptake of nutrients, ions, and water. Membrane transporters that use potential energy stored in concentration gradients are called *secondary active transporters*.

Secondary Active Transport Secondary active transport couples the kinetic energy of one molecule moving down its concentration gradient to the movement of another molecule against its concentration gradient. The cotransported molecules may go in the same direction across the membrane (symport) or one may leave the cell in exchange for the other (antiport). The most common secondary active transport systems are driven by the sodium concentration gradient. As Na^+ moves into the cell, it either brings one or more molecules with it or trades places with molecules exiting the cell. The major Na^+-dependent transporters are listed in Table 5-4. Notice that the cotransported molecules may be either other ions or uncharged molecules, such as glucose.

The mechanism of the Na^+-glucose symporter is illustrated in Figure 5-25 ●. Notice its similarity to both the facilitated diffusion carrier (Fig. 5-19) and the Na^+-K^+-ATPase (Fig. 5-24). Both Na^+ and glucose bind to the carrier on the extracellular fluid side. Sodium binds first and causes a conformational change in the protein that creates a high-affinity binding site for glucose. When glucose binds, the protein changes conformation again and opens its channel to the intracellular fluid side. Sodium is released as it moves down its concentration gradient. The loss of Na^+ from the protein changes the binding site for glucose back to a low-affinity site, so glucose is released and follows Na^+ into the cytoplasm. The net result is the entry of glucose into the cell against its concentration gradient, coupled to the movement of Na^+ into the cell down its concentration gradient.

Table 5-4: Typical Secondary Active Transporters

SYMPORT CARRIERS	ANTIPORT CARRIERS
Sodium-dependent transporters	
Na^+-glucose	Na^+-H^+
Na^+-amino acids (several types)	Na^+-Ca^{2+}
Na^+-K^+-2 Cl^-	
Na^+-bile salts (small intestine)	
Na^+-choline uptake (nerve cells)	
Na^+-neurotransmitter uptake (nerve cells)	
Nonsodium-dependent transporters	
	HCO_3^--Cl^-
	H^+-K^+

● **Figure 5-25** *Sodium-glucose symporter* This transporter uses the potential energy of the Na^+ concentration gradient to move glucose against its concentration gradient.

Why does the body have both a facilitated diffusion GLUT carrier and a Na^+-glucose symporter? The simple answer is that GLUT carriers are found on all cells but the Na^+-glucose symporter is restricted to cells that are bringing glucose into the body from the external environment. GLUT transporters move glucose into or out of the cells depending on the concentration gradient.

For example, when blood glucose levels are high, GLUT transporters on liver cells bring glucose into the cell. During times

Energy is imported into the cell as energy stored in chemical bonds of nutrients such as glucose.

Metabolism

The chemical bond energy is converted into high-energy bonds of ATP through the process of metabolism.

Primary active transport

The energy in the high-energy phosphate bond of ATP is used to move K^+ and Na^+ against their concentration gradients. This creates potential energy stored in the ion concentration gradients.

Secondary active transport

The energy of the Na^+ gradient can be used to move other molecules across the cell membrane against their concentration gradients.

● **Figure 5-26**   *Energy transfer in living cells*

of fasting, when blood glucose levels fall, the liver cell converts its glycogen stores to glucose. When the glucose concentration inside the liver cell builds up and exceeds the glucose concentration in the plasma, glucose leaves the cell on the reversible GLUT transporters. In contrast, the Na^+-glucose transporter can only move substrate into the cell because glucose must follow the Na^+ gradient. Consequently, the Na^+-glucose transporter is found only on transporting epithelial cells that bring glucose into the body from the external environment.

The use of energy to drive secondary active transport is summarized in Figure 5-26●. The organism imports energy from the environment in the chemical bonds of nutrients such as glucose. This energy is transferred to the high-energy bonds of ATP through oxidative phosphorylation in the mitochondria [◈ p. 106]. ATP is used to create ion gradients. For example, the Na^+-K^+-ATPase pushes sodium out of the cell and potassium into it. In doing so, the energy from ATP is transformed into potential energy stored in the ion concentration gradients. This potential energy can be harnessed by various sodium-dependent cotransporters such as the Na^+-K^+-2 Cl^- transporter. The kinetic energy of Na^+ moving into the cell down its concentration gradient is linked to the uphill movement of K^+ and Cl^- into the cell. As you study the different systems of the body, you will find these

secondary active transporters taking part in many physiological processes.

Vesicles Move Large Molecules Across Membranes

What happens to the many macromolecules that are too large to enter or leave cells through protein channels or membrane transporters? They move in and out of the cell with the aid of bubble-like vesicles created from the cell membrane. Cells use two basic mechanisms to import large molecules and particles: phagocytosis and endocytosis. Phagocytosis can be considered a type of endocytosis, but as scientists learned more about the mechanisms behind the two processes, they decided that phagocytosis was fundamentally different. Material leaves cells by the process known as exocytosis, similar to endocytosis in reverse.

Phagocytosis   If you studied *Amoeba* in your biology laboratory, you may have watched these one-cell creatures ingest their food by surrounding it and enclosing it into a vesicle that is brought into the cytoplasm. **Phagocytosis** [*phagein*, to eat + *cyte*, cell + *-sis*, process] is the process by which a cell engulfs a bacterium or other particle into a large membrane-bound **vesicle** called a **phagosome** (Fig. 5-27 ●). The phagosome pinches off from the

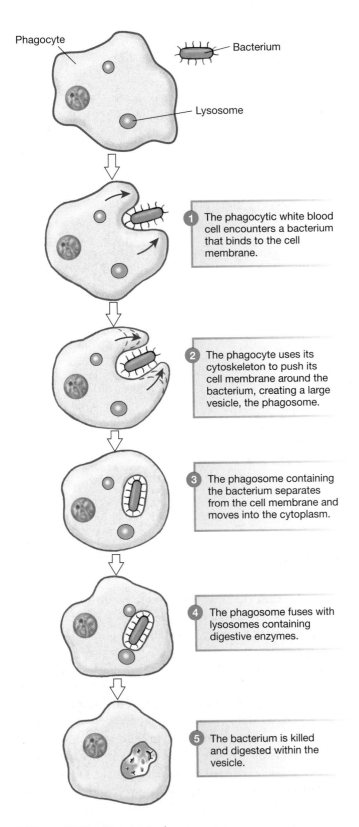

Phagocyte

Bacterium

Lysosome

1 The phagocytic white blood cell encounters a bacterium that binds to the cell membrane.

2 The phagocyte uses its cytoskeleton to push its cell membrane around the bacterium, creating a large vesicle, the phagosome.

3 The phagosome containing the bacterium separates from the cell membrane and moves into the cytoplasm.

4 The phagosome fuses with lysosomes containing digestive enzymes.

5 The bacterium is killed and digested within the vesicle.

● **Figure 5-27** *Phagocytosis*

cell membrane and moves to the interior of the cell, where it fuses with a lysosome [● p. 61], whose digestive enzymes will destroy the bacterium. Phagocytosis requires energy from ATP for the movement of the cytoskeleton and for the intracellular trans-

port of the vesicles. In humans, this phagocytosis occurs only in certain types of white blood cells called *phagocytes* that specialize in "eating" bacteria and other foreign particles.

Endocytosis Another means by which molecules or particles can move into cells is called **endocytosis**. It differs from phagocytosis in that the membrane surface indents rather than pushes out, and the vesicles formed are much smaller. In addition, some endocytosis is *constitutive*, that is, it is an essential function that is always taking place. Phagocytosis must be triggered by the presence of a substance to be ingested.

Endocytosis is an active process that requires ATP. It can be nonselective, allowing extracellular fluid to enter the cell (a process called **pinocytosis** [*pino-*, drink]) or it can be highly selective, allowing only specific molecules to enter the cell. Two types of endocytosis require a ligand to bind to a membrane receptor: receptor-mediated endocytosis and potocytosis.

Receptor-mediated endocytosis Receptor-mediated endocytosis takes place in regions of the cell membrane known as **clathrin-coated pits**, indentations where the cytoplasmic side of the membrane has high concentrations of the protein *clathrin* (Fig. 5-28 ●). In the first step of the process, extracellular ligands that will be brought into the cell bind to their membrane receptors

---

➕ CLINICAL FOCUS

LDL—The Lethal Lipoprotein

Cholesterol is a molecule that is essential for membrane structure and for the synthesis of steroid hormones, such as the sex hormones. On the other hand, elevated cholesterol can lead to heart disease. Why is it that some people have too much cholesterol in their blood (*hypercholesterolemia*)? Cholesterol is an insoluble nonpolar molecule, which means it must be bound to a lipoprotein carrier molecule for transport in the blood. There are several forms of the carrier, the most common of which is *low-density lipoprotein* (*LDL*). LDL-cholesterol is taken into cells by endocytosis when LDL receptors on the cell surface bind to the protein portion of the molecule. People who inherit a genetic defect that decreases the number of LDL receptors on their cell membranes cannot transport cholesterol normally into their cells. As a result, LDL-cholesterol remains in the plasma. Abnormally high blood levels of LDL-cholesterol predispose these people to the development of **atherosclerosis**, commonly known as hardening of the arteries [*atheroma*, a tumor + *skleros*, hard + *-sis*, condition]. In this condition, the accumulation of cholesterol in blood vessels blocks blood flow and contributes to heart attacks.

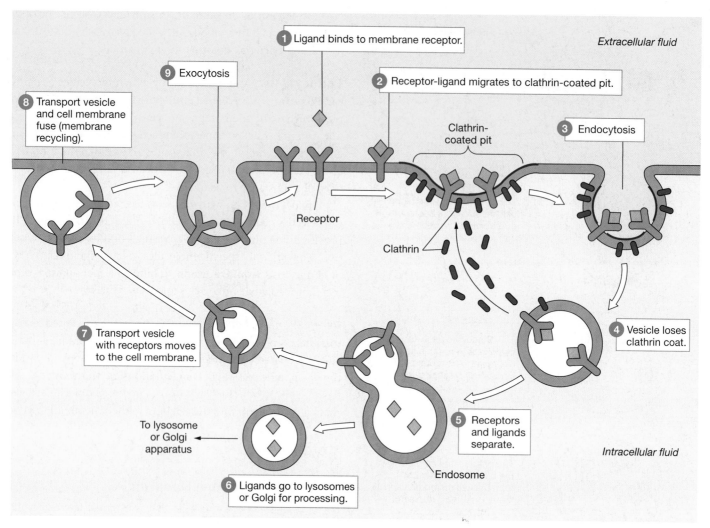

● **Figure 5-28** *Receptor-mediated endocytosis and exocytosis* Steps 1 through 4 illustrate endocytosis, while steps 7 through 9 show exocytosis.

①. The receptor-ligand complex migrates along the cell surface until it encounters a coated pit ②. Once the receptor-ligand complex is in the coated pit, the membrane draws inward, or *invaginates* ③, then pinches off from the cell membrane, and becomes a cytoplasmic vesicle. The clathrin molecules are released and recycle back to the membrane ④. In the vesicle, the receptor and ligand separate, leaving the ligand in an *endosome* ⑤. The ligand goes to a lysosome if it is to be destroyed or to the Golgi apparatus for processing ⑥. The receptors may then be reused in a process known as **membrane recycling** ⑦. The vesicle fuses with the cell membrane ⑧ and its membrane is incorporated back into the cell membrane ⑨.

Receptor-mediated endocytosis transports a variety of substances into the cell including protein hormones, growth factors, antibodies, and plasma proteins that serve as carriers for iron and cholesterol. Abnormalities in receptor-mediated removal of cholesterol from the blood are associated with elevated plasma cholesterol levels and cardiovascular disease.

Potocytosis and caveolae Potocytosis is distinguished from receptor-mediated endocytosis because it uses **caveolae** ("little caves") rather than clathrin-coated pits to concentrate and bring receptor-bound molecules into the cell. Caveolae are membrane regions that are rich in lipids, membrane receptor proteins, and a family of unique membrane proteins named *caveolins*. The receptors in caveolae are linked to the membrane by a glycolipid anchor and are not integral proteins. In many cells, caveolae appear as small, flask-shaped invaginations of the cell membrane, which is how they acquired their name.

Caveolae have several functions: to concentrate and internalize small molecules, to help in the transfer of macromolecules across the capillary endothelium (see Fig. 5-31), and to participate in cell signaling. Caveolae appear to be involved in some disease processes including viral and parastic infections. Two forms of the disease *muscular dystrophy* are associated with abnormalities in the protein caveolin. Scientists are currently trying to discover more details about the role of caveolae in normal physiology and pathophysiology.

Exocytosis Exocytosis is the opposite of endocytosis. In exocytosis, intracellular vesicles move to the cell membrane, fuse with it (Figure 5-28, step 8), and then release their contents to the extracellular fluid (step 9). Cells use exocytosis to export large lipophobic molecules, such as proteins synthesized within the cell, and to get rid of wastes left in lysosomes from intracellular digestion.

The process by which the cell and vesicle membranes fuse is similar in a variety of cell types, from neurons to endocrine cells. Exocytosis involves a series of proteins with names such as SNARE, SNAP, rab, and VAMP. Although the details are still not fully understood, the process usually begins with an increase in intracellular Ca^{2+} concentration. The Ca^{2+} ions interact with one of the proteins, causing the secretory vesicles to attach to the membrane, then fuse with it. The fused area opens, and the vesicle membrane becomes part of the cell membrane as the vesicle contents diffuse into the extracellular space. Exocytosis, like endocytosis, requires energy in the form of ATP.

Exocytosis takes place continually in some cells, making it a *constitutive process*. For example, goblet cells [● p. 70] in the intestine continually release mucus by exocytosis, and fibroblasts in connective tissue release collagen [● p. 72]. In other cell types, exocytosis is an intermittent process that is initiated by a signal. In many endocrine cells, hormones are stored in secretory vesicles in the cytoplasm and released in response to a signal from outside the cell. Exocytosis is also the means by which membrane proteins can be inserted into the cell membrane, as shown in Figure 5-28. You will encounter many examples of exocytosis in your study of physiology.

CONCEPT CHECK   Answers: p. 169

19. How does phagocytosis differ from endocytosis?

20. Name the two membrane proteins associated with regions of endocytosis.

21. How do cells move large proteins into the cell? Out of the cell?

Molecules Entering or Leaving the Body Must Cross an Epithelium

The transport processes described in the previous sections all deal with the movement of molecules across a single membrane, that of the cell. However, any molecules entering and leaving the body across an epithelium must cross two cell membranes. Molecules cross the first membrane when they move into an epithelial cell from the external environment and the second when they leave the epithelial cell to enter the extracellular fluid. Movement across epithelial cells, **transepithelial transport**, uses a combination of active and passive transport.

The transporting epithelia of the intestine and kidney are specialized to selectively transport molecules into and out of the body. These epithelial cells are connected to each other by adhesive junctions and tight junctions [● p. 64] that act as barriers to minimize the unregulated diffusion of material between the cells. The tight junctions also mark the separation of the cell membrane into two regions, or poles. The surface of the epithelial cell that faces the lumen of an organ is called the **apical** (or *mucosal*) membrane (Fig. 5-29 ●). It is often folded into microvilli that increase its surface area. The sides of the cell below the tight junctions and the surface of the cell membrane in contact with the extracellular fluid are collectively called the **basolateral** (or *serosal*) membrane.

Transporting epithelial cells are said to be *polarized* because their apical and basolateral membranes have very different properties. Certain transport proteins, such as the Na^+-K^+-ATPase, are almost always found only on the basolateral membrane, whereas others, like the Na^+-glucose symporter, are localized to the apical membrane. This polarized distribution of transporters results in the one-way movement of certain molecules across the epithelium. Transport of material from the lumen of an organ to the extracellular fluid is called **absorption**. Movement of material from the ECF to the lumen is called **secretion**.

Lumen of intestine or kidney

Apical membrane faces the lumen.

Tight junction prevents movement of substances between the cells.

Transporting epithelial cell

Basolateral membrane faces the *ECF*.

Extracellular fluid

Transport proteins

Polarized epithelia have different transport proteins on apical and basolateral membranes. This allows selective directional transport across the epithelium. Transport from lumen to *ECF* is called **absorption**. Transport from *ECF* to lumen is called **secretion**.

● **Figure 5-29** *Polarized cells of transporting epithelia*

147

The cells of transporting epithelia can alter their permeability by selectively inserting or withdrawing membrane proteins. Transporters pulled out of the membrane may be destroyed in lysosomes or they may be stored in vesicles inside the cell, ready to be reinserted into the membrane in response to a signal (another example of membrane recycling).

The transepithelial movement of molecules requires that the molecules first enter an epithelial cell and then leave it to enter the extracellular fluid. If the molecule can be transported through a protein channel or on transport carriers, then the two-step process usually has one uphill step that requires energy and one downhill step in which the molecule moves passively down its concentration gradient. Molecules that are too large to be moved by membrane proteins can be transported across the cell in vesicles.

Transepithelial Transport Using Membrane Proteins The movement of glucose from the lumen of the kidney tubule or intestine to the extracellular fluid is an important example of directional movement across a transporting epithelium. Transepithelial movement of glucose involves three different transport systems: the secondary active transport of glucose with Na^+ from the lumen into the cell at the apical membrane, followed by the active transport of Na^+ and the facilitated diffusion of glucose out of the cell and into the extracellular fluid on the basolateral side.

Figure 5-30 ● shows the process in detail. The glucose concentration in the transporting epithelial cell is higher than the glucose concentration in either the extracellular fluid or the lumen of the intestine. To move glucose from the lumen into the cell is an uphill process and requires the input of energy—in this case, energy stored in the Na^+ concentration gradient. Sodium

ions in the lumen bind to the symporter, as previously described (see Fig. 5-25), and bring glucose with them into the cell. The energy needed to move glucose against its concentration gradient comes from the kinetic energy of Na^+ moving down its gradient.

Once glucose is in the cell, it leaves by moving down its concentration gradient through the facilitated diffusion GLUT transporter in the basolateral membrane (step 2 of Fig. 5-30). The Na^+ is pumped out of the cell on the basolateral side using the Na^+-K^+-ATPase (step 3). Sodium is more concentrated in the extracellular fluid than in the cell, therefore this step requires energy provided by ATP.

The removal of Na^+ is essential if glucose is to continue to be absorbed, because the Na^+-glucose symporter depends on low intracellular concentrations of Na^+. If the basolateral Na^+-K^+-ATPase is poisoned with *ouabain* ("wah-bane"—a compound related to the heart drug digitalis), Na^+ that enters the cell cannot be pumped out. The Na^+ concentration inside the cell then increases until it is equal to that outside the cell. Without a sodium gradient, there is no energy source to run the Na^+-glucose symporter, and the movement of glucose across the epithelium stops.

Transepithelial transport can use ion movement through channels in addition to carrier-mediated transport. For example, the apical membrane of a transporting epithelium may use the Na^+-K^+-2Cl^- symporter to bring K^+ into the cell against its concentration gradient, using energy from the Na^+ gradient. The K^+ concentration inside the cell is higher than in the extracellular fluid, therefore the K^+ can move out of the cell on the basolateral side through open K^+ leak channels. The Na^+ must be pumped out by the Na^+-K^+-ATPase. By this simple mechanism the body can absorb Na^+ and K^+ at the same time from the lumen of the intestine or the kidney.

CONCEPT CHECK
Answers: pp. 169–170

22. Why does Na^+ movement from the cytoplasm to the extracellular fluid require energy?

23. Ouabain, an inhibitor of the Na^+-K^+-ATPase, cannot pass through cell membranes. What would happen to the transepithelial glucose transport shown in Figure 5-30 if ouabain were applied to the apical side of the epithelium? To the basolateral side of the epithelium?

24. Which GLUT transporter is illustrated in Figure 5-30? (Hint: See the Emerging Concepts box on GLUT transporters.)

Transcytosis and Vesicular Transport Some molecules, such as proteins, are too large to cross epithelia on membrane transporters. Instead they are moved across epithelia by **transcytosis**, which is a combination of endocytosis, vesicular transport across the cell, and exocytosis (Fig. 5-31 ●). In this process, the molecule is brought into the cell using receptor-mediated endocytosis or potocytosis. The resulting vesicle attaches to microtubules in the cytoskeleton and is transported across the

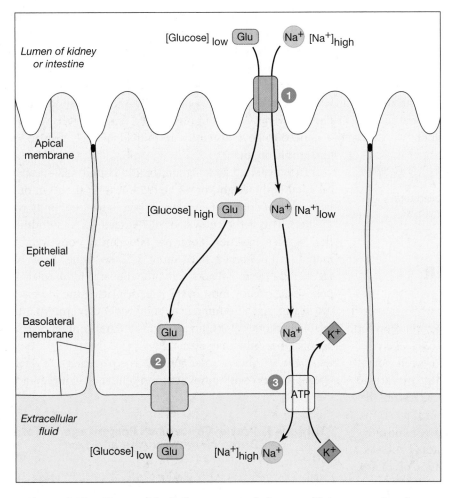

● **Figure 5-30** *Transepithelial transport of glucose* This process involves indirect active transport of glucose across the apical membrane and glucose diffusion across the basolateral membrane.

① **Na⁺- glucose symporter** brings glucose into cell against its gradient using energy stored in the Na⁺ concentration gradient.

② **GLUT transporter** transfers glucose to *ECF* by facilitated diffusion.

③ **Na⁺-K⁺- ATPase** pumps Na⁺ out of the cell, keeping *ICF* Na⁺ concentration low.

FIGURE QUESTIONS

• Match each transporter to its location.
  1. GLUT   a) apical membrane
  2. Na⁺-glucose   b) basolateral membrane
     symporter
  3. Na⁺-K⁺-ATPase
• Is glucose movement across the basolateral membrane active or passive? Explain.
• Why doesn't Na⁺ movement at apical membrane require ATP?

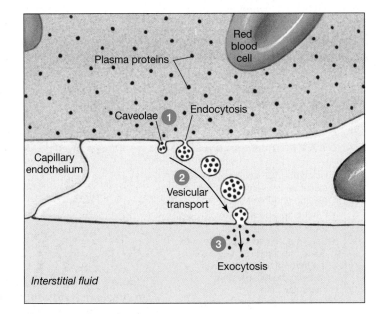

① Plasma proteins are concentrated in caveolae, which then undergo endocytosis and form vesicles.

② Vesicles cross the cell with help from the cytoskeleton.

③ Vesicle contents are released into interstitial fluid by exocytosis.

● **Figure 5-31** *Transcytosis across the capillary endothelium*

149

cell by a process known as **vesicular transport**. At the opposite side of the epithelium, the contents of the vesicle are expelled into the interstitial fluid by exocytosis.

Transcytosis makes it possible for large proteins to move across an epithelium and remain intact. It is the means by which infants absorb maternal antibodies in breast milk. The antibodies are absorbed on the apical surface of the infant's intestinal epithelium and then released into the extracellular fluid.

CONCEPT CHECK   Answers: p. 170

25. If a poison that disassembles microtubules is applied to a capillary endothelial cell, what happens to transcytosis?

THE DISTRIBUTION OF WATER AND SOLUTES IN THE BODY

The distribution of solutes in the body depends on whether a substance can cross the cell membrane by simple diffusion or whether it needs a membrane protein or vesicle to move across. Cell membranes are **selectively permeable**, that is, a cell can select which molecules will enter and which will leave by varying the lipid composition and protein transporters of its membrane. In the remainder of this chapter, we will discuss how the selective permeability of cell membranes creates a body in which the intracellular and extracellular compartments are chemically and electrically different but have the same total concentration of solutes.

The Body Is Mostly Water

Water is the most important molecule in the human body. It is the solvent for all living matter and moves freely between the cells and the extracellular fluid. If we look at the relative volumes of the body compartments (Fig. 5-32 ●), we find that the intracellular compartment contains about 67% of the body's water. The remaining 33% is split between the interstitial fluid (which contains three-quarters of the extracellular water) and the plasma (which contains about one-quarter of the extracellular water).

How much water is in the body? Because one individual differs from the next, there is no single answer. However, in human physiology we often speak of standard values for physiological functions, based on "the 70-kg man." These standard values are derived from data obtained by studying 25-year-old white males whose average weight was 70 kg. Thus, when we speak of standard or average values in physiology, remember that these numbers need to be adjusted for an individual's age, gender, weight, and ethnic origin.

The "average" 70-kilogram, or 154-pound, male has 60% of his total body weight, or 42 kg (92.4 lbs), in the form of water. Each kilogram of water equals one liter, so his **total body water** is 42 liters. This is the equivalent of fourteen 3-liter soft drink bottles! Women have less water per kilogram of body weight than men because women have more adipose tissue. Look back at Figure 3-24 [● p. 74] and note how the large fat droplets in adipose tissue occupy most of the cell, displacing the more aqueous cytoplasm. Age also influences body water content. Infants have relatively more water than adults, and water content decreases as people grow older than 60.

Table 5-5 shows water content as a percentage of total body weight in people of various ages and gender. In clinical practice, it

Table 5-5: Water Content as Percentage of Total Body Weight by Age and Sex

AGE	MALE	FEMALE
Infant	65%	65%
1–9	62%	62%
10–16	59%	57%
17–39	61%	51%
40–59	55%	47%
60+	52%	46%

Adapted from Edelman and Leibman, *American Journal of Medicine* 27: 256–277, 1959.

GRAPH QUESTIONS

- Using the ECF volume shown, calculate the volumes of the plasma and interstitial fluid.
- What is this person's total body water volume?
- Use your answers from the two questions above to calculate the percentage of total body water in the plasma and interstitial fluid.

● **Figure 5-32** *Distribution of volume in the body fluid compartments* This figure shows the compartment volumes for a 70-kg man.

is necessary to allow for the variability of body water content when prescribing drugs. Because women and older people have less body water, they will have a higher concentration of a drug in the plasma than young men will if all are given an equal dose per kilogram of body weight.

CLINICAL FOCUS

Estimating Body Water

Clinicians estimate a person's fluid loss in dehydration by equating weight loss to water loss. Because 1 liter of pure water weighs 1 kilogram, a decrease in body weight of 1 kilogram is considered the equivalent of loss of 1 liter of body fluid. A baby with diarrhea can easily be weighed to estimate its fluid loss. A decrease of 1.1 pounds (0.5 kg) of body weight is assumed to mean loss of 500 mL of fluid. This calculation provides a quick estimate of how much fluid needs to be replaced.

CONCEPT CHECK   Answers: p. 170

26. A mother brings her baby to the Emergency Room because he has had diarrhea and vomiting for two days. The staff weighs the baby and finds that he has lost 2 pounds. If you assume that all the weight loss is water loss, what volume of water has the baby lost? (2.2 pounds = 1 kilogram)

Living Cells Use Energy to Maintain a State of Chemical Disequilibrium

Although water is able to distribute itself freely throughout the body's compartments, most solutes are restricted by the transport properties of the cell membrane. Earlier in this chapter, you learned that diffusion proceeds until the concentration of the diffusing molecule is equal in all parts of the system, a condition known as *chemical equilibrium*. However, few substances in the body are maintained in a state of chemical equilibrium. This is partly because the cell membrane and the capillary endothelium act as selective barriers that prevent diffusion. In addition, the active transport of solutes helps create or maintain differences in solute concentrations.

The **chemical disequilibrium** of the body is illustrated in Figure 5-33 ●. Sodium, chloride, and bicarbonate (HCO_3^-) ions are more concentrated in the extracellular fluid. Potassium concentrations are higher inside the cell. Proteins and other large anions are found primarily in intracellular fluid and in the plasma but are almost absent from the interstitial fluid. Calcium (not shown in the figure) is more concentrated in the extracellular fluid than in the cytosol, although many cells store calcium ions inside membranous organelles such as the endoplasmic reticulum and mitochondria.

Only the constant input of energy keeps the body in this state of chemical disequilibrium. If solutes leak across the cell membrane, active transporters return them to the compartments

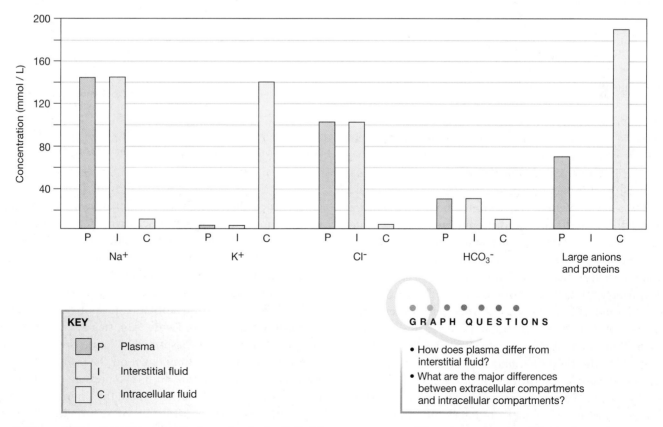

KEY

	P	Plasma
	I	Interstitial fluid
	C	Intracellular fluid

GRAPH QUESTIONS

- How does plasma differ from interstitial fluid?
- What are the major differences between extracellular compartments and intracellular compartments?

● **Figure 5-33** *Distribution of solutes in the body fluid compartments* The compartments of the body are in a state of chemical disequilibrium.

they left. For example, the Na⁺-K⁺-ATPase returns K⁺ that leaks out of the cell to the cytoplasm and pumps Na⁺ that leaks into the cell back to the extracellular fluid. The concentration differences of chemical disequilibrium are a hallmark of the living organism. If cells die and cannot use energy, they obey the second law of thermodynamics [p. 87] and return to a state of randomness.

The Body Is in Osmotic Equilibrium

Water is essentially the only molecule that moves freely between cells and the extracellular fluid and thus can reach a state of equilibrium. The equal distribution of water across body compartments is known as **osmotic equilibrium**. When water moves across a semipermeable membrane in response to a concentration gradient, the process is called **osmosis**. In osmosis, water always moves into the compartment with more concentrated solute.

Look at the example shown in Figure 5-34●, step 1, where two compartments of equal volume are separated by a membrane that is freely permeable to water but does not allow glucose to cross. Compartment A is partly filled with a dilute glucose solution, while compartment B is filled with an equal volume of concentrated glucose solution. This creates a concentration gradient for both glucose and water. However, the membrane is not permeable to glucose, so glucose cannot diffuse to equalize its distribution.

Water, on the other hand, can cross the membrane freely. It will move by osmosis from compartment A, which contains the dilute glucose solution, to compartment B, which contains the more concentrated glucose solution. Thus, water moves to dilute the region of more concentrated solute. Osmosis will continue until the glucose concentrations in the two compartments are equal (Fig. 5-34, step 2). At this point the system has reached osmotic equilibrium, and net water movement stops.

The same situation exists in the body as in the example just discussed. If you drink a glass of water, the water is absorbed through the intestinal epithelium and enters the extracellular fluid. The result is the dilution of solutes in the extracellular fluid. Now the cells have a higher solute concentration than the extracellular fluid. Because there is a concentration gradient, water flows by osmosis across the cell membranes into the cells. Cell volume increases and concentration decreases until the extracellular and intracellular fluid compartments return to equilibrium.

Some people like to think of osmosis as the diffusion of water because water moves from an area of higher *water* concentration to an area of lower *water* concentration. However, this definition is likely to lead to confusion, because water is the solvent and most people use the word *concentration* to refer to the dissolved solute. Look at the two compartments of the system shown in Figure 5-34, step 1. Without counting glucose molecules, which compartment appears to contain the more concentrated solution? Compartment B has more solute for a given volume of solution and therefore has the more concentrated solution. Which

Step 1: Two compartments are separated by a membrane that is permeable to water but not glucose.

Step 2: Water moves by osmosis into the more concentrated solution until concentrations are equal.

Force is applied to B to exactly oppose osmosis from A to B.

Step 3: Osmotic pressure is the pressure that must be applied to B to oppose osmosis.

● **Figure 5-34** *Osmosis and osmotic pressure*

compartment has more water? Compartment A, with less glucose, has more water. Water concentration and solute concentration are inversely related: a concentrated solution has less water; a dilute solution has more water. To avoid confusion over water concentration and solute concentration, remember the rule: *in osmosis, water moves to dilute the area of more concentrated solute.*

How can we make quantitative measurements of osmosis? One method is shown in Figure 5-34, step 3. Place a piston into compartment B, which has a higher solute concentration than compartment A. By pushing down on the piston, you can keep water from flowing from A to B. The pressure that must be applied to the piston to exactly oppose the osmotic movement of water into compartment B is known as **osmotic pressure**. The units for osmotic pressure, like other pressures in physiology, are *atmospheres* (atm) or *millimeters of mercury* (mm Hg).

Osmolarity Describes the Number of Particles in Solution Another way to predict the osmotic movement of water quantitatively is to know the concentrations of the solutions with which we are dealing. In chemistry, concentrations are usually expressed as *molarity* (*M*) or number of moles per liter of solution (mol/L). A *mole* is 6.02×10^{23} (Avogadro's number) of molecules [◉ p. 32]. Therefore, the molarity of the solution describes the number of molecules per liter of solution.

However, using molarity to describe biological concentrations can be misleading. The important factor for osmosis is the number of *particles* in a given volume of solution, not the number of molecules. Some molecules dissociate into ions when they dissolve in a solution, therefore the number of particles in solution is not always the same as the number of molecules. For example, one glucose molecule dissolved in water yields one particle, but one NaCl molecule can yield two ions (particles): Na^+ and Cl^-. Water will move osmotically in response to the total concentration of *particles* in the solution. The particles may be ions, uncharged molecules, or a mixture of both.

Consequently, we express the concentration of biological solutions as **osmolarity**, the number of particles (ions or intact molecules) per liter of solution. Osmolarity is expressed in *osmoles per liter* (osmol/L or OsM) or, for the very dilute physiological solutions, milliosmoles/liter (mOsM). To convert between molarity and osmolarity, use the following equation:

molarity (mol/L) × number of particles/molecule in solution = osmolarity (osmol/L)

Let us look at two examples, glucose and sodium chloride, and compare their molarity with their osmolarity.

One mole of glucose molecules dissolved in enough water to create one liter of solution yields a one molar solution (1 M). Because glucose does not dissociate in solution, the solution has only one mole of osmotically active particles:

1 M glucose × 1 particle/glucose molecule in solution = 1 OsM glucose

Unlike a glucose molecule, a sodium chloride molecule can dissociate into two ions when placed in solution.* Thus, one mole of NaCl molecules dissociates in solution to yield two moles of particles: one mole of Na^+ ions and one mole of Cl^- ions. The result is a 2 OsM solution:

1 M NaCl × 2 ions/NaCl molecule in solution = 2 OsM NaCl

*For the purposes of this discussion, we will assume that all molecules that can dissociate will do so (complete dissociation).

Measuring Osmolarity

In the laboratory, the osmolarity of a solution is measured using an instrument called an osmometer. Many osmometers work on the principle of freezing-point depression. Pure water freezes at 0° C. If solutes are present in the aqueous solution, they will interfere with the crystalline lattice formation of ice and lower the temperature at which the solution freezes. The amount the freezing point is depressed is directly related to the amount of solute present. Osmometers freeze a sample of solution and automatically calculate osmolarity from the freezing-point depression of the solution.

Osmolarity only describes the number of particles in the solution. It says nothing about what the particles are. A 1 OsM solution could be composed of pure glucose or pure Na^+ and Cl^- or a mixture of the three.

The normal osmolarity of the human body ranges from 280 to 296 milliosmoles per liter (mOsM). In this book, to simplify calculations, we will round that number up slightly to 300 mOsM. A term related to osmolarity that you may hear used is osmolality. **Osmolality** is concentration expressed as milliosmoles solute per kilogram of water. Because biological solutions are dilute and little of their weight comes from solute, physiologists sometimes use the terms *osmolarity* and *osmolality* interchangeably. Osmolality is usually used in clinical situations because it is easy to estimate people's body water content by weighing them.

Comparing Osmolarities of Two Solutions Osmolarity is a property of every solution. You can compare the osmolarities of different solutions as long as the concentrations are expressed in the same units, for example, as milliosmoles per liter. If two solutions have the same number of solute particles per unit volume, we say that the solutions are **isosmotic** [*iso-*, equal]. If solution A has a higher osmolarity (has more particles per liter, is more concentrated) than solution B, then we say that solution A is **hyperosmotic** to solution B. In the same example, solution B, with fewer osmoles per liter, is **hyposmotic** to solution A. Table 5-6 shows some examples of comparative osmolarities.

Osmolarity is a *colligative* property of solutions, meaning it depends strictly on the *number* of particles per liter in the

Table 5-6: Comparing Osmolarities

SOLUTION A = 1 OsM GLUCOSE	SOLUTION B = 2 OsM GLUCOSE	SOLUTION C = 1 OsM NaCL
A is hyposmotic to B	B is hyperosmotic to A	C is isosmotic to A
A is isosmotic to C	B is hyperosmotic to C	C is hyposmotic to B

Daniel's medical history tells a frightening story of almost constant medical problems since birth: recurring bouts with respiratory infections, digestive ailments, and, for the past six months, a history of weight loss. Then, last week, when Daniel began having trouble breathing, his mother rushed him to the hospital. A culture taken from Daniel's lungs raised a red flag for cystic fibrosis: the mucus from his airways was unusually thick and dehydrated. In cystic fibrosis, this thick mucus causes life-threatening respiratory congestion and provides a perfect breeding ground for infection-causing bacteria. The thickened mucus is a direct result of faulty CFTR channels. Normal CFTR channels in the lungs transport chloride ions out of the epithelial cells and into the airways.

Question 3: Why would failure to transport chloride ions into the lumen of the airways cause the secreted mucus to be thick? (Hint: Remember that water moves to dilute the more concentrated region.)

solution. Osmolarity says nothing about what the particles are and how they behave. Before we can predict whether osmosis will take place between any two solutions divided by a membrane, we must know the properties of the membrane and of the solutes on each side of the membrane.

If the membrane is permeable only to water and not to any solutes, water will move by osmosis from a less concentrated (hyposmotic) solution into a more concentrated (hyperosmotic) solution, as illustrated in Figure 5-34. This movement follows the saying "water moves to dilute the more concentrated solution." Most biological systems are not this simple, however. Biological membranes are selectively permeable and allow some solutes to cross in addition to water. To predict the movement of water into and out of cells, you must know the *tonicity* of the solution, explained in the next section.

The Tonicity of a Solution Describes How the Size of a Cell Would Change If It Were Placed in the Solution

Tonicity is a physiological term, used to describe solutions, that tells what would happen to cell volume if a cell were placed in the solution [*tonikos*, pertaining to stretching]. If a cell gains water and swells when placed in a solution and allowed to come to equilibrium, we say that the solution is **hypotonic** to the cell. If the cell loses water and shrinks when placed in the solution, the solution is said to be **hypertonic**. If the cell does not change size at equilibrium, the solution is **isotonic** (Table 5-7). By convention, we always describe the tonicity of the solution relative to the cell.

How, then, does tonicity differ from osmolarity? Osmolarity describes the number of particles in a solution. It can be used to describe a single solution (in osmoles/liter) or to compare two so-

Table 5-7: Tonicity of Solutions

SOLUTION	CELL BEHAVIOR WHEN PLACED IN THE SOLUTION	DESCRIPTION OF THE SOLUTION RELATIVE TO THE CELL
A	Cell swells	Solution A is hypotonic
B	Cell doesn't change size	Solution B is isotonic
C	Cell shrinks	Solution C is hypertonic

lutions (solution 1 is hyperosmotic to solution 2, therefore solution 2 is hyposmotic to solution 1). Tonicity, on the other hand, is a comparative term for a solution that describes what would happen to cell volume if a cell were placed in the solution. For example, "Solution 1 is hypotonic to red blood cells" tells you that red blood cells would swell at equilibrium if the cells were placed in solution 1. Tonicity has no units. The tonicity of a solution is characterized relative to a particular cell.

Furthermore, the tonicity of a solution depends not only on its osmolarity, but on the *nature* of the solutes in the solution. By nature of the solutes, we mean whether the solute particles can cross the cell membrane. If the solute particle (ion or molecule) can enter the cell, we call it a **penetrating solute**. We call particles that cannot cross the cell membrane **nonpenetrating solutes**.

The most important nonpenetrating solute in physiology is NaCl. If a cell is placed in a solution of NaCl, the Na^+ and Cl^- will not cross the membrane into the cell. (In reality, a few Na^+ ions may leak across, but they are immediately transported back to the extracellular fluid by the Na^+-K^+-ATPase. NaCl is therefore considered a *functionally* nonpenetrating solute.) If a *saline* solution of water and NaCl is put into the plasma, the Na^+ and Cl^- will stay in the extracellular fluid because they cannot enter the cells.

Suppose you know the composition and osmolarity of a solution. How can you figure out the tonicity of the solution without putting a cell in it? *The key lies in knowing the relative concentrations of nonpenetrating solutes in the cell and in the solution.* By convention, we always assume that cells are filled with nonpenetrating solutes. In other words, the solutes inside the cell are unable to leave as long as the cell membrane remains intact.

To determine tonicity of a solution relative to a cell, you must consider the relative concentrations of nonpenetrating solutes in the solution and in the cell. Here are the rules:

1. *If the cell has a higher concentration of nonpenetrating solutes than the solution*, there will be net movement of water into the cell. The cell swells, and the solution is *hypotonic*.
2. *If the cell has a lower concentration of nonpenetrating solutes than the solution*, there will be net movement of water out of the cell. The cell shrinks, and the solution is *hypertonic*.
3. *If the concentrations of nonpenetrating solutes are the same in the cell and the solution*, there is no net movement of water at equilibrium. The solution is *isotonic* to the cell.

The starting osmolarities of the cell and solution usually cannot be used to determine the tonicity of the solution. One reason is that penetrating solutes, which contribute to osmolarity, must be ignored when determining tonicity. When a cell is placed in a solution with penetrating solutes, these solutes move freely into the cell, as if the cell membrane did not exist, and therefore do not contribute to the solution's tonicity. To understand why this is true, look at the following example.

In Figure 5-35a ●, the cell contains 6 particles of nonpenetrating solute in 1 liter of volume. The 1 liter of solution also contains 6 particles of solute per liter: 3 nonpenetrating particles and 3 penetrating particles. Because cell and solution have the same concentrations (6 particles per liter), they are isosmotic.

To determine tonicity of the solution, compare the concentrations of nonpenetrating solutes in the solution and cell. The solution has 3 nonpenetrating particles per liter, but the cell has 6 nonpenetrating particles per liter. By rule 1 above, this means that water will move from the solution into the cell. The net movement of water into the cell will increase its volume, so the solution is hypotonic to the cell, even though the solution was isosmotic to the cell before they were put together.

If the solution was isosmotic to the cell, what caused osmosis? The answer lies with the gradient of the nonpenetrating solutes. When the cell was first placed in the solution, there was no osmotic gradient (and no water movement) across the cell membrane because cell and solution both had concentrations of 6 particles per liter (Fig. 5-35b). However, there were chemical concentration gradients: the nonpenetrating solutes were more concentrated in the cell (6 versus 3), while the penetrating solutes were more concentrated in the solution (3 versus 0). The nonpenetrating solutes could not cross the membrane, but the penetrating solutes were free to diffuse down their concentration gradient and move into the cell. As soon as one particle moved in, an osmotic gradient was created (Fig. 5-35c). The cell contained 6 + 1 particles, while the solution had 6 − 1 particles. Once an osmotic gradient was created, water followed the solute into the cell and the cell volume increased (solute movement has no significant effect on cell volume). Thus, in this example, an isosmotic solution is hypotonic.

Table 5-8 lists some rules for dealing with osmolarity and tonicity. Understanding the difference between the two properties is critical to making good clinical decisions about intravenous (IV) fluid therapy. In medicine, the tonicity of a solution is an important consideration. One purpose of IV fluids is to get water into dehydrated cells (in which case, a hypotonic solution is used) or to keep fluid in the extracellular fluid to replace blood loss (in which case, an isotonic solution is used). The choice of fluid depends on how the clinician wants the solutes and water to distribute between the extracellular and intracellular fluid compartments. Table 5-9 shows the most common IV solutions and their approximate osmolarity and tonicity relative to the normal human cell.

(a) Cell and solution have equal concentrations, i.e., are isosmotic.

(b) When the cell is placed in the solution, no water moves initially because the cell and solution are in osmotic equilibrium.

(c) However, there is a concentration gradient for the penetrating solute Ⓟ, which diffuses into the cell. This disrupts the osmotic equilibrium, so water follows the solute into the cell.

(d) Equilibrium is restored when the concentrations of nonpenetrating solutes Ⓝ, are equal in the cell and the solution. The cell gained volume at equilibrium, therefore the solution was hypotonic.

FIGURE QUESTION

Using the same cell, give the relative osmolarity and the tonicity of the following solutions, if the solution has (a) 6 nonpenetrating particles, (b) 6 nonpenetrating and 3 penetrating particles, or (c) 3 nonpenetrating and 6 penetrating particles.

● **Figure 5-35** *Tonicity depends on the relative concentrations of nonpenetrating solutes* Water and penetrating solutes Ⓟ can cross the cell membrane but nonpenetrating solutes Ⓝ cannot cross.

Three days after Daniel's sweat test, the lab returns the grim results: salt levels in his sweat are over twice the normal concentration. Daniel is diagnosed with cystic fibrosis. Now, along with antibiotics to prevent lung infections and therapy to loosen the mucus in his airways, Daniel must begin a regimen of enzymes to be taken whenever he eats, for the rest of his life. In cystic fibrosis, thick mucus in the pancreatic ducts blocks the secretion of digestive enzymes into the intestine. Without artificial enzymes, he would starve. Yet even with these treatments, Daniel will probably die before the age of 30.

Question 4: Why will Daniel starve if he does not take artificial pancreatic enzymes?

Table 5-8: Rules for Osmolarity and Tonicity

1. By convention, we assume all intracellular solutes are nonpenetrating.

2. When a cell is exposed to a solution, water and penetrating solutes will move and the system will go to equilibrium. At equilibrium the total concentration of solute particles in the cell and the solution will be equal; that is, the osmolarities will be the same.

3. Compare the osmolarity of a cell and a solution before they are put together, because at equilibrium, the osmolarities will be the same.

4. The tonicity of a solution is determined by the volume of the cell at equilibrium (see Table 5-7).

5. A solution that is hyposmotic to a cell will always be hypotonic, no matter what the nature of the solutes in the solution.

6. You cannot predict the tonicity of isosmotic and hyperosmotic solutions unless you know the concentration of nonpenetrating solutes relative to the cell. Net water movement will be into the compartment that has the higher concentration of nonpenetrating solutes.

 Answers: p. 170

27. Which solution(s) has/have the most water per unit volume? 1 M glucose, 1 M NaCl, or 1 OsM NaCl?

28. Two compartments are separated by a membrane that is permeable to water and urea but not to NaCl. Which way will water move when the solutions below are placed in the two compartments?

Compartment A	Membrane	Compartment B
(a) 1 M NaCl		1 OsM NaCl
(b) 1 M urea		2 M urea
(c) 1 OsM NaCl		1 OsM urea

29. You have a patient who lost 1 liter of blood and you need to restore volume quickly while waiting for a blood transfusion to arrive from the blood bank.

    (a) Would it be better to administer 5% dextrose (glucose) in water or 0.9% NaCl in water? (Hint: Think about how these solutes distribute in the body.) Defend your choice.

    (b) How much of your solution of choice would you have to administer to return blood volume to normal?

The Body Is in a State of Electrical Disequilibrium

Thus far, we have seen that the compartments of the body are in osmotic equilibrium but in chemical disequilibrium. Because many body solutes are ions that carry either positive or negative electrical charges, we must also consider the distribution of electrical charge between the intracellular and extracellular compartments. Although the body as a whole is electrically neutral, it has a few excess negative ions in the intracellular fluid, while their matching positive ions are found in the extracellular fluid.

Potassium (K^+) is the major cation of the cells, and sodium (Na^+) dominates the extracellular fluid (see Fig. 5-33, p. 151). Chloride ions (Cl^-) remain with Na^+ in the extracellular fluid, whereas phosphate ions and negatively charged proteins are the anions of the intracellular fluid. However, there are some protein anions inside the cell that do not have matching cations and some K^+ ions in the extracellular fluid that do not have matching

Table 5-9: Intravenous Solutions

SOLUTION	ALSO KNOWN AS	OSMOLARITY	TONICITY
0.9% saline*	Normal saline	Isosmotic	Isotonic
D_5—0.9% saline	5% dextrose** in normal saline	Hyperosmotic	Isotonic
D_5W	5% dextrose in water	Isosmotic	Hypotonic
0.45% saline	Half-normal saline	Hyposmotic	Hypotonic
D_5—0.45% saline	5% dextrose in half-normal saline	Hyperosmotic	Hypotonic

*Saline = NaCl.   **Dextrose = glucose.

anions. One consequence of this uneven distribution of ions is that the intracellular and extracellular compartments are not in electrical equilibrium. The **electrical disequilibrium** in the body is commonly known as the *resting membrane potential difference*.

Electrical Signals Are Generated in Many Types of Cells The concept of resting membrane potential has traditionally been taught in the chapters on nerve and muscle function because those tissues generate electrical signals known as action potentials. Yet one of the most exciting recent discoveries in physiology is the realization that other kinds of cells also use electrical signals for communication. This section reviews the basic principles of electricity and discusses what creates the resting membrane potential. The chapter ends with a look at how the beta cells of the pancreas use changes in their membrane potential to trigger insulin secretion.

Electricity review Atoms are electrically neutral [❧ p. 21]. They are composed of positively charged protons, negatively charged electrons, and uncharged neutrons, but in balanced proportions, so that an atom is neither positive nor negative. The removal or addition of electrons to an atom creates a charged particle known as an ion. We have discussed several ions that are important in the human body, such as Na^+, K^+, and H^+. For each of these positive ions, somewhere there is a matching electron, usually found as part of a negative ion. For example, when Na^+ in the body enters in the form of NaCl, the "missing" electron from Na^+ can be found on the Cl^-.

The following principles are important to remember when dealing with electricity in physiological systems:

1. The **law of conservation of electric charge** states that the net amount of electric charge produced in any process is zero. This means that, for every positive charge on an ion, there is an electron on another ion. Overall, the human body is electrically neutral.
2. Opposite charges (+ and −) are attracted to each other, but two charges of the same type (+ and + or − and −) repel. The protons and electrons in an atom exhibit this attraction.
3. To separate positive and negative charges, it is necessary to use energy. For example, energy is needed to separate the protons and electrons of an atom.
4. If separated positive and negative charges can move freely toward each other, the material through which they are moving is called a **conductor**. Water is a good conductor of electricity. If separated charges are unable to move through the material that separates them, the material is known as an **insulator**. The phospholipid bilayer of the cell membrane is a good insulator, as is the plastic coating on electrical wires.

The word *electricity* comes from the Greek word *elektron*, for amber, the fossilized resin of trees. The Greeks discovered that if they rubbed a rod of amber with cloth, the amber acquired the ability to attract hair and dust. This attraction (called static elec-

tricity) arises from the separation of electrical charge that occurs when electrons move from the amber atoms to the cloth. To separate these charged particles, energy (work) must be put into the system. In the case of the amber, work was done by rubbing the rod. In the case of biological systems, the work is usually done by energy stored in ATP and other chemical bonds.

The Cell Membrane Allows Separation of Electrical Charge in the Body

In the body, separation of electrical charge takes place across the cell membrane. This process is shown in Figure 5-36●. The diagram shows an artificial cell filled with molecules that dissociate into positive and negative ions, represented by the plus and minus signs. Because the molecules were electrically neutral to begin with, there are equal numbers of positive and negative ions inside the cell. The cell is placed in a similar solution, also electrically neutral. The phospholipid bilayer of the artificial cell, like the bilayer of a real cell, is not permeable to ions. Water can freely cross this cell membrane, making the extracellular and intracellular ion concentrations equal. In Figure 5-36a, the system is at osmotic, chemical, and electrical equilibrium.

In the next step (Figure 5-36b), an active transport protein is inserted into the membrane. This carrier uses energy to move positive ions (+) out of the cell against their concentration gradient. The negative ions (−) in the cell attempt to follow the positive ions because of the attraction of positive and negative charges. But the membrane is impermeable to negative ions, so they remain trapped within the cell. Positive ions outside the cell might try to move into the cell, attracted by the negative charge of the intracellular fluid, but the membrane does not allow ions to leak across it.

As soon as the first positive ion leaves the cell, the electrical equilibrium between the extracellular fluid and intracellular fluid is disrupted: the cell's interior has a net charge of −1, while the cell's exterior has a net charge of +1. The input of energy to transport ions across the membrane has created an electrical gradient: in this example, the inside of the cell became negative relative to the outside.

The active transport of positive ions out of the cell also creates a concentration, or chemical, gradient: there are more positive ions outside the cell than inside. The combination of electrical and concentration gradients is called an **electrochemical gradient**. The cell remains in osmotic equilibrium because water can move freely across the membrane in response to solute movement.

An **electrical gradient** between the extracellular fluid and the intracellular fluid is known as the **resting membrane potential difference** or **membrane potential**, for short. Although the name sounds intimidating, we can break it apart to see what it means.

1. The *resting* part of the name comes from the fact that this electrical gradient is seen in all living cells, even those that

appear to be without electrical activity. The membrane potential has reached a steady state and is not changing.

2. The *potential* part of the name comes from the fact that the electrical gradient created by active transport of ions across the cell membrane is a source of stored, or potential, energy, just as the concentration gradient is a form of potential energy. When oppositely charged molecules come back together, they release energy that can be used to do work, in the same way that molecules moving down their concentration gradient can perform work. The work done by electrical energy includes opening voltage-gated membrane channels and sending electrical signals.

3. The *difference* part of the name is to remind you that the membrane potential represents a difference in the electrical charge inside and outside the cell. The word *difference* is often dropped.

In living systems, we measure electrical gradients on a relative scale rather than an absolute scale. Figure 5-36c compares the two scales. On the absolute scale, the extracellular fluid in our simple example has a charge of +1 from the positive ion it gained, while the intracellular fluid has a charge of −1 for the negative ion that was left behind. However, in real life we cannot measure the charges as numbers of electrons gained or lost. Instead we use a device that measures the difference in electrical charge between two points. This device artificially sets one side of the membrane at zero and measures the second side relative to the first. In our example, resetting the extracellular fluid to zero gives the intracellular fluid a charge of −2, the cell's resting membrane potential difference.

The actual equipment for measuring a cell's membrane potential difference is shown in Figure 5-37 ●. *Electrodes* are created from hollow glass tubes drawn to very fine points. These *micropipets* are filled with fluid that conducts electricity. They are connected to a *voltmeter* that measures the electrical difference between two points in volts (V) or millivolts (mV). A *recording electrode* is inserted through the cell membrane into the cytoplasm of the cell. A *reference electrode* is placed in the external fluid bath, which represents the extracellular fluid.

In living systems, by convention, the extracellular fluid is designated as the *ground* and assigned a charge of 0 mV. When the recording electrode is placed inside a living cell, the voltmeter measures the membrane potential (V_m), the electrical difference between the intracellular fluid and the extracellular fluid. A chart recorder connected to the voltmeter can make a recording of the membrane potential difference versus time.

For nerve and muscle cells, the voltmeter will record a resting membrane potential difference between −40 and −90 mV, indicating that intracellular fluid is negative relative to the extracellular fluid (0 mV). As you saw in Figure 5-36c, the extracellular fluid

(a) Cell and solution are electrically and chemically at equilibrium.

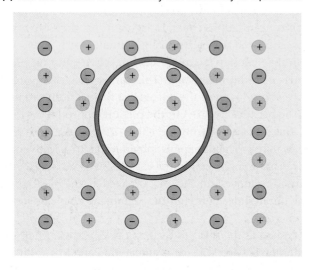

(b) Cell and solution in chemical and electrical disequilibrium. Energy is used to pump one cation out of the cell, leaving a net charge of -1 in the cell and +1 outside the cell.

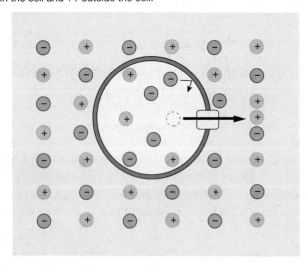

● **Figure 5-36** *Separation of electrical charge*

(c) On an absolute charge scale, the extracellular fluid *(ECF)* would be at +1 and the intracellular fluid *(ICF)* at -1.

Physiological measurements are always on a relative scale, on which the extracellular fluid is assigned a value of zero. This shifts the scale to the left and gives the inside of the cell a relative charge of -2.

A recording electrode is placed inside the cell.

Input

The voltmeter measures the difference in electrical charge between the inside of a cell and the surrounding solution. This value is the **membrane potential difference**, or V_m.

-70   -30   0   +30

Output

Cell

Saline bath

The ground ($=$) or reference electrode is placed in the bath and given a value of 0 millivolts (mV).

The chart recorder measures changes in membrane potential over time.

● **Figure 5-37** *Measuring membrane potential difference*   In the laboratory, a voltmeter measures the difference in electrical charge between the inside of a cell and the surrounding solution.

is not really neutral because it has excess positive charges that exactly balance the excess negative charges inside the cell. The total body remains electrically neutral at all times.

The Resting Membrane Potential Is Due Mostly to Potassium

Which ions create the resting membrane potential difference in animal cells? The artificial cell shown in Figure 5-36b used an active transport protein to move an unspecified ion with a positive charge across a membrane that was otherwise impermeable to ions. But what processes go on in living cells to create an electrical gradient?

Real cells are not completely impermeable to all ions. They have open channels and protein transporters that allow ions to move between the cytoplasm and the extracellular fluid. We can use a different artificial cell to show how the resting membrane potential arises in living cells.

An artificial cell whose membrane is impermeable to ions (Fig. 5-38a ●) contains K^+ ions and large negatively charged proteins, represented by Pr^-. The cell is placed in a solution of Na^+ and Cl^-. Both the cell and solution are electrically neutral and the system is in electrical equilibrium. However, it is not in chemical equilibrium. There are concentration gradients for all four types of ions in the system, and they would all diffuse down their respective concentration gradients if they could cross the cell membrane.

In Figure 5-38b, a K^+ leak channel is inserted into the membrane, making it permeable only to K^+. Because there is no K^+ in the extracellular fluid initially, some K^+ ions will leak out of the cell, moving down their concentration gradient. As the K^+ leave, the negatively charged proteins, Pr^-, are unable to follow. The

proteins gradually build up a negative charge inside the cell as K^+ diffuses out of the cell.

If the only force acting on K^+ were the concentration gradient, K^+ would leak out of the cell until the K^+ concentration inside the cell equaled the K^+ concentration outside. But the loss of positive ions from the cell creates an electrical gradient. Because opposite charges attract, the negative Pr^- inside the cell try to pull K^+ back into the cell. Thus, when K^+ ions move out of the cell down their concentration gradient, the negative charge left inside the cell causes other K^+ to move into the cell.

At some point in this process the electrical force exerted by the negative charge inside the cell becomes equal in magnitude to the chemical concentration gradient driving K^+ out of the cell. At that point, net movement of K^+ across the membrane stops (Fig. 5-38c). The rate at which K^+ move out of the cell down the concentration gradient is exactly equal to the rate at which K^+ move into the cell down the electrical gradient.

In a cell that is permeable to only one ion, such as the artificial cell just described, the membrane potential difference that exactly opposes the concentration gradient of the ion is known as the **equilibrium potential**, or E_{ion}. For example, when the concentration gradient is 150 mM K^+ inside and 5 mM K^+ outside the cell, the equilibrium potential for potassium, or E_K, is -90 mV. The equilibrium potential for any ion at 37° C (human body temperature) can be calculated using the Nernst equation:

$$E_{ion} = \frac{61}{z} \log \frac{[ion]_{out}}{[ion]_{in}}$$

where 61 is a combination of several constants plus temperature

z is the electrical charge on the ion ($+1$ for K^+).

[ion] are the ion concentrations inside and outside the cell

(a) An artificial cell whose membrane is impermeable to ions is filled with K⁺ and large protein anions. It is placed in a solution of Na⁺ and Cl⁻. Both cell and solution are electrically neutral.

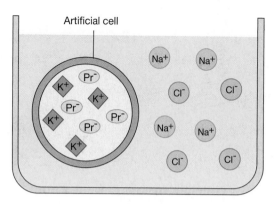

Artificial cell

(b) A K⁺ leak channel is inserted into the membrane. K⁺ leaks out of the cell because there is a K⁺ concentration gradient.

K+ leak channel

FIGURE QUESTION

In the white boxes write the net electrical charge of the intracellular and extracellular compartments as shown.

(c) The negative membrane potential of the cell creates an electrical gradient that attracts K⁺ back into the cell. The resting membrane potential difference at which the electrical gradient exactly opposes the K⁺ concentration gradient is called the **equilibrium potential** for K⁺.

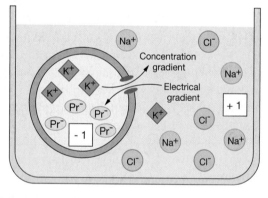

FIGURE QUESTION

Why don't Na⁺, Cl⁻, and the proteins (Pr⁻) cross the membrane?

● **Figure 5-38** *Potassium equilibrium potential*

Now we will use the same artificial cell (K⁺ and Pr⁻ inside, Na⁺ and Cl⁻ outside), but this time we will make the membrane permeable only to Na⁺ (Fig. 5-39 ●). Because Na⁺ is more concentrated outside the cell, it will begin to move into the cell. As it does so, it causes positive charges to accumulate inside the cell. Meanwhile Cl⁻ ions left behind in the extracellular fluid make that compartment negative. This creates an electrical gradient that moves Na⁺ back out of the cell. When the Na⁺ concentration is 150 mM outside and 15 mM inside, the equilibrium potential for Na⁺ (E_{Na}) is +60 mV. In other words, the concentration gradient moving Na⁺ into the cell (mM 150 → 15 mM) is exactly opposed by a positive membrane potential of +60 mV.

In reality, living cells are not permeable to only one ion. The situation in real cells is similar to a combination of the two artifi-

cial systems just described. If a cell is permeable to several ions, we cannot use the Nernst equation to calculate membrane potential. Instead we must use a related equation called the Goldman equation that considers concentration gradients of the permeable ions and the relative permeability of the cell to each ion. For more detail on the Goldman equation, see Chapter 8.

The cell illustrated in Figure 5-40 ● has a resting membrane potential of −70 mV. Most cells are about 40 times more permeable to K⁺ than to Na⁺, and as a result, a cell's resting membrane potential is closer to the E_K of −90 mV than to the E_{Na} of +60 mV. A small amount of Na⁺ leaks into the cell, making the inside of the cell less negative than it would be if Na⁺ were totally excluded. Additional Na⁺ that leaks in is promptly pumped out by the Na⁺-K⁺-ATPase as described earlier. At the same time, K⁺ ions that

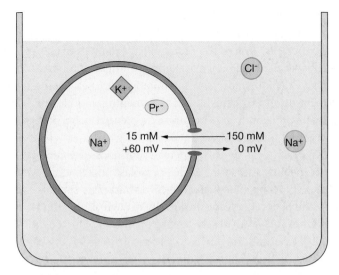

The Na$^+$ concentration gradient shown is exactly opposed by a membrane potential of +60 mV. Therefore +60 mV is the E$_{Na}$ or Na$^+$ equilibrium potential.

● **Figure 5-39** *Sodium equilibrium potential*

leak out of the cell are pumped back in. The pump contributes to the membrane potential by pumping 3 Na$^+$ out for every 2 K$^+$ pumped in. Because the Na$^+$-K$^+$-ATPase helps create the electrical gradient, it is called an *electrogenic* pump.

Not all ion transport creates an electrical gradient. Some transporters, like the Na$^+$-organic anion cotransporter, move one negative charge for each positive charge and are therefore considered electrically neutral. Other transporters make an even exchange of ions: for each charge that enters the cell, one of the same charge leaves. An example is the HCO$_3$$^-$-Cl$^-$ antiporter of red blood cells, which transports these ions in a one-for-one, electrically neutral exchange. Electrically neutral transporters have little effect on the resting membrane potential difference of the cell.

CONCEPT CHECK   Answers: p. 170

30. Starting with the artificial cell shown in Figure 5-38a, add a Cl$^-$ leak channel to the cell. Figure out which way Cl$^-$ will move along concentration and electrical gradients. Will the Cl$^-$ equilibrium potential be positive or negative?

31. What would happen to the resting membrane potential of a cell poisoned with ouabain, an inhibitor of the Na$^+$-K$^+$-ATPase?

Changes in Ion Permeability Change the Membrane Potential

As you have just learned, two factors influence the cell's membrane potential: (1) the concentration gradients of different ions across the membrane and (2) the permeability of the membrane to those ions. If the cell's permeability to an ion changes, the cell's membrane potential changes. We monitor changes in membrane potential using the same intracellular recording electrodes that we use to record resting membrane potential (see Fig. 5-37).

Q
● ● ● ● ● ● ● ●
FIGURE QUESTIONS

• What force(s) promotes Na$^+$ leak into the cell?
• What force(s) promotes K$^+$ leak out of the cell?

● **Figure 5-40** *Resting membrane potential difference in an actual cell*   Most cells in the human body are about 40 times more permeable to K$^+$ than to Na$^+$, and the resting membrane potential difference is about -70 mV.

Terminology for Changes in Membrane Potential

Figure 5-41 ● shows a recording of membrane potential plotted against time. The membrane potential (V$_m$) begins at a steady resting value of -70 mV. When the trace moves upward (becomes less negative), the potential difference between the inside of the cell (-70 mV) and the outside (0 mV) decreases, and

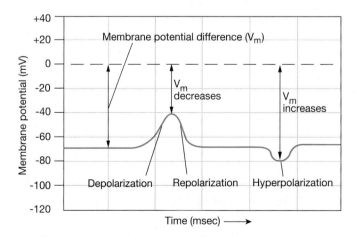

● **Figure 5-41** *Terminology associated with changes in membrane potential*   If the membrane potential becomes less negative, the cell depolarizes. If the membrane potential becomes more negative, the cell hyperpolarizes.

the cell is said to have depolarized. A return to the resting membrane potential is termed repolarization. If the trace moves downward, the membrane potential becomes more negative, the potential difference has increased, and the cell has hyperpolarized.

A major point of confusion when talking about changes in membrane potential is the use of the phrases "the membrane potential decreased" or "the membrane potential increased." Normally, in English we associate "increase" with becoming more positive and "decrease" with becoming more negative—the opposite of what is happening in those cases. One way to avoid confusion is to add the word *difference* after *membrane potential*. If the membrane potential difference is *increasing*, the V_m must be moving away from the ground value of zero and becoming *more negative*. If the membrane potential difference is *decreasing*, the V_m is moving closer to the ground value of 0 mV and is becoming *less negative*.

Changes in Ion Permeability Change Membrane Potential

Four ions usually contribute the most to changes in membrane potential: Na^+, Ca^{2+}, Cl^-, and K^+. The first three are more concentrated in the extracellular fluid than in the cytosol, and the resting cell is minimally permeable to them. If the cell suddenly becomes more permeable to any one of these ions, ions will move across the membrane into the cell. Entry of Ca^{2+} or Na^+ will depolarize the cell (make the membrane potential more positive). Entry of Cl^- will hyperpolarize the cell (make the membrane potential more negative).

Resting cells are fairly permeable to K^+, but making them more permeable will allow more K^+ to leak out. The cell will hyperpolarize until it approaches the equilibrium potential for K^+. Making the cell *less* permeable to K^+ will allow less K^+ to leak out of the cell. When the cell retains K^+, it becomes more positive and depolarizes. You will encounter instances of all these permeability changes as you study physiology.

It is important to remember that a significant change in membrane potential requires the movement of very few ions. *You do not need to reverse the concentration gradient to change the membrane potential.* For example, to change the membrane potential by 100 mV, only one out of every 100,000 K^+ must enter or leave the cell. This is such a tiny fraction of the total number of K^+ in the cell that the concentration of K^+ remains essentially unchanged.

▶ INTEGRATED MEMBRANE PROCESSES: INSULIN SECRETION

The movement of Na^+ and K^+ across cell membranes has been known to play a role in generating electrical signals in excitable tissues for many years. You will study these processes in detail in the chapters on the nervous and muscular systems. Recently,

however, we have come to understand that small changes in membrane potential act as signals in nonexcitable tissues such as endocrine cells. One of the best-studied examples of this process is the beta cell of the pancreas. Release of the hormone insulin by beta cells demonstrates how membrane processes, such as facilitated diffusion, exocytosis, and the opening and closing of ion channels by ligands and membrane potential, work together to regulate cell function.

The beta cells of the pancreas synthesize the protein hormone insulin and store it in cytoplasmic secretory vesicles [● p. 61]. When blood glucose levels increase, such as after a meal, the beta cell releases insulin by exocytosis. Insulin then directs other cells of the body to take up and use glucose, bringing blood concentrations down to pre-meal levels. A key question about the process that went unanswered until recently was "How does the beta cell 'know' that glucose levels have gone up and that it needs to release insulin?" The answer, we have now learned, links the beta cell's metabolism to its electrical activity.

Figure 5-42a ● shows a beta cell at rest. Recall from earlier sections in this chapter that the gates of membrane channels can be opened or closed by chemical or electrical signals. The beta cell has two such channels that help control insulin release. One is a **voltage-gated Ca^{2+} channel**. This channel is closed at the cell's resting membrane potential (Fig. 5-42a, ⑥). The other is a K^+ leak channel (i.e., the channel is usually open) that closes when ATP binds to it. It is called an **ATP-gated K^+ channel**, or **K_{ATP} channel**. In the resting cell, when glucose concentrations are low, the cell makes less ATP (Fig. 5-42a, ①–③). There is little ATP to bind to the K_{ATP} channel and the channel remains open, allowing K^+ to leak out of the cell (Fig. 5-42a, ④). At the resting membrane potential, the voltage-gated Ca^{2+} channels are closed and there is no insulin secretion ⑤–⑦.

Figure 5-42b shows a beta cell secreting insulin. Following a meal, plasma glucose levels increase as glucose is absorbed from the intestine ①. Glucose reaching the beta cell diffuses into the cell with the aid of a GLUT transporter. Increased glucose in the cell stimulates the metabolic pathways of glycolysis and the citric acid cycle [● p. 103], and ATP production increases ②, ③. When ATP binds to the K_{ATP} channel, the gate to the channel closes, preventing K^+ from leaking out of the cell ④, ⑤. Retention of K^+ depolarizes the cell ⑥, which then causes the voltage-sensitive Ca^{2+} channels to open ⑦. Calcium ions enter the cell from the extracellular fluid, moving down their electrochemical gradient. The Ca^{2+} binds to proteins that initiate exocytosis of the insulin-containing vesicles, and insulin is released into the extracellular space ⑧.

The discovery that cells other than nerve and muscle cells use changes in membrane potential as signals for physiological responses has been an exciting one that promises to change our traditional thinking about the role of the resting membrane potential. In the next chapter, we will describe some other types of signals that the body uses for communication and coordination.

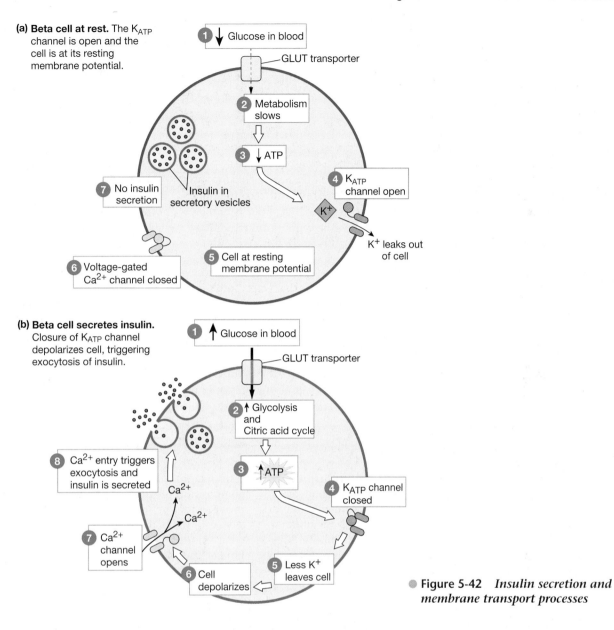

(a) **Beta cell at rest.** The K_{ATP} channel is open and the cell is at its resting membrane potential.

1 ↓ Glucose in blood

GLUT transporter

2 Metabolism slows

3 ↓ ATP

4 K_{ATP} channel open

K^+

K^+ leaks out of cell

7 No insulin secretion

Insulin in secretory vesicles

6 Voltage-gated Ca^{2+} channel closed

5 Cell at resting membrane potential

(b) **Beta cell secretes insulin.** Closure of K_{ATP} channel depolarizes cell, triggering exocytosis of insulin.

1 ↑ Glucose in blood

GLUT transporter

2 ↑ Glycolysis and Citric acid cycle

3 ↑ATP

4 K_{ATP} channel closed

8 Ca^{2+} entry triggers exocytosis and insulin is secreted

Ca^{2+}

Ca^{2+}

7 Ca^{2+} channel opens

6 Cell depolarizes

5 Less K^+ leaves cell

● **Figure 5-42** *Insulin secretion and membrane transport processes*

RUNNING PROBLEM CONCLUSION

Cystic Fibrosis

In this running problem, you learned that cystic fibrosis is caused by a defect in the CFTR channel, which regulates the transport of chloride ions out of epithelial cells. Because the CFTR channel is found in the epithelial cell membranes of several organs—the sweat glands, lungs, and pancreas—cystic fibrosis affects many different body processes. Learn more about the disease in the Explorations section and test your understanding by comparing your answers to those in the summary table.

	QUESTION	FACTS	INTEGRATION AND ANALYSIS
1	Is CFTR a chemically gated, voltage-gated, or mechanically gated channel protein?	Chemically gated channels open when a ligand binds to them. Voltage-gated channels open with a change in the cell's membrane potential. Mechanically gated channels open when a physical force opens the channel. CFTR channels open in response to the binding of nucleotides.	Nucleotides are chemical ligands; therefore, the CFTR channel is a chemically gated channel.

(continues on next page)

	QUESTION	FACTS	INTEGRATION AND ANALYSIS
2	Based on the information given, is the CFTR channel on the apical or basolateral surface of the sweat gland epithelium?	In normal people, the CFTR channels move Cl^- from sweat into epithelial cells.	The epithelial surface that faces the lumen of the sweat gland, which contains sweat, is the apical membrane. Therefore, the CFTR channel is on the apical surface.
3	Why would failure to transport chloride ions into the lumen of the airways cause the secreted mucus to be thick? (Hint: Remember that "water moves to dilute the more concentrated region.")	If Cl^- (followed by Na^+) is secreted into the lumen of the airways, the solute concentration of the airway fluid increases. Water moves in response to concentration gradients.	Normally, the movement of Cl^- and Na^+ creates an osmotic gradient so that water also enters the airway lumen, creating a saline solution that thins the thick mucus. If Cl^- cannot be secreted into the airways, there will be no fluid movement to thin the mucus.
4	Why will Daniel starve if he does not take artificial pancreatic enzymes?	The pancreas secretes mucus and digestive enzymes into ducts that empty into the small intestine. In cystic fibrosis, mucus in the ducts will be thick due to lack of Cl^- and fluid secretion. This blocks the ducts and prevents digestive enzymes from reaching the small intestine.	Without digestive enzymes, Daniel cannot digest the food he eats. His weight loss over the past six months suggests that this has already been a problem. Taking artificial enzymes will allow him to digest his food.

124 131 148 154 156

CHAPTER SUMMARY

This chapter repeats the theme of *compartmentation* that was introduced in Chapters 3 and 4. The dividing walls for body compartments can be tissues such as the peritoneal membrane, or they may be the phospholipid bilayers of cell membranes. Movement of material from one compartment to another illustrates a new theme, *mass flow*. Mass flow across cell membranes occurs in response to osmotic, chemical (concentration), or electrical gradients. The selectively permeable cell membrane creates resistance to mass flow that can be overcome by changing the composition of the membrane lipids or by inserting membrane proteins that act as channels or transporters. *Movement of substances* in the body requires energy from different sources: molecular motion, concentration gradients, or chemical bonds. Finally, the theme of *molecular interactions* is exemplified in the binding of substrates to transporters and in the control of channel opening by ligands.

Membranes in the Body

1. The word **membrane** is used for both cell membranes and for epithelial tissues that line a cavity or separate two compartments. (p. 125; Fig. 5-1)

2. The cell membrane is a barrier that restricts free exchange between the cell and the interstitial fluid. The movement of a substance across the membrane depends on the **permeability** of the membrane to the substance. (p. 125)

Cell Membranes

3. The cell membrane is a barrier between the intracellular and extracellular fluid. (p. 125)

4. The cell membrane provides structural support and regulates exchange and communication between the cell and its environment. (p. 126)

5. The **fluid mosaic model** of membranes says that membranes are **phospholipid bilayers** with proteins inserted into the bilayer. Carbohydrates attach to the extracellular surface. (pp. 126–128; Fig. 5-2)

6. **Membrane-spanning** and other **integral proteins** are tightly bound to the phospholipid bilayer. **Peripheral proteins** attach less tightly to either side of the membrane. (p. 128)

7. **Structural proteins** maintain cell shape and form cell junctions that hold tissues together. (p. 129)

8. **Membrane-associated enzymes** catalyze chemical reactions and help transfer signals across the membrane. (p. 129)

9. **Receptor proteins** on the cell membrane are part of the body's chemical signaling system. A molecule that binds to a receptor is called its **ligand**. (p. 129)

10. **Transport proteins** are the means by which many molecules enter or leave the cell. **Channel proteins** form water-filled channels that link the intracellular and extracellular compartments. **Gated channels** regulate movement of substances through them by opening and closing. Gated channels are reg- ulated by ligands, by the electrical state of the cell, or by physical changes such as pressure. (p. 130; Fig. 5-9)

11. **Carrier proteins** never form a direct connection between the intracellular and extracellular fluid. They bind to substrates, then change shape. (p. 130; Fig. 5-10)

Body Fluid Compartments

IP Fluids & Electrolytes: Introduction to Body Fluids

12. Most cells of the body are surrounded by extracellular fluid (ECF), which can be subdivided into **interstitial fluid** bathing the cells and **plasma**, the fluid portion of the blood. (p. 133; Fig. 5-13)

Movement Across Membranes

13. Lipid-soluble substances can diffuse through the phospholipid bilayer. Less lipid-soluble molecules require the assistance of a membrane protein to cross the membrane. (p. 134)

14. **Passive transport** does not require the input of energy. (p. 134)

15. **Diffusion** is the passive movement of molecules from an area of higher concentration to an area of lower concentration. Diffusion stops when the system reaches **equilibrium**, although molecular movement continues. (pp. 134–135)

16. Diffusion rate depends on the magnitude of the concentration gradient. Diffusion is slow over long distances, is directly related to temperature, and is inversely related to molecular size. (p. 135)

17. **Simple diffusion** is diffusion across the phospholipid bilayer of a membrane. The rate of simple diffusion is directly proportional to the surface area of the membrane, inversely proportional to the thickness of the membrane, and depends on the lipid solubility of the molecule that is diffusing. (p. 136; Fig. 5-16)

18. **Mediated transport** is movement of molecules with the aid of a carrier protein. Passive mediated transport is called **facilitated diffusion**. (p. 138)

19. All mediated transport demonstrates **specificity**, **competition**, and **saturation**. Specificity refers to the ability of a transporter to move only one molecule or a group of closely related molecules. Related molecules will compete for a single transporter. Saturation occurs when a group of membrane transporters are working at their maximum rate. (pp. 138–140)

20. **Active transport** moves molecules against their concentration gradient and requires an outside source of energy. In **primary (direct) active transport**, the energy comes directly from ATP. **Secondary (indirect) active transport** uses the potential energy stored in a concentration gradient and is indirectly driven by energy from ATP. (pp. 134,141)

21. The most important primary active transporter is the **sodium-potassium ATPase** (Na^+-K^+-ATPase), which pumps Na^+ out of the cell and K^+ into the cell. (p. 141; Fig. 5-23)

22. Most secondary active transport systems are driven by the sodium concentration gradient. (p. 143; Tbl. 5-4)

23. Large macromolecules and particles are brought into cells by **phagocytosis** and **endocytosis**. Material leaves cells by **exocytosis**. When vesicles that come into the cytoplasm by endocytosis are returned to the cell membrane, the process is called **membrane recycling**. (pp. 144–146; Figs. 5-27, 5-28)

24. In **receptor-mediated endocytosis**, ligands bind to membrane receptors that concentrate in **clathrin-coated pits**, the site of endocytosis. In **potocytosis**, receptors are located in **caveolae** that have a nonclathrin protein coating. (pp. 145–146; Fig. 5-28)

25. In exocytosis, the vesicle membrane fuses with the cell membrane in a complex process that involves a series of proteins. Exocytosis requires ATP and an increase in cytosolic Ca^{2+} concentration. (pp. 146–147)

26. Transporting epithelia in the intestine and kidney have different membrane proteins on their **apical** and **basolateral** surfaces. This polarization allows one-way movement of molecules across the epithelium. Larger molecules cross epithelia by **transcytosis**, which includes **vesicular transport**. (pp. 147–150; Figs. 5-30, 5-31)

The Distribution of Water and Solutes in the Body

IP Nervous I: The Membrane Potential

27. The body is in a state of **chemical disequilibrium** because cell membranes prevent solutes from diffusing freely and because of the active transport of solutes. (p. 151; Fig. 5-33)

28. Water moves freely between body compartments, so the body is in a state of **osmotic equilibrium**. (p. 152)

29. The movement of water across a membrane in response to a concentration gradient is called **osmosis**. (p. 152)

30. To compare solutions, we express concentration in terms of **osmolarity**, the number of particles (ions or intact molecules)

per liter of solution, expressed as milliosmoles (mosmol) per liter. **Osmolality** is concentration expressed as milliosmoles solute per kilogram of water. (p. 153)

31. **Tonicity** describes what a solution would do to cell volume if the cell were placed in the solution. Cells swell in **hypotonic solutions** and shrink in **hypertonic solutions**. If the cell does not change size at equilibrium, the solution is **isotonic**. (p. 154)

32. The starting osmolarities of a cell and solution cannot be used to determine the tonicity of a solution. The relative concentrations of **nonpenetrating solutes** in the cell and the solution determine the tonicity of the solution. **Penetrating solutes** contribute to the osmolarity of a solution but not to its tonicity. (p. 154; Fig. 5-35)

33. Although the total body is electrically neutral, it is in a state of **electrical disequilibrium** because diffusion and active transport of ions across the cell membrane create an electrical gradient, with the inside of the cell negative relative to the outside. (p. 157)

34. The electrical gradient between the extracellular fluid and the intracellular fluid is known as the **resting membrane potential difference.** (p. 157)

35. The movement of an ion across the cell membrane is influenced by the **electrochemical gradient** for that ion. (p. 157)

36. The membrane potential difference that exactly opposes the concentration gradient of an ion is known as the **equilibrium potential**, or E_{ion}. The equilibrium potential for any ion can be calculated using the Nernst equation. (p. 159; Fig. 5-38)

37. In living cells, K^+ is the primary ion that determines the resting membrane potential. (p. 162)

38. Changes in membrane permeability to ions such as K^+, Na^+, Ca^{2+}, or Cl^- will alter membrane potential and create electrical signals. (p. 162)

Integrated Membrane Processes: Insulin Secretion

39. We have recently learned that the use of electrical signals to initiate a cellular response is not limited to the excitable tissues of nerve and muscle. The beta cell of the pancreas releases insulin in response to a change in membrane potential. (p. 162; Fig. 5-42)

QUESTIONS

LEVEL ONE Reviewing Facts and Terms

1. List the four general functions of the cell membrane.

2. In 1972, Singer and Nicolson proposed a model of the cell membrane known as the _____ model. According to this model, the membrane is composed of a bilayer of _____ and a variety of embedded _____, with _____ on the extracellular surface.

3. List the four functions of membrane proteins and give an example of each.

4. Arrange these compartments in the order a glucose molecule entering the body at the intestine would encounter them: interstitial fluid, plasma, intracellular fluid. Which of these fluids is/are considered to be extracellular fluid(s)?

5. Distinguish between active and passive transport.

6. Which of the following processes are examples of active transport and which of passive transport? Simple diffusion, phagocytosis, facilitated diffusion, exocytosis, osmosis, endocytosis, potocytosis.

7. List four factors that will improve the rate of diffusion.

8. Match the membrane channels with the appropriate description(s). Answers may be used once, more than once, or not at all.

(a) chemically gated channel	1. channel that spends most of its time in the open state
(b) open pore	2. channel that opens in response to a signal
(c) voltage-gated channel	3. channel opens when resting membrane potential changes
(d) mechanically gated channel	4. channel that opens when a ligand binds to it
	5. channel that opens in response to membrane stretch
	6. channel through which water can pass

9. List the three physical methods by which materials enter cells.

10. A cotransporter is a protein that moves more than one molecule at a time. If the molecules are moved in the same direction, the process is called _____; if the molecules are transported in opposite directions, it is called _____. A transport protein that moves only one substrate is called a _____.

11. The two types of active transport are _____, which derives energy directly from ATP, and _____, which couples the kinetic energy of one molecule moving down its concentration gradient to the movement of another molecule against its concentration gradient.

12. A molecule that moves freely between the intracellular and extracellular compartments is said to be _____. A molecule that is not able to enter cells is called a _____ solute.

13. Based on percentage of body water, rank these individuals from highest to lowest: (a) a 25-year-old 70-kg male, (b) a 25-year-old 50-kg female, (c) a 65-year-old 50-kg female, and (d) a 1-year-old male toddler.

14. What determines the osmolarity of a solution? In what units is body osmolarity usually given?

15. What does it mean if we say that a solution is hypotonic to a cell? Hypertonic to the same cell? What determines the tonicity of a solution relative to a cell?

16. In your own words, state the four principles of electricity important in physiology.

17. Match each structure with its primary role in cellular activity.
    (a) Na^+-K^+-ATPase
    (b) proteins
    (c) unit of measurement for membrane potential
    (d) K^+
    (e) Cl^-
    (f) ATP
    (g) Na^+

    1. ion channel
    2. extracellular cation
    3. source of energy
    4. intracellular anion
    5. intracellular cation
    6. millivolts
    7. electrogenic pump
    8. extracellular anion
    9. milliosmoles

18. The membrane potential difference at which the electrical gradient exactly opposes the concentration gradient for an ion is known as the _____.

19. A material that allows free movement of electrical charges is called a(n) _____, while one that prevents this movement is called a(n) _____.

LEVEL TWO *Reviewing Concepts*

20. (a) Create a map of transport across cell membranes using the following terms. You may add additional terms if you wish.

    active transport
    carrier
    caveolae
    channel
    clathrin-coated pit
    concentration gradient
    electrochemical gradient
    exocytosis
    facilitated diffusion
    glucose
    GLUT
    ion
    large polar molecule
    ligand
    osmosis

    Na^+-K^+-ATPase
    passive transport
    phospholipid bilayer
    potocytosis
    receptor
    receptor-mediated endocytosis
    secondary active transport
    simple diffusion
    small polar molecule
    transcytosis
    vesicle
    vesicular transport
    water

    (b) Draw a large rectangle to represent the total body volume. Using the information in Figure 5-32, divide the box proportionately into compartments to represent the different body compartments. Use the information in Figure 5-33 and add solutes to the compartments. Use large size letters for solutes with higher concentrations and small letters for solutes with low concentrations. Label the cell membranes and the endothelial membrane.

21. Explain how inserting cholesterol into the phospholipid bilayer of the cell membrane decreases membrane permeability.

22. Compare and contrast chemically, mechanically, and voltage-gated channels.

23. What factors influence the rate or speed of diffusion across a membrane? Briefly explain each one.

24. Define the following terms and explain how they differ: specificity, competition, saturation. Apply these terms in a short explanation of facilitated diffusion of glucose.

25. Red blood cells are suspended in a solution of NaCl. The cells have an osmolarity of 300 mOsM and the solution has an osmolarity of 250 mOsM. (a) The solution is (hypertonic, isotonic, or hypotonic) to the cells. (b) Water would move (into the cells, out of the cells, or not at all).

26. Two compartments are separated by a membrane that is permeable to glucose. Each compartment is filled with 1 M glucose. After 6 hours, compartment A contains 1.5 M glucose and compartment B contains 0.5 M glucose. What kind of transport occurred?

27. A 2 M NaCl solution and a 2 M glucose solution are separated by a membrane that is permeable to the water, but not to the NaCl or glucose. Are the following statements true or false? Defend each answer with a short statement. (a) The salt solution is isosmotic to the glucose solution. (b) The salt solution is hyperosmotic to the glucose solution. (c) Water will move from the salt solution to the sugar solution. (d) The volume will increase on the glucose side of the membrane. (e) If the salt solution were inside a cell, the cell would swell.

28. Referring to the previous question, change the molarity of the salt and sugar solutions so that each false statement would become true.

29. Explain the difference between a chemical gradient, an electrical gradient, and an electrochemical gradient.

LEVEL THREE *Problem Solving*

30. Sweat glands secrete a fluid into their lumen that is identical to interstitial fluid. As the fluid moves through the lumen on its way to the surface of the skin, the cells of the sweat gland's epithelium make the fluid hypotonic by removing Na^+ and leaving water behind. Design an epithelial cell that will reabsorb Na^+ but not water. You may place water pores, Na^+ leak channels, K^+ leak channels, and the Na^+-K^+-ATPase in the apical membrane, basolateral membrane, or both.

31. Insulin is a hormone that promotes the movement of glucose into many types of cells, thereby lowering blood glucose concentration. Propose a mechanism that explains how this occurs, using your knowledge of cell membrane transport.

32. The following terms have been applied to membrane transport molecules: specificity, competition, saturation. These terms can also be applied to enzymes. How has the application of these terms changed in the two situations? What chemical characteristics do enzymes and transport molecules share that allow these terms to be applied to both?

33. NaCl is a nonpenetrating solute and urea is a penetrating solute. Red blood cells are placed in each of the solutions below. The intracellular concentration of nonpenetrating solute is 300 mOsM. What will happen to the cell volume in each solution? Label each solution with all the terms that apply: hypertonic, isotonic, hypotonic, hyperosmotic, hyposmotic, isosmotic.
    (a) 150 mM NaCl plus 150 mM urea
    (b) 100 mM NaCl plus 50 mM urea
    (c) 100 mM NaCl plus 100 mM urea
    (d) 150 mM NaCl plus 100 mM urea
    (e) 100 mM NaCl plus 150 mM urea

34. Membrane glycoproteins have their sugars added as the proteins pass through the endoplasmic reticulum and Golgi [● p. 116]. Based on this information, would you predict that you would find the sugar "tails" of the proteins on the cytoplasmic side of the membrane, the extracellular side, or both? Explain your reasoning.

LEVEL FOUR *Quantitative Problems*

35. The addition of dissolved solutes to water lowers the freezing point of water. A 1 OsM solution depresses the freezing point of water 1.86° C. If a patient's plasma shows a freezing-point depression of 0.550° C, what is her plasma osmolarity? (Assume that 1 kg water = 1 L.)

36. The patient in the previous question is found to have total body water volume of 42 L, ECF volume of 12.5 L, and plasma volume of 2.7 L.
    (a) What is her intracellular fluid (ICF) volume? Her interstitial fluid volume?
    (b) How much solute (osmoles) exists in her whole body? ECF? ICF? plasma? (Hint: concentration = solute amount/ volume)

37. What is the osmolarity of half-normal saline (= 0.45% NaCl)? [● p. 33] Assume that all NaCl molecules dissociate into two ions.

38. If you give 1 L of half-normal saline to the patient in question 36, what happens to each of the following at equilibrium? (Hint: NaCl is a nonpenetrating solute.)
    (a) her total body volume
    (b) her total body osmolarity
    (c) her ECF and ICF volumes
    (d) her ECF and ICF osmolarities

39. The graph below shows the results of an experiment in which a cell was placed in a solution of glucose. The cell has no glucose in it at the beginning and its membrane can transport glucose. Which of the following processes is/are illustrated by this experiment?
    (a) simple diffusion
    (b) saturation
    (c) competition
    (d) active transport

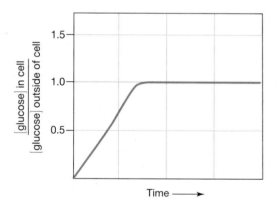

Explorations ⊕

Explorations are also available at *www.physiologyplace.com*

In this chapter you learned about the intricacies and importance of cell membranes. The cell membrane acts both as a barrier and a line of communication, making movement of substances across the membrane an important theme in physiology.

Exploration 1: Caveolae and Caveolin

Caveolae are an interesting example of the progress of science. These flask-like invaginations of the cell membrane were first described and named about 50 years ago by some of the first scientists to study cells with the electron microscope. In those days, before the advent of molecular biology, research on cell organelles relied mostly on morphological characteristics, making the tiny caveolae difficult to study. A MEDLINE search shows that in the 20 years from 1960 to 1980, a total of only 73 scientific papers on caveolae were published—about 2 papers per year. Once molecular biology made it easier to study the biochemical characteristics of the "little caves," however, interest began to grow. In 1995 alone, 59 papers were published. By 2001, once investigators realized that caveolae participated in cell signaling, the number of published papers per year had increased to 191. The newest line of research on these organelles is using knockout mice, animals that have been genetically altered so that they are unable to manufacture caveolin, the protein coat of caveolae.

To find out what happens in caveolin-deficient mice, read "Life without caveolae," *Science* 293:2404–2405, 2001, September 28. (*www.sciencemag.org*). For a history and detailed description of these little membrane formations, read "The role of caveolae and the caveolins in mammalian physiology," *Reviews in Undergraduate Research* 1:44–50, 2002. Available online at *http://www.ruf.rice.edu/~rur/about.html*.

Exploration 2: Cystic Fibrosis

Cystic fibrosis is one of the most common inherited diseases in the United States. It is also one of the diseases where clinical trials are underway to see if gene therapy can successfully restore the normal function of the affected CFTR channels. To learn more about gene therapy and the status of current research, go to the Cystic Fibrosis Foundation website (*www.cff.org*) and click on the link to Research and Clinical Trials. You might also be interested in reading the 1997 consensus statement on "Genetic Testing for Cystic Fibrosis"

published by the U.S. National Institutes for Health. Go to *http://consensus.nih.gov* and select the link for the list of all consensus papers. You will find this paper under the 1997 header.

Some of the most interesting research currently being done uses genetically altered mice. These animal models can be bred to have totally nonfunctional CFTR channels or CFTR channels with altered function corresponding to the mutations of the CFTR gene in humans. An excellent review paper, "The CF mouse: an important tool for studying cystic fibrosis," can be found in the online journal *Expert Reviews in Molecular Medicine* at *http://www-ermm.cbcu.cam.ac.uk/01002551h.htm* (12 March 2001).

Exploration 3: A Nobel Prize for the Na$^+$-K$^+$-ATPase

In 1997 half of the Nobel prize for chemistry was awarded to Dr. Jens C. Skou for his discovery of the sodium-potassium pump, the transporter that helps cells remove Na$^+$ and retain K$^+$. To learn more about Dr. Skou and his work, go to the Nobel website at *www.nobel.se*. Select the tab for chemistry and see what you can learn. To find animations of the pump in action, try a Google search (*www.google.com*) and search for *Na K ATP-ase animation*.

ANSWERS

Answers to Concept Check Questions

Page 129

1. Two types of lipids found in cell membranes are phospholipid and cholesterol.

2. Integral proteins are tightly bound to the membrane. Peripheral proteins are loosely bound to polar regions of the membrane components.

3. The tails of phospholipids are hydrophobic and a single layer would put them in direct contact with aqueous body fluids.

Page 130

4. Positive ions are cations and negative ions are anions.

Page 132

5. Membrane proteins serve as structural elements, receptors, enzymes, and transporters.

6. Ions and water move through open channels.

7. Channels form continuous connections between the two sides of the membrane, have faster transport, and bind more tightly to their substrates.

8. A channel lined with positive charges will attract anions such as Cl$^-$.

9. Glucose is too large to pass through a channel.

Page 135

10. If distance doubles, diffusion takes four times as long.

Pages 137–138

11. Energy for diffusion comes from molecular motion.

12. Lipophilic fatty acids are more likely to cross by simple diffusion.

13. (a) decreases, (b) increases, (c) decreases

14. Compartment A will remain yellow.

15. The skin's thick extracellular matrix is generally impermeable to oxygen. Also, oxygen needs a moist surface to effectively diffuse across a membrane.

Page 140

16. An antiporter moves substrates in opposite directions.

17. If you made the doors larger, you could move more people. This would be analogous to creating a different form of the transporter that could move more substrate per second.

Page 140

18. Facilitated diffusion of glucose reverses and glucose leaves the cell.

Page 147

19. In phagocytosis, the cell pushes its membrane out to engulf the particle in a large vesicle. In endocytosis, the membrane surface indents and the vesicle is much smaller.

20. The proteins are clathrin and caveolin.

21. Proteins move in by endocytosis or out by exocytosis.

Page 148

22. Sodium movement out of the cell is against the concentration gradient.

23. Ouabain applied to the apical side would have no effect because there are no

Na^+-K^+-ATPase molecules on that side. Ouabain applied to the basolateral side will stop the pump. Glucose transport will continue for a time until the Na^+ gradient between the cell and the lumen disappears due to Na^+ entry.

24. The GLUT2 transporter is illustrated.

Page 150

25. Vesicular transport of vesicles from one side of the cell to the other depends on microtubules. If they are not functioning, transcytosis will stop.

Page 151

26. The baby lost 0.91 kg or 0.91 liter.

Page 156

27. 1 M NaCl = 2 OsM. Therefore the 1 M glucose and 1 OsM NaCl have the most water.

28. (a) into A, (b) into B, (c) no net movement

29. (a) The NaCl. Both solutions are isosmotic. Blood is lost from the extracellular compartment, so the best replacement remains in the ECF. Glucose slowly enters cells, taking water with it. (b) If 1 L has been lost, you should replace at least 1 L.

Page 161

30. Cl^- would move into the cell down its concentration gradient, which would make the inside of the cell negative. The positive charges left outside would attract Cl^- back outside. The equilibrium potential would be negative.

31. Over time, Na^+ would leak into the cell, and the resting membrane potential will become more positive.

Answers to Figure Questions

Page 139

Fig. 5-18: (a) Glucose is binding to some of the carriers and displacing galactose. (b) The line is zero from the origin until the point at which glucose is added. It then increases by the same amount that galactose decreases until it becomes constant.

Page 140

Fig. 5-20: The mark on the *x*-axis is at the point where the curve levels off.

Page 149

Fig. 5-30: (a) 1 = b; 2 = a; 3 = b (b) Basolateral glucose transport is passive because it is moving down the concentration gradient. (c) Na^+ movement across the apical membrane does not require ATP because Na^+ is moving down its concentration gradient.

Page 150

Fig. 5-32: (a) 25% of 14 L = 3.5 L plasma. 75% = 10.5 L interstitial fluid. (b) Total body water = 42 L. (c) 3.5 L/42 L = 8.3% plasma; 10.5 L/42 L = 25% interstitial volume

Page 151

Fig. 5-33: (a) Plasma has proteins and large anions not present in interstitial fluid. (b) ECF is higher in Na^+, Cl^-, and bicarbonate and lower in K^+.

Page 155

Fig. 5-35: (a) isosmotic and isotonic (b) hyperosmotic and isotonic (c) hyperosmotic and hypotonic

Page 160

Fig. 5-38: ICF = -1 and ECF = $+1$. The cell membrane is not permeable to Na^+, Cl^-, or proteins.

Page 161

Fig. 5-40: (a) Na^+ leak in is promoted by concentration and electrical gradients. (b) K^+ leak out is promoted by the concentration gradient.

Answers to Level Four Quantitative Problems

Page 168

35. Plasma osmolarity = 296 mOsM

36. (a) ICF volume = 29.5 L; interstitial fluid volume = 9.8 (b) Total solute = 12.432 osmoles; ECF solute = 3.7 osmoles; ICF solute = 8.732 osmoles; plasma solute = 0.799 osmoles.

37. Half-normal saline is approximately 154 mOsM.

38. (a) increases (b) decreases (c) increase (d) decrease

39. The graph shows simple diffusion (a). You know that it is not active transport because the concentration inside the cell does not increase past equilibrium (concentration in = concentration out). The graph is the same shape as the saturation graph in Figure 5-20 but the axes are not the same. There is only one solute being transported, so there is no evidence for competition.

Communication, Integration, and Homeostasis

6

Many chronic diseases represent disorders in communication between cells.

—Howard Rasmussen, in *Biology and Medicine into the 21st Century*, 1991

RUNNING PROBLEM

Diabetes Mellitus

It is 8:00 A.M., and Marvin Garcia, age 20, is hungry. As part of his routine physical examination before school starts, he is having a fasting blood glucose test. In this test, blood is drawn after an overnight fast, and the glucose concentration in the blood is measured. Marvin knows that he is in good physical condition, so he doesn't worry about the results. He is surprised, then, when the nurse practitioner at his family physician's office calls two days later. "Your fasting blood sugar is a bit elevated, Marvin. It is 150 milligrams per deciliter, and normal is 110 or less. Does anyone in your family have diabetes?" "Well, yeah—my dad has it. What exactly is diabetes?"

BACKGROUND BASICS

Homeostasis: **7** | Nucleotides: **41** | Cell junctions: **63** | Extracellular matrix: **64** | Endocrine glands: **70** | Membrane structure: **132** | Membrane proteins: **134** | Diffusion: **134** | Exocytosis: **146–147** |

Complex living organisms are collections of molecules arranged into organelles, cells, tissues, and organs. The sheer size of the human body means that individual cells may be within micrometers of each other or meters apart. How then do the cells manage to communicate and coordinate with one another? Diffusion is only adequate for information exchange between neighboring cells, because as you learned in the last chapter, diffusion is ineffective over long distances. Multicellular organisms require additional methods for moving signals from one part of the body to another.

To meet the challenge of coordinating body functions, various anatomical and physiological mechanisms evolved. Blood flowing through our efficient circulatory system makes a complete circuit about once a minute, moving material from one part of the body to another. The nervous system takes care of more rapid communication, with electrical signals traveling as fast as 120 m/sec (268 mi/hr). The combination of simple diffusion across small distances, widespread distribution of materials through the circulatory system, and rapid, specific delivery of messages by the nervous system enables each cell in the body to communicate with most other cells.

In the first half of this chapter we examine the mechanisms of cell-to-cell communication. In the second half we look at why efficient cell-to-cell communication is important to the body as a whole. We return to the central theme introduced in the first chapter, *homeostasis*, the process by which the body maintains a relatively stable internal environment in the face of either an external or an internal disturbance [💿 p. 7]. In this chapter and those that follow, we will see how the control of homeostasis resides primarily in the nervous and endocrine systems, with their combination of chemical and electrical signals.

▶ CELL-TO-CELL COMMUNICATION

In recent years the amount of information available about cell-to-cell communication has mushroomed due to advances in research technology. Signal pathways that once seemed fairly simple and direct have turned into incredibly complex networks and webs of information transfer. In the sections that follow, we have tried to distill what is known about cell-to-cell communica-

tion into some basic patterns that you can learn and recognize when you encounter them again in your study of physiology.

By most estimates the human body is composed of about 75 *trillion* cells. Those cells face a daunting task—to communicate with each other in a manner that is rapid yet also conveys a tremendous amount of information. Surprisingly, there are only two basic types of physiological signals: electrical and chemical. Electrical signals are changes in a cell's membrane potential [💿 p. 157]. Chemical signals are molecules secreted into the extracellular fluid (ECF) by cells, and they are responsible for most communication within the body. The cells that receive electrical or chemical signals are called **target cells**, or **targets** for short.

Our bodies use four basic methods of cell-to-cell communication: (1) **gap junctions** that allow direct cytoplasmic transfer of electrical and chemical signals between adjacent cells; (2) **contact-dependent signals** that occur when surface molecules on one cell bind to surface molecules on another cell; (3) **local communication** by chemicals that diffuse through the extracellular fluid; and (4) **long-distance communication** through a combination of **electrical signals** carried by nerve cells and **chemical signals** transported in the blood. A given molecule can function as a signal by more than one method. For example, a molecule can act close to the cell that releases it as well as act on cells in distant parts of the body.

Gap Junctions Transfer Chemical and Electrical Signals Directly Between Cells

The simplest form of cell-to-cell communication is the direct transfer of electrical and chemical signals through **gap junctions**, protein channels that create cytoplasmic bridges between adjacent cells (Fig. 6-1a ●) [💿 p. 64]. A gap junction forms from the union of membrane-spanning proteins, called *connexins*, on two adjacent cells. The united connexins create a protein channel (*connexon*) that can open and close. When the channel is open, the connected cells function like a single cell with multiple nuclei (a *syncytium*).

When gap junctions are open, ions and small molecules such as amino acids, ATP, and cyclic AMP diffuse directly from the cytoplasm of one cell to the cytoplasm of the next. As with other membrane channels, larger molecules are excluded. In addition,

(a) Gap junctions form direct cytoplasmic connections between adjacent cells.

(b) Contact-dependent signals require interaction between membrane molecules on two cells.

(c) Autocrine signals act on the same cell that secreted them. **Paracrine signals** are secreted by one cell and diffuse to adjacent cells.

Receptor

● **Figure 6-1** *Direct and local cell-to-cell communication* gap junctions are the only means by which electrical signals can pass *directly* from cell to cell. Movement of molecules through gap junctions can be modulated or shut off completely.

Gap junctions are not all alike. Scientists have discovered more than 20 different variants of connexins that may mix or match to form gap junctions. When a protein such as connexin has several similar forms, the variants are known as subtypes or *isoforms* [*iso-*, equal]. (Isozymes, enzymes that come in related subtypes [🔵 p. 92], are another example of isoforms.) The existence of connexin isoforms allows gap junction selectivity to vary from tissue to tissue. In mammals, gap junctions are found in almost every cell type, including heart muscle, some types of smooth muscle, lung, liver, and neurons of the brain.

Contact-Dependent Signals Require Cell-to-Cell Contact

Some cell-to-cell communication requires that surface molecules on one cell bind to a surface molecule of another cell (Fig. 6-1b). This type of direct contact signaling occurs in the immune system and during growth and development, such as when nerve cells send out long extensions that must grow from the central axis of the body to the *distal* (distant) ends of the limbs. **CAMs**, cell adhesion molecules, were first known for their role in cell-to-cell adhesion [🔵 p. 64] but are now known to participate in cell-cell signaling as well.

Paracrines and Autocrines Are Chemical Signals Distributed by Diffusion

Local communication is accomplished by paracrines and autocrines. A **paracrine** [*para-*, beside + *krinen*, to secrete] is a chemical that is secreted by a cell to act on cells in its immediate vicinity. If this signal molecule acts on the cell that secreted it, the chemical is called an **autocrine** [*auto-*, self]. In some cases a chemical may act both as an autocrine and as a paracrine.

Paracrines and autocrines reach their target cells by diffusing through the interstitial fluid (Fig. 6-1c). Because distance is a limiting factor for diffusion, the effective range of the chemical signal is restricted to adjacent cells. A good example of a paracrine is *histamine*, a chemical released from damaged cells. When you scratch yourself with a pin, the red, raised *wheal* that results is due in part to the local release of histamine from the injured tissue. The histamine acts as a paracrine, diffusing over to the capillaries in the immediate area of the injury and making them more permeable to white blood cells and antibodies in the plasma. Fluid also leaves the blood vessels and collects in the interstitial space, causing swelling around the area of injury.

Several important groups of molecules act as paracrines. *Cytokines* are regulatory peptides that usually act close to the site where they are secreted. *Eicosanoids* [🔵 p. 37] are lipid-derived paracrines and autocrines. Cytokines and eicosanoids will be discussed in more detail below.

Electrical Signals, Hormones, and Neurohormones Carry Out Long-Distance Communication

All cells in the body can release paracrines, but most long-distance communication between cells is the responsibility of the nervous and endocrine systems. The endocrine system communicates using **hormones** [*hormon*, to excite], chemical signals that are secreted into the blood and distributed all over the body by the circulation. Hormones come in contact with most cells of the body but only those cells with receptors for the hormone are target cells (Fig. 6-2a ●).

(a) Hormones are secreted by endocrine glands or cells into the blood. Only target cells with receptors for the hormone will respond to the signal.

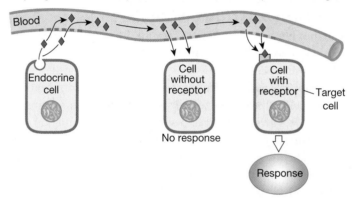

(b) Neurotransmitters are chemicals secreted by neurons that diffuse across a small gap to the target cell. Neurons use electrical signals as well.

(c) Neurohormones are chemicals released by neurons into the blood for action at distant targets.

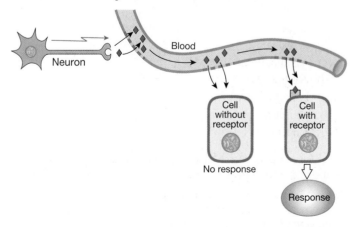

● **Figure 6-2** *Long distance cell-to-cell communication*

The nervous system uses a combination of chemical signals and electrical signals to communicate. An electrical signal travels along a nerve cell (*neuron*) until it reaches the very end of the cell, where it is translated into a chemical signal secreted by the neuron (a **neurocrine**). If the chemical signal diffuses from the neuron across a narrow extracellular space to a target cell and has a rapid effect, it is called a **neurotransmitter** (Fig. 6-2b). If a neurocrine acts more slowly as an autocrine or paracrine, it is called a **neuromodulator**. If the neurocrine released by the neuron diffuses into the blood for distribution, it is called a **neurohormone** (Fig. 6-2c). The similarities between neurohormones and hormones secreted by the endocrine system blur the distinction between the nervous and endocrine systems, making them a continuum rather than two distinct systems (see Fig. 6-30, p. 199).

Cytokines Act as Both Local and Long-Distance Signals

Cytokines are among the most recently identified communication molecules. Initially the term **cytokine** referred only to proteins that modulate immune responses, but in the past few years it has been broadened to include a variety of regulatory peptides. All nucleated cells synthesize and secrete cytokines in response to stimuli. Cytokines control cell development, differentiation, and the immune response. In development and differentiation, cytokines function as autocrines or paracrines. In stress and inflammation, cytokines may act on relatively distant targets and may be transported through the circulation just as hormones are.

How do cytokines differ from hormones? In general, they act on a broader spectrum of target cells. In addition, cytokines are not produced by specialized glands and they are made on demand. In contrast, most protein or peptide hormones are made in advance and stored in the endocrine cell until needed. However, the distinction between cytokines and hormones is sometimes blurry. For example, erythropoietin, the molecule that controls synthesis of red blood cells, is by tradition considered a hormone but functionally fits the definition of a cytokine.

CONCEPT CHECK   Answers: p. 206

1. Match the physiological signal mechanisms on the left with the signal's properties on the right.

   (a) autocrine
   (b) cytokine
   (c) gap junctions
   (d) hormone
   (e) neurohormone
   (f) neurotransmitter
   (g) paracrine

   1. electrical signals
   2. chemical signals
   3. both electrical and chemical signals

2. Which methods of cell-to-cell communication on the list above involve transport through the circulatory system?

3. A cat sees a mouse and pounces on it. Do you think the internal signal to pounce could have been transmitted by a paracrine? Give two reasons to explain why or why not.

▶ SIGNAL PATHWAYS

Chemical signals in the form of paracrines, autocrines, and hormones are released from cells into the extracellular compartment. This is not a very specific way for them to find their targets because substances that travel through the blood reach nearly every cell in the body. Yet cells do not respond to every signal that reaches them. Why do some cells respond to a chemical while others ignore it? The answer lies in the **receptor** proteins to which chemical signals bind [● p. 129]. *A cell cannot respond to a chemical signal if it lacks the appropriate receptor for that signal.*

If a cell has a receptor for a signal molecule, binding of the signal to the receptor will initiate a response pathway. All signal pathways share common features (Fig. 6-3 ●):

1. The **signal molecule**, or *ligand*, binds to its receptor. The signal molecule is also known as a *first messenger* because it brings information to its target cell.
2. Ligand-receptor binding activates the **receptor**.
3. The receptor in turn activates one or more **intracellular signal molecules**.
4. The last signal molecule in the pathway initiates synthesis of a **target protein** or modifies an existing target protein to create a **response**.

In the following sections, we will describe some basic signal-receptor-response pathways. They may seem complex at first, but they follow patterns that you will encounter over and over as you study the systems of the body. Most physiological processes, from the beating of your heart to learning and memory, use some varia-

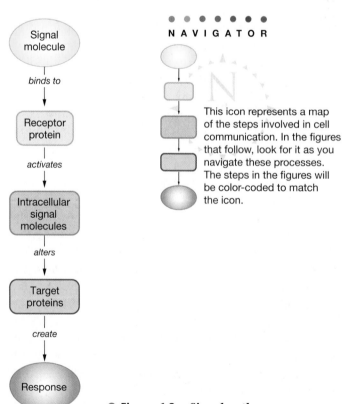

● **Figure 6-3** *Signal pathways*

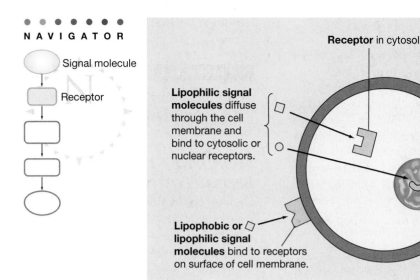

NAVIGATOR

- Signal molecule
- Receptor

Receptor in cytosol

Receptor in nucleus

Lipophilic signal molecules diffuse through the cell membrane and bind to cytosolic or nuclear receptors.

Lipophobic or lipophilic signal molecules bind to receptors on surface of cell membrane.

● **Figure 6-4** *Target cell receptors*

tion of these pathways. One of the wonders of physiology is the fundamental importance of these signal pathways and the way they have been conserved in animals ranging from worms to humans.

Receptors Are Located Inside the Cell or on the Cell Membrane

Chemical signal molecules fall into two broad categories based on their lipid solubility: lipophilic or lipophobic. Target cell receptors may be found in three locations: in the nucleus, in the cytosol, or on the cell membrane as integral proteins (Fig. 6-4 ●). The general category of intracellular pathway for a signal molecule can be predicted by knowing where the molecule's receptor is located.

Lipophilic signal molecules diffuse through the phospholipid bilayer of cell membranes [● p. 136] and bind to *cytosolic receptors* or *nuclear receptors*. In these cases, receptor activation often turns on a gene and directs the nucleus to make new mRNA (*transcription*, [● p. 112]). The mRNA then provides a template

for synthesis of new proteins (*translation*, [● p. 114]). This is a relatively slow process and the cell's response may not be noticeable for an hour or longer. In some instances the activated receptor can also turn off, or repress, gene activity. Many lipophilic signal molecules that follow this pattern are hormones, so we will defer further discussion of their action to the next chapter.

Lipophobic signal molecules are unable to diffuse through the phopholipid bilayer of the cell membrane. Instead, these signal molecules bind to receptors on the target cell membrane. (Some lipophilic signal molecules also bind to cell membrane receptors.) In general, the response time for pathways linked to membrane receptors is very rapid, and responses can be seen within milliseconds to minutes.

Receptors for signal molecules play an important role in physiology and medicine. About half of all drugs currently in use act on receptor proteins. We can group membrane receptors into four major categories, illustrated in Figure 6-5 ●. The simplest

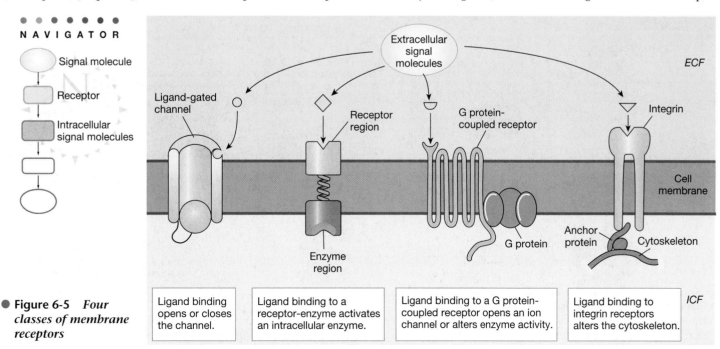

NAVIGATOR

- Signal molecule
- Receptor
- Intracellular signal molecules

| Ligand binding opens or closes the channel. | Ligand binding to a receptor-enzyme activates an intracellular enzyme. | Ligand binding to a G protein-coupled receptor opens an ion channel or alters enzyme activity. | Ligand binding to integrin receptors alters the cytoskeleton. |

● **Figure 6-5** *Four classes of membrane receptors*

Later that day in the physician's office, the nurse practitioner explains diabetes to Marvin. Diabetes mellitus is a family of metabolic diseases caused by defects in the homeostatic pathways that regulate glucose metabolism. Several forms of diabetes exist, and some can be inherited. One form, called type 1 diabetes mellitus, is caused by deficient production of insulin, a protein hormone made in the pancreas. In another form, called type 2 diabetes mellitus, insulin is often present in normal or above normal levels. However, the insulin-sensitive cells of the body do not respond normally to the hormone.

Question 1: In which type of diabetes is the signal pathway for insulin more likely to be defective?

receptors are chemically gated (*ligand-gated*) ion channels [❧ p. 131]. Signal ligand binding opens or closes the channel and alters ion flow across the membrane.

Three other receptor types are shown in Figure 6-5: *receptor-enzymes, G protein-linked receptors*, and *integrin receptors*. For all three, information from the signal molecule must be passed across the membrane to initiate an intracellular response. This transmission of information from one side of a membrane to the other using membrane proteins is known as *signal transduction*.

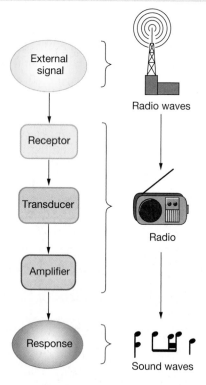

Signal transduction converts one form of signal into a different form.

● **Figure 6-6** *Signal transduction* An example of signal transduction in the physical world is the radio, a transducer that converts radio waves into sound waves.

We will take a closer look at signal transduction before returning to the three receptor types that participate in it.

4. List four components of signal pathways.

5. Name three cellular locations of receptors.

Membrane Proteins Facilitate Signal Transduction

Signal transduction is the process in which an extracellular signal molecule activates a membrane receptor that in turn alters intracellular molecules to create a response. The extracellular signal molecule is the first messenger, and the intracellular molecules form a *second messenger system*. The term *signal transduction* comes from the verb, to transduce, meaning "to lead across" [*trans*, across + *ducere*, to lead].

A **transducer** is a device that converts a signal from one form into a different one. For example, the transducer in a radio converts radio waves into sound waves (Fig. 6-6 ●). In biological systems, transducers convert the message of extracellular signal molecules into intracellular messages that trigger a response.

In biological systems, as in a radio, the original signal is not only transformed but also amplified [*amplificare*, to make larger]. **Signal amplification** turns one signal molecule into multiple second messenger molecules. The process begins with the signal molecule, or ligand, that combines with its receptor (Fig. 6-7 ●). The receptor-ligand complex then turns on an **amplifier enzyme**, an enzyme that activates several more molecules. By the end of the process, the effects of the ligand have been amplified much more than if there were a 1 : 1 ratio between each step. Amplification gives the body "more bang for the buck" by enabling a small amount of signal to create a large effect.

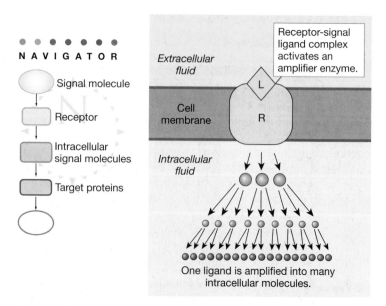

● **Figure 6-7** *Signal amplification*

● **Figure 6-8** *Biological signal transduction*

N A V I G A T O R

- Signal molecule
- Receptor
- Intracellular signal molecules
- Target proteins
- Response

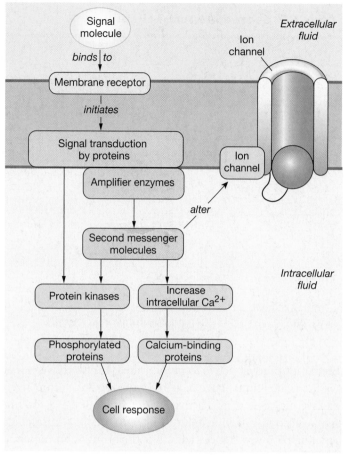

The basic pattern of a biological signal transduction pathway is shown in Figure 6-8● and can be broken down into the following steps.

1. An extracellular *signal molecule* binds to and activates a protein or glycoprotein *membrane receptor*.
2. The activated membrane receptor turns on its associated proteins. These proteins then may:
   (a) activate protein kinases. **Protein kinases** are enzymes that transfer a phosphate group from ATP to a protein [● p. 99]. *Phosphorylation* (addition of a phosphate group) is an important biochemical method of regulating cellular processes.
   (b) activate *amplifier enzymes* that create intracellular **second messengers**. The most common amplifier enzymes and second messengers are listed in Tables 6-1 and 6-2.
3. Second messenger molecules in turn
   (a) alter the open state of ion channels. Changing the open state of ion channels creates electrical signals by altering the cell's membrane potential [● p. 157].
   (b) increase intracellular calcium. Calcium binding to proteins changes their function, creating a cellular response.
   (c) change enzyme activity, especially of protein kinases (which add a phosphate group) or protein **phosphatases**, enzymes that remove a phosphate group. The phosphorylation or *dephosphorylation* of a protein can change its configuration and create a response. Examples of changes that occur with phosphorylation include increased or decreased enzyme activity or changes in the open state of gated ion channels.

4. The proteins modified by calcium binding and phosphorylation control one or more of the following:
   (a) metabolic enzymes,
   (b) motor proteins for muscle contraction and cytoskeletal movement,
   (c) proteins that regulate gene activity and protein synthesis, and
   (d) membrane transport and receptor proteins.

If you think this list includes almost everything a cell does, you're right!

You can see from the pathway above that the steps of a signal transduction pathway form a **cascade** (Fig. 6-9 ●) that starts when a stimulus (the signal molecule) converts inactive molecule A

Table 6-1: Amplifier Enzymes

AMPLIFIER ENZYME	CELLULAR LOCATION	ACTIVATED BY	CONVERTS	TO
Adenylyl cyclase	Membrane	G protein-coupled receptor	ATP	cAMP
Guanylyl cyclase	Membrane Cytosol	Receptor-enzyme Nitric oxide (NO)	GTP	cGMP
Phospholipase C	Membrane	G protein-coupled receptor	Membrane phospholipids	IP$_3$ and DAG*

*IP$_3$ = Inositol triphosphate and DAG = diacylglycerol

Table 6-2: Second Messenger Pathways

SECOND MESSENGER	ACTION	EFFECTS
Ions		
Ca^{2+}	Binds to calmodulin	Alters enzyme activity
	Binds to other proteins	Exocytosis, muscle contraction, cytoskeleton movement, channel opening
Nucleotides		
cAMP	Activates protein kinases, especially protein kinase A	Phosphorylates proteins
	Binds to ion channels	Alters channel opening
cGMP	Activates protein kinases, especially protein kinase G	Phosphorylates proteins
	Binds to ion channels	Alters channel opening
Lipid-derived		
IP_3	Releases Ca^{2+} from intracellular stores	See Ca^{2+} effects above
DAG	Activates protein kinase C	Phosphorylates proteins

(the receptor) to an active form. Activated A then converts inactive molecule B into active B, active molecule B in turn converts inactive molecule C into active C, and so on, until at the final step a substrate is converted into a product. Many intracellular signaling pathways are cascades. Blood clotting is an important example of an extracellular cascade.

In the sections that follow, we will examine in more detail the four major types of membrane receptors: receptor-enzymes, G protein-linked receptors, integrin receptors, and ligand-gated ion channels (see Fig. 6-5). Keep in mind that these receptors may be responding to any of the different kinds of signal molecules, including hormones, neurohormones, neurotransmitters, cytokines, paracrines, or autocrines.

CONCEPT CHECK   Answers: p. 206

6. What are the steps of signal transduction?

7. What happens during amplification?

8. Why do steroid hormones not require signal transduction and second messengers to exert their action? (Hint: Are steroids lipophobic or lipophilic? [● p. 28])

Steps of a cascade

- Initial stimulus — Initial stimulus activates the first step in the cascade.
- Inactive A → Active A
- Inactive B → Active B
- Inactive C → Active C
- Substrate → Product — Conversion of substrate to product is the final step of the cascade.

● **Figure 6-9** *Steps of a cascade*

Receptor-Enzymes Have Protein Kinase or Guanylyl Cyclase Activity

Receptor-enzymes have receptors on the extracellular side of the membrane that activate enzymes on the cytoplasmic side (see Fig. 6-5). In some instances, the receptor and enzyme are parts of the same protein molecule. In other cases, the enzyme is a separate protein. The enzymes of receptor-enzymes are either protein kinases, such as the **tyrosine kinase** illustrated in Figure 6-10 ●, or guanylyl cyclase. **Guanylyl cyclase** is the amplifier enzyme that converts GTP to **cyclic GMP (cGMP)** [● p. 41]. Ligands for receptor-enzymes include many growth factors and cytokines as well as the hormone insulin.

Most Signal Transduction Uses G Proteins

The **G protein-coupled receptors** are a large and complex family of membrane-spanning proteins that cross the phospholipid bilayer seven times. The cytoplasmic tail of the receptor protein is linked to a three-part membrane transducer molecule known as a

N A V I G A T O R

● **Figure 6-10** *Tyrosine kinase, an example of a receptor-enzyme* Tyrosine kinase transfers a phosphate group from ATP to a tyrosine (amino acid) of a protein.

G protein (see Fig. 6-5). Hundreds of G protein-coupled receptors have been identified, and the list continues to grow. The types of ligands that bind to G protein-coupled receptors include hormones, growth factors, olfactory molecules, visual pigments, and neurotransmitters. In 1994 Alfred G. Gilman and Martin Rodbell received a Nobel prize for the discovery of G proteins and their role in cell signaling.

G proteins get their name from the fact that they bind guanosine nucleotides. Inactive G proteins are bound to guanosine diphosphate (GDP). Exchanging the GDP for guanosine triphosphate (GTP) activates the G protein. When G proteins are activated, they either (1) open an ion channel in the membrane or (2) alter enzyme activity on the cytoplasmic side of the membrane.

CLINICAL FOCUS

Insulin's Signal Transduction Pathway

In people with type 2 diabetes mellitus, insulin binds to its receptor but the cell fails to respond normally to the signal. As a result, researchers have been investigating the insulin receptor and its pathways in an effort to uncover the causes of type 2 diabetes. Insulin does not use the well-studied cAMP second messenger system, so for many years the insulin receptor-pathway was a "black box." In 1982 scientists discovered that the insulin receptor is a tyrosine kinase receptor-enzyme. Their next step was to discover what substrates the tyrosine kinase was phosphorylating. This search is still going on. At least eight different substrates for the insulin receptor kinase have been identified, but the downstream steps of their cascades are still under investigation. Although we know what effects insulin has on glucose transport and metabolism, we still do not fully understand the pathways through which these effects take place.

G proteins linked to amplifier enzymes make up the bulk of all known signal transduction mechanisms. The two most common amplifier enzymes for G protein-coupled receptors are adenylyl cyclase and phospholipase C. The pathways for these amplifier enzymes are described below.

Adenylyl Cyclase-cAMP Is the Signal Transduction System for Many Lipophobic Hormones

The **G protein-coupled adenylyl cyclase-cAMP system** was the first identified signal transduction system (Fig. 6-11 ●). It was discovered in the 1950s by Dr. Earl Sutherland when he was studying the effects of hormones on carbohydrate metabolism. This discovery proved so significant to our understanding of signal transduction that Sutherland was awarded a 1971 Nobel prize for his work.

In the adenylyl cyclase-cAMP system, *adenylyl cyclase* is the amplifier enzyme that converts ATP to the second messenger molecule *cyclic AMP* (cAMP). Cyclic AMP then activates *protein kinase A*, which in turn phosphorylates other intracellular proteins as part of the signal cascade. The G protein-coupled adenylyl cyclase-cAMP system is the signal transduction system for many protein hormones.

G Protein-Linked Receptors Also Use Lipid-Derived Second Messengers

G protein-linked receptors are also linked to another amplifier enzyme, phospholipase C (Fig. 6-12 ●). When a signal molecule activates this G protein-linked receptor pathway, **phospholipase C (PL-C)** converts a membrane phospholipid (*phosphatidylinositol bisphosphate*) into two different second messenger molecules, diacylglycerol and inositol triphosphate.

Diacylglycerol (DAG) is a nonpolar diglyceride that remains in the lipid portion of the membrane and interacts with **protein kinase C** (PK-C), a Ca^{2+}-activated enzyme associated with the cytoplasmic face of the cell membrane. Protein kinase C phosphorylates cytosolic proteins that continue the signal cascade.

1. Signal molecule binds to G protein-linked receptor, which activates the G protein.

2. G protein turns on adenylyl cyclase, an amplifier enzyme.

3. Adenylyl cyclase converts ATP to cyclic AMP.

4. cAMP activates protein kinase A.

5. Protein kinase A phosphorylates other proteins, leading ultimately to a cellular response.

● **Figure 6-11** *The G protein-coupled adenylyl cyclase-cAMP system* This was the first second messenger system described.

1. Signal molecule activates receptor and associated G protein.

2. G protein activates phospholipase C (PL-C), an amplifier enzyme.

3. PL-C converts membrane phospholipids into diacylglycerol (DAG) which remains in the membrane, and IP$_3$, which diffuses into the cytoplasm.

4. DAG activates protein kinase C (PK-C), which phosphorylates proteins.

5. IP$_3$ causes release of Ca^{2+} from organelles, creating a Ca^{2+} signal.

KEY
PL-C = phospholipase C
DAG = diacylglycerol
PK-C = protein kinase C
IP$_3$ = inositol triphosphate
ER = endoplasmic reticulum

● **Figure 6-12** *The phospholipase C system* Another G protein-linked second messenger system. Phospholipase C (PL-C) converts membrane lipids into two second messengers, diacylglycerol (DAG) and inositol triphosphate (IP$_3$).

● **Figure 6-13** *Ions create electrical signals*

1 Receptor-channels open or close in response to signal molecule binding.

2 Some channels are directly linked to G proteins.

3 Other ligand-gated channels respond to intracellular second messengers.

Inositol triphosphate (IP₃) is a water-soluble messenger molecule that enters the cytoplasm. There it binds to a calcium channel on the endoplasmic reticulum (ER). IP₃ binding opens the Ca^{2+} channel, allowing Ca^{2+} to diffuse out of the ER and into the cytosol. Calcium is itself an important signal molecule, as discussed below.

Integrin Receptors Transfer Information from the Extracellular Matrix

The membrane-spanning proteins called **integrins** (see Fig. 6-5) mediate blood clotting, wound repair, cell adhesion and recognition in the immune response, and cell movement during development. On the extracellular side of the membrane, integrin receptors bind to proteins of the extracellular matrix [● p. 64] or to ligands such as antibodies and molecules involved in blood clotting. Inside the cell, integrins attach to the cytoskeleton via *anchor proteins*. Ligand binding to the receptor causes integrins to activate intracellular enzymes or to alter the organization of the cytoskeleton.

The importance of integrin receptors is illustrated by two inherited conditions. In the first, absence of β2 integrin receptors results in white blood cells that cannot release bacteria-destroying chemicals, leading to recurrent bacterial infections. In the second condition, platelets, the cell fragments that play a key role in blood clotting, lack an integrin receptor. As a result, blood clotting is defective in these patients.

The Most Rapid Signal Pathways Change Ion Flow Through Channels

Ligand-gated ion channels are the simplest receptors and their activation initiates the most rapid intracellular responses. For that reason these receptors are often located in the excitable tissues of nerve and muscle. When an extracellular signal molecule binds to the receptor-channel protein, a channel gate opens or closes, changing the cell's permeability to an ion. One example of a receptor-channel is the acetylcholine-gated cation channel of skeletal muscle. The neurotransmitter *acetylcholine* diffuses across the extracellular fluid from an adjacent neuron and opens the channel, creating a cascade that initiates muscle contraction.

Increasing or decreasing the permeability of an ion channel changes the cell's membrane potential [● p. 159]. For example, opening the acetylcholine receptor-channel described above allows Na^+ to enter the cytosol down its electrochemical gradient. Net entry of cations depolarizes the cell, creating an electrical signal that alters voltage-sensitive proteins (Fig. 6-13●).

Note that not all ligand-gated ion channels are receptors for extracellular signal molecules. Some ligand-gated channels are controlled by intracellular second messengers such as cAMP. Other ion channels are indirectly linked to receptors by G proteins.

CONCEPT CHECK Answers: p. 207

9. Name the four classes of membrane receptors.

10. What is the difference between a first messenger and a second messenger?

11. Organize the following terms into the correct order for a signal transduction cascade:
    (a) cell response, receptor, second messenger, signal ligand
    (b) amplifier enzyme, cell response, phosphorylated protein, protein kinase, second messenger

12. Will a cell depolarize or hyperpolarize in each of the following situations?
    (a) Cl^- channel opens
    (b) K^+ channel opens
    (c) Na^+ channel opens

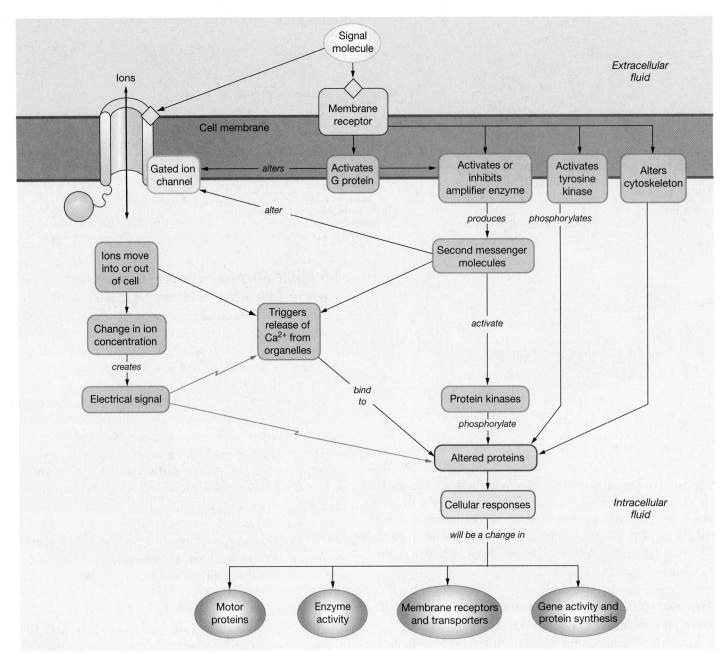

● **Figure 6-14** *Summary of signal transduction systems*

Figure 6-14 ● is a summary map of basic signal transduction, showing the general relationships between first messengers, membrane receptors, second messengers, and the cell's response.

▶ NOVEL SIGNAL MOLECULES

In Chapters 7 and 8, you will learn more about hormones and neurotransmitters and their signal pathways. The sections below introduce you to some unusual signal molecules that are important in physiology and medicine. They include an ion (Ca^{2+}), two gases, and a family of lipid-derived messengers. The processes controlled by these signal molecules have been known for years but the control signals themselves were discovered relatively recently.

Calcium Is an Important Intracellular Signal

Calcium is the most versatile ionic messenger (Fig. 6-15 ●). Calcium enters the cytosol from two sources. It can enter from the extracellular fluid through voltage-gated Ca^{2+} channels or ligand-gated or mechanically gated channels. Calcium can also be released from intracellular compartments by second messengers, such as IP_3. Most intracellular Ca^{2+} is stored in the endoplasmic reticulum [● p. 60], where it is concentrated by active transport.

Release of Ca^{2+} into the cytosol (from any of the sources just mentioned) creates a Ca^{2+} signal or Ca^{2+} "spark" that can be recorded using special Ca^{2+}-imaging techniques (see Biotechnology

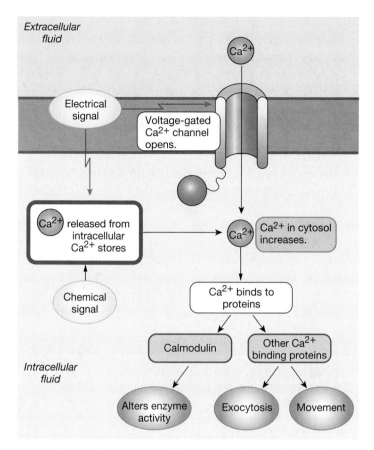

● **Figure 6-15** *Calcium as an intracellular messenger*
Calcium signals occur in the cytosol when Ca^{2+} enters the cell or is released from intracellular stores.

box, Measuring Calcium Signals). The calcium ions combine with cytoplasmic calcium-binding proteins to exert various effects. There are several types of calcium-dependent events in the cell:

1. Ca^{2+} binding to the protein **calmodulin**, found in all cells, alters enzyme activity or the open state of ion channels.
2. Calcium binding to other regulatory proteins alters movement of contractile or cytoskeletal proteins such as microtubules. For example, Ca^{2+} binding to the regulatory protein *troponin* initiates muscle contraction in a skeletal muscle cell.
3. Ca^{2+} binding to regulatory proteins triggers exocytosis of secretory vesicles [● p. 61]. This was illustrated in Chapter 5 in the description of pancreatic beta cell release of insulin.
4. Ca^{2+} binds directly to ion channels to alter their open state. An example of this target is the Ca^{2+}-activated K^+ channel of nerve cells.
5. Ca^{2+} entry into a fertilized egg initiates development of the embryo.

13. Extracellular calcium averages 2.5 mmol/L and free cytosolic Ca^{2+} is about 0.001 mmol/L. If a cell is going to move calcium from its cytosol to the extracellular fluid, will it use passive or active transport? Explain.

Measuring Calcium Signals

If you've ever run your hand through a tropical ocean at night and seen the glow of bioluminescent jellyfish, you've seen a calcium signal. Aequorin, a protein complex isolated from jellyfish, is one of the molecules that scientists use to monitor the presence of calcium during a cellular response. When aequorin combines with calcium, it releases light that can be measured by electronic detection systems. Since the first use of aequorin in 1967, researchers have been designing better and better indicators that allow them to follow calcium signals in cells. With the help of molecules called fura, Oregon green, BAPTA, and chameleons, we can now watch calcium diffuse through gap junctions and flow out of intracellular organelles.

Gases Are Ephemeral Signal Molecules

Not all signal molecules work through receptors. **Nitric oxide (NO)**, a soluble gas, is a novel short-acting paracrine/autocrine that acts close to where it is produced. Nitric oxide took years to identify because it is rapidly broken down, with a half-life of only 2 to 30 seconds. (*Half-life* is the time required for the signal to lose half of its activity.) In tissues, NO is synthesized by the action of the enzyme *nitric oxide synthase* (NOS) on the amino acid arginine:

$$Arginine + O_2 \xrightarrow{\text{nitric oxide synthase}} NO + citrulline \text{ (an amino acid)}$$

NO diffuses into target cells where it activates the cytosolic form of guanylyl cyclase and causes formation of the second messenger cGMP.

Nitric oxide in the brain acts as a neurotransmitter and neuromodulator. In blood vessels, NO is produced by endothelial cells lining the vessel. It then diffuses into adjacent smooth muscle cells, causing them to relax and dilate the blood vessel. In 1998 the Nobel prize for physiology and medicine was awarded jointly to Robert Furchgott, Louis Ignarro, and Ferid Murad for their work on NO as a signal molecule in the cardiovascular system.

Carbon monoxide (CO), a gas known mostly for its toxic effects, is also produced in minute amounts to be a signal molecule in certain cells. Like NO, it activates guanylyl cyclase and cGMP but it may also work independently to exert its effects. Carbon monoxide targets smooth muscle and neural tissue.

Some Lipids Are Important Paracrines

One of the interesting developments from sequencing the genome and using genes to find proteins has been the identification of *orphan receptors*, receptors that have no known ligand.

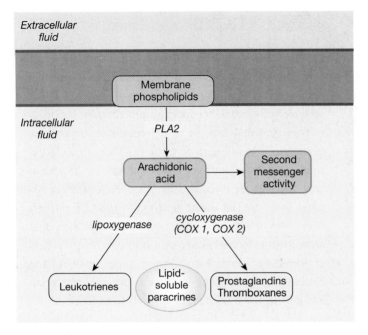

● **Figure 6-16** *The arachidonic acid cascade produces lipid messengers*

Scientists are trying to work backwards through signal pathways to find the ligands that bind to these orphan receptors. As a result of this type of research, investigators have recently come to realize the importance and universality of *eicosanoids,* lipid-derived paracrines that play important roles in many physiological processes.

Eicosanoid signal molecules are all derived from **arachidonic acid** (AA), a 20-carbon fatty acid. The synthesis process is a network called the *arachidonic acid cascade* (Fig. 6-16 ●). For simplicity, we will break the cascade into steps.

Arachidonic acid is produced from membrane phospholipids by the action of an enzyme, **phospholipase A$_2$ (PLA2)**. PLA2 activity is controlled by hormones and other signals. Arachidonic acid itself may act directly as a second messenger, altering ion channel activity and intracellular enzymes. It may also be converted into one of several classes of eicosanoid paracrines. These lipid-soluble molecules can diffuse out of the cell and combine with receptors on neighboring cells to exert their action.

There are two major groups of AA-derived paracrines to be aware of:

1. **Leukotrienes** are molecules produced by the action of the enzyme *lipoxygenase* on arachidonic acid. Leukotrienes are secreted by certain types of white blood cells [*leuko-*, white + *triene,* a molecule with three double bonds between carbon atoms]. They play a significant role in asthma, a lung condition where the smooth muscle of the airways constricts, making it difficult to breathe, and in the severe allergic reaction known as *anaphylaxis.* For this reason, pharmaceutical companies have been actively developing drugs to block leukotriene synthesis or action.

2. *Prostanoids* are molecules produced when the enzyme **cycloxygenase (COX)** acts on arachidonic acid. Prostanoids include **prostaglandins** and **thromboxanes**. These eicosanoids act on many tissues of the body, including smooth muscle in various organs, platelets (the cell fragments involved in blood clotting), kidney, and bone. In addition, prostaglandins are involved in sleep, inflammation, pain, and fever.

The nonsteroidal anti-inflammatory drugs (NSAIDs) such as aspirin and ibuprofen help prevent inflammation by inhibiting COX enzymes and decreasing prostaglandin synthesis. However, the NSAIDs are not specific and may have serious unwanted side effects, such as bleeding in the stomach. The discovery that there are two COX isozymes, COX1 and COX2, allowed the design of drugs that target a specific COX isozyme. By inhibiting only COX2, the enzyme that produces inflammatory prostaglandins, we can treat inflammation with significantly fewer side effects.

CONCEPT CHECK   Answers: p. 207

14. Based on what you have learned about signal molecules, where might a drug that blocks leukotriene action act? Where would a drug that blocks leukotriene synthesis act?

▶ **MODULATION OF SIGNAL PATHWAYS**

As you have just learned, signal pathways in the cell can be very complex. To complicate matters, different cells may respond differently to a single signal molecule. How can one molecule trigger response A in tissue 1 and response B in tissue 2? *For most signal molecules, the target cell response is determined by the receptor and its associated intracellular pathways, not by the ligand.* Because of the importance of signaling pathways, cells use receptors to maintain flexibility in their responses.

Receptors Exhibit Saturation, Specificity, and Competition

Receptors are proteins; therefore, receptor-ligand binding exhibits the characteristics of specificity, competition, and saturation, just like the enzymes discussed in Chapter 4 [● p. 97] and the transporters in Chapter 5 [● p. 138].

Specificity and Competition: Multiple Ligands for One Receptor   Receptors have binding sites for their ligands, just as enzymes and transporters do. As a result, different molecules with similar structures may be able to bind to the same receptor. A classic example of this principle involves two neurocrines that are responsible for the fight-or-flight response: the neurotransmitter *norepinephrine* and its cousin *epinephrine* (also called *adrenaline*), a neurohormone. Both molecules bind to a class of receptors called *adrenergic receptors.* (*Adrenergic* is the adjective relating to adrenaline.) The ability of adrenergic receptors

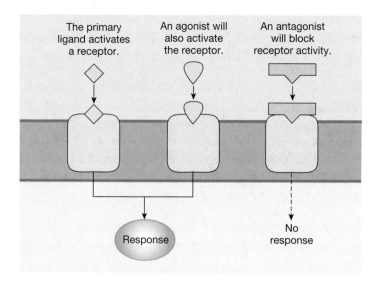

● **Figure 6-17** *Agonists and antagonists*

to bind these neurocrines but not others demonstrates *specificity* of the receptors.

Epinephrine and norepinephrine also exhibit *competition* for a single receptor type. Both neurocrines bind to subtypes of adrenergic receptors designated alpha (α) and beta (β). However, α receptors have a higher binding affinity for norepinephrine. The β receptor subtype designated β_2 has a higher affinity for epinephrine. And a different β adrenergic receptor, designated β_1, has equal affinity for both signal molecules, so they compete for its binding sites on equal footing.

Agonists and Antagonists When a ligand combines with a receptor, one of two events follows. The ligand either turns on the receptor and elicits a response, or the ligand occupies the binding site and prevents the receptor from responding (Fig. 6-17 ●). Ligands that turn on receptors are known as *agonists* and ligands that block receptor activity are called *antagonists*.

Agonists and antagonists may occur in nature. *Nicotine*, the chemical found in tobacco, mimics the activity of the neurotransmitter *acetylcholine*. Agonists can also be synthesized using what scientists learn from the study of receptor and ligand bind-

ing sites. The ability of agonists to mimic the activity of naturally occurring ligands has led to the development of many drugs.

As pharmacologists learn how particular ligands interact with their receptors and with the enzymes that break them down, they are able to design drugs that are longer acting and more resistant to enzymatic degradation. One example is the family of estrogens (female sex hormones) that are found in birth control pills. These drugs are agonists of naturally occurring estrogens but have chemical groups added to protect them from breakdown and extend their active life.

Antagonist ligands behave in the same fashion as competitive inhibitors of enzymes [● p. 95]. They bind to the receptor and block the normal ligand from binding without turning the receptor on themselves. As a result, the cell cannot respond. Antagonist drugs have proven useful for treating many conditions. For example, *tamoxifen*, an antagonist to the estrogen receptor, and *flutamide*, an androgen-receptor blocker, are used in the treatment of hormone-dependent cancers of the breast or prostate gland, respectively.

Multiple Receptors for One Ligand For many years physiologists were unable to explain their observation that a single signal molecule could have different effects in different tissues. For example, epinephrine, the neurohormone described above, dilates blood vessels in skeletal muscle but constricts blood vessels in the intestine. How can one chemical have opposite effects? The answer became clear when scientists discovered that many receptors come in variants known as **isoforms**. When a signal molecule binds to different isoforms of its receptor, the cellular response will depend on the receptor isoform.

For example, the α and β adrenergic receptors for epinephrine described earlier are examples of receptor isoforms. When epinephrine binds to β-receptors on intestinal blood vessel smooth muscle, blood vessels constrict (Fig. 6-18 ●). When epinephrine binds to β-receptors on skeletal muscle blood vessels, they dilate. In other words, the response of the blood vessel to epinephrine depends on the receptor isoform, not on the ligand that activates the receptor. Many drugs now are designed so that they are specific for one receptor isoform and not others (see Emerging Concepts box: Selective Estrogen Receptor Modulators).

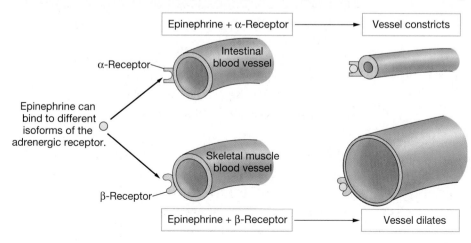

● **Figure 6-18** *Target response depends on the target receptor* In this example, epinephrine will either dilate or constrict blood vessels depending on the receptor found on the blood vessel.

Selective Estrogen Receptor Modulators

For many years scientists believed that estrogen, one of the female sex hormones, had only one receptor isoform and that all estrogen agonists had similar tissue-specific effects. But as pharmacologists developed the drug tamoxifen, they noticed that it was an estrogen agonist in some tissues but an estrogen antagonist in others. Tamoxifen belongs to a class of drugs known as *selective estrogen receptor modulators*, or SERMs. The action of SERMs is tissue specific. For example, SERMs are estrogen agonists in bone, where they are used to prevent osteoporosis. In other tissues, such as breast and uterus, SERMs are antagonists and are used for the treatment of hormone-dependent cancers. The observation that SERMs have tissue-specific effects is consistent with the 1996 discovery that there are two estrogen receptor isoforms, designated ERα and ERβ.

CONCEPT CHECK Answers: p. 207

15. What common property of receptors, enzymes, and transporters allows them all to exhibit saturation, specificity, and competition?

16. Insulin increases the number of glucose transporters on a skeletal muscle cell but not on the membrane of a liver cell. Think of at least two possible mechanisms that would allow this one hormone to have these two different effects.

Up- and Down-Regulation of Receptors Enables Cells to Modulate Cellular Response

Saturation of enzymes or transport proteins refers to the fact that enzymatic or transport activity can be "maxed out" with the presence of excess substrate because cells have limited numbers of protein molecules. The same phenomenon can occur with receptors. A cell's ability to respond to a chemical signal can be limited by the finite number of receptors for that signal.

A single cell contains between 500 and 100,000 receptors on the surface of its cell membrane, with additional receptors in the cytosol and nucleus. In any given cell, the number of receptors may change over time. Old receptors are withdrawn from the membrane by endocytosis and are broken down in lysosomes. New receptors are inserted into the membrane by exocytosis. Intracellular receptors are also made and broken down. This flexibility permits a cell to vary its responses to chemical signals depending on the extracellular conditions and the internal needs of the cell.

What happens if a signal molecule is present in the body in abnormally high concentrations for a sustained period of time? Initially the increased signal creates an enhanced response. As this continues, the target cells may attempt to bring their response back to normal by **down-regulation** of the receptors for the signal.

Down-regulation takes two forms, either a decrease in receptor number or a decrease in binding affinity. In both cases, the result is a lessened response of the target cell even though the concentration of the signal molecule remains high. Down-regulation is partially responsible for the development of *drug tolerance*, a condition in which the response to a given dose decreases despite continuous exposure to the drug. The development of tolerance to opiates such as morphine and codeine occurs when the receptors for these drugs down-regulate.

Up-regulation is the opposite of down-regulation. If the concentration of a ligand decreases, the target cell may insert more receptors into the cell membrane in an attempt to keep its response at a normal level. For example, if a nerve cell is damaged and unable to release normal amounts of neurotransmitter, its target cell will up-regulate its receptors. This up-regulation makes the target cell more sensitive to whatever neurotransmitters are present.

RUNNING PROBLEM

"My dad takes insulin shots for his diabetes," Marvin says. "What does insulin do?" The nurse practitioner replies that normally, insulin helps many cells take up and utilize glucose. But in all types of diabetes, fasting blood glucose concentrations are elevated because the cells are not taking up and using glucose normally. If people with type 1 diabetes are given shots of insulin, their blood glucose levels decline. If people with type 2 diabetes are given insulin, blood glucose levels may change very little.

Question 2: In which form of diabetes are the insulin receptors more likely to be up-regulated?

171 176   190 193 197 201

CONCEPT CHECK Answers: p. 207

17. To down-regulate a receptor's binding affinity, a cell might (circle all that apply):
    (a) synthesize a new isoform of the receptor
    (b) withdraw receptors from the membrane
    (c) insert new receptors into the membrane
    (d) use a covalent modulator (Hint: 🔖 p. 96)

Cells Must Be Able to Terminate Signal Pathways

Signals turn on and off, so cells must be able to tell when a signal is over. This requires that signaling processes have built-in termination mechanisms. For example, to stop the response to a calcium signal, the cell removes Ca^{2+} from the cytosol by pumping it back into the endoplasmic reticulum or out into the extracellular fluid.

Receptor activity can be stopped in a variety of ways. The extracellular first messenger can be degraded by enzymes in the extracellular space. An example of this is the breakdown of the neurotransmitter acetylcholine. Other first messengers, particularly neurotransmitters, can be removed from the extracellular fluid by transporting them into neighboring cells. A widely used class of antidepressant drugs called *selective serotonin reuptake inhibitors*, or SSRIs, extends the active life of the neurotransmitter serotonin by slowing its removal from the extracellular fluid.

Once a receptor is bound to its ligand, activity can also be terminated by endocytosis of the entire receptor-ligand complex. This process was illustrated in Figure 5-28 [● p. 146]. After the vesicle is in the cell, the ligand is removed and the receptors can be returned to the membrane by exocytosis.

Many Diseases and Drugs Target the Proteins of Signal Transduction

As we learn more about cell signaling, scientists are realizing how many diseases are linked to problems with signaling pathways. Disease processes can be caused by alterations in receptors or by problems with G proteins or second messenger pathways (see Table 6-3 for some examples). A single change in the amino acid sequence of a receptor protein can alter the shape of the receptor's binding site and destroy or modify its activity.

Pharmacologists are using information about signaling mechanisms to design drugs to treat disease. Some of the alphabet soup of drugs in widespread use include ARBs (angiotensin receptor blockers), beta-adrenergic receptor blockers, and calcium-channel blockers for treating high blood pressure; SERMs (selective estrogen receptor modulators) for treating estrogen-dependent cancers; H_1 (histamine type 1) receptor antagonists for

CLINICAL FOCUS

G Proteins and Disease

A variety of diseases have now been traced to disturbances in cell-signaling mechanisms (Table 6-3). Abnormalities in G protein function have been linked to certain oncogenes [*onkos*, mass]. An **oncogene** is an abnormal gene that plays a role in the growth and development of cancerous cells. One gene known as *ras* codes for G proteins linked to the phospholipase C pathway. When an abnormal *ras* gene becomes active in tumor cells, the resulting proteins affect cell division, leading to abnormal growth and division of the cancerous cells. Many researchers are studying oncogenes and the signal paths that they control, hoping that a clearer understanding of these mechanisms will lead them to better treatments for various forms of cancer.

decreasing acid secretion in the stomach; and COX2 (cycloxygenase 2) inhibitors that prevent the production of inflammatory prostaglandins. We will encounter many of these drugs again when we study the systems where they are effective.

▶ HOMEOSTASIS

In Chapter 1, you learned that homeostasis is the ability of the body to maintain a relatively stable internal environment [● p. 7]. In this section we examine homeostatic control mechanisms and compare the role of the nervous and endocrine systems in maintaining homeostasis.

Table 6-3: Diseases or Conditions Linked to Abnormal Signaling Mechanisms

Genetically inherited abnormal receptors

RECEPTOR	PHYSIOLOGICAL ALTERATION	DISEASE OR CONDITION THAT RESULTS
Vasopressin receptor (X-linked defect)	Shortens half-life of the receptor	Congenital diabetes insipidus
Calcium sensor in parathyroid gland	Fails to respond to increase in plasma Ca^{2+}	Familial hypercalcemia
Rhodopsin receptor in retina of eye	Improper protein folding	Retinitis pigmentosa

Toxins affecting signal pathways

TOXIN	PHYSIOLOGICAL EFFECT	CONDITION THAT RESULTS
Bordetella pertussis	Blocks inhibition of adenylate cyclase (i.e., keeps it active)	Whooping cough
Cholera toxin	Blocks enzyme activity of G proteins; cell keeps making cAMP	Ions secreted into lumen of intestine cause massive diarrhea

The Development of the Concept of Homeostasis

The idea of an internal environment was developed by the French physician Claude Bernard in the mid-1800s. During his studies of experimental medicine, he noted the stability of various physiological **parameters*** such as body temperature, heart rate, and blood pressure. As chairman of physiology at the University of Paris, he wrote of "la fixité du milieu intérieur" (the constancy of the internal environment). This was an idea that applied to many of the experimental observations of his day, and it became the subject of discussion among physiologists and physicians.

In 1929, an American physiologist named Walter B. Cannon created the name *homeostasis* to describe the regulation of this internal environment. In his essay,[†] Cannon explained that he selected the prefix *homeo-* meaning *like* or *similar* rather than the prefix *homo-* meaning *same* because the internal environment is maintained within a range of values rather than at an exact fixed value. He also pointed out that the suffix *-stasis* in this instance means *a condition*, not a state that is static and unchanging. Thus, Cannon's homeostasis is a state of maintaining "a similar condition," also described as "a relatively constant internal environment."

Some contemporary physiologists feel a literal interpretation of *stasis* [a state of standing] makes the word *homeostasis* imply a static, unchanging state. They argue that we should use the word *homeodynamics* instead, to reflect the small but constant changes always taking place in our internal environment [*dynamikos*, force or power]. Whether the process is called homeostasis or homeodynamics, the important concept to remember is that the body monitors its internal state and takes action to correct disruptions that threaten its normal function.

Cannon's Postulates Describe Regulated Variables and Physiological Control Systems

Cannon proposed a list of parameters that are under homeostatic control based on observations made by numerous physiologists and physicians during the nineteenth and early twentieth centuries. We now know that his list of regulated variables was both accurate and complete. Cannon divided his parameters into what he called environmental factors that affect cells (osmolarity, temperature, and pH) and "materials for cell needs"—nutrients, water, sodium, calcium, other inorganic ions, oxygen, and "internal secretions having general and continuous effects." Cannon's "internal secretions" are the hormones and other chemicals that our cells use to communicate with each other.

Cannon also postulated a number of properties of homeostatic control systems that were validated in the succeeding years. You will encounter these properties repeatedly as you study the various organ systems of the body. Cannon's four postulates are:

1. **The role of the nervous system in preserving the "fitness" of the internal environment.** Fitness in this instance means conditions that are compatible with normal function. The nervous system coordinates and integrates blood volume, osmolarity, blood pressure, and body temperature, among other parameters.

2. **The concept of *tonic level* of activity [*tonos*, tone].** To quote Cannon, "an agent may exist which has . . . a moderate activity which can be varied up and down." This is like the volume control on a radio, which enables you to make the sound level louder or softer by turning a single knob. A physiological example of a tonically active control system is the nervous regulation of diameter in certain blood vessels, where increased input from the nervous system decreases diameter, while decreased input from the nervous system causes the diameter to increase (Fig. 6-19 ●). Tonic control is one of the more difficult concepts in physiology because we have a tendency to think of things as off or on rather than as more or less.

3. **The concept of antagonistic controls.** Cannon wrote, "When a factor is known which can shift a homeostatic state in one direction, it is reasonable to look for . . . a factor or factors having an opposing effect." Systems that are not under tonic control are usually under antagonistic control, either by hormones or the nervous system. For example, insulin and glucagon are antagonistic hormones. Insulin decreases the concentration of glucose in the blood, while glucagon increases it. In pathways controlled by the nervous system, the sympathetic and parasympathetic divisions often have opposing effects. For example, chemical signals from a sympathetic neuron increase heart rate, while chemical signals from a parasympathetic neuron decrease it (Fig. 6-20 ●).

4. **The concept that chemical signals can have different effects in different tissues of the body.** Cannon postulated that "Homeostatic agents, antagonistic in one region of the body, may be cooperative in another region." This theory was subsequently shown to be true. However, it was not until we learned about cell receptors that the basis for the seemingly contradictory actions of some hormones or nerves became clear. As you learned in the first part of this chapter, a single chemical signal can have different effects depending on the receptor at the target cell. For example, epinephrine constricts or dilates blood vessels, depending on whether the vessel has alpha or beta adrenergic receptors (see Fig. 6-18).

CONCEPT CHECK   Answers: p. 207

18. Name three environmental factors that affect cells.

19. What are Canon's "internal secretions"?

20. What is the difference between tonic control and antagonistic control?

21. How can one chemical signal have opposite effects in two different tissues?

---

*A parameter [*para-*, beside + *meter*, measure] is one of the variables in a system.
[†]Walter B. Cannon, "Organization for Physiological Homeostasis," *Physiological Reviews* 9: 399–443, 1929.

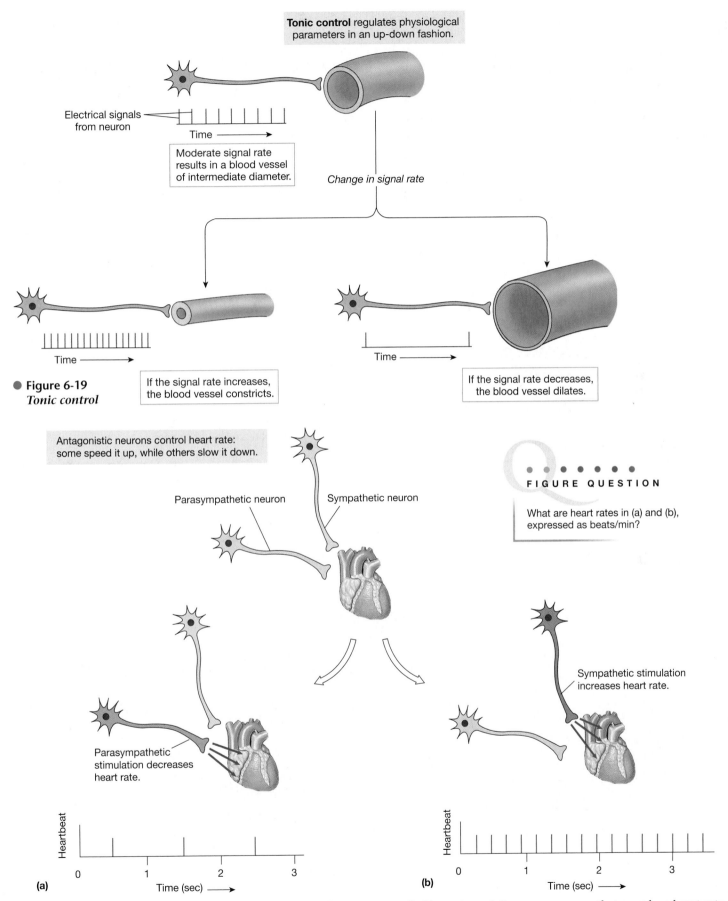

Tonic control regulates physiological parameters in an up-down fashion.

Electrical signals from neuron

Time

Moderate signal rate results in a blood vessel of intermediate diameter.

Change in signal rate

Time

● **Figure 6-19**
Tonic control

If the signal rate increases, the blood vessel constricts.

If the signal rate decreases, the blood vessel dilates.

Antagonistic neurons control heart rate: some speed it up, while others slow it down.

Parasympathetic neuron

Sympathetic neuron

FIGURE QUESTION

What are heart rates in (a) and (b), expressed as beats/min?

Sympathetic stimulation increases heart rate.

Parasympathetic stimulation decreases heart rate.

Heartbeat

Time (sec)

(a)

Heartbeat

Time (sec)

(b)

● **Figure 6-20** *Antagonistic control of heart rate* The heart is controlled by antagonistic neurons: some that speed up heart rate, some that slow down heart rate.

"Why is an elevated blood glucose concentration bad?" Marvin asks. "The elevated blood glucose itself is not bad," says the nurse practitioner. "But when it is high after an overnight fast, it suggests that there is something wrong with the way your body is regulating its glucose metabolism." When a normal person absorbs a meal with carbohydrates, blood glucose levels increase and stimulate insulin release. When cells have taken up the glucose from the meal and blood glucose levels fall, secretion of another pancreatic hormone, glucagon, increases. Glucagon raises blood glucose and helps keep the level within the homeostatic range.

Question 3: The homeostatic regulation of blood glucose levels by the hormones insulin and glucagon is an example of which of Cannon's postulates?

171 176 186 193 197 201

▶ CONTROL PATHWAYS: RESPONSE AND FEEDBACK LOOPS

Homeostasis is a continuous process that involves monitoring multiple parameters, then coordinating appropriate responses to minimize any disturbance. Homeostatic responses may take place in small, localized regions of the body or they may be bodywide, or *systemic*, responses. In either case, the process always has three components: a stimulus or change in a regulated variable, a cell or tissue that evaluates the stimulus and initiates a response, and the cells or tissues that carry out the response.

Homeostasis May Be Maintained by Local or Long-Distance Pathways

The simplest control takes place strictly at the tissue or cell involved. In **local control**, a relatively isolated change occurs in the vicinity of a cell or tissue and evokes a paracrine or autocrine response (Fig. 6-21 ●). More complicated **reflex control pathways** respond to changes that are more widespread or systemic in nature. In a reflex control pathway, an integrating center located away from the affected cell or tissue receives information, evaluates it, and decides whether to send a chemical or electrical signal to initiate a response.

Long-distance reflex pathways are traditionally considered to involve two control systems, the nervous system and the endocrine system. However, cytokines [● p. 174] are now considered the third control system. These proteins, secreted by all nucleated cells, can work by either local or reflex pathways depending on the circumstances. Acting as paracrines and autocrines, cytokines control the white blood cells that defend the body from outside invaders. During stress and systemic inflammatory responses, cytokines work together with the nervous and endocrine systems to integrate information from all over the body into coordinated reflex responses.

Local Control   Paracrines and autocrines are responsible for the simplest control systems. In local control a cell or tissue

● **Figure 6-21**   *Comparison of local and reflex control*

senses a change in its immediate vicinity and responds. The response is restricted to the region where the change took place, hence the term *local control*.

One example of local control can be observed when oxygen concentration in a tissue decreases. The cells lining the small blood vessels bringing blood to that area sense the fall in oxygen concentration and respond by secreting a paracrine. The paracrine relaxes muscles in the blood vessel wall, dilating the blood vessels and bringing more blood and oxygen to the area. Paracrines that are probably involved in this response include carbon dioxide and metabolic products such as lactic acid.

Reflex Control In a reflex pathway, control of the reaction lies outside the organ that carries out the response. We will use the term **reflex** to mean any long-distance pathway that uses the nervous system, endocrine system, or both to receive input about a change, integrate the information, and react appropriately. Reflex pathways can be broken down into two parts: a response loop and a feedback loop (Fig. 6-22 ●). The response loop begins with a stimulus and ends with the response of the target cell. We will discuss response loops first, and then consider how they interact with feedback loops.

A **response loop** has three primary components: an *input signal*, *integration of the signal*, and an *output signal*. These three components can be broken down into the following sequence of

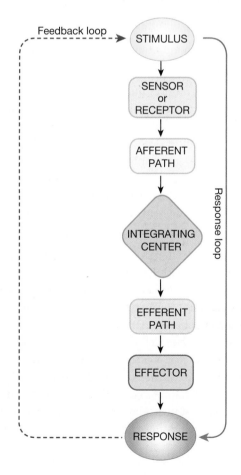

● **Figure 6-22** *Steps of a reflex*

seven steps to form a pattern that is found with slight variations in all reflexes:

$$\text{stimulus} \longrightarrow \text{receptor} \longrightarrow \text{afferent pathway} \longrightarrow \text{integrating}$$
$$\text{center} \longrightarrow \text{efferent pathway} \longrightarrow \text{effector} \longrightarrow \text{response}$$

The input signal of a homeostatic reflex pathway consists of a stimulus, its receptor, and an afferent or incoming pathway. (1) A **stimulus** is the disturbance or change that sets the pathway in motion. The stimulus may be a change in temperature, oxygen content, blood pressure, or any one of myriad parameters. The stimulus is sensed by (2) a *sensor*, or **receptor**, that is continually monitoring its environment. When alerted to a change, the receptor sends out (3) a signal. The signal, or **afferent** (incoming) **pathway**, links the receptor to (4) an integrating center. The **integrating center** is the control center that evaluates the incoming signal, compares it with the **setpoint**, or desired value, and decides on an appropriate response. The integrating center then initiates (5) the output signal or the **efferent** (outgoing) **pathway**. This is an electrical and/or chemical message that travels to (6) the **effector** [*effectus*, the carrying out of a task]. The effector or *target* is the cell or tissue that carries out (7) the appropriate **response** to bring the situation back to within normal limits.

Sensory receptors The first step in a biological response loop is activation of a sensory receptor by the stimulus. NOTICE! This is a new and different application of the word *receptor*. Like many terms in physiology, *receptor* can have different meanings (Fig. 6-23 ●). The receptors of a biological reflex are not protein receptors that bind to signal molecules, like those involved in signal transduction. Reflex, or *sensory*, receptors are specialized cells, parts of cells, or complex multicellular receptors such as the eye that respond to changes in the environment around them.

There are many sensory receptors in the body, each located anatomically where it is in the best position to monitor the parameter it detects. The eyes, ears, and nose are specialized receptors that sense light, sound, motion, and odors. Your skin is covered with less complex receptors that sense touch, temperature, vibration, and pain. Other receptors are internal: receptors in the joints of the skeleton that send information to the brain about body position or blood pressure and oxygen sensors in blood vessels that monitor conditions in the circulatory system. Biological reflex receptors are divided into *central receptors*, located in or closely linked to the brain, and *peripheral receptors* that reside elsewhere in the body.

All receptors have a **threshold**, a minimum stimulus that must be achieved to set the reflex response in motion. If a stimulus is below the threshold, no response loop will be initiated. You can demonstrate threshold easily by touching the back of your hand with a sharp, pointed object such as a pin. If you touch the point to your skin lightly enough, you can see the contact between the point and your skin even though you do not feel anything. In this case, the stimulus (pressure from the point of the

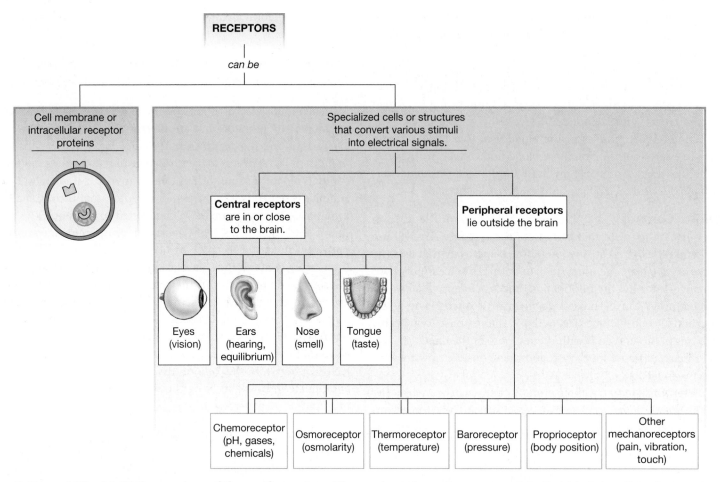

● **Figure 6-23** *Multiple meanings of the word* **receptor**   The word *receptor* may mean a protein that binds to a ligand or a sensory receptor, a specialized cell or structure for transduction of stimuli into electrical signals.

pin) is below threshold and the pressure receptors of the skin are not responding. As you press harder, the stimulus reaches threshold, and the receptors respond by sending a signal through the afferent pathway, causing you to feel the pin.

Afferent pathway   The afferent pathway in a reflex varies depending on the type of reflex. In a nervous reflex, such as the pin touch above, the afferent pathway is the electrical and chemical signal carried by a nerve cell. In endocrine reflexes, there is no afferent path because the stimulus comes directly into the endocrine cell, which serves as both sensor and integrating center.

Integrating center   The integrating center in a reflex is the location that receives information about the change and decides on an appropriate response. In endocrine reflexes, the integrating center is the endocrine cell. In nervous reflexes, the integrating center lies within the central nervous system, which is composed of the brain and the spinal cord.

If information is coming from a single stimulus, it is a relatively simple task for the integrating center to compare that information with the setpoint and initiate a response if necessary. Integrating centers really "earn their pay," however, on occasions when two or more conflicting signals come in from different re-

ceptors. The center must evaluate each signal on the basis of its strength and importance and must come up with an appropriate response that integrates information from all the receptors. This is similar to the kind of decision making you must do when in one evening your parents want to take you to dinner, your friends are having a party, there is a television program you want to watch, and you have a major physiology test in three days. It is up to you to rank those items in order of importance and decide how you will act upon them.

Efferent pathway   Efferent pathways are relatively simple. In the nervous system, the efferent path is always the electrical and chemical signal transmitted by an efferent neuron. Because electrical signals traveling through the nervous system are identical, the distinguishing characteristic of the signal is the anatomical route taken by the nerve cell through which the signal goes. For example, the vagus nerve carries a nervous signal to the heart, and the phrenic nerve carries one to the diaphragm. Because the nature of the electrical message is always the same and there are relatively few types of neurotransmitters, nervous system efferent pathways are named using the anatomical description of the nerve that carries the signal.

In the case of a hormonal signal, the anatomical routing is always the same because all hormones travel in the blood to get to their target. Hormonal efferent pathways are distinguished by the chemical nature of the signal and are therefore named for the hormone that carries the message. For example, the efferent path for a reflex integrated through the pancreas will be either insulin or glucagon depending on the stimulus and the appropriate response.

Effectors The effectors of reflex pathways are the cells or tissues that carry out the response. The targets of nervous reflexes are muscles, glands, and some adipose tissue. The targets of endocrine pathways are any cells that have the proper receptor for the hormone.

Responses There are two levels of response that can be given for any reflex. One is the very specific *cell* or *tissue response* that results from combination of the signal ligand with the receptor: opening a channel, initiating protein synthesis, or modifying enzyme activity. The more general *systemic response* describes what those specific events mean to the tissue or the organism as a whole. For example, when the hormone epinephrine combines with β₂-adrenergic receptors on the walls of certain blood vessels, the cellular response is relaxation of the smooth muscle. The systemic response to relaxation of the blood vessel wall is increased blood flow through the vessel.

RUNNING PROBLEM

Marvin is fascinated by the ability of the body to keep track of glucose. "How does the pancreas know which hormone to secrete?" Special cells in the pancreas called beta cells sense an increase in blood glucose concentrations after a meal, and they release insulin in response. Insulin then acts on many tissues of the body so that they take up and utilize glucose.

Question 4: In the insulin reflex that regulates blood glucose levels, what is the stimulus? the receptor? the integrating center? the efferent pathway? the effector(s)? the response(s)?

CONCEPT CHECK   Answers: p. 207

22. What is the difference between local control and reflex control?

23. Name, in order from the start, the seven basic steps in a simple reflex.

Response Loops Begin with a Stimulus and End with a Response

To illustrate response loops we will now apply them to nonbiological and biological examples. A simple nonbiological analogy to a homeostatic reflex pathway is an aquarium whose heater is set to maintain water temperature at 30° C in a room whose temperature is 25° C (Fig. 6-24 ●). The temperature at which the heater is set is the *setpoint* for the parameter.

Assume that initially the aquarium water is at room temperature, 25° C. When you turn the control box on, you set the response loop in motion. The thermometer (receptor) in the water senses a temperature of 25° C. It sends this information via a wire (afferent path) to the control box (integrating center). The control box evaluates the incoming temperature signal, compares it with the setpoint for the system (30° C), and "decides" that a response is needed to bring the water temperature up to the setpoint. The control box sends a signal via another wire (efferent path) to the heater (effector), which turns on and starts heating the water (the response). This sequence from stimulus to response is the response loop.

The simple example given above demonstrates a parameter (temperature) that is under *tonic control* (see p. 188) by a single control system (the heater). We can also create a nonbiological system that obeys Cannon's postulate of *antagonistic control*. For example, think of a house that has both heating and air-conditioning. The owner would like the house to remain at 70° F (about 21° C). During the autumn when the mornings are chilly, the heater turns on to warm the house. Then, as the day warms up, the heater is no longer needed. As the sun heats the house and the temperature reaches 80° F, the air-conditioner turns on to cool the house back to 70° F. The heater and air-conditioner have antagonistic control over the house temperature. A similar physiological example would be the hormones insulin and glucagon, which exert antagonistic control over glucose metabolism.

CONCEPT CHECK   Answers: p. 207

24. Can you think of a situation where having only a single control system in the aquarium (the heater) would mean the water temperature could not be maintained in the desired range?

Setpoints Can Be Varied

In physiological systems, the **setpoint**, or normal value for any given parameter, can vary from person to person, or even for the same individual over a period of time. Factors that influence an individual's setpoint for a given parameter include inheritance as well as the conditions to which the person has become accustomed. The adaptation of physiological processes to a given set of environmental conditions is known as **acclimatization** if it occurs naturally or **acclimation** if the process is induced artificially in a laboratory setting. Each winter, northerners go south in February, hoping to escape the bitter subzero temperatures and snows of the northern climate. As they walk around in 40° F weather in short-sleeve shirts, the southerners think they are crazy: the weather is cold! The locals are bundled up in coats and gloves. The difference in behavior is due to different temperature acclimatization, a difference in the setpoint for body temperature regulation that is a result of prior conditioning.

① Water temperature decreases to 25° C

② Thermometer

⑦ Water temperature increases to 30° C

Wire ③

④ Control box

⑥ ⑤

Heater   Wire to heater

● **Figure 6-24** *Response loops* The control box of the aquarium is set to maintain a temperature of 30° ± 1° C in the water.

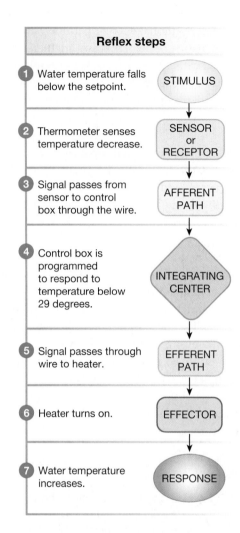

Reflex steps

① Water temperature falls below the setpoint.   STIMULUS

② Thermometer senses temperature decrease.   SENSOR or RECEPTOR

③ Signal passes from sensor to control box through the wire.   AFFERENT PATH

④ Control box is programmed to respond to temperature below 29 degrees.   INTEGRATING CENTER

⑤ Signal passes through wire to heater.   EFFERENT PATH

⑥ Heater turns on.   EFFECTOR

⑦ Water temperature increases.   RESPONSE

Physiological setpoints also vary within individuals in response to external cues such as the daily light-dark cycles and the seasons. These changing setpoints cause certain parameters to vary in predictable ways over a period of time, forming patterns of change known as biorhythms (see Biological Rhythms section later in this chapter).

Feedback Loops Modulate the Response Loop

The response loop is only part of a reflex. For example, in the aquarium described above, the sensor sends temperature information to the control box, which recognizes that the water is too cold. The control box responds by turning on the heater to warm the water. But once the response starts, what keeps the heater from sending the temperature up to 50° C?

The answer is a **feedback loop**, where the response feeds back to influence the input side of the pathway. In the aquarium example, turning on the heater increases the temperature of the water. The sensor continuously monitors the temperature and sends that information to the control box. When the temperature warms up to the maximum acceptable, the control box shuts off the heater, thus ending the reflex response.

Negative Feedback Loops Are Homeostatic For most reflexes, feedback loops are homeostatic, that is, designed to keep the system at or near a setpoint so that the parameter will be relatively stable. How well an integrating center succeeds in maintaining stability depends on the *sensitivity* of the system, the range of normal values. In the case of our aquarium, the control box is programmed to have a sensitivity of plus or minus 1° Celsius. If the water temperature drops from 30° C to 29.5° C, it is still within the acceptable range and no response is triggered. If the water temperature drops below 29° C (30° − 1°), the control box turns the heater on (Fig. 6-25 ●). As the water heats up, the control box constantly receives information about the water temperature from the sensor. When the water reaches 31° C (30° + 1°), the upper limit for the acceptable range, the feedback loop causes the control box to turn the heater off. The water then gradually cools off until the cycle starts all over again. The end result is a physiological variable that *oscillates* [*oscillare*, to swing] around the setpoint.

In biological systems, some receptors are more sensitive than others. For example, the sensors for osmolarity trigger reflexes to conserve water when blood concentration increases only 3% above normal, but the sensors for low oxygen in the blood will not respond until oxygen has decreased by 40%.

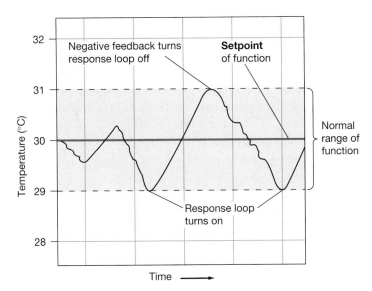

• **Figure 6-25** *Oscillation around the setpoint* Most functions that are controlled homeostatically have a setpoint, or normal value for that function. The reflexes that control the function will operate when the function gets outside a predetermined range.

A pathway where the response opposes or removes the signal is known as **negative feedback**. Negative feedback loops *stabilize* the physiological variable that is being regulated. In the aquarium example, the heater warms the water (the response) and removes the stimulus (low water temperature). With loss of the stimulus for the pathway, the response loop shuts off. All homeostatic reflexes are controlled by negative feedback so that the parameter being controlled will stay within a normal range. *Negative feedback loops can restore the normal state but cannot prevent the initial disturbance out of the normal range.*

Positive Feedback Loops Are Not Homeostatic A few reflexes are not homeostatic. In a **positive feedback loop**, the response *reinforces* the stimulus rather than decreasing it or re-

moving it. In positive feedback, the response destabilizes the variable, triggering a vicious cycle of ever-increasing response and sending the system temporarily out of control (Fig. 6-26 ●). Because the response enhances the loop in positive feedback, these reflexes require some intervention or event outside the loop to stop them.

One example of a positive feedback loop involves the hormonal control of uterine contractions during childbirth (Fig. 6-27 ●). When the baby is ready to be delivered, it drops lower in the uterus and begins to put pressure on the *cervix*, the opening of the uterus. Sensory signals from the cervix to the brain cause release of the hormone *oxytocin*, which causes the uterus to contract and push the baby's head even harder against the cervix, stretching it. The increased stretch causes more oxytocin release, which causes more contractions that push the baby harder against the cervix. This cycle continues until finally the baby is delivered, releasing the stretch on the cervix and stopping the positive feedback loop.

25. Will the system in Figure 6-24 operate with positive feedback or negative feedback? Draw in the appropriate feeback loop on the figure, using an arrowhead with a plus sign for positive feedback or a double parallel "stop" line for negative feedback, as illustrated in Figure 6-26.

Feedforward Control Allows the Body to Anticipate Change and Maintain Stability

Negative feedback loops can stabilize a function and maintain it within a normal range but are unable to prevent the change that triggered the reflex in the first place. A few reflexes have evolved that allow the body to predict that a change is about to occur and

(a) Negative feedback: the response counteracts the stimulus shutting off the response loop.

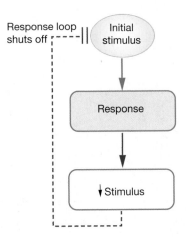

(b) Positive feedback: the response reinforces the stimulus sending the parameter farther from the setpoint.

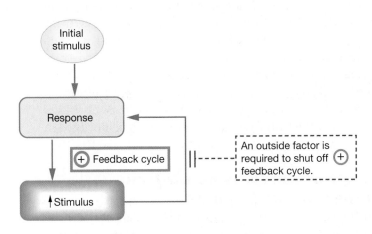

• **Figure 6-26** *Negative and positive feedback*

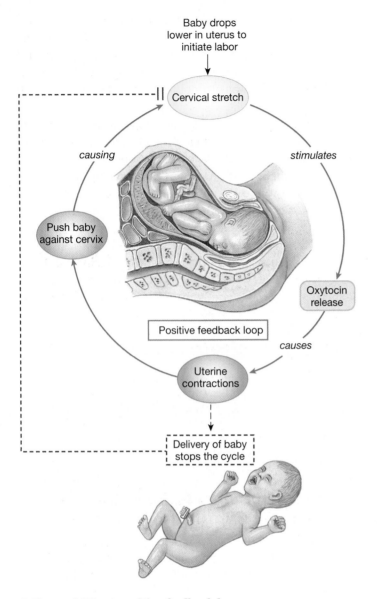

Baby drops lower in uterus to initiate labor

Cervical stretch

causing

stimulates

Push baby against cervix

Oxytocin release

Positive feedback loop

causes

Uterine contractions

Delivery of baby stops the cycle

● **Figure 6-27** *A positive feedback loop*

start the response loop in anticipation of the change. These anticipatory responses are called **feedforward control**.

An easily understood physiological example of feedforward control is the reflex of salivation. The sight, smell, or even the thought of food is enough to start our mouths watering. The saliva is present in expectation of the food that will soon be eaten. This reflex extends even further, because the same stimuli can start the secretion of hydrochloric acid as the stomach anticipates food on the way. One of the most complex feedforward reflexes appears to be the body's response to exercise, to be discussed in Chapter 25.

Biological Rhythms Result from Changes in the Setpoint

In many reflex pathways, the triggers or stimuli are obviously related to the function of the reflex. In the aquarium example, a change in temperature is the trigger to ensure that temperature is

maintained within the desired range. But this is not true of all reflexes. Many hormones, for example, are secreted continuously, with levels that rise and fall throughout the day. Most examples of these apparently spontaneous reflexes occur in a predictable manner and are often timed to coincide with a predictable environmental change such as light-dark cycles or the seasons.

All animals exhibit some form of daily biological rhythm, called a **circadian rhythm** [*circa*, about + *dies*, day]. Humans have circadian rhythms for many body functions, including blood pressure, body temperature, and metabolic processes. Body temperature peaks in the late afternoon and declines dramatically in the wee hours of the morning (Fig. 6-28b ●). Have you ever been studying late at night and noticed that you feel cold? This is not because of a drop in environmental temperature but because your thermoregulatory reflex has turned down your internal thermostat.

Many hormones in humans are secreted so that their concentration in the blood fluctuates predictably through a 24-hour cycle as their setpoints change. Cortisol, growth hormone, and the sex hormones are among the most noted examples. If an abnormality in hormone secretion is suspected, it is important to know the time of day that the test for hormone level is made. A level that is normal in a morning sample may be abnormally high or low in an afternoon sample (Fig. 6-28b). One method used to avoid this error involves collecting all urine for a 24-hour period and getting an average value for the hormone metabolites in the sample.

(a)

(b)

● **Figure 6-28** *Circadian rhythms*

What is the adaptive significance of functions that vary with a circadian rhythm? Our best answer is that biological rhythms create an anticipatory response to a predictable environmental variable. There are seasonal rhythms of reproduction in many mammalian and nonmammalian vertebrates and invertebrates, timed so that the offspring have food and other favorable conditions to maximize survival. Circadian rhythms cued by the light-dark cycle may correspond to our rest-activity cycles. These rhythms allow our bodies to anticipate behavior and coordinate body processes accordingly. You may hear someone who is accustomed to eating dinner at 6 P.M. say that he cannot digest his food if he waits until 10 P.M. to eat because his digestive system has "shut down" in anticipation of going to bed.

One of the interesting correlations between circadian rhythms and behavior involves body temperature. Researchers found that self-described "morning people" have temperature rhythms that begin to climb before they awake in the morning, so that they get out of bed prepared to face the world. On the other hand, "night people" may be forced by school and work schedules to get out of bed while their body temperatures are still at their lowest point, before their bodies are prepared for activity. These same night people are still going strong and working productively into the wee hours of the morning when the morning people's body temperature is dropping and they are ready for bed.

RUNNING PROBLEM

"OK, just one more question," says Marvin. "You said that people with diabetes have high blood glucose levels. If glucose is so high, why can't it just leak into the cells?"

Question 5: Why can't glucose always diffuse into cells when the blood glucose concentration is higher than the intracellular glucose concentration?

Question 6: What do you think happens to the rate of insulin secretion when blood glucose levels fall? What kind of feedback pathway is taking place?

Circadian rhythms arise from special groups of cells in the brain and are reinforced by information about the light-dark cycle that comes in through the eyes. Research in simpler animals such as flies is beginning to explain the molecular basis for biological rhythms. We will discuss the cellular and molecular basis for circadian rhythms further in Chapter 9.

Now that you have been introduced to response loops and feedback loops, we turn to an analysis of the different control systems.

Control Systems Vary in Their Speed and Specificity

Biological reflexes are mediated by the nervous system, the endocrine system, or a combination of the two (Fig. 6-29 ●). A re-

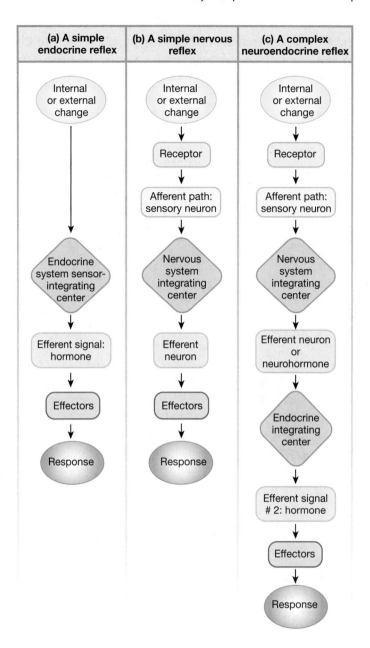

● **Figure 6-29** *Control pathways*

flex mediated solely by the nervous system or solely by the endocrine system is relatively simple, but combination reflexes can be quite complex. In the most complex reflexes, signals pass through three different integrating centers before finally reaching the target tissue. With so much overlap between reflexes controlled by the nervous and endocrine systems, it makes sense to consider them a continuum rather than two discrete systems.

Why does the body need different types of control systems? To answer that question, let us compare endocrine control with nervous control and see what the differences are. Five major differences are summarized in Table 6-4 and discussed below.

Specificity Nervous control is very specific because each nerve cell has a specific target cell or cells to which it sends its

Table 6-4: Comparison of Nervous and Endocrine Control

PROPERTY	NERVOUS REFLEX	ENDOCRINE REFLEX
Specificity	Each neuron terminates on a single target cell or on a limited number of adjacent target cells.	Most cells of the body are exposed to a hormone. The response depends on which cells have receptors for the hormone.
Nature of the signal	Electrical signal through neuron, then chemical neurotransmitters pass the signal from cell to cell. In a few cases, cell-to-cell communication takes place through gap junctions.	Chemical signals that are secreted in the blood for distribution throughout the body.
Speed	Very rapid.	Distribution of the signal and onset of action are much slower than in nervous responses.
Duration of action	Usually very rapid. Slower responses are mediated by neuromodulators.	Duration of action is usually much longer.
Coding for stimulus intensity	Each signal is identical in strength. Stimulus intensity reflected by increased frequency of signaling.	Stimulus intensity reflected by amount of hormone secreted.

message. Anatomically, we can isolate a neuron and trace it from its origin to where it terminates on its target cell(s). Endocrine control is more general because the chemical messenger is released into the blood and can reach virtually every cell in the body. As you learned in the first half of this chapter, the body's response to a specific hormone will depend on which cells have receptors for that hormone. Multiple tissues in the body can respond to a hormone simultaneously.

Nature of the Signal   The nervous system uses both electrical and chemical signals to send information throughout the body. Electrical signals travel long distances through nerve cells while chemical signals, the neurotransmitters, diffuse across the small gap between nerve cells or between nerve cell and target (Fig. 6-30●, ①). In a limited number of instances, electrical signals pass directly from cell to cell through gap junctions.

The endocrine system uses only chemical signals: hormones secreted by endocrine glands or cells into the blood (Fig. 6-30, ⑥). Neurohormones represent a hybrid of the two systems. A nerve cell creates an electrical signal, but the chemical released goes into the blood for general distribution (Fig. 6-30, ②).

CONCEPT CHECK   Answers: p. 207

26. In the simple nervous reflex shown in Figure 6-29, which boxes represent (a) the brain and spinal cord, and (b) the central and peripheral sense organs? In the figure, add a dashed line connecting boxes to show how a negative feedback loop would shut off the reflex.

Speed   Nervous reflexes are much faster than endocrine reflexes. The electrical signals of the nervous system cover great distances very rapidly, with speeds of up to 120 m/sec. Neurotransmitters also create very rapid responses, on the order of milliseconds.

Hormones are much slower than nervous reflexes. Their distribution through the circulatory system and diffusion from the capillary to their receptor takes considerably longer than signals through nerve cells. In addition, hormones have a slower onset of action. In target tissues, the response may take minutes to hours before it can be measured.

Why do we need the speedy reflexes of the nervous system? Consider this example. A mouse ventures out of his hole and sees a cat ready to pounce on him and eat him. A signal must go from the eyes and brain of the mouse down to his feet, telling him to run back into the hole. If his brain and feet were only 5 micrometers (5 μm = 1/200 millimeter) apart, it would take a chemical signal 20 milliseconds (msec) to diffuse across the space. The mouse could escape. If the cells were 50 μm (1/20 millimeter) apart, diffusion would take 2 seconds: the mouse might get caught. But because the head and tail of a mouse are centimeters apart, at this rate of diffusion, it would take a chemical signal *three weeks* to diffuse from the head to the feet of the mouse. Poor mouse! Even if the distribution of the chemical were speeded by help from the circulatory system, the chemical message would still take 10 seconds to get to the feet, and the mouse would become cat food. The moral of this tale is that reflexes requiring a speedy response are mediated by the nervous system because it is so much more rapid.

Duration of Action   Nervous control is of shorter duration than endocrine control. The neurotransmitter released by a nerve cell combines with a receptor on the target cell and initiates a response. But the response is usually very brief because the neurotransmitter is rapidly removed from the vicinity of the receptor by various mechanisms. To get a sustained response, multiple repeating signals must be sent through the nerve cell.

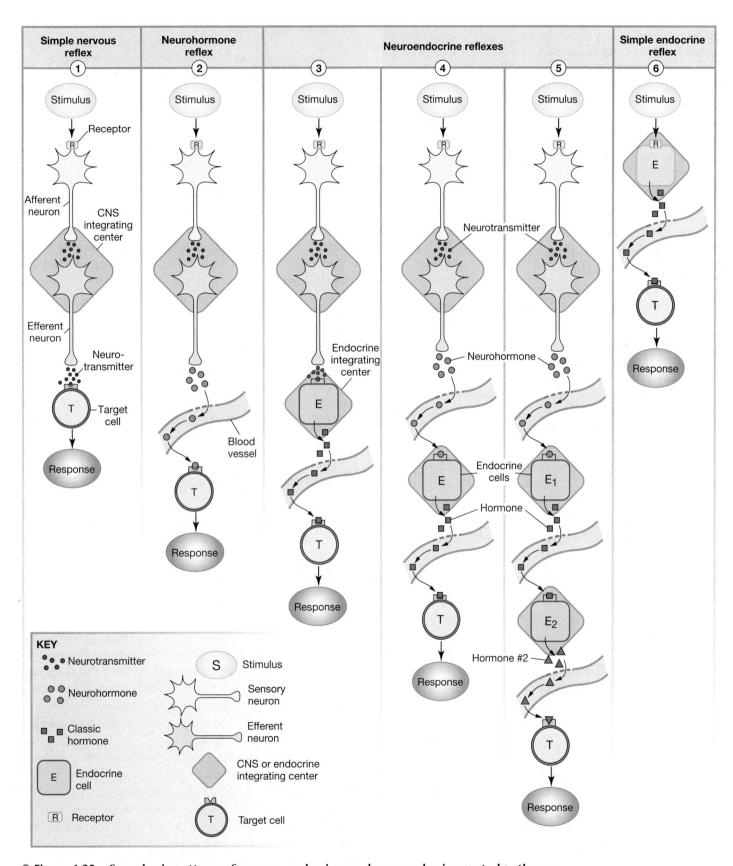

● **Figure 6-30** *Some basic patterns of nervous, endocrine, and neuroendocrine control pathways*

Endocrine reflexes are slower to start but they are of longer duration. This means that most of the ongoing, long-term functions of the body such as metabolism and reproduction fall under the control of the endocrine system.

Coding for Stimulus Intensity   As a stimulus increases in intensity, control systems must have a mechanism for conveying this information to the integrating center. The signal from any one neuron is constant in magnitude and therefore cannot reflect stimulus intensity. Instead, the frequency of signaling through the afferent neuron increases. In the endocrine system, stimulus intensity is reflected by the amount of hormone that is released: the stronger the stimulus, the more hormone.

Complex Reflexes Have Several Integrating Centers

Figure 6-30 summarizes some different variations of endocrine, neuroendocrine, nervous, and mixed reflexes. The simplest reflexes are the simple nervous reflex, the simple endocrine reflex, and the neurohormone reflex.

In a simple endocrine reflex (Fig. 6-29a; Fig. 6-30 ⑥), the endocrine cell acts both as sensor and integrating center and there is no afferent pathway. These endocrine reflexes usually do not have a distinct sensory receptor in the same way that neural and neurohormone reflexes do. Instead, the endocrine cell itself acts as the sensor, monitoring the stimulus parameter directly. The efferent pathway is the hormone, and the target is any cell with the appropriate receptor.

An example of a simple endocrine reflex is the secretion of the hormone insulin in response to changes in blood glucose level. The endocrine cells that secrete insulin monitor blood glucose concentrations by using ATP production within the cell as an indicator, as Chapter 5 described [❀ Fig. 5-42, p. 163]. When blood glucose increases, intracellular ATP production exceeds the threshold level and the endocrine cells respond by secreting insulin into the blood. Any target cell in the body with insulin receptors will respond to the hormone and initiate processes that take glucose out of the blood. The removal of the stimulus acts in a negative feedback manner, and the response loop shuts off when blood glucose levels fall below a certain concentration.

In a simple nervous reflex, all the steps of a reflex are present, from receptor to target (Fig. 6-29b, Fig. 6-30 ①). The nervous reflex is represented in its simplest form by the knee jerk, or patellar tendon reflex (see Fig. 13-7, p. 437). A blow to the knee (the stimulus) activates a stretch receptor. An electrical and chemical signal travels through an afferent neuron to the spinal cord (the integrating center). If the blow is strong enough (exceeds threshold), a signal travels from the spinal cord through an efferent neuron to the muscles of the thigh (the target or effector). In response, the muscles contract, causing the lower leg to kick outward (the knee jerk).

27. On the simple nervous reflex of Figure 6-30, match these terms to the parts of the reflex: blow to knee, leg muscles, neuron to leg muscles, sensory neuron, brain and spinal cord, sensory receptor, muscles contract.

The neurohormone reflex shown in Figure 6-30 ② is identical to the nervous reflex except that the chemical released by the neuron travels in the blood to its target, just like a hormone. A simple neurohormone reflex is the release of breast milk in response to a baby's suckling. The baby's mouth on the nipple stimulates sensory signals that travel through sensory neurons to the brain (integrating center). An electrical signal in the efferent neuron triggers the release of the neurohormone oxytocin from the brain into the circulation. Oxytocin is carried to the breast, where it causes contraction of smooth muscles in the breast (effectors) with resultant ejection of milk.

In complex pathways, there may be more than one integrating center. Figure 6-30 shows three different neuroendocrine pathways. The simplest pattern ③ combines a nervous reflex with an endocrine reflex. The target of the nervous reflex is an endocrine cell that releases a hormone. An example of this pattern can be found in the control of insulin release by the nervous system, where both excitatory and inhibitory neurons terminate on the endocrine cells of the pancreas. The endocrine cells must integrate information from three sources: the two antagonistic inputs from the nervous system and their direct detection of blood glucose levels.

28. Match the following terms with the appropriate parts of the simple neuroendocrine reflex in Figure 6-30 ③ (terms may be used more than once): food in stomach following a meal, brain and spinal cord, endocrine cells of pancreas, stretch receptors, efferent neuron to pancreas, insulin, adipose cell, blood, sensory neuron.

The middle neuroendocrine reflex (Fig. 6-30 ④) combines a neurohormone reflex with an endocrine reflex. The secretion of growth hormone is an example of this pathway. The most complex neuroendocrine pathway, shown on the right, starts with a neurohormone and continues with two sequential endocrine reflexes. This pattern is typical of several hormones released by the anterior pituitary, an endocrine gland located just below the brain (see Chapter 7 for details).

In describing complex reflex pathways, we will assume only one receptor and afferent path, those that start the reflex. For example, in the most complex neuroendocrine pathway (Fig. 6-30 ⑤) the brain is the first integrating center and the neurohormone is the first efferent path. The endocrine target (E_1) of the neurohormone becomes the second integrating center, and its hormone becomes the second efferent path. The second endocrine gland in the pathway (E_2) is the third integrating center, and its hormone is the third

Table 6-5: Comparison of Nervous, Neuroendocrine, and Endocrine Reflexes

	NERVOUS	NEUROENDOCRINE	ENDOCRINE
Receptor	Special and somatic sensory receptors	Special and somatic sensory receptors	Endocrine cell
Afferent path	Afferent sensory neuron	Afferent sensory neuron	None
Integrating center	Brain or spinal cord	Brain or spinal cord	Endocrine cell
Efferent path	Efferent neuron (electrical signal and neurotransmitter)	Efferent neuron (electrical signal and neurohormone)	Hormone
Effector(s)	Muscles and glands, some adipose tissue	Most cells of the body	Most cells of the body
Response	Contraction and secretion primarily. May have some metabolic effects.	Change in enzymatic reactions, membrane transport, or cell proteins	Change in enzymatic reactions, membrane transport, or cell proteins

efferent path. The target of the last signal in the sequence is the effector.

Table 6-5 compares the various steps in nervous, neuroendocrine, and endocrine reflexes. In the remainder of the text, we will use the general patterns shown in Figure 6-30 as a tool for classifying complex pathways. The next chapter looks in detail at some of the reflexes of the endocrine system and the roles they play in homeostasis.

RUNNING PROBLEM CONCLUSION

Diabetes Mellitus

Marvin underwent further tests and was diagnosed with type 2 diabetes. With careful attention to his diet and with a regular exercise program, he has managed to keep his blood glucose levels under control.

In this running problem, you learned about glucose homeostasis and how it is maintained by insulin and glucagon. The disease diabetes mellitus is an indication that glucose homeostasis has been disrupted. Check your understanding of this running problem by comparing your answers to the questions against those in the summary table.

	QUESTION	FACTS	INTEGRATION AND ANALYSIS
1	In which type of diabetes is the signal pathway for insulin more likely to be defective?	Insulin is a peptide hormone that uses membrane receptors linked to second messengers to transmit its signal to cells (● p. 174). People with type 1 diabetes lack insulin; people with type 2 have normal to elevated levels of insulin.	Normal or high insulin levels suggest that the problem is not with amount of insulin but with the action of the insulin at the cell. The problem in type 2 diabetes could be a defective signal transduction mechanism.
2	In which form of diabetes are the insulin receptors more likely to be up-regulated?	Up-regulation of receptors usually occurs if a signal molecule is present in unusually low concentrations (● p. 186). In type 1 diabetes, insulin is not secreted by the pancreas.	In type 1 diabetes, insulin levels are low. Therefore, type 1 is most likely to cause up-regulation of the insulin receptors.
3	The homeostatic regulation of blood glucose levels by the hormones insulin and glucagon is an example of which of Cannon's postulates?	Cannon's postulates describe the role of the nervous system in maintaining homeostasis, and the concepts of tonic activity, antagonistic control, and different effects of signals in different tissues.	Insulin decreases blood glucose levels while glucagon increases them. Therefore, the two hormones are an example of an antagonistic control.

(continues on next page)

	QUESTION	FACTS	INTEGRATION AND ANALYSIS
4	In the insulin reflex that regulates blood glucose levels, what is the stimulus? the receptor? the integrating center? the efferent pathway? the effector(s)? the response(s)?	See the steps of reflex pathways (p. 191).	*Stimulus:* increase in blood glucose levels. *Receptor:* beta cells of the pancreas that sense the change. *Integrating center:* beta cells. *Efferent pathway* (the signal): insulin. *Effectors:* any tissues of the body that respond to insulin. *Responses:* cellular uptake and use of glucose.
5	Why can't glucose always diffuse into cells when the blood glucose concentration is higher than the intracellular glucose concentration?	Glucose is lipophobic. Simple diffusion goes across the phospholipid bilayer. Facilitated diffusion uses protein carriers (p. 138).	Because glucose is lipophobic, it must cross the membrane by facilitated diffusion. If a cell lacks the necessary protein carriers, facilitated diffusion cannot take place.
6	What do you think happens to the rate of insulin secretion when blood glucose levels fall? What kind of feedback pathway is taking place?	The stimulus for insulin release is an increase in blood glucose levels. In negative feedback, the response offsets the stimulus. In positive feedback, the response enhances the stimulus.	An increase in blood glucose concentration stimulates insulin release; therefore, a decrease in blood glucose should decrease insulin release. In this example, the response (lower blood glucose) offsets the stimulus (increased blood glucose), so the pathway exhibits negative feedback.

(171) (176) (186) (190) (193) (197)

CHAPTER SUMMARY

This chapter introduced two of the major themes in physiology: *homeostasis and control systems*, and *communication*. The sensors, integrating centers, and effectors of physiological control systems were described in the context of reflex pathways, which vary from simple to complex. Functional control systems require efficient communication using a combination of chemical and electrical signals. Those signals that cannot cross cell membranes must use membrane receptors and signal transduction to transfer their information into the cell. The interaction of signal molecules with receptors illustrates another fundamental theme of physiology, *molecular interactions*.

Cell-to-Cell Communication

1. There are two basic physiological signals: chemical and electrical. Chemical signals are the basis for most communication within the body. (p. 172)

2. There are four methods of cell-to-cell communication: (1) direct cytoplasmic transfer through gap junctions, (2) contact-dependent signaling, (3) local chemical communication, and (4) long-distance communication. (pp. 172–173; Figs. 6-1 and 6-2)

3. **Gap junctions** are protein channels that connect two adjacent cells. When they are open, ions and small molecules diffuse directly from one cell to the next. (p. 172)

4. **Contact-dependent signals** require direct contact between surface molecules of two cells. (pp. 172–173)

5. Local communication is accomplished by **paracrines**, chemicals that act on cells in the immediate vicinity of the cell that secreted the paracrine. A chemical that acts on the cell that secreted it is called an **autocrine**. The action of paracrines and autocrines is limited by diffusion distance. (p. 173)

6. Long-distance communication is accomplished by **neurocrines** and electrical signals in the nervous system and by **hormones** in the endocrine system. Only cells that possess receptors for a hormone will be **target cells.** (pp. 172–174)

7. **Cytokines** are regulatory molecules that control cell development, differentiation, and the immune response. They function as both local and long-distance signals. (p. 174)

Signal Pathways

8. Chemical signals bind to **receptors** and change intracellular signal molecules that direct the response. (p. 174; Fig. 6-3)

9. Lipophilic signal molecules enter the cell and combine with cytoplasmic or nuclear receptors. Lipophobic signal molecules and some lipophilic molecules combine with membrane receptors. (p. 175; Fig. 6-4)

10. **Signal transduction** pathways use membrane proteins and intracellular second messenger molecules to translate signal information into an intracellular response. (p. 176)

11. Some signal transduction pathways activate **protein kinases**. Others activate **amplifier enzymes** that create **second messenger** molecules. (pp. 176–177)

12. Signal pathways create intracellular **cascades** that amplify the original signal. (p. 177; Figs. 6-7 and 6-9)

13. **Receptor-enzymes** activate protein kinases, such as **tyrosine kinase** (Fig. 6-10), or the amplifier enzyme **guanylyl cyclase** that produces the second messenger **cGMP**. (p. 178)

14. **G proteins** linked to amplifier enzymes are the most prevalent signal transduction system. G protein-coupled receptors also alter ion channels. (p. 179)

15. The **G protein-coupled adenylyl cyclase-cAMP-protein kinase A** pathway is the most common pathway for protein and peptide hormones. (p. 179; Fig. 6-11)

16. The amplifier enzyme **phospholipase C** creates two second messengers, **IP$_3$** and **diacylglycerol**. IP$_3$ causes Ca^{2+} release from intracellular stores. Diacylglycerol activates **protein kinase C**. (pp. 179–181; Fig. 6-12)

17. **Integrin** receptors link the extracellular matrix to the cytoskeleton. (p. 181; Fig. 6-5)

18. **Ligand-gated ion channels** open or close to create electrical signals. (p. 181; Fig. 6-13)

Novel Signal Molecules

19. Calcium is an important signal molecule that binds to **calmodulin** to alter enzyme activity. It also binds to other cell proteins to alter movement and initiate exocytosis. (p. 183; Fig. 6-15)

20. **Nitric oxide (NO)** and **carbon monoxide (CO)** are short-lived gaseous signal molecules. NO activates guanylyl cyclase directly. (p. 183)

21. The arachidonic acid cascade creates lipid paracrines such as **leukotrienes**, **prostaglandins**, and **thromboxanes**. (p. 184; Fig. 6-16)

Modulation of Signal Pathways

22. The response of a cell to a signal molecule is determined by the cell's receptor for the signal. (p. 184)

23. Receptor proteins exhibit specificity, competition, and saturation. (p. 184)

24. A receptor may have multiple ligands. **Agonists** mimic the action of a signal molecule. **Antagonists** block the signal pathway. (p. 185; Fig. 6-17)

25. Receptors come in related forms called **isoforms**. One ligand may have different effects when binding to different isoforms. (p. 185; Fig. 6-18)

26. Cells exposed to abnormally high concentrations of a signal for a sustained period of time attempt to bring their response back to normal by decreasing the number of receptors or decreasing the binding affinity of the receptors. This is known as **down-regulation** of the receptors. **Up-regulation** is the opposite of down-regulation. (p. 186)

27. Cells have mechanisms for terminating signal pathways, such as removing the signal molecule or breaking down the receptor-ligand complex. (p. 186)

28. Many diseases have been linked to defects with various aspects of signal pathways, such as missing or defective receptors. (p. 187)

Homeostasis

29. Walter Cannon first described four basic postulates of homeostasis. (1) The nervous system plays an important role in maintaining homeostasis. (2) Some parameters are under **tonic control**, which allows the parameter to be increased or decreased by a single signal (Fig. 6-19). (3) Other parameters are under **antagonistic control**, in which one hormone or neuron increases the parameter while another decreases it (Fig. 6-20). (4) Chemical signals can have different effects in different tissues of the body, depending on the type of receptor present at the target cell. (p. 188)

Control Pathways: Response and Feedback Loops

30. The simplest homeostatic control takes place at the tissue or cell level and is known as **local control**. (p. 190)

31. In **reflex** control pathways, the decision that a response is needed is made away from the cell or tissue. A chemical or electrical signal to the cell or tissue then initiates the response. Long-distance reflex pathways involve the nervous and endocrine systems and cytokines. (p. 191; Fig. 6-21)

32. Reflex pathways can be broken down into **response loops** and **feedback loops**. A response loop begins when a **stimulus** is sensed by a **receptor**. The receptor is linked by an **afferent pathway** to an **integrating center** that decides on an appropriate response. An **efferent pathway** travels from the integrating center to an **effector** that carries out the appropriate **response**. (pp. 191,194; Fig. 6-22)

33. In **negative feedback**, a homeostatic response is turned off when the response of the system opposes or removes the original stimulus. (p. 195; Fig. 6-26a)

34. In **positive feedback** loops, the response reinforces the stimulus rather than decreasing it or removing it. This destabilizes the system until some intervention or event outside the feedback loop stops the response. (p. 195; Fig. 6-26b)

35. **Feedforward control** allows the body to predict that a change is about to occur and start the response loop in anticipation of the change. (p. 196)

36. Apparently spontaneous reflexes that occur in a predictable manner are called **biological rhythms**. Those that coincide with light-dark cycles are called **circadian rhythms**. (p. 196)

37. Nervous control is faster and more specific than endocrine control but is usually of shorter duration. Endocrine control is less specific and slower to start but is longer lasting and is usually amplified. (pp. 199–200; Tbl. 6-4)

38. Many reflex pathways are combinations of nervous and endocrine control mechanisms. (p. 200; Fig. 6-30)

QUESTIONS

LEVEL ONE *Reviewing Facts and Terms*

1. What are the two routes for long-distance signal delivery in the body?

2. Which two body systems are charged with maintaining homeostasis by responding to changes in the environment?

3. What two types of physiological signals are used to send messages through the body? Of these two types, which is available to all cells?

4. The process of maintaining a relatively stable internal environment is called _____.

5. List at least three parameters maintained by homeostasis.

6. Distinguish between the "target" and the "receptor" in physiological systems.

7. In a signal pathway, the signal ligand, also called a _____ _____, binds to a _____, which activates and changes intracellular _____.

8. The three main amplifier enzymes are (a) _____, which forms cAMP; (b) _____, which forms cGMP; and (c) _____, which converts a phospholipid from the cell's membrane into two different second messenger molecules.

9. An enzyme known as protein kinase adds the functional group _____ to its substrate, by transferring it from a(n) _____ molecule.

10. Put these parts of a reflex in the correct order for a physiological response loop: efferent path, afferent path, effector, stimulus, response, integrating center.

11. Distinguish between central and peripheral receptors.

12. Match each term with its description:

(a) threshold
(b) setpoint
(c) effector
(d) oscillation
(e) sensitivity

1. the desired target value for a parameter
2. the distance allowed away from the setpoint before a response starts
3. the minimum stimulus to trigger a response
4. the organ or gland that performs the change
5. movement of a parameter within the desired range

13. Receptors for signal pathways may be found in the _____, _____, or _____ of the cell.

14. The name for the daily fluctuations of body functions, including blood pressure, temperature, and metabolic processes, is _____ _____. These cycles arise in special cells in the _____.

15. Down-regulation results in a(n) (increased or decreased?) sensitivity to a prolonged signal, resulting in a condition called _____, often seen when a drug is administered over a long term.

16. List the two ways that down-regulation may be achieved.

17. In a negative feedback loop, the effector moves the system in the (same/opposite) direction as the stimulus.

LEVEL TWO *Reviewing Concepts*

18. Explain the relationship among the terms in each of the following sets of terms. Give a physiological example or location if applicable.
    (a) gap junctions, connexins, syncytium, connexon
    (b) autocrine, paracrine, cytokine, neurocrine, hormone
    (c) agonist, antagonist
    (d) transduction, amplification, cascade

19. List and compare the four classes of membrane receptors for signal pathways. Give an example of each.

20. Who was Walter Cannon? Restate his four postulates in your own words.

21. Briefly define each term in question 10 and give an anatomical example where applicable.

22. Explain the differences among positive feedback, negative feedback, and feedforward mechanisms. Under what circumstances would each one be advantageous?

23. Compare and contrast the advantages and disadvantages of nervous versus endocrine control mechanisms.

24. Label each system below as positive or negative feedback.
    (a) glucagon secretion in response to declining blood glucose
    (b) increasing milk letdown and secretion in response to more suckling
    (c) urgency in emptying one's urinary bladder
    (d) sweating in response to rising body temperature

25. Go back to the preceding question and identify the effector organ for each example.

26. Now go back to question 24 again and try to identify the integrating center for each example.

LEVEL THREE *Problem Solving*

27. In each of the following situations, identify the components of the reflex.
    (a) You are sitting quietly at your desk, studying, aware of the bitterly cold winds blowing about at 30 mph, and you begin to feel a little chilly. You start to turn up the thermostat, remember last month's bill, and reach for an afghan to pull around you instead. Pretty soon you are toasty warm again.

    (b) You are strolling through the shopping district and the aroma of cinnamon sticky buns reaches you. You inhale appreciatively, but reassure yourself you're not hungry, since you just had lunch an hour ago. You go about your business, but 20 minutes later you're back at the bakery, sticky bun in hand, ravenously devouring its sweetness, saliva moistening your mouth.

LEVEL FOUR *Quantitative Problems*

28. In a signal cascade for rhodopsin, a photoreceptor molecule, one rhodopsin activates 1000 molecules of transducin, the next molecule in the signal cascade. Each transducin activates one phosphodiesterase, and each phosphodiesterase converts 4000 cGMP to GMP.
    (a) What is the name of the phenomenon described in this paragraph?
    (b) Activation of one rhodopsin will result in the production of how many GMP molecules?

29. The graph shown at right represents the binding of oxygen molecules (O_2) to two different proteins, myoglobin and hemoglobin, over a range of oxygen concentrations. Based on the graph, which protein has the higher affinity for oxygen? Explain your reasoning.

Explorations

Explorations are also available at *www.physiologyplace.com*

In this chapter you learned how the body carries out internal communication, coordinates body functions, and maintains a consistent environment. Without communication, integration, and homeostasis, large complicated organisms such as humans could not survive. The following Explorations will help you gain an appreciation of these coordinating functions.

Exploration 1: The Importance of Proteins

By one estimate, nearly one-third of the proteins in the human genome are associated with membranes as receptors, transporters, enzymes, or structural elements. Combine that with another estimate that nearly half of the drugs in use today act on receptors, and you can see why receptors and proteins are so crucial to understanding physiology. For a discussion of cell proteins and proteomics, see "Proteins Rule," *Scientific American* 286(4):40–47, April 2002 and "Prioritizing the proteome: Identifying pharmaceutically relevant targets," *Drug Discovery Today* 7(9):516–521, 2002.

One complicating factor in keeping track of cellular receptors is the diverse and often conflicting terminology. Up to now, there has been no uniform method of naming receptors, leaving the designations up to scientists working on the proteins. In coming months, however, the Nomenclature Committee of the International Union of Pharmacology will develop a standardized nomenclature and a searchable database of receptors to be housed at the European Bioinformatics Institute (*http://www.ebi.ac.uk/*).

Exploration 2: Communication and Signal Transduction

To fully appreciate how complex signal transduction pathways are, you might look at the series of articles on "Mapping Cellular Signaling" in the May 31, 2002 issue of *Science* magazine. The articles focus on 14 signal pathways, including G proteins and integrins. If your institution has a subscription to *Science*, check out the *Signal Transduction Knowledge Environment (STKE)*, an extensive annotated resource for researchers in the field. STKE has dynamic "connections maps" that show just how complicated second messenger systems can get. Two articles of interest include "Cellular communication via gap junctions," *Science & Medicine* 7(2):18–27, Mar/Apr, 2000 and "Signal transduction pathways in caveolae," *Science & Medicine* 7(1):44–53, Jan/Feb 2000.

Exploration 3: Walter B. Cannon and *The Wisdom of the Body*

Walter B. Cannon, the creator of the term *homeostasis*, was a prolific investigator and writer. He was particularly interested in the mind-body connection and in the physiological effects of stress. To learn more about Cannon, read what his family and colleagues wrote about him at *www.harvardsquarelibrary. org/unitarians/cannon_walter.html*. Or see if you can find his classic book, *The Wisdom of the Body* (New York: W.W. Norton & Company, 1939) in your school library.

Exploration 4: Biological Rhythms and Chronotherapy

Scientists have known for a number of years that an organism's response to chemicals (drugs or toxins) will vary with the time of day or season of the year. For example, a dose of pesticide that is harmless to flies at one time of the day might kill 90% of the same population at a later time of day. The idea that our cells are not equally responsive at all times is now being applied to medical treatments, a field called *chronotherapy*. Run a Google search (*www.google.com*) for chronotherapy to learn about some of the interesting work that is being done in this field. For a discussion of the latest theories on biological clocks, see "Times of our lives," *Scientific American* 287(3):58–65, September 2002.

ANSWERS

Answers to Concept Check Questions

Page 174

1. All the signal mechanisms are chemical signals except for (c) gap junctions, which transfer both chemical and electrical signals. Neurohormones (e) and neurotransmitters (f) are associated with electrical signaling in neurons but are themselves chemicals.

2. Cytokines, hormones, and neurohormones travel through the blood.

3. The signal to pounce could not have been a paracrine because the eyes are too far away from the legs and because the response was too rapid for it to have taken place by diffusion.

Page 176

4. Extracellular signal or messenger, receptor, intracellular messenger(s), target protein

5. Cell membrane, cytosol, nucleus

Page 178

6. Steps are (1) signal binds to receptor that (2) activates a protein that (3) creates second messengers that (4) create a response.

7. Amplification turns one signal molecule into multiple second messenger molecules.

8. Steroids are lipophilic and enter cells to bind to intracellular receptors.

Page 181

9. Receptors are ion channels, receptor-enzymes, G protein-coupled receptors, or integrins.

10. First messengers are extracellular; second messengers are intracellular.

11. (a) signal ligand, receptor, second messenger, cell response (b) amplifier enzyme, second messenger, protein kinase, phosphorylated protein, cell response

12. (a) hyperpolarize (b) hyperpolarize (c) depolarize

Page 183

13. The cell must use active transport to move Ca^{2+} against the concentration gradient.

Page 184

14. A drug that blocks action could act at the receptor or any downstream steps. A drug that blocks leukotriene synthesis might inhibit lipoxygenase.

Page 186

15. The molecules are all proteins.

16. Insulin could be using different second messenger systems or binding to different receptors.

Page 186

17. (a), (d). Changing receptor number would not affect binding affinity.

Page 188

18. Environmental factors are osmolarity, temperature, and pH.

19. "Internal secretions" are hormones, cytokines, paracrines, and autocrines.

20. Tonic control usually uses one control system but antagonistic uses two.

21. A signal can have opposite effects by using different receptors or different signal pathways.

Page 193

22. Local control takes place in or very close to the target cell. Reflex control is mediated by a distance integrating center.

23. stimulus, receptor, afferent path, integrating center, efferent path, effector, response (tissue and systemic)

Page 193

24. If the aquarium water became overheated, there is no control mechanism for bringing it back into the desired range.

Page 195

25. Negative feedback will shut off the heater. The loop goes from the response back to the stimulus, with a double "stop" line at the stimulus end.

Page 198

26. brain and spinal cord = integrating center, sense organs = receptor. Negative feedback would go from the response back to the stimulus.

Page 200

27. blow to knee = stimulus, leg muscles = target (effector), neuron to muscles = efferent neuron, sensory neuron = afferent neuron, brain and spinal cord = integrating center, sensory receptor = receptor, muscles contract = response.

Page 200

28. food in stomach = stimulus, brain and spinal cord = integrating center, endocrine cells = E, stretch receptors = receptor, efferent neuron = efferent neuron to E, insulin = hormone, adipose cell = target, sensory neuron = afferent neuron.

Answers to Figure Questions

Page 189

Fig. 6-19: (a) 60 beats/min (b) 280 beats/min

Answers to Level Four Quantitative Problems

Page 205

28. (a) This paragraph describes amplification and a cascade. (b) One rhodopsin creates (1000×4000) or 4,000,000 GMP.

29. At any given oxygen concentration shown, more myoglobin than hemoglobin is bound to oxygen. Therefore myoglobin has a higher binding affinity.

Introduction to the Endocrine System

Radioactive scan of the thyroid gland

> The separation of the endocrine system into isolated subsystems must be recognized as an artificial one, convenient from a pedagogical point of view but not accurately reflecting the interrelated nature of all these systems.
>
> —Howard Rasmussen, in *Williams' Textbook of Endocrinology*, 1974

RUNNING PROBLEM

Graves' Disease

The ball slid by the hole and trickled off the green: another bogey. Ben Crenshaw's golf game was falling apart. The 33-year-old professional had won the Masters Tournament only a year ago but now something was not right. He was tired and weak, had been losing weight, and felt hot all the time. He attributed his symptoms to stress but his family thought otherwise. At their urging, he finally saw a physician. The diagnosis? Graves' disease, a form of overactive thyroid gland.

BACKGROUND BASICS

Receptors: **129** | Peptides and proteins: **38–39** | Comparison of endocrine and nervous systems: **197** | Signal transduction: **176** | Steroids: **36–37** | Specificity: **91**

David was seven years old when the symptoms first appeared. His appetite at meals increased, and he always seemed to be in the kitchen looking for food. But despite eating more, he was losing weight. When he started asking for water instead of soft drinks, David's mother became concerned. And when he wet the bed three nights in a row, she knew something was wrong. The doctor confirmed the suspected diagnosis after running tests to determine the concentration of glucose in David's blood and urine. David had diabetes mellitus. In David's case, the disease was due to lack of insulin, a hormone produced by the pancreas. David was placed on insulin injections, a treatment that he would continue for the rest of his life.

One hundred years ago, David would have died not long after the onset of symptoms. The field of **endocrinology**, the study of hormones, was then in its infancy. Most hormones had not been discovered, and the functions of known hormones were not well understood. There was no treatment for diabetes, no birth control pill for contraception. Babies born with inadequate secretion of thyroid hormone did not grow or develop normally.

Today, all that has changed. We have identified a long and growing list of hormones. The endocrine diseases that once killed or maimed are under control with synthetic hormones and sophisticated medical procedures. Although physicians do not hesitate to use these treatments, we are still learning exactly how hormones act on their target cells. This chapter provides an introduction to the basic principles of hormone structure and function. You will learn more about individual hormones as you encounter them in your study of the various systems.

▶ HORMONES

As you learned in Chapter 6, hormones are chemical messengers secreted into the blood by specialized cells. Hormones are responsible for many functions that we think of as long-term, ongoing functions of the body. Processes that fall mostly under hormonal control include growth and development, metabolism, regulation of the internal environment (temperature, water balance, ions), and reproduction. Hormones act on their target cells in one of three basic ways. They (1) control the rates of enzymatic reactions, (2) control transport of ions or molecules across cell membranes, or (3) control gene expression and the synthesis of proteins.

Hormones Have Been Known Since Ancient Times

Although the scientific field of endocrinology is relatively young, diseases of the endocrine system have been documented for more than a thousand years. Evidence of endocrine abnormalities can even be seen in ancient art. For example, one pre-Colombian statue of a woman shows a mass on the front of her neck (Fig. 7-1 ●). This is an enlarged thyroid gland, or *goiter*, a common condition high in the Andes, where there was a lack of the dietary iodine needed to make thyroid hormones.

● **Figure 7-1** *Endocrine disorders in ancient art* This pre-Colombian stone carving of a woman shows a mass at her neck. This mass is an enlarged thyroid gland, a condition known as goiter. It was considered a sign of beauty among the people who lived high in the Andes mountains.

The first association of endocrine structure and function was probably the link between the testes and male sexuality. Castration of animals and men was a common practice in both Eastern and Western cultures because it was known to decrease the sex drive and render the male infertile.

In 1849, A. A. Berthold performed the first classic experiment in endocrinology. He removed the testes from roosters and observed that the castrated birds had smaller combs and less aggressiveness or sex drive than normal males. If the testes were surgically placed back into the donor cock or into another castrated male, normal male behavior and comb development resumed. Because the reimplanted testes were not connected to nerves, Berthold concluded that the glands must be secreting something into the blood that affected the entire body.

Experimental endocrinology did not receive much attention, however, until 1889, when the 72-year-old French physician Charles Brown-Séquard made a dramatic announcement of his sexual rejuvenation after injecting himself with extracts made from bull testes ground in water. An international uproar followed, and physicians on both sides of the Atlantic began to inject their patients with extracts of many different endocrine organs, a practice known as *organotherapy*.

We now know that the increased virility Brown-Séquard reported was most likely a placebo effect because testosterone is a hydrophobic steroid that cannot be extracted by an aqueous preparation. But the door to hormone therapy was opened, and in 1891 organotherapy had its first true success: a woman was treated for low thyroid hormone levels with glycerin extracts of sheep thyroid glands.

As the study of "internal secretions" grew, Berthold's experiments became a template for endocrine research. Once a gland or structure was suspected of secreting hormones, the classic steps for identifying an endocrine gland became:

1. Remove the suspected gland and monitor the animal for anatomical, behavioral, or physiological abnormalities. This is equivalent to inducing a state of *hormone deficiency*.
2. Replace the gland or an extract of the gland and see if the abnormalities disappear. *Replacement therapy* should eliminate the symptoms of hormone deficiency.
3. Implant the gland or administer an extract from the gland to a normal animal, and see if symptoms characteristic of *hormone excess* appear.

● **Figure 7-2** *ANATOMY SUMMARY* *Hormones*

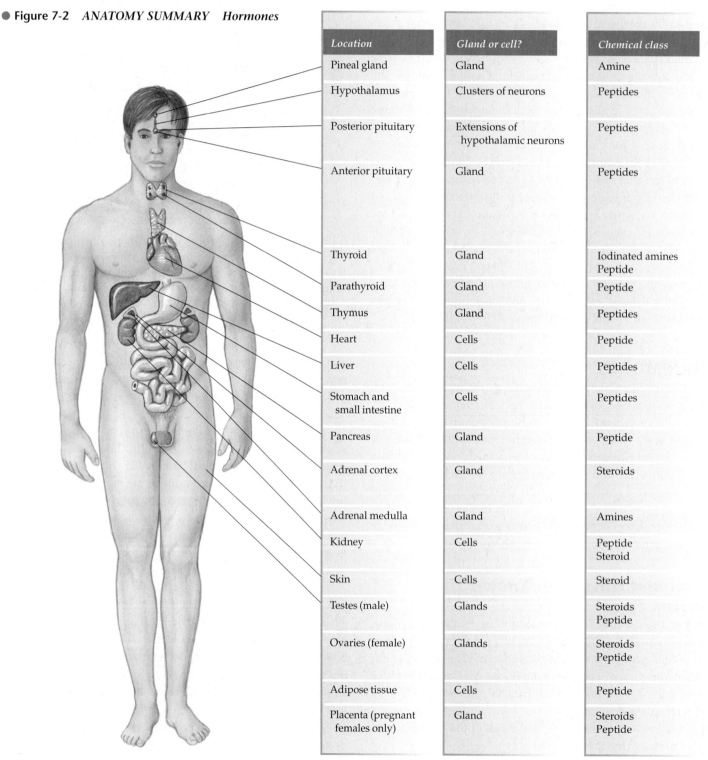

Location	Gland or cell?	Chemical class
Pineal gland	Gland	Amine
Hypothalamus	Clusters of neurons	Peptides
Posterior pituitary	Extensions of hypothalamic neurons	Peptides
Anterior pituitary	Gland	Peptides
Thyroid	Gland	Iodinated amines Peptide
Parathyroid	Gland	Peptide
Thymus	Gland	Peptides
Heart	Cells	Peptide
Liver	Cells	Peptides
Stomach and small intestine	Cells	Peptides
Pancreas	Gland	Peptide
Adrenal cortex	Gland	Steroids
Adrenal medulla	Gland	Amines
Kidney	Cells	Peptide Steroid
Skin	Cells	Steroid
Testes (male)	Glands	Steroids Peptide
Ovaries (female)	Glands	Steroids Peptide
Adipose tissue	Cells	Peptide
Placenta (pregnant females only)	Gland	Steroids Peptide

4. Once a gland is identified as a potential source of hormones, purify extracts of the gland to isolate the active substance. The test for hormone activity is usually a biological assay in which an animal is injected with the purified extract and monitored for a response.

Hormones identified by this technique are sometimes called the *classic hormones*. They include hormones of the pancreas, thyroid, adrenal glands, pituitary, and gonads, all discrete endocrine glands that could be easily identified and surgically removed. Not all hormones come from identifiable glands, however, and we have been slower to discover them.

The Anatomy Summary (Fig. 7-2●) shows the major hormones of the body and the glands or cells that secrete them, along with the major properties of each hormone.

What Makes a Chemical a Hormone?

In 1905, the term *hormone* was coined from the Greek verb meaning "to excite or arouse." The traditional definition of a **hormone**

Hormone	Target	Main Effect
Melatonin	Unclear in humans	Circadian rhythms. Other effects uncertain
Trophic hormones (see Fig. 7-13) See posterior pituitary	Anterior pituitary	Release or inhibit pituitary hormones
Oxytocin (OT)	Breast and uterus	Milk ejection; labor and delivery; behavior
Vasopressin (ADH)	Kidney	Water reabsorption
Prolactin (PRL)	Breast	Milk production
Growth hormone (GH, somatotropin)	Many tissues	Growth and metabolism
Corticotropin (ACTH)	Adrenal cortex	Cortisol release
Thyrotropin (TSH)	Thyroid gland	Thyroid hormone synthesis and release
Follicle stimulating hormone (FSH)	Gonads	Egg or sperm production; sex hormone production
Luteinizing hormone (LH)	Gonads	Sex hormone production; egg or sperm production
Triiodothyronine and thyroxine (T_3, T_4)	Many tissues	Metabolism, growth and development
Calcitonin (CT)	Bone	Plasma calcium levels (minimal effect in humans)
Parathyroid hormone (PTH)	Bone, kidney	Regulate plasma calcium and phosphate levels
Thymosin, thymopoietin	Lymphocytes	Lymphocyte development
Atrial natriuretic peptide (ANP)	Kidneys	Increase sodium excretion
Angiotensinogen	Adrenal cortex, blood vessels, brain	Aldosterone secretion, increase blood pressure
Insulin-like growth factors (IGF)	Many tissues	Growth
Gastrin, cholecystokinin (CCK), secretin, and others	GI tract and pancreas	Assist digestion and absorption of nutrients
Insulin, glucagon, somatostatin (SS), pancreatic polypeptide	Many tissues	Metabolism of glucose and other nutrients
Aldosterone	Kidney	Na^+ and K^+ homeostasis
Cortisol	Many tissues	Stress response
Androgens	Many tissues	Sex drive in females
Epinephrine, norepinephrine	Many tissues	Fight-or-flight response
Erythropoietin (EPO)	Bone marrow	Red blood cell production
1,25 Dihydroxy-vitamin D_3 (calciferol)	Intestine	Increase calcium absorption
Vitamin D_3	Intermediate form of hormone	Precursor of 1,25 dihydroxy-vitamin D_3
Androgen	Many tissues	Sperm production, secondary sex characteristics
Inhibin	Anterior pituitary	Inhibit FSH secretion
Estrogens and progesterone	Many tissues	Egg production, secondary sex characteristics
Ovarian inhibin	Anterior pituitary	Inhibit FSH secretion
Relaxin (pregnancy)	Uterine muscle	Relaxes muscle
Leptin	Hypothalamus, other tissues	Food intake, metabolism, reproduction
Estrogens and progesterone (P)	Many tissues	Fetal and maternal development
Chorionic somatomammotropin (CS)	Many tissues	Metabolism
Chorionic gonadotropin (CG)	Corpus luteum of ovary	Hormone secretion

211

CLINICAL FOCUS

The Discovery of Insulin

Diabetes mellitus, the metabolic disease associated with pathologies of insulin function, is a condition that has been known since ancient times. Detailed clinical descriptions of insulin-deficient diabetes were available for physicians, but they had no means of treating the disease, and patients invariably succumbed. However, in a series of classic experiments in endocrine physiology, researchers finally identified the cause of diabetes. In 1889, Oscar Minkowksi at the University of Strasbourg (Germany) pinpointed the relationship between diabetes and the pancreas. Minkowski surgically removed the pancreas from dogs (*pancreatectomy*) and noticed that they developed symptoms that mimicked diabetes. He also found that implanting pieces of pancreas under the dogs' skin would prevent development of diabetes. Subsequently, in 1921 Fredrick G. Banting and Charles H. Best (Toronto, Canada) identified an antidiabetic substance in pancreas extracts. Banting and Best and others injected pancreatic extracts into diabetic animals and found that the extracts reversed the elevated blood glucose levels of diabetes. From there, it was a relatively short process until, in 1922, purified insulin was used in the first clinical trials.

is a chemical secreted by a cell or group of cells into the blood for transport to a distant target, where it exerts its effect at very low concentrations. However, as scientists learn more about chemical communication in the body, this definition is continually being challenged.

Hormones Are Secreted by a Cell or Group of Cells

Traditionally, the field of endocrinology has focused on chemical messengers secreted by endocrine *glands*, the discrete and readily identifiable tissues derived from epithelial tissue [● p. 70]. However, we now know that molecules that act as hormones are secreted not only by classical endocrine glands but also by isolated endocrine cells (hormones of the *diffuse endocrine system*), by neurons (*neurohormones*), and by cells of the immune system (*cytokines*).

Hormones Are Secreted into the Blood
Secretion, first presented in Chapter 3 [● p. 69], is the movement of a substance from the intracellular compartment to the extracellular compartment or to the external environment. According to the traditional definition, hormones are secreted into the blood. However, the term *ectohormone* [*ektos*, outside] has been given to signal molecules secreted into the external environment.

Pheromones [*pherein*, to bring] are specialized ectohormones that act on other organisms of the same species to elicit a physiological or behavioral response. The existence of pheromones in humans and whether pheromones can be considered hormones is still a matter of debate (see Emerging Concepts: Do humans secrete pheromones?). Some recent studies have shown that human *axillary* (armpit) sweat glands secrete volatile steroids related to sex hormones that appear to serve as human pheromones. Pheromones will be discussed in more detail in Chapter 10.

Hormones Are Transported to a Distant Target
By the traditional definition, a hormone must be transported by the blood to a distant target cell. Experimentally, this property is sometimes difficult to demonstrate. Molecules that are suspected to be hormones but not fully accepted as such are called *candidate hormones*. They are usually identified by the word *factor*. For example, in the early 1970s, the hypothalamic regulating hormones were known as "releasing factors" and "inhibiting factors" rather than releasing and inhibiting hormones.

Currently, **growth factors**, a large group of substances that influence cell growth and division, are being studied to determine if they meet all the criteria for hormones. Many growth factors have been shown to act locally as *autocrines* or *paracrines* [● p. 173], however, evidence is lacking for their widespread distribution via the circulation. A similar situation exists with the lipid-derived signal molecules called *eicosanoids* [● p. 173].

EMERGING CONCEPTS

Do Humans Secrete Pheromones?

Pheromones, like other signal molecules, are used to communicate information. For example, sea anemones secrete alarm pheromones when danger threatens, and ants release trail pheromones to attract fellow workers to food sources. Pheromones are also used to attract members of the opposite sex for mating purposes. Sex pheromones can be found throughout the animal kingdom, in animals from fruit flies to dogs. But do humans have pheromones? Some recent studies have shown that human *axillary* (armpit) sweat glands secrete volatile steroids related to sex hormones that appear to serve as human pheromones. In one study, when female students were asked to rate the odors of T-shirts worn by male students, they preferred the odor of men who were genetically dissimilar. In another study, axillary secretions rubbed on the upper lip altered the timing of menstrual cycles in young women. The selling of putative human pheromones as perfume is becoming the latest fad in the mating game, as you will see if you do a Google search for *human pheromone* (*www.google.com*).

Complicating the classification of signal molecules is the fact that a molecule may act as a hormone when secreted from one location but as a paracrine or autocrine when secreted from a different location. For example, in the 1920s scientists discovered that *cholecystokinin* (CCK) in extracts of intestine caused contraction of the gall bladder. For many years thereafter, CCK was known only as an intestinal hormone. Then in the mid-1970s, CCK was found in neurons of the brain, where it acts as a neurotransmitter or neuromodulator. In recent years, CCK has become famous for its possible role in controlling hunger, a behavior controlled by the brain.

Hormones Exert Their Effect at Very Low Concentrations   One hallmark of a hormone is its ability to act at concentrations in the nanomolar (10^{-9} M) to picomolar (10^{-12} M) range. Some chemical signals transported in the blood to distant targets are not considered to be hormones because they must be present in relatively high concentration before an effect is noticed. For example, histamine released during allergic reactions may act on cells throughout the body but its concentration is higher than the accepted range for a hormone.

As researchers discover new signal molecules and new receptors, the boundary between hormones and nonhormonal signal molecules continues to be challenged, just as the distinction between the nervous and endocrine systems has blurred. Many *cytokines* [◈ p. 174] seem to meet the definition of a hormone as stated above. However, experts in cytokine research do not consider them hormones because peptide cytokines are synthesized and released on demand, in contrast to classic peptide hormones, which are made in advance and stored in the parent endocrine cell. A few cytokines, such as *erythropoietin*, the molecule that controls red blood cell production, were classified as hormones before the term *cytokine* was coined, contributing to overlap between the two groups of signal molecules.

Hormones Act by Binding to Receptors

All hormones bind to target cell receptors and initiate biochemical responses. These responses are known as the **cellular mechanism of action** of the hormone. As you can see from the table in Figure 7-2, one hormone may act on multiple tissues. To complicate matters, the effects may vary in different tissues or at different stages of development. Or a hormone may have no effect at all in a particular cell. Insulin is an example of a hormone with varied effects. In muscle and adipose tissues it alters glucose transport proteins and enzymes for glucose metabolism. In the liver, it modulates enzyme activity but has no direct effect on glucose transport proteins. In brain and certain other tissues, glucose metabolism is totally independent of insulin.

CONCEPT CHECK   Answers: p. 238

1. Name the membrane transport process by which glucose moves from the extracellular fluid into cells.

The variable responsiveness of a cell to a hormone depends primarily on the cell's receptor and signal transduction pathways [◈ p. 174]. If there are no hormone receptors in a tissue, its cells cannot respond. If tissues have different receptors and receptor-linked pathways for the same hormone, they will respond differently.

Hormone Action Must Be Terminated

Signal activity by hormones and other chemical signals must be of limited duration if the body is to respond to changes in its internal state. For example, insulin is secreted when blood glucose concentrations increase following a meal. As long as insulin is present, glucose leaves the blood and enters cells. However, if insulin activity continues too long, blood glucose levels could become so low that the nervous system is unable to function properly—a potentially fatal situation. Normally this situation is avoided in several ways: by limiting insulin secretion, by removing or inactivating insulin circulating in the blood, and by terminating insulin activity in the target cells.

In general, hormones in the bloodstream are *degraded* (broken down) into inactive *metabolites* by enzymes found primarily in the liver and kidney. The metabolites are then excreted in the bile or urine. The rate of hormone breakdown is indicated by a hormone's **half-life** in the circulation, the amount of time required to reduce the concentration of hormone by one-half. Half-life is one indicator of how long a hormone is active in the body.

BIOTECHNOLOGY

Immunocytochemistry

Traditionally, physiologists thought that each hormone came from a single gland (although some glands secrete multiple hormones). But new techniques like immunocytochemistry have turned that view of endocrinology upside down. Immunochemical laboratory tests rely on antibodies made when an animal is injected with a foreign molecule (antigen) such as a hormone. The animal's immune system recognizes the molecule as "not self" and produces specific antibodies to it. The antibodies are extracted from the host animal and combined with a radioactive, fluorescent, or enzymatic marker. The tagged antibody is then applied to cultured cells or sections of tissue and examined under the microscope. If the original antigen is present in the tissue, the antibody binds to it and becomes visible under the microscope. By screening multiple tissues with the antibodies for different hormones, investigators discovered that many hormones are made in tissues far from the site where they were originally isolated.

Hormones bound to target cell receptors have their activity terminated in several ways. Enzymes that are always present in the plasma can degrade peptide enzymes bound to cell membrane receptors. In some cases, the receptor-hormone complex is brought into the cell by endocytosis and the hormone digested within lysosomes [🔖 Fig. 5-28, p. 146]. Intracellular enzymes may also metabolize hormones that enter cells.

CONCEPT CHECK   Answers: p. 238

2. What is the suffix in a chemical name that tells you the molecule is an enzyme? (Hint: 🔖 p. 91) Use that suffix to name an enzyme that digests peptides.

▶ THE CLASSIFICATION OF HORMONES

Hormones can be grouped or classified according to several different schemes: their source, their receptor type, or their chemical structure. One scheme divides hormones into those whose release is controlled by the brain and those that are not. Another scheme groups hormones according to whether they bind to G protein-linked receptors, tyrosine kinase-linked receptors, intracellular receptors, etc.

A third scheme divides hormones into three main chemical classes: peptide/protein hormones, steroid hormones, and amine hormones (Table 7-1). The peptide hormones are composed of linked amino acids. The steroid hormones are all derived from cholesterol [🔖 p. 37]. The amine hormones are all derivatives of a single type of amino acid.

CONCEPT CHECK   Answers: p. 238

3. What is the classic definition of a hormone?

4. Based on what you know about the organelles involved in protein and steroid synthesis [🔖 p. 60], what would be the major differences between the organelle composition of a steroid-producing cell and a protein-producing cell?

Most Hormones Are Peptides or Proteins

The peptide/protein hormones range from small peptides of only three amino acids to large proteins and glycoproteins. Despite the size variability among hormones in this group, for simplicity, they are usually called peptide hormones. You can remember which hormones fall into this category by exclusion: if the hormone is not a steroid and not an amine, then it is probably a peptide or protein.

Table 7-1: Comparison of Peptide, Steroid, and Amine Hormones

| | PEPTIDE HORMONES | STEROID HORMONES | AMINES | |
			CATECHOLAMINES	THYROID HORMONES
Synthesis and storage	Made in advance; stored in secretory vesicles	Synthesized on demand from precursors	Made in advance; stored in secretory vesicles	Made in advance; precursor stored in secretory vesicles
Release from parent cell	Exocytosis	Simple diffusion	Exocytosis	Simple diffusion
Transport in blood	Dissolved in plasma	Bound to carrier proteins	Dissolved in plasma	Bound to carrier proteins
Half-life	Short	Long	Short	Long
Location of receptor	On cell membrane	Cytoplasm or nucleus; some have membrane receptors also	On cell membrane	Nucleus
Response to receptor-ligand binding	Activation of second messenger systems. May activate genes	Activate genes for transcription and translation. May have nongenomic actions	Activation of second messenger systems	Activate genes for transcription and translation
General target response	Modification of existing proteins and induction of new protein synthesis	Induction of new protein synthesis	Modification of existing proteins	Induction of new protein synthesis
Examples	Insulin, parathyroid hormone	Estrogen, androgens, cortisol	Epinephrine, norepinephrine	Thyroxine (T_4)

Peptide Hormone Synthesis, Storage, and Release

The synthesis and packaging of peptide hormones into membrane-bound secretory vesicles is similar to that of other proteins. The initial peptide that comes off the ribosome is a large inactive protein known as a preprohormone (Fig. 7-3 ●). **Preprohormones** contain one or more copies of a peptide hormone, a *signal sequence* that directs the protein into the lumen of the rough endoplasmic reticulum, and other peptide sequences that may or may not have biological activity.

As the inactive peptide moves through the endoplasmic reticulum and Golgi apparatus, the signal sequence is removed, creating a smaller, still inactive molecule called a **prohormone**. In the Golgi apparatus, the prohormone is packaged into secretory vesicles along with *proteolytic* [*proteo-*, protein + *lysis*, rupture] enzymes that chop the prohormone into active hormone and other fragments. This process is called *post-translational modification* [● p. 116].

The secretory vesicles containing peptides are stored in the cytoplasm of the endocrine cell until the cell receives a signal for secretion. At that time, the vesicles move to the cell membrane and release their contents by calcium-dependent exocytosis [● p. 146]. All of the peptide fragments created from the prohormone are released together into the extracellular fluid, in a process known as *co-secretion*.

Post-Translational Modification of Prohormones

Studies of prohormone processing have led to some interesting discoveries. Some prohormones, such as that for *thyrotropin-releasing hormone* (TRH), have multiple copies of the hormone (Fig. 7-4a ●). Another interesting prohormone is called *pro-opiomelanocortin* (Fig. 7-4b). This prohormone processes into three active peptides plus an inactive fragment. In some instances, even the fragments are clinically useful. For example, proinsulin is cleaved into active insulin and an inactive fragment known as *C-peptide* (Fig. 7-4c). Clinicians measure the levels of C-peptide in the blood of diabetics to monitor how much insulin the patient's pancreas is producing.

Transport in the Blood and Half-Life of Peptide Hormones

Peptide hormones are water-soluble and therefore generally dissolve easily in the extracellular fluid for transport throughout the body. The half-life for peptide hormones is usually quite short, in the range of several minutes. If the response to a peptide hormone must be sustained for an extended period of time, the hormone must continue to be secreted.

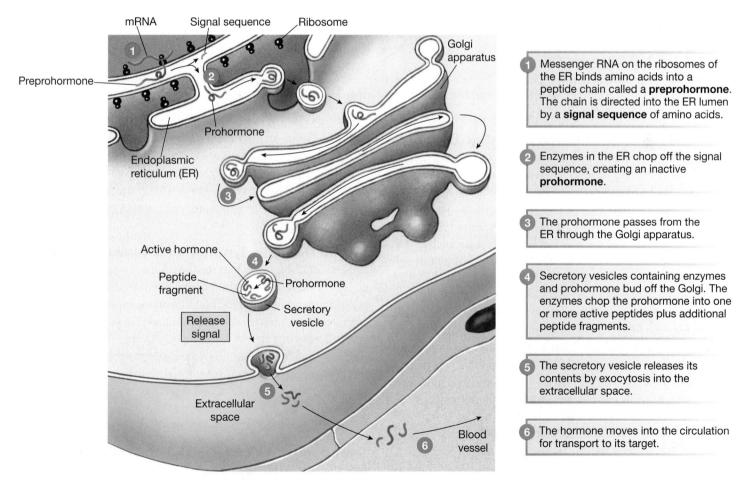

1 Messenger RNA on the ribosomes of the ER binds amino acids into a peptide chain called a **preprohormone**. The chain is directed into the ER lumen by a **signal sequence** of amino acids.

2 Enzymes in the ER chop off the signal sequence, creating an inactive **prohormone**.

3 The prohormone passes from the ER through the Golgi apparatus.

4 Secretory vesicles containing enzymes and prohormone bud off the Golgi. The enzymes chop the prohormone into one or more active peptides plus additional peptide fragments.

5 The secretory vesicle releases its contents by exocytosis into the extracellular space.

6 The hormone moves into the circulation for transport to its target.

● **Figure 7-3** *Peptide hormone synthesis, packaging, and release*

(a) PreproTRH (thyrotropin-releasing hormone) has six copies of the 3-amino acid hormone TRH.

(b) Prohormones, such as pro-opiomelanocortin, the prohormone for ACTH, may contain several peptide sequences with biological activity.

(c) The peptide chain of insulin's prohormone folds back on itself with the help of disulfide (S-S) bonds. The prohormone cleaves to insulin and C-peptide.

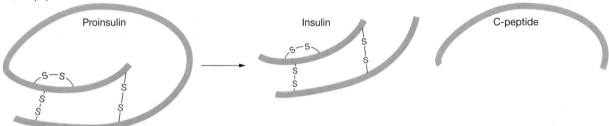

● **Figure 7-4** *Peptide hormone processing* Peptide hormones are made as large, inactive preprohormones that include a signal sequence, one or more copies of hormone, and additional peptide fragments.

Cellular Mechanism of Action of Peptide Hormones

Because peptide hormones are lipophobic, they are usually unable to enter the target cell. Instead, they bind to surface membrane receptors. The hormone-receptor complex initiates the cellular response by means of a *signal transduction* system (Fig. 7-5 ●). Many peptide hormones work through cAMP second messenger systems [● p. 178]. A few peptide hormone receptors, such as that of insulin, have tyrosine kinase activity [● p. 179] or work through other signal transduction pathways.

The response of cells to peptide hormones is usually rapid because second messenger systems modify existing proteins. Some changes triggered by peptide hormones include opening or closing membrane channels and modulating metabolic enzymes or transport proteins. We have recently discovered that some peptide hormones also have longer-lasting effects when their second messenger systems activate genes and direct the synthesis of new proteins.

Steroid Hormones Are Derived from Cholesterol

Steroid hormones have similar chemical structures because they are all derived from cholesterol (Fig. 7-6 ●). Unlike peptide hormones, which are made in tissues all over the body, steroid hor-

mones are made in only a few organs. Three types of steroid hormones are made in the adrenal cortex, the outer portion of the adrenal glands. One adrenal gland sits on top of each kidney [*ad-*, upon + *renal*, kidney; *cortex*, bark]. The gonads produce the sex steroids (estrogens, progesterone, and androgens). In pregnant women, the placenta is also a source of steroid hormones.

Steroid Hormone Synthesis and Release

Cells that secrete steroid hormones have unusually large amounts of smooth endoplasmic reticulum, the organelle where steroids are synthesized. Steroids are lipophilic and diffuse easily across membranes, both out of their parent cell and into their target cell. This property also means that steroid-secreting cells cannot store hormones in secretory vesicles. Instead, they synthesize their hormones as hormone is needed. When a stimulus activates the endocrine cell, precursors in the cytoplasm are rapidly converted to active hormone. The hormone concentration in the cytoplasm rises, and the hormones move out of the cell by simple diffusion.

Transport in the Blood and Half-Life of Steroid Hormones

Like their parent cholesterol, steroid hormones are not very soluble in plasma and other body fluids. For this reason, most of the steroid hormone molecules found in the blood

Peptide hormones (H) cannot enter their target cells and must combine with membrane receptors (R) that initiate signal transduction processes.

Opens ion channel

Second messenger system

phosphorylate

Proteins

Cellular response

KEY

TK = Tyrosine kinase

AE = Amplifier enzyme

G = G protein

● **Figure 7-5** *Membrane receptors for peptide hormones*

● **Figure 7-6** *Steroid hormones are derived from cholesterol*

are bound to protein carrier molecules. Some hormones have specific carriers, such as *corticosteroid-binding globulin*. Others simply bind to general plasma proteins such as *albumin*.

The binding of steroid hormone to a carrier protein protects the hormone from enzymatic degradation and results in an extended half-life. For example, **cortisol**, a hormone produced by the adrenal cortex, has a half-life of 60 to 90 minutes. (Compare this with epinephrine, an amine hormone whose half-life is measured in seconds.)

Although binding steroid hormones to protein carriers extends their half-life, it also blocks their entry into target cells. The carrier-steroid complex remains outside the cell since the carrier proteins are lipophobic and cannot diffuse through the membrane. Only unbound hormone molecules can diffuse into the target cell (Fig. 7-7 ●).

Fortunately, hormones are active in minute concentrations, so a tiny amount of unbound steroid is enough to produce a response. As unbound hormone leaves the blood and enters cells, the carriers release bound steroid so that a small fraction of hormone remaining in the blood is always unbound.

Cellular Mechanism of Action of Steroid Hormones

The best-studied steroid hormone receptors are found within cells, either in the cytoplasm or in the nucleus. The ultimate destination of steroid receptor-hormone complexes is the nucleus, where the complex acts as a *transcription factor*, binding to DNA and activating or *repressing* (turning off) one or more genes (Fig. 7-7). Activated genes create new mRNA that directs the synthesis of new proteins. Any hormone that alters gene activity is said to have a *genomic effect* on the target cell.

Cholesterol is the parent compound for all steroid hormones.

CH₃

modified by enzymes to make steroid hormones such as

Adrenal cortex

In adrenal cortex

CH₂OH

C=O

CH

HO

CH₃

Aldosterone

CH₂OH

C=O

CH₃ OH

HO

CH₃

Cortisol

In ovary

Ovary

OH

CH₃

HO

Estradiol (an estrogen)

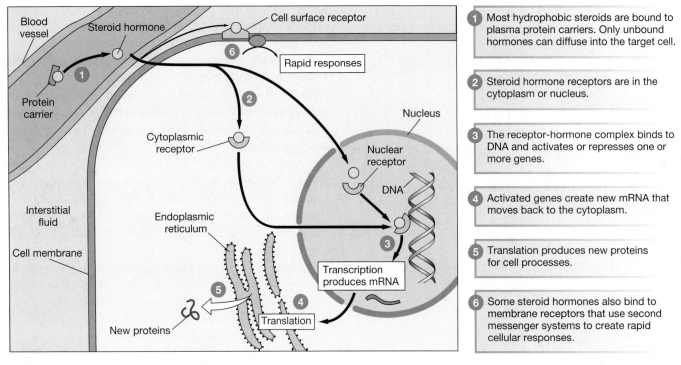

① Most hydrophobic steroids are bound to plasma protein carriers. Only unbound hormones can diffuse into the target cell.

② Steroid hormone receptors are in the cytoplasm or nucleus.

③ The receptor-hormone complex binds to DNA and activates or represses one or more genes.

④ Activated genes create new mRNA that moves back to the cytoplasm.

⑤ Translation produces new proteins for cell processes.

⑥ Some steroid hormones also bind to membrane receptors that use second messenger systems to create rapid cellular responses.

● **Figure 7-7** *Steroid hormone action*

When steroid hormones activate genes to direct the production of new proteins, there is usually a lag time between hormone-receptor binding and the first measurable biological effects. This lag can be as much as 90 minutes. Consequently, steroid hormones do not mediate reflex pathways that require rapid responses.

In recent years researchers have discovered that several steroid hormones, including estrogen and aldosterone, have cell membrane receptors linked to signal transduction pathways, just as peptide hormones do. These receptors enable those steroid hormones to initiate rapid **nongenomic responses** in addition to their slower genomic effects. With the discovery of nongenomic effects of steroid hormones, the differences between steroid and peptide hormone mechanisms of action seem to have almost disappeared.

Answers: p. 238

5. What are the three chemical classes of hormones?

6. The steroid hormone aldosterone has a short half-life for a steroid hormone, only about 20 minutes. What would you predict about the degree to which aldosterone is bound to blood proteins?

Amine Hormones Are Derived from a Single Type of Amino Acid

The amine hormones are small molecules created from one or two amino acids. **Melatonin** is derived from the amino acid tryptophan (see *Focus on the Pineal Gland*, Fig. 7-22). The other amine hormones, the catecholamines and thyroid hormones, are derived from the amino acid tyrosine, notable for its ring structure

(Fig. 7-8 ●). Catecholamines have one tyrosine molecule, the thyroid hormones have two tyrosines plus iodine atoms.

Despite a common precursor, the two groups of tyrosine-based amines have little in common. The **catecholamines** (epinephrine, norepinephrine, and dopamine) are neurohormones that bind to cell membrane receptors like typical peptide hormones. The **thyroid hormones**, produced by the butterfly-shaped thyroid gland in the neck, behave more like steroid hormones, with intracellular receptors that activate genes. Thyroid hormones will be discussed in detail in Chapter 23.

Shaped like a butterfly, the thyroid gland straddles the trachea just below the Adam's apple. Responding to hormonal signals from the hypothalamus and anterior pituitary, the thyroid gland concentrates iodine, an element found in food (most notably as an added ingredient to salt) and combines it with the amino acid tyrosine to make two thyroid hormones, thyroxine and triiodothyronine. These thyroid hormones perform many important functions in the body, including the regulation of growth and development, oxygen consumption, and the maintenance of body temperature.

Question 1:

a. *To which of the three classes of hormones do the thyroid hormones belong?*

b. *If a person's diet is low in iodine, predict what happens to thyroxine production.*

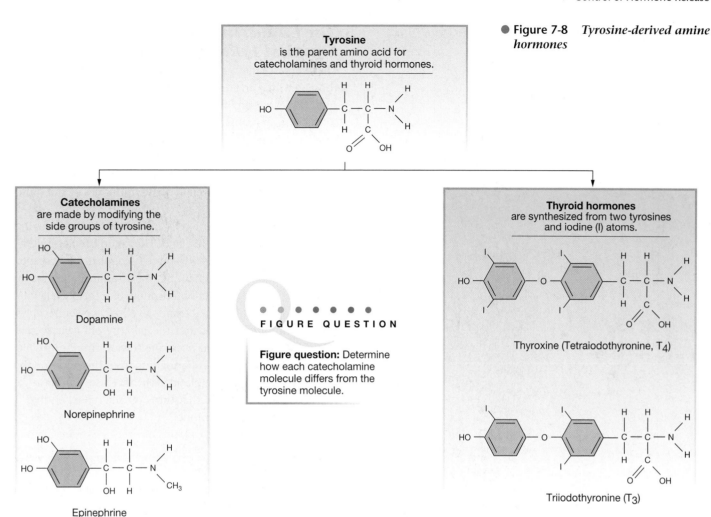

● **Figure 7-8** *Tyrosine-derived amine hormones*

Tyrosine
is the parent amino acid for catecholamines and thyroid hormones.

Catecholamines
are made by modifying the side groups of tyrosine.

Dopamine

Norepinephrine

Epinephrine

FIGURE QUESTION

Figure question: Determine how each catecholamine molecule differs from the tyrosine molecule.

Thyroid hormones
are synthesized from two tyrosines and iodine (I) atoms.

Thyroxine (Tetraiodothyronine, T$_4$)

Triiodothyronine (T$_3$)

▶ CONTROL OF HORMONE RELEASE

Reflex pathways that help maintain homeostasis were introduced in Chapter 6. The sections that follow apply the basic patterns of reflexes to the control pathways for hormones. This discussion is not all-inclusive, and you will encounter a few hormones in later chapters that do not fit exactly into these patterns.

Hormones Can Be Classified by Their Reflex Pathways

Reflex pathways are a convenient method by which to classify hormones and simplify learning the pathways that regulate their secretion. All reflex pathways have similar components: a stimulus, an input signal, integration of the signal, an output signal, and a response [● Fig. 6-22, p. 191]. In endocrine and neuroendocrine reflexes, the output signal is a hormone or neurohormone.

In simple reflex pathways, the response of the pathway usually serves as the *negative feedback* signal that turns off the reflex [● Fig. 6-26a, p. 195]. For example, an increase in blood glucose concentration initiates insulin secretion (Fig. 7-9 ●). Insulin travels through the blood to its target tissues, which increase their

glucose uptake and metabolism. The resultant decrease in blood glucose concentration acts as a negative feedback signal and turns off the reflex, ending release of insulin.

Hormones are not restricted to following only one reflex pathway pattern, however. For example, insulin secretion can also be triggered by input signals from the nervous system (Fig. 7-9) or by a hormone secreted from the digestive tract (not shown). The pancreatic endocrine cells, the integrating center for this reflex, thus must potentially evaluate input signals from three different sources when "deciding" whether to secrete insulin.

CONCEPT CHECK   Answers: p. 238

7. In Figure 6-22 (p. 191), which reflex step(s) represents the input signal? Integration of the signal? The output signal?

8. In the blood glucose example, which step of the reflex is the increase in blood glucose? Insulin secretion? The decrease in blood glucose?

9. Which three pathways in Figure 6-30 (p. 199) match insulin release by the three different mechanisms described above?

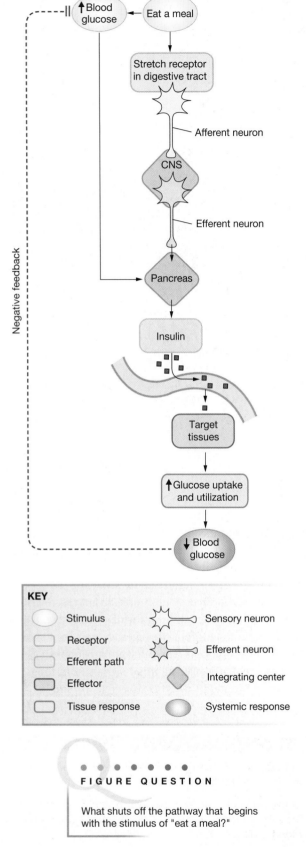

KEY

Stimulus
Sensory neuron

Receptor
Efferent neuron

Efferent path

Effector
Integrating center

Tissue response
Systemic response

FIGURE QUESTION

What shuts off the pathway that begins with the stimulus of "eat a meal?"

● **Figure 7-9** *Hormones may have multiple stimuli for their release* Insulin can be released directly by an increase in blood glucose levels or through nervous stimulation triggered by ingestion of a meal.

The Endocrine Cell Is the Sensor in the Simplest Endocrine Reflexes

The simplest reflex control pathways in the endocrine system are those in which an endocrine cell directly senses a stimulus and responds by secreting its hormone [● Fig. 6-30, pattern 6, p. 199]. In this type of pathway, the endocrine cell acts as both sensor (receptor) and integrating center. **Parathyroid hormone** (PTH) is an example of this simple endocrine reflex. Other hormones that follow a simple reflex pattern include the classic hormones insulin and glucagon, as well as some hormones of the diffuse endocrine system.

Parathyroid endocrine cells are clustered in four small glands that lie behind the thyroid gland. They monitor plasma Ca^{2+} concentration with the aid of G protein-coupled Ca^{2+} receptors on their cell membrane. When a certain number of receptors are bound to Ca^{2+}, PTH secretion is inhibited. If the plasma Ca^{2+} concentration falls below a certain level and fewer receptors are bound, inhibition ceases and the parathyroid cells secrete PTH (Fig. 7-10 ●). Parathyroid hormone travels through the blood to act on its target tissues, initiating responses that increase the concentration of Ca^{2+} in the plasma. The increase in plasma Ca^{2+} is a negative feedback signal that turns off the reflex, ending the release of parathyroid hormone.

CONCEPT CHECK	Answers: p. 238

10. Draw a reflex pathway for parathyroid hormone as described in the paragraph above.

Many Endocrine Reflexes Involve the Nervous System

The nervous system and the endocrine system overlap in their structure and function. Stimuli that are integrated by the central nervous system influence the release of many hormones through efferent neurons, as described above for insulin. In addition, specialized groups of neurons secrete neurohormones, and two endocrine structures are incorporated in the anatomy of the brain: the pineal gland (discussed in a *Focus* box on p. 232) and the pituitary gland.

One of the most fascinating links between the brain and the endocrine system is the influence of emotions over hormone secretion and function. Physicians for centuries have recorded instances in which emotional state has influenced health or normal physiological processes. Women today know that the timing of their menstrual periods may be altered by stress such as travel or final exams. The condition known as "failure to thrive" in infants can often be linked to environmental or emotional stress that increases secretion of some pituitary hormones and decreases production of others. The interaction between stress, the endocrine system, and the immune system is receiving intense study and we will discuss it further in Chapter 24.

● **Figure 7-10**
A simple endocrine reflex: parathyroid hormone

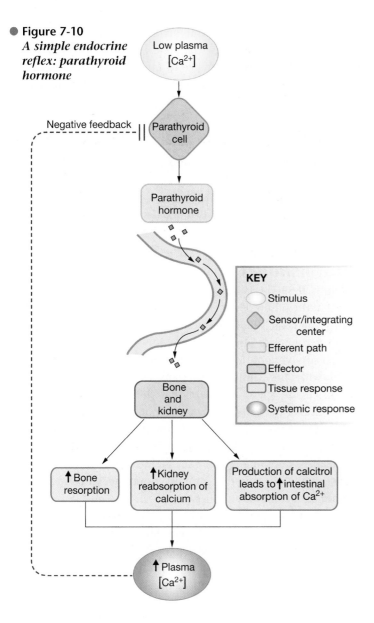

KEY

◯ Stimulus

◇ Sensor/integrating center

▭ Efferent path

▭ Effector

▭ Tissue response

◯ Systemic response

Neurohormones Are Secreted into the Blood by Neurons

Neurohormones are chemical signals released into the blood by a neuron [● Fig. 6-30, pattern 2]. The human nervous system produces three major groups of neurohormones: (1) catecholamines made by modified neurons in the adrenal medulla, (2) hypothalamic neurohormones secreted from the posterior pituitary, and (3) hypothalamic neurohormones that control hormone release from the anterior pituitary. Because the latter two groups of neurohormones are associated with the pituitary gland, we will first describe that important endocrine structure.

11. Catecholamines belong to what chemical class of hormone?

The Pituitary Gland Is Actually Two Fused Glands

The **pituitary gland** is a lima bean-sized structure that extends downward from the brain, connected to it by a thin stalk and cradled in a protective pocket of bone (Fig. 7-11 ●). The first accurate description of the function of the pituitary gland came from Richard Lower (1631–1691), an experimental physiologist at Oxford University. Using observations and some experiments, he theorized that substances produced in the brain passed down the stalk into the gland and from there into the blood.

Lower did not realize that the pituitary gland is actually two different tissue types that merged during embryonic development. The **anterior pituitary** is a true endocrine gland of epithelial origin derived from embryonic tissue that formed the roof of the mouth [● Fig. 3-21, p. 71]. It is also called the *adenohypophysis* [*adeno-*, gland + *hypo-*, beneath + *phyein*, to grow] and its hormones are *adenohypophyseal* secretions. The **posterior pituitary**, or *neurohy-*

Anterior ◀——▶ Posterior

HYPOTHALAMUS

Infundibulum is the stalk that connects the pituitary to the brain.

Posterior pituitary is an extension of the neural tissue.

Sphenoid bone

Anterior pituitary is a true endocrine gland of epithelial origin.

● **Figure 7-11** *Pituitary gland anatomy* The pituitary gland sits in a protected pocket of bone, connected to the brain by a thin stalk.

pophysis, is an extension of the neural tissue of the brain. It secretes neurohormones made in the hypothalamus.

The Posterior Pituitary Stores and Releases Two Neurohormones

The posterior pituitary is the storage and release site for two neurohormones, oxytocin and vasopressin. These small peptide hormones are synthesized in the cell bodies of neurons in the hypothalamus, a region of the brain that controls many homeostatic functions (Fig. 7-12 ●). Each hormone is made in a separate cell type.

Hypothalamic neurohormones follow the standard pattern for peptide synthesis, storage, and release described earlier in this chapter. However, the secretory vesicles containing hormone are transported down long extensions of the cells into the posterior pituitary, where they are stored in the cell terminals. When a stimulus reaches the hypothalamus, an electrical signal passes from the cell body to the posterior pituitary, and the vesicle contents are released into the circulation.

The two posterior pituitary neurohormones are composed of nine amino acids each. **Vasopressin** (also known as antidiuretic hormone) regulates water balance in the body. In women, **oxytocin** released from the posterior pituitary controls the ejection of milk during breast-feeding and contractions of the uterus during labor and delivery.

A few neurons release oxytocin as a neurotransmitter or neuromodulator onto neurons in other parts of the brain. A number of animal and a few human experiments suggest that oxytocin

plays an important role in social, sexual, and maternal behaviors. Some investigators postulate that *autism*, a developmental disorder in which patients are unable to form normal social relationships, may be related to defects in the normal oxytocin-modulated pathways of the brain.

CONCEPT CHECK   Answers: p. 238

12. What organelle is used for transport of secretory vesicles within the cytoplasm?

13. Name the membrane process by which the contents of secretory vesicles are released into the extracellular fluid.

The Anterior Pituitary Secretes Six Hormones

As late as 1889, it was being said in reviews of physiological function that the pituitary was of little or no use to higher vertebrates! But by the early 1900s, researchers had discovered that animals with their anterior pituitary glands surgically removed were unable to survive more than a day or two. This observation, combined with the clinical symptoms associated with pituitary tumors, made scientists realize that the anterior pituitary is a major endocrine gland that secretes not one but six physiologically significant hormones: prolactin (PRL), thyrotropin (TSH), adrenocorticotropin (ACTH), growth hormone (GH), follicle-stimulating hormone (FSH), and luteinizing hormone (LH). Secretion of all the anterior pituitary hormones is controlled by hypothalamic neurohormones.

The anterior pituitary hormones, their associated hypothalamic neurohormones, and their targets are illustrated in Figure 7-13 ●. Notice that all but one of the anterior pituitary hormones have another endocrine gland or cell as one of their targets. A hormone that controls the secretion of another hormone is known as a **trophic** (tropic) **hormone**.

The adjective *trophic* comes from the Greek word *trophikós*, which means "pertaining to food or nourishment" and refers to the manner in which the trophic hormone "nourishes" the target cell. Trophic hormones often have names that end with the suffix *-tropin*, as in *gonadotropin*.* The root word to which the suffix is attached is the target tissue: the gonadotropins are hormones that are trophic to the gonads. The hypothalamic neurohormones that control release of the anterior pituitary hormones are also trophic hormones but for historical reasons they are described as *releasing hormones* (e.g., thyrotropin-releasing hormone) or *inhibiting hormones* (e.g., growth hormone-inhibiting hormone).

One complication you should be aware of is that many of the hypothalamic and anterior pituitary hormones have multiple names as well as standardized abbreviations. For example, **growth hormone-inhibiting hormone** (GHIH) is also called

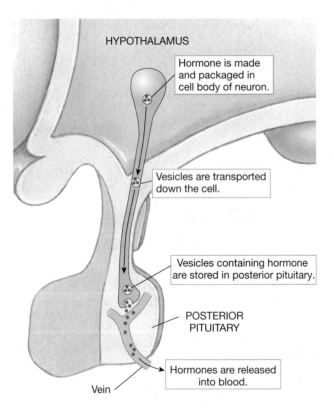

HYPOTHALAMUS

Hormone is made and packaged in cell body of neuron.

Vesicles are transported down the cell.

Vesicles containing hormone are stored in posterior pituitary.

POSTERIOR PITUITARY

Hormones are released into blood.

Vein

● **Figure 7-12** *Synthesis, storage, and release of posterior pituitary hormones*

---

*A few hormones whose names end in *-tropin* do not have endocrine cells as their targets. For example, melanotropin acts on pigment-containing cells in many animals.

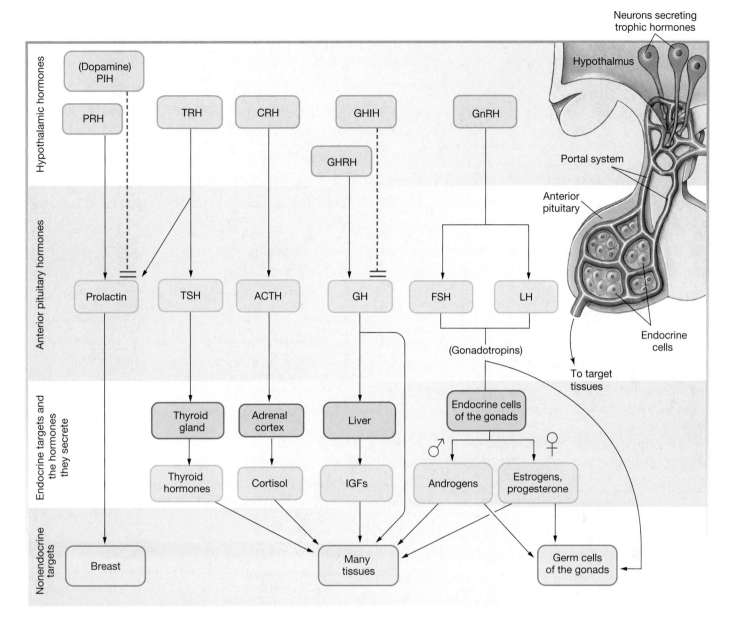

● **Figure 7-13** *Hormones of the hypothalamic-anterior pituitary pathway*   The hypothalamus secretes releasing and inhibiting hormones that act on endocrine cells of the anterior pituitary to influence secretion of their hormones. Anterior pituitary hormones act on additional endocrine glands or act directly on target cells. PIH = prolactin-inhibiting hormone; PRH = prolactin-releasing hormone; TRH = thyrotropin-releasing hormone; CRH = corticotropin-releasing hormone; GHIH = growth hormone-inhibiting hormone (somatostatin); GHRH = growth hormone-releasing hormone; GnRH = gonadotropin-releasing hormone; TSH = thyroid-stimulating hormone (thyrotropin); ACTH = adrenocorticotropic hormone (corticotropin); GH = growth hormone; FSH = follicle-stimulating hormone; LH = luteinizing hormone; IGFs = insulin-like growth factors.

somatostatin. The abbreviations and alternate names are listed in the caption of Figure 7-13.

Feedback Loops Are Different in the Hypothalamic-Pituitary Pathway

The pathways in which anterior pituitary hormones act as trophic hormones are among the most complex endocrine reflexes because they involve three different integrating centers: the hypothalamus, the anterior pituitary, and the endocrine target of the pituitary hormone (Fig. 7-14 ●). Feedback in these complex

pathways follows a different pattern than that described previously. Instead of the response acting as the negative feedback signal, the hormones themselves are the signal. Each hormone in the pathway feeds back to suppress hormone secretion by integrating centers earlier in the reflex.

When secretion of one hormone in a complex pathway increases or decreases, secretion of other hormones changes as well because of the feedback loops that link the hormones. In pathways with two or three hormones in sequence, the "downstream" hormone usually feeds back to suppress the hormone(s) that

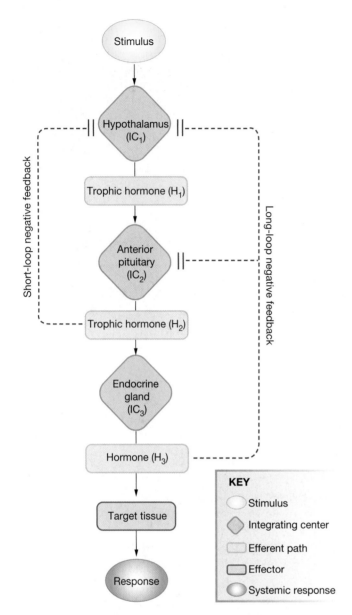

KEY

- Stimulus
- Integrating center
- Efferent path
- Effector
- Systemic response

● **Figure 7-14** *Negative feedback loops in the hypothalamic-anterior pituitary pathway* In complex endocrine pathways, the hormones of the pathway serve as negative feedback signals.

controlled its secretion. (A major exception to this is feedback by ovarian hormones, as you will learn in Chapter 26.)

For example, cortisol secreted from the adrenal cortex feeds back to suppress secretion of the trophic hormones CRH and ACTH (Fig. 7-15 ●). This relationship is called **long-loop negative feedback**. In **short-loop negative feedback**, pituitary hormones feed back to decrease hormone secretion by the hypothalamus. We see this in the same example, where ACTH exerts short-loop negative feedback on the secretion of CRH. With this system of negative feedback, the hormones normally stay within the range needed for an appropriate response. Feedback patterns are important in the diagnosis of endocrine pathologies, to be discussed later in the chapter.

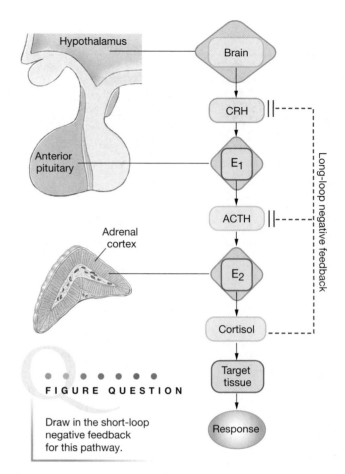

FIGURE QUESTION

Draw in the short-loop negative feedback for this pathway.

● **Figure 7-15** *Control pathway for cortisol secretion* Cortisol is a steroid hormone secreted by the adrenal cortex. ACTH = corticotropin; CRH = corticotropin-releasing hormone.

CONCEPT CHECK   Answers: p. 238

14. Which pathway in Figure 6-30 (p. 199) fits the hypothalamic neurohormone-anterior pituitary hormone-endocrine target of the pituitary hormone pattern described above?

RUNNING PROBLEM

Thyroid hormone production is regulated by thyroid-stimulating hormone, or TSH, a hormone secreted by the anterior pituitary. The production of TSH is in turn regulated by the neurohormone TRH, or thyrotropin-releasing hormone, from the hypothalamus.

Question 2:
a. *In a normal person, when thyroid hormone levels in the blood increase, will negative feedback increase or decrease the secretion of TSH?*

b. *In a person with a hyperactive gland that is producing too much thyroid hormone, would you expect TSH to be higher or lower than in the normal person?*

The Hypothalamic-Hypophyseal Portal System Directs Trophic Hormone Delivery

The hypothalamic trophic hormones that regulate secretion of anterior pituitary hormones are transported directly to the pituitary through a special set of blood vessels known as the **hypothalamic-hypophyseal portal system** (Fig. 7-16 ●). A **portal system** is a specialized region of the circulation consisting of two sets of capillaries directly connected by a set of blood vessels. There are three portal systems in the body: one in the kidneys, one in the digestive system, and this one in the brain.

Hormones secreted into a portal system have a distinct advantage over hormones secreted into the general circulation because a much smaller amount of hormone can be secreted to elicit the same response. A dose of hormone secreted into the general circulation will be rapidly diluted by the total blood volume, which is typically more than 5 L. The same dose secreted into the tiny volume of blood flowing through the portal system remains concentrated while it is taken directly to its target. Thus, a small number of neurosecretory neurons in the hypothalamus can effectively control the anterior pituitary.

The minute amounts of hormone secreted into the hypothalamic-hypophyseal portal system posed a great challenge to the researchers who first isolated these hormones. Because such tiny quantities of hypothalamic-releasing hormones are secreted, Roger Guillemin and Andrew Shalley had to work with tremendous amounts of tissue to obtain enough hormone to analyze. Guillemin and his colleagues processed more than 50 tons of sheep hypothalami, and a major meat packer donated more than one million pig hypothalami to Shalley and his associates. For the final analysis, they needed 25,000 hypothalami to isolate and determine the amino acid sequence of just 1 mg of thyrotropin-releasing hormone, a tiny peptide made of three amino acids (see Fig. 7-4). For their discovery, Guillemin and Shalley shared a Nobel Prize in 1977.

Anterior Pituitary Hormones Control Growth, Metabolism, and Reproduction

The hormones of the anterior pituitary control so many vital functions that the pituitary is often called the "master gland" of the body. In general, we can say that the anterior pituitary hormones control metabolism, growth, and reproduction, all very complex processes.

One anterior pituitary hormone, **prolactin** (PRL), controls milk production in the female breast. In both sexes, prolactin appears to play a role in regulation of the immune system. **Growth hormone** (GH; also called *somatotropin*) affects metabolism of many tissues in addition to stimulating hormone production by the liver (Fig. 7-17 ●). Prolactin and growth hormone are the only two anterior pituitary hormones whose secretion is controlled by both releasing and inhibiting hormones. We will discuss these hormones in detail in Chapters 26 and 23, respectively.

CONCEPT CHECK Answers: p. 238

15. Which pathway in Figure 6-30 (p. 199) fits:
    (a) the hypothalamic trophic hormone-prolactin-breast pattern described above?
    (b) the growth hormone pathway shown in Figure 7-17?

The remaining four anterior pituitary hormones all have another endocrine gland as their primary target. The two hormones known collectively as the **gonadotropins** act on the gonads. **Follicle-stimulating hormone** (FSH) and **luteinizing hormone** (LH) were originally named for their effects on the ovaries, but both hormones are trophic on the male testes as well. **Thyroid-stimulating hormone** (TSH, or *thyrotropin*) controls hormone synthesis and secretion in the thyroid gland. **Adreno-corticotrophic hormone** (ACTH, or *corticotropin*) acts on certain cells of the adrenal cortex to control synthesis and release of the steroid hormone *cortisol*.

The hormones of the anterior and posterior pituitary will be discussed in more detail in later chapters.

CONCEPT CHECK Answers: p. 238

16. What is the target tissue of a hypothalamic hormone secreted into the hypothalamic-hypophyseal portal system?

17. Look at the pathway for insulin release as the result of eating a meal (Fig. 7-9). What event serves as the negative feedback signal to shut off insulin release?

● **Figure 7-16** *The hypothalamic-hypophyseal portal system*

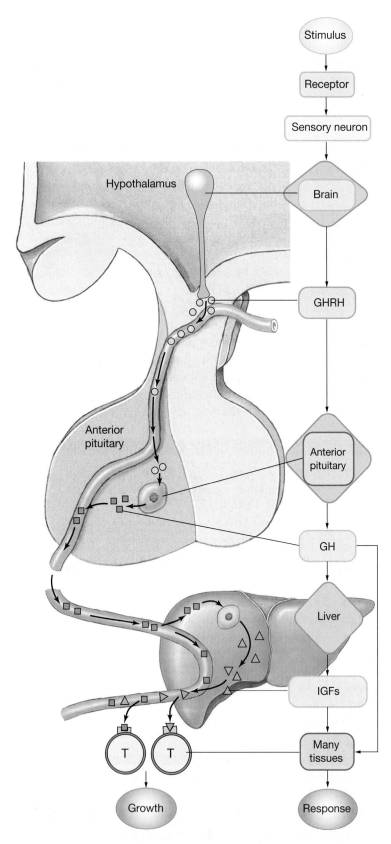

● **Figure 7-17** *A complex endocrine pathway* Growth hormone acts directly on many tissues of the body but also influences liver production of insulin-like growth factors (IGFs or *somatomedins*), another group of hormones that regulate growth.

▶ HORMONE INTERACTIONS

One of the most complicated and confusing aspects of endocrinology is the way that hormones interact at their target cells. It would be simple if each endocrine reflex were a separate entity and if each cell were under the influence of only a single hormone. But in many instances, cells and tissues are controlled by multiple hormones that may be present at the same time. Complicating the picture is the fact that multiple hormones acting on a single cell can interact in ways that cannot be predicted by knowing the individual effects of the hormone. In this section, we examine three types of hormone interaction: synergism, permissiveness, and antagonism.

In Synergism, the Effect of Interacting Hormones Is More Than Additive

Sometimes different hormones have the same effect on the body, although they may accomplish that effect through different cellular mechanisms. One example is the hormonal control of blood glucose levels. Glucagon from the pancreas is the hormone primarily responsible for elevating blood glucose levels. But it is not the only hormone that has that effect. Cortisol raises blood glucose concentration, as does epinephrine.

What happens if two hormones are present, or if all three hormones are secreted at the same time? You might expect their effects to be additive. In other words, if a given amount of epinephrine elevates blood glucose 5 mg/100 mL blood, and glucagon elevates blood glucose 10 mg/100 mL blood, you might expect both hormones acting at the same time to elevate blood glucose 15 mg/100 mL blood (5 + 10).

Frequently, however, two hormones interact at their targets so that the combination of the two yields a result that is more than additive. This type of interaction is called **synergism**. Using the same example, a synergistic reaction would be:

epinephrine	elevates blood glucose	5 mg/100 mL blood
glucagon	elevates blood glucose	10 mg/100 mL blood
epinephrine + glucagon	elevate blood glucose	22 mg/100 mL blood

In other words, the combined effect of the two hormones is greater than the sum of the effects of the two hormones taken individually. An example of synergism between epinephrine, glucagon, and cortisol is shown in Figure 7-18 ●. The cellular mechanisms that underlie synergistic effects are not always clear, but with peptide hormones, synergism is often linked to overlapping effects on second messenger systems.

Synergism is not limited to hormones. It can occur with any two chemicals (or more) in the body. Pharmacologists have developed drugs with synergistic components. For example, the effectiveness of the antibiotic penicillin is enhanced by the presence of clavulanic acid in the same pill.

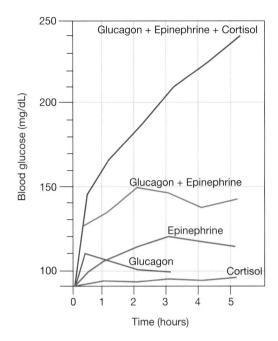

● **Figure 7-18** *Synergism* This graph shows the effect of hormone infusions on blood glucose levels. The effects of combined hormones are greater than the summed effects of the individual hormones, indicating synergistic relationships. (Data adapted from Eigler et al., *J. Clin. Invest.* 63: 114, 1979.)

RUNNING PROBLEM

Ben was diagnosed with Graves' disease, one form of hyperthyroidism. The goal of treatment is to reduce thyroid hormone activity, and Ben's physician offered him several alternatives. One treatment involves use of drugs that prevent the thyroid gland from using iodine. Another treatment is a single dose of radioactive iodine that destroys the thyroid tissue. A third treatment is surgical removal of all or part of the thyroid gland. Ben elected initially to use the thyroid-blocking drug. Several months later he was given radioactive iodine.

Question 3: Why is radioactive iodine rather than some other radioactive element such as cobalt used to destroy thyroid tissue?

A Permissive Hormone Allows Another Hormone to Exert Its Full Effect

In permissiveness, one hormone cannot fully exert its effects unless a second hormone is present. For example, maturation of the reproductive system is controlled by gonadotropin-releasing hormone from the hypothalamus, gonadotropins from the anterior pituitary, and steroid hormones from the gonads. However, if thyroid hormone is not present in sufficient amounts, maturation of the reproductive system is delayed. Because thyroid hormone by itself cannot stimulate maturation of the reproductive

system, thyroid hormone is considered to have a permissive effect on sexual maturation.

The results of this interaction can be summarized as follows:

thyroid hormone alone	no development of the reproductive system
reproductive hormones alone	delayed development of the reproductive system
reproductive hormones with adequate thyroid hormone	normal development of the reproductive system

Antagonistic Hormones Have Opposing Effects

In some situations, two molecules work against each other, one diminishing the effectiveness of the other. This tendency of one substance to oppose the action of another is called antagonism. Recall from Chapter 6 that antagonism may result when two molecules compete for the same receptor [● p. 185]. When one molecule binds to the receptor but does not activate it, that molecule acts as a competitive inhibitor or antagonist to the other molecule. This type of antagonism has been put to use in the development of pharmaceutical compounds, such as the estrogen receptor antagonist *tamoxifen*.

In endocrinology, two hormones are considered antagonists if they have opposing physiological actions. For example, glucagon and growth hormone, both of which raise the concentration of glucose in the blood, are antagonistic to insulin, which lowers the concentration of glucose in the blood. Hormones with antagonistic actions do not necessarily compete for the same receptor. Instead, they may act through different metabolic pathways. Or one hormone may decrease the number of receptors for the opposing hormone. For example, evidence suggests that growth hormone decreases the number of insulin receptors, providing part of its antagonistic effects on blood glucose concentration.

The synergistic, permissive, and antagonistic interactions of hormones make the study of endocrinology both challenging and intriguing. With this brief survey of hormone interactions, you have built a solid foundation for learning more about hormone interactions in later chapters.

► ENDOCRINE PATHOLOGIES

As one endocrinologist said, "There are no good or bad hormones. A balance of hormones is important for a healthy life.... Unbalance leads to diseases."* We can learn much about the normal functions of a hormone by studying the diseases caused by hormone imbalances. There are three basic patterns of endocrine pathology: hormone excess, hormone deficiency, and abnormal responsiveness of target tissues to a hormone.

---

*W. König, preface to *Peptide and Protein Hormones* (New York: VCH Publishers, 1993).

To illustrate the endocrine pathologies, we will use a single example, that of cortisol production by the adrenal cortex (see Fig. 7-15). This is a complex reflex that starts with the secretion of corticotropin-releasing hormone (CRH) from the hypothalamus. CRH stimulates release of adrenocorticotropin (ACTH) from the anterior pituitary. ACTH in turn controls the synthesis and release of cortisol from the adrenal cortex. As in all homeostatic reflexes, negative feedback shuts off the pathway. As cortisol increases, it acts as a negative feedback signal, causing the pituitary and hypothalamus to decrease their output of ACTH and CRH, respectively.

Hypersecretion Exaggerates a Hormone's Effects

If a hormone is present in excessive amounts, the normal effects of the hormone are exaggerated. Most instances of hormone excess are due to **hypersecretion**. There can be multiple causes for hypersecretion including benign tumors (*adenomas*) and cancerous tumors of the endocrine glands. Occasionally, nonendocrine tumors secrete hormones.

Sometimes an individual may exhibit symptoms of hypersecretion as the result of medical treatment with a hormone or agonist. In this case, the condition is said to be *iatrogenic*, or physician-caused [*iatros*, healer + *-gen*, to be born]. It seems simple enough to correct the hormone imbalance by stopping the *exogenous* [*exo-*, outside + *-gen*, to be born] hormone, but this is not always the case.

In our example, exogenous cortisol in the body acts as a negative feedback signal, just as cortisol produced within the body would, shutting off the production of CRH and ACTH (Fig. 7-19 ●). Without the trophic "nourishing" influence of ACTH, the body's own cortisol production shuts down. If the pituitary remains suppressed and the adrenal cortex is deprived of ACTH long enough, the cells of both glands shrink and lose their ability to manufacture ACTH and cortisol. This condition is known as **atrophy** [*a-*, without + *trophikós*, nourishment].

If the cells of an endocrine gland atrophy because of exogenous hormone administration, they may be very slow or totally unable to regain normal function when the treatment with exogenous hormone is stopped. As you may know, steroid hormones can be used to treat poison ivy and severe allergies. However, when treatment is complete, the dosage must be tapered off gradually to allow the pituitary and adrenal gland to work back up to normal hormone production. As a result, packages of steroid pills direct patients ending treatment to take six pills one day, five the day after that, and so on. Low-dose, over-the-counter steroid creams usually do not pose a risk of feedback suppression when used as directed.

Hyposecretion Diminishes or Eliminates a Hormone's Effects

Symptoms of hormone deficiency occur when too little hormone is secreted (**hyposecretion**). Hyposecretion may occur anywhere along the endocrine control pathway, in the hypothalamus, pituitary, or other endocrine glands. The most common cause of hyposecretion syndromes is atrophy of the gland due to some disease process. Hyposecretion of thyroid hormone may occur if there is insufficient dietary iodine for the thyroid gland to manufacture the iodinated hormone.

Negative feedback pathways are affected in hyposecretion, but in the opposite direction from hypersecretion. The absence of feedback causes trophic hormone levels to rise as the trophic hormones attempt to make the defective gland increase its hormone output. For example, if the adrenal cortex atrophies as a result of tuberculosis, cortisol production diminishes. The hypothalamus and anterior pituitary sense that cortisol levels are below normal, so they increase secretion of CRH and ACTH, respectively, in an attempt to stimulate the adrenal gland into making more cortisol.

CONCEPT CHECK   Answers: p. 238

18. Draw a reflex pathway like the one in Figure 7-19 to illustrate what happens to hormone levels and feedback when the adrenal cortex atrophies and cortisol secretion is below normal.

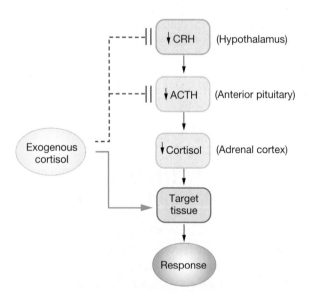

● **Figure 7-19** *Negative feedback by exogenous cortisol* Hormone that is administered to a person for medical reasons will have the same negative feedback effect as the endogenous hormone.

Receptor or Second Messenger Problems Cause Abnormal Tissue Responsiveness

Endocrine diseases do not always arise from problems with endocrine glands. They may also be triggered by changes in the responsiveness of the target tissues to the hormones. In these situations, the target tissues show abnormal responses although the hormone levels may be within normal range. Changes in the target tissue response are usually caused by abnormal interaction between the hormone and its receptor or by alterations in signal transduction pathways. These concepts have been covered for signal molecules in general in Chapter 6 [● p. 174], so we will restrict this discussion to some typical examples of abnormal tissue responsiveness in the endocrine system.

Down-Regulation If hormone secretion is abnormally high for an extended period of time, target cells will *down-regulate* (decrease the number of) their receptors in an effort to diminish their responsiveness to excess hormone. **Hyperinsulinemia** [*hyper-*, elevated + insulin + *-emia*, in the blood] is a classic example of down-regulation in the endocrine system. In this disorder, sustained high levels of insulin cause target cells to remove insulin receptors from the cell membrane. These patients may show symptoms of diabetes despite their high blood insulin levels.

Receptor and Signal Transduction Abnormalities
Many forms of inherited endocrine pathologies can be traced to problems with hormone action in the target cell. Endocrinologists used to believe that these problems were rare, but they are being recognized more and more as scientists increase their understanding of receptors and signal transduction mechanisms.

Some pathologies are due to problems with the hormone receptor. If a mutation alters the protein sequence of the receptor, the cellular response to receptor-hormone binding may be altered. In other mutations the receptors may be absent or completely nonfunctional. For example, in *testicular feminizing syndrome*, androgen receptors are nonfunctional in the male fetus because of a genetic mutation. As a result, androgens produced by the developing fetus are unable to influence development of the genitalia. The result is a child who appears to be female but who lacks a uterus and ovaries.

Genetic alterations in signal transduction pathways can lead to symptoms of hormone excess or deficiency. In the disease called *pseudohypoparathyroidism* [*pseudo-*, false + *hypo-*, decreased + parathyroid + *-ism*, condition or state of being], patients show signs of low parathyroid hormone even though blood levels of the hormone are normal to elevated. These patients have inherited a defect in the G protein that links the hormone receptor to the cAMP amplifier enzyme, adenylyl cyclase. Because the signal transduction pathway does not function, the cells are unable to respond to parathyroid hormone, and symptoms of hormone deficiency appear.

Diagnosis of Endocrine Pathologies Depends on the Complexity of the Reflex

Diagnosis of endocrine pathologies may be simple or complicated, depending on the complexity of the reflex. In the simplest endocrine reflex such as that for parathyroid hormone, if there is too much or too little hormone, there is only one location where the problem can arise, the parathyroid glands (see Fig. 7-10).

However, with complex hypothalamic-pituitary-endocrine gland reflexes, the diagnosis can be much more difficult.

If a pathology (deficiency or excess) arises in the last endocrine gland in a reflex, the problem is considered to be a **primary pathology**. For example, if a tumor in the adrenal cortex begins to produce excessive amounts of cortisol, the resulting condition is called *primary hypersecretion*. If dysfunction occurs in one of the tissues producing trophic hormones, the problem is a **secondary pathology**. For example, if the pituitary is damaged because of head trauma and ACTH secretion diminishes, the resulting cortisol deficiency is considered to be *secondary hyposecretion* of cortisol.

The diagnosis of pathologies in complex endocrine pathways depends on understanding negative feedback in the control pathway. Figure 7-20 ● shows three possible causes of excess cortisol secretion. To determine which is the correct *etiology* (cause) of the disease in a particular patient, the clinician must assess the levels of the three hormones in the control pathway. If the problem is overproduction of CRH by the hypothalamus (Fig. 7-20a), CRH levels will be higher than normal. High CRH in turn causes high ACTH, which in turn causes high cortisol. This is, therefore, secondary hypersecretion arising from a problem in the hypothalamus. In clinical practice, this type of pathology is very rare.

Figure 7-20b shows a secondary hypersecretion of cortisol due to an ACTH-secreting tumor of the pituitary. Once again, the high levels of ACTH cause high cortisol production. But in this example, high cortisol has a negative feedback effect on the hypothalamus, decreasing production of CRH. The combination of low CRH and high ACTH isolates the problem to the pituitary. This pathology is responsible for about two-thirds of cortisol hypersecretion syndromes [*syn-*, together + *-drome*, running; a combination of symptoms characteristic of a particular pathology].

If cortisol levels are high but both trophic hormones are low, the problem must be a primary disorder (Fig. 7-20c). There are two possible explanations: *endogenous* [*endo-*, within + *-gen*, to be born] cortisol hypersecretion or the exogenous administration of cortisol for therapeutic reasons (see Fig. 7-19). In either case, high levels of cortisol act as a negative feedback signal that shuts off production of CRH and ACTH. The pattern of high cortisol with low trophic hormone levels points to a primary disorder.

When the problem is an adrenal tumor that is secreting cortisol in an unregulated fashion, the normal control pathways are totally ineffective. Although negative feedback shuts off the trophic hormones, the tumor is not dependent on them for cortisol production and cortisol secretion continues in their absence. The tumor must be removed or suppressed before cortisol secretion can be controlled.

CONCEPT CHECK   Answers: p. 238

19. Figure 7-21 ● shows two possible etiologies for hyposecretion of cortisol. Using your understanding of negative feedback in the hypothalamic-pituitary control pathway, decide whether the levels of CRH, ACTH, and cortisol will be high or low in each example.

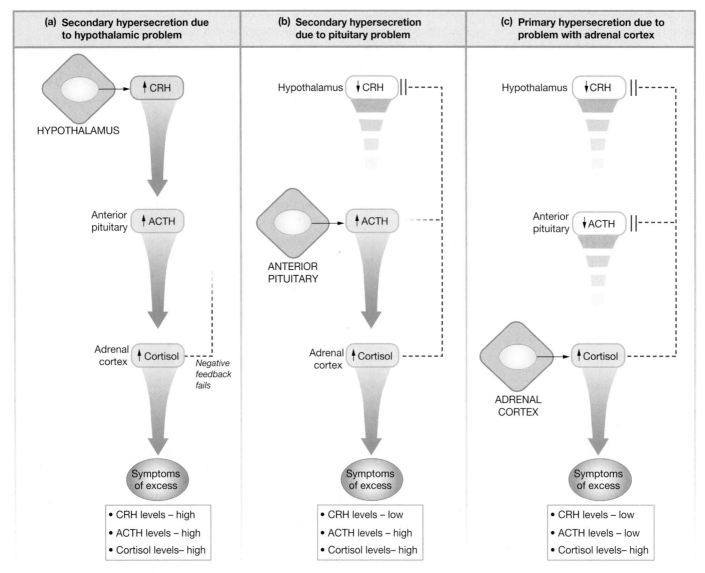

Figure 7-20 *Primary and secondary hypersecretion of cortisol* When there is a pathology in an endocrine gland, negative feedback fails.

Graves' disease is one form of thyroid gland hyperactivity. People with Graves' disease therefore have elevated thyroxine levels in the blood. Their TSH levels are very low.

Question 4: If levels of TSH are low and thyroxine levels are high, is Graves' disease a primary disoder or a secondary disorder (one that arises as a result of a problem with the anterior pituitary or hypothalamus)? Explain your answer.

▶ HORMONE EVOLUTION

Hormone structure and function in many cases have changed amazingly little from the most primitive vertebrates through the mammals. This situation is termed the *evolutionary conservation of* *hormone activity* and is demonstrated by the fact that some hormones from other organisms have biological activity when administered to humans. By studying which portions of a hormone molecule do not change from species to species, scientists have acquired important clues to aid in the design of agonist and antagonist drugs.

The ability of nonhuman hormones to work in humans was a critical factor in the birth of endocrinology. When Best and Banting discovered insulin in 1921 and the first diabetic patients were treated with the hormone, the insulin was extracted from cow, pig, or sheep pancreases. Until recently, slaughterhouses were the major source of insulin for the medical profession. Now, with genetic engineering, the human gene for insulin has been inserted into bacteria, which then synthesize the hormone, providing us with a plentiful source of human insulin.

Although many hormones have the same function in most vertebrates, a few hormones that play a significant role in the phys-

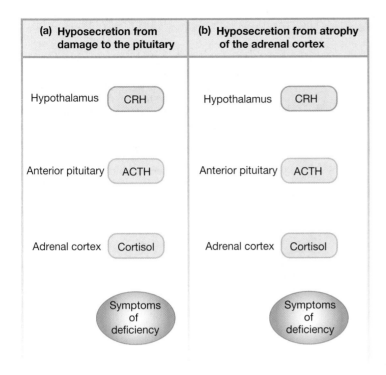

(a) Hyposecretion from damage to the pituitary	(b) Hyposecretion from atrophy of the adrenal cortex
Hypothalamus — CRH	Hypothalamus — CRH
Anterior pituitary — ACTH	Anterior pituitary — ACTH
Adrenal cortex — Cortisol	Adrenal cortex — Cortisol
Symptoms of deficiency	Symptoms of deficiency

FIGURE QUESTION

For each condition, use arrows to indicate whether levels of the three hormones in the pathway will be increased, decreased, or unchanged. Draw in negative feedback loops or indicate where feedback has failed.

● **Figure 7-21** *Patterns of hormone secretion in hypocortisolism*

iology of lower vertebrates seem to be evolutionarily "on their way out" in humans. Calcitonin is a good example of such a hormone. Although it plays a major role in calcium metabolism in fish, calcitonin apparently has no significant influence on daily calcium balance in adult humans. Neither calcitonin deficiency nor excess is associated with any pathological condition or symptom.

Although calcitonin is not a significant hormone in humans, the calcitonin gene does code for a biologically active protein. Cells in the brain process calcitonin gene mRNA to make a peptide known as *calcitonin gene-related peptide* (CGRP) that apparently acts as a neurotransmitter. The ability of one gene to produce multiple peptides is one reason that research is shifting from genomics to proteomics and physiology.

Some endocrine structures that are important in lower vertebrates are vestigial [*vestigium*, footprint] in humans. For example, **melanocyte-stimulating hormone** (MSH) from the intermediate lobe of the pituitary controls pigmentation in reptiles and am-

RUNNING PROBLEM

Researchers have learned that Graves' disease is an autoimmune disorder in which the body fails to recognize its own tissue. In this condition, the body produces antibodies that mimic TSH and bind to the TSH receptor, turning it on. This false signal fools the thyroid gland into overproducing thyroid hormone. More women than men are diagnosed with Graves' disease, perhaps because of the influence of female hormones on thyroid function. Stress and other environmental factors have also been implicated in hyperthyroidism.

Question 5: Antibodies are proteins that bind to the TSH receptor. From that information, what can you conclude about the cellular location of the TSH receptor?

Question 6: In Graves' disease, why doesn't negative feedback shut off thyroid hormone production before it becomes excessive?

 208 218 224 227 230 233

phibians. However, adult humans have only a rudimentary intermediate lobe and normally do not have measurable levels of MSH in their blood.

In the research arena, *comparative endocrinology*, the study of endocrinology in nonhuman organisms, has made significant contributions to our quest to understand the human body. Many of our models of human physiology are based on research carried out in fish or frogs or rats, to name a few. For example, the pineal gland hormone **melatonin** was discovered through research using tadpoles (Fig. 7-22 ●). Many small nonhuman vertebrates have short life cycles that facilitate studying aging or reproductive physiology. Genetically altered mice (transgenic or knockout mice) have provided researchers with valuable information about proteomics, the role of various proteins in physiological function.

Opponents of animal research argue that scientists should not experiment with animals at all and should only use cell cultures and computer models. Although those are valuable tools and can be helpful in the initial stages of medical research, at some point new drugs and procedures must be tested on intact organisms prior to clinical trials in humans. Responsible scientists follow guidelines for appropriate animal use and limit the number of animals killed to the minimum needed to provide valid data.

In this chapter we have examined how the endocrine system with its hormones helps regulate the slower processes in the body. In the next chapter, you will learn how the nervous system can take care of the more rapid responses needed to maintain homeostasis.

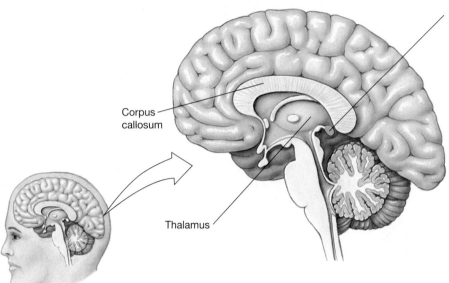

The pineal gland is a pea-sized structure buried deep in the brain of humans. Nearly 2000 years ago, this "seat of the soul" was thought to act as a valve that regulated the flow of vital spirits and knowledge into the brain. By 1950, however, scientists had decided that it was an evolutionary remnant with no known function.

About 1957 one of the wonderful coincidences of scientific research occurred. An investigator heard about a factor in beef pineal glands that could lighten the skin of amphibians. Using the classical methodology of endocrinology, he obtained pineal glands from a slaughterhouse and started making extracts. His biological assay consisted of dropping pineal extracts into bowls of live tadpoles to see if their skin color blanched. Several years and hundreds of thousands of pineal glands later, he had isolated a small amount of melatonin.

Melatonin is an amine hormone derived from the amino acid tryptophan.

Melantonin is the "darkness hormone," secreted at night as we sleep. It is the chemical messenger that transmits information about light-dark cycles to the brain center that governs the body's biological clock.

(Adapted from J. Arendt, *Clin. Endocrinol.* 29:205-229, 1988.)

● **Figure 7-22** *The pineal gland*

Ten years ago scientists and the popular press were trying to link melatonin to sexual function, the onset of puberty, seasonal affective depressive disorder (SADD) in the darker winter months, and sleep-wake cycles. At that time, the only function supported with scientific evidence was the hormone's ability to help shift timing of the body's internal clock, making it useful in overcoming jet lag. Now there is evidence that melatonin is a powerful antioxidant that has the potential to protect the body from the damage caused by free radicals. For a recent review, see "Melatonin: lowering the high price of free radicals," *News in Physiological Sciences* 15: 246-250, Oct. 2000 (*http://nips.physiology.org*).

Graves' Disease

In this running problem, you learned that in Graves' disease, thyroid hormone levels are high due to an immune system protein that mimics TSH. You also learned that the thyroid gland concentrates iodine for synthesis of thyroid hormones, and that radioactive iodine can concentrate in the gland and destroy the thyroid cells.

Graves' disease is the most common form of hyperthyroidism. Other famous people who have suffered from it include former U.S. President George H. W. Bush and First Lady Barbara Bush. To learn more about Graves' disease and other thyroid conditions, check out the Thyroid Foundation of America website at *www.tsh.org*. Check your answers to the problem questions by comparing them to those in the summary table below.

	QUESTION	FACTS	INTEGRATION AND ANALYSIS
1a	To which of the three classes of hormones do thyroid hormones belong?	The three classes of hormones are peptides, steroids, and amines.	Thyroid hormones are made from the amino acid tyrosine, therefore they are amines.
1b	If a person's diet is low in iodine, predict what happens to thyroxine production.	The thyroid gland concentrates iodine and combines it with the amino acid tyrosine to make thyroid hormones.	If iodine is lacking in the diet, a person will be unable to make thyroid hormone.
2a	In a normal person, when thyroid hormone levels in the blood increase, will negative feedback increase or decrease the secretion of TSH?	Negative feedback shuts off response loops.	Normally negative feedback will decrease TSH secretion.
2b	In a person with a hyperactive gland that is producing too much thyroid hormone, would you expect TSH to be higher or lower than in the normal person?		If thyroid hormone is high, you would expect strong negative feedback and even lower levels of TSH.
3	Why is radioactive iodine rather than some other radioactive element such as cobalt used to destroy thyroid tissue?	The thyroid gland concentrates iodine to make thyroid hormones.	Radioactive iodine will be concentrated in the thyroid gland and therefore selectively destroys that tissue. Other radioactive elements would distribute more widely throughout the body and might harm normal tissues.
4	If levels of TSH are low and thyroxine levels are high, is Graves' disease a primary disorder or a secondary disorder (one that arises as a result of a problem with the anterior pituitary or hypothalamus)? Explain your answer.	In secondary hypersecretion disorders, you would expect the levels of the hypothalamic and/or anterior pituitary trophic hormone levels to be elevated.	In Graves' disease, TSH from the anterior pituitary is very low. Therefore, the oversecretion of thyroid hormones is not the result of elevated TSH. This means that Graves' disease is a primary disorder that is caused by a problem in the thyroid gland itself.
5	Antibodies are proteins that bind to the TSH receptor. From that information, what can you conclude about the cellular location of the TSH receptor?	Receptors may be membrane receptors or intracellular receptors. Proteins cannot cross the cell membrane.	The TSH receptor is a membrane receptor. It uses the cAMP second messenger pathway for signal transduction.
6	In Graves' disease, why doesn't negative feedback shut off thyroid hormone production before it becomes excessive?	In normal negative feedback, as thyroid hormone increases, it shuts off TSH secretion. Without TSH stimulation, the thyroid stops producing thyroid hormone.	In Graves' disease, high levels of thyroid hormone have shut off TSH production. However, the thyroid gland still produces hormone in response to the abnormal immune protein. In this situation, negative feedback has failed.

208 218 224 227 230 231

CHAPTER SUMMARY

This chapter introduced you to the endocrine system and the role it plays in *communication* and *control* of physiological processes. As you've seen before, *compartmentation* of the body into intracellular and extracellular compartments requires special mechanisms to allow signals to pass from one compartment to another. The chapter also presented basic patterns that you will encounter again as you study various organ systems: differences between the three chemical classes of hormones, reflex pathways for hormones, types of hormone interaction, and endocrine pathologies.

Hormones

IP Endocrine System: Endocrine System Review

1. The specificity of a hormone depends on its receptors and their associated signal transduction pathways. (p. 213)

2. A **hormone** is a chemical secreted by a cell or group of cells into the blood that acts on a distant target and is effective at very low concentrations. (p. 211)

3. **Pheromones** are chemical signals secreted into the external environment. (p. 212)

4. Hormones bind to receptors to initiate their **cellular mechanism of action**. (p. 213)

5. Hormone activity is limited by ending secretion, removing hormone from the blood, or terminating activity at the target cell. (p. 213)

6. The rate of hormone breakdown is indicated by a hormone's **half-life**. (p. 213)

The Classification of Hormones

IP Endocrine System: Biochemistry, Secretion and Transport of Hormones, and the Actions of Hormones on Target Cells

7. There are three types of hormones: **peptide and protein hormones** composed of three or more amino acids, **steroid hormones** derived from cholesterol, and **amine hormones** derived from one type of amino acid. (p. 214)

8. Peptide hormones are made as inactive **preprohormones** and processed to **prohormones**. Prohormones are chopped into active hormone and peptide fragments that are co-secreted. (p. 215; Figs. 7-3, 7-4)

9. Peptide hormones dissolve in the plasma and have a short half-life. They bind to surface receptors and initiate rapid cellular responses through signal transduction. In some instances, peptide hormones also initiate synthesis of new proteins. (pp. 215–216; Fig. 7-5)

10. Steroid hormones are synthesized as they are needed. They are hydrophobic, and most steroid hormones in the blood are bound to protein carriers. Steroids have an extended half-life. (p. 216)

11. Traditional steroid receptors are inside the target cell, where they turn genes on or off and direct the synthesis of new proteins. Cell response is slower than with peptide hormones. Steroid hormones may bind to membrane receptors and have nongenomic effects. (p. 218; Fig. 7-7)

12. Amine hormones are derived from a single type of amino acid. They may behave like typical peptide hormones or like a combination of steroid and peptide hormone. (p. 218; Fig. 7-8)

Control of Hormone Release

IP Endocrine System: The Hypothalamic-Pituitary Axis

13. Classic endocrine cells act as both sensor and integrating center in the simple reflex pathway. (p. 219; Fig. 7-10)

14. Many endocrine reflexes involve the nervous system, either through **neurohormones** or by neurons that influence hormone release. (p. 221)

15. The pituitary gland is composed of the anterior pituitary, a true endocrine gland, and the posterior pituitary, an extension of the brain. (p. 221; Fig. 7-11)

16. The posterior pituitary releases two neurohormones, oxytocin and vasopressin, that are made in the hypothalamus. (p. 222; Fig. 7-12)

17. **Trophic hormones** control the secretion of other hormones. (p. 222)

18. Anterior pituitary hormones are controlled by releasing and inhibiting hormones from the hypothalamus. (p. 223; Fig. 7-13)

19. In complex endocrine reflexes, hormones of the pathway act as negative feedback signals. (p. 224; Fig. 7-14)

20. The hypothalamic trophic hormones reach the pituitary through the **hypothalamic-hypophyseal portal system**. (p. 225; Fig. 7-16)

21. There are six anterior pituitary hormones: prolactin, growth hormone, follicle-stimulating hormone, luteinizing hormone, thyroid-stimulating hormone, and adrenocorticotrophic hormone. (p. 222; Fig. 7-13)

Hormone Interactions

22. If the combination of two or more hormones yields a result that is more than additive, the interaction is **synergism**. (p. 226; Fig. 7-18)

23. If one hormone cannot exert its effects fully unless a second hormone is present, the second hormone is said to be **permissive** to the first. (p. 227)

24. If one hormone opposes the action of another, they are **antagonistic**. (p. 227)

Endocrine Pathologies

25. Diseases of hormone excess are usually due to **hypersecretion**. Symptoms of hormone deficiency occur when too little hormone is secreted (**hyposecretion**). **Abnormal tissue responsiveness** may result from problems with hormone receptors or signal transduction pathways. (p. 228)

26. **Primary pathologies** arise in the last endocrine gland in a reflex. A **secondary pathology** is a problem with one of the tissues producing trophic hormones. (p. 229; Fig. 7-20)

Hormone Evolution

27. Many human hormones are similar to hormones found in other vertebrate animals. (p. 230)

QUESTIONS

LEVEL ONE Reviewing Facts and Terms

1. The study of hormones is called _____.

2. List the three basic ways hormones act on their target cells.

3. List five endocrine glands and give one example of a hormone each secretes. Give one effect of each hormone you listed.

4. Match the researchers with their experiments:
   (a) Lower
   (b) Berthold
   (c) Guillemin and Shalley
   (d) Brown-Séquard
   (e) Banting and Best
   1. isolated trophic hormones from the hypothalami of pigs and sheep
   2. claimed sexual rejuvenation after he injected himself with testicular extracts
   3. isolated insulin
   4. accurately described the function of the pituitary gland
   5. studied comb development in roosters that had been castrated

5. Put these steps for identifying an endocrine gland in order:
   (a) Purify the extracts and separate the active substances.
   (b) Perform replacement therapy with the gland or its extracts and see if the abnormalities disappear.
   (c) Implant the gland or administer the extract from the gland to a normal animal and see if symptoms characteristic of hormone excess appear.
   (d) Put the subject into a state of hormone deficiency by removing the suspected gland. Monitor the development of abnormalities.

6. For a chemical to be defined as a hormone, it must be secreted into the _____ for transport to a _____ and take effect at _____ concentrations.

7. What is meant by the term *half-life* in connection with the activity of hormone molecules?

8. Metabolites are inactivated hormone molecules, broken down by enzymes found primarily in the _____ and _____, to be excreted in the _____ and _____, respectively.

9. Candidate hormones often have the word _____ as part of their name.

10. List and define the three chemical classes of hormones. Give an example of one hormone in each class.

11. Decide if each statement below applies best to peptide hormones, steroid hormones, amine hormones, all of these, or none of these.
    (a) are lipophobic and must use a signal transduction system
    (b) short half-life, measured in minutes
    (c) often have a lag time of 90 minutes before effects are noticeable
    (d) water-soluble, therefore easily dissolved in the extracellular fluid for transport
    (e) most hormones belong to this class
    (f) all hormones in this group are derived from cholesterol
    (g) consist of three or more amino acids linked together
    (h) released into the blood to travel to a distant target organ
    (i) transported in the blood bound to protein carrier molecules
    (j) all are lipophilic, so diffuse easily across membranes

12. Why do steroid hormones take so much longer to act than peptide hormones?

13. When steroid hormones act on the nucleus, the hormone-receptor complex acts as a/an _____ factor, binds to DNA, and activates one or more _____, which create mRNA to direct the synthesis of new _____.

14. Researchers have discovered that some cells have additional steroid hormone receptors on their _____, enabling a faster response.

15. The amino acids that give rise to the amine hormones are _____, the basis for melatonin, and _____, from which the catecholamines and thyroid hormones are made.

16. A hormone that controls the secretion of another hormone is known as a _____ hormone.

17. In reflex control pathways involving trophic hormones and multiple integrating centers, the hormones themselves act as _____ _____ signals, suppressing trophic hormone secretion earlier in the reflex.

18. What characteristic defines neurohormones?

19. List the two hormones secreted by the posterior pituitary gland. To what chemical class do they belong?

20. What is the hypothalamic-hypophyseal portal system? Why is it important?

21. List the six hormones of the anterior pituitary gland; give an action of each. Which ones are considered to be trophic hormones?

22. How do long- and short-loop negative feedback differ? Give an example.

23. When two hormones work together to create a result that is greater than additive, that interaction is called _____. When two hormones must both be present to achieve full expression of an effect, that interaction is called _____. When hormone activities oppose each other, that effect is called _____.

LEVEL TWO *Reviewing Concepts*

24. Map the following groups of terms. Add additional terms if you like.

List 1	List 2
co-secretion	ACTH
endoplasmic reticulum	anterior pituitary
exocytosis	blood
Golgi apparatus	endocrine cell
hormone receptor	gonadotropins
peptide hormone	growth hormone
preprohormone	hypothalamus
prohormone	inhibiting hormone
secretory vesicle	neurohormone
signal sequence	neuron
synthesis	oxytocin
target cell response	peptide/protein
	peripheral endocrine gland
	portal system
	posterior pituitary
	prolactin
	releasing hormone
	trophic hormone
	TSH
	vasopressin

25. Compare and contrast the following sets of terms:
    (a) paracrine, hormone, cytokine
    (b) primary and secondary endocrine pathologies
    (c) hypersecretion and hyposecretion
    (d) anterior and posterior pituitary

26. Compare and contrast the three chemical classes of hormones.

LEVEL THREE *Problem Solving*

27. You encountered the terms *specificity*, *receptors*, and *down-regulation* in other chapters in this text. Do their meanings change when applied to the endocrine system? What chemical and physical characteristics do hormones, enzymes, transport proteins, and receptors have in common that makes specificity important?

28. Dexamethasone is a drug that is used to suppress the secretion of adrenocorticotrophic hormone (ACTH) from the anterior pituitary. Two patients with hypersecretion of cortisol are given dexamethasone. Patient A's cortisol secretion level falls to normal levels as a result, but patient B's cortisol secretion remains elevated. Draw maps of the reflex pathways for these two patients (see Fig. 7-15 for a template) and use the maps to determine which patient has primary hypercortisolism. Explain your reasoning.

29. Some early experiments for male birth control pills used drugs that suppressed gonadotropin (FSH and LH) release. However, men given these drugs stopped taking them because they decreased testosterone secretion, which decreased their sex drive and caused impotence.

    (a) Use the information given in Figure 7-13 to draw a reflex pathway of the GnRH-FSH/LH-testosterone pathway. Use the pathway to show how suppressing gonadotropins decreases sperm production and testosterone secretion.

    (b) Researchers subsequently suggested that a better treatment would be to give men extra testosterone. Draw another copy of the reflex pathway to show how testosterone could suppress sperm production without the side effect of impotence.

LEVEL FOUR *Quantitative Problems*

30. The graph below represents the disappearance of a drug from the blood as the drug is metabolized and excreted. Based on the graph, what is the half-life of the drug?

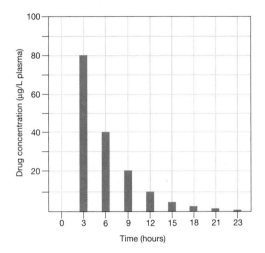

31. The graph in the next column shows plasma TSH in three groups of subjects. Pattern A is the normal subjects. Which

pattern would be consistent with the following pathologies? Explain your reasoning.

(a) primary hypothyroidism
(b) primary hyperthyroidism
(c) secondary hyperthyroidism

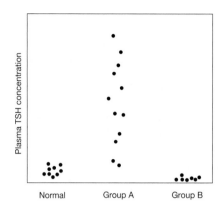

32. Based on what you have learned about the pathway for insulin secretion, draw and label a graph showing the effect of plasma glucose concentration on insulin secretion.

Explorations

Explorations are also available at *www.physiologyplace.com*

Exploration 1: Endocrinology Past and Present

Endocrinology is a relatively young area of physiology, dating only from the end of the nineteenth century. Some scientists wonder if it will survive the changes taking place as we learn more about different forms of chemical communication in the body. For one viewpoint, see "Endocrinology: Survival as a discipline in the twenty-first century?" *Annual Review of Physiology* 62: 947–950, 2000.

For a fascinating history of endocrinology, see the American Physiological Society's book *Endocrinology: People and Ideas* (S. M. McCann, editor. Oxford University Press, 1988).

And to see what's current in endocrinology research and medicine, go to the Endocrine Society homepage at *www.endo-society.org*. Features of interest include "Endocrinology in the News," information about endocrine clinical trials, and links to endocrine-related sites, including patient education.

Exploration 2: Pheromones in Humans

Are we sending chemical signals to attract a mate? See what *Time* magazine says in "The chemistry of love," February 18, 2002, p. 79.

Read some of the research studies on human pheromones for yourself to see if you believe they exist.

- "MHC-dependent mate preferences in humans," *Proceedings of the Royal Society of London, Series B: Biological Sciences* 260 (1359): 245–249, 22 June 1995. This is the study where women rated men's T-shirts.

- "Regulation of ovulation by human pheromones," *Nature* 392: 177–179, 1998. This is the study on changes in the menstrual cycles of young women.

- "The scented brain: pheromonal responses in humans," *Neuron* 31(4): 512–514, 30 Aug 2001. An editorial comment and review of a paper showing changes in brain imaging in response to smelling putative pheromones.

- "The scent of fear," *Neuroendocrinology Letters* 23(2): 79–84, Apr. 2002. This study showed that subjects were able to discriminate between the odors of underarm axillary pads worn when watching a terrifying movie and watching a "neutral" movie.

ANSWERS

Page 213

1. Glucose enters cells by facilitated diffusion (GLUT transporters).

Page 214

2. The suffix *-ase* indicates an enzyme. A *peptidase* would digest peptides.

Page 214

3. A hormone is a chemical secreted into the blood where it acts on a distant target at very low concentrations.

4. Steroid-producing cells would have extensive smooth endoplasmic reticulum; protein-producing cells would have lots of rough endoplasmic reticulum.

Page 218

5. The three classes are peptide, steroid, and amine.

6. The shorter half-life suggests that aldosterone is not bound to plasma proteins as much as other steroid hormones.

Page 219

7. The input signal is the stimulus, receptor, and afferent path. Integration takes place in the integrating center. The output signal is the efferent path.

8. Increased blood glucose is the stimulus. Insulin secretion is the efferent path (output signal). Decrease in blood glucose is the response.

9. Insulin release by blood glucose is #6. Insulin release following a meal is #3. Insulin release in response to a GI hormone is closest to #4 although you are not told if the GI hormone is neurosecretory. If it is not, there is no matching path.

Page 220

10. Stimulus: decreased Ca^{2+}. Sensor/integrating center: PTH cells. Efferent path: PTH. Effector: target tissues. Response: increased Ca^{2+}.

Page 221

11. Catecholamines are amine hormones.

Page 222

12. Microtubules of the cytoskeleton move secretory vesicles (● p. 58).

13. Contents are released by exocytosis.

Page 224

14. Pathway #5 fits the complex pattern described.

Page 225

15. (a) #4, (b) #4 for GH acting directly on targets and #5 for GH acting on the liver.

Page 225

16. The target is endocrine cells of the anterior pituitary.

17. The pathway shuts off when the meal is digested and absorbed. A decrease in blood glucose also will serve as a negative feedback signal.

Page 228

18. Cortisol is low; low negative feedback makes CRH and ACTH high.

Page 229

19. Secondary problem at pituitary: CRH up, ACTH low, cortisol low.

    Primary problem: CRH up, ACTH up, cortisol low.

Page 219

Fig. 7-8: The conversion of tyrosine to dopamine adds a hydroxyl (—OH) group to the 6-carbon ring and changes the carboxyl (—COOH) group to a hydrogen. Norepinephrine is made from dopamine by changing one hydrogen to a hydroxyl group. Epinephrine is made from norepinephrine by changing one hydrogen attached to the nitrogen to a methyl (—CH_3) group.

Page 220

Fig. 7-9: The pathway begun by eating a meal shuts off when the meal is digested and absorbed from the digestive tract.

Page 224

Fig. 7-15: In short-loop negative feedback, ACTH inhibits CRH.

Page 231

Fig. 7-21: See answer for Concept Check question on page 229, shown above.

Page 237

30. Half-life is 6 hours.
31. (a) Group A (b) Group B (c) Group A

32. X-axis is plasma glucose concentration. Y-axis is insulin secretion. As X increases, Y increases.

Neurons:
Cellular and Network Properties

8

Vesicles filled with neurotransmitter (green) cluster near the end of a neuron as it synapses on its target.

The future of clinical neurology and psychiatry is intimately tied to that of molecular neural science.

—Eric R. Kandel, James H. Schwartz, and Thomas M. Jessell,
in the preface to their book, *Principles of Neural Science*, 2000

RUNNING PROBLEM

Mysterious Paralysis

"Like a polio ward from the 1950s" is how Guy McKhann, M.D., a neurology specialist at the Johns Hopkins School of Medicine, describes a section of Beijing Hospital that he visited on a trip to China in 1986. Dozens of paralyzed children—some attached to respirators to assist their breathing—filled the ward to overflowing. The Chinese doctors thought the children had Guillain-Barré syndrome (GBS), an unusually rare paralytic condition. But Dr. McKhann wasn't as convinced. There were simply too many stricken children for the illness to be the rare Guillain-Barré syndrome. Was it polio—as some of the Beijing staff feared? Or was it another illness, perhaps one that had not yet been discovered?

241 244 261 262 267 274 276

BACKGROUND BASICS

Reflex pathways: **191** | Cell organelles: **53** | Neurohormones: **174** | Gated channels: **130–131** | Extracellular matrix: **181** | Gap junctions: **172** | Positive feedback loops: **195** | Phagocytosis: **144** | Antagonistic control: **188** | Neurocrine secretions: **174** | Resting membrane potential: **157** | Exocytosis: **146–147** | Equilibrium potential: **159** | Biological electricity: **157** |

In an eerie scene from a science fiction movie, white-coated technicians move quietly through a room filled with bubbling cylindrical fish tanks. As the camera zooms in on one tank, no fish are seen darting through aquatic plants. The lone occupant of the tank is a gray mass with a convoluted surface like a walnut and a long tail that appears to be edged with beads. Floating off the beads are hundreds of fine fibers, waving softly as the oxygen bubbles weave through them on their way to the surface. This is no sea creature…it is a brain and spinal cord, removed from its original owner and awaiting transplantation into another body. Can this be real? Is this scenario possible? Or is it just the creation of an imaginative movie screenwriter?

The brain is regarded as the seat of the soul, the mysterious source of those traits that we think of as setting humans apart from animals. The brain and spinal cord are also integrating centers for homeostasis, movement, and many other body functions. They are the control center of the **nervous system**, a network of billions or trillions of nerve cells linked together in a highly organized manner to form the rapid control system of the body.

Nerve cells, or **neurons**, are designed to carry electrical signals rapidly over long distances. They are uniquely shaped cells, and most have long, thin extensions, or *processes*, that can extend up to a meter in length. In most pathways, neurons communicate with chemical signals called **neurotransmitters** across a small gap known as a *synapse*. In a few pathways, neurons are linked by *gap junctions* [● p. 172], allowing electrical signals to pass directly from cell to cell.

Reflex pathways in the nervous system do not necessarily follow a straight line from one neuron to the next. One neuron may influence multiple neurons, or many neurons may affect the function of a single neuron. The intricacy of neural networks underlies the **emergent properties** of the nervous system: complex processes such as consciousness, intelligence, and emotion, which cannot be predicted from what we know about the properties of individual nerve cells. The search to explain emergent properties makes neurophysiology one of the most active research areas in physiology today.

Neurophysiology, like many other areas of science, has its own specialized language. In many instances, multiple terms describe one structure or function, which potentially can lead to confusion. Table 8-1 lists some neurophysiology terms used in this book along with their common synonyms.

▶ ORGANIZATION OF THE NERVOUS SYSTEM

Information flow from the internal and external environments through the nervous system follows the basic reflex pattern described in Chapter 6 [● p. 190]. Sensory receptors throughout the body continuously monitor conditions in both the internal and external environments (Fig. 8-1 ●). These receptors send information along **afferent**, or **sensory**, **neurons** to the central nervous

Table 8-1: Synonyms in Neuroscience

TERM USED IN THIS BOOK	SYNONYM
Action potential	Spike, nerve impulse, conduction signal
Autonomic nervous system	Visceral nervous system
Axon	Nerve fiber
Axonal transport	Axoplasmic flow
Axon terminal	Synaptic knob, synaptic bouton, presynaptic terminal
Axoplasm	Cytoplasm of the axon
Cell body	Cell soma
Cell membrane of the axon	Axolemma
Glial cells	Neuroglia, glia
Interneuron	Association neuron
Rough endoplasmic reticulum	Nissl substance, Nissl body
Sensory neuron	Afferent neuron, afferent

system. Afferent pathways that carry incoming information and their associated sensory receptors will be discussed in Chapter 10.

The **central nervous system**, or **CNS**, is the integrating center for neural reflexes. CNS neurons in the **brain** and **spinal cord** integrate incoming information and determine whether a response is needed. Chapter 9 examines the complex structure and functions of the CNS.

The CNS sends output signals directing an appropriate response (if any) through **efferent neurons** to the effector cells of the body. Efferent neurons are subdivided into the **somatic motor division**,* which controls skeletal muscles, and the **autonomic division**, which controls smooth and cardiac muscles, exocrine and some endocrine glands, and some types of adipose tissue. These efferent neurons together with the afferent sensory neurons form the **peripheral nervous system** (sometimes called the PNS). The PNS will be discussed in Chapter 11.

The autonomic division is also called the *visceral nervous system* because it controls contraction and secretion in the various internal organs [*viscera*, internal organs]. Autonomic neurons are further divided into **sympathetic** and **parasympathetic branches**, which can be distinguished by their anatomical organization and by the chemicals they use to communicate with their target cells. Many internal organs receive innervation from both types of autonomic neurons, and it is a common pattern to find that the two divisions exert *antagonistic control* over a single target [● p. 188].

---

*The expression *motor neuron* can be used to refer to all efferent neurons. However, clinically, the term *motor neuron* (or *motoneuron*) is usually used to describe somatic motor neurons that control skeletal muscles.

● **Figure 8-1** *Organization of the nervous system* The peripheral nervous system (PNS) sends information to the central nervous system (CNS) through afferent (sensory) neurons and takes information from the CNS to target cells via efferent neurons. The enteric nervous system can act autonomously or can be controlled by the CNS through the autonomic division of the PNS.

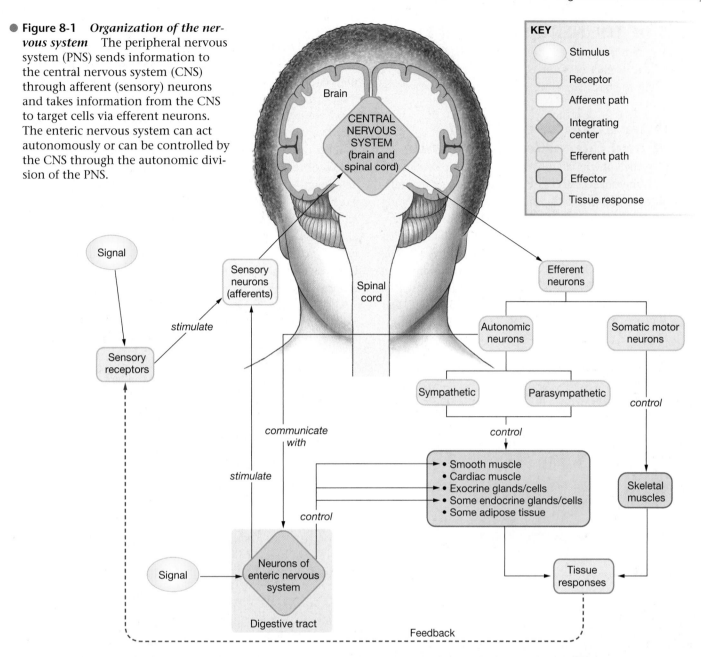

In recent years, a third division of the nervous system has received considerable attention. The **enteric nervous system** is a network of neurons in the walls of the digestive tract. It is frequently controlled by the autonomic division of the nervous system, but it is also able to function autonomously as its own integrating center. We will discuss the enteric nervous system further in Chapter 21 on the digestive system.

Although most of this book will focus on the role of neural reflexes in communication, coordination, and homeostasis in the body, it is important to note that significant processes in the central nervous system can take place without input or output from the peripheral nervous system. The CNS has the ability to initiate activity without sensory input and it need not create any measurable output. Two examples are thinking and dreaming, complex higher brain function that can take place totally within the CNS.

CONCEPT CHECK Answers: p. 282

1. Organize these terms describing functional types of neurons into a map or outline: afferent, autonomic, brain, central, efferent, enteric, parasympathetic, peripheral, sensory, somatic motor, spinal, sympathetic.

RUNNING PROBLEM

Guillain-Barré syndrome is a rare paralytic condition that strikes after a viral infection or immunization. There is no cure, but usually the paralysis and lost sensation slowly disappear. In classic Guillain-Barré, patients can neither feel sensations nor move their muscles.

Question 1: Which division(s) of the nervous system may be involved in Guillain-Barré syndrome?

239   244 261 262 267 274 276

▶ CELLS OF THE NERVOUS SYSTEM

The nervous system is composed primarily of two cell types: support cells known as glial cells (or glia or neuroglia) and nerve cells, or neurons, the basic signaling units of the nervous system.

Neurons Are Excitable Cells That Generate and Carry Electrical Signals

The neuron is the functional unit of the nervous system. (A *functional unit* is the smallest structure that can carry out the functions of a system.) Neurons are uniquely shaped cells with long appendages that extend outward from the cell body. These appendages, or **processes**, are usually classified as either *dendrites* (which receive incoming signals) or *axons* (which carry outgoing information). The shape, number, and length of axons and dendrites vary from one neuron to the next, but they are an essential feature that allows neurons to communicate with each other and with other cells.

Traditionally we use efferent neurons similar to the model neurons shown in Figure 8-2 ● and Figure 8-3e ● to teach how a neuron functions. In other neuron types, the processes shown in the model may be missing or modified. However, if you understand these model neurons, you will understand how most neurons work.

Neurons in the nervous system may be classified either structurally or functionally. Structurally, neurons are classified by the number of processes that originate from the cell body. They may be described as *pseudounipolar* (axon and dendrites fuse during development to create one long process; Fig. 8-3a), *bipolar* (single axon and single dendrite; Fig. 8-3b), *multipolar* (many dendrites and branched axons; Fig. 8-3d), or *anaxonic* (lacking an identifiable axon; Fig. 8-3c). Because physiology is concerned chiefly with function, we will classify neurons according to their function: afferent (sensory) neurons, interneurons, and efferent (somatic motor and autonomic) neurons.

Afferent **sensory neurons** carry information about temperature, pressure, light, and other stimuli from sensory receptors to

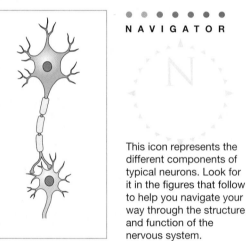

NAVIGATOR

This icon represents the different components of typical neurons. Look for it in the figures that follow to help you navigate your way through the structure and function of the nervous system.

● **Figure 8-2** *Model neuron* A typical neuron has a cell body and numerous extensions. The region where axon terminals communicate with their target cells is known as a synapse.

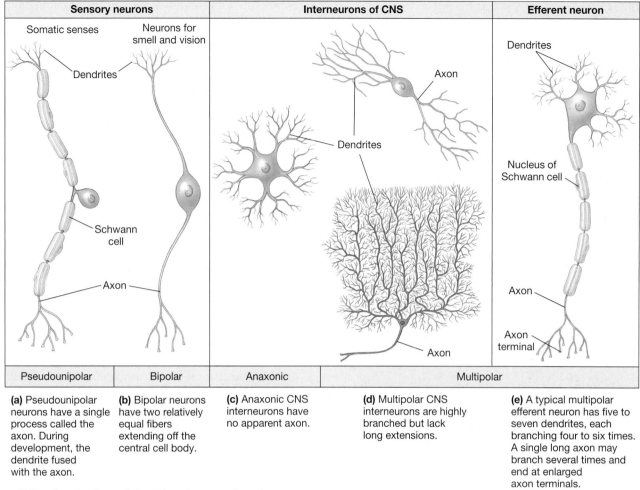

Sensory neurons		Interneurons of CNS		Efferent neuron
Pseudounipolar	Bipolar	Anaxonic	Multipolar	

(a) Pseudounipolar neurons have a single process called the axon. During development, the dendrite fused with the axon.

(b) Bipolar neurons have two relatively equal fibers extending off the central cell body.

(c) Anaxonic CNS interneurons have no apparent axon.

(d) Multipolar CNS interneurons are highly branched but lack long extensions.

(e) A typical multipolar efferent neuron has five to seven dendrites, each branching four to six times. A single long axon may branch several times and end at enlarged axon terminals.

● **Figure 8-3** *Anatomic and functional categories of neurons*

the CNS. They vary from the model neuron in the length and organization of their processes. For example, somatic sensory neurons have cell bodies that are found close to the CNS, with very long processes that extend to receptors in the limbs. In these neurons, the cell body is out of the direct path of signals originating at the dendrites (Fig. 8-3a). In contrast, sensory neurons for smell and vision have two long extensions, one an axon and one a dendrite (Fig. 8-3b). Signals that begin at the dendrites travel through the cell body to the axon.

Neurons that lie entirely within the CNS are known as **interneurons** (short for *interconnecting neurons*). They come in a variety of forms but often have quite complex branching processes that allow them to form synapses with many other neurons (Fig. 8-3c). Some interneurons are quite small compared to the model neuron.

The **efferent neurons** of the somatic motor and autonomic divisions are generally very similar to the model neuron in Figure 8-2. In the autonomic division, some neurons have enlarged regions along the axon called **varicosities** (see Fig. 11-6). The varicosities store and release neurotransmitter.

The long axons of peripheral neurons are bundled together with connective tissue into cordlike fibers called **nerves** that ex-

tend from the CNS to the targets of the component neurons. Nerves may carry afferent signals only (**sensory nerves**), efferent signals only (**motor nerves**), or signals in both directions (**mixed nerves**). Many nerves are large enough to be seen with the naked eye and have been given anatomical names. For example, the *phrenic nerve* runs from the spinal cord to the muscles of the diaphragm.

The Cell Body Is the Control Center of the Neuron

The nerve **cell body** (*cell soma*) resembles the typical cell described in Chapter 3, with a nucleus and all organelles needed to direct cellular activity [● p. 53]. An extensive cytoskeleton extends outward into the axon and dendrites. The position of the cell body varies in different types of neurons, but in most neurons the cell body is small, generally making up one-tenth or less of the total cell volume.

Despite its small size, the cell body with its nucleus is essential to the well-being of the cell. If a neuron is cut apart, any sections separated from the cell body are likely to degenerate slowly and die because they lack the cellular machinery to make essential proteins. With injured somatic motor neurons whose axon has been severed, the degeneration of the distal (distant) portions

In classic Guillain-Barré syndrome, the disease affects both sensory and somatic motor neurons. Dr. McKhann observed that, although the Beijing children could not move their muscles, they could feel a pin prick.

Question 2: Do you think that the paralysis found in the Chinese children affected both sensory and somatic motor neurons? Why or why not?

239 241   261 262 267 274 276

can cause permanent paralysis of the muscles that are *innervated* by the neuron. (The term *innervated* means "controlled by a neuron.") If the damaged neuron is a sensory neuron, the person may experience loss of sensation (numbness or tingling) in the region previously innervated by the neuron. Scientists are trying to understand the cellular events that occur when neurons are damaged so that they can develop procedures to repair the functional loss that accompanies spinal cord and other neurological injuries.

Dendrites Receive Incoming Signals Dendrites [*dendron*, tree] are thin, branched processes that receive incoming information from neighboring cells. Dendrites increase the surface area of a neuron, allowing it to communicate with multiple other neurons. The simplest neurons have only a single dendrite. At the other extreme, neurons in the brain may have multiple dendrites with incredibly complex branching (Fig. 8-3d). A dendrite's surface area can be expanded even more by the presence of **dendritic spines** that vary from thin spikes to mushroom-shaped knobs.

The primary function of dendrites in the peripheral nervous system is to receive incoming information and transfer it to an integrating region within the neuron. Within the CNS, dendrite function is more complex. Dendritic spines can function as independent compartments, sending signals back and forth with other neurons in the brain. Many dendritic spines contain polyribosomes and can make their own proteins. Changes in spine morphology are associated with learning and memory as well as with various pathologies, such as genetic disorders that cause mental retardation and degenerative diseases such as Alzheimer's disease. Because of these associations, dendritic spines are a hot topic in neuroscience research.

Axons Carry Outgoing Signals to the Target Most peripheral neurons have a single **axon**, a projection of the cell body that originates from a special region of the cell body called the **axon hillock** (Fig. 8-2). Axons vary from over a meter in length to only a few micrometers long.

They often branch sparsely along their length, forming **collaterals** [*col-*, with + *lateral*, something on the side]. In our model neuron, each collateral ends in a swelling called an **axon**

terminal. The axon terminal contains mitochondria and membrane-bound vesicles filled with *neurocrine* molecules [p. 174].

Functionally, axons transmit outgoing electrical signals from the integrating center of the neuron to the end of the axon. At the distal end of the axon, the electrical signal is usually translated into a chemical message: secretion of a neurotransmitter, neuromodulator, or neurohormone. Neurons that secrete neurotransmitters and neuromodulators terminate near their target cells, which are other neurons, muscles, or glands.

The region where an axon terminal meets its target cell is called a **synapse** [*syn-*, together + *hapsis*, to join]. The neuron that delivers the signal to the synapse is known as the **presynaptic cell**, and the cell that receives the signal is called the **postsynaptic cell**. The narrow space between the two cells is called the **synaptic cleft**.

2. Where do neurons that secrete neurohormones terminate?

3. Draw a chain of three neurons that synapse on each other in sequence. Label the presynaptic and postsynaptic ends of each neuron, the cell bodies, dendrites, axons, and axon terminals.

Axons are designed to convey chemical and electrical signals. Their cytoplasm is filled with many types of fibers and filaments but lacks ribosomes and endoplasmic reticulum. Therefore, any proteins destined for the axon or the distant axon terminal must be synthesized on the rough endoplasmic reticulum in the cell body. The proteins are then packaged into vesicles and moved down the axon by a process known as **axonal transport**.

Slow axonal transport moves material by **axoplasmic** (cytoplasmic) **flow** from the cell body to the axon terminal. Material moves only at a rate of 0.2–2.5 mm/day; therefore, slow transport is used for components that are not consumed rapidly by the cell, such as enzymes and cytoskeleton proteins.

Fast axonal transport (Fig. 8-4) moves organelles at rates of up to 400 mm (about 15.75 in.) per day. The neuron uses stationary microtubules as tracks along which transported particles "walk" with the aid of attached footlike proteins. These motor proteins alternately bind and unbind to the microtubules with the help of ATP, stepping their attached organelles along the axon in a stop-and-go fashion. The role of motor proteins in axonal transport is similar to their role in muscle contraction and in the movement of chromosomes during cell division.

Fast axonal transport goes in two directions. Forward, or *anterograde*, transport moves synaptic and secretory vesicles and mitochondria from the cell body to the axon terminal. Backward, or *retrograde*, transport returns old membrane components from the axon terminal to the cell body for recycling. There is evidence that nerve growth factors and some viruses also reach the cell body by fast retrograde transport.

● **Figure 8-4** *Axonal transport of membranous organelles*

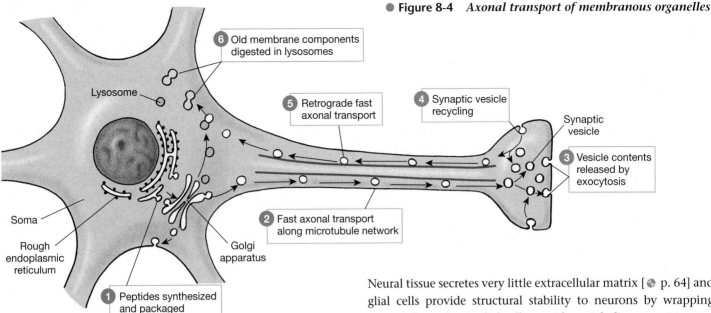

6 Old membrane components digested in lysosomes

Lysosome

5 Retrograde fast axonal transport

4 Synaptic vesicle recycling

Synaptic vesicle

3 Vesicle contents released by exocytosis

Soma

Rough endoplasmic reticulum

Golgi apparatus

2 Fast axonal transport along microtubule network

1 Peptides synthesized and packaged

Glial Cells Are the Support Cells of the Nervous System

Glial cells [*glia*, glue] are the "unsung heroes" of the nervous system, outnumbering neurons in the nervous system by 10–50:1 (Fig. 8-5 ●). Although glial cells do not participate directly in the transmission of electrical signals over long distances, they provide important physical and biochemical support to neurons.

Neural tissue secretes very little extracellular matrix [❀ p. 64] and glial cells provide structural stability to neurons by wrapping around them. Some glial cells provide metabolic support to neurons and they help maintain homeostasis of the brain's extracellular fluid by taking up excess metabolites and K⁺.

Glial cells communicate with neurons and with each other primarily through chemical signals. Glial-derived growth and *trophic* (nourishing) factors help maintain neurons and guide them during repair and development. Research suggests that glial cells in turn respond to neurotransmitters and neuromodulators secreted by neurons. Glial cell function is an active area of neurophysiology

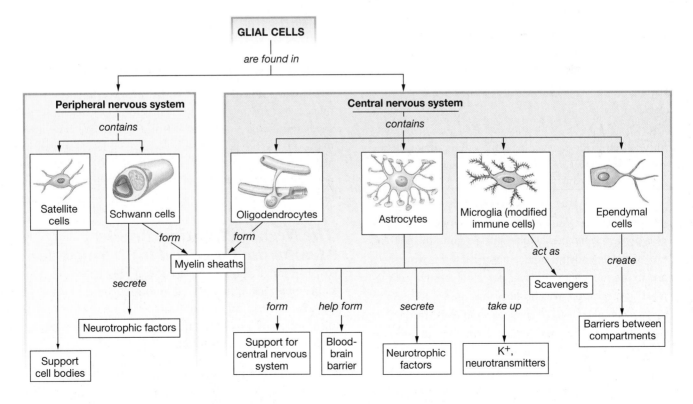

● **Figure 8-5** *Glial cells and their functions* Oligodendrocytes form myelin around portions of several interneuron axons. Astrocytes contact both neurons and blood vessels, but do not form myelin. Microglia are wandering scavenger cells. Ependymal cells create an epithelial barrier.

245

research, and we still do not fully understand all the roles these important cells play in the nervous system.

The peripheral nervous system has two types of glial cells: Schwann cells and satellite cells. The CNS has four types of glial support cells: oligodendrocytes, astrocytes, microglia, and ependymal cells.

Astrocytes [*astron*, a star] are highly branched cells that contact neurons and blood vessels and may transfer nutrients between the two. In addition, they take up K^+ and neurotransmitters from the extracellular fluid to help maintain homeostasis in the extracellular fluid around the neurons. Astrocytes communicate with each other through gap junctions.

Microglia are specialized immune cells that reside permanently within the CNS. When activated, they remove damaged cells and foreign invaders. **Ependymal cells** are epithelial cells that create a selectively permeable barrier between compartments in the brain. In addition, it now appears the ependymal layer is a source of neural stem cells to replace neurons that die or are damaged (see Emerging Concepts box: Neural stem cells).

Schwann cells in the peripheral nervous system and **oligodendrocytes** in the CNS support and insulate axons by creating **myelin**, multiple concentric layers of phospholipid membrane (Fig. 8-6a,c ●). Myelin forms when these glial cells wrap around an axon, squeezing out the glial cytoplasm so that each wrap becomes two membrane layers (Fig. 8-6b). As an analogy, think of wrapping a deflated balloon tightly around a pencil. Some neurons have as many as 150 wraps (300 membrane layers) in the myelin sheath that surrounds their axons. Gap junctions connect the membrane layers and allow flow of nutrients and information from layer to layer.

In the CNS, one oligodendrocyte forms myelin around portions of several axons (Fig. 8-6a). In the peripheral nervous system, each Schwann cell associates with a single axon. One axon may have as many as 500 Schwann cells, each wrapped around a 1–1.5 mm segment of the axon (Fig. 8-6c). Between the insulated areas, a tiny region of axon membrane remains in direct contact with the extracellular fluid. These gaps, called the **nodes of Ranvier**, play an important role in the transmission of electrical signals along the axon.

The final type of glial cell is a nonmyelinating Schwann cell known as a **satellite cell**. These cells form supportive capsules around nerve cell bodies located in ganglia. A **ganglion** [cluster or knot] is a cluster of nerve cell bodies found outside the CNS. Ganglia appear as knots or swellings along a nerve. The equivalent structure inside the CNS is known as a **nucleus** [plural, *nuclei*].

---

4. What is the primary function of the following: myelin, microglia, ependymal cells?

5. Name the two glial cell types that form myelin. How do they differ?

▶ ELECTRICAL SIGNALS IN NEURONS

The unique property of nerve and muscle cells that characterizes them as *excitable tissues* is their ability to propagate electrical signals when ions move across the cell membrane. We now know that many other cell types generate electrical signals to initiate intracellular processes (see insulin secretion, p. 157), but the ability of nerve and muscle cells to send a constant electrical signal over long distance is characteristic of electrical signaling in these tissues.

The Nernst Equation Predicts Membrane Potential for a Single Ion

Recall from Chapter 5 that all living cells have a resting membrane potential difference [● p. 157]. The *membrane potential* (V_m) is the electrical disequilibrium that results from the uneven distribution of ions across the cell membrane. Two factors influence the membrane potential:

1. **concentration gradients** of ions across the membrane. Normally, sodium (Na^+), chloride (Cl^-), and calcium (Ca^{2+}) are more concentrated in the extracellular fluid than in the cytosol. Potassium (K^+) is more concentrated in the cells than in the extracellular fluid.

Neural Stem Cells

Mature neurons are so highly specialized in their structure that they are unable to undergo mitosis and reproduce themselves. Scientists used to believe that once a neuron died, it could never be replaced. However, one of the newest breakthroughs in science is the discovery of neural *stem cells* [● p. 76], immature cells that can differentiate into neurons. During early development a primitive tissue called *neuroepithelium* surrounds the neural tube, which will later become the cavities of the brain and spinal cord. As development proceeds, some cells migrate out of the neuroepithelium and differentiate into neurons. Other cells bordering the neural tube specialize into the epithelium of the ependyma. And among the ependymal cells and in the subependymal layer, some neural stem cells remain unspecialized, waiting until they are called upon to replace damaged cells. In some exciting work, researchers have been able to coax neural stem cells to produce dopamine, a neurotransmitter that is deficient in patients suffering from Parkinson's disease. It now appears possible that in the coming years we may see stem cell transplants designed to reverse the loss of function that comes with a number of different degenerative neurological diseases.

(a) Glial cells of the central nervous system

(b) Myelin formation in the peripheral nervous system

(c) Each Schwann cell forms myelin around a small segment of one axon.

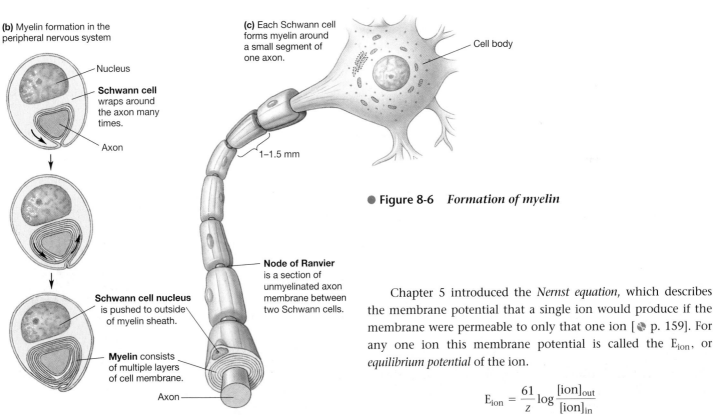

● **Figure 8-6**   *Formation of myelin*

Chapter 5 introduced the *Nernst equation*, which describes the membrane potential that a single ion would produce if the membrane were permeable to only that one ion [● p. 159]. For any one ion this membrane potential is called the E_{ion}, or *equilibrium potential* of the ion.

$$E_{ion} = \frac{61}{z} \log \frac{[ion]_{out}}{[ion]_{in}}$$

where:

2. **membrane permeability** to those ions. The resting cell membrane is much more permeable to K^+ than to Na^+ or Ca^{2+}. This makes K^+ the major ion contributing to the resting membrane potential.

z is the electrical charge on the ion ($+1$ for K^+)
[ion] are the ion concentrations inside and outside the cell
61 is a combination of several constants plus body temperature of 37° C

Using estimated concentrations for K^+ in the neuron and extracellular compartment (Table 8-2), the Nernst equation predicts a membrane potential of -90 mV. However, an average value for the V_m of neurons is -70 mV (inside the cell relative to outside), more positive than predicted by the potassium equilibrium potential. This means that other ions must be contributing to the membrane potential. Neurons at rest are slightly permeable to Na^+ and the leak of positive Na^+ into the cell makes the resting membrane potential slightly more positive than it would be if the cell were only permeable to K^+.

CONCEPT CHECK   Answers: p. 282

6. Given the values in Table 8-2, use the Nernst equation to calculate the equilibrium potential for Ca^{2+}. Use the information about logarithms given in Appendix A and try the calculations without a calculator.

The GHK Equation Predicts Membrane Potential Using Multiple Ions

The **Goldman-Hodgkin-Katz (GHK) equation** is used to calculate the resting membrane potential that results from the contribution of all ions that can cross the membrane. The GHK equation includes membrane permeability values because the permeability of an ion will influence its contribution to the membrane potential. If the membrane is not permeable to an ion, it will not affect the membrane potential.

For mammalian cells, we assume that Na^+, K^+, and Cl^- are the three ions that influence membrane potential in resting cells. Each ion's contribution to the membrane potential is proportional to its ability to cross the membrane. The GHK equation for cells that are permeable to Na^+, K^+, and Cl^- is

$$V_m = 60 \log \frac{P_K[K^+]_{out} + P_{Na}[Na^+]_{out} + P_{Cl}[Cl^-]_{in}}{P_K[K^+]_{in} + P_{Na}[Na^+]_{in} + P_{Cl}[Cl^-]_{out}}$$

where:

V_m = resting membrane potential at $37°$ C

P = relative permeability of the membrane to the ion

[ion] = ion concentration (in or out of cell)

Although this equation looks quite intimidating, it can be simplified into words to say:

Resting membrane potential =
   combined contributions of the (concentration gradient) × (membrane permeability) for each ion

If the membrane is not permeable to an ion, the permeability term in the equation will be zero and the ion drops out of the picture. For example, cells at rest normally are not permeable to Ca^{2+} and therefore Ca^{2+} is not part of the GHK equation.

The GHK equation predicts resting membrane potentials based on given ion concentrations and membrane permeabilities, and it explains how the cell's slight permeability to Na^+ makes the resting membrane potential more positive than the E_K. The equation can also be used to predict what happens to the membrane potential when ion concentrations or membrane permeabilities change. This makes the GHK equation a valuable tool for studying changes in membrane potential that are the basis for electrical signaling in the body.

Ion Movement Across the Cell Membrane Creates Electrical Signals

The resting membrane potential of living cells is determined primarily by the K^+ concentration gradient and the cell's resting permeability to K^+, Na^+, and Cl^-. However, a change in either the K^+ concentration gradient or ion permeabilities will change the membrane potential. An increase in Na^+ permeability plays a key role in generating electrical signals in excitable tissues.

At rest, the cell membrane of a neuron is only slightly permeable to Na^+. However, if the membrane suddenly increases its Na^+ permeability, Na^+ will enter the cell, moving down its electrochemical gradient [● p. 157]. The addition of positive Na^+ to the intracellular fluid *depolarizes* the cell membrane [● Fig. 5-41, p. 161] and creates an electrical signal.

The movement of ions across the membrane can also *hyperpolarize* a cell. If the cell membrane suddenly becomes more permeable to K^+, positive charge is lost from inside the cell and the cell becomes more negative (hyperpolarizes). A cell may also hyperpolarize if negatively charged ions such as Cl^- enter the cell from the extracellular fluid.

Table 8-2: Ion Concentrations and Equilibrium Potentials

ION	EXTRACELLULAR FLUID (mM)	INTRACELLULAR FLUID (mM)	E_{ION} AT $37°$ C
K^+	5 mM (normal range 3.5–5)	150 mM	-90 mV
Na^+	145 mM (normal range 135–145)	15 mM	$+60$ mV
Cl^-	108 mM (normal range 100–108)	10 mM (range 5–15)	-63 mV
Ca^{2+}	1 mM	0.0001 mM	see Concept Check question above

7. Would a cell with a resting membrane potential of -70 mV depolarize or hyperpolarize in the following cases? You must consider both the concentration and electrical gradients of the ion to determine net ion movement.

   (a) Cell becomes more permeable to Ca^{2+}.

   (b) Cell becomes less permeable to K^+.

8. Would the cell membrane depolarize or hyperpolarize if a small amount of Na^+ leaked into the cell?

It is important to understand that a change in membrane potential from -70 mV to a positive value such as $+30$ mV *does not mean that the ion concentration gradients have reversed!* A significant change in membrane potential occurs with the movement of very few ions. For example, to change the membrane potential by 100 mV, only one out of every 100,000 K^+ must enter or leave the cell. This is such a tiny fraction of the total number of K^+ in the cell that the intracellular concentration of K^+ remains essentially unchanged.

An equivalent situation is moving one grain of sand from a beach into your eye. There are so many grains of sand on the beach that the loss of one grain is not significant, just as the movement of one K^+ across the cell membrane does not significantly alter the concentration of K^+. However, the electrical signal created by moving a few K^+ across the membrane has a significant effect on the cell's membrane potential, just as getting that one grain of sand in your eye will create significant discomfort.

Gated Channels Control the Ion Permeability of the Neuron

How does a cell change its ion permeability? The simplest way is to open or close existing channels in the membrane. Neurons contain a variety of gated ion channels that alternate between open and closed states, depending on the intracellular and extracellular conditions [● p. 130]. A slower method is for the cell to insert new channels or remove ion channels in the membrane.

Ion channels are usually named according to the primary ion(s) that they allow to pass through them. For example, Na^+ channels permit Na^+ to move across the membrane. There are four major types of selective ion channels in the neuron: (1) Na^+ channels, (2) K^+ channels, (3) Ca^{2+} channels, and (4) Cl^- channels. Other channels are less selective, such as the *monovalent cation channels* that allow both Na^+ and K^+ to pass.

Ion channels may spend most of their time in an open state (*leak channels*) or they may have gates that open or close in response to particular stimuli:

1. **Mechanically gated ion channels** are found in sensory neurons and open in response to physical forces such as pressure or stretch.
2. **Chemically gated ion channels** in most neurons respond to a variety of ligands, such as extracellular neurotransmitters and neuromodulators or intracellular signal molecules.

Channelopathies

Ion channels are proteins and like any protein, they may lose or change function if their amino acid sequence is altered. **Channelopathies** [*pathos*, suffering] are inherited diseases caused by mutations in ion channel proteins. Because ion channels are so intimately linked to the electrical activity of cells, many channelopathies manifest themselves as disorders of the excitable tissues, nerve, and muscle. One significant contribution of molecular biology to medicine was the discovery that one disease can actually be a family of related diseases with different causes but similar symptoms. For example, the condition known as *long Q-T syndrome* (LQTS; named for changes in the electrocardiogram test) is a cardiac problem characterized by an irregular heart beat (*arrhythmia; a-* without), fainting, and sometimes sudden death. Research studies determined that LQTS is actually three different conditions, two caused by mutations in a K^+ channel in heart muscle and one caused by a change in a Na^+ channel. Other well-known channelopathies include some forms of epilepsy and malignant hyperthermia. The most common channelopathy is cystic fibrosis, which results from defects in chloride channel function (see Chapter 5 Running Problem).

3. **Voltage-gated ion channels** play an important role in the initiation and conduction of electrical signals. Some channels we think of as leak channels are actually voltage-gated channels that remain open in the voltage range of the resting membrane potential.

When ion channels open, ions may move into or out of the cell. The direction of ion movement depends on the *electrochemical* (combined concentration and electrical) gradient for the ion. Potassium ions usually move out of the cell. Na^+, Cl^-, and Ca^{2+} usually flow into the cell. The net movement of electrical charge across the membrane depolarizes or hyperpolarizes the cell, creating an electrical signal.

These electrical signals can be classified into two basic types: graded potentials and action potentials (Table 8-3). **Graded potentials** are variable-strength signals that travel over short distances and lose strength as they travel through the cell. Graded potentials are used for short distance communication. If a depolarizing graded potential is strong enough when it reaches an integrating region within the cell, it will initiate an action potential. **Action potentials** are large, uniform depolarizations that can travel for long distances through the neuron without losing strength. Their function is rapid signaling over long distances.

Table 8-3: Comparison of Graded Potential and Action Potential

	GRADED POTENTIAL	ACTION POTENTIAL
Type of signal	Input signal	Conduction signal
Where occurs	Usually dendrites and cell body	Trigger zone through axon
Types of gated ion channels involved	Mechanically, chemically, or voltage-gated channels	Voltage-gated channels
Ions involved	Usually Na^+, Cl^-, Ca^{2+}	Na^+ and K^+
Type of signal	Depolarizing (e.g., Na^+) or hyperpolarizing (e.g., Cl^-)	Depolarizing
Strength of signal	Depends on initial stimulus; can be summed	Is always the same; (all-or-none phenomena); cannot be summed
What initiates the signal	Entry of ions through channels	Above-threshold graded potential at the trigger zone
Unique characteristics	No minimum level required to initiate	Threshold stimulus required to initiate
	Two signals coming close together in time will sum	Refractory period: two signals too close together in time cannot sum
	Initial stimulus strength is indicated by frequency of a series of action potentials	

Graded Potentials Reflect the Strength of the Stimulus That Initiates Them

Graded potentials are depolarizations or hyperpolarizations that occur in the dendrites, cell body, or less frequently, near the axon terminals. Graded potentials occur when ion channels open or close, causing ions to enter or leave the neuron. These changes in membrane potential are called "graded" because their size or *amplitude* [*amplitudo*, large] is directly proportional to the strength of the triggering event. A large stimulus will cause a strong graded potential, and a small stimulus will result in a weak graded potential.

CONCEPT CHECK
Answers: p. 282

9. Match the ion movement with the type of graded potential it will create.

   (a) Na^+ entry   1. depolarizing
   (b) Cl^- entry   2. hyperpolarizing
   (c) K^+ exit
   (d) Ca^{2+} entry

Graded potentials typically occur in the cell body and dendrites of a neuron. In sensory neurons, a mechanical stimulus may open Na^+ channels. In neurons of the central and efferent nervous systems, graded potentials occur when chemical signals from other neurons open chemically gated ion channels.

In the example shown in Figure 8-7 ●, a chemical or mechanical stimulus opens Na^+ channels. Sodium ions move into the neuron, bringing in electrical energy. The positive charge carried by the Na^+ spreads as a wave of depolarization through the

cytoplasm, just as a stone thrown into water creates ripples or waves that spread outward from the point of entry. The wave of depolarization that moves through the cell is known as **local current flow**. By convention, current flow in biological systems is the net movement of positive electrical charge.

The strength of the initial depolarization in a graded potential is determined by how much charge enters the cell, just as the size of waves in the water is determined by the size of the stone. If more Na^+ channels open, more Na^+ enters and the graded potential has a higher initial amplitude. The stronger the initial amplitude, the farther the graded potential can spread through the neuron before it dies out.

Why do graded potentials lose strength as they move through the cytoplasm? There are two reasons:

1. *Current leak*. Some of the positive charges leak back across the membrane as the depolarization wave moves through the cell. The membrane in this region of the neuron is not a good insulator and has open ion channels that allow positive charge to flow out into the extracellular fluid.
2. *Cytoplasmic resistance*. The cytoplasm itself provides resistance to the flow of electricity, just as water creates resistance that diminishes the ripples from the stone.

The combination of current leak and cytoplasmic resistance means that the strength of the signal inside the cell decreases over distance.

Graded potentials that are strong enough eventually reach the region of the neuron known as the **trigger zone**. In efferent neurons and interneurons, the trigger zone is the axon hillock and the

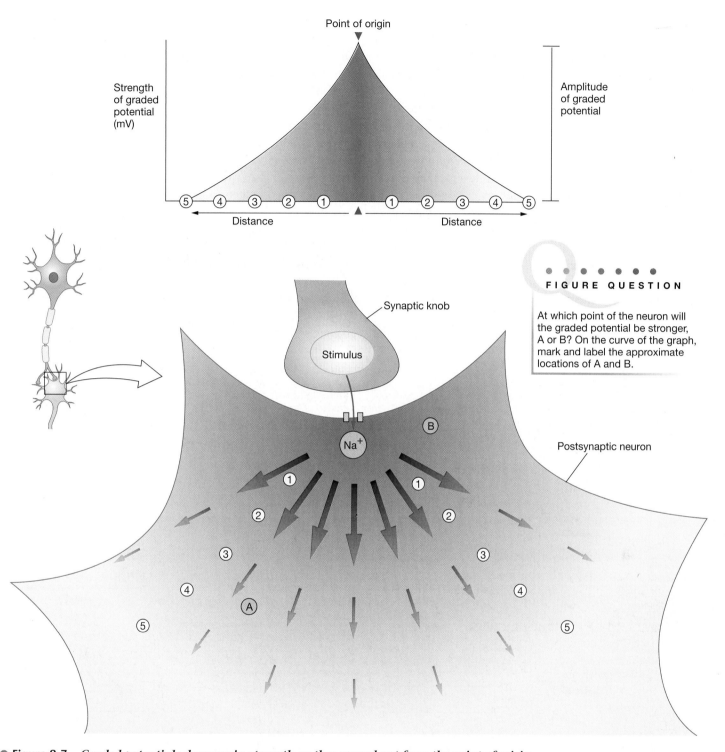

Point of origin

Strength
of graded
potential
(mV)

Amplitude
of graded
potential

⑤ ④ ③ ② ①   ① ② ③ ④ ⑤

Distance   Distance

Synaptic knob

Stimulus

Ⓑ

Na⁺

Postsynaptic neuron

① ①
② ②
③ ③
Ⓐ
④ ④
⑤ ⑤

● **Figure 8-7** *Graded potentials decrease in strength as they spread out from the point of origin*

very first part of the axon, a region known as the **initial segment** (see Fig. 8-2). In sensory neurons, the trigger zone is immediately adjacent to the receptor, where the dendrites join the axon.

CONCEPT CHECK   Answers: p. 282

10. Identify the trigger zones of the neurons illustrated in Figure 8-3, if possible.

The trigger zone is the integrating center of the neuron and contains a high concentration of voltage-gated Na⁺ channels in its membrane. If graded potentials reaching the trigger zone depolarize the membrane to a minimum level known as the **threshold voltage**, the voltage-gated Na⁺ channels open and an *action potential* is initiated. If the depolarization does not reach threshold, no action potential is begun, and the graded potential simply dies out as it moves into the axon.

Because depolarization is necessary to *excite* the cell to fire an action potential, a depolarizing graded potential is known as an **excitatory postsynaptic potential** (EPSP). Any stimulus that makes a neuron more likely to fire an action potential is considered to be *excitatory*.

A hyperpolarizing graded potential moves the membrane potential farther from the threshold value and makes the neuron less likely to fire an action potential. Consequently, hyperpolarizing graded potentials are called **inhibitory postsynaptic potentials** (IPSPs). Any stimulus that makes a cell less likely to fire an action potential is considered to be *inhibitory*.

Figure 8-8 ● shows a neuron that has several recording electrodes placed at intervals along the cell body and trigger zone. In the first example (Fig. 8-8a), a single stimulus triggers a *subthreshold* excitatory postsynaptic potential, one that is below threshold by the time it reaches the trigger zone. Although the cell is depolarized to −40 mV at the site where the graded potential begins, the current decreases as it travels through the cell body. As a result, the graded potential is below threshold by the time it reaches the axon hillock. (For the average mammalian neuron, threshold is about −55 mV.) The stimulus is not strong enough to depolarize the cell to threshold at the trigger zone, and the graded potential dies out without triggering an action potential.

The second example (Fig. 8-8b) shows a stronger initial stimulus that initiates a stronger EPSP and ultimately causes an action potential. Although this excitatory graded potential also diminishes with distance through the neuron, its higher initial strength ensures that it is above threshold at the axon hillock. In this example, an action potential is triggered.

Action Potentials Travel Long Distances without Losing Strength

Action potentials, also known as *spikes*, differ from graded potentials in several ways. Most notably, (1) all action potentials are identical, and (2) they do not diminish in strength as they travel through the neuron.

(a) A graded potential starts above threshold (T) at its initiation point, but decreases in strength as it travels through the cell body. At the trigger zone it is below threshold and therefore does not initiate an action potential.

(b) A stronger stimulus at the same point on the cell body creates a graded potential that is still above threshold by the time it reaches the trigger zone, so an action potential results.

● **Figure 8-8** *Subthreshold and suprathreshold graded potentials in a neuron*

Recordings of action potentials show them to be identical depolarizations of about 100 mV amplitude. The strength of the graded potential that initiates an action potential has no influence on its amplitude. Action potentials are sometimes called **all-or-none** phenomena because they either occur as a maximal depolarization (if the stimulus reaches threshold) or do not occur at all (if the stimulus is below threshold).

The mechanism through which action potentials are generated and conducted along the axon allows them to stay constant. An action potential measured at the distal end of an axon is identical to the action potential that started at the trigger zone. This property is essential for the transmission of signals over long distances, such as from a fingertip to the spinal cord. Table 8-3 compares the key properties of action potentials and graded potentials.

Action Potentials Represent Movement of Na^+ and K^+ Across the Membrane

Action potentials are changes in membrane potential that occur when voltage-gated ion channels open, altering membrane permeability to Na^+ and K^+. Figure 8-9 ● shows the voltage and ion permeability changes that take place during an action potential. The graph can be divided into three phases: the rising phase of the action potential, the falling phase, and the after-hyperpolarization phase. Before and after the action potential, ① and ⑨, the neuron is at its resting membrane potential of −70 mV.

1. *The rising phase of the action potential* is due to a sudden temporary increase in the cell permeability to Na^+. Action potentials begin when a graded potential reaching the trigger zone ② depolarizes the membrane to threshold (−55 mV) ③.

1. Resting membrane potential
2. Depolarizing stimulus
3. Membrane depolarizes to threshold. Voltage-gated Na^+ channels open and Na^+ enters cell. Voltage-gated K^+ channels begin to open slowly.
4. Rapid Na^+ entry depolarizes cell.
5. Na^+ channels close and slower K^+ channels open.
6. K^+ moves from cell to extracellular fluid.
7. K^+ channels remain open and additional K^+ leaves cell, hyperpolarizing it.
8. Voltage-gated K^+ channels close, some K^+ enters cell through leak channels.
9. Cell returns to resting ion permeability and resting membrane potential.

● **Figure 8-9** *The action potential*

As the cell depolarizes, voltage-gated Na$^+$ channels open, making the membrane much more permeable to Na$^+$. Because Na$^+$ is more concentrated outside the cell, and because the negative membrane potential inside the cell attracts these positively charged ions, Na$^+$ flows into the cell.

The addition of positive charge to the intracellular fluid depolarizes the cell membrane, making it progressively more positive ④. On the graph, this is shown by the steep rising phase of the action potential. In the top third of the rising phase, the membrane potential has reversed polarity; that is, the inside of the cell has become more positive than the outside. This reversal is represented on the graph by the **overshoot**, that portion of the action potential above 0 mV.

As soon as the cell membrane potential becomes positive, the electrical driving force moving Na$^+$ into the cell disappears. However, the Na$^+$ concentration gradient remains, so Na$^+$ continues to move in. As long as Na$^+$ permeability remains high, the membrane potential moves toward the Na$^+$ *equilibrium potential* (E$_{Na}$) of +60 mV. (Recall from Chapter 5 that E$_{Na}$ is the membrane potential at which the movement of Na$^+$ into the cell down its concentration gradient is exactly opposed by the positive membrane potential [● p. 159].) However, before the E$_{Na}$ is reached, the Na$^+$ channels in the axon close. Sodium permeability decreases dramatically, and the action potential peaks at +30 mV ⑤.

2. *The falling phase of the action potential* corresponds to an increase in K$^+$ permeability. Voltage-gated K$^+$ channels, like Na$^+$ channels, start to open in response to depolarization. But the K$^+$ channel gates are much slower to open, and peak K$^+$ permeability occurs later than peak Na$^+$ permeability (Fig. 8-9, lower graph). By the time the K$^+$ channels are open, the membrane potential of the cell has reached +30 mV because of Na$^+$ influx through faster-opening Na$^+$ channels.

When the Na$^+$ channels close at the peak of the action potential, the K$^+$ channels have just finished opening, making the membrane very permeable to K$^+$. At a positive membrane potential, the concentration and electrical gradients for K$^+$ favor movement of K$^+$ out of the cell. As K$^+$ moves out of the cytoplasm, the membrane potential rapidly becomes more negative, creating the falling phase of the action potential and sending the cell toward its resting potential ⑥.

When the falling membrane potential reaches −70 mV, the K$^+$ channels have not yet closed. Potassium continues to leave the cell and the membrane hyperpolarizes, approaching the E$_K$ of −90 mV. This after-hyperpolarization is also called the *undershoot* ⑦. Once the slow voltage-gated K$^+$ channels finally close, K$^+$ leakage into the cell exceeds movement out ⑧, bringing the membrane potential back to its normal resting value of −70 mV ⑨.

To summarize, the action potential is a change in membrane potential that occurs when voltage-gated ion channels in the membrane open, increasing the cell's permeability first to Na$^+$

and then to K$^+$. The *influx* (movement into the cell) of Na$^+$ depolarizes the cell. This depolarization is followed by K$^+$ *efflux* (movement out of the cell), which restores the cell to the resting membrane potential.

Na$^+$ Channels in the Axon Have Two Gates

One question that puzzled scientists for many years was how the voltage-gated Na$^+$ channels could close when the cell was depolarized, because depolarization was the stimulus for Na$^+$ channel opening. After many years of study, they found the answer. The voltage-gated Na$^+$ channel has two gates to regulate ion movement rather than a single gate. These two gates, known as the *activation* and *inactivation gates*, flip-flop back and forth to open and shut the Na$^+$ channel.

When the neuron is at its resting membrane potential, the activation gate is closed and no Na$^+$ moves through the channel (Fig. 8-10a ●). At the same time, the inactivation gate, apparently a ball-and-chain–like amino acid sequence on the cytoplasmic side of the channel protein, is open. When the cell membrane close to the channel protein depolarizes, the activation gate

(a) At the resting membrane potential, the activation gate closes the channel.

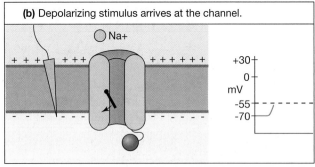

(b) Depolarizing stimulus arrives at the channel.

(c) With activation gate open, Na$^+$ enters the cell.

● **Figure 8-10** *Model of the voltage-gated Na$^+$ channel*

(d) Inactivation gate closes and Na⁺ entry stops.

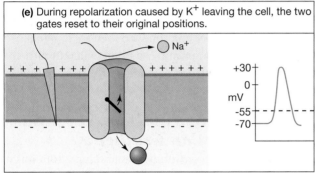

(e) During repolarization caused by K⁺ leaving the cell, the two gates reset to their original positions.

● **Figure 8-10** *continued*

opens (Fig. 8-10b). This opens the channel pore and allows Na⁺ to move into the cell down its electrochemical gradient (Fig. 8-10c). The addition of positive charge further depolarizes the inside of the cell and starts a positive feedback loop [🔖 p. 195] (Fig. 8-11 ●). More Na⁺ channels open and more Na⁺ enters, further depolarizing the cell. As long as the cell remains depolarized, the activation gates remain open.

As happens in all positive feedback loops, outside intervention is needed to stop the escalating depolarization of the cell. This outside intervention is the role of the second gate. Both gates respond to depolarization, but the inactivation gate delays for 0.5 msec. During that time the Na⁺ channel is open, allowing enough Na⁺ influx to create the rising phase of the action potential. When the

slower inactivation gate of the Na⁺ channel finally closes (Fig. 8-10d), Na⁺ influx stops, and the action potential peaks.

As the neuron repolarizes during K⁺ efflux, the Na⁺ channel gates reset to their original positions so they can respond to the next depolarization (Fig. 8-10e). Thus, the voltage-gated Na⁺ channel uses a two-step process for opening and closing rather than a single gate that swings back and forth. This important property allows electrical signals along the axon to be conducted in only one direction, as you will see in the next section.

CONCEPT CHECK   Answers: p. 282

11. The pyrethrin insecticides, derived from chrysanthemums, disable inactivation gates of Na⁺ channels so that the channels remain open. In neurons poisoned with pyrethrins, what would you predict would happen to the membrane potential? Explain your answer.

Action Potentials Will Not Fire During the Absolute Refractory Period

The double gating of the Na⁺ channel plays a major role in the phenomenon known as the **refractory period**. The adjective *refractory* comes from a Latin word meaning "stubborn." The "stubbornness" of the neuron refers to the fact that once an action potential has begun, for about 1 msec, a second action potential cannot be triggered, no matter how large the stimulus. This period is called the **absolute refractory period** (Fig. 8-12 ●) and represents the time required for the Na⁺ channel gates to reset to their resting positions. The absolute refractory period ensures that a second action potential will not occur before the first has finished. *Action potentials cannot overlap and cannot travel backward because of their refractory periods.*

CONCEPT CHECK   Answers: p. 282

12. When the Na⁺ channel gates are resetting, is the activation gate opening or closing? Is the inactivation gate opening or closing?

● **Figure 8-11** *Ion movements during the action potential* Sodium entry creates a positive feedback loop that stops when the Na⁺ channel inactivation gates close.

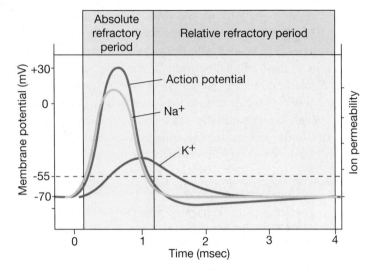

● **Figure 8-12** *Refractory periods* During the absolute refractory period, no stimulus can trigger another action potential. During the relative refractory period, a larger-than-normal stimulus can initiate a new action potential.

A **relative refractory period** follows the absolute refractory period. During this phase, a stronger-than-normal depolarizing graded potential is needed to bring the cell up to the threshold for an action potential. During this time, many but not all Na^+ channel gates have reset to their original positions and a threshold-level depolarization will open them. Those Na^+ channels that have not quite returned to their resting position can be opened by a higher-than-normal graded potential.

In addition, during the relative refractory period, K^+ channels are still open. Although Na^+ can enter through newly re-opened Na^+ channels, depolarization due to Na^+ entry will be offset by K^+ loss. The opposing flow of charge balances out, and the membrane potential does not reach threshold.

The refractory period is a key characteristic that distinguishes action potentials from graded potentials. If two stimuli reach the dendrites of a neuron within a short time, the successive graded potentials can be added to each other. But if two *suprathreshold* (above-threshold) graded potentials reach the action potential trigger zone within the absolute refractory period, the second stimulus will be ignored because the Na^+ channels are inactivated and are incapable of being opened again so soon.

Refractory periods limit the rate at which signals can be transmitted down a neuron. The absolute refractory period also assures one-way travel of an action potential from cell body to axon terminal by preventing the action potential from traveling backward.

Stimulus Intensity Is Coded by the Frequency of Action Potentials

One distinguishing characteristic of action potentials is that every action potential in a given neuron will be identical to every other action potential in that neuron. But if this is so, then how does the neuron transmit information about the strength and dura-

tion of the stimulus that started the action potential? The answer lies not in the amplitude of the action potential, but in the frequency of action potential propagation.

A stimulus in the form of a graded potential reaching the trigger zone does not usually trigger a single action potential. Instead, even a minimal graded potential that is above threshold will trigger a burst of action potentials (Fig. 8-13a ●). If the graded potential increases in strength (amplitude), the frequency of action potentials fired (in action potentials per second) increases (Fig. 8-13b).

The amount of neurotransmitter released at the axon terminal is directly related to the total number of action potentials that arrive at the terminal per unit time. An increase in signal strength will increase the neurotransmitter output, which in turn changes the magnitude of the graded potential in the postsynaptic cell.

One Action Potential Does Not Alter Ion Concentration Gradients

As you just learned, the action potential results from ion movements across the neuron membrane. First, Na^+ moves into the cell and then K^+ moves out. However, it is important to understand that very few ions move across the membrane in a single action potential, so that *the relative Na^+ and K^+ concentrations remain essentially unchanged.*

For example, only one of every 100,000 K^+ must leave the cell to shift the membrane potential from +30 to −70 mV, the falling phase of the action potential. The tiny amounts of ions that cross the membrane during an action potential do not disrupt the Na^+ and K^+ concentration gradients.

Normally, the ions that do move into or out of the cell during action potentials are rapidly restored to their original compartments by the **Na^+-K^+-ATPase** (Na^+-K^+ pump). The pump uses energy from ATP to exchange Na^+ that enters the cell for K^+ that leaked out of it [● p. 141]. *But this exchange does not need to happen before the next action potential fires because the ion concentration gradient was not significantly altered by one action potential!* A neuron without a functional Na^+-K^+ pump must fire a thousand or more action potentials before there will be a significant change in the ion gradients.

13. If you put ouabain, a specific inhibitor of the Na^+-K^+ pump, on a neuron and stimulate it repeatedly, what do you expect to see happen to action potentials generated by that neuron?

  (a) They cease immediately.
  (b) There is no immediate effect but they diminish over time and eventually disappear.
  (c) They get smaller immediately, then stabilize with smaller amplitude.
  (d) Ouabain has no effect on action potentials.

(a) Weak stimulus releases little neurotransmitter

(b) Strong stimulus causes more action potentials and releases more neurotransmitter

Coding for stimulus intensity
All action potentials in a neuron are identical; therefore, the strength of the stimulus is indicated by the frequency of action potential firing. Stronger stimuli release more neurotransmitter into the synapse.

Action Potentials Are Conducted from the Trigger Zone to the Axon Terminal

The movement of an action potential through the axon at high speed is called **conduction**. Conduction represents the flow of electrical energy from one part of the cell to another in a process that ensures that any energy lost to friction or leakage out of the cell is immediately replenished. Thus, an action potential does not lose strength over distance as a graded potential does. To see why the action potential that reaches the end of the axon is identical to the action potential that started at the trigger zone, we must examine conduction at the cellular level.

When a graded potential reaches threshold in the trigger zone, voltage-gated Na^+ channels open, Na^+ enters, and the cell depolarizes (Fig. 8-14a ●). The positive current spreads through the cytoplasm in all directions, just as the depolarization of a graded potential spreads through the cell by local current flow (Fig. 8-14b). Current flow in the forward direction, down into the initial portions of the axon, is critical for conduction of the action potential.

The axon membrane is lined by the same type of voltage-sensitive Na^+ channels found in the membrane of the trigger zone. Positive charge from the depolarized trigger zone spreads to adjacent sections of membrane, repelled by the Na^+ that entered the cytoplasm and attracted by the negative charge of the resting membrane potential. When the depolarization wave reaches voltage-sensitive Na^+ channels adjacent to the trigger zone, the channels open, allowing Na^+ into the cell (Fig. 8-14c). The membrane in

this segment of the axon is now depolarized, and the positive feedback loop of depolarization begins: Na^+ channels open, Na^+ enters, causing depolarization and current flow, which opens more Na^+ channels in the adjacent membrane. The continuous entry of Na^+ down the axon means that the strength of the signal does not diminish as the action potential propagates itself. (Contrast this with graded potentials in Fig. 8-7, where Na^+ enters only at the point of stimulus, resulting in a membrane potential change that loses strength over distance.)

As each segment of axon reaches the peak of the action potential, its Na^+ channels inactivate. During the action potential's falling phase, K^+ channels are open, allowing K^+ to leave the cytoplasm. Finally, the K^+ channels close and the membrane in that segment of axon returns to its resting potential.

Although positive charge from distal, depolarizing segments of membrane may flow backwards by local current flow, the depolarization in that direction has no effect. That part of the axon is in the absolute refractory period, with its Na^+ channel inactivated, so the action potential cannot move backwards.

What happens to current that flows backward from the trigger zone into the cell body? We used to believe that there were few voltage-gated ion channels in that region, so it could be ignored. However, we now know that the cell body and dendrites do have voltage-gated ion channels and may respond to the retrograde flow of current from action potentials. These retrograde signals are able to influence and modify the next signal that reaches the cell.

When we talk about action potentials, it is important to realize that there is no single action potential that moves through the cell. The action potential that occurs at the trigger zone is like the movement in the first domino of a series of dominos standing on end. As the first domino falls, it strikes the next, passing on its kinetic energy. As the second domino falls, it passes kinetic energy to the third domino, and so on. If you could take a snapshot of the line of falling dominos, you would see that as the first domino is coming to rest in the fallen position, the next one is almost down, the third one most of the way down, etc., until you reach the domino that has just been hit and is starting to fall (Fig. 8-15a ●).

In the same way, a wave of action potentials moves down the axon. An action potential is simply a representation of the membrane potential in a given segment of cell membrane at any given moment in time. As the electrical energy of the action potential passes from one part of the axon to the next, the energy state is re-

flected in the membrane potential of that region. If we were to insert a series of recording electrodes along the length of an axon and start an electrical signal at the trigger zone, we would see a series of overlapping action potentials at different parts of the wave form, just like the dominos that are frozen in different positions (Fig. 8-15b).

CONCEPT CHECK   Answers: p. 282

14. If you place an electrode in the middle of an axon and artificially depolarize the cell above threshold, in which direction will an action potential travel: to the axon terminal, to the cell body, or to both? Explain your answer.

Larger Neurons Conduct Action Potentials Faster

Two key physical parameters influence the speed of action potential conduction in mammalian neurons: (1) the diameter of the neurons and (2) the resistance of the neuron membrane to current leak out of the cell. The larger the diameter of the axon or the more leak-resistant the neuron membrane, the faster an action potential will move.

To understand the relationship between diameter and conduction, think of a water pipe with water flowing through it. The water that touches the walls of the pipe encounters resistance due to friction between the flowing water molecules and the station-

● **Figure 8-14**   *Conduction of action potentials*

(a) Initiation of the action potential at the trigger zone

2 Voltage-gated Na⁺ channels open and Na⁺ enters the axon.

Trigger zone

Na⁺

1 A graded potential above threshold reaches the trigger zone.

(b) Local current flow

3 Na⁺ entry depolarizes the membrane, which opens additional Na⁺ channels.

Na⁺

4 Positive charge flows into adjacent sections of the axon by local current flow.

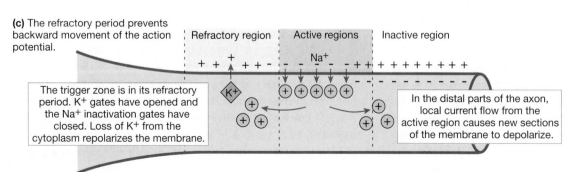

(c) The refractory period prevents backward movement of the action potential.

Refractory region | Active regions | Inactive region

Na⁺

The trigger zone is in its refractory period. K⁺ gates have opened and the Na⁺ inactivation gates have closed. Loss of K⁺ from the cytoplasm repolarizes the membrane.

In the distal parts of the axon, local current flow from the active region causes new sections of the membrane to depolarize.

(a) The transmission of action potentials can be compared to a "snapshot" of dominos falling, where each domino is in a different position.

(b) Simultaneous recordings show that each section of axon is experiencing a different phase of the action potential.

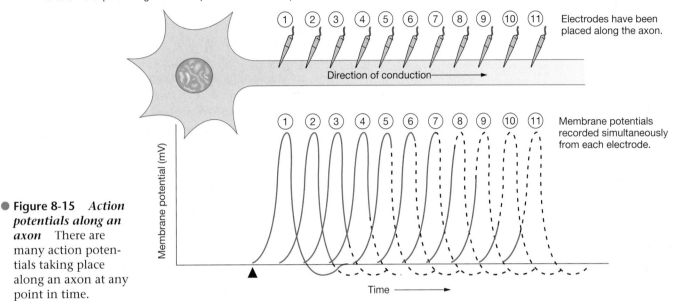

Electrodes have been placed along the axon.

Direction of conduction →

Membrane potentials recorded simultaneously from each electrode.

Membrane potential (mV)

Time →

● **Figure 8-15** *Action potentials along an axon* There are many action potentials taking place along an axon at any point in time.

ary walls. The water in the center of the pipe meets no direct resistance from the walls and therefore flows faster. In a large diameter pipe, a smaller fraction of the water flowing through the pipe is in contact with the walls, making the total resistance lower. In the same way, current flowing inside an axon meets resistance from the membrane. Thus, the larger the diameter of the axon, the lower its resistance to current flow.

The connection between axon diameter and speed of conduction is especially evident in the nerve cells of invertebrate animals such as the squid and the earthworm. The giant axons of these organisms may be up to 1 mm in diameter and have been very important to the development of our understanding of electrical signaling (Fig. 8-16a ●). Larger-diameter axons for rapid conduction are found mostly in animals with small nervous systems, however.

(a) Large axons offer less resistance to current flow, but occupy space.

Squid giant axon

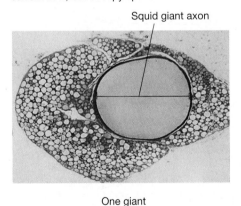

One giant axon from a squid

↓

0.8 mm diameter

(b) Small myelinated axons conduct action potentials as rapidly as large unmyelinated axons.

Nerve   Axons

If the 20 myelinated mammalian axons in the nerve above were each the size of a squid giant axon, the nerve would have to be this size.

↓

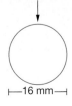

|—16 mm—|

● **Figure 8-16** *Axon diameter and speed of conduction*

If you compare a cross section of a squid giant axon with a cross section of a mammalian nerve, you will find that the mammalian nerve contains about 400 fibers in the same cross-sectional area (Fig. 8-16b). If each of those 400 axons was as large as a squid giant axon, that nerve would have to be 1.6 cm in diameter, much too large for our bodies to have many of them! Complex nervous systems pack more axons into a small nerve by trading large diameter axons for smaller diameter axons wrapped in insulating membranes of myelin.

Conduction Is Faster in Myelinated Axons

The conduction of action potentials down the axon is faster in high-resistance axons where current leak out of the cell is minimized. The unmyelinated axon illustrated in Figure 8-14 has low resistance to current leak because the entire axon membrane is in contact with the extracellular fluid and has ion channels through which current can leak. In contrast, myelinated axons limit the amount of membrane in contact with the extracellular fluid. In these axons, small sections of bare membrane, the nodes of Ranvier, alternate with longer segments wrapped in multiple layers of membrane, the myelin sheath. The myelin sheath creates a high-resistance wall that prevents current flow out of the cytoplasm. The myelin membranes are analogous to heavy coats of plastic surrounding electrical wires, as they increase the effective thickness of the axon membrane by as much as 100-fold.

When an action potential starts at the trigger zone, current flows down the myelinated axon to the first unmyelinated node of Ranvier (Fig. 8-17a ●). In the node regions where the axon lacks a myelin sheath, the cell membrane has low resistance to current flow. Each node has a high concentration of voltage-gated Na⁺ channels, which open and allow Na⁺ into the axon.

To keep the action potential going down the axon undiminished, it must periodically be "refreshed" to compensate for current loss from the cell. The process is identical to that described previously for the unmyelinated axon, except that it occurs only at the nodes in myelinated axons.

When the action potential's depolarization reaches a node, voltage-gated Na⁺ channels open. Sodium entry reinforces the de-

● **Figure 8-17** *Saltatory conduction* (a) Action potentials appear to jump from one node of Ranvier to the next. (b) In demyelinating diseases, conduction slows due to current leaks out of the previously insulated regions between the nodes.

polarization and helps keep the amplitude of the action potential constant as it passes from node to node. The apparent jump of the action potential from node to node is called **saltatory conduction**, from the Latin word *saltare*, meaning "to leap."

What makes conduction more rapid in myelinated axons? The answer is that channel opening slows conduction slightly. In unmyelinated axons, channels must open all the way down the axon membrane to maintain the amplitude of the action potential. In myelinated axons, only the nodes need Na^+ channels because of the insulating properties of the myelin membrane. As the action potential passes along myelinated segments, conduction is not slowed by channel opening.

Saltatory conduction thus is an effective alternative to large-diameter axons and allows rapid action potentials through small axons. A myelinated frog axon 10 μm in diameter conducts action potentials at the same speed as an unmyelinated 500-μm squid axon. Thus, electrical current flows through different axons at different rates, depending on the two parameters of axon diameter and myelination.

CONCEPT CHECK   Answers: p. 282

15. Rank the following neurons for their speed of conduction from fastest to slowest:

    (a) myelinated 20 μm diameter axon

    (b) unmyelinated 20 μm diameter axon

    (c) unmyelinated 200 μm diameter axon

In *demyelinating diseases*, the loss of myelin from vertebrate neurons can have devastating effects on neural signaling. In the central and peripheral nervous systems, the loss of myelin slows the conduction of action potentials. In addition, when current leaks out of the uninsulated regions of membrane between the channel-rich nodes of Ranvier (Fig. 8-17b), the depolarization that reaches a node may not be above threshold and conduction may fail.

Multiple sclerosis is the most common and best-known demyelinating disease. It is characterized by a variety of neurological complaints, including fatigue, muscle weakness, difficulty walking, and loss of vision. At this time, we can treat some of the symptoms but not the causes of demyelinating dis-

RUNNING PROBLEM

Like multiple sclerosis, Guillain-Barré syndrome is an illness in which the myelin that insulates axons is destroyed. One way that multiple sclerosis and other demyelinating illnesses are diagnosed is through the use of a nerve conduction test. This test measures the combined strength of action potentials and the rate at which they are conducted as they travel down neurons.

Question 3: In Guillain-Barré syndrome, what would you expect the results of a nerve conduction test to be?

239 241 244 🏃 262 267 274 276

Mutant Mouse Models

The use of animal models to study human diseases has become a valuable part of biomedical research. Animal models allow scientists to explore causes of disease and test potential treatments. Sometimes, natural mutations produce animal diseases that resemble human diseases. Two examples of such mutants are the twitcher mouse, in which normal myelin degenerates owing to an inherited metabolic problem, and the wobbler mouse, in which somatic motor neurons controlling the limbs die. In other cases, scientists have used biotechnology techniques to create mice that lack specific genes (**knock-out mice**) or to breed mice that contain extra genes that were inserted artificially (**transgenic mice**). By comparing the genetically engineered mice with normal mice, we have expanded our understanding of how the nervous system responds during normal development and to disease. To learn more about mutant mouse models, visit the website for The Jackson Laboratory, *www.jax.org*, and click on the *JAX® Mice and Services* button. In the *JAX® Mice* database you will find descriptions and photos of some of the different knockout and transgenic mice being used for research.

eases, which are either inherited or autoimmune disorders. Currently, researchers are using recombinant DNA technology to study demyelinating disorders in mice.

Electrical Activity Can Be Altered by a Variety of Chemical Factors

A large variety of chemicals alter the conduction of action potentials by binding to Na^+, K^+, or Ca^{2+} channels in the neuron membrane. For example, some *neurotoxins* bind to and block Na^+ channels. Local anesthetics such as procaine, which block sensation, function the same way. If the Na^+ channel is not functional, Na^+ cannot enter the axon. Thus, a depolarization that begins at the trigger zone loses strength as it moves down the axon, much like a normal graded potential. If the wave of depolarization manages to reach the axon terminal, it may be too weak to release neurotransmitter. As a result, the message of the presynaptic neuron is not passed on to the postsynaptic cell.

Alterations in the extracellular fluid concentrations of K^+ and Ca^{2+} are also associated with abnormal electrical activity in the nervous system. The relationship between extracellular fluid K^+ levels and the conduction of action potentials is the most straightforward and easiest to understand.

The concentration of K^+ in the blood and interstitial fluid is the major determinant of the resting potential of all cells. If K^+

BIOTECHNOLOGY

Of Snakes, Snails, Spiders, and Sushi

What do snakes, marine snails, and spiders have to do with neurophysiology? They all provide neuroscientists with compounds for studying synaptic transmission, extracted from the neurotoxic venoms that they use to kill their prey. The Asian snake *Bungarus multicinctus* provides us with α-bungarotoxin, a long-lasting poison that binds tightly to nicotinic acetylcholine receptors. The fish-hunting cone snail, *Conus geographus*, and the funnel web spider, *Agelenopsis aperta*, use toxins that block different types of voltage-gated Ca^{2+} channels. But one of the most potent poisons known comes from the Japanese puffer fish, a highly prized delicacy whose flesh is consumed as *sushi*. The puffer has tetrodotoxin, or TTX, in its gonads. This neurotoxin blocks Na^+ channels on axons and prevents the transmission of action potentials, so ingestion of only a tiny amount can be fatal. The chefs who prepare the puffer fish, known as *fugu*, for consumption are carefully trained to avoid contaminating the fish's flesh as they remove the toxic gonads. But there's always some risk involved in eating *fugu*, one reason that the youngest person at the table is the first to sample the dish.

concentrations in the blood move out of the normal range of 3.5–5 millimoles/liter (mmol/L), the result will be a change in the resting membrane potential of the cells. This change is not important to most cells, but it can have serious consequences to the body as a whole because of the relationship between resting potential and the excitability of nervous and muscle tissue.

An increase in blood K^+ concentration, **hyperkalemia** [*hyper-*, above + *kalium*, potassium + *-emia*, in the blood], will shift the resting membrane potential of a neuron closer to threshold and cause the cells to fire action potentials in response to smaller graded potentials (Fig. 8-18c ●). If blood K^+ concentrations fall too low (**hypokalemia**), the resting membrane potential of the cells hyperpolarizes, moving farther from threshold and requiring a larger-than-normal stimulus to fire an action potential (Fig. 8-18d). This condition shows up as muscle weakness because the neurons that control skeletal muscles are not firing normally.

Hypokalemia and its resultant muscle weakness are one reason that sport drinks supplemented with Na^+ and K^+, such as Gatorade™, were developed. When people sweat excessively, they lose both salts and water. If they replace this fluid loss with pure water, the K^+ remaining in the blood will be diluted, causing hypokalemia. By replacing sweat loss with a dilute salt solution, they can prevent potentially dangerous drops in blood K^+ levels.

Because of the importance of K^+ to normal function of the nervous system, the body regulates blood K^+ levels within a narrow range. We will discuss the important role of the kidneys in maintaining ion balance in Chapter 19.

RUNNING PROBLEM

Dr. McKhann decided to perform nerve conduction tests on some of the paralyzed children. He found that, although the rate of conduction along the children's neurons was normal, the action potentials were greatly diminished.

Question 4: Is the paralytic illness that affected the Chinese children a demyelinating condition? Why or why not?

(a) When blood K^+ is in the normal range (normokalemia), a subthreshold graded potential does not fire an action potential.

(b) In normokalemia, a suprathreshold (above-threshold) stimulus will fire an action potential.

(c) Hyperkalemia, increased blood K^+ concentration, brings the membrane closer to the threshold. Now a stimulus that would normally be subthreshold can trigger an action potential.

(d) Hypokalemia, decreased blood K^+ concentration, hyperpolarizes the membrane and makes the neuron less likely to fire an action potential in response to a stimulus that would normally be above the threshold.

● **Figure 8-18** *Effect of changing extracellular potassium concentration on the excitability of neurons* If the extracellular concentration of K^+ changes, the resting membrane potential of cells changes.

▶ CELL-TO-CELL COMMUNICATION IN THE NERVOUS SYSTEM

Information flow through the nervous system using electrical and chemical signals is one of the most active areas of neurophysiology research today. The specificity of neural communication depends on several factors: the signal molecules secreted by neurons, the target cell receptors for these chemicals, and the anatomical connections between neurons and their targets, which occur in regions known as synapses.

Information Passes from Cell to Cell at the Synapse

A *synapse* is the point at which a neuron meets its target cell (another neuron or a non-neuronal cell). Each synapse has three parts: (1) the axon terminal of the *presynaptic cell*, (2) the **synaptic cleft**, or space between the cells, and (3) the membrane of the *postsynaptic cell* (Fig. 8-19 ●). In a neural reflex, information moves from presynaptic cell to postsynaptic cell. In most neuron-to-neuron synapses, the presynaptic axon terminals are next to the dendrites or cell body of the postsynaptic neuron.

In general, cells with many dendrites will also have many synapses. A moderate number of synapses is 10,000, but some cells in the brain are estimated to have 150,000 or more synapses!

● **Figure 8-19** *A chemical synapse* The axon terminal contains mitochondria and synaptic vesicles filled with neurotransmitter. The postsynaptic membrane has receptors for neurotransmitter that diffuses across the synaptic cleft.

Axon of presynaptic neuron

Mitochondrion
Axon terminal
Synaptic vesicles
Synaptic cleft
Receptor sites
Neurotransmitter
Postsynaptic membrane
Postsynaptic neuron

Synapses can also occur on the axon and even at the axon terminal. Synapses are classified as electrical or chemical depending on the type of signal that passes between cells.

Electrical Synapses **Electrical synapses** pass an electrical signal or current directly from the cytoplasm of one cell to another through gap junctions. Information can flow in both directions through gap junctions. Electrical synapses are uncommon and occur mainly in the CNS. Electrical synapses are also found in glial cells, in cardiac and smooth muscle, and in nonexcitable cells that use electrical signals, such as the pancreatic beta cell. The primary advantage of electrical synapses is rapid conduction of signals from cell to cell.

Chemical Synapses The vast majority of synapses in the nervous system are **chemical synapses** that use neurotransmitters to carry information from one cell to the next. Much of what we know about chemical synapses comes from the study of the *neuromuscular junction*, formed by the synapse of a somatic motor neuron on a skeletal muscle fiber.

Neurotransmitter synthesis can take place either in the nerve cell body or in the axon terminal. However, axon terminals do not have the organelles needed for protein synthesis. Consequently, polypeptide neurotransmitters and protein enzymes needed for metabolism in the axon terminal must be made in the cell body. The dissolved enzymes are brought to axon terminals by slow axonal transport but the neurotransmitters, which are consumed more rapidly than enzymes, are moved in vesicles by fast axonal transport.

When we examine the axon terminal of a presynaptic cell with an electron microscope, we find many small **synaptic vesicles** and large mitochondria in the cytoplasm (Fig. 8-19). Some vesicles are "docked" at active zones along the membrane closest to the synaptic cleft, waiting for a signal to release their contents. Other vesicles act as a reserve pool, clustering close to the docking sites. Each vesicle contains neurotransmitter that is released on demand.

CONCEPT CHECK Answers: p. 282

16. Which organelles are needed to synthesize proteins and package them into vesicles?
17. What is the function of mitochondria in a cell?
18. How do mitochondria get to the axon terminals?

Calcium Is the Signal for Neurotransmitter Release at the Synapse

The release of neurotransmitters into the synapse takes place by exocytosis. From what we can tell, the basic process is similar in neurons and in other types of secretory cells, such as the pancreatic beta cell described in Chapter 5 [● p. 162]. Neurotoxins that block neurotransmitter release, including tetanus and botulinum toxins, exert their action by inhibiting specific proteins of the cell's exocytotic apparatus.

● **Figure 8-20** *Events at the synapse*

1. An action potential depolarizes the axon terminal.

2. The depolarization opens voltage-gated Ca^{2+} channels and Ca^{2+} enters the cell.

3. Calcium entry triggers exocytosis of synaptic vesicle contents.

4. Neurotransmitter diffuses across the synaptic cleft and binds with receptors on the postsynaptic cell.

Figure 8-20● shows how neurotransmitters are released by exocytosis at a synapse. When the depolarization wave of an action potential reaches the axon terminal ①, it sets off a sequence of events. The axon terminal membrane has voltage-gated Ca^{2+} channels that open in response to depolarization ②. Calcium ions are more concentrated outside the cell than in the cytosol, so they move into the cell. The Ca^{2+} ions bind to regulatory proteins and initiate exocytosis ③. The membrane of the synaptic vesicle fuses with the cell membrane, aided by multiple membrane proteins. The fused area opens and neurotransmitter inside the synaptic vesicle moves into the synaptic cleft ④. The neurotransmitter molecules diffuse across the gap to bind with membrane receptors on the postsynaptic cell. When neurotransmitters bind to their receptors, a response is initiated in the postsynaptic cell.

In the classic model of exocytosis, the membrane of the vesicle becomes part of the axon terminal membrane (● Fig. 5-28, p. 146). This increase in membrane surface area is countered by recycling of the vesicles at regions away from the active sites (see Fig. 8-4, p. 245). However, a second model of secretion is emerging. In this model, called the **kiss-and-run pathway**, synaptic vesicles fuse to the membrane at a complex called the **fusion pore**. A small channel opens that is just large enough for neurotransmitter to pass through. Then, instead of opening the fused area wider and incorporating the vesicle membrane into the cell membrane, the vesicle pulls back from the fusion pore and returns to the pool of vesicles in the cytoplasm.

CONCEPT CHECK   Answers: p. 282

19. In an experiment on synaptic transmission, a synapse was bathed in a Ca^{2+}-free medium that was otherwise equivalent to extracellular fluid. An action potential was triggered in the presynaptic neuron. Although the action potential reached the axon terminal at the synapse, the usual response of the postsynaptic cell did not occur. What conclusion did the researchers make on the basis of these results?

EMERGING CONCEPTS

Synaptic Vesicles Kiss and Run

Most of what we know about synaptic vesicle release at the axon terminal comes from a combination of electrophysiology studies and microscopic examination. It has been difficult to follow the dynamic process of exocytosis by microscopy, however. Electron microscopy is performed in a vacuum and the tissue must be killed and preserved. Living cells can be examined by light microscopy, but the resolution of even the best light microscope is too low to show fine details. However, a new technique called **atomic force microscopy** (AFM) is allowing scientists to watch secretion in live cells. With AFM they have found a new membrane structure, the fusion pore, that appears to be made of actin and docking proteins such as t-SNAREs. Fusion pores are key players in the kiss-and-run pathway, in which synaptic vesicles join the cell membrane only long enough to release their contents. Fusion pores have been observed in endocrine and exocrine cells as well as neural tissue, suggesting that their presence may be essential in all forms of secretion. To learn more about fusion pores and to see computerized images of them from AFM, read "Fusion pore in live cells," *News in Physiological Sciences* 17: 219–222, December 2002 (*www.nips.org*).

Neurocrines Convey Information from Neurons to Other Cells

What are these signal molecules that neurons release? Their chemical composition is varied, and they may function as neurotransmitters, neuromodulators, or neurohormones. Neurotransmitters and neuromodulators act as *paracrines*, with target cells located close to the neuron that secretes them. Neurohormones, on the other hand, are secreted into the blood and distributed throughout the body.

Generally, neurotransmitters act at a synapse and elicit a rapid response. Neuromodulators act at both synaptic and nonsynaptic sites and are slower acting. Some neuromodulators and neurotransmitters also act on the cell that secretes them, making them *autocrine* signals as well as paracrines. The number of molecules identified as neurotransmitters and neuromodulators is large and growing daily.

The Nervous System Secretes a Variety of Neurotransmitters

The array of neurotransmitters in the body is truly staggering. They can be informally grouped into seven classes according to their structure: (1) acetylcholine, (2) amino acids, (3) amino acid–derived amines, (4) polypeptides, (5) purines, (6) gases, and (7) lipids. The largest variety of neurotransmitters is found within the CNS (Table 8-4). These neurotransmitters include many polypeptides known mostly for their hormonal activity. In contrast, three primary

Table 8-4: Major Neurocrines

CHEMICAL	RECEPTOR	TYPE*	RECEPTOR LOCATION	KEY RECEPTOR AGONISTS/ANTAGONISTS**
Acetylcholine (ACh)	Cholinergic			
	Nicotinic	ICR (Na^+, K^+)	Skeletal muscles, autonomic neurons, CNS	Nicotine—agonist Curare—antagonist
	Muscarinic	GPCR	Smooth and cardiac muscle, endocrine and exocrine glands, CNS	Muscarine—agonist Atropine—antagonist
Amines				
Norepinephrine (NE)	Adrenergic (α, β)	GPCR	Smooth and cardiac muscle, endocrine and exocrine glands, CNS	α agonist—phenylephrine β antagonist—propranolol
Dopamine (DA)	Dopamine (D)	GPCR	CNS	Antipsychotic drugs—antagonists Bromocriptine—agonist
Serotonin (5-hydroxy-tryptamine, 5-HT)	Serotonergic (5-HT)	ICR (Na^+, K^+) GPCR	CNS	Sumatriptan—agonist LSD—antagonist
Histamine	Histamine (H)	GPCR	CNS	Ranitidine (Zantac®), cimetidine (Tagamet®)—antagonists
Amino Acids				
Glutamate	Glutaminergic			
	AMPA	ICR (Na^+, K^+)	CNS	
	NMDA	ICR (Na^+, K^+, Ca^{2+})	CNS	
GABA (γ-aminobutyric acid)	GABA	ICR (Cl^-) GPCR	CNS	
Glycine	Glycine	ICR (Cl^-)	CNS	
Purines				
Adenosine	Purine (P)	GPCR	CNS	
Gases				
Nitric oxide (NO)	None	N/A	N/A	

* ICR = ion channel-receptor, GPCR = G protein-coupled receptor. AMPA = α-amino-3-hydroxy-5-methylisoxazole-4-proprionic acid. NMDA = N-methyl-D-aspartate. LSD = lysergic acid diethylamine. N/A = Not applicable.

** This list does not include many chemicals that are used as agonists and antagonists in physiological research.

neurotransmitters are used within the peripheral nervous system: acetylcholine, norepinephrine, and epinephrine.

Acetylcholine Acetylcholine (ACh), in a chemical class by itself, is synthesized from choline and acetyl coenzyme A (acetyl CoA). Choline is a small molecule also found in membrane phospholipids. Acetyl CoA is the metabolic intermediate that links glycolysis to the citric acid cycle [● p. 105]. The synthesis of ACh from these two precursors is a simple enzymatic reaction that takes place in the axon terminal (Fig. 8-21 ●). Neurons that secrete ACh and receptors that bind ACh are described as **cholinergic**.

Amines Amine neurotransmitters, like the amine hormones [● p. 218], are derived from single amino acids. The amino acid tyrosine is converted to **dopamine**, **norepinephrine**, and **epinephrine**. All three of these neurocrines also function as neurohormones when secreted by the adrenal medulla [● Fig. 11-10, p. 377].

Neurons that secrete norepinephrine are called **adrenergic neurons**, or, more properly, noradrenergic neurons. The adjective *adrenergic* does not have the same obvious link to its neurotransmitter as *cholinergic* does to *acetylcholine*. Instead, the adjective derives from the British name for epinephrine, *adrenaline*. In the early part of the twentieth century, British researchers thought that sympathetic neurons secreted adrenaline (epinephrine), hence the modifier *adrenergic*. Although our understanding changed, the name persists. Whenever you see reference to "adrenergic control" of a function, you must make the connection to a neuron secreting norepinephrine.

Myasthenia Gravis

What would you think was wrong if suddenly your eyelids started drooping, you had difficulty watching moving objects, and it became difficult to chew, swallow, and talk? What disease attacks these skeletal muscles but leaves the larger muscles of the arms and legs alone? The answer is **myasthenia gravis** [*myo-*, muscle + *asthenes*, weak + *gravis*, severe], an autoimmune disease in which the body fails to recognize the acetylcholine (ACh) receptors on skeletal muscle as part of "self." The immune system then produces molecules called antibodies to attack the receptors. The antibodies bind to the ACh receptor protein and change it in some way that causes the muscle cell to withdraw the receptors from the membrane and destroy them. This destruction leaves the muscle with fewer ACh receptors in the membrane. Even though neurotransmitter release is normal, the muscle target has a diminished response that is exhibited as muscle weakness. Currently medical science does not have an effective treatment for myasthenia gravis, although various drugs can help control its symptoms. To learn more about this disease, visit the website for the Myasthenia Gravis Foundation of America at *www.myasthenia.org*.

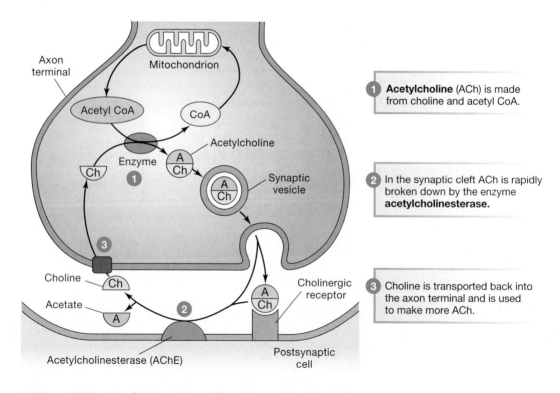

1 **Acetylcholine** (ACh) is made from choline and acetyl CoA.

2 In the synaptic cleft ACh is rapidly broken down by the enzyme **acetylcholinesterase.**

3 Choline is transported back into the axon terminal and is used to make more ACh.

● **Figure 8-21** *Synthesis and recycling of acetylcholine at the synapse*

Other amine neurotransmitters include **serotonin** (also called *5-hydroxytryptamine* or **5-HT**), made from the amino acid tryptophan, and histamine, made from histidine. The amine neurotransmitters are all active in the CNS. In addition, norepinephrine is the major neurotransmitter of the peripheral autonomic sympathetic division.

Amino Acids At least four amino acids function as neurotransmitters in the CNS. **Glutamate** is the primary excitatory neurotransmitter of the CNS, and **aspartate** serves the same function in selected regions of the brain. The main inhibitory neurotransmitter is **gamma-aminobutyric acid (GABA)**. The amino acid **glycine** is also considered inhibitory, although it can bind to one type of glutamate receptor and be excitatory.

Polypeptides The nervous system secretes a variety of peptides that act as neurotransmitters and neuromodulators in addition to functioning as neurohormones. These peptides include **substance P**, involved in some pain pathways, and the **opioid peptides** (**enkephalins** and **endorphins**) that mediate *analgesia* [*an-*, without + *algos*, pain]. Peptides that function as both hormones and neurotransmitters include *cholecystokinin (CCK)*, *vasopressin*, and *atrial natriuretic peptide*. Many peptide neurotransmitters are cosecreted with other neurotransmitters.

Purines *Adenosine*, *adenosine monophosphate* (AMP), and *adenosine triphosphate* (ATP) can all act as neurotransmitters. These molecules, known collectively as *purines* [◐ p. 41], bind to *purinergic* receptors in the CNS and on other excitable tissues such as the heart.

Gases One of the most interesting neurotransmitters is *nitric oxide* (NO), an unstable gas synthesized from oxygen and the amino acid arginine. Nitric oxide is an unusual neurotransmitter that diffuses freely into the target cell rather than binding to a membrane receptor [◐ p. 183]. Once inside the target cell, nitric oxide binds to proteins. With a half-life of only 2–30 seconds, nitric oxide is elusive and difficult to study. It is also released from cells other than neurons and often acts as a paracrine.

Recent work suggests that *carbon monoxide* (CO), best known as a toxic gas, is produced by the body in tiny amounts to serve as a neurotransmitter.

RUNNING PROBLEM

Dr. McKhann then asked to see autopsy reports on some of the children who had died of their paralysis at Beijing Hospital. In the reports, pathologists noted that the patients had normal myelin but damaged axons. In some cases, the axon had been completely destroyed, leaving only a hollow shell of myelin.

Question 5: Do the results of Dr. McKhann's investigation suggest that the Chinese children had Guillain-Barré syndrome? Why or why not?

Multiple Receptor Types Amplify the Effects of Neurotransmitters

All neurotransmitters except nitric oxide have one or more receptor types with which they bind. Each receptor type may have multiple subtypes, a property that allows one neurotransmitter to have different effects in different tissues. Receptor subtypes are distinguished by combinations of letter and number subscripts. For example, serotonin (5-HT) has at least 14 different receptor subtypes that have been identified, such as 5-HT_{1A} or 5-HT_4.

Neurotransmitter receptors fall into two of the membrane receptor categories that we discussed in Chapter 6 [◐ p. 175]: ligand-gated ion channels and G protein–linked receptors. Receptors that alter ion channel function are called *inotropic* receptors, while those that exert their actions through second messenger systems are called *metabotropic* receptors. Some metabotropic G protein–linked receptors regulate the opening or closing of ion channels.

The study of neurotransmitters and their receptors has been greatly simplified by two advances in molecular biology. The genes for many receptor subtypes have been cloned, allowing researchers to create mutant receptors and study their properties. In addition, researchers have discovered or synthesized a variety of agonists and antagonist molecules that mimic or inhibit neurotransmitter activity by binding to the receptors [◐ p. 174]. Table 8-4 includes descriptions of receptor types and some of their agonists or antagonists.

Cholinergic Receptors Cholinergic receptors come in two main subtypes: **nicotinic**, named because *nicotine* is an agonist, and **muscarinic**, for which *muscarine*, a compound found in some fungi, is an agonist. **Cholinergic nicotinic receptors** are found on skeletal muscle, in the autonomic division, and in the CNS. Nicotinic receptors are monovalent cation channels through which both Na^+ and K^+ can pass. Sodium entry into cells exceeds K^+ exit because the electrochemical gradient for Na^+ is stronger. As a result, net Na^+ entry depolarizes the postsynaptic cell and makes it more likely to fire an action potential.

Cholinergic muscarinic receptors come in five related subtypes. They are all coupled to G proteins and are linked to second messenger systems. The tissue response to activation of a muscarinic receptor varies with the receptor subtype. These receptors occur in the CNS and in the autonomic parasympathetic division.

Adrenergic Receptors Adrenergic receptors come in two varieties: α (alpha) and β (beta). Like cholinergic muscarinic receptors, adrenergic receptors are linked to G proteins and initiate second messenger cascades. Indeed, it was the action of epinephrine on β receptors in dog liver that led E. W. Sutherland to the discovery of cyclic AMP and the concept of second messenger systems as transducers of extracellular messengers [◐ p. 176]. The two types of adrenergic receptors work through different second messenger pathways.

Answers: p. 282

20. When pharmaceutical companies design drugs, they try to make drugs as specific as possible for the particular receptor subtype they are targeting. For example, a drug might target adrenergic β_1 receptors rather than all adrenergic α and β receptors. What would be the advantage to this?

Glutaminergic Receptors Glutamate is the main excitatory neurotransmitter in the CNS and also acts as a neuromodulator. The action of glutamate at a particular synapse depends on which of the two receptor types occurs on the cell. The NMDA receptors are named for the glutamate agonist N-*methyl*-D-*aspartate*, and the AMPA receptors are named for their agonist α-*amino-3-hydroxy-5-methylisoxazole-4-proprionic acid*.

AMPA receptors are ligand-gated monovalent cation channels, similar to nicotinic acetylcholine channels. Glutamate binding opens the channel and the cell depolarizes due to net Na^+ influx.

NMDA receptors are unusual for several reasons. First, they are cation channels that allow Na^+, K^+, and Ca^{2+} to pass through the channel. Second, channel opening requires both glutamate binding and a change in membrane potential. The NMDA receptor channel is blocked by a magnesium ion (Mg^{2+}) at resting membrane potentials. Glutamate binding opens the ligand-activated gate but ions cannot flow past the Mg^{2+} ion. However, if the cell depolarizes, the Mg^{2+} blocking the channel is expelled and ions flow through the pore (see Fig. 8-29).

Not All Postsynaptic Responses Are Rapid and of Short Duration

The combination of a neurotransmitter with its receptor sets in motion a series of responses in the postsynaptic cell (Fig. 8-22●). In the simplest response, the neurotransmitter opens a chemically gated ion channel, leading to ion movement between the cell and the extracellular fluid. The resulting change in membrane potential is called a **fast synaptic potential** because it begins quickly and lasts only a few milliseconds. These synaptic potentials are the *excitatory* (depolarizing) and *inhibitory* (hyperpolarizing) *postsynaptic potentials* (EPSPs and IPSPs), previously described.

In slow responses, neurotransmitters bind to receptors linked to G proteins and activate second messenger systems. The second messengers may act from the cytoplasmic side of the cell membrane to open or close ion channels. (Fast synaptic potentials always open ion channels.) Membrane potentials resulting from this process are called **slow synaptic potentials** because the second messenger method takes longer to create a response. In addition, the response itself lasts longer, usually seconds to minutes.

Slow responses are not limited to altering the open state of ion channels. Neurotransmitter activation of second messenger systems may also modify existing cell proteins or regulate the production of new cell proteins. This type of slow response has been linked to the growth and development of neurons and to the mechanisms underlying long-term memory.

Neurotransmitter Activity Is Rapidly Terminated

A key feature of nervous signaling is its short duration, which is achieved by the rapid removal or inactivation of neurotransmitter in the synaptic cleft. This can be accomplished various ways (Fig. 8-23●). Some neurotransmitter molecules simply diffuse away from the synapse, separating them from their receptors. Other neurotransmitters are inactivated by enzymes in the synaptic cleft. Many neurotransmitters are removed from the extracellular fluid by transport back into the presynaptic cell or into adjacent neurons or glial cells.

Acetylcholine ACh in the extracellular fluid is rapidly broken down by the enzyme **acetylcholinesterase** (AChE) in the extracellular matrix and in the membrane of the postsynaptic cell (see Fig. 8-21). Choline from degraded ACh is actively transported back into the presynaptic axon terminal and used to make new acetylcholine to refill recycled synaptic vesicles.

Norepinephrine Norepinephrine action at the target tissue is terminated when the neurotransmitter is actively transported back into the presynaptic axon terminal. Once back in the axon terminal, norepinephrine is either repackaged into vesicles or broken down by intracellular enzymes such as *monoamine oxidase (MAO)*, found in mitochondria.

CNS Neurotransmitters Once released into the synaptic cleft, these amine, polypeptide, amino acid neurotransmitters move into the circulation to be carried away or are actively transported back into the axon terminal to be metabolized or used again.

Answers: p. 282

21. One class of antidepressant drugs is called selective serotonin reuptake inhibitors (SSRIs). What do these drugs do to serotonin activity at the synapse?

▶ INTEGRATION OF NEURAL INFORMATION TRANSFER

Traditionally, scientists viewed chemical synapses as sites of one-way communication, with all messages moving from a presynaptic cell to a postsynaptic cell. We now know that this is not always the case. In the brain, there are some synapses where cells on both sides of the synaptic cleft release neurotransmitter that acts on the opposite cell. Perhaps more important, we have learned that many postsynaptic cells "talk back" to their presynaptic neurons by sending neuromodulators that bind to presynaptic receptors. Let's examine some of the ways that complex neural pathways communicate.

Neural Pathways May Involve Many Neurons Simultaneously

Communication between neurons is not always a one-to-one event. Sometimes a single neuron branches, and its collaterals synapse on multiple target neurons. This pattern is known as

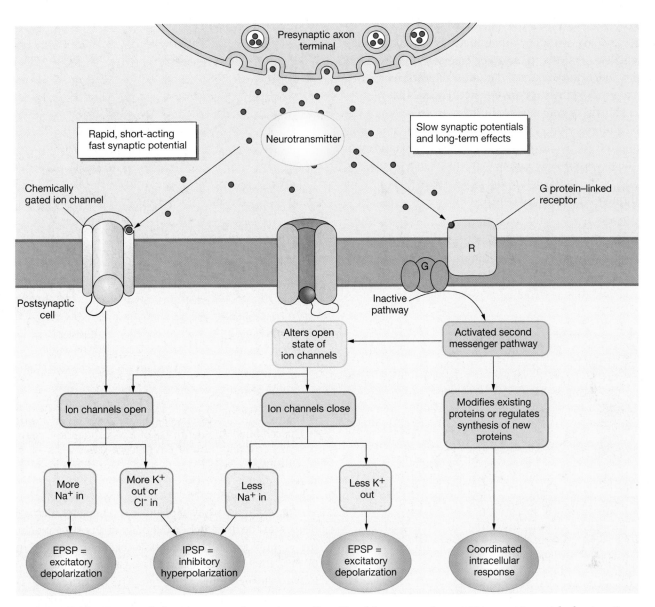

● Figure 8-22 *Fast and slow responses in postsynaptic cells* Many neurotransmitters create rapid, short-acting responses by directly opening ion channels. Some neurotransmitters create slower, longer-lasting responses by activating second messenger systems.

● Figure 8-23 *Inactivation of neurotransmitters*

1. Neurotransmitters can be returned to axon terminals for reuse or transported into glial cells.

2. Enzymes inactivate neurotransmitters.

3. Neurotransmitters can diffuse out of the synaptic cleft.

divergence (Fig. 8-24a ●). On the other hand, a single postsynaptic neuron may have synapses with as many as 10,000 presynaptic neurons (Fig. 8-25 ●). If a larger number of presynaptic neurons provides input to a smaller number of postsynaptic neurons, the pattern is known as **convergence** (Fig. 8-24b).

One advantage of convergence in the nervous system is that input from multiple sources can influence the output of a single postsynaptic cell. When two or more presynaptic neurons converge on the dendrites or cell body of a single postsynaptic cell, the response of the postsynaptic cell will be determined by the summed input from the presynaptic neurons.

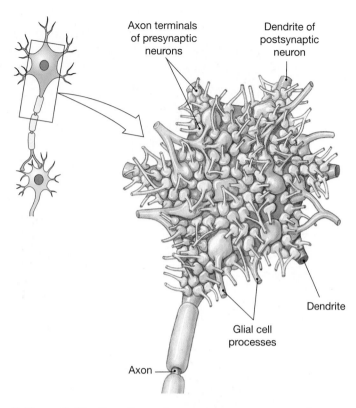

Figure 8-25 *Locations of synapses on a postsynaptic neuron* The cell body and dendrites of a somatic motor neuron are nearly covered in hundreds of axon terminals from other neurons.

(a) In a **divergent pathway**, one presynaptic neuron branches to affect a larger number of postsynaptic neurons.

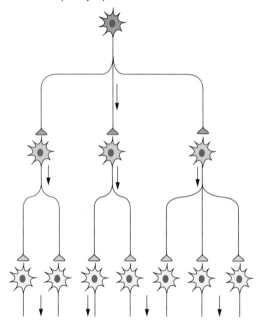

(b) In a **convergent pathway**, many presynaptic neurons converge to influence a smaller number of postsynaptic neurons.

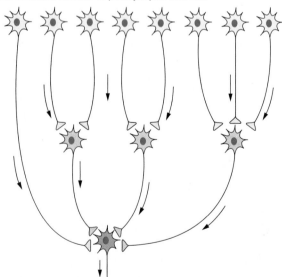

● Figure 8-24   *Convergence and divergence*

For example, what happens if several stimuli arrive simultaneously at the dendrites or cell body of a postsynaptic neuron? If the stimuli are all subthreshold excitatory graded potentials (EPSPs), they can add together to create a *suprathreshold* potential [*supra-*, above] at the trigger zone. The initiation of an action potential from several simultaneous graded potentials is an example of **spatial summation**. The word *spatial* refers to the fact that the graded potentials originate at different locations on the neuron.

Figure 8-26a ● illustrates spatial summation when three excitatory presynaptic neurons converge on one postsynaptic neuron. The graded potentials from the presynaptic neurons are too weak to trigger an action potential by themselves, but if the neurons fire together, the sum of the three stimuli is above threshold and creates an action potential.

Postsynaptic inhibition may occur when a neuron releases an inhibitory neurotransmitter onto a postsynaptic cell and alters its response. Figure 8-26b shows three neurons, two excitatory and one inhibitory, converging on a postsynaptic cell. The neurons fire, creating one inhibitory graded potential (IPSP) and two excitatory graded potentials (EPSPs) that sum as they reach the trigger zone. The IPSP counteracts the EPSPs, creating an integrated signal that is below threshold. As a result, no action potential leaves the trigger zone.

Summation of graded potentials does not always require input from more than one presynaptic neuron. Two subthreshold

● **Figure 8-26** *Spatial summation*

(a)

1 Three excitatory neurons fire. Their graded potentials separately are all below threshold.

2 Graded potentials arrive at trigger zone together and sum to create a suprathreshold signal.

3 An action potential is generated.

(b)

1 One inhibitory and two excitatory neurons fire.

2 The summed potentials are below threshold, so no action potential is generated.

graded potentials from the same presynaptic neuron can be summed if they arrive at the trigger zone close enough together in time. Let's see how this can happen.

Figure 8-27a ● shows recordings from an electrode placed in the trigger zone of a neuron. A stimulus (X) starts a subthreshold graded potential on the cell body at time X_1. The graded potential reaches the trigger zone and depolarizes it as shown on the graph (A_1). A second stimulus occurs at time X_2, and its subthreshold graded potential reaches the trigger zone a short time after the first (A_2). The interval between the two stimuli is so long that the two graded potentials do not overlap. Neither potential by itself is above threshold, so no action potential is triggered.

In Figure 8-27b, the two stimuli are given closer together in time. As a result, the two subthreshold graded potentials arrive at the trigger zone at almost the same time. The second graded potential adds its depolarization to that of the first, allowing the trigger zone to depolarize to threshold. Summation that occurs from graded potentials overlapping in time is called **temporal summation** [*tempus*, time].

In many situations, graded potentials in a neuron will incorporate both temporal and spatial summation. The summation of graded potentials demonstrates a key property of neurons: *postsynaptic integration*. When multiple signals reach a neuron, postsynaptic integration allows the neuron to evaluate the strength and duration of the signals. If the resultant integrated signal is above threshold, the neuron fires an action potential.

CONCEPT CHECK   Answers: p. 282

22. In Figure 8-26b, assume the postsynaptic neuron has a resting membrane potential of −70 mV and threshold of −55 mV. If the inhibitory presynaptic neuron creates an IPSP of 5 mV and the two excitatory presynaptic neurons have EPSPs of 10 and 12 mV, respectively, will the postsynaptic neuron fire an action potential?

23. In the graphs of Figure 8-27, why doesn't the membrane potential (A_1, A_2) change at the same time as the stimulus (X_1, X_2)?

Synaptic Activity Can Be Modulated at the Axon Terminal

Modulation of neuronal pathways can also occur at the axon terminal. Modulation at axon terminals, called **presynaptic modulation**, occurs when a modulatory neuron (inhibitory or excitatory) terminates on or close to an axon terminal of a presynaptic cell (Fig. 8-28a ●). The modulatory neuron secretes a chemical that

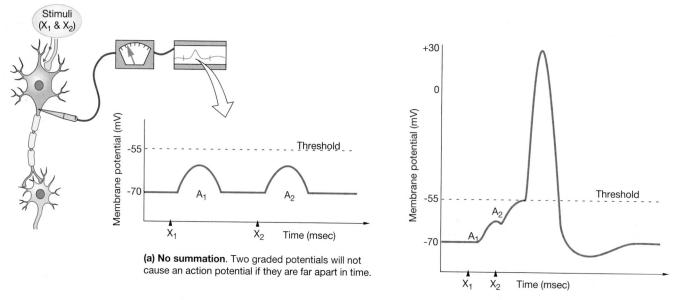

(a) **No summation**. Two graded potentials will not cause an action potential if they are far apart in time.

(b) **Summation causing action potential**. If two subthreshold potentials arrive at the trigger zone within a short period of time, they may sum and create an action potential.

● **Figure 8-27** *Temporal summation*

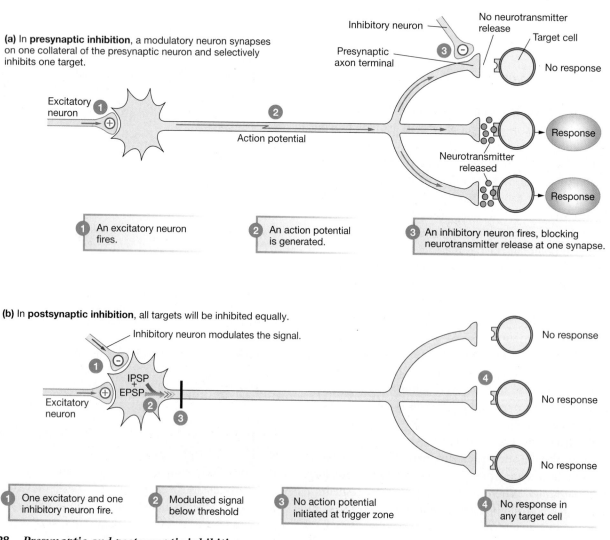

(a) In **presynaptic inhibition**, a modulatory neuron synapses on one collateral of the presynaptic neuron and selectively inhibits one target.

1 An excitatory neuron fires.

2 An action potential is generated.

3 An inhibitory neuron fires, blocking neurotransmitter release at one synapse.

(b) In **postsynaptic inhibition**, all targets will be inhibited equally.

Inhibitory neuron modulates the signal.

1 One excitatory and one inhibitory neuron fire.

2 Modulated signal below threshold

3 No action potential initiated at trigger zone

4 No response in any target cell

● **Figure 8-28** *Presynaptic and postsynaptic inhibition*

influences the amount of neurotransmitter released by the presynaptic cell. If activity in the modulatory neuron decreases neurotransmitter release, the modulation is called *presynaptic inhibition*. In *presynaptic facilitation*, modulatory input increases neurotransmitter release by the presynaptic cell.

Presynaptic modulation provides a more precise means of control than postsynaptic modulation. If the responsiveness of a neuron is changed at the dendrites and cell body (**postsynaptic modulation**), all target cells of the neuron will be affected equally (Fig. 8-28b). In contrast, presynaptic inhibition on a divergent neuron allows selective modulation of collaterals and their targets (Fig. 8-28a). One collateral can be inhibited while others remain unaffected.

Synaptic activity can also be altered by changing the responsiveness of the target cell to neurotransmitter. This may be accomplished by changing the identity, affinity, and number of neurotransmitter receptors. Modulators can alter all of these parameters by influencing the synthesis of enzymes, membrane transporters, and receptors. Most neuromodulators act through second messenger systems that alter existing proteins, and their effects last much longer than do those of neurotransmitters.

24. Why are axon terminals sometimes called "biological transducers"?

Long-Term Potentiation Alters Synaptic Communication

One of the "hot topics" in neurobiology today is **long-term potentiation** (LTP) (*potentia*, power), a process in which activity at a synapse induces sustained changes in the quality or quantity of synaptic connections. Long-term potentiation is believed to be related to the neural processes for learning and memory. Many scientists are working to discover the molecular details of the mechanism.

A key element in long-term potentiation is the amino acid glutamate, the main excitatory neurotransmitter in the CNS. As you learned above, glutamate has two types of receptors: NMDA receptors and AMPA receptors. The NMDA channel has an unusual property: the channel is blocked both by a gate and by a Mg^{2+} ion. Channel opening requires both neurotransmitter binding and depolarization of the cell.

When presynaptic neurons release glutamate, the chemical binds to both types of receptors on the postsynaptic cell

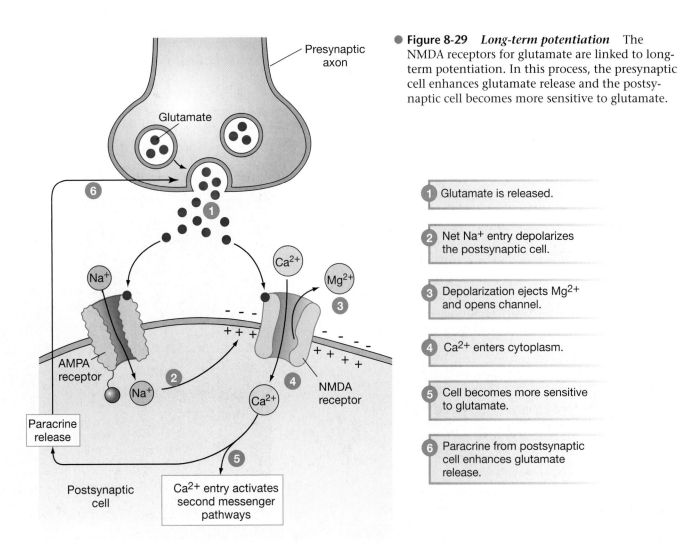

● **Figure 8-29** *Long-term potentiation* The NMDA receptors for glutamate are linked to long-term potentiation. In this process, the presynaptic cell enhances glutamate release and the postsynaptic cell becomes more sensitive to glutamate.

1 Glutamate is released.

2 Net Na^+ entry depolarizes the postsynaptic cell.

3 Depolarization ejects Mg^{2+} and opens channel.

4 Ca^{2+} enters cytoplasm.

5 Cell becomes more sensitive to glutamate.

6 Paracrine from postsynaptic cell enhances glutamate release.

(Fig. 8-29 ●). Activation of the AMPA receptor opens a cation channel and net Na^+ entry depolarizes the cell. The gate of the NMDA receptor opens with glutamate binding but the channel remains blocked by the Mg^{2+} ion until the cell depolarizes. Electrical repulsion then knocks the Mg^{2+} out of the channel, allowing Ca^{2+} to enter the cytosol.

The Ca^{2+} signal initiates second messenger pathways. As a result of these pathways, the postsynaptic cell releases a paracrine that acts on the presynaptic cell to enhance neurotransmitter release. The postsynaptic cell also becomes more sensitive to glutamate, possibly by inserting more glutamate receptors in the postsynaptic membrane.

CONCEPT CHECK   Answers: p. 282

25. Why would depolarization of the membrane repulse a Mg^{2+} ion into the extracellular fluid?

Disorders of Synaptic Transmission Are Responsible for Many Diseases

Synaptic transmission is the most vulnerable step in the process of signaling through the nervous system. It is the point at which many things go wrong, leading to disruption of normal function. Yet, at the same time, the receptors at synapses are exposed to the extracellular fluid, making them more accessible to drugs than intracellular receptors are. In recent years, we have discovered that a variety of nervous system disorders are related to problems with synaptic transmission. These include Parkinson's disease, schizophrenia, and depression. The best understood diseases of the synapse are those that involve the neuromuscular junction. Diseases resulting from synaptic transmission problems within the CNS have proved more difficult to study because, anatomically, they are more difficult to isolate.

Drugs that act on synaptic activity, particularly synapses in the CNS, are the oldest known and most widely used of all pharmacological agents. Caffeine, nicotine, and alcohol are common drugs in many cultures. Medically, some drugs that we use to treat conditions such as schizophrenia, anxiety, and epilepsy act by in-

---

RUNNING PROBLEM

Currently, Dr. McKhann believes that the disease afflicting the Chinese children—which he has named acute motor axonal polyneuropathy (AMAN)—may be caused by a bacterial infection. He also believes that the disease initiates its damage of axons at the neuromuscular junctions.

Question 6: Based on information provided in this chapter, can you name another disease that alters synaptic transmission?

---

fluencing events at the synapse. In many disorders arising in the CNS, we do not yet fully understand either the cause of the disorder or the mechanism of action of the drug. This subject is one major area of pharmacological research, and new classes of drugs are being formulated and approved every year.

Development of the Nervous System Depends on Chemical Signals

During the development of a system as complex as the nervous system, how can more than 100 billion neurons in the brain find their correct targets and make synapses among more than 10 times that many glial cells? How can a somatic motor neuron in the spinal cord find the correct pathway to form a synapse with its target muscle in the big toe? The answer is found in the chemical signals used by the developing embryo, ranging from factors that control differentiation of stem cells to those that direct the elongating axon to its target.

The axons of embryonic nerve cells send out special tips called **growth cones** that extend through the extracellular compartment until they find their target cell (Fig. 8-30 ●). In experiments in which target cells are moved to an unusual location in the embryo, the axons in many instances are still able to find their targets by "sniffing out" their chemical scent. Growth cones depend on many different types of signals to find their way: growth factors, molecules in the extracellular matrix, and membrane proteins on the growth cones and cells along the path. For example, integrins (● p. 66) on the growth cone membrane bind to laminins, protein fibers in the extracellular matrix. Nerve-cell adhesion molecules (N-CAMs; ● p. 64) interact with membrane proteins of other cells.

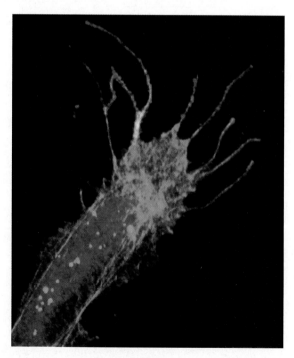

● **Figure 8-30** *Growth cones of a developing axon*

Once an axon reaches its target cell, a synapse forms. But synapse formation must be followed by electrical and chemical activity, or the synapse will disappear. The survival of neurons seems to depend on **neurotrophic factors** [*trophikos*, nourishment] secreted by neurons and glial cells. There is still much we do not understand about this complicated process, and it is an active area of physiological research.

This "use it or lose it" scenario is most dramatically reflected by the fact that the infant brain is only about one-fourth the size of the adult brain. Further brain growth is due not to an increase in cell number but to an increase in size and number of axons, dendrites, and synapses. Development is dependent on electrical activity in the brain, on action potentials moving through sensory pathways and interneurons.

Babies who are neglected or deprived of sensory input may have delayed development because of the lack of stimulation of the nervous system. On the other hand, there is no evidence that extra stimulation in infancy will enhance intellectual development, despite a popular movement to expose babies to art, music, and foreign languages before they can even walk. Once synapses form, they are not fixed for life. Variations in electrical activity can cause rearrangement of the connections. This process, known as the **plasticity** (changeability) of synapses, continues throughout life. It is one reason that older adults are urged to keep learning new skills and information.

When Neurons Are Injured, Segments Separated from the Cell Body Die

We can grow neurons when young, but what happens when neurons are injured? The responses of mature neurons to injury are similar in many ways to the growth of neurons during development, relying on a combination of chemical and electrical signals. This is another area of active research because spinal cord injuries and brain damage from illnesses and accidents disable many people each year. The loss of one neuron from a reflex can have drastic consequences on an entire reflex pathway.

When a neuron is injured, if the cell body dies, the entire neuron dies. If the cell body is intact and only the axon is severed, most of the neuron will survive. At the site of injury, the cytoplasm leaks out until membrane is recruited to seal the opening. The proximal segment (closest to the cell body) swells as organelles and filaments carried into the axon by axonal transport accumulate. Chemical factors produced by Schwann cells near the injury site move by retrograde transport to the cell body, telling it that an injury has occurred. The distal segment of axon, deprived of its source of protein, begins to degenerate slowly (Fig. 8-31 ●). The death of this part of the neuron may take a month or longer, although synaptic transmission ceases almost immediately. The myelin sheath around the distal axon begins to unravel and the axon itself collapses. The fragments are cleared away by scavenger microglia or phagocytes that ingest and digest the debris.

Under some conditions, axons in the peripheral nervous system can regenerate and reestablish their synaptic connections. The Schwann cells of the damaged neuron secrete certain neurotrophic factors that keep the cell body alive and that stimulate regrowth of the axon. The growing tip of a regenerating axon behaves much like that of a developing axon, following chemical signals in the extracellular matrix along its former path until the axon rejoins its target cell. But sometimes the loss of the distal axon is permanent, and the pathway is destroyed.

Regeneration of axons in the central nervous system is less likely to occur naturally. CNS glial cells tend to seal off and scar the damaged region, and damaged CNS cells secrete factors that actually inhibit regrowth of axons. Many scientists are studying these mechanisms of axon growth and inhibition in the hopes of finding treatments that will restore function to victims of spinal cord injury and degenerative neurological disorders.

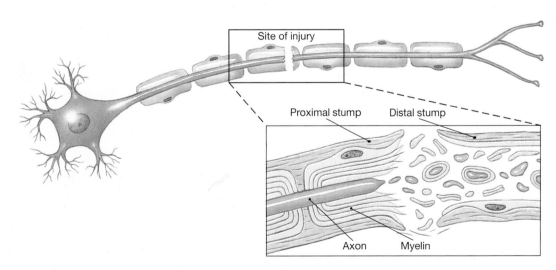

● **Figure 8-31** *Injury to neurons* When an axon is cut, the section attached to the cell body continues to live, but the section of the axon distal to the cut will begin to disintegrate and die. Under some circumstances, the stump of the axon may regrow through the existing sheath of Schwann cells and reform a synapse with the proper target.

Mysterious Paralysis

In this running problem, you learned about a baffling paralytic illness that appears to be a new disease. Although it bears some resemblance to Guillain-Barré syndrome, upon closer investigation, it actually has some significantly different features. To read Dr. McKhann's original report, see "Clinical and electrophysiological aspects of acute paralytic disease of children and young adults in northern China," *Lancet* 338(8767): 593–597, 1991 Sep 7. Check your understanding of this running problem by comparing your answers to those in the summary table.

	QUESTION	FACTS	INTEGRATION AND ANALYSIS
1	Which division(s) of the nervous system may be involved in Guillain-Barré syndrome?	The nervous system is divided into peripheral (sensory and efferent) and central nervous system (CNS). Efferent neurons include somatic motor that control skeletal muscles and autonomic that control glands and smooth and cardiac muscle.	Patients with GBS can neither feel sensations nor move their muscles. This suggests a problem in both sensory and somatic motor divisions. However, it is also possible that there is a problem in the CNS integrating center. You do not have enough information to determine which division is affected.
2	Do you think the paralysis found in the Chinese children affected both sensory and somatic motor neurons? Why or why not?	The Chinese children can feel a pin prick but cannot move their muscles.	Sensory function is normal if they can feel the pin prick. Paralysis of the muscles suggests a problem with somatic motor neurons, the CNS centers controlling movement, or with the muscles themselves.
3	In Guillain-Barré syndrome, what would you expect the results of a nerve conduction test to be?	Nerve conduction tests measure conduction speed and strength of action potentials. In GBS, myelin around neurons is destroyed.	Myelin insulates axons and increases conduction speed. Without myelin, current leaks out of the axon. Thus, in GBS you would expect decreased conduction speed and strength of action potentials.
4	Is the paralytic illness that affected the Chinese children a demyelinating condition? Why or why not?	Nerve conduction tests showed normal conduction speed but decreased strength of the action potentials.	Myelin loss should decrease conduction speed as well as action potential strength, therefore the Chinese illness is probably not a demyelinating disease.
5	Do the results of Dr. McKhann's investigation suggest that the Chinese children had Guillain-Barré syndrome? Why or why not?	Autopsy reports on children who died from the disease showed that the axons were damaged but the myelin was normal.	GBS is a demyelinating disease that affects both sensory and motor neurons. The Chinese children had normal sensory function, and nerve conduction tests and histological studies indicated normal myelin. Therefore it is reasonable to conclude that the disease is not GBS.
6	Based on information provided in this chapter, can you name another disease that alters synaptic transmission?	Synaptic transmission can be altered by blocking neurotransmitter release, action on the target cell, or removal from the synapse.	Parkinson's disease, depression, schizophrenia, and myasthenia gravis are related to problems with synaptic transmission (p. 274).

239 241 244 261 262 267 274

CHAPTER SUMMARY

This chapter introduces the nervous system, one of the major control systems responsible for maintaining *homeostasis*. The divisions of the nervous system correlate with the different steps in a reflex pathway. Sensory receptors monitor regulated variables and send input signals to the central nervous system through sensory neurons. Output signals, both electrical and chemical, travel through the efferent divisions to their targets throughout the body. Information transfer and *communication* depend on electrical signals that pass along neurons, *molecular interactions* between signal molecules and their receptors, and on signal transduction in the target cells. In Chapters 9 through 11 and 13 you will learn more details about the basic processes introduced here.

1. The **nervous system** is a complex network of neurons that form the rapid control system of the body. (p. 240)
2. **Emergent properties** of the nervous system include consciousness, intelligence, and emotion. (p. 240)

Organization of the Nervous System

3. The nervous system is divided into the **central nervous system (CNS)**, composed of the **brain** and **spinal cord**, and the **peripheral nervous system**. (p. 240; Fig. 8-1)

4. The peripheral nervous system has **afferent (sensory) neurons** that bring information into the CNS and **efferent neurons** that carry information away from the CNS back to various parts of the body. (p. 240)

5. The efferent neurons include **somatic motor neurons** that control skeletal muscles and **autonomic neurons** that control smooth and cardiac muscles, glands, and some adipose tissue. (p. 240)

6. Autonomic neurons are subdivided into **sympathetic** and **parasympathetic** neurons. (p. 240)

Cells of the Nervous System

IP Nervous System I: Anatomy Review

7. Neurons have a **cell body** with a nucleus and organelles to direct cellular activity, **dendrites** to receive incoming signals, and an **axon** to transmit electrical signals from the cell body to the **axon terminals**. (pp. 243–244; Fig. 8-2)

8. **Interneurons** are neurons that lie entirely within the CNS. (p. 243; Fig. 8-3)

9. The region where an axon terminal meets its target cell is called a **synapse**. The target cell is called the **postsynaptic cell**, and the neuron that releases the chemical signal is known as the **presynaptic cell**. (p. 244)

10. Material is transported between the cell body and axon terminal by **axonal transport**. (p. 244; Fig. 8-4)

11. **Glial cells** provide physical support and direct the growth of neurons during repair and development. **Schwann cells** and **satellite cells** are associated with the peripheral nervous system. **Oligodendrocytes, astrocytes,** and **ependymal cells** are found in the CNS. **Microglia** are modified immune cells that act as scavengers. (pp. 245–246; Fig. 8-5)

12. Schwann cells and oligodendrocytes form insulating **myelin sheaths** around neurons. The **nodes of Ranvier** are sections of uninsulated membrane between Schwann cells. (p. 246; Fig. 8-6)

13. **Neural stem cells** are found in the ependymal layer. (p. 246)

Electrical Signals in Neurons

IP Nervous System I: The Membrane Potential; Ion Channels; The Action Potential

14. Membrane potential is influenced by the concentration gradients of ions across the membrane and by the permeability of the membrane to those ions. (pp. 246–247)

15. The **Goldman-Hodgkin-Katz (GHK) equation** predicts membrane potential based on ion concentration gradients and membrane permeability. (p. 248)

16. The permeability of a cell to ions changes when ion channels in the membrane open and close. Movement of only a few ions will significantly change the membrane potential. (p. 249)

17. Gated ion channels in neurons open or close in response to chemical or mechanical signals or in response to depolarization of the cell membrane. (p. 249)

18. **Graded potentials** are depolarizations or hyperpolarizations whose strength is directly proportional to the strength of the triggering event. Graded potentials lose strength as they move through the cell. (p. 249; Tbl. 8-3; Fig. 8-7)

19. The wave of depolarization that moves through the cell with a graded potential is known as **local current flow**. (p. 250)

20. **Action potentials** are rapid electrical signals that travel undiminished in amplitude from the cell body to the axon terminals. (p. 252)

21. Action potentials begin in the **trigger zone** if a single graded potential or the sum of multiple graded potentials exceeds a minimum depolarization known as the **threshold**. (pp. 250–251; Fig. 8-8)

22. **Excitatory postsynaptic potentials (EPSPs)** are depolarizing graded potentials that make a neuron more likely to fire an action potential. **Inhibitory postsynaptic potentials (IPSPs)** are hyperpolarizing graded potentials. (p. 252)

23. Action potentials are uniform, **all-or-none** depolarizations. (p. 253)

24. The rising phase of the action potential is due to increased Na^+ permeability. The falling phase of the action potential is due to increased K^+ permeability. (pp. 253–254; Fig. 8-9)

25. The voltage-gated Na^+ channels of the axon have a fast **activation gate** and a slower **inactivation gate**. (p. 254; Fig. 8-10)

26. Once an action potential has begun, there is a brief period of time known as the **absolute refractory period** during which a second action potential cannot be triggered, no matter how large the stimulus. Because of this, action potentials cannot be summed. (p. 255; Fig. 8-12)

27. During the **relative refractory period**, a higher-than-normal graded potential is required to trigger an action potential. (p. 256)

28. Information about the strength and duration of a stimulus is conveyed by the frequency of action potential propagation. (p. 256; Fig. 8-13)

29. Very few ions cross the membrane during an action potential. The Na^+-K^+-ATPase eventually restores Na^+ and K^+ to their original compartments. (p. 256)

30. The movement of an action potential through the axon at high speed is called **conduction**. (p. 257; Fig. 8-14)

31. Larger axon diameter and increased membrane resistance speed up action potential conduction. The myelin sheath speeds up conduction by increasing membrane resistance and decreasing current leak. (pp. 258–260)

32. The apparent jumping of action potentials from node to node is called **saltatory conduction**. (p. 261; Fig. 8-17)

33. Changes in blood K^+ concentration affect resting membrane potential and the conduction of action potentials. (p. 262; Fig. 8-18)

Cell-to-Cell Communication in the Nervous System

IP Nervous System II: Anatomy Review; Synaptic Transmission; Ion Channels

34. In **electrical synapses**, an electrical signal passes directly from the cytoplasm of one cell to another through gap junctions. **Chemical synapses** use neurotransmitters to carry information from one cell to the next. (p. 263)

35. **Neurotransmitters** are synthesized in the cell body or in the axon terminal. They are stored in **synaptic vesicles** and are released by exocytosis when an action potential reaches the axon terminal. Neurotransmitters combine with receptors on target cells. (p. 263; Fig. 8-20)

36. Neurotransmitters come in a variety of forms. **Cholinergic** neurons secrete **acetylcholine**. **Adrenergic** neurons secrete **norepinephrine**. **Glutamate**, **GABA**, **serotonin**, **adenosine**, and **nitric oxide** are other major neurotransmitters. (pp. 266–267; Tbl. 8-4)

37. Neurotransmitter receptors are receptor-channels or G protein–coupled receptors. Ion channels create **fast synaptic potentials**. G protein–coupled receptors create **slow synaptic potentials** or modify cell metabolism. (p. 268; Fig. 8-22)

38. Neurotransmitter action is rapidly terminated by reuptake into cells, diffusion away from the synapse, or enzymatic breakdown. (p. 268; Fig. 8-23)

Integration of Neural Information Transfer

IP Nervous System II: Synaptic Potentials & Cellular Integration

39. If a presynaptic neuron synapses on a larger number of postsynaptic neurons, the pattern is known as **divergence**. If several presynaptic neurons provide input to a smaller number of postsynaptic neurons, the pattern is known as **convergence**. (p. 270; Fig. 8-24)

40. The summation of simultaneous graded potentials from different neurons is known as **spatial summation**. The summation of graded potentials that follow each other sequentially is called **temporal summation**. (pp. 270–271; Figs. 8-26, 8-27)

41. **Presynaptic modulation** of an axon terminal allows selective modulation of collaterals and their targets. **Postsynaptic modulation** occurs when a modulatory neuron, usually inhibitory, synapses on a postsynaptic cell. (pp. 271–273; Fig. 8-28)

42. **Long-term potentiation** is one mechanism by which neurons change the quality or quantity of their synaptic connections. (p. 273)

43. Developing neurons find their way to their targets by using chemical signals. (p. 274)

QUESTIONS

LEVEL ONE *Reviewing Facts and Terms*

1. List the three functional classes of neurons, and explain how they differ from one another structurally and functionally.

2. Match the term with its description:
   (a) axon
   (b) dendrite
   (c) afferents
   (d) efferents
   (e) axon hillock

   1. process of a neuron that receives incoming signals
   2. sensory neurons, transmit information to CNS
   3. long process that transmits signals to the target cell
   4. region of neuron where action potential begins
   5. neurons that transmit information from CNS to the rest of the body

3. Somatic motor neurons control _____. _____ neurons control smooth and cardiac muscles, exocrine and some endocrine glands, and some types of adipose tissue.

4. Autonomic neurons are classified as either _____ or _____ neurons.

5. Name the two primary cell types found in the nervous system.

6. Draw a typical neuron and label the cell body, axon, dendrites, nucleus, trigger zone, axon hillock, collaterals, and axon terminals. Draw mitochondria, rough endoplasmic reticulum, Golgi bodies, and vesicles in the appropriate sections of the neuron.

7. Axonal transport refers to:
   (a) the release of neurotransmitter molecules into the synaptic cleft.
   (b) the use of microtubules to send secretions from the cell body to the axon terminal.
   (c) the transport of vesicles containing proteins down the length of the axon.
   (d) the movement of the axon terminal to synapse with a new postsynaptic cell.
   (e) none of these

8. Match the numbers of the appropriate characteristics with the two types of potentials. Characteristics may apply to one or both types.
   (a) action potential
   (b) graded potential

   1. all-or-none
   2. can be summed
   3. amplitude decreases with distance
   4. exhibits a refractory period
   5. amplitude depends on strength of stimulus
   6. has no threshold

9. A typical neuron synapses with how many other neurons?
   (a) 1
   (b) fewer than 10
   (c) 1000
   (d) 10,000
   (e) more than 100,000

10. List the four major types of ion channels found in neurons. Are they chemically gated, mechanically gated, or voltage-gated?

11. Arrange these events in the proper sequence:
    (a) Efferent neuron reaches threshold and fires an action potential.
    (b) Afferent neuron reaches threshold and fires an action potential.
    (c) Effector organ responds by performing output.
    (d) Integrating center reaches decision about response.
    (e) Sensory organ detects change in the environment.

12. An action potential is (circle all correct answers):
    (a) a reversal of the Na^+ and K^+ concentrations inside and outside the neuron.
    (b) the same size and shape at the beginning and end of the axon.
    (c) initiated by inhibitory postsynaptic graded potentials.
    (d) transmitted to the distal end of a neuron and causes release of neurotransmitter.

In questions 13–17, choose the correct ion to fill in the blanks: Na^+, K^+, Ca^{2+}, Cl^-.

13. The resting cell membrane is more permeable to _____ ions than to _____ ions. Although _____ ions contribute little to the resting membrane potential, they play a key role in generating electrical signals in excitable tissues.

14. _____ is 12 times more concentrated outside the cell than inside.

15. _____ is 30 times more concentrated inside the cell than outside.

16. An action potential occurs when _____ enters the cell.

17. The resting membrane potential is due to the high _____ permeability of the cell.

18. What is the myelin sheath?

19. List two factors that enhance conduction speed.

20. List three ways neurotransmitters are removed from the synapse.

21. Draw and label a graph of an action potential. Below the graph, draw the positioning of the K^+ and Na^+ channel gates during each phase.

LEVEL TWO *Reviewing Concepts*

22. Create a map showing organization of the nervous system using the following terms plus any terms you choose to add:

afferent signals   glial cells   peripheral division
astrocyte   integration   satellite cell
autonomic division   interneuron   Schwann cell
brain   microglia   sensory division
CNS   muscles   somatic motor
efferent neuron   neuron   division
efferent signals   neurotransmitter   spinal cord
ependymal cell   oligodendrocyte   stimulus
glands   parasympathetic   sympathetic
     division   division

23. What causes the depolarization phase of the action potential? (Circle all that apply.)
    (a) K^+ leaving the cell through voltage-gated channels
    (b) K^+ being pumped into the cell by the Na^+-K^+-ATPase
    (c) Na^+ being pumped into the cell by the Na^+-K^+-ATPase
    (d) Na^+ entering the cell through voltage-gated channels
    (e) opening of the Na^+ channel inactivation gate

24. List four different types of glial cells, and briefly explain the role of each type.

25. Arrange the following terms to describe the sequence of events that happens after a neurotransmitter binds to a receptor on a postsynaptic neuron. Terms may be used more than once or not at all.
    (a) action potential fires at axon hillock
    (b) axon hillock reaches threshold
    (c) cell depolarizes
    (d) exocytosis
    (e) graded potential occurs
    (f) ligand-gated ion channel opens
    (g) local current flow
    (h) saltatory conduction
    (i) voltage-gated Ca^{2+} channels open
    (j) voltage-gated K^+ channels open
    (k) voltage-gated Na^+ channels open

26. Match the best term (hyperpolarize, depolarize, repolarize) to the events described. The cell in question has a resting membrane potential of −70 mV.
    (a) membrane potential changes from −70 mV to −50 mV
    (b) membrane potential changes from −70 mV to −90 mV
    (c) membrane potential changes from +20 mV to −60 mV
    (d) membrane potential changes from −80 mV to −70 mV

27. A neuron has a resting membrane potential of −70 mV. Will the neuron hyperpolarize or depolarize when each event below occurs? More than one answer may apply; list all those that are correct. What would happen if:
    (a) Na^+ enters the cell
    (b) K^+ leaves the cell
    (c) Cl^- enters the cell
    (d) Ca^{2+} enters the cell

28. Define, compare, and contrast the following concepts:
    (a) threshold, subthreshold, suprathreshold, spatial summation, temporal summation, all-or-none, overshoot, undershoot
    (b) graded potential, EPSP, IPSP, absolute refractory period, relative refractory period
    (c) afferent neuron, efferent neuron, interneuron
    (d) sensory, somatic motor, sympathetic, autonomic, parasympathetic neuron
    (e) fast and slow synaptic potentials
    (f) temporal and spatial summation
    (g) convergence and divergence

29. If all action potentials within a given neuron are identical, how does the neuron transmit information about the strength and duration of the stimulus?

30. The presence of myelin allows an axon to:
    (a) produce more frequent action potentials.
    (b) conduct impulses more rapidly.
    (c) produce action potentials of larger amplitude.
    (d) produce action potentials of longer duration.

LEVEL THREE *Problem Solving*

31. If human babies' muscles and neurons are fully developed and functional at birth, why can't they focus their eyes, sit up, or learn to crawl within hours of being born? (Hint: Muscle strength is not the problem.)

32. The voltage-gated Na^+ channels of the neuron open when the neuron depolarizes. If depolarization opens the channels, why do they close when the neuron is maximally depolarized?

33. One of the pills that Jim takes for high blood pressure caused his blood K^+ to decrease from 4.5 mM to 2.5 mM. What happens to the resting membrane potential of his liver cells? (Circle all that are correct.)
    (a) it decreases
    (b) it increases
    (c) it does not change

    (d) it becomes more negative
    (e) it becomes less negative
    (f) it fires an action potential
    (g) it depolarizes
    (h) it hyperpolarizes
    (i) it repolarizes

34. Match the example to the proper type of stimulus:
    (a) mechanical   1. bath water at 106° F
    (b) chemical   2. acetylcholine
    (c) thermal   3. a hint of perfume
         4. epinephrine
         5. lemon juice
         6. a punch on the arm

35. In each scenario below, will an action potential be produced? The postsynaptic neuron has a resting membrane potential of −70 mV.

    (a) Fifteen neurons synapse on one postsynaptic neuron. At the trigger zone, 12 of the neurons produce EPSPs of 2 mV each, and the other three produce IPSPs of 3 mV each. The threshold for the postsynaptic cell is −50 mV.

    (b) Fourteen neurons synapse on one postsynaptic neuron. At the trigger zone, 11 of the neurons produce EPSPs of 2 mV each, and the other three produce IPSPs of 3 mV each. The threshold for the postsynaptic cell is −60 mV.

    (c) Fifteen neurons synapse on one postsynaptic neuron. At the trigger zone, 14 of the neurons produce EPSPs of 2 mV each, and the other one produces an IPSP of 9 mV. The threshold for the postsynaptic cell is −50 mV.

36. The GHK equation is sometimes abbreviated to exclude chloride, which plays a minimal role in membrane potential for most cells. In addition, because it is difficult to determine absolute membrane permeability values for Na^+ and K^+, the equation is revised to use the ratio of the two ion permeabilities, α. In the revised equation shown below, $\alpha = P_{Na}/P_K$.

$$V_m = 60 \log \frac{[K^+]_{out} + \alpha[Na^+]_{out}}{[K^+]_{in} + \alpha[Na^+]_{in}}$$

Thus, if you know the relative membrane permeabilities of the two ions and their intracellular and extracellular concentrations, you can predict the membrane potential for a cell.

    (a) Looking at the equation above, can you predict what will happen to V_m (in millivolts) if the only change is an increase in sodium permeability?

    (b) A resting cell has an alpha value of 0.025 and the following ion concentrations:

	INTRACELLULAR	EXTRACELLULAR
Na^+	5 mM	135 mM
K^+	150 mM	4 mM

What is its membrane potential?

Explorations

Explorations are also available at *www.physiologyplace.com*

Exploration 1: Dynamic Action Potentials

In this chapter you learned how neurons communicate over long distances using electrical signals called action potentials. Action potentials are a dynamic physiological process that is difficult to present as a series of static drawings. Many excellent simulations showing action potentials have been developed and are available on the Web. To find some of them, go to Google (*www.google.com*) and search for *action potential animation*. As you view these animations, think critically about how they present the action potential. Does the animation accurately reflect your understanding of the process?

Exploration 2: Sequencing the Mouse Genome: A Gateway to Understanding Human Disease

Looking at a mouse, you might not realize how much we have in common with that furry four-legged creature. On the heels of the release of the human genome sequence, the mouse genome sequencing effort tells us that 99% of the approximately 30,000 mouse genes have direct human homologs (equivalents). What this means is that we can use the mouse to understand what genes and the proteins they encode do in healthy and sick humans. The mouse is the ideal organism to conduct these experiments. It is small, relatively inexpensive, has a short lifespan, its biological processes are similar to those of humans, and the mouse genome can be easily manipulated by genetic engineering technologies. Scientists can put human genes in the mouse and cause diseases that mimic human diseases. Alternatively, a gene can be knocked out or simply turned off or on at specific times during the life of a mouse. It is also possible to turn a mouse gene off or on in a single tissue like the brain or liver. Researchers from universities, private research institutes, and biotechnology and pharmaceutical companies are scavenging the mouse and human genome sequences, available on public databases, and are poised dramatically to accelerate our understanding of human disease, creating new treatments and cures.

To learn more, you might read the articles in a special Mouse Genome issue of the journal *Nature* (*Nature* 420:510–590, 2002). For an introduction, see the article written by the Mouse Genome Sequencing Consortium: "Initial sequencing and comparative analysis of the mouse genome," *Nature* 420:520–562, 2002. You can obtain additional information from the Mouse Genome Informatics databases at The Jackson Laboratory (*www.informatics.jax.org*).

To learn more about how animal models are used in the development of therapeutic drugs, read "Knockouts model the 100 best-selling drugs—will they model the next 100?" *Nat Rev Drug Discov* 2:38–51, 2003.

(continues on next page)

This exploration was written by Carol C. Linder, Ph.D., Assistant Director of Genetic Resources at The Jackson Laboratory, Bar Harbor, Maine. Dr. Linder received her Ph.D. from the University of Texas and was an award-winning graduate teaching assistant in Dr. Dee's physiology course.

Exploration 3: Defining Search Terms

When you decide to search for information, whether on the Web or through one of the many databases that catalog peer-reviewed and lay publications, your first step should be to define your search terms. Databases like MEDLINE, published by the U.S. National Library of Medicine, have a standardized list of indexing terms or you can define the term and do a *keyword search*. For example, suppose you want to learn more about channelopathies, the diseases caused by mutations in ion channels. If you type "channelopathy"

into PubMed, the free public access version of MEDLINE (*www.ncbi.nlm.nih.gov/PubMed/*), the basic search asks you to type in a keyword. If you access MEDLINE through a search engine such as Ovid, the program will also try to match your keyword to the database indexing terms, called MeSH headings for MEDLINE. For channelopathies, the related MeSH headings are various types of ion channels, "nervous system diseases," and pathologies related to ion channels, such as "arrhythmia" or "paralyses, familial periodic."

When you do a web search using a search engine like Google, you must think more carefully about the search terms you use. Small wording changes can make a big difference. To see this for yourself, type CHANNELOPATHY into Google and see how many hits you get. Then, make the term plural—CHANNELOPATHIES—and see the difference in what and how many results you get.

ANSWERS

Answers to Concept Check Questions

Page 241

1. Compare your answer to the map in Figure 8-1, p. 241.

Page 244

2. Neurons that secrete neurohormones terminate close to blood vessels so that neurohormones can enter the circulation.

3. See Figure 8-2.

Page 246

4. Myelin insulates axon membranes. Microglia are scavenger cells. Ependymal cells form barriers between compartments.

5. Schwann cells are in the peripheral nervous system and each cell forms myelin around a small portion of one axon. Oligodendrocytes are in the CNS and one cell forms myelin around axons of several different neurons.

Page 248

6. For Ca^{2+}, $1/0.0001 = 10,000$. $10,000 = 10^4$. Log of 10^4 is 4 (see Appendix A). $(4 \times 61)/+2 = 122$ mV. The charge on Ca^{2+} is +2, therefore z = +2.

Page 249

7. (a) depolarizes   (b) depolarizes

8. depolarize

Page 250

9. (a) 1, (b) 2, (c) 2, (d) 1

Page 251

10. The trigger zone for the sensory neurons will be close to where the dendrites converge. You cannot tell where the trigger zone is for the anaxonic neuron. For the other interneurons and the efferent neuron, the trigger zone will be at the junction of the cell body and the axon.

Page 255

11. It will depolarize and stay depolarized.

Page 255

12. Activation gate is closing, inactivation gate is opening.

Page 256

13. (b)

Page 258

14. The action potential will go in both directions because the Na^+ channels around the stimulation site have not been inactivated by a previous depolarization.

Page 261

15. a, c, b

Page 263

16. Proteins are synthesized on the ribosomes of the rough endoplasmic reticulum, then are directed into the Golgi apparatus.

17. Mitochondria are the primary sites of ATP synthesis.

18. Mitochondria reach the axon terminal by fast axonal transport along microtubules.

Page 264

19. They concluded that some event between arrival of the action potential at the terminal and depolarization of the postsynaptic cell is dependent on extracellular Ca^{2+}. We now know it is neurotransmitter release.

Page 268

20. Different receptor subtypes work through different signal transduction pathways. Targeting drugs therefore decreases the likelihood of unwanted side effects.

Page 268

21. SSRIs decrease reuptake of serotonin into the axon terminal, thereby increasing the time it is active in the synapse.

Page 271

22. Yes, because the net effect would be a 17 mV depolarization, just above the threshold of -55 mV.

23. The timing is different because the depolarization must travel from the point of the stimulus to the recording point.

Page 273

24. Axon terminals convert (transduce) the electrical action potential signal into a chemical neurotransmitter signal.

Page 274

25. Membrane depolarization makes the inside of the membrane more positive in respect to the outside. Like charges repel, so the more positive membrane potential tends to repel Mg^{2+}.

Answers to Figure Questions

Page 251

Fig. 8-7: The graded potential is stronger at B.

Answers to Level Four Quantitative Problems

Page 281

35. (a) $(12 \times 2 \text{ mV} = +24) + (3 \times -3 \text{ mV} = -9)$ = signal strength of +15. Potential $-70 + 15 = -55$. Threshold is -50, so no action potential. (Potential must be equal to or more positive than threshold.)
    (b) $(11 \times 2 = +22) + (3 \times -3 = -9) = +13$. Potential $-70 + 13 = -57$. Threshold is -60, so action potential will fire.

(c) $(14 \times 2 = +28) + (-9) = +19$. Potential $-70 + 19 = -51$. Threshold is -50, so no action potential.

36. (a) An increase in sodium permeability will make alpha larger. That in turn will increase the numerator and make the membrane potential more positive.
    (b) Membrane potential is -78.5 mV.

The Central Nervous System

PET scan of brain

Neuronal assemblies have important properties that cannot be explained by the additive qualities of individual neurons.

—O. Hechter, in *Biology and Medicine into the 21st Century,* **1991**

RUNNING PROBLEM

Infantile Spasms

At four months of age Ben could roll over, hold up his head, and reach for things. At seven months Ben was nearly paralyzed and lying listlessly in his crib. He had lost his abilities so gradually that it was hard to remember when each one had slipped away, but his mother could remember exactly when it began. She was preparing to feed him lunch one day when she heard a cry from the highchair where Ben was sitting. As she watched, Ben's head dropped to his chest, came back up, then went hurtling toward his lap, smacking into his highchair table. Ben's mother snatched him up into her arms, and she could feel him still convulsing against her shoulder. This was the first of many such spells that came with increasing frequency and duration.

BACKGROUND BASICS

Directions of the body: **842** | Up-regulation: **186** | Diffusion through membranes: **136** | Cell-to-cell junctions: **64** | Synapses: **263** | Glial cells: **245** | Transporting epithelium: **67** | Ionic basis of action potentials: **253** | Neurotransmitters and neuromodulators: **265** | Posterior pituitary: **222** | Fast and slow synaptic potentials: **268** | Long-term potentiation: **273** |

Mary Lou awoke to the sound of her 76-year-old husband crying in the darkness. When she turned on the bedside light, she saw Harold standing in his closet, facing the wall. Concerned, Mary Lou got out of bed and went to him. "Harold, what's wrong?" "I can't find my way back to bed!" her husband wailed. Harold suffers from *Alzheimer's disease*, a progressive neurodegenerative disease that robs people of their memories and leaves them unable to recognize familiar places and people. We still do not fully understand the cause of Alzheimer's disease, but we do know that normal communication among brain neurons is lost.

▶ EMERGENT PROPERTIES OF NEURAL NETWORKS

Neurons within the nervous system link together to form circuits with specific functions. The most complex circuits are those of the brain, in which thousands of millions of neurons are linked into intricate networks that converge and diverge, creating an infinite number of possible pathways. Signaling within these pathways creates thinking, language, feeling, learning, and memory, the complex behaviors that make us human. Some neuroscientists have proposed that the functional unit of the nervous system be changed from the individual neuron to neural networks because even the most basic functions require circuits of neurons.

How is it that combinations of neurons linked together into chains or networks collectively possess properties that are not found in any single neuron? We do not yet have an answer to this question. Some scientists seek to answer it by looking for parallels between the nervous system and the integrated circuits of computers.

Computer programs have been written that attempt to mimic the thought processes of humans. This field of study, called *artificial intelligence*, has created some interesting programs such as the "psychiatrist" that is programmed to respond to typed complaints with appropriate comments and suggestions. But we are nowhere near creating a brain as complex as that of a human, or even as complex as that of Hal, the computer in the classic movie *2001*.

Probably one reason that computers cannot yet accurately model brain function is that they lack **plasticity**, the ability to change circuit connections and function in response to sensory input and past experience. Although some computer programs can change their output under specialized conditions, they cannot begin to approximate the plasticity of human brain networks that easily restructure themselves as the result of sensory input, learning, emotion, and creativity. In addition, we now know the brain can add new connections when neural stem cells differentiate. Computers cannot add new circuits to themselves.

How can simply linking neurons together create **affective behaviors**, which are related to feeling and emotion, and **cognitive behaviors** [*cognoscere*, to get to know] related to thinking? In their search for the organizational principles that lead to these behaviors, scientists seek clues in the simplest animal nervous systems.

▶ EVOLUTION OF NERVOUS SYSTEMS

All animals have the ability to sense and respond to changes in their environment. Even single-cell organisms like the *Paramecium* are able to carry out the basic tasks of life—finding food, avoiding being food, finding a mate. Yet they have no obvious brain or integrating center, simply the resting membrane potential that exists in living cells.

The first multicellular animals to develop neurons were members of the phylum Cnidaria, the jellyfish and sea anemones. Their nervous system is a *nerve net* (Fig. 9-1a ●) composed of sensory neurons, connective interneurons, and motor neurons that innervate muscles and glands. These animals respond to stimuli with complex behaviors, yet without input from an identifiable control center. If you watch a jellyfish swim or a sea anemone stuff a piece of shrimp in its mouth, it is hard to imagine how a diffuse network of neurons can create such complex coordinated movements. However, the same basic principles that you learned in the last chapter apply to jellyfish and humans. Electrical signals in the form of action potentials and chemical signals passing across synapses are the same in all animals. It is only in the number and organization of the neurons that we differ.

CONCEPT CHECK   Answers: p. 319

1. Match the terms with the appropriate neuron type(s).

   (a) Afferent neuron   1. Interneuron
   (b) Efferent signal   2. Motor neuron
   (c) Integrating center   3. Sensory neuron
   (d) Input signal
   (e) Output signal

In the primitive flatworms, we see the beginnings of a nervous system as we know it in higher animals, although in these animals the distinction between central nervous system and peripheral nervous system is not clear. Flatworms have a rudimentary brain, a cluster of nerve cell bodies concentrated in the head (*cephalic*) region. A large nerve called a *nerve cord* comes off the primitive brain and leads to a nerve network that innervates distal regions of the flatworm body (Fig. 9-1b).

The segmented worms, or annelids, such as the earthworm, have a more advanced central nervous system (Fig. 9-1c). Clusters of cell bodies are no longer restricted to the primitive brain but also occur in fused pairs, called **ganglia** [singular *ganglion*, a swelling], along a nerve cord. Because each segment of the worm contains a ganglion, simple reflexes can be integrated within a segment without input from the brain. Reflexes that do

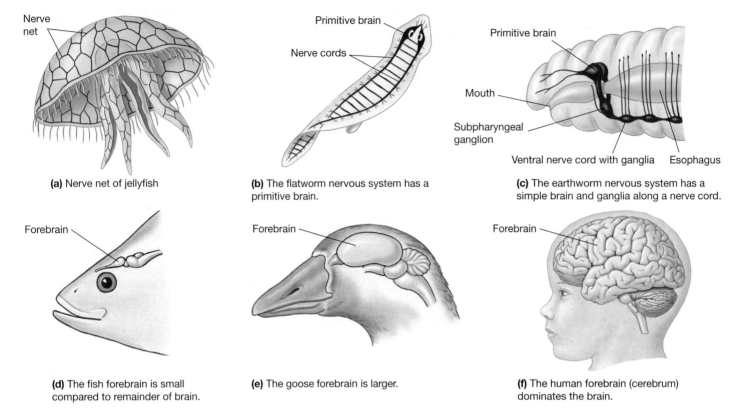

Nerve net

(a) Nerve net of jellyfish

Primitive brain

Nerve cords

(b) The flatworm nervous system has a primitive brain.

Primitive brain

Mouth

Subpharyngeal ganglion

Ventral nerve cord with ganglia   Esophagus

(c) The earthworm nervous system has a simple brain and ganglia along a nerve cord.

Forebrain

(d) The fish forebrain is small compared to remainder of brain.

Forebrain

(e) The goose forebrain is larger.

Forebrain

(f) The human forebrain (cerebrum) dominates the brain.

● **Figure 9-1** *Evolution of the nervous system*

not require integration in the brain also occur in higher animals and are called **spinal reflexes** in humans and other vertebrates.

Annelids and higher invertebrates have complex reflexes controlled through neural networks. Researchers use leeches (a type of annelid) and *Aplysia*, a type of shell-less mollusk, to study neural networks and synapse formation because their neurons can be easily seen and traced. The neural function of these animals provides a simple model system for more complex vertebrate networks.

Nerve cells bodies clustered into brains persist throughout the more advanced phyla and become increasingly more complex. One advantage to cephalic brains is that in most animals, the head is the part of the body that first contacts the environment as the animal moves. In the higher arthropods, such as insects, specific regions of the brain begin to be associated with particular functions and to receive input from specific sensory structures. More sophisticated brains are associated with specialized cephalic receptors such as eyes or chemoreceptors for smell and taste.

In vertebrate brain evolution, the most dramatic change is seen in the *forebrain* region [*fore*, in front], which includes the **cerebrum** [*cerebrum*, brain; adjective *cerebral*]. In fish, the forebrain is a small bulge, dedicated mainly to processing *olfactory* information about odors in the environment (Fig. 9-1d). In birds and rodents, the forebrain has enlarged but still has a smooth surface (Fig. 9-1e). In humans, the cerebrum is the largest and most distinctive part of the brain, with deep grooves and folds

(Fig. 9-1f). More than anything else, the cerebrum is what makes us human. All evidence indicates that it is the part of the brain that allows reasoning and cognition.

BIOTECHNOLOGY

Tracing Neurons in a Network

One of the challenges facing neurophysiologists and neuroanatomists is tracing the networks of neurons that control specific functions, a task that can be likened to following one tiny thread through a tangled mass the size of a beach ball. In 1971, K. Kristensson and colleagues introduced the use of *horseradish peroxidase* (HRP), an enzyme that acts in the presence of its substrate to produce a visible product. When injected into the extracellular fluid near axon terminals, HRP is brought into the neuron by endocytosis. The vesicles containing HRP are then transported by fast retrograde (backward) axonal transport to the cell body and dendrites [● p. 244]. Once the enzyme-substrate reaction is completed, the neuron becomes visible, allowing the researcher to trace the entire neuron from its target back to its origin. Now, with the help of fluorescent antibodies, scientists can even tell what neurotransmitters and receptors the neuron uses.

In this chapter we begin with an overview of central nervous system anatomy and functions. We then look at how neural networks create the higher brain functions of thought and emotion.

▶ ANATOMY OF THE CENTRAL NERVOUS SYSTEM

The vertebrate **central nervous system** (CNS) consists of the brain and the **spinal cord**. As you learned in the previous section, brains increase in complexity and degree of specialization as we move up the phylogenetic tree from fish to humans. However, if we look at the vertebrate nervous system during development, a basic anatomical pattern emerges. In all vertebrates, the central nervous system consists of layers of neural tissue surrounding a fluid-filled central cavity lined with epithelium.

The Central Nervous System Develops from a Hollow Tube

In the very early embryo (c. day 20 of human development), the cells that will become the nervous system lie in a flattened region called the **neural plate** (Fig. 9-2 ●). As development proceeds,

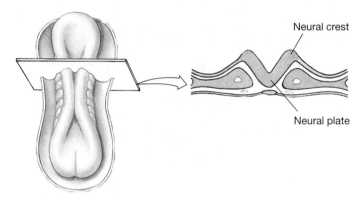

(a) 20 day embryo (dorsal view)

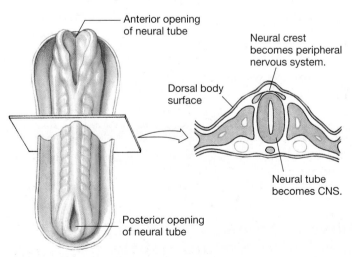

(b) 23 day embryo (dorsal view)

● **Figure 9-2** *The embryonic nervous system develops into a hollow tube.*

neural crest cells along the edge migrate toward the midline and fuse, creating a **neural tube** (Fig. 9-2b, c. day 23 of human development). The lumen of the neural tube will remain hollow and become the central cavity of the CNS. The cells lining the neural tube will differentiate into the epithelial *ependyma* [● p. 246] or remain as undifferentiated *neural stem cells*. The outermost cells of the neural tube will become the neurons and glia of the central nervous system, and the neural crest cells will migrate to become the peripheral nervous system.

By four weeks of human development, the anterior portion of the neural tube has begun to specialize into the regions of the brain (Fig. 9-3a ●). Three divisions are obvious: a forebrain, a midbrain, and a hindbrain. The tube posterior to the hindbrain will become the spinal cord. At this stage, the area that will become the cerebrum is not much larger than the other regions of the brain.

As development proceeds, the growth of the cerebrum begins to outpace that of the other regions (Fig. 9-3b). By six weeks of development, the central nervous system has formed the seven major divisions that are present at birth: (1) the cerebrum, (2) the diencephalon, (3) the midbrain, (4) and (5) the cerebellum and pons, (6) the medulla oblongata, and (7) the spinal cord. The central cavity has begun to enlarge into the hollow **ventricles** [*ventriculus*, belly] of the brain. There are two *lateral ventricles* (the first and second) and two descending ventricles (the third and fourth).

By eleven weeks the cerebrum is noticeably enlarged (Fig. 9-3c), and at birth the cerebrum is the largest and most obvious structure we see when looking at a human brain (Fig. 9-3d). The fully developed cerebrum surrounds the diencephalon, midbrain, and pons, leaving only the cerebellum and medulla visible below it. Because of the flexion of the neural tube early in development (see Fig. 9-3b), some directional terms have different meanings when applied to the brain (Fig. 9-3e).

The Central Nervous System Is Divided into Gray and White Matter

The central nervous system, like the peripheral nervous system, is composed of neurons and supportive glial cells. *Interneurons* are completely contained within the central nervous system. Sensory and efferent neurons link interneurons to peripheral receptors and effectors.

When viewed on a macroscopic level, the tissues of the central nervous system are divided into gray matter and white matter. **Gray matter** consists of unmyelinated nerve cell bodies, dendrites, and axon terminals (Fig. 9-4c ●). The cell bodies are assembled in an organized fashion within both the brain and the spinal cord. They form layers in some parts of the brain or cluster into groups of neurons with a similar function. Clusters of cell bodies in the brain and spinal cord are known as **nuclei**, and they are usually identified by specific names, such as the *lateral geniculate nucleus*, where visual information is processed.

White matter is made up mostly of myelinated axons and contains very few cell bodies. Its pale color comes from the

287

(a) A 4-week human embryo showing the anterior end of the neural tube which has specialized into three brain regions.

(b) At 6 weeks, the neural tube has differentiated into the brain regions present at birth. The hollow spaces shown in the cross section will become the ventricles of the brain. (see Fig. 9-5)

(c) By 11 weeks of embryonic development, the growth of the cerebrum is noticeably more rapid than that of the other divisions of the brain.

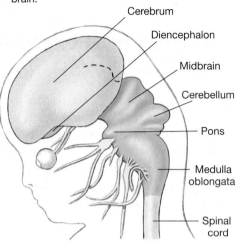

(d) At birth, the cerebrum has covered most of the other brain regions. Its rapid growth within the rigid confines of the cranium forces it to develop a convoluted, furrowed surface.

(e) The directions "dorsal" and "ventral" are different in the brain because of flexion in the neural tube during development.

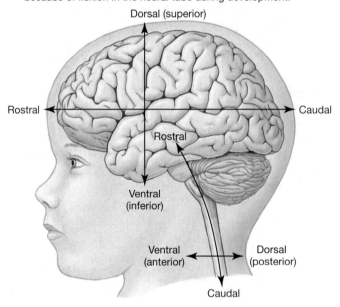

● **Figure 9-3** *The neural tube specializes into the six major regions of the nervous system.*

myelin sheaths that surround the axons. Bundles of axons that connect different regions of the central nervous system are known as **tracts**. Tracts in the central nervous system are equivalent to nerves in the peripheral nervous system.

The consistency of the brain and spinal cord is soft and jelly-like. Although individual neurons and glial cells have highly organized internal cytoskeletons that maintain cell shape and orientation, nervous tissue has minimal extracellular matrix and must rely on external support to help protect it from trauma. This support comes in the form of an outer casing of bone, three layers of connective tissue membrane, and fluid between the membranes.

CONCEPT CHECK   Answers: p. 319

2. Name and give the functions of the four groups of glial cells found in the CNS.

Bone and Connective Tissue Support the Central Nervous System

In vertebrates, the brain is encased in a bony **skull**, or **cranium** (Fig. 9-4a), and the spinal cord runs through a canal in the **vertebral column**. The body segmentation that is characteristic

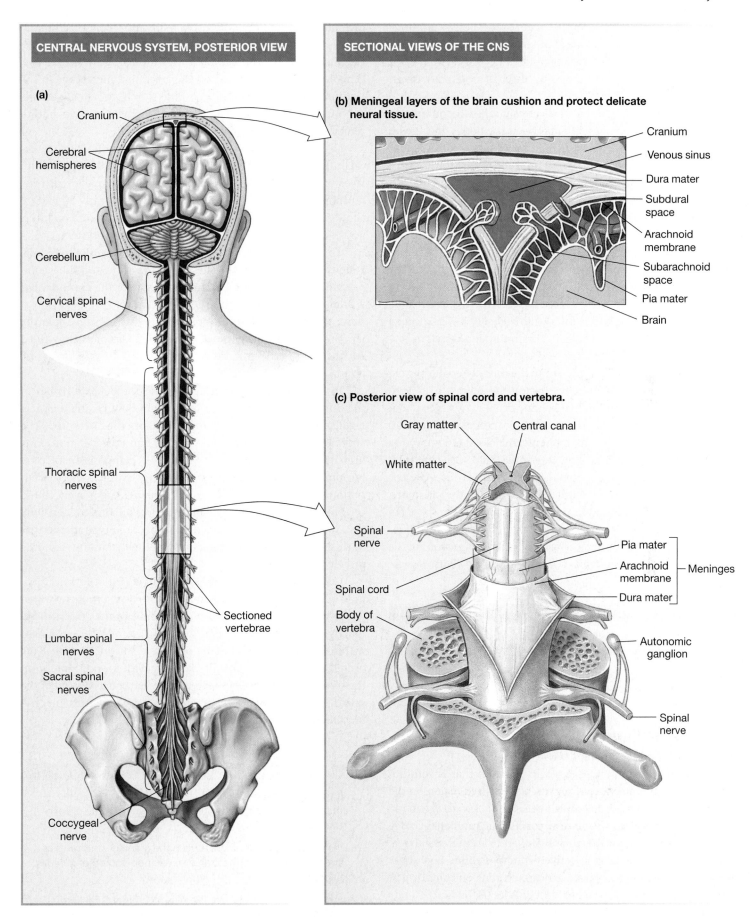

CENTRAL NERVOUS SYSTEM, POSTERIOR VIEW

(a)

- Cranium
- Cerebral hemispheres
- Cerebellum
- Cervical spinal nerves
- Thoracic spinal nerves
- Sectioned vertebrae
- Lumbar spinal nerves
- Sacral spinal nerves
- Coccygeal nerve

SECTIONAL VIEWS OF THE CNS

(b) Meningeal layers of the brain cushion and protect delicate neural tissue.

- Cranium
- Venous sinus
- Dura mater
- Subdural space
- Arachnoid membrane
- Subarachnoid space
- Pia mater
- Brain

(c) Posterior view of spinal cord and vertebra.

- Gray matter
- Central canal
- White matter
- Spinal nerve
- Spinal cord
- Body of vertebra
- Pia mater
- Arachnoid membrane — Meninges
- Dura mater
- Autonomic ganglion
- Spinal nerve

● **Figure 9-4** *ANATOMY SUMMARY The Central Nervous System* of many invertebrates can still be seen in the bony **vertebrae**, which are stacked on top of one another and separated by disks of connective tissue. Nerves of the peripheral nervous system enter and leave the spinal cord by passing through notches between the stacked vertebrae (Fig. 9-4c).

Three layers of membrane, collectively called the **meninges** [singular *meninx*, membrane], lie between the bones and tissues of the central nervous system. These membranes help stabilize the neural tissue and protect it from bruising against the bones of the skeleton. Starting from the bones and moving toward the nervous tissue, the membranes are (1) the dura mater, (2) the arachnoid membrane, and (3) the pia mater (Fig. 9-4b, c). The **pia mater** [*pius*, pious + *mater*, mother] is a thin membrane that adheres to the surface of the brain and spinal cord. Arteries that supply blood to the brain are associated with this layer. The **arachnoid** [*arachnoides*, cobweblike] **membrane** is loosely tied to the pia, leaving a *subarachnoid space* between the two layers. The **dura mater** [*durare*, to last] is thick (think *durable*) and associated with veins that drain blood from the brain through vessels or cavities called *sinuses*.

The final protective component of the central nervous system is extracellular fluid, which helps cushion the delicate neural tissue. The cranium has an internal volume of 1.4 L, of which about 1 L is occupied by the cells. The remaining volume is divided into two distinct extracellular compartments: the blood (100–150 mL) and the cerebrospinal fluid and interstitial fluid (250–300 mL). The cerebrospinal and interstitial fluids form the extracellular environment for neurons. Interstitial fluid lies inside the pia mater; cerebrospinal fluid is found in the ventricles and in the space between the pia mater and the arachnoid membrane. These two fluids communicate with each other across the leaky junctions of the pial membrane and ependymal cell layer lining the ventricles.

CONCEPT CHECK   Answers: p. 319

3. What is a ganglion? What is the equivalent structure in the central nervous system?

4. Peripheral nerves are equivalent to what organizational structure in the central nervous system?

The Brain Floats in Cerebrospinal Fluid

Cerebrospinal fluid, or CSF, is a salty solution that is continuously secreted by the **choroid plexus**, a specialized region on the walls of the ventricles (Fig. 9-5b ●). The choroid plexus is remarkably similar to kidney tissue and consists of capillaries and a transporting epithelium [● p. 67] derived from the ependyma. The choroid plexus cells selectively pump sodium and other solutes from plasma into the ventricles, creating an osmotic gradient that draws water along with the solutes (Fig. 9-5c).

From the ventricles, cerebrospinal fluid flows into the **subarachnoid space** between the pia mater and the arachnoid

membrane, surrounding the entire brain and spinal cord in fluid (Fig. 9-5b). The cerebrospinal fluid flows around the nervous tissue and is finally absorbed back into the blood by special **villi** [singular *villus*, shaggy hair] on the arachnoid membrane in the cranium (Fig. 9-5d). The rate of fluid flow through the central nervous system is sufficient to replenish the entire volume of cerebrospinal fluid about three times a day.

Cerebrospinal fluid serves two purposes: physical and chemical protection. The brain and spinal cord actually float in the thin layer of fluid between the membranes. The buoyancy of cerebrospinal fluid reduces the weight of the brain nearly 30-fold, to about 50 g. Lighter weight translates into less pressure on blood vessels and nerves attached to the central nervous system.

The cerebrospinal fluid also provides protective padding. When there is a blow to the head, the fluid must be compressed before the brain can hit the inside of the cranium. For a dramatic example of the protective power of cerebrospinal fluid, shake a block of tofu representing the brain in an empty jar. Then shake another block of tofu in a jar completely filled with water to see how the cerebrospinal fluid safeguards the brain.

In addition to physically protecting the delicate tissues of the central nervous system, cerebrospinal fluid creates a closely regulated extracellular environment for the neurons. The choroid plexus is selective about what substances it transports into the ventricles, and, as a result, the composition of cerebrospinal fluid is different from that of the plasma. The concentrations of K^+, Ca^{2+}, HCO_3^-, and glucose are lower in the cerebrospinal fluid, and the concentration of H^+ is higher. Only the concentration of Na^+ in cerebrospinal fluid is similar to that in the blood. In addition, cerebrospinal fluid contains very little protein and no blood cells.

Cerebrospinal fluid exchanges solutes with the interstitial fluid of the central nervous system and provides a route by which wastes can be removed. Clinically, a sample of the cerebrospinal fluid is presumed to be an indicator of the chemical environment within the brain. This sampling procedure, known as a *spinal tap* or *lumbar puncture*, is generally done by withdrawing fluid from the subarachnoid space between vertebrae at the lower end of the spinal cord. The presence of proteins or blood cells in cerebrospinal fluid suggests an infection.

CONCEPT CHECK   Answers: p. 319

5. If the concentration of H^+ in cerebrospinal fluid is higher than that in the blood, what can you say about the pH of the CSF?

6. If a blood vessel running between the meninges ruptures as a result of a blow to the cranium, what do you predict might happen?

7. Is cerebrospinal fluid more like plasma or interstitial fluid? Defend your answer.

(a) Ventricles of the brain
The lateral ventricles consist of the first and second ventricles. The third and fourth ventricles extend through the brain stem and connect to the central canal that runs through the spinal cord. Compare the frontal view to the cross-section in Fig. 9-3b.

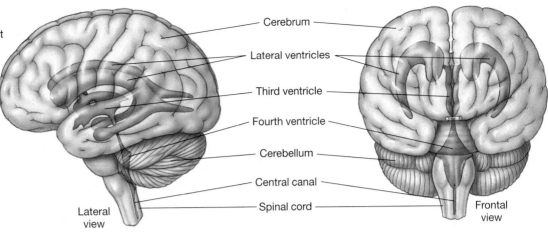

Cerebrum
Lateral ventricles
Third ventricle
Fourth ventricle
Cerebellum
Central canal
Spinal cord
Lateral view
Frontal view

(b) Cerebrospinal fluid is secreted into the ventricles and flows throughout the subarachnoid space where it cushions the central nervous system.

(d) Cerebrospinal fluid is reabsorbed back into the blood at fingerlike projections of the arachnoid membrane called villi

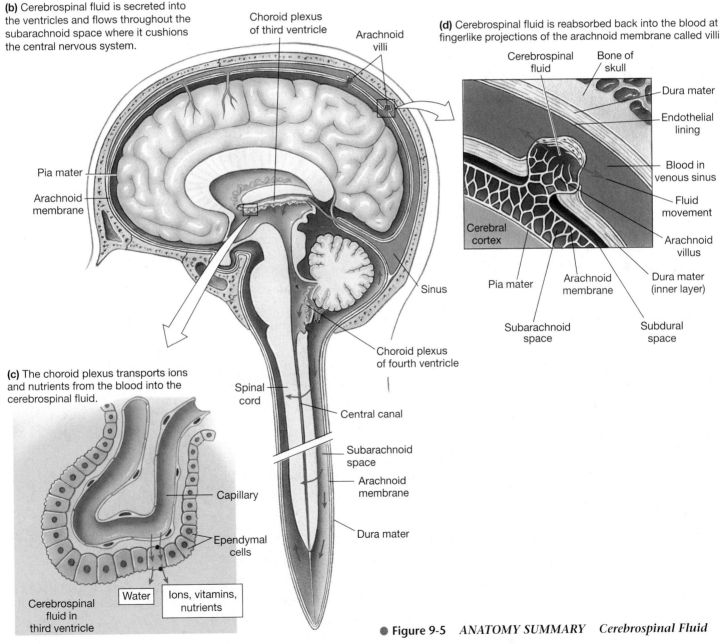

Choroid plexus of third ventricle
Arachnoid villi
Pia mater
Arachnoid membrane
Sinus
Choroid plexus of fourth ventricle
Spinal cord
Central canal
Subarachnoid space
Arachnoid membrane
Dura mater

Cerebrospinal fluid
Bone of skull
Dura mater
Endothelial lining
Blood in venous sinus
Fluid movement
Cerebral cortex
Arachnoid villus
Pia mater
Arachnoid membrane
Dura mater (inner layer)
Subarachnoid space
Subdural space

(c) The choroid plexus transports ions and nutrients from the blood into the cerebrospinal fluid.

Capillary
Ependymal cells
Water
Ions, vitamins, nutrients
Cerebrospinal fluid in third ventricle

● **Figure 9-5** *ANATOMY SUMMARY* *Cerebrospinal Fluid*

The Blood-Brain Barrier Protects the Brain from Harmful Substances in the Blood

The final layer of protection for the brain is a functional barrier between the interstitial fluid and the blood. This barrier is necessary to isolate the body's main control center from potentially harmful substances in the blood and from blood-borne pathogens such as bacteria. To achieve this protection, most of the 400 miles of brain capillaries create a functional **blood-brain barrier** (Fig. 9-6 ●). Although not a literal barrier, the highly selective permeability of brain capillaries shelters the brain from toxins and from fluctuations in hormones, ions, and neuroactive substances such as neurotransmitters in the blood.

Why are brain capillaries so much less permeable than other capillaries? In most capillaries, leaky cell-cell junctions and pores allow free exchange of solutes between the plasma and interstitial fluid [● p. 67]. However, in brain capillaries, the endothelial cells form tight junctions with each other that prevent solute movement between the cells. Tight junction formation apparently is induced by paracrines from adjacent astrocytes whose foot processes surround the capillary. Thus, it is the brain tissue itself that creates the blood-brain barrier.

The selective permeability of the blood-brain barrier can be attributed to its specific transport properties. The capillary endothelium uses selected membrane carriers and channels to move nutrients and other useful material from the blood into the brain interstitial fluid. Other transporters move wastes from the interstitial fluid into the plasma. If a water-soluble molecule is not transported on one of these carriers, it cannot cross the blood-brain barrier.

One interesting illustration of how the blood-brain barrier works is seen in *Parkinson's disease*, a neurological disorder in which brain levels of the neurotransmitter *dopamine* are too low. Dopamine administered in a pill or injection is ineffective because it is unable to cross the blood-brain barrier. But the dopamine precursor, L-*dopa*, is transported across the cells of the blood-brain barrier on an amino acid transporter [● p. 143]. Once neurons have access to L-dopa in the interstitial fluid, they metabolize it to dopamine, thereby correcting the deficiency.

Although the blood-brain barrier excludes many water-soluble substances, smaller lipid-soluble molecules can simply diffuse through the cell membranes [● p. 136]. This is one reason some antihistamines make you sleepy but others do not. Older antihistamines were lipid-soluble amines that readily crossed the blood-brain barrier and acted on brain centers controlling alertness. The newer drugs are much less lipid-soluble and as a result do not have the same sedative effect.

A few areas of the brain lack a functional blood-brain barrier: their capillaries have leaky endothelium like most of the body. In these areas, the function of adjacent neurons depends in some way on direct contact with the blood. For instance, the posterior pituitary releases neurosecretory hormones that must pass into the capillaries for distribution to the body [● p. 222]. Another region that lacks the blood-brain barrier is the vomiting center in the medulla oblongata. These neurons monitor the blood for possibly toxic foreign substances, such as drugs. If they sense something harmful, they initiate a vomiting reflex. Vomiting removes the contents of the digestive system and eliminates ingested toxins.

Nervous Tissue Has Special Metabolic Requirements

One final property of the central nervous system is its specialized metabolism. On one hand, the blood-brain barrier separates plasma from the neural tissue. But at the same time, neurons require a constant supply of oxygen and glucose to make ATP for active transport of ions and neurotransmitters. Because of the brain's high demand for oxygen, it receives about 15% of the blood pumped by the heart. Oxygen passes freely across the blood-brain barrier. Neurons consume oxygen at such high rates that interruption of blood flow to the brain can have devastating results, and brain damage occurs after only a few minutes without oxygen.

(a)

Astrocyte foot processes secrete paracrines that promote tight junction formation.

Astrocyte

Tight junction prevents solute movement between endothelial cells.

(b)

Astrocyte foot processes

Tight junction

Basement membrane

Capillary lumen

● **Figure 9-6** *The blood-brain barrier* The movement of material across the blood-brain barrier is regulated to protect neurons from potentially harmful substances in the blood.

Ben was diagnosed with *infantile spasms*, a form of epilepsy characterized by the onset of head-drop seizures at four to seven months and by arrested or deteriorating mental development. Ben was started on a month-long regimen of adreno-corticotropin (ACTH; ◈ p. 224] shots to control the seizures. Scientists are unsure why this hormone is so effective in controlling this type of seizure, but they have found that among its effects, it increases myelin formation, increases blood-brain barrier integrity, and enhances binding of the neurotransmitter GABA at synapses. As expected, Ben's seizures disappeared completely before the month of treatment ended, and his development began to return to a normal level.

Question 1: How might a leaky blood-brain barrier lead to a cascade of action potentials that trigger a seizure?

Question 2: GABA opens Cl⁻ channels on the postsynaptic cell. What does this do to the cell's membrane potential? Does GABA make the cell more or less likely to fire action potentials?

Question 3: Why is it important to limit the time span of ACTH therapy particularly in very young patients? [◈ p. 228]

Under normal circumstances the only energy source for neurons is glucose. Blood-brain barrier cells have membrane transporters to move the sugar from plasma to interstitial fluid. By some estimates, the brain is responsible for about half of the body's glucose consumption. Consequently, the body uses several homeostatic pathways to ensure that glucose concentrations in the blood always remain adequate to meet the brain's demand. If homeostasis fails, progressive **hypoglycemia** (low blood glucose levels) leads to confusion, unconsciousness, and eventually death.

Hypoglycemia and the Brain

Neurons are picky about their food. Under most circumstances, the only biomolecule that neurons use for energy is glucose. Surprisingly, this can present a problem for diabetic patients, whose problem is too much glucose in the blood. In the face of sustained hyperglycemia (elevated blood glucose), the cells of the blood-brain barrier down-regulate [◈ p. 186] their glucose transporters. Then, if the patient's blood glucose level decreases due to insulin or other therapy, the neurons of the brain may be out of luck. They will be unable to transport glucose fast enough to sustain their electrical activity. The result is impaired CNS function, including confusion, irritability, and slurred speech. Unless the neurons are provided promptly with glucose, they can sustain permanent damage. In extreme cases, hypoglycemia can cause coma or even death.

Now that you have a broad overview of the central nervous system, we will examine the structure and function of the spinal cord and brain in more detail.

8. Oxidative phosphorylation takes place in what organelle?

9. Name the two metabolic pathways for aerobic metabolism of glucose. What happens to $NADH_2$ produced in these pathways?

10. In the late 1800s the scientist Paul Ehrlich injected blue dye into the bloodstream of animals. He noticed that all tissues except the brain stained blue. He was not aware of the blood-brain barrier and drew a different conclusion from his results. What might be some other explanations for his observation?

11. In a subsequent experiment a student of Ehrlich's injected the dye into the cerebrospinal fluid of the same animals. What do you think he observed about staining in the brain and in other body tissues?

▶ THE SPINAL CORD

The spinal cord is the major pathway for information flowing back and forth between the brain and the skin, joints, and muscles of the body. In addition, the spinal cord contains neural networks responsible for locomotion. If the spinal cord is severed, there is loss of sensation from the skin and muscles as well as *paralysis*, loss of the ability to voluntarily control muscles.

The spinal cord is divided into four regions (*cervical, thoracic, lumbar,* and *sacral*), named to correspond to the adjacent vertebrae (see Fig. 9-4a). Each spinal region is subdivided into segments, and each segment gives rise to a bilateral pair of **spinal nerves**. Just before a spinal nerve joins the spinal cord, it divides into two branches called **roots** (Fig. 9-7a ●). The **dorsal root** of each spinal nerve is specialized to carry incoming sensory information. The **dorsal root ganglia**, swellings found on the dorsal roots just before they enter the cord, contain cell bodies of sensory neurons. The **ventral root** carries information from the central nervous system to muscles and glands.

In cross section, the spinal cord has a butterfly- or H-shaped core of gray matter and a surrounding rim of white matter. Sensory fibers from the dorsal roots synapse with interneurons in the **dorsal horns** of the gray matter. The dorsal horn cell bodies are organized into two distinct nuclei, one for somatic information and one for visceral information (Fig. 9-7b).

The **ventral horns** of the gray matter contain cell bodies of motor neurons that carry efferent signals to muscles and glands. The ventral horns are organized into somatic motor and autonomic nuclei. Efferent fibers leave the spinal cord via the ventral root.

The white matter of the spinal cord is the biological equivalent of fiber-optic cables that phone companies use to carry our communications systems. White matter can be divided into a number of **columns** composed of tracts of axons that transfer information

(a) One segment of spinal cord, ventral view, showing its pair of nerves.

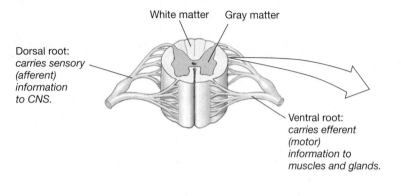

(b) Gray matter consists of sensory and motor nuclei.

(c) White matter in the spinal cord consists of axons carrying information to and from the brain.

Ascending tracts carry sensory information to the brain.

Descending tracts carry commands to motor neurons.

● **Figure 9-7** *Specialization in the spinal cord*   The figures represent different regions of the spinal cord.

up and down the cord. **Ascending tracts** take sensory information to the brain. They occupy the dorsal and external lateral portions of the spinal cord (Fig. 9-7c). **Descending tracts** carry mostly efferent (motor) signals from the brain to the cord. They occupy the ventral and interior lateral portions of the white matter. **Propriospinal tracts** [*proprius*, one's own] remain within the cord.

The spinal cord itself can function as a self-contained integrating center for simple *spinal reflexes*, with signals passing from the sensory neuron through the gray matter to an efferent neuron (Fig. 9-8 ●). In addition, spinal interneurons may route sensory information to the brain through ascending tracts or bring commands from the brain to motor neurons. In many cases, the interneurons also modify information as it passes through them. Reflexes play a critical role in the coordination of movement and will be discussed in more detail in Chapter 13.

CONCEPT CHECK   Answers: p. 320

12. What are the differences between horns, roots, tracts, and columns of the spinal cord?

13. If a dorsal root of the spinal cord is cut, what function will be disrupted?

▶ THE BRAIN

Thousands of years ago, Aristotle declared that the heart was the seat of the soul. However, most people now agree that the brain is the organ that gives the human species its unique attributes. The challenge facing scientists is to understand how circuits formed by millions of neurons result in complex behaviors such as speaking, writing a symphony, or creating imaginary worlds for an interactive computer game. Brain function may be the ultimate *emergent property*, where the characteristics of the whole cannot be predicted by knowing the properties of the component parts. The question remains whether we will ever be able to decipher how emotions such as happiness and love arise from the chemical and electrical signals passing along circuits of neurons.

It is possible to study the brain at many different levels of organization. The most reductionist view looks at the individual neurons and what happens to them in response to chemical or electrical signals. A more integrative study might look at groups of neurons and how they interact with each other in *circuits* or *pathways*. The most complicated approach starts with a behavior or physiological response and works backward to dissect apart the neural circuits that create the behavior or response.

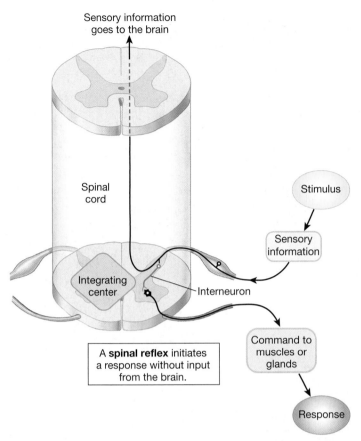

Sensory information goes to the brain

Spinal cord

Integrating center

Interneuron

A **spinal reflex** initiates a response without input from the brain.

Stimulus

Sensory information

Command to muscles or glands

Response

● **Figure 9-8** *The spinal cord is an integrating center* In a spinal reflex, sensory information entering the spinal cord is acted upon without input from the brain.

For centuries, studies of brain function were restricted to anatomical descriptions. However, there is no tidy 1:1 relationship between structure and function when looking at the brain. An adult human brain weighs about 1400 g and contains an estimated 10^{12} neurons. When you consider that each one of these millions of neurons may receive as many as 200,000 synapses, the number of possible neuronal connections is mind-boggling.

A basic principle to remember when studying the brain is that one function, even an apparently simple one such as bending your finger, will involve multiple brain regions (as well as the spinal cord). At the same time, one brain region may be involved in several other functions. In other words, understanding the brain is not simple and straightforward. This section introduces you to major brain structures. The remainder of the chapter then describes how these structures communicate and coordinate for complex integrated functions such as learning and memory.

The Brain Has Six Major Divisions

The brain has six major divisions that are present at birth: (1) the cerebrum, (2) the diencephalon, (3) the midbrain, (4) the cerebel-

lum, (5) the pons, and (6) the medulla oblongata. When the brain is viewed intact in profile, only the medulla, cerebellum, and cerebrum are visible. The remaining three divisions are covered by the cerebrum. Figure 9-9 ● is an anatomy summary to follow as we discuss brain regions, moving from the most primitive to the most complex.

The Brain Stem Is the Transition Between Spinal Cord and Midbrain

The **brain stem** is the oldest and most primitive region of the brain and consists of structures that derive from the embryonic midbrain and hindbrain. The brain stem can be divided into white matter and gray matter, and in some ways its anatomy is similar to that of the spinal cord. Some ascending tracts from the spinal cord pass through the brain stem, while others synapse there. Descending tracts from higher brain centers also travel through the brain stem on their way to the spinal cord.

Pairs of peripheral nerves branch off the brain stem, similar to spinal nerves along the spinal cord (Fig. 9-9d). Eleven of the twelve **cranial nerves** (numbers II–XII) originate along the brain stem. (The first cranial nerve, the olfactory nerve (I), enters the forebrain.) Cranial nerves carry sensory and motor information for the head and neck (Table 9-1), and are described according to whether they include sensory fibers, efferent fibers, or both (mixed nerves). For example, cranial nerve X, the **vagus nerve** [*vagus*, wandering], is a mixed nerve that carries both sensory and motor fibers for many internal organs. An important component of the clinical neurological examination is testing the functions controlled by these nerves.

Located throughout the brain stem are discrete groups of nerve cell bodies (*nuclei*). Many of these nuclei are associated with the **reticular formation**, a diffuse collection of neurons that extends throughout the brain stem. The name *reticular* literally means "network" and comes from the crisscrossed axons that branch profusely up into superior sections of the brain and down into the spinal cord. Nuclei in the brain stem are involved in many basic processes, including arousal and sleep, muscle tone and stretch reflexes, coordination of breathing, blood pressure regulation, and modulation of pain.

Answers: p. 320

14. Are the following white matter or gray matter? (a) ascending tracts (b) reticular formation (c) descending tracts

15. Using the information from Table 9-1, describe the types of activities you might ask a patient to perform if you wished to test the function of each cranial nerve.

16. In anatomical terminology, the cerebrum, which is located next to the top of the skull, is said to be _____ to the brain stem.

LATERAL VIEW OF THE CNS

(a)

Cerebrum

Vertebrae

Spinal cord

ANATOMY OF THE BRAIN

(b) Lateral view of brain

Frontal lobe

Parietal lobe

Occipital lobe

Temporal lobe

Pons

Medulla oblongata

Cerebellum

(e) The skull

Frontal bone

Parietal bone

Temporal bone

Occipital bone

(c) Mid-saggital view of brain

Frontal lobe

Cingulate gyrus

Parietal lobe

Corpus callosum

Occipital lobe

Cerebellum

Temporal lobe

Pons

Medulla oblongata

(d) Lateral view of brain stem

Thalamus

Cut edge of ascending tracts to cerebrum

Optic tract

Midbrain

Pons

Cranial nerves

Cut edges of tracts leading to cerebellum

Medulla oblongata

Spinal cord

● **Figure 9-9** *ANATOMY SUMMARY* *The Brain*

FUNCTIONS OF THE BRAIN	
REGION	**FUNCTION**
Cerebrum (Frontal ●, Parietal ●, Occipital ●, and Temporal ● lobes)	
• Cerebral cortex (See Fig. 9-15)	
Sensory fields	Perception
Motor areas	Skeletal muscle movement
Association areas	Integration of information and direction of voluntary movement
● Basal nuclei (See Fig. 9-11)	Movement
● Limbic system (See Fig. 9-13)	
● Amygdala	Emotion and Memory
● Hippocampus	Learning and Memory
● **Diencephalon** (See Fig. 9-10)	
● Thalamus	Integrating center and relay station for sensory and motor information
Hypothalamus	Homeostasis and behavioral drives (See Table 9-2)
Pituitary	Hormone secretion
● Pineal gland	Melatonin secretion
● **Cerebellum**	Movement coordination
● **Brain stem**	
● Midbrain	Eye movement
● Pons	Relay station between cerebrum and cerebellum; coordination of breathing
● Medulla oblongata	Control of involuntary functions
Reticular formation (See Fig. 9-19)	Arousal, sleep, muscle tone, pain modulation

The Brain Stem Consists of Medulla, Pons, and Midbrain

Starting at the spinal cord and moving toward the top of the skull, the brain stem includes the medulla oblongata, the pons, and the midbrain (Fig. 9-9d). Some authorities include the cerebellum as part of the brain stem. The diamond-shaped fourth ventricle, visible in the early embryo (see Fig. 9-3b), runs through the interior and connects to the central canal of the spinal cord (see Fig. 9-5).

The **medulla oblongata**, frequently just called the *medulla* [*medulla*, marrow; adjective *medullary*], is the transition from the spinal cord into the brain proper (Fig. 9-9c, d). Its white matter includes ascending **somatosensory tracts** [*soma*, body] that bring sensory information to the brain and descending **corticospinal tracts** that convey information from the cerebral cortex to the spinal cord. About 90% of corticospinal tracts cross the midline to the opposite side of the body in a region of the medulla known as the **pyramids**. As a result of this crossover, each side of the brain controls the opposite side of the body. Gray matter in the medulla includes nuclei that control many involuntary functions, such as blood pressure, breathing, swallowing, and vomiting.

The **pons** [*pons*, bridge; adjective *pontine*] is a bulbous protrusion on the ventral side of the brain stem above the medulla and below the midbrain. Because its primary function is to act as a relay station for information transfer between the cerebellum and cerebrum, the pons is often grouped with the cerebellum. The pons also coordinates the control of breathing along with centers in the medulla.

The **midbrain**, or *mesencephalon* [*mesos*, middle], is a relatively small area between the lower brain stem and the diencephalon. Its primary function is the control of eye movement, but it also relays signals for auditory and visual reflexes.

The Cerebellum Coordinates Movement

The cerebellum is the second largest structure in the brain (Fig. 9-9a–c). It is located inside the base of the skull, just above the nape of the neck. The name **cerebellum** [adjective *cerebellar*] means "little brain," and, indeed, most of the nerve cells in the brain are in the cerebellum. The specialized function of the cerebellum is to process sensory information and coordinate the execution of movement. Sensory input into the cerebellum comes from somatic receptors in the periphery of the body and from receptors for equilibrium and balance located in the inner ear. The cerebellum also receives motor input from neurons in the cerebral cortex. Cerebellar function is described in more detail in Chapter 13.

Table 9-1: The Cranial Nerves

NUMBER	NAME	TYPE	FUNCTION
I	Olfactory	Sensory	Olfactory (smell) information from nose
II	Optic	Sensory	Visual information from eyes
III	Oculomotor	Motor	Eye movement, pupil constriction, lens shape
IV	Trochlear	Motor	Eye movement
V	Trigeminal	Mixed	Sensory information from face, mouth; motor signals for chewing
VI	Abducens	Motor	Eye movement
VII	Facial	Mixed	Sensory for taste; efferent signals for tear and salivary glands, facial expression
VIII	Vestibulocochlear	Sensory	Hearing and equilibrium
IX	Glossopharyngeal	Mixed	Sensory from oral cavity, baro- and chemoreceptors in blood vessels; efferent for swallowing, parotid salivary gland secretion
X	Vagus	Mixed	Sensory and efferents to many internal organs, muscles, and glands
XI	Spinal accessory	Motor	Muscles of oral cavity, some muscles in neck and shoulder
XII	Hypoglossal	Motor	Tongue muscles

The Diencephalon Contains the Centers for Homeostasis

The **diencephalon**, or "between-brain," lies between the brain stem and the cerebrum (Fig. 9-9d). It is composed of two main sections, the thalamus and the hypothalamus, and two endocrine structures, the pituitary and pineal glands (Fig. 9-10 ●).

Most of the diencephalon is occupied by many small nuclei that make up the **thalamus** [*thalamus*, bedroom; adjective *thalamic*]. The thalamus receives sensory fibers from the optic tract, ears, and spinal cord, and motor information from the cerebellum. It projects fibers to the cerebrum where the information is processed. The thalamus is often described as a relay station because almost all sensory information from lower parts of the central nervous system passes through the thalamus. But, like the spinal cord, the thalamus can modify information passing through it, making it an integrating center as well as a relay station.

The **hypothalamus** lies beneath the thalamus. Although this structure occupies less than 1% of the total brain volume, it is the center for homeostasis and contains centers for various behavioral drives, such as hunger and thirst. Output from the hypothalamus also influences many functions of the autonomic division of the nervous system as well as a variety of endocrine functions (Table 9-2). The hypothalamus receives input from multiple sources including the cerebral cortex, the reticular formation, and various sensory receptors. Output from the hypothalamus goes first to the thalamus and eventually to multiple effector pathways.

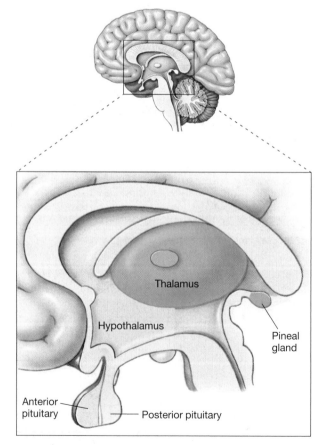

● **Figure 9-10** *The diencephalon* The diencephalon lies between the brain stem and the cerebrum.

Table 9-2: Functions of the Hypothalamus

1. Activation of sympathetic nervous system
   - Control of catecholamine release from adrenal medulla (as in fight-or-flight reaction)
   - Helps maintain blood glucose concentrations through effects on endocrine pancreas

2. Maintenance of body temperature
   - Shivering and sweating

3. Control of body osmolarity
   - Thirst and drinking behavior
   - Secretion of vasopressin

4. Control of reproductive functions
   - Secretion of oxytocin (uterine contractions and milk release)
   - Trophic hormone control of anterior pituitary hormones FSH and LH

5. Control of food intake
   - Satiety center
   - Feeding center

6. Interacts with limbic system to influence behavior and emotions

7. Influences cardiovascular control center in medulla oblongata

8. Secretes trophic hormones that control release of hormones from anterior pituitary gland

Two important endocrine structures are located in the diencephalon: the pituitary gland and the pineal gland. The **pituitary gland** was described in Chapter 7 [❧ p. 221]. The posterior pituitary (*neurohypophysis*) is a downgrowth of the hypothalamus and secretes neurohormones that are synthesized in hypothalamic nuclei. The anterior pituitary (*adenohypophysis*) is a true endocrine gland. Its hormones are regulated by hypothalamic neurohormones secreted into the *hypothalamic-hypophyseal portal system*. The **pineal gland**, which secretes the hormone *melatonin*, was also introduced in Chapter 7 [❧ p. 232] and will be discussed later in this chapter.

CONCEPT CHECK   Answers: p. 320

17. Starting at the spinal cord and moving up, name the subdivisions of the brain stem.

18. What are the four primary structures of the diencephalon?

The Cerebrum Is the Site of Higher Brain Functions

The cerebrum is the largest and most distinctive part of the human brain and fills most of the cranial cavity. It is composed of two hemispheres connected primarily at the **corpus callosum** (Fig. 9-9c), a distinct structure formed by axons passing from one side of the brain to the other. This connection ensures that the two hemispheres communicate and cooperate with each other. Each cerebral hemisphere is divided into four lobes, named for the bones of the skull under which they are located: *frontal*, *parietal*, *temporal*, and *occipital* (Fig. 9-9b, c, e).

The surface of the cerebrum of humans and other primates has a furrowed walnut-like appearance, with grooves called *sulci* [singular *sulcus*, a furrow] dividing convolutions called *gyri* [singular *gyrus*, a ring or circle]. During development, the cerebrum grows faster than the surrounding cranium, causing the tissue to fold back on itself to fit into a smaller volume. The degree of folding is directly related to the level of processing of which the brain is capable; less-advanced mammals, such as rodents, have brains with a relatively smooth surface. The human brain is so convoluted that if it were inflated enough to smooth the surfaces, it would be three times as large and would need a head the size of a beach ball.

The Cerebrum Has Distinct Regions of Gray and White Matter

The outer layer of the cerebrum, only a few millimeters thick, forms the gray matter of the **cerebral cortex** [*cortex*, bark or rind; adjective *cortical*, plural *cortices*] (Fig. 9-11 ●). Neurons of the cerebral cortex

● **Figure 9-11** *The basal nuclei*   The frontal view shown here is similar to the sectional view obtained using modern diagnostic scanning techniques.

(a) Cell layers within the gray matter

Layers
1
2
3
4
5
6

Gray matter

White matter

(b) Photomicrograph of cerebral cortex

● **Figure 9-12** *Cell bodies in the cerebral cortex form distinct layers.*

are arranged in anatomically distinct columns and horizontal layers, or *lamina* (Fig. 9-12 ●). It is within these layers that our higher brain functions such as reasoning arise.

The other gray matter in the cerebrum is found in two interior groups of neurons, the basal nuclei and the limbic system. The **basal nuclei*** (Fig. 9-11) are involved in the control of movement and will be discussed in Chapter 13. The **limbic system** [*limbus*, a border], which surrounds the brain stem (Fig. 9-13 ●), probably represents the most primitive region of the cerebrum. It acts as

---

*The basal nuclei are also called the basal ganglia. However, neuroanatomists prefer to reserve the term *ganglion* for clusters of nerve cell bodies outside the CNS.

the link between higher cognitive functions such as reasoning and more primitive emotional responses such as fear. The major structures in the limbic system are the **amygdala** and **cingulate gyrus**, which are linked to emotion and memory, and the **hippocampus**, which is associated with learning and memory.

White matter in the cerebrum is found primarily in the interior (Fig. 9-11). Bundles of fibers allow different regions of the cortex to communicate with each other and transfer information from one hemisphere to the other primarily through the *corpus callosum*. According to some estimates, the corpus callosum may have as many as 200 million axons passing through it! Information entering and leaving the cerebrum goes along tracts that pass through the thalamus.

19. Name the anatomical location in the brain where neurons from one side of the body cross to the opposite side.

20. Name the divisions of the brain in anatomical order, starting from the spinal cord.

Chemical Communication Influences Neural Pathways

The meaning of signals in the nervous system depends on two factors: (1) where the signals originate and terminate and (2) the physical and chemical connections between neurons in the pathway. The anatomy of the central nervous system just described determines the physical pathway for information transfer. The other factor we must consider in any discussion of the nervous system is chemical communication between neurons in a pathway. This type of signaling has two components: (1) the chemical signals themselves and (2) the target cell receptors for these molecules. Various combinations of receptor and ligand create a large number of possible communication signals.

Chemical signals in the nervous system have three roles. *Neurotransmitters* are responsible for fast synaptic communica-

Cingulate gyrus plays a role in emotion.

Thalamus

Hippocampus is involved in learning and memory.

Amygdala is involved in emotion and memory.

● **Figure 9-13** *The limbic system* Anatomically, the limbic system is part of the cerebrum.

tion. *Neuromodulators* act more slowly, often by using second messenger systems [♦ p. 265]. *Neurohormones* are released into the circulation for action at more distant targets. A single chemical can take on more than one of these roles. For example, *norepinephrine* acts as a neurotransmitter in the central nervous system and peripheral autonomic division but is also released in small amounts to act as a neurohormone.

Neurotransmitters and modulators in the central nervous system include amino acids, acetylcholine (ACh), an assortment of peptides, and amines (Table 9-3). Two of the amino acids, **gamma-aminobutyric acid (GABA)** and **glycine**, are the most common inhibitory neurotransmitters of the central nervous system. They act by opening Cl^- channels, hyperpolarizing their postsynaptic targets, and making them less likely to fire an action potential. GABA acts primarily in the brain; glycine acts chiefly in the spinal cord. The amino acid **glutamate** is the primary excitatory neurotransmitter in the central nervous system. It depolarizes cells by opening channels that allow Na^+ and Ca^{2+} to enter.

► BRAIN FUNCTION

From a simplistic view, the brain is an information processor, much like a computer. It receives sensory input from the internal and external environments, integrates and processes the information, and, if appropriate, creates a response (Fig. 9-14●). This is the basic reflex pathway that was introduced earlier in Chapter 6 [♦ p. 191]. What makes the brain more complicated than this

Table 9-3: Neurotransmitters and Neuromodulators in the CNS

Acetylcholine (ACh)

Adenosine

Amino acids
  γ-aminobutyric acid (GABA)
  Glycine
  Glutamate

Amines
  Dopamine
  Norepinephrine
  Epinephrine
  Serotonin
  Histamine

Nitric Oxide

Peptides (partial list)
  Substance P
  Opioids (enkephalins, β-endorphin)
  Cholecystokinin
  Vasoactive intestinal peptide (VIP)
  Angiotensin II
  Neuropeptide Y

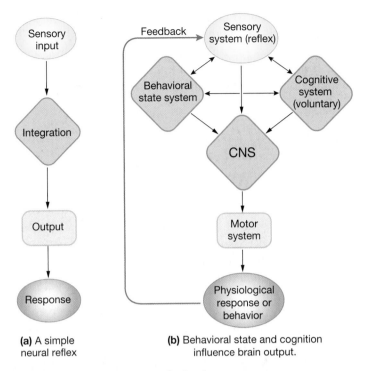

(a) A simple neural reflex

(b) Behavioral state and cognition influence brain output.

● **Figure 9-14** *Reflexes in the brain*

simple reflex, however, is its ability to generate information and output signals *in the absence of external input*. Modeling this intrinsic input requires a more complex diagram than the simple reflex illustrated in Figure 9-14a.

Dr. Larry Swanson of the University of Southern California presents one approach to modeling brain function in his book, *Brain Architecture: Understanding the Basic Plan* (Oxford University Press, 2003). He describes three different systems that influence output by the motor systems of the body: (1) the **sensory system** that monitors the internal and external environment and initiates reflex responses, (2) a **cognitive system** that resides in the cerebral cortex and is able to initiate voluntary responses, and (3) a **behavioral state system**, which also resides in the brain, governing sleep-wake cycles and other intrinsic behaviors. Information about the physiological or behavioral responses created by motor output feeds back to the sensory system, which in turn communicates with the cognitive and behavioral state systems (Fig. 9-14b).

In most of the physiological organ systems of the body that you will study in later chapters of this book, simple reflex pathways initiated through the sensory system are adequate to explain homeostatic control mechanisms. However, the other two systems remain potential sources of influence. At its simplest, this influence may take the form of voluntary behaviors, such as breath-holding, that override automatic functions. More subtle and complicated interactions include the effect of emotions on normal physiology, such as stress-induced heart palpitations, or the role of circadian rhythms in jet lag and shift work.

In the sections that follow we take a brief look at sensory and motor systems in the brain. The receptors and pathways of the sensory system will be described in more detail in Chapter 10, and neural control of motor systems is discussed in Chapters 11 and 13. This chapter concludes with a discussion of some aspects of the behavioral state and cognitive systems, such as circadian rhythms, sleep-wake cycles, emotion, learning, and memory.

The Cerebral Cortex Is Organized into Functional Areas

The cerebral cortex serves as an integrating center for sensory information and a decision-making region for many types of motor output. If we examine the cortex from a functional viewpoint, it can be divided into three specializations: (1) **sensory areas** (fields) that receive sensory input and translate it into perception, (2) **motor areas** that direct skeletal muscle movement, and (3) **association areas** (association cortices), regions that integrate information from multiple sensory and motor areas and that can direct voluntary behaviors (Fig. 9-15 ●). Information passing along a pathway is usually processed in multiple areas of the cortex.

The functional areas of the cerebral cortex do not necessarily correspond to the anatomical lobes of the brain. For one thing, functional specialization is not symmetrical across the cerebral cortex: each lobe has special functions not shared by the other lobe. This **cerebral lateralization** of function is sometimes referred to as *cerebral dominance*, more popularly known as left brain–right brain dominance. Language and verbal skills tend to be concentrated on the left side of the brain, with spatial skills concentrated on the right side. The left brain is the dominant hemisphere for right-handed people, and it appears that the right brain is the dominant hemisphere for many left-handed people (Fig. 9-16 ●).

Even these generalizations are subject to change, however. Neural connections in the cerebrum, like those in other parts of the nervous system, exhibit a certain degree of *plasticity*, an ability to change neuronal connections on the basis of experience. For example, if a person loses a finger, the regions of motor and sensory cortex previously devoted to control of the finger do not go dormant. Instead, adjacent regions of the cortex extend their functional fields and take over the parts of the cortex that are no longer used by the absent finger. Similarly, skills normally associated with one side of the cerebral cortex can be developed in the other hemisphere, as when a right-handed person with a broken hand learns to write with the left hand.

Much of what we know about functional areas of the cerebral cortex comes from study of patients with inherited neurological defects or wounds from accidents or war. In some instances, surgical lesions made to treat some medical condition, such as uncontrollable epilepsy, have revealed functional relationships in particular brain regions. Imaging techniques such as positive emission tomography (PET) scans and magnetic resonance imaging (MRI) provide noninvasive ways for us to watch the human brain at work (Fig. 9-17 ●).

Sensory Information Is Integrated in the Spinal Cord and Brain

The sensory system monitors the internal and external environments and sends information to neural integrating centers, which in turn initiate appropriate responses. In its simplest form,

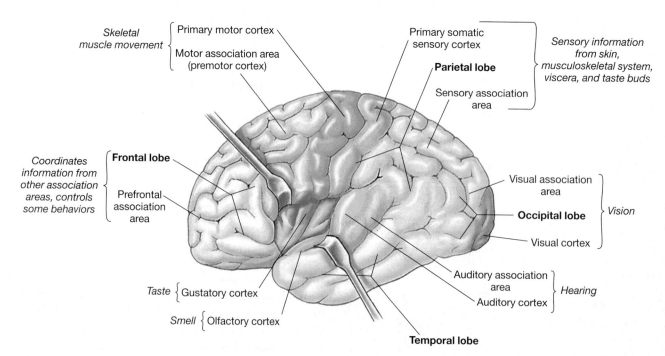

● **Figure 9-15** *Functional areas of the cerebral cortex*   The cerebral cortex is specialized into sensory areas for perception, motor areas that direct movement, and association areas that integrate information.

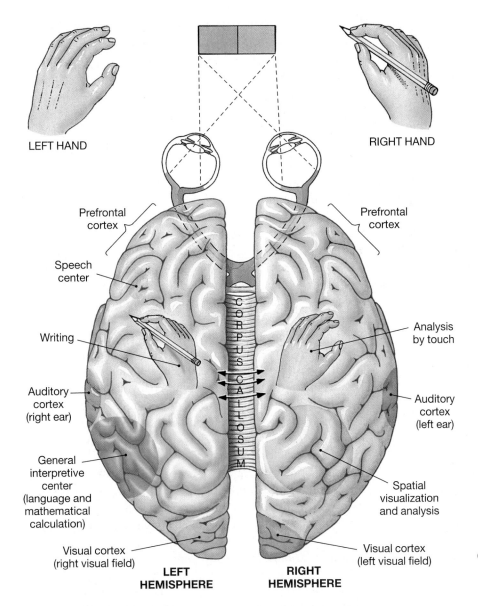

Prefrontal cortex

Prefrontal cortex

Speech center

Writing

Analysis by touch

Auditory cortex (right ear)

Auditory cortex (left ear)

General interpretive center (language and mathematical calculation)

Spatial visualization and analysis

Visual cortex (right visual field)

Visual cortex (left visual field)

LEFT HAND

RIGHT HAND

LEFT HEMISPHERE

RIGHT HEMISPHERE

CORPUS CALLOSUM

Q
● ● ● ● ● ● ●
F I G U R E   Q U E S T I O N S

• What would a person see if a stroke destroyed all function in the right visual cortex?
• What is the function of the corpus callosum?

● **Figure 9-16** *Cerebral lateralization* The distribution of functional areas in the two cerebral hemispheres is not symmetrical.

Hearing words

Seeing words

Speaking words

Generating words

Max

Min

● **Figure 9-17** *PET scan of the brain at work*

this pathway is the classic reflex, illustrated in Figure 9-14a. The simplest reflexes can be integrated in the spinal cord, without input from higher brain centers (see Fig. 9-8). However, even simple spinal reflexes usually send sensory information to the brain, creating *perception*, or awareness, of the stimulus. Brain functions dealing with perception are the most difficult to study because they require communication between the subject and the investigator—the subject must be able to tell the investigator what he or she is seeing, hearing, or feeling.

Sensory information from the body travels in ascending pathways to the brain. Information about muscle and joint position and movement goes to the cerebellum as well as to the cerebral cortex, allowing the cerebellum to assist with automatic subconscious coordination of movement. Most sensory information continues on to the cerebral cortex, where five sensory areas process information. One area handles information from peripheral receptors. The other four areas are each associated with a specific specialized sense organ.

303

The **primary somatic sensory cortex** (*somatosensory cortex*) in the parietal lobe is the termination point of pathways from the skin, musculoskeletal system, and viscera. The somatosensory pathways carry information about touch, temperature, pain, itch, and body position. Damage to this part of the brain leads to reduced sensitivity of the skin on the opposite side of the body because sensory fibers cross to the opposite side of the midline as they ascend through the spine or medulla.

The special senses of vision, hearing, taste, and olfaction (smell) each have different brain regions devoted to processing their sensory input (see Fig. 9-15). The **visual cortex**, located in the occipital lobe, receives information from the eyes. The **auditory cortex**, located in the temporal lobe, receives information from the ears. The **olfactory cortex**, a small region in the temporal lobe, receives input from chemoreceptors in the nose. The **gustatory cortex** deeper in the brain near the edge of the frontal lobe receives sensory information from taste buds. The somatic and special senses are described in detail in the next chapter.

Sensory Information Is Processed into Perception

Once sensory information reaches the appropriate cortical area, its processing has just begun. Neural pathways extend from sensory areas to appropriate association areas, which integrate somatic, visual, and auditory and other stimuli into perception. **Perception** is the brain's interpretation of sensory stimuli.

Often the perceived stimulus is very different from the actual stimulus. For instance, photoreceptors of the eye receive light waves of different frequencies, but we perceive the different wave energies as different colors. Similarly, the brain translates pressure waves hitting the ear into sound, or interprets chemicals binding to chemoreceptors as taste or smell.

One interesting aspect of perception is the way that our brain fills in missing information to create a complete picture or translates a two-dimensional drawing into a three-dimensional shape (Fig. 9-18 ●). Our perceptual translation of sensory stimuli allows the information to be acted upon and used in voluntary motor control or complex cognitive functions such as language.

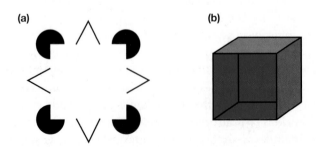

(a) **(b)**

● **Figure 9-18** *Perception* The brain has the ability to interpret sensory information to create the perception of shapes (a) or a three-dimensional object (b).

The Motor System Governs Output from the Central Nervous System

The motor component of the nervous system is associated with the efferent division shown in Figure 8-1 [● p. 241]. Motor output can be divided into three major types: (1) skeletal muscle movement, controlled by the somatic motor division, (2) neuroendocrine signals, neurohormones secreted into the blood by neurons primarily in the hypothalamus and adrenal medulla, and (3) *visceral* responses [*viscera*, internal organs], the actions of smooth and cardiac muscle or endocrine and exocrine glands. Visceral responses are governed by the autonomic division of the nervous system.

Information about skeletal muscle movement is processed in several different regions of the central nervous system. Simple stimulus-response pathways such as the knee jerk reflex are processed in the spinal cord or in the brain stem. Although these reflexes do not require integration in the cerebral cortex, they can be modified or overridden by input from the cognitive system.

Voluntary movements are initiated by the cognitive system and originate in the **primary motor cortex** and **motor association area** in the frontal lobes of the cerebrum (see Fig. 9-15). These regions receive input from sensory areas as well as from the cerebellum and basal ganglia. Long output neurons called *pyramidal cells* project axons from the motor areas through the brain stem to the spinal cord. Other pathways go from the cortex to the basal ganglia and lower brain regions. Descending motor pathways cross to the opposite side of the body; therefore, damage to a motor area will be manifested by paralysis or function loss on the opposite side of the body. A detailed discussion of the coordination of body movement can be found in Chapter 13.

Neuroendocrine and visceral responses are coordinated in subcortical regions of the brain, primarily the hypothalamus and medulla. The brain stem is responsible for maintaining many of the automatic life functions that we take for granted, such as breathing, blood pressure, and control of body movement. It receives sensory information from the body and relays motor commands to peripheral muscles and glands. The hypothalamus contains centers for temperature regulation, eating, and control of body osmolarity, among others. The responses to stimulation of these centers may be neural or hormonal reflexes or a behavioral response. Stress, reproduction, and growth are also mediated by the hypothalamus by way of multiple hormones. You learn more about these reflexes in later chapters as we discuss the various systems of the body.

Sensory input is not the only factor determining motor output by the brain. The behavioral state system can modulate reflex pathways, and the cognitive system exerts both voluntary and involuntary control over motor functions.

The Behavioral State System Modulates Motor Output

The behavioral state system is an important modulator of sensory and cognitive processing. Many neurons in the behavioral state

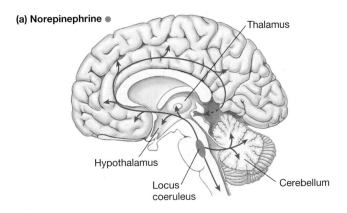

(a) Norepinephrine ●

Thalamus

Hypothalamus

Locus coeruleus

Cerebellum

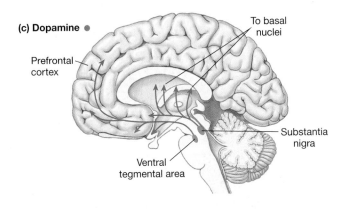

(b) Serotonin ●

To basal nuclei

Raphe nuclei

(c) Dopamine ●

To basal nuclei

Prefrontal cortex

Substantia nigra

Ventral tegmental area

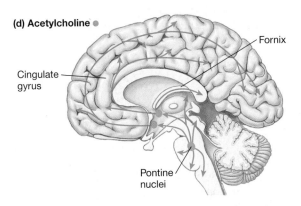

(d) Acetylcholine ●

Fornix

Cingulate gyrus

Pontine nuclei

● **Figure 9-19 *The diffuse modulatory systems modulate brain function*** The neurons collectively known as the diffuse modulatory systems originate in the reticular formation of the brain stem and project their axons to large areas of the brain. The four systems are named for their neurotransmitters.

system are found in regions of the brain outside the cerebral cortex. They include parts of the reticular formation in the brain stem, the hypothalamus, and the limbic system.

The neurons collectively known as the **diffuse modulatory systems** originate in the reticular formation in the brain stem and project their axons to large areas of the brain (Fig. 9-19 ●, Table 9-4). They are generally classified according to the neurotransmitter they secrete: *noradrenergic* (norepinephrine), *serotonergic* (serotonin), *dopaminergic* (dopamine), and *cholinergic* (acetylcholine). The diffuse modulatory systems influence attention, motivation, wakefulness, memory, motor control, mood, and metabolic homeostasis.

The Reticular Activating System Influences States of Arousal

One function that resides in the behavioral state system is control of levels of consciousness as well as sleep-wake cycles. **Consciousness** is the body's state of arousal or awareness of self and environment.

Experimental evidence shows that the **reticular activating system**, a diffuse collection of neurons in the reticular formation, plays an essential role in keeping the "conscious brain" awake. If connections between the reticular formation and the cortex are disrupted surgically, an animal becomes comatose for long periods of time. Other evidence for the importance of the reticular formation in states of arousal comes from studies showing that general anesthetics depress synaptic transmission in that region of the brain. Presumably, blocking ascending pathways between the reticular formation and the cortex creates a state of unconsciousness.

Physiologically, what distinguishes being awake from various stages of sleep? One way to define arousal states is by the pattern of electrical activity created by the cortical neurons. The measurement of brain activity is recorded by a procedure known as **electroencephalography**. Surface electrodes placed on or in the scalp pick up depolarizations of the cortical neurons in the region just under the electrode.

In awake states, many neurons are firing but not in a coordinated fashion. Presumably the desynchronization of electrical activity in waking states is produced by ascending signals from the reticular formation. An *electroencephalogram*, or **EEG**, of the waking-alert (eyes open) state shows a rapid, irregular pattern with no dominant waves (Fig. 9-20 ●). In awake-resting (eyes closed) states, sleep, or coma, electrical activity of the neurons begins to synchronize into waves with characteristic patterns. As the person's state of arousal lessens, the frequency of the waves slows down. The more synchronous the firing of cortical neurons, the larger the amplitude of the waves. Thus, the awake-resting state is characterized by small high-frequency waves, and deep sleep is marked by large-amplitude, slow-frequency waves. The complete cessation of brain waves is one of the clinical criteria for determining death.

Table 9-4: The Diffuse Modulatory Systems

SYSTEM (NEUROMODULATOR)	NEURONS ORIGINATE	NEURONS INNERVATE	FUNCTIONS MODULATED BY THE SYSTEM
Noradrenergic (norepinephrine)	Locus coeruleus of the pons	Cerebral cortex, thalamus, hypothalamus, olfactory bulb, cerebellum, midbrain, spinal cord	Attention, arousal, sleep-wake cycles, learning, memory, anxiety, pain, and mood
Serotonergic (serotonin)	Raphe nuclei along brain stem midline	Lower nuclei project to spinal cord	Pain; locomotion
		Upper nuclei project to most of brain	Sleep-wake cycle, mood and emotional behaviors such as aggression and depression
Dopaminergic (dopamine)	Substantia nigra in midbrain	Cortex	Motor control
	Ventral tegumentum in midbrain	Cortex and parts of limbic system	"Reward" centers linked to addictive behaviors
Cholinergic (acetylcholine)	Base of cerebrum; pons and midbrain	Cerebrum, hippocampus, thalamus	Sleep-wake cycles, arousal, learning, memory, sensory information passing through thalamus

(a)

FIGURE QUESTIONS

Which EEG pattern has the fastest frequency? The greatest amplitude?

(b)

● **Figure 9-20** *Electroencephalograms (EEGs) and the sleep cycle* sleep periods show characteristic patterns.

Recordings of electrical activity in the brain during waking and

Physiological Functions Exhibit Circadian Rhythms

All organisms, even plants, have alternating daily patterns of rest and activity. Sleep-wake rhythms, like many other biological cycles, generally follow a 24-hour light-dark cycle like our day and are known as **circadian rhythms** [*circa*, about + *dies*, a day]. When an

organism is placed in conditions of constant light or dark, these activity rhythms persist, apparently cued by an internal clock.

In mammals, the "clock" resides in networks of neurons located in the **suprachiasmatic nucleus** of the hypothalamus. A very simple interpretation of recent experiments on the molecular basis of the clock is that clock cycling is the result of a complex feedback loop in which specific genes turn on and direct protein

RUNNING PROBLEM

About six months after the start of ACTH treatment, Ben's head-drop seizures returned and his development began to decline once again. In an EEG following Ben's relapse, the doctor found that this EEG did not demonstrate the erratic wave patterns specific to infantile spasms, but did show abnormal activity in the right cortex. The neurologist ordered a positron emission tomography (PET) scan to determine the focus of Ben's seizure activity.

Ben received an injection of radioactively labeled glucose. He was then placed in the center of a PET machine lined with radiation detectors that created a map of his brain showing areas of high and low radiation. Those parts of his brain that were more active absorbed more glucose and thus emitted more gamma radiation when the radioactive compound began to decay.

Question 4: What is the rationale for using radioactively labeled glucose for the PET scan and not some other nutrient?

 284 293 310 312 314

synthesis. The proteins accumulate, turn off the genes, then are themselves degraded. In the absence of the proteins, the genes turn back on and the cycle begins again. The clock has intrinsic activity that is modified by sensory information about light cycles received from the eyes.

Circadian rhythms in humans can be found in most physiological functions and usually correspond to the phases of our sleep-wake cycles. For example, body temperature and cortisol secretion both cycle on a daily basis [❧ Fig. 6-28, p. 196]. Melatonin secretion by the pineal gland also is strongly linked to light-dark cycling and appears to feed back to the suprachiasmatic nucleus to modulate clock cycling.

Why Do We Sleep?

In humans, our major rest period is marked by a behavior known as sleep. **Sleep** is an easily reversible state of inactivity characterized by lack of interaction with the external environment. Why we need to sleep is one of the unsolved mysteries in neurophysiology.

Until the 1960s, sleep was thought to be a passive state that resulted from withdrawal of stimuli to the brain. Then, experiments showed that neuronal activity in ascending tracts from the brain stem to the cortex was required for sleep. From other studies, we know that the sleeping brain consumes as much or more oxygen than the brain awake. As a result, we now consider sleep to be an active state. Each stage of sleep is marked by identifiable, predictable events associated with characteristic EEG patterns and somatic changes (Fig. 9-20). In the deepest stages of sleep, waves of synchronous depolarization sweep across the cerebral cortex.

Sleep is divided into two major phases known as **REM (rapid eye movement) sleep** and **slow-wave sleep (non-REM or deep sleep)**. During a sleep cycle, a person alternates between these two phases, spending variable periods of time in each phase, with transition stages in between. Slow-wave sleep is marked on the EEG by the presence of *delta waves*, low-frequency waves of long duration. During this phase of the sleep cycle, sleepers adjust body position without conscious commands from the brain to do so.

In contrast, REM sleep is marked by an EEG pattern closer to that of an awake person, with rapid, low-amplitude waves. During REM sleep, brain activity inhibits motor neurons to skeletal muscles, paralyzing them. Exceptions to this pattern are the muscles that move the eyes and those that control breathing. The control of homeostatic functions is depressed during REM sleep, and body temperature falls toward ambient temperature.

REM sleep is the period of time during which most dreaming takes place. The eyes move behind closed lids, as if following the action of the dream. Sleepers are most likely to wake up spontaneously from periods of REM sleep.

If sleep is a neurologically active process, what is it that makes us sleepy? The possibility of a sleep-inducing factor was first proposed back in 1913, when scientists found that cerebrospinal fluid from sleep-deprived dogs could induce sleep in normal animals. Since then, a variety of sleep-inducing factors have been identified. Curiously, many of them are also substances that enhance the immune response, such as interleukin-1, interferon, serotonin, and tumor necrosis factor. As a result of this finding, some investigators have suggested that one answer to the puzzle of the biological reason for sleep is that we need to sleep to enhance our immune response. Whether or not that is a reason for why we sleep, the link between the immune system and sleep induction may help explain why we tend to sleep more when we are sick.

Sleep disorders are relatively common, as you can tell by looking at the variety of sleep-promoting agents available over the counter in drugstores. Among the more common sleep disorders are *insomnia*, the inability to go to sleep or remain asleep long enough to awake refreshed, and *sleep apnea* [*apnoos*, breathless], a condition in which sleepers awake when their airway muscles relax to the point of obstructing normal breathing. To learn more about sleep disorders, see the U.S. National Institutes of Health website for the National Center for Sleep Disorder Research (*www.nhlbi.nih.gov/about/ncsdr*).

CONCEPT CHECK   Answers: p. 320

21. During sleep, relay neurons in the thalamus reduce information reaching the cerebrum by altering their membrane potential. Are these neurons more likely to depolarize or hyperpolarize? Explain your reasoning.

Sleepwalking

Sleepwalking, or *somnambulism* [*somnus*, sleep + *ambulare*, to walk], is a sleep behavior disorder that for many years was thought to represent the acting out of dreams. However, most dreaming occurs during REM sleep, whereas sleepwalking takes place during deep sleep. During sleepwalking episodes, which may last from 30 seconds to 30 minutes, the subject's eyes are open and registering the surroundings. The subject is able to avoid bumping into objects, can negotiate stairs, and in some cases is reported to perform tasks such as preparing food or folding clothes. The subject usually has little if any conscious recall of the sleepwalking episode upon awakening. Sleepwalking is most common in children, and the frequency of episodes declines with age. There is also a genetic component, as the tendency to sleepwalk runs in families.

Adenosine and That "Java Jolt"

Caffeine and its *methylxanthine* cousins found in chocolate and tea, *theobromine* and *theophylline*, are probably the most widely consumed psychoactive drugs, known since ancient times for their stimulant effect. Molecular research has revealed that the methylxanthines are receptor antagonists for *adenosine*, a molecule composed of the nucleotide adenine plus the sugar ribose [❀ p. 41]. Adenosine acts as an important neuromodulator in the central nervous system. Four subtypes of adenosine receptor have been identified and they are all G protein-coupled, cAMP-dependent membrane proteins. The discovery that the stimulant effect of caffeine comes from its blockade of adenosine receptors has led scientists to investigate adenosine's role in sleep-wake cycles. Evidence suggests that adenosine accumulates in the extracellular fluid during waking hours, increasingly suppressing activity of the neurons that promote wakefulness. Other roles for adenosine in the brain include its possible involvement in the addiction/reward system and in the development of depression.

Emotion and Motivation Are Complex Neural Pathways

Emotion and motivation are two aspects of brain function that probably represent an overlap of the behavioral state and cognitive systems. The pathways are complex and form closed circuits that cycle information between various parts of the brain, including the hypothalamus, limbic system, and cerebral cortex. We still do not understand the underlying neural mechanisms and this is a large and active area of neuroscience research.

Emotions are difficult to define. We know what they are and can name them, but in many ways they defy description. One characteristic of emotions is that they are difficult to voluntarily turn on or off. The most commonly described emotions that arise in different parts of the brain are anger, aggression, sexual feelings, fear, pleasure, and contentment or happiness.

The limbic system, particularly the region known as the amygdala, is the center of emotion in the human brain. We learned about the role of this brain region through experiments in humans and animals. If the amygdala is artificially stimulated in humans as it might be during surgery for epilepsy, they report experiencing feelings of fear and anxiety. Experimental lesions that destroy the amygdala in animals cause the animals to become tamer and to display hypersexuality. As a result, neurobiologists feel that the amygdala is the center for basic instincts such as fear and for learned emotional states.

The pathways for emotions are complex (Fig. 9-21 ●). Sensory stimuli feeding into the cerebral cortex are constructed

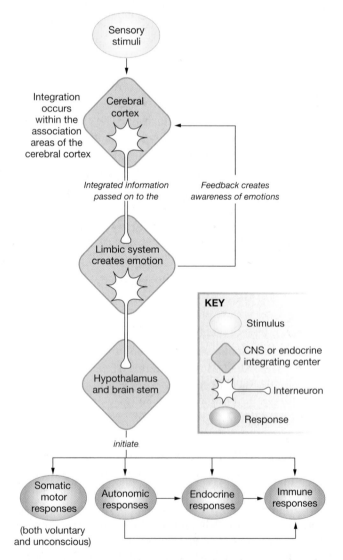

● **Figure 9-21** *The link between emotions and physiological functions* The association between stress and increased susceptibility to viruses is an example of an emotionally linked immune response.

within the brain to create a representation (perception) of the world. After information is integrated by the association areas, it is passed on to the limbic system. Feedback from the limbic system to the cortex creates awareness of the emotion, while descending pathways to the hypothalamus and brain stem initiate voluntary behaviors and unconscious responses mediated by autonomic, endocrine, immune, and somatic motor systems. The physical result of emotions can be as dramatic as the pounding heart of a fight-or-flight reaction or as insidious as the development of an irregular heartbeat. The links between mind and body are difficult to study and will take many years of research to understand.

Motivation is defined as internal signals that shape voluntary behaviors. Some of these behaviors, such as eating, drinking, and sex, are related to survival; others, such as curiosity and sex (again), are linked to emotions. Some motivational states are known as **drives** and generally have three properties in common: (1) they create an increased state of central nervous system arousal or alertness; (2) they create goal-oriented behavior; and (3) they are capable of coordinating disparate behaviors to achieve that goal.

Motivated behaviors often work in parallel with autonomic and endocrine responses in the body, as you might expect with behaviors originating in the hypothalamus. For example, if you eat salty popcorn, your body osmolarity increases. This stimulus acts on the thirst center of the hypothalamus, motivating you to seek something to drink. Increased osmolarity also acts on an endocrine center in the hypothalamus, releasing a hormone that increases water retention by the kidneys. Thus, one stimulus triggers both a motivated behavior and a homeostatic endocrine response.

Some motivated behaviors can be activated by internal stimuli that may not even be obvious to the person in whom they are occurring. Eating, curiosity, and sex drive are three examples of behaviors with complex stimuli underlying their onset. We may eat because we are hungry or because the food looks good or because we do not want to hurt someone's feelings. Many motivated behaviors stop when the person has reached a certain level of satisfaction, or **satiety**, but they may also continue *despite* feeling satiated.

Pleasure is a motivational state that is being intensely studied because of its relationship to *addictive behaviors* such as drug use. Animal studies have shown that pleasure is a physiological state that is accompanied by increased activity of the neurotransmitter dopamine in certain parts of the brain. Drugs that are addictive, such as cocaine and possibly nicotine, act by enhancing the effectiveness of dopamine, thus increasing the pleasurable sensations perceived by the brain. As a result, use of these drugs rapidly becomes a learned behavior. Interestingly, not all behaviors that are addictive are pleasurable. For example, there are a variety of compulsive behaviors that involve self-mutilation, such as pulling out hair by the roots. Fortunately, many behaviors that are learned can also be modulated, given motivation.

Moods Are Long-Lasting Emotional States

Moods are similar to emotions but are longer-lasting, relatively stable subjective feelings related to sense of well-being. Moods are difficult to define at a neurobiological level but evidence from studying and treating mood disorders suggests that what was once thought to be purely psychological in origin is actually a central nervous system function. Depression and other mood disorders appear to result from abnormal neurotransmitter release or reception in different brain regions.

Mood disorders are estimated to be the fourth leading cause of illness in the world today. **Depression** is a mood disturbance that affects nearly 10% of the United States population each year. It is characterized by sleep and appetite disturbances and alterations of mood and libido that may seriously impact the person's ability to function at school or work or in personal relationships. Many people do not realize that depression is not a sign of mental or moral weakness or that it can be treated successfully with drugs. (For detailed information about depression, go to *http://www.nlm.nih.gov/medlineplus/ depression.html*).

The drug therapy for depression has changed in recent years, but the major categories of antidepressant drugs all alter some aspect of synaptic transmission. The older *tricyclic antidepressants*, such as amitriptyline, block reuptake of norepinephrine into the presynaptic neuron, thus extending the active life of the neurotransmitter. The newer class of *selective serotonin reuptake inhibitors*, or SSRIs, slow down the removal of serotonin (and possibly also norepinephrine) from the synapse. As a result of uptake inhibition, serotonin lingers in the synaptic cleft longer than usual, increasing serotonin-dependent activity in the postsynaptic neuron.

Interestingly, patients need to take these drugs for several weeks before they experience their full effect. This suggests that the changes taking place in the brain are long-term modulation of pathways rather than simply enhanced fast synaptic responses. The search to uncover the biological basis of disturbed brain function is a major focus of neuroscience research.

Some research into brain function has become quite controversial, particularly that dealing with sexuality and the degree to which behavior in general is genetically determined in humans. We will not delve deeply into any of these subjects because they are complex and would require lengthy explanations to do them justice. Instead, we will look briefly at some of the recent models proposed to explain the mechanisms that are the basis for higher cognitive functions.

Learning and Memory Change Synaptic Connections in the Brain

For many years, motivation, learning, and memory, all aspects of the cognitive state, were considered to be in the realm of psychology rather than biology. Neurobiologists in decades past were more concerned with the network and cellular aspects of neuronal

RUNNING PROBLEM

Ben's halted development is a feature unique to infantile spasms. The abnormal portions of the brain send out continuous action potentials during frequent seizures and ultimately change the interconnections of the brain neurons. The damaged portions of the brain harm normal portions to such an extent that if some intervention such as medication or surgery is not put into place early in the onset of the disease, the brain is permanently damaged and developmental recovery will never occur.

Question 5: Changes in synaptic connections as a result of neuronal activity are an example of what process?

function. However, in recent years, the two fields have overlapped more and more. Scientists have discovered that the underlying basis for cognitive function seems to be explainable in terms of cellular events, such as long-term potentiation [p. 273] and plasticity. The ability of neuronal connections to change with experience is fundamental to the two cognitive processes of learning and memory.

Learning Is the Acquisition of Knowledge

How do you know when you have learned something? **Learning** can be demonstrated by behavioral changes, but behavioral changes are not required for learning to occur. Learning can be internalized and is not always reflected by overt behavior. Would someone watching you read your textbook be able to tell if you had learned anything?

Learning can be classified into two broad types: associative and nonassociative. **Associative learning** occurs when two stimuli are associated with each other, such as Pavlov's classic experiment in which he simultaneously presented dogs with food and rang a bell. After a period of time, the dogs came to associate the sound of a bell with food and began to salivate in anticipation of food whenever the bell was rung. Another form of associative learning occurs when an animal associates a stimulus with a given behavior. An example would be a mouse that gets a shock each time it touches a certain part of its cage. It soon associates that part of the cage with an unpleasant experience and avoids that area.

Nonassociative learning includes imitative behaviors such as learning a language. This type of learning includes habituation and sensitization, two adaptive behaviors that allow us to filter out and ignore background stimuli while responding more sensitively to potentially disruptive stimuli. In **habituation**, an animal shows a decreased response to an irrelevant stimulus that is repeated over and over. For example, a sudden loud noise may startle you, but if the noise is repeated over and over again, your brain begins to ignore it. Habituated responses allow us to filter out stimuli that we have evaluated and found to be insignificant.

Sensitization is the opposite of habituation and the two behaviors combined help increase an organism's chances for survival. In sensitization learning, exposure to a noxious or intense stimulus causes an enhanced response upon subsequent exposure. For example, people who become ill while eating certain foods may find that they lose their desire to eat that food again. Sensitization is adaptive since it helps us to avoid potentially harmful stimuli.

Memory Is the Ability to Retain and Recall Information

Memory is the ability to retain and recall information. We have several different types of memory: short-term and long-term, reflexive (automatic) and declarative (explicit). Memory processing for different types of memory appears to take place through different pathways. With noninvasive imaging techniques such as MRI and PET scans, researchers have been able to track brain activity as individuals learned to perform tasks.

Memories are stored throughout the cerebral cortex in pathways known as **memory traces**. Declarative memories involve the temporal lobes of the brain; reflexive memory storage requires the amygdala and the cerebellum. Some components of memories are stored in the sensory cortices where they are processed. For example, pictures are stored in the visual cortex and sounds in the auditory cortex.

Learning or recall of a single task may involve multiple circuits in the brain that work in parallel. This parallel processing helps provide backup in case one of the circuits is damaged. It is also believed to be the means by which specific memories are generalized, allowing new information to be processed and matched to stored information. For example, a person who has never seen a volleyball will recognize it as a ball because the volleyball has the same general characteristics as all other balls the person has seen.

In humans, the hippocampus seems to be an important structure in both learning and memory. For example, patients who had part of the hippocampus destroyed to relieve a certain type of epilepsy also had trouble remembering new information to which they were exposed. When given a list of words to repeat, they could remember the words as long as their attention stayed focused on the task. But if they were distracted, the memory of the words disappeared and they had to learn the list again. Information stored in long-term memory before the operation was not affected. This inability to remember newly acquired information is a defect known as **anterograde amnesia** [*amnesia*, oblivion].

Memory has multiple levels of storage, and our memory bank is constantly changing. When a stimulus comes into the central nervous system, it first goes into **short-term memory**, a limited store of information that can hold only about 7 to 12 pieces of information at a time. Items in short-term memory will

Alzheimer's Disease

Alzheimer's disease is a disease of cognitive impairment that accounts for about half the cases of progressive dementia [*demens*, out of one's mind] in the elderly. By one estimate, it affects about 4 million Americans, with the number expected to rise as the Baby Boomers age. Alzheimer's is characterized by memory loss that progresses to a point where the patient doesn't even recognize family members. Over time, even the personality changes and in the final stages, other cognitive functions fail so that patients cannot communicate with caregivers.

Diagnosis of Alzheimer's is difficult, although a study published in 2001 had a 95% success rate predicting early Alzheimer's using PET scans to find regions of decreased brain activity. The only definitive diagnosis comes after death, when brain tissue can be examined for neuronal degeneration, extracellular plaques made of β-*amyloid protein*, and intracellular tangles of *tau*, a protein that is normally associated with microtubules. Although the presence of plaques and tangles is diagnostic, the underlying cause of Alzheimer's is uncertain. One theory says that the β-amyloid plaques are the cause of the pathology (for a review, see "The amyloid hypothesis of Alzheimer's disease: progress and problems on the road to therapeutics," *Science* 297(5580):353–356, 2002 July 19). There is a known genetic component and other theories include oxidative stress and chronic inflammation. Currently there is no proven prevention or treatment, although some patients benefit from therapy with drugs that are either acetylcholine agonists or acetylcholinesterase inhibitors. The forecast of 14 million people with Alzheimer's by the year 2030 has put this disease in the forefront of neurobiological research.

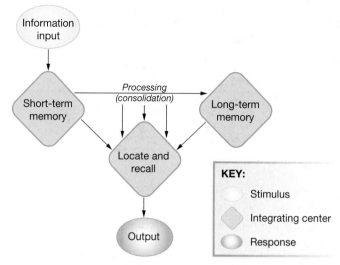

● **Figure 9-22   *Memory processing*** New information goes into short-term memory but will be lost unless it is processed and stored in long-term memory.

disappear unless an effort, such as repetition, is made to put them into a more permanent form (Fig. 9-22 ●).

Working memory is a special form of short-term memory processed in the prefrontal lobes. This region of the cerebral cortex is devoted to keeping track of bits of information long enough to put them to use in a task that takes place after the information has been acquired. Working memory in these regions is linked to long-term memory stores, so that newly acquired information can be integrated with stored information and acted upon.

For example, you are trying to cross a busy road. You look to the right and see that there are no cars coming for several blocks. You then look left and see that there are no cars coming from that direction either. Working memory has stored the information

that the road to the right is clear, so using stored knowledge about safety, you are able to conclude that there is no traffic from either direction and it is safe to cross the road.

In people with damage to the prefrontal lobes of the brain, this task becomes more difficult because they are unable to recall whether the road is clear from the right once they have looked away to assess traffic coming from the left. Working memory allows us to collect a series of facts from short- and long-term memory and connect them together in logical order to solve problems or plan actions.

The processing of information that converts short-term memory into **long-term memory** is known as **consolidation**. Consolidation can take varying periods of time, from seconds to minutes. Information passes through many intermediate levels of memory during consolidation, and in each of these stages, information can be located and recalled.

As scientists studied the consolidation of short-term memory into long-term memory, they discovered that the process involves changes in synaptic connections of the circuits involved in learning. In some cases, new synapses form; in others, the effectiveness of synaptic transmission is altered. These changes are evidence of plasticity and show us that the brain is not "hard-wired."

Long-term memory has been divided into two types that are consolidated and stored using different neuronal pathways (Table 9-5). **Reflexive (implicit) memory** is automatic and does not require conscious processes for its creation or recall. Information stored in this kind of memory is acquired slowly through repetition. Motor skills fall into the category of reflexive memory, as do procedures and rules. For example, you do not need to think about putting a period at the end of each sentence or about how to pick up a fork. Reflexive memory has also been called *procedural memory* because it generally concerns how to do things. Reflexive memories can be acquired through either associative or nonassociative learning processes.

Table 9-5: Types of Long-Term Memory

REFLEXIVE (IMPLICIT) MEMORY	DECLARATIVE (EXPLICIT) MEMORY
Recall is automatic and does not require conscious attention	Recall requires conscious attention
Acquired slowly through repetition	Depends on higher-level thinking skills such as inference, comparison, and evaluation
Includes motor skills and rules and procedures	Memories can be reported verbally
Memories can be demonstrated	

Declarative (explicit) memory, on the other hand, requires conscious attention for its recall. Its creation generally depends on the use of higher-level cognitive skills such as inference, comparison, and evaluation. Declarative memories deal with knowledge about ourselves and the world around us that can be reported or described verbally.

Sometimes information can be transferred from declarative memory to reflexive memory. The quarterback on a football team is a good example. When he learned to throw the football as a small boy, he had to pay close attention to gripping the ball and coordinating his muscles to throw the ball accurately. At that point of learning to throw the ball, the process was in declarative memory and required conscious effort as the boy analyzed his movements. With repetition, however, the mechanics of throwing the ball were transferred to reflexive memory: they became a reflex that could be executed without conscious thought. That transfer allowed the quarterback to use his conscious mind to analyze the path and timing of the pass while the mechanics of the pass became automatic. Athletes often refer to this automaticity of learned body movements as *muscle memory*.

Memory is an individual thing. We process information on the basis of our experiences and perception of the world. Because people have widely different experiences throughout their lives, it follows that no two people will process a piece of information in the same way. If you ask a group of people about what happened during a particular event such as a lecture, a robbery, or an accident, no two descriptions will be identical. Different people will have processed the event according to their own perceptions and experiences. Experiential processing is important to remember when studying in a group situation, because it is unlikely that all group members will learn or recall information the same way.

Language Is the Most Elaborate Cognitive Behavior

One of the hallmarks of an advanced nervous system is the ability of a species to exchange complex information with other members of the same species. Although found predominantly in birds and mammals, this ability also occurs in certain insects that convey amazingly detailed information by means of sound (crickets), touch and sight (bees), and odor (pheromones, also found in many other animals). In humans, communication takes place pri-

RUNNING PROBLEM

The PET scan revealed two abnormal spots, or *loci* (plural of *locus*), on Ben's right hemisphere, one on the parietal lobe and one overlapping a portion of the primary motor cortex. Because the loci triggering Ben's seizures were located on the same hemisphere and were in the cortex, Ben was a candidate for a *hemispherectomy*, removal of the cortex of the affected hemisphere. Surgeons removed eighty percent of his right cerebral cortex, sparing areas crucial to vision, hearing, and sensory processing. Normally the motor cortex would be spared as well, but in Ben's case a seizure locus overlapped much of the region.

Question 6: In which lobes are the centers for vision, hearing, and sensory processing located?

Question 7: Which of Ben's abilities might have suffered if his left hemisphere had been removed instead?

Question 8: By taking only the cortex of the right hemisphere, what parts of the cerebrum did surgeons leave behind?

Question 9: Why were the surgeons careful to spare Ben's right ventricle?

marily through spoken and written language. Because language is considered the most elaborate cognitive behavior, it has received considerable attention from neurobiologists.

Language skills require the input of sensory information (primarily from hearing and vision), processing in various centers in the cerebral cortex, and the coordination of motor output for vocalization and writing. In most people the centers for language ability are found in the left hemisphere of the cerebrum. Even 70% of people who are left-handed (right-brain dominant) or ambidextrous use their left brain for speech. The ability to communicate through speech has been divided into two processes: the combination of different sounds to form words (vocalization) and the combination of words into grammatically correct and meaningful sentences.

The integration of spoken language in the human brain involves two regions in the cortex: **Wernicke's area** in the temporal lobe and **Broca's area** in the frontal lobe close to the motor cortex (Fig. 9-23 ●). Most of what we know about these areas comes

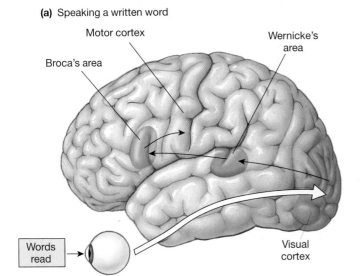

(a) Speaking a written word

Motor cortex

Broca's area

Wernicke's area

Words read

Visual cortex

(b) Speaking a heard word

Motor cortex

Broca's area

Wernicke's area

Spoken words

Auditory cortex

● **Figure 9-23** *Cerebral processing of spoken and visual language* Spoken and written language are processed by different pathways. People with damage to Wernicke's area do not understand spoken or written communication. Those with damage to Broca's area understand, but are unable to respond appropriately.

from studies of people with brain lesions because nonhuman animals are not capable of speech. Even primates that communicate on the level of a small child through sign language and other visual means do not have the physical ability to vocalize the sounds of human language.

Input into the language areas comes from either the visual cortex (reading) or the auditory cortex (listening). Sensory input goes first to Wernicke's area, then to Broca's area. After integration and processing, output to the motor cortex initiates a spoken or written response. If damage occurs to Wernicke's area, a person

is unable to understand any spoken or visual information. The person's own speech, as a result, is nonsense, since the person is unaware of his or her own errors. This condition is known as **receptive aphasia** [*a-*, not + *phatos*, spoken] because the person is unable to understand sensory input.

On the other hand, people with damage to Broca's area understand spoken and written language but are unable to express their response in normal syntax. Their response to a question often consists of appropriate words strung together in random order. These patients may have a difficult time dealing with their disability because they are aware of their mistakes but are powerless to correct them. Damage to Broca's area causes an **expressive aphasia**.

Mechanical forms of aphasia occur as a result of damage to the motor cortex. These patients find themselves unable to physically shape the sounds that make up words or unable to coordinate the muscles of their arm and hand to write.

Personality Is a Combination of Experience and Inheritance

One of the most difficult aspects of brain function to translate from the abstract realm of psychology into the physical circuits of neurobiology is the combination of attributes that we call personality. What is it that makes us individuals? The parents of more than one child will tell you that their offspring were different from birth, and even before that in the womb. If we all have the same brain structure, what makes us different?

This question fascinates many people. The answer that is evolving from neurobiology research is that we are a combination of our experiences and the genetic constraints that we inherit. One complicating factor is the developmental aspect of "experience," as scientists are showing that exposure of developing embryos to hormones while still in the womb can alter brain pathways.

What we learn or experience and what we store in memory create a unique pattern of neuronal connections in our brains. Sometimes these circuits malfunction, creating depression, schizophrenia, or any number of personality disturbances. Psychiatrists for many years attempted to treat these disorders as if they were due solely to events in the person's life, but now we know that there is a genetic component to many of these disorders.

Schizophrenia is an example of a brain disorder that has both a genetic and environmental basis. In the American population as a whole, the risk of developing schizophrenia is about 1%. However, if one parent has schizophrenia, the risk increases to 10%, indicating that people can inherit a susceptibility to develop the disease. Currently the cause of schizophrenia is not known. However it, like many other conditions with altered mental states, can be treated with drugs that influence neurotransmitter release and activity in the brain. To learn more about diagnosis and treatment of schizophrenia, see the National Institutes of Health website *http://www.nlm.nih.gov/medlineplus/schizophrenia.html.*

We still have much to learn about repairing damage to the central nervous system. One of the biggest tragedies in life is the personality change that sometimes accompanies head trauma. Physical damage to the delicate circuits of the brain can not only alter intelligence and memory storage but also can create a new personality. The person who exists after the injury may not seem to be the same person who inhabited that body before the injury. Although the change may not be noticeable to the injured person, it can be devastating to the family and friends of the victim. Perhaps as we learn more about how neurons link to each other, we will be able to find a means of restoring damaged networks and preventing the lasting effects of head trauma and brain disorders.

RUNNING PROBLEM CONCLUSION

Infantile Spasms

Ben has remained seizure-free since the surgery and shows normal development in all areas except motor skills. He remains somewhat weaker and less coordinated on his left side, the side opposite (*contralateral*) to the surgery, but the weakness should subside over time with the aid of physical therapy. Ben's recovery stands as a testament to the incredible plasticity of the brain.

Apart from the physical damage caused to the brain, a number of epileptics have developmental delays that stem from social aspects of their disorder. Young children with frequent seizures often have difficulty socializing with their peers due to overprotective parents, missed school days, and the fear of people who do not understand epilepsy. Their problems can extend into adulthood, when epileptics may have difficulty finding employment or driving if their seizures are not controlled. There are numerous examples of adults who undergo successful epilepsy surgery, but are still unable to fully enter society due to their lack of social and employment skills. Not surprisingly, the rate of depression is much higher among epileptics. To learn more about epilepsy, start with the Epilepsy Foundation (*www.epilepsyfoundation.org*).

This Running Problem was written by Susan E. Johnson, an undergraduate student at the University of Texas at Austin who is studying for a career in the biomedical sciences.

	QUESTION	FACTS	INTEGRATION AND ANALYSIS
1	How might a leaky blood-brain barrier lead to action potentials that trigger a seizure?	Neurotransmitters and other chemicals circulating freely in the blood are normally separated from the brain tissue by the blood-brain barrier.	Negatively charged molecules and neurotransmitters entering the brain might depolarize neurons and trigger action potentials.
2	What does GABA do to the cell's membrane potential? Does GABA make the cell more or less likely to fire action potentials?	GABA opens Cl^- channels.	Cl^- entering a neuron hyperpolarizes the cell and makes it less likely to fire action potentials.
3	Why is it important to limit the time span of ACTH therapy?	Exogenous ACTH acts in a short negative feedback loop, decreasing the output of CRH from the hypothalamus and ACTH production by the anterior pituitary. See Figure 7-15, p. 224.	Long-term suppression of endogenous hormone secretion by ACTH can cause CRH- and ACTH-secreting neurons to atrophy, resulting in a lifelong cortisol deficiency.
4	What is the rationale for using radioactively labeled glucose for the PET scan and not some other nutrient?	Glucose is the primary energy source for the brain.	Glucose usage is more closely correlated to brain activity than any nutrient in the body. Areas of abnormally high glucose usage are suggestive of overactive cells.
5	Changes in synaptic connections as a result of neuronal activity are an example of what process?	Changes in synaptic connections as a result of neuronal activity are an example of plasticity.	N/A*
6	In which lobes are the centers for vision, hearing, and sensory processing located?	Vision is processed in the occipital lobe, hearing in the temporal lobe, and sensory information in the parietal lobe.	N/A
7	Which of Ben's abilities might have suffered if his left hemisphere had been removed instead?	In most people, the left hemisphere contains Wernicke's area and Broca's area, two centers vital to speech. The left brain controls right-sided sensory and motor functions.	Patients who have undergone left hemispherectomies have difficulty with speech (abstract words, grammar, and phonetics). They show loss of right-side sensory and motor functions.

	QUESTION	FACTS	INTEGRATION AND ANALYSIS
8	By taking only the cortex of the right hemisphere, what parts of the cerebrum did surgeons leave behind?	The cerebrum consists of gray matter in the cortex and interior nuclei, white matter, and the ventricles.	The surgeons left behind the white matter, interior nuclei, and ventricles.
9	Why were the surgeons careful to spare Ben's right ventricle?	The walls of the ventricles contain the choroid plexus, which secretes cerebrospinal fluid (CSF). CSF plays a vital protective role by cushioning the brain.	CSF protection is particularly important following removal of portions of brain tissue because the potential damage from jarring of the head is much greater.

* N/A—not applicable

CHAPTER SUMMARY

The brain is the primary control center of the body and as you will learn, homeostatic responses in many organ systems are designed to maintain brain function. The ability of the brain to create complex thoughts and emotions in the absence of external stimuli is one of its *emergent properties*. *Communication* within the central nervous system and between the CNS and the periphery of the body uses a combination of electrical and chemical signals. Neurotransmitters binding to their receptors are an example of *molecular interactions*.

Emergent Properties of Neural Networks

1. Neural networks create **affective** and **cognitive behaviors**. (p. 285)

2. The brain exhibits **plasticity**, the ability to change connections as a result of experience. (p. 285)

Evolution of Nervous Systems

3. Nervous systems evolved from a simple network of neurons to complex brains. (p. 285; Fig. 9-1)

4. The **cerebrum** is responsible for thought and emotion. (p. 286)

Anatomy of the Central Nervous System

5. The central nervous system consists of layers of cells around a fluid-filled central cavity and develops from the **neural tube** of the embryo. (p. 287; Fig. 9-2)

6. The **gray matter** of the CNS consists of nerve cell bodies, dendrites, and axon terminals. The cell bodies form layers in parts of the brain or cluster into groups known as **nuclei**. (p. 287)

7. Myelinated axons form the **white matter** of the CNS and run in bundles called **tracts**. (pp. 287–288)

8. The brain and spinal cord are encased in the **meninges** and the bones of the **cranium** and vertebrae. The meninges are the **pia mater**, **arachnoid membrane**, and the **dura mater**. (pp. 288, 290; Fig. 9-4)

9. The **choroid plexus** secretes **cerebrospinal fluid** (CSF) into the **ventricles** of the brain. Cerebrospinal fluid cushions the tissue and creates a chemically controlled environment. (pp. 287, 290; Fig. 9-5)

10. Tight junctions in brain capillaries create a **blood-brain barrier** that prevents possibly harmful substances in the blood from entering the interstitial fluid. (p. 292; Fig. 9-6)

11. The normal fuel source for neurons is glucose, and the body closely regulates blood glucose concentrations. (p. 293)

The Spinal Cord

12. Each segment of the spinal cord is associated with a pair of **spinal nerves**. (p. 293)

13. The **dorsal root** of each spinal nerve carries incoming sensory information. The **dorsal root ganglia** contain the nerve cell bodies of sensory neurons. (p. 293; Fig. 9-7)

14. The **ventral roots** carry information from the central nervous system to muscles and glands. (p. 293)

15. **Ascending tracts** of white matter carry sensory information to the brain, and **descending tracts** carry efferent signals from the brain. **Propriospinal tracts** remain within the cord. (p. 294)

16. **Spinal reflexes** are integrated within the spinal cord. (p. 294; Fig. 9-8)

The Brain

17. The brain has six major divisions: cerebrum, diencephalon, midbrain, cerebellum, pons, and medulla oblongata. (p. 295; Fig. 9-9)

18. The **brain stem** is divided into medulla oblongata, pons, and midbrain (mesencephalon). **Cranial nerves** II to XII originate here. (pp. 295, 297; Tbl. 9-1)

19. The **reticular formation** is a diffuse group of neurons that play a role in many basic processes. (p. 295)

20. The **medulla oblongata** contains **somatosensory** and **corticospinal tracts** that convey information between the cerebrum and spinal cord. Most tracts cross the midline in the **pyramid** region. The medulla contains control centers for many involuntary functions. (p. 297)

21. The **pons** acts as a relay station for information between the cerebellum and cerebrum. (p. 297)

22. The **midbrain** controls eye movement and relays signals for auditory and visual reflexes. (p. 297)

23. The **cerebellum** processes sensory information and coordinates the execution of movement. (p. 297)

24. The **diencephalon** is composed of the thalamus and hypothalamus. The **thalamus** relays and modifies sensory and motor information going to and from the cerebral cortex. (p. 298; Fig. 9-10)

25. The **hypothalamus** contains centers for behavioral drives and plays a key role in homeostasis by its control over endocrine and autonomic function. (p. 298; Tbl. 9-2)

26. The **pituitary** and **pineal glands** are endocrine glands in the hypothalamus. (p. 299)

27. The cerebrum is composed of two hemispheres connected at the **corpus callosum**. Each cerebral hemisphere is divided into **frontal**, **parietal**, **temporal**, and **occipital lobes**. (p. 299)

28. Cerebral gray matter includes the **cerebral cortex**, basal nuclei, and limbic system. (p. 299; Fig. 9-11)

29. The **basal nuclei** help control movement. (p. 300)

30. The **limbic system** acts as the link between cognitive functions and emotional responses. The **amygdala** and **cingulate gyrus** are linked to emotion and memory. The **hippocampus** is associated with learning and memory. (p. 300; Fig. 9-13)

31. The brain uses a variety of neurotransmitters and neuromodulators. **GABA** and **glycine** are inhibitory neurotransmitters. **Glutamate** is the primary excitatory neurotransmitter. (p. 301; Tbl. 9-3)

Brain Function

32. Three brain systems influence motor output: a **sensory system**, a **cognitive system**, and a **behavioral state system**. (p. 301; Fig. 9-14)

33. Our higher brain functions such as reasoning arise within the cerebral cortex. The cerebral cortex contains three functional specializations: **sensory areas**, **motor areas**, and **association areas**. (p. 302; Fig. 9-15)

34. Each hemisphere of the cerebrum has developed special functions not shared by the other hemisphere, a specialization known as **cerebral lateralization**. (p. 302; Fig. 9-16)

35. Sensory areas receive information from sensory receptors. The **primary somatic sensory cortex** processes information about touch, temperature, and other somatic senses. The **visual cortex**, **auditory cortex**, **gustatory cortex**, and **olfactory cortex** receive information about vision, sound, taste, and odors, respectively. (p. 304)

36. Association areas integrate sensory information into perception. **Perception** is the brain's interpretation of sensory stimuli. (p. 304)

37. Motor output includes skeletal muscle movement, neuroendocrine secretion, and visceral responses. (p. 304)

38. Motor areas direct skeletal muscle movement. Each cerebral hemisphere contains a **primary motor cortex** and **motor association area**. (p. 304)

39. The behavioral state system controls states of arousal. (pp. 304–305)

40. The **diffuse modulatory systems** of the reticular formation influence attention, motivation, wakefulness, memory, motor control, mood, and metabolic homeostasis. (p. 305; Fig. 9-19, Tbl. 9-4)

41. The **reticular activating system** keeps the brain **conscious**, or aware of self and environment. Electrical activity in the brain varies with levels of arousal and can be recorded by **electroencephalography**. (p. 305; Fig. 9-20)

42. **Circadian rhythms** are controlled by an internal clock in the **suprachiasmatic nucleus**. (p. 306)

43. **Sleep** is an easily reversible state of inactivity with characteristic stages. The two major phases of sleep are **REM (rapid eye movement) sleep** and **slow-wave sleep** (non-REM sleep). The physiological reason for sleep is unknown. (p. 307)

44. The limbic system is the center of **emotion** in the human brain. Emotional events influence physiological functions. (p. 308; Fig. 9-21)

45. **Motivation** arises from internal signals that shape voluntary behaviors related to survival or emotions. Motivational **drives** create goal-oriented behaviors. (p. 309)

46. **Moods** are long-lasting emotional states. Many mood disorders can be treated by altering neurotransmission in the brain. (p. 309)

47. **Learning** is the acquisition of knowledge about the world around us. **Associative learning** occurs when two stimuli are associated with each other. **Nonassociative learning** includes imitative behaviors such as learning a language. (p. 310)

48. In **habituation**, an animal shows a decreased response to a stimulus that is repeated over and over. In **sensitization**, exposure to a noxious or intense stimulus creates an enhanced response upon subsequent exposure. (p. 310)

49. **Memory** has multiple levels of storage and is constantly changing. Information is first stored in **short-term memory**, but will disappear unless consolidated into long-term memory. (p. 310; Fig. 9-22)

50. **Long-term memory** includes **reflexive memory**, which does not require conscious processes for its creation or recall, and **declarative memory**, which uses higher-level cognitive skills for formation and requires conscious attention for its recall. (pp. 311–312; Tbl. 9-5)

51. The **consolidation** of short-term memory into long-term memory appears to involve changes in the synaptic connections of the circuits involved in learning. (p. 311)

52. Language is considered the most elaborate cognitive behavior. The integration of spoken language in the human brain involves information processing in **Wernicke's area** and **Broca's area**. (p. 312; Fig. 9-23)

QUESTIONS

LEVEL ONE Reviewing Facts and Terms

1. The ability of human brains to change circuit connections and function in response to sensory input and past experience is known as _____.

2. _____ behaviors are related to feeling and emotion. _____ behaviors are related to thinking.

3. The part of the brain called the _____ is what makes us human, allowing human reasoning and cognition.

4. In vertebrates, the central nervous system is protected by the bones of the _____ and _____.

5. Name the meninges, beginning with the layer next to the bones.

6. List and explain the purposes of cerebrospinal fluid (CSF). Where is CSF made?

7. Compare the concentrations of these substances in CSF and plasma.
   (a) HCO_3^- (b) Ca^{2+}
   (c) glucose   (d) H^+
   (e) Na^+ (f) K^+

8. The only fuel source for neurons under normal circumstances is _____. Low concentrations of this fuel in the blood is termed _____. To synthesize enough ATP to continually transport ions, the neurons also exhibit high rates of _____ consumption. To supply these needs, about _____% of the blood pumped by the heart goes to the brain.

9. Match each area with its function.
   (a) medulla oblongata
   (b) pons
   (c) midbrain
   (d) reticular formation
   (e) cerebellum
   (f) diencephalon
   (g) thalamus
   (h) hypothalamus
   (i) cerebrum

   1. coordinates execution of movement
   2. composed of the thalamus and hypothalamus
   3. controls arousal and sleep
   4. composed of two hemispheres and fills most of the cranium
   5. contains control centers for blood pressure and breathing
   6. relays and modifies information going to and from cerebrum
   7. transfers information to cerebellum
   8. contains integrating centers for homeostasis
   9. relays signals and visual reflexes, eye movement

10. What is the blood-brain barrier and what is its function?

11. How are gray matter and white matter different, both anatomically and functionally?

12. Name the cerebral cortex areas that (a) direct perception, (b) direct movement, and (c) integrate information and direct voluntary behaviors.

13. What does cerebral lateralization refer to? What functions tend to be centered in each hemisphere?

14. Name the two most common inhibitory neurotransmitters of the CNS. Where do they act? Which ion channels do they open?

15. List and define the two major phases of sleep. How are they different from each other?

16. List several homeostatic reflexes and behaviors influenced by output from the hypothalamus. What is the source of emotional input into this area?

17. The _____ region of the limbic system is believed to be the center for basic instincts such as fear and learned emotional states.

18. What are the broad categories of learning? Define habituation and sensitization. What anatomical structure of the cerebrum is important in both learning and memory?

19. What two centers of the cortex are involved in integrating spoken language?

LEVEL TWO Reviewing Concepts

20. Map the following terms describing CNS anatomy. You may draw pictures or add additional terms if you wish.

arachnoid membrane
ascending tracts
blood-brain barrier
brain
capillaries
cell bodies
cerebrospinal fluid
cervical nerves
choroid plexus
cranial nerves
descending tracts
dorsal root
dorsal root ganglion
dura mater
ependyma
gray matter
lumbar nerves
meninges
nuclei
pia mater
propriospinal tracts
sacral nerves
spinal cord
thoracic nerves
ventral root
ventricles
vertebral column
white matter

21. Trace the pathway that the cerebrospinal fluid follows through the nervous system.

22. What are the three brain systems that regulate motor output by the CNS?

23. How can a single molecule act as a neurotransmitter at one time and a neuromodulator at another?

24. Compare and contrast the following concepts:
    (a) diffuse modulatory systems, reticular formation, limbic system, and reticular activating system
    (b) different forms of memory
    (c) nuclei and ganglia
    (d) tracts, nerves, horns, nerve fibers, and roots

25. Fill in the question marks in the table below:

CEREBRAL AREA	LOBE	FUNCTIONS
Primary somatic sensory cortex	?	Receives sensory information from peripheral receptors
?	Occipital	Processes information from the eyes
Auditory cortex	Temporal	?
?	Temporal	Receives input from chemoreceptors in the nose
Motor cortices	?	?
Association areas	NA	?

26. Given the wave below, draw examples with (a) lower frequency, (b) higher amplitude, (c) higher frequency. (Hint: see Figure 9-20, p. 306)

27. What properties do motivational states have in common?

28. What changes occur at synapses as memories are formed?

LEVEL THREE Problem Solving

29. Mr. Andersen, a stroke patient, experiences expressive aphasia. His savvy therapist, Cheryl, teaches him to sing to communicate his needs. What symptoms did he exhibit before therapy? How do you know he did not have receptive aphasia? Using what you have learned about cerebral lateralization, hypothesize why singing worked for him.

30. A study was done in which 40 typical citizens were taught about using seatbelts in their cars. At the end of the presentations, all participants scored at least 90% on a comprehensive test. However, the people were secretly videotaped entering and leaving the parking lot. Twenty subjects entered wearing their seatbelts; 22 left wearing them. Did learning occur? What is the relationship between learning and actually buckling the seatbelts?

31. In 1913, Henri Pieron took a group of dogs and kept them awake for several days. Before allowing them to sleep, he withdrew cerebrospinal fluid from the sleep-deprived animals. He then injected this CSF into normal, rested dogs. The recipient dogs promptly went to sleep for periods ranging from two to six hours. What conclusion can you draw about the possible source of a sleep-inducing factor? What controls should Pieron have run?

Explorations

Explorations are also available at *www.physiologyplace.com*

This chapter introduced you to the organization of the brain and spinal cord. Regions of brain tissue that look the same can function quite differently. Their function depends on the origins and types of signals received. The following Explorations will help you gain an appreciation of the complexity of brain function.

Exploration 1: Does Hormone Replacement Therapy Decrease Alzheimer's Risk?

In a study published in November 2002,* a group of scientists presented the results of a prospective study done in Cache County, Utah. The study began in 1995 with the cognitive assessment of 1,889 women whose mean age was 74.5 years. Investigators also asked about the women's history of taking calcium and multivitamin supplements and postmenopausal hormone replacement therapy (estrogen or estrogen/progesterone). Follow-up interviews three years later looked for the development of Alzheimer's disease in the study population. Data showed 58/800 women who had not used hormone replacement therapy developed Alzheimer's, compared to 26/1,066 women who had used hormones.

1. Can the researchers conclude from the data given that hormone replacement therapy decreases the risk of developing Alzheimer's? Is there other information gathered in the study that should be included in the data analysis?

2. How applicable are these findings to American women as a whole? What other information might you want to know about the study subjects before you draw any conclusions?

*"Hormone replacement therapy and incidence of Alzheimer disease in older women: The Cache County study," *JAMA* 288:2123–2129, 2002 Nov. 6.

Exploration 2: Spinal Cord Injuries

Most victims of spinal cord injury are young men in their teens and twenties. The equestrian accident that paralyzed actor Christopher Reeve brought this type of paralysis to the public's attention, but most spinal cord injuries are caused by motor vehicle accidents, falls, or sports-related accidents. The victim's functional loss will depend on several factors, including the location of the injury along the spine and the amount of damage to the spinal cord. To learn more about spinal cord injuries and the research into treating them, do a Google search (*www.google.com*) for *spinal cord injury*.

Exploration 3: Sleep, Dreams, and Naps

Students are known for pulling "all-nighters," going without sleep so they can cram in more studying. Does that really impair performance? Do short "power naps" during the day help? Does sleep really help your memory? For some scientists' views on these subjects, read the following articles.

Sleep deprivation can decrease learning and recall, according to research published in the article, "Altered brain response to verbal learning following sleep deprivation," *Nature* 403:655–657, 2000 Feb 10 and discussed in "Images of lost sleep," *Nature* 403:605–606, 2000 Feb 10. And short naps of an hour or less do seem to help learning ("The restorative effect of naps on perceptual deterioration," *Nature Neuroscience* 5(7):677–681, 2002 July). Despite these and other studies, there is still some disagreement about the role sleep and dreaming play in memory consolidation. For more on that subject, see the November 2, 2001 *Science* collection of articles on "Sleep, dreams, and memory." (Some articles in *Science* are available free online after one year at *www.sciencemag.org*.)

ANSWERS

Answers to Concept Check Questions

Page 285

1. (a) 3; (b) 2; (c) 1; (d) 3; (e) 2

Page 288

2. Glial cells are astrocytes (support, secretion, absorption), oligodendrocytes (myelin), microglia (scavengers), and ependyma (line ventricles).

Page 290

3. A ganglion is a cluster of nerve cell bodies. The CNS equivalent is a nucleus.

4. Tracts are the CNS equivalent of nerves.

Page 290

5. When H^+ concentration increases, pH decreases.

6. Blood will collect in the space between the membranes, pushing on the soft brain tissue under the skull. (This is called a *subdural hematoma*.)

7. Cerebrospinal fluid is more like interstitial fluid because it has little protein and no blood cells.

Page 293

8. Oxidative phosphorylation takes place in mitochondria.

9. Glucose is metabolized to pyruvate through glycolysis, and then enters the citric acid cycle (also called tricarboxylic acid cycle). $NADH_2$ passes high-energy electrons to the electron transport system for ATP synthesis.

10. Another explanation would be that some property of brain tissue made it resistant to staining by the dye. This is in fact what Ehrlich concluded.

11. The brain stained blue this time but none of the other body tissues stained because the dye was unable to enter the bloodstream across the blood-brain barrier.

Page 294

12. Roots are sections of spinal nerves just before they enter the spinal cord. Horns are areas of gray matter in the spinal cord. Tracts are long projections of white matter (axons) that extend up and down the spinal cord. Columns are groups of tracts carrying similar information.

13. Cutting a dorsal root will disrupt sensory function.

Page 295

14. (a) and (c)—white matter; (b) gray matter

15. Activities would include moving eyes, jaw, or tongue and testing taste, smell, and hearing.

16. The cerebrum is dorsal or superior to the brain stem.

Page 299

17. medulla, pons, and midbrain

18. The diencephalon structures are thalamus, hypothalamus, pituitary, and pineal glands.

Page 300

19. Neurons cross from one side of the body to the other at the pyramids in the medulla.

20. Spinal cord, medulla, pons, cerebellum, midbrain, diencephalon, cerebrum

Page 307

21. Neurons that are sending fewer signals have probably hyperpolarized because they would then require a larger stimulus to initiate an action potential.

Answers to Figure Questions

Page 303

Figure 9-16: (a) Losing function in the right visual hemisphere would mean that the person could see nothing in the left visual field. (b) The corpus callosum consists of tracts that exchange information between the two sides of the cerebrum.

Page 306

Figure 9-20: The alpha waves have the fastest frequency and the delta waves have the greatest amplitude.

Sensory Physiology

10

Stereocilia on hair cells of the ear

Nature does not communicate with man by sending encoded messages.

—Oscar Hechter, in *Biology and Medicine into the 21st Century,* 1991

RUNNING PROBLEM

Ménière's Disease

After a visit to a brothel in Arles, France, on December 23, 1888, Vincent Van Gogh, the nineteenth-century French painter, returned to his room in a boardinghouse, picked up a knife, and cut off his own ear. A local physician, Dr. Felix Ray, examined Van Gogh that night and wrote that the painter had been "assailed by auditory hallucinations" and in an effort to relieve them, "mutilated himself by cutting off his ear." A few months later, Van Gogh committed himself to a lunatic asylum. By 1890, Van Gogh was dead by his own hand. Historians have assumed that Van Gogh suffered from epilepsy. But the painter's strange attacks of dizziness, nausea, and overwhelming tinnitus (ringing or other sounds in the ears), which he recorded in desperate letters to his relatives, are more consistent with Ménière's disease, a condition that affects the inner ear. Today, Anant, a 20-year-old college student, will be examined by an otolaryngologist (ear-nose-throat specialist) to see if his periodic attacks of severe dizziness and nausea are caused by the same condition that drove Van Gogh to suicide.

325 343 349 352 354 361

BACKGROUND BASICS

Summation: **270–271** | Second messenger systems: **177** | Threshold: **324** | G proteins: **179** | Plasticity: **285** | Tonic control: **188–189** | Membrane potential: **250** | Graded potentials: **250** | Neurotransmitter release: **264** |

Imagine floating in the dark in a tank of buoyant salt water: there is no sound, no light, no breeze. The air and water are the same temperature as your body. You are in a sensory deprivation tank, and the only sensations you are aware of come from your own body. Your limbs are weightless, your breath moves in and out, and you feel your heart beating. In the absence of external stimuli, you turn your awareness inward to hear what your body has to say.

In decades past, sensory deprivation tanks were a popular way to counter the stress of a busy world. These facilities are hard to find now, but they illustrate the role of the afferent division of the nervous system: to provide us with information about the environment outside and inside our bodies. Sometimes we perceive sensory signals when they reach a level of conscious awareness, but other times they are processed completely at the subconscious level (Table 10-1). Stimuli that your central nervous system (CNS) monitors without your conscious awareness include changes in muscle stretch and tension, as well as a variety of internal parameters that the body monitors to maintain homeostasis, such as blood pressure and pH. The responses to these stimuli comprise many of the subconscious reflexes of the body.

In this chapter we are concerned primarily with the sensory stimuli that reach the conscious level of perception. We divide these stimuli into the **special senses** of vision, hearing, taste, smell, and equilibrium and the **somatic senses** of touch, temperature, pain, and proprioception. **Proprioception** is mediated by sensory receptors in the muscles and joints of the body that make us aware of our body's movement and position in space. If you close your eyes and raise your arm above your head, you are aware of its position because of the activation of proprioceptors in the muscles and joints.

We will first consider general properties of sensory pathways. We then look at the unique receptors and pathways that distinguish the different sensory systems from each other.

▶ GENERAL PROPERTIES OF SENSORY SYSTEMS

All sensory pathways have certain elements in common. They begin with a stimulus, internal or external, that acts on a sensory receptor. The receptor is a *transducer* that converts the stimulus into an intracellular signal, usually a change in membrane potential. If the stimulus is above threshold, action potentials pass along an afferent sensory neuron to the central nervous system, where incoming signals are integrated. Some stimuli pass upward to the cerebral cortex, where they reach conscious perception, but others are acted upon without our awareness. At each synapse along the way, the nervous system can modulate and shape the sensory information.

Sensory systems in the human body vary widely in complexity. The simplest systems are single sensory neurons with branched dendrites that function as receptors. The most complex systems include multicellular **sense organs** such as the ear and the eye. The cochlea of the ear contains about 16,000 sensory receptors and more than a million associated parts, and one human eye has about 126 million sensory receptors.

In this section we discuss the way sensory signals are transduced, coded, and processed.

Table 10-1: Information Processing by the Sensory Division of the Nervous System

CONSCIOUS	SUBCONSCIOUS
Special senses	**Somatic stimuli**
Vision	Muscle length and tension
Hearing	Proprioception
Taste	**Visceral stimuli**
Smell	Blood pressure
Equilibrium	pH and oxygen content of blood
Somatic senses	pH of cerebrospinal fluid
Touch	Lung inflation
Temperature	Osmolarity of body fluids
Pain	Temperature
Itch	Blood glucose
Proprioception	Distension of gastrointestinal tract

Receptors Are Sensitive to Particular Stimuli

Receptors in the sensory system vary widely in complexity, ranging from the branched endings of a single sensory neuron to complex, highly organized cells such as photoreceptors. The simplest receptors are **somatosensory** (somatic senses) receptors that consist of a neuron with naked ("free") nerve endings (Fig. 10-1a ●). In more complex somatosensory receptors, the nerve endings are encased in connective tissue capsules (Fig. 10-1b). The axons of somatosensory receptors may be myelinated or unmyelinated.

The special senses have the most highly specialized receptors. The receptors for smell are neurons, but the other special senses use non-neural receptor cells that synapse onto sensory neurons. The hair cell of the ear, shown in Figure 10-1c, is an example of a non-neural receptor. When activated, the hair cell releases a neurotransmitter that initiates an action potential in the associated sensory neuron. Taste buds and complex photoreceptors are also non-neural receptors.

Non-neural accessory structures are critical to the operation of many sensory systems. For example, the lens and cornea of the eye help focus incoming light onto the photoreceptors. The hairs

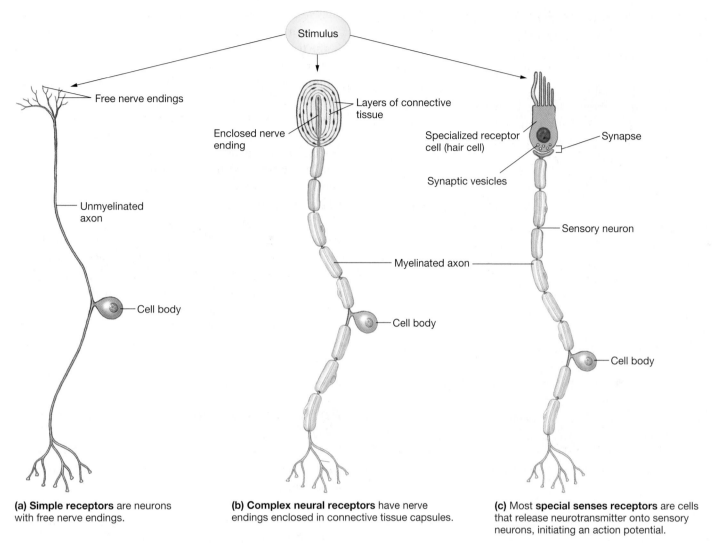

(a) **Simple receptors** are neurons with free nerve endings.

(b) **Complex neural receptors** have nerve endings enclosed in connective tissue capsules.

(c) Most **special senses receptors** are cells that release neurotransmitter onto sensory neurons, initiating an action potential.

● **Figure 10-1** *Sensory receptors* (a) Simple receptors may have myelinated or unmyelinated axons. (b) This illustration is a Pacinian corpuscle, which senses touch. (c) The cell illustrated is a hair cell, found in the ear.

on our arms sense movement in the air millimeters above the skin surface. Non-neural accessories often enhance the information-gathering capability of the sensory system.

Receptors are divided into four major groups, based on the type of stimulus to which they are most sensitive (Table 10-2).

Table 10-2: Types of Sensory Receptors

TYPE OF RECEPTOR	EXAMPLES OF STIMULI
Chemoreceptors	Oxygen, pH, various organic molecules such as glucose
Mechanoreceptors	Pressure (baroreceptors), cell stretch (osmoreceptors), vibration, acceleration, sound
Photoreceptors	Photons of light
Thermoreceptors	Varying degrees of heat

Chemoreceptors respond to chemical ligands that bind to the receptor. **Mechanoreceptors** respond to various forms of mechanical energy, including pressure, vibration, gravity, acceleration, and sound. **Thermoreceptors** respond to temperature and **photoreceptors** to light. The specificity of a receptor for a particular type of stimulus is known as the **law of specific nerve energies**.

CONCEPT CHECK Answers: p. 367

1. What is the advantage of having myelinated axons?

2. What accessory role does the outer ear (the pinna) play in the auditory system?

3. Look at the somatic and visceral stimuli listed in Table 10-1 and decide which of the following receptor types is the appropriate transducer: mechano-, chemo-, photo-, or thermo-.

Sensory Transduction Converts Stimuli into Graded Potentials

How do receptors convert diverse physical stimuli such as light or heat into electrical signals? The first step is **transduction**, the conversion of stimulus energy into information that can be processed by the nervous system.* In many receptors, mechanical, chemical, thermal, or light energy is transformed directly into a change in membrane potential. Some sensory transduction, particularly that mediated by non-neural receptor cells, includes an intermediate step and uses second messenger systems to initiate the change in membrane potential.

Each receptor type has an **adequate stimulus**, a particular form of energy to which it is most responsive. For example, thermoreceptors are more sensitive to temperature than to pressure. Mechanoreceptors respond preferentially to stimuli that deform the cell membrane. Although receptors are specific for one form of energy, they can respond to most forms if the intensity is high enough. Photoreceptors of the eye respond most readily to light, but a blow to the eye may cause us to "see stars," an example of mechanical energy of sufficient force to stimulate the photoreceptors.

Sensory receptors can be incredibly sensitive to their preferred form of stimulus. For example, a single photon of light stimulates certain photoreceptors, and a single *odorant* molecule may activate some olfactory chemoreceptors. The minimum stimulus required to activate a receptor is known as the **threshold**, just as the minimum depolarization required to trigger an action potential is called the threshold [● p. 251].

---

*This type of transduction is not to be confused with signal transduction that involves the transfer of information from an extracellular chemical messenger to an intracellular second messenger [● p. 174].

How is a physical or chemical stimulus converted into a change in membrane potential? The stimulus opens ion channels in the receptor membrane, either directly or indirectly (through a second messenger). In most cases, channel opening results in net influx of Na^+ or other cations into the receptor, depolarizing the membrane. In a few cases, the response to the stimulus is hyperpolarization when K^+ leaves the cell.

The change in receptor membrane potential goes by two different names, depending on the cell type in which it occurs. For somatosensory receptors and smell, the stimulus creates a **generator potential**. Generator potentials are graded potentials whose amplitude is proportional to the strength of the stimulus (see Fig. 10-7). If the generator potential reaches threshold, it initiates an action potential that travels along the sensory fiber to the CNS. This process is similar to a suprathreshold stimulus that reaches the trigger zone in other neurons.

Receptor potentials occur in non-neural receptors such as the hair cell, photoreceptor, or taste receptor. Receptor potentials are changes in membrane potential that influence neurotransmitter secretion by the receptor cell, which in turn alters electrical activity in the associated sensory neuron.

Each Sensory Neuron Has a Receptive Field

A sensory neuron is activated by stimuli that fall within a specific area known as the neuron's **receptive field**. In the simplest scenario, one receptive field is associated with one sensory neuron, which in turn synapses on one CNS neuron (the *secondary neuron* in the pathway). However, receptive fields frequently overlap with neighboring receptive fields (Fig. 10-2 ●).

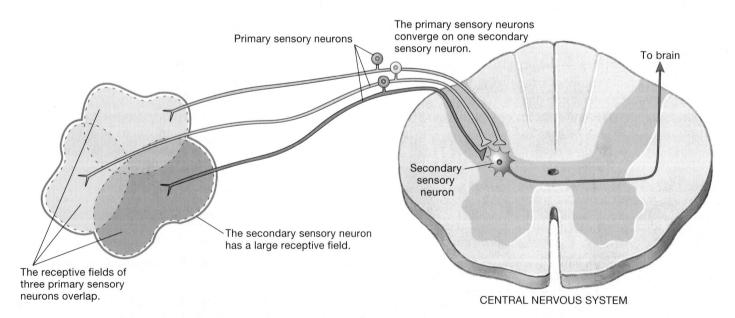

Primary sensory neurons

The primary sensory neurons converge on one secondary sensory neuron.

To brain

Secondary sensory neuron

The secondary sensory neuron has a large receptive field.

The receptive fields of three primary sensory neurons overlap.

CENTRAL NERVOUS SYSTEM

● **Figure 10-2** *Receptive fields of sensory neurons*   Convergence of primary sensory neurons allows simultaneous subthreshold stimuli to sum at the secondary sensory neuron and initiate an action potential. (In this illustration, only part of the secondary sensory neuron is shown.)

In addition, sensory neurons linked to neighboring receptive fields may exhibit *convergence* [✸ p. 270], where multiple presynaptic neurons provide input to a smaller number of postsynaptic neurons. Convergence allows multiple simultaneous subthreshold stimuli to sum at the postsynaptic, or secondary, neuron. When multiple neurons converge on a single secondary neuron, their individual receptive fields merge into a single, large *secondary receptive field*.

The size of secondary receptive fields determines the sensitivity of an area to a stimulus. For example, sensitivity to touch is demonstrated by a **two-point discrimination test**. In some regions of skin, such as that on the arms and legs, two pins placed within 20 mm of each other will be interpreted by the brain as a single pinprick (Fig. 10-3a ●). In these areas, many primary neurons converge on a single secondary neuron, so the secondary receptive field is very large. On the other hand, more sensitive areas of skin have smaller receptive fields, with perhaps a 1 : 1 relationship between primary and secondary sensory neurons. In these regions, two pins separated by as little as 2 mm can be perceived as two separate touches (Fig. 10-3b).

The Central Nervous System Integrates Sensory Information

Sensory information from much of the body enters the spinal cord and travels through ascending pathways to the brain. Some sensory information goes directly into the brain stem via the cranial nerves (✸ p. 296). Sensory information that initiates visceral reflexes is integrated in the brain stem or spinal cord and usually does not reach conscious perception. An example of an unconscious visceral reflex is the control of blood pressure by centers in the brain stem.

Each major division of the brain processes one or more types of sensory information (Fig. 10-4 ●). For example, the midbrain receives visual information and the medulla oblongata receives input for sound and taste. Information about balance and equilibrium is processed primarily in the cerebellum. These pathways, along with those carrying somatosensory information, project to the thalamus, which acts as a relay and processing station before passing the information on to the cerebrum.

Only *olfactory* [*olfacere*, to sniff] information is not routed through the thalamus. The sense of smell, a type of chemoreception, is considered to be the one of the oldest senses, and the most primitive vertebrate brains have well-developed regions for processing olfactory information. Information about odors travels directly from the nose to the cerebrum. Perhaps it is because of this direct input to the cerebrum that odors are so closely linked to memory and emotion. Most people have experienced encountering a smell that suddenly brings back a flood of memories of times or people from the past.

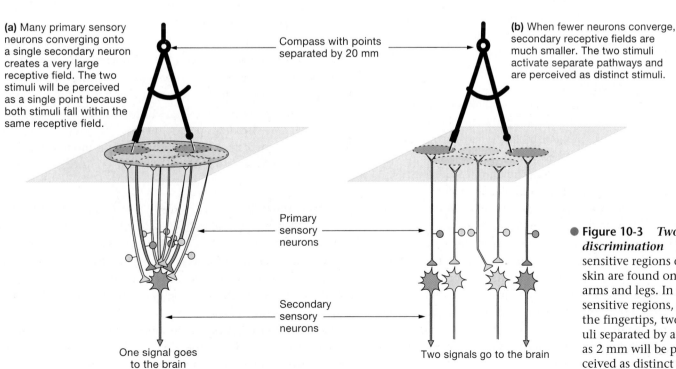

(a) Many primary sensory neurons converging onto a single secondary neuron creates a very large receptive field. The two stimuli will be perceived as a single point because both stimuli fall within the same receptive field.

Compass with points separated by 20 mm

(b) When fewer neurons converge, secondary receptive fields are much smaller. The two stimuli activate separate pathways and are perceived as distinct stimuli.

Primary sensory neurons

Secondary sensory neurons

One signal goes to the brain

Two signals go to the brain

● **Figure 10-3** *Two-point discrimination* Less-sensitive regions of the skin are found on the arms and legs. In more sensitive regions, such as the fingertips, two stimuli separated by as little as 2 mm will be perceived as distinct stimuli.

325

● **Figure 10-4** *Sensory pathways* All pathways except the olfactory pathway pass through the thalamus.

① Olfactory pathways from the nose project to the olfactory cortex.

② Most sensory pathways project to the thalamus. The thalamus modifies and relays information to cortical centers.

③ Equilibrium pathways project to the cerebellum.

The cerebral cortex processes consciously perceived sensory information in ways that we still do not fully understand. One interesting aspect of CNS processing is the **perceptual threshold**, a level of stimulus intensity necessary for us to be aware of a particular sensation. Stimuli bombard our sensory receptors constantly, but our brain can "turn off" some stimuli to avoid being overwhelmed with information.

Decreased perception of a stimulus is accomplished by *inhibitory modulation* [● p. 271] that decreases a suprathreshold stimulus until it is below the perceptual threshold. Inhibitory modulation often occurs in secondary and tertiary neurons of a sensory pathway. If we suddenly need to pay attention to a stimulus, we can consciously focus our attention and overcome the inhibitory modulation.

You have experienced a change in perceptual threshold when you "tune out" the radio while studying or when you "zone out" during a lecture. In both cases, the noise is adequate to stimulate sensory neurons in the ear, but neurons higher in the pathway dampen the perceived signal so that it does not reach the conscious brain. The threshold can be modified by an event such as a sudden question from the professor. At that point, the modulating influence is removed, and your conscious brain seeks to retrieve and recall recent sound input from the subconscious so that you can answer the question.

Coding and Processing Distinguish Stimulus Modality, Location, Intensity, and Duration

If all stimuli are converted to action potentials in the sensory neuron and all action potentials are identical, how can the central nervous system tell the difference between heat and pressure, or between a pinprick to the toe and one to the hand? The attributes of the stimulus must somehow be preserved once the stimulus enters the nervous system for processing. The CNS must distinguish four properties of a stimulus: (1) the nature, or **modality**, of the stimulus, (2) its location, (3) its intensity, and (4) its duration.

Sensory Modality The modality of a signal is indicated by which sensory neurons are activated and where those pathways terminate in the brain. Each receptor type is most sensitive to a particular type of stimulus, according to the law of specific nerve energies mentioned earlier. For example, some neurons respond most strongly to touch; others respond to changes in temperature. Each sensory modality can be subdivided into receptors with varying sensitivities. For instance, color vision is divided into red, blue, and green according to the wavelengths that most strongly stimulate the different receptor subtypes.

In addition, the brain associates a signal coming from a specific group of receptors with a specific modality or submodality. This 1:1 association of a receptor with a sensation is called **labeled line coding**. Stimulation of a cold receptor will always be perceived as cold, whether the actual stimulus was cold or whether an artificial depolarization of the receptor was the source of the stimulus. The blow to the eye that causes us to "see" a flash of light is another example of labeled line coding.

Location of the Stimulus   The location of a stimulus is also coded according to which group of neurons is active. The sensory regions of the cerebrum are highly organized in respect to incoming signals, and input from adjacent sensory receptors is processed in adjacent columns of the cortex [❧ p. 300]. This arrangement preserves the topographical organization of receptors on the skin, eye, or other regions in the processing centers of the brain.

For example, touch receptors in the hand project to a specific area of the cerebral cortex. Experimental stimulation of that area of the cortex during brain surgery is interpreted as a touch to the hand, although there may have been no actual contact. Similarly, the *phantom limb pain* reported by some amputees occurs when secondary sensory neurons in the spinal cord become hyperactive, resulting in the sensation of pain in a limb that is no longer there.

Auditory information is an exception to the localization rule, however. Neurons in the ears are sensitive to different frequencies of sound, but their activation provides no information about the location of the sound. Instead, the brain uses the timing of receptor activation to compute a location, as shown in Figure 10-5 ●. A

sound coming from directly in front of a person will reach both ears simultaneously. A sound placed off to one side will reach the closest ear several milliseconds before it reaches the other ear. The difference in the time it takes for the sound stimuli to reach the two sides of the auditory cortex is registered by the brain and used to compute the sound's source.

Lateral inhibition is another way the location of a stimulus can be isolated. Figure 10-6 ● shows this process for a pressure stimulus to the skin. The pin pushing on the skin activates three primary sensory neurons, each of which releases neurotransmitters onto its corresponding secondary neuron. However, the three secondary neurons do not all respond in the same fashion.

The secondary neuron closest to the actual stimulus (pathway B) suppresses the response of the secondary neurons to its sides (lateral), where the stimulus is weaker. It simultaneously allows its own pathway to proceed without interference. The inhibition of neurons farther from the stimulus enhances the contrast between the center and sides of the receptive field, making the sensation more easily localized.

The pathway in Figure 10-6 also demonstrates **population coding**, the way multiple receptors can function together to send the CNS more information than would be possible from a single receptor. By comparing the input from multiple receptors, the CNS can make complex calculations about the spatial and temporal characteristics of a stimulus.

CONCEPT CHECK   Answers: p. 367

4. What kind of ion channel do you think that secondary neuron B (Fig. 10-6) is opening in neurons A and C to depress their responsiveness: Na^+, K^+, Cl^-, or Ca^{2+}?

Intensity and Duration of the Stimulus   The intensity of a stimulus cannot be directly calculated from a single sensory neuron action potential because a single action potential is "all-or-none," with constant amplitude and duration. Instead, stimulus intensity is coded in two types of information: by the number of receptors activated (another example of *population coding*) and by the frequency of action potentials coming from those receptors (*frequency coding*).

Population coding for intensity occurs because not all receptors have the same threshold to their preferred stimulus. Only the most sensitive receptors with the lowest thresholds will respond to a low-intensity stimulus. As a stimulus increases in intensity, additional receptors will be activated. The CNS then translates the number of active receptors into a measure of stimulus intensity.

For individual sensory neurons, intensity discrimination begins at the receptor. If a stimulus is below threshold, the primary sensory neuron does not respond. Once stimulus intensity exceeds threshold, the amplitude (strength) of the generator potential

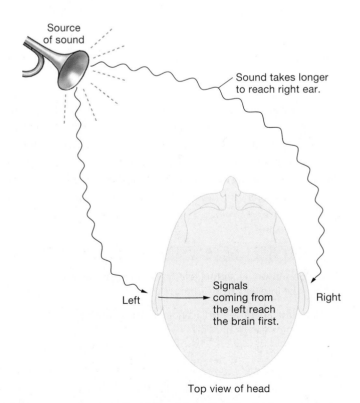

● **Figure 10-5** *Localization of sound*   The brain uses timing differences rather than specific neurons to localize sound.

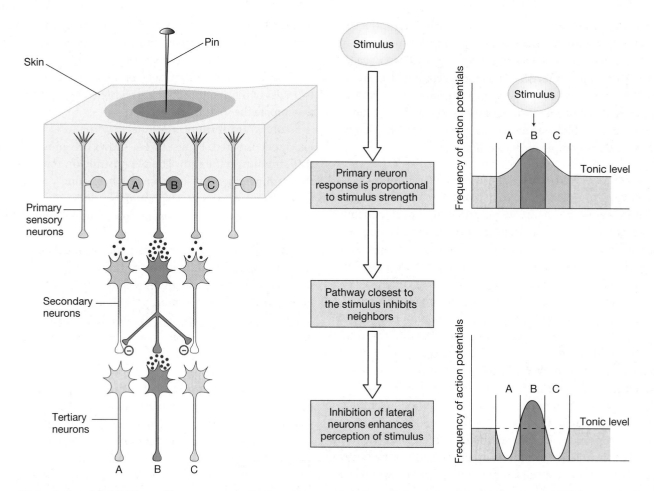

● **Figure 10-6** *Lateral inhibition* The responses of primary neurons A, B, and C are proportional to the intensity of the stimulus in each receptive field. Secondary neurons A and C are inhibited by neuron B, creating greater contrast between B and its neighbors.

increases in proportion to the intensity of the stimulus (Fig. 10-7 ●). As the strength of the generator potential increases, the frequency of action potentials in the primary sensory neuron increases.

Similarly, the duration of a stimulus is coded by the duration of action potentials in the sensory neuron. In general, the longer a stimulus persists, the longer a series of action potentials will be generated in the primary sensory neuron. However, as a stimulus persists, some receptors turn off and cease to respond. Receptors fall into one of two classes, depending on how they react to continuous stimulation.

Tonic receptors are slowly adapting receptors that continue to transmit signals to the CNS as long as the stimulus is present (Fig. 10-8 ●). Pressure-sensitive baroreceptors, nociceptors, and some tactile and proprioreceptors are among the receptors that fall into this category. In general, the stimuli that activate tonic receptors are parameters that must be monitored continuously by the body.

In contrast, **phasic receptors** are rapidly adapting receptors that fire when they first receive a stimulus but cease firing if the strength of the stimulus remains constant. Phasic receptors are attuned to changing conditions. Once a stimulus reaches a steady intensity, phasic receptors **adapt** to the new steady state and turn off. This type of response allows the body to ignore information

that has been evaluated and found not to threaten homeostasis or well-being.

Our sense of smell is an example of a sense that uses phasic receptors. For example, when you put on cologne in the morning, you can smell it, but as the day goes on, your olfactory receptors adapt and are no longer stimulated by the cologne molecules. You no longer smell the fragrance, yet others may comment on it. Adaptation of phasic receptors allows us to filter out extraneous sensory information and concentrate on what is new, different, or essential.

CONCEPT CHECK Answers: p. 367

5. How do sensory receptors tell the central nervous system about the intensity of a stimulus?

6. What is the adaptive significance of having nociceptors that are tonic instead of phasic?

What happens at a receptor during **adaptation**? There is more than one explanation, depending on the receptor type. In some cases, K^+ channels in the receptor membrane open, causing the membrane to repolarize. In other cases, accessory structures

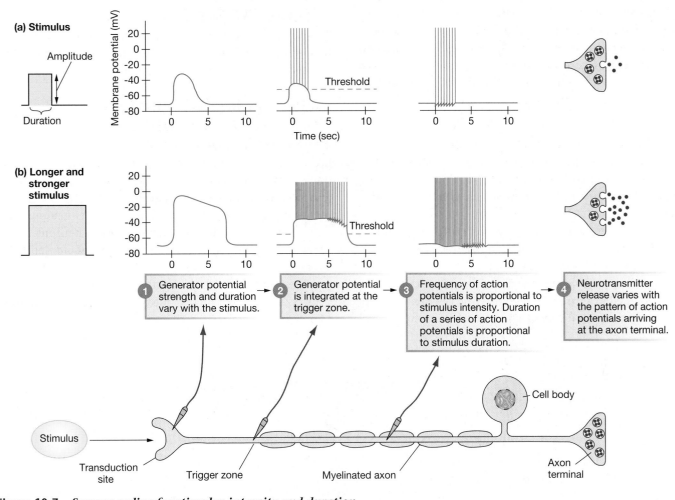

Figure 10-7 *Sensory coding for stimulus intensity and duration*

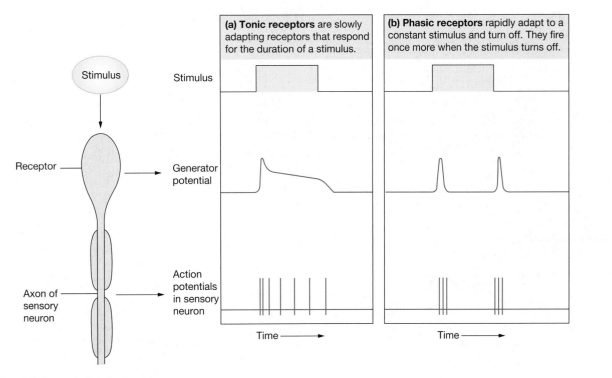

Figure 10-8 *Tonic and phasic receptors*

decrease the amount of stimulus reaching the receptor. This occurs in the ear, for example, where muscles contract and decrease the vibration of small bones in response to loud noises. In general, once adaptation of a phasic receptor has occurred, the only way to create a new signal is to change the intensity of the excitatory stimulus.

To summarize, the specificity of sensory pathways is established in several ways:

1. Each receptor is most sensitive to a particular type of stimulus.
2. A stimulus above threshold initiates action potentials in a sensory neuron that projects to the CNS.
3. Stimulus intensity and duration are coded in the pattern of action potentials reaching the CNS.
4. Stimulus location and modality are coded according to which receptors are activated or, in the case of sound and smell, by the timing of receptor activation.
5. Each sensory pathway projects to a region of the cerebral cortex dedicated to a particular receptive field (see Fig. 10-2). The brain can then tell the exact origin of each incoming signal.

▶ SOMATIC SENSES

There are four somatosensory modalities: touch, proprioception, temperature, and nociception, which includes pain and itch. We will discuss details of proprioception in Chapter 13.

Pathways for Somatic Perception Project to the Somatosensory Cortex and Cerebellum

Receptors for the somatic senses are found both in the skin and in the viscera. Receptor activation triggers action potentials in the associated **primary sensory neuron** that travels from the receptor to the CNS. In the spinal cord, many sensory neurons synapse onto interneurons called **secondary sensory neurons**. (Primary and secondary neurons are also known as *first-order* and *second-order neurons*.)

The location of the synapse between primary and secondary neurons varies according to the type of receptor (Fig. 10-9 ●). Neurons associated with receptors for nociception, temperature,

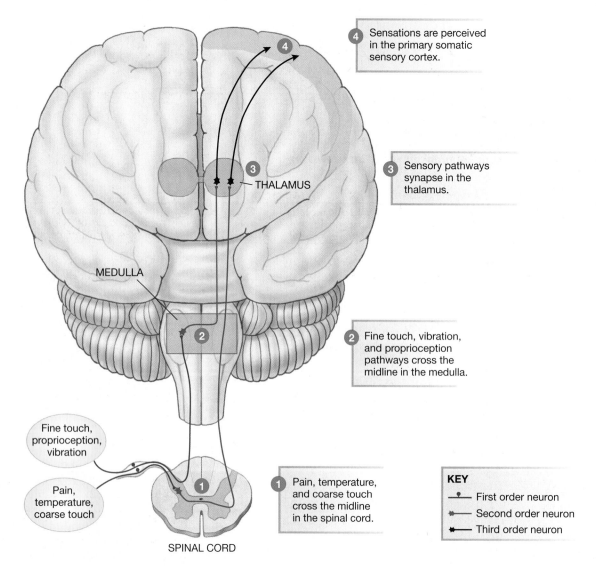

4 Sensations are perceived in the primary somatic sensory cortex.

3 Sensory pathways synapse in the thalamus.

THALAMUS

MEDULLA

2 Fine touch, vibration, and proprioception pathways cross the midline in the medulla.

Fine touch, proprioception, vibration

Pain, temperature, coarse touch

1 Pain, temperature, and coarse touch cross the midline in the spinal cord.

KEY
— First order neuron
— Second order neuron
— Third order neuron

SPINAL CORD

● **Figure 10-9** *Sensory pathways cross the body's midline*

Table 10-3: Sensory Pathways

	Stimulus	
	FINE TOUCH, PROPRIOCEPTION	IRRITANTS, TEMPERATURE, TOUCH (SECONDARY)
Primary sensory neuron terminates in:	Medulla	Dorsal horn of spinal cord
	Path crosses midline of body.	Path crosses midline of body.
Secondary sensory neuron terminates in:	Thalamus	Thalamus
Tertiary sensory neuron terminates in:	Somatosensory cortex	Somatosensory cortex

and coarse touch synapse onto their secondary neurons shortly after entering the spinal cord. In contrast, most fine touch and proprioceptive neurons have very long axons that project up the spinal cord all the way to the medulla (Table 10-3).

All secondary sensory neurons cross the midline of the body at some point, so that sensations from the left side of the body are processed in the right hemisphere of the brain and vice versa. The secondary neurons for nociception, temperature, and coarse touch cross the midline in the spinal cord, then ascend to the brain. Fine touch and proprioceptive neurons cross the midline in the medulla.

In the thalamus, secondary sensory neurons synapse onto **tertiary sensory neurons**, which in turn project to the somatosensory region of the cerebral cortex. In addition, most sensory pathways send branches to the cerebellum so that it can use the information to coordinate balance and movement.

The **somatosensory cortex** is the part of the brain that recognizes where ascending sensory tracts originate. Each sensory tract has a corresponding region of the cortex, so that all sensory pathways for the left hand terminate in one area and so on (Fig. 10-10 ●). Within the cortical region for a particular body part, columns of neurons are devoted to particular types of receptors. For example, a cortical column activated by cold receptors in the left hand may be found next to a column activated by pressure receptors in the skin of the left hand. This columnar arrangement creates a highly organized structure that maintains the association between specific receptors and the sensory modality they transmit.

Some of the most interesting research about the somatosensory cortex has been done on patients during brain surgery for epilepsy. Because brain tissue has no pain fibers, this type of surgery can be performed with the patient awake under local anesthesia. The surgeon stimulates a particular region of the brain and asks the patient about sensations that occur. The ability of the patient to communicate with the surgeon during this process has expanded our knowledge of brain regions tremendously.

Experiments can also be done on animals by stimulating peripheral receptors and monitoring electrical activity in the cortex. We have learned from these experiments that the more sensitive a region of the body is to tactile and other stimuli, the larger the corresponding region in the cortex. Interestingly, the size of the regions is not fixed. If a particular body part is used more exten-

sively, its topographical region in the cortex will expand. For example, people who are visually handicapped and learn to read Braille with their fingertips develop an enlarged region of the somatosensory cortex devoted to the fingertips. This plasticity [❧ p. 285] shows the amazing versatility of the brain.

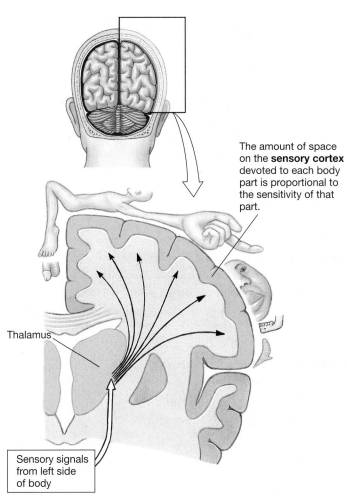

The amount of space on the **sensory cortex** devoted to each body part is proportional to the sensitivity of that part.

Thalamus

Sensory signals from left side of body

Cross section of the right cerebral hemisphere and sensory areas of the cerebral cortex

● **Figure 10-10** *The somatosensory cortex* In this illustration, representations of body parts are shown adjacent to the area of the sensory cortex that processes their stimuli.

Merkel receptors sense steady pressure and texture.

Meissner's corpuscle responds to flutter and stroking movements.

Hair

Free nerve ending

Free nerve ending of hair root senses hair movement.

Hair root

Pacinian corpuscle senses vibration.

Ruffini corpuscle responds to skin stretch.

Sensory nerves carry signals to spinal cord.

Free nerve ending of nociceptor responds to noxious stimuli.

● **Figure 10-11** *Touch-pressure receptors* Different types of touch-pressure receptors are scattered throughout the superficial and deep layers of the skin.

Touch Receptors Respond to Many Different Stimuli

Touch receptors are one of the most common receptors in the body. These receptors respond to many forms of physical contact, such as stretch, steady pressure, fluttering or stroking movement, vibration, and texture. They are found both in superficial layers of the skin (Fig. 10-11 ●) and in deeper regions of the body. Receptors associated with muscles, joints, or internal organs will be discussed in later chapters.

Touch receptors come in many forms. Some are free nerve endings, such as those that wrap around the base of hairs; others are complex receptors (Table 10-4). Most are difficult to study because of their small size. However, the Pacinian corpuscle is one of the largest receptors in the body, and much of what we know about somatosensory receptors comes from studies on these structures.

Pacinian corpuscles are composed of nerve endings encapsulated in layers of connective tissue (see Fig. 10-1b). They are found in the subcutaneous layers of skin, muscles, joints, and internal organs. The concentric layers of connective tissue in the corpuscle create large receptive fields.

Pacinian corpuscles respond best to high-frequency vibrations whose energy is transferred through the connective tissue capsule to the nerve ending, where it opens mechanically gated ion channels [● p. 131]. Pacinian corpuscles are rapidly adapting phasic receptors that turn off even if the stimulus continues (see Fig. 10-8). This property allows them to respond to a change in touch but then ignore it.

Temperature Receptors Are Free Nerve Endings

Temperature receptors are free nerve endings that terminate in the subcutaneous layers of the skin. **Cold receptors** are sensitive primarily to temperatures lower than body temperature. **Warm receptors** are stimulated by temperatures from body temperature

Table 10-4: Touch Receptors

RECEPTOR	STIMULUS	LOCATION	STRUCTURE	ADAPTATION
Free nerve endings	Various touch and pressure stimuli	Around hair roots and under surface of skin	Unmyelinated nerve endings	Variable
Meissner's corpuscle	Flutter, stroking	Superficial layers of skin	Encapsulated in connective tissue	Rapid
Pacinian corpuscle	Vibration	Deep layers of skin	Encapsulated in connective tissue	Rapid
Ruffini corpuscle	Stretch on skin	Deep layers of skin	Enlarged nerve endings	Slow
Merkel receptors	Steady pressure, texture	Superficial layers of skin	Enlarged nerve endings	Slow

(37° C) to about 45° C. Above that temperature, pain receptors are activated, creating a sensation of painful heat.

The receptive field for a thermoreceptor is about 1 mm in diameter, and the receptors are scattered across the body. There are considerably more cold receptors than warm. Temperature receptors adapt between 20° and 40° C, but outside that range, where the likelihood of tissue damage is greater, the receptors do not adapt. Temperature receptors play an important role in thermoregulation, which is discussed in Chapter 22.

Nociceptors Initiate Protective Responses

Nociceptors [*nocere*, to injure] are receptors that respond to a variety of strong noxious stimuli (chemical, mechanical, or thermal) that cause or have the potential to cause tissue damage. Nociceptors are sometimes called pain receptors, even though pain is a perceived sensation rather than a stimulus. Sensory input from nociceptors initiates adaptive, protective responses. For example, if we did not feel joint discomfort with overuse, we would damage our joints more quickly.

Nociceptors are free nerve endings that respond to a variety of chemical, mechanical, and thermal stimuli with the help of cell membrane proteins. For example, *vanilloid receptors* on nociceptors respond to damaging heat as well as to *capsaicin*, the chemical that makes hot chili peppers burn your mouth. At the opposite end of the temperature spectrum, researchers recently identified a membrane protein that responds both to cold and menthol, one reason mint-flavored foods feel cool.

One theory of nociceptor activation states that mechanical and thermal stimuli are mediated through local chemicals. We do know that a variety of molecules released with tissue injury either activate nociceptors or sensitize them by lowering their activation threshold. Chemical mediators of the pain response include K^+, histamine and prostaglandins released from damaged cells, serotonin released from platelets activated by injury, and a peptide known as **substance P** secreted by primary sensory neurons.

Nociceptors may activate two different pathways: (1) reflexive protective responses that are integrated at the level of the spinal cord and (2) ascending pathways to the cortex that become conscious sensation (pain or itch). Primary sensory neurons from nociceptors terminate in the dorsal horn of the spinal cord (see Fig. 10-9). There they synapse onto secondary sensory neurons that project to the brain or onto interneurons for local circuits.

Irritant responses that are integrated in the spinal cord initiate rapid unconscious protective reflexes that automatically remove a stimulated area from the source of the stimulus. For example, if you accidentally touch a hot stove, an automatic **withdrawal reflex** will cause you to pull back your hand even before you are aware of the heat. Reflexes that do not require integration in the brain are called **spinal reflexes**.

The lack of brain involvement in many protective reflexes has been demonstrated in the classic "spinal frog" preparation, in which the animal's brain has been destroyed by pithing. If the frog's foot is placed in a beaker of hot water or strong acid, the withdrawal reflex causes the leg to contract and move the foot away from the stimulus. The frog is unable to feel pain because the brain, which translates sensory input into perception, has been destroyed.

Pain and Itching Are Mediated by Nociceptors

Afferent signals from nociceptors are carried to the CNS in different types of primary sensory fibers (Table 10-5). Two sensations may be perceived when nociceptors are activated: pain and itch. The most common sensation is pain, but activation of a subtype of *C fiber* by histamine and other stimuli creates the sensation we call **itch** (*pruritis*). Itch comes only from nociceptors in the skin, and is characteristic of many rashes and other skin conditions. However, itch can also be a symptom of a number of systemic diseases, including multiple sclerosis, hyperparathyroidism, and diabetes mellitus. The higher pathways for itch are not as well understood as the pathways for pain.

Pain is the brain's interpretation of sensory information transmitted along pathways that begin at nociceptors. **Fast pain**, described as sharp and localized, is rapidly transmitted to the CNS by small, myelinated $A\delta$ (*A-delta*) fibers. **Slow pain**, described as duller and more diffuse, is carried on small, unmyelinated *C fibers*. The timing distinction between the two is most obvious when the stimulus originates far from the CNS, such as when you stub your toe. You first experience a quick stabbing sensation (fast pain), followed shortly by a dull throbbing (slow pain).

The ascending pathways for pain cross the midline in the spinal cord and ascend to the thalamus and sensory areas of the cortex. The pathways also send branches to the limbic system and hypothalamus. As a result, pain may be accompanied by emotional distress and a variety of autonomic reactions such as nausea, vomiting, or sweating.

Table 10-5: Classes of Somatosensory Nerve Fibers

FIBER TYPE	FIBER CHARACTERISTICS	SPEED OF CONDUCTION	ASSOCIATED WITH
Aβ (beta)	Large, myelinated	30–70 m/sec	Mechanical stimuli
Aδ (delta)	Small, myelinated	12–30 m/sec	Cold, fast pain, mechanical stimuli
C	Small, unmyelinated	0.5–2 m/sec	Slow pain, temperature, mechanical stimuli

Our perception of pain is subject to modulation at several levels in the nervous system. It can be magnified by past experiences or suppressed in emergencies when survival depends on ignoring injury. In the latter situation, descending pathways travel through the thalamus and inhibit nociceptor neurons in the spinal cord. Artificial stimulation of these inhibitory pathways is one of the newer techniques being used to control chronic pain.

Pain can also be suppressed in the dorsal horn of the spinal cord, before the stimuli are sent to ascending spinal tracts. Normally, tonically active interneurons within the spinal cord inhibit ascending pathways for pain (Fig. 10-12a●). C fibers from nociceptors synapse on the inhibitory neurons. When activated

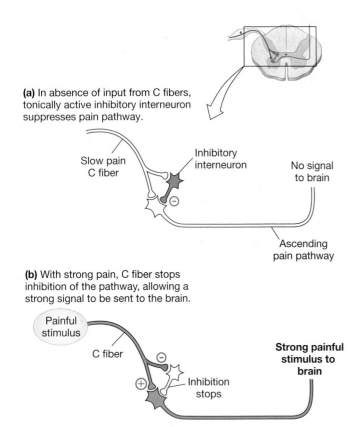

(a) In absence of input from C fibers, tonically active inhibitory interneuron suppresses pain pathway.

Slow pain C fiber

Inhibitory interneuron

No signal to brain

Ascending pain pathway

(b) With strong pain, C fiber stops inhibition of the pathway, allowing a strong signal to be sent to the brain.

Painful stimulus

C fiber

Inhibition stops

Strong painful stimulus to brain

(c) Pain can be modulated by simultaneous somatosensory input.

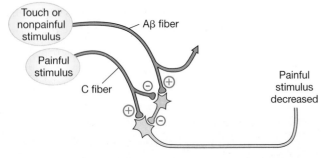

Touch or nonpainful stimulus

Aβ fiber

Painful stimulus

C fiber

Painful stimulus decreased

● **Figure 10-12** *The gate control theory of pain modulation* In the gate control model, nonpainful stimuli can diminish the pain signal.

by a painful stimulus, the C fibers simultaneously excite the ascending path and block the tonic inhibition (Fig. 10-12b). This action allows the pain signal from the C fiber to travel unimpeded to the brain.

In the **gate control theory** of pain modulation, *Aβ fibers* carrying sensory information about mechanical stimuli help block pain transmission. The Aβ fibers synapse on the inhibitory neurons and *enhance* the interneuron's inhibitory activity. If simultaneous stimuli reach the inhibitory neuron from the Aβ and C fibers, the integrated response is partial inhibition of the ascending pain pathway. The pain perceived by the brain is lessened (Fig. 10-12c). You use the gate control theory when you rub a bumped elbow or shin. The tactile stimulus of rubbing activates Aβ fibers and helps decrease the sensation of pain.

Pain can be felt in skeletal muscles (*deep somatic pain*) as well as in the skin. Muscle pain during exercise is associated with the onset of anaerobic metabolism. It is often perceived as a burning sensation within the muscle (as in "go for the burn!"), apparently created by a metabolite released during anaerobic metabolism. Some investigators have suggested that the metabolite is K^+, known to enhance the pain response. Muscle pain from **ischemia** (lack of adequate blood flow that reduces oxygen supply) also occurs during *myocardial infarction* (heart attack).

Pain in the heart and other internal organs (*visceral pain*) is often poorly localized and may be felt in areas far removed from the site of the stimulus (Fig. 10-13a●). For example, the pain of cardiac ischemia may be felt in the neck and down the left shoulder and arm. This **referred pain** apparently occurs because multiple primary sensory neurons converge onto a single ascending tract (Fig. 10-13b). According to this model, when painful stimuli arise in visceral receptors, the brain is unable to distinguish visceral signals from the more common signals arising from somatic receptors. As a result, it interprets the pain as coming from the somatic regions rather than the viscera.

Chronic pain of one sort or another affects millions of people in this country every year, and the alleviation of pain is of considerable interest to health professionals. **Analgesic drugs** [*analgesia*, painlessness] range from aspirin to potent opiates such as morphine. Aspirin inhibits prostaglandins and presumably slows the transmission of pain signals from the site of injury.

The opiate drugs such as *morphine* act directly on *opioid receptors* in the brain and spinal cord. The opioid receptors are part of an analgesic system that responds to endogenous opiate molecules. *Enkephalins* and *dynorphin* are found in neurons associated with the pain pathways. β-endorphin is produced from the same prohormone as ACTH (adrenocorticotropin) in neuroendocrine cells of the hypothalamus [● Fig. 7-4c, p. 216]. Activation of opioid receptors blocks pain perception by decreasing neurotransmitter release from primary sensory neurons and by postsynaptic inhibition of the secondary sensory neurons.

Although opiates are effective at relieving pain, they are also highly addictive. In cases of severe chronic pain, it is sometimes possible to get surgical relief by severing afferent nerves at the

(a) Pain in internal organs is often sensed on the surface of the body, a sensation known as **referred pain**.

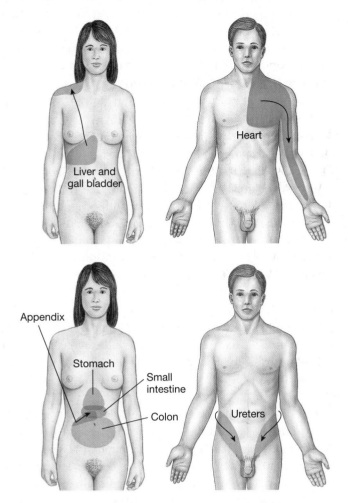

● **Figure 10-13**   *Referred pain*

(b) One theory of referred pain says that nociceptors from several locations converge on a single ascending tract in the spinal cord. Pain signals from the skin are more common than pain from internal organs, and the brain associates activation of the pathway with pain in the skin. Adapted from H.L. Fields, *Pain* (McGraw Hill, 1987).

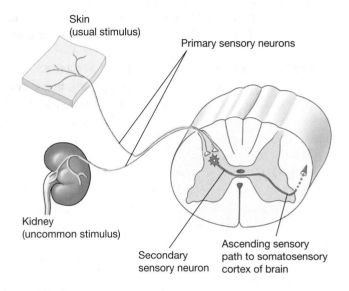

dorsal root or by stimulating the inhibitor pathways of the brain. Acupuncture can also be effective, although the physiological reason for its effectiveness is not clear.

7. What is the adaptive advantage of a spinal reflex?

8. From an adaptive standpoint, why would touch and proprioception move through faster sensory pathways than would pain and temperature information?

9. Our sense of smell uses phasic receptors. Can you think of other receptors (senses) that adapt to ongoing stimuli?

▶ CHEMORECEPTION: SMELL AND TASTE

The five special senses of smell, taste, hearing, equilibrium, and vision are concentrated in the head region. Like somatic senses, the special senses rely on receptors to transform information about the environment into patterns of action potentials that can

CLINICAL FOCUS

Pain Control

Acupuncture is a centuries-old Chinese therapy that is based on the manipulation of needles inserted in special locations on the body. In the Chinese tradition, acupuncture enhances Qi (pronounced *chee*), a form of energy that promotes healing. Western medicine for many years scoffed at acupuncture, claiming that its results were due to the placebo effect: if the patient believes that a treatment will work, in many cases, it does. However, in recent years, evidence has been accumulating that acupuncture has a physiological basis. The leading theory on how acupuncture works proposes that properly placed acupuncture needles trigger the release of endorphins (morphine-like pain suppressants) by the brain. Several leading medical schools have conducted or are participating in controlled clinical trials on the effectiveness of acupuncture. Although much of this research is directed at pain relief, acupuncture may also be useful for repairing neural damage following strokes and for treating drug and alcohol addictions. To learn more about acupuncture research and theories, go to the website for the National Center for Complementary and Alternative Medicine *http://nccam.nih.gov/health/acupuncture/*

be interpreted by the brain. *Chemoreception* is one of the oldest senses from an evolutionary perspective. Primitive animals without formalized nervous systems use chemoreception to locate food and mates. Research on bacteria even indicates that they use chemoreception as an integral part of sensing their environment. Chemoreception is probably the basis for chemical synaptic communication. In humans, chemoreceptive senses have been refined into the special senses of smell (olfaction) and taste (gustation).

Olfaction Is One of the Oldest Senses

Imagine waking up one morning and discovering a whole new world around you, a world filled with odors that you never dreamed existed—scents that told you more about your surroundings than you ever imagined from looking at them. This is exactly what happened to a young patient of Dr. Oliver Sacks (an account is in *The Man Who Mistook His Wife for a Hat and Other Clinical Tales*). Or imagine skating along the sidewalk without a helmet, only to fall and hit your head. When you regain consciousness, the world has lost all odor: no smell of grass or perfume or garbage. Even your food has lost much of its taste, and you now eat only to survive because eating has lost its pleasure.

We do not realize the essential role that our sense of smell plays in our lives until a head cold or injury robs us of the ability to smell. **Olfaction** [*olfacere*, to sniff] allows us to discriminate between thousands of different odors. Even so, our noses are not nearly as sensitive as those of many other animals whose survival depends on olfactory cues. The **olfactory bulb** (Fig. 10-14b●), the extension of the forebrain that receives input from the primary olfactory neurons, is much better developed in vertebrates whose survival is more closely linked to chemical monitoring of their environment (see Fig. 9-1, p. 286).

The human olfactory system consists of primary sensory neurons (olfactory cells) that synapse with secondary sensory neurons in the olfactory bulb. Secondary and higher-order neurons project from the olfactory bulb through the *first cranial nerve* to the *olfactory cortex*. In addition, olfactory pathways lead to the amygdala and hippocampus, parts of the limbic system involved with emotion and memory.

One amazing aspect of olfaction is the link between smell, memory, and emotion. A special cologne or aroma of food can trigger memories and create a wave of nostalgia for the time, place, or people with whom the aroma is associated. In some way that we do not understand, the processing of odors through the limbic system creates deeply buried olfactory memories. Particular combinations of olfactory receptors become linked to other patterns of sensory experience so that stimulating one pathway stimulates them all.

An accessory olfactory structure in the nasal cavity, the **vomeronasal organ** (VNO), is known in rodents to be involved in behavioral responses to sex pheromones. A *pheromone* is an air- or waterborne signaling molecule used for communication between individual organisms. The existence of pheromones in humans

has been postulated but never proved [● p. 212]. Anatomical studies in humans indicate that a VNO is present in the early embryo but it appears to regress during development. A VNO-like pit can be found in some adult humans, but it has no evident sensory cells and no discernable neural connection to the brain.

Olfactory Cells Are Neurons

Olfactory cells in humans are concentrated in a 3-cm^2 patch of **olfactory epithelium** high in the nasal cavity (Fig. 10-14a). The cells are neurons whose processes extend on one side to the surface of the olfactory epithelium and on the other to the olfactory bulb, located on the underside of the frontal lobe. Olfactory cells, unlike other neurons in the body, are continuously dividing cells, with a turnover time of about two months (Fig. 10-14c). This means that the axon of each new neuron must find its way to the olfactory bulb and make the proper synaptic connection. Discovering how these neurons manage to repeat the same connection each time will give us insight on how developing neurons find their targets.

The surface of the olfactory epithelium is composed of the knobby terminals of the olfactory cells, each knob sprouting multiple nonmobile cilia (Fig. 10-14c). The cilia are embedded in a layer of mucus, and odorant molecules must first dissolve in and penetrate the mucus before they can bind to an odorant receptor protein.

Odorant receptors are G protein-cAMP-linked membrane receptor proteins [● p. 179]. Studies in rats suggest that there are 1000 or more related receptors in the odorant receptor family. The combination of an odorant with its receptor activates a special G protein, G_{olf}, which in turn increases intracellular cAMP. This opens cAMP-gated cation channels, depolarizing the cell and triggering a signal that travels along the olfactory pathway to the brain.

What is occurring at the cellular and molecular level that allows us to discriminate between thousands of different odors? Current research suggests olfactory coding is accomplished by thousands of odorant receptors with sensitivity to a variety of substances. The brain uses information from hundreds of different olfactory cells in different combinations to create the perception of many different smells. This is another example of *population coding* in the nervous system.

CONCEPT CHECK   Answers: p. 367

10. Create a map or diagram of the olfactory pathway from the olfactory cell to the olfactory cortex.

11. Create a map or diagram that starts with an odorant binding to its membrane receptor and ends with neurotransmitter release from the primary olfactory neuron.

12. What part(s) of an olfactory cell would be the dendrites?

13. Are olfactory neurons pseudounipolar, bipolar, or multipolar? (Hint: Fig. 8-3, p. 243)

STRUCTURE OF THE OLFACTORY SYSTEM

(a) The olfactory epithelium lies high within the nasal cavity and its olfactory cells project to the olfactory bulb.

Olfactory tract

Olfactory bulb

(b) The olfactory cells synapse with secondary sensory neurons in the olfactory bulb.

Olfactory bulb

2° Sensory neurons

Bone

1° Sensory neurons (olfactory cells)

Olfactory epithelium

(c) Olfactory cells in the olfactory epithelium live only about two months. They are replaced by new cells whose axons must find their way to the olfactory bulb.

Olfactory cell axons carry information to olfactory bulbs.

Lamina propria

Regenerative basal cell divides to replace worn-out olfactory cells.

Developing olfactory cell

Olfactory cell

Supporting cell

Olfactory cilia contain chemoreceptors.

Mucus layer
Odorant molecules must dissolve in this layer.

FIGURE QUESTION

Multiple primary neurons in the epithelium synapsing on one secondary neuron in the olfactory bulb is an example of what principle?

● **Figure 10-14** *ANATOMY SUMMARY* *Olfaction*

Taste Is a Combination of Five Basic Sensations

Taste is closely linked to olfaction. Indeed, much of what we call the taste of food is actually the aroma, as you know if you have ever had a bad cold. Although smell is sensed by hundreds of different types of receptors, taste is a combination of five sensations: sweet, sour, salty, bitter, and *umami*, a taste associated with the amino acid glutamate and some nucleotides. **Umami**, a name derived from the Japanese word for "deliciousness," is a basic taste that enhances the flavor of foods. It is the reason that monosodium glutamate (MSG) is used as a food additive in some countries.

Interestingly, although we have only a limited number of taste sensations, each one is associated with an essential body function. Sour taste is triggered by H^+ and salty by Na^+, two ions whose concentrations in body fluids are closely regulated. The other three taste sensations result from organic molecules. Sweet and umami are associated with nutritious food. Bitter taste is recognized by the body as a warning of possibly toxic components. If something tastes bitter, our first reaction is often to spit it out.

The receptors for taste are located primarily on **taste buds** clustered together on the surface of the tongue (Fig. 10-15a ●). These sensory organs are composed of 50–150 taste cells, support cells, and regenerative basal cells (Fig. 10-15b, c). Taste receptors are also scattered through other regions of the oral cavity, such as the palate.

Taste cells are non-neural polarized epithelial cells [● p. 129], tucked down into the epithelium so that only a tiny tip protrudes through the *taste pore*. Tight junctions link the apical ends of the cells together, limiting movement of molecules between the cells. The apical membrane that extends into the oral cavity is modified into microvilli to increase surface area in contact with the environment. The basal side of the cell forms a synapse with the primary sensory neuron.

For a substance to be tasted, it must first dissolve in the saliva and mucus of the mouth. Taste ligands then interact with an apical membrane protein (Fig. 10-15c). Although the details differ for the five different taste sensations, they all follow the same basic pattern. Ligand-protein creates a calcium signal, either by allowing Ca^{2+} influx from the extracellular fluid or Ca^{2+} release from intracellular stores. The Ca^{2+} signal within the cell triggers exocytosis of neurotransmitter [● p. 263] into the synapse and initiates a series of action potentials in the primary sensory neuron.

The details of taste transduction are somewhat controversial. They are a good example of how our models of physiological

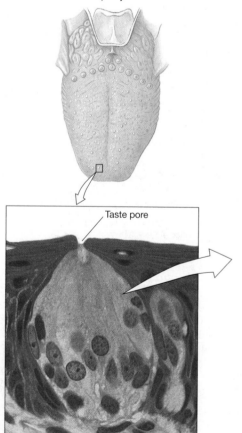

(a) Taste buds are located on the dorsal surface of the tongue. The umami receptors (not shown) are in the back of the pharynx.

Taste pore

(b) A light micrograph of a taste bud. Each taste bud is composed of taste cells and support cells, joined near the apical surface with tight junctions.

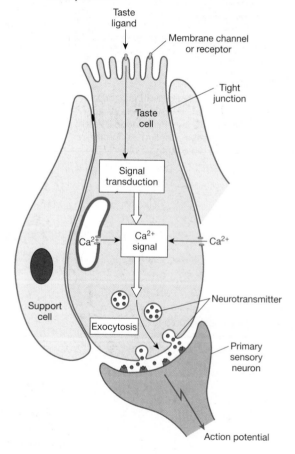

(c) Taste ligands bind to receptors and create Ca^{2+} signals that release neurotransmitters onto primary sensory neurons.

Taste ligand

Membrane channel or receptor

Tight junction

Taste cell

Signal transduction

Ca^{2+}

Ca^{2+} signal

Ca^{2+}

Support cell

Neurotransmitter

Exocytosis

Primary sensory neuron

Action potential

● **Figure 10-15** *Taste buds*

function must periodically be revised as new research data are published. Until recently, the widely held view of taste transduction was that an individual taste cell could sense more than one taste, with cells differing in their sensitivities. However, recently published research using molecular biology and knockout mice suggests that each taste cell is sensitive to one taste. Bitter, sweet, and umami ligands bind to membrane receptors and activate second messenger systems. Ions indicating salty (Na^+) and sour (H^+) taste enter the taste cell through ion channels (Fig. 10-16●).

The details of signal transduction for bitter, sweet, and umami, which once appeared to be relatively straighforward, are apparently more complex than scientists initially thought. The three ligands bind to different G protein-linked receptors, including about 30 variants of bitter receptors. The G proteins appear to be linked to several different signal transduction pathways. Some release Ca^{2+} from intracellular stores while others ultimately open cation channels and allow Ca^{2+} to enter the cell. In any case, the final step in the pathways is a Ca^{2+} signal that causes exocytosis of neurotransmitter, just as occurs in an axon terminal.

Salty and sour transduction is mediated by ion channels rather than by G protein-linked receptors. The channels allow Na^+ or H^+ to enter and depolarize the taste cell. The sour ligand H^+ also binds to and closes K^+ channels, which depolarizes the cell. Depolarization opens a voltage-gated Ca^{2+} channel and creates a Ca^{2+} signal as previously described.

Neurotransmitters from taste cells activate primary sensory neurons whose axons run through cranial nerves VII, IX, and X to the medulla, where they synapse. Sensory information then passes through the thalamus to the *gustatory cortex* (see Fig. 10-4). Central processing of sensory information compares

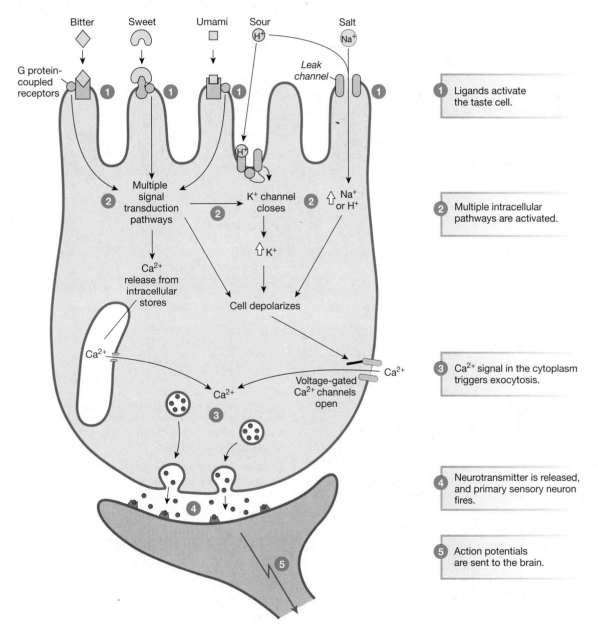

● **Figure 10-16** *Summary of taste transduction* Bitter, sweet, and umami ligands bind to G protein–coupled membrane receptors. Ionic ligands for salt and sour taste enter through ion channels. One taste cell senses one type of ligand, not all five as shown in this figure.

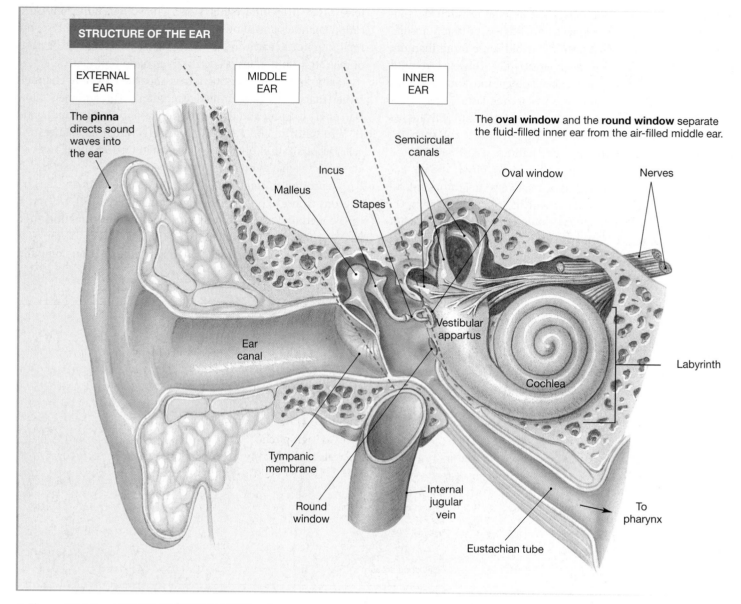

STRUCTURE OF THE EAR

EXTERNAL EAR

The **pinna** directs sound waves into the ear

MIDDLE EAR

INNER EAR

The **oval window** and the **round window** separate the fluid-filled inner ear from the air-filled middle ear.

Semicircular canals

Incus

Malleus

Stapes

Oval window

Nerves

Ear canal

Vestibular appartus

Cochlea

Labyrinth

Tympanic membrane

Round window

Internal jugular vein

To pharynx

Eustachian tube

● **Figure 10-17** *ANATOMY SUMMARY The Ear*

the input from multiple taste cells and interprets the taste sensation based on which populations of neurons are responding most strongly.

An interesting psychological aspect of taste is the phenomenon named **specific hunger**. Humans and other animals who are lacking a particular nutrient may develop a craving for that substance. **Salt appetite**, representing a lack of Na^+ in the body, has been recognized for years. Hunters have used their knowledge of this specific hunger to stake out salt licks because they know that animals will seek them out. Salt appetite is directly related to Na^+ concentration in the body and cannot be assuaged by ingestion of other cations such as Ca^{2+} or K^+. Other appetites, such as craving chocolate, are more difficult to relate to specific nutrient needs and are probably complex mixtures of physical, psychological, environmental, and cultural influences.

CONCEPT CHECK   Answers: p. 367

14. With what essential body function is the umami taste sensation associated?

15. Map or diagram the neural pathway from the taste cell to the gustatory cortex.

▶ THE EAR: HEARING

The ear is a sense organ that is specialized for two distinct functions: hearing and equilibrium. It can be divided into external, middle, and inner sections, with the neurological elements housed in and protected by structures in the inner ear. The vestibular complex of the inner ear is the primary sensor for equilibrium. The remainder of the ear is used for hearing.

The *external ear* consists of the outer ear, or **pinna**, that leads into the **ear canal** (Fig. 10-17 ●). The pinna is another example of an important accessory structure to a sensory system, and it varies in shape and location from species to species, depending on the animals' survival needs. The ear canal is sealed at its internal end by the **tympanic membrane**, or *eardrum*, a thin membranous sheet of tissue that separates the external ear from the middle ear.

The *middle ear* is an air-filled cavity that connects with the nasopharynx through the **eustachian tube**. The eustachian tube is normally collapsed, sealing off the middle ear, but it opens transiently to allow middle ear pressure to equilibrate with atmospheric pressure during chewing, swallowing, and yawning. Colds or infections that cause swelling can block the eustachian tube and result in fluid buildup in the middle ear. If bacteria are trapped in the middle ear fluid, the ear infection known as *otitis media* [*oto-*, ear + *-itis*, inflammation + *media*, middle] results.

Three small bones of the middle ear conduct sound for hearing: the **malleus** [hammer], the **incus** [anvil], and the **stapes** [stirrup]. The three bones are connected to one another with the biological equivalent of hinges. One end of the malleus is attached to the tympanic membrane, and the stirrup end of the stapes is attached to a membrane that separates the middle ear from the inner ear.

The inner ear consists of two major sensory structures. The *vestibular apparatus* with its *semicircular canals* is the sensory transducer for our sense of equilibrium. It will be described in the following section. The **cochlea** of the inner ear contains sensory receptors for hearing. On external view the cochlea is a membranous tube that lies coiled like a snail shell within a bony cavity called the **labyrinth**. Two membranous disks, the *oval window* and the *round window*, separate the fluid-filled cochlea from the air-filled middle ear. Branches of cranial nerve VIII, the *vestibulocochlear nerve*, lead from the inner ear to the brain.

Hearing Is Our Perception of Sound

Hearing is our perception of the energy carried by sound waves. *Sound waves* are alternating pressure waves with peaks of compressed air (or water) and valleys where the molecules are farther apart (Fig. 10-18a ●). *Sound* is our interpretation of the amplitude, frequency, and duration of those waves. The classic question about hearing is, "If a tree falls in the forest with no one to hear, will it make a noise?" The physiological answer is no. The falling tree will emit sound waves, but there is no noise unless someone or something is present to process and perceive the wave energy as sound.

We characterize sound waves by their pitch and their loudness, two properties that are psychophysiological interpretations of what we hear. We interpret **pitch** as high and low sounds, such as the screech of a fingernail on a blackboard (high pitch) or the rumble of distant thunder (low pitch). Physically, pitch is a function of the frequency of sound waves. Our brain interprets low-frequency waves as low-pitched sounds and high-frequency waves as high-pitched sounds.

Sound wave *frequency* is measured in waves per second, or **hertz (Hz)** (Figure 10-18b). The average human ear can hear

(a) Sound waves are alternating pressure waves, with peaks of compressed air and valleys where the molecules are farther apart.

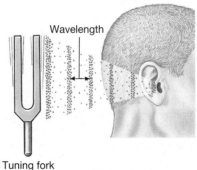

Wavelength

Tuning fork

Sound waves are distinguished by their amplitude, measured in decibels (dB), and frequency, measured in hertz (Hz).

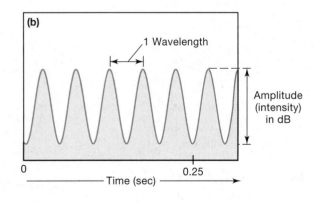

(b)

1 Wavelength

Amplitude (intensity) in dB

0   0.25
⟵ Time (sec) ⟶

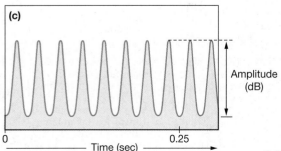

(c)

Amplitude (dB)

0   0.25
⟵ Time (sec) ⟶

●●●●●●●●

FIGURE QUESTIONS

• What are the frequencies of the sound waves in graphs (b) and (c) in Hz (waves/second)?
• Which set of sound waves would be interpreted as having lower pitch?

● **Figure 10-18** *Sound waves*

sounds over the range of 20–20,000 Hz, with the most acute hearing between 1000–3000 Hz. Our hearing is not as acute as that of many other animals, just as our sense of smell is less acute. Bats listen for ultra–high-frequency sound waves (in the kilohertz range) that bounce off objects in the dark. Elephants and some birds can hear sounds in the infrasound (very low frequency) range.

Loudness is our interpretation of sound intensity and will be influenced by the sensitivity of an individual's ear. The *intensity* of a sound wave is a function of the wave *amplitude*. Intensity is measured with a logarithmic scale in units called **decibels (dB)** (Fig. 10-18b). Each 10-dB increase represents a 10-fold increase in intensity. Normal conversation has a typical noise level of about 60 dB. Sounds of 80 dB or more can damage the sensitive hearing receptors of the ear, resulting in hearing loss. A typical heavy metal rock concert will have noise levels around 120 dB, an intensity that puts listeners in immediate danger of damage to their hearing. The amount of damage depends on the duration and frequency of the noise as well as its intensity.

16. What is a kilohertz?

Sound Transduction Is a Multistep Process

Hearing is a complex sense that involves the transduction of energy from air waves into (1) mechanical vibrations, then (2) fluid waves, (3) chemical signals, and finally (4) action potentials. This section summarizes the process. We will then look in detail at signal transduction in the cochlea.

Sound waves striking the outer ear are directed down the ear canal until they hit the tympanic membrane and are converted into vibrations (transduction 1; Fig. 10-19 ●). One end of the malleus, the first of the middle ear bones, is attached to the tympanic membrane so that vibrations are transferred to the malleus, the incus, and the stapes in that sequence. The arrangement of the three middle ear bones creates a lever that multiplies the force

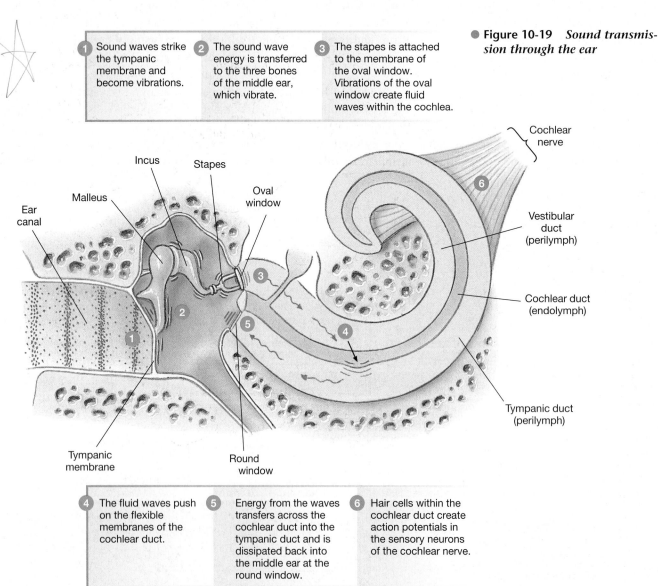

1 Sound waves strike the tympanic membrane and become vibrations.

2 The sound wave energy is transferred to the three bones of the middle ear, which vibrate.

3 The stapes is attached to the membrane of the oval window. Vibrations of the oval window create fluid waves within the cochlea.

● **Figure 10-19** *Sound transmission through the ear*

4 The fluid waves push on the flexible membranes of the cochlear duct.

5 Energy from the waves transfers across the cochlear duct into the tympanic duct and is dissipated back into the middle ear at the round window.

6 Hair cells within the cochlear duct create action potentials in the sensory neurons of the cochlear nerve.

Labels in figure: Cochlear nerve, Incus, Stapes, Malleus, Oval window, Ear canal, Vestibular duct (perilymph), Cochlear duct (endolymph), Tympanic duct (perilymph), Tympanic membrane, Round window of the vibration (*amplification*) so that very little sound energy is lost due to friction. If noise levels are so high that there is danger of damage to the inner ear, small muscles in the middle ear can pull on the bones to decrease their movement and dampen sound transmission to some degree.

When the last bone in the series, the stapes, vibrates, it pulls and pushes on the thin tissue of the **oval window**, to which it is attached. Vibrations at the oval window are converted into fluid waves in the channels of the cochlea (transduction 2). As fluid waves move through the cochlea, they push on the flexible membranes of the *cochlear duct* and activate sensory *hair cells* inside. Because fluid is not compressible, energy of the fluid waves dissipates back into the air of the middle ear at the **round window**.

Activated hair cells release neurotransmitter onto primary sensory neurons inside the cochlea (the chemical signal, transduction 3). The neurotransmitter initiates action potentials in the sensory neurons (transduction 4). Coded information about sound then travels from the cochlea through the *cochlear nerve* to cranial nerve VIII and the brain.

The Cochlea of the Inner Ear Is Filled with Fluid

The transduction of fluid waves into action potentials takes place in the cochlea of the inner ear. Uncoiled, the cochlea can be seen to be composed of three parallel, fluid-filled channels: (1) the **vestibular duct**, or *scala vestibuli* [*scala*, stairway; *vestibulum*, entrance]; (2) the central **cochlear duct**, or *scala media* [*media*, middle]; and (3) the **tympanic duct**, or *scala tympani* [tympanon, drum] (Fig. 10-20 ●). The vestibular and tympanic ducts are continuous, and they connect at the tip of the cochlea through a small opening known as the **helicotrema** [*helix*, a spiral + *trema*, hole]. The cochlea is a dead-end tube but it connects to the vestibular apparatus through a small opening.

The fluid contained within the vestibular and tympanic ducts is similar in ion composition to plasma and is known as **perilymph**. The cochlear duct is filled with **endolymph** secreted by epithelial cells in the duct. Endolymph is unusual because it is more like intracellular fluid than extracellular fluid in its composition, with high concentrations of K^+ and low concentrations of Na^+.

The cochlear duct contains the **organ of Corti**, composed of **hair cell** receptors and support cells. The organ of Corti sits on the **basilar membrane** and is partially covered by the **tectorial membrane**, both flexible tissues that move in response to fluid waves in the vestibular duct (see Fig. 10-20). As fluid waves travel through the cochlea, they displace basilar and tectorial membranes, creating up and down oscillations that bend the hair cells.

Hair cells, like taste cells, are non-neural receptor cells. The apical surface of hair cells is modified into 50–100 stiffened cilia known as **stereocilia**, arranged in ascending height (chapter opening photo and Fig. 10-21a ●). The longest stereocilia of the hair cells are embedded in the overlying tectorial membrane. If the tectorial membrane moves, the underlying cilia do also.

When hair cells move, their stereocilia flex, first one way, then the other. The stereocilia are attached to each other by protein bridges. These bridges act like little springs and are connected to "trap doors" that open and close ion channels in the cilia membrane. When the basilar membrane and cilia are in a neutral position, about 10% of the ion channels are open and there is a low level of tonic neurotransmitter release onto the primary sensory neuron.

When waves deflect the tectorial membrane so that the cilia bend toward the tallest members of a bundle, the bridges pop more channels open, cations enter the cell, and it depolarizes (Fig. 10-21b). Voltage-gated Ca^{2+} channels open, neurotransmitter release increases, and the sensory neuron increases its firing rate.

When the basilar membrane pushes the cilia away from the tallest members, the springy bridges relax and all the ion channels close. Cation influx slows, the membrane hyperpolarizes, less transmitter is released, and the sensory neuron firing decreases (Fig. 10-21c).

The vibration pattern of waves reaching the inner ear is thus transformed into a pattern of action potentials going to the CNS. Because tectorial membrane vibrations reflect the sound wave frequency, the hair cells and sensory neurons must be able to respond to sounds of nearly 20,000 waves per second, the highest frequency audible by a human ear.

Sounds Are Processed First in the Cochlea

The auditory system processes sound waves so that they can be discriminated by pitch, loudness, duration, and location. Localization of sound is a higher function that requires sensory input from both ears coupled with sophisticated computation by the brain (see Fig. 10-5). Initial processing for pitch, loudness, and duration takes place in the cochlea of each ear.

RUNNING PROBLEM

Anant reports to the otolaryngologist that his attacks of dizziness strike without warning and last from 10 minutes to an hour. They often cause him to vomit. He also reports that he has a persistent low buzzing sound in one ear and that he does not seem to hear low tones as well as he could before. The buzzing sound (tinnitus) often gets worse during his dizzy attacks.

Question 2: Subjective tinnitus occurs when an abnormality somewhere along the anatomical pathway for hearing causes the brain to perceive a sound that does not exist outside the auditory system. Starting from the ear canal, name the auditory structures in which problems may arise.

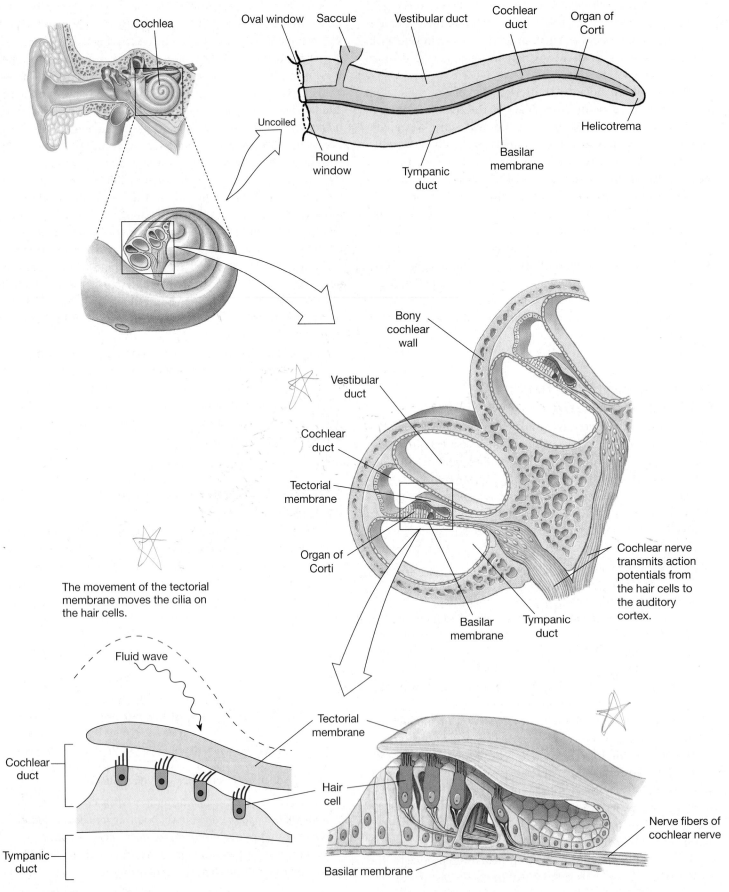

Cochlea

Oval window Saccule Vestibular duct Cochlear duct Organ of Corti

Uncoiled

Round window Tympanic duct Basilar membrane Helicotrema

Bony cochlear wall

Vestibular duct

Cochlear duct

Tectorial membrane

Organ of Corti

Basilar membrane Tympanic duct

Cochlear nerve transmits action potentials from the hair cells to the auditory cortex.

The movement of the tectorial membrane moves the cilia on the hair cells.

Fluid wave

Cochlear duct

Tympanic duct

Tectorial membrane

Hair cell

Basilar membrane

Nerve fibers of cochlear nerve

● **Figure 10-20** *The cochlea*

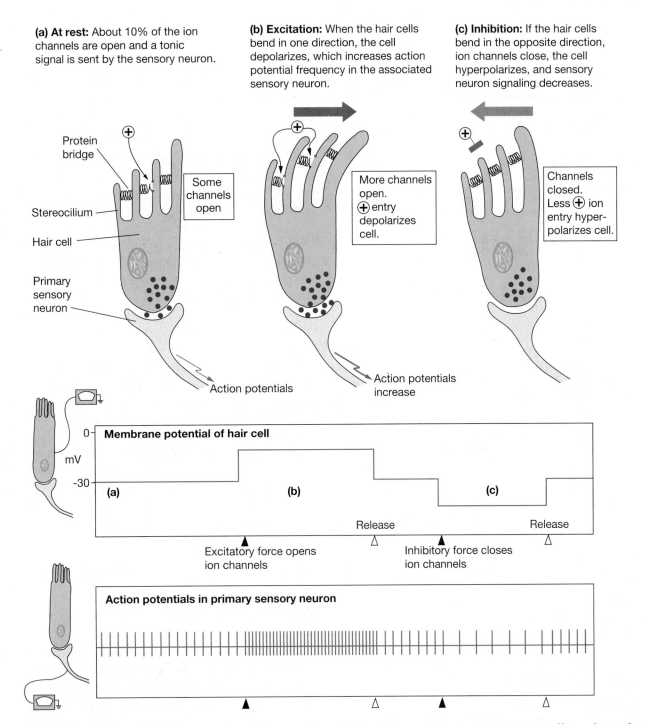

(a) At rest: About 10% of the ion channels are open and a tonic signal is sent by the sensory neuron.

(b) Excitation: When the hair cells bend in one direction, the cell depolarizes, which increases action potential frequency in the associated sensory neuron.

(c) Inhibition: If the hair cells bend in the opposite direction, ion channels close, the cell hyperpolarizes, and sensory neuron signaling decreases.

Protein bridge

Stereocilium

Hair cell

Primary sensory neuron

Some channels open

More channels open. ⊕ entry depolarizes cell.

Channels closed. Less ⊕ ion entry hyperpolarizes cell.

Action potentials

Action potentials increase

Membrane potential of hair cell

0

mV

-30

(a)

(b)

Release

(c)

Release

Excitatory force opens ion channels

Inhibitory force closes ion channels

Action potentials in primary sensory neuron

● **Figure 10-21** *Signal transduction in hair cells* The stereocilia of hair cells have "trap doors" that close off ion channels.

Coding sound for pitch is primarily a function of the basilar membrane. This membrane is stiff and narrow near its attachment by the round and oval windows but widens and becomes more flexible near its distal end (Fig. 10-22 ●). High-frequency waves entering the vestibular duct create maximum displacement of the basilar membrane closer to the oval window and are not transmitted very far along the cochlea. Low-frequency waves travel along the length of the basilar membrane and create their maximum displacement near the flexible distal end.

This differential response to frequency transforms the temporal aspect of frequency (number of sound waves per second) into spatial coding by location along the basilar membrane. This spatial coding is preserved in the auditory cortex as neurons project from the basilar membrane to corresponding regions in the brain.

Loudness is coded by the ear in the same way that signal strength is coded in somatic receptors. The louder the noise, the more rapidly action potentials fire in the sensory neuron.

(a) The basilar membrane has variable sensitivity to sound wave frequency along its length.

(b) The frequency of sound waves determines the displacement of the basilar membrane. The location of active hair cells creates a code that the brain translates as information about the pitch of sound.

● **Figure 10-22** *Sensory coding for pitch*

Auditory Pathways Project to the Auditory Cortex

Once the cochlea transforms sound waves into electrical signals, primary sensory neurons transfer this information to the brain. The cochlear nerve becomes part of cranial nerve VIII, and its neurons project to nuclei in the medulla oblongata. At that point, sound from each ear is projected to both *ipsilateral* (same side) and *contralateral* (opposite side) nuclei. As a result, each side of the brain gets information from both ears. Ascending tracts from the medulla divide and go in three directions. The main auditory pathway synapses in nuclei in the midbrain and thalamus before projecting to the auditory cortex (see Fig. 10-4). Collateral pathways take information to the reticular formation and the cerebellum.

The localization of a sound source is an integrative task for the CNS that requires simultaneous input from both ears. Unless sound is coming from directly in front of a person, it will not reach both ears at the same time (see Fig. 10-5). The brain records the time differential for sound arriving at the ears and uses complex computational skills to create a three-dimensional representation of the sound source.

Hearing Loss May Result from Mechanical or Neural Damage

There are three forms of hearing loss: conductive, sensorineural, and central. In *conductive hearing loss*, sound is unable to be transmitted through the external or middle ear. The causes of conductive hearing loss range from an ear canal plugged with ear wax (*cerumen*), to fluid in the middle ear from an infection, to diseases or trauma that impede vibration of the malleus, incus, or stapes. Correction of conductive hearing loss includes new microsurgical techniques in which the bones of the middle ear can be reconstructed.

Sensorineural hearing loss arises from damage to the structures of the inner ear, including degeneration of hair cells as a result of loud noises. Ninety percent of hearing loss in the elderly (*presbycusis*) is sensorineural. This form of hearing loss is more difficult to treat, but amazing results have been obtained with cochlear implants attached to tiny computers (see Biotechnology box). *Central hearing loss* results from damage to the neural pathways between the ear and cerebral cortex or from damage to the cortex itself. This form of hearing loss is relatively uncommon.

Hearing is probably our most important social sense. Suicide rates are higher among deaf people than among those who have lost their sight. More than any other sense, hearing connects us to other people and the world around us.

BIOTECHNOLOGY

Cochlear Implants

One technique used to treat sensorineural hearing loss is the cochlear implant. The newest generation of cochlear implants has three components: a computerized speech processor, a transmitter, and 8–24 electrodes. The speech processor is a transducer that converts sound into electrical impulses. The smallest speech processors now fit behind the ear like a conventional hearing aid, thanks to microchip technology. The transmitter converts the processor's electrical impulses into radio waves that activate the electrodes. The electrodes are surgically placed under the skin and run directly into the cochlea where they stimulate the sensory nerves. Cochlear implants have been remarkably successful for many profoundly deaf people, allowing them to hear loud noises and modulate their own voices. In the most successful cases, individuals can even use the telephone.

17. Map or diagram the pathways followed by sound, starting in the air at the outer ear and ending on the auditory cortex.

18. Why is somatosensory information projected to only one hemisphere of the brain but auditory information projected to both hemispheres? (Hint: see Figs. 10-5 and 10-9.)

19. Would a cochlear implant help a person who suffers from nerve deafness? From conductive hearing loss?

▶ THE EAR: EQUILIBRIUM

Equilibrium is a state of balance, whether the word is used to describe ion concentrations in body fluids or the positioning of the body in space. In body positioning, sensory information from the inner ear and from joint and muscle proprioceptors tells our brain the location of different body parts in relation to each other and to the environment. Visual information also plays an important role in equilibrium, as you know if you have ever gone to one of the 360° movie theaters where the scene tilts suddenly to one side and the audience tilts with it!

Our sense of equilibrium is mediated by hair cells lining the fluid-filled vestibular apparatus and semicircular canals of the inner ear. These receptors respond to changes in rotational, vertical, and linear acceleration. The hair cells function just like those of the cochlea, but gravity and acceleration rather than sound waves provide the force that moves the stereocilia.

The Vestibular Apparatus Is Filled with Endolymph

The **vestibular apparatus** consists of two saclike **otolith organs**, the **utricle** and the **saccule**, along with three **semicircular canals** (Fig. 10-23a ●). The semicircular canals connect to the utricle at their bases. They are oriented at right angles to each other, like three flat planes that come together to form the corner of a box (Fig. 10-23b).

At one end of each canal is an enlarged chamber, the **ampulla** [bottle], that contains a sensory receptor known as a **crista** [a crest; plural *cristae*]. The crista consists of a gelatinous mass, the **cupula** [small tub], that stretches from floor to ceiling of the ampulla, closing it off (Fig. 10-23c). Embedded in the cupula are the cilia of hair cells. The basal membranes of the hair cells synapse on the sensory neurons of the **vestibular nerve** (a branch of the VIII cranial nerve).

The sensory receptors of the utricle and saccule, the **maculae**, are different from those of the cristae (Fig. 10-23d). Each macula consists of a gelatinous mass known as the **otolith membrane** in which small crystals of calcium carbonate called **otoliths** are embedded [*oto*, ear + *lithos*, stone]. Underneath the otolith membrane are hair cells whose cilia are connected to the membrane.

The vestibular apparatus, like the cochlear duct, is filled with fluid endolymph secreted by epithelial cells. Like cerebrospinal fluid, endolymph is secreted continuously and drains from the inner ear into the venous sinus in the dura mater of the brain. If endolymph production exceeds the drainage rate, buildup of fluid in the inner ear may cause *Ménière's disease*. This condition is marked by episodes of dizziness and nausea, apparently as a result of changes in fluid pressure within the vestibular apparatus. It may also cause hearing loss if the organ of Corti in the cochlear duct is damaged.

The Vestibular Apparatus Provides Information About Movement and Position in Space

The special sense of equilibrium has two components: a *dynamic component* that tells us about our movement through space and a *static component* that tells us if our head is displaced from its normal upright position. The three semicircular canals sense *rotational acceleration* in various directions. The otolith organs tell us about *linear acceleration* and head position.

The receptor cells of both sets of sense organs, the hair cells, function just like the hair cells of the organ of Corti. When the cilia bend in one direction, they depolarize, and when they bend in the opposite direction, they hyperpolarize. Vestibular hair cells have one long cilium called a **kinocilium** that is located at one side of the bundle, creating a reference point for the direction of bending.

The Semicircular Canals Sense Rotational Acceleration

How is rotation sensed? As the head turns, the bony structure of the skull and the membranous labyrinth move, but the endolymph within the canals cannot keep up because of inertia. The drag of the fluid bends the cupula and its hair cells in the direction opposite to the way the head is turning.

For an analogy, think of pulling a paintbrush (the crista attached to the wall of the canal) through sticky wet paint (the endolymph) on a board. If you pull the brush to the right, the drag of the paint on the bristles bends them back to the left (Fig. 10-24 ●). In the same way, the inertia of the fluid in the semicircular canal pulls the cupula and the cilia of the hair cells to the left when the head turns right.

If rotation continues, the moving endolymph finally catches up. Then if head rotation stops suddenly, the fluid has built up momentum and cannot stop as fast. The fluid continues to rotate in the direction of the head rotation, leaving the person with a turning sensation. If the sensation is strong enough, the person may throw his or her body in the direction opposite the rotation in a reflexive attempt to compensate for the apparent loss of equilibrium.

STRUCTURE OF THE VESTIBULAR APPARATUS

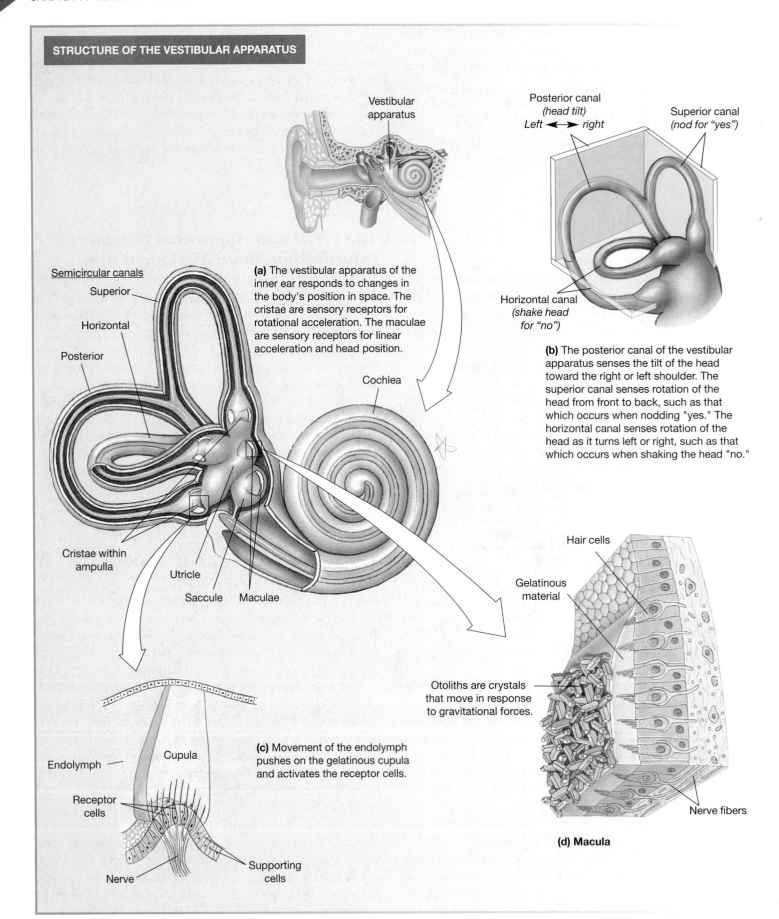

Vestibular apparatus

Posterior canal
(head tilt)
Left ◄──► right

Superior canal
(nod for "yes")

Horizontal canal
(shake head
for "no")

Semicircular canals
Superior
Horizontal
Posterior

(a) The vestibular apparatus of the inner ear responds to changes in the body's position in space. The cristae are sensory receptors for rotational acceleration. The maculae are sensory receptors for linear acceleration and head position.

Cochlea

(b) The posterior canal of the vestibular apparatus senses the tilt of the head toward the right or left shoulder. The superior canal senses rotation of the head from front to back, such as that which occurs when nodding "yes." The horizontal canal senses rotation of the head as it turns left or right, such as that which occurs when shaking the head "no."

Cristae within ampulla

Utricle

Saccule   Maculae

Hair cells

Gelatinous material

Otoliths are crystals that move in response to gravitational forces.

Endolymph

Cupula

(c) Movement of the endolymph pushes on the gelatinous cupula and activates the receptor cells.

Receptor cells

Nerve

Supporting cells

Nerve fibers

(d) Macula

● **Figure 10-23** *ANATOMY SUMMARY Vestibular Apparatus*

When the head turns right, endolymph pushes the cupula to the left.

● **Figure 10-24** *Transduction of rotational forces in the cristae* When the head turns, inertia keeps endolymph inside the ampulla from moving as rapidly as the surrounding cranium.

The Otolith Organs Sense Linear Acceleration and Head Position

The otolith organs are arranged to sense linear forces. The maculae of the utricle are horizontal when the head is in its normal upright position. If the head tips back, the otoliths in the gelatinous membrane slide backward owing to gravity acting on the dense particles (Fig. 10-25 ●). The cilia of the hair cells bend and set off a signal.

The crystalline otoliths are attached to gelatinous material in the maculae.

When the head tilts, gravity causes the otoliths to slide. This bends the stereocilia and increases action potentials in the sensory neurons.

Macula

Gravity

Otolith

● **Figure 10-25** *Otolith organs*

The maculae of the saccule are oriented vertically when the head is erect and are sensitive to vertical forces, such as dropping downward in an elevator. The brain analyzes the pattern of depolarized and hyperpolarized hair cells in order to compute head position and direction of movement.

Equilibrium Pathways Project Primarily to the Cerebellum

The hair cells of the vestibular apparatus stimulate primary sensory neurons in the vestibular nerve. Those neurons either synapse in the vestibular nuclei of the medulla or run without synapsing to the cerebellum (Fig. 10-26 ●). Collateral pathways run from the medulla to the cerebellum or upward through the reticular formation and thalamus. There are some poorly defined pathways to the cerebral cortex, but most integration for equilibrium comes from the cerebellum. Descending pathways from the vestibular nuclei go to certain motor neurons involved in eye movement. These pathways help keep the eyes locked on an object as the head turns.

CONCEPT CHECK   Answers: p. 367

20. The stereocilia of hair cells are bathed in endolymph, which has a very high concentration of K^+ and a low concentration of Na^+. When ion channels in the stereocilia open, what ions move in what direction to cause depolarization?

21. Why does hearing decrease if you have an ear infection with fluid buildup in the middle ear?

22. When dancers perform multiple turns, they try to keep their vision fixed on a single point ("spotting"). How does spotting keep the dancer from getting dizzy?

● **Figure 10-26** *Central nervous system pathways for equilibrium*

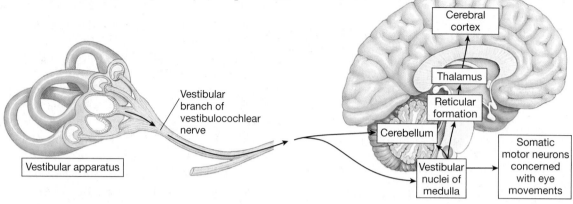

THE EYE AND VISION

The eye is a sensory receptor that functions much like a camera. It focuses light on a light-sensitive surface (the retina) using a lens and an aperture or opening (the pupil), whose size can be adjusted to change the amount of entering light. **Vision**, the process through which light reflected from objects in our environment is translated into a mental image, can be divided into three steps. (1) In the first step, light enters the eye and is focused by a lens on the retina. (2) In the second step, the photoreceptors of the retina transduce light energy into an electrical signal. (3) The third step is processing of the electrical signals through neural pathways.

The Eye Is Protected by the Skull

The external anatomy of the eye is shown in Figure 10-27 ●. Like sensory elements of the ear, the eyes are protected in a bony cav-

ity, the *orbit*, which is formed by facial bones of the skull. Accessory structures associated with the eye include six *extrinsic eye muscles*, skeletal muscles that attach to the outer surface of the sphere and control eye movements. Cranial nerves III, IV, and VI innervate these muscles. The *eyelids* close over the anterior surface of the eye, and the *lacrimal apparatus*, a system of glands and ducts, keeps a continuous flow of tears washing across the cornea so that it remains moist and free of debris.

The eye itself is a hollow sphere divided into two compartments separated by a lens. The compartment in front of the lens and its attachment is filled with the **aqueous humor**, a low-protein plasma-like fluid secreted by the ciliary epithelium supporting the lens (Fig. 10-28a ●). Behind the lens is a much larger chamber filled

---

CLINICAL FOCUS

Glaucoma

The eye disease *glaucoma*, characterized by degeneration of the optic nerve, is the leading cause of blindness worldwide. Many people associate glaucoma with increased *intraocular* (within the eyeball) pressure, but scientists have discovered that increased pressure is only one risk factor for the disease. A significant number of people with glaucoma have normal intraocular pressure and not everyone with elevated pressure develops glaucoma. Many cases of elevated eye pressure are associated with excess aqueous humor. Normally this fluid is secreted by the ciliary epithelium near the lens and drains out through the canal of Schlemm in the anterior chamber of the eye. If outflow is blocked, the aqueous humor accumulates, causing fluid and pressure buildup inside the eyeball. Treatments to decrease intraocular pressure include drugs that inhibit aqueous humor production or surgery to reopen the canal. Research suggests that the optic nerve degeneration in glaucoma may be due to nitric oxide or apoptosis-inducing factors, and studies in these areas are under way.

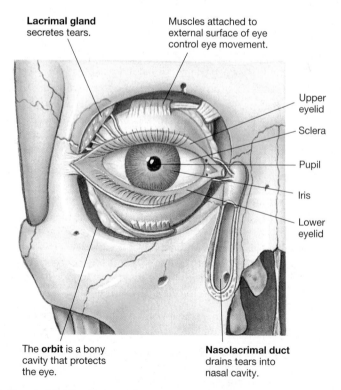

Lacrimal gland secretes tears.

Muscles attached to external surface of eye control eye movement.

Upper eyelid

Sclera

Pupil

Iris

Lower eyelid

The **orbit** is a bony cavity that protects the eye.

Nasolacrimal duct drains tears into nasal cavity.

● **Figure 10-27** *External anatomy of the eye*

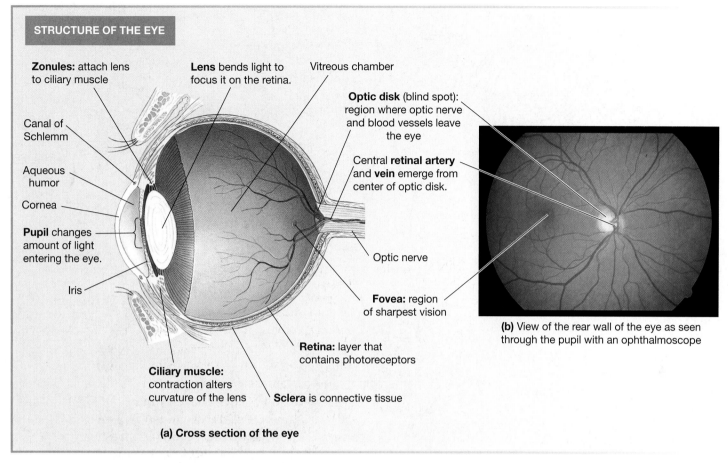

STRUCTURE OF THE EYE

Zonules: attach lens to ciliary muscle

Lens bends light to focus it on the retina.

Vitreous chamber

Optic disk (blind spot): region where optic nerve and blood vessels leave the eye

Canal of Schlemm

Central **retinal artery** and **vein** emerge from center of optic disk.

Aqueous humor

Cornea

Pupil changes amount of light entering the eye.

Iris

Optic nerve

Fovea: region of sharpest vision

(b) View of the rear wall of the eye as seen through the pupil with an ophthalmoscope

Retina: layer that contains photoreceptors

Ciliary muscle: contraction alters curvature of the lens

Sclera is connective tissue

(a) Cross section of the eye

● **Figure 10-28** *ANATOMY SUMMARY The Eye*

mostly with the **vitreous body** [*vitrum*, glass], a clear, gelatinous matrix that helps maintain the shape of the eyeball. The outer wall of the eyeball, the **sclera**, is composed of connective tissue.

Light enters the eye through the **cornea**, a transparent disk of tissue on the anterior surface that is a continuation of the sclera. After crossing the aqueous humor, light passes through the opening of the pupil and strikes the **lens**, a transparent structure with two convex surfaces. The cornea and lens together bend incoming light rays so that they focus on the **retina**, the light-sensitive lining of the eye that contains the *photoreceptors*.

When viewed through the pupil of the eye with an ophthalmoscope [*ophthalmos*, eye], the retina is crisscrossed with small arteries and veins that radiate out from one spot, the **optic disk** (Fig. 10-28b). The optic disk is the location where neurons of the visual pathway form the **optic nerve** (cranial nerve II) and exit the eye. Lateral to the optic disk is a small dark spot, the *fovea*. The fovea and a narrow area of tissue surrounding it, the *macula*, are the regions of the retina with the most acute vision.

Neural pathways for the eyes are illustrated in Figure 10-29 ●. The optic nerves from each eye go to the **optic chiasm** in the brain, where some of the fibers cross to the opposite side. After synapsing in the **lateral geniculate body** (*lateral geniculate nucleus*) of the thalamus, the vision neurons of the tract terminate in the occipital lobe at the **visual cortex**. Collateral pathways go

from the thalamus to the midbrain, where they synapse with efferent neurons that control diameter of the pupils.

CONCEPT CHECK   Answers: p. 367

23. What function(s) might the aqueous humor serve?

Light Enters the Eye Through the Pupil

In the first step of the visual pathway, light from the environment enters the eye. Before it strikes the retina, however, the light is modified two ways. First, the amount of light that reaches photoreceptors is modulated by changes in the size of the pupil. Second, the light is focused by changes in the shape of the lens.

The human eye functions over a 100,000-fold range of light intensity. Most of this ability comes from the sensitivity of the photoreceptors, but the pupils of the eyes assist by regulating the amount of light that falls on the retina. In bright sunlight, the pupils narrow to about 1.5 mm diameter when a parasympathetic pathway constricts the circular pupillary muscles. In the dark, the opening of the pupil dilates to 8 mm, a 28-fold increase. Dilation occurs when a set of radial muscles lying perpendicular to the circular muscles contract under the influence of sympathetic neurons.

Testing **pupillary reflexes** is a standard part of a neurological examination. Light hitting the retina in one eye activates the

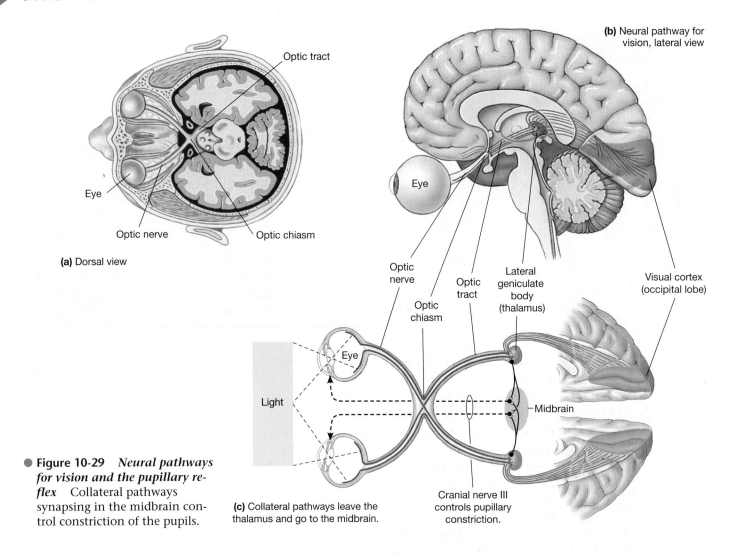

(b) Neural pathway for vision, lateral view

Optic tract

Eye

Optic nerve   Optic chiasm

(a) Dorsal view

Optic nerve

Optic chiasm

Optic tract

Lateral geniculate body (thalamus)

Visual cortex (occipital lobe)

Eye

Light

Midbrain

Cranial nerve III controls pupillary constriction.

● **Figure 10-29** *Neural pathways for vision and the pupillary reflex* Collateral pathways synapsing in the midbrain control constriction of the pupils.

(c) Collateral pathways leave the thalamus and go to the midbrain.

reflex. Signals travel through the optic nerve to the thalamus, then to the midbrain, where efferent neurons constrict the pupils in *both* eyes (Fig. 10-29c). This response is known as the **consensual reflex** and is mediated by parasympathetic fibers running through cranial nerve III.

CONCEPT CHECK   Answers: p. 367

24. Use the neural pathways in Figure 10-29c to answer the following questions.

   (a) Why does shining light into one eye cause pupillary constriction in both eyes?

   (b) If you shine a light in the left eye and get pupillary constriction in the right eye but not in the left eye, what can you conclude about the afferent path from the left eye to the brain? About the efferent pathways to the pupils?

25. Parasympathetic fibers constrict the pupils and sympathetic fibers dilate them. This is an example of what kind of control?

In addition to regulating the amount of light that hits the retina, the pupils create what is known as **depth of field**. A simple example comes from photography. Imagine a picture of a puppy sitting in the foreground of a field of wildflowers. If only the puppy and the flowers immediately around her are in focus, the picture is said to have a shallow depth of field. But if the puppy and the wildflowers all the way back to the horizon are in focus, the picture has full depth of field. Depth of field is created by constricting the pupil (or shutter on the camera) so that only a narrow beam of light enters the eye. In this way, more of the depth of the image is focused on the retina.

RUNNING PROBLEM

The otolaryngologist strongly suspects that Anant has Ménière's disease, with excessive endolymph within the vestibular apparatus and cochlea. Many treatments are available, from simple dietary changes to surgery. For now, the physician suggests that Anant limit his salt intake and take diuretics, drugs that cause the kidneys to remove excess fluid from the body.

Question 5: Why is limiting salt (NaCl) intake suggested as a treatment for Ménière's disease? (Hint: What is the relationship between salt, osmolarity, and volume?)

(a) A **concave lens** scatters light rays.

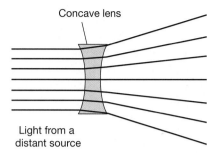

Concave lens

Light from a distant source

(b) A **convex lens** causes light rays to converge.

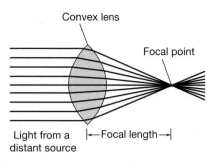

Convex lens

Focal point

Light from a distant source |←—Focal length—→|

The focal length is the distance from the center of the lens to the focal point.

● **Figure 10-30** *Refraction of light*

The Lens Focuses Light on the Retina

When light enters the eye, it passes through the cornea and lens prior to striking the retina. When light waves pass from air into a medium of different density such as glass or water, they will bend, or **refract**. Light entering the eye is refracted twice, most strongly when it passes through the cornea and again when it passes through the lens. To simplify this discussion, we will consider only the refraction that occurs as light passes through the lens.

When light passes from one medium into another, the angle of refraction is influenced by two factors: (1) the difference in density of the two media, and (2) the angle at which light meets the face of the object into which it is passing. For light passing through the lens of the eye, we will consider density to be constant. The angle at which light meets the face of the lens depends on the curvature of the lens surface and the direction of the light beam.

If parallel light waves strike a surface perpendicular to them, the light will pass through without bending. But if the surface they strike is not perpendicular, the light waves will bend. Parallel light waves striking a **concave lens** such as that shown in Figure 10-30a ● will be refracted into a wider beam. A light beam striking a spherical **convex lens** will bend inward and focus to a point—*convex* lenses *converge* light waves (Fig. 10-30b). You can demonstrate the properties of a convex lens by using a magnifying glass to focus sunlight onto a piece of paper or other surface.

When parallel light waves pass through a convex lens, the single point where the waves converge is called the **focal point** (Fig. 10-30b). The distance from the center of a lens to its focal point is known as the **focal length** (or *focal distance*). For any given lens, the focal length and focal point are fixed. To change the focal length, the shape of the lens must change.

When light passes through the lens of the eye, the focal point must fall precisely on the retina of the eye if we are to see an object in focus. In Figure 10-31a ●, parallel light rays from a distant source strike a flattened lens, resulting in a focal point that falls on the retina. The object is therefore in focus. For the

(a) Parallel light rays pass through a flattened lens and the focal point falls on the retina.

(b) For close objects, the light rays are no longer parallel.

(c) Rounding a lens shortens its focal length.

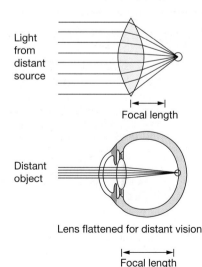

Light from distant source

Focal length

Distant object

Lens flattened for distant vision

Focal length

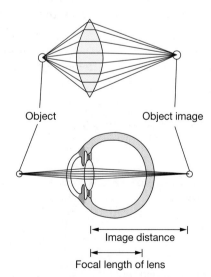

Object

Object image

Image distance

Focal length of lens

The lens and its focal length have not changed but the object is seen out of focus because the light beam is not focused on the retina.

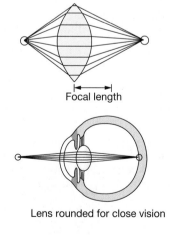

Focal length

Lens rounded for close vision

To keep an object in focus as it moves closer, the lens becomes more rounded.

● **Figure 10-31** *Optics*

human eye, any object that is 20 feet or more away from the eye will create parallel light rays.

But what happens if an object is closer than 20 feet to the lens? In that case, the light rays reflected from the object are not parallel and strike the lens at an oblique angle (Fig. 10-31b). The light rays falling on the retina no longer converge to a point and the object will appear fuzzy and out of focus.

To keep a near object in focus, the lens must round to increase the angle of refraction (Fig. 10-31c). This adjustment causes the light rays to converge *on* the retina instead of behind it, and the object comes into focus. The process by which the eye adjusts the shape of the lens to keep objects in focus is known as **accommodation**. If an object is so close to the lens that it falls within the focal length, no amount of accommodation can bring the object into focus. The closest distance at which you can focus on an object is known as the **near point of accommodation**.

You can demonstrate the accommodation reflex easily by closing one eye and holding your hand up about 8 in. in front of your open eye, fingers spread widely apart. Focus your eye on some object in the distance that is between your fingers. Notice that when you do so, your fingers are visible but out of focus. Your lens is flattened for distance vision, so the focal point for near objects falls behind the retina. Those objects appear out of focus.

Now shift your gaze to your fingers and notice that they come into focus. The light waves coming off your fingers have not changed their angle, but your lens has become more rounded and the light rays now converge on the retina.

How can the lens, which is clear and does not have any muscle fibers in it, change shape? The lens is attached to the ciliary muscle of the eye by inelastic ligaments, or *zonules* (Fig. 10-32 ●). If no tension is placed on the lens by the ligaments, the lens assumes its natural rounded shape because of the elasticity of its capsule. If the ligaments pull on the lens, it flattens out and assumes the shape required for distance vision.

Tension on the ligaments is controlled by the **ciliary muscle**, a ring of smooth muscle that surrounds the lens. When the circular ciliary muscles contract, the muscle ring gets smaller, releasing tension on the ligaments so that the lens rounds. When the cil-

iary muscles are relaxed, the ring is more open and the lens is pulled into a flatter shape.

Young people can focus on items as close as 8 cm, but the accommodation reflex diminishes from the age of 10 on. By age 40, accommodation is only about half of what it was at age 10, and by age 60, many people lose the reflex completely because the lens has lost flexibility. The loss of accommodation, **presbyopia**, is the reason that most people begin to wear reading glasses in their 40s.

Two other common vision problems, nearsightedness (**myopia**) and farsightedness (**hyperopia**), occur when the focal point falls either in front of or behind the retina, respectively (Fig. 10-33 ●). These conditions can be corrected by placement of a lens with the appropriate curvature in front of the eye to change the focal length. A third common vision problem, **astigmatism**, is usually caused by a cornea that is not a perfectly shaped dome, resulting in distorted images.

CONCEPT CHECK   Answers: p. 367

26. What kind of lens is needed to correct myopic vision? Explain.

RUNNING PROBLEM

Anant's condition does not improve with the low-salt diet and diuretics, and he continues to suffer from disabling attacks of vertigo with vomiting. As a last resort, surgery is sometimes performed to treat severe cases of Ménière's disease. In one surgical procedure for the disease, the vestibular nerve is severed. This surgery is difficult to perform, as the vestibular nerve lies near many other important nerves, including facial nerves and the auditory nerve. Patients who undergo this procedure are advised that the surgery can result in deafness if the cochlear nerve is inadvertently severed.

Question 6: Why would severing the vestibular nerve alleviate Ménière's disease?

321 325 343 349 352 361

(a) The lens is attached to the ciliary muscles by inelastic ligaments.

(b) When ciliary muscle is relaxed, the ligaments pull on and flatten the lens.

(c) When ciliary muscle contracts, it releases tension on the ligaments and the lens becomes more rounded.

● **Figure 10-32** *Accommodation*

(a) **Hyperopia,** or far-sightedness, occurs when the focal point falls behind the retina.

Hyperopia (corrected with a convex lens)

(b) **Myopia,** or near-sightedness, occurs when the focal point falls in front of the retina.

Myopia (corrected with a concave lens)

FIGURE QUESTION

Explain (1) what the corrective lenses do to the refraction of light, and (2) why each lens type corrects the visual defect.

● **Figure 10-33**   *Visual defects*

● **Figure 10-34**   *The electromagnetic spectrum*

Phototransduction Occurs at the Retina

The transduction of light into action potentials takes place when light hits the retina, the sensory organ of the eye. Actually, only a small portion of the light that enters our eyes is perceived by our brain. **Visible light** is that electromagnetic energy whose waves have a frequency of $4.0–7.5 \times 10^{14}$ waves per second (hertz) and a wavelength of 400–750 nanometers (nm).

The full electromagnetic spectrum extends from high-energy, very short wavelength waves such as X-rays and gamma rays to low-energy, lower-frequency microwaves and radio waves (Fig. 10-34 ●). The waves of ultraviolet and infrared light border the ends of the visible light spectrum. Although our eyes are unable to perceive these wavelengths, some animals can see light in the infrared and ultraviolet range. For example, bees use ultraviolet "runways" on flowers to guide them to pollen and nectar.

In the human eye, **photoreceptors** in the retina transduce light energy into electrical signals. The retina develops from the same embryonic tissue as the brain, and like the cortex of the brain, its neurons are organized into layers (Fig. 10-35c ●). Backing the photosensitive portion of the human retina is a dark

pigment epithelium layer. Its function is to absorb any light rays that escape the photoreceptors, preventing distracting light from reflecting inside the eye and distorting the visual image. The black color of these epithelial cells comes from granules of the pigment *melanin*.

The layers of retinal neurons are in reverse order from what you might expect. Instead of the photoreceptors facing the pupil where photons of light will strike them first, they are at the back, with their photosensitive tips against the pigment epithelium. Most light entering the eye must pass through several relatively transparent layers of neurons before striking the photoreceptors.

One exception to this pattern occurs in a small region of the retina known as the **fovea** [pit]. In this area, photoreceptors receive light directly because the intervening neurons are pushed off to the side (Fig. 10-36 ●). The fovea and the **macula** immediately surrounding it are the areas of most acute vision. The fovea is the point on which light focuses when you look at an object.

For example, look at Figure 10-37 ●. The eye is focused on the edge of the green bar. Light from that section of the visual field falls on the fovea and will be in sharp focus. Notice that the image falling on the retina is upside down. Later visual processing reverses the image again so that we perceive it in the correct orientation.

Sensory information about light passes from the photoreceptors to a layer of interneurons that includes **bipolar neurons,** then to a layer of **ganglion cells.** The axons of ganglion cells form the *optic nerve* that leaves the eye at the optic disk. Because the optic disk has no photoreceptors, images projected onto this region will not be seen, creating the so-called **blind spot.**

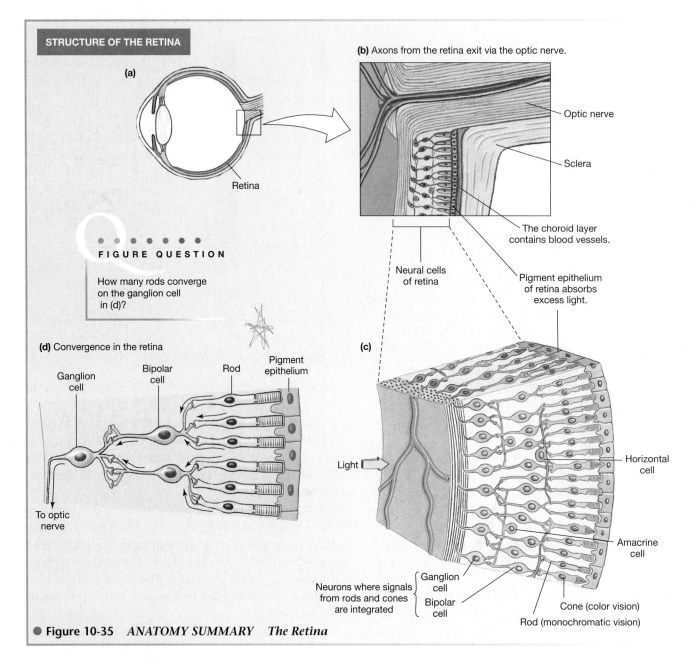

STRUCTURE OF THE RETINA

(a)

Retina

(b) Axons from the retina exit via the optic nerve.

Optic nerve

Sclera

The choroid layer contains blood vessels.

Neural cells of retina

Pigment epithelium of retina absorbs excess light.

FIGURE QUESTION

How many rods converge on the ganglion cell in (d)?

(d) Convergence in the retina

Ganglion cell

Bipolar cell

Rod

Pigment epithelium

To optic nerve

(c)

Light

Horizontal cell

Amacrine cell

Neurons where signals from rods and cones are integrated

Ganglion cell

Bipolar cell

Cone (color vision)

Rod (monochromatic vision)

● **Figure 10-35   ANATOMY SUMMARY   The Retina**

Ganglion cell

Fovea

Light

Photoreceptor

Bipolar neuron

Retina

Light strikes the photoreceptors in the fovea directly because overlying neurons are pushed aside.

Retina

Lens

Fovea

Light

Macula

Fixation point

Pigment epithelium

● **Figure 10-36   *Photoreceptors in the fovea*   Adapted from E. R. Kandel et al., *Principles of Neural Science*, 3rd edition, (New York: McGraw Hill, 2000).**

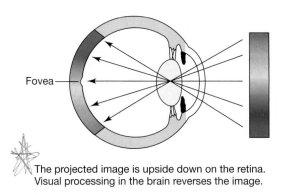

Fovea

The projected image is upside down on the retina.
Visual processing in the brain reverses the image.

● **Figure 10-37** *Image projection onto the retina*

One hallmark of signal processing in the retina is *convergence* [● p. 270], in which multiple neurons synapse onto a single postsynaptic cell (see Fig. 10-35d). As many as 15 to 45 photoreceptors may synapse onto one bipolar neuron. Multiple bipolar neurons in turn innervate a single ganglion cell, so that the information from hundreds of millions of photoreceptors is condensed down to a mere 1 million axons leaving the eye in each optic nerve. Convergence is minimal in the fovea, where some photoreceptors have a 1:1 relationship with their bipolar neurons, and greatest at the outer edges of the retina.

CONCEPT CHECK — Answers: p. 367

27. Animals that see well in very low light levels lack pigment epithelium and instead have a layer called the *tapetum lucidum* behind the retina. What property might this layer have that would enhance vision in low light?

28. Why is the difference in visual acuity between the fovea and the edge of the visual field similar to the difference in touch discrimination between the fingertips and the skin of the arm?

Photoreceptors Transduce Light into Electrical Signals

There are two types of photoreceptors in the eye: rods and cones. Rods are the most numerous by a 20:1 ratio, except in the fovea, where cones are concentrated. **Rods** are responsible for monochromatic nighttime vision and function well in low light levels. **Cones** are responsible for *high-acuity* vision and color vision during the daytime, when light levels are higher. *Acuity* means keenness and is derived from the Latin *acuere*, meaning "to sharpen."

The two types of photoreceptors have the same basic structure: (1) an outer segment whose tip touches the pigment epithelium of the retina, (2) an inner segment that contains the cell nucleus and organelles for ATP and protein synthesis, and (3) a base segment that synapses with the bipolar cells (Fig. 10-38 ●). In the outer segment, the cell membrane has deep folds that form disklike layers, like candy mints stacked in a wrapper. In rods, toward the tip of the outer segments, these layers actually separate

Melanopsin—A New Photosensitive Pigment

Circadian rhythms in mammals are cued by light entering through the eyes. For many years, scientists believed that pathways from the rods and cones of the retina led to the suprachiasmatic nucleus (SCN), the brain center for circadian rhythms. However, in 1999 researchers found that transgenic mice who lacked both rods and cones still had the ability to respond to changing light cues, suggesting that some other photoreceptor must exist in the retina. Now scientists believe they have found it: a subset of retinal ganglion cells that contain a pigment called melanopsin. Axons from these cells project to the SCN, and it appears that these cells join rods and cones as the light-sensing cells of the mammalian retina.

from the cell membrane and form free-floating disks. In the cones, they stay attached.

The light-sensitive **visual pigments** are bound to cell membranes in the outer segments of the photoreceptors. These pigments are the transducers that convert light energy into a change in membrane potential. Rods have one type of visual pigment, **rhodopsin**. Cones have three different pigments that are closely related to rhodopsin.

Cones are responsible for color vision and contain pigments that are excited by different wavelengths of light. White light is a combination of colors, as you see when you separate white light by passing it through a prism. The eye contains cones for red, green, and blue light. Each cone type is stimulated by a range of light wavelengths but is most sensitive to a particular wavelength (Fig. 10-39 ●). Red, green, and blue are the three primary colors for light, just as red, blue, and yellow are the three primary colors for paints.

The color of objects in our environment depends on the wavelengths of light reflected by the object. Green leaves reflect green light and bananas reflect yellow light. White objects reflect most wavelengths. Black objects absorb most wavelengths, one reason they heat up in sunlight while white objects stay cool.

Our brain recognizes the color of an object by interpreting the combination of signals coming to it from the three different color cones. This seems like a simple process. However, the details of color vision are still not well understood and there is some controversy about how color is processed in the cerebral cortex. **Color-blindness** is a condition in which a person inherits a defect in one or more of the three types of cones and has difficulty distinguishing certain colors. Probably the best-known form of color-blindness is red-green, in which people cannot tell red and green apart.

PIGMENT
EPITHELIUM

Melanin
granules

OUTER SEGMENT

Visual pigments in
membrane disks

INNER SEGMENT

Location of major organelles,
metabolic operations such as
photopigment synthesis and
ATP production

SYNAPTIC TERMINAL

Synapses with bipolar cells

Disks

Connecting
stalk

Mitochondria

Cone   Rods

LIGHT

Retinal

Rhodopsin
molecule

Opsin

● **Figure 10-38** *Photoreceptors: rods and cones*
Light transduction takes
place in the outer segment
of the photoreceptor. De-
polarization of the pho-
toreceptor releases
neurotransmitter onto
bipolar cells. The dark pigment epithelium absorbs
extra light and prevents it
from reflecting back and
distorting vision.

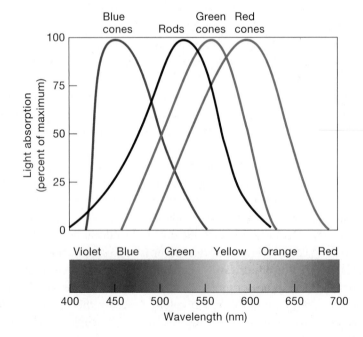

GRAPH QUESTION

Which pigment absorbs light over
the broadest spectrum of wavelengths?
Over the narrowest? Which pigment
absorbs the most light at 500 nm?

● **Figure 10-39** *Light absorption of visual pigments*
There are three types of cone pigment, each with a
characteristic light absorption spectrum.

29. Why is our vision in the dark in black and white rather than color?

Phototransduction The process of phototransduction is similar for rhodopsin and the three color pigments. The visual pigment of the rods, rhodopsin, is composed of two sections: **opsin**, a protein embedded in the membrane of the rod disks, and **retinal**, a vitamin A derivative that is the light-absorbing portion of the pigment. In the absence of light, retinal binds snugly into a binding site on the opsin (Fig. 10-40a ●). When activated by as little as one photon of light, retinal changes shape to a new configuration. The activated retinal no longer binds to opsin and is released from the pigment in the process known as **bleaching**.

But how does rhodopsin bleaching lead to action potentials traveling through the optical pathway? In order to understand the link, we must look at other properties of the rods. As you learned in Chapters 5 and 8, electrical signals in cells occur as a result of the movement of ions between the intracellular and extracellular compartments. A rod contains many cation channels that allow Na$^+$ to enter the rod and K$^+$ channels that allow K$^+$ to leak out of the cell.

When a rod is in darkness and rhodopsin is not active, cyclic GMP levels in the rod are high and both sets of channels are open. Sodium ion influx is greater than K$^+$ efflux, so the rod stays depolarized to an average membrane potential of -40 mV, instead of the more usual -70 mV. At this membrane potential, there is tonic (continuous) release of neurotransmitter from the synaptic portion of the rod onto the adjacent bipolar cell.

When light activates rhodopsin, a second messenger cascade is initiated through **transducin**, the same G protein found in bitter taste receptors (Fig. 10-40b). The second messenger cascade decreases the concentration of cGMP, and, in turn, the cation channels close. As a result, Na$^+$ influx slows or stops.

(a) In darkness, rhodopsin is inactive, cGMP is high, and ion channels are open.

(b) Light bleaches rhodopsin. Opsin decreases cGMP, closes Na$^+$ channels, and hyperpolarizes the cell.

(c) In the recovery phase, retinal recombines with opsin.

Pigment epithelium cell

Inactive rhodopsin (opsin and retinal)

Transducin (G protein)

Na$^+$

cGMP levels high

Disks

K$^+$

Membrane potential in dark = -40mV

Tonic release of neurotransmitter onto bipolar neurons

Activated retinal

Activates transducin

Opsin (bleached pigment)

Cascade

↓cGMP

Na$^+$

Na$^+$ channel closes

K$^+$

Light

Membrane hyperpolarizes to -70 mV

Neurotransmitter release decreases in proportion to amount of light.

Retinal converted to inactive form

Retinal recombines with opsin to form rhodopsin.

● **Figure 10-40** *Phototransduction in rods*

Less Na$^+$ influx but continued K$^+$ efflux causes the inside of the cell to hyperpolarize. Neurotransmitter release onto the bipolar neurons consequently decreases. Bright light will close all Na$^+$ channels and stop all neurotransmitter release. Dimmer light will cause a response that is graded in proportion to the light intensity.

Because a change in neurotransmitter release by rods is mediated through a second messenger system, the response of rods to changes in light intensity is much slower than you might expect for a process associated with the nervous system. When you move from bright light into the dark, or vice versa, the time required for your eyes to adapt to the new light level is partially determined by the slow change in rod membrane potential as ion channels open or close. The recovery of rhodopsin from bleaching also can take some time.

After activation, retinal diffuses out of the rod and is transported into the pigment epithelium. There, it reverts to its inactive form before moving back into the rod and being reunited with opsin (Fig. 10-40c).

We now turn from the cellular mechanism of light transduction to the processing of light signals in the bipolar and ganglion cells of the retina.

	On / Off	Off / On
Light shone onto center	Excites ganglion cell	Inhibits ganglion cell
Light shone onto surround	Inhibits ganglion cell	Excites ganglion cell
Diffuse light on both center and surround	Weak response from ganglion cell	Weak response from ganglion cell

The retina uses contrast rather than absolute light intensity for better detection of weak stimuli.

● **Figure 10-41**   *Ganglion cell receptive fields*

CONCEPT CHECK   Answers: p. 367

30. Draw a map or diagram to explain the processes of phototransduction. Start with when bleaching occurs and end with when neurotransmitter is released.

Signal Processing Begins in the Retina

Release of the neurotransmitter glutamate from photoreceptors onto bipolar neurons completes the first step in the signal pathway. From this point on, signal processing in the retina becomes quite complex. For one thing, glutamate excites some bipolar neurons but inhibits others, depending on the type of glutamate receptor on the bipolar neuron. In this fashion, one stimulus (light) creates two responses with a single neurotransmitter.

The second synapse in the pathway occurs between the bipolar neurons and the ganglion cells. These connections can also be either excitatory or inhibitory. We know more about ganglion cells because they lie on the surface of the retina, and their axons are the most accessible to researchers. Extensive studies have been done in which researchers stimulated the retina with carefully placed light and evaluated the response of the ganglion cells.

Each ganglion cell has a particular area of retina to which it is linked. These areas, known as **visual fields**, have two interesting properties. First of all, they are circular, unlike the irregular shape of receptive fields in the somatic sensory system. The visual field of a ganglion cell near the fovea is quite small. Only a few receptors are associated with each ganglion, so visual acuity is greatest

in these areas. At the edge of the visual field, a single ganglion cell may have multiple receptors, so vision is not as sharp.

The second property of visual fields is that they are divided into two sections: a round center and its doughnut-shaped **surround** (Fig. 10-41 ●). There are two basic types of visual fields. In the "on-center/off-surround" field, the associated ganglion cell will respond most strongly with a series of action potentials when light is brightest in the center of the field. If light is brightest in the "off" surround region of the field, the ganglion cell will be inhibited and stop firing action potentials.

The reverse happens with off-center/on-surround visual fields. If light is uniform across the visual field, the ganglion cell responds weakly. Thus, the retina uses *contrast* rather than absolute light intensity to recognize objects in the environment. One advantage of using contrast is that it allows better detection of weak stimuli.

Scientists have now identified two types of ganglion cells in the retina. Large *magnocellular* ganglion cells, or **M cells**, carry information about movement, location, and depth perception. Smaller *parvocellular* ganglion cells, or **P cells**, transmit signals that pertain to color, form, and texture of objects in the visual field. Once action potentials leave these cells, they travel along the optic nerves to the CNS for further processing.

The optic nerves enter the brain in a region known as the **optic chiasm**. This is also the point at which some fibers from each eye cross to the other side of the brain for processing. Figure 10-42 ● shows how information from the right side of the visual field is processed on the left side of the brain, while information from the left side of the field is processed on the right.

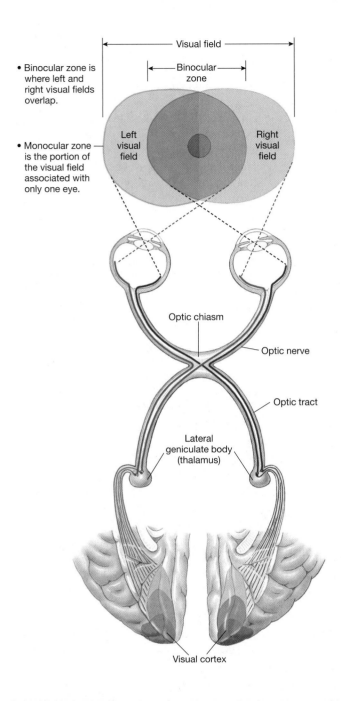

Binocular zone is where left and right visual fields overlap.

Monocular zone is the portion of the visual field associated with only one eye.

Visual field

Binocular zone

Left visual field

Right visual field

Optic chiasm

Optic nerve

Optic tract

Lateral geniculate body (thalamus)

Visual cortex

● **Figure 10-42** *Visual fields and binocular vision* The left side of the visual field in each eye is projected to the right side of the brain, while the right side of the field projects to the left visual cortex. Objects in the binocular zone will be perceived in three dimensions. Objects falling outside the binocular zone will only be perceived in two dimensions.

The central portion of the visual field, where left and right visual fields overlap, is the **binocular zone**. The two eyes have slightly different views of objects in this region, so the brain processes and integrates the two views to create three-dimensional representations of the objects. Our sense of depth perception—that is, whether one object is in front of or behind another—depends on binocular vision. Objects that fall within the visual field of only one eye are in the **monocular zone** and are viewed in two dimensions.

Once axons leave the optic chiasm, a few fibers project to the midbrain, where they participate in control of eye movement or coordinate with somatosensory and auditory information for balance and movement. Most axons, however, project to the *lateral geniculate body* of the thalamus, where the optic fibers synapse onto neurons leading to the visual cortex in the occipital lobe.

The lateral geniculate body is organized in layers that correspond to the different parts of the visual field. Thus, information from adjacent objects is processed together. This **topographical organization** is maintained in the visual cortex, with the six layers of neurons grouped into vertical columns. Within each portion of the visual field, information is further sorted by form, color, and movement.

Visual processing and our perception of the world around us are exceedingly complex subjects whose details are beyond the scope of this book. The cortex merges monocular information from the two eyes to give us a binocular view of our surroundings. Information from on/off combinations of ganglion cells is translated into sensitivity to line orientation in the simplest pathways or into color, movement, and detailed structure in the most complex. Each of these attributes of visual stimuli is processed through a separate pathway, creating a network whose complexity we are just beginning to unravel.

RUNNING PROBLEM CONCLUSION

Ménière's Disease

Anant was told about the surgical options but elected to continue medical treatment rather than risk deafness. Over the next two months, his Ménière's disease gradually resolved. The cause of it was never pinpointed.

In this running problem, you learned that vestibular problems can result in severe dizziness. You also learned about some treatments that are available to alleviate Ménière's disease. Now check your understanding of this running problem by comparing your answers to those in the summary table.

	QUESTION	FACTS	INTEGRATION AND ANALYSIS
1	In what part of the brain is sensory information about equilibrium processed?	The major equilibrium pathways project to the cerebellum. Some information is also processed in the cerebrum.	[Not applicable]

(continues on next page)

	QUESTION	FACTS	INTEGRATION AND ANALYSIS
2	Subjective tinnitus occurs when an abnormality somewhere along the anatomical pathway for hearing causes the brain to perceive a sound that does not exist outside the auditory system. Starting from the ear canal, name the auditory structures in which problems may arise.	The middle ear consists of malleus, incus, and stapes bones that vibrate with sound. The hearing portion of the inner ear consists of hair cells in the fluid-filled cochlea. The cochlear nerve leads to the brain.	Subjective tinnitus could arise from a problem with any of the structures named. Abnormal bone growth can affect the middle ear bones. Excessive fluid accumulation in the inner ear will affect the hair cells. Neural defects may cause the cochlear nerve to fire spontaneously, creating the perception of sound.
3	When a person with BPV changes position, the displaced crystals float toward the semicircular canals. Why would this cause dizziness?	The ends of the semicircular canals contain sensory cristae, each consisting of a cupula with embedded hair cells. Displacement of the cupula creates a sensation of rotational movement.	If the floating crystals displace the cupula of the cristae, the brain will perceive movement that is not matched to sensory information coming from the eyes. The result is vertigo, an illusion of movement.
4	Compare the symptoms of BPV and Ménière's disease. On the basis of Anant's symptoms, which condition do you think he has?	The primary symptom of BPV is brief dizziness following a change in position. Ménière's disease combines vertigo with tinnitus and hearing loss.	Anant complains of dizzy attacks typically lasting up to an hour. BPV dizziness is very brief because it is caused by a change in position. It is more likely that Anant has Ménière's disease.
5	Why is limiting salt (NaCl) intake suggested as a treatment for Ménière's disease?	Ménière's disease is characterized by too much endolymph in the inner ear. Endolymph is an extracellular fluid.	Reducing salt intake should also reduce the amount of fluid in the extracellular compartment because the body will not retain water. Reduction of ECF volume may decrease fluid accumulation in the inner ear.
6	Why would severing the vestibular nerve alleviate Ménière's disease?	The vestibular nerve transmits information about balance and rotational movement from the vestibular apparatus to the brain.	Severing the vestibular nerve prevents false information about body rotation from reaching the brain, thus alleviating the vertigo of Ménière's disease.

321 325 343 349 352 354

CHAPTER SUMMARY

In this chapter you explored sensory receptors in the human body and learned how each type is designed to allow us to perceive different aspects of the world around us. Despite the unique characteristics of each sense, basic patterns emerge for sensory transduction and perception. Membrane proteins in the form of ion channels or G protein-linked receptors convert chemical, mechanical, thermal, and light energy into electrical signals that pass along sensory neurons to CNS integrating centers. The brain processes and filters incoming signals, sometimes acting upon

sensory information without it ever reaching conscious awareness. Many of the visceral reflexes that you will study in later chapters are unconscious responses to sensory input.

1. Sensory stimuli are divided into the **special senses** of vision, hearing, taste, smell, and equilibrium, and the **somatic senses** of touch, temperature, pain, itch, and proprioception. (p. 322; Tbl. 10-1)

General Properties of Sensory Systems

2. Sensory pathways begin with a stimulus that is converted by a receptor into an electrical potential. (p. 322)

3. If the stimulus is above threshold, action potentials pass along a sensory neuron to the central nervous system. We become aware of some stimuli but are never conscious of others. (p. 322)

4. Sensory receptors vary from free nerve endings to encapsulated nerve endings to specialized receptor cells. (p. 322; Fig. 10-1)

5. There are four types of sensory receptors, based on the stimulus to which they are most sensitive: **chemoreceptors, mechanoreceptors, thermoreceptors, photoreceptors.** (p. 323; Tbl. 10-2)

6. The specificity of a receptor for a particular type of stimulus is known as the **law of specific nerve energies.** (p. 323)

7. Each receptor type has an **adequate stimulus**, a particular form of energy to which it is most responsive. (p. 324)

8. If a stimulus is above **threshold**, it will create a graded potential in the receptor. (p. 324)

9. Multiple sensory neurons may converge on one secondary neuron and create a single large **receptive field**. (p. 324; Figs. 10-2, 10-3)

10. Sensory information from the spinal cord projects to the thalamus, then on to the sensory areas of the cortex. Only olfactory information is not routed through the thalamus. (p. 325; Fig. 10-4)

11. The central nervous system is able to change our level of awareness of sensory input. The **perceptual threshold** is the level of stimulus intensity necessary for us to be aware of a particular sensation. (p. 326)

12. The **modality** of a signal and its location are indicated by which sensory neurons are activated. The association of a receptor with a specific sensation is called **labeled line coding**. (pp. 326–327)

13. Localization of auditory information depends on the timing of receptor activation in each ear. (p. 327; Fig. 10-5)

14. **Lateral inhibition** enhances the contrast between the center and side of the receptive field. In **population coding** the brain uses input from multiple receptors to calculate location and timing of a stimulus. (p. 327; Fig. 10-6)

15. Stimulus intensity is coded by the number of receptors that are activated and by the frequency of their action potentials. (p. 327; Fig. 10-7)

16. For **tonic receptors**, the sensory neuron fires action potentials as long as the **generator potential** is above threshold. **Phasic receptors** respond to a change in stimulus intensity, but adapt if the strength of the stimulus remains constant. (pp. 324, 328; Fig. 10-8)

Somatic Senses

17. There are four somatosensory modalities: touch, proprioception, temperature, and nociception. (p. 330)

18. **Secondary sensory neurons** cross the midline so that one side of the brain processes information from the opposite side of the body. Ascending sensory tracts terminate in the **somatosensory cortex**. (pp. 330–331; Fig. 10-9, Tbl. 10-3)

19. Touch receptors come in many varieties. Temperature receptors sense heat and cold. (p. 332; Fig. 10-11, Tbl. 10-4)

20. **Nociceptors** are free nerve endings that respond to chemical, mechanical, or thermal stimuli that are perceived as pain and itch. (p. 333)

21. Some responses to irritants are protective **spinal reflexes**. (p. 333)

22. **Fast pain** is transmitted rapidly by small, myelinated fibers. **Slow pain** is carried on small, unmyelinated fibers. Pain may be modulated by descending pathways from the brain or by **gating** mechanisms in the spinal cord. (pp. 333–334; Fig. 10-12, Tbl. 10-5)

23. **Referred pain** from internal organs occurs when multiple primary sensory neurons converge onto a single ascending tract. (p. 334; Fig. 10-13)

Chemoreception: Smell and Taste

24. Chemoreception is divided into the special senses of smell (**olfaction**) and taste (**gustation**). (p. 336)

25. **Olfactory cells** in the nasal cavity are bipolar neurons whose pathways project directly to the olfactory cortex. They are the only neurons in the body that divide continuously. (p. 336; Fig. 10-14)

26. **Odorant receptors** are G protein-coupled membrane proteins. (p. 336)

27. Taste is a combination of five sensations: sweet, sour, salty, bitter, and **umami**. (p. 338)

28. **Taste cells** are non-neural cells with membrane proteins that interact with taste ligands. This interaction creates an intracellular Ca^{2+} signal that triggers exocytosis of neurotransmitter onto the primary sensory neuron. (p. 333; Figs. 10-15, 10-16)

The Ear: Hearing

29. **Hearing** is our perception of the energy carried by sound waves. Sound transduction turns air waves into mechanical vibrations, then fluid waves, chemical signals, and finally action potentials. (p. 341; Fig. 10-19)

30. The **cochlea** of the inner ear contains three parallel, fluid-filled ducts. The **cochlear duct** contains the **organ of Corti**, which contains **hair cell** receptors. (pp. 341, 343; Fig. 10-20)

31. When sound bends the hair cell cilia, the cells depolarize and release neurotransmitter onto the sensory neurons. (p. 343; Fig. 10-21)

32. The initial processing for pitch, loudness, and duration of sound takes place in the cochlea. Localization of sound is a higher function that requires sensory input from both ears and sophisticated computation by the brain. (pp. 343–346; Fig. 10-22)

The Ear: Equilibrium

33. **Equilibrium** is mediated through hair cells in the **vestibular apparatus** and **semicircular canals** of the inner ear. Gravity and acceleration provide the force that moves the cilia. (p. 347; Figs. 10-23, 10-24, 10-25)

The Eye and Vision

34. **Vision** is the translation of reflected light into a mental image. Photoreceptors of the **retina** transduce light energy into an electrical signal that passes to the visual cortex for processing. (pp. 350–351; Figs. 10-28, 10-29)

35. The amount of light entering the eye is altered by changing the size of the pupil. (p. 351)

36. Light waves are focused by the lens, whose shape is adjusted by contracting or relaxing the **ciliary muscle**. (p. 354; Figs. 10-31, 10-32)

37. Light is converted into electrical energy by the **photoreceptors** of the retina. Signals pass through bipolar neurons to ganglion cells, whose axons form the optic nerve. (p. 355; Fig. 10-35)

38. The **fovea** has the most acute vision because it has the smallest receptive fields. (p. 355; Fig. 10-36)

39. **Rods** are responsible for monochromatic nighttime vision. **Cones** are responsible for high-acuity vision and color vision during the daytime. (p. 357; Fig. 10-38)

40. Light-sensitive **visual pigments** in photoreceptors convert light energy into a change in membrane potential. The visual pigment in rods is **rhodopsin**. Cones have three related visual pigments. (p. 357; Fig. 10-39)

41. Rhodopsin is composed of **opsin** and **retinal**. In the absence of light, retinal binds snugly to opsin. (p. 359; Fig. 10-40)

42. When light bleaches rhodopsin, **retinal** is released and **transducin** begins a second messenger cascade that hyperpolarizes the rod and releases less neurotransmitter onto the bipolar neurons. (p. 359; Fig. 10-40)

43. Signals pass from photoreceptors through bipolar neurons to ganglion cells. These cells use contrast rather than absolute light intensity to code visual information (p. 360; Fig. 10-41)

44. **M ganglion cells** carry information about movement, location, and depth perception. **P ganglion cells** transmit signals that pertain to color, form, and texture of objects in the visual field. (p. 360)

45. Information from one side of the visual field is processed on the opposite side of the brain. Three-dimensional objects fall within the visual fields of both eyes. (pp. 360–361; Fig. 10-42)

QUESTIONS

LEVEL ONE Reviewing Facts and Terms

1. What is the role of the afferent division of the nervous system?

2. Define proprioception.

3. What are the common elements of all sensory pathways?

4. List and briefly describe the four major types of somatic receptors based on the type of stimulus to which they are most sensitive.

5. The receptors of each primary sensory neuron pick up information from a specific area, known as the _____.

6. Match the brain area with the sensory information processed there:
   (a) sound
   (b) odors
   (c) visual information
   (d) taste
   (e) equilibrium

   1. midbrain
   2. cerebrum
   3. medulla
   4. cerebellum
   5. none of the above

7. The conversion of stimulus energy into a change in membrane potential is called _____. The form of energy to which a

receptor responds is called its _____. The minimum stimulus required to activate a receptor is known as the _____.

8. When a sensory receptor membrane depolarizes (or hyperpolarizes in a few cases), the change in membrane potential is called the _____ or _____ potential. Is this a graded potential or an all-or-none potential?

9. Explain what is meant by adequate stimulus to a receptor.

10. The organization of sensory regions in the _____ of the brain preserves the topographical organization of receptors on the skin, eye, or other regions. However, there are exceptions to this rule. In what two senses does the brain rely on the timing of receptor activation to determine the location of the initial stimulus?

11. What is lateral inhibition?

12. Define tonic receptors and list some examples. Define phasic receptors and give some examples. Which type adapts? How is adaptation useful?

13. When heart pain is perceived as coming from the neck and down the left arm, this is an example of _____ pain.

14. What are the five basic tastes? What is the adaptive significance of each taste sensation?

15. The unit of sound wave measurement is _____, which measures the frequency of the sound waves per second. The loudness or intensity of a sound is a function of the _____ of the sound waves and is measured in _____. The range of hearing for the average human ear is from _____ to _____ [units], with the most acute hearing in the range of _____ to _____ [units].

16. Which structure of the inner ear codes sound for pitch? Define spatial coding.

17. Loud noises cause action potentials to:
    (a) fire more frequently
    (b) have higher amplitudes
    (c) have longer refractory periods

18. Once sound waves have been transformed into electrical signals in the cochlea, sensory neurons transfer information to the _____, from which collaterals then take the information to the _____ and _____. The main auditory pathway synapses in the _____ and _____ before finally projecting to the _____ in the _____.

19. The parts of the vestibular apparatus that tell our brain about our movements through space are the _____, which sense rotation, and the _____, which responds to linear forces. Movement is the _____ component, whereas the otolith organs also function in telling us about head position while standing, the _____ component.

20. Put these structures in the correct order for a beam of light entering the eye:
    (a) pupil,
    (b) cornea,
    (c) retina,
    (d) lens, and
    (e) aqueous humor.

21. The three primary colors of vision are _____, _____, and _____. White light containing these colors stimulates photoreceptors called _____. Lack of the ability to distinguish some colors is called _____.

LEVEL TWO *Reviewing Concepts*

22. Compare and contrast the following:
    (a) the special senses with the somatic senses
    (b) different types of touch receptors with respect to structure, size, and location
    (c) transmission of sharp localized pain with the transmission of dull and diffuse pain (include the particular fiber types involved as well as the presence or absence of myelin in your discussion)
    (d) the forms of hearing loss

23. Describe the neural pathways that link pain with emotional distress, nausea, and vomiting.

24. Trace the neural pathways involved in olfaction. What is G_{olf}?

25. Compare the process of signal transduction in the taste buds for salty/sour and sweet/bitter/umami ligands.

26. Put these structures in the order in which a sound wave would encounter them: (a) pinna, (b) cochlear duct, (c) stapes, (d) ion channels, (e) oval window, (f) hair cells/stereocilia, (g) tympanic membrane, (h) incus, (i) vestibular duct, and (j) malleus.

27. Sketch the structures and receptors of the vestibular apparatus for equilibrium. Label the components. Briefly describe how they function to notify the brain of movement.

28. Explain how accommodation by the eye occurs. What is the loss of accommodation called?

29. List four common visual problems and explain how they occur.

30. Explain how the intensity and duration of a stimulus are coded so that the stimulus can be interpreted by the brain. (Remember, action potentials are all-or-none phenomena.)

31. Map the following terms related to vision. Add additional terms if you wish.

accommodation reflex	lens
binocular vision	melanin
bipolar cells	opsin
bleaching	optic chiasm
blind spot	phototransduction
ciliary muscle	pupillary reflex
cones	retina
depth of field	retinal
field of vision	rhodopsin
focal point	rods
fovea	transducin
ganglion cells	visual field
lateral geniculate	zonules

32. Make a table of the special senses. In the first row, start with these stimuli: sound, standing on the deck of a rocking boat, light, a taste, an aroma. Make rows for the location of the receptor for the sense and the structure or properties of each receptor. Last, name the cranial nerve(s) that convey(s) each sensation to the brain. [✎ p. 298]

33. You are prodding your blindfolded lab partner's arm with two needle probes (with her permission). Sometimes she can tell you are using two probes, but when you probe less-sensitive areas, she thinks there is just one probe. What sense are you testing? Which receptors are being stimulated? Explain why she sometimes only feels one probe.

34. Consuming alcohol depresses the nervous system and vestibular apparatus. In a sobriety check, police officers use this information to determine if an individual is inebriated.

What kinds of tests can you suggest that would show evidence of this inhibition?

35. Often, children are brought to medical attention because of speech difficulties. If you were a clinician, which sense would you test first, and why?

36. A clinician shines a light into a patient's left eye and neither pupil constricts. Shining the light into the right eye elicits a normal consensual reflex. What problem in the reflex pathway could explain these observations?

Explorations

Explorations are also available at *www.physiologyplace.com*

We all live in the same world but different animals perceive the world differently. Dogs hear sounds we can't, nocturnal animals have better night vision than we do, and a stallion can smell if a mare is in estrus (ready to breed). An animal can only perceive stimuli for which it has sensory receptors. For example, radio waves fill the air around us, yet we only hear them if the radio is on and tuned to the proper frequency.

Exploration 1: Pheromones and the Human VNO

Do humans use pheromones for communication? There is tantalizing evidence to suggest that we do. But if we do, what sensory receptors are being activated? Anatomical evidence indicates that we do not have a functional vomeronasal organ (VNO).

On the other hand, one group of investigators using surface recordings of electrical activity from the VNO in conscious humans claims that sex-hormone-like steroids create a VNO response, whereas "non-pheromone" odorant molecules do not. Critics of these VNO electrical activity studies point out that (1) the work has never been duplicated, and (2) the investigators' research was partially funded by commercial groups who had potential financial gain from marketing pheromones discovered by these studies.

- Do you think that the research should be discounted for these reasons?

- What additional information would you like to know about the research before coming to a conclusion on its validity?

In 2001, a different group of investigators in Sweden used PET scans to monitor brain activity while subjects smelled sex-hormone-like compounds. Odorants related to male hormones activated the hypothalamus in female subjects, and

female hormone-like molecules activated the hypothalamus in male subjects.

- Do these findings change your interpretation of the earlier VNO research? What conclusions about pheromones and their receptors might you draw from the 2001 PET studies?

Here are citations for the studies cited, as well as some other articles about pheromones and the VNO. To learn more about the investigator who did the VNO research, try a Google search (*www.google.com*) for *Monti-Bloch* or do a MEDLINE author search for Monti-Bloch, L. by going to PubMed (*http://www.ncbi.nlm.nih.gov/PubMed/*) and type *"Monti-Bloch" [au]* in the SEARCH box.

Readings

"Human vomeronasal organ function: a critical review of best and worst cases." *Chemical Senses* 26:433–445, 2001.

"Effect of putative pheromones on the electrical activity of the human vomeronasal organ and olfactory epithelium." *Journal of Steroid Biochemistry and Molecular Biology* 39(4B):573–582, 1991.

"Smelling of odorous sex hormone-like compounds causes sex-differentiated hypothalamic activations in humans." *Neuron* 31:661–668, 2001 Aug 30.

"The scented brain: pheromonal responses in humans." *Neuron* 31:512–514, 2001 Aug 30.

"Histological definition of the vomeronasal organ in humans and chimpanzees, with a comparison to other primates." *The Anatomical Record* 267:166–176, 2002.

Exploration 2: Keeping Up with Changing Models

The sensory system is a good example of how scientists are constantly revising their models of how things work. For example,

(continues on next page)

the Emerging Concepts box on p. 357 describes a new photoreceptive pigment, melanopsin, that apparently plays a role in light reception for circadian rhythms. In the field of taste transduction, recently published research has made us revise our model of what taste cells sense and how they do it. To read about some of this research, see the articles below or try a Google search (www.google.com) for *melanopsin* or *taste transduction*.

"Circadian photoreception." *Science* 299:213–214, 2003 Jan 10.

"Diminished pupillary light reflex at high irradiances in melanopsin-knockout mice." *Science* 299:245–247, 2003 Jan 10.

"Bitter-sweet solution in taste transduction." *Cell* 112:283–284, 2003 Feb 7.

"Making sense of taste." *Scientific American*, 2001 Mar 18. Access through *www.sciam.com*.

"Sweet and salty: transduction in taste." *News in Physiological Sciences* 10:166–170, 1995 Aug. (free access at *http://nips. physiology.org*).

"Salt enhances flavour by suppressing bitterness." *Nature* 387:563, 1997 June 5.

ANSWERS

Answers to Concept Check Questions

Page 323

1. Myelinated axons have a faster conduction velocity than unmyelinated axons.

2. The pinna funnels sound into the ear canal.

3. Muscle length/tension, proprioception = mechano-. Pressure, inflation, distension = mechano-. Osmolarity = mechano-. Temperature = thermo-. Oxygen, glucose, pH = chemo-.

Page 327

4. Neuron B is probably hyperpolarizing neurons A and C by opening K^+ and Cl^- channels.

Page 328

5. Sensory neurons signal intensity of a stimulus by the rate at which they fire action potentials.

6. Nociceptors are telling the body that it is in danger. If possible, the body's response should be something that stops the harmful stimulus. Therefore it is important that signals continue as long as the stimulus is present.

Page 335

7. The adaptive advantage of a spinal reflex is a rapid reaction.

8. Responses to temperature and pain usually do not require rapid responses as balance and touch do.

9. There are many examples: taste receptors; touch receptors that adapt to pressure of clothes against skin.

Page 336

10. Olfactory cell (neuron) to secondary neuron in olfactory bulb to cranial nerve I to olfactory cortex in temporal lobe.

11. Use information in Figure 10-16 to create this map.

12. The knobby terminals of olfactory cells function as dendrites.

13. Olfactory neurons are bipolar neurons.

Page 340

14. Umami is associated with ingestion of the amino acid glutamate.

15. Taste cell to primary sensory neuron through cranial nerves VII, IX, or X to medulla (synapse with secondary neuron) to thalamus to gustatory cortex in parietal lobe.

Page 342

16. A kilohertz is 1000 hertz or waves per second.

Page 347

17. Follow pathway shown in Figure 10-19. Cochlear nerve to cranial nerve VIII to medulla. Main pathway goes to midbrain, thalamus, and auditory cortex in temporal lobe. Collateral pathways to reticular formation in brain stem and to cerebellum.

18. The location of somatosensory information is coded according to the side of the brain where the sensory pathway terminates. The location of sound is coded by the time a stimulus arrives in each hemisphere.

19. A cochlear implant would not help people with nerve deafness or conductive hearing loss. It can only help people with sensorineural hearing loss.

Page 349

20. Depolarization in hair cells is caused by K^+ entry into the cell.

21. When fluid builds up in the middle ear, the eardrum is unable to move freely and therefore cannot transmit sound through the bones of the middle ear as well.

22. When dancers spot, the endolymph in the ampulla moves with each head rotation, but then stops as the dancer holds his/her head still. This results in less inertia than if the head were continuously turning.

Page 351

23. The aqueous humor supports the cornea and lens. It also brings nutrients to and removes wastes from the epithelial layer of the cornea, which has no blood supply.

Page 352

24. (a) The sensory pathway from one eye diverges to activate motor pathways for both pupils. (b) Afferent path and its integration must be functioning because there is an appropriate response on the right. Motor path to left eye must not be functioning.

25. This is antagonistic control.

Page 354

26. In myopia, the lens and cornea bend light too sharply for the length of the eyeball, and the focal point falls in front of the retina. Myopia is corrected with a concave lens that scatters the light and increases the focal distance.

Page 357

27. The tapetum lucidum enhances light hitting the photoreceptors by reflecting it.

28. In both the retina and skin, the finest discrimination occurs in the region with the smallest visual or receptive fields.

Page 359

29. Our dark vision is black and white because only rods (black and white vision), not cones (color vision), are sensitive enough to be stimulated by such low levels of photons.

Page 360

30. Use the information in Figure 10-40 to create your map.

Answers to Figure Questions

Page 337

Figure 10-14: Multiple neurons synapsing on one is an example of convergence.

Page 341

Figure 10-18: Graph (b) shows 20 Hz waves. Graph (c) shows 32 Hz waves. The waves in (b) have lower pitch.

Page 355

Figure 10-33: (1) Convex lenses focus a beam of light; concave lenses scatter a beam of light passing through them. (2) In hyperopia, the focal point lies behind the retina, so the convex lens shortens the focal length. In myopia, the focal point lies in front of the retina, and a concave lens increases the focal length.

Page 356

Figure 10-35: Six rods converge on the ganglion cell.

Page 358

Figure 10-39: Red cones absorb light over the broadest spectrum, blue cones over the narrowest. At 500 nm, blue and green cones absorb light equally.

Efferent Division:
Autonomic and Somatic Motor Control

11

The neuromuscular junction, showing terminal Schwann cells covering axon terminals (red), somatic motor neuron (green), and nicotinic ACh receptors in the postsynaptic muscle membrane (blue). Can you find Schwann cells associated with the axon and a node of Ranvier?

RUNNING PROBLEM

A Powerful Addiction

Every day, more than 50 million Americans intentionally absorb a toxin that has been linked to the deaths of over 6 million people since 1964. Why would people knowingly poison themselves? If you've guessed that the toxin is nicotine, you already know part of the answer. One of more than four thousand chemicals found in tobacco, nicotine is highly addictive. So powerful is this addiction that less than 10% of tobacco users are able to quit smoking the first time they try. Shanika, a smoker for six years, is attempting for the second time to quit smoking. But the odds are in her favor because, this time, Shanika has sought help. She has decided to try the nicotine patch.

BACKGROUND BASICS

Neuron structure: **242** | Membrane receptors: **129** | Synapses: **244** | Neurotransmitters: **174** | Second messenger systems: **177–178** | Catecholamines: **218** | Action potentials: **253** | Slow synaptic potentials: **268** | Up- and down-regulation: **186** | Tonic and antagonist control: **188–189** |

Because a number of cells in the autonomic nervous system act in conjunction, they have relinquished their independence to function as a coherent whole.

—Otto Appenzeller and Emilio Oribe, in *The Autonomic Nervous System*, 1997

The picnic lunch was wonderful. You are now dozing in the warm spring sunlight on the grass as you let the meal digest. Suddenly you feel something moving across your lower leg. You open your eyes, and as they adjust to the bright light, you see a four-foot-long snake slithering over your foot. More by instinct than reason, you fling the snake into the grass while scrambling to a safe perch on top of the nearby picnic table. You are breathing heavily and your heart is pounding.

In less than a second, your body has gone from a state of quiet rest and digestion to a state of panic and frantic activity. This reflex reaction is integrated and coordinated through the central nervous system (CNS), then carried out by the efferent division of the peripheral nervous system. The fibers of efferent neurons are bundled together into nerves that carry commands from the CNS to the muscles and glands of the body. Some nerves, called *mixed nerves*, also carry sensory information through afferent fibers.

The efferent side of the peripheral nervous system has two divisions: (1) the **somatic motor neurons**, which control skeletal muscles, and (2) the **autonomic neurons**, which control smooth muscle, cardiac muscle, many glands, and some adipose tissue. The somatic and autonomic divisions are sometimes called the voluntary and involuntary divisions of the nervous system. Although it is true that movement controlled by somatic pathways usually requires conscious thought and that autonomic reflexes are mainly involuntary, this distinction does not always hold true.

For example, some skeletal muscle reflexes, such as swallowing and the knee jerk reflex, are involuntary. And a person can learn to modulate some involuntary autonomic functions, such as heart rate and blood pressure, through biofeedback training. Thus, the classification of efferent divisions into voluntary and involuntary branches is a broad generalization that is not accurate in all situations.

We begin by looking at the autonomic division. Then we will consider the somatic motor division, as preparation for learning about muscles in Chapter 12.

▶ THE AUTONOMIC DIVISION

The autonomic division of the efferent nervous system (or *autonomic nervous system* for short) is also known in older writings as the *vegetative nervous system*, reflecting the observation that its functions were not under voluntary control. The name *autonomic* comes from the same roots as *autonomous*, meaning *self-governing*. You will also find the autonomic division called the *visceral nervous system* because of its control over internal organs and functions such as blood pressure and heart rate.

The autonomic division is subdivided into **sympathetic** and **parasympathetic branches**. Some parts of the sympathetic branch were first described by the Greek physician Claudius Galen (ca. 130–200), who is famous for his compilation of anatomy, physiology, and medicine as they were known up to his time. As a result of his dissections, Galen proposed that "animal spirits" flowed from the brain to the tissues through hollow nerves, creating "sympathy" between the different parts of the body. Galen's "sympathy"

later gave rise to the name for the sympathetic branch. The prefix *para-*, for the parasympathetic branch, means *beside*, or *alongside of*.

Although the sympathetic and parasympathetic branches can be distinguished anatomically, there is no simple way to separate the actions of the two branches on their targets. The best generalization is to characterize them according to the type of situation in which they are most active. The picnic scene that began the chapter typifies the two extremes at which the sympathetic and parasympathetic branches function. If you are resting quietly after a meal, the parasympathetic branch is dominant, taking command of the routine, quiet activities of day-to-day living such as digestion. Consequently, parasympathetic neurons are sometimes said to control "rest and digest" functions.

In contrast, the sympathetic branch is dominant in stressful situations, such as the potential threat from the snake. One of the most dramatic examples of sympathetic action is the **fight-or-flight** response, in which the brain triggers massive simultaneous sympathetic discharge throughout the body. As the body prepares to fight or flee, the heart speeds up, blood vessels to muscles of the arms, legs, and heart dilate, and the liver starts to produce glucose to provide energy for muscle contraction. Digestion becomes a low priority when life and limb are threatened, so blood is diverted from the gastrointestinal tract to skeletal muscles.

The massive sympathetic discharge that occurs in fight-or-flight situations is mediated through the hypothalamus and is a total-body response to a crisis. If you have ever been scared by the squealing of brakes or a sudden sound in the dark, you know how rapidly the nervous system can influence multiple systems of the body. Most sympathetic responses are not the all-out response of a fight-or-flight reflex, however, and *activating one sympathetic pathway does not automatically activate them all*.

The role of the sympathetic nervous system in mundane daily activities is as important as a fight-or-flight response. For example, one key function of the sympathetic branch is control of blood flow to the tissues, as you will learn in Chapter 15. Most of the time, autonomic control of body function shifts back and forth between the sympathetic and parasympathetic branches, like the seesaw in Figure 11-1 ●, as the two branches cooperate to fine-tune various processes. Only occasionally, as in the fight-or-flight example above, does the seesaw move to one extreme or the other.

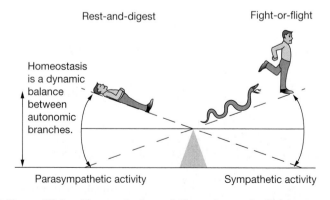

● **Figure 11-1** *Homeostasis and the autonomic division*

CONCEPT CHECK   Answers: p. 388

1. The afferent division of the nervous system consists of _____ and _____.

2. The central nervous system consists of _____ and _____.

Autonomic Reflexes Are Important for Homeostasis

The autonomic nervous system works closely with the endocrine system and the behavioral state system [p. 301] to maintain homeostasis in the body. Sensory information from somatosensory and visceral receptors goes to homeostatic control centers in the hypothalamus, pons, and medulla of the brain (Fig. 11-2). These centers monitor and regulate important functions such as blood pressure, temperature regulation, and water balance (Fig. 11-3). The hypothalamus itself also contains neurons that act as sensors, such as osmoreceptors that monitor osmolarity and thermoreceptors that monitor body temperature. Motor output from the hypothalamus and brain stem creates autonomic responses, endocrine responses, and behavioral responses such as drinking,

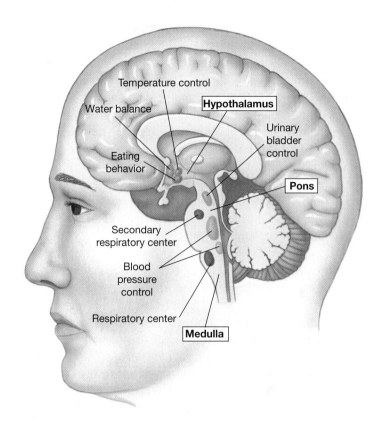

● **Figure 11-3** *Autonomic control centers in the brain*

food-seeking, and temperature regulation (getting out of the heat, putting on a sweater). The behavioral responses are integrated in brain centers responsible for motivated behaviors and control of movement.

In addition, sensory information integrated in the cerebral cortex and limbic system can create emotions that influence autonomic output. Blushing, fainting at the sight of a hypodermic needle, and "butterflies in the stomach" are all examples of emotional influences on autonomic functions. Understanding the autonomic and hormonal control of organ systems is the key to understanding maintenance of homeostasis in virtually every system of the body.

Some autonomic reflexes are capable of taking place without input from the brain. These *spinal reflexes* [Fig. 9-8, p. 295] include urination, defecation, and penile erection, bodily functions that can be influenced by descending pathways from the brain but that do not require this input. For example, people with spinal cord injuries that disrupt communication between the brain and spinal cord may retain some spinal reflexes but lose the ability to sense or control them.

Antagonistic Control Is a Hallmark of the Autonomic Division

The sympathetic and parasympathetic branches of the autonomic nervous system together display all four of Walter Cannon's properties of homeostasis: (1) maintenance of the internal environment,

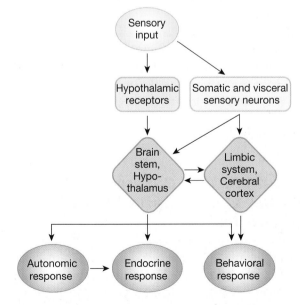

● **Figure 11-2** *The hypothalamus and brain stem initiate autonomic, endocrine, and behavioral responses*

(2) "up-down" regulation by varying tonic control, (3) antagonistic control, and (4) variable tissue responses that result from differences in membrane receptors [◈ p. 188].

Most internal organs are under *antagonistic control*, where one autonomic branch is excitatory and the other branch is inhibitory (see table in Fig. 11-5). For example, sympathetic innervation increases heart rate and parasympathetic stimulation decreases it. Consequently, heart rate can be regulated by altering the relative proportion of sympathetic and parasympathetic control.

Exceptions to dual antagonistic innervation include the sweat glands and smooth muscle in most blood vessels. These tissues are innervated only by the sympathetic branch and rely strictly on tonic (up-down) control.

Although the two autonomic branches are usually antagonistic in their control of a single target tissue, they sometimes work cooperatively on different tissues to achieve a common goal. For example, in the male sex act, blood flow for penile erection is under control of the parasympathetic branch and muscle contraction for sperm ejaculation is directed by the sympathetic branch.

In some autonomic pathways, the neurotransmitter receptor determines the response of the target tissue. For instance, most blood vessels contain one type of *adrenergic receptor* that causes smooth muscle contraction (vasoconstriction). But some blood vessels contain a different subtype of adrenergic receptor that causes smooth muscle to relax (vasodilation). Both receptors are activated by catecholamines. In this example, the receptor rather than the chemical signal determines the response.

CONCEPT CHECK   Answers: p. 388

3. Define homeostasis.

Autonomic Pathways Have Two Efferent Neurons in Series

The sympathetic and parasympathetic branches of the autonomic division share the same general structure (Fig. 11-4 ●). All autonomic pathways consist of two neurons in series. The first neuron, called the **preganglionic neuron**, originates in the central nervous system and projects to an **autonomic ganglion** outside the CNS. There the preganglionic neuron synapses with the second neuron in the pathway, the **postganglionic neuron**. This

Autonomic pathways consist of two neurons that synapse in an autonomic ganglion.

Preganglionic neuron   Postganglionic neuron

CNS   Autonomic ganglion   Target tissue

● **Figure 11-4** *Autonomic pathways*

EMERGING CONCEPTS

Ganglia as Integrating Centers

According to the traditional view of the autonomic division, a signal that originates in the CNS passes directly along a two-neuron chain to the target tissue. In that model, the ganglion is simply a way-station for transfer of signals from preganglionic neurons to postganglionic neurons. But in recent years, scientists have discovered that ganglia can act as mini-integrating centers, receiving sensory input from the periphery and modulating the outgoing responses. Chemical studies show that autonomic interneurons, particularly in sympathetic ganglia, secrete dopamine, which acts as an inhibitory neurotransmitter. The pathways modified by these neurons are still being investigated. It has recently been shown that some sensory receptors in the abdomen are linked directly to postganglionic sympathetic neurons by ganglionic interneurons. Presumably this means that a sensory signal could influence the efferent output of the sympathetic neuron without integration by the CNS.

neuron has its cell body in the ganglion and projects its axon to the target tissue. (A *ganglion* is a cluster of nerve cell bodies that lie outside the CNS. The equivalent in the CNS is a *nucleus*.)

Autonomic ganglia, where the two efferent neurons synapse, are more than a simple collection of axon terminals and nerve cell bodies. The ganglia also contain interneurons that lie completely within them. These ganglionic interneurons can receive sensory information and modulate messages moving through the autonomic neurons, making the ganglia integrating centers for autonomic reflexes.

Divergence [◈ p. 270] is an important feature of autonomic pathways. On average, one preganglionic neuron entering a ganglion synapses with eight or nine postganglionic neurons. Some synapse on as many as 32 neurons! Each postganglionic neuron may innervate a different target, meaning that a single signal from the CNS can affect a large number of target cells simultaneously.

Sympathetic and Parasympathetic Branches Exit the Spinal Cord in Different Regions

How then do the two autonomic branches differ anatomically? The main anatomical differences are (1) where the pathways originate in the CNS and (2) the location of the autonomic ganglia. As Figure 11-5 ● shows, most sympathetic pathways (red) originate in the thoracic and lumbar regions of the spinal cord. *Sympathetic ganglia* are found in a chain that runs close to the

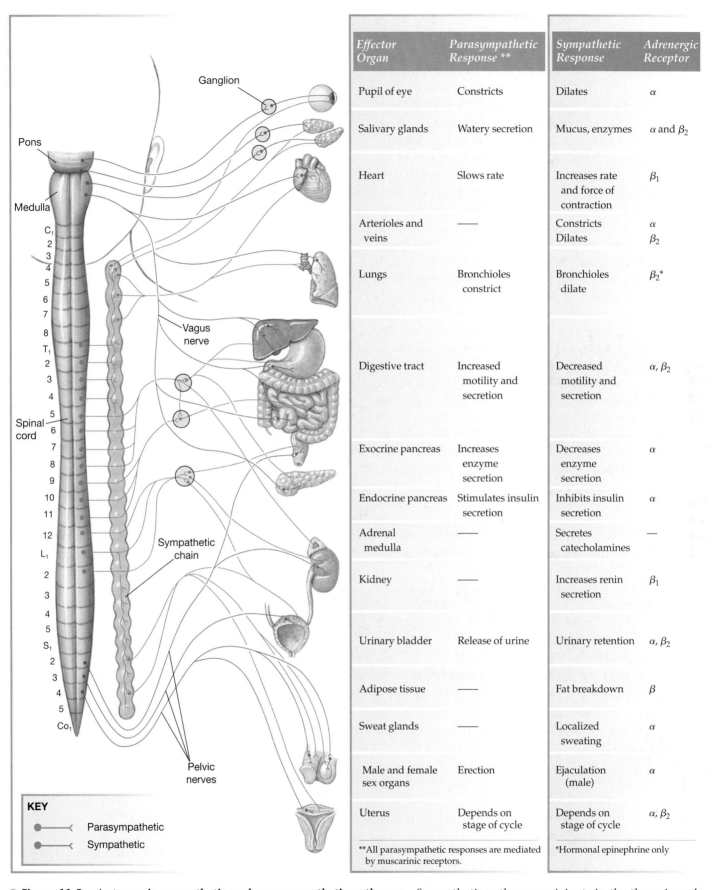

Effector Organ	Parasympathetic Response **	Sympathetic Response	Adrenergic Receptor
Pupil of eye	Constricts	Dilates	α
Salivary glands	Watery secretion	Mucus, enzymes	α and β_2
Heart	Slows rate	Increases rate and force of contraction	β_1
Arterioles and veins	——	Constricts / Dilates	α / β_2
Lungs	Bronchioles constrict	Bronchioles dilate	β_2*
Digestive tract	Increased motility and secretion	Decreased motility and secretion	α, β_2
Exocrine pancreas	Increases enzyme secretion	Decreases enzyme secretion	α
Endocrine pancreas	Stimulates insulin secretion	Inhibits insulin secretion	α
Adrenal medulla	——	Secretes catecholamines	——
Kidney	——	Increases renin secretion	β_1
Urinary bladder	Release of urine	Urinary retention	α, β_2
Adipose tissue	——	Fat breakdown	β
Sweat glands	——	Localized sweating	α
Male and female sex organs	Erection	Ejaculation (male)	α
Uterus	Depends on stage of cycle	Depends on stage of cycle	α, β_2

**All parasympathetic responses are mediated by muscarinic receptors.

*Hormonal epinephrine only

KEY
- Parasympathetic
- Sympathetic

● **Figure 11-5** *Autonomic sympathetic and parasympathetic pathways* Sympathetic pathways originate in the thoracic and lumbar regions and the ganglia are located close to the spinal column or along the descending aorta. Parasympathetic pathways originate in the brain stem or sacral region and the ganglia are located on or near their target organs.

spinal column or along the descending aorta. Long nerves (axons) project to the target tissues. Because most sympathetic ganglia lie close to the spinal cord, sympathetic pathways generally have short preganglionic neurons and long postganglionic neurons.

Many parasympathetic pathways (shown in blue in Figure 11-5) originate in the brain stem and their axons leave the brain in several cranial nerves [● p. 298]. Other parasympathetic pathways originate in the sacral region (near the lower end of the spinal cord) and control pelvic organs. In general, parasympathetic ganglia are located on or near their target organs. Consequently, parasympathetic preganglionic neurons have long axons and postganglionic neurons have short axons.

Parasympathetic innervation goes primarily to the head, neck, and internal organs. The major parasympathetic tract is the **vagus nerve** (cranial nerve X), which contains about 75% of all parasympathetic fibers. The vagus nerve carries both sensory neurons from and parasympathetic fibers to most of the internal organs (Fig. 11-6 ●). *Vagotomy*, a procedure in which the vagus nerve is surgically cut, was an experimental technique used in the nineteenth and early twentieth centuries to study the impact of the autonomic nervous system on various organs. For a time, vagotomy was the preferred treatment for stomach ulcers because removal of parasympathetic innervation to the stomach decreased the secretion of stomach acid. However, this procedure had many unwanted side effects and has been abandoned in favor of drug therapies that treat the problem more specifically.

CONCEPT CHECK   Answers: p. 388

4. A nerve that carries both sensory and motor information is called a _____ nerve.

5. Name the four regions of the spinal cord in order, starting from the brain stem.

The Autonomic Nervous System Uses a Variety of Neurotransmitters and Modulators

Chemically, the sympathetic and parasympathetic branches can be distinguished by their neurotransmitters and receptors, using the following rules:

1. All autonomic preganglionic neurons release acetylcholine (ACh) onto *cholinergic nicotinic receptors* [● p. 267].
2. Most postganglionic sympathetic neurons secrete norepinephrine onto *adrenergic receptors*.
3. Most postganglionic parasympathetic neurons secrete acetylcholine onto *cholinergic muscarinic receptors*.

These rules are illustrated in Figure 11-7 ●. However, there are some exceptions. A few sympathetic neurons, such as those

that terminate on sweat glands, secrete ACh rather than norepinephrine. These neurons are therefore called *sympathetic cholinergic neurons*.

A small number of autonomic neurons secrete neither norepinephrine nor acetylcholine and are known as *non-adrenergic, non-cholinergic neurons*. Some of the chemicals they use as neurotransmitters include substance P, somatostatin, vasoactive intestinal peptide (VIP), adenosine, nitric oxide, and ATP. We will

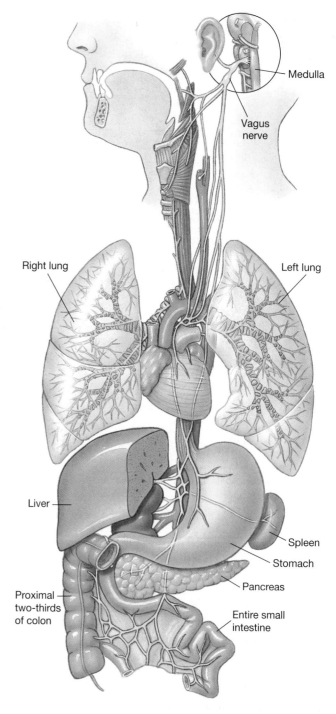

● **Figure 11-6** *The vagus nerve* The vagus nerve carries parasympathetic fibers to most of the internal organs.

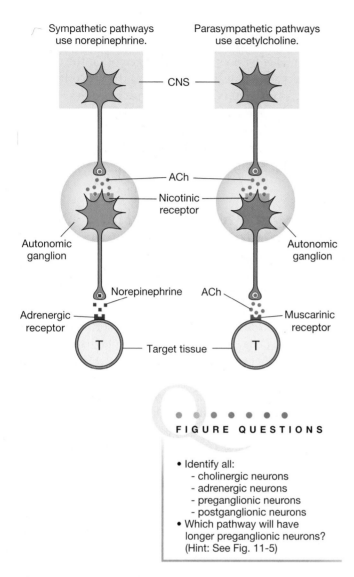

Sympathetic pathways use norepinephrine.

Parasympathetic pathways use acetylcholine.

CNS

ACh

Nicotinic receptor

Autonomic ganglion

Autonomic ganglion

Norepinephrine

ACh

Adrenergic receptor

Muscarinic receptor

T

T

Target tissue

FIGURE QUESTIONS

• Identify all:
  - cholinergic neurons
  - adrenergic neurons
  - preganglionic neurons
  - postganglionic neurons
• Which pathway will have longer preganglionic neurons? (Hint: See Fig. 11-5)

● **Figure 11-7** *Sympathetic and parasympathetic pathways*

Autonomic Pathways Control Smooth Muscle, Cardiac Muscle, and Glands

The targets of autonomic neurons are smooth muscle, cardiac muscle, many exocrine glands, a few endocrine glands, and some adipose tissue. The synapse between a postganglionic autonomic neuron and its target cell is called the **neuroeffector junction**.

The structure of an autonomic synapse differs from the model synapse shown in Chapter 8 [◉ p. 242]. Autonomic axons end with a series of swollen areas at their distal ends, like beads spaced out along a string (Fig. 11-8 ◉). Each of these swellings, known as a **varicosity** [*varicosus*, abnormally enlarged or swollen], contains vesicles filled with neurotransmitter.

The branched ends of the axon lie across the surface of the target tissue, but the underlying membrane does not possess clusters of neurotransmitter receptors in specific sites. Instead, the neurotransmitter is simply released into the interstitial fluid to diffuse to wherever the receptors are, often some distance from the site of release. The result is a less-directed form of communication than that which occurs between a somatic motor neuron and a skeletal muscle. The diffuse release of autonomic neurotransmitter means that a single postganglionic neuron can affect a large area of target tissue.

The release of autonomic neurotransmitters is subject to modulation from a variety of sources. For example, sympathetic varicosities contain receptors for hormones and for paracrines such as histamine. These modulators may either facilitate or inhibit neurotransmitter release. Some preganglionic neurons cosecrete neuropeptides along with acetylcholine. The peptides act as neuromodulators, producing slow synaptic potentials that modify the activity of postganglionic neurons [◉ p. 268].

mention these unusual neurotransmitters in later chapters when they play a significant role. The non-adrenergic, non-cholinergic neurons are assigned to either the sympathetic or parasympathetic branch according to the location where their preganglionic fibers leave the nerve cord.

RUNNING PROBLEM

Nicotinic cholinergic receptors (nAChR) are found throughout the peripheral nervous system and the CNS. Researchers speculate that nicotinic receptors in certain regions of the brain may play a key role in nicotine addiction.

Question 2: Cholinergic receptors are classified as either nicotinic or muscarinic, on the basis of the agonist molecules that bind to them. What happens to the postsynaptic cell when nicotine rather than ACh binds to a nicotinic cholinergic receptor?

369 371 🏃 379 382 383

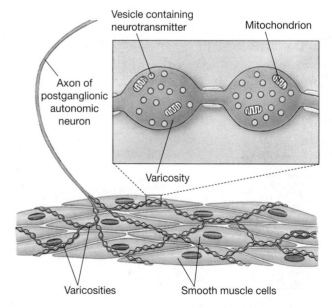

Vesicle containing neurotransmitter

Mitochondrion

Axon of postganglionic autonomic neuron

Varicosity

Varicosities

Smooth muscle cells

● **Figure 11-8** *Varicosities of autonomic neurons*
Varicosities are found along the distal end of the postganglionic axon.

Neurotransmitter Activity at Target Tissues Is Limited

In the autonomic division, neurotransmitter synthesis takes place in the axon varicosities. The primary autonomic neurotransmitters are acetylcholine (ACh) and norepinephrine, both small molecules that are easily synthesized by cytoplasmic enzymes. Neurotransmitter made in the varicosities is packaged into synaptic vesicles for storage.

The process of neurotransmitter release follows the same pattern you learned in Chapter 8. When an action potential arrives at the varicosity, voltage-gated Ca^{2+} channels open, Ca^{2+} enters the neuron, and the synaptic vesicle contents are released by exocytosis. Once neurotransmitters are released into the synapse, they diffuse through the interstitial fluid until they encounter a receptor on the effector cell or drift away from the synapse.

The concentration of neurotransmitter in the synapse is a major factor in the control that an autonomic neuron exerts on its target: the more neurotransmitter, the longer or stronger the response. The concentration of neurotransmitter in a synapse is influenced by its rate of breakdown or removal. Neurotransmitter action terminates when the neurotransmitter (1) diffuses away, (2) is metabolized by enzymes in the extracellular fluid, or (3) is actively transported into cells around the synapse. The return of neurotransmitter to varicosities allows neurons to reuse the chemicals.

These steps were illustrated for acetylcholine in Figure 8-21 [⬗ p. 266] and are shown for norepinephrine in Figure 11-9 ●.

Norepinephrine is synthesized in the varicosity from the amino acid tyrosine. Once released into the synapse, the neurotransmitter may combine with an adrenergic receptor on the target cell, diffuse away, or be transported back into the varicosity. Back in the neuron, norepinephrine is either repackaged into vesicles or broken down by **monoamine oxidase** (MAO), the main enzyme responsible for degradation of catecholamines.

Table 11-1 compares metabolism of the two primary autonomic neurotransmitters.

The Adrenal Medulla Secretes Catecholamines

The adrenal medulla is a specialized neuroendocrine tissue associated with the sympathetic division. During development, the neural tissue destined to secrete the catecholamines norepinephrine and epinephrine splits into two functional entities: the sympathetic branch of the nervous system, which secretes norepinephrine, and the adrenal medulla, which secretes epinephrine. The adrenal medulla forms the core of the adrenal glands, which sit on top of the kidneys (Fig. 11-10a ●). Like the pituitary gland, each *adrenal gland* is actually two glands of different embryological origin that fused during development (Fig. 11-10b). The outer portion, the *adrenal cortex*, is a true endocrine gland of epidermal origin that secretes steroid hormones [⬗ p. 70]. The **adrenal medulla** [*ad-*, upon + *renal*, kidney; *medulla*, marrow] forms the small core of the gland and is a neurosecretory structure.

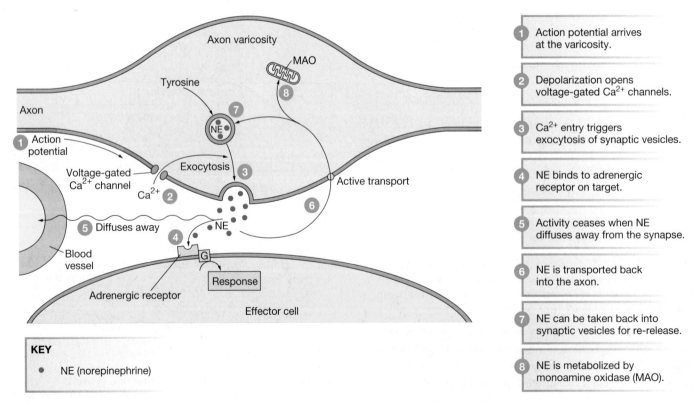

KEY
● NE (norepinephrine)

1. Action potential arrives at the varicosity.
2. Depolarization opens voltage-gated Ca^{2+} channels.
3. Ca^{2+} entry triggers exocytosis of synaptic vesicles.
4. NE binds to adrenergic receptor on target.
5. Activity ceases when NE diffuses away from the synapse.
6. NE is transported back into the axon.
7. NE can be taken back into synaptic vesicles for re-release.
8. NE is metabolized by monoamine oxidase (MAO).

● **Figure 11-9** *Norepinephrine release at a varicosity of a sympathetic neuron*

Table 11-1: Autonomic Neurotransmitters

	SYMPATHETIC DIVISION	PARASYMPATHETIC DIVISION
Neurotransmitter	Norepinephrine	Acetylcholine (ACh)
Synthesized from	Tyrosine	Acetyl CoA + choline
Inactivated by (enzyme)	Monoamine oxidase (MAO)	Acetylcholinesterase (AChE)
Enzyme located in	Mitochondria of varicosity	Synaptic cleft
Varicosity reuptake	Norepinephrine	Choline

The adrenal medulla is often described as a *modified sympathetic ganglion.* Preganglionic sympathetic neurons project from the spinal cord to the medulla, where they synapse. However, the postganglionic neurons lack axons that would normally project to target cells (Fig. 11-10c). Instead, the cell bodies, called *chromaffin cells,* secrete the neurohormone epinephrine directly into the blood. In response to alarm signals from the CNS, the adrenal medulla releases large amounts of epinephrine for general distribution throughout the body as part of a fight-or-flight response.

Most Sympathetic Pathways Secrete Norepinephrine onto Adrenergic Receptors

Norepinephrine from sympathetic neurons and epinephrine from the adrenal medulla bind to adrenergic receptors on their target cells. Adrenergic receptors come in two varieties: α (alpha) and β (beta), with several subtypes of each. Alpha receptors are the most common sympathetic receptor. They respond strongly to norepinephrine and only weakly to epinephrine (Table 11-2).

Beta receptors have two main subtypes, β_1 and β_2, that differ in their affinity for catecholamines. **β_1 receptors** respond equally strongly to norepinephrine and epinephrine. **β_2 receptors** are more sensitive to epinephrine than to norepinephrine. Interestingly, the β_2 receptors are not innervated (no sympathetic neurons terminate near them), which limits their exposure to the neurotransmitter norepinephrine.

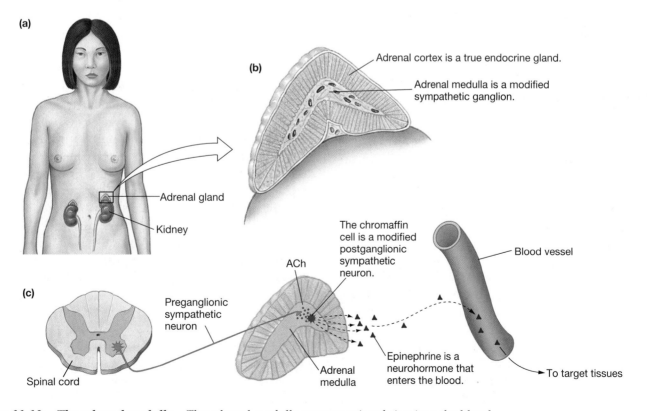

(a)

Adrenal gland

Kidney

(b)

Adrenal cortex is a true endocrine gland.

Adrenal medulla is a modified sympathetic ganglion.

(c)

Spinal cord

Preganglionic sympathetic neuron

ACh

The chromaffin cell is a modified postganglionic sympathetic neuron.

Adrenal medulla

Epinephrine is a neurohormone that enters the blood.

Blood vessel

To target tissues

● **Figure 11-10**   *The adrenal medulla*   The adrenal medulla secretes epinephrine into the blood.

Table 11-2: Sensitivity of Peripheral Adrenergic Receptors to Catecholamines

RECEPTOR	FOUND IN	SENSITIVITY	SECOND MESSENGER
α_1	Most sympathetic target tissues	NE > E	Activates phospholipase C
α_2	Gastrointestinal tract and pancreas	NE > E	Inhibits cAMP
β_1	Heart muscle, kidney	NE = E	Activates cAMP
β_2	Certain blood vessels and smooth muscle of some organs	E > NE	Activates cAMP

NE = norepinephrine, E = epinephrine.

All adrenergic receptors are linked to G proteins but the different subtypes use different second messenger pathways (Table 11-2). Catecholamine binding to α_1 receptors activates phospholipase C and IP_3. This creates intracellular Ca^{2+} signals by opening Ca^{2+} channels in the cell membrane, releasing Ca^{2+} from intracellular stores, or doing both. In general, activation of α_1 receptors causes muscle contraction or secretion by exocytosis. Activated α_2 receptors decrease cyclic AMP and cause smooth muscle relaxation (gastrointestinal tract) or decreased secretion (pancreas).

Catecholamine binding to β receptors activates cyclic AMP and triggers the phosphorylation of intracellular proteins. Activation of β_1 receptors by norepinephrine or epinephrine enhances cardiac muscle contraction and speeds up electrical conduction in cardiac muscle. Activation of β_2 receptors by epinephrine results in smooth muscle relaxation in many tissues.

For all adrenergic receptors, second messenger activity within the target tissue can persist for a longer time than is normally associated with the rapid action of the nervous system. These ongoing metabolic effects result from modification of existing proteins or from the synthesis of new proteins. We will discuss the specific effects of catecholamines on various tissues in subsequent chapters.

CONCEPT CHECK
Answers: p. 388

8. In what organelle is most intracellular Ca^{2+} stored?

9. What enzyme (a) converts ATP to cAMP? (b) does cAMP activate? [Fig. 6-10, p. 179]

Parasympathetic Pathways Secrete Acetylcholine onto Muscarinic Receptors

As a rule, parasympathetic neurons release ACh at their targets. *Cholinergic muscarinic receptors* [p. 267] are found at the neuroeffector junctions of the parasympathetic branch. Muscarinic receptors are all G protein-coupled receptors linked either to second messenger systems or to gated K^+ channels. The tissue response to activation of a muscarinic receptor varies with the receptor subtype, of which there are at least five.

Autonomic Agonists and Antagonists Are Important Tools in Research and Medicine

The study of the autonomic branches has been greatly simplified by advances in molecular biology. The genes for many autonomic receptors and their subtypes have been cloned, allowing researchers to create mutant receptors and study their properties. In addition, researchers have discovered or synthesized a variety of agonists and antagonist molecules (Table 11-3). Direct agonists and antagonists combine with the target receptor to mimic or block neurotransmitter action. Indirect agonists and antagonists act by altering secretion, reuptake, or degradation of neurotransmitters.

For example, cocaine blocks the reuptake of norepinephrine into adrenergic nerve terminals, thereby extending norepinephrine's stimulatory effect on the target. The drug *neostigmine* is a cholinesterase inhibitor that is used to treat *myasthenia gravis* [p. 266], a condition in which skeletal muscles have fewer-than-normal ACh receptors. Neostigmine blocks ACh degradation and extends the active life of each ACh molecule. The toxic *organophosphate insecticides* are also cholinesterase inhibitors.

Many of the drugs used to treat depression act either on membrane transporters for neurotransmitters (tricyclic antidepressants and selective serotonin reuptake inhibitors) or on their metabolism (monoamine oxidase inhibitors). The older antidepressant drugs that act on norepinephrine transport and metabolism (tricyclics and MAO inhibitors) may have side effects related to their actions in the autonomic nervous system, including cardiovascular problems, constipation, urinary difficulty, and sexual dysfunction [*dys-*, abnormal or ill]. The newer serotonin reuptake inhibitors have fewer autonomic side effects because serotonin is primarily a CNS neurotransmitter.

Many new drugs have been developed from our studies of agonists and antagonists. The discovery of α and β adrenergic receptors led to the development of drugs that block only one of the two receptor types. Antagonists of α receptors proved to have little clinical significance, but the drugs known as beta-blockers have given physicians a powerful tool for treating high blood pressure, one of the most common disorders in the United States today.

Table 11-3: Agonists and Antagonists of Neurotransmitter Receptors

RECEPTOR	AGONISTS	ANTAGONISTS	INDIRECT AGONISTS/ANTAGONISTS
Cholinergic	Acetylcholine		*AChE* inhibitors*: neostigmine, parathion *Inhibit ACh release*: botulinus toxin
Muscarinic	Muscarine	Atropine, scopolamine	
Nicotinic	Nicotine	α-bungarotoxin (muscle only), tetraethylammonium (TEA) (ganglia only), curare	
Adrenergic	Norepinephrine, epinephrine		*Stimulate NE release*: ephedrine, amphetamines *Prevent NE uptake*: cocaine
Alpha (α)	Phenylephrine	"Alpha-blockers"	
Beta (β)	Isoproterenol	"Beta-blockers": propranolol (β_1 and β_2), metoprolol (β_1 only)	

* AChE = acetylcholinesterase.

Primary Disorders of the Autonomic Nervous System Are Relatively Uncommon

Diseases and malfunction of the autonomic nervous system are relatively rare. Direct damage (trauma) to hypothalamic control centers may cause abnormalities in water balance or temperature regulation. Generalized sympathetic dysfunction may result from systemic diseases such as cancer and diabetes mellitus. There are also some conditions, such as *multiple system atrophy*, in which the CNS control centers for autonomic functions degenerate. In many cases of sympathetic dysfunction, the symptoms are mani-

fested most strongly in the cardiovascular system, when diminished sympathetic input to blood vessels results in abnormally low blood pressure. Other prominent symptoms of sympathetic pathology include urinary *incontinence* [*in-*, unable + *continere*, to contain], the loss of bladder control, and *impotence*, the inability to obtain or sustain a penile erection.

Occasionally, patients suffer from primary autonomic failure when sympathetic neurons degenerate. In the face of continuing diminished sympathetic input, target tissues *up-regulate* [◈ p. 186], putting more receptors into the cell membrane to maximize the cell's response to available norepinephrine. This increase in receptor number leads to *denervation hypersensitivity*, a state in which the administration of exogenous adrenergic agonists causes a greater-than-expected response.

CLINICAL FOCUS

Diabetes

Diabetic Autonomic Neuropathy

Primary disorders of the autonomic division are rare, but the condition known as **diabetic autonomic neuropathy** is quite common. This complication of diabetes often begins as a sensory neuropathy, with tingling and loss of sensation in the hands and feet. In some patients, pain is the primary symptom. About 30% of diabetic patients go on to develop autonomic neuropathies, manifested by dysfunction of cardiovascular, gastrointestinal, urinary, and reproductive systems (abnormal heart rate, constipation, incontinence, impotence). The cause of diabetic neuropathies is not clear. Patients who have chronically elevated blood glucose levels are more likely to develop neuropathies but the underlying metabolic pathway has not been identified. Other contributing factors for neuropathy include decreased blood flow and autoimmune reactions. At this time there is no prevention for diabetic neuropathies other than controlling blood glucose levels, and no cure, only drugs that treat the symptoms.

RUNNING PROBLEM

One research study that examined brains at autopsy found that smokers have a greater number of receptors that bind nicotine than do nonsmokers. Normally, chronic exposure to an agonist would cause cells to down-regulate their receptors. Up-regulation, an increase in receptor numbers, is usually seen when cells are chronically exposed to receptor antagonists. Molecular studies of the nicotinic acetylcholine receptor (nAChR) show that initial binding of ACh or agonists opens the receptor's channel but continued exposure to ACh inactivates and closes the channel.

Question 3: Although nicotine has been shown in short-term studies to be an agonist at nicotinic receptors, chronic exposure to nicotine causes up-regulation of the receptors. Speculate on why up-regulation might occur.

Question 4: Name another ion channel you have studied that opens in response to a stimulus but inactivates and closes shortly thereafter.

369 371 375 382 383

Summary of Sympathetic and Parasympathetic Divisions

As you have seen in this discussion, the branches of the autonomic nervous system share some features but are distinguished by others. Many of these features are summarized in Figure 11-11 ● and compared in Table 11-4.

1. Both sympathetic and parasympathetic pathways are composed of two neurons. One exception to this rule is the adrenal medulla, in which postganglionic sympathetic neurons have been modified into a neuroendocrine organ.

2. All preganglionic autonomic neurons secrete acetylcholine onto nicotinic receptors. Most sympathetic neurons secrete norepi-

AUTONOMIC PATHWAYS

FIGURE QUESTION

Using the figure, compare:
(a) number of neurons in somatic motor and autonomic pathways
(b) receptors on target cells of somatic motor, sympathetic, and parasympathetic pathways
(c) neurotransmitters used on target cells of somatic motor, sympathetic, and parasympathetic pathways
(d) receptor subtypes for epinephrine to subtypes for norepinephrine
(e) ganglion location for sympathetic and parasympathetic pathways.

● **Figure 11-11** *Summary of efferent pathways* ACh = acetylcholine, NE = norepinephrine, E = epinephrine.

Table 11-4: Comparison of Sympathetic and Parasympathetic Branches

	SYMPATHETIC	PARASYMPATHETIC
Point of CNS origin	1st thoracic to 2nd lumbar segments	Midbrain, medulla, and 2nd–4th sacral segments
Location of peripheral ganglia	Primarily in paravertebral sympathetic chain; 3 outlying ganglia located alongside descending aorta	On or near target organs
Structure of region from which neurotransmitter is released	Varicosities	Boutons and varicosities
Neurotransmitter at target synapse	Norepinephrine (adrenergic neurons)	ACh (cholinergic neurons)
Inactivation of neurotransmitter at synapse	Uptake into varicosity, diffusion	Enzymatic breakdown, diffusion
Types of neurotransmitter receptors	α and β	Muscarinic
Ganglionic synapse	ACh on nicotinic receptor	ACh on nicotinic receptor
Neuron-target synapse	NE on α or β receptor	ACh on muscarinic receptor

nephrine onto adrenergic receptors. Most parasympathetic neurons secrete acetylcholine onto muscarinic receptors.

3. Sympathetic pathways originate in the thoracic and lumbar regions of the spinal cord. The sympathetic ganglia are located close to the spinal cord (*paravertebral*). Parasympathetic pathways leave the CNS at the brain stem and in the sacral region of the spinal cord. Parasympathetic ganglia are located close to or in the target tissue.

4. The sympathetic branch controls functions that are useful in stress or emergencies (fight-or-flight). The parasympathetic branch is dominant during rest-and-digest activities.

▶ THE SOMATIC MOTOR DIVISION

Somatic motor pathways, which control skeletal muscles, differ from autonomic pathways both anatomically and functionally (Table 11-5). Somatic pathways have a single neuron that origi-

nates in the CNS and projects its axon to the target tissue, which is always a skeletal muscle. Unlike autonomic pathways, which may be either excitatory or inhibitory, somatic pathways are always excitatory.

A Somatic Motor Pathway Consists of One Neuron

The cell bodies of somatic motor neurons are located within the gray matter of the spinal cord or brain, with a long single axon projecting to the skeletal muscle target (Fig. 11-11). These myelinated axons may be up to a meter or more in length, like the somatic motor neurons that innervate the muscles of the foot and hand.

Somatic motor neurons branch close to their targets. Each branch divides into a cluster of enlarged axon terminals, or *boutons* [from the French word for "knob" or "button"], that lie on the surface of the skeletal muscle cells (Fig. 11-12 ●). This

Table 11-5: Comparison of Somatic and Autonomic Divisions

	SOMATIC	AUTONOMIC
Number of neurons in efferent path	1	2
Neurotransmitter/receptor at neuron-target synapse	ACh (nicotinic)	ACh (muscarinic) or NE (α or β)
Target tissue	Skeletal muscle	Smooth and cardiac muscle; some endocrine and exocrine glands; some adipose tissue
Structure of axon terminal regions	Boutons	Boutons and varicosities
Effects on target tissue	Excitatory only: muscle contracts	Excitatory or inhibitory
Peripheral components found outside the CNS	Axons only	Preganglionic axons, ganglia, postganglionic neurons
Summary of function	Posture and movement	Visceral function, including movement in internal organs and secretion; control of metabolism

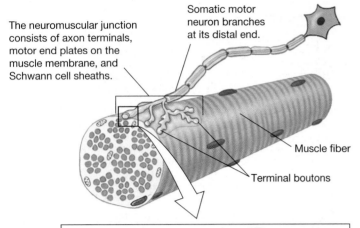

The neuromuscular junction consists of axon terminals, motor end plates on the muscle membrane, and Schwann cell sheaths.

Somatic motor neuron branches at its distal end.

Muscle fiber

Terminal boutons

Schwann cell sheath

Mitochondria

The **motor end plate** is a region of muscle membrane that contains high concentrations of ACh receptors.

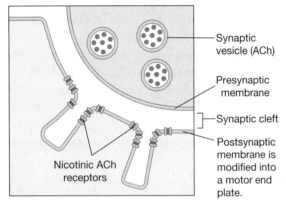

Synaptic vesicle (ACh)

Presynaptic membrane

Synaptic cleft

Postsynaptic membrane is modified into a motor end plate.

Nicotinic ACh receptors

● **Figure 11-12** *Anatomy of the neuromuscular junction*

branching structure allows a single motor neuron to control many muscle fibers at one time.

The synapse of a somatic motor neuron on a muscle fiber is called the **neuromuscular junction** (Fig. 11-12). Like all synapses, this region has three components: (1) the motor neuron's presynaptic axon terminal filled with synaptic vesicles and mitochondria, (2) the synaptic cleft, and (3) the postsynaptic membrane of the skeletal muscle fiber.

In addition, the neuromuscular junction includes extensions of Schwann cells that form a thin layer covering the top of the terminal boutons. For years, it was thought that this cell layer simply provided insulation to speed up the conduction of the ac-

tion potential. But it has now been discovered that the Schwann cells play a critical role in the formation of neuromuscular junctions during development and that they continue to secrete supportive growth factors throughout their life.

On the postsynaptic side of the neuromuscular junction, the muscle cell membrane that lies opposite the axon terminal is modified into a **motor end plate**, a series of folds that look like shallow gutters. Along the upper edge of each gutter, nicotinic ACh receptor channels cluster together into an active zone. Between the axon and the muscle, the synaptic cleft is filled with a fibrous matrix whose collagen fibers hold the axon terminal and the motor end plate in the proper alignment. The matrix also contains **acetylcholinesterase** (AChE), the enzyme that rapidly deactivates ACh by degrading it into acetyl and choline.

The elements of the neuromuscular junction can be seen in color in the chapter opening photo on page 369.

CONCEPT CHECK　　　　　　　Answers: p. 388

10. Compare gating and ion selectivity of ion channels in the motor end plate with the ion channels along the axon of a somatic motor neuron.

The Neuromuscular Junction Contains Nicotinic Receptors

As in all neurons, action potentials arriving at the axon terminal open voltage-gated Ca^{2+} channels in the membrane. Calcium diffuses into the cell down its electrochemical gradient, triggering the release of ACh-containing synaptic vesicles. Acetylcholine diffuses across the synaptic cleft and combines with nicotinic receptor channels on the skeletal muscle membrane (Fig. 11-13 ●). These receptors are similar but not identical to the nicotinic ACh receptors found in ganglia (and those found in the CNS). This difference is evidenced

RUNNING PROBLEM

One of the hallmarks of addiction is the withdrawal symptoms that occur when a person stops ingesting a substance. Nicotine withdrawal symptoms include lethargy, hunger, and irritability. To avoid withdrawal symptoms, people who are trying to stop smoking may use a nicotine patch. These adhesive patches deliver a small, continuous dose of nicotine that is usually enough to keep withdrawal symptoms in check. For Shanika, the first 24 hours without cigarettes have not been easy. Although she wears a nicotine patch, all she wants is a cigarette to smoke. When she reads the nicotine package insert, she notices a warning to keep the patches away from children. An overdose of nicotine (highly unlikely when the patch is used as directed) could result in complete paralysis of the respiratory muscles (the diaphragm and the skeletal muscles of the chest wall).

Question 5: Why might excessive levels of nicotine cause respiratory paralysis?

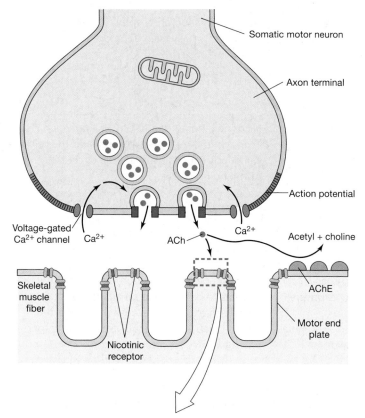

(a) Acetylcholine (ACh) combines with nicotinic receptors or is metabolized by acetylcholinesterase (AChE).

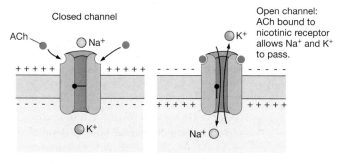

(b) The nicotinic cholinergic receptor binds two ACh molecules, opening a nonspecific monovalent cation channel.

● **Figure 11-13** *Events at the neuromuscular junction*

by the fact that the snake toxin α-bungarotoxin binds to nicotinic skeletal muscle receptors but not to those in autonomic ganglia.

Nicotinic receptors are chemically gated ion channels with two binding sites for ACh (Fig. 11-13b). When ACh binds to the receptor, the channel gate opens and allows monovalent cations to flow through. Net Na^+ entry into the muscle fiber depolarizes it, triggering an action potential that causes contraction of the skeletal muscle cell.

Acetylcholine acting on the motor end plate is always excitatory and creates muscle contraction. There is no antagonistic innervation to relax skeletal muscles. Instead, relaxation occurs when the somatic motor neurons are inhibited, preventing ACh release. In Chapter 13 you will learn more about how inhibition of somatic motor pathways controls body movement.

CONCEPT CHECK   Answers: p. 388

11. A nonsmoker who chews nicotine-containing gum might notice an increase in heart rate, a function controlled by sympathetic neurons. If postganglionic sympathetic neurons secrete norepinephrine, why would nicotine affect heart rate?

12. Patients with myasthenia gravis have a deficiency of ACh receptors on their skeletal muscles and have weak muscle function as a result. Why would administration of an anticholinesterase drug that inhibits acetylcholinesterase improve their muscle function?

Somatic motor neurons do more than simply create contractions: they are necessary for muscle health. "Use it or lose it" is a cliché that is very appropriate to the dynamics of muscle mass, because disrupting synaptic transmission at the neuromuscular junction has devastating effects on the entire body. Without communication between the motor neuron and the muscle, the skeletal muscles for movement and posture weaken, as do the skeletal muscles used for breathing. In the severest cases, loss of respiratory function can be fatal unless the patient is placed on artificial ventilation. Myasthenia gravis, the loss of ACh receptors discussed earlier, is the most common disorder of the neuromuscular junction.

RUNNING PROBLEM CONCLUSION

A Powerful Addiction

Shanika is determined to stop smoking this time because her grandfather, a smoker for many years, was just diagnosed with lung cancer. She finds that the patch alone does not stop her craving for a cigarette, so she attends behavioral modification classes as well. After six months, she proudly informs her family that she thinks she has kicked the habit.

Nicotine is an agonist for nicotinic ACh receptors, but with chronic exposure, nicotine inactivates the ACh receptors. With fewer active receptors, target cells up-regulate their receptor number. When smokers stop smoking, the removal of nicotine from the system allows all receptors to return to an active state. A larger than usual number of active receptors is probably responsible for some of the withdrawal symptoms noted by people who are trying to stop smoking. The nicotine patch allows the former smoker to gradually decrease nicotine levels in the body, preventing withdrawal symptoms during the time the cells are down-regulating their receptors back to the normal number.

In this running problem, you learned that nicotine is an addictive agonist of ACh. You also learned how cells respond to long-term exposure to nicotine. Further check your understanding of this running problem by comparing your answers to those in the summary table.

(continues on next page)

	QUESTION	FACTS	INTEGRATION AND ANALYSIS
1	What is the usual response of cells that are chronically exposed to increased concentrations of a signal molecule? (Hint: This topic was discussed in Chapter 6.)	A cell that is exposed to increased concentrations of a signal molecule will decrease its receptors for that molecule (● p. 186).	Down-regulation of receptors allows a cell to respond normally even if the concentration of ligand is elevated.
2	Cholinergic receptors are classified as either nicotinic or muscarinic, on the basis of the agonist molecules that bind to them. What happens to the postsynaptic cell when nicotine rather than ACh binds to a nicotinic cholinergic receptor?	Nicotine is an agonist of ACh. Agonists mimic the activity of a ligand.	Nicotine binding to a nAChR will open ion channels in the postsynaptic cell and the cell will depolarize. This is the same effect that ACh binding creates.
3	Although nicotine has been shown in short-term studies to be an agonist at nicotinic receptors, chronic exposure to nicotine causes up-regulation of the receptors. Speculate on why up-regulation might occur.	Chronic exposure to an agonist usually causes down-regulation. Chronic exposure to an antagonist usually causes up-regulation. nAChR channels open with initial exposure to agonists but close with continued exposure.	Although nicotine is a short-term agonist, it appears to be acting as an antagonist during long-term exposure because that closes the ion channel. Cells up-regulate their receptors when exposed to antagonists.
4	Name another ion channel you have studied that opens in response to a stimulus but inactivates and closes shortly thereafter.	The voltage-gated Na^+ channel of the axon opens, then closes when the inactivation gate shuts.	Not applicable
5	Why might excessive levels of nicotine cause respiratory paralysis?	Nicotinic receptors are found at the neuromuscular junction that controls skeletal muscle contraction. The diaphragm and chest wall muscles that regulate breathing are skeletal muscles.	The nicotinic receptors of the neuromuscular junction are not as sensitive to nicotine as are those of the CNS and autonomic ganglia. However, excessively high amounts of nicotine will activate the nAChR of the motor end plate, causing the muscle fiber to depolarize and contract. The continued presence of nicotine keeps these ion channels open, and the muscle remains depolarized. In this state, it is unable to contract again, resulting in paralysis.

369 371 375 379 382

CHAPTER SUMMARY

This chapter completes our survey of the nervous system: the central nervous system, which acts as the integrating center for nervous reflexes; the sensory system, which is the input division; and the autonomic and somatic motor divisions, which are the output pathways. *Communication* between the three divisions depends primarily on chemical signaling and *molecular interactions* between neurotransmitters and their receptors. *Homeostasis* requires constant surveillance of body parameters by the nervous system, working in conjunction with the endocrine and immune systems. As you learn about the function of other body systems, you will continue to revisit the principles of communication and coordination introduced in Chapters 6 through 11.

The Autonomic Division

1. The efferent division of the nervous system is divided into **somatic motor neurons**, which control skeletal muscles, and **autonomic neurons**, which control smooth muscle, cardiac muscle, many glands, and some adipose tissue. (p. 370)

2. The autonomic division is subdivided into a **sympathetic branch** and a **parasympathetic branch**. (p. 370; Tbl. 11-4)

3. The maintenance of homeostasis within the body is a balance between autonomic control, endocrine control, and behavioral responses. (p. 371)

4. The autonomic division is controlled by centers in the hypothalamus, pons, and medulla. Some autonomic reflexes are spinal reflexes. Many of these can be modulated by input from the brain. (p. 371; Figs. 11-2, 11-3)

5. The two autonomic branches demonstrate Cannon's properties of homeostasis: maintenance of the internal environment, tonic control, antagonistic control, and variable tissue responses. (p. 371)

6. An autonomic pathway is composed of a **preganglionic neuron** from the CNS that synapses with a **postganglionic neuron** in an **autonomic ganglion**. Autonomic ganglia can modulate and integrate information passing through them. (p. 372; Fig. 11-4)

7. Most sympathetic pathways originate in the thoracic and lumbar regions of the spinal cord. Most sympathetic ganglia lie close to the spinal cord or along the descending aorta. (p. 372; Fig. 11-5)

8. Parasympathetic pathways originate in the brain stem or in sacral region. Parasympathetic ganglia are located on or near their target organs. (p. 374; Fig. 11-5)

9. The primary autonomic neurotransmitters are **acetylcholine** and **norepinephrine**. All preganglionic neurons secrete ACh. As a rule, sympathetic neurons secrete norepinephrine, and parasympathetic neurons secrete ACh. (p. 374; Fig. 11-7, Tbl. 11-1)

10. The synapse between an autonomic neuron and its target cells is called the **neuroeffector junction**. (p. 375)

11. Autonomic axons end with **varicosities** from which neurotransmitter is released. (p. 375; Figs. 11-8, 11-9)

12. The **adrenal medulla** secretes epinephrine and is controlled by sympathetic preganglionic neurons. (p. 376; Fig. 11-10)

13. Adrenergic receptors are G protein-linked receptors. Alpha receptors respond most strongly to norepinephrine. β_1 **receptors** respond equally to norepinephrine and epinephrine. β_2 **receptors** are not associated with sympathetic neurons and respond most strongly to epinephrine. (p. 377; Fig. 11-11, Tbl. 11-2)

14. **Cholinergic nicotinic receptors** are found in autonomic ganglia. ACh binding to a nicotinic receptor opens cation channels and depolarizes the postsynaptic cell. Muscarinic receptors are found at the neuroeffector junctions of the parasympathetic branch. (pp. 374–378; Fig. 11-11)

The Somatic Motor Division

15. Somatic motor pathways, which control skeletal muscles, have a single neuron that originates in the CNS and terminates on a skeletal muscle. Somatic motor neurons are always excitatory and cause muscle contraction. (p. 381; Fig. 11-11)

16. A single **somatic motor neuron** controls many muscle fibers at one time. (p. 381)

17. The synapse of a somatic motor neuron on a muscle fiber is called the **neuromuscular junction**. The muscle cell membrane is modified into a **motor end plate** that contains a high concentration of nicotinic ACh receptors. (p. 381; Fig. 11-12)

18. ACh binding to the receptor opens cation channels. Net Na^+ entry into the muscle fiber, depolarizes the fiber. Acetylcholine in the synapse is broken down by the enzyme **acetylcholinesterase**. (p. 382; Fig. 11-13)

QUESTIONS

LEVEL ONE Reviewing Facts and Terms

1. Name the two efferent divisions of the peripheral nervous system. What effectors does each control?

2. The autonomic nervous system is sometimes called the _____ nervous system. Why is this an appropriate name? List some functions controlled by the autonomic nervous system.

3. What are the two branches of the autonomic nervous system? How are these branches distinguished anatomically and physiologically?

4. Which neurosecretory endocrine gland is closely allied to the sympathetic division?

5. Neurons that secrete acetylcholine are described as _____, whereas those that secrete norepinephrine are called _____ or _____ neurons.

6. List four things that can happen to autonomic neurotransmitters after they are released into the synapse.

7. The main enzyme responsible for catecholamine degradation is _____, abbreviated _____.

8. Somatic motor pathways
   (a) are excitatory or inhibitory?
   (b) are composed of a single neuron or a preganglionic and a postganglionic neuron?
   (c) synapse with glands or with smooth, cardiac, or skeletal muscle?

9. What is acetylcholinesterase? Describe its action.

10. What kind of receptor is found on the postsynaptic cell in the neuromuscular junction?

LEVEL TWO *Reviewing Concepts*

11. **Concept map:** Use the following terms to make a map comparing the somatic, sympathetic, and parasympathetic branches. You may add additional terms.

acetylcholine	muscarinic receptor
adipose tissue	nicotinic receptor
alpha receptor	norepinephrine
autonomic division	one-neuron pathway
beta receptor	parasympathetic branch
cardiac muscle	skeletal muscle
cholinergic receptor	smooth muscle
efferent division	somatic motor division
endocrine gland	sympathetic branch
exocrine gland	two-neuron pathway
ganglion	

12. What is the advantage of divergence of neural pathways in the autonomic nervous system?

13. Compare and contrast the following:
    (a) neuroeffector junctions and neuromuscular junctions
    (b) alpha, beta, muscarinic, and nicotinic receptors. Describe where each is found and the ligands that bind to them.

14. Compare and contrast:
    (a) autonomic ganglia and CNS nuclei
    (b) the adrenal medulla and the posterior pituitary gland
    (c) boutons and varicosities

15. If the target cell's receptor is _____ (use items in left column), the neuron(s) releasing neurotransmitter onto the receptor must be _____. (Use all appropriate items in right column)
    (a) cholinergic nicotinic
    (b) adrenergic α
    (c) cholinergic muscarinic
    (d) adrenergic β

    1. somatic motor neuron
    2. autonomic preganglionic neuron
    3. sympathetic postganglionic neuron
    4. parasympathetic postganglionic neuron

16. Ganglia contain the cell bodies of (choose all that apply):
    (a) somatic motor neurons
    (b) preganglionic autonomic neurons
    (c) interneurons
    (d) postganglionic autonomic neurons
    (e) sensory neurons

LEVEL THREE *Problem Solving*

17. If nicotinic receptor channels allow both Na^+ and K^+ to flow through, why does Na^+ influx exceed K^+ efflux? (Hint: ● p. 160)

18. You have discovered a neuron that innervates an endocrine cell in the intestine. To learn more about this neuron, you place a marker substance at the endocrine cell synapse. The marker is taken into the neuron and transported in a vesicle by retrograde axonal transport to the nerve cell body.
    (a) By what process is the marker probably taken into the axon terminal?

    (b) The nerve cell body is found in a ganglion very close to the endocrine cell. To what branch of the peripheral nervous system does the neuron probably belong? (Be as specific as you can.)
    (c) What neurotransmitter would you predict that the neuron secretes on the endocrine cell?

19. The Huaorani Indians of South America use blowguns to shoot darts poisoned with curare at monkeys. Curare is a plant toxin that binds to and inactivates nicotinic ACh receptors. What happens to the monkey if struck by these darts?

LEVEL FOUR *Quantitative Problems*

20. The American Centers for Disease Control and Prevention (CDC) conduct biennial Youth Risk Behavior Surveys (YRBS) in which they ask high school students to self-report risky behaviors such as alcohol consumption and smoking. The graphs on the next page were created from data in the latest report on cigarette smoking among American high school students (*http://www.cdc.gov/mmwr/preview/mmwrhtml/mm5119a1.htm*).

Current smoking is defined as smoking cigarettes at least one day in the 30 days preceding the survey.

(a) What can you say about cigarette smoking among high school students in the ten-year period from 1991 to 2001?
(b) Which high school students are most likely to be smokers? Least likely to be smokers?

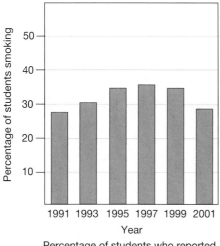

Percentage of students who reported current smoking (1991-2001)

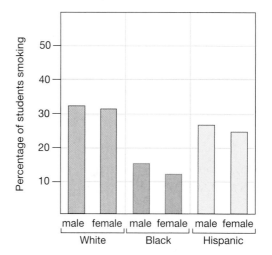

Percentage of students in 2001 who reported current smoking, separated by sex and race/ethnicity*

*Other race/ethnic groups are not shown because the numbers of students reporting were too small for meaningful statistical analysis.

Explorations

Explorations are also available at *www.physiologyplace.com*

The nervous system provides many fascinating examples of the integrative nature of physiology and of the importance of molecular biology in organ system function. In these Explorations, we examine two contemporary issues: cigarette smoking and nicotine addiction, and the role of botulinum toxin in our world.

Exploration 1: Nicotine Addiction

In the Running Problem for this chapter you were introduced to nicotine addiction. Neuroscientists have learned that addictive behaviors such as cigarette smoking develop because nicotine and other addictive chemicals such as cocaine act as *positive reinforcers* in the brain. Nicotine enhances dopamine release in the brain's reward centers and creates pleasurable sensations. Over time, the brain also begins to associate the social aspects of cigarette smoking with pleasure. The combination of physical dependence on nicotine and the conditioned response makes quitting even more difficult. Some people find that they are more successful at quitting if they hold an unlighted cigarette and mimic the social and physical aspects of smoking. To learn more about nicotine addiction and smoking cessation programs, try a Google search (*www.google.com*) for *nicotine addiction*.

Exploration 2: *Clostridium botulinum*—Friend or Foe?

The newspaper advertisement reads "Botox® Party! Erase wrinkles and fine lines. Look great ... Feel great!" Next to it is a news item on botulism and bioterrorism. Both items are referring to the same chemical: botulinum toxin, a polypeptide produced by the bacterium *Clostridium botulinum*. For many years *botulism*, the serious paralytic illness caused by this

toxin, was a rarity, with about 110 cases per year reported in the United States. Now, however, people wanting to erase frown lines and wrinkles are paying to have botulinum toxin injected into their faces. And at the same time, the same toxin is being mentioned as a possible weapon for bioterrorism.

Botulinum toxin paralyzes skeletal muscles by preventing neurotransmitter release at the neuromuscular junction. It binds to receptors on the axon terminal membrane and is taken into the terminal, where it disrupts proteins that facilitate exocytosis of synaptic vesicles. In the absence of neurotransmitter release, the postsynaptic muscle is unable to contract. See if you can apply what you have learned about the neuromuscular junction to answer the following questions:

1. By what mechanism is botulinum toxin taken into the axon terminal?

2. What neurotransmitter is not being released from the terminal?

3. In the absence of neurotransmitter release, what will happen to the number of neurotransmitter receptors on the motor end plate?

4. In the absence of muscle contraction, what will happen to the muscle?

To learn more about botulinum toxin as a cosmetic treatment, do a Google search for *Botox*. To learn more about botulinum toxin as a potential bioterrorism weapon, see "Botulinum toxin as a biological weapon," *JAMA* 285(8):1059–1070, 2001 Feb 28, or go to the website for the CDC at *www.cdc.gov* and click on *Terrorism and Public Health*.

ANSWERS

Answers to Concept Check Questions

Page 371

1. Afferent division consists of sensory receptors and sensory neurons.

2. The CNS consists of brain and spinal cord.

Page 372

3. Homeostasis is the maintenance of a relatively stable internal environment.

Page 374

4. Mixed nerves carry sensory and motor signals.

5. The regions are cervical, thoracic, lumbar, and sacral.

Page 377

6. The adrenal medulla is neurosecretory and therefore like the posterior pituitary.

7. Chromaffin cells are modified postganglionic neurons and therefore have nicotinic receptors.

Page 378

8. Ca^{2+} is stored in the endoplasmic reticulum.

9. Adenylyl cyclase converts ATP to cAMP, and cAMP activates protein kinase A.

Page 382

10. The motor end plate has ligand-gated monovalent cation (Na^+ and K^+) channels.

The axon contains voltage-gated channels, with separate channels for Na^+ and K^+.

Page 383

11. The postganglionic sympathetic neurons are activated by acetylcholine acting on nicotinic receptors. This means nicotine will also excite sympathetic neurons, such as those that increase heart rate.

12. Anticholinesterase drugs decrease the breakdown of ACh at the motor end plate. This allows ACh to remain active at the motor end plate for a longer duration and helps offset the decrease in active receptors.

Answers to Figure Questions

Page 375

Figure 11-7: (a) All preganglionic neurons and the parasympathetic postganglionic neuron are cholinergic. The sympathetic postganglionic neuron is adrenergic. Preganglionic neurons have cell bodies in the CNS; postganglionic have cell bodies in the ganglion. (b) Parasympathetic pathways have the longer preganglionic neurons.

Page 380

Figure 11-11: (a) Somatic has one, autonomic has two. (b) Somatic motor targets have nicotinic cholinergic. Parasympathetic has muscarinic cholinergic. Sympathetic has adrenergic. (c) Somatic motor and parasympathetic use ACh. Sympathetic uses norepinephrine. (d) Epinephrine is most active on α receptors of both subtypes. Nor-epinephrine is most active on β_1 and α receptors. (e) Sympathetic ganglia are close to the CNS. Parasympathetic ganglia are closer to their target tissues.

Answers to Level Four Quantitative Problems

Page 386

(a) Cigarette smoking among high school students increased from 1991 to 1997, then began to decrease. (b) Males are more likely to be smokers than females, and whites smoke more than Hispanics, who smoke more than blacks. Therefore, white males are the most likely to smoke and black females are the least likely to smoke.

Muscles

12

A muscle is … an engine, capable of converting chemical energy into mechanical energy. It is quite unique in nature, for there has been no artificial engine devised with the great versatility of living muscle.

—**Ralph W. Stacy and John A. Santolucito,**

in *Modern College Physiology*, 1966

One somatic motor neuron branches to innervate several skeletal muscle fibers.

RUNNING PROBLEM

Periodic Paralysis

This morning, Paul, six years old, gave his mother the fright of her life. One minute he was happily playing in the backyard with his new beagle puppy. The next minute, after sitting down to rest, he could not move his legs. In answer to his screams, his mother came running and found her little boy unable to walk. Panic-stricken, she scooped him up, brought him in the house, and dialed 9-1-1. But as she hung up the phone and prepared to wait for the paramedics, Paul got to his feet and walked over to her. "I'm OK now, Mom. I'm going outside."

401 **403** **412** **414** **422**

BACKGROUND BASICS

Glucose and glycogen: **35** ⏐ Gap junctions: **172** ⏐ Tendons: **72** ⏐ Kinases and phosphatases: **99** ⏐ Isozymes: **92** ⏐ Anaerobic and aerobic metabolism: **104** ⏐ Glycolysis: **103** ⏐ Fatty acid metabolism: **108** ⏐ Tonic control: **188** ⏐ Nitric oxide: **267** ⏐ Threshold: **251** ⏐ Summation: **270–271** ⏐ Autonomic neurons: **372** ⏐ Somatic motor neurons: **370** ⏐ Neuromuscular junction: **381** ⏐

I t was his first time to be the starting pitcher. As he ran from the bullpen onto the field, his heart was pounding and his stomach felt as if it were tied in knots. He stepped onto the mound and gathered his thoughts before throwing the first practice pitch. Gradually, as he went through the familiar routine of throwing and catching the baseball, his heart slowed and his stomach relaxed. It was going to be a good game.

The pitcher's pounding heart, queasy stomach, and movements as he runs and throws the ball are all the results of muscle contraction. Our muscles have two common functions: (1) to generate motion and (2) to generate force. Our skeletal muscles also generate heat and contribute significantly to the homeostasis of body temperature. When homeostasis is threatened by cold conditions, our brain may direct our muscles to shiver, creating additional heat.

Three types of muscle tissue occur in the human body: skeletal muscle, cardiac muscle, and smooth muscle. Most **skeletal muscles** are attached to the bones of the skeleton, enabling them to control body movement. **Cardiac muscle** [*kardia*, heart] is found only in the heart and is responsible for moving blood through the circulatory system. Skeletal and cardiac muscles are classified as **striated muscles** [*stria*, groove] because of their alternating light and dark bands under the light microscope (Fig. 12-1a, 1b●).

(a) Skeletal muscle

- Nucleus
- Muscle fiber (cell)
- Striations

(b) Cardiac muscle

- Striations
- Muscle fiber
- Intercalated disk
- Nucleus

(c) Smooth muscle

- Muscle fiber
- Nucleus

● **Figure 12-1** *Three types of muscles*

Smooth muscle is the primary muscle of internal organs and tubes such as the stomach, urinary bladder, and blood vessels. Its primary function is to influence the movement of material into, out of, or within the body. An example is the passage of food through the gastrointestinal tract. Smooth muscle has a homogeneous appearance under the microscope, without obvious crossbands (Fig. 12-1c). Its lack of banding results from the less organized arrangement of contractile fibers within the muscle cells.

Skeletal muscles are often described as the voluntary muscles, and smooth and cardiac muscle are described as involuntary. However, this is not a precise classification. Skeletal muscles can contract without conscious direction, and we can learn a certain degree of conscious control over some smooth and cardiac muscle [● p. 370].

Skeletal muscles are unique in that they only contract in response to a signal from a somatic motor neuron. They cannot initiate their own contraction, nor is their contraction influenced directly by hormones.

In contrast, cardiac and smooth muscle have multiple levels of control. Although their primary extrinsic control arises through autonomic innervation, some types of smooth and cardiac muscle can contract spontaneously, without signals from the central nervous system. In addition, the activity of cardiac and some smooth muscle is subject to modulation by the endocrine system. Despite these differences, smooth and cardiac muscle share many properties of skeletal muscle. The following section examines skeletal muscle anatomy and physiology in detail. We then compare properties of smooth and cardiac muscle to those of skeletal muscle.

▶ SKELETAL MUSCLE

Skeletal muscles make up the bulk of muscle in the body, about 40% of total body weight. They are responsible for positioning and moving the skeleton, as their name suggests. Skeletal muscles are usually attached to bones by **tendons** made of collagen [● p. 72]. The **origin** of a muscle is the end of the muscle that is attached closest to the trunk or to the more stationary bone. The **insertion** of the muscle is the more *distal* [*distantia*, distant] or mobile attachment.

When the bones attached to a muscle are connected by a flexible **joint**, contraction of the muscle moves the skeleton. If the centers of the connected bones are brought closer together when the muscle contracts, the muscle is called a **flexor**. If the bones move away from each other when the muscle contracts, the muscle is called an **extensor**.

Most joints in the body have both flexor and extensor muscles, because a contracting muscle can pull a bone in one direction but cannot push it back. Flexor-extensor pairs are called **antagonistic muscle groups** because they exert opposite effects. Figure 12-2● shows a pair of antagonistic muscles in the arm: the *biceps brachii* [*brachion*, arm], which acts as the flexor, and the *triceps brachii*, which acts as the extensor. When the biceps muscle

(a) Flexion

Triceps muscle relaxes

Biceps muscle contracts (flexor)

(b) Extension

Biceps muscle relaxes

Triceps muscle contracts (extensor)

● **Figure 12-2** *Antagonistic muscle groups*
Muscle contraction can pull on a bone, but cannot push a bone away.

Each skeletal muscle fiber is a long, cylindrical cell with up to several hundred nuclei on the surface of the fiber (Fig. 12-3b). Skeletal muscle fibers are the largest cells in the body, created by the fusion of many individual embryonic muscle cells.

The individual fibers of a skeletal muscle are bound with connective tissue and arranged with their long axes in parallel (Fig. 12-3a). Collagen, elastic fibers, nerves, and blood vessels are found between the bundles of muscle fibers. The entire muscle is enclosed in a connective tissue sheath that is continuous with the connective tissue around the muscle fibers and with the tendons holding the muscle to underlying bones.

Muscle Fiber Anatomy Muscle physiologists, like neurophysiologists, have adopted a specialized vocabulary (Table 12-1). The cell membrane of a muscle fiber is called the **sarcolemma** [*sarkos*, flesh + *lemma*, shell], and the cytoplasm is called the **sarcoplasm**. A muscle fiber contains little cytosol. The main intracellular structures are the **myofibrils** [*myo-*, muscle], bundles of contractile and elastic proteins that carry out the work of contraction.

Skeletal muscle fibers contain an extensive **sarcoplasmic reticulum** (SR), a form of modified endoplasmic reticulum that wraps around each myofibril like a piece of lace (Figs. 12-3b, 12-4 ●). The sarcoplasmic reticulum consists of longitudinal tubules, which release Ca^{2+}, and the **terminal cisternae**, which concentrate and sequester Ca^{2+} [*sequestrare*, to put in the hands of a trustee].

A branching network of **transverse tubules**, also known as **t-tubules**, is closely associated with the terminal cisternae. One t-tubule with its flanking terminal cisternae is known as a *triad*. The membranes of t-tubules are a continuation of the muscle fiber membrane. This makes the lumen of the t-tubules continuous with the extracellular fluid.

To understand how this network of tubules in the heart of the muscle fiber communicates with the outside, take a lump of soft clay and stick your finger into the middle of it several times. Notice how the outside surface of the clay (analogous to the cell membrane of the muscle fiber) is now continuous with the sides of the holes that you have poked in the clay (the membrane of the t-tubules).

T-tubules allow action potentials that originate at the neuromuscular junction on the cell surface to move rapidly into the interior of the fiber. Without t-tubules, the action potential could reach the center of the fiber only by diffusion of positive charge

contracts, the hand and forearm move toward the body. Contraction of the triceps extends the flexed arm. In each case, when one muscle contracts and shortens, the antagonistic muscle must relax and lengthen.

CONCEPT CHECK   Answers: p. 427

1. Identify as many pairs of antagonistic muscle groups in the body as you can. If you cannot name them, point out the probable location of the flexor and extensor of each group.

Skeletal Muscles Are Composed of Muscle Fibers

Muscles function together as a unit. A skeletal **muscle** is a collection of muscle cells, or **muscle fibers** (see Anatomy Summary, Fig. 12-3 ●), just as a nerve is a collection of neurons. Groups of muscle fibers that function together and the motor neuron that controls them are called a **motor unit**. A small muscle, such as one of those attached to the bones of the middle ear, may have several hundred muscle fibers organized into a few motor units. Larger muscles, such as the quadriceps muscles of the thigh, may have several thousand fibers organized as many motor units.

Table 12-1: Muscle Terminology

GENERAL TERM	MUSCLE EQUIVALENT
Muscle cell	Muscle fiber
Cell membrane	Sarcolemma
Cytoplasm	Sarcoplasm
Modified endoplasmic reticulum	Sarcoplasmic reticulum

ORGANIZATION OF SKELETAL MUSCLE

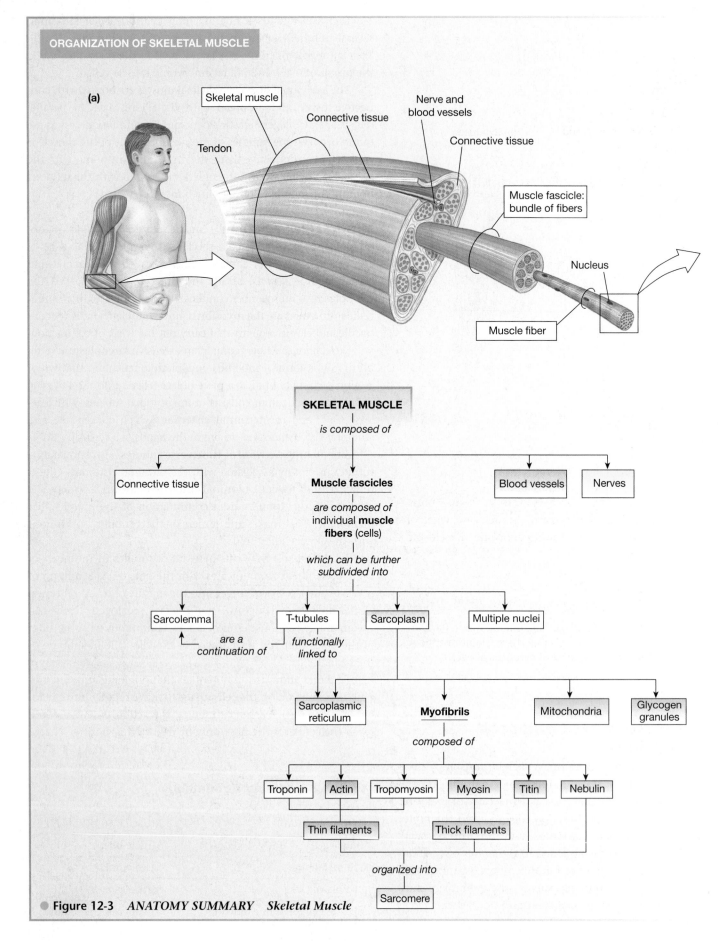

(a) Skeletal muscle · Connective tissue · Nerve and blood vessels · Connective tissue · Tendon · Muscle fascicle: bundle of fibers · Nucleus · Muscle fiber

SKELETAL MUSCLE

is composed of

- Connective tissue
- **Muscle fascicles**

  are composed of individual **muscle fibers** (cells)

  which can be further subdivided into
- Blood vessels
- Nerves

Sarcolemma · T-tubules · Sarcoplasm · Multiple nuclei

are a continuation of · *functionally linked to*

Sarcoplasmic reticulum · **Myofibrils** · Mitochondria · Glycogen granules

Myofibrils *composed of*

Troponin · Actin · Tropomyosin · Myosin · Titin · Nebulin

Thin filaments · Thick filaments

organized into

Sarcomere

● **Figure 12-3** *ANATOMY SUMMARY Skeletal Muscle*

ULTRASTRUCTURE OF MUSCLE FIBER

(b)

Sarcoplasmic reticulum

Sarcolemma

Mitochondria

Nucleus

Thick filament

Myofibril

T-tubules

Thin filament

(c)

A band

Sarcomere

Z disk

Z disk

Myofibril

M line

I band

H zone

(d)

Titin

Z disk

M line

Z disk

Thick filaments

M line

Thin filament

Titin

(e)

Myosin head

Hinge region

Myosin tail

Myosin molecule

(f)

Troponin

Nebulin

Tropomyosin

G-actin molecule

Actin chain

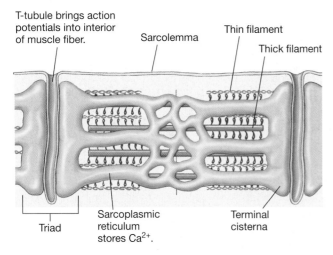

T-tubule brings action potentials into interior of muscle fiber.

Sarcolemma

Thin filament

Thick filament

Sarcoplasmic reticulum stores Ca²⁺.

Triad

Terminal cisterna

● **Figure 12-4** *T-tubules and the sarcoplasmic reticulum*

through the cytosol, a slower process that would delay the response time of the muscle fiber.

The cytosol between the myofibrils contains many glycogen granules and mitochondria. Glycogen, the storage form of glucose found in animals, is a reserve source of energy. Mitochondria provide much of the ATP for muscle contraction through oxidative phosphorylation of glucose and other biomolecules.

Myofibrils Are the Contractile Structures of a Muscle Fiber

Each muscle fiber contains a thousand or more myofibrils that occupy most of the intracellular volume, leaving little space for cytosol and organelles (Fig. 12-3b). Each myofibril is composed of several types of proteins: the contractile proteins *myosin* and *actin*, the regulatory proteins *tropomyosin* and *troponin*, and the giant accessory proteins *titin* and *nebulin*.

Myosin [*myo-*, muscle] is the protein that makes up the thick filaments of the myofibril. Various *isoforms* [*iso-*, equal], or different types, of myosin occur in different types of muscle. They are related to the muscle's speed of contraction.

Each myosin molecule is composed of two heavy protein chains that intertwine to form a long coiled tail and a pair of tadpolelike heads (Fig. 12-3e). Two lightweight protein chains are associated with the heavy chain of each head. In skeletal muscle, about 250 myosin molecules join to create a **thick filament**.

The thick filament is arranged so that the myosin heads are clustered at the ends, and the central region of the filament is a bundle of myosin tails. The rodlike portion of the thick filament is stiff, but the protruding myosin heads have an elastic hinge region where the heads join the rods. This hinge region allows the heads to swivel around their point of attachment.

Actin [*actum*, to do] is the protein that makes up the thin filaments of the muscle fiber. One actin molecule is a globular protein (*G-actin*), represented in Figure 12-3f by a round ball. Usually, multiple globular actin molecules polymerize to form long chains

or filaments (*F-actin*). In skeletal muscle, two F-actin polymers twist together like a double strand of beads, creating the **thin filaments** of the myofibril.

Most of the time, the parallel thick and thin filaments of the myofibril are connected by crossbridges that span the space between the filaments. The **crossbridges** are the myosin heads that loosely bind to the actin filaments. Each G-actin molecule has a single binding site for a myosin head. Under a light microscope, the arrangement of thick and thin filaments in a myofibril creates a repeating pattern of alternating light and dark bands (Figs. 12-1a, 12-3c). One repeat of the pattern forms a **sarcomere** [*sarkos*, flesh + -*mere*, a unit or segment], which has the following elements (Fig. 12-5 ●):

1. **Z disks**: These zigzag structures are made of proteins that serve as the attachment site for the thin filaments. One sarcomere is composed of two Z disks and the filaments found between them. The abbreviation Z comes from *zwischen*, the German word for "between."

2. **I band**: These are the lightest color bands of the sarcomere and represent a region occupied only by thin filaments. The abbreviation *I* comes from *isotropic*, a description from early microscopists meaning that this region reflects light uniformly under a polarizing microscope. A Z disk runs through the middle of an I band, so each half of the I band belongs to a different sarcomere.

3. **A band**: This is the darkest of the bands in a sarcomere and encompasses the entire length of a thick filament. At the outer edges of the A band, the thick and thin filaments overlap. The center of the A band is occupied by thick filaments only. The abbreviation *A* comes from *anisotropic* [*an-*, not], meaning that the protein fibers in this region scatter light unevenly.

4. **H zone**: This central region of the A band is lighter than the outer edges of the A band because the H zone is occupied by thick filaments only. The *H* comes from *helles*, the German word for "clear."

5. **M line**: This band represents the attachment site for the thick filaments, equivalent to the Z disk for the thin filaments. One A band is divided in half by an M line. *M* is the abbreviation for *mittel*, the German word for "middle."

In three-dimensional array, the actin and myosin molecules form a lattice of parallel, overlapping thin and thick filaments (Fig. 12-5b). When viewed end-on, each thin filament is surrounded by three thick filaments, and six thin filaments encircle each thick filament (Fig. 12-5c).

The proper alignment of filaments within a sarcomere is ensured by two types of proteins, titin and nebulin (Fig. 12-6 ●). **Titin** is a huge elastic molecule and the largest known protein, with more than 25,000 amino acids. A single titin molecule stretches from one Z disk to the next M line. To get an idea of the immense size of titin, imagine that one titin molecule is an 8-foot-long piece of the very thick rope used to tie ships to a wharf.

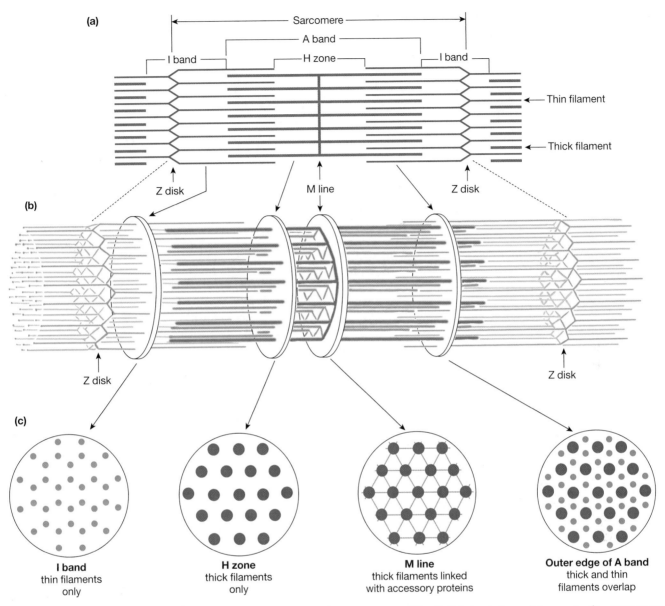

● **Figure 12-5** *The two- and three-dimensional organization of a sarcomere* The Z disk (not shown) has a similar arrangement of accessory proteins that link the thin filaments.

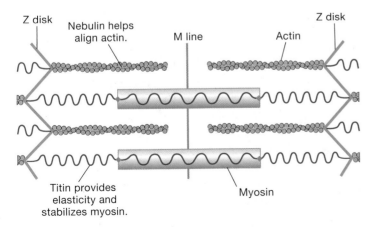

● **Figure 12-6** *Titin and nebulin* Titin spans the distance from one Z disk to the next M line.

By comparison, a single actin molecule would be about the length and weight of a single eyelash.

Titin has two functions: (1) it stabilizes the position of the contractile filaments, and (2) its elasticity returns stretched muscles to their resting length. Titin is helped by **nebulin**, an inelastic giant protein that lies alongside thin filaments and attaches to the Z disk. Nebulin helps align the actin filaments of the sarcomere.

CONCEPT CHECK   Answers: p. 427

2. Why are the ends of the A band the darkest region of the sarcomere when viewed under the light microscope?

3. What is the function of t-tubules?

4. Why are skeletal muscles described as striated?

Muscle Contraction Creates Force

The contraction of muscle fibers is a remarkable process that allows us to create force to move or resist a load. In muscle physiology, the force created by the contracting muscle is called the **muscle tension**. The **load** is a weight or force that opposes contraction of a muscle. The creation of tension in a muscle is an active process that requires energy input from ATP.

Figure 12-7 ● summarizes the major steps leading up to skeletal muscle contraction.

1. **Events at the neuromuscular junction** convert a chemical signal from a somatic motor neuron into an electrical signal in the muscle fiber. These events were described in Chapter 11 [● p. 382].
2. **Excitation-contraction coupling** is the process in which muscle action potentials initiate *calcium signals*. The calcium signals then activate the contraction-relaxation cycle.
3. At the molecular level, **the contraction-relaxation cycle** can be explained by the *sliding filament theory of contraction*. In intact muscles, a muscle *twitch* is one contraction-relaxation cycle.

In the sections that follow, we will examine each of these steps. We start with the molecular basis for muscle contraction that is the heart of the process. From there, we go to the integrated function of a muscle fiber as it undergoes excitation-contraction coupling. The skeletal muscle section ends with a discussion of the organization of intact muscles and how they move bones around joints.

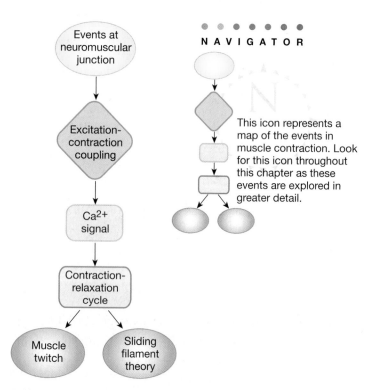

● **Figure 12-7** *Summary of muscle contraction*

Muscles Shorten When They Contract

In previous centuries, scientists observed that when muscles move a load, they shorten. This observation led to early theories of contraction, which proposed that muscles were made of molecules that curled up and shortened when active, then relaxed and stretched at rest, like elastic in reverse. The theory received support when myosin was found to be a helical molecule that shortened upon heating (the reason meat shrinks when you cook it).

In 1954, however, Sir Andrew Huxley and Rolf Niedeigerke discovered that the length of the A band of a myofibril remains constant during contraction. Because the A band represents the myosin filament, they realized that shortening of the myosin molecule could not be responsible for contraction. Subsequently, they proposed an alternative model, the **sliding filament theory of contraction**. In this model, overlapping muscle fibers of fixed length (the thick and thin filaments) slide past each other in an energy-requiring process, resulting in muscle contraction.

The sliding filament theory takes into account the observation that when a muscle contracts, it does not always shorten. A muscle can contract and create force without creating movement. For example, if you push on a wall, you are creating tension in many muscles of your body without shortening the muscles and without moving the wall. According to the sliding filament theory, tension generated in a muscle fiber is directly proportional to interaction between the thick and thin filaments.

Sliding Filament Theory of Contraction If you examined a myofibril at its resting length, you would see that within each sarcomere, the ends of the thick and thin filaments overlap slightly. As the muscle contracts, the thick and thin filaments slide past each other, moving the Z disks of the sarcomere closer together.

This phenomenon can be seen in light micrographs of relaxed and contracted muscle (Fig. 12-8 ●). In the relaxed state, a sarcomere has a large I band (thin filaments only) and an A band whose length is the length of the thick filament. As contraction occurs, the sarcomere shortens. The two Z disks at each end move closer together while the I band and H zone almost disappear. These are regions where actin and myosin do not overlap in resting muscle.

Despite the shortening of the sarcomere, the length of the A band remains constant. These changes are consistent with the sliding of thin actin filaments along the thick myosin filaments as the actin filaments move toward the M line in the center of the sarcomere. It is from this process that the sliding filament theory of contraction derives its name.

The force that pushes the actin filament is the movement of myosin crossbridges that link actin and myosin. Each myosin head has two binding sites on it: one for an ATP molecule and one for actin. The actin molecules serve as the "rope" to which the myosin heads bind.

During the **power stroke** that is the basis for muscle contraction, movement of the flexible myosin crossbridges pushes actin filaments toward the center of the sarcomere. At the end of a

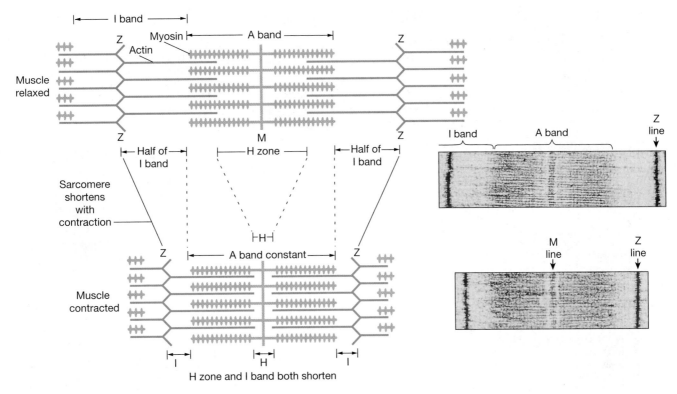

● Figure 12-8 *Changes in a sarcomere during contraction* During contraction the thick and thin filaments do not change length but slide past each other.

power stroke, the myosin head releases its bound actin, then swings back and binds to a new actin molecule, ready to start another cycle.

This process repeats many times as a muscle fiber contracts. The myosin heads repeatedly bind and release actin molecules as myosin pushes the thin filaments toward the center of the sarcomere. An analogy that may help you visualize this is to think of a number of people holding onto one end of a long rope. When the order comes, each person begins pulling on the rope, hand over hand, grabbing, pulling, and releasing as the rope moves past.

In a muscle fiber, what causes movement of the myosin molecules? The answer is energy from ATP. Myosin is a **motor protein** that converts the chemical bond energy of ATP into the mechanical energy of motion.

Each myosin molecule is an **ATPase** (*myosin ATPase*) that binds ATP and hydrolyzes it to ADP and inorganic phosphate (P_i), releasing energy. This energy is stored as potential energy in the angle between the myosin molecule and the long axis of the myosin filament. The potential energy then becomes the power stroke that moves actin.

Figure 12-9 ● shows the molecular events of a contractile cycle in detail.

1. **The rigor state.** Myosin heads are tightly bound to a G-actin molecule. In this state, no nucleotide (ATP or ADP) occupies the second binding site on the myosin head. In living muscle, the rigor state [*rigere*, to be stiff] occurs for only a very brief period.

2. **ATP binds and myosin detaches.** An ATP molecule binds to the myosin head. This changes the actin-binding affinity of myosin and the head releases from actin.

3. **ATP hydrolysis.** The binding site on myosin closes around ATP and hydrolyses it to ADP and inorganic phosphate (P_i). Both products remain bound to myosin.

4. **Myosin reattaches: weak binding.** The energy released from ATP rotates the myosin molecule, which binds weakly to a new G-actin molecule, one or two positions away from the G-actin to which it was previously bound. At this point, the myosin has potential energy, like a stretched spring, and is ready to execute the power stroke that will move the actin filament past it. ADP and P_i are still bound to myosin.

5. **P_i release and the power stroke.** The power stroke begins when inorganic phosphate is released from the myosin binding site. As myosin rotates, it pushes the attached actin filament toward the center of the sarcomere. The power stroke is also called *crossbridge tilting* because the myosin head and neck tilt from a 90° angle relative to the two filaments to a 45° angle. Our current model of the process also suggests that part of the long myosin arm acts as a lever.

6. **Release of ADP.** In the last step of the contractile cycle, myosin releases ADP, the second product of ATP hydrolysis. At this point, the myosin head is again tightly bound to actin in the rigor state. The cycle is ready to begin again when a new ATP binds to myosin.

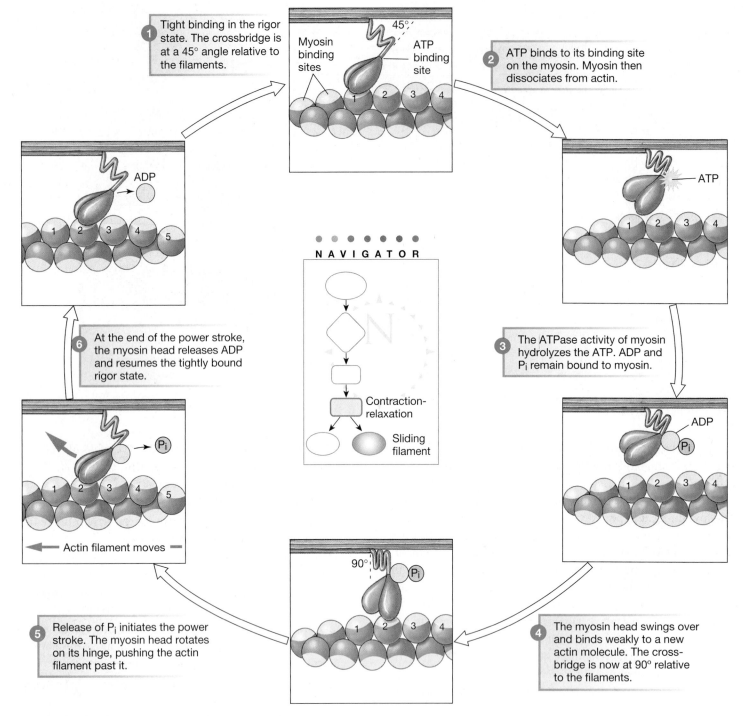

① Tight binding in the rigor state. The crossbridge is at a 45° angle relative to the filaments.

Myosin binding sites

ATP binding site

45°

② ATP binds to its binding site on the myosin. Myosin then dissociates from actin.

ATP

NAVIGATOR

Contraction-relaxation

Sliding filament

③ The ATPase activity of myosin hydrolyzes the ATP. ADP and P_i remain bound to myosin.

ADP

⑥ At the end of the power stroke, the myosin head releases ADP and resumes the tightly bound rigor state.

ADP

P_i

Actin filament moves

④ The myosin head swings over and binds weakly to a new actin molecule. The crossbridge is now at 90° relative to the filaments.

90°

P_i

⑤ Release of P_i initiates the power stroke. The myosin head rotates on its hinge, pushing the actin filament past it.

● **Figure 12-9** *The molecular basis of contraction*

Although our contractile cycle began with the rigor state in which no ATP or ADP was bound to myosin, relaxed muscle fibers remain mostly in step 4 above. The rigor state in living muscle is normally brief because the muscle fiber has a sufficient supply of ATP that binds to myosin when ADP is released in step 6.

After death, however, when metabolism stops and ATP supplies are exhausted, muscles are unable to bind more ATP, so they remain in the tightly bound state described in step 1. In the condition known as *rigor mortis*, the muscles "freeze" owing to immovable crossbridges. The tight binding of actin and myosin persists for a day or so after death, until enzymes within the decaying fiber begin to break down the muscle proteins.

Actin-myosin binding and crossbridge tilting explain how actin filaments slide past myosin filaments during contraction. But what regulates this process? If ATP is always available in the living muscle fiber, what keeps the filaments from continuously interacting? The answer lies with the two regulatory proteins tropomyosin and troponin.

5. Each myosin molecule has binding sites for what molecules?

6. What is the difference between F-actin and G-actin?

Contraction Is Regulated by Troponin and Tropomyosin

In skeletal muscle, the thin actin filaments of the myofibril are associated with two regulatory proteins that prevent myosin heads from completing their power stroke, just as the safety latch on a gun keeps the cocked trigger from being pulled. **Tropomyosin** [*tropos*, to turn] is an elongated protein polymer that wraps around the actin filament and partially blocks the myosin-binding sites (Fig. 12-10a ●). **Troponin** is a calcium-binding protein that controls the position of tropomyosin.

When tropomyosin is in its blocking, or "off," position, weak actin-myosin binding can take place, but myosin is blocked from completing its power stroke. For contraction to occur, tropomyosin

must be shifted to an "on" position that uncovers the remainder of the binding site. This allows the power stroke to take place.

The "off-and-on" positioning of tropomyosin is regulated by troponin, a complex of three proteins associated with tropomyosin. As contraction begins, one protein of the complex, **troponin C**, binds reversibly to Ca^{2+} (Fig. 12-10b). Calcium binding pulls tropomyosin toward the groove of the actin filament and unblocks the myosin-binding sites. This "on" position enables the myosin heads to carry out their power strokes and move the actin filament. Contractile cycles repeat as long as the binding sites are uncovered.

For relaxation to occur, Ca^{2+} concentrations in the cytosol must decrease so that Ca^{2+} unbinds from troponin. Without Ca^{2+}, the troponin-tropomyosin complex returns to its "off" position, covering most of the myosin-binding site. During the portion of the relaxation phase when actin and myosin are not bound to each other, the filaments of the sarcomere slide back to their original positions with the aid of titin and elastic connective tissues within the muscle.

Although the preceding discussion sounds as if we know everything there is to know about the molecular basis of muscle contraction, in reality this is simply our current model. The process is more complex than we have presented here, and it now appears that myosin influences Ca^{2+}-troponin binding, depending on whether the myosin is bound to actin in a strong (rigor) state, a weak state, or not bound at all. The details of this influence are still being worked out.

Studying contraction and the movement of molecules within a myofibril has proved very difficult. Many research techniques rely on crystallized molecules, electron microscopy, and other tools that cannot be used with living tissues. Often we can see filaments only

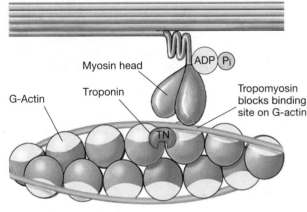

(a) Relaxed state

Myosin head

ADP Pi

G-Actin

Troponin

Tropomyosin blocks binding site on G-actin

TN

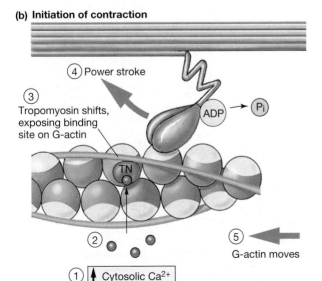

(b) Initiation of contraction

④ Power stroke

③ Tropomyosin shifts, exposing binding site on G-actin

ADP → Pi

TN

②

⑤ G-actin moves

① ↑ Cytosolic Ca^{2+}

● **Figure 12-10** *Regulatory role of tropomyosin and troponin*

BIOTECHNOLOGY

The *In Vitro* Motility Assay

One big step forward in understanding the power stroke of myosin was the development of the *in vitro* motility assay in the 1980s. In this assay, isolated myosin molecules are randomly bonded to a specially coated glass coverslip. A fluorescently labeled actin molecule is placed on top of the myosin molecules. With ATP as a source of energy, the myosin heads bind to the actin and move it across the coverslip, marked by a fluorescent trail as it goes. In even more ingenious experiments, developed in 1995, a single myosin molecule is bound to a tiny bead that elevates it above the surface of the cover slip. An actin molecule is placed on top of the myosin molecule, like the balancing pole of a tightrope walker. As the myosin "motor" moves the actin molecule, lasers measure the nanometer movements and pico-newton forces created with each cycle of the myosin head. Because of this technique, researchers can now measure the mechanical work being done by a single myosin molecule!

at the beginning and end of contraction. Progress is being made, however, and perhaps in the next decade you will see a "movie" of muscle contraction, constructed from photos of sliding filaments.

CONCEPT CHECK   Answers: p. 427

7. Name an elastic fiber in the sarcomere that aids relaxation.

8. In the sliding filament theory of contraction, what prevents the filaments from sliding back to their original position each time they release to bind to the next binding site? (Hint: What would happen if all crossbridges released simultaneously?)

Acetylcholine Initiates Excitation-Contraction Coupling

Now that you have been introduced to the molecular basis of muscle contraction, we will start at the neuromuscular junction and follow the contraction process as it begins at that point. It can be broken down into the following basic steps:

1. Acetylcholine (ACh) is released from the somatic motor neuron.
2. ACh initiates an action potential in the muscle fiber.
3. The muscle action potential triggers calcium release from the sarcoplasmic reticulum.
4. Calcium combines with troponin and initiates contraction.

This combination of electrical and mechanical events in a muscle fiber is called **excitation-contraction coupling**. Now let's look at the process in detail.

Acetylcholine released into the synapse at a neuromuscular junction binds to ACh receptor-channels on the motor end plate (Fig. 12-11●). When these channels open, they allow both Na^+ and K^+ to cross the membrane. However, Na^+ influx exceeds K^+

● **Figure 12-11** *Excitation-contraction coupling*

Paul had experienced mild attacks of muscle weakness in his legs before, usually in the morning. Twice, the weakness had come on after exposure to cold. Each attack had disappeared within minutes, and Paul seemed to suffer no lasting effects. On the advice of Paul's family doctor, Mrs. Leong took her son to see a specialist in muscle disorders. She diagnosed a condition called periodic paralysis. This condition is caused by a genetic defect in the double-gated Na^+ channels in the membranes of his skeletal muscle fibers. Under certain conditions, the defective Na^+ channels fail to inactivate after they open and continuously admit Na^+ into the interior of the muscle cells.

Question 1: What effect does the continuous influx of Na^+ have on the membrane potential of Paul's muscle fibers?

efflux because the electrochemical driving force is greater for Na^+ [⊕ p. 157]. The addition of net positive charge to the muscle fiber depolarizes the membrane, creating an **end-plate potential (EPP)**. Normally, end-plate potentials always reach threshold and initiate a muscle action potential.

The action potential is conducted across the surface of the muscle fiber and into the t-tubules by the opening of voltage-gated Na^+ channels. The process is similar to the conduction of action potentials in axons, although action potentials in skeletal muscle are conducted more slowly than the action potentials of neurons [⊕ p. 257].

The action potential that moves across the membrane and down the t-tubules causes Ca^{2+} release from the sarcoplasmic reticulum (Fig. 12-11b). The t-tubule membrane contains voltage-sensing receptors (the **dihydropyridine**, or DHP, receptors) that are mechanically linked to Ca^{2+} **release channels** in the adjacent sarcoplasmic reticulum (also called **ryanodine receptors** or **RyR**). When a wave of depolarization reaches a DHP receptor, its conformation changes, opening Ca^{2+} release channels in the sarcoplasmic reticulum. Stored Ca^{2+} then moves down its electrochemical gradient into the cytosol, where it initiates contraction.

Free cytosolic Ca^{2+} levels in a resting muscle are normally quite low, but after an action potential, they increase about 100-fold. When cytosolic Ca^{2+} levels are high, Ca^{2+} binds to troponin, tropomyosin moves to the "on" position, and contraction occurs.

Relaxation (the release of muscle tension) occurs when the sarcoplasmic reticulum pumps Ca^{2+} back into its lumen using a Ca^{2+}-**ATPase**. As the free cytosolic Ca^{2+} decreases, Ca^{2+} releases from troponin, tropomyosin slides back to block the myosin-binding site, and the fiber relaxes.

The discovery that Ca^{2+}, not the action potential, is the signal for contraction was the first piece of evidence suggesting that calcium acts as a messenger inside cells. For years, it was thought that calcium signals occurred only in muscles, but we now know that calcium is an almost universal second messenger [⊕ p. 182].

Figure 12-12 ● shows the timing of electrical and mechanical events during excitation-contraction coupling. The somatic motor neuron action potential is followed by the skeletal muscle action potential, which in turn is followed by contraction. A single contraction-relaxation cycle in a skeletal muscle fiber is known as a **twitch**. Notice that there is a short delay, the **latent period**, between the muscle action potential and the beginning of muscle tension development. This delay represents the time required for excitation-contraction coupling to take place.

Once contraction begins, muscle tension increases steadily to a maximum value as crossbridge interaction increases. Tension then decreases in the relaxation phase of the twitch. During relaxation, elastic elements of the muscle return the sarcomeres to their resting length.

A single action potential in a muscle fiber evokes a single twitch (Fig. 12-12, ③). But muscle twitches vary from fiber to fiber in the speed with which they develop tension (the rising slope of the twitch curve), the maximum tension they achieve (the height of the twitch curve), and the duration of the twitch (the width of the twitch curve). We will discuss factors that affect these parameters in upcoming sections. First we will discuss how muscles produce ATP to provide energy for contraction and relaxation.

9. What aspect of contraction requires ATP? Does relaxation require ATP?

10. What events are taking place during the latent period before contraction begins?

Skeletal Muscle Contraction Requires a Steady Supply of ATP

The muscle fiber's use of ATP is a key feature of muscle physiology. Muscles require energy constantly: during contraction for crossbridge movement and release, during relaxation to pump Ca^{2+} back into the sarcoplasmic reticulum, and after excitation-contraction coupling to restore Na^+ and K^+ to the extracellular and intracellular compartments, respectively. Where do muscles get the ATP they need for this work?

The amount of ATP within the muscle fiber at any one time is sufficient for only about eight twitches. As a backup energy source, muscles contain **phosphocreatine**, a molecule whose high-energy phosphate bonds are created from creatine and ATP when muscles are at rest (Fig. 12-13 ●). When muscles become active, such as during exercise, the high-energy phosphate group of phosphocreatine is transferred to ADP, creating more ATP to power the muscles.

The enzyme responsible for transferring the phosphate group is **creatine kinase** (CK), also known as *creatine phosphokinase* (CPK). Muscle cells contain large amounts of this enzyme. Consequently, elevated blood levels of creatine kinase usually indicate damage to

Figure 12-12 *Electrical and mechanical events in muscle contraction* Action potentials in the axon terminal and the muscle fiber are followed by the muscle twitch.

FIGURE QUESTION

- Compare the resting membrane potential of the motor neuron and the muscle fiber.
- Movement of what ion(s) in what direction(s) creates the muscle action potential?

Figure 12-13 *Phosphocreatine* Resting muscle stores energy from ATP in high-energy phosphate bonds of phosphocreatine.

skeletal or cardiac muscle. Because the two muscle types contain different isozymes [● p. 92], clinicians can distinguish tissue damage during a heart attack from skeletal muscle damage.

Energy stored in high-energy phosphate bonds is very limited, so muscle fibers must use metabolism to transfer energy from the chemical bonds of nutrients to ATP. Carbohydrates, particularly glucose, are the most rapid and efficient source of energy for ATP production. Glucose is metabolized through glycolysis to pyruvate [● p. 103]. In the presence of adequate oxygen, pyruvate goes into the citric acid cycle, producing about 30 ATP for each molecule of glucose.

When oxygen concentrations are too low to maintain aerobic metabolism, the muscle fiber shifts to *anaerobic glycolysis*. In this pathway, glucose is metabolized to lactic acid with a yield of only 2 ATP per glucose. Anaerobic metabolism of glucose is a quicker source of ATP but produces many fewer ATP per glucose. When muscle energy demands outpace the amount of ATP that can be produced through anaerobic metabolism of glucose, muscles can function for only a short period of time without fatiguing.

Muscle fibers also obtain energy from fatty acids, although this process always requires oxygen [● p. 108]. During rest and light exercise, skeletal muscles burn fatty acids along with glucose, one reason that modest exercise programs of brisk walking are an effective way to reduce body fat. But *beta-oxidation*, the process by which fatty acids are converted to acetyl CoA, is a slow process and cannot produce ATP rapidly enough to meet the energy needs of muscle fibers during heavy exercise. Under these conditions, muscle fibers rely more on glucose.

Proteins normally are not a source of energy for muscle contraction. Most amino acids found within muscle fibers are used to synthesize proteins rather than to produce ATP.

Do muscles ever run out of ATP? You might think so if you have ever exercised to the point of fatigue, where you feel that you cannot continue or your limbs refuse to obey the commands from your brain. But most studies show that even intense exercise uses only 30% of the ATP within a muscle fiber. The condition we call fatigue must come from other changes in the exercising muscle.

Muscle Fatigue Has Multiple Causes

The term **fatigue** describes a condition in which a muscle is no longer able to generate or sustain the expected power output. Fatigue is highly variable. It is influenced by the intensity and duration of the contractile activity, by whether the muscle fiber is using aerobic or anaerobic metabolism, by the composition of the muscle, and by the fitness level of the individual.

Multiple factors have been proposed to play a role in fatigue (Fig. 12-14 ●). They can be classified into **central fatigue** mechanisms that arise in the central nervous system, and **peripheral fatigue** mechanisms that arise anywhere between the neuromuscular junction and the contractile elements of the muscle. Most experimental evidence suggests that muscle fatigue arises from excitation-contraction failure within the muscle fiber itself rather than from failure of control neurons or neuromuscular transmission.

Central fatigue includes subjective feelings of tiredness and a desire to cease activity. Several studies have shown that this fatigue actually precedes physiological fatigue in the muscles and

Types of fatigue	Process map	Proposed mechanisms
Central fatigue	CNS	• Psychological effects • Protective reflexes
	Somatic motor neuron	
Peripheral fatigue	Neuromuscular junction	• ↓ Neurotransmitter release • ↓ Receptor activation
	Excitation-contraction coupling	• Change in muscle membrane potential
	Ca²⁺ signal	• ↓ Ca²⁺ release • ↓ Ca²⁺-troponin interaction
	Contraction-relaxation	• Depletion theories: PCr, ATP, glycogen • Accumulation theories: H⁺, Pᵢ, lactate

● **Figure 12-14** *Locations and possible causes of muscle fatigue*

RUNNING PROBLEM

Two forms of periodic paralysis exist. One form, called *hypokalemic periodic paralysis*, is characterized by decreased blood levels of K^+ during paralytic episodes. The other form, called *hyperkalemic periodic paralysis*, is characterized by normal or increased blood levels of K^+ during episodes. Results of a blood sample revealed that Paul has the hyperkalemic form.

Question 2: In people with hyperkalemic periodic paralysis, attacks tend to occur after a period of exercise (i.e., repeated muscle contractions). What ion is responsible for the repolarization phase of the muscle action potential, and in which direction does this ion move across the muscle fiber membrane?

---

therefore may be a protective mechanism. *Lactic acid* production from anaerobic metabolism is often mentioned as a possible cause for fatigue. Some evidence suggests that acidosis caused by lactic acid dumped into the blood may influence the sensation of fatigue perceived by the brain. However, because the homeostatic mechanisms for pH balance maintain blood pH at normal levels until exertion is nearly maximal, pH is not a factor in central fatigue in most instances of submaximal exertion.

Neural causes of fatigue could arise either from communication failure at the neuromuscular junction or from failure of the central nervous system command neurons. For example, if ACh is not synthesized in the axon terminal fast enough to keep up with the firing rate of the neuron, neurotransmitter release at the synapse will decrease. Consequently, the muscle end-plate potential may fail to reach the threshold value needed to trigger a muscle fiber action potential, resulting in contraction failure. This type of fatigue is associated with some neuromuscular diseases, but it is probably not a factor in normal exercise.

Fatigue within the muscle fiber itself can occur in any one of several sites. In extended submaximal exertion, fatigue is associated with the depletion of glycogen stores within the muscle. Because most studies show that lack of ATP is not a limiting factor, glycogen depletion may be affecting some other aspect of contraction, such as the release of Ca^{2+} from the sarcoplasmic reticulum.

In short-duration maximal exertion, fatigue apparently has a different cause. It apparently results from increased production of inorganic phosphate (P_i) from ATP and phosphocreatine breakdown. Elevated inorganic phosphate may slow P_i release from myosin and thereby alter the power stroke (Fig. 12-9, ⑤). A new theory suggests that elevated phosphate decreases Ca^{2+} release because the phosphate combines with Ca^{2+} to become calcium phosphate. Some investigators feel that phosphate alters the dynamics of Ca^{2+} release from the sarcoplasmic reticulum.

Potassium is another factor implicated in fatigue. During maximal exercise, K^+ leaves the cell with each action potential, and K^+ concentrations rise in the extracellular fluid of the t-tubules. The shift in K^+ alters the membrane potential of the muscle fiber and is believed to decrease Ca^{2+} release from the sarcoplasmic reticulum.

There appears to be no single cause for muscle fatigue during exercise, and the study of this phenomenon is quite complex. For additional information on this subject, see the Explorations at the end of the chapter.

CONCEPT CHECK   Answers: p. 427

11. If K^+ concentration increases in the extracellular fluid but does not change significantly in the cytoplasm, the cell membrane potential will (select all that apply) *depolarize/hyperpolarize* and become *more/less* negative.

Skeletal Muscle Fibers Are Classified by Contraction Speed and Resistance to Fatigue

Skeletal muscle fibers can be classified into groups on the basis of their speed of contraction and their resistance to fatigue with repeated stimulation. The groups include fast-twitch glycolytic fibers, fast-twitch oxidative fibers, and slow-twitch (oxidative) fibers.

Fast-twitch muscle fibers develop tension two to three times as fast as **slow-twitch fibers**. The speed with which a muscle fiber contracts is determined by the isoform of myosin ATPase present in the fiber's thick filaments. Fast-twitch fibers split ATP more rapidly and can therefore complete multiple contractile cycles more rapidly than slow-twitch fibers. This speed translates into faster tension development in the muscle fiber.

The duration of contraction also varies according to the fiber type. Twitch duration is determined largely by how fast the sarcoplasmic reticulum removes Ca^{2+} from the cytosol. As cytosolic Ca^{2+} concentrations fall, Ca^{2+} unbinds from troponin, allowing tropomyosin to move back and partially block the myosin-binding sites. With the power stroke inhibited, the muscle fiber relaxes.

Fast-twitch fibers pump Ca^{2+} into their sarcoplasmic reticulum more rapidly than slow-twitch fibers, so fast-twitch fibers have quicker twitches. Fast-fiber twitches last only about 7.5 milliseconds (msec), making these muscles useful for fine, quick movements such as playing the piano.

Contractions in slow-twitch muscle fibers may last more than 10 times as long. These muscles are used for strong, sustained movements such as lifting heavy loads. Fast fibers are used occasionally, but the slow-fiber types are used almost constantly for maintaining posture, standing, or walking.

Another major difference between fast-twitch glycolytic fibers and slow-twitch fibers is their ability to resist fatigue. Fast-twitch glycolytic muscle fibers rely primarily on anaerobic glycolysis to produce ATP. However, the accumulation of lactic acid contributes to acidosis, a condition implicated in the development of fatigue. As a result, fast-twitch fibers fatigue more easily than do slow-twitch fibers, which do not produce large amounts of lactic acid.

Slow-twitch fibers, on the other hand, depend primarily on oxidative phosphorylation for production of ATP, hence their descriptive name as *oxidative fibers*. Slow-twitch fibers have more mitochondria, the site of enzymes for the citric acid cycle and oxidative phosphorylation. They also have more blood vessels in their connective tissue to bring oxygen to the muscle fibers (Fig. 12-15 ●).

The efficiency with which muscle fibers obtain oxygen is a factor in their preferred route of glucose metabolism. Slow-twitch fibers are described as "red muscle" because large amounts of **myoglobin**, a red oxygen-binding pigment, give them their characteristic color.

Oxygen brought from the lungs to the muscles must diffuse from the interstitial fluid into the interior of the fiber to reach the mitochondria. In muscles with myoglobin, this diffusion process is faster because myoglobin has a high binding affinity for oxygen. This affinity allows myoglobin to act as a transfer molecule, bringing oxygen more rapidly to the interior of the fiber.

In addition, slow-twitch fibers have a smaller diameter, so the distance through which oxygen must pass before reaching the mitochondria is less. Because slow-twitch muscle fibers have more capillaries to bring blood to the cells, are smaller, and have a high myoglobin content, they maintain a better supply of oxygen and are more likely to use oxidative phosphorylation for ATP production.

Fast-twitch fibers, on the other hand, are described as "white muscle" because of their lower myoglobin content. These muscle fibers are also larger in diameter. The combination of larger size, less myoglobin, and fewer blood vessels means that fast-twitch fibers are more likely to run out of oxygen with repeated contractions.

Fast-twitch fibers have been divided into two subcategories based on their relative diameters and fatigue resistance. **Fast-twitch glycolytic fibers** are the largest in diameter and rely primarily on anaerobic glycolysis for ATP synthesis. They fatigue most rapidly of the three fiber types. **Fast-twitch oxidative fibers** are smaller, contain some myoglobin, and use a combination of oxidative and glycolytic metabolism to produce ATP. Because of their intermediate size and the use of oxidative phosphorylation for ATP synthesis, fast-twitch oxidative fibers are more fatigue-resistant than their fast glycolytic cousins.

● **Figure 12-15** *Fast-twitch glycolytic and slow-twitch muscle fibers*   Large amounts of red myoglobin, numerous mitochondria (M), and extensive capillary blood supply (cap) distinguish slow-twitch oxidative muscle (R) from fast-twitch glycolytic muscle (W).

Human muscles are a mixture of muscle types. The ratio of the different fiber types varies from muscle to muscle and between individuals. For example, who would you think would have more fast-twitch fibers in leg muscles, a marathon runner or a high-jumper? Characteristics of the three muscle fiber types are compared in Table 12-2.

Tension Developed by Individual Muscle Fibers Is a Function of Fiber Length

Within a single muscle fiber, the tension of a twitch is a direct reflection of the length of individual sarcomeres before contraction

begins (Fig. 12-16●). Each sarcomere will contract with optimum force if it is an optimum length, neither too long nor too short, before the contraction begins. Fortunately, the normal resting length of skeletal muscles usually ensures that sarcomeres within the individual muscle fibers are at optimum length when they begin a contraction.

At the molecular level, sarcomere length reflects the overlap between the thick and thin filaments. The sliding filament theory predicts that *the tension a muscle fiber can generate is directly proportional to the number of crossbridges formed between the thick and thin filaments*. If the fibers start a contraction at a very

Table 12-2: Characteristics of Muscle Fiber Types

	SLOW-TWITCH OXIDATIVE; RED MUSCLE	FAST-TWITCH OXIDATIVE; RED MUSCLE	FAST-TWITCH GLYCOLYTIC; WHITE MUSCLE
Time to development of maximum tension	Slowest	Intermediate	Fastest
Myosin ATPase activity	Slow	Fast	Fast
Diameter	Small	Medium	Large
Contraction duration	Longest	Short	Short
Ca^{2+}-ATPase activity in SR	Moderate	High	High
Endurance	Fatigue-resistant	Fatigue-resistant	Easily fatigued
Use	Most used: posture	Standing, walking	Least used: jumping
Metabolism	Oxidative; aerobic; numerous large mitochondria	Glycolytic but becomes more oxidative with endurance training	Glycolytic; more anaerobic than fast-twitch oxidative type
Color	Dark red (myoglobin)	Red	Pale

● **Figure 12-16** *Length-tension relationships in contracting skeletal muscle* The graph compares tension generated by a skeletal muscle with its resting length before contraction starts.

long sarcomere length, the thick and thin filaments are barely overlapped, forming few crossbridges (Fig. 12-16b). This means that in the initial part of the contraction, the sliding filaments can interact only minimally and therefore cannot generate much force.

At the optimum sarcomere length, the filaments begin contracting with more crossbridge linkage between the thick and thin filaments, allowing the fiber to generate optimum force in that twitch (Fig. 12-16a). If the sarcomere is shorter than optimum length at the beginning of the contraction, the thick and thin fibers were too well overlapped before the contraction began. Consequently, the thick filaments can move the thin filaments only a short distance before the thin actin filaments from opposite ends of the sarcomere start to overlap (Fig. 12-16c). This overlap prevents crossbridge formation. If the sarcomere is so short that the thick filaments run into the Z disks, myosin is unable to find new binding sites for crossbridge formation, and tension decreases rapidly (Fig. 12-16d). Thus, the development of single-twitch tension within a muscle fiber is a passive property that depends on filament overlap and sarcomere length.

Force of Contraction Increases with Summation of Muscle Twitches

Although we have just shown that single-twitch tension is determined by the length of the sarcomere, it is important to note that a single twitch does not represent the maximum force that the muscle fiber can develop. The force generated by the contraction of a single muscle fiber can be increased by increasing the rate (frequency) at which muscle action potentials stimulate the muscle fiber.

A typical muscle action potential lasts between 1 and 3 msec, while the muscle contraction itself may last 100 msec (see Fig. 12-12). If repeated action potentials are separated by long intervals of time, the muscle fibers have time to relax completely between stimuli (Fig. 12-17a ●). If the interval of time between action potentials is shortened, the muscle fiber will not have relaxed completely at the time of the second stimulus, resulting in a more forceful contraction (Fig. 12-17b). This process is known as **summation** and is similar to the summation of graded potentials that takes place in neurons [● p. 270].

If action potentials continue to stimulate the muscle fiber repeatedly at short intervals (high frequency), relaxation between contractions diminishes until the muscle fiber achieves a state of maximal contraction known as **tetanus**. There are two types of tetanus. In *incomplete*, or *unfused*, tetanus, the stimulation rate of the muscle fiber is slower, and the fiber relaxes slightly between stimuli (Fig. 12-17c).

In *complete*, or *fused*, tetanus, the stimulation rate is fast enough that the muscle fiber does not have time to relax. Instead, it reaches maximum tension and remains there (Fig. 12-17d). Thus, it is possible to increase the tension developed in a single muscle fiber by changing the rate at which action potentials occur in the fiber. Muscle action potentials are initiated by the somatic motor neuron that controls the muscle fiber.

CONCEPT CHECK   Answers: p. 427

12. Summation in muscle fibers means that the _____ of the fiber increases with repeated action potentials.

13. Temporal summation in neurons means that the _____ of the neuron increases when two depolarizing stimuli occur close together in time.

(a) Single twitches: Muscle relaxes completely between stimuli (▲).

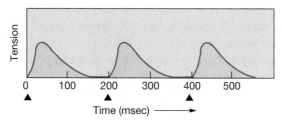

(b) Summation: Stimuli closer together do not allow muscle to relax fully.

(c) Summation leading to unfused tetanus: Stimuli are far enough apart to allow muscle to relax slightly between stimuli.

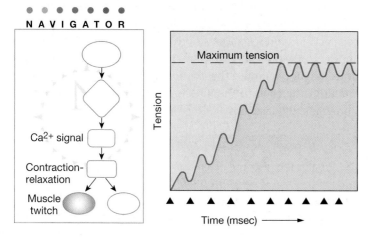

(d) Summation leading to complete tetanus: Muscle reaches steady tension.

● **Figure 12-17** *Summation of contractions*

A Motor Unit Is One Somatic Motor Neuron and the Muscle Fibers It Innervates

The basic unit of contraction in an intact skeletal muscle is a **motor unit**, composed of a group of muscle fibers and the somatic motor neuron that controls them (see chapter opener photo and Fig. 12-18). When the somatic motor neuron fires an action potential, all muscle fibers in the motor unit contract. Note that although one somatic motor neuron innervates multiple fibers, each muscle fiber is innervated by only a single neuron.

The number of muscle fibers in a motor unit varies. In muscles used for fine motor actions, such as the muscles that move the eyes or the muscles of the hand, a motor unit contains as few as three to five muscle fibers. If one motor unit is activated, only a few fibers contract, and the muscle response is quite small. If additional motor units are activated, the response increases by small increments since only a few more muscle fibers contract with the addition of each motor unit. This arrangement allows fine gradations of movement.

In muscles used for gross motor actions, such as standing or walking, a single somatic motor neuron may innervate hundreds or even thousands of muscle fibers. The gastrocnemius muscle in the calf of the leg, for example, has about 2000 muscle fibers in each motor unit. Each time an additional motor unit is activated in these muscles, many more muscle fibers contract, and the muscle response jumps by correspondingly greater increments.

All muscle fibers in a single motor unit are of the same fiber type. Thus, there are fast-twitch motor units and slow-twitch motor units. The determination of which kind of fiber associates with a particular neuron appears to lie with the neuron itself. During embryological development, the somatic motor neuron secretes a growth factor that directs the differentiation of all muscle fibers in its motor unit so that they develop into the same fiber type.

Intuitively, it would seem that people who inherit a predominance of one fiber type over another would excel in certain sports. They do, to some extent. Endurance athletes such as distance runners and cross-country skiers have a predominance of aerobic slow-twitch fibers, whereas sprinters, ice hockey players, and weight lifters tend to have larger percentages of fast-twitch fibers.

But inheritance is not the only determining factor for fiber composition in the body. The metabolic characteristics of muscle fibers can be changed to some extent. With endurance training, the aerobic capacity of some fast-twitch fibers can be enhanced until they are almost as fatigue-resistant as a slow-twitch fiber. Since the conversion occurs only in those muscles that are being trained, a neuromodulator chemical is probably involved. In addition, endurance training increases the number of capillaries and mitochondria in the muscle tissue, allowing more oxygen-carrying blood to reach the contracting muscle and contributing to the increased aerobic capacity of the muscle fibers.

CONCEPT CHECK *Answers: p. 427*

14. Which type of runner would you expect to have more slow-twitch fibers, a sprinter or a marathoner?

Contraction in Intact Muscles Depends on the Types and Numbers of Motor Units

Within a skeletal muscle, each motor unit contracts in an all-or-none manner. How then can muscles create graded contractions of varying force and duration? The answer lies in the fact that intact muscles are composed of multiple motor units of different types (Fig. 12-18 ●). This diversity allows the muscle to vary contraction by (1) changing the types of motor units that are active or (2) changing the number of motor units that are responding at any one time.

The force of contraction within a skeletal muscle can be increased by recruiting additional motor units. **Recruitment** is controlled by the nervous system and proceeds in a standardized sequence. A weak stimulus directed onto a pool of somatic motor neurons in the central nervous system activates only the neurons with the lowest thresholds [● p. 251]. Studies have shown that these low-threshold neurons control fatigue-resistant slow-twitch fibers that generate minimal force.

As the stimulus onto the motor neuron pool increases in strength, additional motor neurons with higher thresholds begin to fire. These neurons in turn stimulate motor units composed of fatigue-resistant oxidative fast-twitch fibers. Because more motor units (and thus, more muscle fibers) are participating in the contraction, greater force is generated.

As the stimulus increases to even higher levels, somatic motor neurons with the highest thresholds begin to fire. These neurons stimulate motor units composed of glycolytic fast-twitch fibers. At this point, the muscle contraction is approaching its maximum force. Because of differences in myosin and crossbridge

One muscle may have many motor units of different fiber types.

● **Figure 12-18** *Motor units*

formation, fast-twitch fibers can generate more force than slow-twitch fibers. However, because fast-twitch fibers fatigue more rapidly, it is impossible to hold a muscle contraction at maximum force for an extended period of time. You can demonstrate this by clenching your fist as hard as you can: how long can you hold it before some of the muscle fibers begin to fatigue?

Sustained contractions within a muscle require a continuous train of action potentials from the central nervous system to the muscle. But, as you learned earlier, increasing the stimulation rate of a muscle fiber results in summation of its contractions. If the muscle fiber is easily fatigued, summation will lead to fatigue and diminished tension.

One way that the nervous system avoids fatigue in a sustained contraction is by **asynchronous recruitment** of motor units. The nervous system modulates the firing rates of the motor neurons so that different motor units take turns maintaining the muscle tension. The alternation of active motor units allows some of the motor units to rest between contractions, preventing fatigue.

Asynchronous recruitment will prevent fatigue only in submaximal contractions, however. In high-tension, sustained contractions, the individual motor units may reach a state of unfused tetanus, in which the muscle fibers cycle between contraction and partial relaxation. In general, we do not notice this cycling, because the different motor units in the muscle are contracting and relaxing at slightly different times. As a result, the contractions and relaxations of the motor units average out and appear to be one smooth contraction. But, as different motor units fatigue, we notice that we are unable to maintain the same amount of tension in the muscle, and the force of the contraction gradually decreases.

CONCEPT CHECK   Answers: p. 427

15. What is the response of a muscle fiber to an increase in the firing rate of the somatic motor neuron?

16. How does the nervous system increase the force of contraction in a muscle composed of many motor units?

▶ MECHANICS OF BODY MOVEMENT

Because one main role of skeletal muscles is to move the body, we now turn to the mechanics of body movement. The term *mechanics* refers to how muscles move loads and how their anatomical relationship to the bones of the skeleton maximizes the work they can do.

Isotonic Contractions Move Loads, but Isometric Contractions Create Force Without Movement

When we described the function of muscles, we noted that they create force to generate movement and that they can also create force without generating movement. You can demonstrate both properties with a pair of heavy weights. Pick up the weights and bend your elbows so that the weights touch your shoulders. You have just performed an **isotonic contraction** [*iso*, equal + *teinein*, to stretch]. Any contraction that creates force and moves a load is an isotonic contraction.

When you bent your arms at the elbows and brought the weights to your shoulders, the biceps muscles shortened in a **concentric action**. If you slowly extend your arms, resisting the tendency of the weights to pull them down, you are performing another type of isotonic contraction known as an **eccentric**, or lengthening, **action**. Eccentric exercise is thought to contribute most to cellular damage after exercise and to lead to delayed muscle soreness.

If you pick up the weights and hold them stationary in front of you, the muscles of your arms are creating tension (force) to overcome the load of the weights but are not creating movement. These contractions that create force without movement are called **isometric** (static) **contractions** [*iso*, equal + *metric*, measurement]. Isotonic and isometric contractions are illustrated in Figure 12-19 ●. To demonstrate an isotonic contraction experimentally, we hang a weight (the load) from the muscle and electrically stimulate the

muscle to contract. The muscle contracts, lifting the weight. The graph on the right shows the development of force throughout the contraction.

To demonstrate an isometric contraction experimentally, we attach a heavier weight to the muscle. When the muscle is stimulated, it develops tension, but the force created is not enough to move the load. In isometric contractions, muscles create force without shortening significantly. For example, when your exercise instructor yells at you to "tuck in your tummy," your response is isometric contraction of abdominal muscles.

How can an isometric contraction create force if the length of the muscle does not change? The elastic elements of the muscle provide the answer. All muscles contain elastic fibers in the tendons and other connective tissues that attach muscles to bone and in the connective tissue between muscle fibers. Within the muscle fibers, elastic cytoskeletal proteins occur between the myofibrils and as part of the sarcomere. Finally, the contractile filaments themselves apparently can stretch even while generating force. All of these elastic components behave collectively as if they were connected in series (one after the other) to the contractile

(a) Isotonic contraction: muscle contracts, shortens, and creates enough force to move the load.

(b) Isometric contraction: muscle contracts but does not shorten. Force cannot move the load.

● **Figure 12-19** *Isotonic and isometric contractions*

Schematic of the series elastic elements

● Figure 12-20 *Series elastic elements in muscle* A muscle has both contractile components (sarcomeres, shown here as a gear and rachet) and elastic components (shown here as a spring).

elements of the muscle. Consequently, they are often called the **series elastic elements** of the muscle (Fig. 12-20 ●).

When the sarcomeres shorten in an isometric contraction, the elastic elements stretch. This stretching of the elastic elements allows the fibers to maintain constant length even though the sarcomeres are shortening and creating tension (Fig. 12-20). Once the elastic elements have been stretched and the force generated by the sarcomeres equals the load, the muscle shortens in an isotonic contraction and lifts the load.

Bones and Muscles Around Joints Form Levers and Fulcrums

The anatomical arrangement of muscles and bones in the body is directly related to how muscles work. The body uses its bones and joints as levers and fulcrums on which muscles exert force to

move or resist a load. A **lever** is a rigid bar that pivots around a point known as the **fulcrum** (Fig. 12-21a ●). In the body, bones form levers, flexible joints form the fulcrums, and muscles attached to bones create force when they contract. The work being performed by the muscle can be expressed as

$$\text{Work} = \text{force} \times \text{distance}$$

In a lever and fulcrum system, distance is measured from the fulcrum to the point along the lever at which force is being exerted. Most lever systems in the body are similar to the one shown where the fulcrum is located at one end of the lever. This arrangement maximizes the distance and speed with which the lever can move the load but also requires that the muscles do more work. We will use the flexion of the forearm to illustrate how this lever system functions.

In the lever and fulcrum system of the forearm, the elbow joint acts as a fulcrum around which movement of the forearm, the lever, takes place (Fig. 12-21a). The biceps muscle is attached at its origin at the shoulder and inserts into the radius bone of the forearm (the lever) about 5 cm away from the elbow joint (the fulcrum). When the biceps contracts, it creates an upward force as it pulls on the bone.

If the biceps muscle is to hold the arm stationary and flexed at a 90° angle, it must do work (W_{up}) that exactly opposes the downward pull of the weight of the forearm and hand (W_{down}). For the arm illustrated in Figure 12-21b, the biceps muscle must create 6 kg of force simply to hold the arm at a 90° angle. Because the muscle is not shortening, this is an isometric contraction.

Now what happens if a 7-kg weight is placed in the hand? This places an additional load on the lever that is even farther from the fulcrum than the center of gravity (Fig. 12-21c). Unless the biceps muscle can create additional upward force to offset the downward force created by the weight, the hand will drop. If you know the added weight and its distance from the elbow, you can calculate the additional force needed to keep the arm from dropping the 7-kg weight.

What happens to the work required of the biceps muscle if the distance between the fulcrum and the muscle insertion changes? Genetic variability in the insertion point could have a dramatic effect on the force required to move or resist a load. For example, if the biceps muscle in Figure 12-21b inserts 6 cm from the fulcrum instead of 5 cm, it would only need to generate 5 kg of force to offset the weight of the arm. Some studies have shown a correlation between the insertion points of muscles and success in certain athletic events.

In the example so far, we have assumed that the load is stationary and that the muscle is contracting isometrically. What happens if we want to flex the arm and lift the load? Then, the biceps muscle must exert a force that exceeds the force created by the load. The disadvantage of a lever system in which the fulcrum is positioned near one end of the lever is that the muscle is required to create large amounts of force to move or resist small loads.

(a)

The work required to move a load on the lever is equal to the force created by the load (*F*) times distance of the load from the fulcrum (*D*).

The human forearm acts as a lever. The fulcrum is close to the elbow.

(b)

The biceps inserts into the lever 5 cm from the fulcrum. Work done by the biceps depends on the force created by the contracting muscle:

Work$_{up}$ = biceps force × 5 cm

The weight of the forearm exerts a downward force of 2 kg at its center of gravity, which is 15 cm from the fulcrum.

Work$_{down}$ = 2 kg × 15 cm

To hold the arm stationary at 90 degrees, the work done by the biceps pulling upward must equal the work created by the downward pull of the arm's weight.

Work$_{up}$ = Work$_{down}$

Biceps force × 5 cm = 2 kg × 15 cm

$$\text{Biceps force} = \frac{30 \text{ kg·cm}}{5 \text{ cm}}$$

Biceps force = 6 kg

(c)

Q · · · · · ·

FIGURE QUESTION

How much additional force must the biceps exert to keep from dropping the weight?

A 7-kg load is added to the hand 25 cm from the elbow.

● **Figure 12-21**   *The arm is a lever and fulcrum system*

Because the insertion of the biceps is close to the fulcrum, a small movement of the biceps becomes a much larger movement of the hand.

When the biceps contracts and shortens 1 cm, the hand moves upward 5 cm.

FIGURE QUESTION

If the biceps shortens 1 cm in 1 second, how fast does the hand move upward?

● **Figure 12-22** *The arm amplifies speed of movement of the load.*

However, this type of lever system maximizes speed and mobility. A small movement of the bone at the point where the muscle inserts becomes a much larger movement at the hand (Fig. 12-22 ●). In addition, the two movements occur in the same amount of time, so the speed of contraction at the insertion point is amplified at the end of the arm. Thus, the lever system of the arm amplifies both the distance and the speed of movement of the load.

In muscle physiology, the speed with which a muscle contracts depends on the type of muscle fiber (fast-twitch versus

RUNNING PROBLEM

Paul's doctor explained to Mrs. Leong that the paralytic attacks associated with hyperkalemic periodic paralysis last only a few minutes to a few hours and generally involve only the muscles of the extremities. "Is there any treatment?" asked Mrs. Leong. The doctor replied that, although the condition is incurable, attacks may be prevented with drugs. Diuretics, for example, increase the body's excretion of water and ions (including Na^+ and K^+), and they have been shown to prevent attacks of paralysis in people with this condition.

Question 3: Why does a Na^+ channel that will not inactivate cause paralysis?

FIGURE QUESTION

- At what point on the line is contraction isometric?
- At what point is the muscle contraction at maximum velocity?

● **Figure 12-23** *Load-velocity relationship in skeletal muscle*

slow-twitch) and on the load that is being moved. Intuitively, you can see that you can flex your arm much faster with nothing in your hand than you can while holding a 7-kg weight in your hand. The relationship between load and velocity of contraction in a muscle fiber, determined experimentally, is graphed in Figure 12-23 ●. Contraction is fastest when the load on the muscle is zero. When the load on the muscle equals the ability of the muscle to create force, the muscle is unable to move the load and the velocity drops to zero. The muscle can still contract, but the contraction becomes isometric instead of isotonic. Because speed is a function of load and muscle fiber type, it cannot be regulated by the body except through recruitment of faster muscle fiber types. However, the arrangement of muscles, bones, and joints allows the body to amplify speed so that regulation at the cellular level becomes less important.

CONCEPT CHECK   Answers: p. 427

17. One study found that many world-class athletes have muscle insertions that are farther from the joint than in the average person. Why would this trait translate into an advantage for a weight lifter?

Muscle Disorders Have Multiple Causes

Dysfunctions of skeletal muscles can arise from a problem with the signal from the nervous system, from miscommunication at the neuromuscular junction, or from defects in the muscle itself. Unfortunately, in many muscle conditions, even the simple ones, we do not fully understand the mechanism of the primary defect. As a result, we can treat the symptoms but may not be able to cure the problem.

One common muscle disorder is a "charlie horse," or *muscle cramp*: a sustained painful contraction of skeletal muscles. As people age, they often develop nighttime cramps in the calves of their legs. Many muscle cramps are caused by hyperexcitability of the somatic motor neurons controlling the muscle. As the neuron fires repeatedly, the muscle fibers of its motor unit go into a state of painful sustained contraction. Sometimes muscle cramps can be relieved by forcibly stretching the muscle. Apparently, stretching sends sensory information to the central nervous system that inhibits the somatic motor neuron, relieving the cramp.

The simplest muscle disorders arise from overuse. Most of us have exercised too long or too hard and suffered from muscle fatigue or soreness as a result. Trauma to muscles can cause muscle fibers, the connective tissue sheath, or the union of muscle and tendon to tear.

Disuse of muscles can be as traumatic as overuse. With prolonged inactivity such as may occur when a limb is immobilized in a cast, the skeletal muscles will **atrophy**, or waste away [*a-*, without + *trephein*, nourishment]. The blood supply to the muscle diminishes and the muscle fibers get smaller. If activity is resumed in less than a year, the fibers usually will regenerate. Atrophy of longer than one year is usually permanent. If the atrophy is occurring because of somatic motor neuron dysfunction, therapists now try to maintain muscle function with the aid of electrical stimulation that directly stimulates the muscle fibers.

Acquired disorders that affect the skeletal muscle system include infectious diseases, such as influenza, that lead to weakness and aching, and poisoning by toxins, such as those produced in botulism (*Clostridium botulinus*) and tetanus (*Clostridium tetani*). Botulinum toxin acts by decreasing the release of acetylcholine from the somatic motor neuron. Clinical investigators have successfully used injections of botulinum toxin as a treatment for writer's cramp, a disabling cramp of the hand that apparently arises as a result of hyperexcitability in the distal portion of the somatic motor neuron. You may also have heard of Botox® injections for cosmetic wrinkle reduction. Botulinum toxin injected under the skin temporarily paralyzes facial muscles that pull the skin into wrinkles.

Inherited muscular disorders are the most difficult to treat. These conditions include various forms of muscular dystrophy as well as biochemical defects in glycogen and lipid storage. In **Duchenne's muscular dystrophy**, the structural protein **dystrophin**, which links actin to proteins in the cell membrane, is absent. Muscle fibers that lack dystrophin have tiny tears in their cell membranes that allow extracellular Ca^{2+} to enter the fiber. Consequently, intracellular enzymes are activated, resulting in breakdown of the fiber components. The major symptom of Duchenne dystrophy is progressive muscle weakness, and patients usually die before age 30 from failure of the respiratory muscles.

McArdle's disease, also known as *myophosphorylase deficiency glycogenosis*, is a condition in which the enzyme that converts glycogen to glucose 6-phosphate is absent in muscles. As a result, muscles lack a usable glycogen energy supply and exercise tolerance is limited.

One way that we are trying to learn more about muscle diseases is by using animal models such as genetically engineered "knockout" mice. These mice lack the genes for certain muscle proteins, and researchers try to correlate the absence of protein with particular disruptions in function.

▶ SMOOTH MUSCLE

Although skeletal muscle has the most muscle mass in the body, cardiac and smooth muscle are more important in the maintenance of homeostasis. Smooth muscle is found predominantly in the walls of hollow organs and tubes, where its contraction changes the shape of the organ. Often smooth muscle generates force to move material through the lumen of the organ. For example, sequential waves of contraction in the intestinal tract move ingested material from the esophagus to the colon.

Smooth muscle is noticeably different from striated muscles in the way that it develops tension (Fig. 12-24 ●). A smooth muscle twitch contracts and relaxes much more slowly than either skeletal or cardiac muscle. At the same time, smooth muscle uses less energy to generate a given amount of force and it can maintain its force for long periods. For example, by one estimate a smooth muscle cell can generate maximum tension with only 25% to 30% of its crossbridges active. We still do not fully understand the mechanisms through which this is accomplished.

In addition, smooth muscle has low oxygen consumption rates yet can sustain contractions for extended periods without fatiguing. This property allows organs such as the bladder to maintain tension despite a continued load. It also allows some smooth muscles to be tonically contracted and maintain tension at most times. The esophageal and urinary bladder **sphincters** [*sphingein*, to close] are examples of tonically contracted muscles whose function is to close off the opening to a hollow organ. These sphincters relax when it is necessary to allow material to enter or leave the organ.

Until recently, smooth muscle had not been studied as extensively as skeletal muscle for many reasons.

1. **Smooth muscle has more variety.** Many types of smooth muscle with widely differing properties are found throughout the animal kingdom, making a single model of smooth muscle function impossible. In humans, smooth muscle can be divided into six major groups: *vascular* (in blood vessel walls), *gastrointestinal* (in walls of the digestive tract and

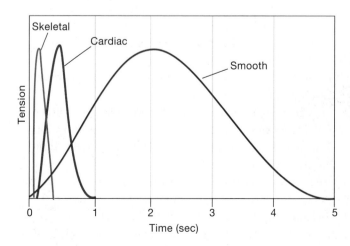

● **Figure 12-24** *Duration of muscle contraction in three types of muscle*

Smooth Muscle and Atherosclerosis

For years, it was believed that mature smooth muscle cells had lost the ability to undergo mitosis. Recent studies, however, have shown otherwise. Not only can smooth muscle fibers reproduce themselves, but their proliferation may also play a significant role in *atherosclerosis*, a disease of the blood vessels in which the vessel lumen narrows because of thickening of the vessel wall. Calcified fatty deposits are found in the extracellular matrix, leading to the hardened state that gives atherosclerosis its popular name, hardening of the arteries. As part of the disease process, smooth muscle cells in the blood vessel wall convert from a contractile state to a proliferative state. During proliferation, the muscle cells migrate toward the lumen, divide, and accumulate cholesterol. This change in smooth muscle function is regulated by growth factors and cytokines released by white blood cells in the region of the atherosclerotic lesion. Researchers are trying to unravel the complex story behind the smooth muscle changes in atherosclerosis so that they can develop drugs to help prevent the disease.

associated organs such as the gall bladder), *urinary* (walls of bladder and ureters), *respiratory* (in the airway passages), *reproductive* (uterus and other reproductive structures in females and males), and *ocular* (in the eye). These muscles have different functions in the body and their physiology is a reflection of their specialized functions. In contrast, skeletal muscle is relatively uniform throughout the body.

2. **Smooth muscle anatomy makes functional studies more difficult.** The contractile fibers of smooth muscle are arranged in oblique bundles rather than in parallel sarcomeres. Consequently, a contraction pulls on the cell membrane in many directions. In addition, within an organ the layers of smooth muscle may run in several directions. For example, the intestine has one layer that encircles the lumen and a perpendicular layer that runs the length of the intestine. It is difficult to measure tension developing in both layers at once.

3. **Smooth muscle contraction is controlled by hormones and paracrines in addition to neurotransmitters.** Unlike skeletal muscle, which is controlled by acetylcholine from somatic motor neurons, smooth muscle activity may be controlled by acetylcholine, norepinephrine, and a variety of other neurotransmitters, hormones, and paracrines.

4. **Smooth muscle fibers exhibit variable electrical properties.** Normal skeletal muscle fibers always respond to an action potential with a twitch. Smooth muscles exhibit a variety of electrical behaviors. They may hyperpolarize as well as depolarize, and they can depolarize without firing action potentials. Contraction may take place after an action potential, a subthreshold graded potential, or without any change in membrane potential.

5. **Multiple pathways influence contraction and relaxation of smooth muscle.** Skeletal muscles contract in response to acetylcholine from a somatic motor neuron and relax when the stimulus for contraction ceases. In marked contrast, multiple neurotransmitters, hormones, and paracrines acting on a smooth muscle fiber can inhibit contraction as well as stimulate it. And, because several different signals might reach the muscle fiber simultaneously, smooth muscle fibers must act as integrating centers.

For example, sometimes blood vessels receive contradictory messages from two different sources: one message signals for contraction, the other for relaxation. The smooth muscle fibers must integrate the two signals and execute an appropriate response. The complexity of overlapping regulatory pathways influencing smooth muscle tone makes the tissue difficult to work with in the laboratory.

Because of the variability in smooth muscle types, we will introduce only their general features in this chapter. Properties that are specific to a certain type will be dealt with when you learn about the different muscles in later chapters.

Smooth Muscles Are Much Smaller than Skeletal Muscle Fibers

Smooth muscles are small spindle-shaped cells. In neurally controlled smooth muscle, neurotransmitter is released from autonomic neuron varicosities [p. 375] close to the surface of the muscle fibers. Smooth muscle lacks specialized receptor regions such as the motor end plates found in skeletal muscle synapses, and the neurotransmitter simply diffuses across the cell surface until it finds a receptor.

Most smooth muscle is **single-unit smooth muscle** (or *unitary smooth muscle*), so called because the individual muscle fibers contract as a single unit. Single-unit smooth muscle is also called **visceral smooth muscle** because it forms the walls of internal organs (viscera) such as the blood vessels and intestinal tract. The fibers of single-unit smooth muscle are all electrically

Three weeks later, Paul had another attack of paralysis, this time at kindergarten after a game of tag. He was rushed to the hospital and given glucose by mouth. Within minutes, he was able to move his legs and arms and asked for his mother.

Question 4: Explain why oral glucose might help bring Paul out of his paralysis. (Hint: Cells use glucose to produce ATP for active transport. The Na$^+$-K$^+$-ATPase actively exchanges K$^+$ and Na$^+$ across the cell membrane. What happens to the extracellular K$^+$ level when Paul is given glucose?)

389 401 403 412   422

connected and an action potential in one cell will spread rapidly through gap junctions to make the entire sheet of tissue contract (Fig. 12-25a ●). Because all fibers contract every time, no reserve units are left to be recruited to increase contraction force. Instead, the amount of Ca^{2+} that enters the cell determines the force of contraction, as you will learn in the discussion that follows.

Multi-unit smooth muscle consists of cells that are not linked electrically. Consequently, each individual muscle fiber must be closely associated with an axon terminal or varicosity and stimulated independently (Fig. 12-25b). This arrangement allows fine control of contractions in these muscles by selective activation of individual muscle fibers. As in skeletal muscle, increasing the force of contraction requires recruitment of additional fibers.

Multi-unit smooth muscle is found in the iris and ciliary body of the eye, in part of the male reproductive tract, and in the uterus, except just prior to labor and delivery. Interestingly, the multi-unit fibers of uterine muscle change and become single-unit during pregnancy. Genes for synthesis of gap junction connexin proteins

turn on, apparently under the influence of pregnancy hormones. The addition of gap junctions to the uterine muscle fibers synchronizes electrical signals, allowing the uterine muscle to contract more effectively while expelling the baby.

CONCEPT CHECK   Answers: p. 427

18. What is the difference in how contraction force is varied in multi-unit and single-unit smooth muscle?

Smooth Muscle Has Longer Actin and Myosin

Smooth muscle uses many of the same contractile elements as skeletal muscle: actin-myosin crossbridges, sarcoplasmic reticulum with Ca^{2+} release channels, and a Ca^{2+} signal that initiates the process. However, details of the structural elements and the contraction process differ in the two muscle types.

Smooth muscles have longer actin and myosin filaments, and its myosin is a different isoform than that of skeletal muscle. Smooth muscle myosin ATPase activity is much slower, decreasing the rate of crossbridge cycling and lengthening the contraction phase. In addition, one of the smaller protein chains in the myosin head serves a regulatory role to control contraction and relaxation. The small regulatory protein chains are called **myosin light chains.**

Actin is more plentiful in smooth muscle, with an actin-to-myosin ratio of 10–15 : 1, compared with 2–4 : 1 in striated muscle. Smooth muscle actin is associated with tropomyosin, as in skeletal muscle. However, unlike skeletal muscle, smooth muscle lacks troponin.

Smooth muscle has less sarcoplasmic reticulum than skeletal muscle, although the amount varies from one type of smooth muscle to another. The primary Ca^{2+} release channel in smooth muscle sarcoplasmic reticulum is an **IP₃-receptor channel.** (Inositol triphosphate (IP_3) is a second messenger created in the phospholipase C pathway [● p. 181].) The calcium-storage function of the sarcoplasmic reticulum may be supplemented by **caveolae** [● p. 146], small vesicles that cluster close to the cell membrane (Fig. 12-26 ●).

(a) **Single-unit smooth muscle** is connected by gap junctions and the cells contract as a single unit.

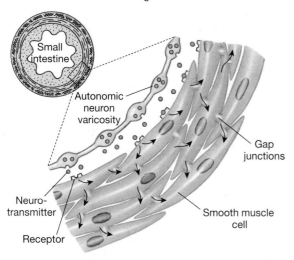

(b) **Multi-unit smooth muscle cells** are not electrically linked and must be stimulated independently.

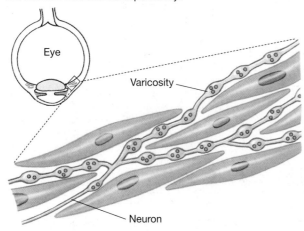

● **Figure 12-25** *Types of smooth muscle*

● **Figure 12-26** *Caveolae in smooth muscle*

Smooth Muscle Contractile Filaments Are Not Arranged in Sarcomeres

Smooth muscle gets its name from its homogeneous appearance under the microscope (see Fig. 12-1c). The contractile fibers are not arranged in sarcomeres, and smooth muscle does not have distinct banding patterns as striated muscle does. Instead, actin and myosin are arranged in long bundles that extend diagonally around the cell periphery, forming a lattice around the central nucleus (Fig. 12-27a ●). The oblique arrangement of contractile elements beneath the cell membrane causes smooth muscle fibers to become globular when they contract, rather than simply shortening as skeletal muscles do (Fig. 12-27b).

The long actin filaments of smooth muscle attach to **dense bodies** of protein in the cytoplasm and terminate at protein plaques in the cell membrane (Fig. 12-27a, c). The less numerous myosin filaments lie bundled between the long actin fibers and are arranged so that their entire surface is covered by myosin

heads (Fig. 12-27d). (Recall that in the sarcomere, the center of the myosin filament is bare of myosin heads.)

The continuous line of myosin heads allows actin to slide along the myosin for longer distances. This unique organization enables smooth muscle to be stretched more, yet still maintain enough overlap to create optimum tension. This is an important property for internal organs such as the bladder, whose volume varies as it alternately fills and empties.

CONCEPT CHECK

Answers: p. 427

19. The dense bodies that anchor smooth muscle actin are analogous to what structure in a sarcomere? (Hint: see Fig. 12-5.)

20. Name three ways smooth muscle myosin differs from skeletal muscle myosin.

21. Name one way actin and its associated proteins differ in skeletal and smooth muscle.

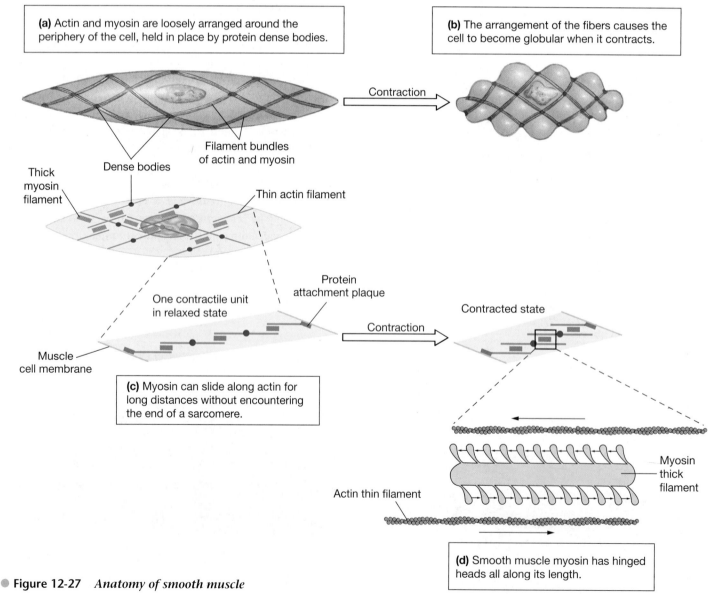

(a) Actin and myosin are loosely arranged around the periphery of the cell, held in place by protein dense bodies.

(b) The arrangement of the fibers causes the cell to become globular when it contracts.

Contraction

Filament bundles of actin and myosin

Dense bodies

Thick myosin filament

Thin actin filament

Protein attachment plaque

One contractile unit in relaxed state

Contracted state

Contraction

Muscle cell membrane

(c) Myosin can slide along actin for long distances without encountering the end of a sarcomere.

Myosin thick filament

Actin thin filament

(d) Smooth muscle myosin has hinged heads all along its length.

● **Figure 12-27** *Anatomy of smooth muscle*

Phosphorylation of Proteins Plays a Key Role in Smooth Muscle Contraction

The molecular events of contraction in smooth muscle are similar in many ways to those in skeletal muscle, but some important differences exist. The primary difference is the role of phosphorylation in the regulation of the contraction process. Here is a summary of the key points of smooth muscle contraction as we currently understand them:

1. An increase in cytosolic Ca^{2+} initiates contraction. Ca^{2+} is released from the sarcoplasmic reticulum and also enters from the extracellular fluid.

2. Ca^{2+} binds to **calmodulin**, a binding protein found in the cytosol. (In skeletal muscle, Ca^{2+} binds to troponin, which smooth muscle lacks.)

3. Ca^{2+} binding to calmodulin is the first step in a cascade that ends with contraction. (In skeletal muscle, Ca^{2+} binding to troponin initiates contraction immediately.)

4. Phosphorylation of proteins is an essential step in smooth muscle contraction.

5. The regulation of myosin ATPase activity is the primary control of contraction.

CONCEPT CHECK   Answers: p. 427

22. Compare points 1, 3, 4, and 5 above to what happens during skeletal muscle contraction. (Hint: see Fig. 12-11.)

Figure 12-28 ● illustrates the steps of smooth muscle contraction. It begins when cytosolic Ca^{2+} concentrations increase following entry from the extracellular fluid and release from the

● **Figure 12-28**   *Smooth muscle contraction*

① Intracellular Ca^{2+} concentrations increase when Ca^{2+} enters cell and is released from sarcoplasmic reticulum.

② Ca^{2+} binds to calmodulin (CaM).

③ Ca^{2+}–calmodulin activates myosin light chain kinase (MLCK).

④ MLCK phosphorylates light chains in myosin heads and increases myosin ATPase activity.

⑤ Active myosin crossbridges slide along actin and create muscle tension.

sarcoplasmic reticulum. The calcium ions bind to calmodulin (CaM), obeying the law of mass action [◉ p. 97]. The Ca^{2+}-calmodulin complex then activates an enzyme called **myosin light chain kinase (MLCK)**. MLCK enhances myosin ATPase activity by phosphorylating light protein chains in the myosin head. When myosin ATPase activity is high, actin binding and cross-bridge cycling increase tension in the muscle. Thus, smooth muscle contraction is primarily controlled through myosin-linked regulatory processes rather than the actin-linked regulation of skeletal muscle.

Relaxation in Smooth Muscle Has Several Steps

Relaxation in a smooth muscle fiber is a multistep process (Fig. 12-29 ◉). As in skeletal muscle, free Ca^{2+} is removed from the cytosol when Ca^{2+}-ATPase pumps it back into the sarcoplasmic reticulum. In addition, smooth muscle extrudes Ca^{2+} from the cell with the help of a Ca^{2+}-Na^+ antiport exchanger and a sarcolemmal Ca^{2+}-ATPase. By the law of mass action, a decrease in free cytosolic Ca^{2+} causes Ca^{2+} to unbind from calmodulin. In the absence of Ca^{2+}-calmodulin, myosin light chain kinase inactivates.

The additional step in smooth muscle relaxation is dephosphorylation of the myosin light chain, which decreases myosin ATPase activity. Removal of myosin's phosphate group is accomplished with the aid of **myosin phosphatase**.

Curiously, dephosphorylation of myosin does not automatically result in relaxation. Under conditions that we do not fully understand, dephosphorylated myosin may remain attached to actin for a period of time in what is known as a **latch state**. This condition maintains tension in the muscle fiber without consuming ATP. It is a significant factor in the ability of smooth muscle to sustain contractions without fatiguing. The hinge muscles of certain bivalve mollusks such as oysters can enter a similar latch state that allows them to remain tightly closed under anaerobic conditions.

CONCEPT CHECK   Answers: p. 427

23. What happens to contraction if a smooth muscle is placed in a saline bath from which all calcium has been removed?

● **Figure 12-29**   *Relaxation in smooth muscle*

1. Free Ca^{2+} in cytosol decreases when Ca^{2+} is pumped out of the cell or back into the sarcoplasmic reticulum.

2. Ca^{2+} unbinds from calmodulin (CaM).

3. Myosin phosphatase removes phosphate from myosin, which decreases myosin ATPase activity.

4. Less myosin ATPase results in decreased muscle tension.

Calcium Entry Is the Signal for Smooth Muscle Contraction

In smooth muscle, an increase in cytosolic Ca^{2+} concentration is the signal to initiate contraction. As you saw in Figure 12-28, Ca^{2+} enters the cell from the extracellular fluid and is released from the sarcoplasmic reticulum. However, because Ca^{2+} stores in smooth muscle are limited, sustained contractions depend on continued influx of Ca^{2+} from the extracellular fluid. The entry of variable amounts of Ca^{2+} into the muscle fiber creates contractions whose force is graded according to the strength of the Ca^{2+} signal.

Recall from the introduction that smooth muscle contraction can be initiated by neurotransmitters, hormones, or paracrines. This is where our model becomes complex. Figure 12-30 ● is a generalized summary of the multiple pathways that can initiate or modulate the contraction process.

Ca^{2+} enters the cell through membrane channels opened by depolarization, membrane stretch, or chemical signals. One recently identified family of Ca^{2+} channels has been named **store-operated Ca^{2+} channels** because these channels open (by an unidentified mechanism) in response to depleted intracellular

stores of Ca^{2+}. Voltage-gated and stretch-activated channels are discussed in more detail below.

Sarcoplasmic reticulum Ca^{2+} release is mediated primarily by an IP_3-activated receptor-channel. The IP_3 receptor-channel opens as a result of signal transduction pathways begun by extracellular signal molecules.

The increase in cytosolic Ca^{2+} initiates contraction through the regulation of myosin ATPase as discussed in the previous section. In some types of smooth muscle, myosin regulation of cross-bridge cycling is supplemented by regulation of actin. Several actin-associated regulatory proteins have been identified, including one called *caldesmon* and another called *calponin*. Second messengers that modulate contraction may act on myosin light chain kinase, myosin phosphatase, or the actin-associated regulatory proteins. The details of this modulation are still being worked out.

24. Compare Ca^{2+} release channels in skeletal and smooth muscle sarcoplasmic reticulum.

Muscle Stretch Opens Ca^{2+} Channels

Smooth muscle cell membranes contain stretch-activated Ca^{2+} channels that open when pressure or other force distorts the cell membrane. Because contraction in this instance originates from a property of the muscle fiber itself, it is known as a **myogenic contraction**. Myogenic contractions are common in blood vessels that maintain a certain amount of tone at all times.

Some smooth muscles adapt if the muscle fiber is stretched for an extended period of time. As the stretch stimulus continues, the Ca^{2+} channels begin to close in a time-dependent fashion. Then, as Ca^{2+} is pumped out of the cell, the muscle relaxes. This adaptation response explains why the bladder develops tension as it fills, then relaxes as it adjusts to the increased volume. (There is a limit to the amount of stretch the muscle can endure, however, and once a critical volume is reached, a urination reflex empties the bladder.)

Some Smooth Muscles Have Unstable Membrane Potentials

A major route of Ca^{2+} influx for smooth muscle is entry through voltage-gated Ca^{2+} channels. However, the role of membrane potentials in smooth muscle contraction is much more complex than in skeletal muscle, where contraction always begins in response to an action potential. In smooth muscle, an action potential is not required to open voltage-gated Ca^{2+} channels. Graded potentials open a few Ca^{2+} channels, allowing small amounts of Ca^{2+} into the cell. This depolarizes the cell and may open additional Ca^{2+} channels.

In addition, many types of smooth muscle display unstable resting membrane potentials that vary between -40 and -80 mV. Cells that exhibit cyclic depolarization and repolarization of their membrane potential are said to have **slow wave potentials** (Fig. 12-31a ●). Sometimes the cell simply cycles through a series of subthreshold slow waves. However, if the peak of the

KEY

IP_3–R = IP_3–activated receptor–channel

* Ligands include norepinephrine, ACh, other neurotransmitters, hormones, and paracrines.

● **Figure 12-30** *Control of smooth muscle contraction*

(a) Slow wave potentials fire action potentials when they reach threshold.

(b) Pacemaker potentials always depolarize to threshold.

(c) Pharmacomechanical coupling occurs when chemical signals change muscle tension without a change in membrane potential.

● **Figure 12-31** *Membrane potentials vary in smooth muscle*

depolarization reaches threshold, action potentials fire, followed by contraction of the muscle.

Other types of smooth muscle with unstable membrane potentials have a regular depolarization that always reaches threshold and fires an action potential (Fig. 12-31b). These depolarizations are called **pacemaker potentials** because they create regular rhythms of contraction. Pacemaker potentials are found in some cardiac muscles as well as in smooth muscle. Both slow wave and pacemaker potentials are due to ion channels in the cell membrane that spontaneously open and close.

Some smooth muscles have action potentials that look very much like action potentials in neurons. At the cellular level, however, they are different from neuronal and skeletal muscle action potentials because the depolarization phase is due to the entry of Ca^{2+} rather than Na^+. The repolarization phase is similar however, and is due to the opening of K^+ channels in all three tissue types.

Smooth muscle function does not depend on firing action potentials. In **pharmacomechanical coupling**, smooth muscle contraction may occur without a significant change in membrane potential (Fig. 12-31c). Chemical signals may also relax muscle tension without a change in membrane potential.

25. How do pacemaker potentials differ from slow wave potentials?

26. Tetrodotoxin (TTX) is a poison that blocks sodium channels. When TTX is applied to certain types of smooth muscle, it does not affect the spontaneous generation of action potentials. What conclusion can you draw about the action potential of this smooth muscle on the basis of this observation?

Smooth Muscle Activity Is Regulated by Chemical Signals

Smooth muscle contraction is controlled by a variety of chemical signals that may be either excitatory or inhibitory. In general, if a stimulus depolarizes the cell, it is more likely to contract. Hyperpolarization of the cell decreases the likelihood of contraction. Contraction can be modulated by second messenger action on myosin or actin regulation.

Autonomic Neurotransmitters   Many smooth muscles have dual innervation and are controlled by both sympathetic and parasympathetic neurons. However, other smooth muscles, such as those found in blood vessels, are controlled by only one of the two autonomic branches. In this type of *tonic control*, the response is graded by increasing or decreasing the amount of neurotransmitter released onto the muscle.

As we have seen, a neurotransmitter can have different effects in different tissues, depending upon the receptors to which it binds. Thus, both the neurotransmitter and its receptor determine the response of a smooth muscle to nervous stimulation. Adrenergic and cholinergic muscarinic receptors act via IP_3 and cAMP second messenger systems. The exact response of the muscle depends on the adrenergic or cholinergic receptor subtype that is activated.

An amazing variety of other neurotransmitters such as vasoactive intestinal peptide and substance P are active in smooth muscle, in addition to norepinephrine and acetylcholine. In many instances, we know how these chemicals affect smooth muscle contraction, but we do not understand the reflex pathways that trigger their release.

27. How can a neuron alter the amount of neurotransmitter it releases? (Hint: ● Fig. 8-13, p. 257)

28. Explain how hyperpolarization decreases likelihood of contraction in smooth muscle.

Hormones and Paracrines Hormones and paracrines also control smooth muscle contraction, unlike skeletal muscle, which is controlled only by the nervous system. Smooth muscles in the cardiovascular, gastrointestinal, urinary, respiratory, and reproductive systems all respond to blood-borne or locally released chemicals. For example, asthma is a condition in which smooth muscle of the airways constricts in response to histamine release. This constriction can be reversed by the administration of a hormone, epinephrine, that relaxes smooth muscle and dilates the airway. Note that not all physiological responses are adaptive or favorable to the body: constriction of the airways triggered during an asthma attack, if untreated, can be fatal.

Another important paracrine that affects smooth muscle contraction is **nitric oxide** [◈ p. 267]. This gas is synthesized by the endothelial lining of blood vessels and relaxes adjacent smooth muscle that regulates the diameter of the blood vessels. For many years, the identity of this *endothelium-derived relaxing factor*, or EDRF, eluded scientists even though its presence could be demonstrated experimentally. We know now that EDRF is nitric oxide, an important paracrine in many systems of the body.

Although smooth muscles in the body do not have nearly the mass of skeletal muscles, they play a critical role in the function of most organ systems. You will learn more about smooth muscle physiology in the chapters to come.

▶ CARDIAC MUSCLE

Cardiac muscle, the specialized muscle of the heart, shares features with both smooth and skeletal muscle (Table 12-3). Like skeletal muscle fibers, cardiac muscle fibers are striated and have a sarcomere structure. However, cardiac muscle fibers are shorter than skeletal muscle fibers, may be branched, and have a single nucleus (unlike multinucleate skeletal muscle fibers).

Like single-unit smooth muscle, cardiac muscle fibers are electrically linked to each other. The gap junctions are contained in specialized cell junctions known as *intercalated disks*. Some cardiac muscle, like some smooth muscle, exhibits pacemaker potentials. In addition, cardiac muscle is under sympathetic and parasympathetic control as well as hormonal control. In Chapter 14, you will learn more about cardiac muscle and how it functions within the heart.

Table 12-3: Comparison of Three Muscle Types

	SKELETAL	SMOOTH	CARDIAC
Appearance under light microscope	Striated	Smooth	Striated
Fiber arrangement	Sarcomeres	Longitudinal bundles	Sarcomeres
Fiber proteins	Actin, myosin; troponin and tropomyosin	Actin, myosin, tropomyosin	Actin, myosin; troponin and tropomyosin
Control	■ Voluntary ■ Ca^{2+} and troponin ■ Fibers independent	■ Involuntary ■ Ca^{2+} and calmodulin ■ Fibers electrically linked via gap junctions	■ Involuntary ■ Ca^{2+} and troponin ■ Fibers electrically linked via gap junctions
Nervous control	Somatic motor neuron	Autonomic neurons	Autonomic neurons
Hormonal influence	None	Multiple hormones	Epinephrine
Location	Attached to bones; a few sphincters close off hollow organs	Forms the walls of hollow organs and tubes; some sphincters	Heart muscle
Morphology	Multinucleate; large, cylindrical fibers	Uninucleate; small spindle-shaped fibers	Uninucleate; shorter branching fibers
Internal structure	T-tubule and sarcoplasmic reticulum	No t-tubules; sarcoplasmic reticulum reduced or absent	T-tubule and sarcoplasmic reticulum
Contraction speed	Fastest	Slowest	Intermediate
Contraction force of single fiber	All-or-none	Graded	Graded
Initiation of contraction	Requires input from motor neuron	Can be autorhythmic	Autorhythmic

RUNNING PROBLEM CONCLUSION

Periodic Paralysis

In this running problem, you were introduced to hyperkalemic periodic paralysis, a condition caused by a genetic defect in the Na^+ channels on muscle cell membranes. To learn more about hyperkalemic periodic paralysis, go to the U.S. National Library of Medicine illustrated medical encyclopedia at *http://www.nlm.nih.gov/medlineplus/encyclopedia.html*. You might also compare the hyperkalemic form to the hypokalemic form of the disease.

Now check your understanding of this running problem by comparing your answers with those in the summary table.

	QUESTION	FACTS	INTEGRATION AND ANALYSIS
1	What effect does the continuous influx of Na^+ have on the membrane potential of Paul's muscle fibers?	The resting membrane potential of cells is negative relative to the extracellular fluid.	The influx of positive charge will depolarize the membrane potential.
2	In people with hyperkalemic periodic paralysis, attacks tend to occur after a period of exercise (i.e., repeated muscle contractions). What ion is responsible for the repolarization phase of the muscle action potential, and in which direction does this ion move across the muscle fiber membrane?	Each muscle twitch results from an action potential in the muscle fiber. In the repolarization phase of the action potential, K^+ leaves the cell.	During the repeated contractions of exercise, K^+ leaves the muscle fiber and accumulates in the t-tubules. This increase in K^+ concentration affects the Na^+ channels and triggers an attack.
3	Why does a Na^+ channel that will not inactivate cause paralysis?	During an attack, the Na^+ channels remain open and continuously admit Na^+, and the cell remains depolarized.	If the muscle fiber is unable to repolarize, it cannot fire additional action potentials. The first action potential causes a twitch, but the muscle then goes into a state of flaccid (uncontracted) paralysis.
4	Explain why oral glucose might help bring Paul out of his paralysis.	Cells use glucose to produce ATP to run the Na^+-K^+-ATPase. This transporter actively moves K^+ and Na^+ across the cell membrane.	Providing glucose to cells may increase ATP available to run the Na^+-K^+-ATPase, which removes K^+ from the extracellular fluid. Insulin, a hormone that enhances glucose uptake into cells, also helps relieve the paralysis.

389 401 403 412 414

CHAPTER SUMMARY

Muscles exhibit many of the physiological themes and concepts that we introduced in Chapter 1. They provide an excellent system for studying *structure-function* relationships at all levels, from actin, myosin, and sliding filaments within the cell to muscles pulling on bones and joints. *Mechanical properties* of muscles that influence contraction include elastic proteins such as titin and the series elastic elements of the intact muscle. *Compartmentation* is essential to muscle function, as demonstrated by the concentration of Ca^{2+} in the sarcoplasmic reticulum and the key role of Ca^{2+} signals in initiating contraction. The *law of mass action* is at work in the dynamics of Ca^{2+}-calmodulin and Ca^{2+}-troponin binding and unbinding. Muscles also show how *biological energy* is transformed from stored energy in ATP's chemical bonds to the movement of motor proteins.

Muscles provide many examples of *communication and control* in the body. Communication is as small as electrical signals spreading among smooth muscle cells via gap junctions or as large as a somatic motor neuron innervating multiple muscle fibers. Skeletal muscles are only controlled by somatic motor neurons, but smooth and cardiac muscle have complex regulation that ranges from neurotransmitters to hormones and paracrines.

1. Muscles generate motion, force, and heat. (p. 390)

2. The three types of muscle are **skeletal muscle**, **cardiac muscle**, and **smooth muscle**. Skeletal and cardiac muscles are **striated muscles**. (p. 390; Fig. 12-1)

3. Skeletal muscles are controlled by somatic motor neurons. Cardiac and smooth muscle are controlled by autonomic innervation, paracrines, and hormones. Some smooth and cardiac muscles are autorhythmic and contract spontaneously. (p. 390)

Skeletal Muscle

IP Muscular Physiology

4. Skeletal muscles are usually attached to bones by **tendons**. The **origin** is the end of the muscle attached closest to the trunk or to the more stationary bone. The **insertion** is the more distal or mobile attachment. (p. 390)

5. At a flexible joint, muscle contraction moves the skeleton. **Flexors** bring bones closer together; **extensors** move bones away from each other. Flexor-extensor pairs are examples of **antagonistic muscle groups**. (p. 390; Fig. 12-2)

6. A skeletal muscle is a collection of **muscle fibers**, large cells with many nuclei. (p. 391; Fig. 12-3, Tbl. 12-1)

7. **T-tubules** allow action potentials to move rapidly into the interior of the fiber and release calcium from the **sarcoplasmic reticulum**. (p. 391; Fig. 12-4)

8. **Myofibrils** are intracellular bundles of contractile and elastic proteins. **Thick filaments** are made of **myosin**. **Thin filaments** are made mostly of **actin**. **Titin** and **nebulin** hold thick and thin filaments in position. (pp. 391–395; Figs. 12-3, 12-6)

9. Myosin binds to actin, creating **crossbridges** between the thick and thin filaments. (p. 394; Fig. 12-3)

10. One **sarcomere** is composed of two **Z disks** and the filaments between them. A sarcomere is divided into **I bands** (thin filaments only), an **A band** that runs the length of the thick filament, and a central **H zone** occupied by thick filaments only. The **M line** and Z disks represent attachment sites for myosin and actin, respectively. (p. 394; Fig. 12-5)

11. The force created by a contracting muscle is called **muscle tension**. The **load** is a weight or force that opposes contraction of a muscle. (p. 396)

12. The **sliding filament theory of contraction** states that during contraction, overlapping thick and thin filaments slide past each other in an energy-dependent manner as a result of actin-myosin crossbridge movement. (p. 396; Fig. 12-8)

13. Myosin converts energy from ATP into motion. **Myosin ATPase** hydrolyses ATP to ADP and P_i. (p. 397; Fig. 12-9)

14. When myosin releases P_i, the myosin molecule moves in the **power stroke**. At the end of the power stroke, myosin releases ADP. The cycle ends in the **rigor state**, with myosin tightly bound to actin. (pp. 396–397; Fig. 12-9)

15. **Tropomyosin** blocks the myosin-binding site on actin. As contraction begins, **troponin** binds to Ca^{2+}. This unblocks the myosin-binding sites and allows myosin to complete its power stroke. (p. 399; Fig. 12-10)

16. During relaxation, the sarcoplasmic reticulum uses a Ca^{2+}-ATPase to pump Ca^{2+} back into its lumen. (p. 401; Fig. 12-11b)

17. In **excitation-contraction coupling** a somatic motor neuron releases ACh, which initiates a skeletal muscle action potential that leads to contraction. (p. 400; Fig. 12-11a)

18. Voltage-sensing **DHP receptors** in the t-tubules open Ca^{2+} **release channels** in the sarcoplasmic reticulum. (p. 401; Fig. 12-11a)

19. A single contraction-relaxation cycle is known as a **twitch**. The **latent period** between the end of the muscle action potential and the beginning of muscle tension development represents the time required for Ca^{2+} release and binding to troponin. (p. 401; Fig. 12-12)

20. Muscle fibers store energy for contraction in **phosphocreatine**. Anaerobic metabolism of glucose is a rapid source of ATP but is not efficient. Aerobic metabolism is very efficient but requires an adequate supply of oxygen to the muscles. (p. 401; Fig. 12-13)

21. **Muscle fatigue** is a condition in which a muscle is no longer able to generate or sustain the expected power output. Fatigue has multiple causes. (p. 403; Fig. 12-14)

22. Skeletal muscle fibers can be classified on the basis of their speed of contraction and resistance to fatigue into **fast-twitch glycolytic fibers**, **fast-twitch oxidative fibers**, and **slow-twitch oxidative fibers**. Oxidative fibers are the most fatigue-resistant. (p. 404; Fig. 12-15; Tbl. 12-2)

23. **Myoglobin** is an oxygen-binding pigment that transfers oxygen to the interior of the muscle fiber. (p. 404)

24. The tension of a skeletal muscle twitch is determined by the length of the sarcomeres before contraction begins. (p. 405; Fig. 12-16)

25. Increasing the stimulus frequency causes summation of twitches with an increase of tension. A state of maximal contraction is known as **tetanus**. (p. 406; Fig. 12-17)

26. A **motor unit** is composed of a group of muscle fibers and the somatic motor neuron that controls them. The number of muscle fibers in a motor unit varies, but all fibers in a single motor unit are of the same fiber type. (p. 407; Fig. 12-18)

27. The force of contraction within a skeletal muscle can be increased by **recruitment** of additional motor units. (p. 408)

Mechanics of Body Movement

28. An **isotonic contraction** creates force and moves a load. An **isometric contraction** creates force without movement. **Concentric actions** are shortening contractions. **Eccentric actions** are lengthening contractions. (p. 409; Fig. 12-19)

29. Isometric contractions occur because **series elastic elements** allow the fibers to maintain constant length even though the sarcomeres are shortening and creating tension. (p. 410; Fig. 12-20)

30. The body uses its bones and joints as **levers** and **fulcrums**. Most lever systems in the body maximize distance and speed but also require that muscles do more work. (p. 410; Figs. 12-21, 12-22)

31. Contraction speed is a function of muscle fiber type and load. Contraction is fastest when the load on the muscle is zero. (p. 412; Fig. 12-23)

Smooth Muscle

32. Smooth muscle is slower than skeletal muscle but can sustain contractions for longer without fatiguing. (p. 413; Fig. 12-24)

33. **Single-unit smooth muscle** contracts as a single unit when depolarizations pass from cell to cell through gap junctions. In **multi-unit smooth muscle**, individual muscle fibers are stimulated independently. (pp. 414–415; Fig. 12-25)

34. Actin and myosin are arranged in smooth muscle cell's periphery. Smooth muscle actin lacks troponin. (p. 416; Fig. 12-27)

35. Smooth muscle has relatively little sarcoplasmic reticulum and two types of Ca^{2+} release channels: a **ryanodine receptor** and an **IP_3-activated channel**. (p. 415)

36. In smooth muscle contraction, Ca^{2+} binds to **calmodulin** and activates **myosin light chain kinase** (MLCK). (pp. 417–418; Fig. 12-28)

37. MLCK phosphorylates **myosin light protein chains**, which activates myosin ATPase. This allows crossbridge power strokes. (p. 415; Fig. 12-28)

38. During relaxation, Ca^{2+} is pumped out of the cytosol and myosin light chains are dephosphorylated by **myosin phosphatase**. (p. 418; Fig. 12-29)

39. Ca^{2+} for contraction enters the cell through Ca^{2+} channels. Additional Ca^{2+} is released from the sarcoplasmic reticulum. (p. 419; Fig. 12-30)

40. In **myogenic contraction**, stretch opens membrane Ca^{2+} channels. (p. 419)

41. Unstable membrane potentials in smooth muscle take the form of **slow wave potentials** or **pacemaker potentials**. (pp. 419–420; Fig. 12-31a, b)

42. The rising phase of smooth muscle action potentials is due to Ca^{2+} entry rather than Na^+ entry. (p. 420)

43. In **pharmacomechanical coupling**, Ca^{2+} entry causes smooth muscle contraction without a significant change in membrane potential. (p. 420; Fig. 12-31c)

44. Smooth muscle is controlled by sympathetic and parasympathetic neurons and a variety of chemical signals. (p. 420)

Cardiac Muscle

45. Cardiac muscle fibers are striated, have a single nucleus, and are electrically linked through gap junctions. Cardiac muscle shares features with both skeletal and smooth muscle. (p. 421; Tbl. 12-3)

QUESTIONS

LEVEL ONE Reviewing Facts and Terms

1. The three types of muscle tissue found in the human body are _____, _____, and _____. Which type is attached to the bones, enabling it to control body movement?

2. Which two muscle types are striated?

3. Which type of muscle tissue is controlled only by somatic motor neurons?

4. Which statement(s) is true about skeletal muscles?
   (a) They constitute about 60% of a person's total body weight.
   (b) They position and move the skeleton.
   (c) The insertion of the muscle is more distal or mobile than the origin.
   (d) They are often paired into antagonistic muscle groups called flexors and extensors.

5. Arrange these skeletal muscle components in order, from outermost to innermost: sarcolemma, connective tissue sheath, myofilaments, myofibrils.

6. The modified endoplasmic reticulum of skeletal muscle is called the _____. Its role is to sequester _____ ions.

7. T-tubules allow _____ to move to the interior of the muscle fiber.

8. List four proteins that make up the myofibrils. Which one creates the power stroke for contraction?

9. List the letters used to label the elements of a sarcomere. Which band has a Z disk in the middle? Which is the darkest band? Why? Which element forms the boundaries of a sarcomere? Name the line that divides the A band in half. What is the function of this line?

10. Briefly explain the functions of titin and nebulin.

11. During contraction, the _____ band remains a constant length. This band is composed primarily of _____ molecules. Which lettered components approach each other during contraction?

12. Explain the sliding filament theory.

13. Match the characteristics below with the appropriate type(s) of muscle.
   (a) largest diameter
   (b) anaerobic metabolism, thus fatigues quickly
   (c) has the most blood vessels
   (d) has some myoglobin
   (e) used for quick, fine movements
   (f) also called red muscle
   (g) uses a combination of oxidative and glycolytic metabolism
   (h) has the most mitochondria

   1. fast-twitch glycolytic fibers
   2. fast-twitch oxidative fibers
   3. slow-twitch oxidative fibers

14. Explain the roles of troponin, tropomyosin, and Ca^{2+} in skeletal muscle contraction.

15. Which neurotransmitter is released by somatic motor neurons?

16. What is the motor end plate, and what kinds of receptors are found there? Explain how neurotransmitter binding to these receptors creates an action potential.

17. A single contraction-relaxation cycle in a skeletal muscle fiber is known as a _____.

18. List the steps of skeletal muscle contraction that require ATP.

19. The basic unit of contraction in an intact skeletal muscle is the _____. The force of contraction within a skeletal muscle is increased by _____ additional motor units.

20. The two functional types of smooth muscle are _____ and _____.

LEVEL TWO *Reviewing Concepts*

21. Make a map of muscle fiber structure using the following words or terms. Add additional terms if you like.

actin
Ca^{2+}
cell
cell membrane
contractile protein
crossbridges
cytoplasm
elastic protein
glycogen
mitochondria
muscle fiber

myosin
nucleus
regulatory protein
sarcolemma
sarcoplasm
sarcoplasmic reticulum
titin
tropomyosin
troponin
t-tubule

22. Arrange the following terms to create a map of skeletal muscle excitation, contraction, and relaxation. Terms may be used more than once. Add any additional terms you like.

acetylcholine
ACh receptor
actin
action potential
ADP
ATP
axon terminal
Ca^{2+}
Ca^{2+}-ATPase
calcium release channels
contraction
crossbridge
DHP receptor
end-plate potential
exocytosis

motor end plate
myosin
Na^+
neuromuscular junction
P_i
power stroke
relaxation
rigor state
sarcoplasmic reticulum
somatic motor neuron
tropomyosin
troponin
t-tubules
voltage-gated Ca^{2+} channels

23. How does an action potential in a muscle fiber trigger a Ca^{2+} signal inside the fiber?

24. Muscle cells depend on a continuous supply of ATP. How do the different types of muscle cells generate ATP? What is used for a backup energy source?

25. Define muscle fatigue. Summarize factors that could play a role in its development. How can muscle cells adapt to resist fatigue?

26. Explain how you vary the strength and effort made by your muscles in picking up a pencil versus a full gallon container of milk.

27. Compare and contrast the cellular anatomy and neural and chemical control of contraction in skeletal and smooth muscle.

28. What is the role of the sarcoplasmic reticulum in muscular contraction? How can smooth muscle contract when it has so little sarcoplasmic reticulum?

29. Compare and contrast:
    (a) fast-twitch oxidative, fast-twitch glycolytic, and slow-twitch muscle fibers
    (b) a twitch and tetanus
    (c) action potentials in motor neurons and action potentials in skeletal muscles
    (d) temporal summation in motor neurons and summation in skeletal muscles
    (e) isotonic contraction, isometric contraction, concentric action, and eccentric action
    (f) slow-wave and pacemaker potentials
    (g) the source and role of Ca^{2+} in skeletal and smooth muscle contraction

30. Explain the different factors that influence Ca^{2+} entry and release in smooth muscle.

LEVEL THREE *Problem Solving*

31. One way that scientists study muscles is to put them into a state of rigor or stiffness by removing ATP. In this condition, the actin and myosin are strongly linked but unable to move. On the basis of what you know about muscle contraction, predict what would happen to these muscles in a state of rigor if you (a) added ATP but no free calcium ions; (b) added ATP with a substantial concentration of calcium ions.

32. When curare, a South American Indian arrow poison, is placed on a nerve-muscle preparation, the muscle will not contract when the nerve is stimulated, even though neuro-transmitter is still being released from the nerve. Name all possible explanations for the action of curare that you can think of.

33. On the basis of what you have learned about muscle fiber types and metabolism, predict what variations in structure you would find among these athletes.
    (a) a 7 foot, 2 inch tall, 325-pound basketball player
    (b) a 5 foot, 10 inch tall, 180-pound steer wrestler
    (c) a 5 foot, 7 inch tall, 130-pound female figure skater
    (d) a 4 foot, 11 inch tall, 89-pound female gymnast

34. Use the arm in Figure 12-21 to answer the following questions.
    (a) How much force would a biceps muscle inserted 4 cm from the fulcrum need to hold the arm stationary at a 90° angle? How does this compare to the arm with insertion at 5 cm?

    (b) If a 7-kg weight band is placed around the wrist 20 cm from the fulcrum, how much force does the biceps muscle inserted 5 cm from the fulcrum need to hold the arm stationary at a 90° angle? How does this compare to placing the same weight in the hand (25 cm from fulcrum)?

Explorations

Explorations are also available at *www.physiologyplace.com*

To understand fitness and exercise, you must first know how muscles work. We will revisit exercise physiology again in Chapter 25 once you have learned more about the cardiovascular and respiratory systems, key players in the integrated process we call exercise. In the meantime, here are some activities to enhance your understanding of muscle physiology.

Exploration 1: Muscle Fatigue

Sports is big business in the United States and around the world, and a considerable amount of money is spent trying to understand how trainers can help athletes perform better. One side benefit of the basic research in exercise physiology is that we also gain a better understanding of how various disease conditions affect the muscular and cardiovascular systems.

In the case of muscle fatigue, the more research is published, the more we realize that muscle fatigue has multiple contributing factors. Look at the graph below, created from data published in "Effect of ambient temperature on human skeletal muscle metabolism during fatiguing submaximal exercise," *Journal of Applied Physiology* 86(3):902–908, 1999. (Free access at *http://jap.physiology.org/*)

What hypotheses might you develop about the cause(s) of muscle fatigue based on these data?

To learn what the experts are saying about muscle fatigue, see the following reviews: "Sarcoplasmic reticulum Ca^{2+} release and muscle fatigue," *Journal of Applied Physiology* 87(2):471–483, 1999 Aug. (free at *http://jap.physiology.org/*)

"Muscle fatigue: Lactic acid or inorganic phosphate the major cause?" *News in Physiological Sciences* 17:17–21, 2002 Feb. (free at *http://nips.physiology.org/*)

Exploration 2: Visualizing Muscle Contraction

The sliding of actin and myosin past each other during muscle contraction is another dynamic physiological process that is tough to visualize from the static, two-dimensional figures in a textbook such as this. Fortunately, many faculty members have developed animations and three-dimensional representations of muscle contraction and made them available on the Web. Here's where the power of a search engine like Google really comes in handy (*www.google.com*). Try searching using any of the following keywords: *actin, myosin, "sliding filament"*, or *titin*. To narrow your choices, you can add the keyword *animation*. Does seeing muscle contraction in action help you remember the steps?

Exploration 3: Neuromuscular Disorders

There are many inherited and acquired diseases that affect skeletal muscle function. Some that you might like to explore include the infectious disease *polio*, which in the last century was a major killer and crippler of children; *multiple sclerosis*, a demyelinating disorder; *malignant hyperthermia*, a channelopathy of the ryanodine receptor in muscle; and *amyotrophic lateral sclerosis* (ALS or Lou Gehrig's disease), a fatal motor neuron disease. Try a Google search (*www.google.com*) to see what you can learn about these diseases. Are there some websites that you would consider to be more reliable sources of information than others? What do you look for when you evaluate a website?

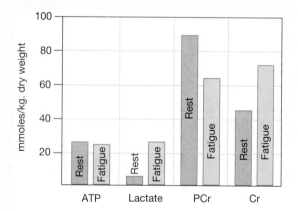

Muscle metabolites in resting muscle and after cycling exercise to fatigue

ANSWERS

Page 391

1. Some examples are biceps and triceps in the upper arm; legs: hamstring (flexor) and quadriceps (extensor).

Page 395

2. Ends of the A bands are darkest because they are where the thick and thin filaments overlap.

3. T-tubules allow action potentials to travel from the surface of the muscle fiber to its interior.

4. The banding pattern of organized filaments in the sarcomere forms striations in the muscle.

Page 399

5. Each myosin molecule has binding sites for ATP and actin.

6. F-actin is a polymer filament of actin made from globular G-actin molecules.

Page 400

7. Titin is an elastic fiber in the sarcomere.

8. The crossbridges do not all unlink at one time, so while some myosin heads are free and swiveling, others are still tightly bound.

Page 401

9. The release of myosin heads from actin requires ATP binding. Energy from ATP is required for the powerstroke. Relaxation does not directly require ATP but relaxation will not occur unless Ca^{2+} is pumped back into the sarcoplasmic reticulum using a Ca^{2+}-ATPase.

10. The latent period includes the muscle action potential, release of Ca^{2+} from the sarcoplasmic reticulum, and diffusion of Ca^{2+} to the contractile filaments.

Page 404

11. Increasing K^+ will cause the cell to depolarize and become less negative.

Page 406

12. Summation in muscle means that the force increases.

13. Temporal summation in neurons means that the graded potential increases.

Page 407

14. A marathoner probably has more slow-twitch muscle fibers, whereas a sprinter probably has more fast-twitch muscle fibers.

Page 408

15. Increased somatic motor neuron firing increases force of contraction due to summation.

16. Force in a muscle can be increased by recruiting additional motor units.

Page 412

17. If the muscle insertion is farther from the joint, the leverage is better and a contraction will create more force for a given increase in distance.

Page 415

18. Multi-unit smooth muscle increases force by recruiting additional muscle fibers. Single-unit smooth muscle increases force by increasing Ca^{2+} entry.

Page 416

19. Dense bodies are analogous to Z disks.

20. Smooth muscle myosin is longer, has heads the entire length of the filament, and has slower ATPase activity.

21. Smooth muscle actin is longer and it lacks troponin.

Page 417

22. Point 1 is similar in skeletal muscle (step 2) except that all Ca^{2+} comes from the sarcoplasmic reticulum. (3) Ca^{2+} binds to troponin in skeletal muscle. (4) Phosphorylation does not regulate skeletal muscle myosin. (5) The primary control point in skeletal muscle is Ca^{2+} binding to troponin.

Page 418

23. Without ECF Ca^{2+}, contraction decreases or stops.

Page 419

24. Skeletal muscle Ca^{2+} release channels are linked to DHP receptors and open with depolarization. Smooth muscle Ca^{2+} release channels are activated by IP_3.

Page 420

25. Pacemaker potentials always reach threshold and create regular rhythms of contraction. Slow wave potentials are variable in magnitude and may not reach threshold each time.

26. The depolarization phase of the action potential must not be due to Na^+ entry.

Page 420

27. More action potentials in the neuron increase neurotransmitter release.

28. Many Ca^{2+} channels open with depolarization, therefore hyperpolarization decreases the likelihood that these channels open. Ca^{2+} is necessary for contraction.

Page 402

Fig. 12-12: (a) Muscle V_m is -90 mV and neuron is -70 mV. This illustrates the fact that V_m is not the same in all cells. (b) Muscle action potential is due to Na^+ entering on the depolarization and K^+ leaving on the repolarization.

Page 411

Fig. 12-21c: Biceps force $\times$ 5 cm = 7 kg $\times$ 25 cm. Biceps force = 35 kg (additional).

Page 412

Fig. 12-22: Hand moves 5 cm/sec.

Page 412

Fig. 12-23: Contraction is isometric at B because muscle does not shorten. Maximal velocity is at A, with zero load on the muscle.

Page 426

(a) The biceps would need to exert 7.5 kg of force to hold the arm stationary, a 125% increase over a biceps inserted at 5 cm. (b) With a 7-kg weight at 20 cm, the biceps needs to exert an additional 28 kg of force. This is less than if the weight is placed in the hand.

Integrative Physiology I:
Control of Body Movement

13

RUNNING PROBLEM

Tetanus

"She has not been able to talk to us. We're afraid she may have had a stroke." That is how her neighbors described 77-year-old Cecile Evans when they brought her to the emergency room. But when a neurological examination revealed no problems other than Mrs. Evans's inability to open her mouth and stiffness in her neck, emergency room physician Dr. Doris Ling began to consider other diagnoses. She noticed some scratches healing on Mrs. Evans's arms and legs and asked the neighbors if they knew what had caused them. "Oh, yes. She told us a few days ago that her dog jumped up and knocked her against the barbed wire fence." At that point, Dr. Ling realized she was probably dealing with her first case of tetanus.

BACKGROUND BASICS

Reflex pathways: **191** | Central nervous system: **287** | Summation of action potentials: **270** | Isometric contraction: **409** | Sensory pathways and receptors: **322** | Graded potentials: **250** | Tonic control: **188** | Tendons: **72** |

Extracting signals directly from the brain to directly control robotic devices has been a science fiction theme that seems destined to become fact.

—**Dr. Eberhard E. Fetz,** *Science News* 156: 142, 8/28/99

Think back to the baseball pitcher in the last chapter. As he stands on the mound, watching the first batter, he receives sensory information from multiple sources: the sound of the crowd, the sight of the batter and the catcher, the feel of the ball within his hand, the alignment of his body as he prepares to throw. Sensory receptors code this information and send it to the central nervous system (CNS), where it is integrated.

The pitcher acts consciously on some of the information: he decides that it is time to throw a fastball. But he processes other information at the subconscious level and acts upon it without conscious thought. As he thinks about starting his pitch, he shifts his weight to offset the impending movement of his arm. The integration of sensory information into an involuntary response is the hallmark of a *reflex* [🔲 p. 191].

In this chapter we first look at the characteristics of neural reflexes. We then discuss how the CNS uses reflexes to control muscle contraction and movement of the body.

▶ NEURAL REFLEXES

All neural reflexes begin with a stimulus that activates a sensory receptor. The receptor sends information in the form of action potentials through sensory neurons to the CNS [🔲 p. 287]. The CNS is the integrating center that evaluates all incoming information and selects an appropriate response. It then initiates action potentials in efferent neurons to direct the response of muscles and glands, the effectors.

A key feature of many reflex pathways is **negative feedback**, a concept introduced in Chapter 6 [🔲 p. 194]. Feedback signals from muscle and joint receptors keep the CNS continuously informed of changing body position. Some reflexes have a **feedforward** component that allows the body to anticipate a stimulus and begin the response [🔲 pp. 195–196]. Bracing yourself in anticipation of a collision would be an example of a feedforward response.

Neural Reflex Pathways Can Be Classified Different Ways

Reflex pathways in the nervous system consist of chains or networks of neurons that link sensory receptors to muscles or glands, the effectors. Neural reflexes can be classified in several ways (Table 13-1):

1. *By the efferent division of the nervous system that controls the response.* Reflexes that involve somatic motor neurons and skeletal muscles are known as **somatic reflexes**. Reflexes whose responses are controlled by autonomic neurons are called **autonomic** or **visceral reflexes**.
2. *By the CNS location where the reflex is integrated.* **Spinal reflexes** are integrated within the spinal cord. These reflexes may be modulated by higher input from the brain, but they can occur without that input. Reflexes integrated in the brain are called **cranial reflexes**.

Table 13-1: Classification of Neural Reflexes

Neural reflexes can be classified by:

1. Efferent division that controls the effector
   a. Somatic motor neurons control skeletal muscles.
   b. Autonomic neurons control smooth and cardiac muscle, glands, and adipose tissue.

2. Integrating region within the central nervous system
   a. Spinal reflexes do not require input from the brain.
   b. Cranial reflexes are integrated within the brain.

3. Time at which the reflex develops
   a. Innate (inborn) reflexes are genetically determined.
   b. Learned (conditioned) reflexes are acquired through experience.

4. The number of neurons in the reflex loop
   a. Monosynaptic reflexes have only two neurons: one afferent (sensory) and one efferent. Only somatic motor reflexes can be monosynaptic.
   b. Polysynaptic reflexes add one or more interneurons between the afferent and efferent neurons. All autonomic reflexes are polysynaptic because they have three neurons: one afferent and two efferent.

3. *By whether the reflex is inborn or learned.* Many reflexes are innate, that is, we are born with them and they are genetically determined. One example is the knee jerk reflex, in which the lower leg kicks out when the patellar tendon at the lower edge of the kneecap is tapped. Other reflexes are acquired through experience. The example of Pavlov's dogs salivating upon hearing a bell is the classic example of a learned (conditioned) reflex.
4. *By the number of neurons in the reflex pathway.* The simplest reflex pathway is a **monosynaptic reflex** involving only two neurons: an afferent sensory neuron and an efferent somatic motor neuron. These two neurons synapse within the spinal cord, allowing a signal initiated at the receptor to go directly from the sensory neuron to the motor neuron (Fig. 13-1a ●). The monosynaptic reflex gets its name from the single synapse between the sensory and somatic motor neurons. (The synapse between the neuron and muscle is ignored.)

Most reflexes have three or more neurons in the pathway, leading to their designation as **polysynaptic reflexes** (Fig. 13-1b). Polysynaptic reflexes may be quite complex, with extensive branching within the CNS to form networks involving multiple interneurons. *Divergence* of pathways allows a single stimulus to affect multiple targets. *Convergence* integrates the input from multiple sources to modulate the response. Recall from Chapter 8 that modulation in polysynaptic pathways may involve excitation or inhibition [🔲 p. 271].

(a) A **monosynaptic reflex** has a single synapse between the afferent and efferent neurons.

(b) Polysynaptic reflexes have two or more synapses.

● **Figure 13-1** *Monosynaptic and polysynaptic somatic motor reflexes* In somatic motor reflexes, the synapses all occur in the central nervous system, as illustrated.

EMERGING CONCEPTS

Visualization Techniques in Sports

Presynaptic facilitation is now believed to be the physiological mechanism that underlies the success of visualization techniques in sports. Visualization, also known as *guided imagery*, enables athletes to maximize their performance by "psyching" themselves, picturing in their minds the perfect vault or the leaping catch for a touchdown. By pathways that we still do not understand, the visual image conjured up by the cerebral cortex is translated into signals that find their way to the muscles. Guided imagery is also being used in medicine as an *adjunct* (supplementary) therapy for cancer treatment and pain management. The ability of the conscious brain to alter physiological function is only one example of many fascinating connections between the higher brain and the body.

▶ **AUTONOMIC REFLEXES**

Autonomic reflexes are also known as *visceral reflexes* because they often involve the internal organs of the body. Some visceral reflexes, such as urination and defecation, are spinal reflexes that can take place without input from the brain. However, spinal reflexes are often modulated by excitatory or inhibitory signals from the brain, carried by descending tracts from higher brain centers.

For example, urination may be voluntarily initiated by conscious thought, or it may be inhibited by emotion or a stressful situation, such as the presence of other people (a syndrome known as "bashful bladder"). Often, the higher control of a spinal reflex is a learned response. Toilet training that we master as toddlers is an example of a learned reflex that our CNS uses to modulate the simple spinal reflex of urination.

Other autonomic reflexes are integrated in the brain, primarily in the hypothalamus, thalamus, and brain stem. These regions contain centers that coordinate body functions needed to maintain homeostasis, such as heart rate, blood pressure, breathing, eating, water balance, and maintenance of body temperature

[Fig. 11-3, p. 371]. The brain stem also contains the integrating centers for autonomic reflexes such as salivating, vomiting, sneezing, coughing, swallowing, and gagging.

An interesting type of autonomic reflex is the conversion of emotional stimuli into visceral responses. The limbic system, the site of primitive drives such as sex, fear, rage, aggression, and hunger, has been called the "visceral brain" because of its role in these emotionally driven reflexes. We speak of "gut feelings" and "butterflies in the stomach"—all transformations of emotion into somatic sensation and visceral function. Other emotion-linked autonomic reflexes include urination, defecation, blushing, blanching, and *piloerection*, in which tiny muscles in the hair follicles pull the shaft of the hair erect ("I was so scared that my hair stood on end!").

Autonomic reflexes are all polysynaptic, with at least one synapse in the CNS between the sensory neuron and the preganglionic autonomic neuron, and an additional synapse in the ganglion between the preganglionic and postganglionic neurons (Fig. 13-2). Many autonomic reflexes are characterized by tonic activity, a continuous stream of action potentials that create ongoing activity in the effector. For example, the tonic control of blood vessels discussed in Chapter 6 is an example of a continuously active autonomic reflex [p. 189]. You will encounter many autonomic reflexes as you continue your study of the systems of the body.

CONCEPT CHECK   Answers: p. 447

1. Name the general steps of a reflex pathway and the anatomical structures in the nervous system that correspond to each step.

2. If the membrane potential hyperpolarizes, does it become more positive or more negative? Move closer to threshold or move farther away from threshold?

RUNNING PROBLEM

Tetanus, also known as lockjaw, is a devastating disease caused by the bacterium *Clostridium tetani*. These bacteria are commonly found in soil and enter the human body through a cut or wound. As the bacteria reproduce in the tissues, they release a protein neurotoxin. This toxin, called *tetanospasmin*, is taken up by motor neurons at the peripheral nerve endings. Tetanospasmin then travels along the axons until it reaches the motor neuron cell body in the spinal cord.

Question 1: Tetanospasmin is a protein. By what process is it taken up into neurons? [Hint: p. 145] By what process does it travel up the axon to the nerve cell body? [Hint: p. 244]

▶ SKELETAL MUSCLE REFLEXES

Although we are not always aware of them, skeletal muscle reflexes are involved in almost everything we do. Receptors that sense changes in joint movements, muscle tension, and muscle length feed this information to the CNS, which responds one of two ways. If muscle contraction is the appropriate response, the CNS activates somatic motor neurons to the muscle fibers. If a muscle needs to be relaxed to achieve the response, sensory input activates inhibitory interneurons in the CNS, and these interneurons *inhibit* activity in somatic motor neurons controlling the muscle.

Recall that somatic motor neurons always cause contraction in skeletal muscle. There is no inhibitory neuron that synapses on skeletal muscles to cause them to relax. Instead, *relaxation results from the absence of excitatory input by the somatic motor neuron.*

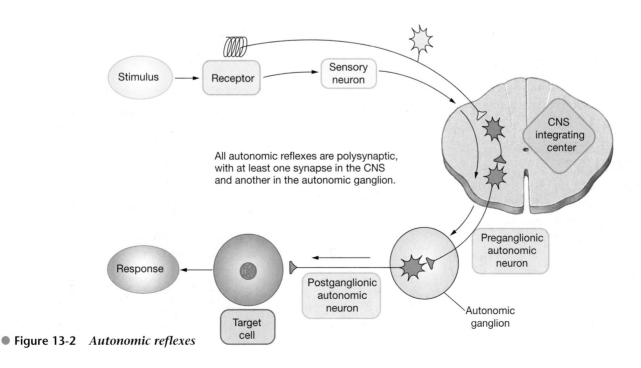

All autonomic reflexes are polysynaptic, with at least one synapse in the CNS and another in the autonomic ganglion.

● **Figure 13-2** *Autonomic reflexes*

Thus, all control of somatic motor neurons and their associated skeletal muscles must occur at synapses within the CNS.

Skeletal muscle reflexes have the following components:

1. *Sensory receptors*, known as **proprioceptors**, are located in skeletal muscles, joint capsules, and ligaments. They monitor the position of our limbs in space, our movements, and the effort we exert in lifting weights.

2. *Sensory neurons* carry the input signal from proprioceptors to the CNS.

3. *The central nervous system* integrates the input signal using networks and pathways of excitatory and inhibitory interneurons. In a reflex, sensory information is integrated and acted upon subconsciously. However, some sensory information may be integrated in the cerebral cortex and become perception, and some reflexes can be modulated by conscious input.

4. *Somatic motor neurons* carry the output signal. The somatic motor neurons that innervate skeletal muscle contractile fibers are called **alpha motor neurons**.

5. The effectors are skeletal muscle *contractile fibers*, also known as **extrafusal muscle fibers**. Action potentials in alpha motor neurons cause contraction in extrafusal fibers.

Three types of proprioceptors are found in the body: muscle spindles, Golgi tendon organs, and joint receptors. *Joint receptors* are found in the capsules and ligaments around every joint in the body. They are stimulated by mechanical distortion that accompanies changes in the relative positioning of bones linked by flexible joints. Sensory information from joint receptors is integrated primarily in the cerebellum.

In the next two sections, we examine the function of muscle spindles and Golgi tendon organs, two interesting and unique receptors. These receptors lie within the skeletal muscles themselves and sense changes in muscle length and tension. Their sensory output activates muscle reflexes.

Muscle Spindles Respond to Muscle Stretch

Muscle spindles are stretch receptors that send information to the spinal cord and brain about muscle length and changes in muscle length. They are small, elongated structures scattered among and arranged parallel to the contractile extrafusal muscle fibers (Fig. 13-3a ●). With the exception of one muscle in the jaw, every skeletal muscle in the body has many muscle spindles. For example, a cat hindlimb muscle has 50 or more spindles.

Each muscle spindle consists of a connective tissue *capsule* that encloses a group of small muscle fibers known as **intrafusal fibers** [*intra-*, within + *fusus*, spindle]. Intrafusal fibers are modified so that the ends are contractile but the central portion lacks myofibrils (Fig. 13-3b). The contractile ends of the intrafusal fibers have their own innervation from **gamma motor neurons**. The noncontractile central portion of the intrafusal fiber is wrapped by sensory nerve endings that are stimulated by stretch. The sensory neurons project to the spinal cord and synapse directly upon alpha motor neurons innervating the same muscle within which the spindles lie.

When a muscle is at its resting length, the central portion of the muscle spindle is stretched enough to activate the sensory fibers (Fig. 13-4a ●). Thus, sensory neurons from the spindles are tonically (continuously) active, sending a steady stream of action potentials to the CNS. Because of this tonic activity, even a muscle at rest maintains a certain level of tension, known as **muscle tone**.

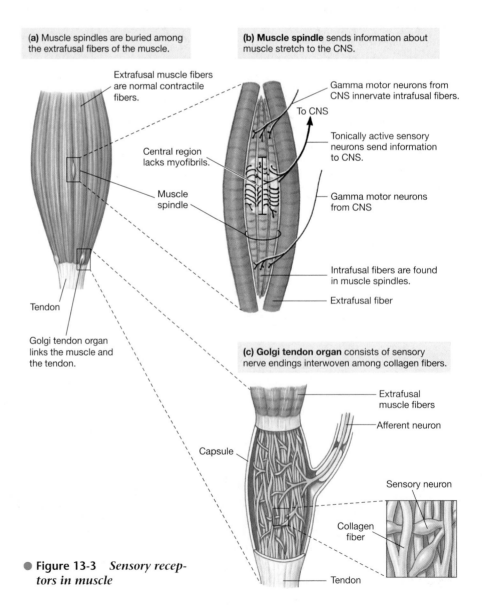

(a) Muscle spindles are buried among the extrafusal fibers of the muscle.

(b) Muscle spindle sends information about muscle stretch to the CNS.

Extrafusal muscle fibers are normal contractile fibers.

Central region lacks myofibrils.

Muscle spindle

Tendon

Golgi tendon organ links the muscle and the tendon.

Gamma motor neurons from CNS innervate intrafusal fibers.

To CNS

Tonically active sensory neurons send information to CNS.

Gamma motor neurons from CNS

Intrafusal fibers are found in muscle spindles.

Extrafusal fiber

(c) Golgi tendon organ consists of sensory nerve endings interwoven among collagen fibers.

Extrafusal muscle fibers

Afferent neuron

Capsule

Sensory neuron

Collagen fiber

Tendon

● **Figure 13-3** *Sensory receptors in muscle*

(a) Spindles are firing even when muscle is relaxed.

Spinal cord

Sensory neuron endings

Intrafusal fibers of muscle spindle

Sensory neuron

Alpha motor neuron

1. Extrafusal muscle fibers at resting length

2. Sensory neuron is tonically active.

3. Spinal cord integrates function.

4. Alpha motor neurons to extrafusal fibers receive tonic input from muscle spindles.

5. Extrafusal fibers maintain a certain level of tension even at rest.

(b) Muscle stretch triggers a stretch reflex.

Muscle stretch → Increased afferent signals to spinal cord → Spinal cord → Increased efferent output through alpha motor neurons → Muscle contracts → Firing rate of afferent sensory neuron decreases

Negative feedback

(1)

(2)

Muscle length

Muscle is stretched.

Muscle returns to initial length.

Action potentials in spindle sensory neuron

(1)

(2)

Time

● **Figure 13-4** *Muscle spindle function*

Muscle spindles are anchored in parallel to the extrafusal muscle fibers. Any movement that increases muscle length also stretches the muscle spindle and causes its sensory fibers to fire more rapidly (Fig. 13-4b). This creates a reflex contraction of the muscle, which prevents damage from overstretching. The reflex pathway in which muscle stretch initiates a contraction response is known as a **stretch reflex**.

CONCEPT CHECK   Answers: p. 447

3. Using the standard steps of a reflex pathway (stimulus, receptor, etc.), draw a reflex map of the stretch reflex.

If stretch activates muscle spindles, what happens to spindle activity when a resting muscle contracts and shortens? You might predict that the release of tension on the center of the intrafusal fibers causes the afferent neurons to slow their firing rate or stop firing altogether (Fig. 13-5a ●). However, in a normal muscle this does not happen because of the process known as **alpha-gamma coactivation**.

The role of gamma motor neurons is to adjust the stretch sensitivity of the muscle spindle so that the spindle is active even when the muscle shortens. Gamma motor neurons innervate the contractile ends of the intrafusal fibers. When gamma motor neurons fire,

(a) If gamma motor axons are cut, the spindle loses activity when muscle contracts.

1. Alpha motor neuron fires.

2. Muscle contracts.

3. Less stretch on center of intrafusal fibers

4. Firing rate of spindle sensory neuron decreases.

(b) **Alpha-gamma coactivation** maintains spindle function when muscle contracts.

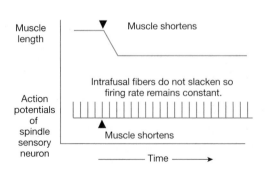

1. Alpha motor neuron fires and gamma motor neuron fires.

2. Muscle contracts.

3. Stretch on centers of intrafusal fibers unchanged. Firing rate of afferent neuron remains constant.

● **Figure 13-5   *Gamma motor neurons***   The gamma motor neurons allow the muscle spindle to maintain function when the muscle is contracting.

the ends of the fibers contract and shorten (Fig. 13-5b). Contracting the ends of the spindle lengthens the central section and maintains stretch on the sensory nerve endings.

In alpha-gamma coactivation, when alpha motor neurons to extrafusal fibers fire, gamma motor neurons to the spindle are activated at the same time. The extrafusal muscle fibers contract and the muscle shortens, releasing tension on the muscle spindle capsule. However, gamma coactivation of the intrafusal fibers stretches the centers of the fibers and counteracts the release of tension on the capsule. Thus, the spindle remains active even during muscle contraction.

An example of how muscle spindles work during a stretch reflex is shown in Figure 13-6 ●. You can try this yourself with an unsuspecting friend. The subject stands with eyes closed, elbows at 90°, and one hand extended with the palm up. A small book or other flat weight is placed in the hand. Now suddenly drop a heavier load, such as another book, onto the subject's hand. The added

weight sends the hand downward, stretching the biceps muscle and activating its muscle spindles. Sensory input into the spinal cord then activates the alpha motor neurons of the biceps muscle. The biceps contracts, bringing the arm back to its original position.

Golgi Tendon Organs Respond to Muscle Tension

A second type of muscle proprioceptor is the **Golgi tendon organ**. These receptors are found at the junction of the tendons and muscle fibers, placing them in series with the muscle fibers. Golgi tendon organs respond primarily to the tension a muscle develops during an isometric contraction, and they cause a relaxation reflex. This is the opposite of muscle spindles, which cause reflex contraction. In contrast to the muscle spindle, the Golgi tendon organ is relatively insensitive to muscle stretch.

Muscle spindle reflex: the addition of a load stretches the muscle and the spindles, creating a reflex contraction.

(a) Add load to muscle.

(b) Muscle and muscle spindle stretch as arm drops.

(c) Reflex contraction initiated by muscle spindle restores arm position.

Golgi tendon reflex protects the muscle from excessively heavy loads by causing the muscle to relax and drop the load.

(d) Muscle contraction stretches Golgi tendon organ.

(e) If excessive load is placed on muscle, Golgi tendon reflex causes relaxation, thus protecting muscle.

1 Neuron from Golgi tendon organ fires.
2 Motor neuron is inhibited.
3 Muscle relaxes.
4 Load is dropped.

● Figure 13-6 *Muscle reflexes*

435

Golgi tendon organs are composed of free nerve endings that wind between collagen fibers inside a connective tissue capsule (Fig. 13-3c). When a muscle contracts, its tendons act as an elastic component during the isometric phase of the contraction [p. 409]. Contraction pulls collagen fibers within the Golgi tendon organ tight, pinching sensory endings of the afferent neurons and causing them to fire.

Afferent input from activation of the Golgi tendon organ excites *inhibitory* interneurons in the spinal cord. The interneurons inhibit alpha motor neurons innervating the muscle, and muscle contraction decreases or ceases. Under most circumstances, this reflex slows muscle contraction as the force of contraction increases. In other instances, the Golgi tendon organs prevent excessive contraction that might injure the muscle.

In the example of the book placed on the outstretched hand, if the added weight requires more tension than the muscle can develop, the Golgi tendon organ will respond as muscle tension nears its maximum. It triggers reflex *inhibition* of the biceps motor neurons, causing the biceps to relax and the arm to fall. Thus, the person will drop the added weight before the muscle fibers can be damaged (Figs. 13-6d, e). Golgi tendon organ input is an important source of inhibition to alpha motor neurons.

CONCEPT CHECK   Answers: p. 447

4. Using the standard steps of a reflex pathway, create a map showing alpha-gamma coactivation and the Golgi tendon reflex. Begin with the stimulus "Alpha motor neuron fires."

Stretch Reflexes and Reciprocal Inhibition Control Movement Around a Joint

Movement around most flexible joints in the body is controlled by groups of synergistic and antagonistic muscles that act in a coordinated fashion. Sensory neurons from muscle receptors and efferent motor neurons that control the muscle are linked together by diverging and converging pathways of interneurons within the spinal cord. The collection of pathways controlling a single joint is known as a **myotatic unit** [*myo-*, muscle + *tasis*, stretching].

The simplest reflex in a myotatic unit is the **monosynaptic stretch reflex** that involves only two neurons: the sensory neuron from the muscle spindle and the somatic motor neuron to the muscle. The knee jerk reflex is an example of a monosynaptic stretch reflex (Fig. 13-7 ●).

To demonstrate the knee jerk reflex, the subject sits on the edge of a table so that the lower leg hangs relaxed. When the patellar tendon below the kneecap is tapped with a small rubber hammer, the tap stretches the quadriceps muscle that runs up the front of the thigh. This stretching activates muscle spindles and sends action potentials via the sensory fibers to the spinal cord. The sensory neurons synapse directly onto the motor neurons that control contraction of the quadriceps muscle (a monosynaptic reflex). The action potentials in the motor neurons

cause motor units in the quadriceps to contract, and the lower leg swings forward.

For muscle contraction to extend the leg, the antagonistic flexor muscles must relax (**reciprocal inhibition**). In the leg, this requires relaxation of the hamstring muscles that run up the back of the thigh. The single stimulus of the tap to the tendon accomplishes both contraction of the quadriceps muscle and reciprocal inhibition of the hamstrings. The sensory fibers branch upon entering the spinal cord. Some of the branches activate motor neurons innervating the quadriceps, while the other branches synapse on inhibitory interneurons. The inhibitory interneurons suppress activity in the motor neurons controlling the hamstrings (a polysynaptic reflex). The result is a relaxation of the hamstrings that allows contraction of the quadriceps to proceed unopposed.

Flexion Reflexes Pull Limbs Away from Painful Stimuli

Flexion reflexes are polysynaptic reflex pathways that cause an arm or leg to be pulled away from a painful stimulus such as a pinprick or a hot stove. These reflexes, like the reciprocal inhibition reflex just described, rely on divergent pathways in the spinal cord. Figure 13-8 ● uses the example of stepping on a tack to illustrate a flexion reflex.

When the foot contacts the point of the tack, nociceptors (pain receptors) in the foot send sensory information to the spinal cord. Here the signal diverges, activating multiple excitatory interneurons. Some of these interneurons excite alpha motor neurons, leading to contraction of the flexor muscles of the stimulated limb. Other interneurons simultaneously activate inhibitory interneurons that cause relaxation of the antagonistic muscle groups. Because of this reciprocal inhibition, the limb is flexed, withdrawing it from the painful stimulus. This type of reflex requires more time than a stretch reflex such as the knee jerk, or patellar tendon reflex, because it is a polysynaptic rather than a monosynaptic reflex.

CONCEPT CHECK   Answers: p. 447

5. Draw a reflex map of the flexion reflex. (Leave space to add additional information.)

RUNNING PROBLEM

Once in the spinal cord, tetanospasmin is released from the motor neuron. It then selectively blocks neurotransmitter release at inhibitory synapses. Patients with tetanus experience muscle spasms that begin in the jaw and may eventually affect the entire body. When the extremities become involved, the arms and legs may go into painful, rigid spasms.

Question 2: Using the reflex pathways diagrammed in Figures 13-7 and 13-8, explain why inhibition of inhibitory interneurons might result in uncontrollable muscle spasms.

428   431   R   439   443   443

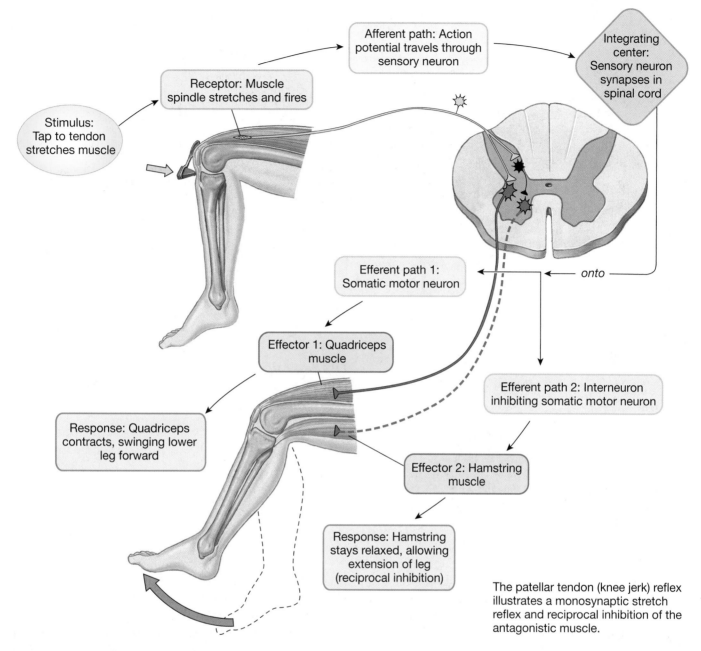

Stimulus:
Tap to tendon
stretches muscle

Receptor: Muscle
spindle stretches and fires

Afferent path: Action
potential travels through
sensory neuron

Integrating
center:
Sensory neuron
synapses in
spinal cord

onto

Efferent path 1:
Somatic motor neuron

Effector 1: Quadriceps
muscle

Response: Quadriceps
contracts, swinging lower
leg forward

Efferent path 2: Interneuron
inhibiting somatic motor neuron

Effector 2: Hamstring
muscle

Response: Hamstring
stays relaxed, allowing
extension of leg
(reciprocal inhibition)

The patellar tendon (knee jerk) reflex
illustrates a monosynaptic stretch
reflex and reciprocal inhibition of the
antagonistic muscle.

● **Figure 13-7** *The knee jerk reflex*

Flexion reflexes, particularly in the legs, are usually accompanied by the **crossed extensor reflex**, a postural reflex that helps maintain balance when one foot is lifted from the ground. In this reflex, also shown in Figure 13-8, the quick withdrawal of the right foot from a painful stimulus (a tack) is matched by extension of the left leg so that it can support the sudden shift in weight. The extensors contract in the supporting left leg and relax in the withdrawing right leg while the opposite occurs in the flexor muscles.

Note in the illustration how the one sensory neuron synapses on multiple interneurons. Divergence of the sensory signal permits a single stimulus to control two sets of antagonistic muscle groups as well as send sensory information to the brain. This type of complex reflex with multiple neuron interactions is more typical of our reflexes than the simple monosynaptic knee jerk stretch reflex.

The most complicated spinal reflex pathways are controlled by networks of neurons in the CNS called **central pattern generators**. Once activated, central pattern generators create spontaneous repetitive movement. In humans, rhythmic movements controlled by central pattern generators include locomotion and the unconscious rhythmicity of quiet breathing.

CONCEPT CHECK   Answers: p. 447

6. Add the extensor reflex in the supporting leg to your map of the flexion reflex.

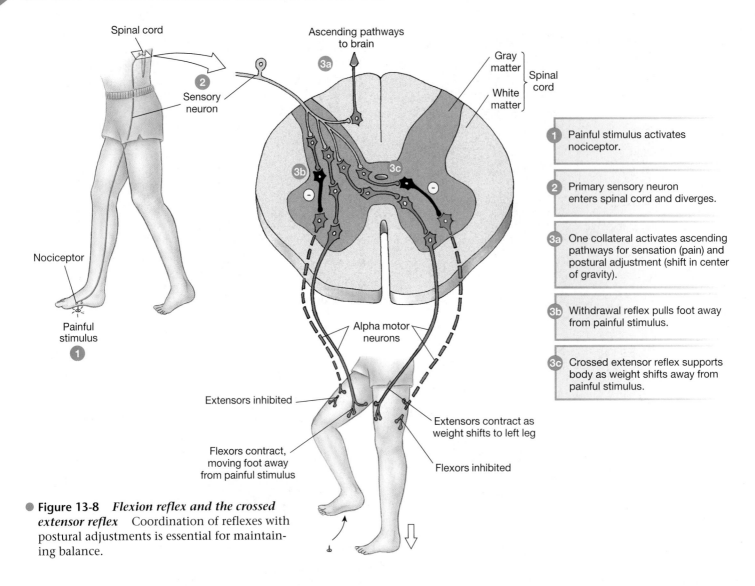

Spinal cord

Ascending pathways to brain

2 Sensory neuron

3a

Gray matter

White matter

Spinal cord

Nociceptor

Painful stimulus **1**

3b **3c**

Alpha motor neurons

Extensors inhibited

Flexors contract, moving foot away from painful stimulus

Extensors contract as weight shifts to left leg

Flexors inhibited

1 Painful stimulus activates nociceptor.

2 Primary sensory neuron enters spinal cord and diverges.

3a One collateral activates ascending pathways for sensation (pain) and postural adjustment (shift in center of gravity).

3b Withdrawal reflex pulls foot away from painful stimulus.

3c Crossed extensor reflex supports body as weight shifts away from painful stimulus.

● **Figure 13-8** *Flexion reflex and the crossed extensor reflex* Coordination of reflexes with postural adjustments is essential for maintaining balance.

In the next section of the chapter, we look at how the CNS controls movements that range from involuntary reflexes to complex, voluntary movement patterns such as dancing, throwing a ball, or playing a musical instrument.

CONCEPT CHECK   Answers: p. 447

7. As you pick up a heavy weight, which of the following are active in your biceps muscle: alpha motor neuron, gamma motor neuron, muscle spindle afferent neurons, Golgi tendon organ afferent neurons?

8. What distinguishes a stretch reflex from a crossed extensor reflex?

▶ THE INTEGRATED CONTROL OF BODY MOVEMENT

Most of us never think about how our body translates thoughts into action. Even the simplest movement requires proper timing so that antagonistic and synergistic muscle groups contract in the appropriate sequence and to the appropriate level. In addition,

the body must continuously adjust its position to compensate for differences between the intended movement and the actual one. For example, the baseball pitcher steps off the mound as he reaches for a ground ball, but in doing so, he slips on a wet patch of grass. His brain quickly compensates for the unexpected change in position through reflex muscle activity, and he stays on his feet to intercept the ball.

Skeletal muscles cannot communicate with each other directly, so they send messages to the CNS, allowing the integrating centers to take charge and direct movement. Most body movements are highly integrated, coordinated responses that require input from multiple regions of the brain. In this section we examine a few of the integrating centers within the CNS that are responsible for control of body movement.

Movement Can Be Classified as Reflex, Voluntary, or Rhythmic

Movement can be loosely classified into three categories: reflex movement, voluntary movement, and rhythmic movement (Table 13-2). **Reflex movements** are the least complex and are

Table 13-2: Types of Movement

	REFLEX	VOLUNTARY	RHYTHMIC
Stimulus that initiates movement	Primarily external via sensory receptors; minimally voluntary	External stimuli or at will	Initiation and termination voluntary
Example	Knee jerk, cough, postural reflexes	Playing piano	Walk, run
Complexity	Least complex; integrated at level of spinal cord or brain stem with higher center modulation	Most complex; integrated in cerebral cortex	Intermediate complexity; integrated in spinal cord with higher center input required
Comments	Inherent, rapid	Learned movements that improve with practice; once learned, may become subconscious ("muscle memory")	Spinal circuits act as pattern generators; activation of these pathways requires input from brain stem

integrated primarily in the spinal cord (for example, see the knee jerk reflex in Fig. 13-7). However, like other spinal reflexes, reflex movements can be modulated by input from higher brain centers. In addition, the sensory input that initiates reflex movements, such as the input from muscle spindles and Golgi tendon organs, goes to the brain and participates in the coordination of voluntary movements and postural reflexes.

Postural reflexes that help us maintain body position as we stand or move through space are integrated in the brain stem. They require continuous sensory input from visual and vestibular (inner ear) sensory systems and from the muscles themselves [● p. 347]. Muscle, tendon, and joint receptors provide information about proprioception, the positions of various body parts relative to one another. You can tell if your arm is bent even when your eyes are closed because these receptors provide information about body position to the brain.

Information from the vestibular apparatus of the ear and visual cues help us maintain our position in space. For example, we use the horizon to tell us our spatial orientation relative to the ground. In the absence of visual cues, we rely on tactile input. People trying to move in a dark room instinctively reach for a wall or piece of furniture to help orient themselves. Without visual and tactile cues, our orientation skills may fail. The lack of cues is what makes flying airplanes in clouds or fog impossible without instruments. The effect of gravity on the vestibular system is such a weak input when compared with visual or tactile cues that pilots may find themselves flying upside down relative to the ground.

Voluntary movements are the most complex type of movement. They require integration at the cerebral cortex, and they can be initiated at will without external stimuli. Learned voluntary movements improve with practice and some even become involuntary, like reflexes. Think about how difficult it was to learn to ride a bicycle. However, once you learned to pedal smoothly and to keep your balance, the movements became automatic. "Muscle memory" is the name dancers and athletes give the ability of the unconscious brain to reproduce voluntary, learned movements and positions.

Rhythmic movements such as walking or running are a combination of reflex movements and voluntary movements. Rhythmic movements are initiated and terminated by input from the cerebral cortex, but once initiated, they can be sustained without further input from the brain. The rhythmic activity of skeletal muscles in the limbs is maintained by networks of spinal interneurons.

RUNNING PROBLEM

Dr. Ling admits Mrs. Evans to the intensive care unit. There Mrs. Evans is given tetanus antitoxin to deactivate any toxin that has not yet entered motor neurons. She also receives penicillin, an antibiotic that kills the bacteria, and drugs to help relax her muscles. Despite these treatments, by the third day Mrs. Evans is having difficulty breathing because of spasms in her chest muscles. Dr. Ling calls in the chief of anesthesiology to administer metocurine, a drug similar to curare. Curare and metocurine induce temporary paralysis of muscles by binding to ACh receptors on the motor end plate. Patients must be placed on respirators that breathe for them while receiving metocurine. For people with tetanus, however, metocurine can temporarily halt the muscle spasms and allow the body to recover.

Question 3: Why does the binding of metocurine to ACh receptors on the motor end plate induce muscle paralysis? (Hint: What is the function of ACh in synaptic transmission?) Is metocurine an agonist or an antagonist of ACh?

These act as central pattern generators, causing alternate contraction and relaxation of muscles in a repetitive, rhythmic fashion until instructed to stop by signals from the brain. As an analogy, think of a battery-operated bunny. When the switch is thrown to "on," the bunny begins to hop. It continues its repetitive hopping until someone turns it off (or until the battery runs down).

The distinction between reflex, voluntary, and rhythmic movement is not always clear-cut. The precision of voluntary movements improves with practice, but so does that of some reflexes. Voluntary movements, once learned, can become reflexive. In addition, most voluntary movements require continuous input from postural reflexes. **Feedforward reflexes** allow the body to prepare for a voluntary movement, and feedback mechanisms are used to create a smooth, continuous motion. Coordination of movement thus requires cooperation from many parts of the brain.

The CNS Integrates Movement

Three levels of the nervous system control movement: (1) the *spinal cord*, which integrates spinal reflexes, (2) the *brain stem and cerebellum*, which control postural reflexes and hand and eye movements, and (3) the *cerebral cortex and basal nuclei*, which are responsible for voluntary movements. The thalamus relays and modifies signals being sent from the spinal cord, basal nuclei, and cerebellum to the cerebral cortex (Table 13-3).

Reflex movements do not require input from the cerebral cortex. Proprioceptors such as muscle spindles, Golgi tendon organs, and joint capsule receptors provide information to the spinal cord, brain stem, and cerebellum (Fig. 13-9 ●). The brain stem is in charge of postural reflexes and hand and eye movements. It also gets commands from the cerebellum, the part of the brain responsible for "fine-tuning" movement. The result is reflex movement. However, some sensory information is sent

EMERGING CONCEPTS

Central Pattern Generators and Spinal Cord Injuries

The ability of CNS pattern generators, or central pattern generators, to sustain rhythmic movement without continued input from the central nervous system has proved important for research on spinal cord injuries. Scientists have artificially stimulated rhythmic movements in experimental animals when the spinal cord between the brain and the motor neurons has been severed. For example, an animal paralyzed by a spinal cord injury that blocks the "start walking" signal from the brain will walk if supported on a moving treadmill and given an electrical stimulus to activate the spinal pattern generator. As the treadmill moves the animal's legs, the spinal pattern generator, reinforced by sensory signals from muscle spindles, drives contraction of the leg muscles. Researchers are trying to take advantage of these rhythmic reflexes in people with spinal cord injuries. Some of the newest therapeutic techniques involve artificially stimulating portions of the spinal cord so that movement is restored to formerly paralyzed limbs.

through ascending pathways to sensory areas of the cortex, where it can be used to plan voluntary movements.

Voluntary movements require coordination between the cerebral cortex, cerebellum, and the basal ganglia. The control of these behaviors can be divided into three steps: (1) decision making and planning, (2) initiating the movement, and (3) executing the

Table 13-3: Neural Control of Movement

LOCATION	ROLE	RECEIVES INPUT FROM:	SENDS INTEGRATIVE OUTPUT TO:
Spinal cord	Spinal reflexes; locomotor pattern generators	Sensory receptors	Brain stem, cerebellum, thalamus/cerebral cortex
Brain stem	Posture, hand and eye	Cerebellum, visual and vestibular sensory receptors	Spinal cord
Motor areas of cerebral cortex	Planning and coordinating complex movement	Thalamus	Brain stem, spinal cord (corticospinal tract), cerebellum, basal ganglia
Cerebellum	Monitors output signals from motor areas and adjusts movements	Spinal cord (sensory), cerebral cortex (commands)	Brain stem, cerebral cortex (Note: All output is inhibitory.)
Thalamus	Contains relay nuclei that modulate and pass messages to cerebral cortex	Basal ganglia, cerebellum, spinal cord	Cerebral cortex
Basal nuclei	Motor planning	Cerebral cortex	Cerebral cortex, brain stem

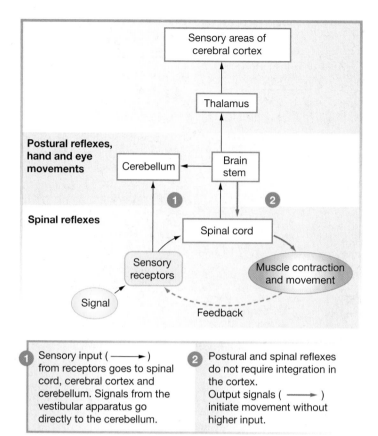

● **Figure 13-9** *Integration of muscle reflexes*

① Sensory input (——▶) from receptors goes to spinal cord, cerebral cortex and cerebellum. Signals from the vestibular apparatus go directly to the cerebellum.

② Postural and spinal reflexes do not require integration in the cortex. Output signals (——▶) initiate movement without higher input.

movement (Fig. 13-10●). The cerebral cortex plays a key role in the first two steps. Behaviors such as movement require knowledge of the body's position in space (where am I?), a decision on what movement should be executed (what shall I do?), a plan for executing the movement (how shall I do it?), and the ability to hold the

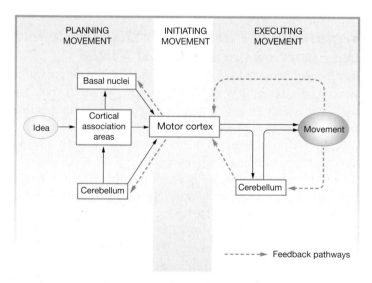

● **Figure 13-10** *CNS control of voluntary movement*
Voluntary movements can be divided into three phases: planning, initiation, and execution. Sensory feedback allows the brain to correct for any deviation between the planned movement and the actual movement.

plan in memory long enough to carry it out. As with reflex movements, sensory feedback is used to continually refine the process.

Let's return to our baseball pitcher and trace the process as he decides whether to throw a fastball or a slow inside curve. Standing out on the mound, the pitcher is acutely aware of his surroundings: the other players on the field, the batter in the box, and the dirt beneath his feet. With the help of visual and somatosensory input to the sensory areas of the cortex, he is aware of his body position as he steadies himself for the pitch (Fig. 13-11●). Deciding which type of pitch to throw and anticipating the consequences occupy many pathways in his prefrontal cortex and association areas. These pathways loop down through the **basal nuclei** (basal ganglia) and thalamus for modulation before cycling back to the cortex.

Once the pitcher makes the decision to throw a fastball, the motor cortex takes charge of organizing the execution of this complex movement. To initiate the movement, descending information travels from the motor association areas and motor cortex to the brain stem, the spinal cord, and the cerebellum. The cerebellum assists in making postural adjustments by integrating feedback from peripheral sensory receptors. The basal nuclei, which assisted the cortical motor areas in planning the pitch, also provide information about posture, balance, and gait to the brain stem.

The pitcher's decision to throw a fastball now is translated into action potentials that travel down through the **corticospinal tract**, a group of interneurons that connect the cortex to the spinal cord (Fig. 13-12●). Most of these descending pathways cross to the opposite side of the body in a region of the medulla known as the *pyramids*, so this pathway is sometimes called the *pyramidal tract*. Descending pathways that do not pass through the pyramids comprise what is sometimes called the *extrapyramidal tract* or the *extrapyramidal system*. However, we now know that the pyramidal and extrapyramidal pathways interact and are not as distinct in their function as was once believed.

As the pitcher begins the pitch, *feedforward postural reflexes* adjust the body position, shifting weight slightly in anticipation of the changes that are about to occur (Fig. 13-13●). Through divergent pathways, action potentials race over appropriate pathways to the pools of somatic motor neurons that control the muscles used for the pitch: some are excited, others are inhibited. The neural circuitry allows precise control over antagonistic muscle groups as the pitcher flexes and retracts his right arm. His weight shifts onto his right foot as his arm moves back.

Each of these movements activates sensory receptors that feed back information to the spinal cord, brain stem, and cerebellum, activating postural reflexes. These reflexes adjust his body position so that the pitcher does not lose his balance and fall over backward. Finally, the pitcher releases the ball, catching his balance on the follow-through: another example of postural reflexes mediated through sensory feedback. His head stays erect and his eyes stay on the ball as it reaches the batter. WHACK... home run. As the pitcher's eyes follow the ball and he evaluates the result of his pitch, his brain is preparing for the next batter, hoping to use what it has learned from these pitches to improve those to come.

● **Figure 13-11** *Control of voluntary movements*

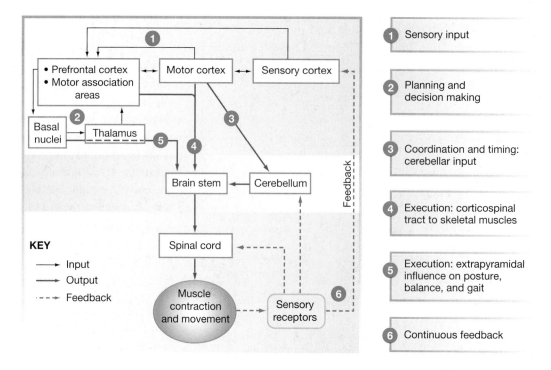

KEY

→ Input
→ Output
---→ Feedback

1 Sensory input

2 Planning and decision making

3 Coordination and timing: cerebellar input

4 Execution: corticospinal tract to skeletal muscles

5 Execution: extrapyramidal influence on posture, balance, and gait

6 Continuous feedback

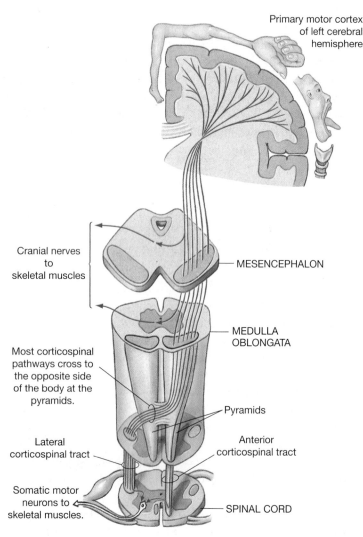

● **Figure 13-12** *The corticospinal tract*

Primary motor cortex of left cerebral hemisphere

Cranial nerves to skeletal muscles

MESENCEPHALON

MEDULLA OBLONGATA

Most corticospinal pathways cross to the opposite side of the body at the pyramids.

Pyramids

Lateral corticospinal tract

Anterior corticospinal tract

Somatic motor neurons to skeletal muscles.

SPINAL CORD

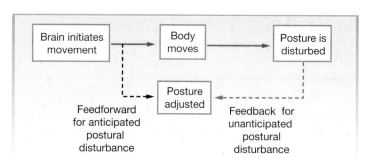

● **Figure 13-13** *Feedforward reflexes and feedback of information during movement*

Symptoms of Parkinson's Disease Reflect the Functions of the Basal Nuclei

Our understanding of the role of the basal nuclei in the control of movement has been slow to develop because, for many years, animal experiments yielded little information. Randomly destroying portions of the basal nuclei did not appear to affect research animals. However, research focusing on **Parkinson's disease** (Parkinsonism) in humans has been more fruitful. From studying patients with Parkinson's, we have learned that the basal nuclei play a role in cognitive function and memory as well as the coordination of movement.

Parkinson's disease is a progressive neurological disorder characterized by abnormal movements, speech difficulties, and cognitive changes. These symptoms are associated with loss of neurons in the basal nuclei that release the neurotransmitter dopamine. One abnormal sign that most Parkinson patients have is tremors in the hands, arm, and legs, particularly at rest. In addition, they have difficulty initiating movement and walk slowly with stooped posture and shuffling gait. They lose facial expression, fail to blink

Four weeks later, Mrs. Evans is ready to go home, completely recovered from her infection and showing no signs of lingering effects. Once she could talk, Mrs. Evans, who was born on the farm where she still lived, was able to tell Dr. Ling that she had never had immunization shots for tetanus or any other diseases. "Well, that made you one of only a handful of people in the United States who will develop tetanus this year," Dr. Ling told her. "You've been given your first two tetanus shots here in the hospital. Be sure to come back in six months for the last one so that this won't happen again." Because of national immunization programs begun in the 1950s, tetanus is now a rare disease in the United States. However, in developing countries without immunization programs, tetanus is still a common and serious condition.

Question 4: On the basis of what you know about who receives immunization shots in the United States, predict the age and background of people who are most likely to develop tetanus this year.

(the reptilian stare), and may develop depression, sleep disturbances, and personality changes.

The cause of Parkinson's disease is usually not known and appears to be a combination of environmental factors and genetic susceptibility. However, a few years ago, a number of young drug users were diagnosed with Parkinsonism. Their disease was traced to the use of homemade heroin that contained a toxic contaminant that destroyed dopaminergic (dopamine-secreting) neurons. This contaminant has been isolated and now enables researchers to induce Parkinson's disease in experimental animals so that we have an animal model on which to test new treatments.

The primary current treatment for Parkinson's is administration of L-dopa, a precursor of dopamine that can cross the blood-brain barrier. Experimental treatments include transplants of dopamine-secreting neurons. Proponents of stem cell research feel that Parkinson's may be one of the conditions that would benefit from the transplant of stem cells into affected brains. For more on Parkinson's disease, see the Explorations at the end of the chapter.

▶ CONTROL OF MOVEMENT IN VISCERAL MUSCLES

Movement created by contracting smooth and cardiac muscles is very different from that created by skeletal muscles, in large part because smooth and cardiac muscle are not attached to bone. In the internal organs, or viscera, muscle contraction usually changes the shape of an organ, narrowing the lumen of a hollow organ or shortening the length of a tube. In many hollow internal organs, muscle contraction pushes material through the lumen of the organ: the heart pumps blood, the digestive tract moves food, the uterus expels a baby.

Visceral muscle contraction is often reflexively controlled by the autonomic nervous system, but not always. Some types of smooth and cardiac muscle are capable of generating their own action potentials, independent of an external signal. Both the heart and digestive tract have spontaneously depolarizing muscle fibers (often called *pacemakers*) that give rise to regular, rhythmic contractions.

Reflex control of visceral smooth muscle varies from that of skeletal muscle. Skeletal muscles are controlled only by the nervous system, but in many types of visceral muscle, hormones are important in regulating contraction. In addition, some visceral muscle cells are connected to each other by gap junctions that allow electrical signals to pass directly from cell to cell.

Because smooth and cardiac muscle have such a variety of control mechanisms, we will discuss their control as we cover the appropriate organ system for each type of muscle. In the next chapter, we examine cardiac muscle and its function in the heart.

Tetanus

In this running problem, you learned about tetanus toxin. Both tetanus toxin and botulinum toxin cause paralysis, but tetanus is a rigid (contracted muscle) paralysis and botulism is a flaccid (relaxed muscle) paralysis (see Exploration 2 for Chapter 11, p. 387). Now check your understanding of this running problem by comparing your answers to those in the summary table.

	QUESTION	FACTS	INTEGRATION AND ANALYSIS
1a	By what process is tetanospasmin taken up into neurons?	Tetanospasmin is a protein.	Proteins are too large to cross cell membranes by mediated transport. Therefore, tetanospasmin must be taken up by endocytosis (p. 145).
1b	By what process does tetanospasmin travel up the axon to the nerve cell body?	Movement of substances from the axon terminal to the cell body is called retrograde axonal transport (p. 244).	Tetanospasmin is taken up by endocytosis, so it will be contained in endocytotic vesicles. These vesicles "walk" along microtubules through retrograde axonal transport.

(continues on next page)

	QUESTION	FACTS	INTEGRATION AND ANALYSIS
2	Using the reflex pathways diagrammed in Figures 13-7 and 13-8, explain why inhibition of inhibitory interneurons might result in uncontrollable muscle spasms.	Muscles often occur in antagonistic pairs. When one muscle is contracting, its antagonist must be inhibited.	If the inhibitory interneurons are not functioning, both sets of antagonistic muscles can contract at the same time. This would lead to muscle spasms and rigidity because the bones attached to the muscles would be unable to move in any one direction.
3a	Why does the binding of metocurine to ACh receptors on the motor end plate induce muscle paralysis?	Acetylcholine (ACh) is the somatic motor neuron neurotransmitter that initiates skeletal muscle contraction.	If metocurine binds to ACh receptors, it prevents ACh from binding. Without ACh binding, the muscle fiber will not depolarize and cannot contract, resulting in paralysis.
3b	Is metocurine an agonist or an antagonist of ACh?	Agonists mimic the effects of a substance. Antagonists block the effects of a substance.	Metocurine blocks ACh action; therefore, it is an antagonist.
4	On the basis of what you know about who receives immunization shots in the United States, predict the age and background of people who are most likely to develop tetanus this year.	Immunizations are required for all children of school age. This practice has been in effect since about the 1950s. In addition, most people who suffer puncture wounds or dirty wounds are given tetanus booster shots when they are treated for those wounds.	Most cases of tetanus in the United States will occur in people over the age of 60 who have never been immunized, in immigrants (particularly migrant workers), and in newborn infants. Another source of the disease is contaminated heroin; injection of the drug under the skin may cause tetanus.

428 431 436 439 443

CHAPTER SUMMARY

This chapter integrates the material you learned in Chapters 8–12 and shows how the nervous *control system* controls body movement. Postural and spinal reflexes follow the basic pattern of a reflex, with sensory input being integrated in the CNS and acted upon when an output signal goes to skeletal muscles. Voluntary movements do not require sensory input to be initiated, but they integrate sensory feedback to ensure smooth execution.

Neural Reflexes

1. A neural reflex consists of the following elements: stimulus → receptor → sensory neurons → efferent neurons → effectors (muscles and glands) → response. (p. 429)

2. Neural reflexes can be classified in several different ways. **Somatic reflexes** involve somatic motor neurons and skeletal muscles. **Autonomic**, or **visceral**, **reflexes** are controlled by autonomic neurons. (p. 429; Tbl. 13-1)

3. **Spinal reflexes** are integrated within the spinal cord. **Cranial reflexes** are integrated in the brain. (p. 429)

4. Many reflexes are inborn. Others are acquired through experience. (p. 429)

5. The simplest reflex pathway is a **monosynaptic reflex** with only two neurons. **Polysynaptic reflexes** have three or more neurons in the pathway. (p. 429; Fig. 13-1)

Autonomic Reflexes

6. Some autonomic reflexes are spinal reflexes that are modulated by input from the brain. Other reflexes needed to maintain homeostasis are integrated in the brain, primarily in the hypothalamus, thalamus, and brain stem. (p. 430)

7. Autonomic reflexes are all polysynaptic, and many are characterized by tonic activity. (p. 431; Fig. 13-2)

Skeletal Muscle Reflexes

8. Skeletal muscle relaxation must be controlled by the CNS because somatic motor neurons always cause contraction in skeletal muscle. (p. 431)

9. The normal contractile fibers of a muscle are called **extrafusal muscle fibers**. Their contraction is controlled by **alpha motor neurons**. (p. 432; Fig. 13-3)

10. **Muscle spindles** send information about muscle length to the CNS. These receptors consist of **intrafusal fibers** with sensory neurons wrapped around the noncontractile center. **Gamma motor neurons** innervate the contractile ends of the intrafusal fibers. (p. 432; Fig. 13-3)

11. Muscle spindles are tonically active stretch receptors. Their output creates tonic contraction of extrafusal muscle fibers. Because of this tonic activity, a muscle at rest maintains a certain level of tension, known as **muscle tone**. (p. 432; Fig. 13-4a)

12. If an intact muscle stretches, the intrafusal fibers of the spindle stretch and initiate reflex contraction of the muscle. The contraction prevents damage from overstretching. This reflex pathway is known as a **stretch reflex**. (p. 434; Fig. 13-4b)

13. When a muscle contracts, **alpha-gamma coactivation** ensures that the muscle spindle remains active. Activation of gamma motor neurons causes contraction of the ends of the intrafusal fibers. This contraction lengthens the central section of the intrafusal fiber and maintains stretch on the sensory nerve endings. (p. 434; Fig. 13-5)

14. **Golgi tendon organs** are found at the junction of the tendons and muscle fibers. They consist of free nerve endings that wind between collagen fibers. Golgi tendon organs respond to both stretch and contraction of the muscle by causing a reflexive relaxation. (p. 435; Figs. 13-3, 13-6)

15. The synergistic and antagonistic muscles that control a single joint are known as a **myotatic unit**. When one set of muscles in a myotatic unit contracts, the antagonistic muscles must relax through a reflex known as **reciprocal inhibition**. (p. 436; Fig. 13-7)

16. **Flexion reflexes** are polysynaptic reflexes that cause an arm or leg to be pulled away from a painful stimulus. Flexion reflexes are usually accompanied by the **crossed extensor reflex**, a postural reflex that helps maintain balance when one foot is lifted from the ground. (pp. 436–437; Fig. 13-8)

17. **Central pattern generators** are networks of neurons in the CNS that function spontaneously to control certain rhythmic muscle movements. (p. 437)

The Integrated Control of Body Movement

18. Movement can be loosely classified into three categories: reflex movement, voluntary movement, and rhythmic movement. (p. 438; Tbl. 13-2)

19. **Reflex movements** are integrated primarily in the spinal cord. **Postural reflexes** are integrated in the brain stem. (pp. 438–439; Fig. 13-9, Tbl. 13-3)

20. **Voluntary movements** are integrated in the cerebral cortex and can be initiated at will. Learned voluntary movements improve with practice and may even become involuntary, like reflexes. (p. 439; Fig. 13-11)

21. **Rhythmic movements** such as walking are a combination of reflexes and voluntary movements. Rhythmic movements can be sustained by central pattern generators. (p. 439)

22. Motor commands travel from cortex to spinal cord through the **corticospinal tract** and from the **basal nuclei** through extrapyramidal pathways. (p. 441; Fig. 13-12)

23. **Feedforward reflexes** allow the body to prepare for a voluntary movement; feedback mechanisms are used to create a smooth, continuous motion. (p. 440; Fig. 13-13)

Control of Movement in Visceral Muscles

24. Contraction in smooth and cardiac muscles may occur spontaneously or may be controlled by hormones or by the autonomic division of the nervous system. (p. 443)

QUESTIONS

LEVEL ONE Reviewing Facts and Terms

1. All neural reflexes begin with a _____ that activates a receptor.

2. Somatic reflexes involve _____ muscles; _____, or visceral, reflexes are controlled by autonomic neurons.

3. The pathway pattern that brings information from many neurons into a smaller number of neurons is known as _____.

4. When the axon terminal of a modulatory neuron (cell M) terminates close to the axon terminal of a presynaptic cell (cell P) and decreases the amount of neurotransmitter released by cell P, the resulting type of modulation is called _____. (Hint: see p. 272)

5. Autonomic reflexes are also called _____ reflexes. Why?

6. Some autonomic reflexes are spinal reflexes; others are integrated in the brain. List examples of each.

7. What part of the brain transforms emotions into somatic sensation and visceral function? List three autonomic reflexes that are linked to one's emotions.

8. How many synapses occur in the simplest autonomic reflexes? Where do the synapses occur?

9. List the three types of sensory receptors that convey information for muscle reflexes.

10. Because of tonic activity in neurons, a resting muscle maintains a low level of tension known as _____.

11. Stretching an intact skeletal muscle causes sensory neurons to (increase/decrease) their rate of firing, causing the muscle to contract, thereby relieving the stretch. Why is this a useful reflex?

12. Match the organ to all correct statements about it.
    (a) muscle spindle
    (b) Golgi tendon organ
    (c) joint capsule mechanoreceptor

    1. strictly a sensory receptor
    2. has afferent neurons that bring information to the CNS
    3. associated with two types of efferent neurons
    4. conveys information about the relative positioning of bones
    5. innervated by gamma motor neurons
    6. modulates activity in alpha motor neurons

13. The Golgi tendon organ responds to both _____ and _____, although _____ elicits the stronger response. Activation (increases/decreases) muscle contraction via the _____ neuron.

14. The simplest reflex requires a minimum of how many neurons? How many synapses? Give an example.

15. List and differentiate the three categories of movement. Give an example of each.

LEVEL TWO *Reviewing Concepts*

16. What is the purpose of alpha-gamma coactivation? Explain how it occurs.

17. Modulatory neuron M synapses on the axon terminal of neuron P, just before P synapses with the effector organ. If M is an inhibitory neuron, what happens to neurotransmitter release by P? What effect does M's neurotransmitter have on the postsynaptic membrane potential of P?

18. At your last physical, your physician checked your patellar tendon reflex by tapping just below your knee while you sat quietly on the edge of the table. (a) What was she checking when she did this test? (b) What would happen if you were worried about falling off the table and were very tense? Where does this additional input to the efferent motor neurons originate? Are these modulatory neurons causing EPSPs or IPSPs at the spinal motor neuron? (c) Your physician notices that you are tense, and asks you to count backwards from 100 by 3's. Why would this enhance your reflex?

LEVEL THREE *Problem Solving*

19. There are several theories about how presynaptic inhibition works at the cellular level. Use what you have learned about membrane potentials and synaptic transmission to explain how each of the following mechanisms could result in presynaptic inhibition:
    (a) Voltage-gated Ca^{2+} channels in axon terminal are inhibited.
    (b) Cl^- channels in axon terminal open.
    (c) K^+ channels in axon terminal open.

20. Andy is working on improving his golf swing. He must watch the ball, swing the club back and then forward, twist his hips, straighten the left arm, then complete the follow-through, where the club arcs in front of him. What specific parts of the brain are involved in adjusting how hard he hits the ball, keeping all his body parts moving correctly, watching the ball, and then repeating these actions once he has verified that this swing is successful?

21. It's Halloween and you are walking through the scariest Haunted House around. As you turn the corner and enter the dungeon, a skeleton reaches out and grabs your arm. You let out a scream. Your heart rate quickens and you feel the hairs on your arm stand on end. (a) What has just happened to you? (b) Where in the brain is fear processed? What are the functions of this part of the brain? Which branch (somatic or autonomic) of the motor output does it control? What will be the target organs for this response? (c) How is it possible for your hair to stand on end when hair is made of proteins that do not contract? (Hint: see Fig. 10-11.) Given that the autonomic nervous system is mediating this reflex response, what type of tissue do you expect to find attached to hair follicles?

22. Using what you've learned about tetanus and botulinum toxins, make a table to compare the two. In what ways are tetanus and botulinum toxin similar? How are they different?

Explorations

Explorations are also available at *www.physiologyplace.com*

How many times have you heard someone say "I did it without thinking?" In effect, they were telling you that their action was just a reflex response. There are many ways to control the functions of muscles and glands of the body, but a neural reflex is the simplest and the fastest.

Exploration 1: Retraining the Spinal Cord

Can patients with spinal cord injuries learn to walk again? Christopher Reeve, the actor paralyzed in an equestrian accident, is hoping so. Already he has regained some sensory function and the ability to wiggle a few fingers. Researchers are using several different techniques to help spinal cord injury patients. In many of these patients, the spinal reflexes controlling their muscles are not damaged but the ability to control the reflexes is lost when higher levels of the spinal cord are damaged. Locomotor training on treadmill or bicycle and *laufband* therapy, which is supported walking on a treadmill, have both been shown to improve function. Functional electrical stimulation, in which the skeletal muscles are electrically stimulated to contract, has also proved helpful. All three forms of therapy help the spinal cord "remember" its intrinsic circuits so that they do not disappear from lack of use. To learn more about retraining the spinal cord, do a Google search (*www.google.com*) for *locomotor training, laufband therapy,* and *functional electrical stimulation.*

To learn more about experiments on spinal cord repair, see "Spinal cord regeneration," *Science* 273: 451, 1996 July 26.

"Spinal cord repair in adult paraplegic rats: partial restoration of hind limb function," *Science* 273: 510–513, 1996 July 26.

Exploration 2: Parkinson's Disease

To many people, Parkinson's is an old person's disease. While it's true that the average age at onset is 60, the award-winning actor Michael J. Fox was only 30 when he was diagnosed with Parkinson's disease. He was able to hide his condition for seven years but finally let the public know. In 2000 Michael quit his popular sit-com *Spin City* and started the Michael J. Fox Foundation for Parkinson's Research (*www.michaeljfox.org*).

Descriptions of Parkinson-like symptoms appear in ancient Chinese and Indian medical writings, but the disease gets its name from Dr. James Parkinson, a London physician who in 1817 published a description of "the shaking palsy." Not all tremors are Parkinson's disease, however. The most common type of tremor in the United States is *essential tremor*, characterized by shaking in the hands and head particularly. In essential tremor, the uncontrollable shaking is often the only symptom. The cause is unknown but there appears to be a strong genetic influence over who will develop this condition.

To learn more about Parkinson's disease and essential tremor, try a Google search (*www.google.com*) or go to the United States National Library of Medicine website (*http://www.nlm.nih.gov/*).

To read more about Parkinson's and other neuromuscular disorders, see "Understanding Parkinson's Disease," *Scientific American*, 1997 January.

"The enigma of Huntington's Disease," *Scientific American*, 2002 December.

"The basal ganglia and disorders of movement: Pathophysiological mechanisms," *News in Physiological Sciences* 17: 51–55, 2002 April (*http://nips.physiology.org*)

To learn more about the role of the cerebellum in coordinating movement, read "The cerebellum: the brain's engine of agility," *Science* 281: 1588–1590, 1998 Sept 11.

ANSWERS

Answers to Concept Check Questions

Page 431

1. Sensor (sensory receptor) → afferent pathway (sensory neuron) → integrating center (central nervous system) → efferent pathway (autonomic or somatic motor neuron) → effector (muscles, glands, some adipose tissue)

2. With hyperpolarization, the membrane potential becomes more negative and moves farther away from threshold.

Page 434

3. Your map of a stretch reflex should match the steps shown in Figure 13-4b.

Page 436

4. The map of alpha-gamma coactivation should match the steps in Figure 13-5b. The stimulus of muscle contraction is the same for the Golgi tendon reflex. The map then branches to show the steps in Figure 13-6d and e.

Page 436

5. The flexion reflex following stepping on a tack has the same steps as the knee jerk in Figure 13-7, with the added contraction of hip flexor muscles in addition to the quadriceps.

Page 437

6. The crossed extensor reflex shares the same initial steps of the flexion reflex until the CNS. There the crossed extensor reflex follows the diagram shown in Figure 13-8, step 3c.

Page 438

7. When you pick up a weight, alpha and gamma neurons, spindle afferents, and Golgi tendon organ afferents are all active.

8. A stretch reflex is initiated by stretch and causes a reflex contraction. A crossed extensor reflex is a postural reflex initiated by withdrawal from a painful stimulus; the extensor muscles contract but the corresponding flexors are inhibited.

Cardiovascular Physiology

Only in the 17th century did the brain displace the heart as the controller of our actions.

—Mary A. B. Brazier, *A History of Neurophysiology in the 19th Century*, 1988

Immunofluorescent staining of cultured cardiac muscle cells. Actin is pink, nebulin is green.

RUNNING PROBLEM

Myocardial Infarction

At 9:06 A.M., the blood clot that had silently formed in Walter Parker's left coronary artery made its sinister presence known. The 53-year-old advertising executive had arrived at the Dallas Convention Center feeling fine, but suddenly a dull ache started in the center of his chest and he became nauseated. At first he brushed it off as aftereffects of the convention banquet the night before. However, when it persisted, he made his way to the Aid Station. "I'm not feeling very well," he told the medic. "I think it may be indigestion." The medic, on hearing Walter's symptoms and seeing his pale, sweaty face, immediately thought of a heart attack. "Let's get you over to the hospital and get this checked out."

452 461 470 475 480 483

BACKGROUND BASICS

Diffusion: **136** | Striated muscle: **390** | Desmosomes: **66** | Excitation-contraction coupling: **400** | Length-tension relationship in muscle: **406** | Tetanus in skeletal muscle: **406** | Muscle contraction: **396** | Gap junctions: **64** | Catecholamines: **218** | Vagus nerve: **374** | Isometric contraction: **409** |

In the classic movie *Indiana Jones and the Temple of Doom*, the evil priest reaches into the chest of a sacrificial victim and pulls out his heart, still beating. This act was not imagined by some Hollywood scriptwriter—it was taken from rituals of the ancient Mayans, who documented their grisly practice in their carvings and paintings. The heart has been an object of fascination for centuries. But how can this workhorse muscle, which pumps 7,200 liters of blood a day, keep beating outside the body? Before we can answer that question, we must first consider the role of hearts in cardiovascular systems.

As life evolved, simple one-celled organisms began to band together, first into cooperative colonies and then into multicelled organisms. In most multicellular animals, only the surface layer of cells is in direct contact with the environment. This body structure presents a problem for the transfer of oxygen and nutrients from the environment to the interior cells. Diffusion is severely limited by distance [● p. 136], and, in larger animals, oxygen consumption in interior cells exceeds the rate at which oxygen can diffuse from the body surface.

One solution to this problem was the evolutionary development of circulatory systems that move fluid between the body surface and its deepest parts. In simple animals, muscular activity creates fluid flow when the animal moves. More complex animals have muscular pumps called hearts to circulate internal fluid.

In the most efficient circulatory systems, the heart pumps blood through a closed system of vessels. This one-way circuit steers blood flow along a specific route and ensures efficient distribution of gases, nutrients, signal molecules, and wastes. A circulatory system with a heart and blood vessels is known as a **cardiovascular system** [*kardia*, heart + *vasculum*, little vessel].

Although the idea of a closed cardiovascular system that cycles blood in an endless loop seems intuitive to us today, it has not always been so. **Capillaries**, the microscopic vessels in which blood exchanges material with the interstitial fluid, were not discovered until Marcello Malpighi, an Italian anatomist, observed them with a microscope in the middle of the seventeenth century. At that time European medicine was still heavily influenced by the ancient belief that the cardiovascular system distributed both blood and air.

Blood was thought to be made in the liver and distributed throughout the body in the veins. Air went from the lungs to the heart, where it was *digested* and picked up *vital spirits*. From the heart, air was distributed to the tissues through vessels called arteries. Anomalies such as the fact that a cut artery squirted blood rather than air were ingeniously explained by unseen links between arteries and veins that opened with injury.

According to this model of the circulatory system, the tissues consumed all blood delivered to them, and the liver had to continuously synthesize new blood. It took the calculations of William Harvey (1578–1657), court physician to King Charles I of England, to show that the weight of the blood pumped by the heart in a single hour is greater than the weight of the entire body! Once it became obvious that the liver could not make blood as rapidly as the heart pumped it, Harvey looked for an anatomical route that would allow the blood to recirculate rather than be consumed in the tissues. He showed that valves in the heart and veins created a one-way flow of blood, and that veins carried blood back to the heart, not out to the limbs. He also showed that blood returning to the right side of the heart from the body had to go to the lungs before it could go to the left side of the heart.

The results of these studies created a furor among Harvey's contemporaries, leading Harvey to say in a huff that no one under the age of 40 could understand his conclusions. Ultimately, Harvey's work became the foundation of modern cardiovascular physiology. Today, we understand the structure of the cardiovascular system at microscopic and molecular levels that Harvey never dreamed existed. Yet some things have not changed. Even now, with our sophisticated technology, we are searching for *spirits* in the blood, although we call them by names such as *hormone* and *cytokine*.

▶ OVERVIEW OF THE CARDIOVASCULAR SYSTEM

In the simplest terms, a cardiovascular system is a series of tubes (the blood vessels) filled with fluid (blood) and connected to a pump (the heart). Pressure generated in the heart propels blood through the system continuously. The blood picks up oxygen at the lungs and nutrients in the intestine, delivering them to cells, while simultaneously removing cellular wastes for excretion. In addition, the cardiovascular system plays an important role in cell-to-cell communication and in defending the body against foreign invaders. This chapter focuses on an overview of the cardiovascular system and on the heart as a pump. We will examine the properties of the blood vessels and the homeostatic controls that regulate blood flow and blood pressure in Chapter 15.

The Cardiovascular System Transports Material Throughout the Body

The primary function of the cardiovascular system is the transport of materials between all parts of the body. Substances transported by the cardiovascular system can be divided into nutrients, water, and gases that enter the body from the external environment, materials that move from cell to cell within the body, and wastes that the cells eliminate (Table 14-1).

Oxygen enters the blood in the lungs. Nutrients and water are absorbed across the intestinal epithelium. Once in the body, the circulation distributes these materials. A steady supply of oxygen for the cells is particularly important because many cells deprived of oxygen become irreparably damaged within a short period of time.

For example, about 5–10 seconds after blood flow to the brain is stopped, a person loses consciousness. If oxygen delivery

Table 14-1: Transport in the Cardiovascular System

SUBSTANCE MOVED	FROM	TO
Materials entering the body		
Oxygen	Lungs	All cells
Nutrients and water	Intestinal tract	All cells
Materials moved from cell to cell		
Wastes	Some cells	Liver for processing
Immune cells, antibodies, clotting proteins	Present in blood continuously	Available for any cell that needs them
Hormones	Endocrine cells	Target cells
Stored nutrients	Liver and adipose tissue	All cells
Materials leaving the body		
Metabolic wastes	All cells	Kidneys
Heat	All cells	Skin
Carbon dioxide	All cells	Lungs

stops for 5–10 minutes, permanent brain damage results. Neurons of the brain have a very high rate of oxygen consumption and are unable to meet their metabolic need for ATP by using anaerobic pathways. Because of the brain's sensitivity to *hypoxia* [*hypo-*, low + *-oxia*, oxygen], homeostatic controls do everything possible to maintain cerebral blood flow, even if it means depriving cells in other tissues of oxygen.

Cell-to-cell communication is a key function of the cardiovascular system. For example, hormones secreted by endocrine glands are carried to their targets. Nutrients such as glucose from the liver or fatty acids released by adipose tissue are transported to metabolically active cells. Finally, the defense team of white blood cells and antibodies patrols the circulation to intercept foreign invaders.

The cardiovascular system also picks up wastes released by cells. The blood transports carbon dioxide and metabolic wastes to the lungs and the kidneys for excretion. Some waste products are transported to the liver for processing before they are excreted in the urine and feces. Heat also circulates through the blood, moving from the body core to the surface, where it dissipates.

The Cardiovascular System Consists of the Heart and Blood Vessels

The cardiovascular system is composed of the heart, the blood vessels (also known as the vasculature), and the cells and plasma of the blood. Blood vessels that carry blood away from the heart are called **arteries**. Blood vessels that return blood to the heart are called **veins**.

As blood moves through the cardiovascular system, a system of **valves** in the heart and veins ensures that blood flows in one direction. Like the turnstiles at an amusement park, the valves keep the blood from reversing its direction of flow. Figure 14-1 ● is a schematic diagram that shows these components and the route that blood flow follows through the body.

The heart is divided by a central wall, or **septum**, into left and right halves. Each half consists of an **atrium** [*atrium*, central room; plural *atria*], which receives blood returning to the heart from the blood vessels, and a **ventricle** [*ventriculus*, belly], which pumps blood out into the blood vessels. The right side of the heart receives blood from the tissues and sends it to the lungs for oxygenation. The left side of the heart receives newly oxygenated blood from the lungs and pumps it to tissues throughout the body.

Starting in the right atrium, trace the path taken by blood as it flows through the cardiovascular system. Note that in Figure 14-1, blood in the right side of the heart is colored blue. This is a convention used to show blood from which the tissues have extracted oxygen. Although this blood is often described as *deoxygenated*, it is not completely devoid of oxygen. It simply has less oxygen than blood going from the lungs to the tissues.

In living people, well-oxygenated blood is bright red; low-oxygen blood becomes darker red. Under some conditions, low-oxygen blood can impart a bluish color to certain areas of the skin, such as around the mouth and under the fingernails. This condition, known as *cyanosis* [*kyanos*, dark blue], is the reason blue is used in drawings to indicate blood with lower oxygen content.

From the right atrium, blood flows into the right ventricle of the heart. From there it is pumped through the **pulmonary arteries** to the lungs, where it is oxygenated. Note the color change from blue to red, indicating higher oxygen content. From the lungs, blood travels back to the left side of the heart through the **pulmonary veins**. The blood vessels that go from the right ventricle to the lungs and back to the left atrium are known collectively as the **pulmonary circulation**.

Blood from the lungs reenters the heart at the **left atrium** and passes into the **left ventricle**. Blood pumped out of the left ventricle enters the large artery known as the **aorta**. The aorta branches into a series of smaller and smaller arteries that finally lead into networks of capillaries.

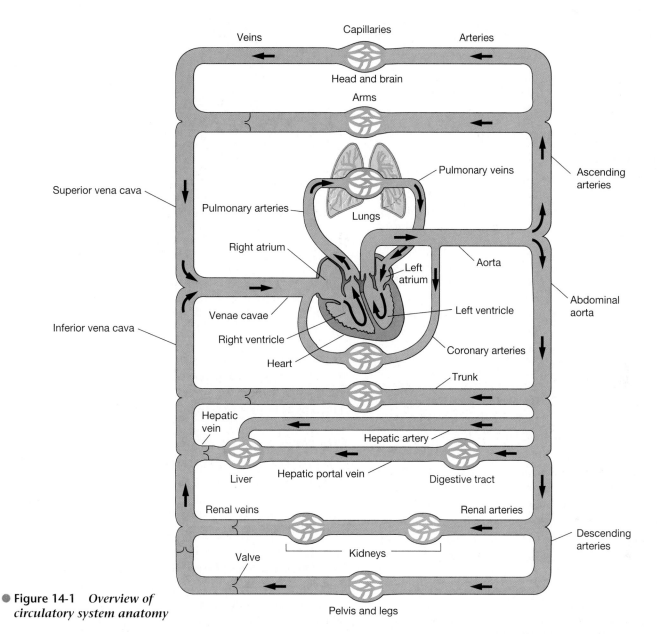

Veins | Capillaries | Arteries

Head and brain

Arms

Pulmonary veins

Ascending arteries

Superior vena cava

Pulmonary arteries

Lungs

Aorta

Abdominal aorta

Right atrium

Left atrium

Venae cavae

Left ventricle

Inferior vena cava

Right ventricle

Coronary arteries

Heart

Trunk

Hepatic vein

Hepatic artery

Hepatic portal vein

Digestive tract

Liver

Renal veins

Renal arteries

Descending arteries

Valve

Kidneys

Pelvis and legs

● **Figure 14-1** *Overview of circulatory system anatomy*

Capillaries are the tiny thin-walled vessels in which material is exchanged between the blood and the tissues. Notice the color change from red to blue in the capillaries, showing that oxygen has left the blood and diffused into the tissues.

Blood in capillaries flows into the venous side of the circulation, moving from small veins into larger and larger veins. The veins from the upper part of the body join together to form the **superior vena cava**. Those from the lower part of the body form the **inferior vena cava**. The two **venae cavae** empty into the right atrium. The blood vessels that go from the left side of the heart to the tissues and back to the right side of the heart are collectively known as the **systemic circulation**.

Return to the aorta and follow its divisions. The first branch represents the coronary arteries that nourish the heart muscle itself. Blood from *coronary arteries* flows into capillaries, then into *coronary veins* that empty directly into the right atrium at the *coronary sinus*. Ascending branches of the aorta go to the head, arms, and brain. The abdominal aorta supplies blood to the

trunk, the legs, and the internal organs such as the kidneys (*renal arteries*), liver (*hepatic artery*), and digestive tract.

Notice two special regions of the circulation. One is the blood supply to the digestive tract and liver. Both regions receive well-oxygenated blood through their own arteries, but, in addition, blood leaving the digestive tract goes directly to the liver by means of the *hepatic portal vein*. The liver is an important site for nutrient processing and plays a major role in the detoxification of foreign substances. Most nutrients absorbed in the intestine are routed directly to the liver, allowing that organ to process the material before it is released into the general circulation.

The two capillary beds of the digestive tract and liver, joined by the hepatic portal vein, are examples of a *portal system*. Another portal system, discussed earlier, is the hypothalamic-hypophyseal portal system that connects the hypothalamus and the anterior pituitary [● p. 225]. The third major portal system is in the kidneys, where two capillary beds are connected in series. You will learn more about this special vascular arrangement in Chapter 19.

▶ PRESSURE, VOLUME, FLOW, AND RESISTANCE

If you ask people why blood flows through the cardiovascular system, many of them respond, "So that oxygen and nutrients can get to all parts of the body." Although this is true, it is a teleological answer, one that describes the purpose of blood flow. In physiology, we are also concerned with how blood flows: the mechanisms or forces that create blood flow.

A simple mechanistic answer to "Why does blood flow?" is that liquids and gases flow down a **pressure gradient (ΔP)** from regions of higher pressure to regions of lower pressure. Therefore, blood can flow in the cardiovascular system only if one region develops higher pressure than other regions.

In humans, high pressure is created in the chambers of the heart when it contracts. Blood flows out of the heart (the region of highest pressure) into the closed loop of blood vessels (a region of lower pressure). As blood moves through the system, pressure is lost because of friction between the fluid and the blood vessel walls. Consequently, pressure falls continuously as blood moves farther from the heart (Fig. 14-2 ●). The highest pressure in the vessels of the circulatory system is found in the aorta and systemic arteries as they receive blood from the left ventricle. The lowest pressure is in the venae cavae, just before they empty into the right atrium.

In this section we review the laws of physics that explain the interaction of pressure, volume, flow, and resistance in the cardio-

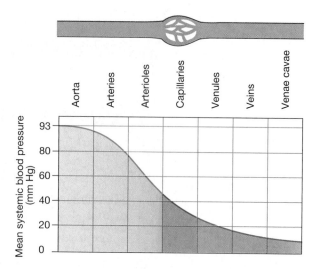

● **Figure 14-2** *Pressure gradient in the blood vessels* The mean blood pressure of the systemic circulation ranges from a high of 93 mm Hg (millimeters of mercury) in the arteries to a low of a few mm Hg in the venae cavae.

vascular system. Many of these principles apply broadly to the flow of all types of liquids and gases, including that of air in the respiratory system. However, in this chapter we will focus on the flow of blood and its relevance to the function of the heart.

The Pressure of Fluid in Motion Decreases over Distance

Pressure in fluids is the force exerted by a fluid on the container holding the fluid. If the fluid is not moving, the pressure is called **hydrostatic pressure** (Fig. 14-3a ●), and force is exerted equally in all directions. For example, a column of fluid in a tube exerts hydrostatic pressure on the floor and sides of the tube. In the heart and blood vessels, pressure is commonly measured in millimeters of mercury (mm Hg, or torr).* One millimeter of mercury pressure is equivalent to the hydrostatic pressure exerted by a 1-mm-high column of mercury on an area 1 cm².

In a system in which fluid is flowing, pressure falls over distance as energy is lost because of friction (Fig. 14-3b). In addition, the pressure exerted by moving fluid has two components: (1) a dynamic, flowing component that represents the kinetic energy of the system and (2) a lateral component that represents the hydrostatic pressure (potential energy) exerted on the walls of the system. Pressure within our cardiovascular system is usually called hydrostatic pressure, although it is a system in which fluid is in motion. Some textbooks are beginning to replace the term *hydrostatic pressure* with the term *hydraulic pressure*. Hydraulics is the study of fluid in motion.

Compressing a Fluid Raises Its Pressure

If the walls of a fluid-filled container contract, the pressure exerted on the fluid in the container increases. You can demonstrate

---

*1 mm Hg = 1 torr. Some physiological literature report pressures in centimeters of water: 1 cm H_2O = 0.74 mm Hg.

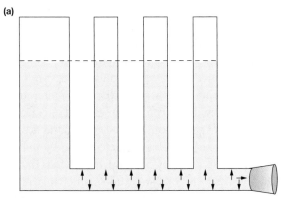

(a)

Hydrostatic pressure is the pressure exerted on the walls of the container by the fluid within the container. Hydrostatic pressure is proportional to the height of the water column.

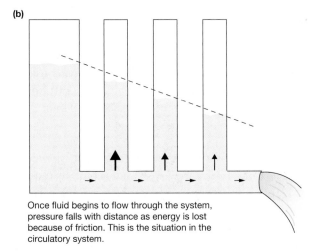

(b)

Once fluid begins to flow through the system, pressure falls with distance as energy is lost because of friction. This is the situation in the circulatory system.

● **Figure 14-3** *Pressure differences of static and flowing fluid*

this principle by filling a balloon with water and squeezing it in your hand. Water is not very compressible, so the pressure applied to the balloon is transmitted throughout the fluid. As you squeeze, higher pressure in the fluid causes parts of the balloon to bulge. If the pressure becomes high enough, the stress on the balloon will cause it to pop. The water volume inside the balloon has not changed, but its pressure has increased.

In the human heart, contraction of the ventricles on blood that fills them is like squeezing a water balloon: the pressure created by the contracting muscle is transferred to the blood. This high-pressure blood then flows out of the ventricle and into the blood vessels, displacing lower-pressure blood already in the vessels. The pressure created within the ventricles is called the **driving pressure** because it is the force that drives blood through the blood vessels.

When the muscular walls of a fluid-filled container relax, the pressure of the fluid inside the container decreases. Thus, when the heart relaxes and expands, pressure in the chambers falls.

Pressure changes can also take place in the blood vessels. If blood vessels dilate, blood pressure inside them falls. If blood vessels constrict, blood pressure increases. Volume changes of the blood vessels and heart are major factors that influence blood pressure within the cardiovascular system.

Blood Flows from an Area of Higher Pressure to One of Lower Pressure

As stated earlier, blood flow through the circulatory system requires a pressure gradient. This pressure gradient, ΔP, is the difference in pressure between two ends of a tube through which fluid flows (Fig. 14-4a ●). Flow through the tube is directly proportional to (∝) the pressure gradient (ΔP):

$$\text{Flow} \propto \Delta P \quad \text{where } \Delta P = P_1 - P_2 \quad (1)$$

This relationship says that the higher the pressure gradient, the greater the flow of fluid.

A pressure gradient is not the same thing as the absolute pressure in the system. For example, the tube in Figure 14-4b has an absolute pressure of 100 mm Hg at each end. However, because there is no pressure gradient between the two ends of the tube, there is no flow through the tube.

On the other hand, two identical tubes can have very different absolute pressures but the same flow (Fig. 14-4c). The first tube has a hydrostatic pressure of 100 mm Hg at one end and 75 mm Hg at the other end, with a pressure gradient between the ends of the tube equal to 25 mm Hg. The identical bottom tube has hydrostatic pressure of 40 mm Hg at one end and 15 mm Hg at the other end. This tube has lower absolute pressure all along its length, but the same pressure gradient as the top tube, 25 mm Hg. Because the pressure difference in the two tubes is identical, the fluid flow through the tubes will be the same.

(a) Fluid flows only if there is a positive pressure gradient.

Higher P ——— Flow ——→ Lower P

Flow →

P_1 P_2

$P_1 - P_2 = \Delta P$

(b) No pressure gradient, so no flow

100 mm Hg   100 mm Hg

ΔP = 0, so no flow

(c) Flow depends on ΔP, not absolute P

100 mm Hg   75 mm Hg

Flow →

ΔP = 100 - 75 = 25 mm Hg

flow is equal

40 mm Hg   15 mm Hg

Flow →

ΔP = 40 - 15 = 25 mm Hg

KEY

P = Pressure
ΔP = Pressure gradient

● **Figure 14-4** *Fluid flow through a tube*

Resistance Opposes Flow

In an ideal system, a substance in motion would remain in motion. However, no system is ideal because all movement creates friction. Just as a ball rolled across the ground loses energy to friction, blood flowing through blood vessels encounters friction from the walls of the blood vessels and from cells within the blood rubbing against each other as they flow.

The tendency of the cardiovascular system to oppose blood flow is called its **resistance** to flow. Resistance (R) is a term that most of us understand already from everyday life. We speak of people being resistant to change or taking the path of least resistance. This concept translates well to the circulatory system because blood flow will also take the path of least resistance. An increase in the resistance of a blood vessel means a decrease in flow through that vessel. We can express that relationship as

$$\text{Flow} \propto 1/R \qquad (2)$$

Flow is inversely proportional to resistance. If resistance increases, flow decreases. If resistance decreases, flow increases.

What parameters determine resistance? For fluid flowing through a tube, resistance is influenced by three components: the length of the tube (L), the radius of the tube (r), and the viscosity (thickness) of the fluid flowing through the tube (η, eta). The following equation, derived by the French physician Jean Leonard Marie Poiseuille and known as **Poiseuille's Law**, shows the relationship between these factors:

$$R = 8L\eta/\pi r^4 \qquad (3)$$

Because the value of $8/\pi$ is a constant, the relationship can be rewritten as

$$R \propto L\eta/r^4 \qquad (4)$$

This expression says that (1) resistance increases as the length of the tube increases, (2) resistance increases as the viscosity of the fluid increases, but (3) resistance decreases as the radius increases.

To remember these relationships, think of drinking through a straw. You do not need to suck as hard on a short straw as on a long one (resistance increases with length). Drinking water through a straw is easier than drinking a thick milkshake (resistance increases with viscosity). And drinking the milkshake through a big fat straw is much easier than through a skinny cocktail straw (resistance increases as radius decreases).

How significant are length, viscosity, and radius to blood flow in a normal individual? The length of the systemic circulation is determined by the anatomy of the system and is essentially constant. The **viscosity** of blood is determined by the ratio of red blood cells to plasma and by how much protein is in the plasma. Normally, viscosity is constant, and small changes in either length or viscosity have little effect on resistance. This leaves changes in the radius of the blood vessels as the main contributor to variable resistance in the systemic circulation.

Let's return to the example of the straw and the milkshake to illustrate how changes in radius affect resistance. If we assume that the length of the straw and the viscosity of the milkshake do not change, this system is similar to the circulatory system—the radius of the tube has the greatest effect on resistance. If we consider only resistance (R) and radius (r) from equation 4, the relationship between resistance and radius can be expressed as follows:

$$R \propto 1/r^4 \qquad (5)$$

If the skinny straw has a radius of 1, its resistance is proportional to $1/1^4$, or 1. If the larger straw has a radius of 2, its resistance is $1/2^4$, or 1/16th, that of the skinny straw. Because flow is inversely proportional to resistance, flow increases 16-fold when the radius doubles (Fig. 14-5 ●).

As you can see from this example, a small change in the radius of a tube will have a large effect on the flow of a liquid through that tube. A small change in the radius of a blood vessel will have a large effect on the resistance to blood flow. A decrease in blood vessel diameter is known as **vasoconstriction** [*vas*, a vessel or duct]. An increase in blood vessel diameter is called **vasodilation**. Vasoconstriction will decrease blood flow through a vessel. Vasodilation will increase blood flow through a vessel.

In summary, by combining equations 1 and 2 above, we get the following equation:

$$\text{Flow} \propto \Delta P/R \qquad (6)$$

The flow of blood within the circulatory system is directly proportional to the pressure gradient in the system and inversely propor-

● **Figure 14-5** *The role of radius in determining resistance to flow*

Radius of B = 2

Radius of A = 1

Volume of B = 16

Volume of A = 1

Resistance $\propto \dfrac{1}{\text{radius}^4}$		Flow $\propto \dfrac{1}{\text{resistance}}$	
Tube A	Tube B	Tube A	Tube B
$R \propto \dfrac{1}{1^4}$	$R \propto \dfrac{1}{2^4}$	$\text{Flow} \propto \dfrac{1}{1}$	$\text{Flow} \propto \dfrac{1}{\frac{1}{16}}$
$R \propto 1$	$R \propto \dfrac{1}{16}$	$\text{Flow} \propto 1$	$\text{Flow} \propto 16$

tional to the resistance of the system to flow. If the driving pressure (ΔP) remains constant, then flow will vary inversely with resistance.

3. Which is more important for determining flow through a tube: absolute pressure or the pressure gradient?

4. The two identical tubes below have the pressures shown at each end. Which tube has the greater flow? Defend your choices.

5. All four tubes below have the same driving pressure. Which tube has the greatest flow? Which has the least flow? Defend your choices.

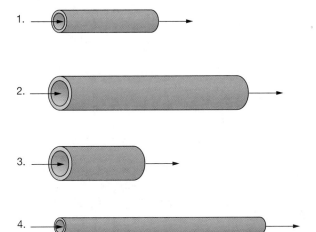

Velocity of Flow Depends on the Flow Rate and the Cross-Sectional Area

The word *flow* is often used imprecisely in cardiovascular physiology, leading to confusion. Flow usually means **flow rate**, the volume of blood that passes one point in the system per unit time. In the circulation, flow is expressed in liters per minute (L/min) or milliliters per minute (mL/min). For instance, blood flow through the aorta in a 70-kg man at rest is about 5 L/min.

Flow rate should not be confused with **velocity of flow**, the distance a fixed volume of blood will travel in a given period of time. Velocity of flow is a measure of *how fast* blood flows past a point. Flow rate measures *how much* (volume) blood flows past a point in a given period of time. For example, look through the open door at the hallway outside your classroom. The number of people passing the door in one minute is the flow rate of people through the hallway. How quickly those people are walking past the door is their velocity.

In a tube of fixed size, velocity of flow is directly related to flow rate. In a tube of variable diameter, if flow rate is constant, blood velocity will vary inversely with the diameter. In other words, velocity will be faster through narrow sections and slower in wider sections. The relationship between flow rate (Q), cross-sectional area of the tube (A), and velocity of flow (v) is expressed by the following equation:

$$\text{Velocity} = \text{flow rate/cross-sectional area} \qquad (7)$$

Figure 14-6 ● shows how the velocity of flow varies as the cross-sectional area of the tube changes. The vessel in the figure has two widths: narrow, with a small cross-sectional area of 1 cm², and wide, with a large cross-sectional area of 12 cm². The flow rate is identical in both parts of the vessel: 12 cm³ per minute.* This flow rate means that in one minute, 12 cm³ of fluid flows past point X in the narrow section, and 12 cm³ flows past point Y in the wide section.

But *how fast* does the fluid need to flow to accomplish that rate? According to equation 7, the velocity of flow at point X is 12 cm/min, but at point Y it is only 1 cm/min. Thus, fluid flows more rapidly through narrow sections of a tube than through wide sections.

To see this principle in action, watch a leaf as it floats down a stream or gutter. Where the stream is narrow, the leaf moves rapidly, carried by the fast velocity of the water. In sections where the stream widens into a pool, the velocity of the water drops and the leaf meanders more slowly.

Table 14-2 summarizes the factors that influence blood flow and velocity of flow. In the remainder of this chapter you will learn how the heart generates pressure gradients that create blood flow.

_____

*1 cm³ = 1 mL

The narrower the vessel, the faster the velocity of flow.

Velocity (v) = $\dfrac{\text{Flow rate (Q)}}{\text{Cross-sectional area (A)}}$	
At point X	**At point Y**
$v = \dfrac{12\ \text{cm}^3/\text{min}}{1\ \text{cm}^2}$	$v = \dfrac{12\ \text{cm}^3/\text{min}}{12\ \text{cm}^2}$
v = 12 cm/min	v = 1 cm/min

● **Figure 14-6** *Flow rate versus velocity of flow*

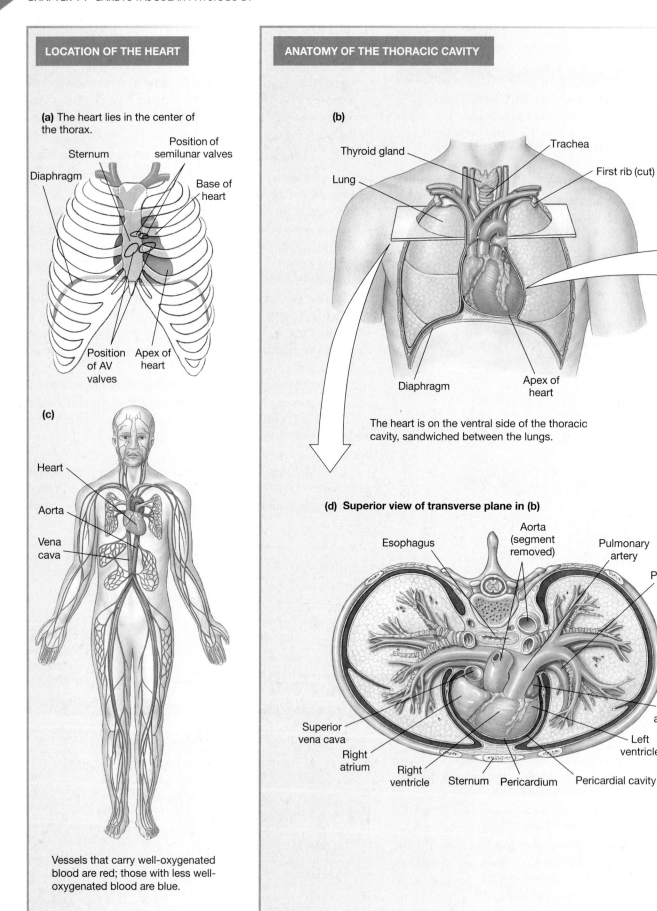

LOCATION OF THE HEART

(a) The heart lies in the center of the thorax.

- Sternum
- Diaphragm
- Position of semilunar valves
- Base of heart
- Position of AV valves
- Apex of heart

(c)

- Heart
- Aorta
- Vena cava

Vessels that carry well-oxygenated blood are red; those with less well-oxygenated blood are blue.

ANATOMY OF THE THORACIC CAVITY

(b)

- Thyroid gland
- Lung
- Trachea
- First rib (cut)
- Diaphragm
- Apex of heart

The heart is on the ventral side of the thoracic cavity, sandwiched between the lungs.

(d) Superior view of transverse plane in (b)

- Esophagus
- Aorta (segment removed)
- Pulmonary artery
- Pulmonary vein
- Superior vena cava
- Right atrium
- Right ventricle
- Sternum
- Pericardium
- Pericardial cavity
- Left atrium
- Left ventricle

● **Figure 14-7** *ANATOMY SUMMARY The Heart*

456

STRUCTURE OF THE HEART

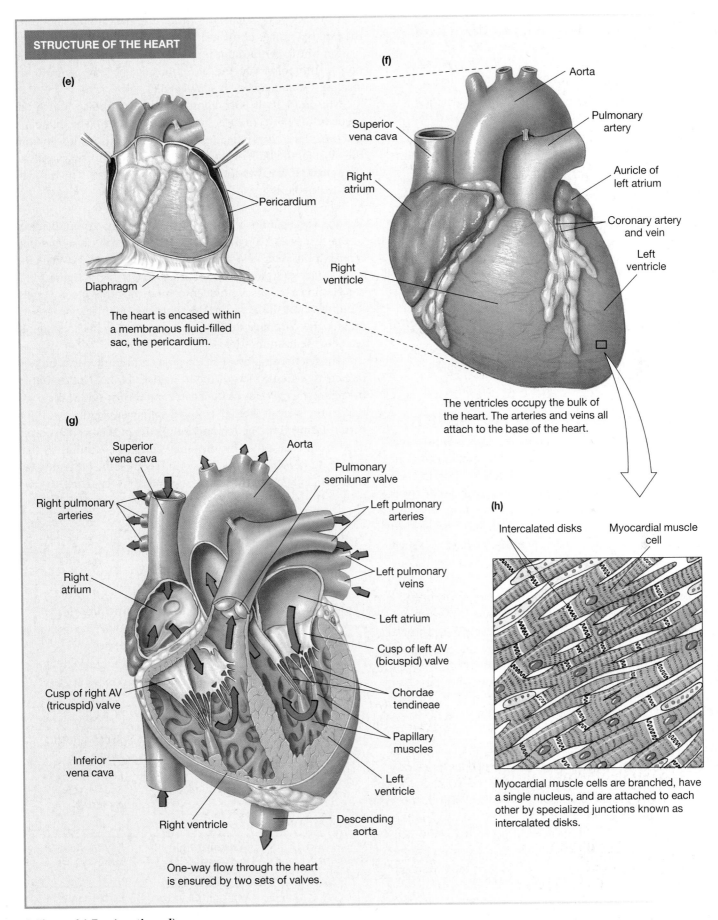

(e)

Pericardium

Diaphragm

The heart is encased within a membranous fluid-filled sac, the pericardium.

(f)

Aorta

Pulmonary artery

Superior vena cava

Right atrium

Auricle of left atrium

Coronary artery and vein

Left ventricle

Right ventricle

The ventricles occupy the bulk of the heart. The arteries and veins all attach to the base of the heart.

(g)

Superior vena cava

Aorta

Pulmonary semilunar valve

Right pulmonary arteries

Left pulmonary arteries

Right atrium

Left pulmonary veins

Left atrium

Cusp of left AV (bicuspid) valve

Cusp of right AV (tricuspid) valve

Chordae tendineae

Papillary muscles

Inferior vena cava

Left ventricle

Right ventricle

Descending aorta

One-way flow through the heart is ensured by two sets of valves.

(h)

Intercalated disks

Myocardial muscle cell

Myocardial muscle cells are branched, have a single nucleus, and are attached to each other by specialized junctions known as intercalated disks.

● **Figure 14-7** *(continued)*

Table 14-2: Rules for the Cardiovascular System

1. Blood flows if a pressure gradient is present. Blood flows from areas of higher pressure to areas of lower pressure. (Fig. 14-4)

2. Blood flow is opposed by the resistance of the system. The three factors that affect resistance are length of the system, radius of the blood vessels, and viscosity of blood. (Fig. 14-5)

3. Flow rate is the volume of blood that passes one point in the system per unit time. It is usually expressed in liters or milliliters per minute (L/min or mL/min). Flow rate is determined by the pressure gradient and by the resistance of the system.

4. Velocity of flow is how far a volume of blood will travel in a given period of time. It is usually expressed as centimeters per minute (cm/min) or millimeters per second (mm/sec). The primary determinant of velocity of flow (when flow rate is constant) is the total cross-sectional area of the vessel. (Fig. 14-6)

CONCEPT CHECK
Answers: p. 488

6. Two canals in Amsterdam are identical in size, but the water flows faster through one than through the other. Which canal has the highest flow rate?

▶ CARDIAC MUSCLE AND THE HEART

To ancient civilizations, the heart was more than a pump—it was *the seat of the mind*. When ancient Egyptians mummified their dead, they removed most of the viscera, but left the heart in place so that the gods could weigh it as an indicator of the owner's worthiness. Aristotle characterized the heart as the most important organ of the body, as well as *the seat of intelligence*. We can still find evidence of these ancient beliefs in modern expressions such as *heartfelt emotions*. The link between the heart and mind is one that is still explored today.

The heart is the workhorse of the body, a muscle that contracts continually, resting only in the milliseconds pause between beats. By one estimate, the heart performs work equivalent to lifting a 5-pound weight up one foot, once a minute. The energy demands of this work require a continuous supply of nutrients and oxygen to the heart muscle.

The Heart Has Four Chambers

The heart is a muscular organ, about the size of a fist, that lies in the center of the *thoracic cavity* (see Anatomy Summary, Fig. 14-7a, b, d ●). The pointed *apex* of the heart angles down to the left side of the body, while the broader *base* lies just behind the breastbone, or *sternum*. Because we usually associate the word *base* with the bottom, remember that the base of a cone is the broad end and the apex is the pointed end. The heart can be thought of as an inverted cone with apex down and base up. Within the thoracic cavity, the heart lies on the ventral side, sandwiched between the two lungs, with its apex resting on the diaphragm (Fig. 14-7b).

The heart is encased in a tough membranous sac, the **pericardium** [*peri*, around + *kardia*, heart] (Fig. 14-7d, e). A thin

layer of clear pericardial fluid inside the pericardium lubricates the external surface of the heart as it beats within the sac. Inflammation of the pericardium (*pericarditis*) may reduce this lubrication to the point that the heart rubs against the pericardium, creating a sound known as a *friction rub*.

The heart itself is composed mostly of cardiac muscle, or **myocardium** [*myo*, muscle + *kardia*, heart], covered by thin outer and inner layers of epithelium and connective tissue. If seen from the outside, the bulk of the heart is the thick muscular walls of the ventricles, the two lower chambers (Fig. 14-7f). The thinner-walled atria lie above the ventricles.

The major blood vessels all emerge from the top of the heart. The aorta and pulmonary trunk (artery) take blood from the heart to the tissues and lungs, respectively. The venae cavae and pulmonary veins return blood to the heart (see Table 14-3). When the heart is viewed from the front (anterior view), as in Figure 14-7f, the pulmonary veins are hidden behind the other blood vessels. Running across the surface of the ventricles are three grooves containing the **coronary arteries** and **coronary veins** that supply blood to the heart muscle.

The relationship between the atria and ventricles can be seen in a cross-sectional view of the heart (Fig. 14-7g). Notice that in the figure the *right* side of the heart is on the *left* side of the page; that is, the heart is labeled as if you were viewing the heart of a person facing you. The left and right sides of the heart are separated from each other by the interventricular septum, so that blood on one side does not mix with blood on the other side. Although blood flow in the left and right heart is separated, the two sides of the heart contract in a coordinated fashion. First the atria contract together, then the ventricles contract.

Blood flows from veins into the atria, thin-walled upper chambers that serve as a reception area. From there, blood passes

Table 14-3: The Heart and Major Blood Vessels

Blue type indicates structures containing blood with lower oxygen content; red type indicates well-oxygenated blood.

	RECEIVES BLOOD FROM	SENDS BLOOD TO
Heart		
Right atrium	Venae cavae	Right ventricle
Right ventricle	Right atrium	Lungs
Left atrium	Pulmonary veins	Left ventricle
Left ventricle	Left atrium	Body except for lungs
Vessels		
Venae cavae	Systemic veins	Right atrium
Pulmonary trunk (artery)	Right ventricle	Lungs
Pulmonary vein	Veins of the lungs	Left atrium
Aorta	Left ventricle	Systemic arteries

through one-way valves into the ventricles, the pumping chambers. Blood leaves the heart via the pulmonary trunk from the right ventricle and via the aorta from the left ventricle. A second set of valves guards the exits of the ventricles so that blood cannot flow back into the heart once it has been ejected.

Notice that blood enters the ventricles at the top of the chamber but also leaves at the top (Fig. 14-7g). During development, the tubular embryonic heart twists back on itself. This puts the arteries (through which blood leaves) close to the top of the ventricles (Fig. 14-8 ●). Functionally, this means that the ventricles must contract from the bottom up so that blood is squeezed out of the top.

(a) Age: embryo, day 25. The heart is a single tube.

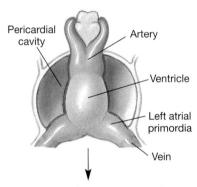

(b) By four weeks of development, the atria and ventricles can be distinguished. The heart begins to twist so that the atria move on top of the ventricles.

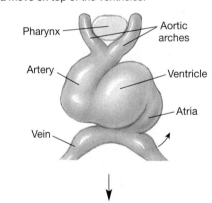

(c) Age: one year (arteries not shown)

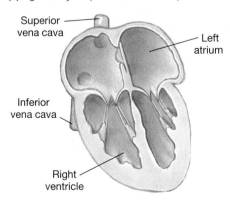

● **Figure 14-8** *Embryological development of the heart*

The heart contains four fibrous connective tissue rings that surround the openings to the major arteries and the openings between the chambers (Fig. 14-7a). These rings form both the origin and insertion for the cardiac muscle, an arrangement that pulls the apex and base of the heart together when the ventricles contract.

In addition, the fibrous connective tissue acts as an insulator, blocking most transmission of electrical signals between the atria and the ventricles. This arrangement ensures that the electrical signal can be directed through special cells to the apex of the heart for the bottom-to-top contraction.

Heart Valves Ensure One-Way Flow in the Heart

As the arrows in Figure 14-7g indicate, blood flows through the heart in one direction. The heart valves ensure one-way flow, with one set between the atria and ventricles and the second set between the ventricles and the arteries. Although the two sets of valves are very different in structure, they serve the same function, preventing the backward flow of blood.

The passageways between the atria and ventricles are guarded by the **atrioventricular**, or **AV**, **valves** (Fig. 14-7g). These valves are thin flaps of tissue attached at their base to a ring of connective tissue. The flaps are slightly thickened at the edge and are attached on the ventricular side to collagenous cords, the **chordae tendineae** (Fig. 14-9b, d ●). The opposite ends of the cords are tethered to moundlike extensions of ventricular muscle known as the **papillary muscles** [*papilla*, nipple]. Most of the chordae attach to the edges of the valve flaps, but some attach to the middle and the base.

When the ventricle contracts, blood pushes against the bottom side of the AV valve and forces it upward into a closed position (Fig. 14-9b). The tendons prevent the valve from being pushed back into the atrium, just as the struts on an umbrella keep the umbrella from turning inside out in a high wind. Occasionally, the tendons fail and the valve is pushed back into the atrium during ventricular contraction, an abnormal condition known as *prolapse*.

The AV valves are not identical. The valve that separates the right atrium and right ventricle has three flaps and is called the **tricuspid valve** [*cuspis*, point] (Fig. 14-9a). The valve between the left atrium and left ventricle has only two flaps and is called the **bicuspid valve**. The bicuspid is also called the **mitral valve** because of its resemblance to the tall headdress, known as a miter, worn by popes and bishops. You can match valves to the proper side of the heart by remembering that the Right Side has the Tricuspid.

The second set of valves lies at the openings from the ventricles into the arteries (aorta and pulmonary trunk). These two valves prevent blood pumped out of the heart from returning to it (Fig. 14-9d). They are called the **semilunar valves** because of their crescent-moon shape in cross section. They are referred to as the **aortic valve** or **pulmonary valve** according to their location. Both semilunar valves have three cuplike leaflets that fill with blood and snap closed when backward pressure is exerted on them (Fig. 14-9c). Because of their shape, the semilunar valves do not need connective tendons as the AV valves do.

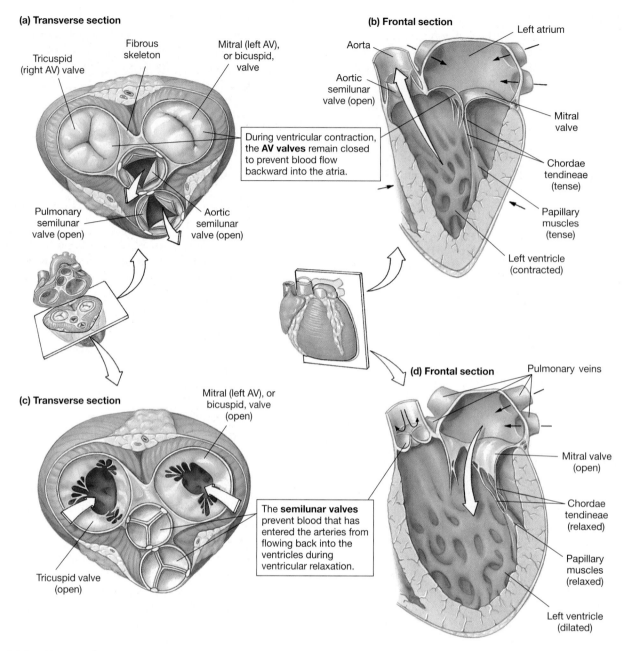

(a) Transverse section

Tricuspid (right AV) valve

Fibrous skeleton

Mitral (left AV), or bicuspid, valve

During ventricular contraction, the **AV valves** remain closed to prevent blood flow backward into the atria.

Pulmonary semilunar valve (open)

Aortic semilunar valve (open)

(b) Frontal section

Aorta

Left atrium

Aortic semilunar valve (open)

Mitral valve

Chordae tendineae (tense)

Papillary muscles (tense)

Left ventricle (contracted)

(c) Transverse section

Mitral (left AV), or bicuspid, valve (open)

Tricuspid valve (open)

The **semilunar valves** prevent blood that has entered the arteries from flowing back into the ventricles during ventricular relaxation.

(d) Frontal section

Pulmonary veins

Mitral valve (open)

Chordae tendineae (relaxed)

Papillary muscles (relaxed)

Left ventricle (dilated)

● **Figure 14-9** *Heart valves* Views (a) and (c) show the valves as viewed from the atria and from inside the arteries.

7. What prevents electrical signals from passing through the connective tissue within the heart?

8. Trace a drop of blood from the superior vena cava to the aorta, naming all structures the drop of blood encounters.

9. What is the function of the AV valves? What happens to blood flow if one of these valves fails?

Cardiac Muscle Cells Contract Without Nervous Stimulation

The bulk of the heart is composed of cardiac muscle cells, or myocardium. Most cardiac muscle is contractile, but about 1% of the myocardial cells are specialized to generate action potentials spon-

taneously. These cells are responsible for a unique property of the heart: its ability to contract without any outside signal.

As mentioned in the introduction to this chapter, Spanish explorers in the New World wrote of witnessing human sacrifices in which hearts that had been torn from the chests of living victims and held aloft continued to beat for minutes. The heart can contract without a connection to other parts of the body because the signal for contraction is *myogenic*, originating within the heart muscle itself.

The signal for myocardial contraction comes not from the nervous system but from specialized myocardial cells known as **autorhythmic cells**. The autorhythmic cells are also called **pacemakers** because they set the rate of the heartbeat. Myocardial autorhythmic cells are anatomically distinct from contractile cells: they are smaller and contain few contractile fibers. Because

RUNNING PROBLEM

When Walter arrived at the University of Texas Southwestern Medical Center emergency room, one of the first tasks was to determine whether he had actually had a heart attack. Walter's vital signs (pulse and breathing rates, blood pressure, temperature) were taken. He was given aspirin and heparin to decrease blood clotting and nitroglycerin to dilate coronary blood vessels. A technician drew blood for enzyme assays to determine the level of cardiac enzymes in Walter's blood. When heart muscle cells die, they release various enzymes that serve as markers of a heart attack. A second tube of blood was sent for an assay of its troponin I level. Troponin I is a good indicator of heart damage following a heart attack.

Question 3: *Some of the enzymes that are used as markers for heart attacks are also found in related forms in skeletal muscle. What are related forms of an enzyme called?* [Hint: 🔖 p. 92]

Question 4: *What is troponin, and why would elevated blood levels of troponin indicate heart damage?* [Hint: 🔖 p. 399]

they do not have organized sarcomeres, the autorhythmic cells do not contribute to the contractile force of the heart.

Most myocardial cells in the heart are typical striated muscle, with contractile fibers organized into sarcomeres [🔖 p. 394]. However, cardiac muscle differs in significant ways from skeletal muscle and shares some properties with smooth muscle. (See Table 12-3, p. 421, for a summary comparison of the three muscle types.)

1. Cardiac muscle fibers are much smaller than skeletal muscle fibers and have a single nucleus per fiber.

2. Individual cardiac muscle cells branch and join neighboring cells end-to-end to create a complex network (Fig. 14-10b ●). The cell junctions are specialized regions known as **intercalated disks**. They consist of interdigitated membranes tightly linked by *desmosomes* that tie adjacent cells together [🔖 p. 66]. These strong connections allow force created in one cell to be transferred to the adjacent cell.

3. The t-tubules of myocardial cells are larger, and they branch inside the myocardial cell.

4. The sarcoplasmic reticulum is smaller than that of skeletal muscle, reflecting the fact that cardiac muscle depends in part on extracellular Ca^{2+} to initiate contraction. In this respect, cardiac muscle resembles smooth muscle.

5. Mitochondria occupy about one-third of the cell volume of a cardiac contractile fiber, a reflection of the high energy demand of these cells. By one estimate, cardiac muscle extracts 70%–80% of the oxygen delivered to it by the blood, more than twice the amount removed by other cells in the body.

   During periods of increased activity, the heart removes almost all oxygen brought to it by the coronary arteries. Thus, the only way to get more oxygen to exercising heart muscle is to increase its blood flow. Reduced myocardial

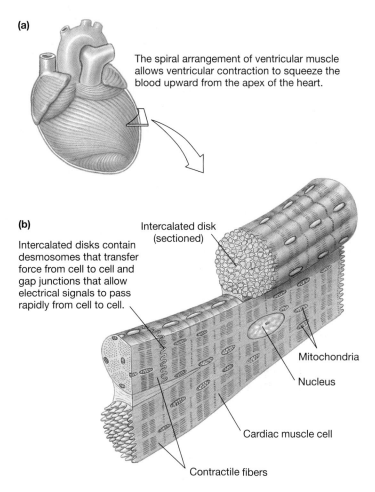

(a)

The spiral arrangement of ventricular muscle allows ventricular contraction to squeeze the blood upward from the apex of the heart.

(b)

Intercalated disks contain desmosomes that transfer force from cell to cell and gap junctions that allow electrical signals to pass rapidly from cell to cell.

Intercalated disk (sectioned)

Mitochondria

Nucleus

Cardiac muscle cell

Contractile fibers

● **Figure 14-10** *Cardiac muscle* For an electron micrograph of gap junctions, see Figure 3-14c on p. 65.

blood flow from narrowing of a coronary vessel by a clot or fatty deposit can damage or even kill myocardial cells.

6. Cardiac muscle cells are electrically connected to each other via *gap junctions* in the intercalated disks. Waves of depolarization spread rapidly from cell to cell, allowing the heart muscle cells to contract almost simultaneously. In this respect, cardiac muscle resembles single-unit smooth muscle.

Cardiac EC Coupling Combines Features of Skeletal and Smooth Muscle

In Chapters 11 and 12 you learned how acetylcholine from a somatic motor neuron causes a skeletal muscle action potential to begin the process of excitation-contraction coupling (EC coupling). In cardiac muscle, an action potential also initiates EC coupling, but the action potential originates spontaneously in the pacemaker cells and spreads into the contractile cells through gap junctions. (Neurotransmitters modulate the pacemaker rate, as you will learn later in this chapter.)

When an action potential enters a contractile cell, it spreads along the sarcolemma and into the t-tubules, where it opens voltage-gated Ca^{2+} channels in the cell membrane. Ca^{2+} enters the cell, and opens *ryanodine receptor-channels (RyR)* in the sarcoplasmic reticulum (Fig. 14-11 ●). Note that these ryanodine receptors are operated by Ca^{2+} binding, not by mechanical linkage as the RyR channels in skeletal muscle are.

The ryanodine receptors are Ca^{2+} channels, and opening them causes **Ca^{2+}-induced Ca^{2+}release**. Stored Ca^{2+} flows out of the sarcoplasmic reticulum and into the cytosol, creating a Ca^{2+} "spark" that can be visualized using special biochemical methods [● p. 183]. Multiple sparks from different RyR channels sum to create a Ca^{2+} signal.

The Ca^{2+} released from the sarcoplasmic reticulum provides about 90% of the Ca^{2+} needed for muscle contraction. Ca^{2+} diffuses through the cytosol to the contractile elements, where the ions bind to troponin and initiate the cycle of crossbridge formation and movement. Contraction takes place by the same type of sliding filament movement that occurs in skeletal muscle [● p. 396]

Relaxation in cardiac muscle occurs when Ca^{2+} unbinds from troponin. Calcium ions are transported back into the sarcoplasmic reticulum with the help of a *Ca^{2+}-ATPase*. They can also be removed from the cell in exchange for sodium ions. The latter process uses a *Na^+-Ca^{2+} antiport protein*. Ca^{2+} moves out of the cell against its electrochemical gradient in exchange for Na^+ entering the cell down its electrochemical gradient. The Na^+ that enters the cell during this transfer is removed by the Na^+-K^+-ATPase.

CONCEPT CHECK   Answers: p. 488

10. If a myocardial contractile cell is placed in interstitial fluid and depolarized, the cell will contract. If Ca^{2+} is removed from the fluid surrounding the myocardial cell, and the cell is depolarized, it will not contract. If the experiment is repeated with a skeletal muscle fiber, the skeletal muscle will contract when depolarized, whether Ca^{2+} is present or not. What conclusion can you draw from the results of this experiment?

Cardiac Muscle Contraction Can Be Graded

A key property of cardiac muscle cells is the ability of a single muscle fiber to execute *graded contractions*, in which the fiber varies the amount of force it generates. (Recall that in skeletal muscle, contraction in a single fiber is all-or-none at any given fiber length.) The force generated by cardiac muscle is proportional to the number of crossbridges that are active. The number of active crossbridges is determined by how much Ca^{2+} is bound to troponin.

If cytosolic Ca^{2+} concentrations are low, some crossbridges will not be activated and force will be low. If additional Ca^{2+} enters the cell from the extracellular fluid, more Ca^{2+} is released from the sarcoplasmic reticulum. This additional Ca^{2+} binds to troponin, en-

1 Action potential enters from adjacent cell.

2 Voltage-gated Ca^{2+} channels open. Ca^{2+} enters cell.

3 Ca^{2+} induces Ca^{2+} release through ryanodine receptor-channels (RyR).

4 Local release causes Ca^{2+} spark.

5 Summed Ca^{2+} sparks create a Ca^{2+} signal.

6 Ca^{2+} ions bind to troponin to initiate contraction.

7 Relaxation occurs when Ca^{2+} unbinds from troponin.

8 Ca^{2+} is pumped back into the sarcoplasmic reticulum for storage.

9 Ca^{2+} is exchanged with Na^+.

10 Na^+ gradient is maintained by the Na^+-K^+-ATPase.

● **Figure 14-11** *Excitation-contraction coupling and relaxation in cardiac muscle*

Foxglove for a Failing Heart

Another mechanism for increasing Ca^{2+} in the cytosol is preventing Ca^{2+} removal from the cell. This mechanism is not active under normal physiological conditions. However, it can be triggered by the administration of cardiac glycosides, a class of molecules first discovered in the plant *Digitalis purpurea* (purple foxglove). The cardiac glycosides include digitoxin and the related compound *ouabain*, a molecule used to inhibit sodium transport in research studies. Because cardiac glycosides depress Na^+-K^+-ATPase activity, they are highly toxic. They poison the pumps on all cells, not just those of the heart. However, if administered in low doses, cardiac glycosides can partially block the removal of Na^+ from the myocardial cell. As Na^+ builds up in the cytosol, the concentration gradient for Na^+ across the cell membrane diminishes. In turn, the potential energy available for indirect active transport diminishes. In the myocardial cell, cardiac glycosides decrease the cell's ability to remove Ca^{2+} by means of the Na^+-Ca^+ exchanger. The resultant increase in cytosolic Ca^{2+} causes stronger myocardial contractions. For this reason, cardiac glycosides have been used since the eighteenth century as a remedy for heart failure, a condition in which the heart is unable to contract forcefully.

hancing the ability of myosin to form crossbridges with actin and creating additional force.

The catecholamines epinephrine and norepinephrine are regulatory molecules that affect the amount of Ca^{2+} available for cardiac muscle contraction [❀ p. 218]. As you may recall, epinephrine is the "fight-or-flight" hormone released by the adrenal medulla. Norepinephrine is the neurotransmitter released by sympathetic neurons. Both molecules modulate contraction by binding to and activating β_1 adrenergic receptors [❀ p. 377] on the membrane of the cardiac muscle cell (Fig. 14-12 ●).

Once activated, β_1 receptors use a cyclic AMP second messenger system to phosphorylate specific intracellular proteins [❀ p. 179]. Phosphorylation of the voltage-gated Ca^{2+} channels increases the probability that they will open. More open channels allow more Ca^{2+} to enter the cell.

In addition, phosphorylation of a regulatory protein known as **phospholamban** enhances Ca^{2+}-ATPase activity in the sarcoplasmic reticulum. This ATPase concentrates Ca^{2+} in the sarcoplasmic reticulum, making more Ca^{2+} available for calcium-induced calcium release. Because the force of contraction increases as the number of active crossbridges increases, the net result of catecholamine stimulation is a stronger contraction.

Catecholamines shorten the duration of cardiac contraction in addition to increasing force. The enhanced Ca^{2+}-ATPase removes Ca^{2+} from the cytosol more rapidly. This in turn shortens

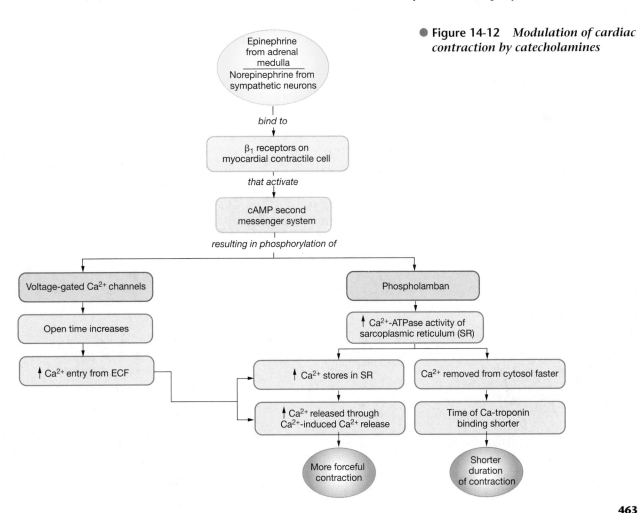

● **Figure 14-12** *Modulation of cardiac contraction by catecholamines*

the time that Ca^{2+} is bound to troponin and the active time of the myosin crossbridges. The muscle twitch is therefore briefer.

When Cardiac Muscle Is Stretched, It Contracts More Forcefully

Another factor that affects the force of contraction in cardiac muscle is the sarcomere length at the beginning of contraction. For both cardiac and skeletal muscle, the tension generated is directly proportional to the initial length of the muscle fiber [● p. 406]. As muscle fiber length and sarcomere length increase, tension increases, up to a maximum (Fig. 14-13 ●).

For years, we believed that the length-tension relationship in cardiac muscle was strictly a function of the overlap between thick and thin filaments. But now evidence suggests the process is much more complex and involves increased intracellular Ca^{2+} concentrations and increased sensitivity of the contractile proteins to Ca^{2+}. Thus, although it is traditional to separate changes in force due to fiber length from changes in force created by catecholamines, there is some overlap between the mechanisms underlying the two.

In the intact heart, stretch on the individual fibers is a function of how much blood is in the chambers of the heart. The relationship between force and ventricular volume is an important property of cardiac function and is discussed in detail later in this chapter.

CONCEPT CHECK   Answers: p. 488

11. A drug that blocks all Ca^{2+} channels in the myocardial cell membrane is placed in the solution around the cell. What will happen to the force of contraction in that cell?

Action Potentials in Myocardial Cells Vary According to Cell Type

Cardiac muscle, like skeletal muscle and neurons, is an excitable tissue with the ability to generate action potentials. Each of the two types of cardiac muscle cells has a distinctive action potential. In both types, Ca^{2+} plays an important role in the action potential, in contrast to the action potentials of skeletal muscle and neurons.

● **Figure 14-13** *Length-tension relationships in skeletal and cardiac muscle* These data represent tension developed during isometric contraction. The physiological range is the sarcomere length within which the muscle normally functions.

Myocardial Contractile Cells The action potentials of myocardial contractile cells are similar in several ways to those of neurons and skeletal muscle [● p. 253]. The rapid depolarization phase of the action potential is due to Na^+ entry, and the steep repolarization phase is due to K^+ leaving the cell. The main difference between the action potential of the myocardial contraction cell and that of a skeletal muscle fiber or a neuron is a lengthening of the action potential due to Ca^{2+} entry. Let's take a look at these longer action potentials (Fig. 14-14 ●).

Phase 4: resting membrane potential. Myocardial contractile cells have a stable resting potential of about −90 mV.

Phase 0: depolarization. When a wave of depolarization moves into a contractile cell through gap junctions, the membrane potential becomes more positive. Voltage-gated Na^+ channels open, allowing Na^+ to enter the cell and rapidly depolarize it. The membrane potential reaches about +20 mV before the Na^+ channels close. These are double-gated Na^+ channels, similar to the voltage-gated Na^+ channels of the axon [● p. 254].

Phase 1: initial repolarization. When the Na^+ channels close, the cell begins to repolarize as K^+ leaves through open K^+ channels.

Phase 2: the plateau. The initial repolarization is very brief. The action potential then flattens into a plateau as the result of two events: a decrease in K^+ permeability and an increase in Ca^{2+} permeability. Voltage-gated Ca^{2+} channels activated by depolarization have been slowly opening during phases 0

● **Figure 14-14** *Action potential of a cardiac contractile cell* The phase numbers are a convention.

and 1. When they finally open, Ca^{2+} enters the cell. At the same time, some K^+ channels close. The combination of Ca^{2+} influx and decreased K^+ efflux causes the action potential to flatten out into a plateau.

Phase 3: rapid repolarization. The plateau ends when Ca^{2+} channels close and K^+ permeability increases once more. The K^+ channels responsible for this phase are similar to those in the neuron: they are activated by depolarization but are slow to open. When the delayed K^+ channels open, K^+ exits rapidly, returning the cell to its resting potential (phase 4).

The influx of Ca^{2+} during phase 2 lengthens the total duration of a myocardial action potential. A typical action potential in a neuron or skeletal muscle fiber lasts between 1 and 5 msec. In a contractile myocardial cell, the action potential is typically 200 msec or more.

The longer action potential in myocardial cells helps prevent the sustained contraction called tetanus. Prevention of tetanus in the heart is important because the muscles must relax between contractions so that the ventricles can fill with blood.

To understand why a longer action potential prevents tetanus, compare the relationship between action potentials and contraction in skeletal and cardiac muscle cells (Fig. 14-15●). As you may recall from Chapter 12, the skeletal muscle action potential is ending as contraction begins (Fig. 14-15a). Thus, a second action potential fired immediately after the refractory period will cause summation of the contractions. If a series of action potentials occurs in rapid succession, the sustained contraction known as tetanus results (Fig. 14-15b).

Tetanus cannot occur in cardiac muscle because the refractory period and the contraction end almost simultaneously due to the longer action potential (Fig. 14-15c). By the time a second action potential takes place, the myocardial cell has almost completely relaxed. Consequently, no summation will occur (Fig. 14-15d).

(a) Skeletal muscle fast-twitch fiber: The refractory period (yellow) is very short compared with the amount of time required for the development of tension.

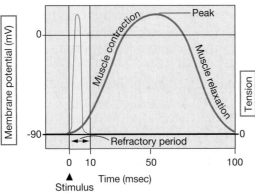

(b) Skeletal muscles that are stimulated repeatedly will exhibit summation and tetanus (action potentials not shown.)

KEY

▲ = Stimulus for action potential

(c) Cardiac muscle fiber: The refractory period lasts almost as long as the entire muscle twitch.

(d) Long refractory period in a cardiac muscle prevents tetanus.

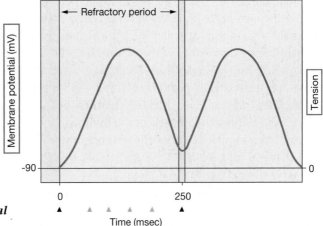

● Figure 14-15 *Refractory periods and summation in skeletal and cardiac muscle*

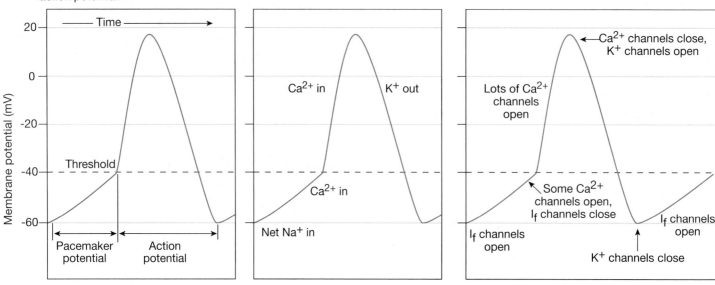

(a) The pacemaker potential gradually becomes less negative until it reaches threshold, triggering an action potential.

(b) Ion movements during an action and pacemaker potential

(c) State of various ion channels

FIGURE QUESTION

Match the appropriate phases of the myocardial contractile cell action potential (Fig.14-14) to the pacemaker action potential above.

● **Figure 14-16** *Action potentials in cardiac autorhythmic cells*

Myocardial Autorhythmic Cells What gives myocardial autorhythmic cells their unique ability to generate action potentials spontaneously in the absence of input from the nervous system? This property results from their unstable membrane potential that starts at $-60\,mV$ and slowly drifts upward toward threshold (Fig. 14-16a ●). Because the membrane potential never "rests" at a constant value, it is called a **pacemaker potential** rather than a resting membrane potential. Whenever the pacemaker potential depolarizes to threshold, the autorhythmic cell fires an action potential.

What causes the membrane potential of these cells to be unstable? Our current understanding is that the autorhythmic cells contain channels that are different from the channels of other excitable tissues. When the cell membrane potential is $-60\,mV$, channels that are permeable to both K^+ and Na^+ open (Fig. 14-16b, c). These channels are called I_f **channels** because they allow current (I) to flow and because of their unusual properties. The researchers who first described the ion current flowing through these channels initially did not understand its behavior and named it *funny* current, hence the subscript *f*.

When I_f channels open at negative membrane potentials, Na^+ influx exceeds K^+ efflux. (This is similar to what happens at the neuromuscular junction when nonspecific cation channels open

[● p. 400].) The net influx of positive charge slowly depolarizes the autorhythmic cell. As the membrane potential becomes more positive, the I_f channels gradually close and some Ca^{2+} channels open. The subsequent influx of Ca^{2+} continues the depolarization, and the membrane potential moves steadily toward threshold.

When the membrane potential reaches threshold, additional Ca^{2+} channels open. Calcium ions rush in, creating the steep depolarization phase of the action potential. Note that this process is different from that in other excitable cells, in which the depolarization phase is due to the opening of voltage-gated Na^+ channels.

When the Ca^{2+} channels close at the peak of the action potential, slow K^+ channels have opened. The repolarization phase of the autorhythmic action potential is due to the resultant efflux of K^+. This phase is similar to repolarization in other types of excitable cells.

Autonomic Neurotransmitters Modulate Heart Rate

The speed with which the pacemaker cells depolarize sets the rate at which the heart contracts (the heart rate). The interval between action potentials can be modified by altering the permeability of the autorhythmic cells to different ions. Increased permeability to Na^+ and Ca^{2+} during the pacemaker potential phase will speed up depolarization and heart rate. Decreased Ca^{2+} permeability or increased K^+ permeability will slow depolarization and slow down the heart rate.

CONCEPT CHECK   Answers: p. 488

15. What would increasing K^+ permeability do to the membrane potential of the cell?

(a) Sympathetic stimulation and epinephrine depolarize the autorhythmic cell and speed up the depolarization rate, increasing the heart rate.

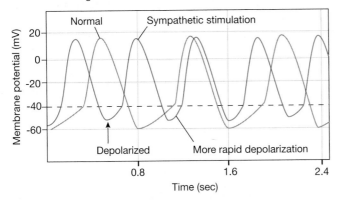

(b) Parasympathetic stimulation hyperpolarizes the membrane potential of the autorhythmic cell and slows depolarization, slowing down the heart rate.

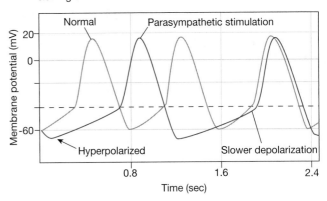

● **Figure 14-17** *Modulation of heart rate by the nervous system*

Sympathetic stimulation of pacemaker cells speeds up heart rate. Norepinephrine from sympathetic neurons and epinephrine from the adrenal medulla increase ion flow through both the I_f and Ca^{2+} channels. More rapid cation entry speeds up the rate of the pacemaker depolarization, causing the cell to reach threshold faster and increasing the rate of action potential firing (Fig. 14-17a ●). When the pacemaker fires action potentials more rapidly, heart rate increases.

The catecholamines exert their effect by binding to and activating β_1 adrenergic receptors on the autorhythmic cells. The β_1 receptors use a cAMP second messenger system to alter the transport properties of the ion channels. In the case of the I_f channels, cAMP itself is the messenger. When cAMP binds to open I_f channels, they remain open longer. The I_f channels belong to a family called the *HCN*

channels, or *hyperpolarization-activated cyclic nucleotide-gated channels*. Other members of the HCN family are found in neurons.

The parasympathetic neurotransmitter acetylcholine (ACh) slows heart rate. Acetylcholine activates muscarinic cholinergic receptors that influence K^+ and Ca^{2+} channels in the pacemaker cell. Potassium permeability increases, hyperpolarizing the cell so that the pacemaker potential begins at a more negative value (Fig. 14-17b). At the same time, Ca^{2+} permeability of the pacemaker decreases. Decreased Ca^{2+} permeability slows the rate at which the pacemaker potential depolarizes. The combination of the two effects causes the cell to take longer to reach threshold, delaying the onset of the action potential in the pacemaker and slowing the heart rate.

Table 14-4 compares action potentials of the two types of myocardial muscle with those of skeletal muscle. Next we will see

Table 14-4: Comparison of Action Potentials in Cardiac and Skeletal Muscle

	SKELETAL MUSCLE	CONTRACTILE MYOCARDIUM	AUTORHYTHMIC MYOCARDIUM
Membrane potential	Stable at $-70\,mV$	Stable at $-90\,mV$	Unstable pacemaker potential; usually starts at $-60\,mV$
Events leading to threshold potential	Net Na^+ entry through ACh-operated channels	Depolarization wave enters via gap junctions	Net Na^+ entry through I_f channels; reinforced by Ca^{2+} entry
Rising phase of action potential	Na^+ entry	Na^+ entry	Ca^{2+} entry
Repolarization phase	Rapid; caused by K^+ efflux	Extended plateau caused by Ca^{2+} entry; rapid phase caused by K^+ efflux	Rapid; caused by K^+ efflux
Hyperpolarization	Due to excessive K^+ efflux at high K^+ permeability when K^+ channels close, leak of K^+ and Na^+ restores potential to resting state	None; resting potential is $-90\,mV$, the equilibrium potential for K^+	None; when repolarization hits $-60\,mV$, the I_f channels open again
Duration of action potential	Short: 1–2 msec	Extended: 200+ msec	Variable; generally 150+ msec
Refractory period	Generally brief	Long because resetting of Na^+ channel gates delayed until end of action potential	None

how action potentials of autorhythmic cells spread throughout the heart to coordinate contraction.

▶ THE HEART AS A PUMP

We now turn from single myocardial cells to the intact heart. How can one tiny noncontractile autorhythmic cell cause the entire heart to beat? And why do those doctors on TV shows shock people with electric paddles when their hearts malfunction? You're about to learn the answers to these questions.

Electrical Conduction in the Heart Coordinates Contraction

The heart is like a group of people around a stalled car. One person can push on the car, but it's not likely to move very far unless everyone pushes together. In the same way, individual myocardial cells must depolarize and contract in a coordinated fashion if the heart is to create enough force to circulate the blood.

Electrical communication in the heart begins with an action potential in an autorhythmic cell. The depolarization spreads rapidly to adjacent cells through gap junctions in the intercalated disks (Fig. 14-18 ●). The depolarization wave is followed by a wave of contraction that passes across the atria, then moves into the ventricles.

The depolarization begins in the **sinoatrial node (SA node)**, autorhythmic cells in the right atrium that serve as the main pacemaker of the heart (Fig. 14-19a ●). The depolarization wave then spreads rapidly through a specialized conducting system of noncontractile autorhythmic fibers. An **internodal pathway** connects the SA node to the **atrioventricular node (AV node)**, a group of autorhythmic cells near the floor of the right atrium.

From the AV node, the depolarization moves into Purkinje fibers in the **atrioventricular bundle (A-V bundle)*** in the wall (septum) between the ventricles. **Purkinje fibers** are specialized conducting cells that transmit electrical signals very rapidly. A short way down the septum, the A-V bundle fibers divide into left and right **bundle branches**. These fibers continue downward to the apex where they divide into smaller Purkinje fibers that spread outward among the contractile cells.

The electrical signal for contraction begins when the SA node fires an action potential and the depolarization spreads to adjacent cells through gap junctions (Fig. 14-19b). Electrical conduction is rapid through the internodal conducting pathways (Fig. 14-19c), but slower through the contractile cells of the atrium (Fig. 14-19d).

As action potentials spread across the atria, they encounter the fibrous skeleton of the heart at the junction of the atria and ventricles. This barricade prevents the transfer of electrical signals

---

*The A-V bundle is also called the **bundle of His** ("hiss").

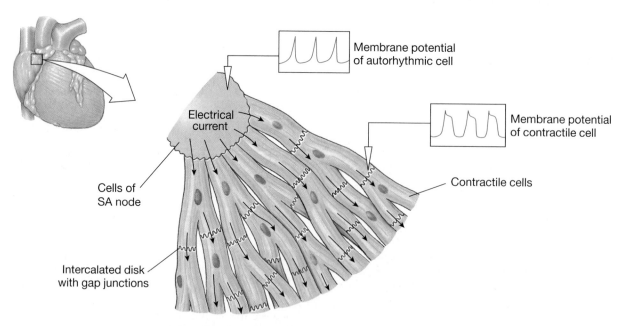

Cells of SA node

Electrical current

Intercalated disk with gap junctions

Membrane potential of autorhythmic cell

Membrane potential of contractile cell

Contractile cells

Depolarizations of autorhythmic cells rapidly spread to adjacent contractile cells through gap junctions.

● **Figure 14-18** *Electrical conduction in myocardial cells*

(b) SA node depolarizes.

SA node

AV node

(c) Electrical activity goes rapidly to AV node via internodal pathways.

(a) The conducting system of the heart

SA node

Internodal pathways

AV node

A-V bundle

Bundle branches

Purkinje fibers

(d) Depolarization spreads more slowly across atria. Conduction slows through AV node.

(e) Depolarization moves rapidly through ventricular conducting system to the apex of the heart.

(f) Depolarization wave spreads upward from the apex.

● **Figure 14-19** *Electrical conduction in the heart* Purple shading in (b)–(f) represents depolarization.

from the atria to the ventricles. Consequently, the AV node is the only pathway through which action potentials can reach the contractile fibers of the ventricles (Fig. 14-19d).

The electrical signal passes from the AV nodes through the A-V bundle and the bundle branches to the apex of the heart (Fig. 14-19e). The Purkinje fibers transmit impulses very rapidly, with speeds up to 4 m/sec, so that all contractile cells in the apex contract nearly simultaneously (Fig. 14-19f).

Why is it necessary to direct the electrical signals through the AV node? Why not allow them to spread downward from the atria? The answer lies in the fact that blood being pumped out of the ventricles leaves through openings at the top of the chambers. If electrical signals from the atria were conducted directly into the ventricles, the ventricles would start contraction at the top. Then blood would be squeezed downward and would be trapped in the bottom of the ventricle (think of squeezing a toothpaste tube at the top). The apex-to-base contraction squeezes blood toward the arterial openings at the base of the heart.

The ejection of blood from the ventricles is aided by the spiral arrangement of the muscles in the walls (see Fig. 14-10a). As these muscles contract, they pull the apex and base of the heart closer together, squeezing blood out openings at the top of the ventricles.

A second function of the AV node is to delay the transmission of action potentials slightly, allowing the atria to complete their contraction before ventricular contraction begins. The **AV node delay** is accomplished by slowing conduction through the AV node cells. Action potentials here move at only 1/20 the rate of action potentials in the atrial internodal pathway.

RUNNING PROBLEM

The results of cardiac enzyme and troponin I assays do not come back for several hours. If a coronary artery was blocked, damage to the heart muscle could be severe by that time. In Walter's case, an electrocardiogram (ECG) showed an abnormal pattern of electrical activity. "He's definitely had an MI," said the ER physician, referring to a myocardial infarction, or heart attack. "Let's start him on a beta-blocker and t-PA." t-PA is short for *tissue plasminogen activator*. t-PA activates plasminogen, a substance produced in the body that dissolves blood clots. When t-PA is given within 1–3 hours of a heart attack, it can help dissolve blood clots that are blocking blood flow to the heart muscle. This will help limit the extent of ischemic damage.

Question 5: How do electrical signals move from cell to cell in the myocardium?

Question 6: What happens to contraction in a myocardial contractile cell if a wave of depolarization passing through the heart bypasses it?

Question 7: A beta-blocker is an antagonist to β_1 adrenergic receptors. What will this drug do to Walter's heart rate? Why is that response helpful following a heart attack?

Pacemakers Set the Heart Rate

The cells of the SA node set the pace of the heartbeat. Other cells in the conducting system, such as the AV node and the Purkinje fibers, have unstable resting potentials and can also act as pacemakers under some conditions. However, because their rhythm is slower than that of the SA node, they do not usually have a chance to set the heartbeat. The Purkinje fibers, for example, can spontaneously fire action potentials, but their rate of firing is very slow, between 25 and 40 beats per minute.

Why does the fastest pacemaker determine the pace of the heartbeat? Consider the following analogy. A group of people are playing "Follow the Leader" as they walk. Initially, everyone is walking at a different pace: some fast, some slow. When the game starts, everyone must match his or her pace to the pace of the person who is walking the fastest. The fastest person in the group is the SA node, walking at 70 steps per minute. Everyone else in the group (autorhythmic and contractile cells) sees that the SA node is fastest, so they pick up their pace and follow the leader. In the heart, the cue to follow the leader is the electrical signal sent from the SA node to the other cells.

Now suppose that the SA node gets tired and drops out of the group. The role of leader defaults to the next fastest person, the AV node, who is walking at a rate of 50 steps per minute. The group slows to match the pace of the AV node, but they are still following the fastest leader.

What happens if the group divides? When they reach the corner, the AV node leader goes left, but a renegade Purkinje fiber decides to go right. Those people who follow the AV node continue to walk at 50 steps per minute, but the people who follow the Purkinje fiber slow down to match his pace of 35 steps per minute. Now there are two leaders, each walking at a different pace.

In the heart, the SA node is the fastest pacemaker and normally sets the heart rate. But if the SA node is damaged and cannot function, one of the slower pacemakers in the heart takes over. Heart rate then matches the rate of the new pacemaker. It is even possible for different parts of the heart to follow different pacemakers, just as the walking group split at the corner.

In a condition known as *complete heart block*, the conduction of electrical signals from the atria to the ventricles through the AV node is disrupted. The SA node fires at its rate of 70 beats per minute, but those signals never reach the ventricles. So the ventricles coordinate with their fastest pacemaker. Because ventricular autorhythmic cells discharge only about 35 times a minute, the ventricles contract at a totally different rate than the atria. In conditions such as complete heart block, in which the rate of ventricular contraction is too slow to maintain adequate blood flow, it may be necessary for the heart's rhythm to be set artificially by a surgically implanted mechanical pacemaker. These battery-powered devices artificially stimulate the heart at a predetermined rate.

Gap Junctions Can Open and Close

The gap junctions between myocardial cells are normally open, allowing free passage of electrical signals. However, they can be closed by unusually high concentrations of Ca^{2+} and H^+ levels in the cytosol. Following a heart attack, cells become oxygen-starved, produce lactic acid (H^+), and are unable to make ATP to pump Ca^{2+} out of the cytoplasm. The resultant increase in H^+ and Ca^{2+} closes gap junctions in the damaged cells. Closure electrically isolates the damaged cells, and they no longer contract. It also forces action potentials to find an alternate route from cell to cell. If the damaged area of myocardium is large, the pumping function of the heart may be severely impaired, even to the point of death.

19. Name two functions of the AV node. What is the purpose of AV node delay?

20. Where is the SA node located?

21. Occasionally an ectopic pacemaker [*ektopos*, out of place] will develop in part of the conducting system. What happens to heart rate if an ectopic atrial pacemaker depolarizes at a rate of 120 times per minute?

The Electrocardiogram Reflects the Electrical Activity of the Heart

In the beginning of the twentieth century, physiologists discovered that they could use electrocardiograms to obtain information about the function of the heart. An **electrocardiogram**, or **ECG**, is a recording of the electrical activity of the heart made from electrodes placed on the surface of the skin. By simply placing electrodes on the arms and legs, we can record electrical activity that is taking place deep within the body.

Salt solutions such as the NaCl-based extracellular fluid are good conductors of electricity and can transfer electrical activity to the skin's surface. The electrical signals are very weak by the time they get to the skin. A ventricular action potential, for example, has a voltage change of 110 mV, but the ECG signal has an amplitude of only 1 mV by the time it reaches the surface of the body. External recordings also allow us to monitor electrical activity in muscles (*electromyogram*, or *EMG*) and in the brain (*electroencephalogram*, or *EEG*).

The first human electrocardiogram was recorded in 1887, but the procedure was not refined for clinical use until the first years of the twentieth century. The father of the modern ECG was a Dutch physiologist named Walter Einthoven. He named the parts of the ECG as we know them today and created "Einthoven's triangle," a hypothetical triangle created around the heart when electrodes are placed on both arms and the left leg (Fig. 14-20 ●).

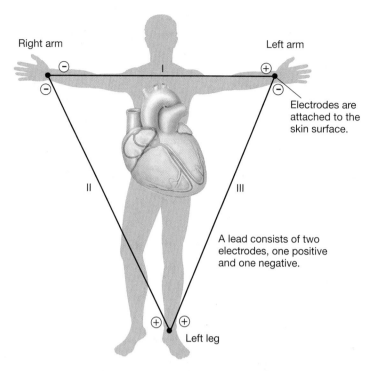

● **Figure 14-20** *Einthoven's triangle* Electrodes attached to both arms and the left leg form a triangle. Two electrodes (a lead) are active at any one time. Net current flow toward the positive electrode of a lead creates an upward deflection on the ECG.

The sides of the triangle are numbered to correspond with the three leads, or pairs of electrodes, used for a recording.

An ECG tracing shows the sum of the electrical potentials generated by all the cells of the heart at any moment. Each component of the ECG reflects depolarization or repolarization of a portion of the heart. Because depolarization is the signal for contraction, the *electrical events* (waves) of an ECG can be associated approximately to contraction or relaxation of the atria or ventricles (collectively referred to as the *mechanical events* of the cardiac cycle). Let's follow an ECG through a single contraction-relaxation, otherwise known as a **cardiac cycle**.

There are two major components of an ECG: waves and segments. Waves appear as deflections above or below the baseline. *Segments* are sections of baseline between two waves. *Intervals* are combinations of waves and segments. Three major waves can be seen on a normal ECG recorded from lead I: the P wave, the combined waves of the QRS complex, and the T wave (Fig. 14-21 ●). The first wave is the **P wave**, which corresponds to depolarization of the atria (Fig. 14-22 ●). The next trio of waves, the **QRS complex**, represents the progressive wave of ventricular depolarization. The final wave, the **T wave**, represents the repolarization of the ventricles. Atrial repolarization is not represented by a special wave but is incorporated into the QRS complex.

The mechanical events of the cardiac cycle lag slightly behind the electrical signals, just as the contraction of a single cardiac muscle cell follows its action potential (see Fig. 14-15c). Atrial contraction begins during the latter part of the P wave

GRAPH QUESTION

If the chart speed is 25 mm /sec, what is the heart rate of the person? (1 little square = 1 mm)

● **Figure 14-21** *The electrocardiogram*
The electrocardiogram (ECG) is divided into waves (P, Q, R, S, and T), segments between the waves, and intervals that include waves. This is a ECG tracing recorded from lead I.

and continues during the P-R segment. Ventricular contraction begins just after the Q wave and continues through the T wave.

One thing that many people find puzzling is that downward deflection may correspond to periods of depolarization in the heart. It is important to remember that the ECG is different from a single action potential (Fig. 14-23 ●). The single action potential is an electrical event in a single cell, recorded using an intracellular electrode. The ECG represents multiple action potentials taking place in the heart muscle at a given time and recorded from the surface of the body.

When recording an ECG, one active surface electrode is the positive electrode, and the other active electrode is the negative electrode. (The third electrode is inactive.) If net electrical current in the heart moves toward the positive electrode, the tracing of the ECG goes up from the baseline. If net current moves toward the negative electrode, the tracing moves downward.

Another important point to remember is that the ECG is an electrical "view" of a three-dimensional object. This is one reason that we use multiple leads to assess heart function. Think of looking at an automobile. From the air, it looks like a rectangle, but from the side and front it has different shapes. Not everything that you see from the front of the car can be seen from its side, and vice versa.

In the same way, ECGs recorded from the different leads look different and give information about different regions of the heart. Clinically, using 12 leads is now common: the three limb electrodes plus nine more electrodes placed on the chest and trunk. These additional "views" provide even more information about electrical conduction in the heart. Electrocardiograms are important diagnostic tools in medicine because they are quick, painless, and noninvasive (that is, do not puncture the skin).

An ECG provides information on heart rate and rhythm, conduction velocity, and even the condition of tissues within the

heart. Thus, although obtaining an ECG is simple, interpreting some of its subtleties can be quite complicated. The interpretation of an ECG begins with the following questions (Fig. 14-24 ●).

1. What is the heart rate? Heart rate is normally timed from the beginning of one P wave to the beginning of the next P wave, or from R wave peak to peak. A normal resting heart rate is 60–100 beats per minute, although trained athletes often have slower heart rates at rest. A faster rate is known as *tachycardia*, and a slower rate is called *bradycardia* [*tachys*, swift; *bradys*, slow].

2. Is the rhythm of the heartbeat regular or irregular? That is, does it occur at regular intervals? An irregular rhythm, or *arrhythmia*, can result from a benign extra beat or from more serious conditions such as atrial fibrillation, in which the SA node has lost control of the pacemaking.

    After determining heart rate and rhythm, the next step in analyzing an ECG is to look at the relationship of the various waves. To help your analysis, you might want to write the letters above the P, R, and T waves.

3. Does a QRS complex follow each P wave, and is the P-R segment constant in length? If not, a problem with conduction of signals through the AV node may exist. In heart block, the conduction problem mentioned earlier, action potentials from the SA node sometimes fail to be transmitted through the AV node to the ventricles. In these conditions, one or more P waves may occur without initiating a QRS complex. In the most severe form of heart block (third-degree), the atria depolarize regularly at one pace while the ventricles contract at a much slower pace (Fig. 14-24b).

The more difficult aspects of interpreting an ECG include looking for subtle changes such as alterations in the shape or duration

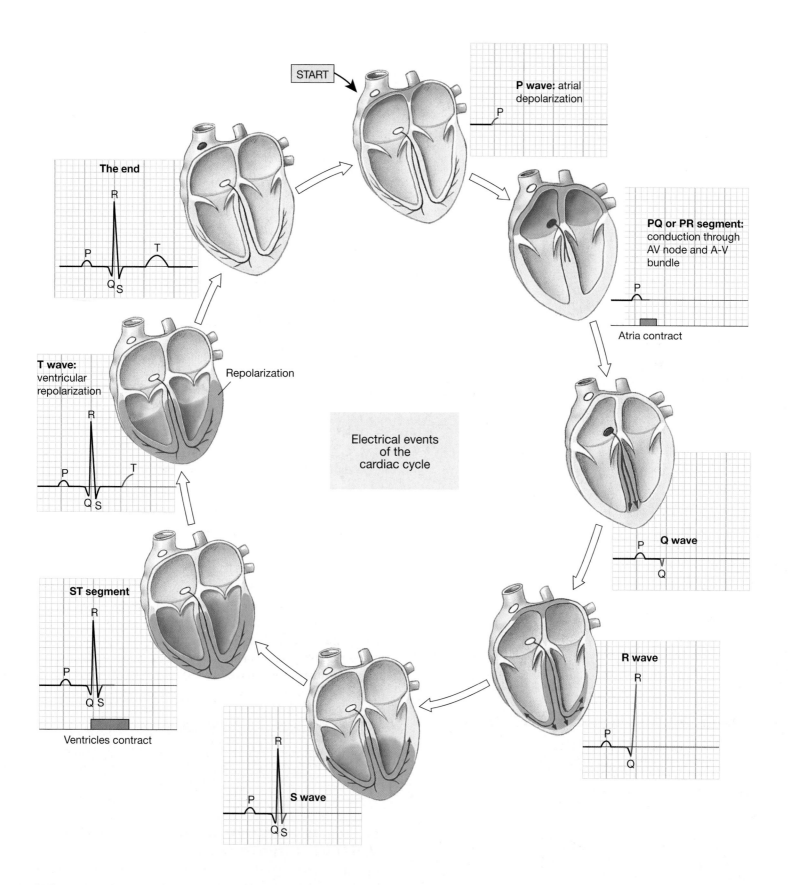

START

P wave: atrial depolarization

P

PQ or PR segment: conduction through AV node and A-V bundle

P

Atria contract

Q wave

P

Q

R wave

R

P

Q

S wave

R

P

Q S

ST segment

R

P

Q S

Ventricles contract

T wave: ventricular repolarization

R

P

Q S

T

Repolarization

Electrical events of the cardiac cycle

The end

R

P

T

Q S

● **Figure 14-22** *Correlation between the ECG and electrical events in the heart*   The depolarized regions are shown in purple, along with the corresponding section of the ECG. The repolarizing regions are shown in peach.

(a)

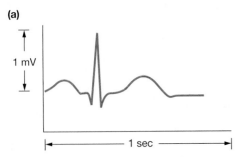

The electrocardiogram represents the summed electrical activity of all cells recorded from the surface of the body.

(b)

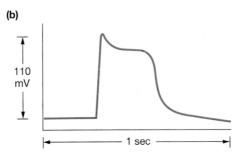

The ventricular action potential is recorded from a single cell using an intracellular electrode. Notice that the voltage change is much greater when recorded intracellularly.

● **Figure 14-23** *Comparison of the ECG and a myocardial action potential*

(a) Normal ECG

(b) Third-degree block

(c) Atrial fibrillation

(d) Ventricular fibrillation

CONCEPT CHECK   Answers: p. 488

22. Identify the waves on the ECG below. Look at the pattern of their occurrence and describe what has happened to electrical conduction in the heart.

(e)

● **Figure 14-24** *Normal and abnormal electrocardiograms*
All tracings represent 10-second recordings.

QUESTIONS TO ASK WHEN ANALYZING ECG TRACINGS:

1. What is the rate? Is it within the normal range of 60-100 beats per minute?

2. Is the rhythm regular?

3. Are all normal waves present in recognizable form?

4. Is there one QRS complex for each P wave? If yes, is the P-R segment constant in length?

5. If there is not one QRS complex for each P wave, count the heart rate using the P waves, then count it according to the R waves. Are the rates the same? Which wave would agree with the pulse felt at the wrist?

of various waves or segments. An experienced clinician can find signs pointing to changes in conduction velocity, enlargement of the heart, or tissue damage resulting from periods of ischemia. An amazing number of conclusions can be drawn about heart function simply by looking at alterations in the electrical activity of the heart as recorded on an ECG.

CONCEPT CHECK   Answers: p. 488

23. Some abnormal ECGs are shown in Figure 14-24. Study them and see if you can relate the ECG changes to disruption of the normal electrical conduction pattern in the heart.

The Heart Contracts and Relaxes Once During a Cardiac Cycle

A **cardiac cycle** is the period of time from the beginning of one heartbeat to the beginning of the next. Each cycle has two phases: **diastole**, the time during which cardiac muscle relaxes, and

RUNNING PROBLEM

The electrocardiogram indicated that Walter suffered a myocardial infarction, or blockage of blood vessels nourishing the left ventricle. The exact location of the damage depends on which artery and which branch or branches have become occluded.

Question 8: If the ventricle of the heart is damaged, in which wave or waves of the electrocardiogram would you expect to see abnormal changes?

448 452 461 470 480 483

systole, the time during which the muscle is contracting [*diastole*, dilation; *systole*, contraction]. The atria and ventricles do not contract and relax at the same time, so we will discuss atrial and ventricular events separately.

In thinking about blood flow during the cardiac cycle, remember that blood flows from an area of higher pressure to one of lower pressure and that contraction increases pressure while relaxation decreases pressure. In this discussion, we divide the cardiac cycle into five phases. Follow the steps below in Figure 14-25 ●:

1. The heart at rest: atrial and ventricular diastole. We begin our cardiac cycle with the brief moment during which both the atria and ventricles are relaxing. The atria are filling with blood from the veins, while the ventricles have just completed a contraction. As the ventricles relax, the AV valves between the atria and ventricles open. Blood flows by gravity from the atria into the ventricles. The relaxing ventricles expand to accommodate the entering blood.

CONCEPT CHECK   Answers: p. 488

24. Is pressure in the atrium higher or lower than pressure in the venae cavae during atrial filling?

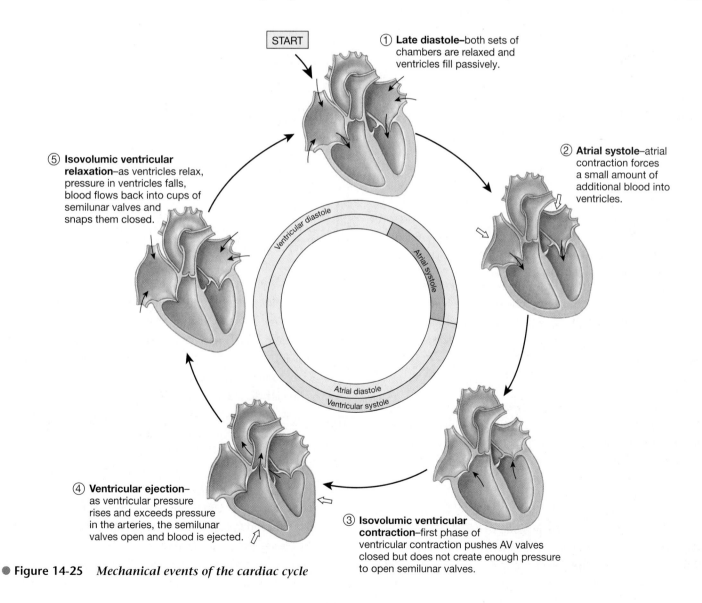

START

① **Late diastole**–both sets of chambers are relaxed and ventricles fill passively.

② **Atrial systole**–atrial contraction forces a small amount of additional blood into ventricles.

③ **Isovolumic ventricular contraction**–first phase of ventricular contraction pushes AV valves closed but does not create enough pressure to open semilunar valves.

④ **Ventricular ejection**– as ventricular pressure rises and exceeds pressure in the arteries, the semilunar valves open and blood is ejected.

⑤ **Isovolumic ventricular relaxation**–as ventricles relax, pressure in ventricles falls, blood flows back into cups of semilunar valves and snaps them closed.

Ventricular diastole
Atrial systole
Atrial diastole
Ventricular systole

● **Figure 14-25** *Mechanical events of the cardiac cycle*

2. Completion of ventricular filling: atrial systole. Although most blood enters the ventricles while the atria are relaxed, the last 20% of filling is accomplished when the atria contract and push blood into the ventricles. (This applies to a normal person at rest. When heart rate increases, as during exercise, atrial contraction plays a greater role in ventricular filling.)

Atrial systole, or contraction, begins following the wave of depolarization that sweeps across the atria. The pressure increase that accompanies contraction pushes blood into the ventricles. A small amount of blood is forced backward into the veins. There are no one-way valves to block backward flow, although the openings of the veins do narrow during contraction. The retrograde movement of blood back into the veins may be observed as a pulse in the jugular vein of a normal person who is lying with the head and chest elevated about 30°. (Look in the hollow formed where the sternocleidomastoid muscle runs under the clavicle.) An observable jugular pulse higher on the neck of a person sitting upright is a sign that pressure in the right atrium is higher than normal.

3. Early ventricular contraction and the first heart sound. As the atria are contracting, the depolarization wave is moving slowly through the conducting cells of the AV node, then rapidly down the Purkinje fibers to the apex of the heart. Ventricular systole begins there as spiral bands of muscle squeeze the blood upward toward the base. Blood pushing against the underside of the AV valves forces them closed so that blood cannot flow back into the atria. Vibrations following closure of the AV valves create the **first heart sound**, the "lub" of "lub-dup."

With both sets of AV and semilunar valves closed, blood in the ventricles has nowhere to go. Nevertheless, the ventricles continue to contract, squeezing on the blood in the same way that you might squeeze a water balloon in your hand. This is similar to an isometric contraction, in which muscle fibers create force without movement [❧ p. 409]. To return to the toothpaste analogy, it is like squeezing the toothpaste tube with the cap still on: high pressure develops within the tube, but the toothpaste has nowhere to go. This phase is called **isovolumic ventricular contraction** [*iso-*, equal], to underscore the fact that the volume of blood in the ventricle is not changing (Fig. 14-25 step 3).

While the ventricles begin to contract, the atrial muscle fibers are repolarizing and relaxing. When atrial pressure falls below that in the veins, blood flows from the veins into the atria again. Closure of the AV valves isolates the upper and lower cardiac chambers, so atrial filling is independent of events taking place in the ventricles.

4. The heart pumps: ventricular ejection. As the ventricles contract, they generate enough pressure to open the semilunar valves and push blood into the arteries (Fig. 14-25 step 4). The pressure created by ventricular contraction becomes the driving force for blood flow. High-pressure blood is forced into the arteries, displacing the low-pressure blood that fills them and

pushing it farther into the vasculature. During this phase the AV valves remain closed and the atria continue to fill.

5. Ventricular relaxation and the second heart sound. At the end of ventricular ejection, the ventricles begin to repolarize and relax. As they do so, ventricular pressure decreases. Once ventricular pressure falls below the pressure in the arteries, blood starts to flow backward into the heart. This backflow of blood fills the cuplike cusps of the semilunar valves, forcing them together into the closed position. The vibrations created by semilunar valve closure are the **second heart sound**, the "dup" of "lub-dup."

Once the semilunar valves close, the ventricles again become a sealed chamber. The AV valves remain closed because ventricular pressure, although falling, is still higher than atrial pressure (Fig. 14-25 step 5). This period is called **isovolumic ventricular relaxation** because the volume of blood in the ventricles is not changing.

When ventricular relaxation causes ventricular pressure to become less than atrial pressure, the AV valves open. Blood that has been accumulating in the atria during ventricular contraction rushes into the ventricles. The cardiac cycle has begun again.

CLINICAL FOCUS

Gallops, Clicks, and Murmurs

The simplest direct assessment of heart function consists of listening to it through the chest wall, a process known as **auscultation** [*auscultare*, to listen to] that has been practiced since ancient times. In its simplest form, auscultation is done by placing an ear against the chest. Today, it is usually performed by listening through a stethoscope placed against the chest and back. Normally, there are two audible heart sounds. One is associated with closure of the AV valves, and the other with the closure of the semilunar valves. Additional heart sounds can be recorded with very sensitive electronic stethoscopes. The third heart sound is caused by turbulent blood flow into the ventricles during ventricular filling, and the fourth heart sound is associated with turbulence during atrial contraction. With certain abnormal conditions, these sounds may become audible through a regular stethoscope. They are called gallops because their timing puts them close to one of the normal heart sounds: "lub—dup-dup," or "lub-lub—dup." Other abnormal heart sounds include clicking sounds caused by abnormal movement of one of the valves and murmurs caused by the "whoosh" of blood leaking through an incompletely closed or excessively narrowed (*stenosed*) valve.

25. Which chamber has higher pressure during the following phases of the cardiac cycle? (a) ventricular ejection, (b) isovolumic ventricular relaxation, (c) atrial and ventricular diastole, (d) isovolumic ventricular contraction

26. Murmurs are abnormal heart sounds caused by blood forced through a narrowed valve opening or by backwards flow (regurgitation) through a valve that has not closed completely. *Valvular stenosis* [*stenos*, narrow] may be inherited or may result from inflammation or other disease processes. At which step(s) in the cardiac cycle (Fig. 14-25) would you expect to hear a murmur caused by the following pathologies? (a) aortic valve stenosis, (b) mitral valve regurgitation, (c) aortic valve regurgitation

Pressure-Volume Curves Represent One Cardiac Cycle

Another way to describe the cardiac cycle is with the use of a pressure-volume graph, shown in Figure 14-26 ●. This figure represents the changes in volume (*x*-axis) and pressure (*y*-axis) that occur during one cardiac cycle. The flow of blood through the heart is governed by the same principle that governs the flow of all liquids and gases: flow proceeds from areas of higher pressure to areas of lower pressure. When the heart contracts, the pressure increases and blood flows out of the heart into areas of lower pressure.

The graph shown in Figure 14-26 represents pressure and volume changes in the left ventricle, which sends blood into the systemic circulation. The left side of the heart creates higher pressures than the right side, which sends blood through the shorter pulmonary circuit.

The cycle begins at point A. The ventricle has completed a contraction and contains the minimum amount of blood that it will hold during the cycle. It has relaxed, and its pressure is also at its minimum value. Blood is flowing into the atrium from the pulmonary veins.

Once pressure in the atrium exceeds pressure in the ventricle, the mitral valve, between the atrium and ventricle, opens (Fig. 14-26, point A). Atrial blood now flows into the ventricle, increasing its volume (point A to point B). As blood flows in, the relaxing ventricle expands to accommodate the entering blood. Consequently, the volume of the ventricle increases, but the pressure in the ventricle goes up very little.

The last portion of ventricular filling is completed by atrial contraction. The ventricles now contain the maximum volume of blood that they will hold during this cardiac cycle (point B). Because maximum filling occurs at the end of ventricular relaxation (diastole), this volume is called the **end-diastolic volume (EDV)**.

In a 70-kg man at rest, end-diastolic volume is about 135 mL. The EDV will vary under different conditions. For instance, during periods of very high heart rate, when the ventricles do not have time to fill completely between beats, the end-diastolic value may be less.

KEY

EDV = End-diastolic volume
ESV = End-systolic volume

● ● ● ● ● ● ●
FIGURE QUESTIONS

Match the following segments to the corresponding ventricular events:

A → B:	(a) Ejection of blood into aorta
B → C:	(b) Isovolumic contraction
C → D:	(c) Isovolumic relaxation
D → A:	(d) Passive filling and atrial contraction

Match the following events to points A – D:
(a) aortic valve opens
(b) mitral valve opens
(c) aortic valve closes
(d) mitral valve closes

● **Figure 14-26** *Left ventricular pressure-volume changes during one cardiac cycle*

When ventricular contraction begins, the mitral valve closes. With both AV and semilunar valves closed, blood in the ventricle has nowhere to go. Nevertheless, the ventricle continues to contract, causing pressure within the ventricle to increase rapidly during isovolumic contraction (B → C). Once ventricular pressure exceeds the pressure in the aorta, the aortic valve opens (point C). Pressure continues to increase as the ventricle contracts further, but ventricular volume decreases as blood is pushed out into the aorta (C → D).

The heart does not empty itself completely of blood each time the ventricle contracts. The amount of blood left in the ventricle at the end of contraction is known as the **end-systolic volume (ESV)**. The ESV is the minimum amount of blood that the ventricle will contain during one cycle (point D). An average value for the ESV in a person at rest is 65 mL, meaning that nearly half of the 135 mL that was in the ventricle at the start of the contraction is still there at the end of the contraction.

At the end of each ventricular contraction, the ventricle begins to relax. As it does so, ventricular pressure decreases. Once pressure in the ventricle falls below aortic pressure, the semilunar valve closes and the ventricle again becomes a sealed chamber. The remainder of relaxation occurs without a change in blood volume, so this phase is called *isovolumic relaxation* (Fig. 14-26, D → A). When ventricular pressure finally falls to the point at which atrial pressure exceeds ventricular pressure, the mitral valve opens and the cycle begins again.

The electrical and mechanical events of the cardiac cycle are summarized together in Figure 14-27 ●, known as a Wiggers diagram after the physiologist who first created it.

CONCEPT CHECK   Answers: p. 488

27. On Figure 14-25, at what points on the cycle do EDV and ESV occur?

28. On the Wiggers diagram in Figure 14-27, match the following events to the lettered boxes.

    (a) end diastolic volume, (b) aortic valve opens, (c) mitral valve opens, (d) aortic valve closes, (e) mitral valve closes, (f) end systolic volume

29. Why does atrial pressure increase following point C? Why does it decrease during the initial part of ventricular systole, then increase? Why does it decrease suddenly at point D?

30. Why does ventricular pressure shoot up suddenly at point C?

Stroke Volume Is the Volume of Blood Pumped by One Ventricle in One Contraction

What is the purpose of leaving blood in the ventricles at the end of each contraction? For one thing, the resting end-systolic volume of 65 mL provides a safety margin. With a more forceful contraction, the heart can decrease its ESV, sending additional blood to the tissues. Like many physiological functions, the heart does not usually work "all out."

The amount of blood pumped by one ventricle during a contraction is known as the **stroke volume**. It is measured in milliliters per beat and can be calculated as follows:

Volume of blood before contraction −
    volume of blood after contraction = stroke volume

EDV − ESV = stroke volume

For the average contraction in a person at rest:

135 mL − 65 mL = 70 mL, the normal stroke volume

Stroke volume is not constant and can increase to as much as 100 mL during exercise. Stroke volume, like heart rate, is homeostatically regulated by mechanisms discussed later in this chapter.

Fibrillation

Coordination of myocardial contraction is essential for normal cardiac function. In extreme cases in which the myocardial cells contract in an extremely disorganized manner, a condition known as *fibrillation* results. Ventricular fibrillation is a life-threatening emergency because without coordinated contraction of the muscle fibers, the ventricles cannot pump enough blood to supply adequate oxygen to the brain. One way to correct this problem is to administer an electrical shock to the heart. The shock creates a depolarization that triggers action potentials in all cells simultaneously, coordinating them again. You have probably seen this procedure on television hospital shows, when the doctors place flat paddles on the patient's chest and tell everyone to stand back ("Clear!") while they pass an electrical current through the body.

Cardiac Output Is a Measure of Cardiac Performance

How can we assess the effectiveness of the heart as a pump? One way is to measure cardiac output, the volume of blood pumped by the heart in a given period of time. All blood that leaves the heart flows through the tissues, therefore cardiac output is an indicator of total blood flow through the body. However, cardiac output does not tell us how blood is distributed to various tissues. That aspect of blood flow is regulated at the tissue level.

Cardiac output is defined as the volume of blood pumped per ventricle per unit time. It can be calculated by multiplying heart rate (beats per minute) by stroke volume (mL/beat, or contraction). For an average resting heart rate of 72 beats per minute and stroke volume of 70 mL per beat:

Cardiac output (CO) = heart rate × stroke volume

CO = 72 beats/min × 70 mL/beat

CO = 5040 mL/min ≈ 5 L/min

Average total blood volume is about 5 liters. This means that at rest, one side of the heart pumps all the blood in the body through it in only one minute!

During exercise, cardiac output may increase to 30–35 L/min. Homeostatic changes in cardiac output are accomplished by varying the heart rate, the stroke volume, or both. Normally, cardiac output is the same for both ventricles. If one side of the heart begins to fail for some reason and is unable to pump efficiently, cardiac output becomes mismatched. In that situation, blood will pool in the circulation behind the weaker side of the heart.

● **Figure 14-27** *The Wiggers diagram* The Wiggers diagram shows the relationship between the electrocardiogram, heart sounds, and pressure and volume changes in the left heart and aorta. The boxed letters refer to Concept Check question 28.

Walter was in the cardiac care unit by 1 P.M., where the cardiologist visited him. "We need to keep an eye on you here for the next week. There is a good chance the damage from your heart attack could cause an irregular heartbeat." Once Walter was stable, he would have a coronary angiogram, a procedure in which an opaque dye visible on X-rays shows regions where the lumen of the vessel narrows. Depending on the results of that test, the physician might recommend either balloon angioplasty, in which a tube passed into the coronary artery is inflated to open up the blockage, or coronary bypass surgery, in which veins from other parts of the body are grafted onto the heart arteries to provide bypass channels around blocked regions.

Question 9: If Walter's heart attack has damaged the muscle of his left ventricle, what do you predict will happen to his left cardiac output?

448 452 461 470 475 483

CONCEPT CHECK
Answers: p. 488

31. If the stroke volume of the left ventricle is 250 mL/beat and the stroke volume of the right ventricle is 251 mL/beat, what will happen to the relative distribution of blood between the systemic and pulmonary circulation after 10 beats?

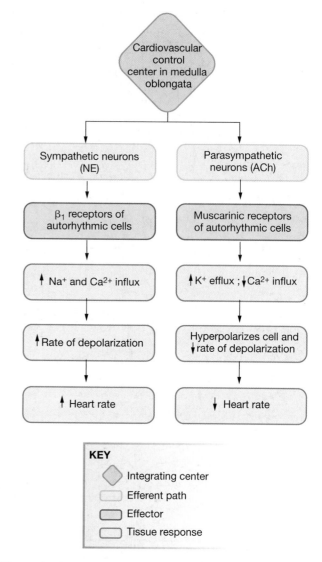

● **Figure 14-28** *Reflex control of heart rate*

Heart Rate Is Varied by Autonomic Neurons and Catecholamines

An average resting heart rate in an adult is 70 beats per minute (bpm). The normal range for heart rate is highly variable, however. Trained athletes may have resting heart rates of 50 bpm or less. Someone who is excited or anxious may have a rate of 125 bpm or higher. Children have higher average heart rates than adults. Although heart rate is initiated by autorhythmic cells in the SA node, it is modulated by neural and hormonal input.

The sympathetic and parasympathetic branches of the autonomic division influence heart rate through antagonistic control (Fig. 14-28●). Parasympathetic activity slows heart rate, while sympathetic activity speeds it up. Normally, the tonic control of heart rate is dominated by the parasympathetic branch.

This can be shown experimentally by blocking all autonomic input to the heart. When all sympathetic and parasympathetic input is blocked, the spontaneous depolarization rate of the SA node is 90–100 times per minute. This means that to achieve a heart rate of 70 beats per minute, tonic parasympathetic activity must slow the intrinsic rate down from 90 beats per minute.

An increase in heart rate can be achieved in two ways. (1) The simplest method to increase rate is to decrease parasympathetic activity. As parasympathetic influence is withdrawn from the autorhythmic cells, they resume their intrinsic rate of depolarization, and heart rate increases to 90–100 beats per minute. (2) To

increase heart rate above the intrinsic rate, sympathetic input is required. As you learned earlier, norepinephrine (or epinephrine) on β_1 receptors speeds up the depolarization rate of the autorhythmic cells and increases heart rate.

Both autonomic branches also alter the rate of conduction through the AV node. Acetylcholine secreted by parasympathetic neurons slows the conduction of action potentials through the AV node, thus increasing AV node delay. In contrast, the catecholamines epinephrine and norepinephrine enhance conduction of action potentials through the AV node and through the conducting system.

Multiple Factors Influence Stroke Volume

Stroke volume, the volume of blood pumped per ventricle per contraction, is directly related to the force generated by cardiac muscle during a contraction. The greater the force of contraction, the greater the stroke volume. Force is affected by two parameters:

the length of the muscle fiber at the beginning of contraction and the contractility of the heart.

Contractility is the intrinsic ability of a cardiac muscle fiber to contract at any given fiber length. Changes in contractility are traditionally considered different from changes in force due to variations in muscle (sarcomere) length. However, it now appears that increasing sarcomere length makes cardiac muscle more sensitive to Ca^{2+}, linking contractility to muscle length.

Length-Tension Relationships and Starling's Law of the Heart

As you learned previously, the force created by a myocardial muscle fiber is directly related to the length of the sarcomere. As sarcomere length increases (up to an optimum length), tension created during contraction increases. In the intact heart, as tension on the ventricular wall increases, so does the stroke volume. If additional blood flows into the ventricles, the muscle fibers stretch, then contract more forcefully, ejecting more blood.

This relationship between stretch and force in the intact heart was first described by a German physiologist, Otto Frank. A British physiologist, Ernest Starling, then expanded upon it. Starling used an isolated heart-lung preparation from a dog and hooked it to a reservoir so that he could regulate the amount of blood returning to the heart. In the absence of any nervous or hormonal control, he found that under normal conditions, the heart would pump all the blood that returned to it.

The relationship between stretch and force in the intact heart is plotted on a Starling curve (Fig. 14-29 ●). The *x*-axis represents the end-diastolic volume (EDV), the volume of blood in the ventricle at the beginning of contraction. Sarcomere length is proportional to the end-diastolic volume. The *y*-axis of the Starling curve represents the stroke volume, or amount of blood ejected from the ventricle per contraction.

The graph shows that stroke volume is proportional to force. As additional blood enters the heart, the heart contracts more forcefully. This relationship is known as the **Frank-Starling law of the heart**. It means that within physiological limits, the heart pumps all the blood that returns to it.

Stroke Volume and Venous Return

According to the Frank-Starling law, the more blood there is in the ventricle at the beginning of contraction (increased end-diastolic volume), the greater the stroke volume is. End-diastolic volume in turn is determined by **venous return**, the amount of blood that enters the heart from the venous circulation. The degree of myocardial stretch created by venous return is sometimes called the *pre-load* on the heart because it represents the load placed on cardiac muscles before contraction.

Three factors affect venous return: (1) contraction or compression of veins returning blood to the heart, (2) pressure changes in the abdomen and thorax during breathing, and (3) sympathetic innervation of veins. The skeletal muscle pump is the name given to skeletal muscle contractions that squeeze veins (particularly in the legs), compressing them and pushing blood toward the heart.

The so-called **respiratory pump** is created by the movement of the thorax during inspiration (breathing in). As the chest expands and the diaphragm moves toward the abdomen, the thoracic cavity, a sealed compartment, enlarges and develops a subatmospheric pressure. This low pressure creates lower pressure in the inferior vena cava as it passes through the thorax. The lower pressure helps draw more blood into the vena cava from veins in the abdomen.

The respiratory pump is aided by higher pressure placed on the outside of abdominal veins when the abdominal contents are compressed during inspiration. The combination of increased pressure in the abdominal veins and decreased pressure in thoracic veins enhances venous return during inspiration.

Constriction of veins by sympathetic activity is a third factor that affects venous return. When the veins constrict, their volume decreases. As a result, more blood flows out of the veins into the heart, and a larger volume of blood is in the ventricles at the beginning of the next contraction (increased end-diastolic volume). In turn, according to the Frank-Starling law, the ventricles contract more forcefully, sending the blood out into the arterial side of the circulation. In this manner, sympathetic innervation of veins allows the body to redistribute some venous blood to the arterial side of the circulation.

GRAPH QUESTION

What is the maximum stroke volume achieved in this experiment? At what end-diastolic volume is maximum stroke volume first achieved?

● **Figure 14-29** *Length-force relationships in the intact heart*
In the intact heart, stroke volume is used as an indicator of contractile force.

Reflex Control of Contractility

Contractility of the heart is controlled by the nervous and endocrine systems. Any chemical that affects contractility is called an *inotropic agent* [*ino*, fiber]. For example, catecholamines and drugs such as digitalis enhance contractility and are therefore considered to have a positive inotropic effect.

Figure 14-30 ● shows a normal Starling curve (the control curve) along with a curve showing how the stroke volume changes with increased contractility due to norepinephrine. Note that contractility is distinct from the length-tension relationship. A muscle can remain at one length (for example, one reflected by the end-diastolic volume marked A) but show increased contractility. Contractility increases as calcium available for contraction increases. As you learned earlier, catecholamines increase Ca^{2+} entry and storage as the basis for their positive inotropic effect.

The factors that affect cardiac output are summarized in Figure 14-31 ●. Cardiac output varies with both heart rate and stroke volume. Heart rate is modulated by the autonomic division of the nervous system and by epinephrine. The determination of stroke volume is more complex, consisting of an intrinsic myocardial response to stretch (Starling's law), interacting with adrenergically mediated changes in contractility. Venous return is a major determinant of end-diastolic volume and stretch. In the next chapter, we will examine how cardiac output plays a key role in blood flow through the circulation.

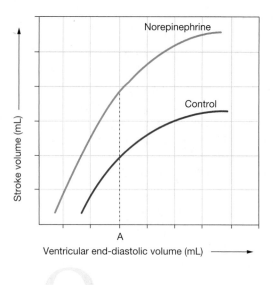

GRAPH QUESTION

At the end-diastolic volume indicated by point A, which heart will create more force: the control heart or the heart under the influence of norepinephrine?

● **Figure 14-30** *The effect of norepinephrine on contractility of the heart* Norepinephrine is a positive inotropic agent.

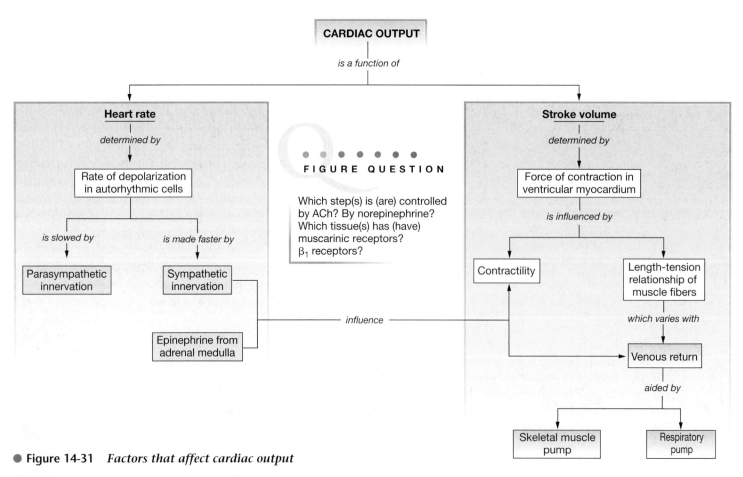

● **Figure 14-31** *Factors that affect cardiac output*

FIGURE QUESTION

Which step(s) is (are) controlled by ACh? By norepinephrine? Which tissue(s) has (have) muscarinic receptors? β_1 receptors?

Myocardial Infarction

Walter's angiogram showed two blocked arteries, which were opened by balloon angioplasty. He returned home with instructions from his doctor for modifying his lifestyle to include a better diet, regular exercise, and no cigarette smoking. As part of his follow-up, Walter had a *myocardial perfusion imaging* test, in which he was administered radioactive thallium. The distribution of thallium is an indicator of blood flow to the heart muscle. Photos from nuclear myocardial perfusion imaging are on the cover of this book.

In this running problem, you learned about some current techniques for diagnosing and treating heart attacks. You also learned that many of these treatments depend on speed to work effectively. Check your understanding of physiology in this running problem by comparing your answers to those in the summary table.

	QUESTION	FACTS	INTEGRATION AND ANALYSIS
1	Why did the medic give Walter oxygen?	The medic suspects that Walter has had a heart attack. Blood flow and oxygen supply to the heart muscle may be blocked.	If the heart is not pumping effectively, the brain may not receive adequate oxygen. Administration of oxygen will raise the amount of oxygen that reaches both the heart and the brain.
2	What effect would the injection of isotonic saline have on Walter's extracellular fluid volume? On his intracellular fluid volume? On his total body osmolarity?	An isotonic solution is one that does not change cell volume (● p. 154). Isotonic saline (NaCl) is isosmotic to the body.	The extracellular volume will increase because all of the saline administered will remain in that compartment. The intracellular volume and total body osmolarity will not change.
3	Some enzymes that are used as markers for heart attacks are also found in related forms in skeletal muscle. What are related forms of an enzyme called?	Related forms of an enzyme are called isozymes (● p. 92).	Although isozymes are variants of the same enzymes, their activity may vary with different conditions, and their structures are slightly different. Cardiac and skeletal muscle isozymes can be distinguished by their different structures.
4	What is troponin, and why would elevated blood levels of troponin indicate heart damage?	Troponin is the regulatory protein bound to tropomyosin (● p. 399). Ca^{2+} binding to troponin uncovers the myosin-binding site of actin to allow contraction.	Troponin is part of the contractile apparatus of the muscle cell. If troponin escapes from the cell and enters the blood, this is an indication that the cell has been damaged or is dead.
5	How do electrical signals move from cell to cell in the myocardium?	Electrical signals can pass through gap junctions in intercalated disks (● p. 64).	The cells of the heart are electrically linked by gap junctions.
6	What happens to contraction in a myocardial contractile cell if a wave of depolarization passing through the heart bypasses it?	Depolarization in a muscle cell is the signal for contraction.	If a myocardial cell is not depolarized, it will not contract. Failure to contract creates a nonfunctioning region of heart muscle and impairs the pumping function of the heart.
7	What will a beta-blocker do to Walter's heart rate? Why is that response helpful following a heart attack?	A beta-blocker is an antagonist to β_1 adrenergic receptors. Activation of β_1 receptors increases heart rate.	A beta-blocker therefore decreases heart rate and lowers oxygen demand. Cells that need less oxygen are less likely to die if their blood supply is diminished.
8	If the ventricle of the heart is damaged, in which wave or waves of the electrocardiogram would you expect to see abnormal changes?	The P wave of the ECG represents atrial depolarization. The QRS complex and T wave represent ventricular depolarization and repolarization, respectively.	The QRS complex and the T wave are most likely to show changes after a heart attack. Changes indicative of myocardial damage include enlargement of the Q wave, shifting of the S-T segment off of the base-line (elevated or depressed), and inversion of the T wave.
9	If Walter's heart attack has damaged the muscle of his left ventricle, what do you predict will happen to his left cardiac output?	Cardiac output equals stroke volume (the amount of blood pumped by one ventricle per contraction) times heart rate.	If the ventricular myocardium has been weakened, stroke volume may decrease. Decreased stroke volume would in turn decrease cardiac output.

448 452 461 470 475 480

CHAPTER SUMMARY

The cardiovascular system exemplifies many of the basic themes in physiology. Blood flows through vessels as a result of high pressure created during ventricular contraction (*mass flow*). The circulation of blood provides an essential route for *cell-to-cell communication*, particularly for hormones and other chemical signals. Myocardial contraction, like contraction in skeletal and smooth muscle, demonstrates the importance of *molecular interactions*, *biological energy*, and the *mechanical properties* of cells and tissues. This chapter also introduced the *control systems* for cardiovascular physiology, a theme that will be expanded in the next chapter.

Overview of the Cardiovascular System

 IP Cardiovascular—Anatomy Review: The Heart

1. The human **cardiovascular system** consists of a **heart** that pumps blood through a closed system of blood vessels. (p. 449; Fig. 14-1)

2. The primary function of the cardiovascular system is the transport of nutrients, water, gases, wastes, and chemical signals among all parts of the body. (p. 449; Tbl. 14-1)

3. Blood vessels that carry blood away from the heart are called **arteries**. Blood vessels that return blood to the heart are called **veins**. **Valves** in the heart and veins ensure unidirectional blood flow. (p. 450; Fig. 14-1)

4. The heart is divided into two **atria** and two **ventricles**. (p. 450; Fig. 14-1)

5. The **pulmonary circulation** goes from the right side of the heart to the lungs and back to the heart. The **systemic circulation** goes from the left side of the heart to the tissues and back to the heart. (pp. 450–451; Fig. 14-1)

Pressure, Volume, Flow, and Resistance

6. Blood flows down a **pressure gradient** (ΔP), from the highest pressure in the arteries to the lowest pressure in the **venae cavae** and **pulmonary veins**. (pp. 450–452; Fig. 14-2)

7. In a system in which fluid is flowing, pressure decreases over distance. (p. 452)

8. The pressure created when the ventricles contract is called the **driving pressure** for blood flow. (p. 453)

9. **Resistance** of a fluid flowing through a tube increases as length of the tube and **viscosity** (thickness) of the fluid increase. Resistance increases as the radius (r) of the tube decreases. Radius has the greatest effect on resistance. (p. 454)

10. If resistance increases, flow decreases. If resistance decreases, flow increases. (p. 454; Fig. 14-5)

11. Fluid flow through a tube is proportional to the pressure gradient (ΔP). A pressure gradient is not the same thing as the absolute pressure in the system. (p. 453; Fig. 14-4)

12. **Flow rate** (flow) is the volume of blood that passes one point in the system per unit time. (p. 455)

13. **Velocity of flow** is the distance a volume of blood will travel in a given period of time. At a constant flow rate, blood will flow faster through a smaller tube than it will through a larger tube. (p. 455; Fig. 14-6)

Cardiac Muscle and the Heart

 IP Cardiovascular—Cardiac Action Potential

14. The heart is composed mostly of cardiac muscle, or **myocardium**. Most cardiac muscle is typical striated muscle. (p. 456; Fig. 14-7)

15. The signal for contraction originates in **autorhythmic cells** within the heart. Autorhythmic cells are noncontractile myocardium. (p. 460)

16. Myocardial cells are linked by **intercalated disks** that contain gap junctions. Gap junctions allow depolarization to spread rapidly from cell to cell. (p. 461; Fig. 14-10)

17. In contractile cell excitation-contraction coupling, an action potential opens Ca^{2+} channels. Ca^{2+} entry triggers the release of Ca^{2+} from the sarcoplasmic reticulum through **calcium-induced calcium release**. (p. 462; Fig. 14-11)

18. The force of cardiac muscle contraction can be graded according to how much Ca^{2+} enters the cell. (p. 462)

19. Epinephrine and norepinephrine increase the force of myocardial contraction when they bind to β_1 adrenergic receptors. They also shorten the duration of cardiac contraction. (p. 463; Fig. 14-12)

20. As initial muscle fiber length increases, force of contraction increases. (p. 464; Fig. 14-13)

21. The action potentials of myocardial contractile cells have a rapid depolarization phase created by Na^+ influx and a steep

repolarization phase due to K⁺ efflux. The action potential also has a plateau phase created by Ca^{2+} influx. (p. 464; Fig. 14-14)

22. Autorhythmic myocardial cells have an unstable membrane potential called a **pacemaker potential**. The pacemaker potential is due to I_f channels that allow net influx of positive charge. (p. 466; Fig. 14-16)

23. The steep depolarization phase of the autorhythmic cell action potential is caused by Ca^{2+} influx. The repolarization phase is due to K⁺ efflux. (p. 466; Fig. 14-16)

24. Norepinephrine and epinephrine act on β₁ receptors to speed up the rate of the pacemaker depolarization and increase heart rate. Acetylcholine activates muscarinic receptors and slows down heart rate. (p. 467; Fig. 14-17)

The Heart as a Pump

 IP Cardiovascular—Intrinsic Conduction System

25. Action potentials originate at the **sinoatrial node** (SA node) and spread rapidly from cell to cell. Action potentials are followed by a wave of contraction. (p. 468; Fig. 14-18)

26. The electrical signal moves from the SA node through the **internodal pathway** to the **atrioventricular node** (AV node), then into the **A-V bundle**, **bundle branches**, terminal **Purkinje fibers**, and myocardial contractile cells. (p. 468; Fig. 14-19)

27. The SA node sets the pace of the heartbeat. If the SA node malfunctions, other autorhythmic cells in the AV node or ventricles will take control of heart rate. (p. 470)

28. An **electrocardiogram**, or ECG, is a surface recording of the electrical activity of the heart. The **P wave** represents atrial depolarization. The **QRS complex** represents ventricular depolarization. The **T wave** represents ventricular repolarization. Atrial repolarization is incorporated in the QRS complex. (p. 471; Figs. 14-21, 14-22)

29. An ECG provides information on heart rate and rhythm, conduction velocity, and the condition of tissues within the heart. (p. 472)

 IP Cardiovascular—The Cardiac Cycle

30. One **cardiac cycle** includes one cycle of contraction and relaxation. **Systole** is the contraction phase. **Diastole** is the relaxation phase. (pp. 474–475; Fig. 14-25)

31. Most blood enters the ventricles while the atria are relaxed. Only 20% of ventricular filling is due to atrial contraction. (p. 476)

32. The **AV valves** prevent backflow of blood into the atria. Vibrations following closure of the AV valves create the **first heart sound**. (pp. 460, 476; Figs. 14-9, 14-25)

33. During **isovolumic ventricular contraction**, the ventricular blood volume does not change but pressure rises. When ventricular pressure exceeds arterial pressure, the **semilunar valves** open and blood is ejected into the arteries. (pp. 460, 476; Figs. 14-9, 14-25)

34. When the ventricles relax and ventricular pressure falls, the semilunar valves close, creating the **second heart sound**. (pp. 460, 476; Fig. 14-9, 14-25)

35. The amount of blood pumped by one ventricle during a contraction is known as the **stroke volume**. (p. 478; Fig. 14-26)

 IP Cardiovascular—Cardiac Output

36. **Cardiac output** is the volume of blood pumped per ventricle per unit time. It is equal to heart rate times stroke volume. The average cardiac output at rest is 5 L/min. (p. 478)

37. Homeostatic changes in cardiac output are accomplished by varying heart rate, stroke volume, or both. (p. 480; Fig. 14-31)

38. Parasympathetic activity slows heart rate. Sympathetic activity speeds it up. (p. 480; Fig. 14-28)

39. The **Frank-Starling law of the heart** says that an increase in **end-diastolic volume** will result in a greater stroke volume. (p. 481; Fig. 14-29)

40. End-diastolic volume is determined by **venous return**. Venous return is affected by skeletal muscle contractions, the respiratory pump, and constriction of veins by sympathetic activity. (p. 481)

41. **Contractility** of the heart is enhanced by catecholamines and certain drugs. (p. 481; Fig. 14-30)

QUESTIONS

LEVEL ONE *Reviewing Facts and Terms*

1. What contributions to understanding the cardiovascular system did each of these people make?
   (a) William Harvey
   (b) Frank and Starling
   (c) Malpighi

2. List three functions of the cardiovascular system.

3. Put these structures in the order blood passes through them, starting and ending with the left ventricle:
   (a) left ventricle
   (b) systemic veins

   (c) pulmonary circulation
   (d) systemic arteries
   (e) aorta
   (f) right ventricle

4. The primary reason blood flows through the body is a _____ gradient. In humans, this value is highest at the _____ and in the _____. It is lowest in the _____. In a system in which fluid is flowing, pressure decreases over distance due to _____.

5. If vasodilation occurs in a blood vessel, pressure (increases/decreases).

6. The specialized cell junctions between myocardial cells are called _____. These areas contain _____ that allow rapid conduction of electrical signals.

7. Trace an action potential from the SA node through the conducting system of the heart.

8. Distinguish between the following pairs:
   (a) end-systolic volume and end-diastolic volume
   (b) sympathetic and parasympathetic control of heart rate
   (c) diastole and systole
   (d) systemic and pulmonary circulation
   (e) AV node and SA node

9. Match the descriptions with the correct anatomic terms. Not all terms are used. Give a definition for the unused terms.
   (a) tough membranous   1. aorta
       sac that encases the heart 2. apex
   (b) valves between ventricles 3. artery
       and the main arteries   4. atria
   (c) a vessel that carries blood 5. atrium
       away from the heart   6. AV valve
   (d) lower chamber of the heart 7. base

   (e) valve between left atrium   8. bicuspid valve
       and left ventricle   9. ventricle
   (f) primary artery of the   10. myocardium
       systemic circulation   11. pericardium
   (g) muscular layer of the heart 12. semilunar valve
   (h) narrow end of the heart;   13. tricuspid valve
       points downward   14. endothelium
   (i) valve with papillary muscles
   (j) the upper chambers of the heart

10. What events cause the two principal heart sounds?

11. What is the proper term for each of the following?
    (a) number of heart contractions per minute
    (b) volume of blood in the ventricle before the heart contracts
    (c) volume of blood that enters the aorta with each contraction
    (d) volume of blood that leaves the heart in one minute
    (e) volume of blood in the entire body

LEVEL TWO *Reviewing Concepts*

12. List the events of the cardiac cycle in sequence, beginning with atrial and ventricular diastole. Note where valves open and close. Describe what happens to pressure and blood flow in each chamber at each step of the cycle.

13. Compare and contrast the structure of a cardiac muscle cell with that of a striated muscle cell. What unique properties of cardiac muscle are essential to its function?

14. Concept maps: You may add any terms you like to the lists given.
    (a) Create a map showing blood flow through the heart and body. Label as many structures as you can.
    (b) Create a map for control of cardiac output using the terms below. You may add additional terms.

ACh	length-tension relationship
adrenal medulla	muscarinic receptor
autorhythmic cells	norepinephrine
Ca^{2+}	parasympathetic neurons
Ca^{2+}-induced Ca^{2+} release	respiratory pump
cardiac output	skeletal muscle pump
contractile myocardium	stroke volume
contractility	sympathetic neurons
force of contraction	venous return
heart rate	β_1 receptor

15. Explain why contractions in cardiac muscle cannot sum or exhibit tetanus.

16. Correlate the waves of the ECG with mechanical events in the atria and ventricles. Why are there only three electrical events but four mechanical events?

17. Match the ion movements with the appropriate phrase. More than one answer may apply to a single phrase. Answers may not all be used.
    (a) slow rising phase of   1. K^+ from ECF to ICF
        autorhythmic cells   2. K^+ from ICF to ECF
    (b) plateau phase of   3. Na^+ from ECF to ICF
        contractile cells   4. Na^+ from ICF to ECF
    (c) rapid rising phase of   5. Ca^{2+} from ECF to ICF
        contractile cells   6. Ca^{2+} from ICF to ECF
    (d) rapid rising phase of
        autorhythmic cells
    (e) rapid falling phase of
        contractile cells
    (f) falling phase of
        autorhythmic cells
    (g) cardiac muscle contraction
    (h) cardiac muscle relaxation

18. List and briefly explain four types of information that the ECG provides about the heart.

19. Define inotropic effect. Give examples of two drugs with a positive inotropic effect. What is the role of Ca^{2+} in this process?

LEVEL THREE *Problem Solving*

20. Two drugs used to reduce cardiac output are calcium channel blockers and beta (receptor) blockers. What effect do these drugs have on the heart that explains how they decrease cardiac output?

21. Police Captain Jeffers has suffered a myocardial infarction.
    (a) Explain to his (nonmedically oriented) family what has happened to his heart.

    (b) When you analyzed his ECG, you referred to several different leads, such as lead I. What are leads?
    (c) Why is it possible to record an ECG on the body surface without direct access to the heart?

22. What might cause a longer-than-normal P-R interval in an ECG?

23. The following paragraph is a summary of a newspaper article:

    A new treatment for atrial fibrillation due to an excessively rapid rate at the SA node involves a high-voltage electrical pulse administered to the AV node to destroy its autorhythmic cells. A ventricular pacemaker is then implanted in the patient.

    Briefly explain the physiological rationale for this treatment. Why is a rapid atrial depolarization rate dangerous, why is the AV node destroyed, and why must a pacemaker be implanted?

LEVEL FOUR *Quantitative Problems*

24. Police Captain Jeffers in question 21 above has an *ejection fraction* (SV divided by EDV) of only 25%. His stroke volume is 40 mL and his heart rate is 100 beats per minute. What are his EDV, ESV, and CO? Show your calculations.

25. If 1 cm water = 0.74 mm Hg:
    (a) Convert a pressure of 120 mm Hg to cm H_2O.
    (b) Convert a pressure of 80 torr to cm H_2O.

26. Calculate cardiac output if stroke volume is 65 mL and heart rate is 80 bpm.

27. Calculate end-systolic volume if end-diastolic volume is 150 mL and stroke volume is 65 mL.

28. A person has a total blood volume of 5 liters. Of this total, assume 4 liters is contained in the systemic circulation and 1 liter is in the pulmonary circulation. If the person has a cardiac output of 5 L/min: (a) How long will it take for a drop of blood leaving the left ventricle to return to the left ventricle? (b) How long will it take for a drop of blood to go from the right ventricle to the left ventricle?

Explorations

Explorations are also available at *www.physiologyplace.com*

The cardiovascular system is a wonderful example of both electrical and mechanical physiology. The heart is made of a variety of tissues that function distinctly yet in cooperation with one another. In the activities below, expand your understanding of cardiac function.

Exploration 1: The Electrocardiogram

Your heart beats about 3,000,000,000 times during your life, each beat arising from an action potential. What is just as amazing is that we can record this electrical activity using electrodes placed on the skin's surface. To learn more about the discovery of "animal electricity" and the development of the electrocardiogram, try a Google search (*www.google.com*) for *animal electricity* and *electrocardiography history*. For some other interesting articles on the history of cardiac physiology, go to *http://nips.physiology.org* to read "From funny current to HCN channels: 20 years of excitation," *News in Physiological Sciences* 17: 32–37, 2002 Feb.

"Who discovered the Frank-Starling mechanism?" *News in Physiological Sciences* 17: 181–184, 2002 Oct.

"William Harvey and the circulation of the blood: The birth of a scientific revolution and modern physiology," *News in Physiological Sciences* 17: 175–180, 2002 Oct.

The term *electrocardiogram* is abbreviated both ECG and EKG, from the Greek word *kardia*, meaning *heart*. To find out more about the clinical use of electrocardiography, do a Google search for ECG or EKG. You will find a number of websites with tutorials and examples of abnormal ECGs so that you can practice your interpretation skills.

Exploration 2: Cardiac Pathologies

The heart is a complex organ with many parts that can malfunction. In the next chapter you will learn about high blood pressure and atherosclerosis, and how these conditions can cause the heart to fail in its role as a pump. Another form of cardiac pathology that can have potentially fatal consequences is the cardiac *arrhythmias* [a-, without + rhythm]. These are electrical problems in the generation or conduction of action potentials through the heart, and they can usually be seen on an ECG. Some arrhythmias are "dropped beats," where the ventricles do not get their usual signal to contract. Other arrhythmias are extra beats that occur when an autorhythmic cell other than the SA node jumps in and fires an action potential out of sequence.

One heart condition called *long QT syndrome* (LQTS) is named for its effect on the QT interval on the ECG. LQTS has several forms. Several are inherited channelopathies, with mutations in myocardial Na^+ or K^+ channels. In another form, the ion channels are normal but the protein *ankyrin-B* that anchors the channels to the cell membrane is defective. An *iatrogenic* (physician-caused) form of LQTS occurred in the 1990s when patients took a nonsedating antihistamine called terfenadine (Seldane®) that binds to K^+ repolarization channels. After at least eight deaths were attributed to the drug, the U.S. Food and Drug Administration removed it from the market.

For more on LQTS, arrhythmias, and some of the treatments for them, try a web search or read: "Sudden cardiac death," *Scientific American* 1980 Nov.

"Molecular biology of K^+ channels and their role in cardiac arrhythmias," *The American Journal of Medicine* 110: 50–59, 2001 Jan.

ANSWERS

Page 452

1. A cardiovascular system has tubes (vessels), fluid (blood), and a pump (heart).

2. (a) Pulmonary circulation goes to lungs, systemic to remainder of the body. (b) Artery carries blood away from heart; vein carries blood to the heart. (c) Atrium is an upper heart chamber that receives blood; ventricle is a lower chamber that pumps blood.

Page 455

3. The pressure gradient determines flow.

4. The bottom tube has the greatest flow because it has the larger pressure gradient (50 mm Hg versus 40 mm Hg).

5. Tube 3 has the greatest flow because it has the larger radius (less resistance) and the shorter length (less resistance). Tube 4, with the greatest resistance due to length and narrow radius, has the least flow.

6. If the canals are identical in size and therefore in cross-sectional area, the canal with the faster flow will have the higher rate of flow.

Page 460

7. Connective tissue is not an excitable tissue and is unable to conduct action potentials.

8. Superior vena cava → right atrium → tricuspid (right AV) valve → right ventricle → pulmonary (right semilunar) valve → pulmonary trunk → right or left pulmonary artery → pulmonary arteriole → pulmonary capillary → pulmonary venule → pulmonary vein → left atrium → mitral (bicuspid, left AV) valve → left ventricle → aortic (left semilunar) valve → aorta

9. The AV valves prevent backward flow of blood. If one fails, blood leaks back into the atrium.

Page 462

10. From this experiment, it is possible to conclude that myocardial cells require extracellular calcium for contraction, but skeletal muscle cells do not.

Page 464

11. If all calcium channels in the muscle are blocked, there will be no contraction. If only some are blocked, the force of contraction will be less.

Page 465

12. Na^+ influx causes neuronal depolarization and K^+ efflux causes neuronal repolarization.

13. The refractory period represents the time required for the Na^+ channel gates to reset (activation gate closes, inactivation gate opens).

14. If cardiac Na^+ channels are blocked with lidocaine, the cell will not depolarize and therefore will not contract.

Page 466

15. Increasing K^+ permeability hyperpolarizes the membrane potential.

Page 468

16. The Ca^{2+} channels in the autorhythmic cells are not the same as the Ca^{2+} channels in the contractile cells. Autorhythmic Ca^{2+} channels open rapidly when the membrane potential reaches about -50 mV and close when it reaches about $+20$ mV. The Ca^{2+} channels in contractile cells are slower and do not open until the membrane has depolarized fully.

17. If TTX is applied to a myocardial autorhythmic cell, nothing will happen as there are no voltage-gated Na^+ channels in those cells.

18. Cutting the nerve caused heart rate to speed up, therefore the nerve must contain parasympathetic fibers that slow the rate.

Page 471

19. The AV node conducts action potentials from atria to ventricles. It also slows down their conduction of those action potentials, allowing atrial contraction to end before ventricular contraction begins.

20. The SA node is in the upper right atrium.

21. The fastest pacemaker sets the heart rate. Therefore, the heart rate increases to 120 beats/min.

Page 474

22. Starting at left, the waves are P, P, QRS, T, P, P, QRS, T, P, P, P, and so on. A P wave that is not followed by a QRS wave suggests an intermittent conduction block at the AV node. See also part (b)

23. In Figure 14-24b, notice that there is no regular association between the P waves and the QRS complexes (the P-R segment varies in length). Notice also that not every P wave has an associated QRS complex. Both sets of waves appear at regular intervals, but the atrial rate (P waves) is faster than the ventricular rate (QRS complexes). In Figure 14-24c, there are identifiable R waves but no P waves. In (d), there are no recognizable waves at all, showing that the depolarizations are not following the normal conduction path.

Page 475

24. The atrium has lower pressure than the venae cavae.

Page 477

25. (a) ventricles, (b) ventricles, (c) atria, (d) ventricles

26. (a) ventricular ejection, (b) isovolumic ventricular contraction and ventricular ejection (c) from isovolumic ventricular relaxation until ventricular contraction begins again

Page 478

27. EDV is step 3 and ESV is step 5.

28. (a) E, (b) A, (c) D, (d) B, (e) C, (f) F

29. Atrial pressure increases because pressure on the mitral valve pushes it back into the atrium, decreasing atrial volume. Atrial pressure decreases during the initial part of ventricular systole as the atria relax. It then increases as the atria fill with blood. Atrial pressure decreases at point D when the mitral valve opens and blood flows down into the ventricles.

30. Ventricular pressure shoots up when the ventricles contract on a fixed volume of blood.

Page 480

31. After 10 beats, the pulmonary circulation will have gained 10 mL of blood and the systemic circulation will have lost 10 mL.

Answers to Figure Questions

Page 472

Figure 14-21: The heart rate is 75–80 beats per minute, depending on how you calculate it. It is more accurate to estimate rate by using several seconds of the ECG tracing rather than using one R-R interval, as beat-to-beat intervals may vary. If you start counting at the first R wave and go right for 3 seconds, there are 4 beats in that time period, which equals 80 bpm.

Page 477

Figure 14-26: Match segments to events: (a) 3, (b) 2, (c) 4, (d) 1. Match points to events: (a) C, (b) A, (c) D, (d) B.

Page 481

Figure 14-29: Maximum stroke volume is 160 mL, first achieved when end-diastolic volume is about 330 mL.

Page 482

Fig. 14-30: At point A, the heart under the influence of norepinephrine has a larger stroke volume and therefore is creating more force.

Page 482

Fig. 14-31: Heart rate is the only parameter controlled by ACh. Heart rate and contractility are both controlled by norepinephrine. The SA node has muscarinic receptors. The SA node and contractile myocardium have β_1 receptors.

Answers to Level Four Quantitative Problems

Page 487

24. SV/EDV = 0.25. If SV = 40 mL, EDV = 160 mL. SV = EDV − ESV, so 40 mL = 160 mL − ESV, therefore ESV = 120 mL. CO = HR × SV = 100 bpm × 40 mL/beat = 4000 mL/min or 4 L/min.

25. (a) 162.2 cm H_2O (b) 108.1 cm H_2O

26. 5200 mL/min or 5.2 L/min

27. 85 mL

28. (a) 1 min (b) 12 sec

Blood Flow and the Control of Blood Pressure

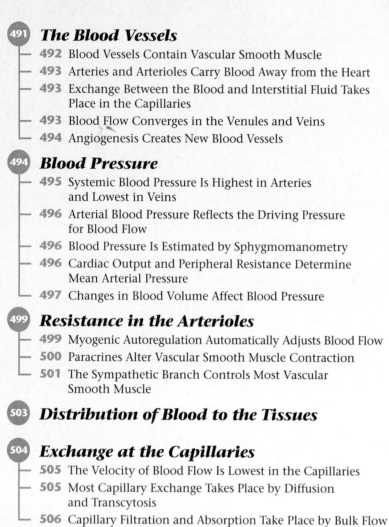

Since 1900, CVD (cardiovascular disease) has been the No. 1 killer in the United States every year but 1918.

—American Heart Association, *Heart Disease and Stroke Statistics—2003 Update*

Colored scan of the heart enclosed in the rib cage

RUNNING PROBLEM

Essential Hypertension

"Doc, I'm as healthy as a horse," says Kurt English, age 56, during his annual physical examination. "I don't want to waste your time. Let's get this over with." But to Dr. Arthur Cortez, Kurt does not appear to be the picture of health: he is about 30 pounds overweight. When Dr. Cortez asks about his diet, Kurt replies, "Well, I like to eat." Exercise? "Who has the time?" replies Kurt. Dr. Cortez slips a blood pressure cuff around Kurt's arm and takes a reading. "Your blood pressure is 164 over 100," says Dr. Cortez. "We'll take it again in 15 minutes. If it's still high, we'll need to discuss it further." Kurt stares at his doctor, flabbergasted. "But how can my blood pressure be too high? I feel fine!" he protests.

BACKGROUND BASICS

Basement membrane: **67** ∣ Nitric oxide: **183, 267** ∣ Transcytosis: **148** ∣ Fight-or-flight response: **370** ∣ Exchange epithelium: **67** ∣ Catecholamines and receptors: **376ff** ∣ Caveolae: **146** ∣ Diffusion: **136** ∣ Tonic control: **188** ∣

Anthony was always sure he was going to be a physician, until the day in physiology laboratory that they studied blood types. When the lancet pierced his fingertip and he saw the drop of bright red blood well up, the room started to spin, then everything went black. He awoke, much embarrassed, to the sight of his classmates and the teacher bending over him.

Anthony suffered an attack of *vasovagal syncope* (fainting), a benign and common reaction to blood, hypodermic needles, or other upsetting sights. Normally, homeostatic regulation of the cardiovascular system maintains adequate blood flow (*perfusion*) to the heart and brain. In vasovagal syncope, signals from the nervous system cause a sudden fall in blood pressure, and the patient faints from lack of oxygen to the brain. In this chapter you will learn how the heart and blood vessels work together most of the time to prevent such problems.

A simplified model of the cardiovascular system (Fig. 15-1 ●) summarizes the key points that we will discuss in this chapter. This model shows the heart as two separate pumps that work in series (one after the other), with the right heart pumping blood to the lungs, then to the left heart. The left heart then pumps blood through the rest of the body.

Blood leaving the left heart enters the systemic arteries, shown here as an expandable, elastic region. Pressure produced by contraction of the left ventricle is stored in the elastic walls of the arteries and slowly released through *elastic recoil*. This mechanism maintains a continuous driving pressure for blood flow during the time when the ventricles are relaxing.

The arterioles create a high-resistance outlet for arterial blood flow. In addition, the arterioles direct distribution of blood flow to individual tissues by selectively constricting and dilating. Arteriolar diameter is regulated both by local factors, such as tissue oxygen, and by homeostatic control.

Once blood flows into the capillaries, a leaky epithelium allows exchange of material between the plasma, the interstitial fluid, and the cells of the body. At the distal end of the capillaries, blood flows into the venous side of the circulation and from there back to the right side of the heart.

Total blood flow through any level of the circulation is equal to the cardiac output. For example, if cardiac output is 5 L/min, blood flow through all the systemic capillaries is 5 L/min. In the same manner, blood flow through the pulmonary side of the circulation is equal to blood flow through the systemic circulation.

▶ THE BLOOD VESSELS

The walls of blood vessels are made from layers of smooth muscle, elastic connective tissue, and fibrous connective tissue (Fig. 15-2 ●). The inner lining of all blood vessels is a thin layer of **endothelium**, a type of epithelium. For years, the endothelium was thought to be simply a passive barrier. However, we now know that endothelial cells play important roles in the regulation of blood pressure, blood vessel growth, and absorption of materials.

Surrounding the endothelium are layers of connective tissue and smooth muscle. The thickness and type of connective tissue vary with different vessels, as does the thickness of the smooth muscle layer. The descriptions below apply to the vessels of the systemic circulation, although those of the pulmonary circulation

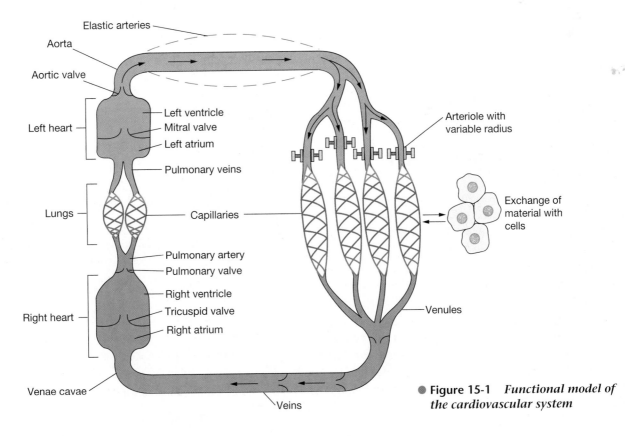

● **Figure 15-1** *Functional model of the cardiovascular system*

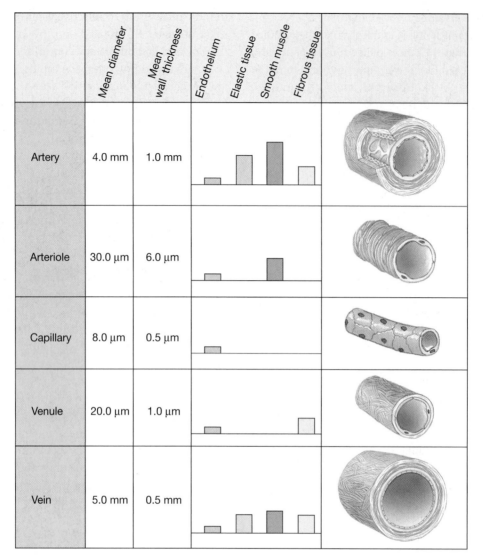

	Mean diameter	Mean wall thickness	Endothelium	Elastic tissue	Smooth muscle	Fibrous tissue	
Artery	4.0 mm	1.0 mm					
Arteriole	30.0 µm	6.0 µm					
Capillary	8.0 µm	0.5 µm					
Venule	20.0 µm	1.0 µm					
Vein	5.0 mm	0.5 mm					

● **Figure 15-2** *Blood vessels* The walls of blood vessels vary in their diameter and composition.

are generally similar. Functional aspects of the different types of blood vessels will be described in detail later in the chapter.

Blood Vessels Contain Vascular Smooth Muscle

The smooth muscle of blood vessels is known as **vascular smooth muscle**. Most blood vessels contain smooth muscle, arranged in either circular or spiral layers. *Vasoconstriction* narrows the diameter of the vessel lumen and *vasodilation* widens it.

In most blood vessels, smooth muscle cells maintain a state of partial contraction at all times, creating the condition known as *muscle tone*. Contraction of smooth muscle, like that of cardiac muscle, depends on the entry of Ca^{2+} from the extracellular fluid through Ca^{2+} channels [● p. 417]. A variety of chemicals influences vascular smooth muscle tone, including neurotransmitters, hormones, and paracrines. Many vasoactive paracrines are secreted by endothelial cells lining the blood vessels or by the tissues surrounding the vessels.

Calcium Channel Blockers

The calcium channels of both vascular smooth muscle and cardiac muscle can be blocked by a class of drugs known as calcium channel blockers. These drugs bind to Ca^{2+} channel proteins, making it less likely that the channels will open in response to depolarization. With diminished Ca^{2+} entry, vascular smooth muscle dilates, while in the heart, the depolarization rate of the SA node decreases. Vascular smooth muscle is more sensitive than cardiac muscle to certain classes of calcium channel blockers, and it is possible to get vasodilation at low doses that have no effect on heart rate. Other tissues with Ca^{2+} channels, such as neurons, are only minimally affected by calcium channel blockers because their Ca^{2+} channels are a different subtype.

Arteries and Arterioles Carry Blood Away from the Heart

The aorta and major arteries are characterized by walls that are both stiff and springy. Arteries have thick smooth muscle layers with large amounts of elastic and fibrous tissue (Fig. 15-2). Because of the stiffness of the fibrous tissue, substantial amounts of energy are required to stretch the walls of an artery outward. This energy comes in the form of high-pressure blood ejected from the left ventricle. Once the artery is distended with blood, the energy stored by stretching elastic fibers is released through elastic recoil.

The arteries and arterioles are characterized by a divergent pattern of blood flow. As the major arteries divide into smaller and smaller arteries, the character of the wall changes, becoming less elastic and more muscular. Finally, the smallest arteries become **arterioles**. The walls of arterioles contain several layers of smooth muscle that contract and relax under the influence of various chemical signals.

Some arterioles branch into vessels known as **metarterioles** [*meta-*, beyond]. Metarterioles have only part of their wall surrounded by smooth muscle, in comparison with true arterioles, whose walls have a continuous muscle layer. Blood flowing through metarterioles either can be directed into adjoining capillary beds (Fig. 15-3 ●) or can bypass the capillaries and go directly to the venous circulation if *precapillary sphincters* (rings of muscle) are contracted.

In addition to regulating blood flow through the capillaries, metarterioles allow white blood cells to go directly from the arterial to the venous circulation. Capillaries are barely large enough to let red blood cells through, much less white blood cells that are twice as large.

The arterioles, along with the capillaries and venules, are called the *microcirculation*. The regulation of blood flow through the microcirculation is an active area of physiological research.

Exchange Between the Blood and Interstitial Fluid Takes Place in the Capillaries

Capillaries are the smallest vessels in the cardiovascular system. They and the postcapillary venules are the site of exchange between the blood and the interstitial fluid. To facilitate exchange of material, capillary walls lack smooth muscle and elastic or fibrous tissue reinforcement (Fig. 15-2). Instead, they consist of a flat layer of endothelium, one cell thick, supported on an acellular matrix called the *basement membrane* or *basal lamina* [p. 67].

Many capillaries are closely associated with cells known as **pericytes** [*peri-*, around]. These highly branched contractile cells surround the capillaries, forming a meshlike outer layer between the endothelium and the interstitial fluid. Pericytes contribute to the "tightness" of capillary permeability: the more pericytes, the less leaky the capillary endothelium. For example, cerebral capillaries are surrounded by pericytes and glial cells and have tight junctions that create the *blood-brain barrier* [p. 292].

Pericytes secrete factors that influence capillary growth, and they can differentiate to become new endothelial or smooth muscle cells. Loss of pericytes around capillaries of the retina is a hallmark of the disease *diabetic retinopathy*, a leading cause of blindness. Scientists are now trying to determine whether pericyte loss is a cause or consequence of the retinopathy.

Blood Flow Converges in the Venules and Veins

Blood flows from the capillaries into small vessels called **venules**. The very smallest venules are similar to capillaries, with a thin exchange epithelium and little connective tissue (Fig. 15-2). They are distinguished from capillaries by their convergent pattern of flow.

Smooth muscle begins to appear in the walls of larger venules. From venules, blood flows into veins that become larger in diameter as they travel toward the heart. Finally, the largest veins, the venae cavae, empty into the right atrium.

Veins are more numerous than arteries and have a larger diameter. As a result of their large volume, the veins hold more than half of the blood in the circulatory system. Veins lie closer to the surface of the body than arteries, forming the bluish blood vessels that you see running just under the skin. Veins have thinner walls than arteries, with less elastic tissue. As a result, they expand easily when they fill with blood.

When you have blood drawn from your arm (*venipuncture*), the technician uses a tourniquet to exert pressure on the blood vessels. Blood flow coming into the arm through high-pressure arteries is not affected, but pressure of the tourniquet stops outflow through

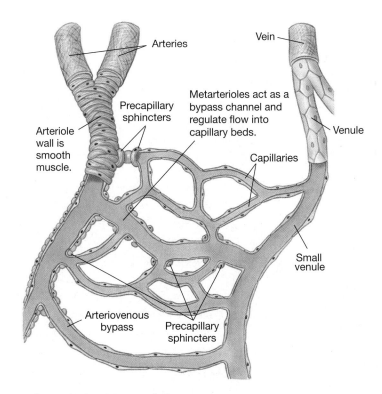

Arteries

Vein

Precapillary sphincters

Metarterioles act as a bypass channel and regulate flow into capillary beds.

Arteriole wall is smooth muscle.

Venule

Capillaries

Small venule

Arteriovenous bypass

Precapillary sphincters

● **Figure 15-3** *Metarterioles*

the low-pressure veins. As a result, blood collects in the surface veins, making them stand out against the underlying muscle tissue.

Angiogenesis Creates New Blood Vessels

One topic of tremendous interest to researchers is **angiogenesis**, the process by which new blood vessels develop, especially after birth [*angeion*, vessel + *gignesthai*, to beget]. In children, capillary growth is necessary for normal development. In adults, angiogenesis takes place during wound healing and during growth of the uterine lining after menstruation. Angiogenesis also occurs with endurance exercise training, enhancing blood flow to the heart muscle and to skeletal muscles.

From studies of normal blood vessels and tumor cells, scientists learned that angiogenesis is controlled by a balance of angiogenic and antiangiogenic cytokines. A number of related growth factors, including *vascular endothelial growth factor* (VEGF) and *fibroblast growth factor* (FGF), promote angiogenesis. These growth factors are *mitogens*, meaning they promote mitosis, or cell division. They are normally produced by smooth muscle cells and pericytes. Cytokines that inhibit angiogenesis include *angiostatin*, made from the blood protein plasminogen, and *endostatin* [*stasis*, a state of standing still].

Scientists are currently using these cytokines to develop new treatments for two major illnesses: cancer and coronary heart disease. As cancer cells invade tissues and multiply, they must grow new blood vessels to maintain a supply of nutrients and oxygen.

Without these vessels, the interior cells of a cancerous mass are unable to get adequate oxygen and nutrients, so they die. Angiostatin and endostatin are being used in clinical trials to see if they can block angiogenesis and literally starve tumors to death.

Coronary heart disease, in contrast, is a condition in which we would like to be able to selectively induce angiogenesis. In coronary heart disease, blood flow to the myocardium is decreased by fatty deposits that narrow the lumen of the coronary blood vessels. If we can induce the growth of new blood vessels to replace the vessels that are becoming blocked, we might be able to prevent the damage caused by lack of oxygen.

▶ BLOOD PRESSURE

The pressure created by ventricular contraction is the driving force for blood flow through the vessels of the system [❀ p. 453]. As blood is ejected from the left ventricle, the aorta and arteries expand to accommodate it (Fig. 15-4●). When the ventricle relaxes and the semilunar valve closes, the elastic arterial walls recoil, propelling the blood forward into smaller arteries and arterioles (Fig. 15-4b).

By sustaining the driving pressure for blood flow during ventricular relaxation, the arteries create continuous blood flow through the blood vessels. Flow in the arterial side of the circulation is *pulsatile*, reflecting the changes in arterial pressure throughout a cardiac cycle. Once past the arterioles, pulse waves disappear.

(a) Ventricular contraction

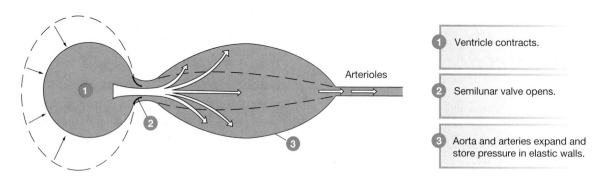

Arterioles

1 Ventricle contracts.

2 Semilunar valve opens.

3 Aorta and arteries expand and store pressure in elastic walls.

(b) Ventricular relaxation

1 Isovolumic ventricular relaxation

2 Semilunar valve shuts.

3 Elastic recoil of arteries sends blood forward into rest of circulatory system.

● **Figure 15-4** *Elastic recoil in the arteries*

Systemic Blood Pressure Is Highest in Arteries and Lowest in Veins

Blood pressure is highest in the arteries and decreases continuously as blood flows through the circulatory system. The decrease in pressure occurs because energy is lost due to the resistance of the vessels to blood flow (Fig. 15-5 ●). Resistance to blood flow also comes from friction between the blood cells.

In the systemic circulation, the highest pressure occurs in the aorta and reflects the pressure created by the left ventricle. Aortic pressure reaches an average high of 120 mm Hg during ventricular systole (**systolic pressure**), then falls steadily to a low of 80 mm Hg during ventricular diastole (**diastolic pressure**). Notice that although pressure in the ventricle falls to nearly 0 mm Hg as the ventricle relaxes, the diastolic pressure in the large arteries remains relatively high. The high diastolic pressure in the arteries reflects the ability of those vessels to capture and store energy in their elastic walls.

The rapid pressure increase that occurs when the left ventricle pushes blood into the aorta can be felt as a **pulse**, or pressure wave, transmitted through the fluid-filled arteries of the cardiovascular system. The pressure wave travels about 10 times as fast as the blood itself. Even so, a pulse felt in the arm is occurring slightly after the ventricular contraction that created the wave.

The amplitude of the pulse decreases over distance because of friction, finally disappearing at the capillaries (Fig. 15-5). **Pulse pressure**, a measure of the strength of the pressure wave, is defined as systolic pressure minus diastolic pressure:

Systolic pressure − diastolic pressure = pulse pressure

For example, in the aorta:

120 mm Hg − 80 mm Hg = 40 mm Hg pulse pressure

By the time blood flow reaches the veins, pressure has fallen because of friction, and a pulse wave no longer exists. Low-pressure blood in veins below the heart must flow "uphill," or against gravity, to return to the heart. Try holding your arm straight down without moving for several minutes and notice how the veins in the back of

Valves in the veins prevent backflow of blood.

Valve closed

When the skeletal muscles compress the veins, they force blood toward the heart (the skeletal muscle pump).

Valve opened

Valve closed

● **Figure 15-6** *Valves create one-way flow in the veins*

your hand begin to stand out as they fill with blood. (This effect may be more evident in older people, whose subcutaneous connective tissue has lost elasticity.) Then raise your hand so that gravity assists the venous flow and watch the bulging veins disappear.

To assist venous flow, some veins have internal one-way valves (Fig. 15-6 ●). These valves, like those in the heart, ensure that blood passing the valve cannot flow backward. Once blood reaches the vena cava, there are no valves. Blood flow is steady, pushed along by the continuous movement of blood into the venous circulation from the capillaries.

Venous return to the heart is aided by the *skeletal muscle* and *respiratory pumps* [◈ p. 481]. When muscles such as those in the calf of the leg contract, they compress the veins, forcing the blood upward past the valves. While your hand is hanging down, try clenching and unclenching your fist to see the effect muscle contraction has on distention of the veins.

CONCEPT CHECK
Answers: p. 522

1. Do you think that veins from the brain to the heart have valves? Defend your answer.
2. If you checked the pulse in a person's carotid artery and left wrist at the same time, would the pressure waves occur simultaneously? Explain.
3. Who has the higher pulse pressure, someone with blood pressure of 90/60 or someone whose pressure is 130/95?

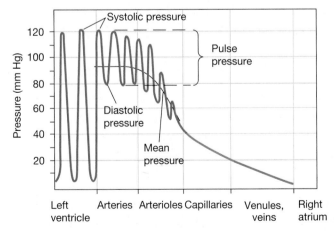

● **Figure 15-5** *Pressure throughout the systemic circulation* Pressure waves created by ventricular contraction are reflected into the blood vessels. They diminish in amplitude with distance and disappear at the capillaries.

Arterial Blood Pressure Reflects the Driving Pressure for Blood Flow

Arterial blood pressure, or simply "blood pressure," reflects the driving pressure created by pumping action of the heart. Because ventricular pressure is difficult to measure, it is customary to assume that arterial blood pressure reflects ventricular pressure. Arterial pressure is pulsatile, so we use a single value for arterial pressure that is representative of the driving pressure. This number is the **mean arterial pressure** (MAP), estimated as the diastolic pressure plus one-third of the pulse pressure:

$$\text{MAP} = \text{diastolic P} + 1/3\,(\text{systolic P} - \text{diastolic P})$$

For a person whose systolic pressure is 120 and diastolic pressure is 80:

$$\text{MAP} = 80\,\text{mm Hg} + 1/3(120 - 80\,\text{mm Hg})$$
$$\text{MAP} = 93\,\text{mm Hg}$$

Mean arterial pressure is closer to diastolic pressure than to systolic pressure because diastole lasts twice as long as systole.

Abnormally high or low arterial blood pressure can be indicative of a problem in the cardiovascular system. If blood pressure falls too low (*hypotension*), the driving force for blood flow will be unable to overcome opposition by gravity. In this instance, blood flow and oxygen supply to the brain are impaired, and the subject may become dizzy or faint.

On the other hand, if blood pressure is chronically elevated (a condition known as *hypertension*, or high blood pressure), high pressure on the walls of blood vessels may cause weakened areas to rupture and bleed into the tissues. If a rupture occurs in the brain, it is called a *cerebral hemorrhage* and may cause the loss of neurological function commonly called a *stroke*. If a weakened area ruptures in a major artery such as the descending aorta, rapid blood loss into the abdominal cavity will cause blood pressure to fall below the critical minimum. Without prompt treatment, rupture of a major artery is fatal.

CONCEPT CHECK   Answers: p. 522

4. The formula given for MAP applies to a typical resting heart rate of 60–80 beats per minute. If heart rate increases, would the contribution of systolic pressure to mean arterial pressure decrease or increase, and would MAP decrease or increase?

5. Peter's systolic pressure is 112 mm Hg and his diastolic pressure is 68 mm Hg (written 112/68). What is his pulse pressure? His mean arterial pressure?

Blood Pressure Is Estimated by Sphygmomanometry

We estimate arterial blood pressure in the radial artery of the arm using a *sphygmomanometer*, an instrument consisting of an inflatable cuff and a pressure gauge [*sphygmus*, pulse + *manometer*, an instrument for measuring pressure of a fluid]. The cuff encircles the upper arm and is inflated until it exerts pressure higher than the

---

RUNNING PROBLEM

Kurt's second blood pressure reading is 158/98. Dr. Cortez asks him to take his blood pressure at home daily for two weeks, then return to the office. When Kurt comes back with his diary, the story is the same: his blood pressure continues to average 160/100. After running some tests, Dr. Cortez concludes that Kurt is one of approximately 50 million adult Americans with high blood pressure, also called hypertension. If not controlled, hypertension can lead to heart failure, stroke, and kidney failure.

Question 1: Why are people with high blood pressure at higher risk for having a hemorrhagic, or bleeding, stroke?

---

systolic pressure driving arterial blood. When cuff pressure exceeds arterial pressure, blood flow into the lower arm stops (Fig. 15-7a ●).

Now pressure on the cuff is gradually released. When cuff pressure falls below systolic arterial blood pressure, blood begins to flow again. As blood squeezes through the still-compressed artery, a thumping noise called a **Korotkoff sound** can be heard with each pressure wave (Fig. 15-7b). Once the cuff pressure no longer compresses the artery, the sounds disappear (Fig. 15-7c).

The pressure at which blood flow is first heard represents the highest pressure in the artery and is recorded as the **systolic pressure**. The point at which sound disappears is the lowest pressure in the artery and is recorded as the **diastolic pressure**. By convention, blood pressure is written as systolic pressure over diastolic pressure.

The average value for blood pressure is considered to be 120/80. Like many average physiological values, however, these numbers are subject to wide variability between people or within a single individual from moment to moment. A systolic pressure that is repeatedly over 130 mm Hg at rest or a diastolic pressure that is chronically over 85 mm Hg is considered to be a sign of *hypertension*, or high blood pressure.

Cardiac Output and Peripheral Resistance Determine Mean Arterial Pressure

Mean arterial pressure, often called simply *blood pressure*, is the driving force for blood flow. But what determines mean arterial pressure? Arterial pressure is a balance between blood flow into the arteries and blood flow out of the arteries to the tissues. If flow in exceeds flow out, blood collects in the arteries and mean arterial pressure increases. If flow out exceeds flow in, mean arterial pressure falls.

Blood flow into the aorta, the main artery, is equal to the cardiac output of the left ventricle. Blood flow out of the arteries is influenced primarily by the resistance of the arterioles (**peripheral resistance**). We can express these relationships as

$$\text{Mean arterial pressure} \propto \\ \text{cardiac output} \times \text{arteriolar resistance}$$

$$\text{MAP} \propto \text{CO} \times \text{R}_{\text{arterioles}}$$

Figure 15-7 *Measurement of arterial blood pressure* Arterial blood pressure is measured with a sphygmomanometer, consisting of an inflatable cuff and a pressure gauge, and a stethoscope. The inflation pressure shown is for a person whose blood pressure is 120/80.

Mean arterial pressure (MAP) is a function of cardiac output (CO) and the resistance (R) of the arterioles (Fig. 15-8 ●).

Let's consider how this works. For example, if cardiac output increases, the heart pumps more blood into the arteries per unit time. If resistance to blood flow out of the arteries does not change, then flow into the arteries is greater than flow out, blood volume in the arteries increases, and arterial blood pressure increases.

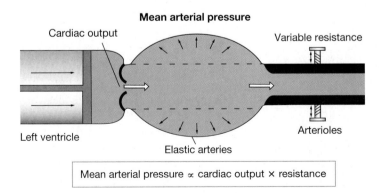

Figure 15-8 *Mean arterial pressure is a function of cardiac output and resistance in the arterioles* In this model, the ventricle is represented by a syringe. The variable diameter of the arterioles is represented by a screw clamp.

In another example, suppose cardiac output remains unchanged but peripheral resistance increases. Flow in is unchanged but flow out is less. Blood again accumulates in the arteries, and the blood pressure increases. Most cases of hypertension are believed to be caused by increased peripheral resistance without changes in cardiac output.

Two additional factors can influence arterial blood pressure: total blood volume and the distribution of blood in the systemic circulation.

Changes in Blood Volume Affect Blood Pressure

Although the volume of the blood within the circulation is usually relatively constant, changes in blood volume can affect arterial blood pressure. If blood volume increases, blood pressure increases. When blood volume decreases, blood pressure decreases.

To understand the relationship between blood volume and pressure, think of the circulatory system as an elastic balloon filled with water. If only a small amount of water is in the balloon, little pressure is exerted on the walls, and the balloon is soft and flabby. As more water is added to the balloon, more pressure is exerted on the elastic walls. If you fill a balloon close to the bursting point, you risk popping the balloon. The best way to reduce this pressure is to remove some of the water.

Small increases in blood volume occur throughout the day due to ingestion of food and liquids, but they usually do not create long-lasting changes in blood pressure because of homeostatic compensations. Adjustments for increased blood volume are primarily the responsibility of the kidneys. If the blood volume increases, the kidneys restore normal volume by excreting excess water in the urine (Fig. 15-9 ●).

Compensation for decreased blood volume is more difficult because it requires an integrated response from the kidneys and cardiovascular system to restore homeostasis. If blood volume decreases, *the kidneys cannot restore the lost fluid*. They can only conserve blood volume to prevent further decreases in blood pressure. The only way to restore lost fluid volume is through drinking or intravenous infusions. This is an example of mass balance: volume lost to the external environment must be replaced from the external environment.

Cardiovascular compensation for decreased blood volume includes vasoconstriction and increased sympathetic stimulation of the heart [● Fig. 14-31 p. 482]. However, there are limits to the effectiveness of cardiovascular compensation, and, if the fluid loss is too great, adequate blood pressure cannot be maintained. Typical events that might cause significant changes in blood volume include dehydration, hemorrhage, and ingestion of a large quantity of fluid. The integrated compensation for these events will be discussed in Chapter 20.

In addition to the absolute volume of blood in the cardiovascular system, the relative distribution of blood between the arter-

ial and venous sides of the circulation can be an important factor in maintaining arterial blood pressure. Arteries are low-volume vessels that usually contain only about 11% of the total blood volume. Veins, in contrast, are high-volume vessels that hold about 60% of the circulating blood volume. Venous blood acts as a reservoir that can be redistributed to the arteries if needed. When arterial blood pressure falls, increased sympathetic activity constricts veins, decreasing their holding capacity and redistributing blood to the arterial side of the circulation.

Figure 15-10 ● summarizes the four key factors that influence mean arterial blood pressure.

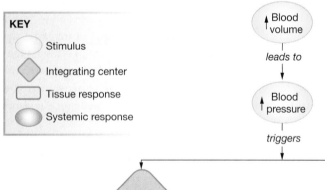

KEY
- Stimulus
- Integrating center
- Tissue response
- Systemic response

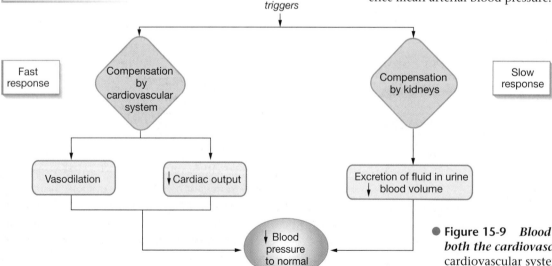

● **Figure 15-9** *Blood pressure control involves both the cardiovascular and renal systems* The cardiovascular system carries out rapid responses, while the renal response is slower.

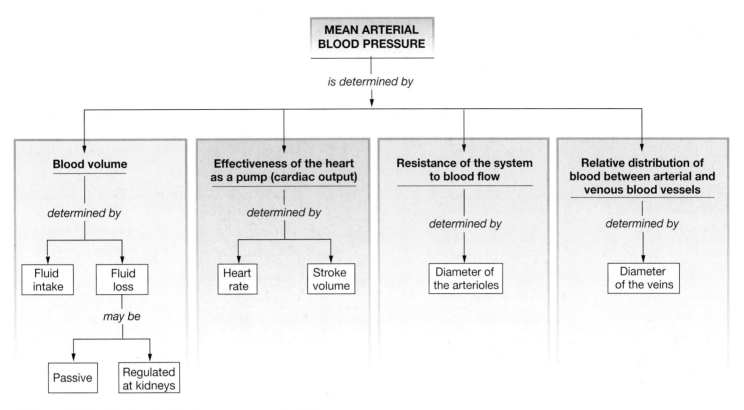

● **Figure 15-10** *Factors that influence mean arterial pressure*

    501 509 516

▶ RESISTANCE IN THE ARTERIOLES

As we saw in Chapter 14 [◉ p. 454], resistance to blood flow (R) is directly proportional to (∝) the length of the tubing through which the fluid flows (L) and the viscosity (η) of the fluid, and inversely proportional to the radius of the tubing (r) to the fourth power:

$$R \propto L\eta/r^4$$

Normally the length of the systemic circulation and the blood's viscosity are relatively constant. That leaves only the radius of the blood vessels as the primary resistance to blood flow:

$$R \propto 1/r^4$$

The arterioles are the main site of variable resistance in the systemic circulation and contribute more than 60% of the total resistance to flow in the system. Resistance in arterioles is variable because of the large amounts of smooth muscle in the arteriolar walls. When the smooth muscle contracts and relaxes, the radius of the arterioles changes.

Arteriolar resistance is influenced by both reflex and local control mechanisms (Table 15-1):

1. **Sympathetic reflexes** mediated by the central nervous system maintain mean arterial blood pressure and govern blood distribution for certain homeostatic needs, such as temperature regulation.
2. **Local control** of arteriolar resistance matches tissue blood flow to the metabolic needs of the tissue. In the heart and skeletal muscle, these local controls actually take precedence over reflex control by the central nervous system.
3. **Hormones,** particularly those that regulate salt and water excretion by the kidney, influence blood pressure by acting directly on the arterioles and by altering autonomic reflex control. The following sections look at the factors that influence blood flow at the tissue level.

Myogenic Autoregulation Automatically Adjusts Blood Flow

Vascular smooth muscle has the ability to regulate its own state of contraction, a process called **myogenic autoregulation.** In the absence of autoregulation, an increase in blood pressure increases blood flow through an arteriole. However, when smooth muscle fibers in the wall of the arteriole stretch because of increased

Table 15-1: Chemicals Mediating Vasoconstriction and Vasodilation

CHEMICAL	PHYSIOLOGICAL ROLE	SOURCE	TYPE
Vasoconstriction			
Norepinephrine (α receptors)	Baroreceptor reflex	Sympathetic neurons	Neurotransmitter
Serotonin	Platelet aggregation, smooth muscle contraction	Neurons, digestive tract, platelets	Paracrine, neurotransmitter
Substance P	Pain, increase capillary permeability	Neurons, digestive tract	Paracrine, neurotransmitter
Endothelin	Paracrine mediator	Vascular endothelium	Paracrine
Vasopressin	Increase blood pressure in hemorrhage	Posterior pituitary	Neurohormone
Angiotensin II	Increase blood pressure	Plasma hormone	Hormone
Vasodilation			
Nitric oxide (NO)	Paracrine mediator	Endothelium	Paracrine
Bradykinin (via NO)	Increase blood flow	Multiple tissues	Paracrine
Adenosine	Increase blood flow to match metabolism	Hypoxic cells	Paracrine
$\downarrow O_2$, $\uparrow CO_2$, $\uparrow H^+$, $\uparrow K^+$	Increase blood flow to match metabolism	Cell metabolism	Paracrine
Histamine	Increase blood flow	Mast cells	Paracrine
Epinephrine (β_2 receptors)	Increase blood flow to skeletal muscle, heart, liver	Adrenal medulla	Neurohormone
Acetylcholine (via NO)	Erection of clitoris or penis	Parasympathetic neurons	Neurotransmitter
Vasoactive intestinal peptide	Digestive secretion, relax smooth muscle	Neurons	Neurotransmitter, neurohormone
Natriuretic peptides (Example—ANP)	Reduce blood pressure	Atrial myocardium, brain	Hormone, neurotransmitter

blood pressure, the arteriole constricts. This vasoconstriction increases the resistance of the arteriole, automatically decreasing its blood flow. With this simple and direct response to pressure, arterioles regulate their own blood flow.

The mechanism responsible for the intrinsic response of vascular smooth muscle is stretch that opens Ca^{2+} channels in the muscle membrane. Calcium entering the smooth muscle cell combines with calmodulin and activates myosin light chain kinase, which in turn increases myosin ATPase activity and cross-bridge activity [👓 p. 417].

Paracrines Alter Vascular Smooth Muscle Contraction

Local control of arteriolar resistance is an important method by which individual tissues regulate their own blood supply. Local regulation is accomplished by paracrines (including the gases O_2, CO_2, and NO) that are secreted by vascular endothelium or by cells that the arterioles are supplying with blood (Table 15-1). The concentrations of many paracrines change as cells become more or less metabolically active.

For example, if aerobic metabolism increases, tissue O_2 levels decrease while CO_2 production goes up. Both low O_2 and high CO_2 dilate arterioles. Vasodilation increases blood flow into the tissue, bringing in additional O_2 to match increased metabolism and removing waste CO_2 (Fig. 15-11a ●). This process is known as **active hyperemia**: an increase in blood flow accompanies an increase in metabolic activity [*hyper-*, above normal + *(h)aimia*, blood].

If blood flow to a tissue stops completely for a few seconds to a few minutes, metabolically produced paracrines such as CO_2 and H^+ accumulate in the interstitial fluid around the cells. When blood flow resumes, the increased concentration of these paracrines immediately triggers significant vasodilation. As the vasodilators are washed away by the restored tissue blood flow, the radius of the arteriole gradually returns to normal. An increase in tissue blood flow following a period of low perfusion is known as **reactive hyperemia** (Fig. 15-11b).

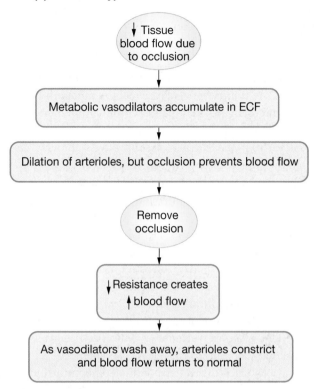

● **Figure 15-11** *Hyperemia*

One vasodilator paracrine being studied is the nucleotide **adenosine**. In heart muscle, if oxygen consumption exceeds the rate of oxygen supply by the blood, myocardial *hypoxia* results [*hypo-*, low + *oxia*, oxygen]. In response to low tissue oxygen, the myocardial cells release adenosine. Adenosine dilates coronary arterioles in an attempt to bring additional blood flow into the muscle.

Not all vasoactive paracrines reflect changes in metabolism. For example, *kinins* and *histamine* are potent vasodilators that play a role in inflammation. *Serotonin*, previously mentioned as a CNS neurotransmitter [● p. 267], is a paracrine released by activated platelets. Serotonin triggers the constriction of arterioles, so if blood vessels are damaged, constriction helps slow blood loss. A family of serotonin agonists called triptans (for example, *sumatriptan*) is used to treat migraine headaches that are caused by inappropriate cerebral vasodilation.

CONCEPT CHECK   Answers: p. 522

6. Resistance to blood flow is determined *primarily* by (a) blood viscosity, (b) blood volume, (c) cardiac output, (d) blood vessel diameter, or (e) blood pressure gradient (ΔP).

7. Extracellular fluid concentration of K^+ increases in exercising skeletal muscles. What effect will the increase in K^+ have on blood flow in the muscle?

RUNNING PROBLEM

After a month, Kurt returns to the office for a checkup. He has lost five pounds and is walking at least a mile daily, but his blood pressure has not changed. "I swear, I'm trying to do better," says Kurt. "But it's difficult." Because lifestyle changes alone have not lowered Kurt's blood pressure, Dr. Cortez prescribes an antihypertensive drug. "This drug, called an ACE inhibitor, blocks production of a chemical called angiotensin II, a powerful vasoconstrictor. This medication should bring your blood pressure back to a normal value."

Question 3: Why would blocking the action of a vasoconstrictor lower blood pressure?

409   496   499   509   516

The Sympathetic Branch Controls Most Vascular Smooth Muscle

Smooth muscle contraction in arterioles is regulated by neural and hormonal signals in addition to locally produced paracrines. Among the hormones with significant vasoactive properties are *atrial natriuretic peptide* and *angiotensin II*. These hormones also have significant effects on the kidney's excretion of ions and water, as you will learn in Chapter 19.

Most systemic arterioles are innervated by sympathetic neurons. A notable exception is arterioles involved in the erection reflex of the penis and clitoris. They are controlled indirectly by parasympathetic innervation that causes paracrine release of nitric oxide, resulting in vasodilation.

Tonic discharge of norepinephrine from sympathetic neurons helps maintain myogenic tone (Fig. 15-12 ●). Norepinephrine binding to α receptors on vascular smooth muscle causes vasoconstriction. If sympathetic release of norepinephrine decreases, the arterioles dilate. If sympathetic stimulation increases, arterioles constrict.

Epinephrine from the adrenal medulla travels through the blood and binds with α receptors, reinforcing vasoconstriction. However, α receptors have a lower affinity for epinephrine and do not respond as strongly to it as they do to norepinephrine [● p. 267].

Epinephrine also binds to β₂ receptors, found on vascular smooth muscle of heart, liver, and skeletal muscle arterioles. β₂ receptors are not innervated and therefore respond primarily to circulating epinephrine [● p. 377]. Activation of vascular β₂ receptors by epinephrine causes vasodilation.

One way to remember which arterioles have β₂ receptors is to think of a fight-or-flight response to a stressful event [● p. 370]. The fight-or-flight response includes a generalized increase in sympathetic activity, along with the release of epinephrine. Blood vessels that have β₂ receptors will respond to epinephrine by vasodilating. β₂-mediated vasodilation enhances blood flow to the heart, skeletal muscles, and liver, tissues that are active during the fight-or-flight response. (The liver produces glucose for muscle contraction.)

From Dynamite to Vasodilation

In 1998 the Nobel prize for physiology and medicine was awarded to Robert Furchgott, Louis Ignarro, and Ferid Murad for their research into the role of nitric oxide as a signal molecule in the cardiovascular system (*www.nobel.se/medicine*). Who would have thought that this component of smog and derivative of dynamite would turn out to be a biological messenger? Certainly not the referees who initially rejected Louis Ignarro's attempts to publish his research findings on the elusive gas. However, the ability of nitrate-containing compounds to relax blood vessels has been known for over 100 years, ever since workers in Alfred Nobel's dynamite factory complained of headaches. Physicians in the 1860s used nitroglycerin to relieve *angina*, heart pain that results from blocked blood vessels, and heart patients even today carry little nitroglycerin tablets to slide under their tongues when angina strikes. Still, it took years of work to find the short-lived gas that is the biologically active molecule. Even today, direct research on NO is difficult, and many studies look at its influence indirectly by studying the location and activity of nitric oxide synthase (NOS), the enzyme that produces NO.

● **Figure 15-12** *Tonic control of arteriolar diameter*

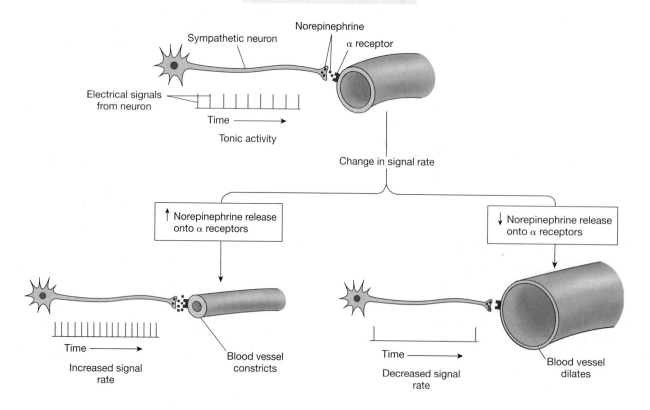

Arteriole diameter is controlled by tonic release of norepinephrine.

Norepinephrine

Sympathetic neuron

α receptor

Electrical signals from neuron

Time

Tonic activity

Change in signal rate

↑ Norepinephrine release onto α receptors

↓ Norepinephrine release onto α receptors

Time

Increased signal rate

Blood vessel constricts

Time

Decreased signal rate

Blood vessel dilates

During fight-or-flight, increased sympathetic activity at arteriolar α receptors causes vasoconstriction. The increase in resistance diverts blood from nonessential organs, such as the gastrointestinal tract, to the skeletal muscles, liver, and heart. The nervous system's ability to selectively alter blood flow to organs is an important aspect of cardiovascular regulation.

CONCEPT CHECK   Answers: p. 522

8. What happens when epinephrine combines with β_1 receptors in the heart? With β_2 receptors in the heart? (Hint: "in the heart" is vague. The heart has multiple tissue types. Which heart tissues possess the different types of β receptors?)

9. Skeletal muscle arterioles have both α and β receptors on their smooth muscle. Epinephrine can bind to both. Will the arterioles constrict or dilate in response to epinephrine? Explain.

▶ DISTRIBUTION OF BLOOD TO THE TISSUES

The distribution of systemic blood varies according to the metabolic needs of individual organs. Distribution is governed by a combination of local control mechanisms and homeostatic reflexes. For example, at rest, skeletal muscles receive about 20% of the cardiac output. During exercise, when the muscles use more oxygen and nutrients, they receive as much as 85%.

Blood flow to individual organs is set to some degree by the number and size of arteries feeding the organ. Figure 15-13 ● shows how blood is distributed to various organs when the body is at rest. Usually, over two-thirds of the cardiac output is routed to the digestive tract, liver, muscles, and kidneys. On a per unit weight basis, the kidneys have the highest flow of all the major organs.

Variations in blood flow to individual tissues are possible because the arterioles in the body are arranged in parallel. That is, all arterioles receive blood at the same time from the aorta (see Fig. 15-1). Total blood flow through *all* the arterioles of the body is always equal to the cardiac output.

However, the flow through individual arterioles depends on their resistance. The higher the resistance in the arteriole, the lower the blood flow through it. If an arteriole constricts and resistance goes up, blood flow through that arteriole goes down (Fig. 15-14 ●):

$$\text{Flow}_{\text{arteriole}} \propto 1/\text{resistance}_{\text{arteriole}}$$

In other words, blood is diverted from high-resistance arterioles to lower-resistance arterioles. You might say that blood traveling through the arterioles takes the path of least resistance.

Within a tissue, blood flow into individual capillaries can be regulated by **precapillary sphincters** [*sphingein*, to hold tight]. These are small bands of smooth muscle found at the junction of the metarteriole and capillary. If precapillary sphincters constrict, they

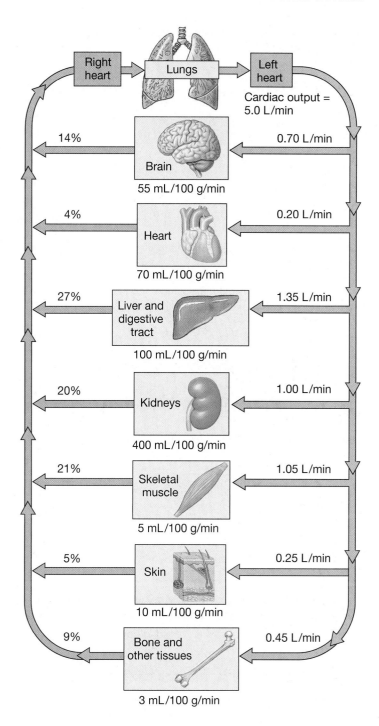

FIGURE QUESTION

What is the rate of blood flow through the lungs?

● **Figure 15-13 *Distribution of blood in the body at rest***
Blood flow to the major organs is represented as a percentage of total flow, on a per 100-g tissue-weight basis, and as an absolute rate of flow in liters per minute.

(a) Blood flow through four identical vessels (A–D) is equal. Total flow into vessels equals total flow out.

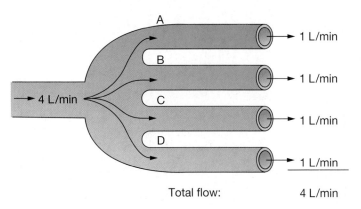

Total flow: 4 L/min

(b) When vessel B constricts, resistance of B increases and flow through B decreases. Flow diverted from B is divided among the lower-resistance vessels A, C, and D.

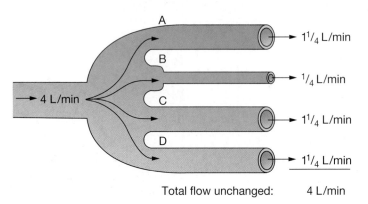

Total flow unchanged: 4 L/min

● **Figure 15-14**   *Blood flow through individual blood vessels is determined by their resistance*

restrict blood flow through them into the capillaries (Fig. 15-15 ●). If the sphincters dilate, blood flow increases. This mechanism provides an additional site for local control of blood flow.

CONCEPT CHECK   Answers: p. 522

10. Use Figure 15-13 to answer these questions. (a) Which tissue has the highest blood flow per unit weight? (b) Which tissue has the least blood flow, regardless of weight?

▶ EXCHANGE AT THE CAPILLARIES

The transport of materials around the body is only part of the function of the cardiovascular system. Once blood reaches the capillaries, the plasma and the cells exchange material across the thin walls. Most cells are located within 0.1 mm of the nearest capillary, so diffusion over this short distance proceeds rapidly.

The capillary density in any given tissue is directly related to the metabolic activity of its cells. Tissues with a higher metabolic

(a) When precapillary sphincters are relaxed, blood flows through all capillaries in the bed.

(b) If precapillary sphincters constrict, blood flow bypasses capillaries completely and flows through metarterioles.

● **Figure 15-15**   *Precapillary sphincters*

rate require more oxygen and nutrients. Those tissues have more capillaries per unit area. Subcutaneous tissue and cartilage have the lowest capillary density. Muscles and glands have the highest. By one estimate, the adult human body has about 50,000 miles of capillaries with a total exchange surface area of more than 6300 m², nearly the surface area of two football fields.

Capillaries have the thinnest walls of all the blood vessels, composed of a single layer of flattened endothelial cells supported on a basal lamina. The diameter of a capillary is just larger than that of a red blood cell, forcing the blood cells to pass through single file. Cell junctions between the endothelial cells vary from tissue to tissue and help determine the "leakiness" of the capillary.

Capillaries can be divided into two categories based on structure. The most common is the **continuous capillary**, whose endothelial cells are closely joined (Fig. 15-16a ●). These capillaries are found in muscle, connective tissues, and neural tissue. The continuous capillaries of the brain form the *blood-brain barrier*, with tight junctions that protect neural tissue from toxins in the bloodstream [● p. 292].

Fenestrated capillaries [*fenestra*, window] have large pores (*fenestrae*) in the capillary that allow high volumes of fluid to pass rapidly between the plasma and interstitial fluid (Fig. 15-16b). These capillaries are found primarily in the kidney and the intestine, where they are associated with absorptive transporting epithelia.

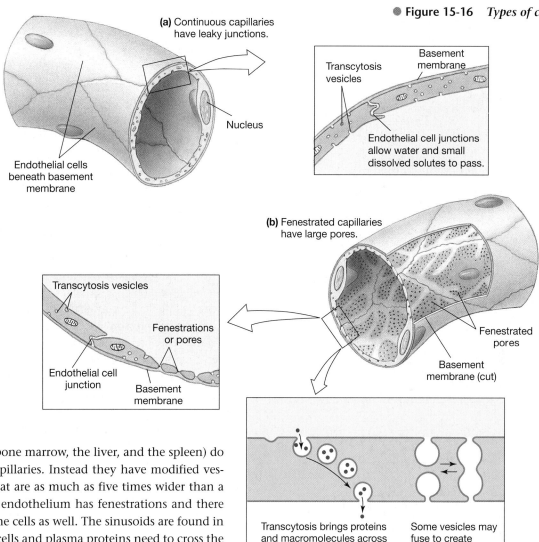

● **Figure 15-16** *Types of capillaries*

(a) Continuous capillaries have leaky junctions.

Nucleus

Endothelial cells beneath basement membrane

Transcytosis vesicles

Basement membrane

Endothelial cell junctions allow water and small dissolved solutes to pass.

(b) Fenestrated capillaries have large pores.

Transcytosis vesicles

Fenestrations or pores

Endothelial cell junction

Basement membrane

Fenestrated pores

Basement membrane (cut)

Transcytosis brings proteins and macromolecules across endothelium.

Some vesicles may fuse to create temporary channels.

Three tissues (the bone marrow, the liver, and the spleen) do not have traditional capillaries. Instead they have modified vessels called **sinusoids** that are as much as five times wider than a capillary. The sinusoid endothelium has fenestrations and there may be gaps between the cells as well. The sinusoids are found in locations where blood cells and plasma proteins need to cross the endothelium to enter the blood.

The Velocity of Blood Flow Is Lowest in the Capillaries

In Chapter 14, you learned that at a constant flow rate, velocity of flow will be higher in a smaller vessel than it will be in a larger vessel [● p. 455]. From this, you might conclude that blood moves very rapidly through a capillary because it is the smallest blood vessel. But the primary determinant for velocity of flow is not the diameter of an individual capillary but the *total cross-sectional area* of all the capillaries.

What is total cross-sectional area? Imagine circles representing cross sections of all the capillaries placed edge to edge, and you have it. For the capillaries, those circles would cover an area much larger than the total cross-sectional areas of all the arteries and veins combined. Therefore, because total cross-sectional area of the capillaries is so large, the velocity of their blood flow is low.

Figure 15-17 ● compares cross-sectional areas of different parts of the systemic circulation with the velocity of blood flow in each part. The fastest flow is in the relatively small-diameter arterial system. The slowest flow is in the capillaries and venules,

which collectively have the largest cross-sectional area. The low velocity of blood flow through capillaries is a useful characteristic that allows diffusion to go to equilibrium [● p. 136].

Most Capillary Exchange Takes Place by Diffusion and Transcytosis

Exchange between the plasma and interstitial fluid takes place by movement between the endothelial cells (the *paracellular pathway*) or through the cells (*transendothelial transport*). Dissolved solutes and gases move by diffusion, while larger solutes and proteins move mostly by vesicular transport.

The diffusion rate for dissolved solutes is determined primarily by the concentration gradient between the plasma and the interstitial fluid. Oxygen and carbon dioxide diffuse freely across the thin endothelium. Their concentrations reach equilibrium with the interstitial fluid and cells by the time blood reaches the venous end of the capillary. In capillaries with leaky cell junctions, most small dissolved solutes can diffuse freely between the cells or through the fenestrae.

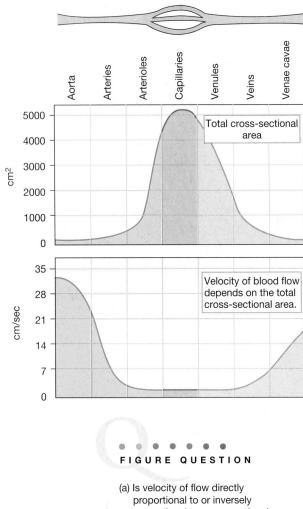

FIGURE QUESTION

(a) Is velocity of flow directly proportional to or inversely proportional to cross-sectional area?
(b) What effect does changing only the cross-sectional area have on flow rate?

● **Figure 15-17** *The velocity of flow depends on the total cross-sectional area.*

Blood cells and most plasma proteins are unable to pass through the junctions between the capillary endothelial cells. But we know that proteins do move from plasma to interstitial fluid and vice versa. In most capillaries larger molecules, including selected proteins, are transported across the endothelium by *transcytosis* [● p. 148]. The endothelial cell surface appears dotted with numerous *caveolae* and noncoated pits that become vesicles for transcytosis. It appears that in some capillaries, chains of vesicles fuse to create open channels that extend across the endothelial cell.

Capillary Filtration and Absorption Take Place by Bulk Flow

A third form of capillary exchange is bulk flow into and out of the capillary. **Bulk flow** refers to the mass movement of fluid between the blood and the interstitial fluid as the result of hydrostatic or osmotic pressure gradients. If the direction of bulk flow is into the capillary, it is called **absorption**. If the direction of flow is out of the capillary, the fluid movement is known as **filtration**. Capillary filtration is caused by hydrostatic pressure that literally forces fluid out of the capillary through leaky cell junctions. As an analogy, think of the garden "soaker" hoses whose perforated walls allow water to ooze out.

Most capillaries show a transition from net filtration at the arterial end to net absorption at the venous end. There are some exceptions to this rule. Capillaries in part of the kidney filter fluid along their entire length. Some capillaries in the intestine are only absorptive, picking up digested nutrients that have just been transported into the interstitial fluid from the lumen.

Two forces regulate bulk flow in the capillaries. One is *hydrostatic pressure*, the lateral pressure component of blood flow that pushes fluid out through the capillary pores [● p. 452]. The second factor is osmotic pressure [● p. 152]. These forces are sometimes called *Starling forces*, after the English physiologist E. H. Starling who first described them (the same Starling of Starling's law of the heart).

Osmotic pressure is determined by solute concentration differences between two compartments. The main solute difference between plasma and interstitial fluid is due to proteins that are present in the plasma but mostly absent from interstitial fluid. Osmotic pressure due to the presence of these proteins is known as **colloid osmotic pressure (π)**. Colloid osmotic pressure is *not* equivalent to the total osmotic pressure. It is simply a measure of the osmotic pressure created by proteins. The capillary endothelium is freely permeable to other solutes in the plasma, and they therefore do not contribute to the osmotic gradient.

Colloid osmotic pressure is higher in the plasma (π_{cap} = 25 mm Hg) than in the interstitial fluid (π_{IF} = 0 mm Hg). Therefore the osmotic gradient favors water movement by osmosis from the interstitial fluid into the plasma (Fig. 15-18a ●). For the purposes of our discussion, we will consider colloid osmotic pressure to be constant along the length of the capillary.

Capillary hydrostatic pressure (P_{cap}), on the other hand, decreases along the length of the capillary as energy is lost to friction. Average values for capillary hydrostatic pressure are 32 mm Hg at the arterial end of a capillary and 15 mm Hg at the venous end. The hydrostatic pressure of the interstitial fluid (P_{IF}) is very low, so we will consider it to be essentially zero. This means that water movement due to hydrostatic pressure will always be directed out of the capillary, with the pressure gradient decreasing from the arterial end to the venous end.

Net fluid flow across the capillary is determined by the difference between the hydrostatic pressure gradient favoring filtration and the colloid osmotic pressure favoring absorption.

Filtration (P_{out}) = hydrostatic pressure gradient
$$= (P_{cap} - P_{IF})$$

Absorption (π_{in}) = colloid osmotic pressure gradient
$$= (\pi_{IF} - \pi_{cap})$$

Net pressure = hydrostatic pressure gradient
+ colloid osmotic pressure gradient

(a) Filtration in systemic capillaries

Net pressure = hydrostatic pressure − colloid osmotic pressure

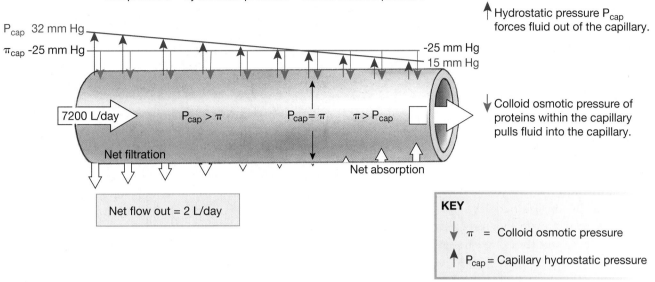

↑ Hydrostatic pressure P_{cap} forces fluid out of the capillary.

↓ Colloid osmotic pressure of proteins within the capillary pulls fluid into the capillary.

P_{cap} 32 mm Hg
π_{cap} -25 mm Hg

-25 mm Hg
15 mm Hg

7200 L/day

$P_{cap} > \pi$ $P_{cap} = \pi$ $\pi > P_{cap}$

Net filtration

Net absorption

Net flow out = 2 L/day

KEY

↓ π = Colloid osmotic pressure

↑ P_{cap} = Capillary hydrostatic pressure

(b) Relationship between capillaries and lymph vessels

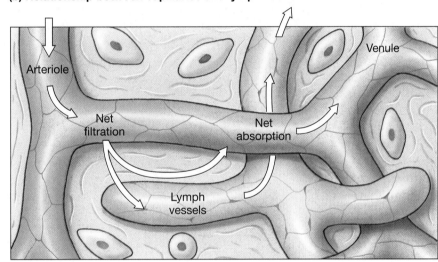

Venule

Arteriole

Net filtration

Net absorption

Lymph vessels

The excess water and solutes that filter out of the capillary are picked up by the lymph vessels and returned to the circulation.

● **Figure 15-18** *Fluid exchange at the capillary* An average of 3 L/day of fluid filters out of the capillaries.

If we assume that the interstitial hydrostatic and colloid osmotic pressures are zero, as discussed above, then we get the following values at the arterial end of a capillary:

$$\text{Net pressure}_{\text{Arterial end}} = (32 \text{ mm Hg} - 0) + (0 \text{ mm Hg} - 25)$$
$$= 32 - 25 \text{ mm Hg} = 7 \text{ mm Hg}$$

P_{out} is greater than π_{in}, therefore the net pressure is 7 mm Hg of filtration pressure. At the venous end, where capillary hydrostatic pressure is less:

$$\text{Net P}_{\text{Venous end}} = (15 \text{ mm Hg} - 0) + (0 \text{ mm Hg} - 25)$$
$$= 15 - 25 \text{ mm Hg} = -10 \text{ mm Hg}$$

π_{in} is greater than P_{out}, therefore the net pressure is 10 mm Hg favoring absorption. (A negative net pressure indicates absorption.)

Fluid movement down the length of a capillary is shown in Figure 15-18a. At the arterial end, there is net filtration, and at the venous end, there is net absorption. If the point at which filtration equals absorption occurred in the middle of the capillary, there would be no net movement of fluid. All volume that was filtered at the arterial end would be absorbed at the venous end.

However, filtration is usually greater than absorption, resulting in bulk flow of fluid out of the capillary into the interstitial space. By most estimates, that bulk flow amounts to about 3 liters per day, or the equivalent of the entire plasma volume! Unless this filtered fluid is returned to the plasma, the blood will turn into a sludge of blood cells and proteins. Restoring fluid lost from the capillaries to the circulatory system is one of the functions of the lymphatic system.

CONCEPT CHECK   Answers: p. 522

11. Hydrostatic pressure in a capillary increases to 35 mm Hg on the arterial end and 15 mm Hg on the venous end. Will net filtration in this capillary decrease, increase, or stay the same?

12. A person with liver disease may lose the ability to synthesize plasma proteins. What will happen to the colloid osmotic pressure of his blood? What will happen to the balance between filtration and absorption in his capillaries?

13. Why did this discussion refer to the osmotic pressure of the plasma rather than the osmolarity of the plasma?

▶ THE LYMPHATIC SYSTEM

The vessels of the lymphatic system interact with three physiological systems: the cardiovascular system, the digestive system, and the immune system. Functions of the lymphatic system include (1) returning fluid and proteins filtered out of the capillaries to the circulatory system, (2) picking up fat absorbed at the small intestine and transferring it to the circulatory system, and (3) serving as a filter to help capture and destroy foreign pathogens. In this discussion, we focus on the role of the lymphatic system in fluid transport. The other two functions will be discussed in connection with digestion (Chapter 21) and immunity (Chapter 24).

The lymphatic system is designed for the one-way movement of interstitial fluid from the tissues into the circulation. Closed-end lymph vessels called *lymph capillaries* lie close to all blood capillaries, except those in the kidney and central nervous system (Fig. 15-18b). The lymph capillaries are a single layer of flattened endothelium, even thinner than that of the blood capillaries.

The walls of lymph capillaries are anchored to the surrounding connective tissue by fibers that hold the thin-walled vessels open. Large gaps between cells allow fluid, interstitial proteins, and particulate matter such as bacteria to be swept into the lymph capillary by bulk flow. Once inside the lymphatic system, this clear fluid is simply called **lymph**.

Lymph capillaries in the tissues join to form larger collecting vessels that progressively increase in size (Fig. 15-19 ●). These vessels have a system of semilunar valves, similar to valves in the venous circulation. Finally, the largest lymph vessels empty into the venous circulation just under the collarbones, where the left and right subclavian veins join the internal jugular veins. At intervals along the way, vessels enter **lymph nodes**, bean-shaped nodules of tissue with a fibrous outer capsule and an internal collection of immunologically active cells, including lymphocytes and macrophages.

The lymphatic system has no single pump like the heart. Lymph flow depends primarily on waves of contraction of smooth muscle in the walls of larger lymph vessels. Flow is aided by contractile fibers in the endothelial cells, by the one-way valves, and by external compression created by skeletal muscles.

The skeletal muscle pump plays a significant role in lymph flow, as you know if you have ever injured a wrist or ankle. An immobilized limb frequently swells from the accumulation of fluid in the interstitial space, a condition known as **edema** [*oidema*, swelling]. Edema is the reason patients are told to elevate an injured limb above the level of the heart: gravity will assist lymph flow back to the blood.

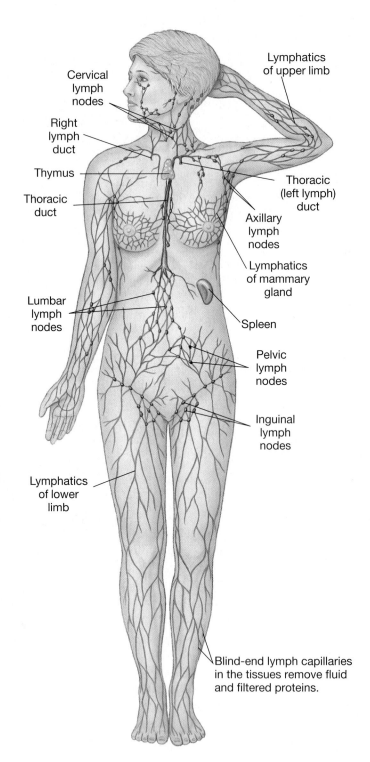

Lymph fluid empties into the venous circulation.

● **Figure 15-19** *The lymphatic system*

An important reason for returning filtered fluid to the circulation is the recycling of plasma proteins. The body must maintain low protein concentration in the interstitial fluid because colloid osmotic pressure is the only significant force that opposes capillary hydrostatic pressure. If proteins move from the plasma to the interstitial fluid, the osmotic pressure gradient that opposes filtration decreases. With less opposition to capillary hydrostatic pressure, additional fluid moves into the interstitial space.

For example, inflammation is a situation in which the balance of colloid osmotic and hydrostatic pressures is disrupted. Histamine released in the inflammatory response makes capillary walls leakier and allows proteins to escape from plasma into the interstitial fluid. Thus, the local swelling that accompanies a region of inflammation is an example of edema caused by redistribution of proteins from the plasma to the interstitial fluid.

Edema Is the Result of Alterations in Capillary Exchange

Edema is a sign that normal capillary-lymph exchange has been disrupted. Edema usually arises from one of two causes: (1) inadequate drainage of lymph or (2) capillary filtration that greatly exceeds capillary absorption.

Inadequate lymph drainage occurs with obstruction of the lymphatic system, particularly at the lymph nodes. Parasites, cancer, or fibrotic tissue growth caused by therapeutic radiation can block the movement of lymph through the system. For example, *elephantiasis* is a chronic condition marked by gross enlargement of the legs and lower appendages when parasites block the lymph vessels. Lymph drainage may also be impaired if lymph nodes are removed during surgery, a common procedure in the diagnosis and treatment of cancer.

Factors that disrupt the normal balance between capillary filtration and absorption include:

1. *An increase in capillary hydrostatic pressure.* Increased capillary pressure is usually indicative of elevated venous pressure. An increase in arterial pressure is generally not noticeable at the capillaries because of autoregulation of pressure in the arterioles.

   One common cause of increased venous pressure is *heart failure*, a condition in which the ventricle on one side of the heart is unable to pump all the blood sent to it by the other ventricle. As blood backs up, pressure rises first in the atrium, then in the veins and capillaries supplying blood to that side of the heart. When capillary hydrostatic pressure increases, filtration greatly exceeds absorption, leading to edema.

2. *A decrease in plasma protein concentration.* Plasma protein concentrations may decrease as a result of severe malnutrition or liver failure. The liver is the main site for plasma protein synthesis.

3. *An increase in interstitial proteins.* As discussed earlier, excessive leakage of proteins out of the blood will decrease plasma colloid osmotic pressure and will increase net capillary filtration.

On occasion, changes in the balance between filtration and absorption help the body maintain homeostasis. For example, if arterial blood pressure falls, capillary hydrostatic pressure also falls. This change increases fluid absorption. If pressure falls low enough, there will be net absorption in the capillaries rather than

net filtration. This passive mechanism helps maintain blood volume in situations in which blood pressure is very low, such as hemorrhage or severe dehydration.

14. If the left ventricle fails to pump normally, blood will back up into what set of blood vessels? Where would you expect edema to occur?

15. Malnourished children who have inadequate protein in their diet often have grotesquely swollen bellies (Fig. 15-20 ●). This condition is known as *ascites* and can be described as edema of the abdomen. Use the information you have just learned about capillary filtration to explain why malnutrition causes ascites.

RUNNING PROBLEM

Another few weeks go by, and Kurt again returns to Dr. Cortez for a checkup. Kurt's blood pressure is finally closer to the normal range and has been averaging 135/87. "But Doc, can you give me something for this dry, hacking cough I've been having? I don't feel bad, but it's driving me nuts." Dr. Cortez explains that a dry cough is an occasional side effect of taking ACE inhibitors. "It is more of a nuisance than anything else. But let's change your medicine. I'd like to try you on a calcium channel blocker this time."

Question 4: How do calcium channel blockers lower blood pressure?

● **Figure 15-20** *Ascites (abdominal edema) in a child with protein malnutrition* The African word for protein malnutrition is *kwashiorkor.*

▶ REGULATION OF BLOOD PRESSURE

The central nervous system coordinates the reflex control of blood pressure. The main integrating center is in the medulla oblongata of the brain. Because of the difficulty of studying neural networks in the brain, however, we still know relatively little about the nuclei, neurotransmitters, and interneurons of the **medullary cardiovascular control centers**.

The Baroreceptor Reflex Is the Primary Homeostatic Control for Blood Pressure

The primary goal of the cardiovascular control center is to maintain adequate blood flow to the brain and heart. Sensory input to this integrating center comes from a variety of peripheral sensory receptors. Stretch-sensitive mechanoreceptors known as **baroreceptors** are located in the walls of the carotid artery and aorta (Fig. 15-21 ●), where they monitor the pressure of blood flowing to the brain (carotid body receptors) and to the body (aortic body receptor). The carotid and aortic baroreceptors are tonically active stretch receptors that fire action potentials continuously at normal blood pressures.

The primary reflex pathway for homeostatic control of blood pressure is the **baroreceptor reflex**. When increased blood pressure in the arteries stretches the baroreceptor membrane, the firing rate of the receptor increases. If blood pressure falls, the firing rate of the receptor decreases.

Action potentials from the baroreceptors travel to the cardiovascular control center via sensory neurons. The cardiovascular control center integrates the sensory input and initiates an appropriate response. The response of the baroreceptor reflex is quite rapid: changes in cardiac output and peripheral resistance occur within two heartbeats of the stimulus.

Efferent output from the medullary cardiovascular control center is carried via both sympathetic and parasympathetic autonomic neurons. Peripheral resistance is under tonic sympathetic control, with increased sympathetic discharge causing vasoconstriction.

Heart function is regulated by antagonistic control. Increased sympathetic activity increases heart rate at the SA node, shortens conduction time through the AV node, and enhances the force of myocardial contraction. Increased parasympathetic activity slows down heart rate but has only a small effect on ventricular contraction.

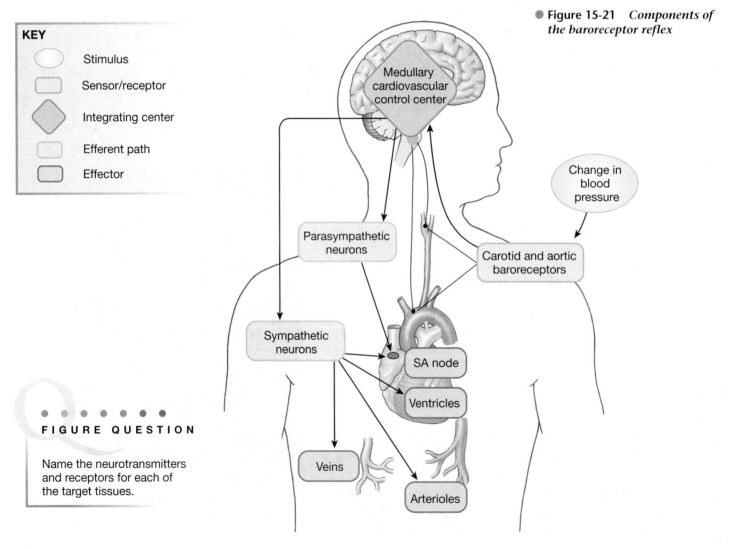

● **Figure 15-21** *Components of the baroreceptor reflex*

KEY

◯ Stimulus

▢ Sensor/receptor

◇ Integrating center

▢ Efferent path

▢ Effector

Q

FIGURE QUESTION

Name the neurotransmitters and receptors for each of the target tissues.

The baroreceptor reflex is summarized in Figure 15-22 ●. Baroreceptors increase their firing rate as blood pressure increases, activating the cardiovascular control center. In response, the cardiovascular control center increases parasympathetic activity and decreases sympathetic activity to slow down the heart. When heart rate falls, cardiac output falls. In the circulation, decreased sympathetic activity causes dilation of the arterioles, allowing more blood to flow out of the arteries. The combination of reduced cardiac output and decreased peripheral resistance lowers the mean arterial blood pressure.

Cardiovascular function can be modulated by input from peripheral receptors other than the baroreceptors. For example, if arterial chemoreceptors are activated by low blood oxygen levels, they increase cardiac output. The cardiovascular control center also has reciprocal communication with centers in the medulla that control breathing.

The integration of function between the respiratory and circulatory systems is adaptive. If tissues require more oxygen, it is supplied by the cardiovascular system working in tandem with the respiratory system. Consequently, increases in breathing rate are usually accompanied by increases in cardiac output.

Blood pressure is also subject to modulation by higher brain centers such as the hypothalamus and cerebral cortex. The hypothalamus is responsible for vascular responses involved in body

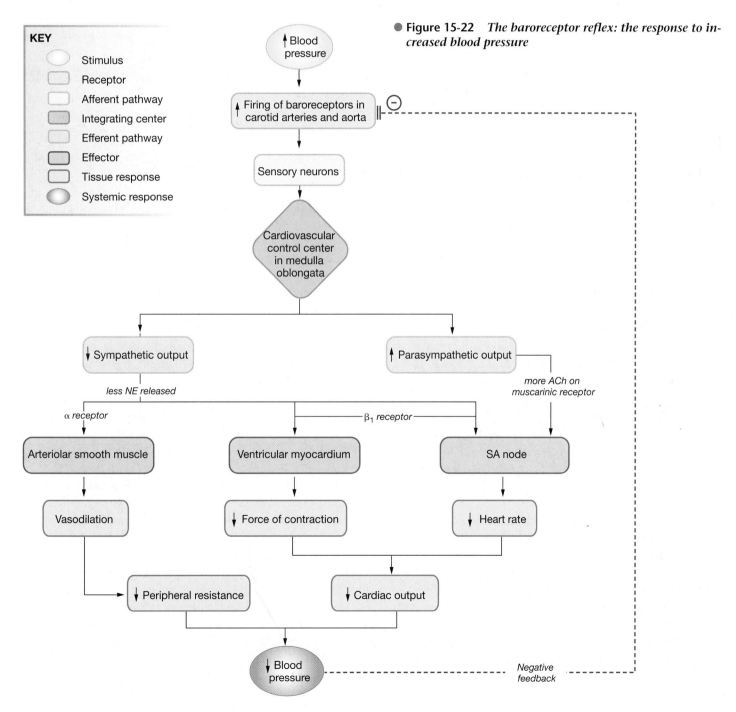

● **Figure 15-22** *The baroreceptor reflex: the response to increased blood pressure*

temperature regulation (see Chapter 22) and for the fight-or-flight response. Learned and emotional responses may originate in the cerebral cortex and be expressed by cardiovascular responses such as blushing and fainting.

One such reflex is the *vasovagal response*, which may be triggered in some people by the sight of blood or a hypodermic needle. (Recall Anthony's experience at the beginning of this chapter.) In this pathway, increased parasympathetic activity and decreased sympathetic activity slow heart rate and cause widespread vasodilation. Cardiac output and peripheral resistance both fall, triggering a precipitous fall in blood pressure. With insufficient blood to the brain, the individual faints.

Regulation of blood pressure in the cardiovascular system is closely tied to the regulation of body fluid balance by the kidneys. Thus, certain hormones from the heart act on the kidneys, while hormones from the kidneys act on the heart and blood vessels. Together, the heart and the kidneys play a major role in maintaining the homeostasis of body fluids, the subject of Chapter 20. The overlap of these systems is an excellent example of the integration of organ system function that must take place in the body in order to maintain homeostasis.

CONCEPT CHECK Answers: p. 522

16. Baroreceptors have stretch-sensitive ion channels in their cell membrane. Increased pressure opens the channels and initiates action potentials. What ion probably flows through these channels and in what direction, into or out of the cell?

Orthostatic Hypotension Triggers the Baroreceptor Reflex

The baroreceptor reflex functions every morning when you get out of bed. When you are lying flat, gravitational forces are distributed evenly up and down the length of the body, so blood also is distributed evenly throughout the circulation. When you stand up, gravity causes blood to pool in the lower extremities. This pooling creates an instantaneous decrease in venous return. As a result, less blood is in the ventricles at the beginning of the next contraction. Cardiac output falls from 5 L/min to 3 L/min, causing arterial blood pressure to decrease. This decrease in blood pressure upon standing is known as *orthostatic hypotension* [*orthos*, upright + *statikos*, to stand].

Orthostatic hypotension in a normal person triggers the baroreceptor reflex. The carotid and aortic baroreceptors respond to the fall in arterial blood pressure by decreasing their firing rate (Fig. 15-23 ●). Diminished sensory input into the cardiovascular control center increases sympathetic activity and decreases parasympathetic activity. As a result of autonomic changes, heart rate and force of contraction increase, while arterioles and veins constrict. The combination of increased cardiac output and increased peripheral resistance increases mean arterial pressure and restores it to normal within two heartbeats. The skeletal muscle pump also contributes to the recovery by enhancing venous re-

turn when abdominal and leg muscles contract to maintain an upright position.

The baroreceptor reflex is not always effective, however. For example, during extended bed rest or in the zero-gravity conditions of space flights, blood from the lower extremities is redistributed more evenly throughout the body. This redistribution raises arterial pressure, triggering the kidneys to excrete what they perceive to be excess fluid. Over the course of three days, excretion of water leads to a 12% decrease in blood volume. When the person finally gets out of bed or returns to earth, gravity again causes blood to pool in the legs. Orthostatic hypotension occurs, and the baroreceptors attempt to compensate. But, in this instance, the cardiovascular system is unable to restore normal pressure because of the loss of blood volume. As a result, the subject may become dizzy or even faint from reduced delivery of oxygen to the brain.

CONCEPT CHECK Answers: p. 522

17. In the movie *Jurassic Park*, Dr. Ian Malcolm must flee from the *T. rex*. Draw a reflex map showing the cardiovascular response to his fight-or-flight situation. (Hints: What is the stimulus? Fear is integrated in the limbic system.)

▶ CARDIOVASCULAR DISEASE

Diseases of the heart and blood vessels, such as heart attacks and strokes, play a role in more than half of all deaths in the United States. According to the American Heart Association, in 2003 all cardiovascular diseases will cost people in the United States more than $351 billion in medical expenses and lost wages. **Coronary heart disease**, also known as *coronary artery disease*, accounts for the majority of cardiovascular disease deaths. This condition, in which the heart's blood vessels become blocked by cholesterol deposits and blood clots, is the single largest killer of Americans.

The prevalence of cardiovascular diseases is reflected in the tremendous amount of research that is being done worldwide. The scientific investigations range from large-scale clinical studies that track cardiovascular disease in thousands of people, such as the Framingham (Massachusetts) Heart Study, to experiments at the molecular level.

Much of the research at the cellular and molecular level is designed to expand our understanding of both normal and abnormal function in the heart and blood vessels. Scientists are studying a virtual alphabet soup of transporters and regulators. Some of these molecules, such as adenosine, endothelin, vascular endothelial growth factor (VEGF), phospholamban, and nitric oxide, you have studied in these chapters.

As we increase our knowledge of cardiovascular function, we begin to understand how drugs that have been used for centuries work. A classic example is the drug *digitalis* [● p. 463], whose mechanism of action we finally understand. It is a humbling thought to realize that for many therapeutic drugs, we know *what* they do without fully understanding *how* they do it.

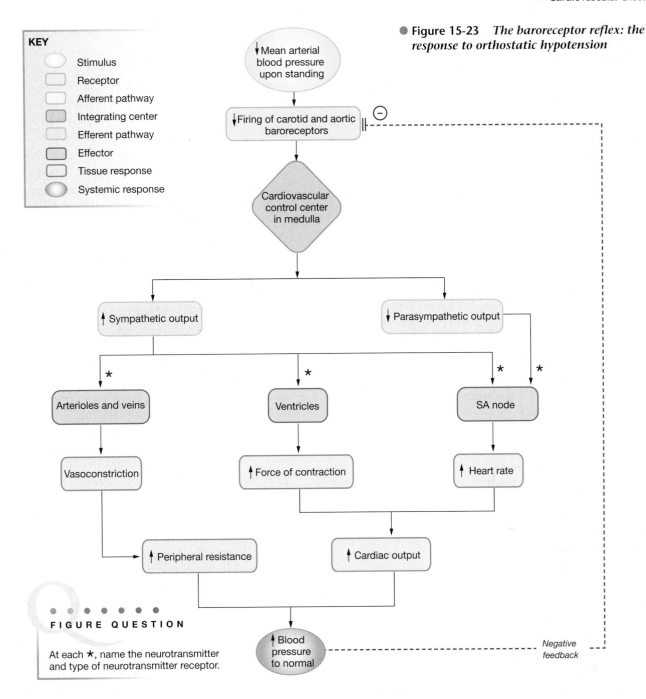

KEY

Stimulus
Receptor
Afferent pathway
Integrating center
Efferent pathway
Effector
Tissue response
Systemic response

● **Figure 15-23** *The baroreceptor reflex: the response to orthostatic hypotension*

↓Mean arterial blood pressure upon standing

↓Firing of carotid and aortic baroreceptors ⊖

Cardiovascular control center in medulla

↑Sympathetic output   ↓Parasympathetic output

Arterioles and veins *   Ventricles *   SA node * *

Vasoconstriction   ↑Force of contraction   ↑Heart rate

↑Peripheral resistance   ↑Cardiac output

↑Blood pressure to normal

Negative feedback

● ● ● ● ● ● ● ●
FIGURE QUESTION

At each ★, name the neurotransmitter and type of neurotransmitter receptor.

Risk Factors for Cardiovascular Disease Include Smoking, Obesity, and Inheritable Factors

We can predict the likelihood that a person will develop cardiovascular disease during his or her lifetime by examining the various risk factors that the person possesses. The list of risk factors is the result of many years following the medical histories of thousands of people in studies such as the famous Framingham Heart Study. As more data become available, additional risk factors may be added.

Risk factors are generally divided into those over which the person has no control and those that can be controlled. Medical intervention is aimed at reducing the risk from the controllable factors.

The risk factors that cannot be controlled include gender, age, and family history of early cardiovascular disease. Men have a 3–4 times higher risk of developing coronary artery disease through middle age than do women. After age 55, when most women have entered menopause, the death rate from coronary artery disease equalizes in men and in women. In general, the risk of coronary artery disease increases as people age.

Heredity also plays an important role. If a person has one or more close relatives with coronary artery disease, his or her risk is elevated. Diabetes mellitus, a metabolic disorder with a strong genetic component, also puts a person at risk for developing coronary artery disease. Diabetes contributes to the development of fatty deposits in the blood vessels.

Diabetes and Cardiovascular Disease

Having diabetes is one of the major risk factors for developing cardiovascular disease, and almost two-thirds of people with diabetes will die from cardiovascular problems. In diabetes, cells that cannot use glucose for energy turn to fats and proteins. The body breaks down fat stores into fatty acids [● p. 108] and dumps them into the blood. Plasma cholesterol levels are also elevated. When LDL-cholesterol remains in the blood, the excess is ingested by macrophages, starting a series of events that lead to atherosclerosis. Because of the pivotal role that LDL-cholesterol plays in atherosclerosis, many forms of therapy, ranging from dietary modification and exercise to drugs, are aimed at lowering LDL-cholesterol. Untreated, blockage of small and medium blood vessels in the lower extremities, combined with loss of sensation, can lead to *gangrene* (tissue death) in the feet. Atherosclerosis in larger vessels causes heart attacks and strokes. To learn more about diabetes and the increased risk of cardiovascular disease, visit the American Diabetes Association website at *www.diabetes.org.*

Risk factors that can be controlled include cigarette smoking, obesity, sedentary lifestyle, elevated serum cholesterol and triglycerides, untreated high blood pressure (hypertension), and diabetes mellitus. Most of these factors contribute directly to the development of **atherosclerosis**, also known as hardening of the arteries.

In atherosclerosis, macrophages ingest cholesterol and other lipids, then form *fatty streaks* in the extracellular matrix just under the endothelial lining of the larger arteries (Fig. 15-24b ●). As the streak grows, it forms a lipid core. Paracrines released by the macrophages attract smooth muscle cells that migrate from the underlying layer of vascular smooth muscle.

As the condition progresses, the lipid core grows and smooth muscle cells reproduce, forming bulging *plaques* that protrude into the lumen of the artery. In the advanced stages of atherosclerosis, the plaques develop hard, calcified regions and fibrous collagen caps.

We used to believe that the *occlusion* (blockage) of coronary blood vessels by large plaques was the primary cause of heart attacks, but that theory has been revised. The new model for the relationship between atherosclerosis and heart attacks divides plaques into two groups: stable plaques and vulnerable (unstable) plaques.

Stable plaques have thick fibrous caps that separate the lipid core from the blood (Fig. 15-24c). *Vulnerable plaques* have thin caps that are likely to rupture, expose collagen, and trigger the formation of blood clots (*thrombi*) (Fig. 15-24d). One theory says that atherosclerosis is an inflammatory process in which macrophages release enzymes that dissolve collagen and convert stable plaques to vulnerable plaques. If a clot blocks blood flow to the heart muscle, a heart attack, or myocardial infarction, results.

Blocked blood flow decreases the heart muscle's oxygen supply. Without oxygen, the cells may be irreversibly damaged, and the normal conduction of electrical signals through the ventricle may be disrupted. This disruption can lead to an irregular heartbeat (*arrhythmia*) and potentially result in cardiac arrest and death.

The role of elevated blood cholesterol levels in the development of coronary heart disease is well established. Cholesterol, like other lipids, is not very soluble in aqueous solutions such as the plasma. Therefore, when cholesterol in the diet is absorbed from the digestive tract, it combines with a lipoprotein molecule so that it dissolves in the plasma. There are multiple types of lipoprotein carriers for cholesterol, but clinicians generally are concerned with two: **high-density lipoprotein**, or **HDL**, and **low-density lipoprotein**, or **LDL**. HDL-cholesterol is the more desirable form because high levels of HDL-cholesterol are associated with lower risk of heart attacks. (Remember, "H" stands for "healthy.")

LDL-cholesterol is sometimes called the bad cholesterol, because elevated plasma LDL levels are associated with coronary heart disease. (Remember this by associating "L" with "lethal.") LDL-cholesterol in normal levels is not bad, however, because the LDL carrier is necessary for uptake of cholesterol into cells. The LDL portion of the complex combines with an LDL receptor on the cell membrane. The entire receptor-LDL-cholesterol complex is brought into the cell by endocytosis. Once freed from its carrier lipoprotein, cholesterol is used to make cell membranes or steroid hormones.

Hypertension Represents a Failure of Homeostasis

The last controllable risk factor for coronary heart disease is hypertension. **Hypertension** is defined as chronically elevated blood pressure, with either systolic pressures greater than 130 mm Hg or diastolic pressures greater than 85 mm Hg. Hypertension is a common disease in the United States and is the single most common reason for visits to physicians and for the use of prescription drugs.

More than 90% of all patients with hypertension are considered to have *essential* or *primary hypertension*, with no clear-cut cause other than heredity. Cardiac output is usually normal in these people, and their elevated blood pressure appears to be associated with increased peripheral resistance. Some investigators have speculated that the increased resistance may be due to a lack of nitric oxide, the locally produced vasodilator that is produced by endothelial cells in the arterioles.

In the remaining 5%–10% of hypertensive cases, the cause is known, and the hypertension is considered to be secondary to an underlying pathology. For instance, the cause might be an endocrine disorder that causes fluid retention.

A key feature of hypertension from all causes is adaptation of the carotid and aortic baroreceptors to higher pressure, with subsequent down-regulation of their activity. Without input from the baroreceptors, the medullary cardiovascular control center interprets the high blood pressure as "normal," so no reflex reduction of pressure occurs.

Hypertension is a risk factor for atherosclerosis because high pressure in the arteries damages the endothelial lining of the vessels and promotes the formation of atherosclerotic plaques. In addition, high arterial blood pressure puts additional strain on the heart. When resistance in the arterioles is high, the myocardium must work harder to push the blood into the arteries.

To understand why this occurs, think of a swinging door (the semilunar valves) between the kitchen (the ventricles) and the dining room (the arteries). A waiter pushing through the door uses a certain amount of force to push the door open each time he passes through. The increased resistance of essential hypertension is equivalent to piling furniture against the back of the door. When that

happens, the waiter must use more force to push through the door because he must push not only the door but also the furniture. The additional resistance created by the furniture's weight is analogous to increased resistance of the arterioles to blood flow.

Amazingly, stroke volume in hypertensive patients remains constant, up to a mean blood pressure of about 200 mm Hg, despite the increasing amount of work that the ventricle must perform as blood pressure increases. The cardiac muscle of the left ventricle responds to chronic high systemic resistance in the same way that skeletal muscle responds to a weight-lifting routine. The heart muscle *hypertrophies*, increasing the size and strength of the muscle fibers.

However, if resistance remains high over time, the heart muscle is unable to meet the work load and begins to fail. At that point, cardiac output by the left ventricle decreases. If the cardiac output of the right heart remains normal while the output from the left side decreases, fluid collects in the lungs, creating *pulmonary edema*.

At this point, a detrimental positive feedback loop begins. Oxygen exchange in the lungs diminishes because of the pulmonary edema, leading to less oxygen in the blood. Lack of oxygen for aerobic metabolism further weakens the heart, so its pumping

● **Figure 15-24** *The development of atherosclerotic plaques*

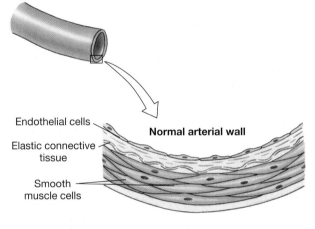

(a) The normal arterial wall consists of smooth muscle and connective tissue with an endothelial cell lining.

(b) In early stages, excess LDL-cholesterol accumulates between the endothelium and connective tissue. There, it is oxidized and phagocytosed. The macrophages produce paracrines that attract smooth muscle cells. At this stage, the lesion is called a fatty streak.

(c) As cholesterol accumulates, fibrous scar tissue forms around it. Migrating smooth muscle cells divide, thickening the arterial wall and narrowing the lumen of the artery. This stage is known as a fibrous plaque.

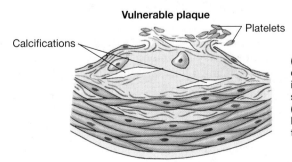

(d) In the advanced stages of atherosclerosis, calcified scar tissue will form. If the endothelium is damaged and collagen is exposed, platelets stick to the damaged area and a blood clot (thrombus) forms. If blood flow in the coronary blood vessel is stopped, a heart attack is the result.

effectiveness diminishes even more. Unless treated, this condition, known as *congestive heart failure*, eventually leads to death.

Many of the treatments for hypertension have their basis in the cardiovascular physiology that you learned in these chapters. Calcium channel blockers decrease Ca^{2+} entry into vascular smooth muscle cells, causing arterioles to dilate and reducing peripheral resistance (see box on *Calcium Channel Blockers*, p. 492). Beta-blocking drugs that target β_1 receptors decrease catecholamine stimulation of cardiac output.

Two other groups of antihypertensive drugs, the ACE inhibitors and the angiotensin receptor blockers, act by decreasing activity of angiotensin, a powerful vasoconstrictor substance. You will learn more about angiotensin later in the book, when you study the integrated control of blood pressure by the cardiovascular and renal systems. In the future, we may be seeing additional treatments for hypertension that are based on the molecular physiology of the heart and blood vessels.

RUNNING PROBLEM CONCLUSION

Essential Hypertension

Kurt remained on the calcium channel blocker, and, after several months, his blood pressure stabilized at 130/85—a significant improvement. Kurt's new diet also brought his total blood cholesterol down below 200 mg/dL plasma. By removing two of his controllable risk factors, Kurt decreased his chances of having a heart attack.

In this running problem, you learned about hypertension and some of the therapies currently used to treat it. Further check your understanding of this running problem by comparing your answers with those in the summary table.

	QUESTION	FACTS	INTEGRATION AND ANALYSIS
1	Why are people with high blood pressure at higher risk for having a hemorrhagic, or bleeding, stroke?	High blood pressure exerts force on the walls of the blood vessels.	If a blood vessel has an area of its wall that is weakened or damaged, high blood pressure may cause that area to rupture, allowing blood to leak out of the vessel into the surrounding tissues.
2	What is the rationale for reducing salt intake to control hypertension?	Salt causes water retention.	Blood pressure will increase if the circulating blood volume increases. By restricting salt in the diet, a person can decrease retention of fluid in the extracellular compartment, which includes the plasma.
3	Why would blocking the action of a vasoconstrictor lower blood pressure?	Peripheral blood pressure is determined by cardiac output and peripheral resistance.	Resistance is inversely proportional to the radius of the blood vessels. Therefore, if blood vessels dilate due to the blocking action of vasoconstrictors, resistance and blood pressure will decrease.
4	How do calcium channel blockers lower blood pressure?	Calcium entry from the extracellular fluid plays an important role in both smooth and cardiac muscle contraction.	Blocking Ca^{2+} entry through Ca^{2+} channels will decrease force of cardiac contraction and decrease contractility of vascular smooth muscle. Both of these effects will lower blood pressure.

CHAPTER SUMMARY

Blood flow through the cardiovascular system is an excellent example of *mass flow* in the body. Cardiac contraction creates high pressure in the ventricles that drives blood through the vessels of the systemic and pulmonary circuits, speeding up cell-to-cell *communication*. Resistance to flow is regulated by *local and reflex control mechanisms* that act on arteriolar smooth muscle and help match tissue perfusion to tissue needs. The *homeostatic* baroreceptor reflex monitors arterial pressure to ensure adequate perfusion of the brain and heart. Capillary *exchange of material* between the plasma and interstitial fluid *compartments* uses several transport mechanisms, including diffusion, transcytosis, and bulk flow.

1. Homeostatic regulation of the cardiovascular system is aimed at maintaining adequate blood flow to the brain and heart. (p. 491)

2. Total blood flow at any level of the circulation is equal to the cardiac output. (p. 491; Fig. 15-1)

The Blood Vessels

IP Cardiovascular—Anatomy Review: Blood Vessel Structure & Function

3. Blood vessels are composed of layers of smooth muscle, elastic and fibrous connective tissue, and **endothelium**. (p. 491; Fig. 15-2)

4. **Vascular smooth muscle** maintains a state of muscle tone. (p. 492)

5. The walls of the aorta and major arteries are both stiff and springy. This property allows them to absorb energy and release it through elastic recoil. (p. 493)

6. **Metarterioles** regulate blood flow through capillaries and allow white blood cells to go directly from arterioles to the venous circulation. (p. 493; Fig. 15-3)

7. Capillaries and postcapillary **venules** are the site of exchange between blood and interstitial fluid. (p. 493)

8. Veins hold more than half of the blood in the circulatory system. Veins have thinner walls with less elastic tissue, so they expand easily when they fill with blood. (p. 493)

9. **Angiogenesis** is the process by which new blood vessels grow and develop, especially after birth. (p. 494)

Blood Pressure

IP Cardiovascular—Measuring Blood Pressure

10. The ventricles create high pressure that is the driving force for blood flow. (p. 494; Fig. 15-4)

11. Blood pressure is highest in the arteries and falls continuously as blood flows through the circulatory system. At rest, average **systolic pressure** is 120 mm Hg, and average **diastolic pressure** is 80 mm Hg. (p. 495; Fig. 15-5)

12. Pressure created by the ventricles can be felt as a **pulse** in the arteries. **Pulse pressure** equals systolic pressure minus diastolic pressure. (p. 495)

13. Blood flow against gravity in the veins is assisted by one-way valves and the respiratory and skeletal muscle pumps. (p. 495; Fig. 15-6)

14. Arterial blood pressure is indicative of the driving pressure for blood flow. **Mean arterial pressure** (MAP) is defined as diastolic pressure+1/3 (systolic pressure − diastolic pressure). (p. 496)

15. Arterial blood pressure is usually measured with a sphygmomanometer. Blood squeezing through a compressed brachial artery makes a **Korotkoff sound**. (p. 496; Fig. 15-7)

16. Arterial pressure is a balance between cardiac output and resistance of arterioles to blood flow (**peripheral resistance**). (p. 496; Fig. 15-8)

17. If blood volume increases, blood pressure increases. When blood volume decreases, pressure decreases. (p. 497; Fig. 15-9)

18. Venous blood volume can be shifted to the arteries if arterial blood pressure falls. (p. 498; Fig. 15-10)

Resistance in the Arterioles

IP Cardiovascular—Factors that Affect Blood Pressure

19. The arterioles are the main site of variable resistance in the systemic circulation. A small change in the radius of an arteriole creates a large change in resistance: $R \propto 1/r^4$. (p. 499)

20. Arterioles regulate their own blood flow through **myogenic autoregulation**. Vasoconstriction increases resistance of an arteriole and decreases its blood flow. (p. 499)

21. Arteriolar resistance is influenced by local control mechanisms that match tissue blood flow to the metabolic needs of the tissue. Vasodilator paracrines include nitric oxide, H^+, K^+, CO_2, prostaglandins, adenosine, and histamine. Low O_2 causes vasodilation. Endothelins are powerful vasoconstrictors. (p. 500; Tbl. 15-1)

22. **Active hyperemia** is a process in which increased blood flow accompanies increased metabolic activity. **Reactive hyperemia** is an increase in tissue blood flow following a period of low perfusion. (p. 500; Fig. 15-11)

23. Most systemic arterioles, except those in the brain, are under tonic sympathetic control. Norepinephrine causes vasoconstriction. Decreased sympathetic stimulation causes vasodilation. (p. 502; Fig. 15-12)

24. Epinephrine binds to arteriolar α receptors and causes vasoconstriction. Epinephrine on β_2 receptors, found in the arterioles of the heart, liver, and skeletal muscle, causes vasodilation. (p. 502)

Distribution of Blood to the Tissues

25. Changing the resistance of the arterioles affects mean arterial pressure and alters blood flow through the arteriole. (p. 503; Fig. 15-14)

26. The flow through individual arterioles depends on their resistance. The higher the resistance in the arteriole, the lower the blood flow in those arterioles: Flow$_{arteriole}$ ∝ 1/R$_{arteriole}$. (p. 503)

27. Blood flow into individual capillaries can be regulated by **precapillary sphincters**. (p. 503; Fig. 15-15)

Exchange at the Capillaries

 IP Cardiovascular—Autoregulation and Capillary Dynamics

28. Exchange of material between the blood and the interstitial fluid is primarily by diffusion. (p. 504)

29. **Continuous capillaries** have leaky junctions between cells but also transport material using transcytosis. Continuous capillaries with tight junctions form the blood-brain barrier. (p. 504; Fig. 15-16)

30. **Fenestrated capillaries** have pores that allow large volumes of fluid to pass rapidly. (p. 504; Fig. 15-16)

31. The velocity of blood flow through the capillaries is slow and diffusion goes to equilibrium. (p. 505; Fig. 15-17)

32. **Bulk flow** refers to the mass movement of fluid between the blood and the interstitial fluid. Fluid movement is called **filtration** if the direction of flow is out of the capillary, or **absorption** if the flow is directed into the capillary. (p. 506; Fig. 15-18)

33. The osmotic pressure difference between plasma and interstitial fluid due to the presence of plasma proteins is the **colloid osmotic pressure**. (p. 506)

The Lymphatic System

 IP Fluids & Electrolytes—Electrolyte Homeostasis, Edema

34. About 3 liters of fluid filter out of the capillaries each day. The lymphatic system returns it to the circulatory system. (p. 508; Fig. 15-19)

35. The lymphatic system also takes fat absorbed at the small intestine to the circulatory system and captures and destroys foreign pathogens. (p. 508)

36. Lymph capillaries accumulate fluid, interstitial proteins, and particulate matter by bulk flow. Lymph flow depends on smooth muscle in vessel walls, one-way valves, and the skeletal muscle pump. (p. 508)

37. Excess fluid in the interstitial space is called **edema**. Factors that disrupt the normal balance between capillary filtration and absorption will cause edema. (p. 508; Fig. 15-20)

Regulation of Blood Pressure

 IP Cardiovascular—Blood Pressure Regulation

38. The reflex control of blood pressure resides in the medulla oblongata. **Baroreceptors** in the carotid and aortic bodies monitor arterial blood pressure and trigger the **baroreceptor reflex**. (p. 510; Fig. 15-21)

39. Efferent output from the **medullary cardiovascular control center** goes to the heart and arterioles. Increased sympathetic activity increases heart rate and force of contraction. Increased parasympathetic activity slows down heart rate. Increased sympathetic discharge at the arterioles causes vasoconstriction. There is no significant parasympathetic control of arterioles. (p. 510; Fig. 15-22)

40. Cardiovascular function can be modulated by input from higher brain centers and from the respiratory control center of the medulla. (p. 511)

41. The baroreceptor reflex functions each time a person stands up. The decrease in blood pressure upon standing is known as orthostatic hypotension. (p. 512; Fig. 15-23)

Cardiovascular Disease

42. **Coronary heart disease** is the leading cause of death in the United States. Risk factors predict the likelihood that a person will develop cardiovascular disease during his or her lifetime. (p. 512)

43. **Atherosclerosis** is a condition in which fatty deposits develop in arteries. If these deposits are unstable, they may block the arteries by triggering blood clots. (p. 514; Fig. 15-24)

QUESTIONS

LEVEL ONE *Reviewing Facts and Terms*

1. The first priority of blood pressure homeostasis is to maintain adequate perfusion to what two organs?

2. Match the blood vessels with their function(s). Each vessel may have more than one match, and matching items may be used more than once.
   - (a) arterioles
   - (b) arteries
   - (c) capillaries
   - (d) veins
   - (e) venules
     1. store pressure generated by the heart
     2. walls are both stiff and elastic
     3. carry low-oxygen blood
     4. thin walls of exchange epithelium
     5. act as a volume reservoir
     6. diameter can be altered by nervous input
     7. blood flow slowest through these vessels
     8. have the lowest blood pressure
     9. main site of variable resistance

3. List the four tissue components of blood vessel walls, in order from inner lining to outer covering. Briefly describe the importance of each tissue.

4. Blood flow to individual tissues is regulated by selective vasoconstriction and dilation of which vessels?

5. Aortic pressure reaches a typical high value of _____ (give units) during _____, or contraction of the heart. As the heart relaxes during the event called _____, aortic pressure declines to a typical low value of _____. This blood pressure reading would be written as _____/_____.

6. The rapid pressure increase that occurs when the ventricles push blood into the aorta can be felt as a pressure wave, or _____. What is the equation used to calculate the strength of this pressure wave?

7. List the factors that aid venous return to the heart.

8. What is hypertension, and why is it a threat to a person's well-being?

9. When are you likely to hear Korotkoff sounds?

10. List at least three paracrines that cause vasodilation. What is the source of each one? In addition to paracrines, list two other ways to control smooth muscle contraction in arterioles.

11. What is hyperemia? How does active hyperemia differ from reactive hyperemia?

12. Most systemic arterioles are innervated by the _____ branch of the nervous system. Increased sympathetic input will have what effect on arteriole diameter?

13. Match the event at left with all appropriate neurotransmitter(s) *and* receptor(s) from the list on the right.
    - (a) vasoconstriction of intestinal arterioles
    - (b) vasodilation of coronary arterioles
    - (c) increased heart rate
    - (d) decreased heart rate
    - (e) vasoconstriction of coronary arterioles
      1. norepinephrine
      2. epinephrine
      3. acetylcholine
      4. β_1 receptor
      5. α receptor
      6. β_2 receptor
      7. nicotinic receptor
      8. muscarinic receptor

14. What organs receive over two-thirds of the cardiac output at rest? Which organs have the highest flow of blood on a per unit weight basis?

15. By looking at the density of capillaries in a tissue, you can make assumptions about what property of the tissue? Which tissues have the lowest and highest capillary densities?

16. What type of transport is used to move each listed substance across the capillary endothelium?
    - (a) oxygen
    - (b) proteins
    - (c) glucose
    - (d) water

17. With which three physiological systems do the vessels of the lymphatic system interact?

18. Define edema. List some ways in which it can arise.

19. Define the following terms and explain their significance to cardiovascular physiology.
    - (a) perfusion
    - (b) colloid osmotic pressure
    - (c) vasoconstriction
    - (d) angiogenesis
    - (e) metarterioles
    - (f) pericytes

20. The two major lipoprotein carriers of cholesterol are _____ and _____. Which type is bad in elevated amounts?

LEVEL TWO *Reviewing Concepts*

21. **Concept map:** Map all the following factors that influence mean arterial pressure. You may add additional terms.

aorta	parasympathetic neuron
arteriole	peripheral resistance
baroreceptor	SA node
blood volume	sensory neuron
cardiac output	stroke volume
carotid artery	sympathetic neuron
contractility	vein
heart rate	venous return
medulla oblongata	ventricle

22. Compare and contrast the following sets of terms:
    - (a) lymphatic capillaries and systemic capillaries
    - (b) roles of the sympathetic and parasympathetic branches in blood pressure control
    - (c) lymph fluid and blood
    - (d) continuous and fenestrated capillaries
    - (e) hydrostatic and colloid osmotic pressure in systemic capillaries

23. Drugs known as calcium channel blockers prevent Ca^{2+} movement through Ca^{2+} channels. Explain two ways that this

519

action can lower blood pressure. Why are neurons and other cells unaffected by these drugs?

24. Define myogenic autoregulation. What mechanisms have been proposed to explain it?

25. Left ventricular failure may be accompanied by edema, shortness of breath, and increased venous pressure. Explain how these symptoms develop.

LEVEL THREE *Problem Solving*

26. Draw a reflex map that would explain Anthony's vasovagal syncope at the sight of blood. Be sure to include all the steps of the reflex and explain if pathways are being stimulated or inhibited.

27. A physiologist placed a section of excised arteriole in a perfusion chamber with saline. When the oxygen content of the saline perfusing (flowing through) the arteriole was reduced, the arteriole dilated. In a follow-up experiment, she placed an isolated piece of arteriolar smooth muscle stripped away from the other layers of the arteriole wall in saline. When the oxygen content of the saline was reduced as in the first experiment, the muscle showed no response. What do these two experiments suggest about how low oxygen exerts local control over arterioles?

28. Robert is a nonsmoker, 52 years old. He weighs 180 lbs, stands 5'9" tall, and his blood pressure averaged 145/95 on three successive visits to his doctor's office. His father, grandfather, and uncle all had heart attacks in their early 50s, and his mother died of a stroke at the age of 71.
    (a) Identify Robert's risk factors for coronary heart disease.
    (b) Does Robert have hypertension? Explain.
    (c) Robert's doctor prescribes a drug called a beta blocker. Explain the mechanism by which a beta-receptor-blocking drug may help lower blood pressure.

29. The figure below is a schematic representation of the systemic circulation. Use it to help answer the questions below. CO = cardiac output, MAP = mean arterial pressure

MAP

CO

Left ventricle

Arteries

Arterioles

Resistance

① ② ③ ④

Flow in vessels downstream

(a) If resistance in vessels 1 and 2 increases due to local paracrines but cardiac output is unchanged, what happens to MAP? What happens to flow through vessels 1 and 2? Through 3 and 4?

(b) Homeostatic compensation occurs within seconds. Draw a reflex map to explain the compensation (stimulus, receptor, etc.).

(c) When vessel 1 constricts, what happens to the filtration pressure in the capillaries downstream from that arteriole?

30. The graphs below are recordings of contractions in an isolated frog heart. The intact frog heart is innervated by sympathetic neurons that increase heart rate and by parasympathetic neurons that decrease heart rate. Based on these four graphs, what conclusion can you draw about the mechanism of action of atropine? (Atropine does not cross the cell membrane.)

A
(add epinephrine)

B
(add epinephrine + atropine)

C
(add ACh)

D
(add ACh + atropine)

LEVEL FOUR *Quantitative Problems*

31. Using the appropriate equation, mathematically explain what happens to blood flow if the diameter of a blood vessel increases from 2 mm to 4 mm.

32. Duplicate William Harvey's experiment that led him to believe that blood circulated in a closed loop.

(a) Take your resting pulse.
(b) Assume that your heart at rest pumps 70 mL/beat and that 1 mL of blood weighs one gram. Calculate how long it would take your heart to pump your weight in blood. (2.2 pounds = 1 kilogram)

33. Calculate the mean arterial pressure (MAP) and pulse pressure for a person with a blood pressure of 115/73.

34. According to the Fick principle, the oxygen consumption rate of an organ is equal to the blood flow through that organ times the amount of oxygen extracted from the blood as it flows through the organ:

oxygen consumption rate = blood flow ×
    (arterial oxygen content − venous oxygen content)

$$\text{mL O}_2 \text{ consumed/min} = \text{mL blood/min} \\ \times \text{ mL O}_2/\text{mL blood}$$

A person has a total body oxygen consumption rate of 250 mL oxygen consumed per minute. The oxygen content of blood in his aorta is 20 mL O_2/mL blood. The oxygen content of pulmonary artery blood is 16 mL O_2/mL blood. What is his cardiac output?

Explorations 🌐

Explorations are also available at *www.physiologyplace.com*

In this chapter you learned about the factors that contribute to the maintenance of blood pressure and the efficient delivery of oxygen and nutrients to the tissues.

Exploration 1: Cardiovascular Disease—Not Just for the Elderly

Do you think of cardiovascular disease (CVD) as an "old folks" problem? Think again. Each year about 40,000 children in the United States are born with some type of congenital cardiovascular defect, and CVD is the second leading cause of death in children under the age of 15. About 10% of babies born with congenital heart defects have a *patent* (open) *ductus arteriosus*. During fetal development, most blood in the pulmonary artery bypasses the lungs and goes into the aorta by way of a channel called the ductus arteriosus. Normally this bypass channel closes during the first day after birth. If it doesn't, the baby will usually require surgery to close off the channel. See if you can apply your understanding of cardiovascular anatomy and physiology to answer the following questions.

1. In the fetus, why does most blood bypass the lungs?

2. Poiseuille's Law [🌐 p. 454] states that the resistance of a system to flow is proportional to the length of the vessels. If the systemic side of circulatory system is longer than the pulmonary side, which circuit has the higher resistance?

3. If flow is equal in the pulmonary and systemic circulation, which side of the heart must generate more pressure to overcome resistance?

4. Using your answer to question 3, figure out which way blood will flow through a patent ductus arteriosus.

To learn more about patent ductus arteriosus, use that phrase as your search term for a Google search (*www.google.com*). To learn more about some of the other CVD that affect children, go to the American Heart Association website at *www.americanheart.org* and look for the link to CVD in children.

Exploration 2: Interpreting Data: Cause or Association?

As you learned in Chapter 1, conducting and interpreting research in humans is a complicated endeavor because of the difficulty of conducting controlled experiments. The economic and social importance of cardiovascular disease (CVD) make it the focus of many studies each year as researchers try to improve treatments and prediction algorithms. (An *algorithm* is a set of rules or sequence of steps used to solve a problem.)

However, sometimes it is difficult to determine whether a positive correlation between a parameter and the incidence of CVD represents a cause-effect relationship or a simple association. For example, two factors associated with higher incidence of heart disease are *C-reactive protein* (CRP) and *homocysteine*. CRP is a molecule involved in the body's response to inflammation, and in one study, women who had elevated CRP in their blood were more than twice as likely to have a serious cardiovascular problem as women with low CRP. Does that mean that CRP is causing cardiovascular disease? Or is it simply a marker that can be used clinically to predict who is more likely to develop cardiovascular problems? Similarly, elevated homocysteine levels are associated with an increased incidence of CVD. (Homocysteine is an amino acid in a complicated metabolic pathway that also requires folate and vitamin B_{12} as cofactors.) Should physicians routinely measure homocysteine along with cholesterol? Currently there is little clinical evidence to show that reducing either CRP or homocysteine reduces a person's risk of developing CVD. If these two markers are not indicators for *modifiable* risk factors, should a patient's insurance be asked to pay for the tests for them? To learn more about these two marker molecules and what health care providers are saying about them, go to the American Heart Association website (*www.americanheart.org*) and enter *homocysteine* or *C-reactive protein* in the SEARCH box.

"Homocysteine and cardiovascular disease: current evidence and future prospects," *American Journal of Medicine* 112:556–565, 2002.

ANSWERS

Page 495

1. Veins from the brain to the heart do not require valves because the flow of blood toward the heart is aided by gravity.

2. The pressure wave at the carotid artery would arrive slightly ahead of the pressure wave to the left wrist because the distance from the heart to the carotid artery is shorter.

3. Pressure of 130/95 has higher pulse pressure (35 mm Hg).

Page 496

4. If heart rate increases, the relative time spent in diastole decreases. In that case, the contribution of systolic pressure to mean arterial pressure increases.

5. Pulse pressure = systolic pressure minus diastolic pressure. $(112 - 68) = 44$ mm Hg.

   Mean arterial pressure = diastolic pressure $+1/3$ pulse pressure. $68 + 1/3(44) = 82.7$ mm Hg

Page 501

6. Resistance is determined primarily by blood vessel diameter.

7. Extracellular K^+ dilates arterioles, thereby increasing blood flow into the tissue.

Page 503

8. β_1 receptors are found on myocardium and their activation increases heart rate and force of contraction. β_2 receptors are found on the vascular smooth muscle of heart arterioles and their activation causes vasodilation.

9. Alpha receptors have lower affinity for epinephrine than β_2 receptors, therefore the muscle will have a stronger-mediated response and will dilate.

Page 504

10. (a) The tissue with the highest blood flow per unit weight is the kidney. (b) The tissue with the least total blood flow is the heart.

Pages 507–508

11. Net filtration out of the capillary will increase due to increased hydrostatic pressure.

12. Loss of plasma proteins will decrease colloid osmotic pressure. As a result, hydrostatic pressure will have a greater effect and increase net filtration.

13. Using osmotic pressure rather than osmolarity allows a direct comparison between absorption force and filtration force, which is also a pressure expressed in mm Hg.

Page 509

14. If the left ventricle fails, blood will back up into the pulmonary veins and then into lung capillaries. Edema in the lungs is known as *pulmonary edema*.

15. Low protein diets result in a low concentration of plasma proteins, the solute responsible for the osmotic pressure difference between the plasma and the interstitial fluid. With lower osmotic pressure inside the capillaries, reabsorption is reduced while filtration remains constant. This results in a greater net outward movement of fluid and leads to edema.

Page 512

16. The most likely ion is Na^+, moving into the receptor cell. Action potentials are created when a cell depolarizes to threshold. Depolarization results from net influx of positive ions or decreased efflux of K^+.

Page 512

17. Stimulus: Sight, sound, and smell of the *T. rex*. Receptors: Eyes, ears, and nose. Integrating center: cerebral cortex, with descending pathways through limbic system. Divergent pathways go to the medullary cardiovascular control center, which increases sympathetic output to heart and arterioles. A second descending spinal pathway goes to the adrenal medulla, which releases epinephrine. Epinephrine on β_2 receptors of liver, heart, and skeletal muscle arterioles causes vasodilation of those arterioles. Norepinephrine from sympathetic innervation onto α receptors in most other arterioles causes vasoconstriction. Both catecholamines increase heart rate and force of contraction.

Page 503

Figure 15-13: Blood flow through the lungs is 5 L/min.

Page 506

Figure 15-17: (a) Velocity of flow is inversely proportional to area. As area increases, velocity decreases. (b) Changing only cross-sectional area has no effect on flow rate.

Page 510

Figure 15-21: The SA node has muscarinic cholinergic receptors for ACh and β_1 receptors for catecholamines. The ventricles have β_1 receptors for catecholamines. Sympathetic neurons terminating on arterioles and veins release norepinephrine onto α receptors.

Page 513

Figure 15-23: Arterioles and veins: norepinephrine, α receptors. Ventricles: norepinephrine, β_1 receptors. SA node: norepinephrine, β_1 receptors, and ACh, muscarinic receptors.

Pages 520–521

31. Flow $\propto \Delta P \times$ radius^4. If diameter goes from 2 to 4, flow increases 16-fold.

32. Answers will vary. For a 50-kg individual with a resting pulse of 70 bpm, her weight in blood will be pumped in approximately 10 minutes.

33. Mean arterial pressure (MAP) = $1/3 (115 - 73) + 73$ or 87 mm Hg. Pulse pressure = $115 - 73 = 42$ mm Hg.

34. His cardiac output is 6250 mL/min or 6.25 L/min.

Blood

Red blood cells trapped in fibrin mesh

RUNNING PROBLEM

Von Willebrand's Disease

Emily Richards, 20, had heard horror stories about getting wisdom teeth pulled, but none could compare with what she was experiencing. In addition to the pain and swelling, the extraction sites would not stop bleeding. After three days of lying in bed with gauze stuffed in her mouth, she finally called her doctor. "I think something weird is going on," she told the receptionist. "Could I have hemophilia?" Emily's doctor ran blood tests and called her the next day with the results: "You do have a bleeding disorder, Emily, but it is not hemophilia. You have von Willebrand's disease."

526 529 536 542

BACKGROUND BASICS

Protein structure: **38** | Connective tissue: **71** | Phagocytosis: **144** | Second messenger cascade: **177** | Viscosity and resistance: **454** | Collagen: **72** | Cell organelles: **53** | Cytokines: **174** |

Who would have thought the old man to have had so much blood in him?

—Macbeth, V, i, 42, *by William Shakespeare*

Blood, the fluid that circulates in the cardiovascular system, has occupied a prominent place throughout history as an almost mystical fluid. Humans undoubtedly made the association between blood and life by the time they began to fashion tools and hunt animals. A wounded animal that lost blood would weaken and die if the blood loss was severe enough. The logical conclusion was that blood was necessary for existence. This observation eventually led to the term *lifeblood*, meaning anything essential for existence.

Ancient Chinese physicians linked blood to energy flow in the body. They wrote about circulation of blood through the heart and blood vessels long before William Harvey described it in seventeenth-century Europe. In China, changes in blood flow were used as diagnostic clues to illness. Chinese physicians were expected to recognize some 50 variations in the pulse. Because blood was considered a vital fluid to be conserved and maintained, bleeding patients to cure disease was not a standard form of treatment.

In contrast, Western civilizations came to believe that disease-causing evil spirits circulated in the blood. The way to remove these spirits was to remove the blood containing them. However, it was recognized that blood was an essential fluid, so bloodletting had to be done judiciously. Veins were opened with knives or sharp instruments (*venesection*), or blood-sucking leeches were applied to the skin. In ancient India, people believed that leeches could distinguish between healthy and infected blood.

There is no written evidence for venesection in ancient Egypt, but Galen of Pergamum in the second century A.D. influenced Western medicine for nearly 2000 years by advocating bleeding as treatment for many disorders. The location, timing, and frequency of the bleeding depended on the condition, and the physician was instructed to remove enough blood to bring the patient to the point of fainting. Over the years, this practice undoubtedly killed more people than it cured.

What is even more remarkable is the fact that as late as 1923, there was an American medical textbook that advocated bleeding for treating certain infectious diseases such as pneumonia! Now that we better understand the importance of blood in the immune response, it is doubtful that modern medicine will ever again turn to blood removal as a nonspecific means of treating disease. It is still used, however, for selected *hematological disorders* [*haima*, blood].

▶ PLASMA AND THE CELLULAR ELEMENTS OF BLOOD

What is this remarkable fluid that flows through the circulatory system? Blood makes up one-fourth of the extracellular fluid, the internal environment that bathes the cells and acts as a buffer between the cells and the external environment. Blood is the circulating portion of the extracellular fluid, responsible for carrying material from one part of the body to another. Total blood volume in the 70-kg man is equal to about 7% of his total body weight. Thus, if we assume that 1 kilogram of blood occupies a volume of 1 liter, then the 70-kg man has about 5 liters of blood

$(0.07 \times 70\,\text{kg} = 4.9\,\text{L})$. Of this volume, about 2 liters is composed of blood cells, while the remaining 3 liters is composed of plasma, the fluid portion of the blood.

This chapter presents an overview of the components of blood, which are summarized in Figure 16-1 ●. We will consider the functions of plasma, red blood cells, and platelets in detail but will reserve detailed discussion of white blood cells for Chapter 24.

Plasma Is Composed of Water, Ions, Organic Molecules, and Dissolved Gases

Plasma is the fluid portion of the blood, within which cellular elements are suspended. Water is the main component of plasma, accounting for about 92% of its weight. Proteins account for another 7%. The remaining 1% is dissolved organic molecules (amino acids, glucose, lipids, nitrogenous wastes), ions (Na^+, K^+, Cl^-, H^+, Ca^{2+}, and HCO_3^-) trace elements and vitamins, and dissolved oxygen (O_2) and carbon dioxide (CO_2).

Plasma is identical in composition to interstitial fluid except for the presence of **plasma proteins**. The liver makes most plasma proteins and secretes them into the blood. **Albumins** are the most prevalent type of protein in the plasma. **Globulins** and the clotting protein **fibrinogen** are also common. Some globulins, known as *immunoglobulins* or *antibodies*, are synthesized and secreted by special cells rather than by the liver.

The presence of proteins in the plasma makes the osmotic pressure of the blood higher than that of the interstitial fluid. This gradient pulls water from the interstitial fluid into the capillaries, as you learned in the last chapter [● p. 506].

Plasma proteins participate in many functions, including blood clotting and defense against foreign invaders. In addition, they act as carriers for steroid hormones, cholesterol, certain ions such as iron (Fe^{2+}), and drugs. Finally, some plasma proteins act as hormones or as extracellular enzymes. Table 16-1 summarizes the functions of plasma proteins.

The Cellular Elements Include Red Blood Cells, White Blood Cells, and Platelets

Three main cellular elements are found in blood: **red blood cells**, also called **erythrocytes** [*erythros*, red]; **white blood cells**, also called **leukocytes** [*leukos*, white]; and **platelets**. Only the white blood cells are fully functional cells in the circulation. Red blood cells have lost their nuclei by the time they enter the bloodstream. Platelets are cell fragments that have split off a parent cell known as a **megakaryocyte**.

Red blood cells play a key role in transporting oxygen and carbon dioxide between the lungs and the tissues. Platelets are instrumental in *coagulation*, the process by which blood clots prevent blood loss in damaged vessels. White blood cells play a key role in defending the body against foreign invaders such as parasites, bacteria, and viruses. Although most white blood cells

● **Figure 16-1** *Composition of blood*

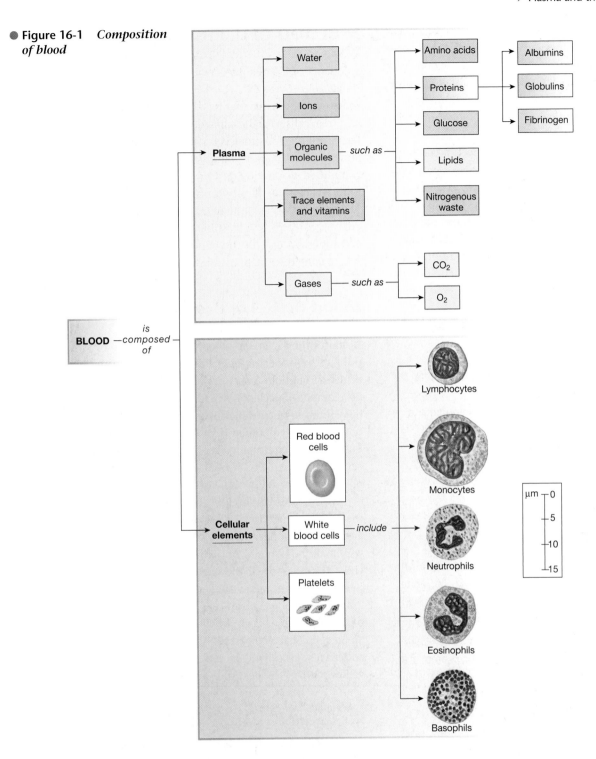

Table 16-1: Functions of Plasma Proteins

NAME	SOURCE	FUNCTION
Albumins (multiple types)	Liver	Major contributors to colloid osmotic pressure of plasma; carriers for various substances
Globulins (multiple types)	Liver and lymphoid tissue	Clotting factors, enzymes, antibodies, carriers for various substances
Fibrinogen	Liver	Forms fibrin threads essential to blood clotting

circulate through the body in the blood, their work is usually carried out in the tissues rather than inside the circulatory system.

Blood contains five types of mature white blood cells: (1) **lymphocytes**, (2) **monocytes**, (3) **neutrophils**, (4) **eosinophils**, and (5) **basophils**. Monocytes that leave the circulation and enter the tissues develop into **macrophages**. Tissue basophils are called **mast cells**.

The types of white blood cells may be grouped according to common morphological or functional characteristics. Neutrophils, monocytes, and macrophages are collectively known as **phagocytes** because they can engulf and ingest foreign particles such as bacteria (phagocytosis; ● p. 144). Lymphocytes are sometimes called **immunocytes** because they are responsible for specific immune responses directed against invaders. Basophils, eosinophils, and neutrophils are called **granulocytes** because they contain cytoplasmic inclusions that give them a granular appearance.

CONCEPT CHECK   Answers: p. 545

1. Name the five major types of leukocytes.

2. Why do we say that erythrocytes and platelets are not fully functional cells?

3. On the basis of what you have learned about the origin and role of plasma proteins, explain why patients with advanced degeneration of the liver frequently suffer from edema.

CLINICAL FOCUS

The Differential White Cell Count

When the body's defense system is called on to fight off foreign invaders, both the absolute number of white blood cells and the relative proportions of the different types of white blood cells in the circulation change. Clinicians often rely on a *differential white cell count* to help them arrive at a diagnosis (Fig. 16-2●). For example, a person with a bacterial infection will usually have a high total number of white blood cells in the blood with an increase in the percentage that are neutrophils. A person with a viral infection may have a high, normal, or low total white cell count but often shows an increase in the percentage of lymphocytes. Identifying the different types of white blood cells is not as simple as you might think from doing it in the student laboratory. In abnormal smears, accurate identification may be complicated by the presence of immature blood cells that are difficult to distinguish from each other. Medical technologists undergo months of special training in hematology, and those who work in laboratories certified by the College of American Pathologists are tested on cell identification three times each year.

▶ BLOOD CELL PRODUCTION

Where do these different blood cells come from? They are all descendants of a single precursor cell type known as the *pluripotent hematopoietic stem cell*. This cell type is found primarily in **bone marrow**, a soft tissue that fills the hollow centers of bones. Pluripotent stem cells have the remarkable ability to develop into many different cell types. As they specialize, they narrow their possible fates and become first *uncommitted stem cells*, then *progenitor cells,* that are committed to developing into one or perhaps two different cell types (Fig. 16-3●). Progenitor cells differentiate into red blood cells, lymphocytes, other white blood cells, and megakaryocytes, the parent cells of platelets. It is estimated that only about one out of every 100 thousand cells in the bone marrow is an uncommitted stem cell, making the isolation and study of these cells very difficult.

Blood Cells Are Produced in the Bone Marrow

Hematopoiesis [*haima*, blood + *poiesis*, formation], the synthesis of blood cells, begins early in embryonic development and continues throughout a person's life. In about the third week of fetal development, special cells in the yolk sac of the embryo form clusters. Some of these cell clusters are destined to become the endothelial lining of blood vessels, while others become blood cells. The common embryological origin of the endothelium and blood cells perhaps explains why many cytokines that control hematopoiesis are released by the vascular endothelium.

As the embryo develops, blood cell production spreads from the yolk sac to the liver, spleen, and bone marrow. By birth, the liver and spleen no longer produce blood cells. Hematopoiesis continues within all the bones of the skeleton until age five. However, as the child continues to age, the active regions of marrow decrease.

In adults, the pelvis, spine, ribs, cranium, and proximal ends of long bones produce blood cells. Active bone marrow is red because of the presence of *hemoglobin*, the oxygen-binding protein of red blood cells. Inactive marrow is yellow because of an abundance of adipocytes (fat cells). You can see the difference between red and yellow marrow the next time you look at bony cuts of meat in the

---

RUNNING PROBLEM

Von Willebrand's disease, named for the Finnish physician who discovered it in the late nineteenth century, is often diagnosed after an individual undergoes a dental or other surgical procedure. Besides prolonged bleeding, other symptoms include nosebleeds and bruising. The disease is caused by a deficiency in von Willebrand factor, a protein found in the blood that acts as a carrier for clotting factor VIII. Von Willebrand factor also promotes adherence of platelets to the endothelium and to each other during coagulation.

Question 1: Why does von Willebrand's disease cause prolonged bleeding after a dental or surgical procedure?

523   529   536   542

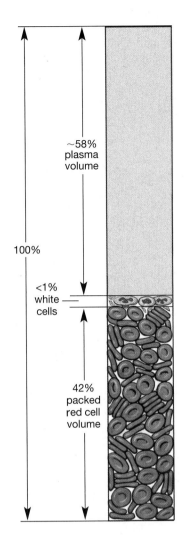

1. The **hematocrit** is the percentage (%) of the total blood volume that is occupied by packed (centrifuged) red blood cells (see figure at left).

2. The **hemoglobin** (Hb) content of erythrocytes. This value is measured as total hemoglobin content of blood (g Hb/dL).

3. **Mean red cell volume** (MCV). In some disease states, the red blood cells may be either abnormally large or abnormally small. For example, they are abnormally small in iron-deficiency anemia.

4. **Red cell count** in millions of cells per microliter. A machine counts the cells as they stream through a beam of light.

5. The **morphology** of the red blood cells also gives clues to diseases. Sometimes the cells lose their flattened disk shape and become spherical (spherocytosis). In sickle cell anemia, the cells are shaped like a sickle or crescent moon.

6. **Total white cell count** tells the total number of all types of white blood cells but does not distinguish between the different types.

7. The **differential white cell** count estimates the relative numbers of the five types of white cells. The differential count is made by a medical technologist from a thin smear of blood that is stained with biological dyes.

8. **Platelet count** is suggestive of the ability of the blood to clot.

	Males	Females
Hematocrit (% of whole blood that is packed red cells)	40%–54%	37%–47%
Hemoglobin (g/dL blood)	14–17	12–16
Red cell count (cells/μL)	4.5–6.5×10^6	3.9–5.6×10^6
Total white cell count (cells/μL)	4–11×10^3	4–11×10^3
Differential white cell count		
Neutrophils	50%–70%	50%–70%
Eosinophils	1%–4%	1%–4%
Basophils	<1%	<1%
Lymphocytes	20%–40%	20%–40%
Monocytes	2%–8%	2%–8%
Platelets (per μL)	200–500×10^3	200–500×10^3

● **Figure 16-2**   *The blood count*   The table at the bottom shows the normal range of values. A deciliter is equal to 100 mL.

grocery store. Although blood synthesis in adults is limited to small areas, the liver, spleen, and inactive (yellow) regions of marrow can resume blood cell production in times of need.

In the regions of marrow that are actively producing blood cells, about 25% of the cells are developing red blood cells, while 75% are destined to become white blood cells. The life span of white blood cells is considerably shorter than that of red blood cells, so they must be replaced more frequently. For example, neutrophils have a six-hour half-life, and the body must make *100 billion* neutrophils each day in order to replace those that die. Red blood cells, on the other hand, live for nearly four months in the circulation.

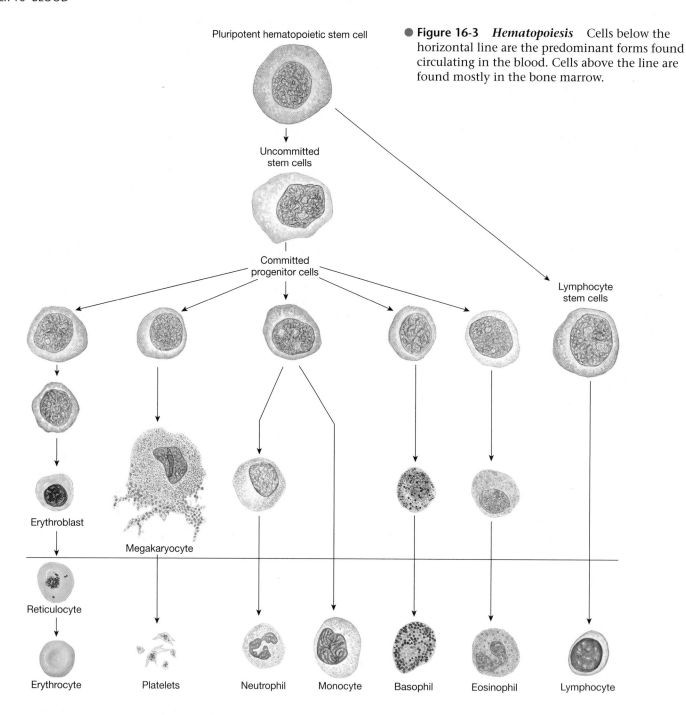

Pluripotent hematopoietic stem cell

● **Figure 16-3** *Hematopoiesis* Cells below the horizontal line are the predominant forms found circulating in the blood. Cells above the line are found mostly in the bone marrow.

Uncommitted stem cells

Committed progenitor cells

Lymphocyte stem cells

Erythroblast

Megakaryocyte

Reticulocyte

Erythrocyte   Platelets   Neutrophil   Monocyte   Basophil   Eosinophil   Lymphocyte

Hematopoiesis Is Controlled by Cytokines, Growth Factors, and Interleukins

What controls the production and development of blood cells? Chemical factors known as cytokines are responsible. *Cytokines* are peptides or proteins released from one cell that affect the growth or activity of another cell [● p. 174]. When cytokines are first discovered, they are often called *factors* and given a modifier that describes their actions: growth factor, differentiating factor, trophic factor.

Some of the best-known cytokines in hematopoiesis are the **colony-stimulating factors**, molecules made by endothelial cells

and white blood cells. Others are the **interleukins** [*inter-*, between + *leuko*, white], such as IL-3. The name *interleukin* was first given to cytokines released by one white blood cell to act on another white blood cell. Numbered interleukin names are given to cytokines once their amino acid sequences have been determined. Interleukins also play important roles in the immune system, as we will discuss in Chapter 24.

Another hematopoietic cytokine is *erythropoietin*, which controls red blood cell synthesis. You will often see erythropoietin called a hormone, but technically it fits the definition of a cytokine because it is made on demand rather than stored in vesicles.

Table 16-2 lists a few of the many cytokines linked to hematopoiesis. In fact, one review on the subject was titled

Table 16-2: Cytokines Involved in Hematopoiesis

NAME	CELL THAT PRODUCES CYTOKINE	INFLUENCES GROWTH OR DIFFERENTIATION OF
Erythropoietin (EPO)	Kidney cells	Red blood cells
Thrombopoietin (TPO)	Liver primarily	Megakaryocytes
Colony-stimulating factors, interleukins, stem cell factor	Endothelium and fibroblasts of bone marrow, leukocytes	All types of blood cells; mobilizes hematopoietic stem cells

"Regulation of hematopoiesis in a sea of chemokine family members with a plethora of redundant activities"!* Because of the complexity of the subject, we will not attempt to discuss hematopoietic cytokines in detail.

Colony-Stimulating Factors Regulate Leukopoiesis

Colony-stimulating factors (CSFs) were identified and named for their ability to stimulate the growth of leukocyte colonies in culture. These cytokines, made by endothelial cells, marrow fibroblasts, and white blood cells, regulate leukocyte production and development (**leukopoiesis**). They are required to induce both cell division (mitosis) and cell maturation in stem cells. Once a leukocyte matures, it loses its ability to undergo mitosis.

One fascinating aspect of hematopoiesis is that leukocyte production is regulated in part by the existing white blood cells. This form of control allows leukocyte development to be very specific and tailored to the body's needs. For example, if you have a bacterial infection, cytokines released by active white blood cells fighting the infection stimulate the production of additional neutrophils and macrophages. The complex process by which leukocyte production is matched to need is still not well understood and is an active area of research.

Scientists are working to create a model for the control of hematopoiesis so that they can develop effective treatments for diseases characterized by either a lack or an excess of white blood cells. The *leukemias* are a group of diseases characterized by the abnormal growth and development of white blood cells. In *neutropenias* [*penia*, poverty], patients have too few white blood cells and are unable to fight off bacterial and viral infections as a result. Researchers hope to find better treatments for both leukemias and neutropenias by unlocking the secrets of how the body regulates cell growth and division.

Thrombopoietin Regulates Platelet Production

Thrombopoietin (TPO) is a glycoprotein that regulates the growth and maturation of megakaryocytes, the parent cells of

platelets. It is produced primarily in the liver but is also present in the kidney. The gene for this cytokine, whose activity was first described in 1958, was cloned in 1994. Within a year, TPO was widely available to researchers through the use of recombinant DNA techniques, and there has been an explosion of papers describing its effects on megakaryocytes and platelet production. Although we still do not understand everything about the basic biology of the process, thrombopoietin synthesis is a huge step forward in the search for answers.

Erythropoietin Regulates Red Blood Cell Production

Red blood cell production (**erythropoiesis**) is controlled by the glycoprotein hormone (cytokine) **erythropoietin** (**EPO**), assisted by several other cytokines. Erythropoietin is made primarily in the kidney of adults. The stimulus for EPO synthesis and release is *hypoxia*, low oxygen levels in the tissues. This pathway, like other endocrine pathways, helps the body maintain homeostasis. By stimulating red blood cell synthesis, EPO puts more hemoglobin into the circulation to carry oxygen.

The existence of a substance controlling red blood cell production was first suggested in the 1950s, but two decades passed before scientists succeeded in purifying the hormone. One reason

---

RUNNING PROBLEM

"Is this disease serious?" Emily asks her doctor. "It can be," he replies. "Several different types of the disease exist, with varying degrees of severity. Since you have not had symptoms before this, you probably have one of the milder forms." Von Willebrand factor is present in megakaryocytes, endothelial cells, platelets, and plasma. In each location, the factor exists in three different "sizes," or multimers. The different types of von Willebrand disease vary according to the presence or absence of different multimers. The least serious type (type I) is characterized by decreased production of all von Willebrand multimers in all tissues. The most serious type (type III) is characterized by the complete absence of von Willebrand multimers in plasma.

Question 2: Why is type I the least serious form of von Willebrand's disease?

---

*H. E. Broxmeyer, and C. H. Kim, *Experimental Hematology* 27(7):1113–1123, 1999, July.

for the delay is that EPO is made on demand and not stored, as in an endocrine cell. Scientists took another nine years to sequence the hormone and to isolate and clone the gene for it. However, an incredible leap in progress was made after cloning the gene for EPO: only two years later, the hormone was produced by recombinant DNA technology and put into clinical use.

Today, physicians can prescribe not only genetically engineered EPO (epoetin alfa) but also several colony-stimulating factors (sargramostim and filgrastim) that stimulate white blood cell synthesis. Other hematopoietic cytokines are being tested in clinical trials. Cancer patients in whom hematopoiesis has been suppressed by chemotherapy have benefited from injections of these hematopoietic hormones.

CONCEPT CHECK   Answers: p. 545

4. Name the cytokine(s) that regulates growth and maturation in (a) erythrocytes, (b) leukocytes, and (c) megakaryocytes.

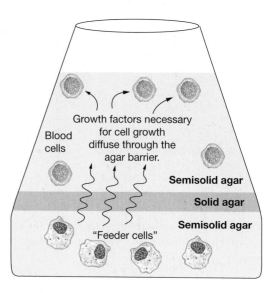

● **Figure 16-4** *Blood cell culture* Feeder cells in the bottom layer secrete cytokine growth factors that diffuse through the solid agar.

BIOTECHNOLOGY

Blood Cell Culture

The successful laboratory culture of isolated blood cells was first reported around 1961, and by 1966, researchers realized that blood cell culture required growth factors secreted by other leukocytes. Blood cells in a culture flask are suspended in a semisolid layer of nutrient material, but in the absence of "feeder cells" (macrophages or other cytokine-secreting cells), the cultured cells fail to develop. If feeder cells are added, the blood cells will develop. Direct cell-to-cell contact is not needed, however. If feeder cells are separated from the immature cells by a solid agar layer, cytokine growth factors can diffuse through the solid agar layer and development proceeds (Fig. 16-4 ●). The soluble growth factors were named *colony-stimulating factors* because they were first discovered in cell culture.

Colony-stimulating factors are secreted in such small amounts that initially it seemed impossible to get enough for clinical use. But with technological advances, the factors were identified, and their genes were located and cloned. We now have commercially available recombinant factors in widespread clinical use. The cost of developing colony-stimulating factors for therapeutic use is extraordinary. It has been estimated that the development cost for one hematopoietic factor from laboratory to bedside is greater than $500 million. Yet the market, which includes anyone who has lost bone marrow function from chemotherapy, is huge, and many patients stand to benefit.

RED BLOOD CELLS

Red blood cells, or *erythrocytes*, are the most abundant cell type in the blood. A microliter of blood contains about 5 million red blood cells, compared with only 4,000–11,000 white blood cells and 20,000–500,000 platelets. The primary function of red blood cells is to facilitate oxygen and carbon dioxide transport between the lungs and cells.

The ratio of red blood cells to plasma is indicated clinically by the **hematocrit** and is expressed as a percentage of the total blood volume (see Fig. 16-2). Hematocrit is determined by drawing a blood sample into a narrow capillary tube and spinning it in a centrifuge so that the heavier red blood cells go to the bottom of the sealed tube, leaving the thin "buffy layer" of lighter white blood cells and platelets in the middle and the plasma on top.

The column of *packed red cells* is measured and compared to the total sample volume to find the hematocrit. The normal range of hematocrit is 40%–54% for a man and 37%–47% for a woman. This test provides a rapid and inexpensive way to estimate a person's red cell count because blood for a hematocrit can be collected by simply sticking a finger.

Mature Red Blood Cells Lack a Nucleus

In the bone marrow, committed stem cells differentiate through several stages into large, nucleated *erythroblasts*. As erythroblasts mature, the nucleus condenses and the cell shrinks in diameter from 20 μm to about 7 μm. In the last stage before maturation, the nucleus is pinched off and phagocytosed by marrow macrophages. At the same time, other membranous organelles such as mitochondria break down and disappear. The final immature cell form, called a *reticulocyte*, leaves the marrow and enters the circulation, where it matures into an erythrocyte in about 24 hours (Fig. 16-5 ●).

(a) The bone marrow, hidden within the bones of the skeleton, is easily overlooked as a tissue, although collectively, it is nearly the size and weight of the liver!

Bone marrow

(b) Marrow is a highly vascular tissue, filled with blood sinuses, widened regions lined with epithelium.

Bone cortex

Central sinus

Stroma of marrow

Nutrient artery

Venous sinuses

Radial artery

(c) Bone marrow consists of blood cells in different stages of development and supporting tissue known as the **stroma** (mattress).

Mature blood cells squeeze through the endothelium to reach the circulation.

Red blood cell maturation

Stem cell

Reticulocyte expelling nucleus

Platelets

Mature neutrophil

Fragments of megakaryocyte break off to become platelets.

Reticular fiber

Venous sinus

Neutrophil maturation

Reticular cell

Stem cell

Macrophage

Monocyte

The stroma is composed of fibroblast-like reticular cells, collagenous fibers, and extracellular matrix.

Lymphocyte

● **Figure 16-5**

Mature mammalian red blood cells are biconcave disks, shaped much like jelly doughnuts with the filling squeezed out of the middle (Fig. 16-6a ●). They lack a nucleus and membranous organelles and are a simple membranous "bag" filled with enzymes and hemoglobin, a red oxygen-carrying pigment.

Because red blood cells contain no mitochondria, they cannot carry out aerobic metabolism. Glycolysis is their primary source of ATP. Without a nucleus and endoplasmic reticulum to carry out protein synthesis, the erythrocyte is unable to make new enzymes or to renew its membrane components. This inability leads to an increasing loss of membrane flexibility, making older cells more fragile and likely to rupture.

The biconcave shape of the red blood cell is one of its most distinguishing features. The membrane is held in place by a complex cytoskeleton composed of filaments linked to transmembrane attachment proteins (Fig. 16-6c). Despite the cytoskeleton, red cells are remarkably flexible, like a partially filled water balloon that can compress into various shapes. This flexibility allows erythrocytes to change shape as they squeeze through the narrow capillaries of the circulation.

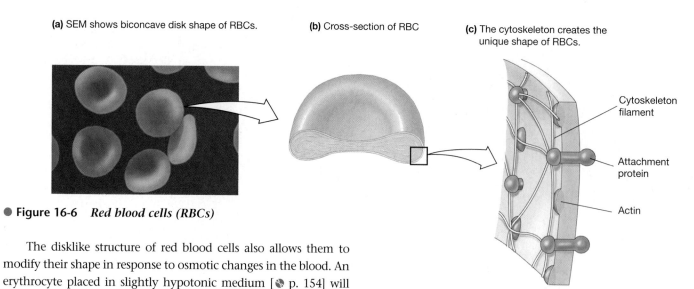

(a) SEM shows biconcave disk shape of RBCs.

(b) Cross-section of RBC

(c) The cytoskeleton creates the unique shape of RBCs.

Cytoskeleton filament

Attachment protein

Actin

● **Figure 16-6** *Red blood cells (RBCs)*

The disklike structure of red blood cells also allows them to modify their shape in response to osmotic changes in the blood. An erythrocyte placed in slightly hypotonic medium [● p. 154] will swell and form a sphere without disruption of its membrane integrity. In hypertonic media, red cells shrink up and develop a spiky surface when the membrane pulls tight against the cytoskeleton.

Hemoglobin Synthesis Requires Iron

Hemoglobin (Hb), the oxygen-carrying pigment of red blood cells, is a large, complex molecule. Its quaternary structure has four globular protein chains, each of which is wrapped around an iron-containing **heme** group (Fig. 16-7 ●). There are several forms of the **globin** protein chain. The most common forms are designated *alpha* (α), *beta* (β), *gamma* (γ), and *delta* (δ), depending on the structure of the chain. Most adult hemoglobin has two alpha and two beta chains (designated *HbA*). However, a small portion of adult hemoglobin (about 2.5%) has two alpha and two delta chains (*HbA₂*).

The human fetus has a different form of hemoglobin adapted to attract oxygen from maternal blood in the placenta. *Fetal hemoglobin (HbF)* has two gamma chains in place of the two beta chains found in adult hemoglobin. Shortly after an infant's birth, fetal hemoglobin is replaced with the adult form as new red blood cells are made.

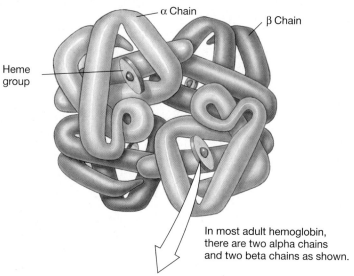

(a) A hemoglobin molecule is composed of four protein globin chains, each centered around a heme group.

α Chain

β Chain

Heme group

In most adult hemoglobin, there are two alpha chains and two beta chains as shown.

(b) Each heme group consists of a porphyrin ring with an iron atom in the center.

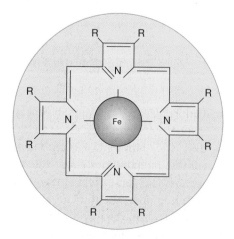

R = additional C, H, O groups

● **Figure 16-7** *Hemoglobin*

Diabetes

CLINICAL FOCUS

DIABETES: Hemoglobin and Hyperglycemia

One of the goals of diabetes treatment is to keep blood glucose concentrations as close to normal as possible. But how can a clinician tell if a patient has been doing this? One way is to measure the patient's hemoglobin. Glucose in the plasma will bind covalently to hemoglobin, producing a glycohemoglobin known as **hemoglobin A₁c**. The amount of hemoglobin A₁c in the plasma is directly related to hemoglobin's exposure to glucose over the preceding 8–12 weeks. By using this assay, a clinician can monitor long-term fluctuations in blood glucose levels and adjust a diabetic patient's therapy appropriately.

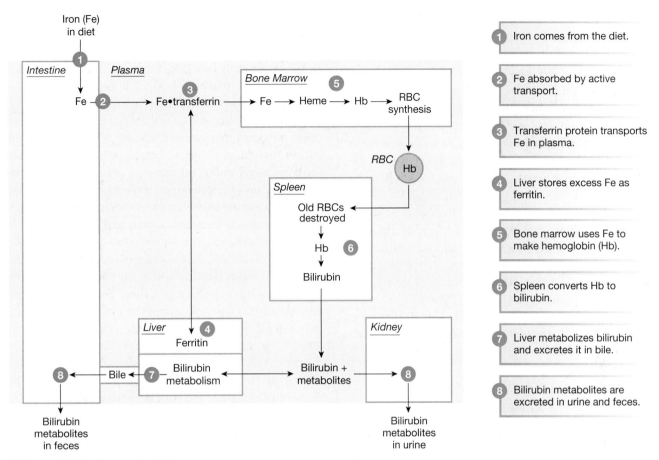

● **Figure 16-8** *Iron metabolism*

The four heme groups in a hemoglobin molecule are identical. Each consists of a carbon-hydrogen-nitrogen *porphyrin ring* with an iron atom (Fe) in the center (Fig. 16-7b). Some 70% of the iron in the body is found in the heme groups of red blood cells.

Hemoglobin synthesis requires an adequate supply of iron in the diet (Fig. 16-8 ●). Iron is absorbed in the small intestine by active transport. It is transported in the blood by a carrier protein called **transferrin**. In the bone marrow, developing red blood cells use iron to make the heme portion of hemoglobin. Excess iron in the body is stored, mostly in the liver, in the protein **ferritin** and its derivatives.

It is possible to ingest too much iron, and poisoning in children sometimes occurs when children ingest too many vitamins with iron. Initial symptoms are gastrointestinal pain, cramping, and bleeding when iron corrodes the digestive epithelium. Later onset problems include liver failure, which can be fatal.

Red Blood Cells Live About Four Months

Red blood cells in the circulation live for 120 days (plus or minus 20 days). The increasingly fragile older erythrocytes may rupture as they try to squeeze through narrow capillaries, or they may be captured by scavenging macrophages as they pass through the spleen. Many components of hemoglobin are recycled. Amino acids from the globin chains become new proteins, while some iron from heme groups may be reused in new heme groups.

The remainder of the old heme is converted by cells of the spleen and liver to a colored pigment called **bilirubin**. Bilirubin is carried by plasma albumin to the liver, where it is metabolized and incorporated into a secretion called **bile** (Fig. 16-8). Bile is secreted into the digestive tract, and the bilirubin metabolites leave the body in the feces. Small amounts of other bilirubin metabolites are filtered from the blood into the kidneys, where they contribute to the yellow color of the urine.

In some circumstances, bilirubin levels in the blood become elevated (*hyperbilirubinemia*). This condition, known as **jaundice**, causes the skin and whites of the eyes to take on a yellow cast. The accumulation of bilirubin can occur from multiple causes. Newborns whose fetal hemoglobin is being broken down and replaced with adult hemoglobin are particularly susceptible to bilirubin toxicity, so doctors monitor babies for jaundice in the first weeks of life. Another common cause of jaundice is liver disease, in which the liver is unable to process or excrete bilirubin.

CONCEPT CHECK   Answers: p. 545

5. Distinguish between (a) heme and hemoglobin, (b) ferritin and transferrin.

6. Is bile an endocrine or an exocrine secretion?

533

Red Blood Cell Disorders Decrease Oxygen Transport

Hemoglobin plays a critical role in oxygen transport, so the red blood cell count and hemoglobin content of the body is important. If hemoglobin content is too low, a condition known as **anemia**, the blood cannot transport enough oxygen to the tissues. People with anemia are usually tired and weak, especially during exercise. The major causes of anemia are summarized in Table 16-3. Anemia usually results either from blood loss, accelerated red blood cell destruction, or defective red blood cell or hemoglobin synthesis.

In the *hemolytic anemias* [*lysis*, rupture], the rate of red blood cell destruction exceeds the rate of red blood cell production. The hemolytic anemias are usually hereditary defects in which the body makes fragile cells. For example, in *hereditary spherocytosis*, the cytoskeleton does not link properly because of defective or deficient cytoskeletal proteins. Consequently, the cells are shaped more like spheres than like biconcave disks. This disruption in the cytoskeleton results in cells that rupture easily and that are not able to withstand osmotic changes as well as normal cells.

Table 16-3: Causes of Anemia

Accelerated red blood cell loss

Blood loss: Cells are normal in size and hemoglobin content but low in number

Hemolytic anemias: Cells rupture at an abnormally high rate

  Hereditary

    Membrane defects (example: hereditary spherocytosis)

    Enzyme defects

    Abnormal hemoglobin (example: sickle cell anemia)

  Acquired

    Parasitic infections (example: malaria)

    Drugs

    Autoimmune reactions

Decreased red blood cell production

Defective red blood cell or hemoglobin synthesis in the bone marrow

  Aplastic anemia: Can be caused by certain drugs or radiation

  Inadequate dietary intake of essential nutrients

  Iron deficiency (iron is required for heme production)

  Folic acid deficiency (folic acid is required for DNA synthesis)

  Vitamin B_{12} deficiency. May be due to lack of intrinsic factor for B_{12} absorption (vitamin B_{12} is required for DNA synthesis)

Inadequate production of erythropoietin

Other anemias result from the failure of the bone marrow to make adequate amounts of hemoglobin. One of the most common examples of an anemia that results from insufficient hemoglobin synthesis is **iron-deficiency anemia**. If iron loss exceeds iron intake, the marrow does not have adequate iron to make heme groups, and hemoglobin synthesis slows.

People with iron-deficiency anemia have a low red blood cell count (reflected in a low hematocrit) or a low hemoglobin content in their blood. Their red blood cells are often smaller than usual (*microcytic* red blood cells), and the lower hemoglobin content may cause the cells to be paler than normal (*hypochromic*). Women who menstruate are likely to suffer from iron-deficiency anemia because of iron loss in menstrual blood.

Finally, some anemias are the result of abnormal hemoglobin molecules. *Sickle cell disease* is a genetic defect in which glutamate, the number 6 amino acid in the 146-amino acid beta chain of hemoglobin, is replaced by valine. The result is abnormal hemoglobin (*HbS*) that crystallizes when it gives up its oxygen. This pulls the red blood cells into a sickle shape, like a crescent moon (Fig. 16-9 ●). The sickled cells become tangled with other sickled red cells as they pass through the smallest blood vessels, causing them to jam up and block blood flow to the tissues. This blockage creates tissue damage and pain from hypoxia.

Although the anemias are common, it is also possible to have too many red blood cells. *Polycythemia vera* [*vera*, true] is a stem cell dysfunction that produces too many blood cells, white as well as red. These patients may have hematocrits as high as 60%–70% (normal is 37%–54%). The increased number of cells causes the blood to become more viscous and thus more resistant to flow through the circulatory system [● p. 454].

In *relative polycythemia*, the person's actual red blood cell number is normal, but the hematocrit is elevated due to loss of plasma volume. This might occur with dehydration, for example. The opposite problem can also occur. If an athlete overhydrates, the hematocrit may decrease temporarily because of increased plasma volume. In both of these situations, there is no actual pathology with the red blood cells.

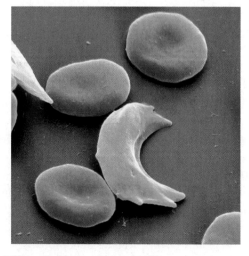

● **Figure 16-9** *Sickled red blood cells*

Sickle Cell Disease and Hydroxyurea

One new treatment developed for sickle cell disease is the administration of hydroxyurea, a compound that inhibits DNA synthesis. In the bone marrow, hydroxyurea causes immature red blood cells to produce fetal hemoglobin (HbF), with alpha and gamma globin chains, instead of adult hemoglobin. The theory for hydroxyurea therapy is that fetal hemoglobin interferes with the crystallization of sickle cell hemoglobin (HbS), so the red blood cells no longer develop a sickle shape. However, some studies show improvements in sickle cell symptoms before HbF levels increase. One theory of why this occurs is based on the observation that hydroxyurea in rats is metabolized to nitric oxide (NO), which causes vasodilation. Inhaled nitric oxide is now being tested as a new treatment for sickle cell disease symptoms. To learn more about the current sickle cell treatments, visit the website for Emory University's Sickle Cell Information Center *http://www.scinfo.org/index.htm*.

CONCEPT CHECK Answers: p. 545

7. Erythropoietin (EPO) was first isolated from the urine of anemic patients who had high circulating levels of the hormone. Although these patients had high concentration of EPO, they were unable to produce adequate amounts of hemoglobin or red cells. Give some possible reasons why the patients' EPO was unable to correct their anemia.

8. A person who goes from sea level to a city that is 5000 feet above sea level will show an increased hematocrit within two to three days. Draw the reflex pathway that links the hypoxia of high altitude to increased red blood cell production.

9. An individual who becomes dehydrated will have an elevated hematocrit. Explain how dehydration leads to an elevated hematocrit although blood cell production is normal.

▶ PLATELETS AND COAGULATION

Because of its fluid nature, blood flows freely throughout the circulatory system. However, if there is a break in the "piping" of the system, blood will be lost unless steps are taken. One of the challenges for the body is to plug holes in damaged blood vessels while still maintaining blood flow through the vessel.

It would be simple to block off a damaged blood vessel completely, like putting a barricade across a street full of potholes. But just as shopkeepers on that street lose business if traffic is blocked, cells downstream from the point of injury die from lack of oxygen and nutrients if the vessel is completely blocked. The body's task is to allow blood flow through the vessel while simultaneously repairing the damaged wall.

This challenge is complicated by the fact that blood in the system is under pressure. If the repair "patch" is too weak, it will simply be blown out by the blood pressure. Thus, stopping blood loss involves several steps. First, the pressure in the vessel must be decreased long enough to create a secure mechanical seal in the form of a blood clot. Once the clot is in place and blood loss has been stopped, the body's repair mechanisms can take over. Then, as the wound heals, enzymes gradually dissolve the clot, while scavenger white blood cells ingest and destroy the debris.

Platelets Are Small Fragments of Cells

Platelets (*thrombocytes*) are cell fragments produced in the bone marrow from huge cells called *megakaryocytes* [*mega*, extremely large + *karyon*, kernel + *-cyte*, cell]. Megakaryocytes develop their formidable size by undergoing mitosis up to seven times without undergoing nuclear or cytoplasmic division. The result is a *polyploid* cell with a lobed nucleus (Fig. 16-10 ●).

The outer edges of marrow megakaryocytes extend through the endothelium into the lumen of marrow blood sinuses, where the cytoplasmic extensions fragment into disklike platelets (see Fig. 16-5). Platelets are smaller than red blood cells, are colorless, and have no nucleus. Their cytoplasm contains mitochondria, smooth endoplasmic reticulum, and many granules filled with clotting proteins and cytokines.

The typical life span of a platelet is about 10 days. Platelets are always present in the blood, but they are not active unless damage has occurred to the walls of the circulatory system.

Hemostasis Prevents Blood Loss from Damaged Vessels

Hemostasis [*haima*, blood + *stasis*, stoppage] is the process of keeping blood within a damaged blood vessel. (The opposite of hemostasis is *hemorrhage* [-*rrhagia*, abnormal flow].) Hemostasis has three major steps: (1) vasoconstriction, (2) temporary blockage of a break by a platelet plug, and (3) blood coagulation or formation of a clot that seals the hole until tissues are repaired.

Figure 16-11 ● summarizes this process. The first step in hemostasis is immediate constriction of damaged vessels caused by vasoconstrictive paracrines released by the endothelium. Vasoconstriction temporarily decreases flow and pressure within the vessel, helping formation of the platelet plug. When you put pressure on a bleeding wound, you also decrease flow within the damaged vessel.

Vasoconstriction is rapidly followed by the second step, mechanical blockage of the hole by a **platelet plug**. Platelets stick to the exposed collagen (**platelet adhesion**) and become activated, releasing cytokines into the area around the injury. Platelet factors reinforce local vasoconstriction and activate more platelets, which stick to each other (**platelet aggregation**) to form a loose platelet plug.

(a)

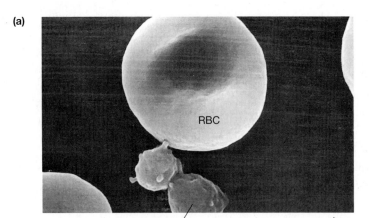

Inactive platelets are small disklike cell fragments.

(b)

Activated platelets (shown enlarged) develop a
spiky outer surface and adhere to each other.

(c)

Megakaryocytes are giant cells with
multiple copies of DNA in the nucleus.

Platelets

Red
blood
cell

Endoplasmic
reticulum

The edges of the megakaryocyte
break off to form cell fragments
called platelets.

● **Figure 16-10** *Megakaryocytes and platelets*

Simultaneously, in the third step, exposed collagen and tissue factor initiate a series of reactions known as the **coagulation cascade**. In the cascade, inactive plasma proteins are converted into active enzymes. In the last step, the enzyme **thrombin** converts **fibrinogen** into **fibrin** fibers that intertwine with and reinforce the platelet plug. The reinforced platelet plug is called a **clot**. Finally, as cell growth and division repair the damaged vessel, the clot retracts and is slowly dissolved by the enzyme **plasmin**.

The body must maintain the proper balance during hemostasis. Too little hemostasis allows excessive bleeding. Too much creates a **thrombus**, or blood clot, that adheres to the undamaged wall of a blood vessel [*thrombos*, a clot or lump]. A large thrombus can block the lumen of the vessel and stop blood flow.

Although hemostasis seems straightforward, there are still unanswered questions at the cellular and molecular levels. Because inappropriate blood clotting plays an important role in strokes and heart attacks, this area of research is very active. Research has led to the development and use of "clot busters," enzymes that can dissolve clots in arteries after heart attacks and strokes.

A detailed study of hemostasis involves many chemical factors, some of which play multiple roles and have multiple names. Thus, learning about hemostasis may be especially challenging. For example, some factors participate in both platelet plug forma-

tion and coagulation. Another factor activates enzymes for both clot formation and clot dissolution. Because of the complexity of the coagulation cascade, we will discuss only a few aspects of hemostasis in additional detail.

Platelet Activation Begins the Clotting Process

When a blood vessel wall is first damaged, exposed collagen and chemicals from endothelial cells activate platelets (Fig. 16-12●;

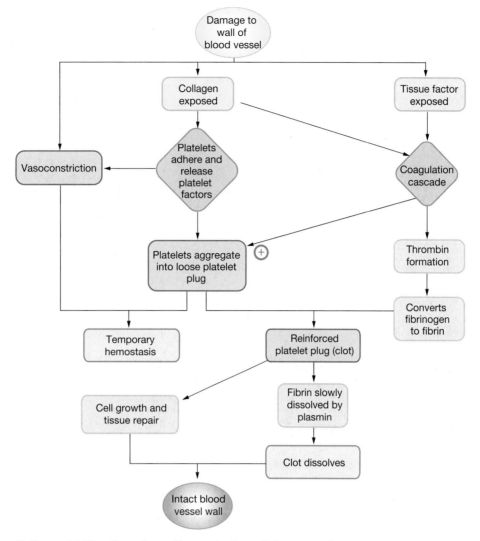

● **Figure 16-11** *Overview of hemostasis and tissue repair*

1 Exposed collagen binds and activates platelets.

2 Release of platelet factors

3 Attracts more platelets

4 Aggregate into platelet plug

● **Figure 16-12** *Platelet plug formation*

Table 16-4: Factors Involved in Platelet Function

CHEMICAL FACTOR	SOURCE	ACTIVATED BY OR RELEASED IN RESPONSE TO	ROLE IN PLATELET PLUG FORMATION	OTHER ROLES AND COMMENTS
Collagen	Subendothelial extracellular matrix	Injury exposes platelets to collagen	Binds platelets to begin platelet plug	——
von Willebrand factor (vWF)	Endothelium, megakaryocytes	Exposure to collagen	Links platelets to collagen	Deficiency or defect causes prolonged bleeding
Serotonin	Secretory vesicles of platelets	Platelet activation	Platelet aggregation	Vasoconstrictor
Adenosine diphosphate (ADP)	Platelet mitochondria	Platelet activation, thrombin	Platelet aggregation	——
Platelet-activating factor (PAF)	Platelets, neutrophils, monocytes	Platelet activation	Platelet aggregation	Plays role in inflammation; increases capillary permeability
Thromboxane A_2	Phospholipids in platelet membranes	Platelet activating factor	Platelet aggregation	Vasoconstrictor; eicosanoid
Platelet-derived growth factor (PDGF)	Platelets	Platelet activation	——	Promotes wound healing by attracting fibroblasts and smooth muscle cells

Table 16-4). Normally, the blood vessel endothelium separates the collagenous matrix fibers from the circulating blood. But if the vessel is damaged, collagen is exposed, and platelets rapidly begin to adhere to it.

Platelets adhere to collagen with the help of *integrins*, membrane receptor proteins that are linked to the cytoskeleton [● pp. 66, 181]. Binding activates platelets so they release the contents of their intracellular granules, including *serotonin* (5-hydroxytryptamine), ADP, and **platelet-activating factor** (PAF). PAF sets up a positive feedback loop by activating more platelets. It also initiates pathways that convert platelet membrane phospholipids into **thromboxane** A_2.

Serotonin and thromboxane A_2 are vasoconstrictors. They also contribute to platelet aggregation, along with ADP and PAF. The net result is a growing platelet plug that seals the damaged vessel wall.

If platelet aggregation is a positive feedback event, what prevents the platelet plug from continuing to form and thus spreading beyond the site of injury to other areas of the blood vessel wall? The answer lies in the fact that platelets will not adhere to normal endothelium.

Intact vascular endothelial cells convert their membrane lipids into **prostacyclin**, an eicosanoid [● p. 37] that blocks platelet adhesion and aggregation. Nitric oxide (NO) released by normal, intact endothelium also inhibits platelets from adhering. The combination of platelet attraction to the injury site and repulsion from the normal vessel wall creates a localized response that limits the platelet plug to the area of damage.

Coagulation Converts the Platelet Plug into a More Stable Clot

The third major step in hemostasis, coagulation, is a complex process in which fluid blood forms a gelatinous clot. The **coagulation cascade** is a series of enzymatic reactions that ends with formation of a protein fiber mesh that stabilizes the platelet plug. Some chemical factors involved in the coagulation cascade also promote platelet adhesion and aggregation to the damaged region.

Coagulation is divided into two pathways (Fig. 16-13 ●). An **intrinsic pathway** (yellow) begins with collagen exposure and uses proteins already present in the plasma. Collagen activates the first enzyme, factor XII, to begin the cascade. An **extrinsic pathway** (blue) starts when damaged tissues expose a protein-phospholipid mixture called **tissue factor**, or *tissue thromboplastin*. Tissue factor activates factor VII to begin the extrinsic pathway. The two pathways unite at the **common pathway** (green) to create thrombin, the enzyme that converts fibrinogen into insoluble fibrin fibers. These fibers become part of the clot.

Coagulation was initially regarded as a cascade similar to the second messenger cascades diagrammed in Chapter 6 [● p. 177]. At each step an enzyme converts an inactive precursor into an active enzyme, often with the help of Ca^{2+}, membrane phospholipids, or additional factors.

We now know, however, that the process is more than a simple cascade. Factors in the intrinsic and extrinsic pathways interact with each other, making coagulation a network rather than a

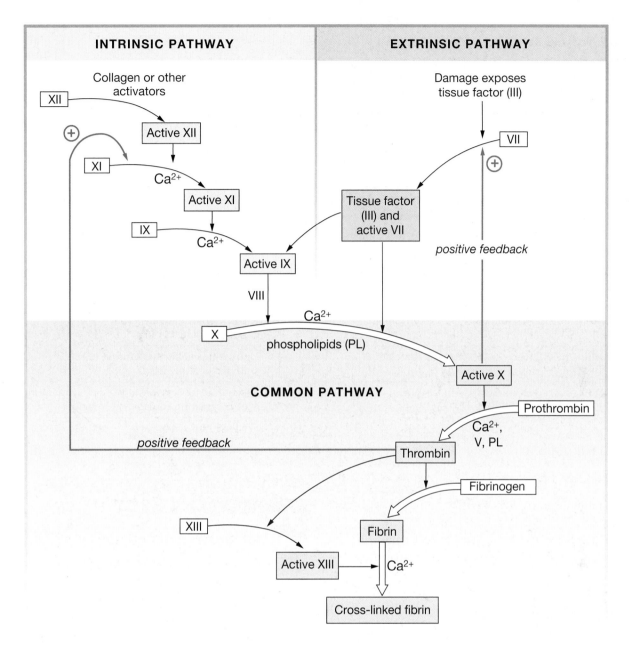

● **Figure 16-13** *The coagulation cascade* Inactive plasma proteins are converted into active enzymes in each step of the pathway.

simple cascade. In addition, there are several positive feedback loops that sustain the cascade until one or more of the participating plasma proteins is completely consumed.

The end result of the coagulation process is the conversion of fibrinogen into fibrin, a reaction catalyzed by the enzyme thrombin (Figure 16-14 ●). The fibrin fibers weave through the platelet plug and trap red blood cells within their mesh (see chapter opening photo on p. 523). Active factor XIII converts fibrin into a cross-linked polymer that stabilizes the clot.

As the clot forms, it incorporates **plasmin** molecules, the seeds of its own destruction. Plasmin is an enzyme created from plasminogen by thrombin and **tissue plasminogen activator** (**t-PA**). Plasmin breaks down fibrin polymers into fibrin fragments. The dissolution

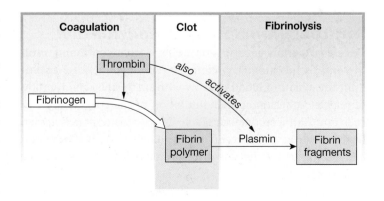

● **Figure 16-14** *Coagulation and fibrinolysis*

Table 16-5: Factors Involved in Coagulation

CHEMICAL FACTOR	SOURCE	ACTIVATED BY OR RELEASED IN RESPONSE TO	ROLE IN COAGULATION	OTHER ROLES AND COMMENTS
Collagen	Subendothelial extracellular matrix	Injury exposes collagen to plasma clotting factors	Starts intrinsic pathway	NA
von Willebrand factor (vWF)	Endothelium, megakaryocytes	Exposure to collagen	Regulates level of factor VIII	Deficiency or defect causes prolonged bleeding
Kininogen and kallikrein	Liver and plasma	Cofactors normally present in plasma pathway	Cofactors for contact activation of intrinsic pathway	Mediate inflammatory response; enhance fibrinolysis
Tissue factor (tissue thromboplastin or factor III)	Most cells except platelets	Damage to tissue	Starts extrinsic pathway	NA
Prothrombin and thrombin (factor II)	Liver and plasma	Platelet lipids, Ca^{2+}, and factor V	Fibrin production	NA
Fibrinogen and fibrin (factor I)	Liver and plasma	Thrombin	Forms insoluble fibers that stabilize platelet plug	NA
Fibrin-stabilizing factor (XIII)	Liver, megakaryocytes	Platelets	Cross-links fibrin polymers to make stable mesh	NA
Ca^{2+} (factor IV)	Plasma ions	NA	Required for several steps of coagulation cascade	Never a limiting factor
Vitamin K	Diet	NA	Needed for synthesis of factors II, VII, IX, X	NA

of fibrin by plasmin is known as **fibrinolysis**. As the damaged vessel wall slowly repairs itself, fibrinolysis helps remove the clot.

The large number of factors involved in coagulation and the fact that a single factor may have many different names can be confusing (Table 16-5). Scientists assigned numbers to the coagulation factors but the factors are not numbered in the order in which they participate in the coagulation cascade. Instead, they are numbered according to the order in which they were discovered.

CONCEPT CHECK

Answers: p. 545

10. On Figure 16-11, which boxes or ovals correspond to the intrinsic and extrinsic pathways of coagulation?

Anticoagulants Prevent Coagulation

Once coagulation begins, what keeps it from continuing until the entire circulation has clotted? Two different mechanisms limit the extent of blood clotting within a vessel: (1) inhibition of platelet adhesion and (2) inhibition of the coagulation cascade and fibrin production (Table 16-6). As mentioned earlier, factors such as prostacyclin in the blood vessel endothelium and plasma ensure that the platelet plug is restricted to the area of damage.

In addition, endothelial cells release chemicals known as **anticoagulants** that prevent coagulation from taking place. Most act by blocking one or more of the reactions in the coagulation

EMERGING CONCEPTS

Clot Busters and Antiplatelet Agents

The discovery of the factors controlling coagulation and fibrinolysis was an important step in developing treatments for heart attacks, more properly called *myocardial infarctions* (MIs). An **infarct** is an area of tissue deprived of its blood (and oxygen) supply because the vessel is blocked by an inappropriate clot (thrombus) or by an atherosclerotic plaque. Unless the blockage is removed promptly, the tissue will die or be severely damaged. If the blockage occurs in a coronary artery, the heart may sustain so much damage that it can no longer function properly. Many clinical trials around the world are studying the best way to treat myocardial infarctions. One option is to use fibrinolytic drugs, such as streptokinase (from bacteria) and tissue plasminogen activator (t-PA), to dissolve the clots. These drugs are now being combined with other agents that prevent further platelet plug and clot formation. Radio ads for aspirin remind consumers that this old standby can help prevent heart attack damage. The newest drugs for treating heart attacks are antiplatelet agents, including some that are antagonists to platelet integrin receptors.

Table 16-6: Endogenous Factors Involved in Fibrinolysis and Anticoagulation

CHEMICAL FACTOR	SOURCE	ACTIVATED BY OR RELEASED IN RESPONSE TO	ROLE IN ANTICOAGULATION OR FIBRINOLYSIS	OTHER ROLES AND COMMENTS
Plasminogen and plasmin	Liver and plasma	t-PA and thrombin	Dissolves fibrin and fibrinogen	NA
Tissue plasminogen activator (t-PA)	Many tissues	Normally present; levels increase with stress, protein C	Activates plasminogen	Recombinant t-PA used clinically to dissolve clots
Antithrombin III	Liver and plasma	NA	Anticoagulant; blocks factors IX, X, XI, XII, thrombin, kallikrein	Facilitated by heparin; no effect on thrombin despite name
Prostacyclin (prostaglandin I, or PGI_2)	Endothelial cells	NA	Blocks platelet aggregation	Vasodilator

cascade. The body produces two anticoagulants, **heparin** and **antithrombin III**, which work together to block active factors IX, X, XI, and XII. **Protein C**, another anticoagulant in the body, inhibits clotting factors V and VIII.

Anticoagulant drugs may be prescribed for people who are in danger of forming small blood clots that could block off critical vessels in the brain, heart, or lungs. The family of *coumarin anticoagulants*, such as *warfarin* (Coumadin®), block the action of vitamin K, a cofactor [🐾 p. 93] in the synthesis of clotting factors II, VII, IX, and X. These anticoagulants were discovered when cattle that developed severe bleeding problems were found to have been eating spoiled sweet clover.

When blood samples are drawn into glass tubes, clotting takes place very rapidly unless the tube contains an anticoagulant. Several of the anticoagulants that are used for this purpose remove free Ca^{2+} from the plasma. Ca^{2+} is an essential clotting factor, so with no Ca^{2+}, no coagulation can occur. In the living body however, plasma Ca^{2+} levels never decrease to levels that interfere with coagulation.

Acetylsalicylic acid (aspirin) is an agent that prevents platelet plug formation. It acts by inhibiting the COX enzymes [● p. 184] that promote synthesis of the platelet activator thromboxane A_2. People who are at risk of developing small blood clots are sometimes told to take one aspirin every other day "to thin the blood."

The aspirin does not actually dilute the blood, but it does prevent clots from forming by blocking platelet aggregation. Aspirin is now given routinely as emergency treatment for a suspected heart attack (see the box on Clot Busters).

Several inherited diseases affect the coagulation process. Patients with coagulation disorders bruise easily. In severe forms, spontaneous bleeding may occur throughout the body. Bleeding into the joints and muscles can be painful and disabling. If bleeding occurs in the brain, it can be fatal.

The best-known coagulation disorder is **hemophilia**, a name given to several diseases in which one of the factors in the coagulation cascade is defective or lacking. Hemophilia A, a factor VIII deficiency, is the most common form, occurring in about 80% of all cases. This disease is a recessive sex-linked trait that usually affects only males.

One exciting development in the treatment of hemophilia was a report on the first patients to be given gene therapy for hemophilia B, a deficiency in clotting factor IX. The patients who were injected with a virus engineered to carry the gene for factor IX started to produce some of the factor on their own, reducing their need for expensive injections of artificial factor IX. To learn more about these clinical trials and the latest treatments for hemophilia, visit the National Hemophilia Foundation website at *www.hemophilia.org*.

RUNNING PROBLEM CONCLUSION

Von Willebrand's Disease

"Can I live a normal life with von Willebrand's disease?" asks Emily. "Yes," her doctor replies. "This coagulation disorder is different from hemophilia, in which a clotting factor is defective. Your clotting factor is normal." Interestingly, people with hemophilia have normal amounts of von Willebrand factor.

In this running problem, you learned about a coagulation disorder called von Willebrand's disease. You also learned how this disease is treated and how it differs from hemophilia. Now compare your answers with those in the summary table.

(continues on next page)

	QUESTION	FACTS	INTEGRATION AND ANALYSIS
1	Why does von Willebrand's disease cause prolonged bleeding after a dental or surgical procedure?	Von Willebrand's disease is caused by a deficiency in von Willebrand factor, a protein found in the blood that promotes adherence of platelets.	Because of a deficiency in von Willebrand factor, platelets do not stick together during coagulation, and blood clots do not form easily.
2	Why is type I the least serious form of von Willebrand's disease?	Type I is characterized by decreased production of von Willebrand factor.	Because at least some von Willebrand factor is present, some clotting does take place. Therefore, this form is the least serious.
3	Explain the rationale for giving an antifibrinolytic agent to individuals with von Willebrand's disease.	Antifibrinolytic agents inhibit the action of plasmin in a blood clot.	Plasmin dissolves blood clots in the latter stages of blood clotting. Inhibiting plasmin blocks this dissolution and allows clot formation to proceed.

CHAPTER SUMMARY

Blood is an interesting tissue, with blood cells and cell fragments suspended in a liquid matrix. The plasma forms one of the two extracellular *compartments*. *Exchange* between the plasma and interstitial fluid takes place only in the capillaries. *Mass flow* of blood through the body depends on the pressure gradient created by the heart. At the same time, high pressure in the blood vessels poses a danger should the wall of the vessel rupture. Collectively the cellular and protein components of blood serve as a functional unit that provides protection against hemorrhage.

Plasma and the Cellular Elements of Blood

1. Blood is the circulating portion of the extracellular fluid. (p. 524)

2. **Plasma** is composed mostly of water, with dissolved proteins, organic molecules, ions, and dissolved gases. (p. 524; Fig. 16-1)

3. The **plasma proteins** include **albumins**, **globulins**, and the clotting protein **fibrinogen**. They function in blood clotting, defense, and as hormones, enzymes, or carriers for different substances. (p. 524)

4. The cellular elements of blood are red blood cells (**erythrocytes**), white blood cells (**leukocytes**), and **platelets**. Platelets are fragments of cells called **megakaryocytes**. (p. 524)

5. Blood contains five types of white blood cells: (1) **lymphocytes**, (2) **monocytes**, (3) **neutrophils**, (4) **eosinophils**, and (5) **basophils**. (p. 526; Fig. 16-1)

Blood Cell Production

6. All blood cells develop from a pluripotent hematopoietic stem cell. (p. 526; Fig. 16-3)

7. **Hematopoiesis** begins early in embryonic development and continues throughout a person's life. Most hematopoiesis takes place in the bone marrow. (p. 526; Fig. 16-5)

8. **Colony-stimulating factors** and other cytokines control white blood cell production. **Thrombopoietin** regulates the growth and maturation of megakaryocytes. Red blood cell production is regulated primarily by **erythropoietin**. (pp. 528–529)

Red Blood Cells

9. Mature mammalian red blood cells are biconcave disks without a nucleus. They contain hemoglobin, a red oxygen-carrying pigment. (p. 531; Fig. 16-6)

10. One **hemoglobin** molecule consists of four **globin** chains, each centered around a **heme** group. (p. 532; Fig. 16-7)

11. Hemoglobin synthesis requires iron in the diet. Iron is transported in the blood on **transferrin** and stored mostly in the liver, on **ferritin**. (p. 533; Fig. 16-8)

12. When hemoglobin is broken down, some heme groups are converted into **bilirubin**, which is incorporated into **bile** and excreted. Elevated bilirubin concentrations in the blood cause **jaundice**. (p. 533; Fig. 16-8)

Platelets and Coagulation

13. Platelets are cell fragments filled with granules containing clotting proteins and cytokines. Platelets are activated by damage to vascular endothelium. (p. 535; Fig. 16-10)

14. **Hemostasis** begins with vasoconstriction and the formation of a **platelet plug**. (p. 535; Fig. 16-12)

15. Exposed collagen triggers **platelet adhesion** and **platelet aggregation**. The platelet plug is converted into a clot when reinforced by **fibrin**. (pp. 535–536; Fig. 16-11)

16. In the last step of the **coagulation cascade**, fibrin is made from **fibrinogen** through the action of **thrombin**. (p. 538; Fig. 16-13)

17. As the damaged vessel is repaired, **plasmin** trapped in the platelet plug dissolves fibrin (**fibrinolysis**) and breaks down the clot. (pp. 539–540; Fig. 16-14)

18. Platelet plugs are restricted to the site of injury by **prostacyclin** in the membrane of intact vascular endothelium. **Anticoagulants** limit the extent of blood clotting within a vessel. (pp. 538, 540; Fig. 16-12)

QUESTIONS

LEVEL ONE *Reviewing Facts and Terms*

1. The fluid portion of the blood, called _____, is composed mainly of _____.

2. List the three types of plasma proteins in order of prevalence. Name at least one function of each type.

3. List the cellular elements found in blood and name at least one function of each.

4. Blood cell production is called _____. When and where does it occur?

5. What role do colony-stimulating factors, cytokines, and interleukins play in blood cell production? How are these chemical signal molecules different? Give two examples of each.

6. List the technical terms for production of red blood cells, platelets, and white blood cells, respectively.

7. The hormone that directs red blood cell synthesis is called _____. Where is it produced and what is the stimulus for its production?

8. How are the terms *hematocrit* and *packed red cells* related? What are normal hematocrit values for men and women?

9. Distinguish between an erythroblast and an erythrocyte. Give three distinct characteristics of erythrocytes.

10. What chemical element in the diet is important for hemoglobin synthesis?

11. Define the following terms and explain their significance in hematology.
    (a) jaundice
    (b) anemia
    (c) transferrin
    (d) hemophilia

12. Chemicals that prevent blood clotting from occurring are called _____.

LEVEL TWO *Reviewing Concepts*

13. **Concept maps:** Combine each list of terms into a map. You may add other terms.

List 1	List 2	List 3
ADP	clot	bile
collagen	coagulation	bilirubin
integrins	fibrin	bone marrow
membrane	fibrinogen	erythropoietin
phospholipids	fibrinolysis	ferritin
platelet activation	infarct	globin
platelet adhesion	plasmin	hematocrit
platelet	plasminogen	heme
aggregation	polymer	hemoglobin
platelet plug	thrombin	intestine
platelet-activating		iron
factor		liver
positive feedback		porphyrin
serotonin		reticulocyte
thromboxane A_2		transferrin
vasoconstriction		

14. Distinguish between the intrinsic, extrinsic, and common pathways of the coagulation cascade.

15. Once platelets are activated to aggregate, what factors halt their activity?

LEVEL THREE *Problem Solving*

16. Rachel is undergoing chemotherapy for breast cancer. She has blood cell counts at regular intervals; below is the computer printout of her results:

	Normal Count	10 Days Post-Chemotherapy	20 Days Post-Chemotherapy
WBC ($\times 10^3$)	4.6–10.2	2.6	4.9
RBC ($\times 10^6$)	4.04–5.48	3.85	4.2
Platelets	142–424	133	151

At the time of the first test, Jen, the nurse, notes that Rachel, although pale and complaining of being tired, does not have any bruises on her skin. Jen tells Rachel to eat foods high in protein, take a multivitamin tablet with iron, and stay home and away from crowds as much as possible. How are Jen's observations and recommendations related to the results of the first and second blood tests?

LEVEL FOUR *Quantitative Problems*

17. If we estimate that total blood volume is 7% of body weight, calculate the total blood volume in a 200-lb man and in a 130-lb woman (2.2 lb/kg). What are their plasma volumes if the man's hematocrit is 52% and the woman's hematocrit is 41%?

18. The total blood volume of an average person is 7% of total body weight. Using this figure and the fact that 1 kg of blood occupies about 1 liter, figure the total erythrocyte volume of a 50 kg woman with a hematocrit of 40%.

Explorations

Explorations are also available at *www.physiologyplace.com*

In this chapter you were introduced to the concept of blood as a tissue. Blood cells are essential for oxygen transport, hemostasis, and defense, as you will see in later chapters.

Exploration 1: Stem Cells from Umbilical Cord Blood

In recent years scientists have been working to isolate and grow uncommitted hematopoietic stem cells to use as replacements for patients whose own stem cells have been killed by cancer chemotherapy. These stem cells were first obtained from bone marrow or peripheral blood. Now umbilical cord blood, collected at birth, has been found to be a rich source of hematopoietic stem cells. Besides being more plentiful, umbilical cord stem cells are less likely to cause a foreign tissue rejection reaction than are stem cells obtained from a donor who is unrelated to the recipient. The potential for a source of easily collected stem cells has led to the development of cord blood banking programs in the United States and Europe. The Placental Blood Program at the New York Blood Center, supported by the National Institutes of Health, currently has more than 13,000 donations of umbilical cord blood. To learn more about stem cells and umbilical cord banking programs, go to *http://health.nih.gov/* and search for *stem cells*, or try a Google search (*www.google.com*) for *umbilical cord stem cell*.

Some recent research suggests that stem cells from umbilical cords may be able to form new liver cells and even new neurons. For example, see:

Matrix cells from Wharton's jelly form neurons and glia, *Stem Cells Online* 21:50–60, 2003. Available at *http://stemcells.alphamedpress.org/*.

Exploration 2: Erythropoietin and the Olympics

Athletes spend hundreds of hours training, trying to build their endurance. But suppose they could increase the oxygen-carrying capacity of their blood by an injection of DNA-recombinant human epoetin-alfa (rhEPO), genetically engineered erythropoietin—wouldn't that be tempting? Tempting, yes; legal in the Olympics, no. Despite this, some estimate that as many as 7% of elite endurance athletes are using rhEPO. The International Olympic Committee tested for rhEPO at the 2002 Salt Lake City Winter Olympics and had over 100 positive results. How can they test for a hormone that the body makes naturally?

To learn more about rhEPO and its use in sports, read "Detection of DNA-recombinant human epoetin-alfa as a pharmacological ergogenic aid," *Sports Medicine* 32(2):125–142, 2002. You can also do a Google search (*www.google.com*) for *epo sports* or *epo olympics* and see what the popular press has to say about the illicit athletic use of this clinically important drug.

Exploration 3: Clinical Applications—The CBC and Chem Tests

Have you ever heard a doctor request a "CBC, Chem" for a patient and wonder what that means? As you learned in this chapter, blood is a complex tissue composed of multiple cell types, fluid, molecules, and proteins that are normally maintained within a specific range of values. The CBC (complete blood count) and the Chem 20 are tests to evaluate the levels of different blood components to determine if they are different from healthy homeostatic levels. Do a Google search (*www.google.com*) for *complete blood count* and another for *Chem 20* to find the list of blood components that are measured with each test. For example, the National Library of Medicine online encyclopedia (*www.nlm.nih.gov/medlineplus/encyclopedia.html*) has complete descriptions of both tests. How many blood components do you recognize? Does the combination of the two tests provide a complete view of everything found in blood? Can you identify which test would reveal ion imbalances and which test would reveal infections that have caused increased white blood cell numbers? Which test allows determination of hematocrit? What are some normal plasma levels for different blood components, such as sodium and potassium ions?

Exploration 4: Reading

Hormones that control hematopoiesis are a focus of contemporary physiological research. The following articles can be accessed free online at *http://nips.physiology.org*.

- "Pleiotropic functions and tissue-specific expression of erythropoietin," *News in Physiological Sciences* 16:110–113, 2001 June. (*Pleiotrophy* is the ability of a molecule to participate in a variety of apparently unrelated processes.)
- "Thrombopoietin: The novel hepatic hormone," *News in Physiological Sciences* 17:6–10, 2002 February.

To learn more about anticoagulant therapy for treating heart disease, see the following articles in the *New England Journal of Medicine* (*www.nejm.org*):

- "COX inhibitors and thromboregulation," *New England Journal of Medicine* 347(13):1025–1026, 2002 Sep. 26.
- "Antithrombocytic therapy after myocardial infarction," *New England Journal of Medicine* 347(13):1019–1021, 2002 Sep. 26.

ANSWERS

Answers to Concept Check Questions

Page 526

1. The five groups of leukocytes are lymphocytes, macrophages/monocytes, basophils, neutrophils, and eosinophils.

2. Erythrocytes and platelets lack nuclei and therefore are unable to carry out protein synthesis.

3. Liver degeneration reduces the total plasma protein concentration, which reduces the colloid osmotic pressure in the capillaries. This increases net capillary filtration, and edema results.

Page 530

4. (a) erythropoietin, (b) colony-stimulating factors and interleukins, (c) thrombopoietin.

Page 533

5. (a) Heme is a porphyrin ring of carbon, hydrogen, and nitrogen with an iron atom in the middle. Hemoglobin contains four protein chains, each centered on a heme group. (b) Ferritin is the liver protein that stores iron. Transferrin is the plasma protein for iron transport in the blood.

6. Bile is an exocrine secretion because it is secreted into the lumen of the intestine, which is the body's external environment.

Page 535

7. Some other factor that is essential for red blood cell synthesis must be lacking. These would include iron for hemoglobin production, folic acid, and vitamin B_{12}.

8. Low atmospheric oxygen at high altitude causes low arterial oxygen that is sensed by kidney cells. The kidney secretes erythropoietin that acts on bone marrow to increase production of red blood cells.

9. A dehydrated person will have a lower plasma volume, but red blood cell volume will remain constant. This results in an elevated hematocrit.

Page 540

10. In Figure 16-11, the intrinsic pathway starts at collagen and the extrinsic pathway starts at tissue factor.

Answers to Level Four Quantitative Problems

Page 544

17. Total blood volume in a 200-lb man is about 6.4 L; his plasma volume is about 3.1 L. A 130-lb woman has total blood volume of 4.1 L and plasma volume of about 2.4 L.

18. Her total blood volume is 3.5 L and her total erythrocyte volume is 1.4 L.

Mechanics of Breathing

17

Airways are lined with ciliated epithelium (green) interspersed with goblet cells covered in microvilli (orange). Two yellow mucus droplets sit on the microvilli.

RUNNING PROBLEM

Emphysema

"Diagnosis: COPD (blue bloater)," reads Edna Wilson's patient chart. COPD—chronic obstructive pulmonary disease—is a name given to diseases in which air exchange is impaired by narrowing of the airways. Most people with COPD have emphysema or chronic bronchitis or a combination of the two. Individuals in whom chronic bronchitis predominates are nicknamed "blue bloaters," owing to the bluish tinge of their skin and a tendency to be overweight. "Pink puffers" suffer more from emphysema. They tend to be thin, have normal (pink) skin coloration, and breathe shallow, rapid breaths. Because most COPD is caused by smoking, most people can avoid the disease simply by not smoking. Unfortunately, Edna has been a heavy smoker for 35 of her 47 years.

551 555 559 563 568

BACKGROUND BASICS

Ciliated and exchange epithelia: **67** | Pressure, volume, flow, and resistance: **454** | Pulmonary circulation: **450** | Hydrogen bonds and surface tension: **31** | Autonomic and somatic motor neurons: **372** | Velocity of flow: **455** |

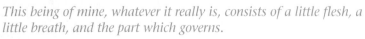

This being of mine, whatever it really is, consists of a little flesh, a little breath, and the part which governs.

—Marcus Aurelius Antoninus (121–180 A.D.)

Imagine covering the playing surface of a racquetball court (about 75 m²) with thin plastic wrap, then crumpling up the wrap and stuffing it into a 3-liter soft drink bottle. Impossible? Maybe so, if you use plastic wrap and a drink bottle. But the lungs of a 70-kg man have a gas exchange surface the size of that plastic wrap, compressed into a volume that is less than the size of the bottle. This tremendous area for gas exchange is needed to supply the trillions of cells in the body with adequate amounts of oxygen.

Aerobic metabolism in animal cells depends on a steady supply of oxygen and nutrients from the environment, coupled with the removal of carbon dioxide and other wastes. In very small aquatic animals, simple diffusion across the body surface meets these needs. Distance limits diffusion rate, however, and as a result, most multicelled animals require specialized respiratory organs associated with a circulatory system. Respiratory organs take a variety of forms, but all possess a large surface area compressed into a small space.

Besides needing a large exchange surface, humans and other terrestrial animals face an additional physiological challenge: dehydration. Their exchange surfaces must be thin and moist to allow gases to pass from air into solution, yet at the same time they must be protected from drying out due to exposure to air. Some terrestrial animals such as the slug, a shell-less snail, meet the challenge of dehydration with behavioral adaptations that restrict them to humid environments and nighttime activities.

However, a more common solution is anatomical: an internalized respiratory epithelium. Human lungs are enclosed within the chest cavity to control their contact with the outside air. Internalization creates a humid environment for the exchange of gases with the blood and protects the exchange surface from damage.

Internalized lungs create another problem, however: how to exchange air between the atmosphere and the exchange surface deep within the body. Air flow requires a muscular pump to create pressure gradients. Thus, in more complex animals, the respiratory system consists of two separate components: a muscle-driven pump and a thin, moist exchange surface. In humans, the pump is the musculoskeletal structure of the thorax. The lungs themselves consist of the exchange epithelium and associated blood vessels.

The primary functions of the respiratory system are:

1. **Exchange of gases between the atmosphere and the blood**. The body brings in oxygen for distribution to the tissues, and eliminates carbon dioxide waste produced by metabolism.
2. **Homeostatic regulation of body pH**. The lungs can alter body pH by selectively retaining or excreting CO_2.
3. **Protection from inhaled pathogens and irritating substances**. Like all epithelia that contact the external environment, the respiratory epithelium is well supplied with mechanisms to trap and destroy potentially harmful substances before they can enter the body.
4. **Vocalization**. Air moving across the vocal cords creates vibrations used for speech, singing, and other forms of communication.

In addition to serving these functions, the respiratory system is also a source of significant losses of water and heat from the body. These losses must be balanced using homeostatic compensations.

In this chapter you will learn how the respiratory system carries out these functions by exchanging air between the environment and the interior air spaces of the lungs. This exchange is the *bulk flow* of air and it follows many of the same principles as the bulk flow of blood through the cardiovascular system that you studied in the previous chapters:

1. Flow takes place from regions of higher pressure to regions of lower pressure.
2. A muscular pump creates pressure gradients.
3. Resistance to air flow is influenced primarily by the diameter of the tubes through which the air is flowing.

The primary difference between air flow in the respiratory system and blood flow in the circulatory system is that air is a compressible mixture of gases while blood is a noncompressible liquid.

▶ THE RESPIRATORY SYSTEM

The word *respiration* can have several meanings (Fig. 17-1 ●). **Cellular respiration** refers to the intracellular reaction of oxygen with organic molecules to produce carbon dioxide, water, and energy in the form of ATP [❧ p. 103]. **External respiration**, the

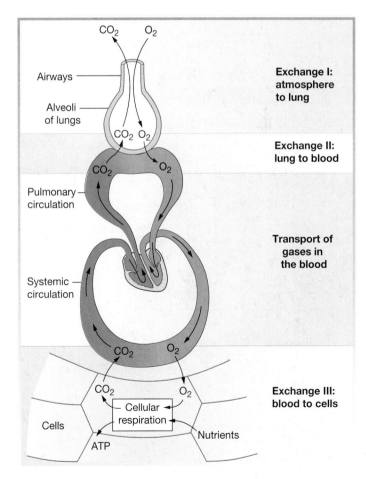

● **Figure 17-1** *Overview of external and cellular respiration*

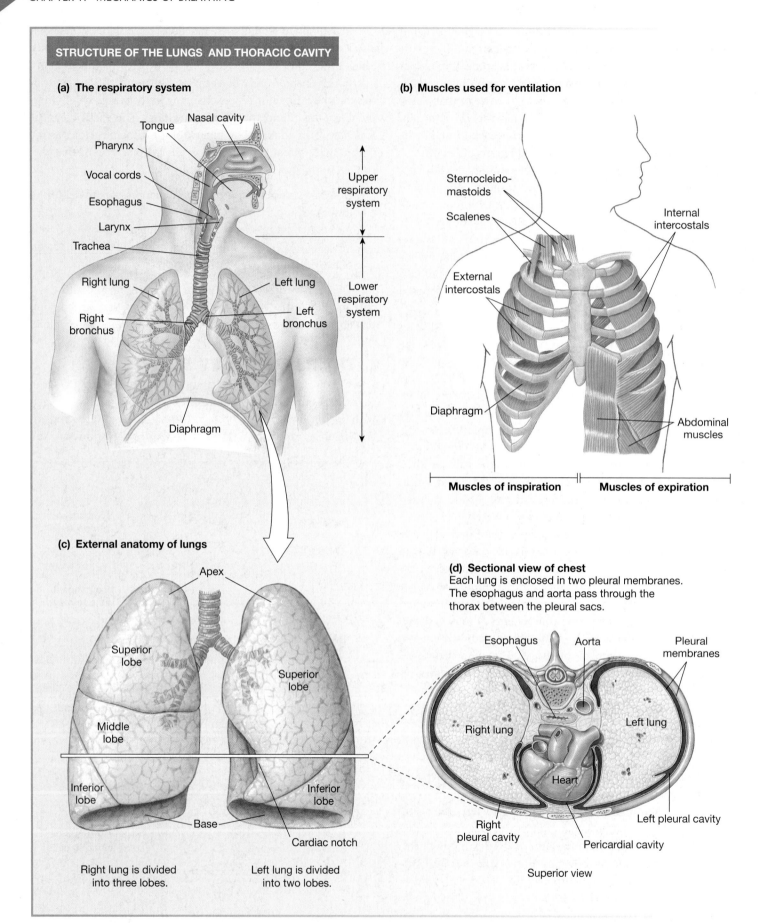

STRUCTURE OF THE LUNGS AND THORACIC CAVITY

(a) The respiratory system

Nasal cavity
Tongue
Pharynx
Vocal cords
Esophagus
Larynx
Trachea
Right lung
Right bronchus
Diaphragm
Left lung
Left bronchus

Upper respiratory system
Lower respiratory system

(b) Muscles used for ventilation

Sternocleido-mastoids
Scalenes
External intercostals
Diaphragm
Internal intercostals
Abdominal muscles

Muscles of inspiration   Muscles of expiration

(c) External anatomy of lungs

Apex
Superior lobe
Middle lobe
Inferior lobe
Superior lobe
Inferior lobe
Base
Cardiac notch

Right lung is divided into three lobes.
Left lung is divided into two lobes.

(d) Sectional view of chest
Each lung is enclosed in two pleural membranes. The esophagus and aorta pass through the thorax between the pleural sacs.

Esophagus   Aorta   Pleural membranes
Right lung
Left lung
Heart
Right pleural cavity
Left pleural cavity
Pericardial cavity

Superior view

● Figure 17-2 *ANATOMY SUMMARY The Respiratory System*

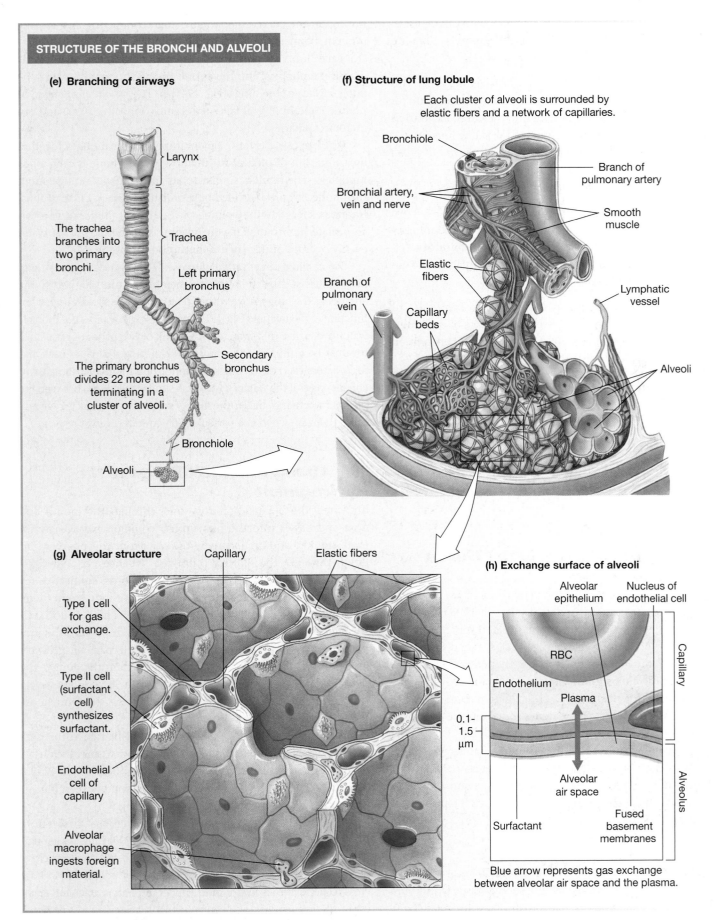

STRUCTURE OF THE BRONCHI AND ALVEOLI

(e) Branching of airways

Larynx

The trachea branches into two primary bronchi.

Trachea

Left primary bronchus

Secondary bronchus

The primary bronchus divides 22 more times terminating in a cluster of alveoli.

Bronchiole

Alveoli

(f) Structure of lung lobule

Each cluster of alveoli is surrounded by elastic fibers and a network of capillaries.

Bronchiole

Branch of pulmonary artery

Bronchial artery, vein and nerve

Smooth muscle

Branch of pulmonary vein

Elastic fibers

Lymphatic vessel

Capillary beds

Alveoli

(g) Alveolar structure

Capillary

Elastic fibers

Type I cell for gas exchange.

Type II cell (surfactant cell) synthesizes surfactant.

Endothelial cell of capillary

Alveolar macrophage ingests foreign material.

(h) Exchange surface of alveoli

Alveolar epithelium

Nucleus of endothelial cell

RBC

Capillary

Endothelium

Plasma

0.1–1.5 μm

Alveolar air space

Alveolus

Surfactant

Fused basement membranes

Blue arrow represents gas exchange between alveolar air space and the plasma.

● **Figure 17-2** *(continued)*

topic of this chapter and the next, is the interchange of gases between the environment and the body's cells. External respiration can be subdivided into four integrated processes:

1. *The exchange of air between the atmosphere and the lungs.* This process is known as **ventilation**, or breathing. **Inspiration** (or inhalation) is the movement of air into the lungs. **Expiration** (or exhalation) is the movement of air out of the lungs.
2. *The exchange of oxygen and carbon dioxide between the lungs and the blood.*
3. *The transport of oxygen and carbon dioxide by the blood.*
4. *The exchange of gases between blood and the cells.*

External respiration requires the coordinated functioning of the respiratory and cardiovascular systems. The **respiratory system** is composed of the structures involved in ventilation and gas exchange (Fig. 17-2 ●). It consists of:

1. The **conducting system** of passages, or **airways**, that lead from the environment to the exchange surface of the **lungs**.
2. The *alveoli*, the exchange surface of the lungs, where oxygen and carbon dioxide transfer between air and blood.
3. The bones and muscles of the **thorax** (chest) that assist in ventilation.

The respiratory system can be divided into two parts. The **upper respiratory tract** consists of the mouth, nasal cavity, pharynx, and larynx. The **lower respiratory tract** consists of the trachea, two bronchi, their branches, and the lungs. It is also known as the *thoracic* portion of the respiratory system because it is enclosed in the thorax.

The Bones and Muscles of the Thorax Surround the Lungs

The thorax, or chest cavity, is bounded by the bones of the spine and rib cage and their associated muscles. Together the bones and muscles are called the *thoracic cage*. The ribs and spine (the *chest wall*) form the sides and top of the cage. A dome-shaped sheet of skeletal muscle, the **diaphragm**, forms the floor (Fig. 17-2a).

Two sets of **intercostal muscles**, internal and external, connect the 12 pairs of ribs (Fig. 17-2b). Additional muscles, the **sternocleidomastoids** and the **scalenes**, run from the head and neck to the sternum and first two ribs.

Functionally, the thorax is a sealed container filled with three membranous bags, or sacs. One, the *pericardial sac*, contains the heart. The other two bags, the **pleural sacs**, contain the lungs [*pleura*, rib, or side]. The esophagus and thoracic blood vessels and nerves pass between the pleural sacs (Fig. 17-2d).

Pleural Sacs Enclose the Lungs

The **lungs** (Fig. 17-2a, c) consist of light, spongy tissue whose volume is mostly occupied by air-filled spaces. These irregular cone-shaped organs nearly fill the thoracic cavity, with their bases resting on the curved diaphragm. Rigid conducting airways, the bronchi, connect the lungs to the main airway, the trachea.

Each lung is surrounded by a double-walled pleural sac whose membranes line the inside of the thorax and cover the outer surface of the lungs (Fig. 17-3 ●). The *pleural membranes*, or **pleura**, contain several layers of elastic connective tissue and numerous capillaries.

The opposing layers of pleural membrane are held together by a thin film of **pleural fluid** whose total volume is only a few milliliters. The result is similar to an air-filled balloon (the lung) surrounded by a water-filled balloon (the pleural sac). Most illustrations exaggerate the volume of the pleural fluid, but you can appreciate its thinness if you imagine spreading 3 mL of water evenly over the surface of a 3-liter soft drink bottle.

Pleural fluid serves several purposes. First, it creates a moist, slippery surface so that the opposing membranes can slide across one another as the lungs move within the thorax. The second important function of pleural fluid is to hold the lungs tight against the thoracic wall. To visualize this arrangement, think of two panes of glass stuck together by a thin film of water. You can slide the panes back and forth across each other, but you cannot pull them apart because of cohesiveness of the water [● p. 31]. A similar fluid bond between the pleural membranes makes the lungs "stick" to the thoracic cage and holds them stretched in a partially inflated state, even at rest.

The Airways Connect Lungs to the Environment

Air enters the upper respiratory tract through the mouth and nose and passes into the **pharynx**, a common passageway for food, liquids, and air [*pharynx*, throat]. From the pharynx, air flows through the **larynx** into the **trachea**, or windpipe (Fig. 17-2a). The larynx contains the **vocal cords**, connective tissue bands that tighten to create sound when air moves past them.

The trachea is a semiflexible tube held open by 15 to 20 C-shaped cartilage rings (Fig. 17-2e). It extends down into the thorax, where it branches (division 1) into a pair of **primary bronchi**, one *bronchus* to each lung. Within the lungs, the bronchi branch repeatedly (divisions 2–11) into progressively smaller bronchi (Fig. 17-2a, e). Like the trachea, the bronchi are semirigid tubes supported by cartilage.

Within the lungs, the smallest bronchi branch to become **bronchioles**, small collapsible passageways with smooth muscle walls. The bronchioles continue branching (divisions 12–23) until the *respiratory bronchioles* form a transition between the airways and the exchange epithelium of the lung.

The diameter of the airways becomes progressively smaller from the trachea to the bronchioles, but as the individual airways get narrower, their numbers increase (Fig. 17-4 ●). As a result, the total cross-sectional area increases with each division of the airways. Total cross-sectional area is lowest in the upper respiratory tract and greatest in the bronchioles, analogous to the increase in cross-sectional area that occurs from the aorta to the capillaries [● p. 505].

The pleural sac forms a double membrane surrounding the lung,
similar to a fluid-filled balloon surrounding an air-filled balloon.

The pleural fluid has a much smaller volume
than is suggested by this illustration.

● **Figure 17-3** *The relationship between the pleural sac and the lung*

		Name	Division	Diameter (mm)	How many?	Cross-sectional area (cm)
Conducting system		Trachea	0	15-22	1	2.5
		Primary bronchi	1	10-15	2	
		Smaller bronchi	2		4	
			3			
			4	1-10		
			5			
			6-11		1 x 10⁴	
Exchange surface		Bronchioles	12-23	0.5-1	2 x 10⁴ ↓ 8 x 10⁷	100 ↓ 5 x 10³
		Alveoli	24	0.3	3-6 x 10⁸	>1 x 10⁶

● **Figure 17-4** *Branching of the airways*

Answers: p. 572

CONCEPT CHECK

1. What is the difference between cellular respiration and external respiration?
2. Name the components of the (a) upper respiratory tract and (b) lower respiratory tract.
3. Based on the total cross-sectional area of different parts of the respiratory system, where do you think velocity of air flow is highest and lowest?
4. Give two functions of pleural fluid.
5. Name the components (including muscles) of the thoracic cage. List the contents of the thorax.
6. Which air passages of the respiratory system are collapsible?

The Alveoli Are the Site of Gas Exchange

The bulk of lung tissue consists of interconnected exchange sacs known as **alveoli** [*alveus*, a concave vessel; singular *alveolus*] that cluster at the ends of terminal bronchioles (Fig. 17-2f, g). Their primary function is the exchange of gases between the alveoli and the blood.

Each tiny alveolus is composed of a single layer of epithelium (Fig. 17-2g). Two types of epithelial cells are found in the alveoli,

RUNNING PROBLEM

Patients with chronic bronchitis have excessive mucus production and general inflammation of the entire respiratory tract. The mucus narrows the airways and makes breathing difficult.

Question 1: What does narrowing of the airways do to resistance of the airways to air flow? (Hint: The relationship between radius and resistance is the same for air flow as it was for blood flow in the circulatory system; p. 454.)

and they occur in roughly equal numbers. The smaller but thicker **type II alveolar cells** synthesize and secrete a type of chemical known as surfactant. Surfactant mixes with the thin fluid lining of the alveoli to ease the expansion of the lungs during breathing, as we will see later in this chapter.

The larger **type I alveolar cells** are very thin so that gases can diffuse rapidly through them to the capillaries underneath (Fig. 17-2h). In much of the exchange area, a layer of basal lamina fuses the alveolar epithelium to the capillary endothelium, and only a small amount of interstitial fluid is present.

The thin walls of the alveoli do not contain muscle, because muscle fibers would block rapid gas exchange. As a result, lung tissue itself cannot contract. However, connective tissue between the alveolar epithelial cells contains many elastin fibers that create elastic recoil when lung tissue is stretched.

The close association of alveoli with an extensive network of capillaries demonstrates the intimate link between the respiratory and cardiovascular systems. Blood vessels cover 80%–90% of the alveolar surface, forming an almost continuous "sheet" of blood in close contact with the air-filled alveoli (Fig. 17-2f). The proximity of capillary blood to air in the alveolus is essential for the rapid exchange of gases.

The Pulmonary Circulation Is a High-Flow, Low-Pressure System

The pulmonary circulation begins with the pulmonary trunk that receives low-oxygen blood from the right ventricle. The pulmonary trunk divides into two pulmonary arteries, one to each lung [● Figure 14-1, p. 451]. Oxygenated blood from the lungs returns to the left atrium via the pulmonary veins.

At any given moment, the pulmonary circulation contains about 0.5 liters of blood, or 10% of the total blood volume. About 75 mL of this amount is found in the capillaries, where gas exchange takes place, with the remainder in the pulmonary arteries and veins. The rate of blood flow through the lungs is quite high when compared with other tissues [● p. 503] because the lungs receive the entire cardiac output of the right ventricle, 5 L/min. This means that as much blood flows through the lungs in one minute as flows through the rest of the body in the same amount of time!

Despite the high flow rate, blood pressure in the pulmonary circulation is low. Pulmonary arterial blood pressure averages 25/8 mm Hg, compared with the average systemic arterial blood pressure of 120/80 mm Hg. The right ventricle does not have to pump as forcefully to create blood flow through the lungs because the resistance of the pulmonary circulation is low. This low resistance can be attributed to the shorter total length of the pulmonary blood vessels and the distensibility and large cross-sectional area of pulmonary arterioles.

Because mean blood pressure is low in the pulmonary capillaries, the net hydrostatic pressure forcing fluid out of the capillary into the interstitial space is also low [● p. 506]. The lymphatic system efficiently removes fluid filtered from the pulmonary capillaries, and normally lung interstitial fluid volume is small. As a result, the distance between alveolus and capillary is short, and gases diffuse rapidly between them.

Answers: p. 572

7. Is blood flow through the pulmonary trunk greater than, less than, or equal to blood flow through the aorta?

8. A person has left ventricular failure but normal right ventricular function. As a result, blood pools in the pulmonary circulation, doubling pulmonary capillary hydrostatic pressure. What happens to net fluid flow across the walls of the pulmonary capillaries?

▶ GAS LAWS

Air flow in the respiratory system is very similar to blood flow in the cardiovascular system in many respects, although blood is a noncompressible liquid and air is a compressible mixture of gases. Blood pressure and environmental air pressure (**atmospheric pressure**) are both reported in millimeters of mercury (mm Hg).* At sea level, atmospheric pressure is 760 mm Hg.

One convention we will follow in this book is the designation of atmospheric pressure as 0 mm Hg. Because atmospheric pressure varies with altitude and because very few people live exactly at sea level, this convention allows us to compare pressure differences that occur during ventilation without correcting for altitude. Negative numbers designate subatmospheric pressures, and positive numbers denote higher-than-atmospheric pressures.

Table 17-1 summarizes the rules that govern the behavior of gases in air. These rules provide the basis for the exchange of air between the environment and the alveoli. The gas laws that govern the solubility of gases in solution will be considered in Chapter 18.

Air Is a Mixture of Gases

The atmosphere surrounding the earth is a mixture of gases and water vapor. **Dalton's law** states that the total pressure of a mixture of gases is the sum of the pressures of the individual gases. Thus, in dry air at an atmospheric pressure of 760 mm Hg, 78% of the total pressure is due to nitrogen molecules, 21% to oxygen molecules, and so on (Table 17-2).

---

*Respiratory physiologists sometimes report gas pressures in units of centimeters of water: 1 mm Hg = 1.36 cm H_2O.

Table 17-1: Gas Laws

1. The total pressure of a mixture of gases is the sum of the pressures of the individual gases (Dalton's law).

2. Gases, singly or in a mixture, move from areas of higher pressure to areas of lower pressure.

3. If the volume of a container of gas changes, the pressure of the gas will change in an inverse manner (Boyle's law).

Table 17-2: Partial Pressures of Some Atmospheric Gases at 25° C and 760 mm Hg

GAS	PARTIAL PRESSURE IN ATMOSPHERIC AIR, DRY	PARTIAL PRESSURE IN ATMOSPHERIC AIR, 100% HUMIDITY
Nitrogen (N_2)	593 mm Hg	575 mm Hg
Oxygen (O_2)	160 mm Hg	152 mm Hg
Carbon dioxide (CO_2)	0.25 mm Hg	0.24 mm Hg
Water vapor	0 mm Hg	23.8 mm Hg

In respiratory physiology, we are concerned not only with total atmospheric pressure but also with the individual pressures of oxygen and carbon dioxide. The pressure of a single gas is known as its **partial pressure**, abbreviated P_{gas}. To find the partial pressure of a gas, multiply the atmospheric pressure (P_{atm}) by the gas's relative contribution (%):

Partial pressure of an atmospheric gas =
P_{atm} × % of gas in atmosphere

Partial pressure of oxygen = 760 mm Hg × 21%

$P_{O_2} = 760 \times 0.21 = 160$ mm Hg

Thus, the P_{O_2}, or partial pressure of oxygen, in dry air at sea level is 160 mm Hg. The pressure of an individual gas is determined only by its relative abundance in the mixture and is independent of the molecular size or weight of the gas.

The actual partial pressure of gases varies slightly depending on how much water vapor is in the air. Water pressure "dilutes" the contribution of other gases to the total pressure. To calculate the partial pressure of a gas in humid air, you must first subtract the water vapor pressure from the total pressure. Table 17-2 compares the partial pressures of some atmospheric gases in dry air and at 100% humidity.

CONCEPT CHECK   Answers: p. 572

9. If nitrogen is 78% of atmospheric air, what is the partial pressure of nitrogen (P_{N_2}) when the dry atmospheric pressure is 720 mm Hg?

Gases Move from Areas of Higher Pressure to Areas of Lower Pressure

Air flow occurs whenever there is a pressure gradient. Air flow, like blood flow, moves from areas of higher pressure to areas of lower pressure. Meteorologists predict the weather by knowing that areas of high atmospheric pressure move in to replace areas of low pressure. In ventilation, flow of air down pressure gradients explains why air exchanges between the environment and the lungs. The movement of the thorax during breathing creates alternating conditions of high and low pressure within the lungs.

Movement down pressure gradients also applies to single gases. For example, oxygen moves from areas of higher oxygen partial pressure (P_{O_2}) to areas of lower oxygen partial pressure.

Boyle's Law Describes Pressure-Volume Relationships of Gases

The pressure exerted by a gas or mixture of gases in a sealed container is created by the collisions of moving gas molecules with the walls of the container and with each other. If the size of the container is reduced, the collisions will become more frequent and the pressure will rise. This relationship can be expressed by the following equation:

$P_1V_1 = P_2V_2$ where P represents pressure
and V represents volume

For example, start with a 1-liter container (V_1) of a gas whose pressure is 100 mm Hg (P_1). One side of the container moves in to decrease the volume to 0.5 L (Fig. 17-5 ●). What happens to the pressure of the gas? According to the equation,

$$P_1V_1 = P_2V_2$$
$$100 \text{ mm Hg} \times 1 \text{ L} = P_2 \times 0.5 \text{ L}$$
$$P_2 = 200 \text{ mm Hg}$$

If the volume is reduced by one-half, the pressure doubles. If the volume were to double, the pressure would be reduced by one-half. This relationship between pressure and volume was first noted by Robert Boyle in the 1600s and has been called **Boyle's law** of gases.

In the respiratory system, changes in volume of the chest cavity during ventilation cause pressure gradients that create air flow. When the chest increases in volume, the alveolar pressure falls and air flows into the respiratory system. When the chest decreases in volume, the pressure rises and air flows out into the atmosphere. This movement of air is called *bulk flow* because the entire gas mixture is moving rather than merely one or two gas species.

Boyle's Law: $P_1V_1 = P_2V_2$

Decreasing volume increases collisions and increases pressure.

$V_1 = 1.0$ L
$P_1 = 100$ mm Hg

$V_2 = 0.5$ L
$P_2 = 200$ mm Hg

● **Figure 17-5** *Boyle's law*   Boyle's law ($P_1V_1 = P_2V_2$) assumes that temperature and number of gas molecules remain constant.

▶ VENTILATION

The first exchange in respiratory physiology is ventilation, or breathing, the movement of air between the environment and the alveoli.

The Airways Warm, Humidify, and Filter Inspired Air

The upper airways and the bronchi do more than simply serve as passageways for air. They play an important role in conditioning air before it reaches the alveoli. Conditioning has three components:

1. *Warming* air to body temperature (37° C) so that core body temperature will not change and alveoli will not be damaged by cold air
2. *Adding water vapor* until the air reaches 100% humidity so that the moist exchange epithelium will not dry out
3. *Filtering out foreign material* so that viruses, bacteria, and inorganic particles will not reach the alveoli

Inhaled air is warmed and moistened by heat and water evaporating from the mucosal lining of the airways. Under normal circumstances, by the time air reaches the trachea, it has been conditioned to 37° C and 100% humidity.

Breathing through the mouth is not nearly as effective at warming and moistening air as breathing through the nose. If you exercise outdoors in very cold weather, you may be familiar with the ache in your chest that results from breathing cold air through your mouth.

Filtration of air takes place in the trachea and bronchi as well. These airways are lined with a ciliated epithelium that secretes both mucus and a dilute saline solution (see chapter opening photo, p. 546). The cilia themselves are bathed in a watery saline layer. On top of them lies a sticky layer of mucus that traps most inhaled particles larger than 2 µm.

The mucus layer is secreted by *goblet cells* in the epithelium (Fig. 17-6●). The underlying cilia beat with an upward motion that moves the mucus continuously toward the pharynx, creating what is called the *mucus escalator*. Once mucus reaches the pharynx, it is swallowed. Mucus contains *immunoglobulins* that can disable many pathogens, and stomach acid and enzymes destroy any remaining microorganisms.

Secretion of the watery fluid layer beneath the mucus is essential for a functional mucus escalator. Without it, the cilia become trapped in thick, sticky mucus. For example, in the disease *cystic fibrosis*, inadequate fluid secretion prevents mucus clearance and allows bacteria to colonize the airways, resulting in recurrent lung infections.

CONCEPT CHECK   Answers: p. 572

10. As inhaled air becomes humidified passing down the airways, what happens to the P_{O_2} of the air?

11. Cigarette smoking paralyzes the cilia in the epithelial lining of the airways. Why would paralysis of the cilia cause smokers to develop a cough?

During Ventilation, Air Flows Because of Pressure Gradients

Air flows into the lungs because of pressure gradients created by a pump, just as blood flows in the cardiovascular system because of

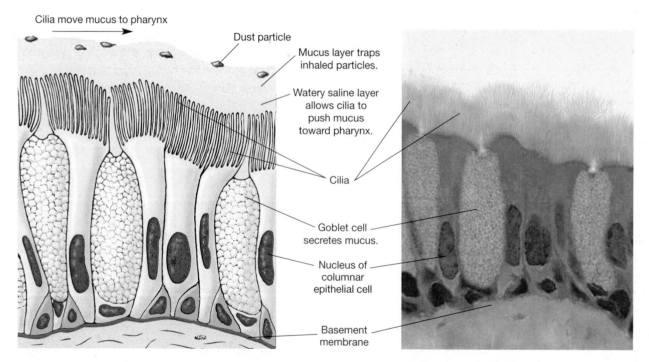

Cilia move mucus to pharynx

Dust particle

Mucus layer traps inhaled particles.

Watery saline layer allows cilia to push mucus toward pharynx.

Cilia

Goblet cell secretes mucus.

Nucleus of columnar epithelial cell

Basement membrane

Ciliated epithelium of the trachea

● **Figure 17-6** *Ciliated respiratory epithelium*

Smokers usually develop chronic bronchitis before they develop emphysema. Cigarette smoke paralyzes the cilia that sweep debris and mucus out of the airways. Without the action of cilia, mucus and debris pool in the airways, leading to a chronic cough. Eventually, breathing becomes difficult.

Question 2: Why do people with chronic bronchitis have a higher-than-normal rate of respiratory infections?

the pumping action of the heart. In the respiratory system, most lung tissue is thin exchange epithelium, so the muscles of the thoracic cage and diaphragm must function as the pump. When these muscles contract, the lungs expand, held to the inside of the chest wall by the pleural fluid.

Breathing is an active process that uses muscle contraction to create a pressure gradient. The primary muscles involved in quiet breathing (breathing at rest) are the diaphragm, the external intercostals, and the scalenes. During forced breathing, other muscles of the chest and abdomen may be recruited to assist. Examples of situations in which breathing is forced include exercise, playing a wind instrument, and blowing up a balloon.

Air flow in the respiratory tract obeys the same rule as blood flow in the cardiovascular system:

$$\text{Flow} \propto \Delta P/R$$

This equation means that (1) air flows in response to a pressure gradient (ΔP) and that (2) flow decreases as the resistance (R) of the system to flow increases. Before we discuss resistance, we will consider how the respiratory system creates a pressure gradient.

Pressures in the respiratory system can be measured either in the air spaces of the lungs (**alveolar pressure**, P_A) or within the pleural fluid (**intrapleural pressure**). Because atmospheric pressure is relatively constant, pressure in the lungs must be higher or lower than atmospheric pressure for air to flow between the environment and the alveoli.

The respiratory system ends in a dead end, so the direction of air flow must reverse. Air moves into the lungs when you inhale (*inspiration*) and out of the lungs when you exhale (*expiration*). A

single **respiratory cycle** consists of an inspiration followed by an expiration. The pressure-volume relationships of Boyle's law provide the basis for pulmonary ventilation.

CONCEPT CHECK   Answers: p. 572

12. Compare the direction of air movement during one respiratory cycle to the direction of blood flow during one cardiac cycle.
13. Explain the relationship between the lungs, the pleura, the pleural fluid, and the thoracic cage.

Inspiration Occurs When Alveolar Pressure Decreases

For air to move into the lungs, the pressure inside the lungs must become lower than the pressure of the atmosphere. According to Boyle's law, an increase in volume will create a decrease in pressure. During **inspiration**, the thoracic volume increases when the skeletal muscles of the rib cage and diaphragm contract.

When the diaphragm contracts, it loses its dome shape and drops down toward the abdomen. In quiet breathing, the diaphragm moves about 1.5 cm. This movement increases the volume of the thoracic cavity by flattening its floor (Fig. 17-7●). Contraction of the diaphragm causes between 60% and 75% of the inspiratory volume change during normal quiet breathing.

Movement of the rib cage creates the remaining 25% to 40% of the volume change. **External intercostal** and **scalene muscles** contract and pull the ribs upward and out. Rib movement during inspiration has been likened to a pump handle lifting up and away from the pump (the ribs moving up and away from the spine) and to the movement of a pair of bucket handles as they lift away from the sides of a bucket (ribs moving outward in a lateral direction). The combination of these two movements broadens the rib cage in all directions (Fig. 17-8●). As the volume of the thoracic cavity increases, pressure decreases and air flows into the lungs.

For many years, quiet breathing was attributed solely to the action of the diaphragm and the external intercostal muscles. It was thought that the scalenes and *sternocleidomastoid muscles* were active only during deep breathing. But in recent years, studies of patients with neuromuscular disorders have changed our understanding of how these accessory muscles contribute to quiet breathing.

  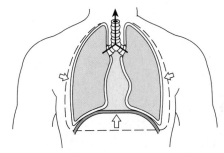

(a) At rest, diaphragm is relaxed.

(b) Diaphragm contracts, thoracic volume increases.

(c) Diaphragm relaxes, thoracic volume decreases.

● **Figure 17-7** *Movement of the diaphragm*

(a) "Pump handle" motion increases anterior-posterior dimension of rib cage.

(b) "Bucket handle" motion increases lateral dimension of rib cage.

● **Figure 17-8** *Movement of the rib cage during inspiration*

It now appears that without lifting of the sternum and upper ribs by the scalenes, contraction of the diaphragm during inspiration pulls the lower ribs inward. This action decreases the volume of the thoracic cage and works against inspiration. Because we know that the lower ribs move up and out in normal quiet breathing, the scalenes must assist quiet inspiration.

New evidence also downplays the role of the external intercostal muscles during quiet breathing. However, the external intercostals play an increasingly important role as respiratory activity increases. Because the exact contribution of the external intercostals and scalenes varies depending upon the type of breathing, we will group these muscles together and call them the *inspiratory muscles*.

Now let's see what happens to alveolar pressure as it changes during a single inspiration. Follow the graphs in Figure 17-9● as you read through the process.

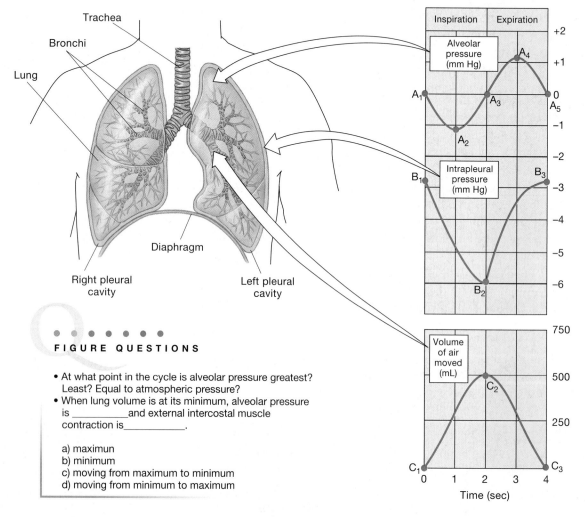

● ● ● ● ● ● ●
FIGURE QUESTIONS

• At what point in the cycle is alveolar pressure greatest? Least? Equal to atmospheric pressure?
• When lung volume is at its minimum, alveolar pressure is _____ and external intercostal muscle contraction is_____.

a) maximun
b) minimum
c) moving from maximum to minimum
d) moving from minimum to maximum

● **Figure 17-9** *Pressure changes during quiet breathing*

Time 0. In the brief pause between breaths, alveolar pressure is equal to atmospheric pressure (point A_1). When pressures are equal, there is no air flow.

Time 0–2 Inspiration. As inspiration begins, the muscles of the thoracic cage contract and thoracic volume increases. With the increase in volume, alveolar pressure falls about 1 mm Hg below atmospheric pressure (-1 mm Hg; point A_2) and air begins to flow into the alveoli (C_1 to C_2). The thoracic volume changes faster than air can flow, so alveolar pressure reaches its lowest value about halfway through inspiration (point A_2).

As air continues to flow into the alveoli, pressure increases until the thoracic cage stops expanding, just before the end of inspiration. Air movement continues for a fraction of a second longer, until the pressure inside the lungs equalizes with atmospheric pressure (point A_3). At the end of inspiration, the volume of air in the lungs is at its maximum for the respiratory cycle (point C_2) and alveolar pressure is equal to atmospheric pressure.

You can demonstrate this phenomenon by taking a deep breath and stopping the movement of your chest at the end of inspiration. (Do not "hold your breath," because doing so causes the opening of the pharynx to close and prevents air flow.) If you do this correctly, you will notice that air flow stops after you freeze the inspiratory movement. This exercise shows that, at the end of inspiration, air pressure in the alveoli is equal to atmospheric pressure.

Expiration Occurs When Alveolar Pressure Exceeds Atmospheric Pressure

At the end of inspiration, impulses from somatic motor neurons to the inspiratory muscles cease and the muscles relax. Elastic recoil of the lungs returns the diaphragm and rib cage to their original relaxed positions, just as a stretched elastic waistband recoils when released. Because expiration during quiet breathing involves passive elastic recoil rather than active muscle contraction, it is called **passive expiration**.

Time 2–4 Expiration. As the volume of the lungs and thorax decreases during expiration, air pressure in the lungs increases, reaching about 1 mm Hg above atmospheric pressure at its maximum (Fig. 17-9, point A_4). Alveolar pressure is now higher than atmospheric pressure, so air flow reverses and air moves out of the lungs.

Time 4. At the end of expiration, air movement ceases when the alveolar pressure is again equal to atmospheric pressure (point A_5). The volume of air in the lungs reaches its minimum for the respiratory cycle (point C_3). At this point, the respiratory cycle has ended and is ready to begin again with the next breath.

The pressure differences shown in Figure 17-9 apply to quiet breathing. During exercise or forced heavy breathing, these values will become proportionately larger. **Active expiration** occurs during voluntary exhalations and when ventilation exceeds 30–40 breaths per minute. (Normal resting ventilation rate is 12–20 breaths per minute for an adult.) Active expiration uses a different set of muscles from those used during inspiration, namely the internal intercostal muscles and the abdominal muscles. These muscles are collectively called the *expiratory muscles*.

The **internal intercostal muscles** line the inside of the rib cage. When they contract, they pull the ribs inward, reducing the volume of the thoracic cavity. To feel this action, place your hands on your rib cage. Forcefully blow as much air out of your lungs as you can, noting the movement of your hands as you do so.

The internal and external intercostals function as antagonistic muscle groups [◈ p. 390] to alter the position and volume of the rib cage during ventilation. The diaphragm, however, has no antagonistic muscles. Instead, the *abdominal muscles* contract during active expiration to supplement the activity of the internal intercostals.

Abdominal contraction during active expiration pulls the lower rib cage inward and decreases abdominal volume. This displaces the intestines and liver upward, pushing the diaphragm into the thoracic cavity and passively decreasing the chest volume even more. The action of abdominal muscles during forced expiration is why aerobics instructors tell you to blow air out as you lift your head and shoulders during abdominal "crunches." The active process of blowing air out helps contract the abdominals, the very muscles you are trying to strengthen.

Any neuromuscular disease that weakens skeletal muscles or damages their motor neurons can adversely affect ventilation. With decreased ventilation, less fresh air enters the lungs. In addition, loss of the ability to cough increases the risk of pneumonia and other infections. Examples of diseases that affect the motor control of ventilation include **myasthenia gravis** [◈ p. 266], an illness in which acetylcholine receptors of the motor end plate of skeletal muscles are destroyed, and **polio** (poliomyelitis), a viral illness that paralyzes skeletal muscles.

CONCEPT CHECK   Answers: p. 572

14. Scarlett O'Hara was trying to squeeze herself into a corset with an 18-inch waist. Was she more successful when she took a deep breath and held it or when she blew all the air out of her lungs? Why?

15. Why would loss of the ability to cough increase the risk of respiratory infections? (Hint: What does coughing do to mucus in the airways?)

Intrapleural Pressure Changes During Ventilation

Ventilation requires that the lungs, which are unable to expand and contract on their own, move in association with the contraction and relaxation of the thorax. As we noted earlier in this chapter, the lungs are "stuck" to the thoracic cage by cohesive forces of the fluid between the two pleural membranes. Thus, if the thoracic cage moves, the lungs move with it.

The *intrapleural pressure*, pressure within the fluid between the pleural membranes, is normally subatmospheric. This subatmospheric pressure arises during development, when the thoracic cage with its associated pleural membrane grows more rapidly than the lung with its associated pleural membrane. The two pleural membranes are held together by the pleural fluid bond, so the elastic lungs are forced to stretch to conform to the larger volume of the thoracic cavity. At the same time, however, elastic recoil of the lungs creates an inwardly directed force that attempts to pull the lungs away from the chest wall (Fig. 17-10a ●). The combination of the outward pull of the thoracic cage and inward recoil of the elastic lungs creates an intrapleural pressure of about −3 mm Hg.

You can create a similar situation with a syringe half-filled with fluid and capped with a plugged-up needle. Begin with the fluid inside the syringe at atmospheric pressure. Now pick up the syringe and hold the barrel (the chest wall) in one hand while you try to withdraw the plunger (the elastic lung pulling away from the chest wall). As you pull on the plunger, the volume inside the barrel increases slightly, but the cohesive forces between the water molecules cause the fluid inside the syringe to resist expansion. The pressure within the barrel, which was initially equal to atmospheric pressure, decreases slightly as you pull on the plunger. If you release the plunger, it snaps back to its resting position, restoring atmospheric pressure inside the syringe.

What happens to subatmospheric intrapleural pressure if an opening is made between the sealed pleural cavity and the atmosphere? A knife thrust between the ribs, a broken rib that punctures the pleural membrane, or any event that opens the pleural cavity to the atmosphere will allow air to flow in down its pressure gradient, just as air enters when you break the seal on a vacuum-packed can.

Air in the pleural cavity breaks the fluid bond holding the lung to the chest wall. The chest wall expands outward while the elastic lung collapses to an unstretched state, like a deflated balloon (Fig. 17-10b). This condition, called **pneumothorax** [*pneuma*, air + *thorax*, chest], results in a collapsed lung that is unable to function normally. Pneumothorax can also occur spontaneously if a congenital *bleb*, or weakened section of lung tissue, ruptures, allowing air from inside the lung to enter the pleural cavity.

Correction of a pneumothorax has two components: removing as much air from the pleural cavity as possible with a suction pump, and sealing the hole to prevent more air from entering. Any air remaining in the cavity will gradually be absorbed into the blood, restoring the pleural fluid bond and reinflating the lung.

Pressures in the pleural fluid vary during a respiratory cycle. At the beginning of inspiration, intrapleural pressure is about −3 mm Hg (Fig. 17-9, point B_1). As inspiration proceeds, the pleural membranes and lungs follow the thoracic cage because of the pleural fluid bond. But the elastic lung tissue resists being stretched. The lungs attempt to pull farther away from the chest wall, causing the intrapleural pressure to become even more negative (Fig. 17-9, point B_2).

Because this process is difficult to visualize, return to the analogy of the fluid-filled syringe with the blocked needle. You can pull the plunger out a small distance without much effort, but if you try to pull it out farther, it is harder to do because of the cohesiveness of the fluid. The increased amount of work you do trying to pull out the plunger is paralleled by the work your muscles must do when they contract for inspiration. The bigger the breath, the more work is required to "stretch" the pleural fluid and the elastic lung.

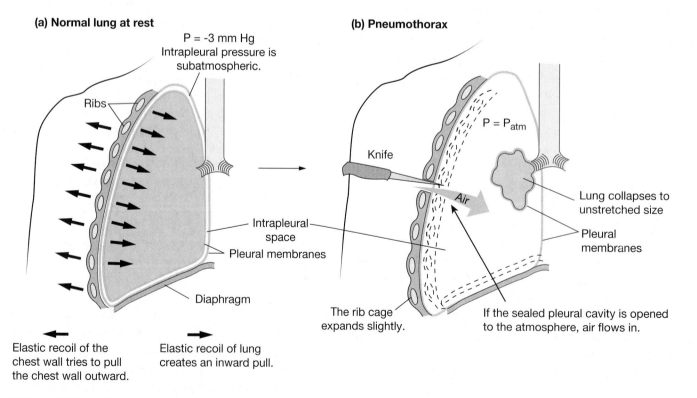

(a) Normal lung at rest

P = -3 mm Hg
Intrapleural pressure is subatmospheric.

Ribs

Intrapleural space
Pleural membranes

Diaphragm

Elastic recoil of the chest wall tries to pull the chest wall outward.

Elastic recoil of lung creates an inward pull.

(b) Pneumothorax

P = P$_{atm}$

Knife

Air

Lung collapses to unstretched size

Pleural membranes

The rib cage expands slightly.

If the sealed pleural cavity is opened to the atmosphere, air flows in.

● **Figure 17-10** *Pressure in the pleural cavity*

By the end of quiet inspiration, when lungs are fully expanded, intrapleural pressure falls to around −6 mm Hg (point B₂). During exercise or other powerful inspirations, intrapleural pressure may reach −8 mm Hg.

Upon expiration, the thoracic cage returns to its resting position. The lungs are released from their extra-stretched position, and the intrapleural pressure returns to its normal value of −3 mm Hg (point B₃). Notice that intrapleural pressure never equilibrates with the atmosphere because the pleural cavity is a closed compartment.

The pressure gradients required for air flow are created by the work of skeletal muscle contraction. Normally, about 3%–5% of the body's energy expenditure is used for quiet breathing. During exercise, the energy required for breathing increases substantially. The two factors that have the greatest influence on the amount of work needed for breathing are (1) the stretchability, or compliance, of the lungs and (2) the resistance of the airways to air flow.

CONCEPT CHECK

Answers: p. 572

16. A person has periodic spastic contractions of the diaphragm, otherwise known as hiccups. What happens to intrapleural and alveolar pressures when a person hiccups?

17. A stabbing victim is brought into the emergency room with a knife wound between the ribs on the left side of his chest. What has probably happened to his left lung? To his right lung? Why does the left side of his rib cage seem larger than the right side?

Lung Compliance and Elastance May Change in Disease States

Adequate ventilation depends on the ability of the lungs to expand normally. Most of the work of breathing goes into overcoming the resistance of the elastic lungs and the thoracic cage to stretch. Clinically, the ability of the lung to stretch is called **compliance**. A high-compliance lung stretches easily, just as a compliant person is easy to persuade. A low-compliance lung requires more force from the inspiratory muscles to stretch it.

Compliance is different from **elastance** (elasticity). The fact that a lung stretches easily (high compliance) does not necessarily mean that it will return to its resting volume when the stretch-

RUNNING PROBLEM

Emphysema is characterized by a loss of elastin, the elastic fibers that help the alveoli recoil during expiration. Researchers believe that elastin is destroyed by proteases released by immune system cells, which work overtime in smokers to rid the lungs of irritants. People with emphysema have more difficulty exhaling air than inhaling. Their alveoli have lost elastic recoil, which makes expiration—normally a passive process—require conscious effort. They literally have to work to push air out of their lungs.

Question 3: Name the muscles that patients with emphysema use to exhale actively.

ing force is released (elastance). You may have experienced this with old gym shorts. After many washings the elastic waistband is easy to stretch (high compliance) but lacking in elasticity, making the shorts impossible to keep around your waist.

Analogous problems occur in the respiratory system. For example, *emphysema* is a disease in which the elastin fibers normally found in lung tissue are destroyed. Destruction of the elastin fibers results in lungs that exhibit high compliance and stretch easily during inspiration. However, these lungs also have decreased elastance, and therefore they do not recoil to their resting position during expiration.

To understand the importance of elastic recoil to expiration, think of an inflated balloon and an inflated plastic bag. The balloon is similar to the normal lung. Its elastic walls squeeze on the air inside the balloon, thereby increasing the internal air pressure. When the neck of the balloon is opened to the atmosphere, elastic recoil causes air to flow out of the balloon. The inflated plastic bag, on the other hand, is like the lung of an individual with emphysema. It has high compliance and is easily inflated, but it has little elastic recoil. If the inflated plastic bag is opened to the atmosphere, most of the air remains inside the bag.

A decrease in lung compliance affects ventilation because more work must be expended to stretch a stiff lung. Pathological conditions in which compliance is reduced are called **restrictive lung diseases**. In these conditions, the energy expenditure required to stretch less-compliant lungs can far exceed the normal range. Two common causes of decreased compliance are inelastic scar tissue formed in *fibrotic lung diseases* and inadequate production of surfactant in the alveoli.

✚ CLINICAL FOCUS

Fibrotic Lung Disease

Fibrotic lung diseases often result from chronic inhalation of fine particulate matter that escapes the mucus lining of the airways and reaches the exchange epithelium of the alveoli. The only protective mechanism in that region of the respiratory system is removal by wandering alveolar macrophages (Fig. 17-2g). These phagocytic cells patrol the alveoli, engulfing any airborne particles that reach the exchange surface. If the particles are organic, the macrophages digest them with lysosomal enzymes. But if the particles cannot be digested or if they accumulate in large numbers, an inflammatory process ensues. Intracellular accumulation of the particles causes the macrophage to secrete growth factors that stimulate fibroblasts in the lung's connective tissue. These fibroblasts produce collagen that forms inelastic, fibrous scar tissue. Large amounts of scar tissue reduce the compliance of the lung. The proliferation of inelastic scar tissue in the lung is called **fibrotic lung disease**, or **fibrosis**. Particles that can trigger fibrosis include asbestos, coal dust, silicon, and even dust, paper particles, and pollutants from urban areas.

Surfactant Decreases the Work of Breathing

For years, physiologists assumed that elastin and other elastic fibers were the primary source of resistance to stretch in the lung. But studies comparing the work required to expand air-filled and saline-filled lungs showed that air-filled lungs are much harder to inflate. From this, researchers concluded that lung tissue itself contributes less to resistance than they had thought. Some other property of the normal air-filled lung, not present in a saline-filled lung, must create most of the resistance to stretch.

This property is **surface tension** [✲ p. 31] created by the thin fluid layer between the alveolar cells and the air. At any air-fluid interface, the surface of a liquid behaves as if it is under tension, like a thin membrane being stretched. This surface tension arises because of the hydrogen bonds between water molecules. The surface molecules are attracted to other water molecules beside and beneath them, but not to the air.

If water is isolated in a drop, it will take on a rounded shape [✲ Fig. 2-10, p. 31]. If water is in the form of a bubble, surface tension exerts a force directed toward the center of the bubble (Fig. 17-11a ●). The **Law of LaPlace** is an expression of this force, or pressure. It states that the pressure within a fluid bubble is a function of two factors: the surface tension of the fluid and the radius of the bubble. This relationship is expressed by the equation

$$P = 2 \times T/r$$

where P is pressure inside the bubble
T is the surface tension of the fluid
r is the radius of the sphere

If two bubbles have different diameters but are lined by fluids with the same surface tension, the pressure inside the smaller bubble will be greater (Fig. 17-11a).

How does this apply to the lung? In physiology, we equate the bubble to a fluid-lined alveolus, although alveoli are not perfect spheres. The presence of fluid lining the alveoli creates surface tension that increases the resistance of the lung to stretch. The greater the lung's resistance, the more work needed to expand the alveoli with each breath. Small alveoli require more work than larger alveoli.

Normally, however, our lungs secrete a chemical surfactant that reduces surface tension. **Surfactants** are molecules that disrupt cohesive forces between water molecules. For example, in dishwasher rinses, surfactants keep rinse water from beading up on the dishes. In the lungs, surfactant decreases the surface tension of the fluid lining the alveoli and decreases the resistance of the lung to stretch.

Surfactant is more concentrated in smaller alveoli, making their surface tension less than that in larger alveoli (Fig. 17-11b). Lower surface tension helps equalize the pressure among the different sizes of alveoli and makes it easier to inflate the smaller alveoli. With lower surface tension, the work needed to expand the alveoli with each breath is also greatly reduced.

Human surfactant is a mixture containing lipoproteins such as *dipalmitoylphosphatidylcholine*. It is secreted into the alveolar air space by the type II alveolar cells (see Fig. 17-2g). Normally, surfactant synthesis begins about the twenty-fifth week of fetal development under the influence of various hormones. Production usually reaches adequate levels by the thirty-fourth week (about six weeks before normal delivery). Babies who are born prematurely without adequate concentrations of surfactant in their alveoli develop *newborn respiratory distress syndrome (NRDS)*. In addition to "stiff" (low-compliance) lungs that require tremendous effort to expand with each breath, babies with NRDS also have alveoli that collapse.

Airway Diameter Is the Primary Determinant of Airway Resistance

The other factor besides compliance that influences the work of breathing is the resistance of the respiratory system to air flow. Resistance in the respiratory system is similar in many ways to resistance in the cardiovascular system [✲ p. 454]. Parameters that contribute to resistance (R) are:

1. The length of the system (L)
2. The viscosity of the substance flowing through the system (η)
3. The radius of the tubes in the system (r)

The equation that expresses the relation of those parameters is:

$$R \propto L\eta/r^4$$

Because the length of the respiratory system is constant, we can ignore that variable in the equation. The viscosity of air is almost constant, although you may have noticed that it feels harder to breathe in a sauna filled with steam than in a room with

---

(a) Pressure is greater in the smaller bubble.

Larger bubble
r = 2
T = 3
P = (2 × 3)/2
P = 3

Smaller bubble
r = 1
T = 3
P = (2 × 3)/1
P = 6

Law of LaPlace: P = 2T/r

P = pressure
T = surface tension
r = radius

According to the law of LaPlace, if two bubbles have the same surface tension, the small bubble will have higher pressure.

(b) Surfactant reduces surface tension (T). Pressure is equalized in the large and small bubbles.

r = 2
T = 2
P = (2 × 2)/2
P = 2

r = 1
T = 1
P = (2 × 1)/1
P = 2

● **Figure 17-11** *Surfactant reduces surface tension*

Newborn Respiratory Distress Syndrome and Surfactant

The clue to what creates additional resistance in an air-filled alveolus came from studies of premature babies with newborn respiratory distress syndrome (NRDS). From their first breath, NRDS babies have difficulty keeping their lungs inflated. With each breath, their lungs collapse, and they must use a tremendous amount of energy to expand them again for the next breath. Unless treatment is initiated rapidly, about 50% of these infants die. When President and Mrs. John Fitzgerald Kennedy's son was born with NRDS in the 1960s, the physicians could do little but administer oxygen and hope that the baby's lungs would mature before hypoxia and exertion overwhelmed him. Now the prognosis for NRDS babies is much better. Amniotic fluid can be sampled to assess if the baby's lungs are producing adequate amounts of surfactant. If they are not, and if delivery cannot be delayed, the NRDS baby can be treated by artificial ventilation that forces air into the lungs and keeps the alveoli open. The current treatment includes the aerosol administration of artificial surfactant until the baby's lungs mature enough to produce their own.

normal humidity. Water droplets in the steam increase the viscosity of the steamy air, thereby increasing its resistance to flow.

Viscosity also changes slightly with atmospheric pressure, decreasing as pressure decreases. A person at high altitude may feel less resistance to air flow than a person at sea level. Despite these exceptions, viscosity plays very little role in resistance to air flow.

Because length and viscosity are essentially constant for the respiratory system, the radius (or diameter) of the airways becomes the primary determinant of airway resistance. Normally, however, the work needed to overcome resistance of the airways to air flow is small compared with the work needed to overcome the resistance of the lungs and thoracic cage to stretch.

Nearly 90% of airway resistance normally can be attributed to the trachea and bronchi, rigid structures with the smallest total cross-sectional area of the airways. These structures are supported by cartilage and bone, so their diameters normally do not change, and their resistance to air flow is constant. However, mucus accumulation from allergies or infections can dramatically increase resistance. If you have ever tried breathing through your nose when you have a cold, you can appreciate how narrowing an upper airway limits air flow!

The bronchioles normally do not contribute significantly to airway resistance because their total cross-sectional area is about 2000 times that of the trachea. But because the bronchioles are collapsible tubes, a decrease in their diameter can suddenly turn them into a significant source of airway resistance. **Bronchoconstriction** increases resistance to air flow and decreases the amount of fresh air that reaches the alveoli.

Bronchioles, like arterioles, are subject to reflex control by the nervous system and by hormones. However, most minute-to-minute changes in bronchiolar diameter occur in response to paracrines. Carbon dioxide in the airways is the primary paracrine that affects bronchiolar diameter. Increased carbon dioxide in expired air relaxes bronchiolar smooth muscle and cause **bronchodilation**.

Histamine is a paracrine that acts as a powerful bronchoconstrictor. This chemical is released by *mast cells* [✎ p. 526] in response to tissue damage or allergic reactions. In severe allergic reactions, large amounts of histamine may lead to widespread bronchoconstriction and difficult breathing. Immediate medical treatment in these patients is imperative.

The primary neural control of bronchioles comes from parasympathetic neurons that cause bronchoconstriction, a reflex designed to protect the lower respiratory tract from inhaled irritants. There is no significant sympathetic innervation of the bronchioles in humans.

However, smooth muscle in the bronchioles is well supplied with β_2 receptors that respond to epinephrine. Stimulation of β_2 receptors relaxes airway smooth muscle and results in bronchodilation. This reflex is used therapeutically in the treatment of asthma and various allergic reactions characterized by histamine release and bronchoconstriction. Table 17-3 summarizes the factors that alter airway resistance.

Table 17-3: Factors That Affect Airway Resistance

FACTOR	AFFECTED BY	MEDIATED BY
Length of the system	Constant; not a factor	
Viscosity of air	Usually constant. Humidity and altitude may alter slightly.	
Diameter of airways		
Upper airways	Physical obstruction	Mucus and other factors
Bronchioles	Bronchoconstriction	Parasympathetic neurons (muscarinic receptors), histamine, leukotrienes
	Bronchodilation	Carbon dioxide, epinephrine (β_2 receptors)

18. In a normal person, which contributes more to the work of breathing: airway resistance or lung and chest wall compliance?

19. Coal miners who spend years inhaling fine coal dust have much of their alveolar surface area covered with scarlike tissue. What happens to their lung compliance as a result?

20. What happens to work required for breathing if surfactant is not present?

21. A cancerous lung tumor has grown into the walls of a group of bronchioles, narrowing their lumens. What has happened to the resistance to air flow in these bronchioles?

22. Name the neurotransmitter and receptor for parasympathetic bronchoconstriction.

Pulmonary Function Tests Measure Lung Volumes During Ventilation

Physiologists and clinicians assess a person's pulmonary function by measuring how much air the person moves during quiet breathing and with maximum effort. These **pulmonary function tests** use a **spirometer**, an instrument that measures the volume of air moving with each breath (Fig. 17-12●). Most spirometers in clinical use today are small computerized machines rather than the traditional spirometer illustrated.

Although pulmonary function tests are relatively simple to perform, they have considerable diagnostic value. For example, in asthma, the bronchioles are constricted. They tend to collapse and close off before a forced expiration is completed, reducing both the amount and rate of air flow. Diseases in which air flow during expiration is diminished due to narrowing of the bronchioles are known as **obstructive lung diseases**. Emphysema and chronic bronchitis are sometimes called *chronic obstructive pulmonary disease*, or *COPD*, because of their ongoing, or *chronic*, nature.

Asthma

Asthma is an obstructive lung disease characterized by bronchoconstriction and airway edema, which increase airway resistance and decrease air flow. Patients complain of shortness of breath (*dyspnea*), and when they are asked to exhale forcefully, a wheezing sound is heard as air whistles through the narrowed lower airways. The severity of asthma attacks ranges from mild to life threatening. Asthma is an inflammatory condition that is often associated with allergies. It can also be triggered by exercise (exercise-induced asthma) and by rapid changes in the temperature or humidity of inspired air. At the cellular level, a variety of chemical signals may be responsible for inducing bronchoconstriction, including acetylcholine, histamine, substance P (a neuropeptide), and leukotrienes secreted by mast cells, macrophages, and eosinophils. *Leukotrienes* are lipid-like bronchoconstrictors that are released during the inflammatory response. Asthma is treated with inhaled and oral medications that include β_2-adrenergic agonists, anti-inflammatory drugs, and new leukotriene antagonists.

Lung Volumes   The air moved during breathing can be divided into four lung volumes: (1) tidal volume, (2) inspiratory reserve volume, (3) expiratory reserve volume, and (4) residual volume. These volumes are diagrammed in Figure 17-13● and described below.

The numerical values given represent average volumes for the 70-kg man. The volumes for women are typically about 20%–25% less. Lung volumes vary considerably with the age, sex, and height of the individual. Each paragraph below begins with the instructions that you would be given if you were being tested for these volumes.

"Breathe quietly." The volume of air that moves in a single inspiration or expiration is known as the **tidal volume** (V_T). Average tidal volume during quiet breathing is about 500 mL.

When the subject inhales, air moves into the lungs. The volume of the bell decreases, and the pen rises on the tracing.

● **Figure 17-12**   *The recording spirometer*   The subject inserts a mouthpiece that is attached to an inverted bell filled with air or oxygen. The volume of the bell and the volume of the subject's respiratory tract create a closed system.

A spirometer tracing showing lung volumes and capacities.

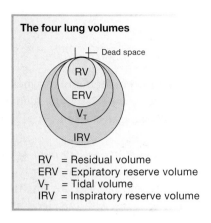

The four lung volumes

RV = Residual volume
ERV = Expiratory reserve volume
V_T = Tidal volume
IRV = Inspiratory reserve volume

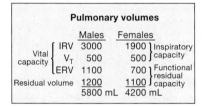

Pulmonary volumes			
		Males	Females
Vital capacity	IRV	3000	1900 } Inspiratory capacity
	V_T	500	500
	ERV	1100	700 } Functional residual capacity
Residual volume		1200	1100
		5800 mL	4200 mL

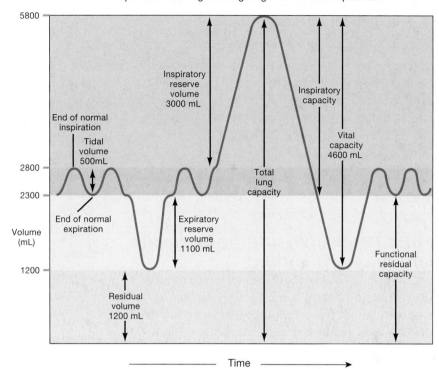

● **Figure 17-13** *Lung volumes and capacities*

Capacities are sums of 2 or more volumes.

"Now, at the end of a quiet inspiration, take in as much additional air as you possibly can." The additional volume you inspire above the tidal volume represents your **inspiratory reserve volume** (IRV). In a 70-kg man, this volume is about 3000 mL, a sixfold increase over the normal tidal volume.

"Now stop at the end of a normal exhalation, then exhale as much air as you possibly can." The amount of air exhaled after the end of a normal expiration is the **expiratory reserve volume** (ERV), which averages about 1100 mL.

The fourth volume is one that cannot be measured directly. Even if you blow out as much air as you can, air still remains in the lungs and the airways. The volume of air in the respiratory system after maximal exhalation, about 1200 mL, is called the **residual volume** (RV). Most of this residual volume exists because the lungs are held stretched against the thoracic wall by the pleural fluid.

Lung Capacities   The sum of two or more lung volumes is called a **capacity**. The **vital capacity** (VC) is the sum of the inspiratory reserve volume, expiratory reserve volume, and tidal volume. Vital capacity represents the maximum amount of air that can be voluntarily moved into or out of the respiratory system with one breath. To measure vital capacity, you would instruct the person to take in as much air as possible, then blow it all out. Vital capacity decreases with age.

The vital capacity plus the residual volume yields the **total lung capacity** (TLC). Other capacities of importance in pulmonary medicine include the **inspiratory capacity** (tidal volume + inspiratory reserve volume) and the **functional residual capacity** (expiratory reserve volume + residual volume).

Auscultation of Breath Sounds   Auscultation of breath sounds is an important diagnostic technique in pulmonary medicine, just as auscultation of heart sounds is used in cardiovascular diagnosis [❧ p. 476]. But breath sounds are more complicated to interpret due to a wide range of normal variation.

In general, breath sounds are distributed evenly over the lungs and resemble a quiet "whoosh" made by flowing air. In conditions in which air flow is reduced, such as pneumothorax, breath sounds may be diminished or absent. Abnormal sounds include a variety of squeaks, pops, wheezes, or bubbling sounds caused by fluid and secretions in the airways or alveoli. Inflammation of the pleural membrane results in a crackling or grating sound known as a *friction rub*. It is caused by swollen, inflamed pleural membranes rubbing against each other, and it disappears when fluid again separates them.

CONCEPT CHECK Answers: p. 572

23. How are lung volumes and lung capacities related?

24. Which lung volume cannot be measured directly?

25. Restrictive lung disease decreases the compliance of the lung. How will the inspiratory reserve volume change in patients with a restrictive lung disease?

26. Chronic obstructive lung disease causes patients to lose the ability to exhale fully. How does residual volume change in these patients?

27. If vital capacity decreases with age but total lung capacity does not change, what volume must be changing? In which direction?

Rate and Depth of Breathing Determine the Efficiency of Breathing

You may recall that the efficiency of the heart is measured by the cardiac output, which is calculated by multiplying heart rate by stroke volume. Likewise, we can estimate the effectiveness of ventilation by calculating the **total pulmonary ventilation**, the volume of air moved into and out of the lungs each minute. Total pulmonary ventilation, also known as the *minute volume*, is calculated as follows:

$$\text{Ventilation rate} \times \text{tidal volume } (V_T) = \text{total pulmonary ventilation}$$

The normal ventilation rate for an adult is 12–20 breaths per minute. Using the average tidal volume of 500 mL and the slowest ventilation rate, we get:

$$12 \text{ breaths/min} \times 500 \text{ mL/breath} = 6000 \text{ mL/min} = 6 \text{ L/min}$$

Total pulmonary ventilation represents the physical movement of air into and out of the respiratory tract. But is it a good indicator of how much fresh air reaches the alveolar exchange surface? Not necessarily.

Some air that enters the respiratory system does not reach the alveoli because part of every breath remains in the conducting airways, such as the trachea and bronchi. Because the conducting airways do not exchange gases with the blood, they are known as the **anatomic dead space**. Anatomic dead space averages about 150 mL.

To illustrate the difference between air that enters the airways and fresh air that reaches the alveoli, let's consider a typical breath that moves 500 mL of air during a respiratory cycle (Fig. 17-14 ●).

1. At the end of an inspiration, lung volume is maximum and fresh air fills the dead space.
2. The tidal volume of 500 mL is exhaled. The first air out is the 150 mL of fresh air that had been in the dead space, followed by 350 mL of "stale" air from the alveoli (150 + 350 = 500 mL exhaled). 500 mL of air left the alveoli, but only

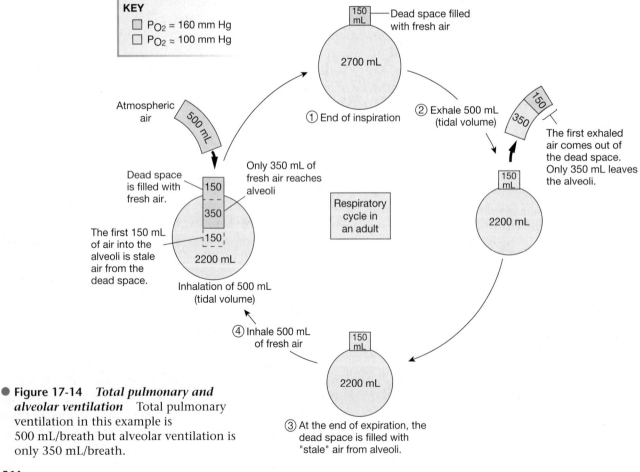

● **Figure 17-14 Total pulmonary and alveolar ventilation** Total pulmonary ventilation in this example is 500 mL/breath but alveolar ventilation is only 350 mL/breath.

Table 17-4: Effects of Breathing Pattern on Alveolar Ventilation

TIDAL VOLUME (mL)	RESPIRATORY RATE (BREATHS/MIN)	TOTAL PULMONARY VENTILATION (mL/MIN)	FRESH AIR TO ALVEOLI (mL) (TIDAL VOLUME − DEAD SPACE VOLUME*)	ALVEOLAR VENTILATION (mL/MIN)
500 (normal)	12 (normal)	6000	350	4200
300 (shallow)	20 (rapid)	6000	150	3000
750 (deep)	8 (slow)	6000	600	4800

*Dead space volume is assumed to be 150 mL.

Table 17-5: Types and Patterns of Ventilation

NAME	DESCRIPTION	EXAMPLES
Eupnea	Normal quiet breathing	
Hyperpnea	Increased respiratory rate and/or volume in response to increased metabolism	Exercise
Hyperventilation	Increased respiratory rate and/or volume without increased metabolism	Emotional hyperventilation; blowing up a balloon
Hypoventilation	Decreased alveolar ventilation	Shallow breathing; asthma; restrictive lung disease
Tachypnea	Rapid breathing; usually increased respiratory rate with decreased depth	Panting
Dyspnea	Difficulty breathing (a subjective feeling sometimes described as "air hunger")	Various pathologies or hard exercise
Apnea	Cessation of breathing	Voluntary breath-holding; depression of CNS control centers

350 mL of that volume left the body. The remaining 150 mL of "stale" alveolar air stays in the dead space.

3. At the end of expiration, lung volume is at its minimum and stale air fills the dead space.

4. With the next inspiration, 500 mL of fresh air enters the airways. The entering air returns the 150 mL of stale air in the anatomic dead space to the alveoli, followed by the first 350 mL of the fresh air. The last 150 mL of inspired fresh air remains in the dead space and never reaches the alveoli.

Thus, although 500 mL of air entered the alveoli, only 350 mL of that volume was fresh air. The fresh air entering the alveoli equals the tidal volume minus the dead space volume.

Because a significant portion of inspired air never reaches an exchange surface, a more accurate indicator of the *efficiency* of ventilation is **alveolar ventilation**, the amount of fresh air that reaches the alveoli each minute. Alveolar ventilation is calculated by multiplying ventilation rate by the volume of fresh air that reaches the alveoli:

Ventilation rate × (tidal volume − dead space volume) = alveolar ventilation

Using the same ventilation rate and tidal volume as above and a dead space volume of 150 mL yields an alveolar ventilation of:

12 breaths/min × (500 mL/breath − 150 mL/breath) = 4200 mL/min

Thus, at 12 breaths per minute, the alveolar ventilation is 4.2 L/min. Although 6 L/min of fresh air entered the respiratory system, only 4.2 L reached the alveoli.

Alveolar ventilation can be drastically affected by changes in the rate or depth of breathing. Table 17-4 shows that three people can have the same total pulmonary ventilation (minute volume) but dramatically different alveolar ventilation. *Maximum voluntary ventilation*, which involves breathing as deeply and quickly as possible, may increase total pulmonary ventilation to as much as 170 L/min. Table 17-5 describes various patterns of ventilation.

Gas Composition in the Alveoli Varies Little During Normal Breathing

How much can a change in alveolar ventilation affect the amount of fresh air and oxygen that reaches the alveoli? Fig. 17-15 ● shows how P_{O_2} and P_{CO_2} in the alveoli vary with hyper- and hypoventilation.

● **Figure 17-15** *Effect of changing alveolar ventilation on* P_{O_2} *and* P_{CO_2} *in the alveoli*

As alveolar ventilation increases above normal levels (**hyperventilation**), alveolar P_{O_2} rises to about 120 mm Hg and alveolar P_{CO_2} falls to around 20 mm Hg. During **hypoventilation**, when less fresh air enters the alveoli, alveolar P_{O_2} decreases and alveolar P_{CO_2} increases.

Although a dramatic change in alveolar ventilation pattern will affect gas partial pressures in the alveoli, the P_{O_2} and P_{CO_2} in the alveoli change surprisingly little during normal quiet breathing. Alveolar P_{O_2} is fairly constant at 100 mm Hg, and alveolar P_{O_2} stays close to 40 mm Hg.

Intuitively, you might think that P_{O_2} would increase when fresh air first enters the alveoli, then decrease steadily as oxygen leaves to enter the blood. Instead, we find only very small swings in P_{O_2}. Why? The reasons are that (1) the amount of oxygen that enters the alveoli with each breath is roughly equal to the amount of oxygen that enters the blood, and (2) the amount of

fresh air that enters the lungs with each breath is only a little more than 10% of the total lung volume at the end of inspiration.

Ventilation and Alveolar Blood Flow Are Matched

Moving oxygen from the atmosphere to the alveolar exchange surface is only the first step in external respiration. Next there must be normal gas exchange across the alveolar-capillary interface. Finally, blood flow (*perfusion*) past the alveoli must be adequate to pick up the available oxygen. Matching ventilation into groups of alveoli with blood flow past those alveoli is a two-part process involving local regulation of both air flow and blood flow.

Alterations in pulmonary blood flow depend almost exclusively on properties of the capillaries and on local factors such as the concentrations of oxygen and carbon dioxide in the lung tissue. Capillaries in the lungs are unusual because they are collapsible. If the pressure of blood flowing through the capillaries falls below a certain point, the capillaries close off, diverting blood to other pulmonary capillary beds with higher pressure.

In a person at rest, some capillary beds found in the apex (top) of the lung are closed off because of low hydrostatic pressure. Capillary beds at the base of the lung have higher hydrostatic pressure because of gravity and remain open. Consequently, blood flow is diverted toward the base of the lung.

During exercise, when blood pressure rises, the closed apical capillary beds open, ensuring that the increased cardiac output will be fully oxygenated as it passes through the lungs. The ability of the lungs to recruit additional capillary beds during exercise is an example of the reserve capacity of the body.

At the local level, the body attempts to match air and blood flow in each section of the lung by regulating the diameters of the arterioles and the bronchioles. Bronchiolar diameter is mediated primarily by CO_2 levels in exhaled air passing through them (Table 17-6). An increase in the P_{CO_2} of expired air causes the bronchioles to dilate. A decrease in the P_{CO_2} of expired air causes bronchioles to constrict.

Table 17-6: Local Control of Arterioles and Bronchioles by Oxygen and Carbon Dioxide

GAS COMPOSITION	BRONCHIOLES	PULMONARY ARTERIOLES	SYSTEMIC ARTERIOLES
P_{CO_2} increases	Dilate	(Constrict)	Dilate
P_{CO_2} decreases	Constrict	(Dilate)	Constrict
P_{O_2} increases	(Constrict)	Dilate	Constrict
P_{O_2} decreases	(Dilate)	Constrict	Dilate

Note: Responses in parentheses indicate weak responses.

Although there is some autonomic innervation of the pulmonary arterioles, there is apparently little neural control of pulmonary blood flow. The resistance of pulmonary arterioles to blood flow is regulated primarily by the oxygen content of the interstitial fluid around the arteriole (Fig. 17-16 ●).

If ventilation of alveoli in one area of the lung is diminished, the tissue P_{O_2} in that area decreases. The arterioles respond to lower P_{O_2} by constricting. Vasoconstriction is adaptive because it diverts blood away from the underventilated region to better-ventilated parts of the lung.

On the other hand, if P_{O_2} in the lung becomes higher than normal, the alveoli in that region are being overventilated relative to the blood flow past them. The arterioles supplying those alveoli dilate to bring in additional blood to pick up the extra oxygen.

Note that constriction of pulmonary arterioles in response to low P_{O_2} is exactly the opposite of what occurs in the systemic circulation [● p. 500]. In the systemic circulation, a decrease in the concentration of oxygen in the tissues causes the arterioles to dilate, bringing more blood to metabolically active tissues that are consuming oxygen.

One important point must be noted here. Local control mechanisms are not effective regulators of air and blood flow under all circumstances. If blood flow is blocked in one pulmonary artery, or if air flow is blocked at the level of the larger airways, local responses that shunt air or blood to other parts of the lung are ineffective because no part of the lung has normal ventilation or perfusion.

This completes our discussion of the mechanics of ventilation. In the next chapter, we shift from bulk flow of air to the diffusion and transport of oxygen and carbon dioxide as they travel between the air spaces of the alveoli and the cells of the body.

CONCEPT CHECK   Answers: p. 573

30. If a tumor in the lung tissue decreases blood flow in one small section of the lung to a minimum, what happens to P_{O_2} in those alveoli and in the surrounding tissue? What happens to P_{CO_2} in the same region? What is the compensatory response of the bronchioles in that region? Will the compensation bring ventilation in that section of the lung back to normal? Explain.

Table 17-7: Normal Ventilation Values in Pulmonary Medicine

Total pulmonary ventilation	6 L/min
Total alveolar ventilation	4.2 L/min
Maximum voluntary ventilation	125–170 L/min
Respiration rate	12–20 breaths/min

(a) Ventilation in alveoli is matched to perfusion through pulmonary capillaries.

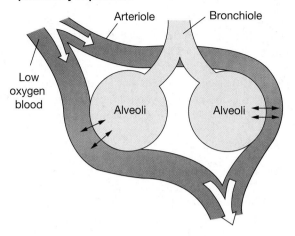

(b) Ventilation-perfusion mismatch.
If ventilation decreases in a group of alveoli (blue), P_{CO_2} increases and P_{O_2} decreases. Blood flowing past those alveoli does not get oxygenated.

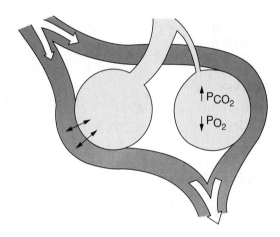

(c) Local control mechanisms try to keep ventilation and perfusion matched.
Decreased tissue P_{O_2} around underventilated alveoli constricts their arterioles, diverting blood to better ventilated alveoli.

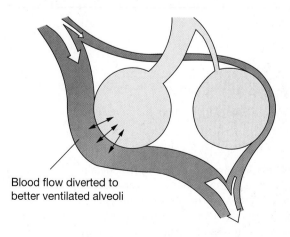

● **Figure 17-16** *Local control matches ventilation and perfusion*

RUNNING PROBLEM CONCLUSION

Emphysema

Edna is discharged from the hospital after three days with prescriptions for bronchodilator and anti-inflammatory drugs to keep her airways as open as possible. Unfortunately, the lung changes that take place with COPD are not reversible, and she will require treatment for the rest of her life. According to the American Lung Association, COPD is the fourth leading cause of death in the United States and costs over $30 billion per year in direct medical costs and indirect costs such as lost wages.

In this running problem, you learned about chronic obstructive pulmonary disease. Now check your understanding of the physiology in the problem by comparing your answers to those in the table below.

	QUESTION	FACTS	INTEGRATION AND ANALYSIS
1	What does narrowing of the airways do to resistance of the airways to air flow?	The relationship between radius and resistance is the same for air flow as it was for blood flow in the circulatory system: As radius decreases, resistance increases (p. 454).	When resistance increases, the body must use more energy to create air flow or blood flow.
2	Why do people with chronic bronchitis have a higher-than-normal rate of respiratory infections?	Cigarette smoke paralyzes the cilia that sweep debris and mucus out of the airways. Without the action of cilia, mucus and trapped particles pool in the airways.	Bacteria trapped in the mucus can multiply and cause respiratory infections.
3	Name the muscles that patients with emphysema use to exhale actively.	Normal expiration depends on elastic recoil of muscles and elastic tissue in the lungs.	Forceful expiration uses the internal intercostal muscles and the abdominal muscles.
4	Why does Edna have an increased hematocrit?	Because of Edna's COPD, her arterial P_{O_2} is low. The major stimulus for red blood cell synthesis is hypoxia.	Low arterial oxygen levels will trigger the synthesis of additional red blood cells. This provides more binding sites for oxygen transport.

546 551 555 559 563

CHAPTER SUMMARY

The process by which air flows into and out of the lungs is another example of the principle of *mass flow*. Like blood flow in the cardiovascular system, air flow in the lungs requires a pump to create the pressure gradient and encounters resistance primarily from changes in the diameter of the tubes through which it flows. The *mechanical properties* of the pleural sac and elastic recoil in the chest wall and lung tissue are essential for normal ventilation.

1. Aerobic metabolism in living cells consumes oxygen and produces carbon dioxide. (p. 547)

2. Gas exchange requires a large, thin, moist exchange surface, a pump to move air, and a circulatory system to transport gases to the cells. (p. 547)

3. Respiratory system functions include gas exchange, pH regulation, vocalization, and protection from foreign substances. (p. 547)

The Respiratory System

IP Respiratory System: Anatomy Review

4. **Cellular respiration** refers to the metabolic processes of the cell that consume oxygen. **External respiration** is exchange of gases between the atmosphere and cells. It includes ventilation, gas exchange at the lung and cells, and transport of gases in the blood. **Ventilation** is the movement of air into and out of the lungs. (pp. 547, 550; Fig. 17-1)

5. The **respiratory system** consists of anatomical structures involved in ventilation and gas exchange. (p. 550)

6. The **upper respiratory tract** includes the mouth, nasal cavity, **pharynx**, and **larynx**. The **lower respiratory tract** includes the **trachea**, **bronchi**, **bronchioles**, and exchange surfaces of the **alveoli**. (pp. 548, 550; Fig. 17-2a)

7. The thoracic cage is bounded by the ribs, spine, and **diaphragm**. Two sets of **intercostal muscles** connect the ribs. (pp. 548, 550; Fig. 17-2b)

8. The **lungs** are contained within a double-walled **pleural sac** that contains a small quantity of **pleural fluid**. (pp. 548, 550; Figs. 17-2d, 17-3)

9. The two **primary bronchi** enter the lungs. Each primary bronchus divides into progressively smaller bronchi and finally into collapsible **bronchioles**. (pp. 548, 550; Figs. 17-2e, 17-4)

10. The alveoli consist mostly of thin-walled **type I alveolar cells** for gas exchange. **Type II alveolar cells** produce surfactant. A network of capillaries surrounds each alveolus. (p. 552; Fig. 17-2f, g)

11. Blood flow through the lungs equals the cardiac output. Resistance to blood flow in the pulmonary circulation is low. Pulmonary arterial pressure averages 25/8 mm Hg. (p. 552)

Gas Laws

IP Respiratory System: Pulmonary Ventilation

12. The total pressure of a mixture is the sum of the pressures of the individual gases in the mixture (**Dalton's law**). **Partial pressure** is the pressure contributed by a single gas such as oxygen. (pp. 552–553; Tbls. 17-1, 17-2)

13. Bulk flow of air occurs down pressure gradients, as does the movement of any individual gas. (p. 533)

14. **Boyle's law** states that as volume increases, pressure decreases. The body creates pressure gradients by changing thoracic volume. (p. 553; Fig. 17-5)

Ventilation

IP Respiratory System: Pulmonary Ventilation

15. The upper respiratory system filters, warms, and humidifies inhaled air. (p. 554)

16. Air flow in the respiratory system is directly proportional to the pressure gradient and inversely related to the resistance of the airways. (p. 555)

17. A single **respiratory cycle** consists of an inspiration and an expiration. (p. 555; Fig. 17-12)

18. During **inspiration**, **alveolar pressure** decreases and air flows into the lungs. Inspiration requires contraction of the inspiratory muscles and the diaphragm. (p. 555; Fig. 17-7)

19. **Expiration** is usually passive, resulting from elastic recoil of the lungs. (p. 557)

20. **Active expiration** requires contraction of internal intercostal and abdominal muscles. (p. 557)

21. **Intrapleural pressures** are always subatmospheric because the pleural cavity is a sealed compartment. (pp. 555–557; Fig. 17-10)

22. **Compliance** is the ease with which the chest wall and lungs expand. Loss of compliance increases the work of breathing. **Elastance** is the ability of a stretched lung to resume its normal volume. (p. 559)

23. **Surfactant** decreases surface tension in the fluid lining the alveoli. This prevents smaller alveoli from collapsing. (p. 560; Fig. 17-11)

24. The diameter of the bronchioles determines resistance to air flow. (p. 561)

25. Increased CO_2 in expired air dilates bronchioles. Parasympathetic neurons cause **bronchoconstriction** in response to irritant stimuli. There is no significant sympathetic innervation of bronchioles, but epinephrine causes **bronchodilation**. (p. 561; Tbl. 17-3)

26. **Tidal volume** is the amount of air taken in during a single normal inspiration. **Vital capacity** is tidal volume plus **expiratory** and **inspiratory reserve volumes**. The air left in the lungs at the end of maximum expiration is the **residual volume**. (pp. 562–563; Fig. 17-13)

27. **Total pulmonary ventilation** = tidal volume × ventilation rate. **Alveolar ventilation** = ventilation rate × (tidal volume − dead space volume). (pp. 564–565; Fig. 17-14)

28. Alveolar gas composition changes very little during a normal respiratory cycle. **Hyperventilation** increases alveolar P_{O_2} and decreases alveolar P_{CO_2}. **Hypoventilation** has the opposite effect. (p. 566; Fig. 17-15)

29. Air flow is matched to blood flow around the alveoli by local mechanisms. CO_2 dilates bronchioles, and decreased O_2 constricts pulmonary arterioles. (p. 566; Fig. 17-16)

QUESTIONS

LEVEL ONE *Reviewing Facts and Terms*

1. List four functions of the respiratory system.

2. Give two different definitions for the word *respiration*.

3. Which sets of muscles are used for normal quiet inspiration? For normal quiet expiration? For active expiration? What kind(s) of muscles are the different muscles?

4. What is the function of pleural fluid?

5. Name the anatomical structures that an oxygen molecule passes on its way from the atmosphere to the blood.

6. Diagram the structure of an alveolus and give the function of each part. How are capillaries associated with an alveolus?

7. Trace the path of the pulmonary circulation. About how much blood is found here at any given moment? What is a typical arterial blood pressure for the pulmonary circuit and how does this compare to that of the systemic circulation?

8. What happens to inspired air as it is conditioned during its passage through the airways?

9. During inspiration, most of the thoracic volume change is due to movement of the _____.

10. Describe the changes in alveolar and intrapleural pressure during one respiratory cycle.

11. What is the function of surfactants in general? In the respiratory system?

12. Of the three factors that contribute to the resistance of air flow through a tube, which plays the largest role in changing resistance in the human respiratory system?

13. Match the following items with their correct effect on the bronchioles:
(a) histamine   1. bronchoconstriction
(b) epinephrine   2. bronchodilation
(c) acetylcholine   3. no effect
(d) increased P_{CO_2}

14. On the spirogram below, label:
(a) tidal volume (V_T), inspiratory and expiratory reserve volumes (IRV and ERV), residual volume (RV), vital capacity (VC), total lung capacity (TLC).
(b) What is the value of each of the volumes and capacities you labeled?
(c) What is this person's ventilation rate?

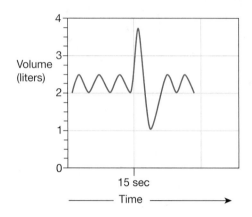

LEVEL TWO *Reviewing Concepts*

15. Compile the following terms into a map of ventilation. Use the modifiers [↑], [↓], greater than (>), less than (<). You may add additional terms.

abdominal muscles	forced breathing	$P_{intrapleural}$
air flow	in, out, from, to	quiet breathing
contract	inspiratory muscles	relax
diaphragm	internal intercostals	scalenes
expiratory muscles	P_{alv}	
external intercostals	P_{atm}	

16. Compare and contrast the following sets of concepts:
(a) compliance and elastance
(b) inspiration, expiration, and ventilation
(c) intrapleural and alveolar pressures
(d) total pulmonary ventilation and alveolar ventilation
(e) type I and type II alveolar cells
(f) pulmonary and systemic circulation

17. List the major paracrines and neurotransmitters that cause bronchoconstriction and bronchodilation. What receptors do they act through? (muscarinic, nicotinic, α, β-1, β-2,)

18. For each of the following parameters, decide if it will increase, decrease, or not change in the situations given.
(a) airway resistance with bronchodilation
(b) intrapleural pressure during inspiration
(c) air flow with bronchoconstriction
(d) bronchiolar diameter with increased P_{CO_2}
(e) tidal volume with decreased compliance
(f) alveolar pressure during expiration

19. Define the following terms: pneumothorax, spirometer, auscultation, hypoventilation, bronchoconstriction, minute volume, partial pressure of a gas.

20. The cartoon coyote is blowing up a balloon as part of an attempt to once more catch a roadrunner. He first breathes in as much air as he can, then blows out all that he can into the balloon.
(a) The volume of air in the balloon is equal to the _____ _____ of the coyote's lungs. This volume can be measured directly, as above, or by adding what respiratory volumes together?
(b) In 10 years, when the coyote is still chasing the roadrunner, will he still be able to put as much air into the balloon in one breath? Explain.

21. Match the descriptions to the appropriate phase(s) of ventilation:
(a) inspiration
(b) expiration
(c) both inspiration and expiration
(d) neither
_____ usually depends on elastic recoil
_____ is easier when lung compliance decreases
_____ due mainly to positive intrapleural pressure generated by muscular contraction
_____ is usually an active process requiring smooth muscle contraction

22. Draw and label a graph showing the P_{O_2} of air in the primary bronchi during one respiratory cycle. (*Hint*: What parameter goes on each axis?)

23. Lung compliance decreases as we age. In the absence of other changes, would the following parameters increase, decrease, or not change?
    (a) the work of breathing
    (b) ease with which lungs
    (c) inflate lung elastance
    (d) airway resistance during inspiration

24. Will pulmonary surfactant increase, decrease, or not change the following?
    (a) the work of breathing
    (b) lung compliance
    (c) surface tension in the alveoli

LEVEL THREE *Problem Solving*

25. Assume a normal female has a resting tidal volume of 400 mL, respiratory rate of 13 breaths/min, and dead space of 125 mL. When she exercises, which of the following scenarios would be most efficient for increasing her oxygen delivery to the lungs?
    (a) increase respiratory rate to 20 breaths/min but no change in tidal volume
    (b) increase tidal volume to 550 mL but no change in respiratory rate
    (c) increase tidal volume to 500 mL and respiratory rate to 15 breaths/min
    Which of the scenarios above is most likely to occur during exercise in real life?

26. A 30-year-old computer programmer has had asthma for 15 years. When she lies down at night, she has spells of wheezing and coughing. Over the years, she has found that she can breathe better if she sleeps sitting nearly upright. Upon examination, she has an enlarged thorax. Her lungs are overinflated on X-ray. Here are the results of her examination and pulmonary function tests. Use the normal data and abbreviations in Figure 17-13 to help answer the questions.

Ventilation rate	16 breaths/min
Tidal volume	600 mL
ERV	1000 mL
RV	3500 mL
Inspiratory capacity	2500 mL
Vital capacity	2800 mL
FRC	4400 mL
TLC	7050 mL

After being given a bronchodilator, her vital capacity increased to 3650 mL.
    (a) What is her minute volume?
    (b) Explain the change in vital capacity with bronchodilators.
    (c) Which other values are abnormal? Can you explain why they might be, given her history and findings?

LEVEL FOUR *Quantitative Problems*

27. A container of gas with a movable piston has a volume of 500 mL and a pressure of 60 mm Hg. The piston is moved, and the new pressure is measured as 150 mm Hg. What is the new volume of the container?

28. You have a mixture of gases in dry air, with an atmospheric pressure of 760 mm Hg. Calculate the partial pressure of each gas for each of the examples below:
    (a) 21% oxygen, 78% nitrogen, 0.3% carbon dioxide
    (b) 40% oxygen, 13% nitrogen, 45% carbon dioxide, 2% hydrogen
    (c) 10% oxygen, 15% nitrogen, 1% argon, 25% carbon dioxide

29. Li is a tiny woman, with a tidal volume of 400 mL and a respiratory rate of 12 breaths per minute at rest. What is her total pulmonary ventilation? Just before a physiology exam, her ventilation increases to 18 breaths per minute from nervousness. Now what is her total pulmonary ventilation? Assuming her dead space is 120 mL, what is her alveolar ventilation in each case?

30. You collected the following data on your classmate Neelesh.

    Minute volume = 5000 mL/min
    3 breaths/15 seconds
    Vital capacity = 4800 mL
    Expiratory reserve volume = 1000 mL

    What are Neelesh's tidal volume and inspiratory reserve volume?

31. Use the figures below to help solve this problem. A spirometer with a volume of 1 liter (V_1) is filled with a mixture of oxygen and helium with a helium concentration of 4 g/L (C_1). Helium does not move from the lungs into the blood or from blood into the lungs. A subject is told to blow out all the air he possibly can. His lung volume at that time is (V_2). He then puts the spirometer tube in his mouth and breathes quietly for several breaths. By that time the helium is evenly dispersed within the spirometer and the subject's lungs. A measurement shows the new concentration of helium is 1.9 g/L. What was the subject's lung volume at the start of the experiment? (Hint: $C_1V_1 = C_2V_2$)

Helium/O_2 mixture V_1

Explorations

Explorations are also available at *www.physiologyplace.com*

Exploration 1: Pneumothorax

Paula Dixon thought the injuries she'd sustained in the motorcycle accident on the way to the airport were minor. But a few hours later, in the middle of a 15-hour flight from Hong Kong to London, her broken rib punctured her left lung and collapsed it. A pneumothorax is easily treated in an emergency room, but treatment in an airplane with only a small medical kit is another matter. Fortunately two physicians on board were able to save Paula's life with the help of tubing from an oxygen mask, a wire coat hanger, and a bottle of cognac.

Pneumothorax occurs as the result of trauma, an infection, or even spontaneously when a weak portion of the lung ruptures. Paula's pneumothorax was life-threatening because with each breath she took, air was sucked into her thorax but was unable to leak back out. This condition is known as a *tension pneumothorax.*

Read the full story of Paula's rescue in "Emergency on Flight BA 032," in the February 1996 issue of *Reader's Digest.* To learn more about the different types of pneumothorax and their treatment, check out the MEDLINEplus medical encyclopedia at *http://www.nlm.nih.gov/medlineplus/encyclopedia.html.*

Exploration 2: Congestive Heart Failure

When is a lung problem not a lung problem? The answer: when it's really a heart problem. Congestive heart failure (CHF) is an excellent example of the interrelationships between body systems and of how a pathology in one system can have a domino effect in the others. The primary symptoms of heart failure are shortness of breath (*dyspnea*), wheezing when breathing, and sometimes a productive cough that may include *phlegm* (pronounced "flem") that is pinkish due to the presence of blood. Congestive heart failure arises when the right heart is a more effective pump than the left heart (see Concept Check question 30 on p. 567). Blood accumulates in the pulmonary circulation, increasing pulmonary blood pressure and capillary hydrostatic pressure. Capillary filtration exceeds the ability of the lymph system to drain interstitial fluid, resulting in pulmonary edema. Treatment of CHF includes increasing urinary output of fluid, which brings yet another organ system into the picture. By current estimates, nearly 5 million Americans suffer from CHF. To learn more about this condition, visit the American Heart Association website (*www.americanheart.org*) or see the Health Information section for the National Heart, Lung, and Blood Institute of the National Institutes of Health (*www.nhlbi.nih.gov*).

ANSWERS

Answers to Concept Check Questions

Page 551

1. Cellular respiration is intracellular metabolism that uses oxygen and organic substrates to produce ATP. External respiration is the exchange and transport of gases between the environment and the cell.

2. Upper respiratory tract includes the mouth, nasal cavity, pharynx, and larynx. Lower respiratory tract includes the trachea, bronchi, bronchioles, and exchange surface of lungs.

3. Velocity of air flow is highest in the trachea and lowest in the bronchioles.

4. Pleural fluid reduces friction between the membranes when they slide past each other, and it holds the lungs tight against the chest wall.

5. The thoracic cage consists of rib cage with intercostal muscles, spinal (vertebral) column, and diaphragm. The thorax contains two lungs in pleural sacs, heart in pericardial sac, esophagus, and major blood vessels entering and leaving the heart.

6. The bronchioles are collapsible.

Page 552

7. Blood flow is equal in the pulmonary trunk and the aorta.

8. Increased hydrostatic pressure results in greater net filtration out of the capillaries.

Page 553

9. $720 \, \text{mm Hg} \times 0.78 \, N_2 = 561.6 \, \text{mm Hg}$

Page 554

10. As air becomes humidified, the partial pressure of oxygen (or any component gas) decreases.

11. If cilia cannot move mucus up the mucus escalator, the collected mucus will trigger a cough reflex in an attempt to clear it out.

Page 555

12. Air flow reverses direction during a cycle but blood flows in a loop and never reverses direction.

13. See Figures 17-2d and 17-3. The lungs are enclosed in a pleural sac. One membrane attaches to the lung and the other membrane lines the thoracic cage. Pleural fluid fills the inside of the pleural sac.

Page 557

14. Scarlett will be more successful if she exhales deeply, as this will decrease her thoracic volume and will pull her lower rib cage inward.

15. Coughing clears the airways of mucus that has captured bacteria and viruses. Inability to cough decreases the ability to expel the potentially harmful material.

Page 559

16. A hiccup results from a quick contraction of the diaphragm that causes a rapid decrease in both intrapleural and intrapulmonary pressures.

17. The knife wound would collapse the left lung if the knife punctured the pleural membrane. Loss of adhesion between lung and chest wall would release the inward pressure exerted on the chest wall, and the rib cage would expand outward. The right side would be unaffected as the right lung is contained within its own pleural sac.

Page 562

18. Normally, lung and chest wall compliance contribute more to the work of breathing.

19. Scar tissue will reduce lung compliance.

20. In the absence of surfactant, the work required for breathing increases.

21. If the bronchiolar lumen (diameter) narrows, the resistance increases.

22. The neurotransmitter is acetylcholine and the receptor is muscarinic.

Page 564

23. A lung capacity is the sum of two or more lung volumes.

24. The residual volume cannot be measured directly.

25. Inspiratory reserve volume will decrease in a lung with lower compliance.

26. Residual volume will increase in a patient who cannot fully exhale.

27. Residual volume increases in aging subjects if total lung capacity does not change.

Page 566

28. An increase in tidal volume will increase alveolar P_{O_2}.

29. An increase in breathing rate also increases alveolar P_{O_2}. Increasing either rate or tidal volume will increase alveolar ventilation.

Page 567

30. If blood flow into one small section of lung decreases, the P_{O_2} in those alveoli will increase because oxygen is not leaving the alveoli and entering the blood. The P_{CO_2} in those alveoli will decrease because new CO_2 is not entering the alveoli from the blood. Bronchioles constrict when P_{CO_2} decreases, shunting air to areas of the lung with better blood flow. The compensation will not bring ventilation in this section of lung back to normal because the original pathology still exists and local control is insufficient to correct the disparity.

Answers to Figure Questions

Page 556

Fig. 17-9 Alveolar pressure is greatest in the middle of expiration and least in the middle of inspiration. It is equal to atmospheric pressure at the beginning and end of inspiration and expiration. When lung volume is at its minimum, alveolar pressure is (c) moving from max to min and external intercostal muscle contraction is (b) minimum.

Answers to Level Four Quantitative Problems

Page 571

27. $P_1V_1 = P_2V_2$ New volume = 200 mL

28. (a) 21% O_2 = 160 mm Hg,
    78% nitrogen = 593 mm Hg,
    0.3% CO_2 = 3 mm Hg.
    (b) 40% O_2 = 304 mm Hg,
    13% nitrogen = 99 mm Hg,
    45% CO_2 = 342 mm Hg,
    2% H_2 = 15 mm Hg.
    (c) 10% O_2 = 76 mm Hg,
    15% nitrogen = 114 mm Hg,
    1% argon = 8 mm Hg,
    25% CO_2 = 190 mm Hg.

29. 400 mL/breath × 12 breaths/min = total pulmonary ventilation of 4800 mL/min. Before an exam, ventilation is 7200 mL/min. Alveolar ventilation is (400−120) × rate, or 3360 and 5040 mL/min.

30. Tidal volume = 471 mL/breath. IRV= 3383 mL.

31. Lung volume is 1.1 L. (Did you forget to subtract the volume of the spirometer?)

Gas Exchange and Transport

18

Model of a hemoglobin molecule with its four subunits. Heme goups are fuchsia with orange iron atoms in the center.

RUNNING PROBLEM

High Altitude

In 1981 a group of 20 physiologists, physicians, and climbers, supported by 42 Sherpa assistants, formed the American Medical Research Expedition to Mt. Everest. The purpose of the expedition was to study human physiology at extreme altitudes, starting with the base camp at 5400 m and continuing on to the summit at 8850 m. From the work of these scientists and others, we now have a good picture of the physiology of high-altitude acclimatization.

578 579 582 586 591 593

BACKGROUND BASICS

Exchange epithelia: **67** | pH and buffers: **33, 34** | Law of mass action: **97** | Cerebrospinal fluid: **290** | Simple diffusion: **136** | Autonomic and somatic motor neurons: **372** | Structure of the brain stem: **297** | Red blood cells and hemoglobin: **532** | Blood-brain barrier: **292** |

The successful ascent of Everest without supplementary oxygen is one of the great sagas of the 20th century.

—John B. West, *Climbing with O's,*
NOVA Online (*www.pbs.org*)

The book *Into Thin Air* by Jon Krakauer chronicles an ill-fated trek to the top of Mt. Everest. To reach the summit of Mt. Everest, climbers must pass through the "death zone" located at about 8000 meters. Of about 4000 people who have attempted the summit, 142 have died, and only 660 have been successful. What are the physiological challenges of climbing Mt. Everest (8850 m or 29,035 ft), and why did it take so many years before humans successfully reached the top? The lack of oxygen at high altitude is part of the answer.

In the last chapter we looked at the mechanics of breathing, the events that create bulk flow of air into and out of the lungs. In this chapter we focus on the two gases that are most significant to physiology, oxygen and carbon dioxide, and look at how they move between the alveolar air spaces and the cells of the body. The process can be divided into two components: the exchange of gases between compartments, which requires diffusion across cell membranes, and the transport of gases in the blood (Fig. 18-1 ●).

▶ DIFFUSION AND SOLUBILITY OF GASES

The simple diffusion of oxygen and carbon dioxide between lung and blood, or between blood and cells, obeys the rules for simple diffusion across a membrane that are summarized in *Fick's law of diffusion* [● p. 137]:

$$\text{Diffusion rate} \propto \frac{\text{surface area} \times \text{concentration gradient}}{\text{membrane thickness} \times \text{membrane resistance}}$$

If we assume that membrane resistance is constant, three factors influence diffusion in the lungs:

1. *Concentration gradient* The rate of diffusion is directly proportional to the concentration (partial pressure) gradient.
2. *Surface area* The rate of diffusion is directly proportional to the available surface area.
3. *Membrane thickness* The rate of diffusion is inversely proportional to the thickness of the membrane.

And from the general rules for diffusion, we can add:

4. *Diffusion distance* Diffusion is most rapid over short distances.

Under most circumstances, diffusion distance, surface area, and membrane thickness are constants and are maximized to facilitate diffusion. The most important factor for gas exchange in normal physiology is the concentration gradient.

When we think of concentrations in physiology, units such as moles per liter or milliosmoles per liter come to mind. However, respiratory physiologists commonly use *partial pressures* to express gas concentration in solution. This measure allows direct comparison with partial pressures of the gases in air, which is important for establishing whether there is a concentration gradient between the alveoli and the blood.

The Solubility of Gases in Liquids Depends on Pressure, Solubility, and Temperature

When a gas is placed in contact with water, if there is a pressure gradient, the gas molecules move from one phase to the other. If gas pressure is higher in the water than in the gaseous phase, gas molecules will leave the water. If the gas pressure is higher in the gaseous phase than in the water, the gas will dissolve into the water.

The movement of gas molecules from air into solution is directly proportional to three factors: (1) the pressure gradient of the individual gas, (2) the solubility of the gas in the given liquid, and (3) temperature. Because temperature is relatively constant in mammals, we will ignore its contribution in this discussion.

The ease with which a gas dissolves in a solution is its **solubility**. If a gas is very soluble, large numbers of gas molecules will go into solution at low partial pressures. With less soluble gases, high partial pressures may cause only a few molecules of the gas to dissolve.

For example, imagine a container of water exposed to air with a P_{O_2} of 100 mm Hg (Fig. 18-2a ●). Initially, the water has no oxygen dissolved in it (P_{O_2} = 0 mm Hg). As the gas phase stays in contact with the water, the moving oxygen molecules in the gas diffuse into the water and dissolve (Fig. 18-2b). This process continues until equilibrium is reached. At equilibrium (Fig. 18-2c), the movement of oxygen into the water is equal to the movement of oxygen back into the air.

We refer to the amount of oxygen that dissolves in the water at any given P_{O_2} as the *partial pressure of the gas in solution*. Thus, if

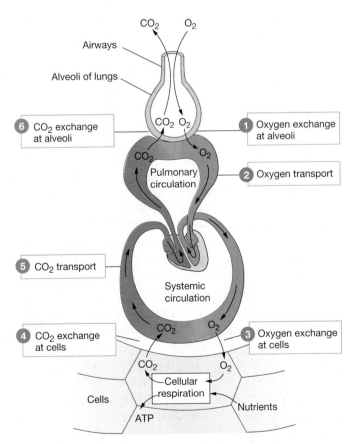

● **Figure 18-1** *Overview of oxygen and CO_2 exchange and transport*

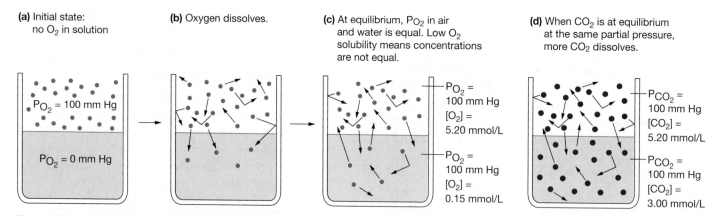

(a) Initial state: no O_2 in solution

(b) Oxygen dissolves.

(c) At equilibrium, P_{O_2} in air and water is equal. Low O_2 solubility means concentrations are not equal.

(d) When CO_2 is at equilibrium at the same partial pressure, more CO_2 dissolves.

P_{O_2} = 100 mm Hg

P_{O_2} = 0 mm Hg

P_{O_2} = 100 mm Hg
$[O_2]$ = 5.20 mmol/L

P_{O_2} = 100 mm Hg
$[O_2]$ = 0.15 mmol/L

P_{CO_2} = 100 mm Hg
$[CO_2]$ = 5.20 mmol/L

P_{CO_2} = 100 mm Hg
$[CO_2]$ = 3.00 mmol/L

● **Figure 18-2   *Gases in solution*** The amount of a gas that dissolves in solution depends on the solubility and partial pressure of the gas if temperature is assumed to be constant.

the gaseous phase has a P_{O_2} of 100 mm Hg, at equilibrium the water will also have a P_{O_2} of 100 mm Hg.

Note that this does *not* mean that the concentration of oxygen is the same in the air and in the water! The concentration of dissolved oxygen also depends on the *solubility* of oxygen. For example, when the P_{O_2} of both air and water is 100 mm Hg, the concentration of oxygen in air is 5.2 mmoles O_2/L air, but the concentration of oxygen in water is only 0.15 mmoles O_2/L water (Fig. 18-2c). As you can see, oxygen is not very soluble in aqueous solutions. Its low solubility is one reason for the evolution of oxygen-carrying molecules in blood.

Now compare oxygen solubility to that of carbon dioxide (Fig. 18-2d). Carbon dioxide is about 20 times as soluble in water as oxygen is. At a P_{CO_2} of 100 mm Hg, the CO_2 concentration in air is 5.2 mmoles CO_2/L air and its concentration in solution is 3.0 mmoles/L water. So although the P_{gas} in the air and water is the same for the two gases, the amount of each gas that dissolves in water is very different.

CONCEPT CHECK   Answers: p. 596

1. A saline solution is exposed to a gas mixture with equal partial pressures of H_2 and N_2. What information do you need to know to predict if equal amounts of H_2 and N_2 will dissolve in the saline?

2. If nitrogen is 78% of atmospheric air, what is the partial pressure of nitrogen (P_{N_2}) when the dry atmospheric pressure is 720 mm Hg?

3. True or false? Plasma with a P_{O_2} of 40 mm Hg and a P_{CO_2} of 40 mm Hg has the same concentrations of oxygen and CO_2.

▶ GAS EXCHANGE IN THE LUNGS AND TISSUES

The gas laws state that individual gases flow from regions of higher partial pressure to regions of lower partial pressure, and this rule governs the exchange of oxygen and carbon dioxide at the lungs and tissues.

Normal alveolar P_{O_2} is 100 mm Hg (Fig. 18-3a ●). The P_{O_2} of systemic venous blood arriving at the lungs is 40 mm Hg. Oxygen therefore moves down its partial pressure (concentration) gradient from the alveoli into the capillaries. Diffusion goes to equilibrium and the P_{O_2} of arterial blood leaving the lungs is the same as in the alveoli, 100 mm Hg.

When arterial blood reaches tissue capillaries, the gradient is reversed. Cells are continuously using oxygen for oxidative phosphorylation. In a cell at rest, intracellular P_{O_2} averages 40 mm Hg. Arterial blood entering the systemic capillaries has a P_{O_2} of 100 mm Hg. Because P_{O_2} is lower in the cells, oxygen diffuses down its gradient from plasma into cells. Once again, diffusion goes to equilibrium and venous blood has the same P_{O_2} as the cells it just passed.

Conversely, P_{CO_2} is higher in tissues than in systemic capillary blood because of CO_2 production during metabolism (Fig. 18-3b). Resting cell P_{CO_2} is about 46 mm Hg, compared to arterial plasma P_{CO_2} of 40 mm Hg. The gradient causes carbon dioxide to diffuse out of cells into the capillaries. Diffusion goes to equilibrium and systemic venous blood has an average P_{CO_2} of 46 mm Hg.

At the lungs, the process reverses. Venous blood bringing waste CO_2 from the cells has a P_{CO_2} of 46 mm Hg. Alveolar P_{CO_2} is 40 mm Hg. The partial pressure is higher in the plasma so CO_2 moves from the blood into the alveoli. By the time blood leaves the alveoli, it has a P_{CO_2} of 40 mm Hg, identical to the partial pressure of the alveoli (Table 18-1).

If the diffusion of gases between alveoli and blood is significantly impaired, **hypoxia**, or too little oxygen in the cells, results. Hypoxia frequently goes hand in hand with **hypercapnia**, elevated concentrations of carbon dioxide. These two conditions are clinical signs, not diseases, and the clinician must gather additional information to pinpoint their cause (Table 18-2).

There are three categories of problems that result in low arterial oxygen content: (1) inadequate oxygen reaching the alveoli, (2) problems with oxygen exchange between the alveoli and the blood, or (3) inadequate transport of oxygen in the blood. We will consider these issues in the sections that follow.

(a) Oxygen diffusion

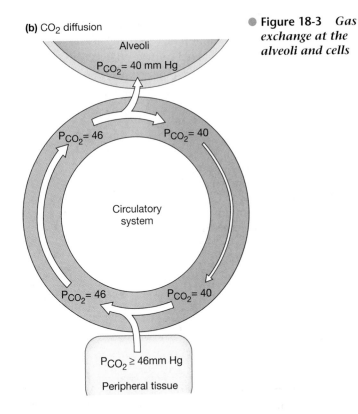

(b) CO_2 diffusion

● **Figure 18-3** *Gas exchange at the alveoli and cells*

Table 18-1: Normal Blood Values in Pulmonary Medicine

	ARTERIAL	VENOUS
P_{O_2}	95 mm Hg (85–100)	40 mm Hg
P_{CO_2}	40 mm Hg (35–45)	46 mm Hg
pH	7.4 (7.38–7.42)	7.37

Table 18-2: Causes of Hypoxia and Examples

- **Hypoxic hypoxia:** low arterial P_{O_2}
  - Physiological: high altitude
  - Alveolar hypoventilation
  - Decreased lung diffusion capacity
  - Abnormal ventilation-perfusion ratio

- **Anemic hypoxia:** decrease in total amount of O_2 bound to hemoglobin
  - Blood loss and anemia
  - Carbon monoxide prevents O_2 binding

- **Ischemic hypoxia:** due to reduced blood flow in tissue
  - General (heart failure), peripheral (shock), or to a single organ (coronary thrombosis)

- **Histotoxic hypoxia:** failure of cell to use O_2 due to poisoning
  - Cyanide and other metabolic poisons

4. Cellular metabolism review: match the items with the metabolic pathway(s) that are *directly* associated with them. (See Chapter 4 if you need help.)
   (a) O_2 consumption
   (b) CO_2 production
   1. glycolysis
   2. citric acid cycle
   3. electron transport system

5. Why doesn't the movement of oxygen from the alveoli into the plasma decrease the P_{O_2} of the alveoli?

Decrease in Alveolar P_{O_2} Decreases Oxygen Uptake at the Lungs

The first requirement for adequate oxygen delivery to the tissues is adequate oxygen intake, as reflected by the P_{O_2} of the alveoli. A decrease in alveolar P_{O_2} will result in less oxygen entering the blood. If alveolar P_{O_2} is low, there are two possible causes: (1) the composition of the inspired air is abnormal or (2) alveolar ventilation is inadequate.

The main factor that affects the oxygen content of inspired air is altitude. The partial pressure of oxygen in air decreases along with total atmospheric pressure as you move from sea level (760 mm Hg) to higher altitudes. For example, Denver, 1609 m above sea level, has a dry atmospheric pressure of about 628 mm Hg. The P_{O_2} of dry air in Denver is 132 mm Hg, down from 160 mm Hg at sea level. Unless a person is traveling, however, altitude remains constant. If alveolar P_{O_2} is low but the composition of inspired air is normal, then the problem lies with alveolar ventilation. Low alveolar ventilation is also known as *hypoventilation* and is characterized by less fresh air entering the alveoli. Pathological factors that cause alveolar hypoventilation include increased airway resistance (asthma; Fig. 18-4e ●), decreased lung compliance (fibrosis; Fig. 18-4c), and overdoses of drugs or alcohol that depress the central nervous system and slow the ventilation rate and depth.

CONCEPT CHECK
Answers: p. 596

6. At the summit of Mt. Everest, an altitude of 8850 m, atmospheric pressure is only 250 mm Hg. What is the P_{O_2} of dry atmospheric air on Everest? If water vapor added to inhaled air is 47 mm Hg, what is the P_{O_2} of the inhaled air when it reaches the alveoli?

RUNNING PROBLEM

Hypoxia is the primary problem that people experience when ascending to high altitude. High altitude is considered anything above 1500 m (5000 ft), but most pathological responses to altitude occur above 2500 m. By one estimate, 25% of people arriving at 2590 m will experience some form of altitude sickness.

Question 1: If water vapor contributes 47 mm Hg to fully humidified air, what is the P_{O_2} of inspired air reaching the alveoli at 2500 m (dry atmospheric pressure is 542 mm Hg)? How does this compare to an equivalent value at sea level?

    586 591

Changes in the Alveolar Membrane Alter Gas Exchange

In situations where the gas composition of the alveoli is normal but P_{O_2} of arterial blood leaving the lungs is low, the exchange process between alveoli and blood is defective. The transfer of gases between alveoli and blood requires diffusion across the barrier created by type I alveolar cells and the capillary endothelium (Fig. 18-5 ●).

Normally, the diffusion distance is small because the cells are thin and there is little or no interstitial fluid between the two cell layers. In addition, oxygen and CO_2 are both water- and lipid-soluble. Gas exchange in the lungs is rapid, blood flow through pulmonary capillaries is slow, and diffusion reaches equilibrium in less than one second.

● **Figure 18-4** *Pulmonary pathologies that affect alveolar ventilation and gas exchange*

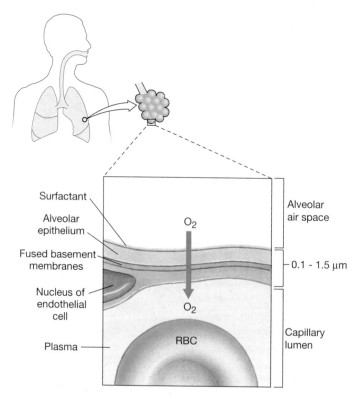

● **Figure 18-5** *Oxygen diffuses across the alveolar and endothelial cells to enter the plasma*

Pathological changes that adversely affect gas exchange include (1) a decrease in the amount of alveolar surface area that is available for gas exchange, (2) an increase in the thickness of the alveolar membrane, or (3) an increase in the diffusion distance between the alveoli and the blood.

Physical loss of alveolar surface area is demonstrated dramatically in *emphysema*, a degenerative lung disease most often caused by cigarette smoking. The irritating effect of smoke in the alveoli activates alveolar macrophages that release proteolytic enzymes. These enzymes destroy the elastic fibers of the lung, as discussed earlier, and kill cells, breaking down the walls of the alveoli. The result is a high-compliance/low-elastic recoil lung with fewer and larger alveoli and less surface area for gas exchange (Fig. 18-4b).

Pathological changes in the alveolar membrane that alter its properties will slow gas exchange. For example, in fibrotic lung diseases, deposition of scar tissue thickens the alveolar membrane (Fig. 18-4c). Diffusion of gases through this scar tissue is much slower than normal. However, the lungs have a built-in reserve capacity. As a result, one-third of the exchange epithelium must be incapacitated before arterial P_{O_2} falls significantly.

One situation in which the diffusion distance between alveoli and blood increases is **pulmonary edema**, characterized by excessive interstitial fluid volume in the lungs. Normally, only small amounts of interstitial fluid are found in the lungs, due to low pulmonary blood pressure and effective lymph drainage. But if pulmonary blood pressure rises for some reason such as left ventricular failure or mitral valve dysfunction, the normal filtration/reabsorption balance at the capillary is disrupted [◉ Fig. 15-18, p. 507].

When capillary hydrostatic pressure increases, more fluid filters out of the capillary. If filtration increases too much, the lym-

phatics are unable to remove all the fluid and excess accumulates in the pulmonary interstitial space, creating *pulmonary edema* (Fig. 18-4d). In severe cases, fluid even leaks across the alveolar membrane, collecting inside the alveoli.

Oxygen has low solubility in body fluids and takes longer to cross the increased diffusion distance, resulting in decreased arterial P_{O_2}. On the other hand, carbon dioxide is relatively soluble in body fluids, so the increased diffusion distance may not significantly affect carbon dioxide exchange. In some cases of pulmonary edema, arterial P_{O_2} is low but arterial P_{CO_2} is normal due to the solubility difference of the two gases.

▶ GAS TRANSPORT IN THE BLOOD

Now that we have described gas exchange in the lungs, we turn our attention to how oxygen and carbon dioxide are transported in the blood. The *law of mass action* [◉ p. 97] plays an important role in this process. Changes in oxygen or CO_2 concentration disturb the equilibrium of reactions, shifting the balance between substrates and products.

Hemoglobin Transports Most Oxygen to the Tissues

Oxygen is transported two ways in the blood: (1) dissolved in the plasma and (2) bound to hemoglobin (Hb). This fact can be summarized in the following statement:

Total blood oxygen content = amount dissolved in plasma
+ amount bound to hemoglobin

Hemoglobin (Hb), the oxygen-binding protein in red blood cells [❀ p. 532], binds reversibly to oxygen, as summarized in the following equation:

$$Hb + O_2 \rightleftharpoons HbO_2$$

In the lungs, where oxygen concentrations are high, hemoglobin binds oxygen. At the cells, where oxygen is being used and concentrations are lower, hemoglobin gives up its oxygen. Because oxygen is only slightly soluble in aqueous solutions, we must have adequate amounts of hemoglobin in our blood to survive.

The importance of hemoglobin to oxygen transport is summarized in Figure 18-6●. More than 98% of the oxygen in a given volume of blood is transported inside red blood cells bound to hemoglobin. The balance remains dissolved in plasma.

To understand just how low the solubility of oxygen is, let's look at the following example. In a situation where blood has no hemoglobin, only 3 mL of O_2 will dissolve in the plasma fraction of 1 liter of arterial blood (Fig. 18-7a●). Thus, with a typical cardiac output of 5 L blood/min, about 15 mL of dissolved oxygen reaches the systemic tissues each minute.

● **Figure 18-6** *Summary of oxygen transport in the blood*

(a) Oxygen transport in blood without hemoglobin. Alveolar P_{O_2} = arterial P_{O_2}

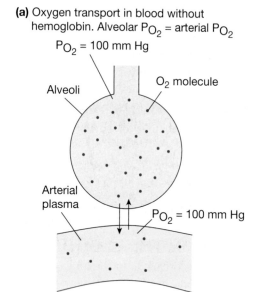

Oxygen dissolves in plasma.

O_2 content of plasma = 3 mL O_2/L blood	
O_2 content of red blood cells = 0	
Total O_2 carrying capacity	3 mL O_2/L blood

(b) Oxygen transport at normal P_{O_2} in blood with hemoglobin

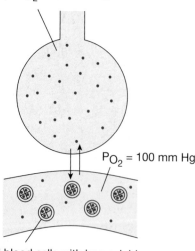

Red blood cells with hemoglobin are carrying 98% of their maximum load of oxygen

O_2 content of plasma = 3 mL O_2/L blood	
O_2 content of red blood cells = 197 mL O_2/L blood	
Total O_2 carrying capacity	200 mL O_2/L blood

(c) Oxygen transport at reduced P_{O_2} in blood with hemoglobin

Red blood cells carrying 50% of their maximum load of oxygen

O_2 content of plasma = 0.8 mL O_2/L blood	
O_2 content of red blood cells = 99.5 mL O_2/L blood	
Total O_2 carrying capacity	100.3 mL O_2/L blood

● **Figure 18-7** *The role of hemoglobin in oxygen transport*

This small amount cannot begin to meet the needs of the tissues, however. Oxygen consumption at rest is about 250 mL O_2 consumed/min, and that figure increases dramatically with exercise. The oxygen brought to cells dissolved in plasma provides less than 10% of what the cells consume. Thus, the body is heavily dependent on oxygen carried by hemoglobin.

At normal hemoglobin levels, oxygen content of red blood cells is about 197 mL/L blood (Fig. 18-7b). Thus:

Total blood oxygen content =

$$3 \text{ mL plasma } O_2 + 197 \text{ mL HbO}_2 = 200 \text{ mL } O_2/\text{L blood}$$

If the cardiac output remains 5 L/min, oxygen delivery to cells jumps to almost 1000 mL/min, nearly four times the oxygen consumption of the tissues at rest.

The amount of oxygen that binds to hemoglobin depends on two factors: the P_{O_2} of the plasma surrounding the red blood cells and the number of potential oxygen-binding sites available within the red blood cells.

1. *The P_{O_2} of plasma* is the primary factor determining how many of the available hemoglobin binding sites are occupied by oxygen. Figure 18-7c shows what happens to oxygen transport when alveolar and arterial P_{O_2} decrease. As you learned in previous sections, arterial P_{O_2} is established by the (1) composition of inspired air, (2) the alveolar ventilation rate, and (3) the efficiency of gas exchange between lung and blood.

2. *The total number of binding sites* depends on the number of hemoglobin molecules in the blood. Clinically, this number can be estimated either by counting the red blood cells and quantifying the amount of hemoglobin per red blood cell (*mean corpuscular hemoglobin*) or by simply determining the blood hemoglobin content (g Hb/dL whole blood). Any pathology that decreases the amount of hemoglobin in the cells or the number of red blood cells will adversely affect the blood's oxygen-transporting capacity.

People who have lost large amounts of blood need to replace hemoglobin for oxygen transport. A blood transfusion is the ideal replacement for blood loss, but in emergencies this is not always possible. Saline infusions can replace lost blood volume, but saline, like plasma, cannot transport sufficient quantities of oxygen to support cellular respiration.

Researchers are currently testing artificial oxygen carriers to replace hemoglobin. In times of large-scale disasters, these hemoglobin substitutes will eliminate the need to establish blood type before giving transfusions for blood loss.

Each Hemoglobin Molecule Binds Up to Four Oxygen Molecules

Why is hemoglobin an effective oxygen carrier? The answer lies in its molecular structure. One hemoglobin molecule is composed of four globin subunits, each centered around a heme group whose central iron atom binds reversibly with an oxygen molecule [Fig. 16-7, p. 532]. Because there are four iron atoms

Blood Substitutes

Physiologists have been attempting to find a substitute for blood ever since 1878, when an intrepid physician named T. Gaillard Thomas transfused a patient with whole milk in place of blood. Although milk seems an unlikely replacement for blood, it has two important properties: proteins to provide colloid osmotic pressure and molecules (emulsified lipids) capable of binding to oxygen. In the development of hemoglobin substitutes, oxygen transport is the most difficult property to mimic. A hemoglobin solution would seem to be the obvious answer, but hemoglobin that is not compartmentalized in red blood cells behaves differently than hemoglobin that is. Investigators are making progress by polymerizing hemoglobin into larger, more stable molecules and loading these hemoglobin polymers into phospholipid liposomes [p. 128]. Perfluorocarbon emulsions are also being tested as oxygen carriers. To learn more about this research, read "Physiological properties of blood substitutes," in *News in Physiological Sciences* 16:38–41, 2001 Feb. (*http://nips.physiology.org*).

per hemoglobin, one hemoglobin molecule has the potential to bind four oxygen molecules. The iron-oxygen interaction is a weak bond, and the Hb-O_2 link can be easily broken without altering either the hemoglobin or the oxygen.

Hemoglobin (Hb) bound to oxygen is known as **oxyhemoglobin**, abbreviated HbO$_2$. (It would be more accurate to show the actual number of oxygen molecules carried on each hemoglobin molecule: $Hb(O_2)_{1-4}$, but we will use the simpler abbreviation because the number bound varies from molecule to molecule.)

Oxygen-hemoglobin binding obeys the *law of mass action*:

$$Hb + O_2 \rightleftharpoons HbO_2$$

If the oxygen concentration increases (left side of the equation above), more oxygen binds to hemoglobin and the balance of the equation shifts to the right. If the concentration of oxygen on the left side of the equation *decreases*, the equation shifts to the left as hemoglobin releases some of its bound oxygen.

Because of the law of mass action, the amount of oxygen bound to hemoglobin depends primarily on the P_{O_2} of the surrounding plasma (Fig. 18-6). In the pulmonary capillaries, dissolved O_2 in plasma diffuses into red blood cells, where it binds to hemoglobin. This removes dissolved O_2 from the plasma, so more oxygen diffuses in from the alveoli. The transfer of oxygen from air to plasma to red blood cells and onto hemoglobin occurs so rapidly that blood leaving the alveoli normally picks up as much oxygen as the P_{O_2} of the plasma and number of red blood cells permit.

Once arterial blood reaches the tissues, the exchange process that took place in the lungs reverses. Dissolved oxygen diffuses out of plasma into the cells and the resultant decrease in plasma P_{O_2} disturbs the equilibrium of the oxygen-hemoglobin binding reaction by removing oxygen from the left side of the equation. The reaction shifts to the left according to the law of mass action, and the hemoglobin molecules release their oxygen stores.

Like oxygen loading at the lungs, this process takes place very rapidly and goes to equilibrium. The P_{O_2} of the cell determines how much oxygen unloads from hemoglobin. As cells increase their metabolic activity, their P_{O_2} decreases and hemoglobin releases more oxygen to them.

Oxygen Binding Is Expressed as a Percentage

The amount of oxygen bound to hemoglobin at any given P_{O_2} is expressed as a percentage: (Actual amount of O_2 bound/maximum that could be bound) $\times$ 100 = percent saturation of hemoglobin.

The **percent saturation of hemoglobin** refers to the percentage of available binding sites that are bound to oxygen. If all binding sites of all hemoglobin molecules are occupied by oxygen molecules, the blood is 100% oxygenated, or *saturated* with oxygen. If half the available binding sites are carrying oxygen, the hemoglobin is 50% saturated, and so on.

The relationship between plasma P_{O_2} and oxygen-hemoglobin binding can be explained with the following analogy. The hemoglobin molecules carrying oxygen are like students moving books from an old library to a new one. Each student (a hemoglobin molecule) can carry a maximum of four books (100% saturation). The librarian in charge controls how many books (O_2 molecules) each student will carry, just as P_{O_2} determines the amount of oxygen that binds to hemoglobin.

At the same time, the total number of books being carried depends on the number of available students, just as the amount of oxygen delivered to the tissues depends on the number of available hemoglobin molecules. For example, if there are 100 students and the librarian gives each of them four books, then 400 books will be carried to the new library.

If the librarian has only three books for each (decreased P_{O_2}), then only 300 books will go to the new library, even though each student can carry four (decreased percent saturation of hemoglobin).

If the librarian is handing out four books per student but only 50 students show up (fewer hemoglobin molecules), then only 200 books will get to the new library, even though the students will be carrying the maximum number of books they can carry.

▶ P_{O_2} DETERMINES HEMOGLOBIN BINDING OF OXYGEN

The physical relationship between P_{O_2} and oxygen binding to hemoglobin can be studied *in vitro* in the laboratory. Researchers expose samples of hemoglobin to varying P_{O_2} levels and quantitatively determine the amount of oxygen that binds. **Oxyhemoglobin** (HbO_2) **dissociation curves**, such as the one shown in Figure 18-8 ●, are the result of these *in vitro* binding studies.

The shape of the HbO_2 dissociation curve reflects the properties of the hemoglobin molecule and its affinity for oxygen. If you look at the curve, you find that at normal alveolar and arterial P_{O_2} (100 mm Hg), 98% of the hemoglobin is bound to oxygen. In other words, as blood passes through the lungs, hemoglobin picks up nearly the maximum amount of oxygen that it can carry.

Notice that the curve is nearly flat at higher P_{O_2} levels (i.e., slope approaches zero). Above 100 mm Hg, large changes in P_{O_2}

GRAPH QUESTIONS

(a) When the P_{O_2} is 20 mm Hg, what is the percent O_2 saturation of hemoglobin?

(b) At what P_{O_2} is hemoglobin 50% saturated with O_2?

● **Figure 18-8** *Oxygen-hemoglobin dissociation curve*

cause only minor changes in the percent saturation. Hemoglobin will not be 100% saturated until the P$_{O_2}$ reaches nearly 650 mm Hg, a partial pressure far higher than anything we encounter in everyday life.

The flattening of the dissociation curve at higher P$_{O_2}$ also means that alveolar P$_{O_2}$ can fall a good bit below 100 mm Hg without having a significant effect on hemoglobin saturation. As long as P$_{O_2}$ in the pulmonary capillaries stays above 60 mm Hg, hemoglobin will be more than 90% saturated and maintain near-normal levels of oxygen transport. However, once the P$_{O_2}$ falls below 60 mm Hg, the curve becomes steeper. The steep slope means that a small decrease in P$_{O_2}$ causes a relatively large release of oxygen.

For example, if P$_{O_2}$ falls from 100 mm Hg to 60 mm Hg, the percent saturation of hemoglobin goes from 98% saturation to about 88% saturation, a decrease of 10%. This is equivalent to a saturation change of 2.5% for each 10 mm Hg change.

If P$_{O_2}$ falls further, from 60 to 40 mm Hg, the percent saturation goes from 88% to 75%, a decrease of 6.5% for each 10 mm Hg. In the 40–20 mm Hg range, the curve is even steeper. Hemoglobin saturation declines from 75% to 35%, a change of 20% for each 10 mm Hg.

What is the physiological significance of the HbO$_2$ curve's shape? At a P$_{O_2}$ of 40 mm Hg, the partial pressure of resting cells, hemoglobin is still 75% saturated and has released only one-fourth of the oxygen it is capable of carrying. The 75% of the oxygen that remains bound serves as a reservoir that cells can draw upon if metabolism increases.

When metabolically active tissues use additional oxygen, their P$_{O_2}$ decreases and additional oxygen is released by hemoglobin. At a P$_{O_2}$ of 20 mm Hg, an average value for exercising muscle, hemoglobin saturation falls to about 35%. With this 20 mm Hg decrease in P$_{O_2}$ (40 mm Hg to 20 mm Hg), hemoglobin releases an additional 40% of the oxygen it is capable of carrying. This is another example of the built-in reserve capacity of the body.

9. Can a person breathing 100% oxygen at sea level achieve 100% saturation of her hemoglobin?

10. What effect does hyperventilation have on the percent saturation of arterial hemoglobin? (Hint: Fig. 17-15)

Temperature, pH, and Metabolites Affect Oxygen-Hemoglobin Binding

Any factor that changes the conformation of the hemoglobin protein may affect its ability to bind oxygen. In humans, physiological changes in pH, P$_{CO_2}$, and temperature of the blood all alter the oxygen-binding capability of hemoglobin. Changes in binding affinity are reflected by changes in the shape of the HbO$_2$ dissociation curve.

Increased temperature and P$_{CO_2}$ or decreased pH decrease the affinity of hemoglobin for oxygen and shift the oxygen-hemoglobin dissociation curve to the right (Fig. 18-9 ●). When these factors change in the opposite direction, binding affinity increases and the HbO$_2$ curve shifts to the left.

Notice that when the curve shifts in either direction, the changes are much more pronounced in the steep part of the curve. This means that oxygen binding at the lungs is not greatly affected, but oxygen delivery at the tissues will be significantly altered.

Let's examine one specific example, the shift that takes place when pH goes from 7.4 (normal) to 7.2 (more acidic). A

FIGURE QUESTIONS

(a) At a P$_{O_2}$ of 20mm Hg, how much more oxygen is released at an exercising muscle cell whose pH is 7.2 than by a cell with a pH of 7.4?
(b) What happens to oxygen release when the exercising muscle cell warms up?

● **Figure 18-9** *Physical factors alter oxygen binding to hemoglobin*

(a) Effect of pH

(b) Effect of temperature

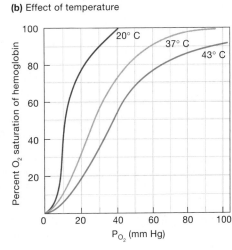

(c) Effect of P$_{CO_2}$

pH of 7.2 is lower than normal for the body (normal range 7.38–7.42) but is compatible with life. At a P_{O_2} of 40 mm Hg and pH of 7.2, hemoglobin molecules release 15% more oxygen than they do at a pH of 7.4.

When does the body undergo shifts in pH? Anaerobic exercise is one example of a situation in which body pH decreases. Anaerobic metabolism in exercising muscle fibers produces lactic acid, which in turn releases H^+ into the cytoplasm and extracellular fluid. As H^+ concentrations increase, pH falls, the affinity of hemoglobin for oxygen decreases, and the HbO_2 dissociation curve shifts to the right. More oxygen is released at the tissues as the blood becomes more acidic (pH decreases). A shift in the hemoglobin saturation curve that results from a change in pH is called the **Bohr effect**.

An additional factor that affects oxygen-hemoglobin binding is **2,3-diphosphoglycerate** (2,3-DPG; also called *2,3-bisphosphoglycerate* or 2,3-BPG), a compound made from an intermediate of the glycolysis pathway. **Chronic hypoxia** (extended periods of low oxygen) triggers an increase in 2,3-DPG production in red blood cells. Like the other factors discussed, 2,3-DPG lowers the binding affinity of hemoglobin and shifts the HbO_2 dissociation curve to the right (Fig. 18-10●). Ascent to high altitude and anemia are two situations that increase 2,3-DPG production.

Changes in hemoglobin structure also will change oxygen binding. For example, fetal hemoglobin has a unique protein chain for two of its subunits. This change enhances the ability of fetal hemoglobin to bind oxygen in the low-oxygen environment of the placenta. The altered binding affinity is reflected by the different shape of the fetal HbO_2 dissociation curve (Fig. 18-11●).

Figure 18-12● summarizes all the factors that influence oxygen transport in the blood.

11. A muscle that is actively contracting may have an intracellular P_{O_2} of 25 mm Hg. What happens to oxygen binding to hemoglobin at this P_{O_2}? What is the P_{O_2} of the venous blood leaving the active muscle?

Carbon Dioxide Is Transported Three Ways

Gas transport in the blood is a two-way process because the removal of carbon dioxide from the body is as essential as the uptake of oxygen. The hemoglobin molecule, as we will see, is as important in CO_2 removal as it was in the transport of oxygen.

Why is removing CO_2 so important? The reason is that elevated P_{CO_2} (*hypercapnia*) causes the pH disturbance known as *acidosis* (see Chapter 19). Abnormally high P_{CO_2} levels also depress central nervous system function, causing confusion, coma, or even death. Thus, CO_2 is a potentially toxic waste product that must be removed by the lungs.

GRAPH QUESTIONS

(a) Because of incomplete gas exchange across the thick membranes of the placenta, hemoglobin in fetal blood leaving the placenta is 80% saturated with oxygen. What is the P_{O_2} of that placental blood?
(b) Blood in the vena cava of the fetus has a P_{O_2} around 10 mm Hg. What is the percent O_2 saturation of maternal hemoglobin at the same P_{O_2}?

GRAPH QUESTION

Blood stored in blood banks loses its normal content of 2,3-DPG. Is this good or bad? Explain.

● **Figure 18-10**   *2,3-DPG alters hemoglobin affinity for oxygen*

● **Figure 18-11**   *Differences in the oxygen-binding properties of maternal and fetal hemoglobin*

● **Figure 18-12** *Factors contributing to the total oxygen content of arterial blood*

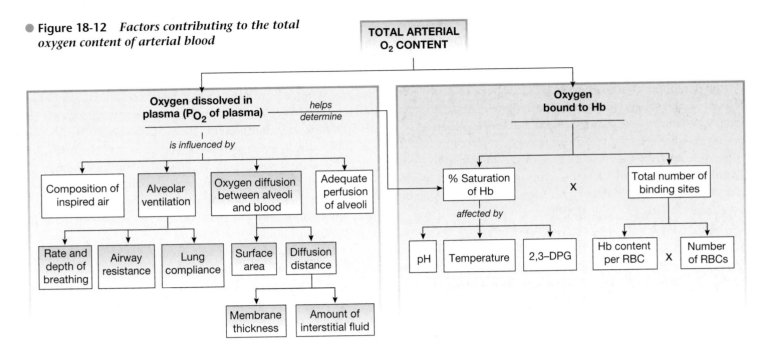

Carbon dioxide is produced as a by-product of cellular respiration [● p. 105]. CO$_2$ is more soluble in body fluids than oxygen, but the cells produce far more CO$_2$ than can dissolve in the plasma. Only about 7% of carbon dioxide carried by venous blood is in the form of dissolved CO$_2$. The remaining 93% diffuses into red blood cells, where 70% is converted to bicarbonate ion, as explained below, and 23% binds to hemoglobin (Hb·CO$_2$). Figure 18-13 ● summarizes these three methods of carbon dioxide transport in the blood.

CO$_2$ and Bicarbonate About 70% of the CO$_2$ that enters the blood is transported to the lungs as bicarbonate ions (HCO$_3^-$) dissolved in the plasma. The conversion of carbon dioxide to

● **Figure 18-13** *Carbon dioxide transport in the blood* CA = carbonic anhydrase

bicarbonate serves two purposes: (1) it provides an additional means by which carbon dioxide can be transported from cells to lungs, and (2) the bicarbonate is available to act as a buffer for metabolic acids [❀ p. 34], thereby helping stabilize the body's pH.

But how does CO_2 turn into bicarbonate? The rapid conversion depends on the presence of **carbonic anhydrase (CA)**, an enzyme found concentrated in red blood cells. Let's see how this happens.

Dissolved CO_2 in plasma diffuses into red blood cells, where it reacts with water in the presence of carbonic anhydrase to form *carbonic acid*. Carbonic acid then dissociates into a hydrogen ion and a bicarbonate ion. The equation for the reaction is written as follows:

$$CO_2 + H_2O \underset{\text{carbonic acid}}{\overset{\text{carbonic anhydrase}}{\rightleftharpoons}} H_2CO_3 \rightleftharpoons H^+ + HCO_3^-$$

Because of the dissociation of carbonic acid, we sometimes ignore the intermediate step and summarize the reaction as:

$$CO_2 + H_2O \rightleftharpoons H^+ + HCO_3^-$$

This reaction is reversible. The rate in either direction depends on the relative concentrations of the substrates and obeys the law of mass action.

When CO_2 moves into red blood cells, carbonic anhydrase converts it into HCO_3^- and H^+ (Fig. 18-13). This reaction will continue until an equilibrium state is reached. (Water is always in excess in the body, so water concentration plays no role in the dynamic equilibrium of this reaction.) To keep the reaction going, the products, H^+ and HCO_3^-, must be removed from the cytoplasm of the red blood cell. If the product concentrations are kept low, the reaction cannot reach equilibrium.

Two separate steps remove free H^+ and HCO_3^-. In the first step, bicarbonate ions produced from CO_2 leave the red blood cell on an antiport protein [❀ p. 138]. This transport process, known as the **chloride shift**, exchanges one HCO_3^- for one Cl^-. The one-for-one exchange maintains electrical neutrality so that the cell's membrane potential is not affected.

The transfer of HCO_3^- into the plasma makes additional buffer available to moderate pH changes caused by production of metabolic acids. Bicarbonate is the most important extracellular buffer in the body.

Hemoglobin and H⁺   The second step removes free H^+ from the red blood cell cytoplasm when hemoglobin acts as a buffer and binds hydrogen ions ($Hb \cdot H$). Hemoglobin's buffering of H^+ is an important step that prevents large changes in the body's pH.

If blood P_{CO_2} is elevated much above normal, the hemoglobin buffer cannot soak up all H^+ produced from the reaction of CO_2 and water. In those cases, excess H^+ accumulates in the plasma, causing the condition known as **respiratory acidosis**. Further information on the role of the respiratory system in maintaining pH homeostasis is found in Chapter 20.

As H^+ and HCO_3^- disappear from the red blood cell cytoplasm, the reaction continues and more CO_2 converts to carbonic acid. This CO_2 in turn is replaced by CO_2 that diffuses into the red blood cell from the plasma. Then, as plasma P_{O_2} decreases, additional CO_2 diffuses out of cells.

Hemoglobin and CO₂   Although most carbon dioxide that enters the red blood cells is converted to bicarbonate ions, about 23% of the CO_2 in venous blood binds directly to hemoglobin. When oxygen leaves its binding sites on the hemoglobin molecule, CO_2 binds with free hemoglobin at exposed amino groups (NH_2), forming **carbaminohemoglobin**. This reaction can be summarized as

$$CO_2 + Hb \rightleftharpoons Hb \cdot CO_2 \text{ carbaminohemoglobin}$$

The formation of carbaminohemoglobin is facilitated by the presence of both CO_2 and H^+. Both factors decrease hemoglobin's binding affinity for oxygen (see Fig. 18-9).

CO₂ Removal at the Lungs   When venous blood reaches the lungs, the processes that took place in the systemic capillaries reverse (Fig. 18-13). The P_{CO_2} of alveoli is lower than that of venous blood in pulmonary capillaries. Carbon dioxide diffuses out of plasma into the alveoli, and the plasma P_{CO_2} begins to fall.

The decrease in plasma P_{CO_2} allows dissolved CO_2 to diffuse out of red blood cells. As CO_2 levels in the red blood cells decrease, the equilibrium of the CO_2-bicarbonate reaction is disturbed.

Removal of CO_2 causes H^+ to leave the hemoglobin molecules. The chloride shift reverses: Cl^- returns to the plasma in exchange for HCO_3^- that moves back into the red blood cells. The HCO_3^- and H^+ re-form into carbonic acid that is converted into water and CO_2. The CO_2 is then free to diffuse out of the red blood cell and into the alveoli.

In summary, Figure 18-14● shows the combined transport of CO_2 and O_2 in the blood. At the alveoli, O_2 diffuses down its pressure gradient into the plasma and, from there, into the red blood cells. Hemoglobin binds O_2, increasing the amount of oxygen that can be transported to the cells.

At the cells, the process reverses. P_{O_2} in the cells is lower than that in the arterial blood, so O_2 diffuses from the plasma to the

---

RUNNING PROBLEM

The usual homeostatic response to high-altitude hypoxia is hyperventilation that begins upon arrival. Hyperventilation enhances alveolar ventilation, but this may not help arterial P_{O_2} significantly when atmospheric P_{O_2} is low. However, hyperventilation does lower plasma P_{CO_2}.

Question 6: What happens to plasma pH during hyperventilation? (Hint: apply the law of mass action to figure out what happens to the balance between CO_2 and $H^+ + HCO_3^-$).

Question 7: How does this change in pH affect oxygen binding at the lungs when P_{O_2} is decreased? How does it affect unloading of oxygen at the cells?

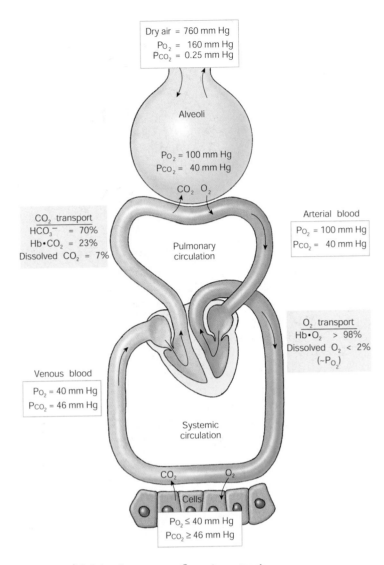

▶ REGULATION OF VENTILATION

Breathing is a rhythmic process that occurs without conscious thought. In that respect, it resembles the rhythmic beating of the heart. However, skeletal muscles, unlike autorhythmic cardiac muscles, are not able to contract spontaneously. Instead, skeletal muscle contraction must be initiated by somatic motor neurons, which in turn are controlled by the central nervous system.

In the respiratory system, contraction of the diaphragm and intercostals is initiated by groups of neurons in the brain stem (Fig. 18-15 ●). These neurons take the form of a network with a **central pattern generator** that has intrinsic rhythmic activity [● p. 437]. Research suggests the rhythmic activity arises from pacemaker neurons with unstable membrane potentials.

Direct study of the brain centers controlling ventilation is difficult because of the complexity of the neuronal network and its anatomical location. Consequently, some of our understanding of the control of ventilation has come from observing patients with brain damage. Other information has come from animal experiments in which the neural connections between major parts of the brain stem were severed.

From these observations and hypotheses, the following model for the control of ventilation has been proposed. Some parts of the model are well supported with experimental evidence. Other aspects are still under investigation. The model states that:

1. Respiratory neurons in the medulla control inspiration and expiration.
2. Neurons in the pons modulate ventilation.
3. The rhythmic pattern of breathing arises from a network of spontaneously discharging neurons.
4. Ventilation is subject to modulation by various chemical factors and by higher brain centers.

Neurons in the Medulla Control Breathing

Classic descriptions of the brain's respiratory control of ventilation divided groups of neurons into various control centers. The most recent descriptions have become less specific about assigning function to particular centers and now simply refer to the network of neurons in the brain stem as the central pattern generator.

The central pattern generator is unique in that it functions automatically throughout a person's life, and yet it can be controlled voluntarily, up to a point. Complicated synaptic interactions between neurons in the network create the rhythmic cycles of inspiration and expiration, influenced continuously by sensory input from receptors for CO_2, O_2, and H^+. Ventilation pattern depends in large part on the levels of those three substances in the blood.

Although there is still much to be learned about the central pattern generator, we do know that respiratory neurons are found concentrated in two nuclei in the medulla oblongata. The **dorsal respiratory group** (DRG) contains mostly *inspiratory neurons*

● **Figure 18-14**   *Summary of gas transport*

cells. The decrease in plasma P_{O_2} causes hemoglobin to release O_2, giving up additional oxygen to enter the cells.

Carbon dioxide from aerobic metabolism simultaneously enters the blood, dissolving in the plasma. From there, it enters red blood cells, where most is converted to HCO_3^- and H^+. The HCO_3^- is returned to the plasma in exchange for a Cl^-, while the H^+ binds to hemoglobin. A fraction of the CO_2 also binds directly to hemoglobin. At the lungs, the process reverses as CO_2 diffuses out into the alveoli.

In order to understand fully how the respiratory system coordinates delivery of oxygen to the lungs with its transport in the circulation, we will now consider the central nervous system control of ventilation.

CONCEPT CHECK   Answers: p. 596

12. How would an obstruction of the airways affect alveolar ventilation, arterial P_{CO_2}, and the body's pH?

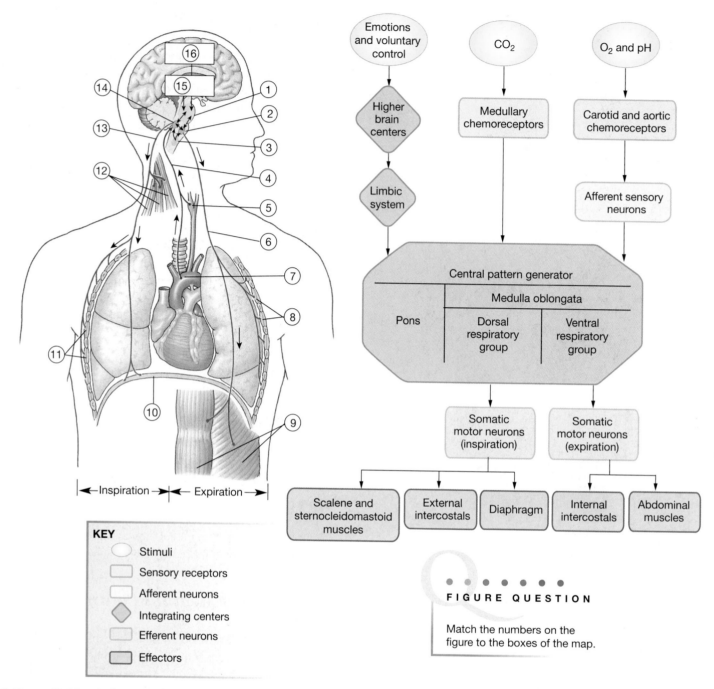

KEY
- ⬭ Stimuli
- ▭ Sensory receptors
- ▢ Afferent neurons
- ◇ Integrating centers
- ▭ Efferent neurons
- ▭ Effectors

FIGURE QUESTION

Match the numbers on the figure to the boxes of the map.

● **Figure 18-15** *Reflex control of ventilation* Chemoreceptors monitor blood gases and pH. Control centers in the brain stem regulate activity in somatic motor neurons leading to respiratory muscles.

(I neurons) that control the external intercostal muscles and the diaphragm (Fig. 18-15). The **ventral respiratory group** (VRG) contains neurons that control the muscles used for active expiration (E neurons) and for greater-than-normal inspiration (I+ neurons), such as occurs during vigorous exercise.

During quiet respiration, the inspiratory neurons of the dorsal respiratory group gradually increase stimulation of the inspiratory muscles for 2 seconds. This increase is sometimes called ramping because of the shape of the graph of inspiratory neuron activity (Fig. 18-16 ●).

A few inspiratory neurons fire to begin the ramp. The firing of these neurons recruits other inspiratory neurons to fire in an apparent positive feedback loop. As more neurons fire, more skeletal muscle fibers are recruited. The rib cage expands smoothly as the diaphragm contracts.

At the end of 2 seconds, the inspiratory neurons abruptly stop firing and the respiratory muscles relax. Over the next 3 seconds, passive expiration occurs because of elastic recoil of the inspiratory muscles and the elastic lung tissue. However, there is some motor neuron activity during passive expiration, suggesting

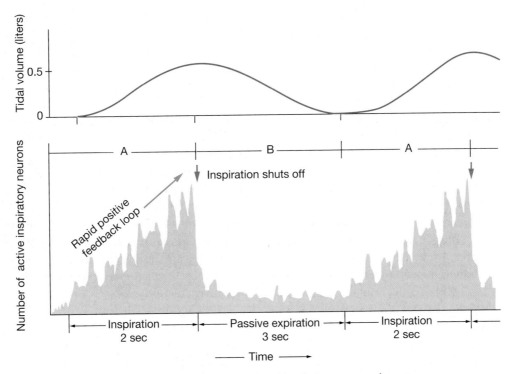

During inspiration, the acitvity of inspiratory neurons increases
steadily, apparently through a positive feedback mechanism.
At the end of inspiration, the activity shuts off abruptly and
expiration takes place through recoil of elastic lung tissue.

● **Figure 18-16** *Rhythmic breathing*

that perhaps muscles in the upper airways contract to slow the
flow of air out of the respiratory system.

The expiratory neurons and I+ neurons of the ventral respiratory group remain mostly inactive during quiet respiration.
They function primarily during forced breathing, when inspiratory movements are exaggerated, or during active expiration.

In forced breathing, the increased activity of DRG inspiratory
neurons in some way activates the I+ neurons of the ventral respiratory group. Those I+ neurons in turn stimulate accessory inspiratory muscles such as the sternocleidomastoids. Contraction
of the accessory inspiratory muscles enhances expansion of the
thorax by raising the sternum and upper ribs.

In active expiration, expiratory (E) neurons from the ventral
respiratory group activate the internal intercostal and abdominal
muscles. There seems to be reciprocal inhibition between the inspiratory and expiratory neurons. The I neurons inhibit the E neurons
during inspiration, and the E neurons inhibit I neurons during active expiration. In addition, the cycling of respiratory activity is
subject to continuous modulation as CO_2 in the blood changes.

Carbon Dioxide, Oxygen, and pH Influence Ventilation

Sensory input from two sets of chemoreceptors modifies the rhythmicity of the central pattern generator. Carbon dioxide is the primary stimulus for changes in ventilation. Oxygen and plasma pH
play lesser roles.

The chemoreceptors for oxygen and carbon dioxide are strate-
gically associated with the arterial circulation. If too little oxygen
is present in the arterial blood destined for the tissues, the rate and
depth of breathing increase. Or, if the rate of carbon dioxide production by the cells exceeds the rate of carbon dioxide removal by
the lungs, ventilation is intensified to match carbon dioxide removal to production. These homeostatic reflexes operate constantly, keeping arterial P_{O_2} and P_{CO_2} within a narrow range.

Peripheral chemoreceptors located in the *carotid* and *aortic
bodies* sense changes in the oxygen concentration, pH, and P_{CO_2}
of the plasma (Fig. 18-15). (These are the same carotid and aortic
bodies that contain the baroreceptors involved in reflex control
of blood pressure [● p. 510].)

Central chemoreceptors monitor cerebrospinal fluid (CSF)
composition and respond to changes in the concentration of CO_2
in the cerebrospinal fluid. These receptors lie on the ventral surface
of the medulla, close to neurons involved in respiratory control.

The Carotid and Aortic Bodies The carotid and aortic
chemoreceptors are sensitive to the P_{O_2}, P_{CO_2}, and pH of arterial
blood. When specialized **glomus cells** [*glomus*, a ball-shaped
mass] are activated by decreases in P_{O_2} or pH, or an increase in
P_{CO_2}, they trigger a reflex increase in ventilation.

How do the glomus cells sense oxygen, CO_2, or H^+ and activate their associated sensory neurons? Some of the details remain
to be worked out, but the basic pattern is similar to one you encountered for insulin release by pancreatic beta cells [● p. 162] or

taste transduction in taste buds [p. 339]. In all three examples, the stimulus inactivates K^+ channels, causing the receptor cell to depolarize (Fig. 18-17●). Depolarization opens voltage-gated Ca^{2+} channels, and Ca^{2+} entry causes exocytosis of neurotransmitter (probably dopamine) onto the sensory neuron. Dopamine initiates action potentials in sensory neurons leading to the brain stem respiratory centers, signaling them to increase ventilation.

Under most circumstances, oxygen is not an important factor in modulating ventilation. Arterial P_{O_2} must be less than 60 mm Hg before ventilation is stimulated. This decrease in P_{O_2} is equivalent to ascending to an altitude of 3000 m. (For reference, Denver is located at an altitude of 1609 m.)

Because the peripheral chemoreceptors respond only to dramatic changes in arterial P_{O_2}, they do not play a role in the everyday regulation of ventilation. However, unusual physiological conditions (such as ascending to high altitude) and pathological conditions such as chronic obstructive pulmonary disease (COPD) can reduce arterial P_{O_2} to low levels that activate the peripheral chemoreceptors.

The peripheral chemoreceptors in the carotid and aortic bodies are more responsive to changes in P_{O_2} than to changes in plasma pH and P_{CO_2}. However, any condition that reduces plasma pH or increases P_{CO_2} stimulates ventilation by way of peripheral chemoreceptors.

Central Chemoreceptors The most important chemical controller of ventilation is carbon dioxide, mediated through central chemoreceptors located in the medulla (Fig. 18-18●). These receptors set the respiratory pace, providing continuous input into the central pattern generator.

When arterial P_{CO_2} increases, carbon dioxide crosses the blood-brain barrier quite rapidly and activates the central chemoreceptors. These receptors signal the central pattern generator to increase the rate and depth of ventilation, thereby increasing alveolar ventilation and removing carbon dioxide from the blood (Fig. 18-19●).

Although we say that the central chemoreceptors respond to carbon dioxide, they actually respond to pH changes in the cerebrospinal fluid. Carbon dioxide that diffuses across the blood-brain barrier into the cerebrospinal fluid is converted to bicarbonate and H^+:

$$CO_2 + H_2O \rightleftharpoons H_2CO_3 \rightleftharpoons H^+ + HCO_3^-$$

Experiments indicate that the H^+ produced by this reaction actually initiates the chemoreceptor reflex.

Note, however, that pH changes in the plasma *do not* influence the central chemoreceptor directly. Free H^+ in the plasma is unable to cross the blood-brain barrier and therefore has no direct effect on the central chemoreceptors.

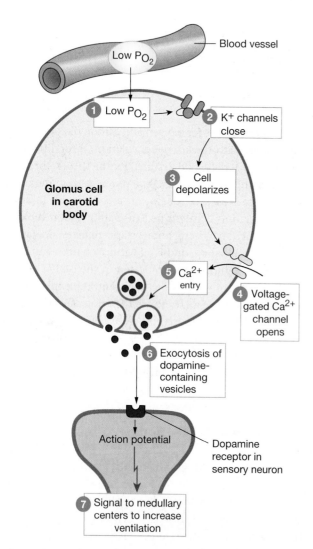

● **Figure 18-17** *Carotid body oxygen sensor releases neurotransmitter when* P_{O_2} *decreases*

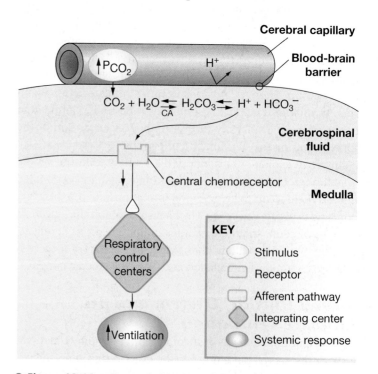

● **Figure 18-18** *Central chemoreceptor monitors* CO_2 *in cerebrospinal fluid* CA = carbonic anhydrase

● **Figure 18-19** *Chemoreceptor response to increased P_{CO_2}*

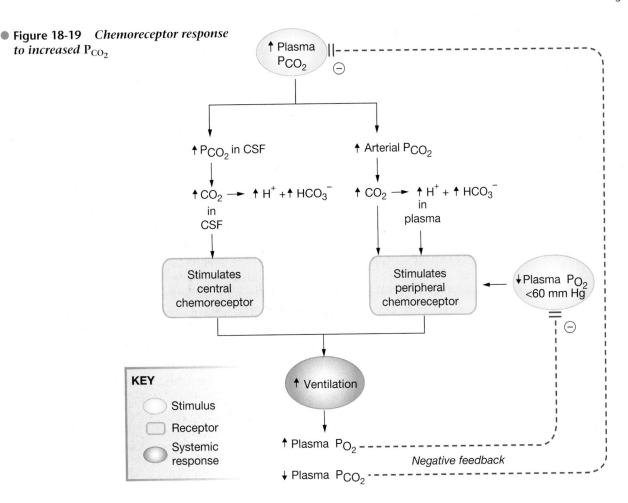

KEY
- Stimulus
- Receptor
- Systemic response

Decreases in arterial P_{CO_2} also influence ventilation. If alveolar P_{CO_2} falls, as it might during hyperventilation, plasma P_{CO_2} and P_{CO_2} in the cerebrospinal fluid decrease. As a result, central chemoreceptor activity declines, and the central pattern generator slows the ventilation rate.

When ventilation decreases, carbon dioxide begins to accumulate in alveoli and the plasma. Eventually, the arterial P_{CO_2} rises above the threshold level for the chemoreceptors. At that point, the receptors fire, and the central pattern generator again increases ventilation.

The central chemoreceptors initially respond strongly to an increase in plasma P_{CO_2} by increasing ventilation. However, if P_{CO_2} remains elevated for several days, ventilation decreases. Over time, the choroid plexus begins to transport bicarbonate ions into the cerebrospinal fluid. The bicarbonate acts as a buffer, removing H^+ from the cerebrospinal fluid and decreasing H^+ stimulation of the central pattern generator.

Fortunately for people with chronic lung diseases, the response of peripheral chemoreceptors to low arterial oxygen remains intact over time, even though central chemoreceptors adapt to high P_{CO_2}. In some situations, low P_{O_2} becomes the primary chemical stimulus for ventilation.

For example, patients with severe chronic lung disease such as emphysema have chronic hypercapnia and hypoxia. Their arterial P_{CO_2} may rise to 50–55 mm Hg (normal 35–45), while their P_{O_2} falls to 45–50 mm Hg (normal 75–100). Because these levels are chronic, the central chemoreceptors gradually adapt to the elevated P_{CO_2}.

Most of the chemical stimulus for ventilation then comes from low P_{O_2}, sensed by the carotid and aortic chemoreceptors. If these patients are given too much oxygen, they may stop breathing because their chemical stimulus for ventilation is eliminated.

Protective Reflexes Guard the Lungs

In addition to the chemoreceptor reflexes that help regulate ventilation, the body has protective reflexes that respond to physical injury or irritation of the respiratory tract and to overinflation of the lungs. The major protective reflex is *bronchoconstriction*, mediated

---

RUNNING PROBLEM

The hyperventilation response to hypoxia creates a peculiar breathing pattern called *periodic breathing*, in which the person goes through a 10–15 second period of breath-holding, followed by a short period of hyperventilation. Periodic breathing occurs most often during sleep.

Question 8: Based on your understanding of the control of ventilation, why do you think periodic breathing occurs most often during sleep?

through parasympathetic neurons that innervate bronchiolar smooth muscle. Inhaled particles or noxious gases stimulate **irritant receptors** in the airway mucosa. The irritant receptors send signals through sensory neurons to control centers in the central nervous system that trigger bronchoconstriction. Protective reflex responses also include coughing and sneezing.

The **Hering-Breuer inflation reflex** is designed to prevent overexpansion of the lungs during strenuous exercise. If tidal volume exceeds 1 liter, stretch receptors in the lung signal the brain stem to terminate inspiration. However, because normal tidal volume is only 500 mL, this reflex does not operate during quiet breathing and mild exertion.

Higher Brain Centers Affect Patterns of Ventilation

Conscious and unconscious thought processes also affect respiratory activities. Higher centers in the hypothalamus and cerebrum can alter the activity of the central pattern generator and change ventilation rate and depth. Voluntary control of ventilation falls into this category. But higher brain center control is not a *requirement* for ventilation. Even if the brain stem above the pons is severely damaged, essentially normal respiratory cycles continue.

Respiration can also be affected by stimulation of portions of the limbic system. As a result, emotional and autonomic activities such as fear and excitement may affect the pace and depth of respiration. In some of these situations, the neural pathway goes directly to the somatic motor neurons, bypassing the central pattern generator in the brain stem.

Although we can temporarily alter our respiratory performance, we cannot override the chemoreceptor reflexes. Holding one's breath is a good example. We can hold our breath voluntarily only until elevated P_{CO_2} in the blood and cerebrospinal fluid activates the chemoreceptor reflex, forcing us to inhale.

Small children having temper tantrums sometimes attempt to manipulate parents by threatening to hold their breath until they die. However, the chemoreceptor reflexes make it impossible for the children to carry out that threat. Extremely strong-willed children can continue holding their breath until they turn blue and pass out from hypoxia, but once they are unconscious, normal breathing will automatically resume.

Breathing is intimately linked to cardiovascular function. The integrating centers for both functions are located in the brain stem, and interneurons project between the two centers, allowing signaling back and forth. In Chapter 20, we will examine the integration of cardiovascular, respiratory, and renal function as the three systems work together to maintain fluid and acid-base homeostasis. But first, in Chapter 19, we examine the physiology of the kidneys.

RUNNING PROBLEM CONCLUSION

High Altitude

On May 29, 1953, Sir Edmund Hillary and Tenzing Norgay of the British Everest Expedition were the first humans to reach the summit of Mt. Everest. They carried supplemental oxygen with them, as it was believed that this feat was impossible without it. But in 1978, Reinhold Messner and Peter Hebeler achieved the impossible. On May 8, they struggled to the summit using sheer willpower and no extra oxygen. In Messner's words, "I am nothing more than a single narrow gasping lung, floating over the mists and summits." Learn more about these Everest expeditions by doing a Google search (www.google.com) for *Hillary Everest* or *Messner Everest*.

To learn more about different types of mountain sickness, see "All about altitude illness," *www.high-altitude-medicine.com*; "High altitude medicine," *American Family Physician* 1998 Apr 15, *www.aafp.org/afp/980415ap/harris.html*; and "High-altitude pulmonary edema," *New England Journal of Medicine* 346(21):1606–1607, 2002 May 23, *www.nejm.org*.

In this running problem you learned about normal and abnormal responses to high altitude. Check your understanding of the physiology behind this challenge by comparing your answers to the ones in the table below.

	QUESTION	FACTS	INTEGRATION AND ANALYSIS
1	What is the P_{O_2} of inspired air reaching the alveoli at 2500 m (dry atmospheric pressure is 542 mm Hg)? How does this compare to an equivalent value at sea level?	Water vapor contributes 47 mm Hg to fully humidified air. Oxygen is 21% of dry air. P_{atm} at sea level is 760 mm Hg.	Correct for water vapor: $542 - 47 = 495$ mm Hg $\times 21\% = P_{O_2}$ of 104 mm Hg. In humidifed air at sea level, $P_{O_2} = 150$ mm Hg.
2	Why would someone with HAPE be short of breath?	Pulmonary edema increases the diffusion distance for oxygen.	The increased diffusion distance worsens the normal hypoxia of altitude.
3	Based on what you learned about mechanisms for matching ventilation and perfusion in the lung, can you explain why patients with HAPE have elevated pulmonary arterial blood pressure?	Low oxygen constricts pulmonary arterioles.	Constriction of pulmonary arterioles would cause blood to collect in the pulmonary arteries behind the constriction. This increases arterial pressure.

	QUESTION	FACTS	INTEGRATION AND ANALYSIS
4	How does adding more erythrocytes to the blood help the acclimatization to high altitude?	98% of arterial oxygen is carried bound to hemoglobin.	Additional hemoglobin increases the oxygen-carrying capacity of the blood.
5	What does adding more erythrocytes do to the viscosity of the blood? What effect will that have on blood flow?	Adding cells increases blood viscosity.	According to Poiseuille's law, increased viscosity increases resistance to flow.
6	What happens to plasma pH during hyperventilation?	Apply the law of mass action to the equation $CO_2 + H_2O \rightleftharpoons H^+ + HCO_3^-$.	If CO_2 decreases, the equation shifts to the left. This decreases H^+ and increases pH (alkalosis).
7	How does this change in pH affect oxygen binding at the lungs when P_{O_2} is decreased? How does it affect unloading of oxygen at the cells?	See graph (a) in Figure 18-9.	The left shift of the curve means that more O_2 binds in the lungs. Less O_2 will unbind at the tissues for a given P_{O_2}, but cell P_{O_2} is probably decreased from normal, and there may be no change in unloading.
8	Why do you think periodic breathing occurs most often during sleep?	Periodic breathing alternates periods of breath-holding (apnea) and hyperventilation.	An awake person is more likely to make a conscious effort to breathe during the breath-holding spells, eliminating the cycle of periodic breathing.

(574) (578) (579) (582) (586) (591)

CHAPTER SUMMARY

The exchange and transport of oxygen and carbon dioxide illustrates the *mass flow* of gases along concentration gradients. The *homeostatic balance* of these blood gases demonstrates *mass balance*: the concentration in the blood varies according to what enters or leaves at the lungs and tissues. The *law of mass action* governs the chemical reactions through which hemoglobin binds oxygen, and carbonic anhydrase catalyzes the conversion of CO_2 and water to carbonic acid.

Diffusion and Solubility of Gases

1. Diffusion of oxygen and CO_2 is influenced by the partial pressure gradients, surface area, membrane thickness, and diffusion distance. (p. 575)

2. The amount of a gas that dissolves in a liquid is proportional to the partial pressure of the gas and to the **solubility** of the gas. Carbon dioxide is 20 times more soluble than oxygen. (p. 575; Fig. 18-2)

Gas Exchange in the Lungs and Tissues

IP Respiratory System: Gas Exchange

3. Normal alveolar and arterial P_{O_2} is 100 mm Hg. Normal arterial P_{CO_2} is 40 mm Hg. (p. 576; Fig. 18-3)

4. Venous P_{O_2} is 40 mm Hg and P_{CO_2} is 46 mm Hg. (p. 576)

5. The composition of inspired air and effectiveness of alveolar ventilation affect P_{O_2} of the alveoli. (p. 577)

6. Changes in alveolar surface area and membrane thickness, and interstitial distance affect gas exchange and arterial P_{O_2}. (p. 579; Fig. 18-4)

Gas Transport in the Blood

Interactive Physiology

IP Respiratory System: Gas Transport

7. Oxygen is transported dissolved in plasma and bound to hemoglobin. More than 98% of blood oxygen is HbO_2. (p. 579; Fig. 18-6)

8. The P_{O_2} of the plasma determines how much oxygen will bind to hemoglobin. (p. 581; Fig. 18-8)

P_{O_2} Determines Hemoglobin Binding of Oxygen

9. Oxygen-hemoglobin binding is affected by pH, temperature, and **2,3-diphosphoglycerate (2,3-DPG)**. (pp. 583–584; Figs. 18-9, 18-10)

10. Venous blood carries 7% of its carbon dioxide dissolved in plasma, 23% as **carbaminohemoglobin**, and 70% as bicarbonate ion in the plasma. (pp. 585–586; Fig. 18-13)

11. **Carbonic anhydrase** inside red blood cells converts to CO_2 to carbonic acid, which dissociates into H^+ and HCO_3^-. H^+ binds to hemoglobin. HCO_3^- enters the plasma using the **chloride shift**. (p. 586)

Regulation of Ventilation

Interactive Physiology

IP Respiratory System: Control of Respiration

12. Respiratory control resides in a **central pattern generator**, a network of neurons in the brain stem. (p. 587)

13. The medullary **dorsal respiratory group** contains inspiratory neurons that control somatic motor neurons to the inspiratory muscles and diaphragm. The **ventral respiratory group** of neurons assists in greater-than-normal inspiration and active expiration. (pp. 587–588; Fig. 18-15)

14. Carbon dioxide is the primary stimulus for changes in ventilation. **Central chemoreceptors** in the medulla respond to changes in P_{CO_2}. (p. 589; Fig. 18-18)

15. **Peripheral chemoreceptors** in the carotid and aortic bodies monitor P_{O_2}, P_{CO_2}, and pH. Ventilation increases when P_{O_2} falls below 60 mm Hg. (p. 589; Fig. 18-17)

16. Protective reflexes monitored by peripheral mechanoreceptors prevent injury to the lungs from overinflation or irritants. (p. 591)

17. Conscious and unconscious thought processes can affect respiratory activity. (p. 592)

QUESTIONS

LEVEL ONE *Reviewing Facts and Terms*

1. List five factors that influence the diffusion of gases from alveolus to blood.

2. More than _____% of the oxygen in arterial blood is transported bound to hemoglobin. How is the remaining oxygen transported to the cells?

3. Name four factors that influence the amount of oxygen that binds to hemoglobin. Which of these four factors is the most important?

4. Describe the structure of a hemoglobin molecule. What element is essential for hemoglobin synthesis?

5. The centers for control of ventilation are found in the _____ and _____ of the brain. What do the dorsal and ventral respiratory groups of neurons control? What is a central pattern generator?

6. Name the chemoreceptors that influence ventilation and explain how they do so. What chemical is the most important controller of ventilation?

7. Describe the protective reflexes of the respiratory system. What does the Hering-Breuer reflex prevent? How is it initiated?

8. What causes the exchange of oxygen and carbon dioxide between alveoli and blood or blood and cells?

9. List five possible physical changes that could result in less oxygen reaching the arterial blood.

LEVEL TWO *Reviewing Concepts*

10. **Concept map:** Construct a map of gas transport using the following terms. You may add additional terms.

P_{CO_2}	carbonic anhydrase
P_{O_2}	chloride shift
alveoli	dissolved CO_2
arterial blood	dissolved O_2
carbaminohemoglobin	hemoglobin

hemoglobin saturation	pressure gradient
oxyhemoglobin	red blood cell
plasma	venous blood

11. In respiratory physiology, it is customary to talk of the P_{O_2} of the plasma. Why is this not the most accurate way to describe the oxygen content of blood?

12. Compare and contrast the following sets of concepts:
    (a) transport of oxygen and carbon dioxide in arterial blood
    (b) partial pressure and concentration of a gas dissolved in a liquid

13. Will HbO_2 binding increase, decrease, or not change with decreased pH?

14. Define the following terms: hypoxia, COPD, hypercapnia.

15. Why did oxygen-transporting pigments need to evolve in animals?

16. Draw and label the following graphs.
    (a) The effect of ventilation on arterial P_{O_2}
    (b) The effect of arterial P_{CO_2} on ventilation

17. As the P_{O_2} of the plasma increases:
    (a) what happens to the amount of oxygen that dissolves in plasma?
    (b) what happens to the amount of oxygen that binds to hemoglobin?

18. If a person is anemic and has less-than-normal hemoglobin content in her red blood cells, what will her arterial P_{O_2} be, compared to normal?

19. Create reflex pathways (stimulus, receptor, afferent path, etc.) for the chemical control of ventilation starting with the following stimuli. Be as specific as possible regarding anatomical locations. Where known, include neurotransmitters and their receptors.
    (a) Increased arterial P_{CO_2}
    (b) Arterial P_{O_2} = 55 mm Hg

LEVEL THREE *Problem Solving*

20. Marco tries to hide at the bottom of a swimming hole by breathing in and out through a garden hose that greatly increases his dead space. What happens to the following parameters in his arterial blood, and why?
    (a) P_{CO_2} (b) P_{O_2}
    (c) bicarbonate ion   (d) pH

21. Which person will carry more oxygen in his blood:
    (a) one with normal hemoglobin of 15 g/dL and an arterial P_{O_2} of 80 mm Hg
    (b) one with a reduced hemoglobin of 12 g/dL and a normal arterial P_{O_2} of 100 mm Hg

22. What would happen to the following in a person suffering from an acute asthma attack (bronchoconstriction)?
    (a) arterial P_{O_2}
    (b) arterial hemoglobin saturation
    (c) alveolar ventilation

23. In early research on the control of rhythmic breathing, scientists made the following observations. What hypotheses might the researchers have formulated from each observation?
    (a) *Observation.* If the brain stem is severed below the medulla, all respiratory movement ceases.
    (b) *Observation.* If the brain stem is cut above the level of the pons, ventilation is normal.
    (c) *Observation.* If the medulla is completely separated from the pons and higher brain centers, ventilation becomes gasping and irregular in depth, but the respiratory rhythm remains.

24. A hospitalized patient with severe chronic obstructive lung disease has a P_{CO_2} of 55 mm Hg and a P_{O_2} of 50 mm Hg. To elevate his blood oxygen, he is given pure oxygen through a nasal tube. The patient immediately stops breathing. Explain why this might occur.

25. You are a physiologist on the manned space flight to a distant planet. You find intelligent humanoid creatures inhabiting the planet, and they willingly submit to your tests. Some of the data you have collected are described below. The first graph shows the oxygen dissociation curve for the oxygen-carrying pigment in the blood of the humanoid named Bzork. Bzork's normal alveolar P_{O_2} is 85 mm Hg. His normal cell P_{O_2} is 20 mm Hg but it drops to 10 mm Hg with exercise.
    (a) What is the percent saturation for Bzork's oxygen-carrying pigment in blood at the alveoli? At an exercising cell?
    (b) What conclusions can you draw about Bzork's oxygen requirement during normal activity and during exercise, based on the graph?

26. The next experiment on Bzork involves his ventilatory response to different conditions. The data from that experiment are graphed below. Line A represents ventilation with an alveolar P_{O_2} of 50 mm Hg. Line B represents ventilation with an alveolar P_{O_2} of 85 mm Hg. Line C represents ventilation with an alveolar P_{O_2} of 85 mm Hg after Bzork has been given beer to drink. Interpret the results of experiments A and C.

LEVEL FOUR *Quantitative Problems*

27. You are given the following information on a patient.

    Blood volume = 5.2 liters

    Hematocrit = 47%

    Hemoglobin concentration = 12 g/dL whole blood

    Total amount of oxygen carried in this person's blood = 1015 mL oxygen

    Arterial plasma P_{O_2} = 100 mm Hg

    You know that at a P_{O_2} of 100 mm Hg, plasma contains 0.3 mL oxygen/dL and that hemoglobin is 98% saturated. Each hemoglobin molecule can bind to a maximum of four molecules of oxygen. Using the information above, calculate the maximum oxygen-carrying capacity of hemoglobin (when hemoglobin is 100% saturated). Units will be mL oxygen/g hemoglobin.

28. The Fick equation says:

$$\text{cardiac output} = \frac{O_2 \text{ consumption}}{\text{difference in oxygen content of arterial and venous blood}}$$

    A person has a cardiac output of 4.5 L/min. Arterial oxygen content is 105 mL O_2/L blood and vena cava oxygen content is 60 mL O_2/L blood. What is this person's metabolic rate (oxygen consumption)?

29. On the oxygen-hemoglobin saturation curve in Figure 18-8, show what happens to the curve when blood hemoglobin falls from 15 g/dL to 10 g/dL.

Explorations

Explorations are also available at *www.physiologyplace.com*

In this chapter, you learned why climbing Mt. Everest is such a challenge for the human body and why people with emphysema experience the same respiratory challenges at sea level.

Exploration 1: Training at High Altitude
A scandal rocked the world of bike racing when testing revealed that at least 14% of competitors in the 1998 Tour de France had taken a man-made form of erythropoietin (EPO) to artificially stimulate red blood cell production. Detection of EPO abuse by athletes is difficult, however, because the effects of EPO abuse are similar to other training regimens, such as training at high altitude. For years top Olympic athletes have naturally stimulated their blood's oxygen-carrying capacity by training at high-altitude training camps, such as the ones in Colorado and Kenya. Is there an ethical difference between taking EPO and training at a high altitude if the results are the same? What if you knew that athletes risk cardiovascular failure and death from EPO abuse? To read more about EPO testing and the ongoing battle between cheating athletes and testers, see the news feature in *Nature* 407:124–127, 2000 and the article by Lasne and de Ceaurriz, *Nature* 405:635, 2000. To read more about altitude training effects try a Google search (*www.google.com*) for *high altitude training*.

Exploration 2: An Interactive Respiratory Physiology Model
The Johns Hopkins School of Medicine has designed a lung model on their Interactive Respiratory Physiology website that you can manipulate to test your understanding of respiratory physiology. You can change multiple conditions, such as tidal volume or altitude, to see their effect on blood gases and other variables. Find this lung model at *http://oac.med.jhmi.edu/LungModel4.2/*. First, note the input and output variables and the diagram of blood flow direction. Start by clicking CALCULATE without making any changes to the input values provided. Examine the output. Note that most variables are grouped according to where in the circulation they are measured, such as venous versus systemic arterial blood. Next, change each input variable independently, examining changes in output variables as you go. To *really* test your understanding, make a prediction about what will happen *before* you look at the model's results. For example, if you decrease hemoglobin amount from 15 to 10 g/dL to mimic anemia, what would happen to arterial O_2 saturation and O_2 content? (See Question 29 above.) Did you accurately predict that only one would change? Another suggestion: check your own calculations of minute ventilation and alveolar ventilation against the model's calculations after increasing the dead space volume from 150 to 200 mL. The lung model was designed and implemented by Johns Hopkins University © 1999. For more respiratory physiology activities from the Johns Hopkins University, see their website at *http://oac.med.jhmi.edu/res_phys/*, © 1996.

ANSWERS

Page 576

1. The other factor that affects how much of each gas will dissolve is their solubilities.

2. $720 \text{ mm Hg} \times 0.78 \text{ N}_2 = 561.6 \text{ mm Hg}$

3. False. CO_2 is more soluble and there will be more CO_2 dissolved.

Page 577

4. (a) O_2 consumption occurs during 3. electron transport system. (b) CO_2 production occurs during 2. citric acid cycle. (The citric acid cycle stops in the absence of oxygen but it does not directly consume oxygen.)

5. The P_{O_2} of the alveoli does not decrease because it is constantly being replenished by fresh air from the environment. [✎ p. 566]

Page 578

6. On Everest, 21% of 250 mm Hg = P_{O_2} of 53 mm Hg. Correct for water vapor pressure: $(250 - 47) \times 21\% = P_{O_2}$ of 43 mm Hg.

Page 579

7. These pathologies cause blood to pool in the lungs because the left heart is unable to pump as efficiently as the right heart. Blood pools in the lungs, increasing pulmonary pressure.

8. When alveolar ventilation increases, arterial P_{O_2} increases because more fresh air is entering the alveoli. Arterial P_{CO_2} decreases because the fresh air has a lower P_{CO_2} and alveolar P_{CO_2} decreases, causing more CO_2 to leave the blood. Venous P_{O_2} and P_{CO_2} will not change because these are determined by oxygen consumption and CO_2 production in the cells.

Page 583

9. Yes. Hemoglobin reaches 100% saturation at 650 mm Hg. At sea level, 100% oxygen equals P_{O_2} of 760 mm Hg.

10. Hyperventilation causes a minimal increase in percent saturation of arterial hemoglobin.

Page 584

11. As the P_{O_2} of the exercising muscle falls, more oxygen is released by hemoglobin. The P_{O_2} of the venous blood leaving the muscle will be 25 mm Hg, the same as the P_{O_2} of the muscle cell.

Page 587

12. An obstruction of the airways would decrease alveolar ventilation and increase arterial P_{CO_2}. This would increase the H^+ concentration and the pH would fall.

Page 582

Fig. 18-8: (a) When P_{O_2} is 20 mm Hg, hemoglobin is about 34% saturated with oxygen. (b) Hemoglobin is 50% saturated with oxygen at a P_{O_2} of 28 mm Hg.

Page 583

Fig. 18-9: (a) When pH falls from 7.4 to 7.2, hemoglobin saturation decreases by 13%, from 37% saturation to 24%. (b) When an exercising muscle cell warms up, it releases more oxygen at any given P_{O_2}.

Page 584

Fig. 18-10: Loss of 2,3-DPG is bad because then hemoglobin binds more tightly to oxygen at P_{O_2} values found in cells.

Page 584

Fig. 18-11: (a) The P_{O_2} of placental blood is about 28 mm Hg. (b) At a P_{O_2} of 10 mm Hg, maternal blood is only about 8% saturated with oxygen.

Page 588

Fig. 18-15: 1. pons; 2. ventral respiratory group; 3. medullary chemoreceptor; 4. sensory neuron; 5. carotid chemoreceptor; 6. somatic motor neuron; 7. aortic chemoreceptor; 8. internal intercostals; 9. abdominal muscles; 10. diaphragm; 11. external intercostals; 12. scalenes and stemocleiodmastoids; 13. somatic motor neuron; 14. dorsal respiratory group; 15. limbic system; 16. higher brain centers (emotions and voluntary control)

Page 596

27. The maximum O_2 carrying capacity is 1.65 mL O_2/gm Hb.

28. Oxygen consumption is 202.5 mL O_2/min.

29. Nothing happens to the curve. The percent of available Hb bound to O_2 is unchanged at any given P_{O_2}. However, with less Hb available, less oxygen will be transported.

The Kidneys

A section of latex-injected kidney showing blood vessels (red) and a Bowman's capsule and afferent arteriole (yellow).

RUNNING PROBLEM

Gout

Michael Moustakakis, 43, had spent the last two days on the sofa from a throbbing, relentless pain in his left big toe. When the pain began, Michael thought he had a mild sprain or perhaps the beginnings of arthritis. But then the pain intensified, and the toe joint became hot and red. Finally Michael hobbled into his doctor's office, feeling a little silly about his problem. But upon hearing his symptoms, the doctor seemed to know instantly what was wrong. "Sounds to me like you have gout," said Dr. Garcia.

603 612 614 616 619

BACKGROUND BASICS

Transepithelial transport: **148** | Mass balance: **9** | Osmolarity and tonicity: **152** | Filtration in capillaries: **506** | Transcytosis: **148** | Saturation: **139** | Autoregulation of vascular resistance: **499** | Transport across membranes: **134** | pH and buffers: **33** | Transporting epithelium: **67** | Fenestrated capillaries: **504** | Phosphocreatine: **401** | Competition: **138** |

Plasma undergoes modification to urine in the nephron.

—**Arthur Grollman**, in *Clinical Physiology: The Functional Pathology of Disease*, 1957

About 100 A.D. Areteaus the Cappadocian wrote, "Diabetes is a wonderful affection, not very frequent among men, being a melting down of the flesh and limbs into urine. . . . The patients never stop making water [urinating], but the flow is incessant, as if from the opening of aqueducts." Physicians have known for centuries that *urine*, the fluid waste produced by the kidneys, reflects the function of the body. To aid them in their diagnosis of illness, they even carried special flasks for collection and inspection of patients' urine.

The first step in examining a urine sample was to determine its color. Was it dark yellow (concentrated), pale straw (dilute), red (indicating the presence of blood), or black (indicating the presence of hemoglobin metabolites)? One form of malaria was called *blackwater fever* because metabolized hemoglobin from abnormal breakdown of red blood cells turned victims' urine black or dark red.

Physicians also inspected urine samples for clarity, smell, taste, and froth (abnormal presence of proteins). Physicians who did not want to taste the urine themselves would allow their students the "privilege" of tasting it for them. A physician without students might expose insects to the urine and study their reaction.

Probably the most famous example of using urine for diagnosis is the taste test for diabetes mellitus, historically known as the honey-urine disease. Diabetes is an endocrine disorder characterized by the presence of glucose in the urine. The urine of diabetics tasted sweet and attracted insects, making the diagnosis clear.

Although today we have much more sophisticated tests for glucose in the urine, the first step of a *urinalysis* is still to examine the color, clarity, and odor of the urine. In this chapter, you will learn why we can tell so much about the body's function by what is present in the urine.

► FUNCTIONS OF THE KIDNEYS

If you asked people on the street, "What is the most important function of the kidney?" they would probably say, "The removal of wastes." Actually, the most important function of the kidney is the homeostatic regulation of the water and ion content of the blood, also called "salt and water balance" or "fluid and electrolyte balance." Waste removal is important, but disturbances in blood volume or ion levels will cause serious medical problems before the accumulation of metabolic wastes reaches toxic levels.

The kidneys maintain normal concentrations of ions and water by balancing intake of those substances with excretion in the urine, obeying the principle of mass balance [● p. 9]. We can divide kidney function into six general areas:

1. **Regulation of extracellular fluid volume.** When extracellular fluid volume decreases, blood pressure also decreases [● p. 497]. If volume and blood pressure fall too low, the body cannot maintain adequate blood flow to the brain and other essential organs. The kidneys work in an integrated fashion with the cardiovascular system to ensure that blood pressure and tissue perfusion remain within an acceptable range.

2. **Regulation of osmolarity.** The body integrates kidney function with behavioral drives such as thirst to maintain blood osmolarity at a value close to 290 mOsM. We will examine the reflex pathways for regulation of volume and osmolarity in Chapter 20.

3. **Maintenance of ion balance**. The kidneys keep concentrations of key ions within a normal range by balancing dietary intake with urinary loss. Sodium (Na^+) is the major ion involved in the regulation of extracellular fluid volume and osmolarity. Potassium (K^+) and calcium (Ca^{2+}) concentrations are also closely regulated.

   We will discuss renal control of sodium and potassium balance in Chapter 20 but defer the discussion of calcium until Chapter 22, when we look at all aspects of calcium homeostasis.

4. **Homeostatic regulation of pH.** The pH of plasma is normally kept within a narrow range [● p. 33]. If extracellular fluid becomes too acidic, the kidneys remove H^+ and conserve bicarbonate ions (HCO_3^-), which act as a buffer [● p. 34]. Conversely, when extracellular fluid becomes too alkaline, the kidneys remove HCO_3^-, and conserve H^+. The kidneys do not correct pH disturbances as rapidly as the lungs do, but they play a significant role in pH homeostasis, as you will learn in the next chapter.

5. **Excretion of wastes and foreign substances.** The kidneys remove two types of wastes: those that are by-products of metabolism, and foreign substances such as drugs and environmental toxins. Metabolic wastes include creatinine from muscle metabolism [● p. 401] and the nitrogenous wastes *urea* [● p. 108] and *uric acid*. A metabolite of hemoglobin called *urobilinogen* gives urine its characteristic yellow color.

   Examples of foreign substances that are actively removed by the kidneys include the artificial sweetener *saccharin* and the anion *benzoate*, part of the preservative *potassium benzoate* that you ingest each time you drink a diet soft drink.

6. **Production of hormones.** Although the kidneys are not endocrine glands, they play important roles in three endocrine pathways. Kidney cells synthesize *erythropoietin*, the cytokine/hormone that regulates red blood cell synthesis [● p. 529]. They also release *renin*, an enzyme that regulates the production of hormones involved in sodium balance and blood pressure homeostasis. Renal enzymes help convert vitamin D_3 into a hormone that regulates Ca^{2+} balance.

The kidneys, like many organs, are supplied with a tremendous reserve capacity. By most estimates, you must lose nearly three-fourths of your kidney function before homeostasis begins to be affected. Many people function perfectly normally with only one kidney, including the 1-in-1000 people born with only one kidney (the mate fails to develop during gestation) or those people who donate a kidney for transplantation.

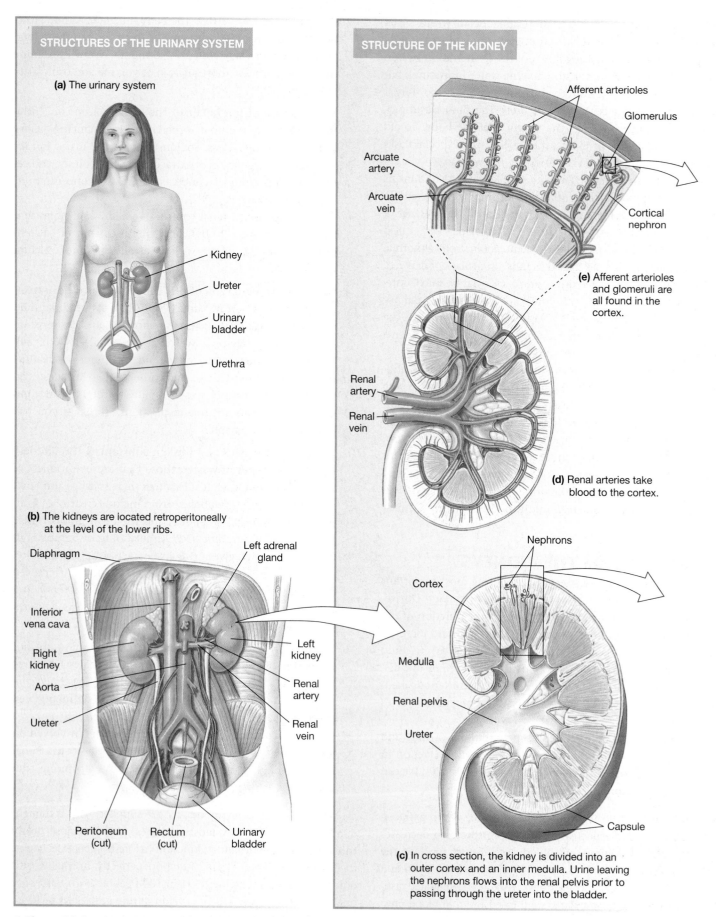

STRUCTURES OF THE URINARY SYSTEM

(a) The urinary system

Kidney

Ureter

Urinary bladder

Urethra

(b) The kidneys are located retroperitoneally at the level of the lower ribs.

Diaphragm

Left adrenal gland

Inferior vena cava

Right kidney

Left kidney

Aorta

Renal artery

Ureter

Renal vein

Peritoneum (cut)

Rectum (cut)

Urinary bladder

STRUCTURE OF THE KIDNEY

Afferent arterioles

Glomerulus

Arcuate artery

Arcuate vein

Cortical nephron

(e) Afferent arterioles and glomeruli are all found in the cortex.

Renal artery

Renal vein

(d) Renal arteries take blood to the cortex.

Nephrons

Cortex

Medulla

Renal pelvis

Ureter

Capsule

(c) In cross section, the kidney is divided into an outer cortex and an inner medulla. Urine leaving the nephrons flows into the renal pelvis prior to passing through the ureter into the bladder.

● **Figure 19-1** *ANATOMY SUMMARY The Urinary System*

STRUCTURE OF THE NEPHRON

(f) The capillaries of the glomerulus form a ball-like mass.

Glomerulus

(g) Each nephron has two arterioles and two sets of capillaries associated with it.

Efferent arteriole
Glomerulus (capillaries)
Peritubular capillaries
Juxtaglomerular apparatus
Afferent arteriole

(j) Parts of a nephron

Proximal tubule
Distal tubule
Collecting tubules

Bowman's capsule
Descending limb of loop begins
Ascending limb of loop ends

Descending limb
Ascending limb

Collecting duct

(i) Some nephrons dip deep into the medulla.

Arterioles
Nephrons
Cortex
Medulla

Loop of Henle

To bladder

(h) Juxtamedullary nephron with vasa recta

Peritubular capillaries
Glomerulus
Vasa recta
Collecting duct
Loop of Henle

● **Figure 19-1** *(continued)*

1. Ion regulation is a key feature of kidney function. What happens to the resting membrane potential of a neuron if extracellular K^+ levels decrease? [● p. 262]

2. What happens to the force of cardiac contraction if plasma Ca^{2+} levels decrease substantially? [● p. 462]

▶ ANATOMY OF THE URINARY SYSTEM

The **urinary system** [*ouron*, urine] is composed of the kidneys and accessory structures. The study of kidney function is called **renal physiology**, from the Latin word *renes*, meaning "kidneys."

The Urinary System Consists of Kidneys, Ureters, Bladder, and Urethra

Let's begin by following the route a drop of water would take on its way to excretion in the urine. In the first step of urine production, water and solutes move from plasma into hollow tubules (*nephrons*) that make up the bulk of the paired **kidneys** (Fig. 19-1a ●). These tubules modify the composition of the fluid as it passes through.

The modified fluid leaves the kidney and passes into a hollow tube called a **ureter**. There are two ureters, one leading from each kidney to the **urinary bladder**. The bladder expands and fills with urine until, by reflex action, it contracts and expels urine through a single tube, the **urethra**.

The urethra in males exits the body through the shaft of the penis. In females, the urethral opening is found anterior to the openings of the vagina and anus. *Micturition*, or urination, is the process by which urine is **excreted**, or removed from the body.

The kidneys are the site of urine formation. They lie on either side of the spine at the level of the eleventh and twelfth ribs, just above the waist (19-1b). Although they are below the diaphragm, they are technically outside the abdominal cavity (**retroperitoneal**), sandwiched between the membranous **peritoneum**, which lines the abdomen, and the bones and muscles of the back.

The concave surface of each kidney faces the spine. The renal blood vessels, nerves, lymphatics, and ureters all emerge from this surface. **Renal arteries**, which branch off the abdominal aorta, supply blood to the kidneys. **Renal veins** carry blood from the kidneys back to the inferior vena cava.

At any given time, the kidneys receive 20%–25% of the cardiac output, even though they are only 0.4% of the total body weight (4.5–6 ounces each). This high rate of blood flow through the kidneys is critical to renal function.

The Nephron Is the Functional Unit of the Kidney

A cross section through a kidney shows that the interior is arranged in two layers: an outer **cortex** and inner **medulla** (Fig. 19-1c). The layers are formed by the organized arrangement of microscopic tubules called **nephrons** (Fig. 19-1i). About 80% of

Urinary Tract Infections

Because of the shorter length of the female urethra and its proximity to bacteria from the large intestine, women are more prone than men to develop bacterial infections of the bladder and kidneys, or **urinary tract infections** (UTIs). The most common cause of UTIs is the bacterium *Escherichia coli*, a normal inhabitant of the human large intestine. *Escherichia coli* is not harmful while restricted to the lumen of the large intestine, but it is pathogenic [*patho-*, disease + *-genic*, causing] if it gets into the urethra. The most common symptoms of a UTI are pain or burning with urination and frequency of urination. A urine sample from a patient with a UTI will often contain many red and white blood cells, neither of which is commonly found in normal urine. UTIs are treated with antibiotics.

our nephrons are almost completely contained within the cortex (*cortical* nephrons) but the others dip down into the medulla (*juxtamedullary* nephrons).

The nephron is the functional unit of the kidney, with about 1 million nephrons in each kidney. (A *functional unit* is the smallest structure that can carry out all the functions of an organ.) Each nephron has both **vascular elements**, the blood vessels of the nephron, and **tubular elements**.

Vascular Elements of the Nephron Blood enters the kidney through the *renal artery*, flowing into smaller arteries and then into arterioles in the cortex (Fig. 19-1d, e). At this point, the arrangement of blood vessels becomes unique. Blood flows from an arteriole into a set of capillaries, then into a *second* arteriole and a second set of capillaries. This arrangement creates a *portal system*, one of three in the body [● p. 225]. Fluid filters out of the blood in the first set of capillaries, then re-enters the blood at the second set.

Arterial blood arriving at a nephron passes from the **afferent arteriole** into a ball-like network of capillaries known as the **glomerulus** [*glomus*, a ball-shaped mass; plural *glomeruli*] (Fig. 19-1f). All glomeruli are located in the cortex of the kidney. The glomerulus is the site where plasmalike fluid filters out of the capillaries and into the lumen of the tubule. This process is similar to the filtration of water and molecules out of systemic capillaries in other tissues, as you will see.

Blood leaving the glomerulus flows into an **efferent arteriole**. It then enters a series of **peritubular capillaries** [*peri-*, around] that surround the tubule (Fig. 19-1g). Long capillaries that dip into the medulla are part of the **vasa recta** (Fig. 19-1h). Fluid moves out of the tubule and back into the blood as it passes through the peritubular capillaries and vasa recta. Finally, renal capillaries join to form venules and small veins, exiting the kidney through the **renal vein**.

Tubular Elements of the Nephron   The tubular portion of the nephron begins with a hollow ball-like structure called **Bowman's capsule** that surrounds the glomerulus (Fig. 19-1j). The capillary endothelium of the glomerulus is fused to the epithelium of Bowman's capsule so that fluid filtering out of the capillaries passes directly into the lumen of the tubule. The combination of glomerulus and Bowman's capsule is called the **renal corpuscle.**

From Bowman's capsule, filtered fluid flows into the **proximal tubule** [*proximal*, close or near], then into the **loop of Henle**, a hairpin-shaped segment that dips down into the medulla and then back up to reenter the cortex. The loop of Henle is divided into two *limbs*, a **descending limb**, in which fluid flows into the medulla, and an **ascending limb**, in which fluid flows back to the cortex.

Fluid leaving the loop of Henle goes into the **distal tubule** [*distal*, distant or far]. Notice in Figure 19-1g how the kidney tubule twists and folds back on itself so that the distal tubule passes between afferent and efferent arterioles. This region is known as the **juxtaglomerular apparatus** [*juxta-*, beside]. The proximity of tubule and arterioles allows paracrine communication between the structures, a key feature of kidney autoregulation.

RUNNING PROBLEM

Gout is a metabolic disease characterized by high blood concentrations of uric acid (*hyperuricemia*). If uric acid concentrations reach a critical level (7.5 mg/dL), crystals of monosodium urate form in peripheral joints, particularly in the feet, ankles, and knees. These crystals trigger an inflammatory reaction and cause periodic attacks of excruciating pain. Uric acid crystals may also form kidney stones in the renal pelvis.

Question 1: Trace the route followed by kidney stones when they are excreted.

 598 · 612 · 614 · 616 · 619

Because the twisted configuration of the kidney tubule makes it difficult to follow fluid flow, we will unfold the nephron in the remaining figures in this chapter so that fluid flows from left to right across the figure.

The distal tubules of up to eight nephrons drain into a single larger tube called the **collecting duct**. (The distal tubules and collecting duct together form the **distal nephron.**) Collecting ducts pass from the cortex through the medulla and drain into the **renal pelvis**. From the renal pelvis, the filtered and modified fluid, now called **urine**, goes into the ureter on its way to excretion.

▶ OVERVIEW OF KIDNEY FUNCTION

Imagine drinking a 12-ounce soft drink every three minutes around the clock: by the end of 24 hours, you would have consumed the equivalent of 90 2-liter bottles. The thought of putting 180 liters into our intestinal tract is staggering, but that is how much plasma moves into the nephrons in a day! Yet the average volume of urine leaving the kidneys is only 1.5 L/day. More than 99% of the fluid that enters nephrons must find its way back into the blood, or the body would rapidly dehydrate.

The Three Processes of the Nephron Are Filtration, Reabsorption, and Secretion

There are three basic processes that take place in the nephron: filtration, reabsorption, and secretion (Fig. 19-2 ●). **Filtration** is the movement of fluid from blood into the lumen of the nephron. Filtration takes place only in the renal corpuscle, where the walls of glomerular capillaries and Bowman's capsule are modified to allow bulk flow of fluid.

Once filtered fluid passes into the lumen of the nephron, it becomes part of the body's external environment, just as substances in the lumen of the intestinal tract are part of the external environment

● **Figure 19-2**   *Filtration, reabsorption, secretion, and excretion*

KEY

> **F** = Filtration: blood to lumen
>
> **R** = Reabsorption: lumen to blood
>
> **S** = Secretion: blood to lumen
>
> **E** = Excretion: lumen to external environment

(Figure 1-2, p. 3). Thus, anything that filters into the nephron is destined for removal in the urine unless it is reabsorbed into the body.

After filtered fluid leaves Bowman's capsule, it is modified by the addition of solute and the removal of solutes and water. **Reabsorption** is the process of moving filtered material from the lumen of the nephron back to the blood as it flows through peritubular capillaries.

Secretion removes selected molecules from the blood and adds them to the fluid in the lumen. Although secretion and filtration both move substances from blood into the lumen, secretion is a more selective process that usually uses membrane proteins to move molecules across the tubule epithelium.

Volume and Osmolarity Change as Fluid Flows Through the Nephron

Now let's follow some fluid through the nephron to learn what happens to it in various segments of the tubule. An average of 180 liters a day filters into Bowman's capsule (Table 19-1). This filtrate is almost identical in composition to plasma and is nearly isosmotic, about 300 mOsM [p. 153].

As the 180 liters of filtrate flow through the proximal tubule, about 70% of the volume is reabsorbed, leaving 54 liters. The proximal tubule cells transport solutes out of the lumen, and water follows by osmosis. Fluid leaving the proximal tubule therefore has the same concentration as filtrate but its composition has changed. Thus, the primary function of the proximal tubule is the *bulk reabsorption of isosmotic fluid*.

Proximal tubule fluid then passes into the loop of Henle, the primary site for creation of dilute fluid. As the filtrate passes through the loop, proportionately more solute is reabsorbed than water. This process leaves hyposmotic fluid in the lumen. By the time filtrate leaves the loop, it averages 100 mOsM and its volume has fallen from 54 L/day to about 18 L/day. Now 90% of the original filtered volume has been reabsorbed.

From the loop of Henle, filtrate passes into the distal tubule and the collecting duct. In these two segments, the fine regulation of salt and water balance takes place under the control of several hormones. Reabsorption and, to a lesser extent, secretion

in the collecting duct determine the final composition and concentration of the fluid.

Once tubule fluid leaves the collecting duct, it can no longer be modified. In humans, the composition of the urine does not change during storage in the bladder.

By the end of the collecting duct, tubule fluid has a volume of 1.5 L/day and an osmolarity that can range from 50 to 1200 mOsM. Both the volume and osmolarity of urine depend on the body's need to conserve or excrete water and solute.

A word of caution here: it is very easy to confuse *secretion* with *excretion*. Try to remember the origins of the two prefixes. *Se-* means *apart*, as in to separate something from its source. In the nephron, secreted solutes are moved from plasma to tubule lumen.

Ex- means *out*, or *away*, as in out of or away from the body. Excretion refers to the removal of a substance from the body. Besides the kidneys, other organs that carry out excretory processes include the lungs (CO_2) and intestines (undigested food, bilirubin).

Figure 19-2 summarizes filtration, reabsorption, secretion, and excretion. Filtration takes place in the renal corpuscle as fluid moves from the capillaries of the glomerulus into Bowman's capsule. Reabsorption and secretion occur along the remainder of the tubule, transferring material between lumen and peritubular capillaries. The quantity and composition of the substances being reabsorbed or secreted vary in different segments of the nephron. Fluid that remains in the lumen at the end of the nephron is excreted as urine.

The amount of any substance excreted in the urine reflects how it was handled during its passage through the nephron (Fig. 19-3). The amount excreted is equal to the amount filtered into the tubule, minus the amount reabsorbed, plus any additional amount of the substance secreted into the lumen:

Amount excreted =
amount filtered − amount reabsorbed + amount secreted

This equation is a useful way to think about renal handling of solutes. In the following sections, we will look in more detail at the important processes of filtration, reabsorption, secretion, and excretion.

Table 19-1: Changes in Volume and Osmolarity Along the Nephron

LOCATION IN NEPHRON	VOLUME OF FLUID	OSMOLARITY OF FLUID
Bowman's capsule	180 L/day	300 mOsM
End of proximal tubule	54 L/day	300 mOsM
End of loop of Henle	18 L/day	100 mOsM
End of collecting duct (final urine)	1.5 L/day (average)	50–1200 mOsM

Amount filtered **F**	−	amount reabsorbed **R**	+	amount secreted **S**	=	Amount of solute excreted **E**

 Figure 19-3 *The excretion of a substance depends on the amount that was filtered, reabsorbed, and secreted*

3. Name one way in which filtration and secretion are alike. Name one way in which they differ.

4. A water molecule enters the renal corpuscle from the blood and ends up in the urine. Name all anatomical structures that the molecule passes through on its trip to the outside world.

5. What would happen to the body if filtration continued at a normal rate but reabsorption dropped to one-half of normal?

▶ FILTRATION

The filtration of plasma into the kidney tubule is the first step in urine formation. This relatively nonspecific process creates a filtrate whose composition is like that of plasma minus the plasma proteins. Under normal conditions, blood cells are not filtered and they remain in the capillary. Filtered fluid is composed only of water and dissolved solutes.

The Renal Corpuscle Consists of the Glomerulus and Bowman's Capsule

Filtration takes place in the renal corpuscle, which consists of the glomerular capillaries surrounded by Bowman's capsule. Substances leaving the plasma must pass through three *filtration barriers* before reaching the tubule lumen: the glomerular capillary endothelium, the basal lamina (basement membrane), and the epithelium of Bowman's capsule (Fig. 19-4d ●).

The first barrier is the capillary endothelium. **Glomerular capillaries** are *fenestrated capillaries* [● p. 504] with large pores that allow most components of plasma to filter out of the blood (Fig. 19-4d). The pores are small enough, however, to prevent blood cells from leaving the capillary. Their negatively charged surface proteins also help repel negatively charged plasma proteins.

Glomerular **mesangial cells** lie between and around the glomerular capillaries (Fig. 19-4c). Mesangial cells have bundles of actinlike filaments in their cytoplasm that allow them to contract

(a) The epithelium around glomerular capillaries is modified into podocytes.

Efferent arteriole
Bowman's capsule
Capsular epithelium
Distal tubule
Podocyte
Proximal tubule
Afferent arteriole
Glomerular capillary
Lumen of Bowman's capsule

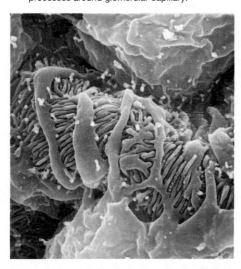

(b) Micrograph showing podocyte foot processes around glomerular capillary.

(c) Podocyte foot processes surround each capillary, leaving slits through which filtration takes place.

Podocyte
Lumen of Bowman's capsule
Glomerular capillary
Podocyte foot processes
Capillary endothelium
Mesangial cell

● **Figure 19-4**
Structure of the renal corpuscle

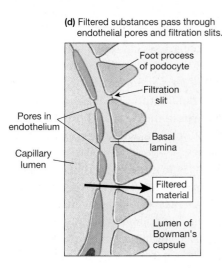

(d) Filtered substances pass through endothelial pores and filtration slits.

Foot process of podocyte
Filtration slit
Pores in endothelium
Basal lamina
Capillary lumen
Filtered material
Lumen of Bowman's capsule

605

and alter blood flow through the capillaries. In addition, mesangial cells secrete cytokines associated with immune and inflammatory processes. Disruptions of mesangial cell function have been linked to several disease processes in the kidney.

The second filtration barrier is an acellular layer of extracellular matrix called the **basal lamina**, which separates the capillary endothelium from the epithelial lining of Bowman's capsule. The basal lamina consists of negatively charged glycoproteins and a collagen-like material that act like a coarse sieve, excluding most plasma proteins from the fluid that filters through it.

The third filtration barrier is the epithelium of Bowman's capsule. The portion of the capsule epithelium that surrounds each capillary consists of specialized cells called **podocytes** [*podos*, foot]. The podocytes have long, fingerlike cytoplasmic extensions that extend from the main cell body (Fig. 19-4a, b). Adjacent "fingers," or **foot processes**, wrap around the capillaries and interlace, leaving narrow **filtration slits** closed by a membrane. Podocytes, like mesangial cells, contain contractile fibers. These are connected to the basal lamina by integrins.

When you visualize plasma filtering out of the glomerular capillaries, it is easy to imagine that all the plasma in the capillary moves into Bowman's capsule. However, total filtration of the plasma would leave behind a sludge of blood cells and proteins that could not flow out of the glomerulus. Normally, only about one-fifth of the plasma that flows through the kidneys at any one time filters into the nephrons. The remaining four-fifths of the plasma, along with the plasma proteins and blood cells, flow into the peritubular capillaries (Fig. 19-5 ●). The percentage of total plasma volume that filters is called the **filtration fraction**.

Filtration Occurs Because of Hydrostatic Pressure in the Capillaries

What drives filtration across the walls of the glomerular capillaries? It is similar in many ways to filtration of fluid out of systemic capillaries [● p. 506]. The following forces influence glomerular filtration:

1. The *hydrostatic pressure of blood flowing through glomerular capillaries* forces fluid through the leaky endothelium. Capillary blood pressure averages 55 mm Hg and favors filtration. Although pressure declines along the length of the capillaries, it remains higher than the opposing pressures. Consequently, filtration takes place along nearly the entire length of glomerular capillaries.

2. The *colloid osmotic pressure inside glomerular capillaries* is higher than that of fluid within Bowman's capsule due to the presence of proteins in the plasma. The osmotic pressure gradient averages 30 mm Hg and favors fluid movement back into the capillaries.

3. Bowman's capsule is an enclosed space (unlike the interstitial fluid), so the presence of fluid within the capsule creates *hydrostatic fluid pressure* that opposes fluid movement into Bowman's capsule. Fluid filtering out of the capillaries must displace the fluid already in the lumen. Hydrostatic fluid pressure in the capsule averages 15 mm Hg opposing filtration.

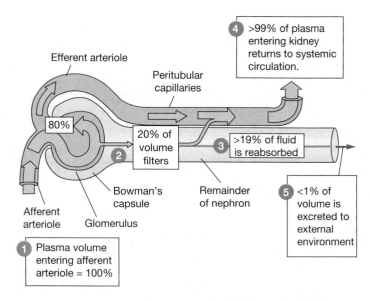

● **Figure 19-5** *The filtration fraction* Only 20% of plasma that passes through the glomerulus is filtered. Less than 1% of filtered fluid is eventually excreted.

The three pressures—capillary blood pressure, colloid osmotic pressure, and capsule fluid pressure—that influence glomerular filtration are summarized in Figure 19-6 ●. The net driving force is 10 mm Hg, favoring filtration. Although this pressure may not seem very high, when combined with the very leaky nature of the fenestrated capillaries, it results in the filtration of 180 L of fluid per day into the nephrons.

6. The osmotic pressure of plasma in the efferent arterioles is higher than that in the afferent. Why?

EMERGING CONCEPTS

Diabetes

Diabetic Nephropathy

End-stage renal failure, where kidney function has deteriorated beyond recovery, is a life-threatening complication in 30%–40% of people with type 1 diabetes and 10%–20% of type 2 diabetics. Like many complications of diabetes, its exact causes are not clear. Diabetic nephropathy usually begins with an increase in glomerular filtration. This is followed by the appearance of proteins in the urine (proteinuria), an indication that the normal filtration barrier has been altered. In later stages, filtration rates decline. This stage is associated with thickening of the glomerular basal lamina and changes in both podocytes and mesangial cells. Abnormal proliferation of mesangial cells compresses the glomerular capillaries and impedes blood flow, contributing to the decrease in glomerular filtration.

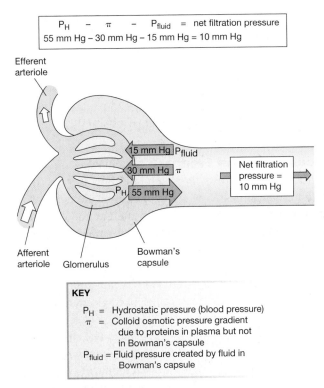

$$P_H - \pi - P_{fluid} = \text{net filtration pressure}$$
$$55 \text{ mm Hg} - 30 \text{ mm Hg} - 15 \text{ mm Hg} = 10 \text{ mm Hg}$$

Efferent arteriole

15 mm Hg P_{fluid}

30 mm Hg π

P_H 55 mm Hg

Net filtration pressure = 10 mm Hg

Afferent arteriole   Glomerulus   Bowman's capsule

KEY

P_H = Hydrostatic pressure (blood pressure)
π = Colloid osmotic pressure gradient due to proteins in plasma but not in Bowman's capsule
P_{fluid} = Fluid pressure created by fluid in Bowman's capsule

● **Figure 19-6**   *Filtration pressure in the renal corpuscle*

Glomerular Filtration Rate Averages 180 Liters per Day

Filtration efficiency is described by the **glomerular filtration rate** (**GFR**), the amount of fluid that filters into Bowman's capsule per unit time. Average GFR is 125 mL/min, or 180 L/day, an incredible rate considering that the total plasma volume is only about 3 liters. This rate means that the kidneys filter the entire plasma volume 60 times a day, or 2.5 times every hour. If most of the filtrate were not reabsorbed during its passage through the nephron, we would run out of plasma in only 24 minutes of filtration!

GFR is influenced by two factors: (1) the net filtration pressure just described and (2) the *filtration coefficient*. Filtration pressure is determined primarily by renal blood flow and pressure. The filtration coefficient has two components: surface area of the glomerular capillaries available for filtration, and permeability of the capillary-Bowman's capsule interface. In this respect, GFR is similar to gas exchange at the alveoli, where the rate of gas exchange depends on partial pressure differences, surface area of the alveoli, and permeability of the alveolar membrane.

Blood Pressure and Renal Blood Flow Influence GFR

Blood pressure provides the hydrostatic pressure that drives glomerular filtration. Therefore it would seem reasonable to assume that if blood pressure rises, GFR goes up, or that if blood pressure falls, GFR goes down. But that is not usually the case. GFR is remarkably constant over a wide range of blood pressures. As long as mean arterial blood pressure remains between 80 and 180 mm Hg, GFR averages 180 L/day (Fig. 19-7 ●).

Autoregulation maintains a nearly constant GFR when mean arterial blood pressure is between 80 and 180 mm Hg.

180

Normal mean blood pressure

40   80   120   160   200
Mean arterial blood pressure (mm Hg)

● **Figure 19-7**   *Autoregulation of glomerular filtration rate*

CONCEPT CHECK   Answers: p. 624

7. If average blood pressure is 120/80 mm Hg and mean blood pressure is diastolic pressure +1/3 the pulse pressure, what is an average value for mean blood pressure?

The control of GFR is accomplished primarily by regulation of blood flow through the renal arterioles (Fig. 19-8 ●). If the resistance of the renal arterioles increases, renal blood flow decreases and blood is diverted to other organs [● Fig. 15-14, p. 504]. The effect of increased resistance on GFR, however, depends upon *where* the resistance change takes place.

If resistance increases in the *afferent* arteriole (Fig. 19-8b), hydrostatic pressure decreases on the downstream side of the constriction. This translates into a decrease in GFR. If resistance increases in the *efferent* arteriole, blood "dams up" in front of the increased resistance and hydrostatic pressure in the glomerular capillaries increases (Fig. 19-8c). GFR goes up. The opposite occurs with decreased resistance in the afferent or efferent arterioles. Most regulation occurs at the afferent arteriole.

GFR Is Subject to Autoregulation

Autoregulation of GFR is a local control process in which the kidney maintains a relatively constant GFR in the face of normal fluctuations in blood pressure. We do not completely understand the process, but several different mechanisms are at work. The **myogenic response** is the intrinsic ability of vascular smooth muscle to respond to pressure changes. **Tubuloglomerular feedback** is a paracrine signaling mechanism through which changes in fluid flow through the distal tubule influence GFR.

Myogenic Response   The myogenic response of afferent arterioles is similar to autoregulation in other systemic arterioles. When smooth muscle in the arteriole wall stretches because of increased blood pressure, stretch-sensitive ion channels open, and the muscle cells depolarize, then contract [● p. 500]. Vasoconstriction increases resistance to flow, so blood flow through the arteriole diminishes. The decrease in blood flow decreases filtration pressure within the glomerulus.

If blood pressure decreases, you might expect the opposite response to occur. The tonic level of arteriolar contraction disappears,

(a) Renal blood flow and GFR change if resistance in the arterioles changes.

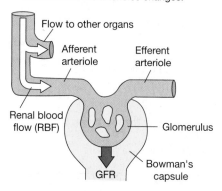

(b) Vasoconstriction of the afferent arteriole increases resistance and decreases renal blood flow, capillary blood pressure (P_H), and GFR.

(c) Increased resistance of efferent arteriole decreases renal blood flow but increases P_H and GFR.

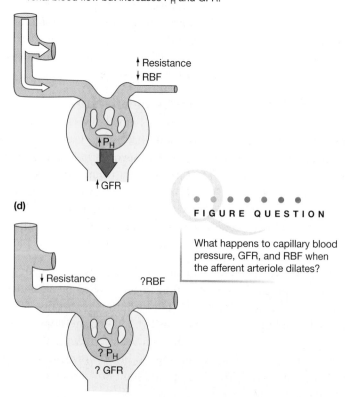

(d)

↓ Resistance   ?RBF

? P_H
? GFR

● ● ● ● ● ● ●
FIGURE QUESTION

What happens to capillary blood pressure, GFR, and RBF when the afferent arteriole dilates?

● **Figure 19-8** *Resistance changes in renal arterioles alter GFR and renal blood flow*

and the vessel becomes maximally dilated. However, vasodilation is not as effective at maintaining GFR as vasoconstriction because the afferent arteriole is normally fairly relaxed.

Consequently, if mean blood pressure drops below 80 mm Hg, GFR will decrease. This decrease is adaptive in the sense that if less plasma is filtered, the potential for fluid loss in the urine is decreased. In other words, a decrease in GFR helps the body conserve blood volume.

Tubuloglomerular Feedback   Tubuloglomerular feedback is a local control pathway in which fluid flow through the distal tubule influences GFR. The configuration of the nephron causes the distal tubule to pass between the afferent and efferent arterioles (Fig. 19-9 ●). The tubule and arteriolar walls are modified in the regions where they contact each other.

The modified portion of the distal tubule epithelium is called the **macula densa**. The adjacent walls of the arterioles have specialized smooth muscle cells called **juxtaglomerular cells (JG cells**; also known as granular cells). The JG cells secrete renin, an enzyme involved in salt and water balance. The complex of macula densa and JG cells is called the *juxtaglomerular apparatus*.

When fluid flow through the distal tubule increases as a result of increased GFR, the macula densa cells send a paracrine message to the neighboring afferent arteriole (Fig. 19-10 ●). The afferent arteriole constricts, increasing resistance and decreasing GFR.

The exact stimulus that signals increased tubule fluid flow is unclear. Experimental evidence indicates that some aspect of NaCl absorption at the macula densa is involved, because inhibitors of NaCl transport decrease the response. Other experiments show that there are several different paracrine signals, such as nitric oxide and adenosine, that pass from the macula densa to the arteriole.

Hormones and Autonomic Neurons Also Influence GFR

Systemic signals such as hormones and the autonomic nervous system also affect GFR by changing resistance in the arterioles or by altering the filtration coefficient. Integrating centers outside the kidney initiate these signals, often in response to changes in systemic blood pressure or volume.

Both the afferent and efferent arterioles are innervated by sympathetic neurons. Norepinephrine binding to α receptors on vascular smooth muscle causes vasoconstriction. If sympathetic activity is moderate, there is little effect on GFR.

If systemic blood pressure drops sharply, as occurs with hemorrhage or severe dehydration, sympathetically induced vasoconstriction of the arterioles decreases GFR and renal blood flow. This is an adaptive response that helps conserve fluid volume.

A variety of hormones influence arteriolar resistance. Among the most important are angiotensin II, a potent vasoconstrictor, and prostaglandins, which act as vasodilators.

These same hormones may affect the filtration coefficient by acting on podocytes or mesangial cells. Podocytes change the size of filtration slits. If the slits widen, more surface area is available for filtration, and GFR increases. Contraction of mesangial cells

(a) **(b)**

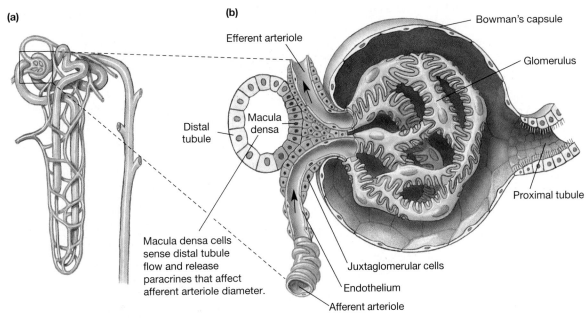

Efferent arteriole

Bowman's capsule

Glomerulus

Distal tubule

Macula densa

Proximal tubule

Macula densa cells sense distal tubule flow and release paracrines that affect afferent arteriole diameter.

Juxtaglomerular cells

Endothelium

Afferent arteriole

● **Figure 19-9**   *The juxtaglomerular apparatus*

Efferent arteriole

Bowman's capsule

Glomerulus

JG cells

Afferent arteriole

Proximal tubule

Macula densa

Distal tubule

Collecting duct

Loop of Henle

1 GFR ↑

2 Flow through tubule ↑

3 Flow past macula densa ↑

4 Paracrine from macula densa to afferent arteriole

5 Afferent arteriole constricts

Resistance in afferent arteriole ↑

Hydrostatic pressure in glomerulus ↓

GFR ↓

● **Figure 19-10**   *Tubuloglomerular feedback*

apparently changes the glomerular capillary surface area available for filtration. We still have much to learn about these processes, and physiologists are actively investigating them.

8. Systemic blood pressure remains constant but the afferent arteriole of a nephron constricts. What happens to renal blood flow and GFR in that nephron?

9. A person with cirrhosis of the liver has lower-than-normal levels of plasma proteins and a higher-than-normal GFR. Explain why a decrease in plasma protein concentration would cause an increase in GFR.

▶ REABSORPTION

Each day, 180 liters of filtered fluid pass from the glomerular capillaries into the nephrons, yet only about 1.5 liters are excreted in the urine. More than 99% of what filters therefore must be reabsorbed as the fluid moves through the nephron. Most reabsorption takes place in the proximal tubule. Finely regulated reabsorption takes place in the distal segments of the nephron and allows the kidneys to return ions and water to the plasma selectively as needed to maintain homeostasis.

One question you might be asking is, "Why bother to filter 180 L/day and then reabsorb 99% of it? Why not simply filter and excrete the 1% that needs to be eliminated?" There are two reasons. First, many foreign substances are filtered into the nephron but not reabsorbed. The high daily filtration rate helps clear such substances from the plasma very rapidly.

Second, filtering ions and water into the tubule simplifies their regulation. If a portion of filtrate that reaches the distal nephron is not needed to maintain homeostasis, it passes into the urine. With a high GFR, this excretion can occur quite rapidly. However, if the ions and water are needed, they are reabsorbed.

Reabsorption May Be Active or Passive

Reabsorption of water and solutes from the tubule lumen to the extracellular fluid depends upon active transport. Filtrate moving out of Bowman's capsule into the proximal tubule has the same solute concentrations as extracellular fluid. Therefore, the tubule cells must use active transport to create concentration gradients. Water osmotically follows solutes as they are reabsorbed.

Most reabsorption involves *transepithelial transport*, in which substances cross both the apical and basolateral membranes of the tubule epithelial cell [※ p. 148]. (A few substances can pass between the cells because this is a leaky epithelium, but we will ignore that route in this discussion.) The concentration gradient of a molecule determines the mechanism required to transport it across a membrane.

Molecules moving down their concentration (or electrochemical) gradient use open leak channels or facilitated diffusion carriers. Molecules that need to be pushed against their concentration gradient are moved by either primary or secondary active transport. Sodium is directly or indirectly involved in many instances.

● **Figure 19-11**   *Sodium reabsorption in the proximal tubule*

Active Transport of Sodium   Filtrate entering the proximal tubule is similar in ion composition to plasma, with a higher Na^+ concentration than is found within cells. Subsequently, Na^+ in the filtrate moves into the proximal tubule cell through open leak channels, moving down its electrochemical gradient (Fig. 19-11 ●). Once inside the tubule cell, Na^+ is actively transported into the extracellular fluid by the Na^+-K^+-ATPase on the basolateral membrane. The result is Na^+ reabsorption across the epithelium.

Secondary Active Transport: Symport with Sodium Sodium-linked secondary active transport in the nephron is responsible for the reabsorption of many substances, including glucose, amino acids, ions, and various organic metabolites. Figure 19-12● shows Na^+-dependent glucose movement across the epithelium [※ Fig. 5-28, p. 146].

The apical membrane contains a Na^+-glucose co-transporter that brings glucose into the cytoplasm against its concentration gradient by harnessing the energy of Na^+ moving down its electrochemical gradient. On the basolateral side of the cell, Na^+ is pumped out by the Na^+-K^+-ATPase, while glucose diffuses out with the aid of a facilitated diffusion carrier. The same basic pattern holds for the other molecules absorbed by Na^+-dependent transport: an apical symport protein and a basolateral facilitated diffusion carrier.

Passive Reabsorption: Urea Reabsorption Urea has no active transporters in the proximal tubule, but it will move across the epithelium if there is a urea concentration gradient.

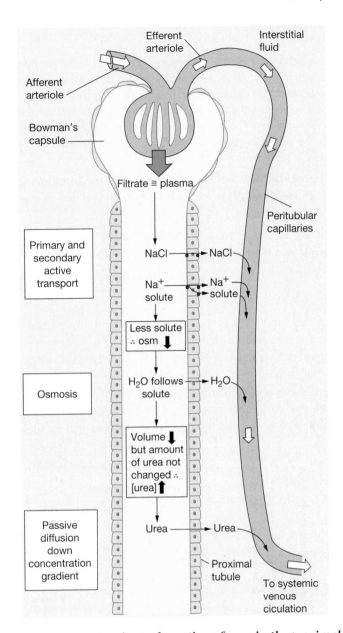

● **Figure 19-12** *Sodium-linked glucose reabsorption in the proximal tubule*

① Na⁺ moving down its gradient pulls glucose into the cell against its gradient.

② Glucose diffuses out basolateral side of cell.

③ Na⁺ is pumped out by Na⁺-K⁺-ATPase.

Initially, urea concentrations in the filtrate and extracellular fluid are equal. However, the active transport of Na⁺ and other solutes creates a urea concentration gradient by the following process.

When Na⁺ and other solutes are reabsorbed from the lumen of the nephron, the transfer of osmotically active particles makes the extracellular fluid more concentrated than the fluid remaining in the lumen (Fig. 19-13 ●). In response to the osmotic gradient, water moves by osmosis across the epithelium. Up to this point, no urea molecules have moved out of the lumen because there has been no urea concentration gradient. Now, however, when water leaves the lumen, the *concentration* of urea goes up because the same amount of urea is contained within a smaller volume. Once a concentration gradient for urea exists, urea diffuses out of the lumen into the extracellular fluid.

Transcytosis: Plasma Proteins Filtration of plasma at the glomerulus normally excludes most plasma proteins, but some smaller protein hormones and enzymes can pass through the filtration barrier. Most filtered proteins are reabsorbed in the proximal tubule. Normally, only trace amounts of protein appear in urine.

● **Figure 19-13** *Passive reabsorption of urea in the proximal tubule*

Proteins are too large to be reabsorbed by carriers or through channels. Instead they enter epithelial cells by endocytosis at the apical membrane. Once in the cell, they may be digested and released as amino acids or delivered intact to the extracellular fluid (transcytosis, [● p. 148]).

Saturation of Renal Transport Plays an Important Role in Kidney Function

Most transport in the nephron is mediated by membrane proteins and exhibits saturation, specificity, and competition, the three characteristics of mediated transport [● p. 138].

Saturation refers to the maximum rate of transport that occurs when all available carriers are occupied with substrate (i.e., have become saturated with substrate). At substrate concentrations below the saturation point, transport rate is directly related to substrate

RUNNING PROBLEM

Uric acid, the molecule that causes gout, is a normal product of purine metabolism. Increased uric acid production may be associated with cell and tissue breakdown, or it may occur as a result of inherited enzyme defects. Uric acid in the blood filters freely into Bowman's capsule but is almost totally reabsorbed in the proximal tubule. Some uric acid that appears in the urine is secreted into the proximal tubule.

Question 2: Why would uric acid levels in the blood go up when cell breakdown increases? (Hint: Purines are part of what biomolecules?)

Question 3: Based on what you have learned about uric acid, predict two different ways a person may develop hyperuricemia.

concentration (Fig. 19-14●). At substrate concentrations equal to or above the saturation point, transport occurs at a maximum rate. The transport rate at saturation is called the **transport maximum**, or T_m.

Glucose reabsorption in the nephron provides us with an excellent example of the consequences of saturation. At normal plasma glucose concentrations, all glucose that enters the nephron is reabsorbed before it reaches the end of the proximal tubule. The

GRAPH QUESTIONS

What is the transport rate at the following plasma substrate concentrations: 3 mg/mL, 5 mg/mL, 8 mg/mL? At what plasma substrate concentration is the transport rate 4 mg/min?

● **Figure 19-14** *Saturation of mediated transport* The transport rate of a substance is proportional to the plasma concentration of the substance, up to the point at which transporters become saturated. Once saturation occurs, transport rate reaches a maximum, known as the transport maximum. The plasma concentration of substrate at which the transport maximum occurs is called the renal threshold.

tubule epithelium is well supplied with carriers to capture the glucose as it flows past.

But what happens if blood glucose concentrations become excessive, as they do in the disease diabetes mellitus? In that case, glucose is filtered faster than the carriers can reabsorb it. The carriers become saturated and are unable to reabsorb all the glucose that flows through the tubule. As a result, some glucose escapes reabsorption and is excreted in the urine.

Consider the following analogy. Assume that the carriers are like seats on a train at Disney World. Instead of boarding the stationary train from a stationary platform, passengers step onto a moving sidewalk that rolls them past the train. As the passengers see an open seat, they grab it. But if more people are allowed onto the moving sidewalk than there are seats in the train, some people will not find seats. And because the sidewalk is moving these people past the train toward an exit, the extra people cannot wait for the next train. Instead, they are transported out the exit.

In the same fashion, glucose molecules entering the kidney tubule as part of the filtrate are like passengers stepping onto the moving sidewalk. In order to be reabsorbed, each glucose molecule must bind to a transporter as the filtrate flows through the proximal tubule. If only a few glucose molecules filter at a time, each one will find a free transporter and be reabsorbed, just as a small number of people on the moving sidewalk all find seats on the train. However, if glucose molecules filter into the tubule faster than the glucose carriers can transport them, some glucose remains in the lumen and is excreted in the urine.

Figure 19-15● is a graphic representation of glucose handling by the kidney. Graph (a) shows that the filtration rate of glucose from plasma into Bowman's capsule is proportional to the plasma concentration of glucose. Because filtration does not exhibit saturation, the graph continues infinitely in a straight line. The concentrations of glucose in plasma and in filtrate are always equal.

Figure 19-15b plots the reabsorption rate of glucose in the proximal tubule against the plasma concentration of glucose. Reabsorption exhibits a maximum transport rate (T_m) when the carriers reach saturation.

Figure 19-15c plots the excretion rate of glucose in relation to the plasma concentration of glucose. When plasma glucose concentrations are low enough that 100% of the filtered glucose is reabsorbed, no glucose is excreted. Once the carriers reach saturation, glucose excretion begins. The plasma concentration at which glucose first appears in the urine is called the **renal threshold**.

Figure 19-15d is a composite graph that compares filtration, reabsorption, and excretion of glucose. Recall from our earlier discussion that

Amount excreted =
  amount filtered − amount reabsorbed + amount secreted

For glucose, which is not secreted, the equation can be rewritten as

Amount glucose excreted =
  amount glucose filtered − amount glucose reabsorbed

(a) Filtration of glucose is proportional to the plasma concentration.

(b) Reabsorption of glucose is proportional to plasma concentration until the transport maximum (T_m) is reached.

(c) Glucose excretion is zero until the renal threshold is reached.

(d) Composite graph shows the relationship between filtration, reabsorption, and excretion of glucose.

● **Figure 19-15** *Glucose handling by the nephron*

Under normal conditions, all filtered glucose is reabsorbed. In other words, filtration is equal to reabsorption.

Notice in Figure 19-15d that the lines representing filtration and reabsorption are identical up to the plasma glucose concentration that equals the renal threshold. If filtration equals reabsorption, the algebraic difference between the two is zero, and there is no excretion.

Once the renal threshold is reached, filtration begins to exceed reabsorption. Notice on the graph that the filtration and reabsorption lines diverge once the transport maximum for reabsorption is reached. The difference between the filtration line and the reabsorption line represents the excretion rate:

$$\text{Excretion} = \underset{\text{(increasing)}}{\text{filtration}} - \underset{\text{(constant)}}{\text{reabsorption}}$$

Excretion of glucose in the urine is called **glucosuria**, or **glycosuria** [*-uria*, in the urine], and usually indicates an elevated blood glucose concentration. Rarely, glucose appears in the urine even though the blood glucose concentrations are normal. This situation is due to a genetic disorder in which the tubule does not make enough carriers.

Peritubular Capillary Pressures Favor Reabsorption

The capillary reabsorption we have just discussed refers to the movement of solutes and water from the tubule lumen to the interstitial fluid. But how does that fluid get into the capillary? The answer is the low hydrostatic pressure along the entire length of the peritubular capillaries, which favors reabsorption.

Peritubular capillaries have an average hydrostatic pressure of 10 mm Hg, in contrast to the glomerular capillaries, where hydrostatic pressure averages 55 mm Hg. Colloid osmotic pressure, which favors movement of fluid into the capillaries, is 30 mm Hg. As a result, the pressure gradient in peritubular capillaries is 20 mm Hg, favoring the reabsorption of fluid. Fluid that is reabsorbed passes from the capillaries to the venous circulation and returns to the heart.

▶ SECRETION

Secretion is the transfer of molecules from the extracellular fluid into the lumen of the nephron. Secretion, like reabsorption, depends mostly on membrane transport systems. The secretion of K^+ and H^+ by the nephron is important in the homeostatic regulation of those ions (see Chapter 20). In addition, many organic compounds are secreted in the kidney. These compounds include both metabolites produced in the body and substances brought into the body.

Secretion enables the nephron to enhance excretion of a molecule. If a substance is filtered and not reabsorbed, it will be excreted very effectively. But if it is filtered, not reabsorbed, *and* the nephron removes more of it from the peritubular capillaries by secretion, excretion will be even more effective.

Secretion is an active process because it requires moving substrates against their concentration gradients. Most organic compounds are transported across the tubule epithelium by secondary active transport.

Competition Decreases Penicillin Secretion

An interesting and important example of an organic molecule secreted by the kidney tubule is the antibiotic **penicillin**. Many people today take antibiotics for granted, but until the early decades of the twentieth century, infections were a leading cause of death.

In 1928, Alexander Fleming discovered a substance in the *Penicillium* bread mold that retarded the growth of bacteria. But the antibiotic was difficult to isolate, so it did not become available for clinical use until the late 1930s. During World War II, penicillin made a major difference in the number of deaths and amputations caused by infected wounds. The only means of producing penicillin, however, was to isolate it from bread mold, and supplies were limited.

Demand for the drug was heightened by the fact that kidneys secrete penicillin. Renal secretion is so efficient at clearing foreign molecules from the blood that within three to four hours of administering a dose of penicillin, about 80% has been excreted in the urine. During the war, the drug was in such short supply that it was common procedure to collect the urine from patients being treated with penicillin so that the antibiotic could be isolated and reused.

RUNNING PROBLEM

Michael finds it amazing that a metabolic problem could lead to pain in his big toe. "There's a way to treat gout, right?" he asks. Dr. Garcia explains that the treatment includes anti-inflammatory agents, lots of water, and avoidance of alcohol, which can trigger gout attacks. "In addition, I would like to put you on a *uricosuric agent*, like probenecid, that will enhance renal excretion of uric acid," replies Dr. Garcia. "By enhancing excretion, we can reduce uric acid levels in your blood and thus provide relief." Michael agrees to try these measures, and on his way home, stops off at a convenience store for a case of his favorite bottled water.

Question 4: Uric acid is reabsorbed into the proximal tubule on a membrane transporter. Uricosuric agents are also organic acids. Knowing those two facts, explain how uricosuric agents might enhance excretion of uric acid.

This was not a satisfactory solution, however, so researchers looked for a way to slow penicillin secretion. They hoped to find a molecule that could compete with penicillin for the organic acid transporter responsible for secretion [p. 138]. That way, when presented with both drugs, the carrier would bind preferentially to the competitor and secrete it, leaving penicillin behind in the blood. A synthetic compound named *probenecid* was the answer. When probenecid is administered concurrently with penicillin, the transporter removes probenecid preferentially, prolonging the activity of penicillin. Once mass-produced synthetic penicillin became available and supply was no longer a problem, the medical use of probenecid declined.

▶ EXCRETION

Urine output is the result of all the processes that take place in the kidney. By the time fluid reaches the end of the nephron, it bears little resemblance to the filtrate that started in Bowman's capsule. Glucose, amino acids, and useful metabolites are gone, having been reabsorbed into the blood. Organic wastes are more concentrated. The concentrations of ions and water in the urine are highly variable depending on the state of the body.

Although excretion tells us what the body is eliminating, excretion by itself cannot tell us the details of renal function. Recall that for any substance,

Amount excreted =
    amount filtered − amount reabsorbed + amount secreted

Simply looking at the excretion rate of a substance tells us nothing about how the kidney handles that substance. The excretion rate of a substance depends on (1) its filtration rate and (2) whether the substance is reabsorbed or secreted as it passes through the nephron.

These three processes in the kidney are often of interest, however. For example, clinicians use information about a person's glomerular filtration rate as an indicator of overall kidney function. Pharmaceutical companies developing drugs must provide the Food and Drug Administration with complete information on how the human kidney handles each new compound.

But how can investigators dealing with living humans assess filtration, reabsorption, and secretion at the level of the individual nephron? They have no way to do this directly: the kidneys are not easily accessible, and the nephrons are microscopic. Scientists therefore had to develop a technique that would allow them to assess renal function using only analysis of the urine and blood. From their work came the concept of clearance.

Clearance Is a Noninvasive Way to Measure GFR

Clearance of a solute describes how many milliliters of plasma passing through the kidneys have been totally cleared of that solute in a given period of time. Before we jump into the mathematical expression of clearance, let's look at an example that shows how clearance relates to kidney function.

598 603 612 616 619

For our example, we will use **inulin**, a polysaccharide isolated from the tuberous roots of dahlia plants. (Careful! Inulin is not the same as *insulin*, the protein hormone that regulates glucose metabolism.) Scientists discovered from experiments with isolated nephrons that inulin injected into the plasma filters freely into the nephron. As it passes through the kidney tubule, inulin is neither reabsorbed nor secreted. Thus, 100% of the inulin that filters is excreted.

How does this relate to clearance? Let's take a look (Fig. 19-16). Assume for the sake of this discussion that 100% of a filtered volume of plasma is reabsorbed. (This is not too far off the actual value, which is more than 99%.) Inulin has been injected so that the plasma concentration is 4 inulin per 100 mL plasma.

If the GFR is 100 mL plasma filtered per minute, we can calculate the filtration rate, or *filtered load*, of inulin using the following equation:

Filtered load of a substance =

plasma concentration of the substance × GFR

Filtered load of inulin =

(4 inulin/100 mL plasma) × 100 mL plasma filtered/min

Filtered load of inulin = 4 inulin filtered/min

As inulin and the filtered plasma pass along the nephron, the plasma is reabsorbed but the inulin remains in the tubule. The reabsorbed plasma has no inulin in it, so it has been totally *cleared* of inulin. The *inulin clearance* therefore is 100 mL of plasma cleared/min. At the same time, the excretion rate of inulin is 4 inulin excreted per minute.

● **Figure 19-16** *Inulin clearance*

What good is this information? For one thing, we can use it to calculate the glomerular filtration rate. Notice from Figure 19-16 that inulin clearance (100 mL plasma cleared/min) is equal to the GFR (100 mL plasma filtered/min).

For any substance (such as inulin) that is freely filtered but neither reabsorbed nor secreted, its clearance is equal to the GFR.

Now let's show mathematically that inulin clearance is equal to the GFR. We saw above that:

Filtered load of inulin =

plasma concentration of inulin × GFR   (1)

But we know that 100% of the inulin that filters is excreted. In other words:

Filtered load of inulin = excretion rate of inulin   (2)

We can substitute excretion rate for filtered load in equation (1) by using algebra (if A = B and A = C, then B = C).

Excretion rate of inulin =

plasma concentration of inulin × GFR   (3)

This equation can be rearranged to read:

$$\text{GFR} = \frac{\text{excretion rate of inulin}}{\text{plasma concentration of inulin}} \quad (4)$$

It turns out that the right side of the equation is identical to the clearance equation for inulin. The equation for clearance is:

$$\text{Clearance of a substance} = \frac{\text{urine excretion rate of the substance (mg/min)}}{\text{plasma concentration of the substance (mg/mL plasma)}} \quad (5)$$

For inulin:

$$\text{Clearance of inulin} = \frac{\text{excretion rate of inulin}}{\text{plasma concentration of inulin}} \quad (6)$$

By using algebra again:

GFR = clearance of inulin   (7)

So why is this important? For one thing, you have just learned how we can measure GFR in a living human by taking only blood and a timed urine sample. Try the example in the concept check box to see if you understand.

Inulin is not practical for routine clinical applications because it does not occur naturally in the body and must be administered by continuous intravenous infusion. As a result, inulin use is restricted to research. Unfortunately, no substance that occurs naturally in the human body is handled by the kidney exactly like inulin.

In clinical settings, physicians use creatinine to estimate GFR. **Creatinine** is the breakdown product of phosphocreatine, an energy-storage compound found primarily in muscles. It is constantly produced by the body and does not need to be administered. Normally, the production and breakdown rates of phosphocreatine are relatively constant, and plasma concentration of creatinine does not vary much.

RUNNING PROBLEM

Three weeks later, Michael is back in Dr. Garcia's office. The anti-inflammatory drugs and probenecid had eliminated the pain in his toe, but last night, he had gone to the hospital with a very painful kidney stone. "We'll have to wait until the analysis comes back," says Dr. Garcia, "But I will guess that it is a uric acid stone. Did you drink as much water as I told you to?" Sheepishly, Michael admits that he had good intentions, but could never find the time at work to get out his bottled water. "You have to drink enough water while on this drug to produce 3000 milliliters or more of urine a day. That's over three quarts. Otherwise, you may end up with another uric acid kidney stone." Michael agrees that this time, he will follow instructions to the letter.

Question 5: Explain why not drinking enough water while taking uricosuric agents may cause uric acid stone formation in the urinary tract.

598   603   612   614   619

Although creatinine is always present in the plasma and is easy to measure, it is not the perfect molecule for estimating GFR because a small amount is secreted into the urine. However, the error is small enough that, in most people, *creatinine clearance* is routinely used to estimate GFR.

10. If plasma creatinine = 1.8 mg/100 mL plasma, urine creatinine = 1.5 mg/mL urine, and urine volume is 1100 mL in 24 hours, what is the creatinine clearance? What is this person's GFR?

Knowing the GFR Helps Us Determine How the Kidney Handles a Solute

Once we know a person's GFR, we can determine how the kidney handles any solute by measuring the solute's plasma concentration and excretion rate. If we assume that the solute is freely filtered at the glomerulus, we know that:

Filtered load of a solute =
plasma concentration of the solute × GFR

If we compare the filtered load of the solute with its excretion rate, then we know how the nephron handled that substance (Table 19-2). For example, if less of the substance appears in the urine than was filtered, then there was net reabsorption (excreted = filtered − reabsorbed). If *more* of the substance appears in the urine than was filtered, there must have been net secretion of the substance into the lumen (excreted = filtered + secreted). If the same amount of the substance is filtered and excreted, then the substance is handled like inulin: neither reabsorbed nor secreted.

For example, suppose glucose is present in the plasma at 100 mg glucose/dL plasma, and the GFR is calculated from creatinine clearance to be 125 mL plasma/min:

Filtered load of glucose =
(100 mg glucose/100 mL plasma) × 125 mL plasma/min

Filtered load of glucose = 125 mg glucose/min

But there is no glucose present in the urine: glucose excretion rate is zero. Because glucose was filtered at a rate of 125 mg/min but excreted at a rate of 0 mg/min, it must have been totally reabsorbed.

Clearance Can Be Used to Determine Renal Handling of a Substance

Clearance values are also used to determine how the nephron handles a solute filtered into it. In this method, researchers calculate creatinine or inulin clearance, then compare the clearance of the solute being investigated with the creatinine or inulin clearance (Table 19-2). If the clearance of the solute is less than the inulin clearance, the solute has been reabsorbed.

On the other hand, if the clearance of the solute is higher than the inulin clearance, some solute has been secreted into the nephron. More plasma was cleared of the solute than was filtered, so the additional material must have been removed from the plasma by secretion.

Figure 19-17 ● shows clearance of three different molecules: glucose, urea, and penicillin. These examples are similar to the earlier figure on inulin clearance (Fig. 19-16), where inulin clearance was 100 mL plasma/min. All solutes have the same concentration, 4 molecules/100 mL plasma. GFR is 100 mL/min, and the entire 100 mL of plasma volume that filters is reabsorbed.

Table 19-2: Renal Handling of Solutes

For any molecule X that is freely filtered at the glomerulus:	
If filtration rate is greater than excretion rate,	there is net reabsorption of X.
If excretion rate is greater than filtration rate,	there is net secretion of X.
If filtration and excretion rate are the same,	X passes through the nephron without net reabsorption or secretion.
If the clearance of X is less than inulin clearance,	there is net reabsorption of X.
If the clearance of X is equal to inulin clearance,	X is neither reabsorbed nor secreted.
If the clearance of X is greater than inulin clearance,	there is net secretion of X.

(a) Glucose clearance

Filtration
(100 mL/min)

① Glucose
molecules

②

③

No glucose
excreted

100 mL,
100% glucose
reabsorbed

④ Clearance of
glucose = 0 mL/min

KEY

= 100 mL of
plasma/filtrate

① Plasma concentration
is 4/100 mL

② GFR = 100 mL /min

③ 100 mL plasma is
reabsorbed.

④ Clearance depends on
renal handling of solute

(b) Urea clearance

Filtration
(100 mL/min)

① Urea
molecules

②

③

50% of urea
excreted

100 mL,
50% of urea
reabsorbed

④ Urea clearance
= 50 mL/min

(c) Penicillin clearance

Filtration
(100 mL/min)

① Penicillin
molecules

②

③

More penicillin
is excreted than
was filtered

100 mL,
0 penicillin
reabsorbed.
Some penicillin
secreted

④ Penicillin
clearance =
150 mL/min

● **Figure 19-17** *The relationship between clearance and excretion* The figure represents the events taking place in one minute. For simplicity, 100% of the filtered volume was reabsorbed.

For any solute, its clearance reflects how the kidney tubule handles it. For example, 100% of the glucose that filters is reabsorbed and glucose clearance is zero (Fig. 19-17a). Urea is partially reabsorbed (Fig. 19-17b, point 3). Four molecules filter, but only two are reabsorbed. Consequently, urea clearance is 50 mL plasma cleared per minute. Urea and glucose clearance are both less than inulin clearance, which tells you that urea and glucose have been reabsorbed.

Penicillin (Fig. 19-17c) is filtered, not reabsorbed, and additional penicillin molecules are secreted from plasma in the per-

itubular capillaries. In this example, an extra 50 mL of plasma have been cleared of penicillin in addition to the original 100 mL that filtered. The penicillin clearance therefore is 150 mL plasma cleared per minute. Penicillin clearance is greater than inulin clearance, which tells you that there is net secretion of penicillin.

Note that a comparison of clearance values tells you only the *net* handling of the solute. It does not tell you if a molecule is both reabsorbed and secreted. For example, nearly all K^+ filtered is reabsorbed in the proximal tubule and loop of Henle, then a small amount is secreted back into the lumen by the distal nephron. On the basis of K^+ clearance, it appears that there was only net reabsorption.

Clearance calculations are simple because all you need to know are urine excretion rates and their plasma concentrations, both easily obtained. If you also know either inulin or creatinine clearance, then you can determine the renal handling of any compound.

▶ MICTURITION

Once fluid leaves the collecting ducts, it flows into the renal pelvis then down the ureter to the bladder with the help of rhythmic smooth muscle contractions. The bladder is a hollow organ whose walls contain well-developed layers of smooth muscle. In the bladder, urine is stored until released in the process known as urination, voiding, or more formally, **micturition** [*micturire*, to desire to urinate].

The bladder is a hollow sac that can expand to hold a volume of about 500 mL. The neck of the bladder is continuous with the urethra, a single tube through which urine passes to reach the external environment. The opening between the bladder and urethra is closed by two rings of muscle called **sphincters** (Fig. 19-18 ●).

(a) Bladder at rest

(b) Micturition

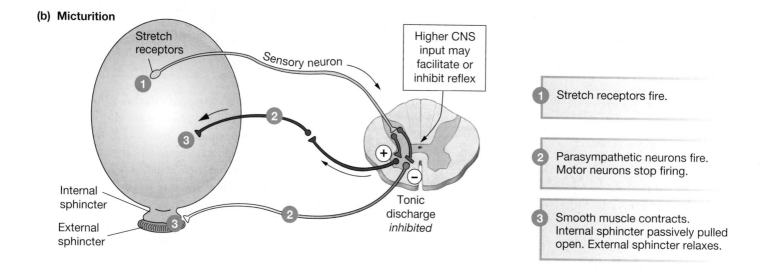

● **Figure 19-18** *The micturition reflex*

1. Stretch receptors fire.

2. Parasympathetic neurons fire. Motor neurons stop firing.

3. Smooth muscle contracts. Internal sphincter passively pulled open. External sphincter relaxes.

The *internal sphincter* is a continuation of the bladder wall and consists of smooth muscle. Its normal tone keeps it contracted. The *external sphincter* is a ring of skeletal muscle controlled by somatic motor neurons. Tonic stimulation from the central nervous system maintains contraction of the external sphincter except during urination.

Micturition is a simple spinal reflex that is subject to both conscious and unconscious control from higher brain centers. As the bladder fills with urine and its walls expand, stretch receptors send signals via sensory neurons to the spinal cord (Fig. 19-18b). There, the information is integrated and transferred to two different sets of neurons. The stimulus of a full bladder *excites* parasympathetic neurons leading to the smooth muscle in the bladder wall. The smooth muscle contracts, increasing the pressure on the bladder contents. Simultaneously, somatic motor neurons leading to the external sphincter are *inhibited*.

Contraction of the bladder occurs in a wave that pushes urine downward toward the urethra. Pressure exerted by urine forces the internal sphincter open while the external sphincter relaxes. Urine passes into the urethra and out of the body, aided by gravity.

This simple micturition reflex occurs primarily in infants who have not yet been toilet trained. A person who has been toilet trained acquires a learned reflex that keeps the micturition reflex inhibited until the person consciously desires to urinate. The learned reflex involves additional sensory fibers in the bladder that signal the degree of fullness. Centers in the brain stem and cerebral cortex receive that information and override the basic micturition reflex by directly inhibiting the parasympathetic fibers and by reinforcing contraction of the external sphincter. When an appropriate time to urinate arrives, those same centers remove the inhibition and facilitate the reflex by inhibiting the external sphincter.

In addition to conscious control of urination, various subconscious factors can affect the micturition reflex. "Bashful bladder" is a condition in which a person is unable to urinate in the presence of other people despite the conscious intent to do so. The sound of running water facilitates micturition and is often used to help patients urinate if the urethra is irritated from insertion of a *catheter*, a tube inserted into the bladder to drain it passively.

Gout

In this running problem, you learned that gout, which often presents as a debilitating pain in the big toe, is a metabolic problem whose cause and treatment may be linked to kidney function. Further check your understanding of this running problem by comparing your answers against those in the summary table. To learn more about the causes, symptoms, and treatments for gout, go to Medline Plus (*http://www.nlm.nih.gov/medicineplus*) and search under *G* for gout.

	QUESTION	FACTS	INTEGRATION AND ANALYSIS
1	Trace the route followed by kidney stones when they are excreted.	Kidney stones often form in the renal pelvis.	From the renal pelvis, a stone passes down the ureter and into the urinary bladder. From there, it leaves the body through the urethra.
2	Why would uric acid levels in the blood go up when cell breakdown increases?	Purines include adenine and guanine, which are components of DNA, RNA, and ATP.	When a cell dies, the nucleus and other components are broken down into biomolecules. Degradation of DNA, RNA, and ATP would increase purine production, which in turn would increase uric acid.
3	Based on what you have learned about uric acid, predict two different ways a person may develop hyperuricemia.	Hyperuricemia is a condition of elevated uric acid levels in the blood. Uric acid is made from purines. It is filtered, reabsorbed, and secreted by the kidney tubule.	Hyperuricemia results from overproduction of uric acid or from failure of the kidneys to excrete uric acid. Because uric acid normally gets into the urine by secretion, a defect in the renal secretory mechanism would lead to hyperuricemia.
4	Uric acid is reabsorbed in the proximal tubule on a membrane transporter. Uricosuric agents are organic acids. Knowing those two facts, explain how uricosuric agents might enhance excretion of uric acid.	Mediated transport exhibits competition, where related molecules compete for one transporter. Usually, one molecule will bind preferentially and therefore inhibit transport of the second molecule.	Uricosuric agents such as probenecid may compete with uric acid for the proximal tubule transporter. This would block uric acid access to the transporter, leaving it in the lumen and increasing its excretion.
5	Explain why not drinking enough water while taking uricosuric agents may cause uric acid stone formation in the urinary tract.	Uric acid crystals form when uric acid concentrations exceed a critical level.	If a person drinks large volumes of water, the excess water will be excreted by the kidneys. Large amounts of water will dilute the urine and thus prevent high concentrations of uric acid.

CHAPTER SUMMARY

The urinary system, like the lungs, uses the principle of *mass balance* to maintain homeostasis. Put simply, to keep the internal environment constant, what comes into the body must go out. The second *portal system* of the body is found in the kidney, where it carries out filtration in the first capillary bed and absorption in the second. The *pressure-flow-resistance* relationship that you encountered in the cardiovascular and pulmonary systems also plays a role in glomerular filtration and urinary excretion. *Compartmentation* is illustrated by the movement of water and solutes between the internal and external environments as filtrate is modified along the nephron. Reabsorption and secretion of solutes depend on *molecular interactions* and the *movement of molecules across membranes* of the tubule cells.

Functions of the Kidneys

1. The kidneys regulate extracellular fluid volume and osmolarity, maintain ion balance, regulate pH, excrete wastes and foreign substances, and participate in endocrine pathways. (p. 599)

Anatomy of the Urinary System

IP Urinary System: Glomerular Filtration

2. The **urinary system** is composed of kidneys, ureters, bladder, and urethra. (p. 600; Fig. 19-1a)

3. Each **kidney** has about 1 million **nephrons**. In cross-section, a kidney is arranged into an outer **cortex** and inner **medulla**. (p. 600; Fig. 19-1c)

4. Blood flow through a nephron goes from **afferent arteriole** to **glomerulus** to **efferent arteriole** to **peritubular capillaries**. The **vasa recta** capillaries dip into the medulla. (p. 600; Fig. 19-1g)

5. Fluid filters from the glomerulus into **Bowman's capsule**. From there, it flows through the **proximal tubule, loop of Henle, distal tubule,** and **collecting duct,** then drains into the **renal pelvis. Urine** goes through the **ureter** to the **urinary bladder.** (pp. 600, 603; Fig. 19-1c, j)

Overview of Kidney Function

6. **Filtration** is the movement of fluid from plasma into Bowman's capsule. **Reabsorption** is the movement of filtered material from nephron to blood. **Secretion** is the movement of selected molecules from blood into the nephron. (pp. 603–604; Fig. 19-2)

7. Average urine volume is 1.5 L/day. Osmolarity varies between 50 and 1200 mOsM. (p. 604; Tbl. 19-1)

8. The amount of a solute excreted equals the amount filtered minus the amount reabsorbed plus the amount secreted. (p. 604; Fig. 19-3)

Filtration

IP Urinary System: Glomerular Filtration

9. Filtered solutes must pass through glomerular capillary endothelium, **basal lamina**, and Bowman's capsule epithelium before reaching the lumen of Bowman's capsule. (pp. 605–606; Fig. 19-4)

10. Filtration allows most components of plasma to enter the tubule but excludes blood cells and most plasma proteins. (p. 605)

11. The capsule epithelium has specialized cells called **podocytes** that wrap around the capillaries and create **filtration slits. Mesangial cells** are associated with glomerular capillaries. (pp. 605–606; Fig. 19-4)

12. One-fifth of renal plasma flow filters into the nephrons. The percentage of total plasma volume that filters is called the **filtration fraction.** (p. 606; Fig. 19-5)

13. Hydrostatic pressure in glomerular capillaries averages 55 mm Hg, favoring filtration. Opposing filtration are colloid osmotic pressure of 30 mm Hg and hydrostatic capsule fluid pressure averaging 15 mm Hg. The net driving force is 10 mm Hg, favoring filtration. (p. 606; Fig. 19-6)

14. The **glomerular filtration rate (GFR)** is the amount of fluid that filters into Bowman's capsule per unit time. Average GFR is 125 mL/min, or 180 L/day. (p. 607)

15. Blood pressure in glomerular capillaries can be altered by changing resistance in renal arterioles. (p. 607; Fig. 19-8)

16. Autoregulation of glomerular filtration is accomplished by a **myogenic response** of vascular smooth muscle to respond to pressure changes and by **tubuloglomerular feedback.** When fluid flow through the distal tubule increases, the **macula densa** cells send a paracrine signal to the afferent arteriole, which constricts. (pp. 607–608; Fig. 19-10)

17. Reflex control of GFR is mediated through systemic signals such as hormones and the autonomic nervous system. (p. 608)

Reabsorption

IP Urinary System: Early Filtrate Processing

18. Most reabsorption takes place in the proximal tubule. Finely regulated reabsorption takes place in the later segments of the nephron. (p. 610)

19. Reabsorption of water and solutes depends upon active transport of solutes from the lumen to the extracellular fluid. Most reabsorption involves transepithelial transport. (p. 610)

20. Na^+ in the lumen moves into tubule cells through apical leak channels, then is actively transported to the extracellular fluid by Na^+-K^+-ATPase on the basolateral membrane. (p. 610; Fig. 19-11)

21. Glucose, amino acids, ions, and various organic metabolites are reabsorbed by Na^+-linked secondary active transport. (p. 610; Fig. 19-12)

22. The active transport of Na^+ and other solutes creates concentration gradients for passive reabsorption of urea and other solutes. (p. 611; Fig. 19-13)

23. Most renal transport is mediated by membrane proteins and exhibits **saturation**, specificity, and competition. The **transport maximum**, or T_m, is the transport rate at saturation. (pp. 611–612; Fig. 19-14)

24. The **renal threshold** is the plasma concentration at which a substance first appears in the urine. (p. 612; Fig. 19-14)

25. Peritubular capillaries reabsorb fluid along their entire length. (p. 613)

Secretion

26. Secretion enhances excretion by removing solutes from the peritubular capillaries. K^+, H^+, and a variety of organic compounds are secreted. (p. 613)

27. Molecules that compete for renal carriers slow the secretion of a molecule. (p. 614)

Excretion

28. The excretion rate of a solute depends on (1) its filtered load and (2) whether it is reabsorbed or secreted as it passes through the nephron. (p. 614)

29. **Clearance** describes how many milliliters of plasma passing through the kidneys have been totally cleared of a solute in a given period of time. (p. 614)

30. **Inulin** clearance is equal to GFR. In clinical settings, **creatinine** is used to measure GFR. (p. 615; Fig. 19-16)

31. Clearance can be used to determine how the nephron handles a solute filtered into it. (p. 616; Fig. 19-17)

Micturition

32. The external sphincter of the bladder is skeletal muscle that is tonically contracted except during urination. (p. 618; Fig. 19-18)

33. Micturition is a simple spinal reflex subject to conscious and unconscious control. Parasympathetic neurons cause contraction of the smooth muscle in the bladder wall. Somatic motor neurons leading to the external sphincter are simultaneously inhibited. (p. 618)

QUESTIONS

LEVEL ONE *Reviewing Facts and Terms*

1. List and explain the significance of the five characteristics of urine that can be found by physical examination.

2. List and explain the six major kidney functions.

3. At any given time, what percentage of the cardiac output goes to the kidneys?

4. List the major structures of the urinary system in sequence from the kidneys to the urine leaving the body. Give a brief function for each structure.

5. Arrange these structures in the order that a drop of water entering the nephron would encounter them:
   (a) afferent arteriole
   (b) Bowman's capsule
   (c) collecting duct
   (d) distal tubule
   (e) glomerulus
   (f) loop of Henle
   (g) proximal tubule
   (h) renal pelvis

6. Name the three filtration barriers that solutes must cross as they move from plasma to the lumen of Bowman's capsule. What components of blood are usually excluded by these layers?

7. What force(s) promotes glomerular filtration? What force(s) opposes it? What is meant by the term *net driving force?*

8. What does the abbreviation GFR stand for? What is a typical numerical value for GFR in milliliters per minute? In liters per day?

9. Identify the following structures, then explain their significance in renal physiology:
   (a) juxtaglomerular apparatus
   (b) macula densa
   (c) mesangial cell
   (d) podocyte
   (e) sphincters in the bladder
   (f) renal cortex

10. In which segment of the nephron does most reabsorption take place? When a molecule or ion is reabsorbed from the lumen of the nephron, where does it go? If a solute is filtered and not reabsorbed from the tubules, where does it go?

11. Match the molecule with its primary mode(s) of transport across the kidney epithelium.
   (a) Na^+
   (b) glucose
   (c) urea
   (d) plasma proteins
   (e) water

   1. transcytosis
   2. primary active transport
   3. secondary active transport
   4. facilitated diffusion
   5. movement through open channels
   6. simple diffusion through the phospholipid bilayer

12. List three solutes secreted by the kidney tubule.

13. What solute, normally present in the body, is used to estimate GFR in humans?

14. What is micturition?

LEVEL TWO *Reviewing Concepts*

15. Map the following terms. You may add additional terms if you like.

α receptor	glomerulus
afferent arteriole	JG cells
autoregulation	macula densa
basal lamina	mesangial cell
Bowman's capsule	myogenic autoregulation
capillary blood pressure	norepinephrine
capsule fluid pressure	paracrine
colloid osmotic pressure	plasma proteins
efferent arteriole	podocyte
endothelium	resistance
epithelium	vasoconstriction
GFR	

16. Define, compare, and contrast the following sets of terms and/or events:
   (a) filtration, secretion, and excretion
   (b) saturation, transport maximum, and renal threshold
   (c) probenecid, creatinine, inulin, and penicillin
   (d) clearance, excretion, and glomerular filtration rate

17. What are the advantages of a kidney that filters a large volume of fluid and then reabsorbs 99% of it?

18. If the afferent arteriole of a nephron constricts, what happens to GFR in that nephron? If the efferent arteriole of a nephron constricts, what happens to GFR in that nephron? Assume that no autoregulation takes place.

19. Diagram the micturition reflex. How is this reflex altered by toilet training? How do higher brain centers influence micturition?

LEVEL THREE *Problem Solving*

20. You have been asked to study kidney function in a new species of rodent found in the Amazonian jungle. You isolate some nephrons and expose them to inulin. The graph shows the results of your studies. (a) How is the rodent nephron handling inulin? Is inulin filtered? excreted? net reabsorption? net secretion? (b) On the graph, accurately draw in a line to show the net reabsorption or secretion. (Hint: Excretion = filtration − reabsorption + secretion)

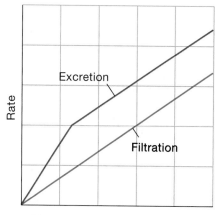

Plasma concentration of inulin

LEVEL FOUR *Quantitative Problems*

21. Darlene weighs 50 kg. Assume that her total blood volume is 8% of her body weight, her heart pumps her total blood volume once a minute, and her renal blood flow is 25% of her cardiac output. Calculate the volume of blood that flows through Darlene's kidneys each minute.

22. Dwight was competing for a spot on the Olympic equestrian team. As his horse, Nitro, cleared a jump, the footing gave way, causing the horse to somersault, landing on Dwight and crushing him. The doctors feared kidney damage and ran the following test. Dwight's serum creatinine level was 2 mg/100 mL. His 24-hour urine specimen contained 1 L and had a creatinine concentration of 20 mg/mL. A second test 24 hours later had the same serum creatinine and urine volume, but a urine creatinine concentration of 4 mg/ml. How many milligrams of creatinine are in each specimen? What is Dwight's creatinine clearance in each test? What is his GFR? Evaluate these results and comment on Dwight's kidney function.

23. You are a physiologist taking part in an archeological expedition to search for Atlantis. One of the deep-sea submersibles has come back with a mermaid, on which you are running a series of experiments. You have determined that the mermaid's GFR is 250 mL/min. The mermaid's kidney reabsorbs glucose with a transport maximum (T_m) of 50 mg/min. What is the mermaid's renal threshold for glucose? When the mermaid's plasma concentration of glucose is 15 mg/mL, what is its glucose clearance?

Explorations 🌐

Explorations are also available at *www.physiologyplace.com*

This chapter introduced you to the kidney. The components of urine are constantly changing and reflect the kidney's functions of regulating ions and water and removing wastes.

Exploration 1: Clearance—Napoleon's Hair and Garlic Breath

The concept of *clearance* that you learned in this chapter is one of the more difficult topics in physiology. Perhaps it is easier to understand clearance if you realize that the kidney is only one organ that clears solutes from the plasma. The liver is the other major site for clearance. Hepatic cells metabolize many different types of molecules, including hormones and drugs. The metabolites are incorporated into bile and secreted into the digestive tract. Some hepatic metabolites are reabsorbed and finally excreted by the kidney. Other drugs, such as acetaminophen, are cleared primarily by the liver.

Saliva, sweat, breast milk, and hair also contain solutes that have been cleared from the plasma. Salivary secretion of the hormone *cortisol* is a simple noninvasive method for monitoring chronic stress. Drugs and alcohol in breast milk are important because a breast-feeding infant will ingest these substances. One scam that you can read about on the Internet is hair analysis, in which people claim to be able to tell you about your nutritional state and health by analyzing a hair sample. A legitimate use of hair analysis is to test for heavy metal poisoning. A 1960s analysis of Napoleon Bonaparte's hair showed significant concentrations of the poison arsenic. The question remains whether Napoleon was murdered or accidentally poisoned.

The lungs clear volatile lipid-soluble materials from the blood when these substances pass into the alveoli and are expelled during breathing. One everyday example of lung clearance is "garlic breath." Ethanol is also cleared by the lungs, and exhaled alcohol is the basis of the "breathalyzer" test used by law enforcement agencies. To learn more about these topics, try a Google search (*www.google.com*) for *hair analysis* or *Napoleon hair arsenic* or *breathalyzer*.

Exploration 2: Artificial Kidneys

Many people with severe renal disease depend on *dialysis*, a medical procedure that supplements or replaces their kidney function. Imagine trying to make a machine or procedure that performs the functions of the kidney. What would it have to do?

Dialysis is based on diffusion through a semipermeable membrane. Solutes and water pass from extracellular fluid across the membrane into a dialysis fluid. *Hemodialysis* routes blood from the arm past a membrane in an external dialysis machine. This technique requires attachment to the machine for about 3–5 hours three days a week and is used for more severe cases of renal failure. *Peritoneal dialysis* is also called *continuous ambulatory peritoneal dialysis* (CAPD) because it takes place while the patient moves about during daily activities. For CAPD, dialysis fluid is injected into the peritoneal cavity. There it accumulates waste products from the blood for 4 to 6 hours before being drained out.

Create a model system that you think would work for dialysis. Draw compartments to represent extracellular fluid and dialysis fluid separated by a semipermeable membrane. List normal extracellular fluid solutes and their concentrations (see table with normal values of blood components inside the back cover of this book). What will happen to the concentrations of these solutes in kidney failure? Which of these solutes should you put in the peritoneal fluid and what should their concentrations be? Here are some things to consider for your model. For each solute, does the body need to maintain a constant concentration or excrete excess? Do you want diffusion into or out of the dialysis fluid or no net movement? What would happen if the patient was retaining too much water? How could you get the artificial kidney to remove the excess?

You can check which functions of the kidney are appropriately regulated by dialysis from a chart shown on the TransPacific Renal Network website: *http://www.network17.org/consumerarticles-/numbc2.htm*. For more information about dialysis, see the website for the National Institute of Diabetes, Digestive, and Kidney Diseases (*http://www.niddk.nih.gov*). Under the Health Information: Kidney topics list you will find information on both *hemodialysis* and *peritoneal dialysis*.

ANSWERS

Answers to Concept Check Questions

Page 602

1. If extracellular K^+ decreases, more K^+ will leave the cell and the membrane potential hyperpolarizes (becomes more negative, increases).

2. If plasma Ca^{2+} decreases, the force of contraction will decrease.

Page 605

3. Filtration and secretion both represent movement from the extracellular fluid into the lumen. Filtration is always passive; secretion may be passive or active.

4. Glomerulus → Bowman's capsule → proximal tubule → loop of Henle → distal tubule → collecting duct → renal pelvis → ureter → urinary bladder → urethra.

5. If reabsorption decreases to half of normal, the body would run out of plasma in under an hour.

Page 606

6. Osmotic pressure is higher in efferent arterioles because fluid volume is decreased, leaving the same amount of protein in a smaller volume.

Page 607

7. Average mean blood pressure is 93 mm Hg.

Page 610

8. If the afferent arteriole constricts, the resistance in that arteriole increases, flow through that arteriole is diverted to lower resistance arterioles, and GFR in that nephron will decrease.

9. The primary driving force for GFR is blood pressure, opposed by fluid pressure in Bowman's capsule and colloid osmotic pressure due to plasma proteins. With fewer plasma proteins, the plasma has lower colloid osmotic pressure. With less colloid osmotic pressure opposing GFR, GFR increases.

Page 616

10. GFR is equal to creatinine clearance, which in this case is about 92 L/day.

Answers to Figure Questions

Page 608

Figure 19-8: Capillary blood pressure, GFR, and RBF all increase.

Page 612

Figure 19-14: The transport rate at 3 mg/mL is 3 mg/min; at 5 and 8 mg/mL, it is 4 mg/min. The transport rate is 4 mg/min at any plasma concentration equal to or higher than 4 mg/mL.

Answers to Level Four Quantitative Problems

Page 622

21. Renal blood flow (25%) is 1 L/min.
22. The first specimen contains 20,000 mg creatinine. Clearance = 1000 L plasma/day. Normally creatinine clearance = GFR. However, this value of 1000 L/day is not at all realistic for GFR (normal average is 180 L/day). The repeat test had 4000 mg of creatinine and gives a clearance of 200 L/day, which is within normal limits. The abnormal values on the first test were probably a laboratory error. Dwight's kidney function is normal.

23. For any solute that filters:

    plasma concentration $\times$ GFR = filtration rate

    At the transport maximum:

    filtration rate = reabsorption rate or T_m

    By substitution:

    plasma concentration $\times$ GFR = T_m

    The renal threshold represents the plasma concentration at which the transporters working at their maximum (T_m). By substitution:

    renal threshold $\times$ GFR = T_m

    Mermaid's GFR is 250 mL/min and T_m is 50 mg/min

    At 15 mg glucose/mL plasma:

    plasma concentration $\times$ GFR = 3750 mg/min glucose filtered

    3750 mg/min glucose filtered minus 50 mg/min reabsorbed = 3700 mg/min excreted

    How many mL of plasma contain 3700 mg?

    1 mL plasma/15 mg = ? mL plasma/3700 mg.

    Clearance = 246.7 mL plasma/min.

Integrative Physiology II:
Fluid and Electrolyte Balance

20

At a 10% loss of body fluid, the patient will show signs of confusion, distress, and hallucinations and at 20%, death will occur.

—**Poul Astrup,** in *Salt and Water in Culture and Medicine,* 1993

RUNNING PROBLEM

Diabetes Insipidus

"Urine without flavor." That is the literal meaning of the term *diabetes insipidus*, a disorder in which the mechanism that concentrates urine is inadequate or absent. Christopher Godell, age 1, had been exhibiting two symptoms of this rare disease: intense thirst, particularly for ice water, and copious excretion of dilute urine. His mother took him to the pediatrician because of problems she noted. "Christopher is always thirsty, and always urinating," she explained. "I know this isn't normal."

632 640 645 653

BACKGROUND BASICS

pH and buffers: **33** | Control of blood pressure: **510** | CO_2 excretion by lungs: **586** | Skin: **78** | Membrane recycling: **146** | Peptide hormones: **214** | Membrane transport: **134** | Juxtaglomerular apparatus: **603** | Role of K^+ in membrane potential: **262** | Carbonic anhydrase: **586** | Osmolarity and tonicity: **153** | Blood pressure and blood volume: **497** | Posterior pituitary hormones: **222** | Body fluid compartments: **150** | Steroid hormones: **216** | Polarized epithelial cells: **147** | Second messenger systems: **177** | Glomerular filtration rate: **607** | Protein structure: **39** |

The American businesswoman in Tokyo finished her workout and stopped at the snack bar of the fitness club to ask for a sports drink. The attendant handed her a bottle labeled "Pocari Sweat®." Although the thought of drinking sweat is not very appealing, the physiological basis for the name is sound.

During exercise, the body secretes sweat, a dilute solution of water and ions, particularly Na^+, K^+, and Cl^-. To maintain homeostasis, the body must replace any substance that it has lost to the external environment. Therefore, the replacement fluid after exercise should resemble sweat.

In this chapter, we explore how the body maintains salt and water balance, also known as fluid and electrolyte balance. The homeostatic control mechanisms for fluid and electrolyte balance are aimed at maintaining four parameters: volume, osmolarity, concentrations of individual ions, and pH of body fluids.

FLUID AND ELECTROLYTE HOMEOSTASIS

The human body is in a state of constant flux. Over the course of a day we ingest about two liters of food and drink that contain from 6 to 15 grams of NaCl. In addition, we bring in varying amounts of other electrolytes, including K^+, H^+, Ca^{2+}, HCO_3^-, and phosphate ions. The body's task is to maintain *mass balance* [● p. 9]: what comes in must be excreted if the body does not need it.

The body has several potential routes for excreting ions and water. The kidneys are the primary route for water loss and for removal of many ions. Under normal conditions, small amounts of these substances are lost in the feces as well. In addition, the lungs help remove H^+ and HCO_3^- by excretion of CO_2.

Although physiological mechanisms that maintain fluid and electrolyte balance are important, behavioral mechanisms also play an essential role. *Thirst* is critical because drinking is the only normal way to replace lost water. *Salt appetite* is a behavior that leads people and animals to seek and ingest salt (sodium chloride, NaCl).

Why are we concerned with homeostasis of these substances? Water and Na^+ are associated with extracellular fluid (ECF) volume and osmolarity. Disturbances in K^+ balance can cause serious problems with cardiac and muscle function by disrupting the membrane potential of excitable cells. Ca^{2+} is involved in a variety of body processes, from exocytosis and muscle contraction to bone formation and blood clotting. And H^+ and HCO_3^- are the ions whose balance determines body pH.

ECF Osmolarity Affects Cell Volume

Why is maintaining osmolarity so important to the body? The answer lies in the fact that water crosses most cell membranes freely. If the osmolarity of the ECF changes, water moves into or out of cells and changes intracellular volume. If ECF osmolarity decreases due to excess water intake, water moves into the cells and they swell. If ECF osmolarity increases due to salt intake, water moves out of the cells and they shrink.

Changes in cell volume, either shrinking or swelling, can impair cell function. For example, when cells swell, ion channels in the membrane open, disrupting membrane potential and cell signaling. The brain, encased in the rigid skull, is particularly vulnerable to damage from swelling. Some cells can regulate their own volume in response to swelling or shrinking, but this capability is limited. In general, maintenance of ECF osmolarity within a normal range is essential to maintain homeostasis.

Fluid and Electrolyte Balance Requires Integration Among Multiple Systems

The process of fluid and electrolyte balance is truly integrative because it involves the respiratory and cardiovascular systems in addition to renal and behavioral responses. Adjustments made by the lungs and cardiovascular system are primarily under nervous control and can be made quite rapidly. Homeostatic compensation by the kidneys occurs more slowly because the kidneys are primarily under endocrine and neuroendocrine control.

For example, small changes in blood pressure due to fluid gain or loss are quickly corrected by the cardiovascular control center in the brain [● p. 497]. If volume changes are persistent or of larger magnitude, the kidneys step in to help maintain homeostasis.

Figure 20-1● summarizes the integrated response of the body to changes in blood volume and blood pressure. Signals from carotid, aortic, and atrial baroreceptors initiate a quick neural response mediated through the cardiovascular control center and a slower response elicited from the kidneys. In addition, low blood pressure stimulates thirst. In both situations, renal function integrates with the cardiovascular system to keep blood pressure within a normal range.

EMERGING CONCEPTS

Regulation of Cell Volume

The regulation of cell volume is so important that many cells have independent mechanisms for volume regulation. For example, renal tubule cells in the medulla of the kidney are constantly exposed to high extracellular fluid osmolarity, yet they maintain normal cell volume. They do this by synthesizing internal organic solutes so that their intracellular osmolarity matches that of the extracellular fluid. The organic solutes used to raise intracellular osmolarity include sugar alcohols such as sorbitol, and certain amino acids such as taurine. In addition, cells can regulate their volume by changing their ionic composition. In some instances, changes in cell volume are believed to act as signals that initiate certain cellular responses. For example, swelling of liver cells activates protein and glycogen synthesis, and shrinking of liver cells causes breakdown of protein and glycogen.

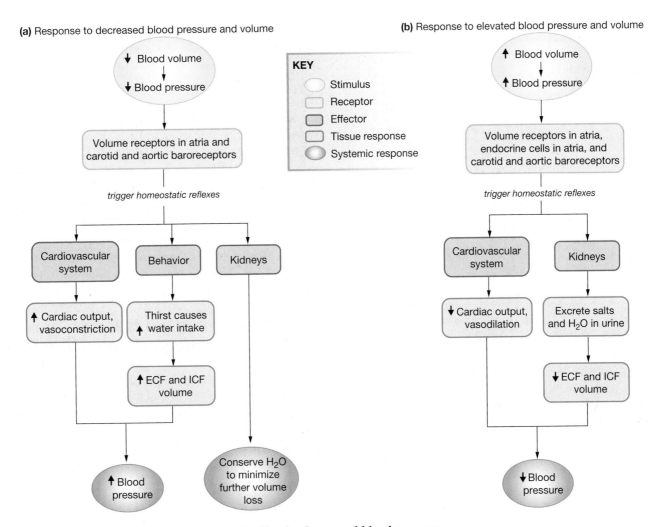

(a) Response to decreased blood pressure and volume

(b) Response to elevated blood pressure and volume

KEY
- ◯ Stimulus
- ☐ Receptor
- ☐ Effector
- ☐ Tissue response
- ◯ Systemic response

● **Figure 20-1** *Integrated responses to changes in blood volume and blood pressure*

Because of the overlap in their functions, a change made by one system is likely to have consequences that affect the other. Endocrine pathways initiated by the kidneys have direct effects on the cardiovascular system. Hormones released by myocardial cells act on the kidneys. Sympathetic pathways from the cardiovascular control center affect not only cardiac output and vasoconstriction but also glomerular filtration and hormone secretion by the kidneys.

Thus, the maintenance of blood pressure, blood volume, and ECF osmolarity forms a network of interwoven control pathways. This integration of function in multiple systems is one of the more difficult concepts in physiology but, at the same time, one of the most exciting areas of physiological research.

▶ WATER BALANCE AND THE REGULATION OF URINE CONCENTRATION

Water is the most abundant molecule in the body, constituting about 50% of total body weight in a 17- to 39-year-old woman and 60% of total body weight in a man of the same age. A 60-kg (132 lb) woman therefore has about 30 liters of body water. The prototype 70-kg man has about 42 liters of water volume. Two-thirds of his water (about 28 liters) is inside the cells, about 3 liters are in the plasma, and the remaining 11 liters are in the interstitial fluid [● Fig. 5-32, p. 150].

Daily Water Intake and Excretion Are Balanced

To maintain constant body water content, we must take in the same amount of water that we excrete: intake must equal output. There are multiple sources for water gain and loss during a day (Fig. 20-2●). On the average, an adult ingests a little more than 2 liters of water in food and drink in a day. Normal cellular respiration (glucose + O_2 ⇌ CO_2 + H_2O) adds about 0.3 liters of water, bringing the total daily intake to approximately 2.5 liters.

Notice that the only means by which water normally enters the body from the external environment is by absorption through the digestive tract. Unlike some animals, we cannot absorb significant amounts of water directly through our skin. If fluids must be rapidly replaced or an individual is unable to eat and drink, fluid can be added directly to the plasma by means of **intravenous (IV) injection**, a medical procedure.

● **Figure 20-2** *Water balance in the body*

The major route of water loss from the body is the urine, which has a daily volume of about 1.5 liters. A small volume of water, about 100 mL, is lost in the feces. We also experience **insensible water loss** (called insensible because we are not normally aware of it). Humans are adapted for life in a terrestrial habitat, and our epidermis is modified with an outer layer of keratin to reduce evaporative water loss [● p. 78]. Despite this epidermal protection, however, we still lose about 900 mL of pure water per day across the skin surface and by exhaling humidified air. Thus, the 2.5 liters of water that we take in are balanced by 2.5 liters that leave the body. Of all the water lost, only water excreted in the urine can be regulated.

Although urine is normally the major route of water loss, in certain situations other routes of water loss can become significant. Excessive sweating is one example. Another example is water loss through diarrhea, a condition that can pose a major threat to the maintenance of water balance, particularly in infants.

Pathological water loss disrupts homeostasis in two ways. Volume depletion from the extracellular compartment decreases blood volume and blood pressure. If blood pressure cannot be maintained, the tissues do not get adequate oxygen. If the fluid lost is hyposmotic to the body, as in excessive sweating, the solutes left behind in the body raise osmolarity, potentially disrupting cell function.

Normally, water balance takes place automatically. Salty food makes us thirsty. Drinking 42 ounces of a soft drink means an extra trip to the bathroom. Salt and water balance is a subtle process that we are only peripherally aware of, like breathing and the beating of the heart.

Now that we have seen *why* regulation of osmolarity is important, let's see *how* the body accomplishes that goal.

The Kidneys Conserve Water

Figure 20-3 ● summarizes the role of the kidneys in water balance. In this figure, the mug with a hollow handle represents the body. The handle represents the kidneys, where body fluid filters into the nephrons and is reabsorbed. Some solutes and water leave in the urine, but the amount that leaves can be regulated, as represented by the little gates.

If additional fluid is added to the mug, normal volume can easily be restored by allowing the extra water to leak out of the handle. If fluid is lost from the mug, however, the only way to replace it is to add water from a source outside the mug. During volume loss, fluid still flows through the handle, but even maximal reabsorption only conserves the water that is in the mug. Reabsorption of water by the kidneys can do nothing to increase the volume. This figure underscores the fact that the kidneys cannot replenish lost water. They can only conserve it.

Urine Concentration Is Determined in the Loop of Henle and Collecting Duct

The concentration, or osmolarity, of urine is a measure of water excretion by the kidneys. When the body needs to eliminate excess water, the kidneys put out copious amounts of dilute urine with osmolarities as low as 50 mOsM. Removal of excess water in dilute urine is known as **diuresis** [*diourein*, to pass in urine]. Drugs that cause diuresis are known as *diuretics*. When the kidneys are conserving water, the urine becomes quite concentrated, up to four times as concentrated as the blood (1200 mOsM vs. 300 mOsM).

● **Figure 20-3** *Role of the kidneys in water balance*

The kidneys control urine concentration by varying the amounts of water and Na^+ reabsorbed in the distal regions of the nephron. (For a review of renal anatomy, see ● Figure 19-1, p. 600.) To produce dilute urine, the kidney must reabsorb solute without allowing water to follow osmotically. This means that the cell membranes across which solute is transported must not be permeable to water.

On the other hand, to produce concentrated urine the nephron must reabsorb water while leaving solute in the lumen. At one time, scientists believed that water was actively transported on carriers, just as Na^+ and other ions are. However, once they developed micropuncture techniques to sample fluid inside kidney tubules, they discovered that water is reabsorbed only by osmosis.

Osmosis will not take place unless a concentration gradient exists. Through an ingenious arrangement of blood vessels and tubules, to be discussed later, the renal medulla maintains a high osmotic concentration in its interstitial fluid. This high *medullary interstitial osmolarity* allows urine to be concentrated.

Let's follow some fluid through the nephron to see where these changes in fluid osmolarity take place (Fig. 20-4 ●). You

may recall from the last chapter that reabsorption in the proximal tubule is isosmotic. Fluid entering the loop of Henle has an osmolarity of about 300 mOsM. Fluid leaving the loop of Henle, on the other hand, is hyposmotic, around 100 mOsM. Cells in the ascending limb of the loop are impermeable to water. When they pump Na^+, K^+, and Cl^- out of the tubule lumen, water cannot follow. This creates a hyposmotic solution in the lumen.

Once the hyposmotic fluid leaves the loop of Henle and passes through the distal segments of the nephron, its concentration will be determined by the permeability of those tubule membranes to water. If the membrane is not permeable, water remains in the tubule and dilute urine will be excreted. A small amount of additional solute can be reabsorbed in the collecting duct. As a result, the most dilute urine we can form is about 50 mOsM.

Conversely, if the urine is to become more concentrated, the tubule epithelium must become permeable to water. When the collecting duct membrane is permeable to water, osmosis draws water out of the lumen and into the interstitial fluid. At maximal water permeability, removal of water from the tubule leaves behind a concentrated urine with an osmolarity of 1200 mOsM.

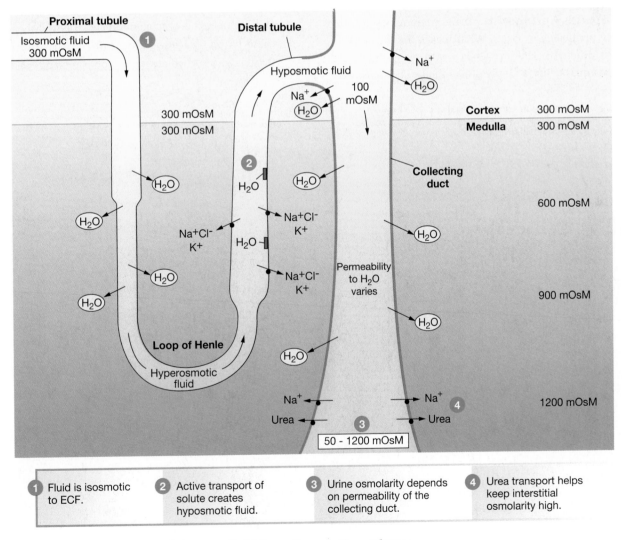

● **Figure 20-4** *Osmolarity changes as fluid flows through the nephron*

Vasopressin Regulates Urine Osmolarity

How does the collecting duct alter its permeability to water and determine the final concentration of urine? The process involves adding or removing water pores in the apical membrane under the direction of the posterior pituitary hormone **vasopressin** [● p. 222]. Because vasopressin causes the body to retain water, it is also known as **antidiuretic hormone** or ADH.

When vasopressin acts on the cell, water pores are present in the apical membrane, and water moves out of the lumen by osmosis (Fig. 20-5a ●). Osmotic movement of water occurs because the cells and interstitial fluid of the kidney medulla are more concentrated than fluid in the tubule.

In the absence of vasopressin, the collecting duct is impermeable to water (Fig. 20-5b). Although a concentration gradient is present across the epithelium, water remains in the tubule, producing dilute urine.

The water permeability of the collecting duct is not an all-or-none phenomenon, as the paragraph above might suggest. Permeability is variable, depending on how much vasopressin is present. The graded effect of vasopressin allows the body to match urine concentration closely to the body's needs.

Vasopressin and Aquaporins   Most membranes in the body are freely permeable to water. What makes the cells of the collecting duct different? The answer lies with its water pores. Water pores are **aquaporins**, a family of membrane channels with

at least ten different isoforms that occur in mammalian tissues. The kidney has at least six different types of aquaporins, including aquaporin-2 (AQP2), the water channel regulated by vasopressin.

AQP2 in collecting duct cells may be found in two locations: on the apical membrane facing the tubule lumen and in the

(a) With maximal vasopressin, the collecting duct is freely permeable to water. Water leaves by osmosis and is carried away by the vasa recta capillaries. Urine is concentrated.

(b) In the absence of vasopressin, the collecting duct is impermeable to water and the urine is dilute.

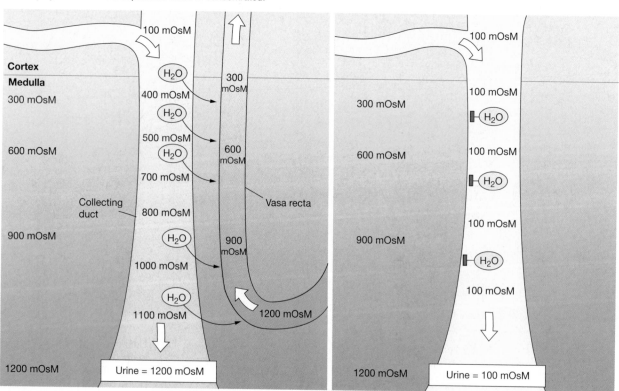

● **Figure 20-5**   *Water movement in the collecting duct in the presence and absence of vasopressin*

membrane of cytoplasmic vesicles. (Two other types of aquaporin water channels are present in the basolateral membrane, but they are not regulated by vasopressin.)

When vasopressin levels and collecting duct water permeability are low, the tubule cell has few water pores in its apical membrane (Fig. 20-6 ●). It stores its AQP2 water pores in cytoplasmic vesicles.

When vasopressin arrives from the posterior pituitary, it binds to its receptor on the basolateral side of the collecting duct cell. Binding activates a G-protein/cAMP second messenger system [● p. 179]. Subsequent phosphorylation of intracellular proteins causes the AQP2 vesicles to move to the apical membrane and fuse with it. Exocytosis inserts the AQP2 water pores into the apical membrane. Now the cell is permeable to water. This process, in which parts of the cell membrane are alternately added by exocytosis and withdrawn by endocytosis, is known as **membrane recycling** [● Fig. 5-26, p. 144].

CONCEPT CHECK   Answers: p. 657

1. Will the apical membrane of a collecting duct cell have more water channels when vasopressin is present or when it is absent?

2. The collecting duct cells are surrounded by extremely high osmolarity on their basolateral sides, yet do not shrivel up. How can they maintain normal cell volume in the face of such high ECF osmolarity? (Hint: read the box on regulation of cell volume, p. 626.)

Changes in Blood Pressure and Osmolarity Trigger Water Balance Reflexes

What stimuli control vasopressin secretion? There are two: ECF osmolarity and blood pressure (Fig. 20-7 ●). The most potent stimulus for vasopressin release is an increase in blood osmolarity. Osmolarity is monitored by **osmoreceptors**, stretch-sensitive cells that activate sensory neurons when osmolarity increases above threshold.

The primary osmoreceptors for vasopressin release are found in the hypothalamus. When osmolarity is below the threshold value of 280 mOsM, the osmoreceptors do not fire, and vasopressin secretion ceases (Fig. 20-8 ●). If osmolarity rises above 280 mOsM, the osmoreceptors stimulate release of vasopressin.

Water reabsorption in the kidneys conserves water and can decrease osmolarity when coupled to excretion of solute. However, vasopressin cannot restore lost fluid volume. Only the ingestion or infusion of water can replace water that has been lost.

Decreases in blood pressure and blood volume are less powerful stimuli for vasopressin release. The primary receptors for volume are stretch-sensitive receptors in the atria. Blood pressure is monitored by the same baroreceptors in the carotid and aortic bodies that initiate cardiovascular responses [● p. 510]. When blood pressure or blood volume is low, these receptors signal the hypothalamus to secrete vasopressin and conserve fluid.

● **Figure 20-6** *The mechanism of action of vasopressin* In the absence of vasopressin, the water pores are withdrawn from the apical membrane and stored in cytoplasmlic vesicles.

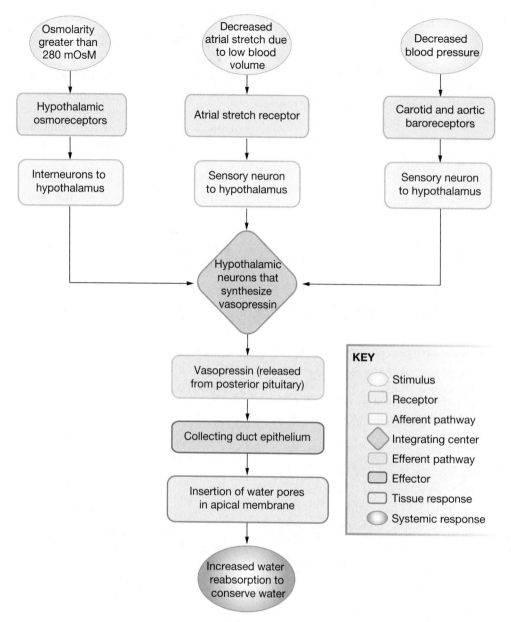

● **Figure 20-7** *Factors affecting vasopressin release*

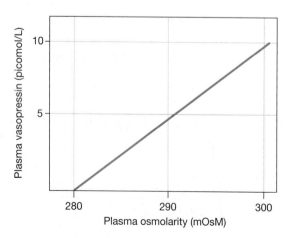

● **Figure 20-8** *The effect of plasma osmolarity on vasopressin secretion*

RUNNING PROBLEM

Diabetes insipidus is caused either by a deficiency in vasopressin or by inability of nephrons to respond to vasopressin. Diabetes insipidus due to vasopressin deficiency (central diabetes insipidus) can be caused by head injury, trauma, or various drugs that damage the pituitary or hypothalamus. In the other form of diabetes insipidus, called *nephrogenic diabetes insipidus*, the collecting duct cells fail to increase their water permeability in response to vasopressin.

Question 1: Christopher's doctor sends him for a test that measures the amount of vasopressin in his blood. The test shows that Christopher has normal levels of vasopressin. Is your diagnosis central or nephrogenic diabetes insipidus?

3. A scientist monitoring the activity of osmoreceptors noticed that infusion of hyperosmotic saline (NaCl) caused increased firing of the osmoreceptors. Infusion of hyperosmotic urea (a penetrating solute) had no effect. If osmoreceptors fire only when cell volume decreases, explain why hyperosmotic urea did not affect them.

4. If vasopressin increases water reabsorption by the nephron, would vasopressin secretion be increased or decreased with dehydration?

5. If vasopressin secretion is suppressed, will urine be dilute or concentrated?

The Loop of Henle Is a Countercurrent Multiplier

Vasopressin is the signal, but the key to the kidney's ability to produce concentrated urine is the high osmolarity of the medullary interstitium. Without it, there would be no concentration gradient for osmotic movement of water out of the collecting duct.

What creates this high ECF osmolarity? And why isn't the high osmolarity diluted by water being reabsorbed from the collecting duct and descending limb of the loop of Henle (see Fig. 20-4, p. 629)? The answer to these questions can be found in the anatomical arrangement of the loop of Henle and its associated blood vessels, the vasa recta, into a countercurrent exchange system.

Countercurrent Systems Exchange Molecules or Heat

Countercurrent exchange systems require arterial and venous blood vessels that pass very close to each other, with fluid flow moving in opposite directions (the countercurrent). This anatomical arrangement allows transfer of heat or molecules from one vessel to the other. Because the countercurrent heat exchanger is

easier to understand, we first examine how it works, then translate the same principle to the kidney.

The countercurrent heat exchanger in mammals and birds evolved to reduce heat loss in flippers, tails, and other limbs that are poorly insulated and that have a high surface-to-volume ratio. Without such a system, warm blood flowing out from the body core into the limb easily loses heat to the surrounding environment (Fig. 20-9a ●).

With the countercurrent heat exchanger (Fig. 20-9b), blood entering the limb in arteries transfers its heat to blood going back into the body in veins. As cooler blood at the tip of the limb flows back into the body, it picks up heat from the warm blood flowing out into the limb. This arrangement reduces heat loss to the environment.

The countercurrent system of the kidney, the loop of Henle, works on the same principle, except that it transfers solute instead of heat from the circulating fluid. However, because the kidney forms a closed system, the solutes are not lost to the environment

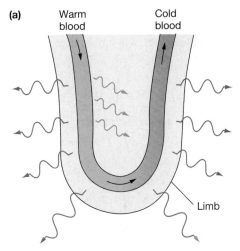

(a) Warm blood / Cold blood / Limb

If blood vessels are not close to each other, heat is dissipated to the external environment.

(b) Warm blood | Warm blood

Countercurrent heat exchanger allows warm blood entering the limb to transfer heat directly to blood flowing back into the body.

● **Figure 20-9** *A countercurrent heat exchanger*

Bed-Wetting and Vasopressin

Bed-wetting, or **nocturnal enuresis**, is a problem that affects about 8% of school-age children. For many years, enuresis was thought to be of psychological origin, but now it is felt that most enuretic children have a physiological cause for their bed-wetting. In normal children, vasopressin secretion shows a circadian rhythm, with levels increasing at night. The result is nocturnal production of a smaller volume of more concentrated urine. Many children with enuresis fail to increase vasopressin secretion at night, with the result that their urine output stays elevated to the point that the bladder fills to its maximum capacity and empties spontaneously during sleep. These children can be successfully treated with a nasal spray of desmopressin, a vasopressin derivative, at bedtime.

as heat is. Instead, the solutes concentrate in the interstitial fluid. This process is aided by active transport of solute out of the ascending limb, which makes the ECF osmolarity even greater. For this reason, the loop of Henle is known as a **countercurrent multiplier**.

The countercurrent multiplier system in the renal medulla is shown in Figure 20-10 ●. Fluid from the proximal tubule flows into the descending limb of the loop of Henle. The descending limb is permeable to water but does not transport ions. As the loop dips into the medulla, water moves by osmosis into the progressively more concentrated interstitial fluid, leaving solutes behind in the lumen.

The tubule fluid becomes progressively more concentrated as it moves deeper into the medulla. At the tips of the longest loops of Henle, the tubule fluid reaches a concentration of 1200 mOsM. Fluid in shorter loops that do not extend into the most concentrated regions of the medulla does not reach such a high concentration.

When the fluid reverses direction and enters the ascending limb of the loop, the properties of the tubule epithelium change. The tubule becomes impermeable to water while actively transporting Na^+, K^+ and Cl^- out of the tubule into the interstitial fluid. The loss of solute from the lumen causes the osmolarity of

the tubule fluid to decrease steadily, from 1200 mOsM at the bottom of the loop to 100 mOsM at the cortex. The net result of the countercurrent multiplier in the kidney is to produce hyperosmotic interstitial fluid in the medulla and hyposmotic fluid leaving the loop of Henle.

The Vasa Recta Removes Water   It is easy to see how transport of solute from the ascending limb dilutes the fluid in the loop of Henle and helps concentrate fluid in the medulla. Still, why doesn't the water leaving the descending limb of the loop dilute the concentrated interstitial fluid of the medulla? The answer lies in the close anatomical association of the loop of Henle and those peritubular capillaries known as the **vasa recta**.

These capillaries, like the loop of Henle, dip down into the medulla and then go back up to the cortex, forming hairpin loops. Although textbooks traditionally show a single nephron with a single loop of capillary, each kidney has millions of collecting ducts and loops of Henle packed between millions of vasa recta capillaries. Thus, functionally, blood flow in the limbs of the vasa recta is in the opposite direction from fluid flow in the loop, as shown in Figure 20-10.

Water or solutes that leave the tubule move into the vasa recta if a concentration or osmotic gradient exists. For example, assume that blood entering the medulla in the vasa recta is 300 mOsM, isosmotic with the cortex. As the blood flows into the medulla, it loses water and picks up solutes transported out of the ascending limb of the loop of Henle, carrying them farther into the medulla. By the time the blood reaches the bottom of the vasa recta loop, it has a high osmotic concentration, similar to that of the surrounding interstitial fluid.

Then, as the blood in the vasa recta flows out of the medulla, its high osmolarity attracts the water that is being lost from the tubule. The movement of water into the vasa recta decreases the osmolarity of the blood while simultaneously preventing the water from diluting the concentrated interstitial fluid.

As blood in the vasa recta leaves the medulla, it removes water reabsorbed from the loop of Henle. Without the vasa recta, water moving out of the descending loops of Henle would eventually dilute the medullary interstitial fluid. The vasa recta thus are an important part of keeping the medullary concentration high.

Urea Increases Osmolarity of the Medullary Interstitium   The high concentration in the medullary interstitium is only partly due to NaCl. Nearly half the solute in the medullary interstitial fluid is urea. Where does it come from? For many years scientists thought urea only crossed cell membranes by passive transport. However, in recent years we have learned that there are several different transporters for urea in the collecting duct. One is a facilitated diffusion carrier, and the other is a Na^+-dependent secondary active transporter. The transporters help concentrate urea in the medullary interstitium, where it contributes to the high interstitial osmolarity.

● **Figure 20-10**   *Countercurrent exchange in the medulla of the kidney*

▶ SODIUM BALANCE AND THE REGULATION OF ECF VOLUME

As noted in the introduction to this chapter, we ingest a lot of NaCl—an average of 9 grams in one day. This is about 2 teaspoons of salt or 155 milliosmoles of Na^+ and 155 milliosmoles of Cl^-. Let's see what would happen to our bodies if the kidneys could not get rid of this Na^+.

Our normal plasma Na^+ concentration is 135–145 milliosmoles Na^+ per liter. Sodium distributes freely between plasma and interstitial fluid, so this value also represents our ECF Na^+ concentration. If we add 155 milliosmoles Na^+ to the ECF, how much water would we have to add to keep ECF sodium at 140 mOsM?

$$155 \text{ mosmol} / ? \text{ liters} = 140 \text{ mosmol/liter}$$

We would have to add over a liter of water to the ECF to compensate for the addition of the Na^+. Normal ECF volume is about 14 liters, so that increase would represent about an 8% gain! Imagine what that volume increase would do to blood pressure.

On the other hand, suppose we add the NaCl and don't drink water. What would happen to osmolarity? If we assume that normal total body osmolarity is 300 mOsM and volume is 42 L, the addition of 155 milliosmoles of Na^+ and 155 milliosmoles of Cl^- would increase total body osmolarity to 307 mOsM—a substantial increase. In addition, because NaCl is a nonpenetrating solute, it would stay in the ECF. Higher osmolarity in the ECF

would draw water from the cells, shrinking them and disrupting normal cell function.

Fortunately, our homeostatic mechanisms maintain mass balance: anything extra that comes into the body will be excreted. Figure 20-11● shows a generalized homeostatic pathway for sodium balance in response to salt ingestion. Here's how it works.

The addition of NaCl to the body raises osmolarity. This stimulus triggers two responses: vasopressin secretion and thirst. Vasopressin release causes the kidneys to conserve water and concentrate the urine. Thirst prompts us to drink water or other fluids.

Ingested water decreases osmolarity, but the combination of salt and water intake increases ECF volume and blood pressure. The increase in ECF volume and pressure then triggers another series of control pathways, which bring blood pressure and osmolarity back into the normal range by excreting extra salt and water.

The kidneys are the primary means of Na^+ excretion. Normally, only a small amount leaves the body in feces and perspiration. However, in situations such as vomiting, diarrhea, and heavy sweating, we may lose significant amounts of Na^+ and Cl^- through nonrenal routes.

Although we speak of ingesting and losing salt (NaCl), the body monitors only Na^+ for homeostasis. (Chloride usually accompanies Na^+.) Sodium regulation is mediated by aldosterone from the adrenal cortex. Interestingly, the stimuli that set the aldosterone pathway in motion are more closely tied to blood volume, blood pressure, and osmolarity than to Na^+ itself.

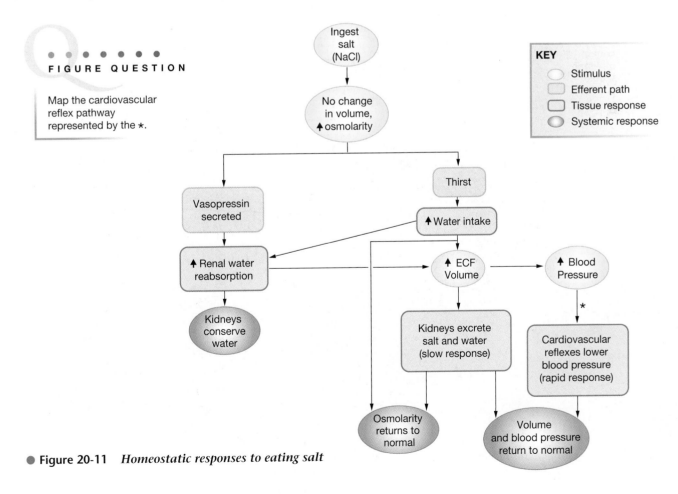

FIGURE QUESTION

Map the cardiovascular reflex pathway represented by the ∗.

● **Figure 20-11** *Homeostatic responses to eating salt*

Aldosterone Controls Sodium Balance

The regulation of blood Na^+ levels takes place through one of the most complicated endocrine pathways of the body. The reabsorption of Na^+ in the distal tubule and collecting duct of the kidney is regulated by the steroid hormone **aldosterone**: the more aldosterone, the more Na^+ reabsorption. Because one target of aldosterone is increased activity of the Na^+-K^+-ATPase, aldosterone also causes K^+ secretion.

Aldosterone is a steroid hormone synthesized in the adrenal cortex, the outer portion of the glands that sit on top of each kidney [⚫ p. 376]. Like typical steroid hormones, aldosterone is secreted into the blood and transported on a protein carrier to its target. The primary site of aldosterone action is the last third of the distal tubule and the portion of the collecting duct that runs through the kidney cortex (the *cortical collecting duct*).

The target cell of aldosterone is the **principal cell**, or **P cell** (Fig. 20-12 ⚫). Principal cells look much like other polarized transporting epithelial cells, with Na^+-K^+-ATPase pumps on the basolateral membrane, and various channels and transporters on the apical membrane [⚫ p. 147]. The apical membranes of principal cells contain leak channels for both Na^+ and K^+.

Aldosterone enters cells by simple diffusion. In target cells, aldosterone combines with a cytoplasmic receptor. In the earliest response phase, apical Na^+ channels increase their open time under the influence of an as-yet-unidentified signal molecule. As intracellular Na^+ levels rise, the Na^+-K^+-ATPase speeds up, trans-

porting cytoplasmic Na^+ into the ECF. The net result is a rapid increase in Na^+ reabsorption that does not require the synthesis of new channel or ATPase proteins. In the slower phase of aldosterone action, newly synthesized channels and pumps are inserted into epithelial cell membranes.

Note that Na^+ and water reabsorption are separately regulated in the distal nephron. Water does not automatically follow Na^+ reabsorption: vasopressin must be present. In contrast, in the proximal tubule, Na^+ reabsorption is automatically followed by water reabsorption because the proximal tubule epithelium is always freely permeable to water.

CONCEPT CHECK   Answers: p. 657

6. In Figure 20-12, what force(s) causes Na^+ and K^+ to cross the apical membrane?

Blood Pressure, Osmolarity, and K^+ Influence Aldosterone Secretion

What controls aldosterone secretion? There are three primary stimuli: increased K^+, increased osmolarity, and a trophic hormone called *angiotensin II* (ANG II).

The simplest stimuli for aldosterone release act directly on the adrenal cortex cells: increased plasma K^+ and increased osmolarity (Table 20-1). An increase in K^+ concentration stimulates aldosterone

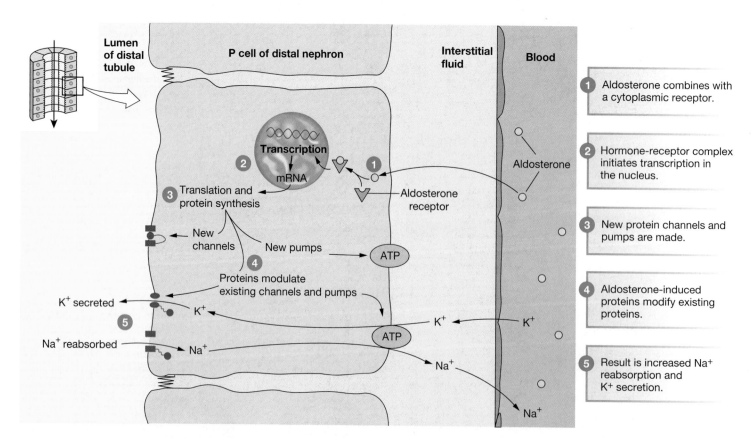

⚫ **Figure 20-12**   *Aldosterone action in principal cells*

Table 20-1: Factors Affecting Aldosterone Release

Direct, at the adrenal cortex	
Increased extracellular K^+ concentration	Stimulates
Increased osmolarity	Inhibits
Indirect, through the RAAS pathway	
Decreased blood pressure	Stimulates
Decreased flow past the macula densa	Stimulates

production, which results in K^+ secretion by the nephron. This reflex protects the body from *hyperkalemia* (elevated blood K^+).

The other stimulus that acts directly on the adrenal gland is osmolarity: an increase in ECF osmolarity inhibits aldosterone secretion. Less aldosterone means less Na^+ reabsorption and therefore increased Na^+ excretion, which helps decrease osmolarity.

The direct inhibitory effect of high osmolarity on the adrenal cortex is an important part of the body's response to dehydration.

The Renin-Angiotensin-Aldosterone Pathway The primary signal for aldosterone release is a trophic hormone known as **angiotensin II**. ANG II is one component of the **renin-angiotensin-aldosterone system (RAAS)**, a complex, multistep pathway for maintaining blood pressure.

The RAAS pathway begins when the *juxtaglomerular cells* (JG cells) in the afferent arterioles of nephrons [p. 603] secrete an enzyme called **renin** (Fig. 20-13 ●). Renin converts an inactive plasma protein, **angiotensinogen**, into **angiotensin I (ANG I)**. (The suffix *-ogen* indicates an inactive precursor.) When ANG I in the blood encounters an enzyme called **angiotensin converting enzyme (ACE)**, ANG I is converted into ANG II.

This conversion was originally thought to take place only in the lungs, but ACE is now known to occur on the endothelium of blood vessels throughout the body. When ANG II in the blood

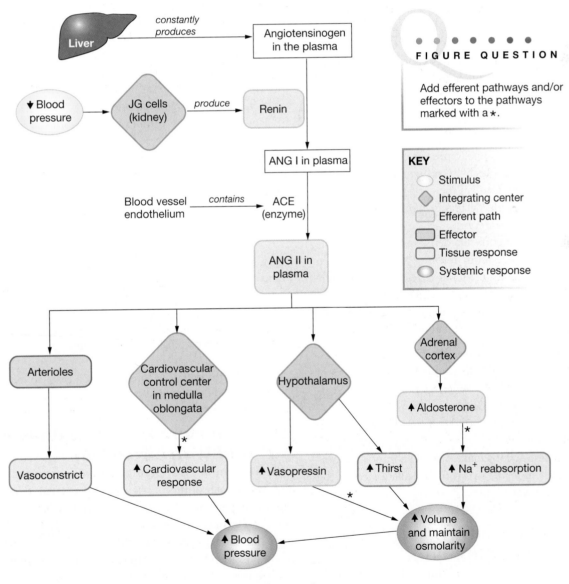

● **Figure 20-13** *The renin-angiotensin-aldosterone pathway*

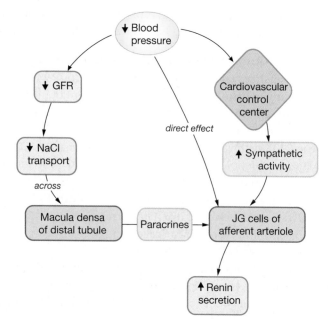

● **Figure 20-14** *Decreased blood pressure stimulates renin secretion*

reaches the adrenal gland, it causes synthesis and release of aldosterone. Finally, at the distal nephron, aldosterone initiates a series of intracellular reactions that cause the tubule to reabsorb Na^+.

The stimuli that begin the RAAS pathway are all related directly or indirectly to low blood pressure (Fig. 20-14 ●):

1. The *JG cells* are directly sensitive to pressure. They respond to low pressure in renal arterioles by secreting renin.
2. *Sympathetic neurons*, activated by the cardiovascular control center when blood pressure decreases, terminate on the JG cells and stimulate renin secretion.
3. *Paracrine feedback* from the macula densa in the distal tubule to the JG cells stimulates renin release [● p. 173]. When fluid flow through the distal tubule is relatively high, the macula densa cells release paracrines which inhibit renin release. If fluid flow in the distal tubule decreases, macula densa cells signal the JG cells to secrete renin.

Sodium reabsorption does not directly raise low blood pressure, but retention of Na^+ does help stimulate fluid intake and volume expansion (see Fig. 20-11). When blood volume increases, blood pressure also increases.

The RAAS pathway does not end here, however. Angiotensin II is a remarkable hormone with additional effects directed at raising blood pressure. These actions make ANG II an important hormone in its own right, not merely an intermediate step in the aldosterone control pathway.

Angiotensin II Influences Blood Pressure Through Multiple Pathways

Angiotensin II has significant effects on fluid balance and blood pressure beyond stimulating aldosterone secretion. ANG II in-

creases blood pressure both directly and indirectly through four additional pathways (see Fig. 20-13):

1. Activation of ANG II receptors in the brain increases vasopressin secretion. Fluid retention in the kidney under the influence of vasopressin helps conserve blood volume.
2. ANG II stimulates thirst. Fluid ingestion is a behavioral response that expands blood volume and raises blood pressure.
3. ANG II is one of the most potent vasoconstrictors known in humans. Vasoconstriction causes blood pressure to increase without a change in blood volume.
4. ANG II receptors are found in the cardiovascular control center. Their activation increases sympathetic output to the heart and blood vessels, which increases cardiac output and vasoconstriction. These responses both increase blood pressure.

Once these blood pressure-raising effects of ANG II became known, it was not surprising that pharmaceutical companies started looking for drugs to block ANG II. Their research produced a new class of antihypertensive drugs called *ACE (angiotensin converting enzyme) inhibitors*. These drugs block the ACE-mediated conversion of ANG I to ANG II, thereby helping to relax blood vessels and lower blood pressure. Less ANG II also means less aldosterone release, a decrease in Na^+ reabsorption and, ultimately, a decrease in ECF volume. All these responses contribute to lowering blood pressure.

However, the ACE inhibitors have not been without side effects. ACE inactivates a cytokine called *bradykinin*. When ACE is inhibited by drugs, bradykinin levels increase and in some patients, this creates a dry, hacking cough. One solution has been the development of drugs called *sartans* that block the blood pressure–raising effects of ANG II by binding to AT_1 *receptors*, a subtype of ANG II receptor.

CONCEPT CHECK   Answers: p. 657

7. A man comes to the doctor with high blood pressure. Tests show that he also has elevated levels of renin in his blood and atherosclerotic plaques that have nearly blocked blood flow through his renal arteries. How does decreased blood flow in his renal arteries cause renin secretion to increase?
8. Describe the pathways through which elevated renin causes high blood pressure in this man.

Atrial Natriuretic Peptide Promotes Na^+ and Water Excretion

Once it was known that aldosterone and vasopressin increase Na^+ and water reabsorption, scientists speculated that other hormones might cause Na^+ loss (**natriuresis**) [*natrium*, sodium + *ourein*, to urinate] and water loss (diuresis) in the urine. If found, these hormones might be used clinically to lower blood volume and blood pressure in patients with essential hypertension [● p. 514]. During years of searching, however, evidence for the other hormones was not forthcoming.

Then, in 1981, a group of Canadian researchers found that injections of homogenized rat atria caused rapid but short-lived excretion of Na$^+$ and water in the urine. They hoped that they had found the missing hormone, one whose activity would complement that of aldosterone and vasopressin. As it turned out, they had discovered the first in an entire family of hormones, some produced in the myocardial cells of the heart and others found in the brain.

Atrial natriuretic peptide (ANP; also known as *atriopeptin*) is a peptide hormone produced in specialized myocardial cells in the atria of the heart. It is synthesized as part of a large prohormone that is cleaved into several active hormone fragments [● p. 215]. Atrial natriuretic peptide is secreted when atrial cells stretch more than is normal, as would occur with increased blood volume.

At the systemic level, ANP enhances Na$^+$ excretion and urinary water loss, but the exact mechanisms by which it does so still are not clear. Atrial natriuretic peptide increases GFR, apparently by making more surface area available for filtration (Fig. 20-15 ●). In addition, ANP has been reported to decrease NaCl and water reabsorption in the collecting duct. The mechanism by which ANP affects tubular reabsorption is not known.

Atrial natriuretic peptide also has several indirect effects on kidney function that alter Na$^+$ and water excretion. It inhibits the release of renin, aldosterone, and vasopressin. These actions all reinforce its natriuretic-diuretic effect. At this time, however, scientists are not convinced that ANP is the primary hormone responsible for regulating Na$^+$ excretion, so the search for other natriuretic factors continues.

Atrial natriuretic peptide and related peptides are also secreted by neurons in the brain. These neurons originate in the hypothalamus and terminate in the cardiovascular control center of the medulla. When secreted as a neurotransmitter or neuromodulator in this location, ANP apparently helps lower blood pressure, the very effect that makes it of interest to drug companies. Unfortunately, ANP has a short half-life in the body, and patients would have to take ANP continuously to maintain its effects.

Researchers are hoping to learn more about the cellular mechanism of ANP action so that they can produce a longer-acting agonist that can be used to treat hypertension.

▶ POTASSIUM BALANCE

The regulation of body potassium levels is essential to maintain a state of well-being. Most K$^+$ is found within the cells: only about 2% is in the ECF. Although only a small fraction of the body's K$^+$ is in the ECF, potassium homeostasis keeps plasma K$^+$ concentrations within a narrow range. As you learned in Chapter 8, changes in the concentration of extracellular K$^+$ affect the resting membrane potential of cells by altering the K$^+$ concentration gradient between the cytoplasm and the ECF (● Fig. 8-18, p. 262).

If plasma (and ECF) K$^+$ concentrations decrease (**hypokalemia**), the concentration gradient becomes larger, more K$^+$ leaves the cell, and the resting membrane potential becomes more negative. If plasma (and ECF) K$^+$ increases (**hyperkalemia**), the concentration gradient decreases and more K$^+$ remains in the cell, depolarizing it.

Because of the effect of plasma K$^+$ on excitable tissues such as the heart, clinicians are always concerned about keeping plasma K$^+$ within the normal range (3.5–5 meq/L). If K$^+$ falls below 3 meq/L or rises above 6 meq/L, the excitable tissues of muscle and nerve begin to show altered function.

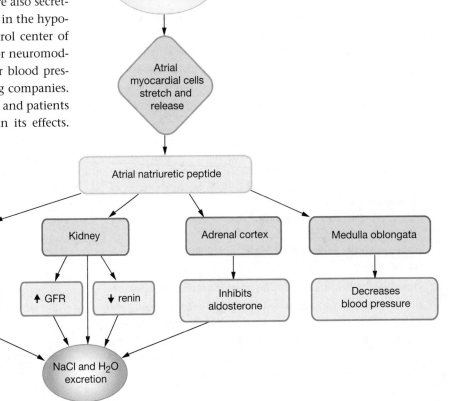

● **Figure 20-15** *Atrial natriuretic peptide*

Hypokalemia causes muscle weakness because it is more difficult for hyperpolarized neurons and muscles to fire action potentials. The danger in this condition lies in the failure of respiratory muscles and the heart. Fortunately, skeletal muscle weakness is usually significant enough to lead the patient to seek treatment before cardiac problems occur. Mild hypokalemia may be corrected by oral intake of K^+ supplements and K^+-rich foods, such as orange juice and bananas.

Hyperkalemia is a more dangerous potassium disturbance because depolarization of excitable tissues makes them more excitable initially. Subsequently, the cells are unable to repolarize fully and actually become *less* excitable. In this state, they have action potentials that are either smaller than normal or nonexistent. Cardiac muscle excitability affected by changes in plasma K^+ and hyperkalemia can lead to life-threatening cardiac arrhythmias.

Under normal conditions, the body very carefully matches K^+ excretion to K^+ ingestion. If plasma K^+ concentration goes up, aldosterone is secreted through the direct effect of hyperkalemia on the adrenal cortex. Aldosterone prompts the kidneys to excrete more K^+ while simultaneously retaining Na^+. (Remember that when plasma K^+ concentrations change, anions such as Cl^- are also added or subtracted to the ECF in a 1:1 ratio, maintaining overall electrical neutrality.)

Disturbances in K^+ balance may result from kidney disease, loss of K^+ in diarrhea, or the use of certain types of diuretics that prevent the kidneys from fully reabsorbing K^+. Inappropriate correction of dehydration can also create K^+ imbalance.

For example, a golfer went out to play a round of golf when the temperature was above 100° F. He was aware of the risk of dehydration, so he took water along to replace his fluid loss. The replacement of sweat loss with pure water kept the golfer's ECF volume normal but dropped his total blood osmolarity and his K^+ and Na^+ concentrations. He was unable to finish the round of golf because of muscle weakness, and he required medical attention that included ion replacement therapy.

RUNNING PROBLEM

Nephrogenic diabetes insipidus is associated with low water permeability in the collecting duct despite adequate levels of vasopressin. Normally, vasopressin stimulates a cAMP second messenger pathway, which prompts insertion of water pores into apical membranes of the tubule cells. In the absence of vasopressin, the cells remove water pores and store them in the cytoplasm.

Question 2: Scientists recently discovered that nephrogenic diabetes insipidus can be caused by two different defects. One defect affects the vasopressin V_2 receptor, leading to the inability of vasopressin to bind to the receptor and activate adenylyl cyclase. In the other defect, the V_2 receptors are normal. What could be the problem in the other defect?

Potassium balance is also closely tied to acid-base balance, as you will learn in the last section of the chapter. Correction of a pH disturbance requires close attention to plasma K^+ levels. Similarly, correction of K^+ imbalance may alter body pH.

▶ BEHAVIORAL MECHANISMS IN SALT AND WATER BALANCE

Although neural, neuroendocrine, and endocrine reflexes play key roles in salt and water homeostasis, behavioral responses are critical in restoring the normal state, especially when ECF volume decreases or osmolarity increases. Drinking water is normally the only way to restore lost water, and eating salt is the only way to raise the body's Na^+ content. Both behaviors are essential for normal salt and water balance. The clinician must recognize the absence of these behaviors in patients who are unconscious or otherwise unable to obey behavioral urges, and must adjust treatment accordingly. The study of the biological basis for behaviors, including drinking and eating, is a field known as *physiological psychology.*

Drinking Replaces Fluid Loss

Thirst is one of the most powerful urges known in humans. In 1952, the Swedish physiologist Bengt Andersson showed that stimulating certain regions of the hypothalamus triggered drinking behavior. This discovery led to the identification of hypothalamic osmoreceptors that initiate drinking when osmolarity rises above 280 mOsM. This is an example of a behavior initiated by an internal stimulus.

Interestingly, although increased osmolarity triggers thirst, the act of drinking is sufficient to relieve thirst. The ingested water does not need to be absorbed for thirst to be quenched. There are as-yet-unidentified receptors in the mouth and pharynx (*oropharynx receptors*) that respond to cold water by decreasing thirst and decreasing vasopressin release, although plasma osmolarity remains high. This oropharynx reflex is one reason that surgery patients are allowed to suck on ice chips: the ice alleviates their thirst without putting significant amounts of fluid into the digestive system.

A similar reflex exists in camels. When led to water, they will drink just enough to replenish their water deficit. The oropharynx receptors apparently act as a feedforward "metering" system that helps prevent wide swings in osmolarity by matching water intake to water need.

In humans, the reflex response of drinking to relieve increased osmolarity is complicated by cultural rituals. Drinking takes place during social events, not just in response to the stimulus of thirst. As a result, our bodies must be capable of eliminating excess fluid ingested in various social situations.

CONCEPT CHECK   Answers: p. 658

9. Incorporate the thirst reflex into Figure 20-7.

Low Na⁺ Stimulates Salt Appetite

Thirst is not the only urge associated with fluid balance. **Salt appetite** is a craving for salty foods that occurs when plasma Na^+ concentrations drop. It can be observed in deer and cattle attracted to salt blocks or naturally occurring salt licks. In humans, salt appetite is linked to aldosterone and angiotensin, hormones that regulate Na^+ balance. The centers for salt appetite are in the hypothalamus close to the center for thirst.

Avoidance Behaviors Help Prevent Dehydration

Other behaviors play a role in fluid balance by preventing or triggering dehydration. Desert animals avoid the heat of the day and become active only at night, when environmental temperatures fall and humidity rises. Humans, especially now that we have air conditioning, are not always so wise.

The midday nap, or *siesta*, is a cultural adaptation in tropical countries that keeps people indoors during the hottest part of the day, thereby helping prevent dehydration and overheating. In the United States, we have abandoned this civilized custom and are active continuously during daylight hours, even when the temperature soars during summer in the South and Southwest. Fortunately, our homeostatic mechanisms usually keep us out of trouble.

▶ INTEGRATED CONTROL OF VOLUME AND OSMOLARITY

The disruption of salt and water balance is corrected using an integrated response from two systems. The cardiovascular system responds to changes in blood volume and the kidneys respond to changes in blood volume or osmolarity. Maintaining homeostasis throughout the day is a continuous process in which the amounts of salt and water in the body shift, according to whether you just drank a soft drink or sweated through an aerobics class.

In that respect, maintaining fluid balance is like driving a car down the highway: small adjustments keep the car in the center of the lane. But just as exciting movies are made from wild car chases, not from sedate driving, the exciting part of fluid homeostasis is the body's response to crisis situations such as severe dehydration or hemorrhage. In this section, we examine challenges to salt and water balance.

Osmolarity and ECF Volume Can Change Independently

Normally, blood volume and osmolarity in the body are homeostatically maintained within a narrow range. However, under some circumstances, fluid loss exceeds fluid gain or vice versa, and the body goes out of balance. Common examples of fluid loss include excessive sweating, vomiting, diarrhea, and **hemorrhage** (excessive blood loss). All of these situations may require medical intervention. In contrast, fluid gain is seldom a medical emergency.

	Osmolarity Decrease	Osmolarity No change	Osmolarity Increase
Volume Increase	Drinking large amount of water	Ingestion of isotonic saline	Ingestion of hypertonic saline
Volume No change	Replacement of sweat loss with plain water	Normal volume and osmolarity	Eating salt without drinking water
Volume Decrease	Incomplete compensation for dehydration	Hemorrhage	Dehydration (e.g., sweat loss or diarrhea)

● **Figure 20-16** *Disturbances of volume and osmolarity*

Volume and osmolarity of the ECF can each have three possible states: normal, increased, or decreased. The relation of volume and osmolarity changes can be represented by a matrix as shown in Figure 20-16 ●. The center box represents the normal state and the surrounding boxes represent the most common examples of the possible changes.

In all cases, the appropriate homeostatic compensation for the change attempts to follow the principle of mass balance: whatever fluid and solute were added to the body must be removed, or what was lost must be replaced. However, perfect compensation is not always possible. Let's begin at the upper right corner and move right to left across each row.

1. **Increased volume and increased osmolarity.** This situation might occur if you ate salty food and drank liquids at the same time, such as popcorn and a soft drink at the movies. The net result may be ingestion of hypertonic saline that increases ECF volume and osmolarity. The appropriate homeostatic response is excretion of hypertonic urine. For homeostasis to be maintained, the osmolarity and volume of the urinary output must match the salt and water input from the popcorn and soft drink.

2. **Increased volume but no change in osmolarity.** If the proportion of salt and water in the ingested food is equivalent to an isotonic NaCl solution, volume will increase but osmolarity will not change. The appropriate response is excretion of isotonic urine whose volume equals that of the ingested fluid.

3. **Increased volume and decreased osmolarity.** This situation would happen if you drank pure water without ingesting any solute. The goal here would be to excrete a very dilute urine to maximize water loss while conserving salts. However, because our kidneys cannot excrete pure water, there will always be some loss of solute in the urine. In this example, urinary output cannot exactly match input, so compensation is imperfect.

4. **Normal volume with increased osmolarity.** This disturbance might occur if you ate salted popcorn without drinking anything. The ingestion of salt without water increases ECF osmolarity and causes some water to shift from cells to the ECF. The homeostatic response is intense thirst, which prompts ingestion of water to dilute the added solute. The kidneys help by creating a highly concentrated urine of minimal volume, conserving water while removing the excess NaCl. Once water is ingested, the disturbance becomes that described in situation 1 or situation 2 above.

5. **Normal volume with decreased osmolarity.** This scenario might occur when a person who is dehydrated (situation 6: decreased volume, increased osmolarity) replaces the fluid loss with pure water, like the golfer described earlier. The decreased volume is corrected, but the replacement fluid has no solute to replace that which was lost. Consequently, a new imbalance is created.

   This situation led to the development of electrolyte-containing sports beverages. If people working out in hot weather replace their sweat loss with pure water, they may restore volume but they run the risk of diluting their plasma K^+ and Na^+ concentrations to dangerously low levels (*hypokalemia* and *hyponatremia*).

6. **Decreased volume and increased osmolarity.** Dehydration is a common example of this fluid and electrolyte disturbance. Dehydration has multiple causes. During prolonged heavy exercise, water loss from the lungs can double, while sweat loss may increase from 0.1 liter to as much as 5 liters! The fluid secreted by sweat glands is hyposmotic, so the fluid left behind in the body becomes hyperosmotic.

   Diarrhea [*diarhein*, to flow through], excessive watery feces, is a pathological condition involving major water and solute loss, this time from the digestive tract. In both sweating and diarrhea, if too much fluid is lost from the circulatory system, the blood volume decreases to the point that the heart can no longer pump blood effectively to the brain. In addition, cell shrinkage caused by increased osmolarity disrupts cell function.

7. **Decreased volume with no change in osmolarity.** This situation occurs with excessive blood loss, or *hemorrhage*. Blood loss represents the loss of isosmotic fluid from the extracellular compartment, similar to scooping a cupful of seawater out of a large bucketful of seawater. If a blood transfusion is not immediately available, the best replacement solution is one that is isosmotic and remains in the ECF, such as isotonic NaCl.

8. **Decreased volume and decreased osmolarity.** This situation might also result from incomplete compensation of dehydration, but it is uncommon, so we will ignore it.

Dehydration Triggers Renal and Cardiovascular Responses

To put together an integrated response to changes in volume and osmolarity, you must first have a clear idea of which pathways become active in response to various stimuli. Table 20-2 is a summary of the many pathways involved in the homeostasis of salt and water balance. For details of individual pathways, refer to the figures cited in Table 20-2.

Severe dehydration is an excellent example of how the body works to maintain blood volume and cell volume in the face of decreased volume and increased osmolarity. It also illustrates the role of neural and endocrine integrating centers. In dehydration, the adrenal cortex receives two opposing signals. One says, "Secrete aldosterone. " The other says, "Do not secrete aldosterone." The rule of thumb that fluid balance integrating centers use when faced with conflicting signals is: *Attempt to restore osmolarity to normal before restoring volume.*

The body has multiple mechanisms to deal with diminished blood volume, but high osmolarity causes cells to shrink and presents a more immediate threat to well-being. Thus, faced with decreased volume and increased osmolarity, the adrenal cortex will not secrete aldosterone. (If secreted, aldosterone would cause Na^+ reabsorption that could worsen the already-high osmolarity associated with dehydration.)

In severe dehydration, compensatory mechanisms are aimed at restoring normal blood pressure, ECF volume, and osmolarity by (1) conserving fluid to prevent additional loss, (2) triggering cardiovascular reflexes to increase blood pressure, and (3) stimulating thirst so that normal fluid volume and osmolarity can be restored. Figure 20-17● maps the interwoven nature of these responses. This figure is complex and intimidating at first glance, so we will discuss it step by step.

At the top of the map (in yellow) are the two stimuli caused by dehydration: decreased volume and increased osmolarity. Decreased ECF volume causes decreased blood pressure.

Blood pressure acts as a stimulus for several pathways, mediated through the baroreceptors of the carotid and aortic bodies, stretch receptors in the atria, and the pressure-sensitive JG cells:

1. The carotid and aortic baroreceptors signal the cardiovascular control center (CVCC) to raise blood pressure. Sympathetic output from the cardiovascular control center increases while parasympathetic output decreases.

   (a) Heart rate goes up as control of the SA node shifts from predominantly parasympathetic to sympathetic.

   (b) The force of ventricular contraction also increases under sympathetic stimulation. Increased ventricular contraction combines with increased heart rate to increase cardiac output.

   (c) Simultaneously, sympathetic input causes arteriolar vasoconstriction, increasing peripheral resistance.

   (d) Sympathetic vasoconstriction of afferent arterioles in the kidneys decreases GFR, helping conserve fluid.

   (e) Increased sympathetic activity at the JG cells of the kidneys increases renin secretion.

2. Decreased peripheral blood pressure directly decreases GFR. A lower GFR conserves ECF by filtering less of it into the nephron.

3. In addition, lower GFR decreases fluid flow past the macula densa. Paracrine feedback causes the JG cells to release renin.

Table 20-2: Reflexes Triggered by Changes in Volume, Blood Pressure, and Osmolarity

STIMULATES	ORGAN OR TISSUE INVOLVED	RESPONSE	FIGURE
Decreased blood pressure			
Direct effects	JG cells	Renin secretion	20-14
	Glomerulus	Decreased GFR	19-6, 19-7
Reflexes			
Carotid and aortic baroreceptors	Cardiovascular control center	Increased sympathetic, decreased parasympathetic output	15-23
Carotid and aortic baroreceptors	Hypothalamus	Thirst stimulation	20-1a
Carotid and aortic baroreceptors	Hypothalamus	Vasopressin secretion	20-7
Atrial baroreceptors	Hypothalamus	Thirst stimulation	20-17
Atrial baroreceptors	Hypothalamus	Vasopressin secretion	20-17
Increased blood pressure			
Direct effects	Glomerulus	Increased GFR (transient)	
	Atrial cells	ANP secretion	20-15
Reflexes			
Carotid and aortic baroreceptors	Cardiovascular control center	Decreased sympathetic, increased parasympathetic output	15-22
Carotid and aortic baroreceptor	Hypothalamus	Thirst inhibition	
Carotid and aortic baroreceptor	Hypothalamus	Vasopressin inhibition	20-7
Atrial baroreceptors	Hypothalamus	Thirst inhibition	
Atrial baroreceptors	Hypothalamus	Vasopressin inhibition	20-7
Increased osmolarity			
Direct effects	Adrenal cortex	Decreased aldosterone secretion	20-17
Reflexes			
Osmoreceptors	Hypothalamus	Thirst stimulation	20-17
Osmoreceptors	Hypothalamus	Vasopressin secretion	20-17
Decreased osmolarity			
Direct effects	None	None	
Reflexes			
Osmoreceptors	Hypothalamus	Decreased vasopressin secretion	

4. The JG cells respond directly to decreased blood pressure. The combination of decreased blood pressure, increased sympathetic input, and signals from the macula densa stimulates renin release and ensures increased production of angiotensin II.

5. Decreased arterial pressure, decreased blood volume, increased osmolarity and ANG II all stimulate vasopressin and thirst centers of the hypothalamus.

The redundancy in the control pathways ensures that all four main compensatory mechanisms are activated: cardiovascular responses, ANG II, vasopressin, and thirst.

1. **Cardiovascular responses** combine enhanced cardiac output and increased peripheral resistance to raise blood pressure. Note, however, that this increase in blood pressure does *not necessarily* mean that blood pressure returns to normal.

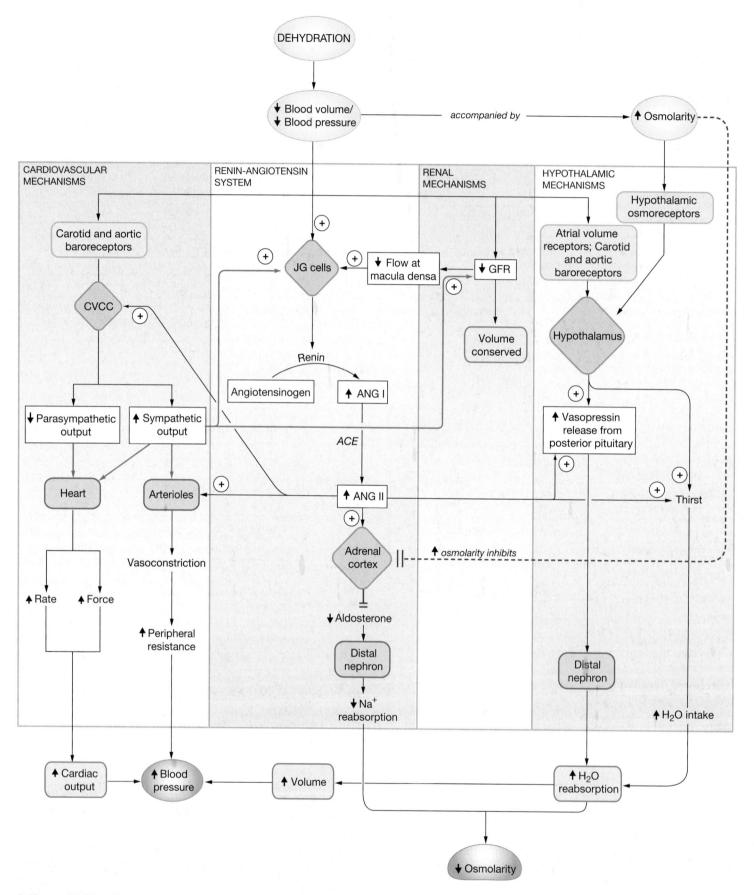

● **Figure 20-17**   *Homeostatic compensation for severe dehydration*

If dehydration is severe, compensation may be incomplete, and blood pressure may remain below normal.

2. **Angiotensin II** has a variety of effects aimed at raising blood pressure including stimulation of thirst, vasopressin release, direct vasoconstriction, and reinforcement of the cardiovascular control center output. ANG II also goes to the adrenal cortex and attempts to stimulate aldosterone release. In dehydration, however, Na^+ reabsorption would worsen the already-high osmolarity. Consequently, high osmolarity at the adrenal cortex directly inhibits aldosterone release, blocking the action of ANG II. The result of the RAAS pathway in dehydration permits the beneficial blood pressure-enhancing effects of ANG II while avoiding the detrimental effects of Na^+ reabsorption. This is a beautiful example of integrated function.

3. **Vasopressin** increases the water permeability of the collecting ducts, allowing water reabsorption to conserve fluid. Without fluid replacement, however, vasopressin cannot bring volume and osmolarity back to normal.

4. **Oral (or intravenous) intake of water** is the only mechanism for replacing lost fluid volume and for restoring osmolarity to normal.

The net result of all four mechanisms is shown in green at the bottom of Figure 20-17. The net result includes: (1) restoration of blood volume by water conservation and fluid intake, (2) maintenance of blood pressure through increased blood volume, enhanced cardiac output, and vasoconstriction, and (3) restoration of normal osmolarity by decreased Na^+ reabsorption and increased water reabsorption and intake.

Using the pathways listed in Table 20-2, see if you can create reflex maps for the other disturbances of volume and osmolarity shown in Figure 20-16.

▶ ACID-BASE BALANCE

Acid-base balance, also called pH homeostasis, is one of the essential functions of the body. The pH of a solution is a measure of its H^+ concentration [p. 33]. A normal arterial plasma sample has an H^+

concentration of 0.00004 meq/L, minute compared with other ions. (For example, the plasma concentration of Na^+ is about 135 meq/L.)

Because H^+ concentration is so low, it is commonly expressed on a logarithmic pH scale of 0–14, where a pH of 7.0 is neutral. If pH falls below 7.0, H^+ concentration has increased above 1×10^{-7} M and the solution is considered acidic. If pH rises above 7.0, the solution has an H^+ concentration lower than 1×10^{-7} M and is considered alkaline (basic).

The normal pH of the body is 7.40, slightly alkaline. A change of 1 pH unit represents a 10-fold change in H^+ concentration. If you need to review pH and the logarithmic scale upon which it is based, please see Appendix A.

Enzymes and the Nervous System Are Particularly Sensitive to Changes in pH

The normal range of pH in the plasma is 7.38–7.42. Extracellular pH usually reflects intracellular pH and vice versa. Because monitoring intracellular conditions is difficult, plasma values are used clinically as an indicator of whole body pH.

Body fluids that are "outside" the body's internal environment, such as those in the lumen of the gastrointestinal tract or kidney tubule, can have a much greater range of pH. Acidic secretions in the stomach may create gastric pH as low as 1. The pH of urine varies between 4.5 and 8.5, depending on the body's need to excrete H^+ or HCO_3^-.

The concentration of H^+ in the body is closely regulated. Intracellular proteins such as enzymes and membrane channels are particularly sensitive to pH because the function of these proteins depends on their three-dimensional shape [p. 39]. Changes in H^+ concentration alter the tertiary structure of proteins by interacting with hydrogen bonds in the molecule, disrupting the protein's three-dimensional structure and activity.

Abnormal pH may significantly affect the activity of the nervous system. If pH is too low, the condition known as **acidosis**, neurons become less excitable and CNS depression results. Patients become confused and disoriented, then slip into a coma. If CNS depression progresses, the respiratory centers cease to function, causing death.

If pH is too high, the condition known as **alkalosis**, neurons become hyperexcitable, firing action potentials at the slightest signal. This condition shows up first as sensory changes, such as numbness or tingling, then as muscle twitches. If alkalosis is severe, muscle twitches turn into sustained contractions (*tetanus*) that paralyze respiratory muscles.

Disturbances of acid-base balance are associated with disturbances in K^+ balance. This is partly due to a renal transporter that moves the two ions in an antiport fashion. In acidosis, the kidneys excrete H^+ and reabsorb K^+ using an H^+-K^+-ATPase. Conversely, when the body goes into a state of alkalosis, the kidneys reabsorb H^+ and excrete K^+. Potassium imbalance usually shows up as disturbances in excitable tissues, especially the heart.

Acids and Bases in the Body Come from Many Sources

In day-to-day functioning, the body is challenged by intake and production of acids more than bases. Hydrogen ions come both from food and from internal metabolism. Maintaining mass balance requires that acid intake and production be balanced by acid excretion. Hydrogen balance in the body is summarized in Figure 20-18 ●.

Acid Input Many metabolic intermediates and foods are organic acids that ionize and contribute H^+ to body fluids. Examples of organic acids include amino acids, fatty acids, intermediates in the citric acid cycle, and lactic acid produced by anaerobic metabolism. Often these organic acids are recognizable by the suffix -*ate*, which denotes the anion (A) half of an acid:

$$HA \rightleftharpoons A^- + H^+$$

Example: pyruvic acid $\rightleftharpoons$ pyruvate + H^+

Metabolic production of organic acids each day creates a significant amount of H^+ that must be excreted to maintain mass balance.

Under extraordinary circumstances, metabolic organic acid production can increase significantly and create a crisis. For example, severe anaerobic conditions, such as circulatory collapse, produce so much lactic acid that normal homeostatic mechanisms cannot keep pace, resulting in a state of *lactic acidosis*. In diabetes mellitus, abnormal fat and amino acid metabolism creates strong acids known as **ketoacids**. These acids cause a state of metabolic acidosis known as *ketoacidosis*.

The biggest source of acid on a daily basis is the production of CO_2 during aerobic respiration. Carbon dioxide is not an acid because it does not contain any hydrogen atoms. However, CO_2 from respiration combines with water to form carbonic acid (H_2CO_3), which dissociates into H^+ and HCO_3^-:

$$CO_2 + H_2O \rightleftharpoons H_2CO_3 \rightleftharpoons H^+ + HCO_3^-$$

This reaction takes place in all cells and in the plasma, but at a slow rate. However, in certain cells of the body, the reaction goes very rapidly due to the presence of large amounts of *carbonic anhydrase* [● p. 586]. This enzyme catalyzes the conversion of CO_2 and H_2O to H_2CO_3.

The production of H^+ from CO_2 and H_2O is the single biggest source of acid input under normal conditions. By some estimates, CO_2 from resting metabolism produces 12,500 meq of H^+ in one day. If this amount of acid were placed in a volume of water equal to the plasma volume, it would create an H^+ concentration of 4167 meq/L, over *100 million* (10^8) times as concentrated as the normal plasma H^+ concentration of 0.00004 meq/L!

These numbers show that CO_2 from aerobic respiration has the potential to seriously affect pH in the body. Fortunately, homeostatic mechanisms normally prevent CO_2 from accumulating in the body.

Base Input If acid-base physiology focuses on acids, it is for good reasons. First, the diet and metabolism have few significant sources of bases. Some fruits and vegetables contain anions that metabolize to HCO_3^-, but their influence is far outweighed by the contribution of acidic fruits, amino acids, and fatty acids. Second, in acid-base disturbances, those due to excess acid are more common than those due to excess base. For these reasons, the body spends most of its time removing excess acid.

pH Homeostasis Depends on Buffers, the Lungs, and the Kidneys

How does the body cope with minute-to-minute changes in pH? There are three mechanisms: (1) buffers, (2) ventilation, and (3) renal regulation of H^+ and HCO_3^-. Buffers are the first line of defense, always present and waiting to prevent wide swings in pH. Ventilation, the second line of defense, is a rapid, reflexively controlled response that can take care of 75% of most pH disturbances. The final defense of pH lies with the kidneys. They are slower than buffers or the lungs but are very effective at coping with any remaining pH disturbance under normal conditions. These three mechanisms help the body balance acid so effectively that normal body pH varies only slightly. Let's take a closer look at each of them.

Buffer Systems Include Proteins, Phosphate Ions, and HCO_3^-

A *buffer* is a molecule that moderates but does not prevent changes in pH by combining with or releasing H^+. In the absence of buffers, the addition of acid to a solution will cause a sharp change in pH. In the presence of a buffer, the pH change will be moderated or may even be unnoticeable. Because acid production

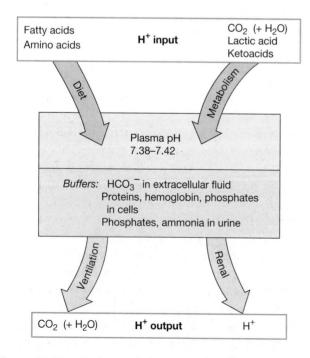

● **Figure 20-18** *Hydrogen balance in the body*

is the major challenge to pH homeostasis, most physiological buffers combine with H^+.

Buffers are found both within the cells and in the plasma. Intracellular buffers include cellular proteins, phosphate ions (HPO_4^{2-}), and hemoglobin. As we have seen, hemoglobin in red blood cells buffers the H^+ produced by the reaction of CO_2 with H_2O (Fig. 17-16, p. 567).

Each H^+ ion buffered by hemoglobin leaves a matching bicarbonate (HCO_3^-) ion inside the red blood cell. The HCO_3^- then leaves the red blood cell by exchanging with a plasma Cl^- ion. This exchange is the *chloride shift* described in Chapter 18 [p. 586].

The large amounts of bicarbonate produced from metabolic CO_2 create the most important extracellular buffer system of the body. Plasma HCO_3^- concentration averages 24 meq/L, which is approximately 600,000 times as concentrated as plasma H^+. Although H^+ and HCO_3^- are created in a 1:1 ratio from CO_2 and H_2O, intracellular buffering of the H^+ by hemoglobin is a major reason why the two ions do not appear in the plasma in the same concentration. The HCO_3^- in plasma can then buffer H^+ that comes from nonrespiratory sources such as metabolism.

The relationship between CO_2, HCO_3^-, and H^+ in the plasma is expressed by the following equation:

$$CO_2 + H_2O \rightleftharpoons \underset{\text{carbonic acid}}{H_2CO_3} \rightleftharpoons H^+ + HCO_3 \qquad (1)$$

According to the law of mass action, any change in CO_2, H^+, or HCO_3^- will cause the reaction to shift until a new equilibrium is reached. (Water is always in excess in the body and does not contribute to the equilibrium state of the reaction.)

For example, if CO_2 increases (shown in blue), the equation shifts to the right, creating one additional H^+ and one additional HCO_3^- from each CO_2 and water:

$$\uparrow CO_2 + H_2O \rightarrow H_2CO_3 \rightarrow \uparrow H^+ + \uparrow HCO_3^- \qquad (2)$$

Once a new equilibrium state is reached, both H^+ and HCO_3^- levels have increased. The addition of H^+ makes the solution more acidic and lowers its pH. It does not matter that a HCO_3^- buffer molecule has also been produced: HCO_3^- acts as a buffer only when it binds to H^+.

Now suppose H^+ is added to the plasma from some metabolic source such as lactic acid. In this case, plasma HCO_3^- will act as a buffer, combining with some of the additional H^+ until the reaction reaches a new equilibrium state. Before equilibrium

$$CO_2 + H_2O \rightleftharpoons H_2CO_3 \rightleftharpoons \uparrow H^+ + HCO_3^- \qquad (3)$$
$$\text{(additional } H^+ \text{ from lactic acid)}$$

The increase in H^+ shifts the equation to the left:

$$CO_2 + H_2O \leftarrow H_2CO_3 \leftarrow \uparrow H^+ + HCO_3^- \qquad (4)$$

The net result at equilibrium is that H^+ is still elevated, but not as much as initially. HCO_3^- is decreased because some has been used as a buffer. The buffered H^+ is converted to CO_2 and H_2O, increasing the amounts of both:

$$\uparrow CO_2 + \uparrow H_2O \rightleftharpoons H_2CO_3 \rightleftharpoons \uparrow H^+ + \downarrow HCO_3^- \qquad (5)$$

The law of mass action is a useful way to think about the relationship between changes in the concentrations of H^+, HCO_3^-, and CO_2 as long as you remember certain qualifications.

First, a change in HCO_3^- concentration according to the reaction may not show up clinically as a HCO_3^- concentration outside the normal range. This is because HCO_3^- is 600,000 times as concentrated in the plasma as H^+ is. If both H^+ and HCO_3^- are added to the plasma, you may observe changes in pH but not in HCO_3^- concentrations because so much HCO_3^- was present initially. Both H^+ and HCO_3^- experience an *absolute* increase, but because so many HCO_3^- were in the plasma to begin with, the *relative increase* in HCO_3^- goes unnoticed.

As an analogy, think of two football teams playing in a packed stadium before 80,000 fans. If 10 more players (H^+) run out onto the field, everyone notices. But if 10 people (HCO_3^-) come into the stands at the same time, no one pays any attention because there were already so many people watching the game that 10 more make no significant difference.

The second qualification for the law of mass action is that when the reaction shifts to the left and increases plasma CO_2, a nearly instantaneous increase in ventilation takes place (in a normal person). If extra CO_2 is ventilated off, arterial P_{CO_2} may remain normal or even fall below normal as a result of hyperventilation.

Ventilation Can Compensate for pH Disturbances

The increase in ventilation just described is a *respiratory compensation* for acidosis. Ventilation and acid-base status are intimately linked, as shown by the now-familiar equation below:

$$CO_2 + H_2O \rightleftharpoons H_2CO_3 \rightleftharpoons H^+ + HCO_3^-$$

Changes in ventilation can correct disturbances in acid-base balance, but they can also cause them. Because of the dynamic equilibrium between CO_2 and H^+, any change in plasma P_{CO_2} will affect both H^+ and HCO_3^- content of the blood.

For example, if a person hypoventilates and P_{CO_2} increases, the equation shifts to the right. More carbonic acid is formed and H^+ goes up, creating a more acidotic state:

$$\uparrow CO_2 + H_2O \rightarrow H_2CO_3 \rightarrow \uparrow H^+ + \uparrow HCO_3^- \qquad (6)$$

If a person hyperventilates, blowing off CO_2 and decreasing the plasma P_{CO_2}, the equation shifts to the left. H^+ combines with HCO_3^-, thereby raising the pH:

$$\downarrow CO_2 + H_2O \leftarrow H_2CO_3 \leftarrow \downarrow H^+ + \downarrow HCO_3^- \qquad (7)$$

In these two examples, you can see that a change in P_{CO_2} will affect the H^+ concentration and therefore the pH of the plasma.

The body uses ventilation as a method for adjusting pH only if a stimulus associated with pH triggers the reflex response. Two stimuli can do so: H^+ and CO_2.

Ventilation is affected directly by H^+ through carotid and aortic chemoreceptors (Fig. 20-19). These are located in the carotid and aortic bodies along with the oxygen and blood pressure

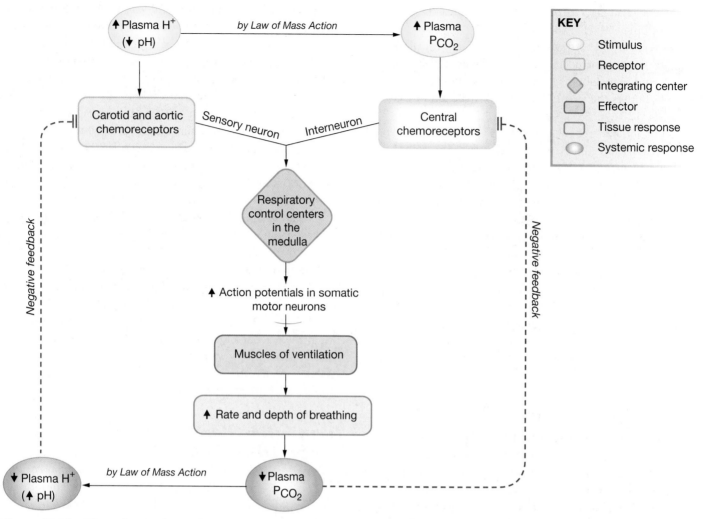

● **Figure 20-19**   *The reflex pathway for respiratory compensation of metabolic acidosis*

sensors we discussed previously [● pp. 510, 589]. An increase in plasma H$^+$ stimulates the chemoreceptors, which in turn signal the medullary control centers to increase ventilation. Increased alveolar ventilation allows the lungs to excrete more CO_2 and convert H$^+$ to carbonic acid.

The central chemoreceptors of the medulla oblongata cannot respond directly to changes in plasma pH because H$^+$ does not cross the blood-brain barrier. However, changes in pH change P_{CO_2}, and CO_2 stimulates the central chemoreceptors [● Fig. 18-18, p. 590]. Dual control of ventilation through the central and peripheral chemoreceptors helps the body respond rapidly to changes in either pH or plasma CO_2.

10. In Figure 20-19, name the muscles of ventilation that might be involved in this reflex.

11. In equation 6 above, HCO$_3^-$ is shown as increasing in amount. Why doesn't the HCO$_3^-$ buffer the increased H$^+$ and prevent acidosis from occurring?

Kidneys Excrete or Reabsorb H$^+$ and HCO$_3^-$

The kidneys take care of the 25% of compensation that the lungs cannot handle. They alter pH two ways: (1) directly by excreting or reabsorbing H$^+$ and (2) indirectly by changing the reabsorption or excretion of HCO$_3^-$ buffer.

In acidosis, the kidney secretes H$^+$ into the tubule lumen using direct and indirect active transport (Fig. 20-20●). Ammonia and phosphate ions in the kidney act as buffers, trapping large amounts of H$^+$ as NH$_4^+$ and H$_2$PO$_4^-$ and allowing more H$^+$ to be excreted. Phosphate ions (HPO$_4^{2-}$) are present in filtered fluid and combine with H$^+$ secreted by the tubule:

$$HPO_4^{2-} + H^+ \rightleftharpoons H_2PO_4^-$$

Ammonia is made from amino acids, as described in the next section.

Even with these buffers, urine can become quite acidic, down to a pH of about 4.5. While H$^+$ is being excreted, the kidneys make new HCO$_3^-$ from CO_2 and H$_2$O. The HCO$_3^-$ is reabsorbed into the blood to act as a buffer and increase pH.

● Figure 20-20 *Overview of renal compensation for acidosis*
The transporters shown in this figure are generic transporters.
The actual transporters are shown in Figures 20-21 and 20-22.

In times of alkalosis, the kidney reverses the general process described above, excreting HCO_3^- and reabsorbing H^+ in an effort to bring pH back into the normal range. Renal compensations are slower than respiratory compensations, and their effect on pH may not be noticed for 24–48 hours. However, once activated, renal compensations effectively handle all but severe acid-base disturbances.

The cellular mechanisms for renal handling of H^+ and HCO_3^- resemble transport processes in other epithelia. However, these mechanisms involve some membrane transporters that you have not encountered before:

1. The **apical Na^+-H^+ antiport** is an indirect active transporter that brings Na^+ into the epithelial cell in exchange for moving H^+ against its concentration gradient into the lumen.
2. The **basolateral Na^+-HCO_3^- symport** moves Na^+ and HCO_3^- out of the epithelial cell and into the interstitial fluid. This indirect active transporter couples the energy of HCO_3^- diffusing down its concentration gradient to the uphill movement of Na^+ from the cell to the ECF.
3. The **H^+-ATPase** uses energy from ATP to acidify the urine, pushing H^+ against its concentration gradient into the lumen of the distal nephron.
4. The **H^+-K^+-ATPase** puts an H^+ into the urine in exchange for a reabsorbed K^+. This exchange contributes to the potassium imbalance that sometimes accompanies acid-base disturbances.
5. A **Na^+-NH_4^+ antiport** moves an ammonium ion from the cell to the lumen in exchange for Na^+.

In addition to these transporters, the renal tubule also uses the ubiquitous Na^+-K^+-ATPase and the same HCO_3^--Cl^- antiport protein that is responsible for the chloride shift in red blood cells.

The Proximal Tubule: Hydrogen Ion Excretion and Bicarbonate Reabsorption The kidneys filter the equivalent of a pound of baking soda bicarbonate each day, and most of it must be reabsorbed to maintain the body's buffer capacity. The proximal tubule reabsorbs most filtered HCO_3^- by indirect methods because there is no membrane transporter to bring HCO_3^- into the tubule cell. In Figure 20-21 ●, the numbers correspond to the steps below. In these figures, you will see how the transporters listed in the previous section function together.

In the proximal tubule, there are two pathways for bicarbonate reabsorption:

1. H^+ is secreted into the lumen in exchange for a filtered Na^+, using the Na^+-H^+ antiport protein.
2. The secreted H^+ then combines with filtered HCO_3^- to form CO_2 in the lumen.
3. CO_2 diffuses into the proximal tubule cell and combines with water to form H_2CO_3, which dissociates to form H^+ and HCO_3^- in the cytoplasm.
4. The H^+ can be secreted again, replacing the H^+ that combined with the filtered HCO_3^-.
5. The HCO_3^- is transported out of the cell on the basolateral side by the HCO_3^--Na^+ symporter.

The net result of this process is reabsorption of filtered Na^+ and HCO_3^- and excretion of H^+.

A second way to reabsorb bicarbonate and excrete H^+ comes from metabolism of the amino acid *glutamine* (Fig. 20-21).

6. Glutamine loses its two amino groups, which become ammonia (NH_3). The ammonia buffers H^+ to become ammonium ion.
7. Ammonium is transported into the lumen in exchange for Na^+.
8. The α-ketoglutarate molecule made from deamination of glutamine is metabolized further to HCO_3^-, which is transported into the blood along with Na^+.

The net result of these pathways is reabsorption of sodium bicarbonate—baking soda.

The Distal Nephron: H^+ and HCO_3^- Handling Depend on the Acid-Base State of the Body The distal nephron plays a significant role in the fine regulation of acid-base balance. Special cells known as **intercalated cells**, or **I cells**, are interspersed among the principal cells in this section of the nephron. They are responsible for acid-base regulation.

Intercalated cells are characterized by high concentrations of carbonic anhydrase in their cytoplasm. This enzyme allows them to convert large amounts of CO_2 into H^+ and HCO_3^-. The H^+ ions are pumped out of the intercalated cell by the H^+-ATPase or by the ATPase that exchanges one H^+ for one K^+. Bicarbonate leaves the cell by means of the HCO_3^--Cl^- antiport exchanger.

Intercalated cells come in two variants whose transporters are found on different faces of the epithelial cell. In times of acidosis, type A intercalated cells secrete H^+ and conserve bicarbonate.

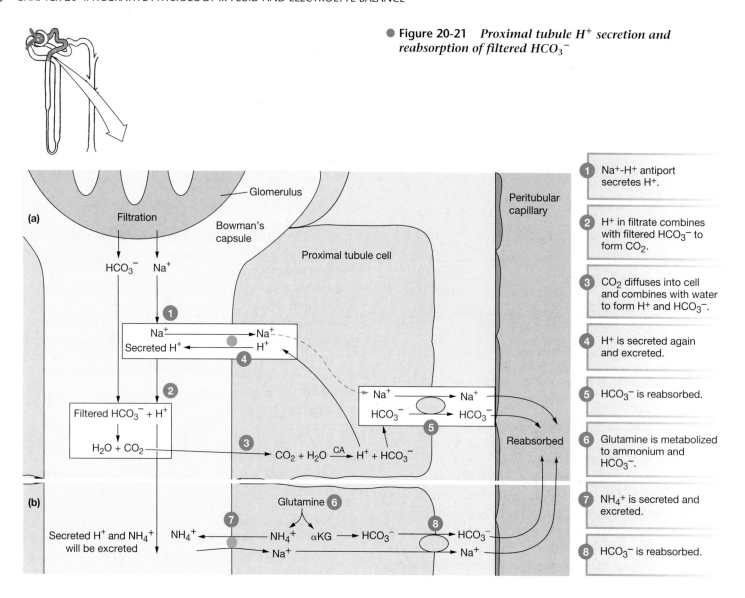

● **Figure 20-21** *Proximal tubule H⁺ secretion and reabsorption of filtered HCO₃⁻*

1. Na⁺-H⁺ antiport secretes H⁺.

2. H⁺ in filtrate combines with filtered HCO_3^- to form CO_2.

3. CO_2 diffuses into cell and combines with water to form H⁺ and HCO_3^-.

4. H⁺ is secreted again and excreted.

5. HCO_3^- is reabsorbed.

6. Glutamine is metabolized to ammonium and HCO_3^-.

7. NH_4^+ is secreted and excreted.

8. HCO_3^- is reabsorbed.

During periods of alkalosis, type B intercalated cells secrete HCO_3^- and reabsorb H⁺.

Fig. 20-22a ● shows how the type A intercalated cell works in times of acidosis, secreting H⁺ and reabsorbing HCO_3^-. The process is similar to H⁺ secretion in the proximal tubule except for the specific H⁺ transporters. The distal nephron uses a H⁺-ATPase and a H⁺-K⁺-ATPase rather than a Na⁺-H⁺ antiport protein.

In times of alkalosis, when the H⁺ concentration of the body is too low, H⁺ is reabsorbed and HCO_3^- buffer is excreted in the urine (Fig. 20-22b). Once again, the ions are produced by the dissociation of H_2CO_3 formed from H_2O and CO_2. Hydrogen ions are reabsorbed into the ECF on the basolateral side, while HCO_3^- is secreted into the lumen. The polarity of the two types of cells is reversed, with the transport proteins found on the opposite sides of the cell.

The H⁺-K⁺-ATPase of the distal nephron helps create parallel disturbances of acid-base and K⁺ balance. In acidosis, when plasma H⁺ is high, the kidney excretes H⁺ and reabsorbs K⁺. Thus, acidosis is often accompanied by hyperkalemia. (There are also nonrenal events that contribute to elevated ECF K⁺ concentrations in acidosis.)

The reverse is true for alkalosis, when blood H⁺ levels are low. The mechanism that allows the distal nephron to retain H⁺ simultaneously causes it to excrete K⁺, so alkalosis goes hand in hand with hypokalemia.

CONCEPT CHECK   Answers: p. 658

12. Why does the K⁺-H⁺ transporter of the distal nephron require ATP to secrete H⁺ but the Na⁺-H⁺ exchanger of the proximal tubule does not?

13. In hypokalemia, the intercalated cells of the distal nephron reabsorb K⁺ from the tubule lumen. What happens to pH in hypokalemia as a result?

● **Figure 20-22** *Role of the intercalated cell in acidosis and alkalosis* Intercalated cells in the collecting duct secrete or reabsorb H^+ and HCO_3^- according to the needs of the body.

(a) Type A intercalated cell function in acidosis. H^+ is excreted; HCO_3^- and K^+ are reabsorbed.

(b) Type B intercalated cell function in alkalosis. HCO_3^- and K^+ are excreted; H^+ is reabsorbed.

Acid-Base Disturbances May Be Respiratory or Metabolic in Origin

The three compensatory mechanisms (buffers, ventilation, and renal excretion) take care of most variations in plasma pH. But under some circumstances, the production or loss of H^+ or HCO_3^- is so extreme that compensatory mechanisms fail to maintain pH homeostasis. In these states, the pH of the blood moves out of the normal range of 7.38–7.42. If the body fails to keep pH within a range of pH 7.00–7.70, acidosis or alkalosis can be fatal.

Acid-base problems are classified both by the direction of the pH change (acidosis or alkalosis) and by the underlying cause (metabolic or respiratory). As you learned earlier, changes in P_{CO_2} due to hyper- or hypoventilation cause pH to shift. These disturbances are said to be of respiratory origin. If the pH problem arises from acids or bases of non-CO_2 origin, the problem is said to be a metabolic problem.

Note that by the time an acid-base disturbance shows up as a change in plasma pH, the body's buffers are ineffectual. The loss of buffering ability leaves the body with only two options: respiratory compensation and renal compensation.

If the problem is of respiratory origin, only one homeostatic compensation is available—the kidneys. On the other hand, in metabolic acid-base disturbances, both respiratory and renal mechanisms can compensate.

The combination of an initial pH disturbance and compensatory changes is one factor that makes analysis of acid-base disorders in the clinical setting so difficult. In this book, we will concentrate on the simple disturbances with a single cause.

Respiratory Acidosis A state of respiratory acidosis occurs when alveolar hypoventilation results in CO_2 retention and elevated plasma P_{CO_2}. Some situations in which this occurs include respiratory depression due to drugs or alcohol, increased airway resistance in asthma, impaired gas exchange in fibrosis or severe pneumonia, or muscle weakness in muscular dystrophy and other muscle diseases. The most common cause of respiratory acidosis is *chronic obstructive pulmonary disease* (COPD), such as emphysema, in which inadequate gas exchange is compounded by loss of alveolar exchange area.

No matter what the cause of respiratory acidosis, plasma CO_2 levels increase (shown in blue), leading to elevated H^+ and HCO_3^-:

$$\uparrow CO_2 + H_2O \rightarrow H_2CO_3 \rightarrow \uparrow H^+ + \uparrow HCO_3^-$$

The hallmark of simple respiratory acidosis is decreased pH and elevated bicarbonate levels (Table 20-3). Because the problem is of respiratory origin, the body cannot carry out respiratory compensation. (However, depending on the problem, mechanical ventilation can be used to assist breathing.)

651

Table 20-3: Plasma P_{CO_2}, Ions, and pH in Acid–Base Disturbances

DISTURBANCE	P_{CO_2}		H^+	pH	HCO_3^-
Acidosis					
Respiratory	↑		↑	↓	↑
Metabolic	Normal* or ↓		↑	↓	↓
Alkalosis					
Respiratory	↓		↓	↑	↓
Metabolic	Normal* or ↑		↓	↑	↑

* These values are different from what you would expect from the law of mass action because almost instantaneous respiratory compensation occurs to keep P_{CO_2} from changing significantly.

Any compensation must be through renal mechanisms that excrete H^+ and reabsorb HCO_3^-. The excretion of H^+ raises plasma pH. Reabsorption of HCO_3^- provides additional buffer that combines with H^+, lowering free H^+ and increasing the pH.

In chronic obstructive pulmonary disease, renal compensation mechanisms can moderate the pH change, although they may not be able to return it to normal. If you look at pH and HCO_3^- in patients with compensated respiratory acidosis, you find that those values are both higher than they were when the acidosis first occurred.

Metabolic Acidosis Metabolic acidosis is a disturbance of mass balance that occurs when the dietary and metabolic input of acid exceeds acid excretion. Metabolic causes of acidosis include lactic acidosis, which is a result of anaerobic metabolism, and ketoacidosis, which occurs when there is excessive breakdown of fats or certain amino acids. The metabolic pathway that produces ketoacids in diabetes mellitus is described in Chapter 21. Ingested substances that cause metabolic acidosis include methanol, aspirin, and ethylene glycol (antifreeze).

Metabolic acidosis can also occur if the body loses HCO_3^-. The most common example is loss of HCO_3^- from the intestines with diarrhea. The exocrine pancreas produces HCO_3^- by a mechanism that is similar to the renal mechanism illustrated in Figure 20-20. The H^+ that is made at the same time is released into the blood. Normally, HCO_3^- is secreted into the small intestine, then reabsorbed, buffering the H^+ made at the same time. However, if HCO_3^- is not reabsorbed due to diarrhea, a state of acidosis results.

Metabolic acidosis is expressed in the equation below. Hydrogen ion concentration increases because of H^+ contributed by the metabolic acids. This increase shifts the equation to the left, increasing CO_2 and using up HCO_3^- buffer:

$$\uparrow CO_2 + H_2O \leftarrow H_2CO_3 \leftarrow \uparrow H^+ + \downarrow HCO_3^-$$

The decrease in HCO_3^- concentration is an important way to distinguish metabolic from respiratory acidosis (Table 20-3).

You would think from looking at the equation at the bottom of the left column that metabolic acidosis would be accompanied by elevated P_{CO_2}. But unless the subject has a lung disease as well, respiratory compensation takes place almost instantaneously. Both elevated CO_2 and H^+ stimulate ventilation through the pathways described earlier. As a result, P_{CO_2} decreases to normal or even below-normal levels due to hyperventilation.

Uncompensated metabolic acidosis is rarely seen clinically. Indeed, a common sign of metabolic acidosis is hyperventilation, evidence of respiratory compensation occurring in response to the acidosis.

The same renal compensations discussed for respiratory acidosis take place in metabolic acidosis: excretion of H^+ and reabsorption of HCO_3^-. Renal compensations take several days to reach full effectiveness, so they are not usually seen in acute disturbances.

Respiratory Alkalosis States of alkalosis are much less common than acidotic conditions. Respiratory alkalosis occurs as a result of hyperventilation, when alveolar ventilation goes up without a matching increase in metabolic CO_2 production. Consequently, plasma P_{CO_2} drops and alkalosis results:

$$\downarrow CO_2 + H_2O \leftarrow H_2CO_3 \leftarrow \downarrow H^+ + \downarrow HCO_3^-$$

The decrease in CO_2 pulls the equation to the left, so both H^+ and HCO_3^- concentrations drop. Low HCO_3^- levels in alkalosis indicate a respiratory disorder.

The primary clinical cause of respiratory alkalosis is excessive artificial ventilation. Fortunately, this condition is easily corrected by adjusting the ventilator.

The most common physiological cause of respiratory alkalosis is hysterical hyperventilation due to anxiety. In these situations, the neurological symptoms caused by alkalosis can be partially reversed by having the patient breathe into a paper bag. In doing so, the patient rebreathes exhaled CO_2, which raises arterial P_{CO_2} and corrects the problem.

Because this problem is of respiratory origin, the body can only carry out renal compensation. Filtered bicarbonate is not reabsorbed in the proximal tubule and is therefore excreted. In the distal nephron, bicarbonate is secreted and H^+ is reabsorbed. These compensations decrease the body's HCO_3^- and increase its H^+, which helps correct the alkalosis.

Metabolic Alkalosis Metabolic alkalosis has two common causes: excessive vomiting of acidic stomach contents and excessive ingestion of bicarbonate-containing antacids. In both cases, the resulting alkalosis reduces H^+ concentration:

$$\downarrow CO_2 + H_2O \rightarrow H_2CO_3 \rightarrow \downarrow H^+ + \uparrow HCO_3^-$$

The decrease in H$^+$ shifts the reaction to the right. Carbon dioxide (P$_{CO_2}$) decreases and HCO$_3^-$ levels go up.

Just as in metabolic acidosis, respiratory compensation is very rapid. The increase in pH and drop in P$_{CO_2}$ depress ventilation. Less CO$_2$ is blown off, raising the P$_{CO_2}$ and creating more H$^+$ and HCO$_3^-$.

The ventilatory compensation helps correct the pH problem but elevates HCO$_3^-$ levels even more. However, the ventilatory compensation is limited because hypoventilation causes hypoxia. Once the arterial P$_{O_2}$ drops below 60 mm Hg, hypoventilation ceases.

The renal response to metabolic alkalosis is the same as that for respiratory alkalosis: HCO$_3^-$ is excreted and H$^+$ is reabsorbed.

This chapter has used fluid and acid-base balance to integrate function between various systems of the body. Changes in body fluid volume, reflected by changes in blood pressure, trigger both cardiovascular and renal homeostatic responses. Disturbances of acid-base balance are met with compensatory responses from both the respiratory and renal systems. Because of the interwoven responsibilities of these three systems, a pathology in one system is likely to cause disturbances in the other two. Recognition of this fact is an important aspect of treatment for many clinical conditions.

RUNNING PROBLEM CONCLUSION

Diabetes Insipidus

The doctor told Mrs. Godell that the treatment for Christopher's nephrogenic diabetes insipidus was to ensure that he had adequate fluid intake at all times. He also prescribed a thiazide diuretic. Giving a diuretic to someone with a dehydrating condition seems paradoxical, because traditionally, diuretics have been used to rid the body of excess water. But in patients with nephrogenic diabetes insipidus, thiazide diuretics actually reduce the volume of dilute urine by indirectly increasing reabsorption of NaCl in the proximal tubule.

In this running problem, you learned about a rare disease characterized by the inability to concentrate urine. You also learned that the inability to concentrate urine can be due to defects in either the brain or the kidney. Further check your understanding of this running problem by comparing your answers with those in the summary table.

	QUESTION	FACTS	INTEGRATION AND ANALYSIS
1	Does Christopher have central or nephrogenic diabetes insipidus?	Central diabetes insipidus is caused by a defect in the hypothalamus that leads to decreased levels of vasopressin. Nephrogenic diabetes insipidus is caused by a defect in the kidneys. Christopher has normal blood levels of vasopressin.	Nephrogenic diabetes insipidus is caused by a defect in the kidneys, therefore vasopressin levels are normal. Christopher has nephrogenic diabetes insipidus.
2	Nephrogenic diabetes insipidus can be caused by two different defects. One defect affects the vasopressin receptor. What could the other defect be?	Vasopressin initiates a cAMP second messenger pathway and prompts insertion of water pores into the apical membranes of tubule cells.	The second defect could be any part of the process that occurs after the vasopressin receptor is activated. For example, problems with the cytoskeleton might prevent membrane recycling, or the water pores themselves could be defective.
3	Why do people with diabetes insipidus become dehydrated even though they are drinking a lot?	In diabetes insipidus, the collecting duct is unable to reabsorb water.	Inability to reabsorb water in the collecting duct leads to excretion of large volumes of dilute urine. If patients do not continuously drink water, they can lose enough fluid to go into a state of severe dehydration.
4	Explain why intense thirst is a main symptom of diabetes insipidus.	Thirst is triggered when osmolarity rises above 280 mOsM or when blood pressure decreases.	Excretion of large volumes of dilute urine decreases blood volume and pressure and increases osmolarity. Decreased blood pressure stimulates ANG II and thirst. Increased blood osmolarity triggers osmoreceptors for thirst.

625 632 640 645

CHAPTER SUMMARY

Homeostasis of body fluid volume, electrolytes, and pH follows the principle of *mass balance*: whatever comes into the body must be excreted. The *control systems* that regulate these parameters are among the more complicated reflexes of the body due to the overlapping functions of the kidneys, lungs, and cardiovascular system. At the cellular level, however, the *movement of molecules across membranes* follows familiar patterns, as transfer of water and solutes from one *compartment* to another depends on osmosis, diffusion, and protein-mediated transport.

Fluid and Electrolyte Homeostasis

1. The renal, respiratory, and cardiovascular systems control fluid and electrolyte balance. Behaviors such as drinking also play an important role. (p. 626; Fig. 20-1)

2. Pulmonary and cardiovascular compensations are more rapid than renal compensation. (p. 626)

Water Balance and the Regulation of Urine Concentration

IP Urinary: Early Filtrate Processing

3. Most water intake comes from food and drink. The major route of water loss is 1.5 liters/day in urine. Smaller amounts are lost in feces, by evaporation from skin, and in exhaled humidified air. (p. 628; Fig. 20-2)

4. Renal water reabsorption conserves water but cannot restore lost volume. (p. 628; Fig. 20-3)

5. To produce dilute urine, the nephron must reabsorb solute without water. To concentrate urine, the nephron must reabsorb water without solute. (p. 629)

6. Fluid leaving the ascending limb of the loop of Henle is dilute. The final concentration of urine depends on the water permeability of the collecting duct. (p. 629; Fig. 20-4)

7. The hypothalamic hormone **vasopressin** controls collecting duct permeability to water in a graded fashion. When vasopressin is absent, water permeability is nearly zero. (p. 630; Fig. 20-5)

8. Vasopressin causes the cell to insert **aquaporin** water pores in the apical membrane of collecting duct cells. (p. 630; Fig. 20-6)

9. An increase in ECF osmolarity or a decrease in blood pressure will stimulate vasopressin release from the posterior pituitary. Osmolarity is monitored by hypothalamic **osmoreceptors.** Blood pressure and volume are sensed by receptors in the carotid and aortic bodies and atria respectively. (p. 631; Figs. 20-7, 20-8)

10. The loop of Henle is a **countercurrent multiplier** that creates high osmolarity in the medullary interstitial fluid by actively transporting Na^+, Cl^-, and K^+ out of the tubule. This high medullary osmolarity is necessary for formation of concentrated urine as fluid flows through the collecting duct. (pp. 633–634; Figs. 20-9, 20-10)

11. The **vasa recta** capillaries carry away water leaving the tubule so that it does not dilute the medullary interstitium. (p. 634; Fig. 20-10)

12. Urea contributes to the high osmolarity in the renal medulla. (p. 634)

Sodium Balance and the Regulation of ECF Volume

IP Urinary: Late Filtrate Processing

13. The total amount of Na^+ in the body is a primary determinant of ECF volume. (p. 635; Fig. 20-11)

14. The steroid hormone **aldosterone** increases Na^+ reabsorption and K^+ excretion. (p. 636; Fig. 20-12)

15. Aldosterone acts on **principal cells** (P cells) of the distal nephron. It enhances activity of the Na^+-K^+-ATPase and increases open time of Na^+ and K^+ leak channels. Aldosterone also stimulates the synthesis of new pumps and channels. (p. 636; Fig. 20-12)

16. Aldosterone secretion can be controlled directly at the adrenal cortex. Increased K^+ stimulates aldosterone secretion but increased ECF osmolarity inhibits it. (p. 636; Tbl. 20-1)

17. Aldosterone secretion is also stimulated by **angiotensin II.** JG cells in the kidney secrete **renin**, which converts **angiotensinogen** in the blood to **angiotensin I.** **Angiotensin converting enzyme (ACE)** converts ANG I to ANG II. (p. 637; Fig. 20-13)

18. The stimuli that release renin are related directly or indirectly to low blood pressure. (p. 638; Fig. 20-14)

19. ANG II has additional effects that raise blood pressure, including increased vasopressin secretion, stimulation of thirst, vasoconstriction, and activation of the cardiovascular control center. (p. 638; Fig. 20-13)

20. **Atrial natriuretic peptide** (ANP) enhances Na^+ excretion and urinary water loss by increasing GFR, inhibiting tubular reabsorption of NaCl, and inhibiting the release of renin, aldosterone, and vasopressin. (pp. 638–639; Fig. 20-15)

Potassium Balance

21. Potassium homeostasis keeps plasma K^+ concentrations in a narrow range. **Hyperkalemia** and **hypokalemia** cause problems with excitable tissues, especially the heart. (p. 639)

Behavioral Mechanisms in Salt and Water Balance

22. **Thirst** is triggered by hypothalamic osmoreceptors and relieved by drinking. (p. 640)

23. **Salt appetite** is triggered by aldosterone and angiotensin. (p. 641)

Integrated Control of Volume and Osmolarity

IP Fluids & Electrolytes: Water Homeostasis

24. Homeostatic compensations for changes in salt and water balance attempt to follow the law of mass balance. Fluid and solute added to the body must be removed, or fluid and solute lost must be replaced. However, perfect compensation is not always possible. (p. 641; Fig. 20-16; Tbl. 20-2)

Acid-Base Balance

IP Fluids & Electrolytes: Acid/Base Homeostasis

25. The body's pH is closely regulated because pH affects intracellular proteins such as enzymes and membrane channels. (p. 645)

26. Acid intake and production are the biggest challenge to body pH. The biggest source of acid is CO_2 from respiration that combines with water to form carbonic acid (H_2CO_3). (p. 646; Fig. 20-18)

27. The body copes with changes in pH by using buffers, ventilation, and renal excretion or reabsorption of H^+ and HCO_3^-. (p. 646; Fig. 20-18)

28. Bicarbonate produced from CO_2 is the most important extracellular buffer system of the body. Bicarbonate buffers organic acids produced by metabolism. (p. 647)

29. Ventilation can correct disturbances in acid-base balance because changes in plasma P_{CO_2} affect both the H^+ and the HCO_3^- content of the blood. An increase in P_{CO_2} stimulates central chemoreceptors. An increase in plasma H^+ stimulates carotid and aortic chemoreceptors. Increased ventilation excretes CO_2 and decreases H^+. (p. 647; Fig. 20-19)

30. In **acidosis**, the kidneys excrete H^+ and reabsorb HCO_3^-. In **alkalosis**, the kidneys excrete HCO_3^- and reabsorb H^+. (pp. 645, 648–649; Figs. 20-20, 20-21)

31. **Intercalated cells** in the collecting duct are responsible for the fine regulation of acid-base balance. (p. 649; Fig. 20-22)

QUESTIONS

LEVEL ONE Reviewing Facts and Terms

1. What is an electrolyte? Name five electrolytes whose concentrations must be regulated by the body.

2. List five organs and four hormones important in maintaining fluid and electrolyte balance.

3. Compare the routes by which water enters the body to the ways the body loses water.

4. List the receptors that regulate osmolarity, blood volume, blood pressure, ventilation, and pH. Where are they located, what stimulates them, and what compensatory mechanisms are triggered by these receptors?

5. How do the two limbs of the loop of Henle differ in their permeability? How is this possible?

6. Which ion is a primary determinant of ECF volume? Which ion is the determinant of extracellular pH?

7. What happens to the resting membrane potential of excitable cells when plasma concentrations of K^+ decrease? Which organ is most likely to be affected by changes in K^+ concentration?

8. Appetite for what two substances is important in regulating fluid volume and osmolarity?

9. Write out the words for the following abbreviations: ADH, ANP, ACE, ANG II, JG cell, P cell, I cell.

10. Make a list of all the different membrane transporters in the kidney. For each transporter, tell (a) which section(s) of the nephron contain(s) the transporter; (b) whether the transporter is on the apical or basolateral membrane, or both; (c) whether it participates in reabsorption, excretion, or both.

11. List and briefly explain three reasons why monitoring and regulating the ECF pH are important. What three mechanisms does the body use to cope with changing pH?

12. Which is more likely to accumulate in the body, acids or bases? List some sources of each one.

13. What is a buffer? List three intracellular buffers. Name the primary extracellular buffer.

14. Name two ways that the kidneys alter the plasma pH. Which compounds serve as urinary buffers?

15. Write the equation that shows how CO_2 is related to pH. What enzyme increases the rate of this reaction? Name two different cell types that possess high concentrations of this enzyme.

16. When ventilation increases, what happens to arterial P_{CO_2}? To pH? To plasma H^+ concentration?

LEVEL TWO *Reviewing Concepts*

17. **Concept map:** Map the homeostatic reflexes that occur in response to each of the following situations:
    (a) decreased volume, normal osmolarity
    (b) increased volume, increased osmolarity
    (c) normal volume, increased osmolarity

18. Figures 20-19 and 20-20 show the respiratory and renal compensations for acidosis. Draw similar maps for alkalosis.

19. Explain how the loop of Henle and vasa recta work together to create dilute fluid.

20. Diagram the mechanism by which vasopressin alters the composition of the urine.

21. Make a table and specify for each substance listed (a–f): hormone or enzyme? steroid or peptide? cell or tissue that produces it? target cell or tissue? response of the target to the substance?
    (a) ANP   (c) renin
    (b) aldosterone   (d) ANG II

    (e) vasopressin   (f) angiotensin-converting enzyme

22. Name the four main compensatory mechanisms for restoring low blood pressure to normal. Why do you think there are so many homeostatic pathways for raising low blood pressure?

23. Compare and contrast the following sets of terms:
    (a) principal and intercalated cells
    (b) renin, angiotensin II, aldosterone, ACE
    (c) respiratory acidosis with metabolic acidosis, including their causes and compensations
    (d) water reabsorption in the proximal tubule, distal tubule, and ascending limb of the loop of Henle
    (e) respiratory alkalosis with metabolic alkalosis, including their causes and compensations

LEVEL THREE *Problem Solving*

24. A 45 year-old man visiting from out of town arrives at the emergency room with an asthma attack caused by pollen. Blood drawn before treatment showed the following: $HCO_3^- = 30$ meq/L (normal 24), $P_{CO_2} = 70$ mm Hg, pH = 7.24.
    (a) What is the man's acid-base state? Is this an acute or chronic situation? The man was treated and made a complete recovery. Over the next ten years he continued to smoke a pack of cigarettes a day and a year ago his family doctor diagnosed chronic obstructive pulmonary disease (emphysema). The man's most recent blood test showed the following: $HCO_3^- = 45$ meq/L, $P_{CO_2} = 85$ mm Hg, and pH = 7.34.
    (b) What is the man's acid-base state now? Is this acute or chronic?
    (c) Explain why in his second illness his plasma bicarbonate level and P_{CO_2} are higher than in the first illness but his pH is closer to normal.

25. Karen has bulimia, in which she induces vomiting to avoid weight gain. When the doctor sees her, her weight is 89 lb

and respiration is 6 breaths/min (normal 12). Her blood HCO_3^- is 62 meq/L (normal 24–29). Her arterial blood pH is 7.61 and her P_{CO_2} is 61 mm Hg.
    (a) What is her acid-base condition called?
    (b) Explain why her plasma bicarbonate level is so high.
    (c) Why is she hypoventilating? What effect will this have on the pH and total oxygen content of her blood? Explain your answers.

26. Hannah, a 31-year-old woman, decided to have colonic irrigation, a procedure during which large volumes of distilled water were infused into her rectum. During the treatment she absorbed 3000 mL of water. About 12 hours later, her roommate found her in convulsions and took her to the emergency room. Her blood pressure was 140/90, her plasma Na^+ concentration was 106 meq/L (normal is 135 meq/L), and her plasma osmolarity was 270 mOsM. In a concept map or flow chart, diagram all the homeostatic responses that her body was using to attempt compensation for the changes in blood pressure and osmolarity.

LEVEL FOUR *Quantitative Problems*

27. The Henderson-Hasselbalch equation is a mathematical expression of the relationship between pH, HCO_3^- concentration, and concentration of dissolved CO_2. One variant of the equation uses P_{CO_2} instead of dissolved CO_2: pH = 6.1 + log $[HCO_3^-]/0.03 \times P_{CO_2}$

    (a) If arterial blood has a P_{CO_2} of 40 mm Hg and its HCO_3^- concentration is 24 mM (normal), what is its pH? (You will need a log table or calculator with a logarithmic function capability.)
    (b) What is the pH of venous blood with the same HCO_3^- but a P_{CO_2} of 46 mm Hg?

28. Hyperglycemia in a diabetic patient leads to osmotic diuresis and dehydration. Given the following information, answer the questions.

Plasma glucose = 400 mg/dL   Normal urine flow = 1 L per day

GFR = 130 mL/min   Normal urine osmolarity = 300 mOsM

Glucose T_m = 400 mg/min   Molecular weight glucose = 180 daltons

Renal plasma flow = 500 mL/min
(a) How much glucose filters/min?
(b) How much glucose is reabsorbed/min?
(c) How much glucose will be excreted in the urine per day?
(d) Assuming that dehydration causes maximal ADH secretion and allows the urine to concentrate to 1200 mOsM, how much additional urine will this diabetic excrete in a day?

Exploration

Explorations are also available at *www.physiologyplace.com*

Exploration 1: Understanding the Countercurrent Multiplier

The countercurrent multiplier created by the loop of Henle is fundamental to kidney function. Although simple in principle, it can initially be more difficult to understand than other kidney mechanisms. The loop of Henle uses two basic processes to create a gradient of increasing solution concentration as it descends deeper in the kidney medulla: (1) fluid flow in opposite directions (countercurrent) in the two limbs of the loop of Henle and, (2) active solute reabsorption from the ascending limb. The high interstitial concentration created by the countercurrent multiplier is essential for the kidney's ability to excrete concentrated urine, and the removal of solute from the ascending limb is essential for the production of dilute urine.

The mechanism of the loop of Henle is perhaps best understood when seen in action. A series of excellent interactive simulations for a countercurrent multiplier can be found at *http://endeavor.med.nyu.edu/courses/physiology/courseware/wwwcc* (W. James Sullivan, M.D. and Felice Aull, Ph.D., January 2000,

New York University School of Medicine). The first activity starts you with a simple model of countercurrent multiplication showing how the gradient becomes established as a result of countercurrent flow and solute transfer. In the second activity you apply the principles to a more realistic model of the loop of Henle and see that the mechanism works basically the same as the simpler model. Next, you can work through a series of problems that show how variations in flow, solute movement, and loop length affect kidney function. You will be surprised at how much you know! Finally, you can work with some more difficult endocrine and clinical problems to really test your understanding.

To learn more, do a Google internet search for *countercurrent multiplier*. This will bring up other models, including those that use similar mechanisms for other physiological processes. You should also search for *countercurrent exchanger*. The difference in the two processes is that the multiplier uses active transport to create and maintain the gradient, whereas the exchanger simply relies on existing gradients for transfer. In the kidney, the tubule in the loop of Henle is a multiplier and the blood vessels of the vasa recta are a passive exchanger.

ANSWERS

Answers to Concept Check Questions

Page 631

1. Apical membranes on collecting duct cells will have more water pores when vasopressin is present.

2. The collecting duct cells concentrate organic solutes that counteract the hypertonic effect of their environment.

Page 633

3. Hyperosmotic NaCl is hypertonic and causes the osmoreceptors to shrink, but hyperosmotic urea is hypotonic and causes them to swell. Thus, they will not fire.

4. In dehydration, the kidneys try to conserve water. Vasopressin enhances water reabsorption, therefore vasopressin levels would increase with dehydration.

5. If vasopressin secretion is suppressed, the urine will be dilute.

Page 636

6. In Figure 20-12, Na^+ and K^+ are moving down their electrochemical gradients.

Page 638

7. Atherosclerotic plaques block blood flow, which decreases GFR and decreases pressure in the afferent arteriole. These are both stimuli for renin release.

8. Renin secretion begins a cascade that produces angiotensin II. ANG II is a powerful vasoconstrictor, acts on the medullary cardiovascular control center to increase blood pressure, increases vasopressin and aldosterone secretion and thirst, resulting in increased fluid volume in the body. These responses may all contribute to increased blood pressure.

Page 640

9. On the left side of Figure 20-7, interneurons lead from osmoreceptors to the hypothalamic thirst centers as well.

Page 648

10. The muscles of ventilation that are involved include the diaphragm, the external intercostals, and the scalenes.

11. The bicarbonate is increasing as the reaction shifts to the right and achieves a new equilibrium. At equilibrium, both H+ and bicarbonate are increased. The bicarbonate cannot act as a buffer when the system is at equilibrium.

Page 650

12. Both K^+ and H^+ are being moved against their concentration gradients, which requires ATP. In the proximal tubule, the Na^+ is moving down its concentration gradient, providing the energy to push H^+ against its gradient.

13. When intercalated cells reabsorb K^+, they secrete H^+; therefore pH will increase.

Answers to Figure Questions

Page 635
Figure 20-11: See Fig. 15-22, p. 511.

Figure 20-13: See Fig. 15-21, p. 510, for the cardiovascular pathway and Fig. 20-12 for aldosterone.

Answers to Level Four Quantitative Problems

Page 656
27. (a) pH = 6.1 + log 24/(.03 × 40) = 7.40 (b) 7.34

Page 657
28. (a) 400 mg glucose/100 mL × 130 mL/min = 520 mg glucose/min filters. (b) Can reabsorb up to T_m therefore 400 mg/min reabsorbed. (c) Excreted = filtered − reabsorbed or 120 mg/min × 1440 min/day = 172.8 g/day excreted. (d) Must convert grams of glucose to milliosmoles: 172.8 g × 180 g/mole × 1000 mosmol/mole = 960 mosmol glucose excreted/day. Concentration = amount/volume. 1200 mOsM = 960 mosmol/ ? liters. To excrete the glucose will require 0.8 additional liters of urine.

Digestion

21

Vibrio cholerae (green), the bacterium that causes the diarrheal disease cholera, occurs in fresh water. Here it is shown attached to the surface of freshwater algae (red).

RUNNING PROBLEM

Peptic Ulcers

Your stomach is one of the most hostile environments on earth. Bathed in acid strong enough to digest nails, its walls protected with a sticky mucus, and constantly churning, the stomach was once thought to be uninhabitable by any life form. However, scientists have discovered a remarkable organism that actually thrives in the stomach's formidable environment. This hardy bacterium, called *Helicobacter pylori*, has evolved a set of defenses so unique that it escaped detection until 1982. But this bacterium is not a silent passenger in the stomach: it is associated with many cases of peptic ulcers worldwide. Proper drug treatment kills *H. pylori* and allows ulcers in the stomach and duodenum to heal. That is good news to Tonya Berry, whose physician has just informed her that she has a peptic ulcer. "I thought it was stress," she said.

BACKGROUND BASICS

Composition of biomolecules: **34ff** ∣ Micelles: **128** ∣ Smooth muscle structure and contraction: **413** ∣ Anatomy of the lymphatics: **508** ∣ Secondary active transport: **143** ∣ Cell-to-cell junctions: **64** ∣ Renal transport of sodium: **610** ∣ Positive feedback loops: **195** ∣ Microvilli: **69** ∣ Endocrine and exocrine gland formation: **71** ∣ Transporting and secretory epithelia: **67, 70** ∣ Apical and basolateral surfaces: **69** ∣ Enzymes: **90** ∣ Micelles: **128** ∣ Protein synthesis and storage: **112** ∣ Acidification of the urine: **648** ∣ Exocytosis and transcytosis: **146-148** ∣ Portal systems: **451** ∣

Give me a good digestion, Lord, and also something to digest.
—Anonymous, *A Pilgrim's Grace*

Ashotgun wound to the stomach seems an unlikely beginning to the scientific study of digestive processes. But in 1822 at Fort Mackinac, young Canadian trapper Alexis St. Martin narrowly escaped death when a gun discharged only three feet from him, tearing open his chest and abdomen and leaving a hole in his stomach wall. American army surgeon William Beaumont attended St. Martin and nursed him back to health over the next two years.

The gaping wound over the stomach failed to heal properly, leaving a *fistula*, or opening, into the lumen. St. Martin was destitute and unable to care for himself, so Beaumont "retained St. Martin in his family for the special purpose of making physiological experiments." In a legal document, St. Martin even agreed to "obey, suffer, and comply with all reasonable and proper experiments of the said William [Beaumont] in relation to ... the exhibiting ... of his said stomach and the power and properties ... and states of the contents thereof."

Beaumont's observations on digestion and the state of St. Martin's stomach under various conditions created a sensation. In 1832, just before Beaumont's observations were published, the nature of gastric juice [*gaster*, stomach] and digestion in the stomach was a subject of much debate. Beaumont's careful observations went far toward solving the mystery.

Like physicians of old who tasted urine when making a diagnosis, Beaumont tasted the mucous lining of the stomach and the gastric juices. He described them both as "saltish," but mucus was not at all acid and gastric fluid was very acid. Beaumont collected copious amounts of gastric fluid through the fistula, and in controlled experiments he confirmed that gastric fluid digested meat, aided by a combination of hydrochloric acid and another active factor that we now know is the enzyme pepsin.

These observations and others about motility and digestion within the stomach became the foundation of what we know about digestive physiology. Although research today is being conducted more at the cellular and molecular level, researchers still create surgical fistulas in experimental animals to observe and sample the contents of the lumen of the digestive tract.

Why is the digestive system of such great interest? Gastrointestinal diseases today account for nearly one-tenth of the money spent on health care. Many of these conditions, such as heartburn, indigestion, gas, and constipation, are troublesome rather than major health risks, but their significance should not be underestimated. Go into any drugstore and look at the number of over-the-counter medications for digestive disorders to get a feel for the impact that gastrointestinal diseases have on our society. In this chapter we examine the digestive system and the remarkable way that it transforms the food we eat into nutrients for the body's use.

FUNCTION AND PROCESSES OF THE DIGESTIVE SYSTEM

The **gastrointestinal tract**, or *GI tract*, is a long tube whose function is to move nutrients, water, and electrolytes from the external to the internal environment. The food we eat is mostly in the form of macromolecules such as proteins and complex carbohydrates, so our digestive systems must secrete powerful enzymes to digest them into smaller molecules, or nutrients, that can be absorbed. At the same time, however, the enzymes must not digest the cells of the GI tract itself. If protective mechanisms against autodigestion fail, we develop raw patches known as *ulcers* on the walls of the gastrointestinal tract.

Another challenge the digestive system faces daily is mass balance matching input with output. Various exocrine glands and cells secrete about 7 liters of fluid per day into the lumen of the tract. These secretions contain digestive enzymes, mucus, and water. They are critical to proper digestive function, but they must be reabsorbed or the body will rapidly dehydrate. *Diarrhea*, or watery stools, can become an emergency if fluid loss from the GI tract depletes the extracellular fluid volume to the point that the circulatory system is unable to maintain adequate blood pressure.

These physiological challenges are met by coordinating the four basic processes of the digestive system: digestion, absorption, motility, and secretion (Fig. 21-1 ●). **Digestion** is the chemical and mechanical breakdown of foods into smaller units that can be taken across the intestinal epithelium into the body. **Absorption** is the active or passive transfer of substances from the lumen of the GI tract to the extracellular fluid (ECF). **Motility** [*movere*, move + *tillis*, characterized by] is movement of material

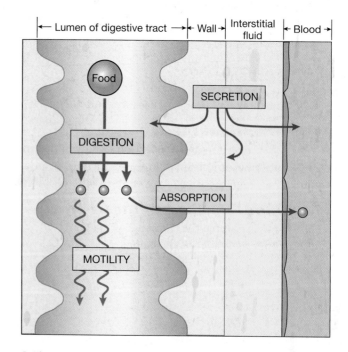

● **Figure 21-1** *Processes of the digestive system*

in the GI tract as a result of muscle contraction. **Secretion** refers to (1) the transfer of water and ions from the ECF to the lumen, and (2) the release of synthesized material by GI epithelial cells.

For most nutrients, absorption is not regulated, so "what you eat is what you get." In contrast, motility and secretion are continuously regulated to maximize the availability of absorbable material. Motility is regulated because if food moves through the system too rapidly, there is not enough time for everything in the lumen to be digested and absorbed. Secretion is regulated because if digestive enzymes are not secreted in adequate amounts, food in the GI tract cannot be broken down into an absorbable form.

When digested nutrients have been absorbed and reach the cells, cellular metabolism directs their use or storage (see Chapter 22). Some of the same chemical signal molecules that alter digestive motility and secretion also participate in the control of metabolism, providing an integrating link between the two steps.

Although we tend to think of the digestive system primarily in terms of its digestive function, repelling foreign invaders is another challenge the system faces. The GI tract passes through the interior of the body, but the lumen and its contents are part of the external environment [Fig. 1-2, p. 3]. Indeed, the digestive system provides the largest amount of contact between the internal environment and the outside world, with a total surface area estimated to be the size of a tennis court.

The GI tract faces daily conflict between the need to absorb water and nutrients and the need to keep bacteria, viruses, and other pathogens from entering the body. As a result, the transporting epithelium of the GI tract is combined with an array of physiological defense mechanisms including mucus, digestive enzymes, acid, and the largest collection of lymphoid tissue in the body, the **gut-associated lymphoid tissue** (GALT). By one estimate, 80% of all lymphocytes [p. 526] in the body are found in the small intestine.

▶ ANATOMY OF THE DIGESTIVE SYSTEM

The digestive system begins with the oral cavity (mouth and pharynx), which serves as a receptacle for food (Fig. 21-2●). The oral cavity is where the first stages of digestion begin with chewing and secretion of saliva by three pairs of **salivary glands**: *sublingual* under the tongue, *submandibular* under the mandible, or jawbone, and *parotid* that lie near the hinge of the jaw.

Once swallowed, food moves into the GI tract, a long tube with muscular walls and a lining of transporting epithelium. The tube is closed off by skeletal muscle sphincters at the ends. At intervals along its length, rings of smooth muscle function as sphincters to separate the tract into segments with distinct functions.

Food moves through the GI tract propelled by waves of muscle contraction. Along the way, secretions are added to the food by secretory epithelium, liver, and pancreas, creating a soupy mixture known as **chyme.**

Digestion takes place primarily in the lumen of the tube. The products of digestion are absorbed across the epithelium of the wall and pass into the extracellular space. From there, they move into the blood or lymph for distribution throughout the body. Any waste remaining at the end of the GI tract leaves the body through the opening known as the anus.

CONCEPT CHECK   Answers: p. 694

1. Define digestion.
2. What is the difference between digestion and metabolism [p. 99]?
3. What is the difference between absorption and secretion?

The Digestive System Consists of the GI Tract and Accessory Glandular Organs

A piece of food that enters the mouth and is swallowed passes into the **esophagus**, a narrow tube that travels through the thorax to the abdomen (Fig. 21-2a). The esophageal walls are skeletal muscle initially but transition to smooth muscle about two-thirds of the way down the length. Just below the diaphragm, the esophagus ends at the **stomach**, a baglike organ that can hold 2–3 liters of food and fluid when fully (but uncomfortably) expanded.

The stomach is roughly divided into three sections: the upper **fundus**, central **body**, and lower **antrum** (Fig. 21-2b). The stomach continues digestion, mixing food with acid and enzymes to create chyme. The opening between the stomach and **small intestine** is guarded by the pyloric valve, or **pylorus**. This thickened band of smooth muscle relaxes to allow only small amounts of chyme into the small intestine at any one time.

In this way, the stomach acts as an intermediary between the behavioral act of eating and the physiological acts of digestion and absorption in the intestine. By regulating the rate at which chyme reaches the intestine, the stomach ensures that the intestine will not be overwhelmed with more than it can digest and absorb.

Most digestion takes place in the small intestine, which is divided into three sections: the **duodenum** (the first 25 cm), **jejunum**, and **ileum** (the last two together are about 260 cm long). Digestion is carried out by intestinal enzymes, aided by exocrine secretions from two accessory glandular organs, the **pancreas** and the **liver**. Their secretions enter the initial section of the duodenum through ducts. A tonically contracted sphincter (the *sphincter of Oddi*) keeps pancreatic fluid and bile from entering the small intestine except during a meal.

Digestion is essentially completed in the small intestine, and nearly all digested nutrients and secreted fluids are absorbed there, leaving about 1.5 liters of chyme per day to pass into the **large intestine** (Fig. 21-2a). In the **colon**, the proximal section of the large intestine, watery chyme is converted into semisolid **feces** [*faeces*, dregs] through the absorption of water and electrolytes.

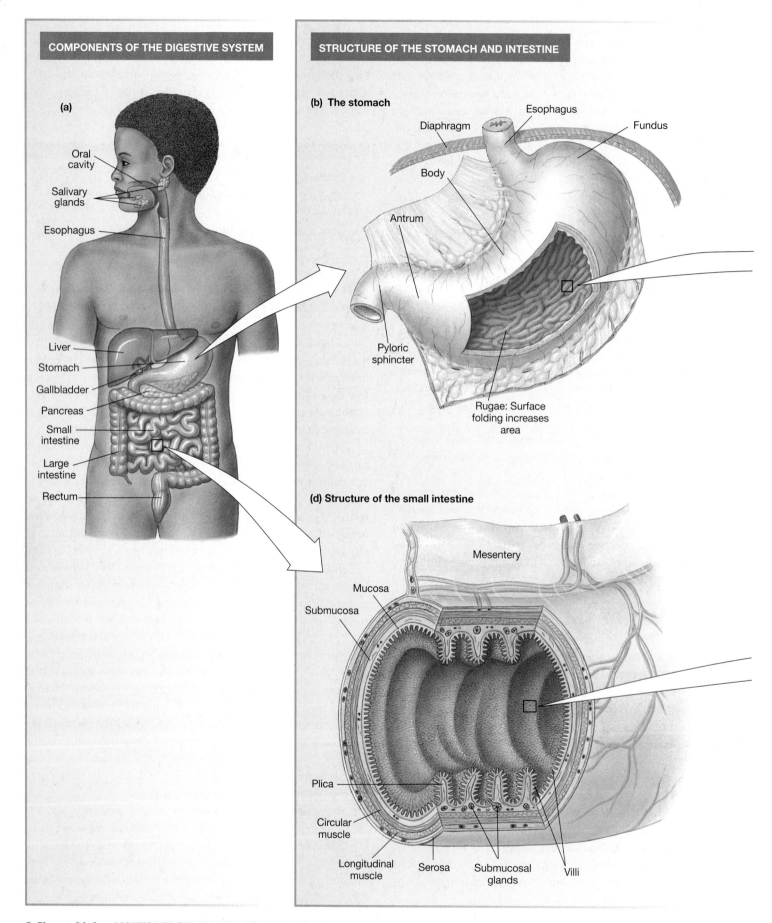

COMPONENTS OF THE DIGESTIVE SYSTEM

(a)

Oral cavity
Salivary glands
Esophagus
Liver
Stomach
Gallbladder
Pancreas
Small intestine
Large intestine
Rectum

STRUCTURE OF THE STOMACH AND INTESTINE

(b) The stomach

Diaphragm
Esophagus
Fundus
Body
Antrum
Pyloric sphincter
Rugae: Surface folding increases area

(d) Structure of the small intestine

Mesentery
Mucosa
Submucosa
Plica
Circular muscle
Longitudinal muscle
Serosa
Submucosal glands
Villi

● Figure 21-2 *ANATOMY SUMMARY* *The Digestive System*

SECTIONAL VIEWS OF THE STOMACH AND INTESTINE

(c) In the stomach, surface area is increased by invaginations called gastric glands.

Opening to gastric gland
Epithelium
Lymph vessel
Lamina propria
Muscularis mucosae
Submucosa
Circular muscle
Longitudinal muscle
Serosa
Artery and vein
Submucosal plexus
Myenteric plexus

(g) Absorptive cell

Brush border
Microvilli

(e) Intestinal surface area is enhanced by finger-like villi.

Villi
Crypt
Mucosa
Muscularis mucosae
Submucosa
Circular muscle
Longitudinal muscle
Serosa
Submucosal artery and vein
Submucosal plexus
Peyer's patch
Myenteric plexus
Lymph vessel

(f)

Absorptive cells
Goblet cells
Crypt lumen
Lamina propria
Endocrine cells
Muscularis mucosae

● **Figure 21-2** *(continued)*

When feces are propelled into the terminal section known as the **rectum**, distension of the rectal wall triggers the *defecation reflex*. The final opening through which gastrointestinal material must pass is the **anus**, with its external anal sphincter of skeletal muscle, which is under voluntary control.

In a living person, the GI tract from mouth to anus is about 450 cm, or nearly 15 feet, long! Of this length, 395 cm, or about 13 feet, consists of the large and small intestines. Try to imagine 13 feet of rope ranging from 1 to 3 inches in diameter all coiled up inside your abdomen from the belly button down. The tight arrangement of the abdominal organs helps explain why you feel the need to loosen your belt after consuming a large meal!

Measurements of intestinal length made during autopsies are nearly double those given here because after death, the longitudinal muscles of the intestinal tract relax. This accounts for the wide variation in intestinal length that you may encounter in different references.

The GI Tract Wall Has Four Layers

The basic structure of the gastrointestinal wall is similar in the stomach and the intestines, although variations exist between different sections of the GI tract (Fig. 21-2c, d, e, f). The gut wall consists of four layers: an inner *mucosa*, a middle layer known as the *submucosa*, and an outer section, the *muscularis externa*, composed of layers of smooth muscle covered with connective tissue (the *serosa*).

The Mucosa   The **mucosa** is the inner lining of the gastrointestinal tract, created from (1) a single layer of epithelial cells, (2) the **lamina propria**, subepithelial connective tissue that holds the epithelium in place, and (3) the **muscularis mucosae**, a thin layer of smooth muscle. Several structural modifications increase the amount of surface area to enhance absorption.

First, the entire wall is crumpled into folds known as *rugae* in the stomach and *plicae* in the small intestine. The intestinal mucosa also projects into the lumen in small fingerlike extensions known as **villi**. Additionally, the surface area of each individual intestinal cell is increased by **microvilli** [◐ p. 69] along the apical membrane (Fig. 21-2g). Microvilli create the *brush border*, a feature also seen in the epithelial cells of the nephron.

Additional surface area is added by tubular invaginations of the surface that extend down into the supporting connective tissue. These invaginations are called **gastric glands** in the stomach and **crypts** in the intestine. Some of the deepest invaginations form secretory **submucosal glands** that open into the lumen through ducts.

Epithelial cells that line the lumen are the most variable feature of the GI tract, changing from section to section. The cells include transporting epithelial cells, endocrine and exocrine secretory cells, and stem cells. Transporting epithelial cells move ions and water into the lumen and absorb ions, water, and nutrients. At the *mucosal* (apical) surface, secretory cells release enzymes, mucus, and paracrines into the lumen of the digestive system. On their *serosal* (baso-

lateral) sides, they secrete hormones or paracrines into the interstitial fluid, where they act on neighboring cells.

GI *stem cells* are rapidly dividing undifferentiated cells in the crypts or glands that continuously produce new epithelium. As stem cells divide, the newly formed cells are pushed up toward the luminal surface of the epithelium. The average life span of a GI epithelial cell is only a few days, a good indicator of the rough life such cells lead.

The cell-to-cell junctions that tie GI epithelial cells together vary [◐ p. 64]. In the stomach, the junctions form a tight barrier so that little can pass between the cells. In the small intestine, junctions are not as tight. This intestinal epithelium is considered "leaky" because some water and solutes can be absorbed by going *between* the cells instead of *through* them. We have recently learned that these junctions have plasticity; that is, their "tightness" can be regulated to some extent by the nutritional status of the body.

The subepithelial connective tissue, the *lamina propria*, contains nerve fibers and small blood and lymph vessels into which absorbed nutrients pass (Fig. 21-2c). The lamina propria also contains wandering immune cells, such as macrophages and lymphocytes, on patrol for invaders that enter through breaks in the epithelium.

In the intestine, collections of lymphoid tissue form small nodules and larger **Peyer's patches** (Fig. 21-2e) that create visible bumps in the mucosa. These lymphoid aggregations are a major part of the gut-associated lymphoid tissue (GALT).

The *muscularis mucosae* separates the mucosa from the submucosa. Contraction of this thin smooth muscle layer may somehow alter the effective surface area for absorption by moving the villi back and forth, like the waving tentacles of a sea anemone.

CONCEPT CHECK   Answers: p. 694

4. Is the lumen of the digestive tract on the apical or basolateral side of intestinal epithelium?

5. Name the structures a piece of food passes as it travels mouth to anus.

The Submucosa   The middle layer of the gut wall is composed of connective tissue with larger blood and lymph vessels (Fig. 21-2c, e). The submucosa also contains the **submucosal plexus** [*plexus*, interwoven], one of the two major nerve networks of the **enteric nervous system** [◐ p. 241]. The enteric nervous system is a unique division of the nervous system that helps coordinate digestive function, as we will see.

The Muscularis Externa and Serosa   The outer wall of the intestinal tract consists mostly of two layers of smooth muscle: an inner circular layer and an outer longitudinal layer (Fig. 21-2c, d, e). Contraction of the circular layer decreases the diameter of the lumen. Contraction of the longitudinal layer shortens the tube.

The stomach has an incomplete third layer of oblique muscle between the circular muscles and the submucosa. The second nerve network of the enteric nervous system, the **myenteric plexus** [*myo-*, muscle + *enteron*, intestine], lies between the two muscle layers of the digestive tract.

A connective tissue membrane called the **serosa** surrounds the entire digestive tract. The serosa is a continuation of the **peritoneal membrane** (*peritoneum*) that lines the abdominal cavity. The peritoneum also forms sheets of **mesentery** that hold the intestines in place so that they do not become tangled as they move.

Now let's take a brief look at the processes of motility, secretion, digestion, and absorption. Gastrointestinal physiology is a rapidly expanding field, and this textbook does not attempt to be all-inclusive. Instead, it focuses on certain broad aspects of digestive physiology, with additional material placed in tables for reference.

CONCEPT CHECK   Answers: p. 694

6. Why is the digestive system associated with the largest collection of lymphoid tissue in the body?

▶ MOTILITY

Motility in the gastrointestinal tract serves two purposes: moving food from the mouth to the anus and mechanically mixing food to break it into uniformly small particles. This mixing maximizes exposure of the particles to digestive enzymes by increasing their surface area. Gastrointestinal motility is determined by the properties of its smooth muscle and modified by chemical input from nerves, hormones, and paracrines.

GI Smooth Muscle Contracts Spontaneously

Most of the gastrointestinal tract is composed of single-unit smooth muscle, with groups of cells electrically connected by gap junctions [❀ p. 64] to create contracting segments. Different re-

gions exhibit different types of contraction. **Tonic contractions** that are sustained for minutes or hours occur in some smooth muscle sphincters as well as in the anterior portion of the stomach. **Phasic contractions** with contraction-relaxation cycles that last only a few seconds occur in the posterior region of the stomach and the small intestine.

Cycles of smooth muscle contraction and relaxation are associated with spontaneous cycles of depolarization and repolarization known as **slow wave potentials**, or simply *slow waves*. Slow wave potentials differ from myocardial pacemaker potentials because they do not reach threshold with each cycle (Fig. 21-3 ●). A slow wave that does not reach threshold will not cause a contraction in the muscle fiber.

When a slow wave potential does reach threshold, voltage-gated Ca^{2+} channels in the muscle fiber open, Ca^{2+} enters, and the cell fires one or more action potentials. The depolarization phase of the action potential, like that of myocardial autorhythmic cells, is due to Ca^{2+} entry into the cell. In addition, Ca^{2+} entry initiates muscle contraction [❀ p. 417].

Contraction of smooth muscle, like that of cardiac muscle, is graded according to the amount of Ca^{2+} that enters the fiber. The higher the amplitude of the slow wave, the more action potentials fire, and the greater the contraction force in the muscle. Similarly, the longer the duration of the slow wave, the longer the duration of contraction. Both amplitude and duration can be modified by neurotransmitters, hormones, or paracrines.

Slow wave frequency varies by region of the digestive tract, ranging from 3 waves/min in the stomach to 12 waves/min in the duodenum. Current research indicates that slow waves originate in modified smooth muscle cells called **interstitial cells of Cajal** (see Emerging Concepts box) and pass into smooth muscle cells through gap junctions. The interstitial cells of Cajal are also associated with enteric neurons and may act as an intermediary between the neurons and smooth muscle.

RUNNING PROBLEM

A peptic ulcer is a raw, inflamed area in the stomach or duodenum that extends from a break in the surface epithelium past the muscularis mucosa. Most peptic ulcers occur in the duodenum (duodenal ulcers). Other peptic ulcers are located within the stomach itself (gastric ulcers). Until 1982, the year *H. pylori* was discovered, peptic ulcers were believed to be caused by high levels of acid and enzyme production in the stomach. However, since the 1980s, scientists have found *H. pylori* in 90% of people with duodenal ulcers and 70% of people with gastric ulcers. To confirm the diagnosis of an *H. pylori*-caused ulcer, Tonya Berry will undergo tests to isolate the bacterium from her stomach.

Question 1: Why are peptic ulcers found in the duodenum rather than in the jejunum or ileum?

● **Figure 21-3**   *Slow wave potentials in GI smooth muscle*

Action potentials fire when slow wave potentials exceed threshold.

The force and duration of muscle contraction is directly related to the amplitude and frequency of action potentials.

Membrane potential (mV)

Slow wave

Action potential

Threshold

Force of muscle contraction

Time →

GI Smooth Muscle Exhibits Different Patterns of Contraction

Muscle contractions in the gastrointestinal tract occur in three general patterns. Between meals, when the tract is largely empty, a series of contractions begins in the stomach and passes slowly from section to section, taking about 90 minutes to reach the large intestine. This pattern, known as the **migrating motor complex**, is a "housekeeping" function that sweeps food remnants and bacteria out of the upper GI tract and into the large intestine.

Muscle contractions during and following a meal fall into one of two other patterns. **Peristaltic contractions** are progressive waves of contraction that move from one section of the GI tract to another, just like the human "waves" that ripple around a football stadium or basketball arena.

In peristalsis, circular muscles contract just behind a mass, or *bolus*, of food. This pushes the bolus forward into a *receiving segment*, where the circular muscles are relaxed (Fig. 21-4a ●). The receiving segment then contracts, continuing the forward movement. Peristaltic contractions push a bolus forward at speeds between 2 and 25 cm/sec.

Peristalsis propels material forward through the esophagus, and mixes food as it digests in the stomach. In normal digestion, intestinal peristaltic waves are limited to short distances. Hormones, paracrines, and the autonomic nervous system influence peristalsis in all regions of the GI tract.

In **segmental contractions**, short (1–5 cm) segments of intestine contract and relax (Fig. 21-4b). In the contracting segment, circular muscles contract while longitudinal muscles relax. These

(a) Peristaltic contractions are responsible for forward movement.

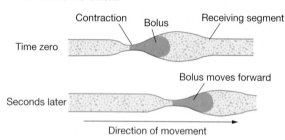

(b) Segmental contractions are responsible for mixing.

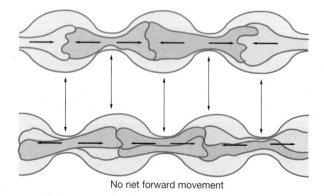

No net forward movement

● **Figure 21-4** *Contractions in the GI tract*

contractions may occur randomly along the intestine or may occur at regular intervals. Alternating segmental contractions of the type shown churn intestinal contents back and forth, mixing them and keeping them in contact with the absorptive epithelium. When segmental contractions occur sequentially, in an oral-to-aboral direction [*ab-*, away], digesting material is propelled short distances.

Motility disorders are among the more common gastrointestinal problems. They range from esophageal spasms and delayed gastric emptying to constipation and diarrhea. *Irritable bowel syndrome* is a chronic functional disorder characterized by altered bowel habits and abdominal pain.

CONCEPT CHECK

Answers: p. 694

7. Why are some sphincters of the digestive system tonically contracted?

▶ SECRETION

In a typical day, 9 liters of fluid pass through the lumen of an adult's gastrointestinal tract—the equivalent of three 3-liter soft drink bottles! Only about 2 liters of that volume enter the system through the mouth. The remaining 7 liters of fluid come from body water secreted along with enzymes and mucus (Fig. 21-5 ●). About half the secreted fluid comes from accessory organs and glands such as the salivary glands, pancreas, and liver. The remaining 3.5 liters are secreted by the epithelial cells of the digestive tract itself.

The significance of the secreted volume is more apparent if you realize that the secreted fluid equals one-sixth of the entire body water volume, or more than twice the plasma volume. In other words, if the fluid secreted into the lumen is not reabsorbed as it passes along the GI tract, the body will rapidly dehydrate.

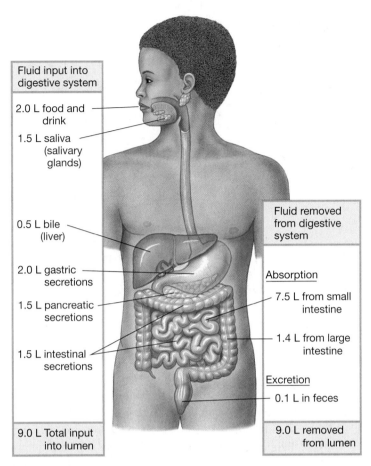

Fluid input into digestive system

2.0 L food and drink

1.5 L saliva (salivary glands)

0.5 L bile (liver)

2.0 L gastric secretions

1.5 L pancreatic secretions

1.5 L intestinal secretions

9.0 L Total input into lumen

Fluid removed from digestive system

Absorption

7.5 L from small intestine

1.4 L from large intestine

Excretion

0.1 L in feces

9.0 L removed from lumen

● **Figure 21-5** *Daily mass balance in the digestive system*

Digestive Enzymes Are Secreted into the Mouth, Stomach, and Intestine

Digestive enzymes are secreted either by exocrine glands (salivary glands and the pancreas) or by epithelial cells in the mucosa of the stomach and small intestine. Enzymes are proteins, which means they are synthesized on the rough endoplasmic reticulum, packaged by the Golgi apparatus into secretory vesicles, and

stored within the cell until needed. On demand, they are released by exocytosis [◐ p. 146].

Some digestive enzymes are secreted in an inactive **proenzyme** form and are known collectively as *zymogens*. These enzymes must be activated in the GI lumen before they can carry out digestion. Late activation prevents the enzymes from digesting the cells in which they are synthesized. It also allows them to be stockpiled within the cells until needed.

The control pathways for enzyme release vary but include a variety of neural, hormonal, and paracrine signals. Usually, stimulation of parasympathetic neurons in the vagus nerve enhances enzyme secretion.

Specialized Cells Secrete Mucus

Mucus is a viscous secretion composed primarily of glycoproteins that collectively are called **mucins**. The primary functions of mucins are to form a protective coating over the GI mucosa and to lubricate the contents of the gut.

Mucus is made in specialized exocrine cells called **mucous cells** in the stomach and **goblet cells** in the intestine. Goblet cells comprise between 10% and 24% of the intestinal cell population. In the mouth, the minor salivary glands secrete about 70% of the salivary mucus.

The release signals for mucus include parasympathetic innervation, a variety of neuropeptides found in the enteric nervous system, and cytokines from immunocytes. Parasitic infections and inflammatory processes in the gut both cause substantial increases in mucus output as the body attempts to increase its protective barrier.

The Digestive System Secretes Ions and Water

A large portion of the 7 liters of fluid secreted by the digestive system each day is composed of water and ions, particularly Na^+, K^+, Cl^-, HCO_3^-, and H^+. The ions are first secreted into the lumen of the tract, then reabsorbed. Water follows osmotic gradients created by the transfer of solutes from one side of the epithelium to another.

Gastrointestinal epithelial cells, like those in the kidney, have distinct apical and basolateral membranes [◐ p. 147], each with proteins for active transport, facilitated diffusion, and ion movement through open channels. Many of these transporters are similar to those of the renal tubule.

The basolateral membrane contains the ubiquitous Na^+-K^+-ATPase. Cotransporters include a Na^+-K^+-$2Cl^-$ symporter, Cl^--HCO_3^- antiporter, and an H^+-K^+-ATPase. Ion channels include Na^+, K^+, and Cl^- channels.

▶ DIGESTION AND ABSORPTION

Digestion of macromolecules into absorbable units is accomplished by a combination of mechanical and enzymatic breakdown [● p. 90]. Chewing and churning create smaller pieces of food, which expose more surface area to digestive enzymes. *Bile*, a complex chemical mixture secreted by the liver, serves a similar purpose by dispersing lipids (more commonly called *fats* in digestive physiology) as fine droplets with greater surface area.

The pH at which different digestive enzymes function best [● p. 94] reflects the location where they are most active. Enzymes that act in the stomach work well at acidic pH. Those that are secreted into the small intestine work best at alkaline pH.

Most nutrient absorption takes place in the small intestine, with additional absorption of water and ions in the large intestine. Like secretion, absorption of nutrients and ions across the GI epithelium uses many of the same transport proteins that the kidney tubule epithelium uses.

Carbohydrates Are Digested to Monosaccharides

About half the calories the average American ingests are in the form of carbohydrates, mainly *starch* and *sucrose* (table sugar). Other dietary carbohydrates include the glucose polymers *glycogen*

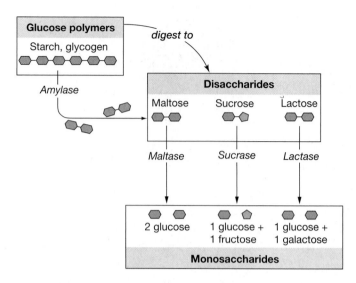

● **Figure 21-6** *Carbohydrate digestion*

and *cellulose*, disaccharides such as *lactose* and *maltose*, and the simple sugars *glucose* and *fructose* [● Fig. 2-13, p. 36].

The complex carbohydrates we eat include cellulose, glycogen, and starch. We are unable to digest cellulose because we lack the necessary enzymes. As a result, the cellulose in plant matter becomes what is known as fiber, or roughage, in our diet. Glycogen, found mainly in meat, is usually degraded during cooking.

Starch, therefore, is the primary complex carbohydrate that we must digest (Fig. 21-6 ●). **Amylase** is the enzyme that breaks long glucose polymers into smaller glucose chains and into the disaccharide maltose (Table 21-1). Maltose and the other disaccharides

Table 21-1: Digestive Enzymes

BIOMOLECULE	DIGESTED BY	INTO	ENZYME LOCATION
Carbohydrate			
Starch	Amylase	Maltose	Saliva, pancreas
Disaccharides			Intestine
Maltose	Maltase	2 Glucose	
Lactose	Lactase	Glucose + galactose	
Sucrose	Sucrase	Glucose + fructose	
Proteins			
Interior peptide bonds	Endopeptidases	Peptides	Stomach, intestine, pancreas
Terminal peptide bonds	Exopeptidases	Amino acids	Stomach, intestine, pancreas
Works at NH_2-terminal end	Aminopeptidase		
Works at COOH-terminal end	Carboxypeptidase		
Fats			
Triglycerides	Lipase	Monoglyceride and free fatty acids	Mouth, stomach, pancreas
Phospholipids	Phospholipase	Lysophospholipid and free fatty acids	Pancreas

must be broken down into their component monosaccharides before they can be absorbed. Enzymes known as **disaccharidases** (maltase, sucrase, and lactase) are responsible for this final step.

Proteins Are Digested into Small Peptides and Amino Acids

Unlike carbohydrates, which are ingested in forms ranging from the simple to the complex, most proteins are ingested as proteins or large polypeptides [◉ Fig. 2-16, p. 39]. Not all proteins are equally digestible, however.

Plant proteins are the least digestible. Egg protein is among the most digestible, with 85%–90% of the protein in a form that can be digested and absorbed. Surprisingly, between 30% and 60% of the protein found in the intestinal lumen comes not from ingested food but from the sloughing of dead cells and from protein secretions such as enzymes and mucus.

The enzymes for protein digestion are classified into two broad groups: endopeptidases (also called *proteases*) and exopeptidases. **Endopeptidases** attack peptide bonds in the interior of the amino acid chain and make smaller peptide fragments from a long chain (Fig. 21-7 ◉). These enzymes are secreted as inactive

proenzymes from epithelial cells in the stomach, intestine, and pancreas. **Exopeptidases** release single amino acids from peptides by chopping them off the ends, one at a time.

Fat Digestion Is Assisted by Bile

Fats and related molecules in the Western diet include triglycerides, cholesterol, phospholipids, long-chain fatty acids, and the fat-soluble vitamins [◉ Figure 2-14, p. 37]. Nearly 90% of our fat calories come from triglycerides because they are the primary form of lipid in both plants and animals.

Fat digestion is complicated by the fact that most lipids are not particularly water-soluble. As a result, they form large clumps in the aqueous chyme solution of the intestinal tract. Enzymatic fat digestion must therefore be aided by nonenzyme secretions that help make fat globules into smaller particles.

Enzymatic fat digestion is carried out by **lipases**. These enzymes remove two fatty acids from triglycerides, resulting in a monoglyceride and two free fatty acids (Fig. 21-8 ◉). Phospholipids are digested by pancreatic *phospholipase*. Free cholesterol does not have to be digested before being absorbed.

Nucleic Acids Are Digested into Bases and Monosaccharides

The nucleic acid polymers DNA and RNA do not make up a significant part of most diets. They are digested by pancreatic and intestinal enzymes, first into their component nucleotides and then into bases and monosaccharides [◉ Figure 2-17, p. 40]. The bases are absorbed by active transport, and the monosaccharides are absorbed by facilitated diffusion and secondary active transport like other simple sugars.

▶ REGULATION OF GI FUNCTION

The digestive system has remarkably complex regulation with overlapping pathways that involve neural, endocrine, and local control mechanisms. One particularly challenging part of mastering GI regulation is learning the terminology. Some of the hormones and neurotransmitters involved in digestive regulation were first

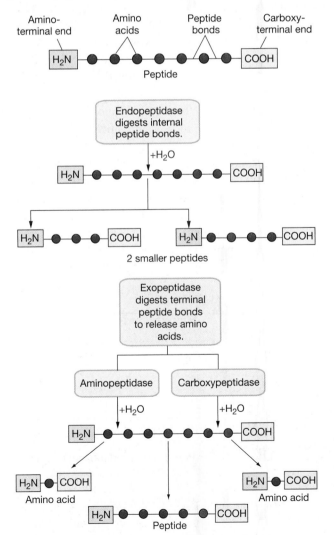

● **Figure 21-7** *Endopeptidases and exopeptidases*

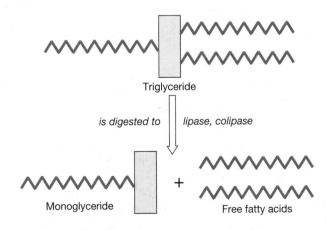

● **Figure 21-8** *Fat digestion*

identified in other systems of the body. As a result, their names have nothing to do with their function in the gastrointestinal system. Here is an overview of the different control mechanisms.

1. **Classic neural reflexes.** The classic neural reflex begins with a stimulus transmitted along a sensory neuron to the CNS, where it is integrated and acted upon. In the digestive system some classic reflexes originate in the GI tract itself but others originate outside. Digestive reflexes integrated in the CNS are called **long reflexes** (Fig. 21-9 ●).

   Gastrointestinal reflexes that originate completely outside the digestive system include feedforward reflexes [● p. 195] and emotional reflexes. These reflexes are called **cephalic reflexes** because they originate in the cephalic brain [*cephalicus*, head]. *Feedforward reflexes* begin with stimuli such as the sight, smell, sound, or thought of food; they prepare the digestive system for food that the brain is anticipating. For example, if you are hungry and smell dinner cooking, your mouth waters and your stomach growls.

   The influence of emotions on the GI tract illustrates another link between the brain and the digestive system. Emotional responses that have GI manifestations range from travelers' constipation to "butterflies in the stomach" to psychologically induced diarrhea. Fight-or-flight reactions also influence GI function.

   In long reflexes, the smooth muscle and glands of the GI tract are under autonomic control. In general, we say that parasympathetic neurons to the GI tract, carried mostly in the vagus nerve, are excitatory and enhance GI functions. Sympathetic neurons usually inhibit GI function.

2. **Short reflexes integrated in the enteric nervous system.** Neural control of the GI tract does not rely strictly on the central nervous system (CNS). Instead, the enteric nerve plexus acts as a "little brain," allowing some local reflexes to begin and end in the gastrointestinal tract. Reflexes that originate within the enteric nervous system and are integrated there without outside input are called **short reflexes** (Fig. 21-9).

   The primary responses controlled by the enteric nervous system are related to motility, secretion, and growth. The *submucosal nerve plexus* contains sensory neurons that receive signals from the lumen. Submucosal neurons control secretion by GI cells. In addition, they integrate sensory input and signal *myenteric motor neurons* to influence motility. Although the enteric nervous system can work in isolation, it also coordinates function with autonomic neurons bringing signals from the CNS.

3. **GI peptides.** Peptides secreted by cells of the digestive tract may act as hormones or paracrines. GI hormones are secreted into the blood. They act on the GI tract, on accessory organs such as the pancreas, or on more distant targets, such as the brain. Paracrines are secreted into the lumen of the GI tract or into the extracellular fluid. Luminal paracrines combine with receptors on the apical membrane to elicit a response. Paracrines act locally, on cells close to where they were secreted.

   In general, the GI peptides excite or inhibit motility and secretion (Fig. 21-10 ●). Motility effects include altering peristaltic activity, contraction of the gallbladder for bile release, and controlling gastric emptying to maximize digestion and absorption. Secretory processes influenced by GI peptides include both endocrine and exocrine functions. Some of the most interesting pathways are those in which GI peptides act on the brain. For example, in experimental studies the GI hormone CCK was found to enhance *satiety*, the state of feeling that hunger has been satisfied. However, these findings are complicated by the fact that CCK is also manufactured by neurons and functions as a neurotransmitter in the brain.

● **Figure 21-9** *The enteric nervous system*

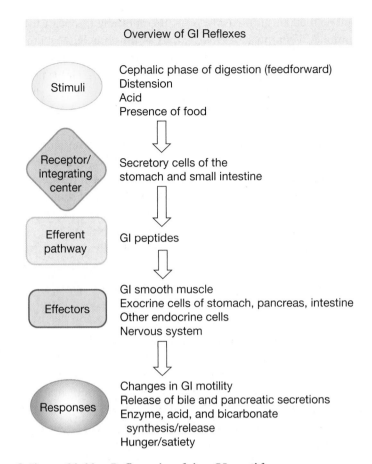

Overview of GI Reflexes

Stimuli
Cephalic phase of digestion (feedforward)
Distension
Acid
Presence of food

Receptor/integrating center
Secretory cells of the stomach and small intestine

Efferent pathway
GI peptides

Effectors
GI smooth muscle
Exocrine cells of stomach, pancreas, intestine
Other endocrine cells
Nervous system

Responses
Changes in GI motility
Release of bile and pancreatic secretions
Enzyme, acid, and bicarbonate synthesis/release
Hunger/satiety

● **Figure 21-10** *Reflexes involving GI peptides*

The Enteric Nervous System Is Known As the Little Brain

The enteric nervous system was first recognized over a century ago, when scientists noted that an increase in intraluminal pressure caused a reflex wave of peristaltic contraction to sweep along sections of isolated intestine removed from the body. What they observed was the ability of the enteric nervous system (ENS) to carry out a reflex independent of control by the CNS.

In this respect, the enteric nervous system is much like the nerve networks of jellyfish and sea anemones (the phylum Cnidaria) that were introduced in Chapter 9 [p. 285]. You might have seen sea anemones being fed at an aquarium. As the piece of shrimp or fish drifts down close to the tentacles, they begin to wave, picking up chemical "odors" through the water. Once the food contacts the tentacles, it is directed toward the mouth, passed from one tentacle to another until it disappears into the digestive cavity.

This purposeful reflex is accomplished without a brain, eyes, or a nose. The anemone's nervous system consists of a network composed of sensory neurons, interneurons, and efferent neurons that control the muscles and secretory cells of the anemone's body. Somehow, the neurons of the network are linked in a way that allows them to integrate information and act upon it. In the same way that the anemone captures its food, the enteric nervous system receives stimuli and acts upon them.

Anatomically and functionally, the enteric nervous system shares many features with the CNS, so much so that it has been nicknamed "the little brain." Some similarities between the ENS and the brain are:

1. **Neurotransmitters and neuromodulators.** The neurons of the enteric nervous system release more than 30 neurotransmitters and neuromodulators, most of which are identical to molecules found in the brain. These neurotransmitters are sometimes called *non-adrenergic, non-cholinergic* to distinguish them from the traditional autonomic neurotransmitters, norepinephrine and acetylcholine. Among the best-known neurotransmitters and neuromodulators in the enteric nervous system are serotonin, vasoactive intestinal peptide (VIP), and nitric oxide (NO).

2. **Support cells.** The support cells of neurons within the enteric nervous system are more similar to astroglia of the brain than to Schwann cells of the peripheral nervous system.

3. **Diffusion barrier.** The capillaries that surround ganglia within the enteric nervous system are not very permeable and create a diffusion barrier that is similar to the blood-brain barrier of cerebral blood vessels.

4. **Integrating center.** Reflexes that originate in the GI tract can be integrated and acted upon without neural signals leaving the enteric nervous system. Thus, the neuron network of the enteric nervous system is its own integrating center, much like the brain and spinal cord.

It was thought that if we could explain how the ENS integrates simple behaviors, we could use it as a simple model system for CNS function. But study of enteric nervous system function is difficult because there is no discrete command center for enteric reflexes.

Instead, in an interesting twist, GI physiologists are applying information gleaned from studies of the brain and spinal cord to the function of the enteric nervous system. The complex interactions between the enteric and central nervous systems, the endocrine system, and the immune system promise to provide scientists with questions to investigate for many years to come.

RUNNING PROBLEM

Tonya undergoes endoscopy, a procedure in which a fiberoptic scope is inserted into her esophagus and maneuvered into her stomach. Small pincers attached at the end of the scope pluck a piece of her stomach lining, which is sent to the pathology laboratory for study. Tonya is awake but groggy from a tranquilizer during the procedure. That afternoon, she feels fine, but her stomach hurts. She chews a tablet of antacid for relief.

Question 3: Some antacids contain sodium bicarbonate (NaHCO₃), while many others contain aluminum hydroxide (Al(OH)₃). Antacids neutralize hydrochloric acid (HCl) secreted by the stomach by acting as buffers. For both types of antacids, write the chemical equation showing how the antacids buffer hydrochloric acid.

GI Peptides Include Hormones, Neurocrines, and Cytokines

The hormones of the gastrointestinal tract occupy an interesting place in the history of endocrinology. In 1902, two Canadian physiologists, W. M. Bayliss and E. H. Starling, discovered that acidic chyme entering the small intestine from the stomach caused the release of pancreatic juices even if all nerves to the pancreas were cut. Because the only communication remaining between the intestine and the pancreas was the blood supply that ran between them, they postulated the existence of some blood-borne (*humoral*) factor that was released by the intestine.

When duodenal extracts applied directly to the pancreas stimulated secretion, they knew they were dealing with a chemical produced by the duodenum. They named the substance *secretin*. Starling further proposed that the general name *hormone*, from the Greek word meaning "I excite," be given to all humoral agents that act at a site distant from their release.

A few years later, in 1905, J. S. Edkins postulated the existence of a gastric hormone that stimulated gastric acid secretion. It took more than 30 years for a relatively pure extract of the gastric hormone to be made successfully, and it was 1964 before the hormone named *gastrin* was finally purified.

Why was research on the digestive hormones so slow to develop? A major reason is the way hormones of the GI tract are secreted by isolated endocrine cells scattered among other cells of the mucosal epithelium. In the early years, the only way to obtain these hormones was to make a crude extract of the entire epithelium, a procedure that also liberated digestive enzymes and paracrines made in adjacent cells. Thus, it was very difficult to tell if the physiological effect elicited by the extract came from one hormone, from more than one hormone, or from a paracrine such as histamine.

Today, although researchers have sequenced over 30 peptides from the GI mucosa, only 5 are widely accepted as hormones. A few peptides have well-defined paracrine effects, but most fall into a long list of candidate hormones. In addition, we know of nonpeptide regulatory molecules, such as histamine, a paracrine. Because of the uncertainty associated with the field, we will restrict our focus to the major regulatory molecules.

The gastrointestinal peptides are generally divided into three groups:

1. The *gastrin family* includes the hormones **gastrin** and **cholecystokinin** (CCK), with several variants of each.
2. The *secretin family* includes **secretin**, the neurocrine [p. 174] molecule **vasoactive intestinal peptide** (VIP), and **GIP**, a hormone known originally as *gastric inhibitory peptide* because it inhibited gastric acid secretion in early experiments. More recent studies, however, found that GIP administered in lower physiological doses does not block acid secretion. A new name was proposed, **glucose-dependent insulinotropic peptide**, which more accurately describes

Table 21-2: The Digestive Hormones

	SECRETED BY	TARGET(S)	EFFECTS	
			ENDOCRINE SECRETION	EXOCRINE SECRETION
Gastrin	G cells in stomach antrum	ECL cells; parietal cells	None	Stimulates gastric acid
Cholecystokinin (CCK)	Endocrine cells of small intestine; neurons of brain and gut	Gallbladder, pancreas, gastric smooth muscle	None	Stimulates pancreatic enzyme secretion; potentiates bicarbonate secretion
Secretin	Endocrine cells in small intestine	Pancreas, stomach	None	Stimulates bicarbonate secretion and pepsin release; inhibits gastric acid
Gastric inhibitory peptide (GIP)	Endocrine cells in small intestine	Beta cells of endocrine pancreas	Stimulates insulin release (feedforward mechanism)	Inhibits acid secretion
Motilin	Endocrine cells in small intestine	Smooth muscle of antrum and duodenum	None	None
Glucagon-like Peptide 1	Endocrine cells in small intestine	Endocrine pancreas	Stimulates insulin release; inhibits glucagon release	Possibly inhibits acid secretion

its actions (see below). Both names are used in scientific publications, but Medline uses "gastric inibitory peptide" as its preferred subject heading. Another member of the secretin family is the candidate hormone, **glucagon-like peptide 1** (GLP-1).

3. The third group of peptides includes those that do not fit into the two families above. The primary member of this group is the hormone **motilin**.

The sources, targets, and effects of some GI peptides are summarized in Table 21-2.

In the remainder of this chapter we will follow some food as it passes through the GI tract. As a preview, look at Figure 21-11 ●, a summary of the main events that occur in each section of the GI tract. Food processing is traditionally divided into three phases: a cephalic phase, a gastric phase, and an intestinal phase.

► THE CEPHALIC PHASE

Digestive processes in the body begin before food ever enters our mouths. Simply smelling, seeing, or even *thinking* about food can make our mouths water and our stomachs rumble. These long reflexes that begin in our cephalic brain create a feedforward response that is known as the **cephalic phase** of digestion (Fig. 21-12 ●).

These anticipatory stimuli and the stimulus of food in the oral cavity activate neurons in the medulla oblongata. The medulla in turn sends an efferent signal through autonomic neurons to the salivary glands and through the vagus nerve to the enteric nervous system. In response to these signals, the stomach, intestine, and accessory glandular organs begin secretion and increase motility in anticipation of the food to come.

Chemical and Mechanical Digestion Begins in the Mouth

When food first enters the mouth, it is met by a flood of the secretion we call **saliva**, a dilute solution of water, ions, and proteins such as mucins, lysozyme, and immunoglobulins. Salivary secretion is under autonomic control and can be triggered by multiple stimuli such as the sight, smell, touch, and even thought of food.

Saliva serves multiple purposes. Its water and mucus soften and lubricate food to make it easier to swallow. You can appreciate this function if you've ever tried to swallow a dry soda cracker without chewing it thoroughly. Saliva also dissolves food so that we can taste it.

Saliva begins chemical digestion with the secretion of *salivary amylase* and a small amount of lipase. Amylase breaks starch into maltose after it is activated by chloride ions, which are abundantly available in saliva. If you chew on an unsalted soda cracker for a long time, you may be able to taste the conversion of flour starch in the cracker into sweeter maltose.

MOTILITY	OTHER	STIMULUS FOR RELEASE	RELEASED INHIBITED BY	OTHER INFORMATION
None	Enhanced mucosal cell growth	Peptides and amino acids peptide in lumen; gastrin releasing peptide and ACh in nervous reflexes	pH < 1.5; somatostatin	NA
Stimulates gallbladder contraction for bile release; inhibits gastric emptying	Stimulates satiety	Fatty acids and some amino acids	NA	Some effects may be due to CCK as a neuropeptide rather than as a hormone
Inhibits gastric emptying	None	Acid in small intestine	Somatostatin	NA
None	None	Glucose, fatty acids, and amino acids in small intestine	NA	Acid inhibition questionable at physiological concentrations
Stimulates migrating motor complex	None	Fasting: periodic release every 1.5–2 hours by neural stimulus	NA	Changes associated with both constipation and diarrhea, but relationship is unclear
Possibly inhibits gastric emptying	None	Mixed meal that includes carbohydrates or fats in the lumen	NA	Related to but not identical to pancreatic glucagon; may act together with GIP

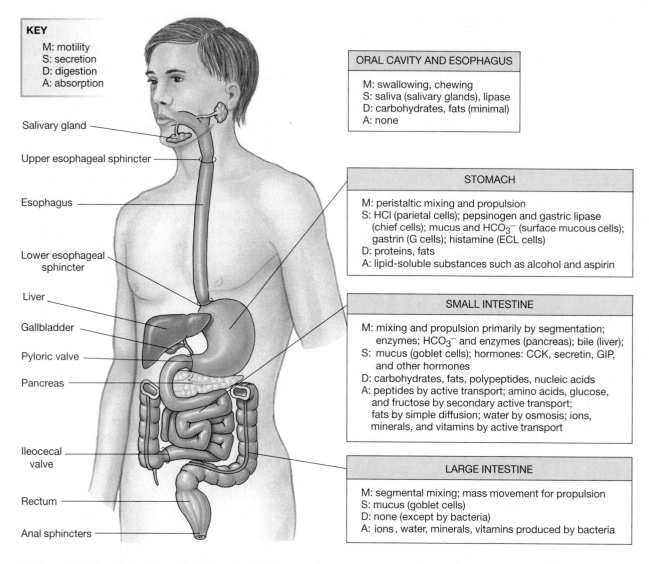

KEY

M: motility
S: secretion
D: digestion
A: absorption

Salivary gland

Upper esophageal sphincter

Esophagus

Lower esophageal sphincter

Liver

Gallbladder

Pyloric valve

Pancreas

Ileocecal valve

Rectum

Anal sphincters

ORAL CAVITY AND ESOPHAGUS

M: swallowing, chewing
S: saliva (salivary glands), lipase
D: carbohydrates, fats (minimal)
A: none

STOMACH

M: peristaltic mixing and propulsion
S: HCl (parietal cells); pepsinogen and gastric lipase (chief cells); mucus and HCO_3^- (surface mucous cells); gastrin (G cells); histamine (ECL cells)
D: proteins, fats
A: lipid-soluble substances such as alcohol and aspirin

SMALL INTESTINE

M: mixing and propulsion primarily by segmentation; enzymes; HCO_3^- and enzymes (pancreas); bile (liver);
S: mucus (goblet cells); hormones: CCK, secretin, GIP, and other hormones
D: carbohydrates, fats, polypeptides, nucleic acids
A: peptides by active transport; amino acids, glucose, and fructose by secondary active transport; fats by simple diffusion; water by osmosis; ions, minerals, and vitamins by active transport

LARGE INTESTINE

M: segmental mixing; mass movement for propulsion
S: mucus (goblet cells)
D: none (except by bacteria)
A: ions, water, minerals, vitamins produced by bacteria

● **Figure 21-11** *Overview of functions in different regions of the digestive system*

Small nonsalivary glands in the mouth secrete a second digestive enzyme, *lingual lipase.* This enzyme does not contribute significantly to fat digestion in the mouth but remains active in the stomach.

The final function of saliva is protection. *Lysozyme* is an antibacterial enzyme, and *immunoglobulins* are proteins that disable bacteria and viruses. In addition, saliva helps wash down the teeth and keep the tongue free of food particles.

Mechanical digestion of food begins in the oral cavity with chewing. The lips, tongue, and teeth all contribute to the **mastication** of food, creating a softened, moistened mass, or *bolus*, that can be easily swallowed.

Swallowing Moves Food from the Mouth to the Stomach

Swallowing, or **deglutition**, is a reflex action that pushes a bolus of food or liquid into the esophagus (Fig. 21-13 ●). The stimulus for swallowing is pressure created when the bolus is pushed

against the soft palate and back of the mouth with the tongue. Sensory input to a swallowing center in the medulla oblongata begins the reflex.

First, the **epiglottis** folds down over the opening of the larynx to prevent food and liquid from entering the airways. At the same time, respiration is inhibited and the upper esophageal sphincter relaxes as the bolus enters the esophagus. Waves of peristaltic contraction then push the bolus toward the stomach, aided by gravity.

The lower end of the esophagus lies just below the diaphragm and is separated from the stomach by the lower esophageal sphincter. This area is not a true sphincter but a region of higher muscle tension that acts as a barrier between the esophagus and the stomach. When food is swallowed, the tension relaxes, allowing the bolus to pass into the stomach.

If the lower esophageal sphincter does not stay contracted, gastric acid and pepsin can irritate the lining of the esophagus, leading to the pain and irritation of heartburn. The walls of the esophagus expand during inspiration, when the intrapleural pressure falls [● p. 557]. Expansion creates subatmospheric pres-

The sight, smell, and taste of food initiate long reflexes that prepare the stomach.

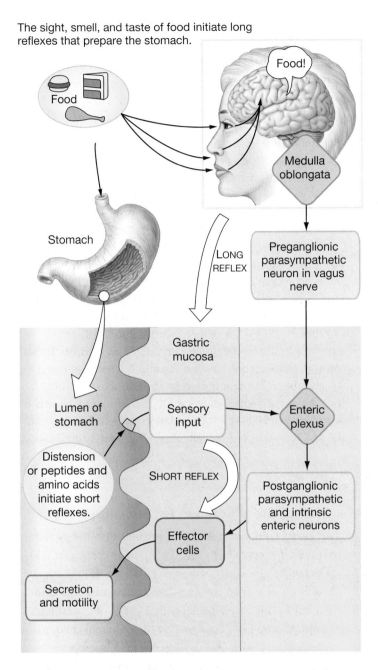

● **Figure 21-12** *Long and short reflexes in the stomach*

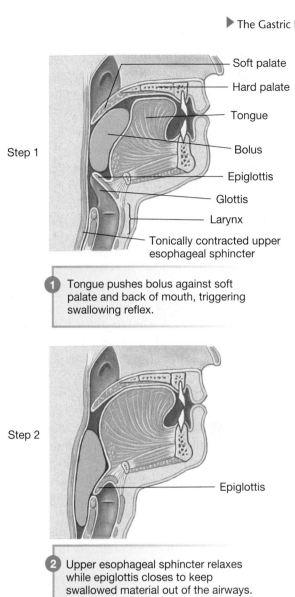

Step 1

- Soft palate
- Hard palate
- Tongue
- Bolus
- Epiglottis
- Glottis
- Larynx
- Tonically contracted upper esophageal sphincter

1 Tongue pushes bolus against soft palate and back of mouth, triggering swallowing reflex.

Step 2

- Epiglottis

2 Upper esophageal sphincter relaxes while epiglottis closes to keep swallowed material out of the airways.

Step 3

3 Food moves downward into the esophagus, propelled by peristaltic waves and aided by gravity.

● **Figure 21-13** *The swallowing reflex*

sure in the esophageal lumen that can literally suck acidic contents out of the stomach if the sphincter is relaxed. The churning action of the stomach when filled with food can also squirt acid back into the esophagus if the sphincter is not contracted fully.

▶ THE GASTRIC PHASE

About 3.5 liters of food, drink, and saliva enter the fundus of the stomach each day. The stomach has three general functions:

1. **Storage.** It stores food and regulates its passage into the small intestine, where most digestion and absorption take place.
2. **Digestion.** It chemically and mechanically digests food into the soupy mixture of uniformly small particles called chyme.

675

3. **Protection.** It protects the body by destroying many of the bacteria and other pathogens that are swallowed with food or trapped in airway mucus. At the same time, the stomach must protect itself from being damaged by its own secretions.

Digestive activity in the stomach begins with the long **vagal reflex** of the cephalic phase, before food even arrives. Then, when food enters the stomach, stimuli within the gastric lumen initiate a series of short reflexes. These short reflexes are the **gastric phase** of digestion.

In gastric phase reflexes, distension of the stomach and the presence of peptides or amino acids in the lumen activate endocrine cells and enteric neurons. Hormones, neurocrines, and paracrines then influence motility and secretion.

The Stomach Stores Food

When food arrives, the stomach relaxes and expands to hold the increased volume. The upper half of the stomach remains relatively quiet, holding food until it is needed. The storage function of the stomach is perhaps the least obvious aspect of digestion. However, at times we ingest more than we need from a nutritional standpoint. In these situations, it falls to the stomach to regulate the rate at which food enters the small intestine.

Without such regulation, the small intestine would not be able to digest and absorb the load presented to it, and significant amounts of unabsorbed chyme would pass into the large intestine. The epithelium of the large intestine is not designed for large-scale nutrient absorption, so most of the chyme would pass out in the feces, resulting in diarrhea. This "dumping syndrome" is one of the less pleasant side effects of surgery that removes portions of either the stomach or small intestine.

While the upper stomach is quietly holding food, the lower stomach is busy with digestion. In the distal half of the stomach, a series of peristaltic waves pushes the food down toward the pyloric valve, mixing it with acid and digestive enzymes. As large food particles are digested to the more uniform texture of chyme, each contractile wave squirts a small amount of chyme through the pyloric valve into the duodenum. Enhanced gastric motility during a meal is primarily under neural control and is stimulated by distension of the stomach.

Digestion in the Stomach   Chemical digestion in the stomach is accomplished by enzymes and hydrochloric acid.

- *Gastric acid* (HCl) kills bacteria and other microorganisms that are swallowed. Acid also *denatures* proteins [● p. 94], destroying their tertiary structure by breaking disulfide and hydrogen bonds. Unfolding protein chains makes the peptide bonds between amino acids accessible to enzymes.
- *Pepsin* is a protease that carries out the initial digestion of proteins. It is particularly effective on collagen and therefore plays an important role in digesting meat.
- *Gastric lipase* is co-secreted with pepsin and begins fat digestion, aided by lingual lipase. About 10% of fat digestion takes place in the stomach.

- The carbohydrate digestion that began in the mouth will continue in the stomach until mixing exposes the amylase to gastric acid. Salivary amylase is inactivated at low pH.

Secretion in the Stomach   The stomach secretes a variety of substances from an assortment of cells. We will discuss the functions and relationships of these secretions in the following section.

- **Parietal cells** deep in the gastric glands secrete hydrochloric acid. Acid secretion in the stomach averages 1–3 liters per day and can create a luminal pH as low as 1. The cytoplasmic pH of the parietal cell is about 7.2, which means the cells are pumping H^+ against a gradient that is 2.5 *million* times as concentrated as the H^+ concentration in the cell.

  The cellular pathway for acid secretion is shown in Figure 21-14●. In an unusual twist, the H^+ comes directly from water by an unknown mechanism. An H^+-K^+-ATPase pumps the H^+ into the lumen. Cl^- moves into the lumen through an open chloride channel. The OH^- from water combines with CO_2 to form HCO_3^-, which is absorbed into the blood.
- Parietal cells also secrete a protein known as **intrinsic factor**. Intrinsic factor complexes with vitamin B_{12} and is essential for B_{12} absorption in the intestine. In the absence of intrinsic factor, vitamin B_{12} deficiency causes the condition known as *pernicious anemia*. In this state, red blood cell synthesis (erythropoiesis), which depends on vitamin B_{12}, is severely diminished.

BIOTECHNOLOGY

Olestra, the No-Calorie Fat Substitute

Although nutritionists have determined that the typical Western diet contains too much fat, many consumers are reluctant to give up fat in their diet. As a result, food manufacturers such as Procter and Gamble have been developing low-calorie fat substitutes. In 1996, a fat substitute named Olestra® was approved by the Food and Drug Administration. Olestra is sucrose polyester: a sucrose molecule with 6–8 fatty acids attached to it. This unusual arrangement of naturally occurring substances tastes and feels like dietary fats but cannot be digested by intestinal enzymes or absorbed across the intestinal epithelium. Foods containing Olestra were introduced with considerable fanfare, but, within weeks, unhappy consumers were reporting unpleasant side effects. Intestinal cramping, gas, and diarrhea were significant enough in some people to cause them to stop using Olestra. In addition, nutritionists are concerned that excessive consumption of Olestra will lead to deficiencies of the fat-soluble vitamins (A, D, E, and K) that are normally absorbed along with dietary fats. The Food and Drug Administration requires a warning label on foods with Olestra and is monitoring its effects.

● Figure 21-14 *Acid secretion by parietal cells*
CA = carbonic anhydrase

Lack of intrinsic factor cannot be remedied directly, but patients with pernicious anemia can be given vitamin B_{12} shots.

- **Chief cells** in the gastric glands secrete the inactive enzyme **pepsinogen**. Pepsinogen is cleaved to active pepsin in the lumen of the stomach by the action of H^+.

- **D cells**, which are closely associated with the parietal cells, secrete the paracrine **somatostatin**.

- **Enterochromaffin-like** (ECL) **cells** secrete the paracrine histamine.

- **G cells**, found deep within the gastric glands, secrete the hormone **gastrin**. The release of gastrin is stimulated by the presence of amino acids and peptides in the stomach, by distension of the stomach, and by nervous reflexes mediated by **gastrin-releasing peptide**. Coffee, even decaffeinated, is also a good stimulant of gastrin release, one reason that people with excess acid secretion syndromes are advised to avoid coffee. Gastrin release is inhibited by somatostatin and by a luminal pH below 1.5.

- **Mucous cells** in the neck of the gastric glands secrete both mucus and bicarbonate. These substances create a barrier that protects the stomach from digesting itself (autodigestion). The mucus forms a physical barrier, while the bicarbonate creates a chemical buffer barrier underlying the mucus (Fig. 21-15 ●).

Researchers using micro pH electrodes have shown that the bicarbonate layer just above the cell surface in the stomach has a pH that is close to 7, even when the pH in the lumen is highly acidic at pH 2. Mucus secretion is increased when the stomach is irritated, such as by the ingestion of aspirin (acetylsalicylic acid) or alcohol.

The various secretions of the stomach, their stimuli for release, and their functions are summarized in Figure 21-16 ●.

The Stomach Balances Digestion and Protection

Now that you have an overview of digestion and secretion in the stomach, let's see how everything functions together when you eat a meal.

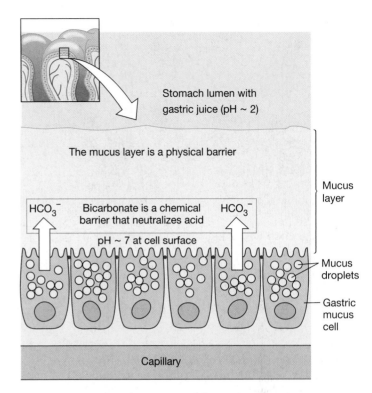

● Figure 21-15 *The mucus-bicarbonate barrier of the gastric mucosa*

When food comes into the mouth, the feedforward cephalic vagal reflex begins secretion in the stomach. Once food enters the stomach, the gastric phase stimuli continue to promote secretion and increased motility. The coordinated function of the gastric secretory cells is shown in Figure 21-17 ●.

- Parasympathetic neurons from the vagus nerve stimulate G cells to release gastrin into the blood. Amino acids or peptides into the lumen do the same. They also trigger a short reflex for gastrin release.

- Gastrin in turn promotes acid release, both directly and indirectly by stimulating histamine release.

- Histamine is released from ECL cells in response to gastrin and acetylcholine from the enteric nervous system. Histamine diffuses to its target, the parietal cells.

- Histamine stimulates acid secretion by combining with histamine H_2 receptors on parietal cells.

- Acid in the lumen stimulates pepsinogen release from chief cells through a short reflex. In the lumen, acid converts pepsinogen into pepsin, and protein digestion begins.

- Acid also triggers somatostatin release from D cells. Somatostatin acts in a negative feedback fashion to inhibit gastric acid, gastrin, and pepsinogen secretion.

Under normal conditions, the gastric mucosa is protected from enzymes and acid by the mucus-bicarbonate barrier. However, in *Zollinger-Ellison syndrome*, patients secrete excessive levels of gastrin, usually from gastrin-secreting tumors in the pancreas. As a result, hyperacidity in the stomach overwhelms the normal protective mechanisms and causes a *peptic ulcer*. In peptic ulcers, acid

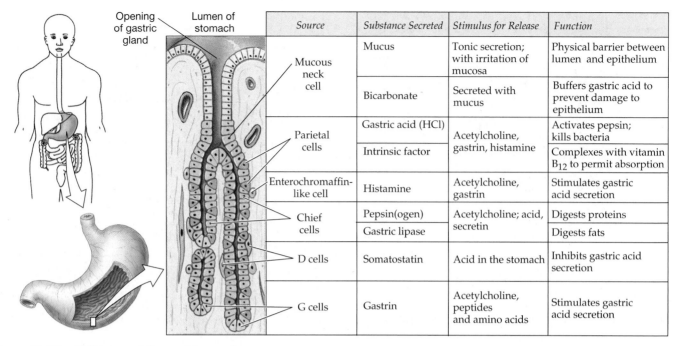

● **Figure 21-16** *Cell types of the gastric mucosa*

Source	Substance Secreted	Stimulus for Release	Function
Mucous neck cell	Mucus	Tonic secretion; with irritation of mucosa	Physical barrier between lumen and epithelium
	Bicarbonate	Secreted with mucus	Buffers gastric acid to prevent damage to epithelium
Parietal cells	Gastric acid (HCl)	Acetylcholine, gastrin, histamine	Activates pepsin; kills bacteria
	Intrinsic factor		Complexes with vitamin B_{12} to permit absorption
Enterochromaffin-like cell	Histamine	Acetylcholine, gastrin	Stimulates gastric acid secretion
Chief cells	Pepsin(ogen)	Acetylcholine; acid, secretin	Digests proteins
	Gastric lipase		Digests fats
D cells	Somatostatin	Acid in the stomach	Inhibits gastric acid secretion
G cells	Gastrin	Acetylcholine, peptides and amino acids	Stimulates gastric acid secretion

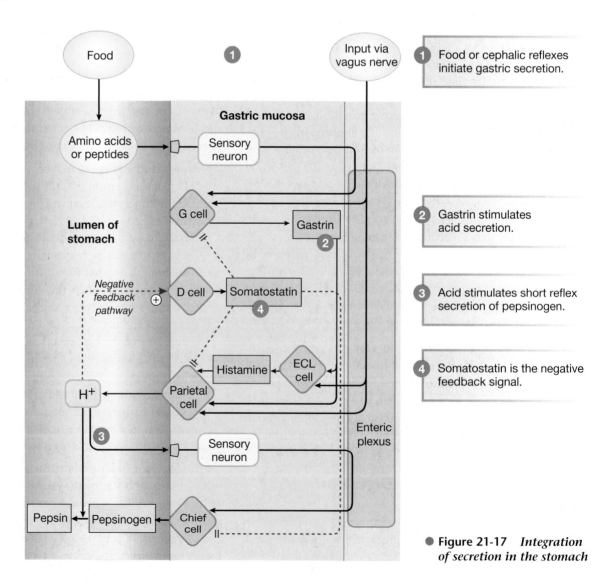

1 Food or cephalic reflexes initiate gastric secretion.

2 Gastrin stimulates acid secretion.

3 Acid stimulates short reflex secretion of pepsinogen.

4 Somatostatin is the negative feedback signal.

● **Figure 21-17** *Integration of secretion in the stomach* and pepsin destroy the mucosa, creating holes that extend into the submucosa and muscularis of the stomach, duodenum, or esophagus.

Excess acid secretion is an uncommon cause of peptic ulcers. By far the most common causes are nonsteroidal anti-inflammatory drugs (NSAIDs) such as aspirin and *Helicobacter pylori*, a bacterium that creates inflammation of the gastric mucosa.

For many years the primary therapy for excess acid secretion, or *dyspepsia*, was the ingestion of *antacids*, agents that could neutralize acid in the gastric lumen. But as molecular biologists explored the mechanism for acid secretion by parietal cells, the potential for new therapies became obvious. Today we have two new classes of drugs to fight hyperacidity: H_2 histamine receptor antagonists (cimetidine and ranitidine, for example) and *proton pump inhibitors* (PPIs) that block the H^+-K^+-ATPase (omeprazole and lansoprazole, for example).

▶ THE INTESTINAL PHASE

As chyme enters the small intestine, the **intestinal phase** of digestion begins. Before we move on to see what takes place in the small intestine itself, we should look at some reflexes from this phase. These reflexes regulate the delivery rate of chyme from the stomach and are both neural and hormonal (Fig. 21-18 ●).

The presence of acidic chyme in the duodenum releases the hormone secretin. Secretin then inhibits acid production and gastric motility, slowing gastric emptying. In addition, secretin stimulates production of pancreatic HCO_3^- to neutralize the acidic chyme that has entered the intestine.

If a meal contains fats, then the hormone cholecystokinin is secreted into the bloodstream. CCK also slows gastric motility and acid secretion. Fat digestion is slower than either protein or carbohydrate digestion, so it is crucial that the stomach only allow small amounts of fat into the intestine at one time.

If the meal contains carbohydrates, then the hormone GIP is released. In experiments, large doses of GIP inhibit gastric acid secretion, but there is some question whether GIP release reaches these levels with a normal meal.

The mixture of acid, enzymes, and digested food in chyme usually forms a hyperosmotic solution. Osmoreceptors in the intestine wall are sensitive to the osmolarity of the entering chyme. When stimulated by high osmolarity, they inhibit gastric emptying in a reflex mediated by some unknown blood-borne substance.

The net result of all three phases of gastric function is the digestion of proteins in the stomach by pepsin; the formation of chyme by the action of pepsin, acid, and peristaltic contractions; and the controlled entry of chyme into the small intestine so that further digestion and absorption can take place.

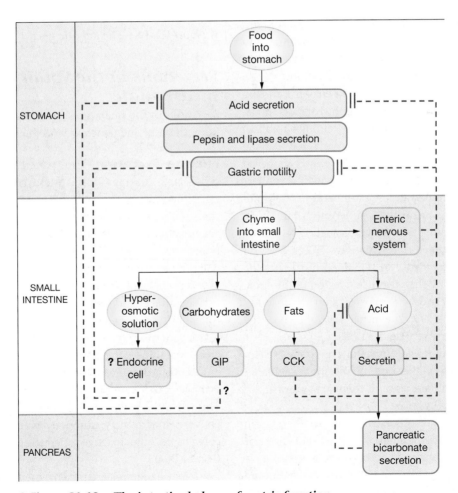

● **Figure 21-18** *The intestinal phase of gastric function*

Most Digestion and Absorption Occurs in the Intestine

As chyme from the stomach passes through the small intestine, additional enzymes and fluid are added by the intestinal epithelium, the exocrine portion of the pancreas, and the liver. Most digestion and absorption take place in the duodenum and jejunum. The small intestine and its accessory organs fulfill these roles:

- Bicarbonate secretions, primarily by the pancreas, neutralize acidic chyme to prevent damage to the intestinal mucosa. Intestinal goblet cells secrete mucus for protection and lubrication.

- The pancreas and intestinal epithelium secrete fluid containing enzymes for the digestion of proteins, carbohydrates, and fats. The liver adds *bile*, a complex mixture that contains wastes to be excreted and agents to aid fat digestion.

- Motility in the intestine mixes chyme with enzymes and exposes chyme to the mucosa for absorption. Forward movement through the intestine must be slow enough to allow digestion and absorption to go to completion.

   Once food enters the small intestine, it is slowly propelled forward by a combination of peristaltic and segmental contractions. Parasympathetic innervation, gastrin, and CCK all promote intestinal motility. Sympathetic innervation inhibits it.

- Absorption in the intestine takes place through transporting **absorptive cells**, also called *enterocytes* (Fig. 21-2g). Most digested nutrients are absorbed into the blood of the hepatic portal system [● p. 451]. This specialized region of the circulation has two capillary beds: one set to pick up absorbed nutrients at the intestine and the other set to deliver the nutrients directly to the liver (Fig. 21-19●).

   Digested fats, however, go into the lymphatic system rather than into the blood, because the intestinal capillaries have a basement membrane around them that most lipids are unable to cross.

- As described in the previous section, endocrine secretions from the intestine coordinate gastric emptying. Other endocrine and neural signals stimulate enzyme and bicarbonate secretion or release of bile from the gallbladder. GIP acts in a feedforward fashion to initiate insulin release from the endocrine pancreas.

● **Figure 21-19** *The hepatic portal system*

Secretions in the Small Intestine Include Bicarbonate

Secretions in the small intestine come from the intestinal epithelium, the exocrine pancreas, and the liver.

Enzyme Secretion The pancreas and intestinal epithelial cells secrete enzymes for the intestinal phase. Pancreatic enzymes enter the intestine in a watery fluid and are active in the lumen. Intestinal enzymes, on the other hand, remain bound to the apical membranes of brush border cells, anchored by transmembrane protein "stalks." **Brush border enzymes** are stationary and will not be swept out of the small intestine along with the chyme as it is propelled forward. The brush border enzymes include peptidases, disaccharidases, and a protease called **enteropeptidase** (previously called *enterokinase*).

A variety of enzymes is secreted by exocrine cells in the pancreas, including pancreatic amylase, lipase, and several proteases and exopeptidases. Some of these enzymes are secreted as inactive zymogens that must be activated upon arrival in the intestine. The activation process is a cascade. Brush border enteropeptidase converts inactive **trypsinogen** to active **trypsin**. Trypsin then converts the other inactive pancreatic enzymes to their active forms (Fig. 21-20●).

What signals release the enzymes? The brush border enzymes are bound to the membrane and are always in place. The pancreas

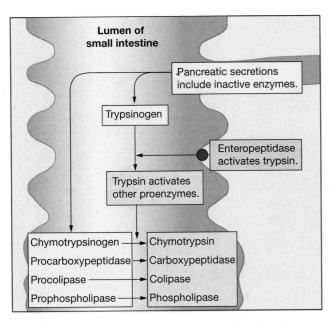

● **Figure 21-20** *Activation of pancreatic secretions*

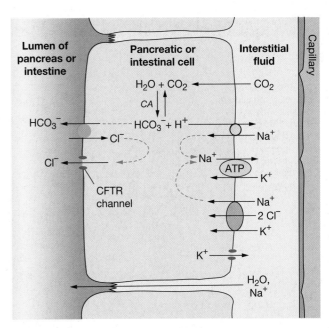

● **Figure 21-21** *Bicarbonate secretion* HCO_3^- is exchanged for Cl^- at the apical membrane while H^+ is exchanged for Na^+ at the basolateral membrane.

releases enzymes in response to distension and the presence of food in the intestine. The pancreatic reflex is mediated both by the nervous system and by the hormone CCK.

Bicarbonate Secretion Bicarbonate secretion into the small intestine neutralizes the highly acidic chyme that enters from the stomach. Most bicarbonate comes from the pancreas, which secretes a watery solution of $NaHCO_3$. The cells that produce bicarbonate are different from the enzyme-secreting cells, and they respond to different signals. Bicarbonate secretions are released in response to neural stimuli and the hormone *secretin*.

In the pancreas, as in the renal tubule cells and red blood cells, bicarbonate production requires high levels of the enzyme *carbonic anhydrase* [● p. 586]. Bicarbonate produced from CO_2 and water is secreted by a Cl^--HCO_3^- antiport protein into the pancreatic duct for delivery to the small intestine (Fig. 21-21 ●). A basolateral Na^+-H^+ antiporter transports H^+ into the extracellular fluid. The H^+ thus reabsorbed helps balance the HCO_3^- put into the blood by parietal cells.

Sodium secretion in these tissues is a passive process, driven by electrochemical gradients. To create these gradients, the epithelial cells transport Cl^- from the extracellular fluid into the lumen. The basolateral membrane contains a Na^+-K^+-$2Cl^-$ symport protein that brings Cl^- into the cell. On the apical membrane, a gated channel known as the **cystic fibrosis transmembrane regulator**, or **CFTR chloride channel**, allows Cl^- to enter the lumen. The Cl^- then re-enters the cell in exchange for HCO_3^-.

The movement of negative charge from the ECF to the lumen attracts positive Na^+, which moves down its electrochemical gradient through leaky junctions *between* the cells. This type of transepithelial movement, which we have not discussed before, is called the *paracellular* pathway. The secretion of Na^+ and

HCO_3^- into the lumen creates an osmotic gradient, so water follows by osmosis. The net result is secretion of a watery sodium bicarbonate solution from the pancreas.

In cystic fibrosis, an inherited defect causes the CFTR channel to be defective or absent from the membrane. As a result, secretion of chloride and fluid ceases, but goblet cells continue to secrete mucus. The result is thickened mucus that clogs the small pancreatic ducts and accumulates in the airways of the respiratory system, where the CFTR channel is also found [● p. 554].

Bile Secretion Bile is a non-enzyme solution secreted from **hepatocytes**, or liver cells (see *Focus on the Liver*, Fig. 21-22 ●). The key components of bile are (1) bile salts for fat digestion, (2) bile pigments, such as bilirubin, that are the waste products of hemoglobin degradation, and (3) cholesterol. **Bile acids** are steroids that combine with amino acids to form **bile salts**. Bile salts are detergents with polar side chains that allow them to interact with both lipids and water, creating water-soluble droplets (Fig. 21-23 ●).

Bile is secreted into hepatic ducts that lead to the **gallbladder**, which stores and concentrates the solution. During a meal, contraction of the gallbladder sends bile into the duodenum through the **common bile duct**, along with a watery solution of bicarbonate and digestive enzymes from the pancreas. Gall bladder contraction is stimulated by CCK.

Carbohydrate Digestion and Absorption Carbohydrate digestion in the small intestine converts digestible polysaccharides and disaccharides into monosaccharides so that they can be absorbed. Amylase from the pancreas continues to digest starch into maltose. Maltose and disaccharides from food, such as sucrose (table sugar) and lactose (milk sugar), must then be digested by the appropriate brush border disaccharidase (maltase, sucrase, and lactase) to

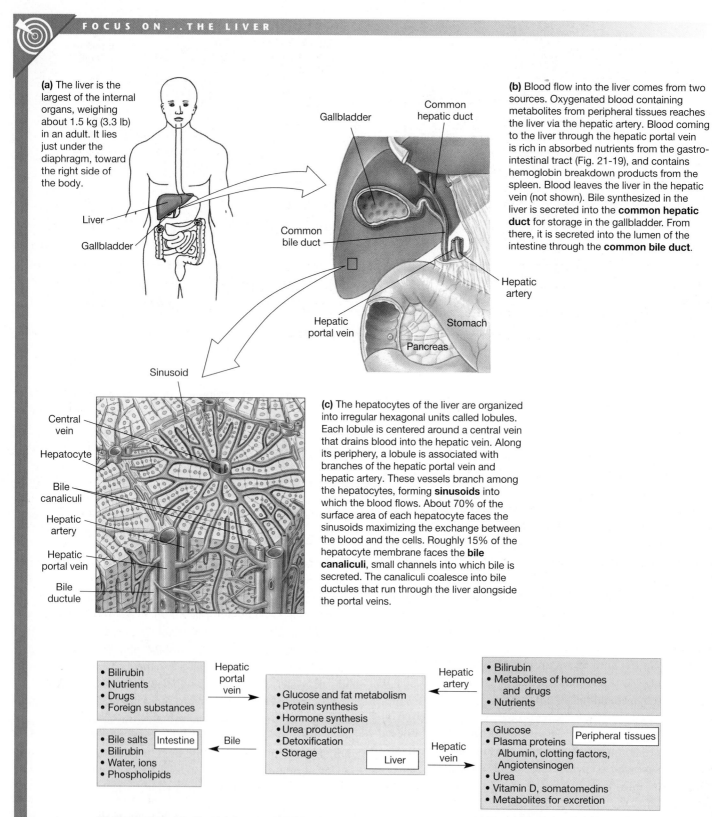

(a) The liver is the largest of the internal organs, weighing about 1.5 kg (3.3 lb) in an adult. It lies just under the diaphragm, toward the right side of the body.

Liver

Gallbladder

Gallbladder

Common hepatic duct

Common bile duct

Hepatic artery

Hepatic portal vein

Stomach

Pancreas

(b) Blood flow into the liver comes from two sources. Oxygenated blood containing metabolites from peripheral tissues reaches the liver via the hepatic artery. Blood coming to the liver through the hepatic portal vein is rich in absorbed nutrients from the gastrointestinal tract (Fig. 21-19), and contains hemoglobin breakdown products from the spleen. Blood leaves the liver in the hepatic vein (not shown). Bile synthesized in the liver is secreted into the **common hepatic duct** for storage in the gallbladder. From there, it is secreted into the lumen of the intestine through the **common bile duct**.

Sinusoid

Central vein

Hepatocyte

Bile canaliculi

Hepatic artery

Hepatic portal vein

Bile ductule

(c) The hepatocytes of the liver are organized into irregular hexagonal units called lobules. Each lobule is centered around a central vein that drains blood into the hepatic vein. Along its periphery, a lobule is associated with branches of the hepatic portal vein and hepatic artery. These vessels branch among the hepatocytes, forming **sinusoids** into which the blood flows. About 70% of the surface area of each hepatocyte faces the sinusoids maximizing the exchange between the blood and the cells. Roughly 15% of the hepatocyte membrane faces the **bile canaliculi**, small channels into which bile is secreted. The canaliculi coalesce into bile ductules that run through the liver alongside the portal veins.

- Bilirubin
- Nutrients
- Drugs
- Foreign substances

Hepatic portal vein

- Glucose and fat metabolism
- Protein synthesis
- Hormone synthesis
- Urea production
- Detoxification
- Storage

Liver

Hepatic artery

- Bilirubin
- Metabolites of hormones and drugs
- Nutrients

- Bile salts | Intestine
- Bilirubin
- Water, ions
- Phospholipids

Bile

Hepatic vein

- Glucose
- Plasma proteins | Peripheral tissues
  Albumin, clotting factors, Angiotensinogen
- Urea
- Vitamin D, somatomedins
- Metabolites for excretion

(d) Blood entering the liver brings nutrients and foreign substances from the digestive tract, bilirubin from hemoglobin breakdown, and metabolites from peripheral tissues of the body. In turn, the liver excretes some of these in the bile and stores or metabolizes others. Some of the liver's products are wastes to be excreted by the kidney; others are essential nutrients, such as glucose. In addition, the liver synthesizes an assortment of plasma proteins.

● **Figure 21-22**

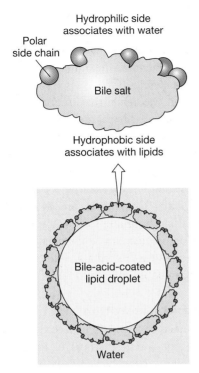

Bile salts coat lipids to make emulsions.

● **Figure 21-23** *Bile salts*

complete the digestion process. The end products of carbohydrate digestion are glucose, galactose, and fructose.

Intestinal glucose and galactose absorption uses the same transporters as the renal proximal tubule: an apical Na^+-glucose symport protein and a basolateral GLUT2-facilitated diffusion transporter (Fig. 21-24 ●). The glucose transporters move galactose as well as glucose. Most fructose absorption, on the other hand, is not Na^+-dependent. Fructose moves by facilitated diffusion on the GLUT5 transporter [● p. 139].

If glucose is the major metabolic substrate for aerobic respiration, why don't intestinal epithelial cells use the glucose they ab-

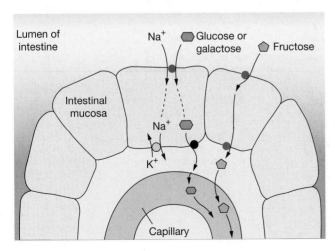

Glucose is absorbed by indirect active transport coupled to facilitated diffusion on a GLUT2 transporter. Fructose uses GLUT5.

● **Figure 21-24** *Carbohydrate absorption*

Lactose Intolerance

Lactose, or milk sugar, is a disaccharide composed of glucose and galactose. Ingested lactose must be digested before it can be absorbed, a task accomplished by the intestinal brush border enzyme *lactase*. Generally lactase is found only in juvenile mammals, except in some humans of northern European descent. Those people inherit a dominant gene that allows them to produce lactase after childhood. Scientists believe the gene provided a selective advantage to their ancestors, who developed a dairy-oriented culture in which milk and milk products played an important role. In non-Western cultures, where dairy products are not part of the diet after weaning, most adults lack the gene and synthesize less intestinal lactase. Decreased lactase activity is associated with a condition known as *lactose intolerance*. If a person with lactose intolerance drinks milk or eats dairy products, diarrhea may result. In addition, bacteria in the large intestine ferment lactose to gas and organic acids, leading to bloating and flatulence. The simplest remedy is to remove milk from the diet, although milk predigested with lactase is available.

sorb for their own metabolism? How can they keep intracellular glucose concentrations high so that facilitated diffusion will move glucose into the extracellular space? From what we can tell, the metabolism of intestinal cells differs from that of most other cells: they do not use glucose as their preferred energy substrate. Current studies suggest that these cells use the amino acid glutamine as their main source of energy, thus allowing absorbed glucose to pass unchanged into the bloodstream.

Protein Digestion and Absorption Bicarbonate in the intestine neutralizes acidic chyme, which increases luminal pH. Pepsin is inactivated at this higher pH, but pancreatic proteases and at least 17 different additional proteases and peptidases in the brush border continue protein digestion. The primary products of protein digestion are free amino acids, dipeptides, and tripeptides, all of which can be absorbed.

Amino acid structure is so variable [● Figure 2-15, p. 38] that there are multiple amino acid transport systems in the intestine. Most free amino acids are carried by Na^+-dependent cotransport proteins similar to those in the proximal tubule of the kidney (Fig. 21-25 ●). A few amino acid transporters are H^+-dependent.

Di- and tripeptides are carried into the mucosal cell using H^+-dependent cotransport. Once inside the epithelial cell, di- and tripeptides have two possible fates. Most are digested by cytoplasmic peptidases into amino acids, which are then transported across the basolateral membrane and into the circulation.

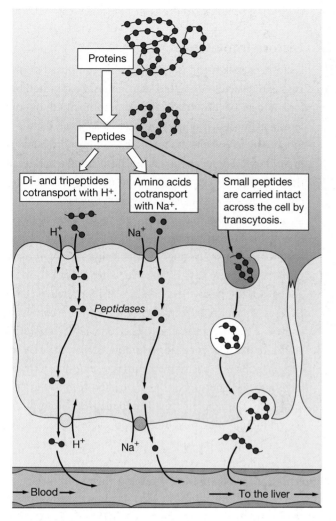

● **Figure 21-25** *Peptide absorption*

Undigested di- and tripeptides are transported intact across the basolateral membrane. Interestingly, the transport system that moves these peptides also is responsible for intestinal uptake of certain drugs, including beta-lactam antibiotics, angiotensin-converting enzyme inhibitors, and thrombin inhibitors.

Some larger peptides are absorbed by transcytosis [● p. 148] after binding to a receptor on the luminal surface of the epithelium. The discovery that ingested proteins can be absorbed as peptides has implications in medicine. Peptide fragments may act as *antigens*, substances that stimulate antibody formation and result in allergic reactions. Consequently, the intestinal absorption of large peptides may be a significant factor in the development of food allergies and intolerances.

It has been shown that in newborns, large peptide absorption takes place primarily in intestinal crypt cells. At birth, intestinal villi are very small, and the crypts are well exposed to the luminal contents. As the villi grow and the crypts have less access to chyme, the high peptide absorption rates present at birth decline steadily. If parents delay the ingestion of allergy-inducing peptides,

the gut has a chance to mature, lessening the likelihood of antibody formation.

One of the most common antigens responsible for food allergies is gluten, a component of wheat. Interestingly, the incidence of childhood gluten allergies has decreased since the 1970s, when the recommendation was made that infants not be fed gluten-based cereals until they are several months of age.

In another medical application, intestinal absorption of intact peptides is being used by pharmaceutical companies to develop undigestible peptide drugs that can be given orally instead of by injection. Probably the best-known example is DDAVP (1-deamino-8-D-arginine vasopressin), the synthetic analog of vasopressin used clinically. If the natural hormone vasopressin is ingested, it is digested rather than absorbed intact. By changing the structure of the hormone slightly, scientists created a synthetic hormone with the same activity that can be absorbed without being digested.

Fat Digestion and Absorption   Fats enter the small intestine from the stomach in the form of a coarse emulsion created by acid lipase digestion and mechanical mixing (Fig. 21-26 ●). An

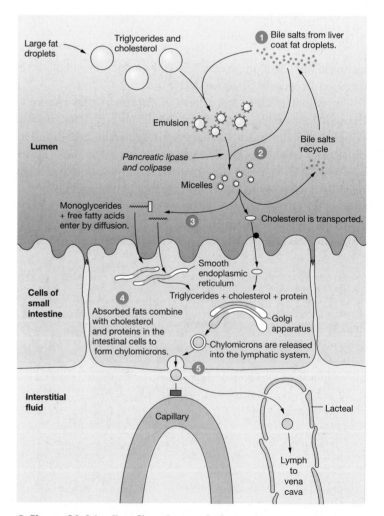

● **Figure 21-26** *Fat digestion and absorption*

emulsion consists of small droplets suspended in a liquid, similar to the emulsion you create when you shake up a bottle of vinegar and oil for salad dressing. In the duodenum, bile salts coat the coarse emulsion to help stabilize the droplets.

The digestion of triglycerides in the emulsion is carried out by pancreatic lipase. Lipase alone is unable to penetrate the bile salt coating and requires the presence of **colipase**, a protein co-factor secreted from the pancreas. Colipase displaces some bile salts, allowing lipase access to fats inside the bile salt coating.

Lipase digests triglycerides trapped in the interior of fat droplets into monoglycerides and free fatty acids. As enzymatic and mechanical digestion proceed, fatty acids, bile salts, monoglycerides, phospholipids, and cholesterol form small cylindrical **micelles** [● p. 128].

Fats are absorbed primarily by simple diffusion across the apical membrane of the intestinal cells. At the brush border surface, fatty acids and monoglycerides move out of micelles and into the epithelial cell. Cholesterol is transported on a specific, energy-dependent membrane transporter.

Once in the cytoplasm of the epithelial cell, monoglycerides and fatty acids move to the smooth endoplasmic reticulum, where they are resynthesized into triglycerides. The triglycerides then combine with cholesterol and proteins into large droplets called **chylomicrons**.

Because of their size, chylomicrons have to be packaged into secretory vesicles and leave the cell by exocytosis. Once in the extracellular space, their size also prevents them from crossing the basement membrane that surrounds the capillaries. Instead, chylomicrons are absorbed into the lymph vessels of the villi, the **lacteals**. They pass through the lymphatic system and finally enter the venous blood just before it flows into the heart [● p. 508].

Some shorter fatty acids (10 or fewer carbons) are not assembled into chylomicrons. They can therefore cross the capillary basement membrane and go directly into the blood.

Bile salts are not altered during fat digestion. When they reach the terminal section of the small intestine, the ileum, they encounter cells that transport them back into the hepatic portal circulation. From there, they return to the liver, are taken back up into the hepatocytes, and resecreted. This recirculation of bile salts is essential to maintain fat digestion because the body's pool of bile salts will cycle from two to five times for each meal.

Bilirubin and other wastes secreted in bile are not reabsorbed and pass into the large intestine for excretion.

Vitamins and Minerals In general, the fat-soluble vitamins (A, D, E, and K) are absorbed along with fats. The water-soluble vitamins are absorbed by mediated transport. The major exception is vitamin B_{12}, also known as *cobalamin*. This vitamin is made by bacteria but we obtain most of our dietary supply from seafood, meat, and milk products. The intestinal transporter for B_{12} recognizes the vitamin only when it is complexed with **intrinsic factor**. This transporter is found only in the ileum.

Mineral absorption usually occurs by active transport. Iron and calcium are two of the few substances whose intestinal absorption is linked to their concentration in the body. For both minerals, a decrease in body concentrations of the mineral leads to enhanced uptake at the intestine. (Calcium balance is described in detail in the next chapter.)

CONCEPT CHECK   Answers: p. 694

8. The recipes for oral rehydration therapy usually include sugar (sucrose) and table salt. The salt enhances intestinal absorption of glucose. Explain how.

9. Does bile digest triglycerides into monoglycerides and free fatty acids?

Absorption of Ions and Water About 5.5 liters of fluid and food per day enter the small intestine. Another 3.5 liters of pancreatic and intestinal secretions are added, for a total of 9 liters in the lumen. Of this total, most (7.5 liters) is reabsorbed from the small intestine.

The absorption of organic nutrients, which takes place mostly in the duodenum and jejunum, creates an osmotic gradient for water absorption. Na^+ is actively reabsorbed across the intestinal epithelium using transporters similar to those in the kidney [● p. 610]. One apical transporter carries Cl^- with Na^+. Another is a Na^+-K^+-$2Cl^-$ symport that brings all three types of ions into the cell. K^+ and Cl^- also pass through the cell junctions in the small intestine.

By the end of the ileum, about 1.5 liters of unabsorbed chyme remain. It passes into the large intestine through the **ileocecal valve**. This is a tonically contracted region of muscularis that narrows the opening between the ileum and the **cecum**, the initial section of the large intestine (Fig. 21-27 ●). The ileocecal valve relaxes each time a peristaltic wave reaches it. It also relaxes at the same time that food leaves the stomach as part of the *gastroileal reflex*.

The Large Intestine Concentrates Waste for Excretion

The large intestine has seven regions. Chyme from the ileum enters the cecum, a blind-end pouch. The *appendix* is a small finger-like projection at the end of the ventral end of the cecum. Material moves from the cecum upward through the **ascending colon**, horizontally across the body through the **transverse colon**, then down through the **descending colon** and **sigmoid colon** [*sigmoeides*, shaped like a sigma, σ]. The **rectum** is the short (about 12 cm) terminal section of the large intestine. It is separated from the external environment by the **anus**, an opening

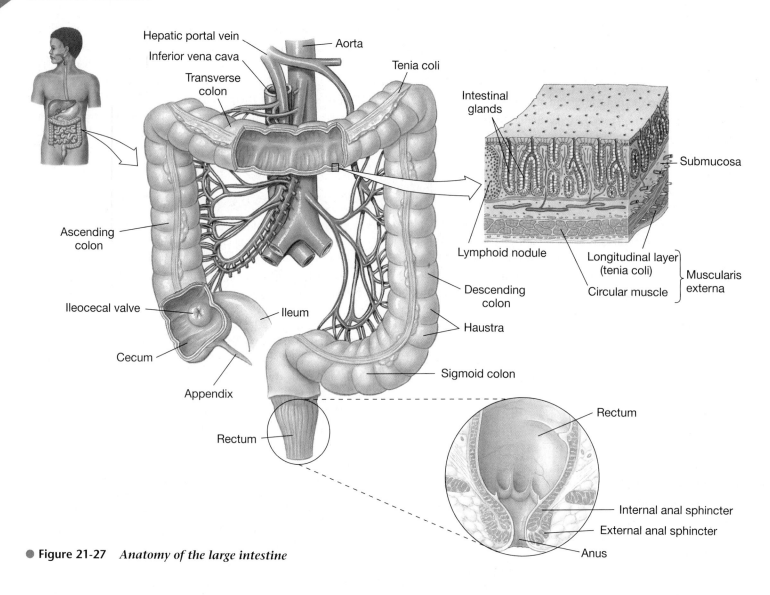

● **Figure 21-27** *Anatomy of the large intestine*

closed by two sphincters, an internal smooth muscle sphincter and external skeletal muscle sphincter.

The wall of the colon differs from that of the small intestine. The muscularis has an inner circular layer but a discontinuous longitudinal muscle layer that is concentrated into three bands called the **tenia coli.** Contractions of the tenia pull the wall into bulging pockets called **haustra.**

The mucosa of the colon has two regions, like that of the small intestine. The luminal surface lacks villi and appears smooth. It is composed of transporting epithelial cells called *colonocytes* and mucus-secreting goblet cells. The crypts contain stem cells that divide to produce new epithelium, goblet cells, endocrine cells, and maturing colonocytes (*crypt cells*).

Motility in the Large Intestine

Chyme that enters the colon continues to be mixed by segmental contractions. Forward movement is minimal during mixing contractions and depends primarily on a unique colonic contraction known as **mass movement.** A wave of contraction decreases the diameter of a segment of colon and sends a substantial bolus of material forward. These contractions occur 3–4 times a day and are associated with eating and distension of the stomach through the *gastrocolic reflex.* Mass movement is responsible for the sudden distension of the rectum that triggers defecation.

The **defecation reflex** removes undigested feces from the body. Defecation resembles urination in that it is a spinal reflex triggered by distension of the organ wall. The movement of fecal material into the normally empty rectum triggers the reflex. The smooth muscle anal sphincter relaxes and peristaltic contractions in the rectum push material toward the anus. At the same time, the external sphincter, which is under voluntary control, is consciously relaxed if the situation is appropriate. Defecation is often aided by conscious abdominal contractions and forced expiratory movements against a closed glottis, the *Valsalva maneuver.*

Defecation, like urination, is subject to emotional influence. Stress may increase intestinal motility and cause psychosomatic diarrhea in some individuals but decrease motility and cause *constipation* in others. When feces are retained in the colon, either through consciously ignoring a defecation reflex or through decreased motility, continued water absorption creates hard, dry feces that are difficult to expel. One treatment for constipation is glycerin suppositories, small bullet-shaped wads that are inserted through the anus into the rectum. Glycerin attracts water and helps soften the feces to promote defecation.

Digestion and Absorption in the Large Intestine

According to the traditional view of the large intestine, no significant digestion of organic molecules takes place there. However, in recent years this view has been revised. We now know that the numerous bacteria that inhabit the colon break down significant amounts of undigested complex carbohydrates and proteins through fermentation. The end products include lactate and short-chain fatty acids such as butyric acid. The fatty acids are used by colonocytes as their preferred energy substrate.

Colonic bacteria also produce significant amounts of absorbable vitamins, especially vitamin K. Intestinal gases, such as hydrogen sulfide, that escape from the gastrointestinal tract are a less useful product. Some starchy foods, such as beans, are notorious for their tendency to produce intestinal gas (**flatus**).

Secretion and Absorption of Fluid and Electrolytes

The colon is responsible for absorbing most of the water that enters it in the form of chyme. An average of 1.5 L/day enter the large intestine, but only about 0.1 liter is lost daily in feces under normal conditions. The colonocytes on the mucosal surface absorb NaCl using a variety of membrane transporters (Fig. 21-28 ●). At the same time, some K^+ is secreted at the apical membrane. This process is influenced by aldosterone, which promotes Na^+ reabsorption and K^+ secretion, just as it does in the kidney [● p. 636].

Crypt cells in the colon (and also the small intestine) secrete Cl into the lumen (Fig. 21-29 ●). Chloride enters the cell by secondary active transport, then exits via the apical CFTR channel. Sodium and water follow by the paracellular pathway. The secreted fluid mixes with mucus secreted by goblet cells to help lubricate the contents of the gut.

Diarrhea Can Cause Dehydration

Diarrhea is a pathological state in which intestinal secretion of fluid is not balanced by absorption, resulting in watery stools. Diarrhea will occur if normal intestinal water absorption mechanisms are disrupted or if there are unabsorbed osmotically active solutes that "hold" water in the lumen. Substances that cause *osmotic diarrhea* include undigested lactose and sorbitol, a sugar alcohol from plants. Sorbitol is used as an "artificial" sweetener in some chewing gums and in foods made for diabetics. Another unabsorbed solute that can cause osmotic diarrhea, intestinal cramping, and gas is Olestra, the "fake fat" made from vegetable oil and sugar [● p. 676].

① Na⁺ enters colonic cells by multiple pathways.

② The Na⁺-K⁺-ATPase pumps Na⁺ into the ECF.

● **Figure 21-28** *NaCl reabsorption by colonocytes*

In clinical settings, patients who need to have their bowels cleaned out before surgery or other procedures are often given four liters of an isotonic solution of unabsorbable polyethylene glycol and electrolytes to drink. The large volume of unabsorbed solution passes into the colon, where it triggers copious diarrhea that removes all solid waste from the GI tract.

Secretory diarrheas occur when bacterial toxins such as cholera toxin from *Vibrio cholerae* and *Escherichia coli* enterotoxin enhance colonic Cl^- secretion (Fig. 21-29). When excessive fluid secretion is coupled with increased motility, diarrhea results. Secretory diarrhea in response to intestinal infection can be viewed as adaptive because it helps flush pathogens out of the lumen. However, it also has the potential to cause dehydration if fluid loss is excessive.

The World Health Organization estimates that in developing countries, four million people a year die from diarrhea. In the United States, diarrhea in children causes about 200,000 hospitalizations a year. Oral replacement fluids for treatment of diarrheal salt and water loss can prevent the *morbidity* (sickness) and *mortality* (death) associated with diarrhea. Oral rehydration solutions usually contain glucose as well as Na^+, K^+, and Cl^- because the inclusion of glucose enhances Na^+ absorption. If dehydration is severe, intravenous fluid therapy may be necessary.

▶ IMMUNE FUNCTIONS OF THE GI TRACT

As you learned in the beginning of the chapter, the GI tract is the largest immune organ in the body. It is continuously exposed to disease-causing organisms, and it must prevent them

1 Na$^+$ and Cl$^-$ enter by cotransport.

2 Cl$^-$ enters lumen through CFTR channel.

3 Na$^+$ is reabsorbed.

4 Negative Cl$^-$ in lumen attracts Na$^+$ by paracellular pathway. Water follows.

● **Figure 21-29**   *NaCl secretion by colonic crypt cells*

from entering the body through delicate absorptive tissues. The first lines of defense are the enzymes and immunoglobulins of saliva and the high acid environment of the stomach. If pathogens or toxic material make it into the small intestine, sensory receptors and the immune cells of the GALT respond. Two common responses are diarrhea, just described, and vomiting.

M Cells Sample the Contents of the Gut

The immune system of the intestinal mucosa consists of immune cells scattered throughout the mucosa, clusters of immune cells in Peyer's patches (see Fig. 21-2e), and specialized epithelial cells called **M cells** that overlie the Peyer's patches. The M cells provide information about the contents of the lumen to the immune cells of the GALT.

M cells have fewer and more widely spaced microvilli than the typical intestinal cell. The apical surface also contains clathrin-coated pits [● p. 145] with membrane receptors. When *antigens*, molecules that activate the immune response, bind to the luminal receptors, the M cell transports them to its basolateral membrane using transcytosis. Macrophages and lymphocytes [● p. 526] are waiting in the extracellular compartment for the M cell to present them with antigens.

If the antigen is a substance that threatens the body, the immune system swings into action. It secretes cytokines to attract additional immune cells that can attack the invader. It also secretes cytokines that trigger an inflammatory response. A third response to cytokines is increased Cl$^-$, fluid, and mucus secretion to flush the invader from the GI tract.

In *inflammatory bowel diseases* (ulcerative colitis and Crohn's disease), the immune response is triggered inappropriately by the normal contents of the gut. One experimental therapy for these diseases that seems to be successful is to block the action of cytokines released by the gut-associated lymphoid tissues.

One question that has puzzled scientists for years was how certain pathogenic bacteria cross the barrier created by the intestinal epithelium. With the discovery of M cells, we may have the answer. It appears that some bacteria, such as *Salmonella* and *Shigella*, have evolved surface molecules that bind to the M cell receptors. The M cells then obligingly transport the bacteria across the epithelial barrier and deposit them inside the body, where the immune system immediately reacts. Both bacteria cause diarrhea. *Salmonella* also causes fever and vomiting.

Vomiting Is a Protective Reflex

Vomiting, or *emesis*, is the forceful expulsion of gastric and duodenal contents from the mouth. It is a protective reflex that is designed to remove toxic material from the GI tract before it can be absorbed. However, excessive or prolonged vomiting with loss of gastric acid can cause metabolic alkalosis [● p. 652].

The vomiting reflex is coordinated through a vomiting center in the medulla. It begins with stimulation of sensory receptors and is often, but not always, accompanied by nausea. A variety of stimuli from all over the body can trigger vomiting. They include chemicals in the blood, such as cytokines and certain drugs, pain, and disturbed equilibrium, such as occurs in a moving car or rocking boat. Tickling the back of the pharynx can induce vomiting.

Efferent signals from the vomiting center initiate a wave of reverse peristalsis that begins in the small intestine and moves upward. The motility wave is aided by abdominal contraction that increases intra-abdominal pressure. The stomach relaxes so that the increased pressure forces gastric and intestinal contents back into the esophagus and out of the mouth.

During vomiting, respiration is inhibited. The epiglottis and soft palate close off the trachea and nasopharynx to prevent the *vomitus* from being inhaled (*aspirated*). Should acid or small food particles get into the airways, they could damage the respiratory system and cause *aspiration pneumonia*.

RUNNING PROBLEM CONCLUSION

Peptic Ulcers

Tonya's lab results are in. Tests performed on her stomach tissue confirm that she has an *H. pylori* infection that is probably the cause of her duodenal ulcer. Tonya's physician gives her a double antibiotic to take for seven days and changes her H_2 blocker to a proton pump inhibitor.

In this running problem, you learned about peptic ulcers and an unusual bacterium that is associated with most duodenal ulcers. You also learned that antacids and histamine-receptor blockers reduce the level of acid in the stomach.

Further check your understanding of this running problem by comparing your answers with those in the summary table.

	QUESTION	FACTS	INTEGRATION AND ANALYSIS
1	Why are peptic ulcers found in the duodenum rather than in the jejunum or ileum?	Peptic ulcers are caused by gastric acid and stomach enzymes. Chyme moving out of the stomach passes into the duodenum.	If protective mechanisms in the intestine have failed, the duodenum is the first part of the small intestine to be exposed to gastric acid and enzymes.
2	How does the conversion of urea into ammonia protect *H. pylori* from the hostile environment of the stomach?	A base is an anion that combines with H^+. Ammonia combines with H^+ to create ammonium ions, NH_4^+. This reduces acidity (increases pH).	*H. pylori* has enzymes that produce ammonia, a base. Ammonia combines with H^+ to reduce acidity. In this way, the bacteria protect themselves from being killed by the acidic environment of the stomach.
3	Write the chemical equation showing how the antacids sodium bicarbonate ($NaHCO_3$) and aluminum hydroxide ($Al(OH)_3$) buffer hydrochloric acid (HCl).	Buffers are molecules that combine with H^+. Bicarbonate (HCO_3^-) and OH^- are buffers.	$NaHCO_3 + HCl \rightarrow NaCl + H_2CO_3$ $Al(OH)_3 + 3\,HCl \rightarrow AlCl_3 + 3\,H_2O$
4	How does blocking H_2 receptors stop acid production in the stomach?	Histamine is a paracrine that stimulates acid secretion by the stomach.	If histamine receptors are blocked by H_2 blockers, histamine cannot combine with them and stimulate acid secretion.

659 665 667 671 680

CHAPTER SUMMARY

The digestive system, like the renal system, plays a key role in *mass balance* in the body. Material that enters the system, whether by mouth or by secretion, is mostly absorbed before it reaches the end of the GI tract. In pathologies such as diarrhea, where absorption and secretion are unbalanced, the loss of material through the GI tract can seriously disrupt *homeostasis*. Absorption and secretion in the GI tract provide lots of examples of *movement across membranes*, most of which follow patterns you have encountered in other systems. Finally, regulation of GI tract function illustrates the complex interactions that take place between endocrine and neural *control systems* and the protective function of the immune system.

Function and Processes of the Digestive System

1. The gastrointestinal (GI) tract moves nutrients, water, and electrolytes from the external to the internal environment. (p. 660)

2. The four processes of the digestive system are digestion, absorption, motility, and secretion. **Digestion** is chemical and mechanical breakdown of foods into absorbable units. **Absorption** is transfer of substances from the lumen of the GI tract to the ECF. **Motility** is movement of material through the GI tract. **Secretion** refers to the transfer of fluid and electrolytes from ECF to lumen or the release of substances from cells. (pp. 660–661; Fig. 21-1)

3. The GI tract contains the largest collection of lymphoid tissue in the body, the **gut-associated lymphoid tissue (GALT)**. (p. 661)

Anatomy of the Digestive System

4. **Chyme** is a soupy substance created as ingested food is broken down by mechanical and chemical digestion. (p. 661)

5. Food entering the digestive system passes through mouth, pharynx, **esophagus, stomach (fundus, body, antrum), small intestine (duodenum, jejunum, ileum), large intestine, rectum**, and **anus**. (pp. 661, 664; Fig. 21-2a)

6. The **salivary glands, pancreas**, and **liver** add exocrine secretions with enzymes and mucus to the lumen. (p. 661; Fig. 21-2a)

7. The wall of the GI tract consists of four layers: an inner mucosa, a middle submucosa, a smooth muscle layer, and an outer **serosa**. (pp. 664–665; Fig. 21-2d)

8. The **mucosa** consists of epithelium, the **lamina propria**, lymphoid tissue, and the **muscularis mucosa**. Small **villi** increase the surface area, as do **microvilli** on the cells. (p. 664; Fig. 21-2e, f, g)

9. The **submucosa** contains blood and lymph vessels and the **submucosal plexus** of the **enteric nervous system**. (p. 664; Fig. 21-2c, d)

10. The outer wall consists of two layers of circular and longitudinal muscle. The **myenteric plexus** lies between the muscle layers. (p. 665; Fig. 21-2c)

Motility

11. Motility moves food from mouth to anus and mechanically mixes food. (p. 665)

12. GI smooth muscle cells are electrically connected by gap junctions. Some segments are **tonically contracted** but others exhibit **phasic contractions**. (p. 665)

13. Intestinal muscle exhibits spontaneous **slow wave potentials**. When a slow wave reaches threshold, it fires action potentials and contracts. (p. 665; Fig. 21-3)

14. Slow waves originate in the **interstitial cells of Cajal**. (p. 665)

15. Between meals, the **migrating motor complex** moves remnants from the upper GI tract to the lower regions. (p. 666)

16. **Peristaltic contractions** are progressive waves of contraction. Peristalsis is mediated by the enteric nervous system and altered by hormones, paracrines, and neurocrines. (p. 666; Fig. 21-4)

17. **Segmental contractions** are primarily mixing contractions. (p. 666; Fig. 21-4)

Secretion

18. About 2 liters of fluid per day enter the GI tract through the mouth. Another 7 liters are secreted by the body. Nearly all of this volume is reabsorbed. (p. 666; Fig. 21-5)

19. Digestive enzymes are secreted by exocrine glands or by gastric and intestinal epithelium. Some enzymes are secreted as zymogens. (p. 667)

20. Mucus from **mucous cells** and **goblet cells** forms a protective coating and lubricates contents of the gut. (p. 667)

21. The digestive system secretes and absorbs water and ions. (p. 667)

Digestion and Absorption

22. Most nutrient absorption takes place in the small intestine. The large intestine absorbs water and ions. (p. 668)

23. **Amylase** digests starch to maltose. **Disaccharidases** digest disaccharides to monosaccharides. (pp. 668–669; Fig. 21-6)

24. **Endopeptidases** (also called proteases) break proteins into smaller peptides. **Exopeptidases** remove amino acids from peptides. (p. 669; Fig. 21-7)

25. Fat digestion requires **lipase** and **colipase**. (pp. 669, 685; Fig. 21-8)

Regulation of GI Function

26. The enteric nervous system is called the little brain because it can integrate information without input from the CNS. (p. 670)

27. **Short reflexes** originate within the enteric nervous system and are integrated there. **Long reflexes** are integrated in the CNS. Long reflexes may originate within or outside the enteric nervous system. (p. 670; Fig. 21-9)

28. Generally, parasympathetic innervation is excitatory for GI function and sympathetic innervation is inhibitory. (p. 670)

29. Digestive hormones are divided into the gastrin family (**gastrin, cholecystokinin**), secretin family (**secretin, vasoactive intestinal peptide, glucose-dependent insulinotropic peptide**, and **glucagon-like peptide 1**), and hormones that do not fit into either of those two families (**motilin**). (pp. 672–673; Tbl. 21-2)

30. GI peptides excite or inhibit motility and secretion. Most stimuli for GI peptide secretion arise from the ingestion of food. (p. 670; Fig. 21-10)

31. GI paracrines, such as histamine, interact with hormones and neural reflexes. (p. 670)

The Cephalic Phase

32. In the **cephalic phase** of digestion, the sight, smell, or taste of food initiates GI reflexes. (p. 673; Fig. 21-12)

33. Mechanical digestion begins with chewing, or **mastication**. **Saliva** moistens and lubricates food. Salivary amylase digests carbohydrates. (pp. 673–674)

34. Swallowing, or **deglutition**, is a reflex integrated by a medullary center. (p. 674; Fig. 21-13)

The Gastric Phase

35. The stomach digests proteins and fats and regulates movement of chyme into the intestine. (p. 676)

36. The stomach secretes mucus and bicarbonate from **mucous cells**, gastric acid from **parietal cells**, pepsinogen from **chief cells**, somatostatin from **D cells**, histamine from **ECL cells**, and gastrin from **G cells**. (pp. 676–677; Fig. 21-16)

The Intestinal Phase

37. Acid in the intestine, CCK, and secretin delay gastric emptying. (p. 679; Fig. 21-18)

38. Intestinal enzymes are part of the **brush border**. Some pancreatic enzymes need to be activated after secretion. (p. 680; Fig. 21-20)

39. The pancreas also secretes a watery $NaHCO_3$ solution that neutralizes gastric acid. (p. 681; Fig. 21-21)

40. Epithelial cells secrete Cl^- using the **cystic fibrosis transmembrane regulator**, or **CFTR chloride channel**. (p. 681; Fig. 21-21)

41. Bile from the liver contains **bile salts**, bilirubin, and cholesterol. (p. 681; Fig. 21-22)

42. Glucose absorption uses the Na^+-glucose symporter and GLUT2 transporter. Fructose uses the GLUT5 transporter. (p. 683; Fig. 21-24)

43. Amino acids are absorbed by Na^+-dependent cotransport. Di- and tripeptides are absorbed using H^+-dependent cotransport. Some larger peptides are absorbed intact by transcytosis. (p. 683; Fig. 21-25)

44. Fat digestion requires bile, which emulsifies fats. As enzymatic and mechanical digestion proceed, fat droplets form **micelles**. (p. 685; Fig. 21-26)

45. Fat absorption occurs primarily by simple diffusion. (p. 685)

46. Monoglycerides and fatty acids reassemble into triglycerides in the intestinal cell, then combine with cholesterol and proteins into **chylomicrons**. Chylomicrons are absorbed into the lymph. (p. 685; Fig. 21-26)

47. Fat-soluble vitamins are absorbed along with fats. Water-soluble vitamins are absorbed by mediated transport. Vitamin B_{12} absorption requires **intrinsic factor** secreted by the stomach. Mineral absorption usually occurs by active transport. (p. 685)

48. Most volume reabsorption takes place in the small intestine. The large intestine concentrates the 1.5 liters of chyme that enter it daily. (p. 685)

49. Undigested material in the colon moves forward by **mass movement**. The **defecation reflex**, a spinal reflex subject to higher control, is triggered by sudden distension of the rectum. (p. 686)

50. Colonic bacteria use fermentation to digest organic material. (p. 687)

51. Cells of the colon can reabsorb or secrete fluid. Excessive fluid secretion or decreased reabsorption causes diarrhea. (p. 687; Fig. 21-28, 21-29)

Immune Functions of the GI Tract

52. Protective mechanisms of the GI tract include acid, mucus, vomiting, and diarrhea. (p. 688)

53. **M cells** sample gut contents and present antigens to cells of the GALT. (p. 688)

54. **Vomiting** is a protective reflex integrated in the medulla. (p. 688)

QUESTIONS

LEVEL ONE Reviewing Facts and Terms

1. Define the four basic processes of the digestive system and give an example of each.

2. For most nutrients, the process of _____ is not regulated, whereas _____ and _____ are continuously regulated. Why do you think these differences exist? Defend your answer.

3. Match each of the following descriptions with the appropriate term(s):

(a) chyme is produced here
(b) organ where most digestion occurs
(c) initial section of small intestine
(d) this adds exocrine secretions to the duodenum via a duct
(e) sphincter between stomach and intestine
(f) enzymes produced here
(g) distension of its walls triggers the defecation reflex

1. colon
2. stomach
3. small intestine
4. duodenum
5. ileum
6. jejunum
7. pancreas
8. pylorus
9. rectum
10. liver

4. List the four layers found in the GI tract walls. What type of tissue predominates in each?

5. Describe the functional types of epithelium found lining the stomach and intestines.

6. What are Peyer's patches?

7. What purposes does motility serve in the gastrointestinal tract? What types of tissue contribute to gut motility? What types of contraction do the tissues undergo?

8. What is a zymogen? What is a proenzyme? List two specific examples.

9. Match each cell with the product(s) it secretes. Items may be used more than once.

(a) parietal cells
(b) goblet cells
(c) brush border cells
(d) pancreatic cells
(e) D cells
(f) ECL cells
(g) chief cells
(h) G cells

1. enzymes
2. histamine
3. mucus
4. pepsinogen
5. gastrin
6. somatostatin
7. HCO_3^-
8. HCl
9. intrinsic factor

10. How does each of these factors affect digestion? Briefly explain how and where the factor exerts its effects.
(a) emulsification
(b) neural activity
(c) pH
(d) size of food particles

11. Most digested nutrients are absorbed into the _____ of the _____ system, delivering nutrients to the _____ (organ). However, digested fats go into the _____ system because intestinal capillaries have a _____ around them that most lipids are unable to cross.

12. What is the enteric nervous system and what is its function?

13. What are short reflexes? What types of responses do they regulate? What is meant by the term *long reflex*?

14. What role do paracrines play in digestion? Give specific examples.

LEVEL TWO *Reviewing Concepts*

15. **Map exercise:** List the three major groups of biomolecules across the top of a large piece of paper. Down the left side of the paper write: mouth, stomach, small intestine, absorption. For each biomolecule, fill in how it is digested in each of the GI tract regions listed and how it is absorbed.

16. Define, compare, and contrast the following terms:
(a) mastication, deglutition
(b) microvilli, villi
(c) peristalsis, segmentation, migrating motor complex, reflux esophagitis, mass movements, defecation, vomiting, diarrhea
(d) chyme, feces
(e) short reflexes, long reflexes

(f) submucosal plexus, myenteric plexus, enteric nervous system, vagus nerve
(g) cephalic, gastric, and intestinal phases of digestion

17. Diagram the mechanisms by which Na^+, K^+, and Cl^- are transported out of the intestine. Diagram the mechanisms by which H^+ and HCO_3^- are secreted into the lumen.

18. Compare the enteric nervous system with the cephalic brain. Give some specific examples concerning neurotransmitters, neuromodulators, and supporting cells.

19. List and briefly describe the actions of the members of each of the three groups of GI hormones.

20. Explain how H_2 receptor antagonists and proton pump inhibitors decrease gastric acid secretion.

LEVEL THREE *Problem Solving*

21. Erica's baby, Justin, has had a severe bout of diarrhea and is now dehydrated. Is his blood more likely to be acidotic or alkalotic? Why?

22. Mary Littlefeather arrives in her physician's office complaining of severe, steady pain in the upper right quadrant of her abdomen. The pain began shortly after she ate a meal of fried chicken, french fries, and peas. Lab tests and an ultra-

sound reveal the presence of gallstones in the common bile duct that drains the liver, gallbladder, and pancreas into the small intestine.
(a) Why was Mary's pain precipitated by the meal she ate?
(b) Which of the following will be affected by the gallstones: micelle formation in the intestine, carbohydrate digestion in the intestine, protein absorption in the intestine. Explain your reasoning.

23. Intestinal transport of the amino acid analog MIT (monoiodotyrosine) can be studied using the "everted sac" preparation. A length of intestine is turned inside out, filled with a solution containing the substrate to be transported (MIT), tied at both ends, then placed in a bath containing nutrients, salts, and an equal concentration of MIT. Changes in the concentration of MIT are monitored in the bath (mucosal or apical side), in the intestinal cells (tissue), and within the sac (serosal or basolateral side) over a 240-minute period. The results are shown in the graph. (Data from Nathans et al., *Biochimica et Biophysica Acta* 41:271–282, 1960)

    (a) Based on the data shown, is the transepithelial transport of MIT a passive process or an active process?

    (b) Which way does MIT move: (1) mucosal to tissue to serosal or (2) serosal to tissue to mucosal? Is this movement absorption or secretion?

    (c) Is transport across the mucosal (apical) membrane active or passive? Explain your reasoning.

    (d) Is transport across the serosal (basolateral) membrane active or passive? Explain your reasoning.

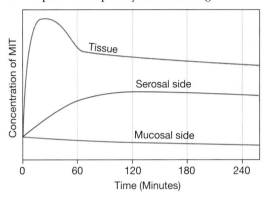

Explorations

Explorations are also available at *www.physiologyplace.com*

Exploration 1: Heartburn and Parietal Cells

"Acid indigestion" and heartburn are two of the most common digestive disorders in the United States. Stroll the aisles of any pharmacy and look at the space devoted to over-the-counter medications for these conditions. Traditional antacids are buffers such as calcium carbonate, sodium bicarbonate, and aluminum and magnesium hydroxide that neutralize H^+ in the stomach lumen. Newer therapies have been developed based on physiological studies of gastric acid secretion. For example, cimetidine, ranitidine, and famotidine are histamine-2 (H_2) receptor antagonists that decrease parietal cell acid secretion. Proton pump inhibitors (PPIs or "prazoles" such as omeprazole) block the H^+-ATPase on the luminal side of the parietal cell. Use Fig. 21-14 and locate the sites of action of these two classes of drugs. To see acid secretion in action, do a Google search (*www.google.com*) for *parietal cell animation*.

To learn more about the role of bacteria in the development of gastric and duodenal ulcers, read "The effect of ranitidine and amoxicillin plus metronidazole on the eradication of *Helicobacter pylori* and the recurrence of duodenal ulcer," *New England Journal of Medicine* 328:308–312, 1993.

Exploration 2: Dry Mouth = Guilty

The digestive tract often reflects our emotions in a very noticeable way. In ancient China, a person under suspicion for a crime might have been given a mouth of dry rice to chew during questioning. If he could produce enough saliva to moisten the rice and swallow it, he went free. But if his nervous state dried up his salivary reflex, he was pronounced guilty. Interestingly, recent research has confirmed that stress, such as that associated with lying, decreases salivary secretion. Based on what you have learned about nervous control of digestive function, what efferent division of the nervous system do you think promotes the production of saliva?

Saliva is a complex fluid that includes water, ions, mucus, and enzymes such as lysozyme to kill bacteria and amylase to begin starch digestion. At resting flow rates saliva is slightly acid, with a pH of 6–7. The ionic composition of saliva is determined in two distinct epithelial transport steps. The salivary glands are exocrine glands, with a secretory epithelium (the *acinus*) that opens to the outside environment through a duct (see Fig. 3-21, p. 71). Fluid secreted into the acinus resembles extracellular fluid in its ionic composition. As saliva passes through the duct, the epithelial cells reabsorb Na^+ and secrete K^+ until their ratio is more like that of intracellular fluid.

- Using what you have learned about epithelial transport in this chapter and the renal chapters, draw a picture of the salivary duct cells and lumen. Arrange membrane channels and transporters on the apical and basolateral membranes so that the duct cell absorbs Na^+ and secretes K^+.

- With neural stimulation, the flow rate of saliva can increase from 0.4 mL/min to 2 mL/min. What do you think happens to the Na^+ and K^+ content of saliva at higher flow rates?

ANSWERS

Page 661

1. Digestion is the chemical and mechanical breakdown of food into absorbable units.

2. Both involve chemical reactions. Digestion takes place in the GI tract lumen, which is external environment. Metabolism takes place inside the body.

3. Absorption moves material from lumen to ECF. Secretion moves substances from the cells or ECF into the lumen.

Page 664

4. The lumen of the digestive tract is on the apical side of the intestinal epithelium.

5. Mouth → pharynx → esophagus → stomach (fundus, body, antrum) → small intestine (duodenum, jejunum, ileum) → large intestine (colon, rectum) → anus

Page 665

6. The GI tract has a large, vulnerable surface area facing the external environment; therefore, it needs the immune cells of lymphoid tissue to combat invaders.

Page 666

7. The sphincters are tonically contracted to close off the GI tract from the outside world and to keep material from passing freely between sections of the tract.

Page 685

8. Na^+ from table salt enhances the intestinal absorption of glucose because glucose must be co-transported with Na^+.

9. Bile does not digest triglycerides. It simply emulsifies them into smaller globules so that lipase can digest them.

Page 693

23. (a) MIT started out with equal concentrations in both solutions but by the end of the experiment, it was more concentrated on the serosal side. Therefore MIT must be moving by active transport.

(b) MIT moves mucosal to tissue to serosal. This is apical to basolateral transport, or absorption. (c) Transport across the mucosal (apical) membrane goes from the bath into the tissue. The tissue has more MIT than the bath, therefore this must be active transport.

(d) Transport across the serosal (basolateral) membrane goes from the tissue into the sac of intestine. The tissue has more MIT than fluid on the serosal side, therefore this must be passive transport.

Page 693

Sodium can enter the duct cell through a Na^+ channel and leave through the basolateral Na^+-K^+-ATPase. Potassium enters through the ATPase and can leave through an apical K^+ channel. You could also use other ion transporters to get the same result. Research suggests that the apical membrane also contains both Na^+-H^+ and K^+-H^+ antiporters.

As flow of saliva increases, the epithelial cells have less time to alter the composition of a bolus of fluid flowing past them. Therefore Na^+ concentration in saliva increases and K^+ secretion decreases at higher flow rates.

Metabolism and Energy Balance

22

PET scans of the brain use glucose metabolism to indicate regions of greater neural activity (red and yellow).

RUNNING PROBLEM

Eating Disorders

Sara Inman and Nicole Baker had been best friends growing up but hadn't seen each other for several semesters. When Sara heard Nicole was in the hospital, she went to visit but wasn't prepared for what she saw. Nicole, who had always been thin, now was so emaciated that Sara hardly recognized her. Nicole had been taken to the Emergency Room when she fainted in a ballet class. When Dr. Ayani saw Nicole, he was as concerned about her weight as her broken wrist. At 5'6", Nicole weighed only 95 pounds (normal healthy range for that height is 118–155 pounds). Dr. Ayani suspected an eating disorder, probably anorexia nervosa, and he ordered blood tests to confirm his suspicion.

BACKGROUND BASICS

Essential amino acids: **37** | Glycogen: **35** | Brown adipose tissue: **73** | Biological work: **85** | Intermediary metabolism: **100** | Beta oxidation of fats: **109** | GLUT transporters: **139** | Receptor-mediated endocytosis: **145** | Tyrosine kinase-linked receptors: **178** | Peptide hormones: **214** | Homeostatic mechanisms for blood pressure: **510** | Renal threshold for glucose: **612** | Exocrine pancreas: **661** | Cholecystokinin: **672** | GIP: **672** | Chylomicrons: **685** |

Probably no single abnormality has contributed more to our knowledge of the intermediary metabolism of animals than has the disease known as diabetes.

—Helen R. Downes, *The Chemistry of Living Cells*, 1955

Magazine covers at the grocery store checkout stands tell visitors to this country a lot about Americans. Headlines screaming "Lose 10 pounds in a week without dieting" or "CCK: the hormone that makes you thin" vie for attention with glossy full-color photos of luscious desserts dripping with chocolate, whipped cream, fat, and calories. As one magazine article put it, we are a nation obsessed with staying trim and with eating, mutually exclusive occupations. But what determines how much, what, and when we eat? This is an area of intense research because the act of eating is the main point at which the body exerts control of its energy input.

The Brain Controls Food Intake

The control of food intake is a complex process. Hunger and *satiety* [*satis*, enough] are affected by psychological, environmental, and social factors as well as by physiological mechanisms. As a result, we still do not fully understand what governs when, what, and how much we eat.

Our current model is based on hypothalamic centers [✦ Fig. 11-3, p. 371] that are influenced by neurotransmitters and hormones. Studies using transgenic and knockout mice show that control of these centers involves multiple brain neuropeptides and "brain-gut" hormones. What follows is an overview of this complex and constantly changing field.

The hypothalamus contains two centers that regulate food intake: a **feeding center** that is tonically active and a **satiety center** that stops food intake by inhibiting the feeding center. Efferent signals from these centers cause changes in eating behavior and create sensations of hunger and fullness. If the feeding center is destroyed, animals will cease to eat. If the satiety center is destroyed, animals will overeat and become obese.

There are two classic theories for regulation of food intake: the glucostatic theory and the lipostatic theory. The **glucostatic theory** states that glucose utilization by hypothalamic centers regulates food intake. When blood glucose concentrations decrease, the satiety center is suppressed and the feeding center is dominant. When glucose utilization is up, the satiety center inhibits the feeding center.

The **lipostatic theory** of energy balance proposes that a signal from the body's fat stores to the brain modulates eating behavior so that the body maintains a particular weight. If fat stores are increased, eating decreases. In times of starvation, eating increases. Obesity results from disruption of this pathway.

The discovery of **leptin** [*leptos,* thin], a protein hormone synthesized in adipocytes, provided evidence for a signal between adipose tissue and the brain. Leptin is synthesized in adipocytes under control of the *obese (ob) gene.* Mice that lack the *ob* gene and leptin protein become obese, as do mice with defective leptin receptors. However, only a small percentage of obese humans are leptin-deficient. The majority of them have *elevated* leptin levels. As we are learning, leptin is only part of the story.

Another important signal molecule is **neuropeptide Y** (NPY), a brain neurotransmitter that seems to be the actual stimulus for food intake. In normal-weight animals, leptin inhibits NPY in a

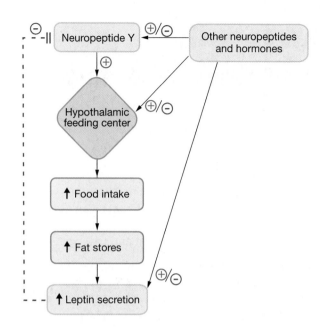

● **Figure 22-1** *Roles of peptides in regulation of food intake*

negative feedback pathway (Fig. 22-1 ●). **Ghrelin** is a peptide secreted by the stomach that increases hunger when infused into human subjects. Other neuropeptides and hormones also influence NPY and the hypothalamic centers (Table 22-1). Interestingly, many of these peptides affect functions in addition to food intake. Ghrelin promotes release of growth hormone and the brain peptides called *orexins* appear to play a role in sleep. Our understanding of how all these factors interact is incomplete.

Table 22-1: Some Peptides That Modulate Food Intake

NAME	SOURCE
Increase food intake	
Neuropeptide Y (NPY)	Hypothalamus
Orexins (also called hypocretins)	Hypothalamus
Galanin	Hypothalamus
Melanin-concentrating hormone (MCH)	Hypothalamus
Ghrelin	Stomach
Decrease food intake	
CCK	Small intestine; neurons
Corticotropin-releasing hormone (CRH)	Hypothalamus
α-Melanocyte-stimulating hormone (α-MSH)	Hypothalamus
CART (cocaine- and amphetamine-regulated transcript)	Hypothalamus
Glucagon-like peptide-1 (GLP-1)	Intestines
PYY_{3-36}	Intestines

Appetite and eating are influenced by sensory input through the nervous system in addition to chemical signals. The simple acts of swallowing and chewing food help create a sensation of fullness. The sight, smell, and taste of food can either stimulate or suppress appetite.

One interesting study tried to determine if chocolate craving could be attributed to psychological factors or physiological stimuli.* Subjects were given either dark chocolate, white chocolate with none of the pharmacological agents of cocoa, cocoa capsules, or placebo capsules. The results showed that white chocolate was the best substitute for the real thing and that aroma plays a significant role in satisfying chocolate craving.

Psychological factors, such as eating or not eating when stressed, are known to play a significant role in regulating food intake. The eating disorder *anorexia nervosa* probably has both a psychological component and a physiological one. The concept of appetite is closely linked to the psychology of eating, and it may explain why dieters who crave smooth, cold ice cream cannot be satisfied by a crunchy carrot stick. A considerable amount of money is directed to research on eating behaviors.

CONCEPT CHECK   Answers: p. 725

1. Explain the role of the satiety and feeding centers. Where are they located?
2. Studies show that most obese humans have elevated leptin levels in their blood. Based on your understanding of endocrine pathologies [❂ p. 227], explain why leptin in these people is not decreasing food intake.

▶ ENERGY BALANCE

Once food has been digested and absorbed, the body's chemical reactions, known collectively as *metabolism*, determine what happens to those nutrients. Are they destined to burn off as heat? Become muscle? Or turn into extra pounds that make it difficult to zip up blue jeans? In this section we examine energy balance in the body.

Energy Input Equals Energy Output

The *first law of thermodynamics* states that energy in the universe is constant. By extension, this statement means that all energy that goes into a biological system such as the human body can be accounted for. According to the concept of *mass balance*:

$$\text{Body energy} = \text{energy intake} + \text{energy output} \qquad (1)$$

Energy intake for humans consists of the nutrients we eat, digest, and absorb. However, because the digestive system itself does not regulate energy intake, we must depend on behavioral mechanisms such as hunger and satiety to tell us when and how much

*"Pharmacological versus sensory factors in the satiation of chocolate craving," *Physiology & Behavior* 56(3):419–422, 1994 Sep.

Novel Approaches for Discovering Peptides

In the early days of molecular biology, scientists took tissues that they knew contained active peptides, isolated and purified the peptides, and then analyzed their amino acid sequence. Now, as we enter the age of proteomics, investigators are using reverse techniques to discover new proteins within the body. For example, the *orexin* (or hypocretin) peptides were found by isolating the mRNA that was expressed most in a particular region of the hypothalamus. Investigators then used the mRNA to create the amino acid sequence of the prepropeptide. The orexin peptides were also discovered simultaneously by a different group of scientists who worked backwards from an "orphan" G protein-coupled receptor [❂ p. 183] to find its peptide ligand. The endogenous hunger hormone *ghrelin* was discovered by a similar method. Pharmacologists who were testing synthetic peptides to stimulate the release of growth hormone found that their peptides were binding to a previously unknown receptor, and from that receptor, they discovered ghrelin.

to eat. Psychological and social aspects of eating, such as parents who say "Clean your plate," complicate the physiological control of food intake.

Energy output by the body either does work (which includes storing energy) or is returned to the environment as heat:

$$\text{Energy output} = \text{heat} + \text{work} \qquad (2)$$

RUNNING PROBLEM

Sara and Nicole first met in ballet class at age 11. They were both serious about dance and worked hard to maintain the perfect thin ballet-dancer's body. Both girls occasionally took diet pills or laxatives, and it was almost a contest to see who could eat the least food. Sara was always impressed with Nicole's will-power, but then again, Nicole was a perfectionist. Two years ago Sara found herself less interested in dance and went away to college. Nicole, on the other hand, was accepted into a prestigious ballet company and remained focused on her dancing—and her body image. Nicole's regimented diet became stricter, and if she felt she'd eaten too much, she would make herself throw up. Her weight kept dropping but each time she looked in the mirror, she saw a fat girl looking back.

Question 1: If you measured Nicole's leptin level, what would you expect to find?

Question 2: Would you expect Nicole to have elevated or depressed levels of NPY?

In the human body, at least half the energy released in chemical reactions is lost as unregulated "waste" heat.

Biological work takes one of three forms [🌑 p. 85]:

1. *Transport work* moves molecules from one side of a membrane to another. Transport processes bring material into and out of the body and transfer them between compartments within the body.

2. *Mechanical work* uses intracellular fibers and filaments to create movement. This includes external work, such as movement created by skeletal muscle contraction, and internal work, such as the cytoplasmic movement of vesicles.

3. *Chemical work* is used for growth, maintenance, and storage of information and energy. Chemical work in the body can be subdivided into synthesis and storage. Storage includes both *short-term energy storage* in high-energy phosphate compounds such as ATP, and *long-term energy storage* in the chemical bonds of glycogen and fat.

Most components of energy output, including synthesis, transport, and heat production, are not under conscious control. Therefore movement such as exercise is the only way people can voluntarily increase energy output. People can control their energy intake, however, by watching what they eat.

Although energy balance is a very simple concept, it is a difficult one for people to accept. Behavior modifications such as eating less and exercising more are among the most frequent instructions health care professionals give to patients. They are also the most difficult instructions to follow, and patient compliance is usually low.

Energy Balance Is Reflected by an Individual's Oxygen Consumption

To compile an energy balance sheet for the human body, we must be able to estimate both the energy content of food (energy intake) and the energy expenditure from heat loss and various types of work.

Oxygen Consumption Can Be Equated to Metabolic Rate

The most direct way to measure the energy content of food is by **direct calorimetry**. In this procedure, food is burned in an instrument called a bomb calorimeter, and the released heat is trapped and measured. The energy content of food is usually measured in **kilocalories** (kcal). One kilocalorie is the amount of heat needed to raise the temperature of 1 liter of water by 1° C. A kilocalorie is the same as a *Calorie* (with a capital C).

Although direct calorimetry is a quick way of measuring total energy content, the metabolic energy content of food is actually slightly less because not all foods can be fully digested and absorbed. The metabolic energy content of proteins and carbohydrates is 4 kcal/g. Fats contain more than twice as much energy at 9 kcal/g.

The caloric content of food can be calculated by multiplying the number of grams of each component by its metabolic energy content (kcal/g): 2 g fat × 9 kcal/g = 18 kcal. In the United States you can find examples of the energy content for various foods on the Nutrition Facts label of food packages.

Estimating an individual's energy expenditure, or **metabolic rate**, is more complex than figuring the caloric content of food. The heat released by the body can be measured by enclosing a person in a sealed compartment. Energy intake in food equals work plus heat, so the difference between the heat produced and the person's caloric intake would be the energy used for chemical, mechanical, and transport work. Practically speaking, however, total body heat release is not a very easy way to measure energy use.

Probably the most common method for estimating energy expenditure is to measure a person's **oxygen consumption**, the amount of oxygen consumed per kilocalorie of food metabolized. Studies have shown that oxygen consumption for different foods is relatively constant at a rate of 1 liter oxygen consumed for each 4.5–5 kcal of energy. Thus, the rate of oxygen consumption is generally equated to metabolic rate. Measurement of oxygen consumption is one form of **indirect calorimetry**.

Another method of estimating metabolic energy use is to measure carbon dioxide production, either alone or in combination with oxygen consumption. Aerobic metabolism both consumes O_2 and produces CO_2. However, the ratio of CO_2 produced to O_2 consumed varies with the composition of the diet.

This ratio of $CO_2 : O_2$, known as the **respiratory quotient**, or **RQ**, varies from a high of 1.0 with a pure carbohydrate diet to 0.8 for pure protein and 0.7 for pure fat. The average American diet has an RQ of about 0.82. Whether measured by oxygen consumption or by carbon dioxide production, metabolic rate can be highly variable from one person to another or from day to day in a single individual.

Many Factors Influence Metabolic Rate An individual's lowest metabolic rate is considered the **basal metabolic rate**, or **BMR**. In reality, metabolic rate would be lowest when an individual is asleep. But because measuring the metabolic rate of a sleeping person is difficult, BMR is measured after a 12-hour fast in

---

RUNNING PROBLEM

When Nicole's lab results came back, Dr. Ayani immediately wrote orders to start an intravenous infusion and heart monitoring. The lab slip showed a plasma potassium of 2.5 mEq/L (normal 3.5–5.0 mEq/L), plasma HCO_3^- of 40 mEq/L (normal 24–29), and plasma pH of 7.52. Dr. Ayani admitted Nicole to the hospital for further treatment and evaluation, hoping she could be convinced that she needed help for her anorexia. *Anorexia*, meaning "no appetite," can have both physiological and psychological origins.

Question 3: What is Nicole's K^+ disturbance called? What effect does her K^+ disturbance have on the resting membrane potential of her cells?

Question 4: Why does Dr. Ayani want to monitor Nicole's cardiac function?

Question 5: Based on her clinical values, what is Nicole's acid-base status?

a person who is awake but resting. Factors that affect metabolic rate in humans include age, gender, amount of lean muscle mass, activity level, diet, hormones, and genetics.

1. **Age and gender.** Adult males have an average BMR of 1.0 kcal per hour per kilogram body weight. Adult females have a lower rate than males: 0.9 kcal/hr/kg. The difference arises because women have a higher percentage of adipose tissue and less lean muscle mass.

   Metabolic rates in both sexes decline with age. Some of this decline is due to decreases in lean muscle mass.

2. **Amount of lean muscle mass.** Muscle has a higher oxygen consumption rate at rest than fat (most of the volume of an adipose tissue cell is occupied by metabolically inactive lipid droplets). This is one reason that advice on how to lose weight often includes weight training in addition to aerobic exercise. By adding to lean muscle mass, people raise their basal metabolic rate and burn off more calories at rest.

3. **Physical activity level.** Physical activity and muscle contraction increase metabolic rate over the basal rate. Sitting and lying down consume relatively little energy. Rowing in a race and bicycling are among the energetically most expensive activities.

4. **Effect of diet.** We now know that resting (nonactive) metabolic rate increases after a meal, a phenomenon termed **diet-induced thermogenesis**. In other words, there is an energetic cost to the digestion and assimilation of food.

   Diet-induced thermogenesis is related to the type and amount of food ingested. Fats cause relatively little diet-induced thermogenesis. Proteins increase heat production the most. This phenomenon may support the claim of some nutritionists that eating one calorie of fat is different than eating one calorie of protein, although they contain the same amount of energy when measured by direct calorimetry.

5. **Effects of hormones.** Basal metabolic rate is increased by thyroid hormones and by catecholamines (epinephrine and norepinephrine). Some of the peptides that regulate food intake also appear to influence metabolism.

In the end, the two factors that a person can control to balance energy expenditure and avoid weight gain are energy intake (eating) and activity. If a person's activity includes strength training that increases lean muscle mass, resting metabolic rate goes up. The addition of lean muscle mass to the body creates additional energy use, which in turn decreases the number of calories that go into storage.

Energy Is Stored in Fat and Glycogen

A person's daily energy requirement, expressed as caloric intake, varies with the needs and activity of the body. For example, large male athletes in an intensive training program may need more than 10,000 kcal per day. On the other hand, a 70-kg man engaged in normal activities may only require 2000 kcal/day.

Assume that our average man's energy requirement could be met by ingesting only glucose. Glucose has an energy content of 4 kcal/g, therefore to get 2000 kcal he would have to consume 500 g, or 1.1 pounds, of glucose in a day. Our bodies cannot ab-

When Nicole was admitted to the hospital, her blood pressure was 80/50 and her pulse was a weak and irregular 90. She weighed less than 85% of the minimal healthy weight for a woman of her height and age. She had an intense fear of gaining weight, even though she was underweight. Her menstrual periods were irregular, she had just suffered a fractured wrist from a fall that normally shouldn't have caused a fracture, and her hair was thinning. When Dr. Ayani questioned Nicole, she admitted that she had been feeling weak during dance rehearsals and had been having difficulty concentrating at times.

Question 6: Based on what you learned in Chapters 14 and 15 about heart rate and blood pressure, speculate on why Nicole has low blood pressure with a rapid pulse.

Question 7: Would you expect Nicole's renin and aldosterone levels to be normal, elevated, or depressed? How might this relate to her K^+ disturbance?

Question 8: Give some possible reasons Nicole had been feeling weak during dance rehearsals.

sorb crystalline glucose, however, so those 500 g of glucose would have to be dissolved in water. If the glucose was made as an isosmotic 5% solution, the 500 g would have to be dissolved in 10 liters of water: a substantial volume to drink during a day!

Fortunately, we do not usually ingest glucose as our primary fuel. Proteins, complex carbohydrates, and fats also provide energy. *Glycogen*, a glucose polymer [p. 35], is a more compact form of energy than an equal number of individual glucose molecules. Glycogen also requires less water for hydration.

Our cells therefore convert glucose to glycogen for storage. Normally we keep about 100 g of glycogen in the liver and 200 g in skeletal muscles. But even 300 g of glycogen can provide only enough energy for 10 to 15 hours of the body's energy needs. The brain alone requires 150 g of glucose per day.

Consequently, the body keeps most of its energy reserves in compact, high-energy fat molecules. One gram of fat has 9 kcal, more than twice the energy content of an equal amount of carbohydrates or proteins. This feature makes adipose tissue a very efficient way for the body to keep large amounts of energy stored in minimal space. Metabolically, however, the energy in fat is harder to access, and the metabolism of fats is slower than that of carbohydrates.

CONCEPT CHECK   Answers: p. 725

3. Name six factors that can influence a person's metabolic rate.

4. Why is most energy stored in fat and not in glycogen?

5. Complete the reaction for aerobic metabolism of one glucose and balance the equation:

$$C_6H_{12}O_6 + O_2 \rightarrow ? + ?$$

6. What is the RQ for the balanced equation in question 5?

▶ METABOLISM

Chapter 4, *Cellular Metabolism*, introduced the basic pathways of cellular metabolism. **Metabolism** is the sum of all chemical reactions in the body. These reactions (1) extract energy from nutrients, (2) use energy for work, and (3) store excess energy so that it can be used later. Some metabolic reactions release energy; others require energy (*exergonic* and *endergonic reactions;* ● p. 88).

Metabolic pathways that synthesize small molecules into larger ones are called **anabolic pathways** [*ana-*, completion + *metabole*, change]. Pathways that break large molecules into smaller components are considered **catabolic** [*cata-*, down or back]. The classification of a pathway is its end result, not what happens in one or two steps.

For example, in the first step of *glycolysis* [● p. 103], glucose adds a phosphate to become a larger molecule, glucose 6-phosphate. This single reaction is anabolic. But by the end of glycolysis the initial 6-carbon glucose molecule has been converted to two 3-carbon pyruvate molecules. This makes glycolysis a catabolic pathway.

In the human body, we divide metabolism into two states. The period of time following a meal, when the products of digestion are being absorbed, used, and stored, is called the **fed**, or **absorptive**, **state**. This is an *anabolic* state in which the energy of nutrient biomolecules is transferred to high-energy compounds or is stored in the chemical bonds of other molecules.

Once nutrients from a recent meal are no longer in the bloodstream and available for use by the tissues, the body enters a **fasted**, or **postabsorptive**, **state**. As the pool of available nutrients in the blood decreases, the body taps into its stored reserves. The postabsorptive state is *catabolic* because cells break down large molecules into smaller molecules. The energy released by breaking chemical bonds is used to do work.

Energy from Ingested Nutrients May Be Used Immediately or Stored

The biomolecules we eat are destined to meet one of three fates:

1. **Energy.** Energy released from chemical bonds of nutrients is trapped in ATP, phosphocreatine, and other high-energy compounds. This energy can then be used to do mechanical or transport work.
2. **Synthesis.** Biomolecules are used for synthesis of basic components needed for growth and maintenance of cells and tissues.
3. **Storage.** If the food ingested exceeds the body's requirements for energy and synthesis, excess nutrients go into storage as glycogen and fat. Storage makes energy available for times of fasting.

The specific fate of an absorbed nutrient depends on whether the biomolecule is a carbohydrate, protein, or fat. Figure 22-2 ● is a schematic diagram that follows these biomolecules from the diet into the three *nutrient pools* of the body: the fatty acid pool, the glucose pool, and the amino acid pool. Nutrient pools are nutrients that are available for immediate use. They are located primarily in the plasma.

Fats are absorbed primarily as *fatty acids* and *glycerol*. Fatty acids form the primary pool of fats in the blood. They can be used as an energy source by many tissues but are also easily stored as fat (*triglycerides*) in adipose tissue.

Carbohydrates are absorbed mostly as glucose. Plasma glucose concentration is the most closely regulated of the three nutrient pools because glucose is the only fuel that the brain can metabolize, except in times of starvation.

Notice in Figure 22-2 that if the glucose pool falls below a certain level, only the brain has access to glucose. This conservation measure ensures that the brain has an adequate energy supply.

● **Figure 22-2** *Summary of metabolism* Adapted from L. L. Langley, *Homeostasis* (New York: Reinhold, 1965).

Just as the circulatory system gives priority to supplying oxygen to the brain, metabolism also gives priority to the brain.

If the body's glucose pool is within the normal range, most tissues use glucose for their energy source. Excess glucose goes into storage as glycogen. The synthesis of glycogen from glucose is known as **glycogenesis**. Glycogen stores are limited, however, and additional excess glucose is converted to fat (**lipogenesis**).

If plasma glucose concentrations decrease, the body converts glycogen back to glucose through the process called **glycogenolysis**. By balancing oxidative metabolism, glycogen synthesis, glycogen breakdown, and fat synthesis, the body maintains plasma glucose concentrations within a narrow range.

If homeostasis fails and plasma glucose exceeds a critical level, as occurs in diabetes mellitus, excess glucose is excreted in the urine. Glucose excretion occurs only when the *renal threshold* for glucose reabsorption is exceeded [❀ p. 612].

The amino acid pool of the body is used primarily for protein synthesis. However, if glucose intake is low, amino acids can be converted into glucose through the pathways known as **gluconeogenesis**. This word literally means "the birth of new glucose" and refers to the synthesis of glucose from a noncarbohydrate precursor.

Amino acids are the main source for glucose through the gluconeogenesis pathways, but glycerol from triglycerides can also be used. Both gluconeogenesis and glycogenolysis are important backup sources for glucose during periods of fasting.

Hormones Control Metabolic Pathways by Changing Enzyme Activity

The most important biochemical pathways for energy production are illustrated in Figure 22-3 ●. This figure does not include all of the metabolic intermediates in each pathway (see Chapter 4 for

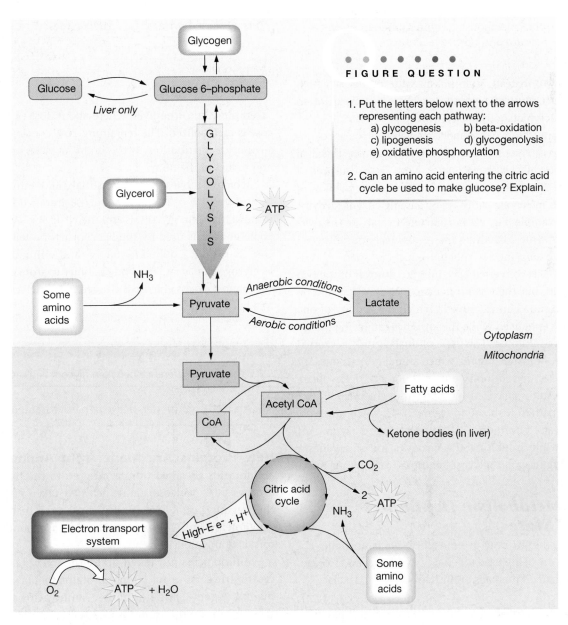

● **Figure 22-3** *Summary of biochemical pathways for energy production*

(a)

Without regulation of enzymatic activity, the pathway will simply cycle back and forth.

(b)

Fed-state metabolism under the influence of insulin

(c)

Fasted-state metabolism under the influence of glucagon

● **Figure 22-4   4 Dual (push-pull) control of metabolism** Push-pull control of metabolism requires different enzymes to catalyze the forward and reverse reactions.

detailed pathways). Instead, it emphasizes the points at which different pathways intersect, because these intersections are often key points at which metabolism is controlled.

One significant feature of metabolic regulation is the use of different enzymes to catalyze forward and reverse reactions. This dual control, sometimes called *push-pull control*, allows close regulation of a reaction. Figure 22-4● shows how hormones regulate the flow of nutrients through metabolic pathways by altering enzyme activity.

In the first example (Fig. 22-4a), enzyme 1 catalyzes the reaction of A to B. Enzyme 2 catalyzes the reverse reaction. If the activity of the two enzymes is roughly equal, as soon as A is converted into B, B is converted back into A. Turnover of the two substrates is rapid, but there is no net production of product B.

Figure 22-4b shows an example in which the activity of one set of enzymes is enhanced while the enzymes catalyzing the reverse reaction are inhibited. This type of dual control is exerted by the pancreatic hormone *insulin*, which stimulates enzymes for glycogen synthesis and inhibits enzymes for glycogen breakdown. The net result is glycogen synthesis from glucose.

The reverse pattern is shown in Figure 22-4c, where *glucagon*, another pancreatic hormone, stimulates the enzymes of glycogenolysis while inhibiting the enzymes for glycogenesis. The net effect of glucagon is glucose synthesis from glycogen.

Anabolic Metabolism Dominates in the Fed State

As we've noted, the *fed state* or *absorptive state* is anabolic: absorbed nutrients are being used for energy, synthesis, and storage. Table 22-2 summarizes the fates of nutrients in the fed state.

Carbohydrates Provide Energy   Glucose absorbed after a meal goes into the circulation of the hepatic portal system and is

Table 22-2: Fate of Nutrients in the Fed State

Carbohydrates, absorbed primarily as glucose

1. Used immediately for energy through aerobic pathways*
2. Used for lipoprotein synthesis in the liver
3. Stored as glycogen in liver and muscle
4. Excess converted to fat and stored in adipose tissue
   (glucose→ pyruvate→ acetyl CoA→ fatty acids)

Proteins, absorbed primarily as amino acids

1. Most amino acids go to tissues for protein synthesis*
2. If needed for energy, amino acids converted in liver to intermediates for aerobic metabolism
3. Excess converted to fat and stored in adipose tissue
   (amino acids→ acetyl CoA→ fatty acids)

Fats, absorbed primarily as triglycerides

1. Stored as fats primarily in liver and adipose tissue*

* Primary fate.

taken directly to the liver, where about 30% of all ingested glucose is metabolized. The remaining 70% continues in the bloodstream for distribution to the brain, muscles, and other organs and tissues of the body.

Glucose moves from interstitial fluid into cells via GLUT transporters [● p. 139]. Most glucose absorbed from a meal goes immediately into glycolysis and the citric acid cycle to make ATP. Some glucose is used by the liver for lipoprotein synthesis. Glucose that is not required for energy and synthesis is stored either as glycogen or as fat. The body's ability to store glycogen is limited, so most excess glucose is converted to triglycerides and stored in adipose tissue.

CONCEPT CHECK   Answers: p. 725

7. What is the difference between glycogenesis and gluconeogenesis?

8. Are GLUT transporters active or passive transport?

New Proteins Are Made from Amino Acids   Most amino acids absorbed from a meal go to the tissues for protein synthesis. Like glucose, amino acids are taken first to the liver by the hepatic portal system. The liver uses them to synthesize lipoproteins and plasma proteins, such as albumin, clotting factors, and angiotensinogen.

Amino acids not taken up by the liver are used by cells to create structural or functional proteins such as cytoskeletal elements, enzymes, and hormones. Amino acids are also incorporated into nonprotein molecules such as amine hormones and neurotransmitters.

If glucose intake is low, amino acids can be used for energy. On the other hand, if more protein is ingested than is needed for synthesis and energy needs, excess amino acids are converted to fat. Some bodybuilders spend large amounts of money on special amino acid supplements that are advertised to build bigger muscles. What the athletes may not realize is that these amino acids do not automatically go into protein synthesis. When amino acid intake exceeds the body's need for protein synthesis, excess amino acids are burned to create energy or are stored as fat.

Fats Store Energy   Most ingested fats are assembled into lipoprotein complexes called *chylomicrons* in the intestinal epithelium [◈ p. 685]. Chylomicrons reach the venous circulation via the lymphatics, bypassing the liver (Fig. 22-5, step 1 ●). As they circulate through the blood, **lipoprotein lipase** bound to the capillary endothelium of muscles and adipose tissue converts triglycerides in the chylomicron to free fatty acids and monoglycerides (step 2). These molecules then diffuse into the cells, where they are used for energy (step 4) or reassembled into triglycerides for storage (step 3).

Chylomicron remnants that remain in the circulation are composed primarily of lipoproteins and cholesterol. The combination of lipid with proteins makes cholesterol more soluble in plasma, but the complexes are unable to diffuse through cell membranes. Instead, lipoprotein complexes must be brought into cells using *receptor-mediated endocytosis* [◈ p. 145]. Chylomicron remnants are taken up and metabolized by the liver.

Once in liver cells, cholesterol from the remnants joins the liver's pool of lipids (step 5). If cholesterol is in excess, some may be excreted in the bile. The remainder is added to newly made cholesterol and fatty acids, then packaged again into lipoprotein complexes for secretion back into the blood.

There are several lipoproteins that form these cholesterol-protein complexes, ranging from *very low-density lipoprotein (VLDL)* to *high-density lipoprotein (HDL)*. The complexes contain varying ratios of protein, triglyceride, phospholipids, and cholesterol. The more protein a complex contains, the heavier its weight.

Most lipoprotein in the blood is **low-density lipoprotein**, or **LDL**. LDL-cholesterol is the form in which most cells get cholesterol through receptor-mediated endocytosis. LDL-cholesterol is sometimes known as the "lethal cholesterol" because elevated concentrations of plasma LDL are associated with the development of atherosclerosis.

The second most common lipoprotein is **high-density lipoprotein**, or **HDL**. HDL-cholesterol is sometimes called the "healthy cholesterol" because HDL takes cholesterol into the liver cells, where it is metabolized or excreted. The ratio of HDL-cholesterol to LDL-cholesterol in the plasma is as important as the absolute concentration. The ratio is used by clinicians to predict a person's risk of developing atherosclerosis.

The protein components of lipoproteins are called **apoproteins**, or *apolipoproteins* [apo-, derived from]. LDL complexes contain *apoprotein B* (apoB), which combines with receptors that bring

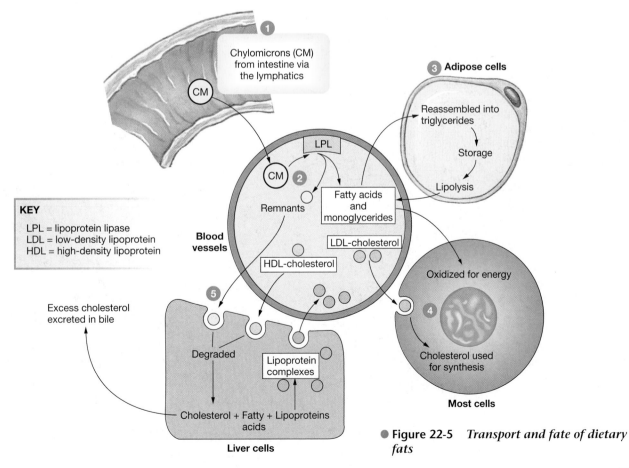

KEY

LPL = lipoprotein lipase
LDL = low-density lipoprotein
HDL = high-density lipoprotein

● **Figure 22-5**   *Transport and fate of dietary fats*

LDL-cholesterol into most cells of the body. HDL complexes contain *apoE*, which activates receptors in the liver. Several inherited forms of *hypercholesterolemia* (elevated plasma cholesterol levels) have been linked to defective forms of either apoE or apoB. These abnormal apoproteins may help explain the accelerated development of atherosclerosis in people with hypercholesterolemia.

Catabolic Metabolism Dominates in the Fasted State

Once all nutrients from a meal have been digested, absorbed, and distributed to various cells, plasma concentrations of glucose begin to fall. This is the signal for the body to shift from fed- (absorptive) state to fasted- (postabsorptive) state metabolism. The goal of the fasted state is to maintain plasma glucose concentrations within an acceptable range so that the brain and neurons have adequate fuel.

The liver is the primary source of glucose production during the fasted state. Liver glycogen can provide enough glucose through glycogenolysis to meet four to five hours of the body's energy needs (Fig. 22-6 ●).

Skeletal muscle glycogen can be metabolized to glucose in the fasted state but not through direct conversion. Muscle cells, like most other cells, lack the enzyme that makes glucose from glucose 6-phosphate, so they metabolize glucose 6-phosphate to pyruvate (aerobic conditions) or lactate (anaerobic conditions). Pyruvate and lactate are then transported to the liver, which uses them to make glucose.

Additional glucose or ATP can be made from amino acids, particularly those in muscle proteins. Enzymes remove amino groups from the amino acids (*deamination*, ● p. 108) and convert the amino groups to *urea*, which is excreted. Some deaminated amino acids become citric acid cycle intermediates and make ATP. This alternative ATP source thus spares plasma glucose for use by the brain. Other amino acids are processed to pyruvate, which goes to the liver and is made into glucose, as previously described.

In the fasted state, adipose tissue breaks down its stores of triglycerides into fatty acids and glycerol. Glycerol goes to the liver and can be converted to glucose. Fatty acids are released into the blood and are used as a source of energy by many tissues.

The long carbon chains of fatty acids are chopped into two-carbon units through the process of **β-oxidation** [● p. 109]. These two-carbon units then feed into the citric acid cycle through acetyl CoA. If there is excess fatty acid breakdown, β-oxidation in

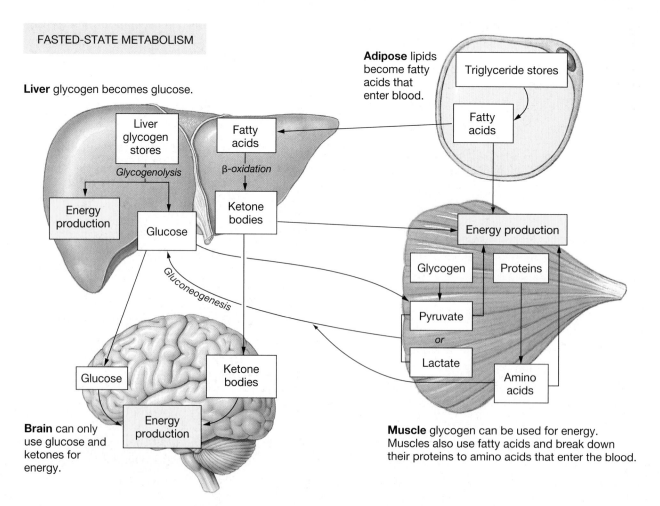

FASTED-STATE METABOLISM

Liver glycogen becomes glucose.

Adipose lipids become fatty acids that enter blood.

Triglyceride stores

Fatty acids

Liver glycogen stores

Fatty acids

Glycogenolysis

β-oxidation

Energy production

Glucose

Ketone bodies

Gluconeogenesis

Energy production

Glycogen

Proteins

Pyruvate

or

Lactate

Amino acids

Glucose

Ketone bodies

Brain can only use glucose and ketones for energy.

Energy production

Muscle glycogen can be used for energy. Muscles also use fatty acids and break down their proteins to amino acids that enter the blood.

● **Figure 22-6**   *Fasted-state metabolism*   The goal of fasted-state metabolism is to maintain adequate plasma glucose concentrations for the brain.

the liver creates molecules known as **ketone bodies**. Ketone bodies are transported from the liver to other cells that convert them to acetyl CoA. Acetyl CoA then enters the citric acid cycle.

Ketone bodies are the only fuel besides glucose that the brain can use, so in cases of prolonged starvation, ketones become a significant source of energy. Unfortunately, the ketones *acetoacetic acid* and *β-hydroxybutyric acid* are moderately strong acids, and excessive ketone production leads to a state of acidosis known as **ketoacidosis**. People in ketoacidosis have a fruity odor to their breath due to acetone, a volatile ketone whose odor you might recognize from nail polish remover.

With this summary of metabolic pathways as background, we now turn to the endocrine and neural regulation of metabolism.

CONCEPT CHECK   Answers: p. 725

9. When amino acids are used for energy, which pathways in Figure 22-3 do they follow?

10. If cholesterol is lipid-soluble, why does plasma cholesterol need a membrane transporter to enter cells?

▶ HOMEOSTATIC CONTROL OF METABOLISM

The endocrine system has primary responsibility for metabolic regulation, although the nervous system does have some influence, particularly in terms of governing food intake. Hour-to-

CLINICAL FOCUS

Ketogenic Diets

Some people intentionally restrict or eliminate their intake of carbohydrates in an effort to lose weight. Diets that are very low in carbohydrate and high in fat and protein shift metabolism to β-oxidation of fats and ketone body production. These diets are known therefore as *ketogenic diets*. People who go on these diets are often pleased by the initial rapid weight loss, but this is due to glycogen breakdown and water loss, not fat use. Sticking with ketogenic diets decreases caloric intake and eventually results in loss of body fat, but any diet that restricts calories will do the same. Ketogenic diets have risks associated with them, including dehydration, electrolyte loss, inadequate intake of calcium and vitamins, gout, and kidney problems. The only recommended use of ketogenic diets is for children under 10 who have epilepsy that is not responding fully to drug therapy. For reasons that we do not understand, maintaining a state of ketosis in these children decreases the incidence of seizures.

hour regulation depends primarily on the ratio of insulin to glucagon, two hormones secreted by endocrine cells of the pancreas. Both hormones have short half-lives and must be continuously secreted if they are to have a sustained effect.

The Pancreas Secretes Insulin and Glucagon

The endocrine cells of the pancreas are less than 2% of the organ's total mass, with most pancreatic tissue devoted to production and exocrine secretion of digestive enzymes and bicarbonate [p. 661]. In 1869, the German anatomist Paul Langerhans described small clusters of cells scattered throughout the body of the pancreas (Fig. 22-7 ●). These clusters, now known as the **islets of Langerhans**, contain four distinct cell types, each associated with secretion of one or more different peptide hormones.

Nearly three-quarters of the islet cells are **beta cells**, which produce **insulin** and a peptide called *amylin*. Another 20% are **alpha cells**, which secrete **glucagon**. Most of the remaining cells are *somatostatin*-secreting **D cells**. A few rare cells called *PP cells* (or *F cells*) produce *pancreatic polypeptide*.

Like all endocrine glands, the islets are closely associated with capillaries into which the hormones are released. Both sympathetic and parasympathetic neurons terminate on the islets, providing a means by which the nervous system can influence metabolism.

The Insulin-to-Glucagon Ratio Regulates Metabolism

Insulin and glucagon act in antagonistic fashion to keep plasma glucose concentrations within an acceptable range. Both hormones are present in the blood most of the time. It is the ratio of the two hormones that determines which hormone dominates.

In the fed state, when the body is absorbing nutrients, insulin dominates and the body undergoes net anabolism (Fig. 22-8a ●). Ingested glucose is used for energy production, with excess glucose stored as glycogen or fat. Proteins go primarily to protein synthesis.

In the fasted state, the goal of metabolic regulation is to prevent low plasma glucose concentrations (*hypoglycemia*). When glucagon predominates (Fig. 22-8b), the liver uses glycogen and nonglucose intermediates to synthesize glucose for release into the blood.

Figure 22-9 ● shows plasma glucose, glucagon, and insulin concentrations over the course of a day. In a normal person, fasting plasma glucose is maintained around 90 mg/dL plasma. After absorption of nutrients from a meal, plasma glucose rises. The increase in glucose stimulates insulin release, which in turn promotes glucose transfer into cells. Plasma glucose concentrations thus fall back toward the fasting level shortly after each meal.

During an overnight fast, plasma glucose concentrations fall to a low but steady value, and insulin secretion also decreases.

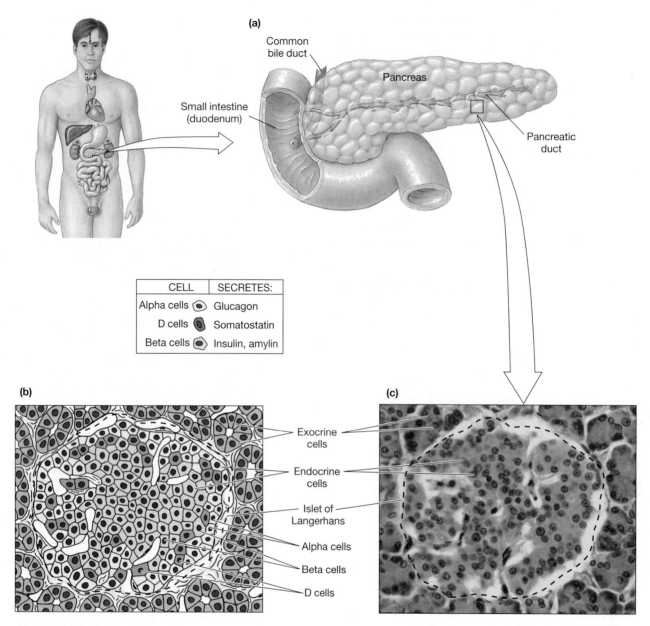

● Figure 22-7   *The endocrine pancreas*

Glucagon secretion remains relatively steady during the 24-hour period, supporting the theory that it is the ratio of insulin to glucagon that determines the direction of metabolism.

Insulin Is the Dominant Hormone of the Fed State

Insulin is a typical peptide hormone (Table 22-3). It is synthesized as an inactive prohormone and activated prior to secretion [● Fig. 7-4c, p. 216]. Glucose is the primary stimulus for insulin secretion but additional factors may stimulate, amplify, or inhibit secretion.

1. **Increased glucose concentrations.** The primary stimulus for insulin release is plasma glucose concentrations greater than 100 mg/dL. Glucose absorbed from the small intestine reach-

es the beta cells, where it is taken up by a GLUT-2 transporter. With more glucose available as substrate, ATP production increases and ATP-gated K^+ channels close. The cell depolarizes, voltage-gated Ca^{2+} channels open, and Ca^{2+} entry initiates exocytosis of insulin [● Fig. 5-42, p. 163].

2. **Increased amino acid concentrations.** Increased plasma amino acid concentrations following a meal also trigger insulin secretion.

3. **Feedforward effects of GI hormones.** Multiple GI hormones, including GIP (*glucose-dependent insulinotropic peptide*), CCK, and gastrin, amplify insulin secretion. The most potent of these hormones is *glucagon-like peptide-1 (GLP-1)*, which is released by glucose in the lumen of the small intestine. The GI hormones travel through the circulation to beta cells and may arrive there even before the first glucose is absorbed. The

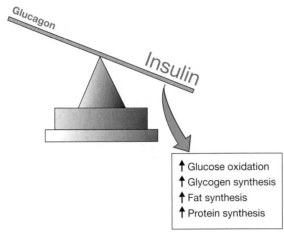

(a) Fed state: insulin dominates

Glucagon

Insulin

↑ Glucose oxidation
↑ Glycogen synthesis
↑ Fat synthesis
↑ Protein synthesis

(b) Fasted state: glucagon dominates

Insulin

Glucagon

↑ Glycogenolysis
↑ Gluconeogenesis
↑ Ketogenesis

● **Figure 22-8** *Metabolism is controlled by insulin and glucagon*

Table 22-3: Insulin

Cell of origin	Beta cells of pancreas
Chemical nature	51-amino acid peptide
Biosynthesis	Typical peptide
Transport in the circulation	Dissolved in plasma
Half-life	5 minutes
Factors affecting release	Plasma [glucose] > 100 mg/dL; ↑ blood amino acids; GI hormones (feedforward reflex) and parasympathetic amplify. Sympathetic inhibits.
Target cells or tissues	Liver, muscle, and adipose tissue primarily; brain, kidney, and intestine not insulin-dependent
Target receptor	Membrane receptor with tyrosine kinase activity; pathway with insulin-receptor substrates
Whole body or tissue action	↓ Plasma [glucose] by ↑ transport into cells or ↑ metabolic use of glucose
Action at cellular level	↑ Glycogen synthesis; ↑ aerobic metabolism of glucose; ↑ protein and triglyceride synthesis
Action at molecular level	Inserts GLUT transporters in muscle and adipose cells; alters enzyme activity. Complex signal transduction pathway.
Feedback regulation	↓ Plasma [glucose] shuts off insulin release
Other information	Growth hormone and cortisol are antagonistic.

Plasma glucagon concentration (ηg/mL)

Plasma glucose concentration (mg/dL)

Plasma insulin concentration (µU/mL)

8 AM   Noon   4 PM   8 PM   2 AM   8 AM

▲ Meal   ▲ Meal   ▲ Meal

● **Figure 22-9** *Glucose, glucagon, and insulin levels in a 24-hour period* Hour-to-hour metabolic regulation depends on the ratio of insulin to glucagon.

anticipatory release of insulin in response to these hormones prevents a sudden surge in plasma glucose concentrations when the meal is absorbed.

4. **Parasympathetic activity.** Parasympathetic activity to the GI tract and pancreas increases during and following a meal. Parasympathetic input to beta cells stimulates insulin secretion.

5. **Sympathetic activity.** Insulin secretion is inhibited by sympathetic neurons. In times of stress, sympathetic input to the endocrine pancreas increases and is reinforced by catecholamine release from the adrenal medulla (Table 22-4). Epinephrine and norepinephrine inhibit insulin secretion and switch metabolism to gluconeogenesis to provide extra fuel for the nervous system and skeletal muscles.

Insulin Promotes Anabolism

Like other peptide hormones, insulin combines with a membrane receptor on its target cells. The insulin receptor has *tyrosine kinase* activity that initiates complex intracellular cascades whose details

Table 22-4: Adrenal Catecholamines (Epinephrine and Norepinephrine)

Origin	Adrenal medulla (90% epinephrine and 10% norepinephrine)
Chemical nature	Amines made from tyrosine
Biosynthesis	Typical peptide
Transport in the circulation	Some bound to sulfate
Half-life	2 minutes
Factors affecting release	Primarily fight-or-flight reaction through CNS and autonomic nervous system; hypoglycemia
Target cells or tissues	Mainly neurons, pancreatic endocrine cells, heart, blood vessels, adipose tissue
Target receptor	G protein-coupled receptors; α and β subtypes
Second messenger	cAMP (α_2 and all β receptors); IP_3 α_1
Whole body or tissue action	↑ Plasma [glucose]; activate fight-or-flight and stress reactions; ↑ glucagon and ↓ insulin secretion
Onset and duration of action	Rapid and brief

are still not completely understood. A general summary of insulin action is shown in Figure 22-10●.

The activated insulin receptor phosphorylates proteins that include a group known as the **insulin-receptor substrates** (IRS). These proteins then act through complicated pathways to influence transport and cellular metabolism. The enzymes that regulate metabolic pathways may be inhibited or activated directly, or their synthesis is influenced indirectly through transcription factors.

The primary target tissues for insulin are the liver, adipose tissue, and skeletal muscles. The usual target cell response is increased glucose metabolism. In some target tissues, insulin also regulates the GLUT transporters. Other tissues, including the brain and transporting epithelia of the kidney and intestine, do not require insulin for glucose uptake and metabolism.

How does insulin lower plasma glucose? It exerts its action in the following four ways:

1. **Insulin increases glucose transport into most, but not all, insulin-sensitive cells.** Adipose tissue and resting skeletal muscle require insulin for glucose uptake. Without insulin, their GLUT-4 transporters are withdrawn from the membrane and stored in cytoplasmic vesicles (Fig. 22-11a●). When the insulin receptor is activated, the resulting signal transduction cascade causes vesicles to move to the cell membrane and insert the transporters by exocytosis (Fig. 22-11b). The cells then take up glucose from the interstitial fluid by facilitated diffusion.

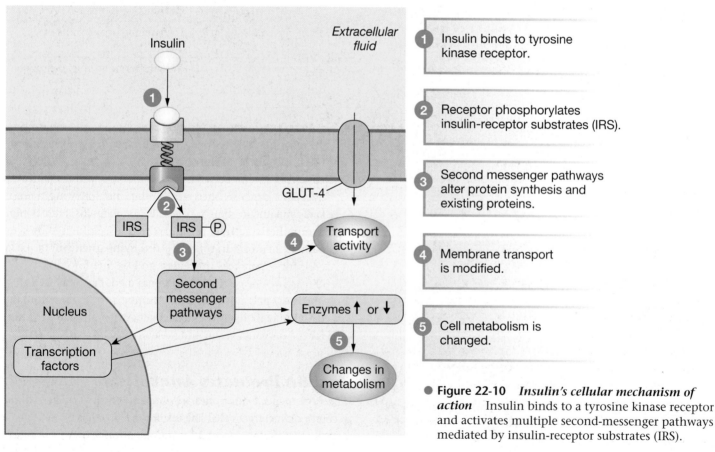

● **Figure 22-10** *Insulin's cellular mechanism of action* Insulin binds to a tyrosine kinase receptor and activates multiple second-messenger pathways mediated by insulin-receptor substrates (IRS).

(a) In the absence of insulin, glucose cannot enter the cell.

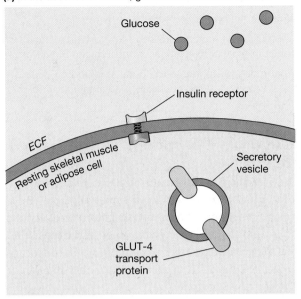

(b) Insulin signals the cell to insert GLUT-4 transporters into the membrane, allowing glucose to enter cell.

● **Figure 22-11** *Insulin alters glucose transport in adipose tissue and resting skeletal muscle, enabling glucose to enter the cells.*

(a) Hepatocyte. In fed state, liver cell takes up glucose.

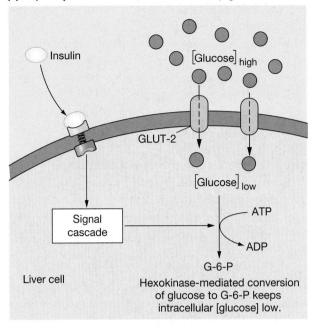

(b) Hepatocyte. In fasted state, liver cell makes glucose and transports it out into the blood.

● **Figure 22-12** *Insulin action on liver cells* Insulin does not directly regulate glucose transport in liver cells (hepatocytes).

Curiously, *exercising* skeletal muscle is not dependent on insulin activity for its glucose uptake. When muscles contract, GLUT-4 transporters are inserted into the membrane and glucose uptake increases. The intracellular signal for this is unclear but appears to involve Ca^{2+} and decreased inorganic phosphate (P_i).

Glucose transport into liver cells (*hepatocytes*) is not *directly* insulin-dependent but will be influenced by the presence or absence of insulin. Liver cells have GLUT-2 transporters that are always present in the cell membrane. In the fed state (Fig. 22-12a ●), insulin activates *hexokinase*, the enzyme that phosphorylates glucose to glucose 6-phosphate.

This reaction keeps free intracellular glucose concentrations low relative to the plasma, so that glucose continuously diffuses into the hepatocyte on the GLUT-2 transporter.

In the fasted state, when insulin is low, glucose moves *out* of the liver and into the blood to help maintain glucose homeostasis. In this process (Fig. 22-12b), hepatocytes are converting glycogen stores and amino acids to glucose. Newly formed glucose moves down its concentration gradient out of the cell using the same GLUT-2 transporters running in the reverse direction. If the GLUT transporters were pulled from the membrane during the fasted state, as they are in muscle and adipose tissue, glucose would have no way to leave the liver cells.

2. **Insulin enhances cellular utilization and storage of glucose** (Fig. 22-13 ●). Insulin activates enzymes for glucose utilization (*glycolysis*) and glycogen and fat synthesis (*glycogenesis* and *lipogenesis*). Insulin simultaneously inhibits enzymes for glycogen breakdown (*glycogenolysis*), glucose synthesis (*gluconeogenesis*), and fat breakdown (*lipolysis*) to ensure that metabolism moves in the anabolic direction. If more glucose has been ingested than is needed for energy and synthesis, the excess is made into glycogen or fatty acids.

3. **Insulin enhances utilization of amino acids.** Insulin activates enzymes for protein synthesis and inhibits enzymes that promote protein breakdown. If a meal includes protein, amino acids are used for protein synthesis by both the liver and muscle. Excess amino acids are converted to fatty acids.

4. **Insulin promotes fat synthesis.** Insulin inhibits beta-oxidation of fatty acids and promotes conversion of excess glucose or amino acids into triglycerides (*lipogenesis*). Excess triglyceride is stored as lipid droplets in adipose tissue.

Thus, insulin is an anabolic hormone because it promotes glycogen, protein, and fat synthesis. When insulin is absent or deficient, cells go into catabolic metabolism.

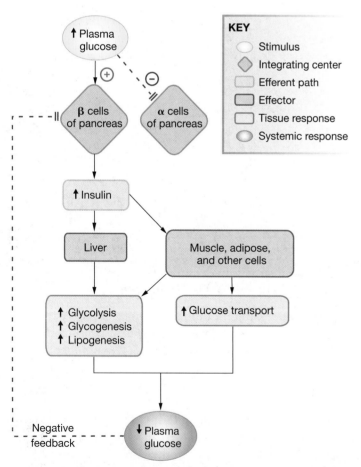

● **Figure 22-13** *Fed-state metabolism* Fed-state metabolism under the influence of insulin promotes glucose metabolism by cells.

11. What are the primary target tissues for insulin?

12. Why is glucose metabolism and transport independent of insulin in renal and intestinal epithelium and in neurons?

13. What is the advantage to the body of inhibiting insulin release during a sympathetically mediated fight-or-flight response?

Glucagon Is Dominant in the Fasted State

Glucagon, secreted by pancreatic alpha cells, is generally antagonistic to insulin in its effects on metabolism (Table 22-5). When plasma glucose concentrations decline after a meal, insulin secretion slows and the effects of glucagon on tissue metabolism take on greater significance. It is the ratio of insulin to glucagon that determines the direction of metabolism rather than an absolute amount of either hormone.

The goal of glucagon action is to prevent hypoglycemia, and the primary stimulus for glucagon release is plasma glucose concentration. When plasma glucose concentrations fall below 100 mg/dL, glucagon secretion rises dramatically. At glucose concentrations above 100 mg/dL, when insulin is being secreted, glucagon secretion is inhibited and remains at a low but relatively constant level (see Fig. 22-9). The strong relationship between insulin secretion and glucagon inhibition has led to

Table 22-5: Glucagon

Cell of origin	Alpha cells of pancreas
Chemical nature	29-amino acid peptide
Biosynthesis	Typical peptide
Transport in the circulation	Dissolved in plasma
Half-life	4–6 minutes
Factors affecting release	Stimulated by plasma [glucose] < 200 mg/dL, with maximum secretion below 50 mg/dL; ↑ blood amino acids.
Target cells or tissues	Liver primarily
Target receptor/second messenger	G protein-coupled receptor linked to cAMP
Whole body or tissue action	↑ Plasma [glucose] by glycogenolysis and gluconeogenesis; ↑ lipolysis leads to ketogenesis in liver
Action at molecular level	Alters existing enzymes and stimulates synthesis of new enzymes
Feedback regulation	↑ Plasma [glucose] shuts off glucagon secretion
Other information	Member of secretin family along with VIP, GIP, and GLP-1

speculation that the alpha cells are regulated by some factor linked to insulin rather than by plasma glucose concentrations directly.

The liver is glucagon's primary target tissue (Fig. 22-14 ●). Glucagon stimulates glycogenolysis and gluconeogenesis to increase glucose output. It is estimated that during an overnight fast, 75% of the glucose produced by the liver comes from glycogen stores and the remaining 25% from gluconeogenesis.

Glucagon release is also stimulated by plasma amino acids. This pathway prevents hypoglycemia after ingestion of a pure protein meal. Let's see how that might occur in the absence of glucagon.

If a meal contains protein but no carbohydrate, amino acids in the food cause insulin secretion. Even though no glucose has been absorbed, insulin-stimulated glucose uptake increases and plasma glucose concentrations fall. Unless something counteracts this process, the brain's fuel supply is threatened by hypoglycemia.

Co-secretion of glucagon in this situation prevents hypoglycemia by stimulating hepatic glucose output. Thus, although only amino acids were ingested, both glucose and amino acids are made available to peripheral tissues.

CONCEPT CHECK   Answers: p. 725

14. How are glycogenolysis and gluconeogenesis similar and different?

Diabetes Mellitus Is a Family of Metabolic Diseases

The most common pathology of the pancreatic endocrine system is the family of metabolic disorders known as **diabetes mellitus**. Diabetes is characterized by abnormally elevated plasma glucose concentrations (*hyperglycemia*) due to inadequate insulin secretion, abnormal target cell responsiveness [● p. 228], or both. Chronic hyperglycemia and associated metabolic abnormalities cause the many complications of the disease, including damage to blood vessels, eyes, kidneys, and the nervous system.

Diabetes has been known to affect humans since ancient times, and written accounts of the disease highlight the calamitous consequences of insulin deficiency. Aretaeus the Cappadocian (A.D. 81–138) wrote of the "wonderful" nature of this disease that consisted of the "melting down of the flesh … into urine," accompanied by terrible thirst that could not be quenched. The copious production of glucose-laden urine gave the disease its name. *Diabetes* refers to the flow of fluid through a siphon, and *mellitus* comes from the word for honey. In the Middle Ages, diabetes was known as "the pissing evil."

The severe type of diabetes described by Aretaeus is **type 1 diabetes**. It is a condition of insulin deficiency due to beta cell destruction. Type 1 diabetes is most commonly an *autoimmune* disease, in which the body fails to recognize the beta cells as "self" and destroys them with antibodies and white blood cells.

The other major variant of diabetes mellitus is **type 2 diabetes mellitus**. Type 2 diabetes is also known as *insulin-resistant*

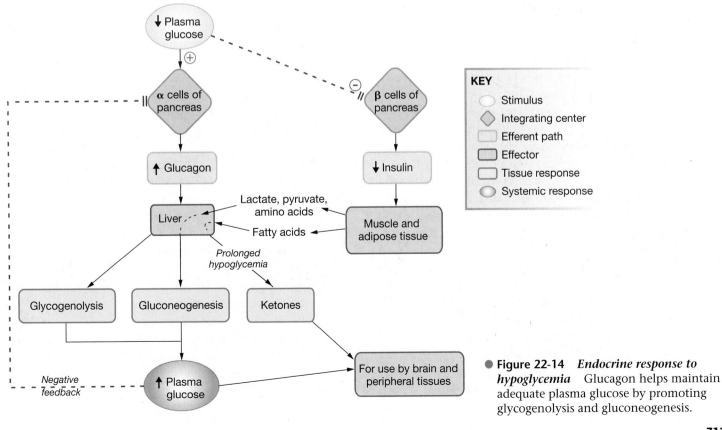

● **Figure 22-14** *Endocrine response to hypoglycemia* Glucagon helps maintain adequate plasma glucose by promoting glycogenolysis and gluconeogenesis.

KEY
- Stimulus
- Integrating center
- Efferent path
- Effector
- Tissue response
- Systemic response

diabetes because in most patients, insulin levels in the blood are normal or even elevated initially. Later in the disease process many type 2 diabetics become insulin-deficient and require insulin injections. Type 2 diabetes is actually a whole family of diseases with a variety of causes.

Type 1 Diabetics Are Ketosis-Prone

Type 1 diabetes is a complex disorder whose onset in genetically susceptible individuals is sometimes preceded by a viral infection. Many type 1 diabetics develop their disease in childhood, giving rise to the old name of *juvenile-onset diabetes*. About 10% of all diabetics have type 1 diabetes.

Because individuals with type 1 diabetes are insulin-deficient, the only treatment is insulin injections. Until the arrival of genetic engineering, most pharmaceutical insulin came from pork, cow, and sheep pancreases. However, once the gene for human insulin was cloned, biotechnology companies began to manufacture artificial human insulin for therapeutic use. In addition, scientists are developing techniques for implanting encapsulated beta cells in the body, in the hope that individuals with type 1 diabetes will no longer need to rely on regular insulin injections.

The events that follow ingestion of carbohydrate in an insulin-deficient diabetic create a picture of what happens to metabolism in the absence of insulin (Fig. 22-15 ●). Following a meal, nutrient absorption by the intestine proceeds normally because this process is insulin-independent. However, insulin-dependent nutrient uptake and metabolism by liver, resting skeletal muscle, and adipose tissue are severely diminished in the absence of insulin. Lacking nutrients to metabolize, the cells go into fasted-state metabolism.

1. *Protein metabolism.* Without glucose and amino acids for energy and protein synthesis, muscles break down their proteins to provide substrate for ATP production. Amino acids leave the muscles and are transported to the liver.
2. *Fat metabolism.* Adipose tissue in fasted-state metabolism breaks down its fat stores. Fatty acids provide substrate for ATP production and enter the blood for transport to the liver. The liver uses β-oxidation to break down fatty acids. However, it is limited in its ability to send fatty acids through the citric acid cycle, and the excess fatty acids are converted to *ketones*. Ketones reenter the circulation and can be used by other tissues, such as muscle and brain, for ATP synthesis.

   The catabolism of muscle and adipose tissue in the absence of insulin leads to tissue loss and the symptom of "melting down of the flesh" described by Arataeus.
3. *Glucose metabolism.* In the absence of insulin, glucose remains in the blood, creating a condition of elevated plasma glucose, or **hyperglycemia**. The liver, unable to metabolize this glucose, initiates fasted-state pathways of glycogenolysis and gluconeogenesis. These pathways produce *additional*

glucose from glycogen, amino acids, and glycerol. When the liver returns this glucose to the blood, hyperglycemia worsens.

4. *Brain metabolism.* Tissues that are not insulin-dependent, such as most neurons in the brain, carry on metabolism as usual. However, glucose use in the brain's satiety center is insulin sensitive. Therefore, in the absence of insulin, the satiety center is unable to take up plasma glucose. It perceives the absence of intracellular glucose as starvation and allows the feeding center to increase food intake. The result is *polyphagia* (excessive eating), a classic symptom associated with untreated type 1 diabetes mellitus.

   These derangements of fat and glucose metabolism create some of the most serious symptoms of uncontrolled insulin deficiency.
5. *Osmotic diuresis and polyuria.* If the hyperglycemia of diabetes exceeds the *renal threshold* for glucose, glucose reabsorption in the proximal tubule of the kidney becomes saturated [● p. 612]. Consequently some filtered glucose is not reabsorbed, and it is excreted in the urine (*glucosuria*).

   The presence of additional solute in the collecting duct lumen causes less water to be reabsorbed and more to be excreted (see problem #28 on p. 657). This creates large volumes of urine (*polyuria*) and, if unchecked, will cause dehydration. The loss of water in the urine due to unreabsorbed solutes is known as **osmotic diuresis**.
6. *Dehydration.* Dehydration resulting from osmotic diuresis leads to decreased circulating blood volume and decreased blood pressure. Falling blood pressure triggers homeostatic mechanisms for maintaining blood pressure, including secretion of ADH, thirst that causes constant drinking (*polydipsia*), and cardiovascular compensations [● Fig. 20-17, p. 644].
7. *Metabolic acidosis.* Metabolic acidosis in diabetes has two potential sources: anaerobic metabolism and ketone production.

   If cardiovascular compensation fails, blood pressure may decrease to the point that perfusion of peripheral tissues becomes inadequate. When the tissues no longer receive enough oxygen to support aerobic metabolism, pathways shift to anaerobic glycolysis, which creates lactic acid. Lactic acid leaves the cells and enters the blood, contributing to a state of metabolic acidosis.

   The primary cause of *metabolic acidosis* in uncontrolled insulin-deficient diabetes is the production of acidic *ketone bodies* by the liver. Patients in *diabetic ketoacidosis* exhibit the symptoms of metabolic acidosis that were discussed in Chapter 20: increased ventilation, acidification of the urine, and hyperkalemia [● p. 652].

   If untreated, the combination of acidosis and hypoxia from circulatory collapse can cause coma and even death. The treatment for a patient in diabetic ketoacidosis is insulin replacement, accompanied by fluid and electrolyte therapy to replenish lost volume and ions.

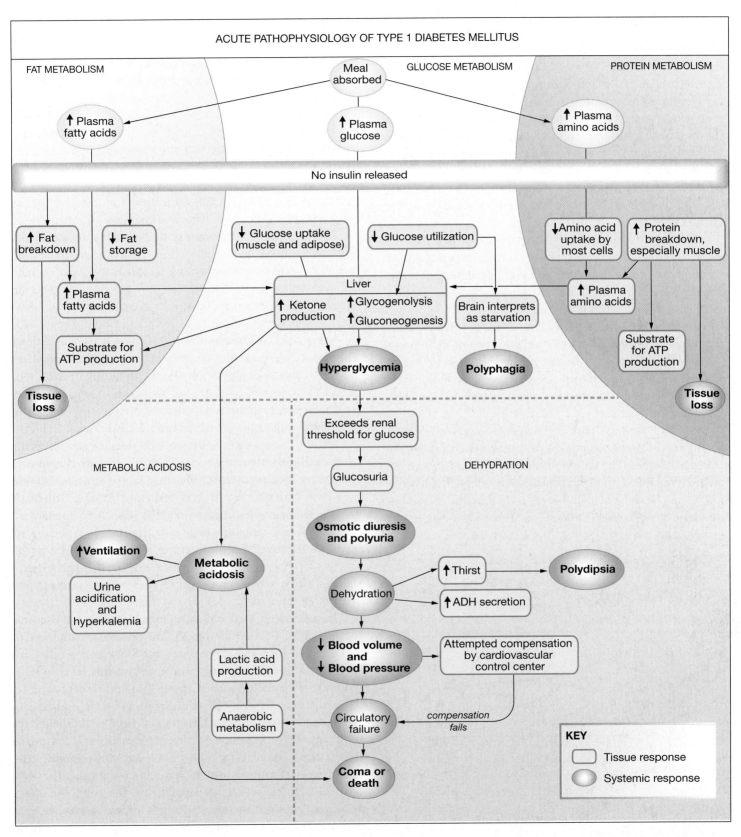

● Figure 22-15 *Acute pathophysiology of type 1 diabetes mellitus* Untreated type 1 diabetes is marked by tissue breakdown, glucosuria, polyuria, polydipsia, polyphagia, and metabolic ketoacidosis.

Type 2 Diabetics Often Have Elevated Insulin Levels

Type 2 diabetes is reaching epidemic proportions in the United States and is responsible for 10% of all health care dollars spent in 2002. According to some projections, one of every four Americans will have diabetes in the year 2028. Experts attribute the cause of the epidemic to our sedentary lifestyle, ample food, and obesity, which affects more than 50% of the population.

Type 2 diabetics comprise 90% of all diabetics. There is a significant genetic predisposition to develop the disease among certain ethnic groups. For example, about 25% of Hispanics over age 45 have diabetes. The disease is more common in people over the age of 40, but there is growing concern about the increased diagnosis of diabetes in children and adolescents. About 80% of type 2 diabetics are obese.

A common hallmark of type 2 diabetes is **insulin resistance**, demonstrated by a delayed response to an ingested glucose load. This response can be demonstrated by a procedure known as a *glucose tolerance test* (Fig. 22-16 ●). First, the subject's fasting plasma glucose concentration is determined (time 0). Then the person consumes either a special glucose drink or a typical meal. Plasma glucose is measured periodically for two or more hours.

Normal subjects have a fasting plasma glucose of 109 mg/dL or less. They show a slight increase in plasma glucose concentration soon after a meal, but the level rapidly returns to normal following insulin secretion.

In diabetic patients, however, fasting plasma glucose concentrations are usually above normal, and they rise even higher as glucose is absorbed. The cells of the body are slow to remove glucose from the blood, so plasma glucose stays elevated for two or more hours. This slow response suggests that insulin, even if present in the blood, is unable to carry out its normal function.

An abnormal glucose tolerance test simply shows that the body's response to an ingested glucose load is not normal. The test cannot distinguish between problems with insulin synthesis, problems with insulin release, or the responsiveness of target tissues to insulin. Some type 2 diabetics have both resistance to insulin action and decreased insulin secretion. Others have normal to high insulin secretion but decreased target cell responsiveness.

In addition, although type 2 diabetics are hyperglycemic, they often have elevated glucagon levels as well. This seems contradictory until you realize that the pancreatic alpha cells, like muscle and adipose cells, require insulin for glucose uptake. Thus, in diabetes, the alpha cells do not take up glucose, which prompts them to secrete glucagon. Glucagon then contributes to the hyperglycemia by promoting glycogenolysis and gluconeogenesis.

In type 2 diabetes, the acute symptoms of the disease are not nearly as severe because insulin is usually present and the cells, although resistant to insulin's action, are able to carry out a certain amount of glucose metabolism. The liver, for example, does not have to turn to ketone production, so ketosis is uncommon in type 2 diabetes.

Nevertheless, metabolism overall is not normal, and patients with this condition develop a variety of diabetes-related problems because of abnormal glucose and fat metabolism. Complications of type 2 diabetes include atherosclerosis, neurological changes, renal failure, and blindness from diabetic retinopathy. As many as 70% of type 2 diabetics die from cardiovascular disease.

Because many people with type 2 diabetes are asymptomatic when diagnosed, they can be very difficult to treat. People who come in for their yearly checkup feeling fine, only to be told that they have diabetes, can be very resistant to making dramatic lifestyle changes when they do not feel sick. Unfortunately, by the time diabetic symptoms appear, damage to tissues and organs is well under way. Patient compliance at that point can slow the progress of the disease but cannot reverse the pathological changes. The goal of treatment is to correct hyperglycemia to prevent complications.

The first therapy recommended for most type 2 diabetics and for those at high risk of developing the disease is to lose weight and exercise. For some patients, losing weight reverses their insulin resistance. Exercise decreases hyperglycemia because exercising skeletal muscle does not require insulin for glucose uptake.

Drugs used to treat type 2 diabetes may (1) stimulate beta cell secretion of insulin, (2) slow the digestion of carbohydrates in the intestine, (3) inhibit hepatic glucose output, or (4) make target tissues more responsive to insulin. For example, the *sulfonylureas* stimulate insulin secretion by closing the K_{ATP} channel of beta cells and depolarizing the cell. The drugs known as *α-glucosidase inhibitors* block intestinal enzymes that digest complex carbohydrates and therefore decrease glucose absorption. Another class of drugs that includes *metformin* reduces hepatic glucose production by an unknown mechanism. Drugs that act at the target cell include the "glitazones" such as

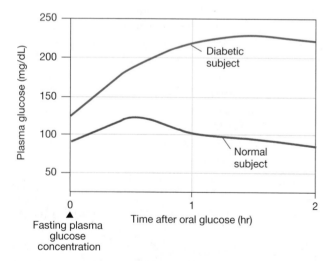

● **Figure 22-16** *Normal and abnormal glucose tolerance tests* At time 0, the subject consumes a glucose drink or meal. In diabetic subjects, glucose concentrations are at or above 200 mg/dL after two hours. A value of 140–199 mg/dL is considered to be impaired glucose tolerance. Below 140 mg/dL is normal.

Diabetes and Amylin

In 1987 researchers discovered that insulin is not the only hormone secreted by pancreatic beta cells. A second peptide named **amylin** is co-secreted with insulin. The functions of amylin are still being investigated, but at this time it appears that the hormone helps regulate glucose homeostasis following a meal. Amylin slows gastric emptying and gastric acid secretion, which delays digestion and absorption of carbohydrates. The combined actions of amylin and GIP and GLP-1 thus set up a self-regulating cycle. Glucose in the intestine causes GIP and GLP-1 release. The hormones initiate insulin and amylin secretion. Amylin then goes back to the GI tract to slow the rate at which food enters the intestine. In rodents, amylin also decreases food intake. Studies are underway with an amylin agonist called *pramlintide* to see if this drug can be used to treat diabetes or obesity.

rosiglitazone, which acts in part by activating a nuclear receptor called PPAR [◈ p. 760]. One of the newest drugs being tested is an analog of **amylin**, a peptide hormone that is co-secreted with insulin (see box Diabetes and Amylin).

CONCEPT CHECK   Answers: p. 725

15. Why must insulin be administered as a shot and not as an oral pill?

16. Patients who are admitted to the hospital with acute diabetic ketoacidosis and dehydration are given insulin and fluids that contain K^+ and other ions. The patient's acidosis is usually accompanied by hyperkalemia, so why is K^+ included in the rehydration fluids? (Hint: Dehydrated patients may have a high *concentration* of K^+, but their total body fluid volume is low. If fluid is replaced without giving K^+, what happens to the concentration of K^+ as the volume returns to normal?)

RUNNING PROBLEM

The medical staff and Nicole's family decided to talk to her about her weight and eating habits. *Anorexia nervosa* (AN) is a disorder that can have serious physiological consequences. The mortality rate for girls age 15–24 suffering from anorexia is 12 times greater than that for the general population. The most common causes of death are cardiac arrest, electrolyte imbalance, and suicide. Successful treatment of AN includes renutrition, psychotherapy, and family therapy. Current research is looking at the role of NPY and other brain peptides for the treatment of anorexia.

Question 9: Why might an NPY agonist help anorexia?

▶ REGULATION OF BODY TEMPERATURE

Type 2 diabetes is an excellent example of the link between body weight and metabolism. The development of *obesity* (excess body fat) may be linked to the efficiency with which the body converts food energy into cell and tissue components. According to one theory, people who are more efficient in transferring energy from food to fat are the ones who put on weight. In contrast, people who are less metabolically efficient can eat the same number of calories and not gain weight because more food energy is released as heat. Much of what we know about the regulation of energy balance comes from studies on body temperature regulation. Body temperature results from a balance between heat production, gain, and loss (Fig. 22-17 ●).

Body Temperature Is a Balance Between Heat Production, Gain, and Loss

Temperature regulation in humans is linked to metabolic heat production (*thermogenesis*). Humans are **homeothermic** animals who regulate their internal body temperatures within a relatively narrow range. Average body temperature is 37° C (98.6° F) with a normal range of 35.5°–37.7° C (96°–99.9° F).

These average values are subject to considerable variation, both from one individual to the next and throughout the day in a single individual. The site at which temperature is measured also makes a difference, because the temperature at the core of the body may be higher than that at the surface of the skin. Oral temperatures are about 0.5° C less than rectal temperatures.

Several factors affect body temperature in a given individual. Body temperature increases with exercise or after a meal because of *diet-induced thermogenesis*. Temperature also cycles throughout the day: the lowest (basal) body temperature occurs in the early morning and the highest in the early evening. Women of reproductive age exhibit a monthly cycle of temperature. Basal body temperatures are about 0.5° higher in the second half of the menstrual cycle, after ovulation, than before ovulation.

Heat Gain and Loss Are Balanced Temperature balance in the body, like energy balance, depends on a dynamic equilibrium between heat input and heat output (Fig. 22-18 ●). *Internal heat production* includes heat from normal metabolism and heat released during muscle contraction. *External heat input* comes from the environment through *radiation* or *conduction*.

All objects with a temperature above absolute zero give off radiant energy (radiation) with infrared or visible wavelengths. This energy can be absorbed by other objects and constitutes **radiant heat gain**. You absorb radiant energy each time you sit in the sun or in front of a fire. **Conductive heat gain** is the transfer of heat between objects that are in contact with each other, such as the skin and a heating pad or the skin and warm water.

We lose heat from the body in four ways: conduction, radiation, convection, and evaporation. **Conductive heat loss** is the

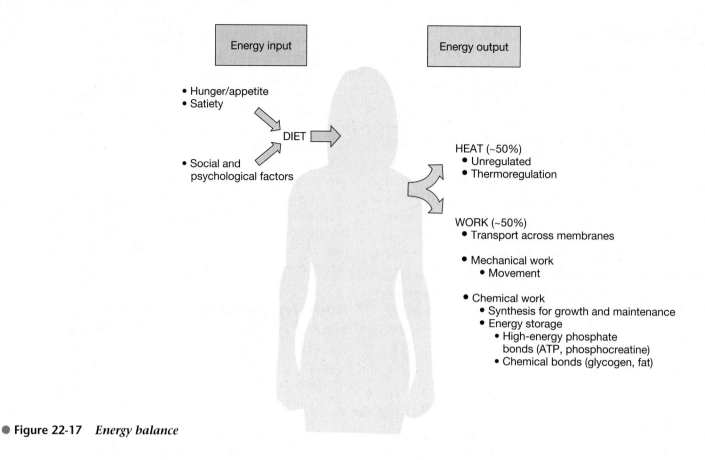

● **Figure 22-17** *Energy balance*

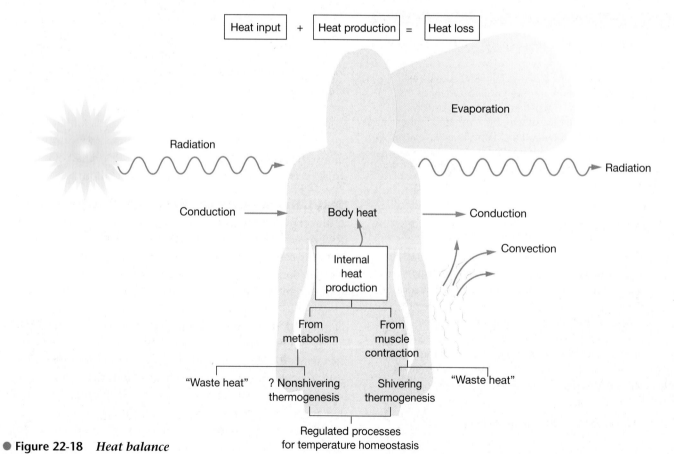

● **Figure 22-18** *Heat balance*

loss of body heat to a cooler object, such as cool water or a cold stone bench. **Radiant heat loss** from the human body is estimated to account for nearly half the heat lost from a person at rest in a normal room. *Thermography* is an imaging technique in diagnostic medicine that measures radiant heat loss. For example, some cancerous tumors can be visually identified because they have higher metabolic activity and give off more heat than surrounding tissues.

Radiant and conductive heat loss is enhanced by **convective heat loss**, heat that is carried away by warm air rising from the surface of the body. Convective air currents are created wherever a temperature difference in the air exists: hot air rises and is replaced by cooler air. Convection helps move warmed air away from the skin's surface. Clothing, which traps air and prevents convective air currents, therefore helps retain heat close to the body.

The fourth type of heat loss in the body is **evaporative heat loss**, which takes place as water evaporates at the skin's surface or in the respiratory tract. The conversion of water from a liquid to a gaseous (vapor) state requires the input of substantial amounts of energy. Thus, when water on the body evaporates, it acts as a "heat sink."

You can demonstrate the effect of evaporative cooling by wetting one arm but not the other. The wet arm will feel much cooler as it dries because heat is drawn from the skin to vaporize the water. Similarly, the half-liter of water vapor that leaves the body through lungs and skin each day takes with it a significant amount of body heat. Evaporative heat loss is affected by the humidity of surrounding air, with less evaporation taking place at higher humidities.

Conductive, convective, and evaporative heat loss from the body are all enhanced by *bulk flow* of air across the body, such as air moved across the skin by a fan or breeze. The effect of wind on body temperature regulation is described in the winter as the "wind chill factor," a combination of absolute environmental temperature and the effect of convective heat loss.

Body Temperature Is Homeostatically Regulated

The human body is usually warmer than its environment and therefore loses heat. However, normal metabolism generates enough heat to maintain body temperature when the environmental temperature stays between 27.8° and 30° C (82° and 86° F). This range is known as the **thermoneutral zone**.

In temperatures above the thermoneutral zone, the body gains heat because heat production exceeds heat loss. Below the thermoneutral zone, heat loss exceeds heat production. In both cases, the body must use homeostatic compensation to maintain internal temperature.

A human without clothing can thermoregulate at environmental air temperatures between 10° and 55° C (50° and 131° F). Because we are seldom exposed to the higher end of that temperature range, the main physiological challenge in thermal regulation comes from cold environments.

Humans have been described by some physiologists as tropical animals because we are genetically adapted for life in warm climates. But we have retained a certain amount of genetic flexibility. The physiological mechanisms by which we thermoregulate show some ability to adapt to changing conditions.

Body temperature regulation is under control of centers in the hypothalamus. Sensory **thermoreceptors** are located peripherally in the skin and in the anterior hypothalamus. These sensors monitor skin temperature and body core temperature and send that information to the hypothalamus.

The hypothalamic "thermostat" compares the input signals with the desired temperature setpoint and coordinates an appropriate physiological response to raise or lower the core temperature (Fig. 22-19 ●). Heat loss is promoted by dilation of blood vessels in the skin and by sweating. Heat is generated by shivering and possibly by nonshivering thermogenesis.

Alterations in Cutaneous Blood Flow Conserve or Release Heat Heat loss across the surface of the skin is regulated by controlling blood flow in *cutaneous* [*cutis*, skin] blood vessels near the skin's surface. These blood vessels can pick up heat by convection and transfer it to the body core, or they can lose heat to the surrounding air.

Blood flow through cutaneous blood vessels varies from close to zero when heat needs to be conserved to nearly one-third of cardiac output when heat needs to be released to the environment. Local control influences cutaneous blood flow to a certain degree, possibly through vasodilators produced by the vascular endothelium. However, neural regulation is the primary determining factor.

Most arterioles in the body are under tonic sympathetic control [● p. 501]. If core body temperature falls, the hypothalamus selectively activates sympathetic neurons innervating cutaneous blood vessels. The vessels constrict, increasing their resistance to blood flow and diverting blood to lower-resistance vessels in the interior of the body. This response keeps warmer core blood away from the cooler skin surface to reduce heat loss.

In warm temperatures, the opposite happens: cutaneous arterioles dilate to enhance heat loss at the skin's surface. In skin at the extremities, vasodilation results from decreased sympathetic input. In skin away from the extremities, vasodilation is mediated through special sympathetic neurons that secrete acetylcholine. It is unclear at this point whether the effect is direct or whether it is mediated by bradykinin or other paracrine vasodilator substances.

CONCEPT CHECK   Answers: p. 725

17. What neurotransmitter and neurotransmitter receptor mediate cutaneous vasoconstriction?

18. What observations might have prompted researchers who discovered sympathetic neurons that secrete ACh to classify them as sympathetic rather than parasympathetic? [Hint: ● p. 372]

● **Figure 22-19** *Thermoregulatory reflexes*

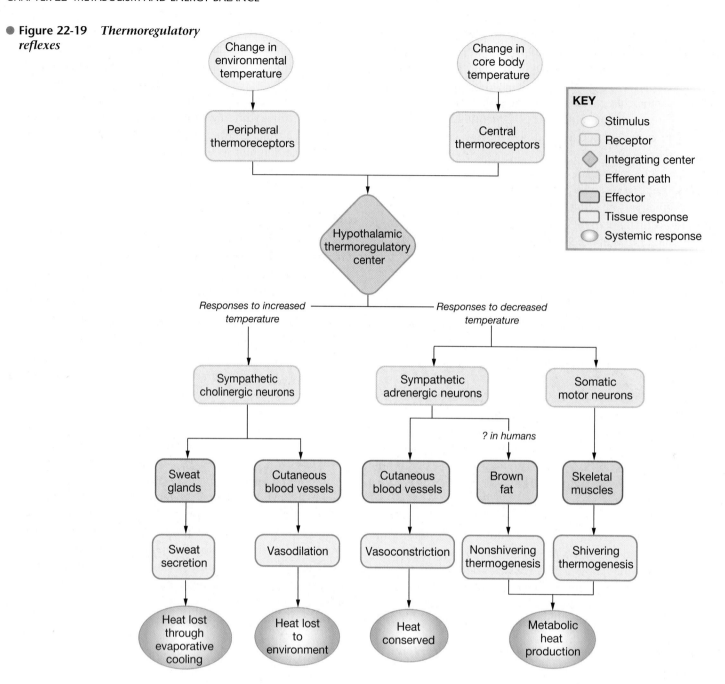

Sweat Contributes to Heat Loss Surface heat loss is enhanced by the evaporation of sweat. By some estimates the human integument has between 2 and 3 million sweat glands. The highest concentrations are found on the forehead, scalp, *axillae* (armpits), palms of hands, and soles of feet.

Sweat glands are made of transporting epithelium. Cells deep in the gland secrete an isotonic solution similar to interstitial fluid into the lumen. As the fluid travels through the duct to the skin, NaCl is reabsorbed, resulting in hypotonic sweat. A typical value for sweat production is 1.5 L/hr. With acclimatization to hot weather, some people sweat at rates of 4–6 L/hr. However, they can only maintain this high rate for short periods unless they are drinking to replace lost fluid vol-

ume. Sweat production is regulated by cholinergic sympathetic neurons.

Cooling by evaporative heat loss depends on the vaporization of water from sweat on the skin's surface. Because water evaporates rapidly in dry environments but slowly or not at all in humid ones, the body's ability to withstand high temperature is directly related to the relative humidity of the air. Evaporative cooling slows in humid environments because sweat does not evaporate from the skin's surface as rapidly as it does in dry environments. The combination of heat and humidity is reported by meteorologists as the *heat index* or *humidex*. Air moving across a sweaty skin surface enhances vaporization even with high humidity, one reason that fans are so useful in hot weather.

The Body Produces Heat Through Movement and Metabolism

Heat production by the body falls into two broad categories. The first is unregulated heat production from voluntary muscle contraction and normal metabolic pathways. The second is regulated heat production used to maintain temperature homeostasis in low ambient temperatures. Regulated heat production is further divided into two subcategories: shivering thermogenesis and nonshivering thermogenesis.

In **shivering thermogenesis**, the body uses shivering (rhythmic tremors caused by skeletal muscle contraction) to generate heat. Signals from the hypothalamic control center initiate these skeletal muscle tremors. Shivering generates five to six times as much heat as resting muscle. Shivering can be partially suppressed by voluntary control.

Nonshivering thermogenesis is metabolic heat production by pathways other than shivering. In laboratory animals such as the rat, cold exposure significantly increases heat production in brown fat. The mechanism for brown fat heat production is *mitochondrial uncoupling*. In this process, energy flowing through the electron transport system [p. 106] is released as heat rather than being trapped in ATP. Mitochondrial uncoupling in response to cold is promoted by thyroid hormones and by increased sympathetic activity.

The importance of nonshivering thermogenesis in adult humans is unsettled. Humans are born with significant amounts of brown fat, found primarily in the *interscapular* area between the shoulder blades. In newborns, nonshivering thermogenesis in this brown fat contributes significantly to raising and maintaining body temperature. However, as the child ages, white fat gradually replaces most brown fat. The brown fat that remains is very difficult to study because the experimental methodology for doing so requires that fat be extracted from the body.

The body's responses to high and low temperature are summarized in Figure 22-20 ●. In cold environments, the body tries to reduce heat loss while increasing heat production. In hot temperatures, the opposite takes place.

Notice that voluntary behavioral responses play a significant role in temperature regulation. We slow our activity level during hot weather, thereby decreasing muscle heat production. In cold weather, we put on extra clothing, tuck our hands in our armpits, or curl up in a ball to slow heat loss.

The Body's Thermostat Can Be Reset

Variations in body temperature regulation can be both physiological and pathological. Examples of physiological variation include the circadian rhythm of body temperature mentioned earlier, menstrual cycle variations in temperature, postmenopausal hot flashes, and fever. All these processes seem to arise from a common mechanism: resetting of the hypothalamic thermostat.

Hot flashes appear to be transient decreases in the thermostat's setpoint caused by the absence of estrogen. When the setpoint lowers, a room temperature that had previously been comfortable suddenly feels too hot. This discomfort triggers the usual thermoregulatory responses to heat, including sweating and cutaneous vasodilation, which leads to flushing of the skin.

● Figure 22-20
Homeostatic responses to environmental extremes

Fever for many years was thought to be a pathological response to infection, but it is now considered part of the body's normal immune response. Toxins from bacteria and other pathogens trigger the release of chemicals known as **pyrogens** [*pyr*, fire] from various immunocytes. Pyrogens are fever-producing cytokines that also have many other effects.

Experimentally, some interleukins (IL-1, IL-6), some interferons, and tumor necrosis factor have all been shown to induce fever. They do so by acting on the hypothalamic thermostat, resetting it to a higher setpoint. Normal room temperature feels too cold and the patient begins to shiver, creating additional heat. Pyrogens may also increase nonshivering thermogenesis, causing body temperature to rise.

The adaptive significance of fever is still unclear, but it seems to enhance the activity of white blood cells involved in the immune response. For this reason, some people question whether patients with a fever should be given aspirin and other fever-reducing drugs simply for the sake of comfort. High fever can be dangerous, however, as a fever of 41° C (106° F) for more than a brief period causes brain damage.

Pathological conditions in which body temperature strays outside the normal range include different states of *hyperthermia* and *hypothermia*. Heat exhaustion and heat stroke are the most common forms of hyperthermia. **Heat exhaustion** is marked by severe dehydration and body core temperatures of 37.5°–39° C (99.5°–102.2° F). Patients may have muscle cramps, nausea, and headache. They are usually pale and sweating profusely. Heat exhaustion often occurs in people who are physically active in hot, humid climates to which they are not acclimatized. It also occurs in the elderly who have lost the ability to thermoregulate.

Heat stroke is a more severe form of hyperthermia, with higher body core temperatures. The skin is usually flushed and dry. Rapid cooling is important, as enzymes and other proteins begin to denature at temperatures above 41° C (106° F). Mortality in heat stroke is nearly 50%.

Malignant hyperthermia, in which body temperature becomes abnormally elevated, is a genetically linked condition. A defective Ca^{2+} channel in skeletal muscle releases too much Ca^{2+} into the cytoplasm. As cell transporters work to move the Ca^{2+} back into mitochondria and the sarcoplasmic reticulum, the heat released from ATP hydrolysis substantially raises the body temperature. Some investigators have suggested that a mild version of this process plays a role in nonshivering thermogenesis in mammals.

Hypothermia, a condition in which body temperature falls abnormally low, is also a dangerous condition. As body core temperature falls, enzymatic reactions slow down and the person loses consciousness. When metabolism slows, oxygen consumption also decreases. Drowning victims in cold water can sometimes be revived without brain damage if they have gone into a state of hypothermia. This observation led to the development of induced hypothermia for certain surgical procedures such as heart surgery. The patient is cooled to 21°–24° C (70°–75° F) so that tissue oxygen demand can be met by artificial oxygenation of the blood as it passes through a bypass pump. After surgery is complete, the patient is gradually rewarmed.

CONCEPT CHECK Answers: p. 725

19. Why do people with water beds have to heat them to sleep comfortably?
20. Will a person who is exercising outside overheat faster if the weather is dry or if it is humid?

RUNNING PROBLEM CONCLUSION

Eating Disorders

Nicole finally agreed to undergo counseling and enter a treatment program for anorexia nervosa. She was lucky—her wrist would heal, and her medical complications could have been much worse. While 90% of AN cases are female, the number of male cases is increasing. AN reportedly affects as many as 3% of females in industrialized nations at some point in their lifetime.

After seeing Nicole and discussing her anorexia, Sara realized that she also needed to see a counselor. Sara still used diet pills, diuretics, and laxatives when she became uncomfortable with her weight, and she had started binging and purging—eating much more than normal when she was stressed, and forcing herself to vomit to keep the food from adding any weight. These are the behavioral patterns of *bulimia nervosa* (BN). BN is an equally serious condition that affects an estimated 4% of females. Its physiological effects and treatments are similar to AN.

To learn more about the pathophysiology and treatment of anorexia and bulimia, look for "eating disorders" on the University of California at Davis Health System website, *http://wellness.ucdavis.edu*. For help finding a support group, see the National Association of Anorexia Nervosa and Associated Disorders website at *www.anad.org*.

	QUESTION	FACTS	INTEGRATION AND ANALYSIS
1	If you measured Nicole's leptin level, what would you expect to find?	Leptin is a hormone secreted by adipose tissue.	Nicole has little adipose tissue and would have low leptin levels.

	QUESTION	FACTS	INTEGRATION AND ANALYSIS
2	Would you expect Nicole to have elevated or depressed levels of NPY?	NPY is inhibited by leptin. NPY stimulates feeding centers.	Leptin is low, so you might predict that NPY would be elevated and feeding stimulated. However, the feeding center is affected by other factors besides NPY (Fig. 22-1). Brain studies of anorexic patients show high levels of CRH (see Table 22-1), which opposes NPY and depresses feeding.
3	What is Nicole's K^+ disturbance called? What effect does her K^+ disturbance have on the resting membrane potential of her cells?	Nicole's K^+ is 2.5 mEq/L and normal is 3.5–5.	Low plasma K^+ is hypokalemia. Hypokalemia causes membrane potential to hyperpolarize [● p. 262].
4	Why does Dr. Ayani want to monitor Nicole's cardiac function?	Cardiac muscle is an excitable tissue whose activity depends on changes in membrane potential.	Hypokalemia can alter the membrane potential of cardiac autorhythmic and contractile cells and cause a cardiac arrhythmia that may be fatal.
5	Based on her clinical values, what is Nicole's acid-base status?	Nicole's pH is 7.52 and her plasma HCO_3^- is elevated at 40 mEq/L.	Normal pH is 7.35–7.45, so she is in alkalosis. Her elevated HCO_3^- indicates a metabolic alkalosis. The cause is probably her induced vomiting and loss of HCl from her stomach.
6	Based on what you learned in Chapters 14 and 15 about heart rate and blood pressure, speculate on why Nicole has low blood pressure with a rapid pulse.	Her blood pressure is 80/50 (low) and pulse is 90 (high).	Normally increasing heart rate would increase blood pressure. In this case, the increased pulse is a compensatory attempt to raise her low blood pressure. The low blood pressure is probably from dehydration.
7	Would you expect Nicole's renin and aldosterone levels to be normal, elevated, or depressed? How might this relate to her K^+ disturbance?	The primary stimuli for renin secretion are all associated with low blood pressure. Renin begins the RAAS pathway that stimulates aldosterone secretion.	Nicole's blood pressure is low and you would expect elevated renin and aldosterone. Aldosterone promotes renal K^+ secretion, which would lower her body load of K^+. She probably also has low dietary K^+ intake, which contributes to her hypokalemia.
8	Give some possible reasons Nicole had been feeling weak during dance rehearsals.	In fasted-state metabolism, the body breaks down skeletal muscle.	Loss of skeletal muscle proteins, hypokalemia, and possibly hypoglycemia could all be causes of Nicole's weakness.
9	Why might an NPY agonist help anorexia?	NPY stimulates the feeding center.	An NPY agonist might stimulate the feeding center and help Nicole want to eat.

(695) (697) (698) (699) (715)

CHAPTER SUMMARY

Energy balance in the body demonstrates *mass balance*: body energy equals input minus output. Metabolism is very similar: the amount of nutrient in the different nutrient pools depends on intake and output. Glucose *homeostasis* is one of the most important goals of regulated metabolism, for without adequate glucose, the brain is unable to function. Flow of material through the biochemical pathways of metabolism depends on the *molecular interactions* of substrates and enzymes.

1. The hypothalamus contains a tonically active **feeding center** and a **satiety center** that inhibits the feeding center. (p. 696)

2. Blood glucose concentrations (the **glucostatic theory**) and body fat content (the **lipostatic theory**) influence food intake. (p. 696)

3. Food intake is influenced by a variety of peptides including **leptin, neuropeptide Y,** and **ghrelin.** (p. 696; Fig. 22-1)

Energy Balance

4. In energy balance, energy intake equals energy output. (p. 697)

5. The body uses energy for transport, movement, and chemical work. (p. 698)

6. The energy content of food is measured by **direct calorimetry** and reported in **kilocalories** (kcal). (p. 698)

7. Measuring **oxygen consumption** is the most common method of estimating energy expenditure by an organism. (p. 698)

8. The **respiratory quotient** (RQ) is the ratio of CO_2 produced to O_2 consumed. RQ varies with diet. (p. 698)

9. **Basal metabolic rate**, or BMR, is an individual's lowest metabolic rate. (p. 698)

10. **Diet-induced thermogenesis** is an increase in heat production after eating. (p. 699)

11. Glycogen and fat are the two primary forms of energy storage in the human body. (p. 699)

Metabolism

12. **Metabolism** is all the chemical reactions that extract, use, or store energy. (p. 700; Figs. 22-2, 22-3)

13. **Anabolic pathways** synthesize small molecules into larger ones. **Catabolic pathways** break large molecules into smaller ones. (p. 700)

14. Metabolism is divided into the **fed** (absorptive) **state** and **fasted** (postabsorptive) **state**. The fed state is anabolic. The fasted state is catabolic. (p. 700)

15. **Glycogenesis** is synthesis of glycogen from glucose. **Glycogenolysis** is glycogen breakdown. (p. 701; Fig. 22-4)

16. **Gluconeogenesis** is synthesis of glucose from noncarbohydrate precursors, especially amino acids. (p. 701)

17. Ingested fats circulate as chylomicrons. **Lipoprotein lipase** removes triglycerides, leaving chylomicron remnants to be taken up and metabolized by the liver. (p. 703; Fig. 22-5)

18. The liver secretes lipoprotein complexes such as **low-density lipoprotein** (LDL) and **high-density lipoprotein** (HDL). (p. 703; Fig. 22-5)

19. The goal of fasted-state metabolism is to maintain adequate plasma glucose concentrations because glucose is normally the only fuel that the brain can metabolize. (p. 704; Fig. 22-6)

20. In the fasted state, the liver produces glucose from glycogen and amino acids. **Beta oxidation** of fatty acids forms acidic **ketones**. (pp. 704–705; Fig. 22-6)

Homeostatic Control of Metabolism

21. Hour-to-hour metabolic regulation depends on the ratio of insulin to glucagon. Insulin dominates the fed state and decreases plasma glucose. Glucagon dominates the fasted state and increases plasma glucose. (p. 705; Figs. 22-8, 22-9)

22. The **islets of Langerhans** secrete insulin and amylin from beta cells, glucagon from alpha cells, and somatostatin from D cells. (p. 705; Fig. 22-7)

23. Increased plasma glucose and amino acids stimulate insulin release. GI hormones and parasympathetic input amplify it. Sympathetic signals inhibit insulin secretion. (p. 706)

24. **Insulin** binds to a tyrosine kinase receptor and activates multiple **insulin-receptor substrates**. (pp. 705, 708; Fig. 22-10)

25. Major insulin target tissues are liver, adipose tissue, and skeletal muscles. Some tissues are insulin-independent. (p. 708)

26. Insulin increases glucose transport into muscle and adipose tissue, as well as glucose utilization and storage of glucose and fat. (p. 708; Figs. 22-11, 22-13)

27. **Glucagon** stimulates glycogenolysis and gluconeogenesis. (pp. 705, 711; Fig. 22-14)

28. **Diabetes mellitus** is a family of diseases marked by abnormal secretion or activity of insulin that causes hyperglycemia. In type 1 diabetes, pancreatic beta cells are destroyed by antibodies. In type 2 diabetes, target tissues fail to respond normally to insulin. (p. 711)

29. **Type 1 diabetes** is marked by catabolism of muscle and adipose tissue, glucosuria, polyuria, and metabolic ketoacidosis. **Type 2 diabetics** have less acute symptoms. In both types, complications include atherosclerosis, neurological changes, and problems with the eyes and kidneys. (p. 711; Fig. 22-15)

Regulation of Body Temperature

30. Body temperature homeostasis is controlled by the hypothalamus. (p. 717)

31. Heat loss takes place by radiation, conduction, convection, and evaporation. Heat loss is promoted by cutaneous vasodilation and sweating. (pp. 715–717; Fig. 22-18)

32. Heat is generated by **shivering** and **nonshivering thermogenesis**. (p. 719; Fig. 22-19)

QUESTIONS

1. Define metabolic, anabolic, and catabolic pathways.

2. List and briefly explain the three forms of biological work.

3. Define a kilocalorie. What is measured with these units? What is direct calorimetry?

4. What is the respiratory quotient (RQ)? What is a typical RQ value for an American diet?

5. Define basal metabolic rate (BMR). Under what conditions is it measured? Why does the average BMR differ in adult males and females? List at least four factors other than gender that may affect BMR in humans.

6. What are the three general fates of biomolecules in the body?

7. What are the main differences between metabolism in the absorptive and postabsorptive states?

8. What is a *nutrient pool?* What are the three primary nutrient pools of the body?

9. What is the primary goal of fasted-state metabolism?

10. Excess energy in the body is stored in what forms?

11. What are the three possible fates for ingested proteins? For ingested fats?

12. Name the two hormones that regulate glucose metabolism and explain what effect each hormone has on blood glucose concentrations.

13. What noncarbohydrate molecules can be made into glucose? What are the pathways called through which these molecules are converted to glucose?

14. Under what circumstances are ketone bodies formed? From what biomolecule are ketone bodies formed? How are they used by the body, and why is their formation potentially dangerous?

15. Name two stimuli that increase insulin secretion and one stimulus that inhibits insulin secretion.

16. What are the two types of diabetes mellitus? How do their causes and basic symptoms differ?

17. What factors release glucagon? What organ is the primary target of glucagon? What effect(s) do(es) glucagon produce?

18. Define the following terms and explain their physiological significance:
    (a) lipoprotein lipase   (e) apoprotein
    (b) amylin   (f) leptin
    (c) ghrelin   (g) osmotic diuresis
    (d) neuropeptide Y   (h) insulin resistance

19. What does insulin do to:
    (a) glycolysis   (d) lipogenesis
    (b) gluconeogenesis   (e) protein synthesis
    (c) glycogenesis

20. Concept map: Draw a concept map that compares the fed state and the fasted state. For each state, compare metabolism in skeletal muscles, the brain, adipose tissue, and the liver. Indicate which hormones are active in each stage and at what points they exert their influence.

21. Examine the graphs of insulin and glucagon secretion in Figure 22-9. Why have some researchers concluded it is the ratio of these two hormones that determines whether glucose will be stored or removed from storage?

22. Define, compare, and contrast or relate the terms within each set below:
    (a) glucose, glycogenolysis, glycogenesis, gluconeogenesis, glucagon, glycolysis
    (b) shivering, nonshivering, and diet-induced thermogenesis
    (c) lipoproteins, chylomicrons, cholesterol, HDL, LDL, apoprotein
    (d) direct and indirect calorimetry

    (e) conductive heat loss, radiant heat loss, convective heat loss, evaporative heat loss
    (f) absorptive and postabsorptive states

23. Describe (or map) the physiological events that lead to the following symptoms in a type 1 diabetic:
    (a) hyperglycemia   (d) ketosis
    (b) glucosuria   (e) dehydration
    (c) polyuria   (f) severe thirst

24. Both insulin and glucagon are released following ingestion of a protein meal that raises plasma amino acid levels. Why is the secretion of both hormones necessary?

25. Explain the current theory of the control of food intake. Use the following terms in your explanation: hypothalamus, feeding and satiety centers, appetite, leptin, NPY, neuropeptides.

26. Compare human thermoregulation in hot environments to thermoregulation in cold environments.

27. Scott is a bodybuilder who consumes large amounts of amino acid supplements in the belief that this will increase his muscle mass. He believes that the amino acids he consumes are stored in his body until he needs them. Is Scott correct? Explain.

28. Draw and label a graph for the following relationship: the effect of insulin secretion on plasma glucose concentration.

29. One of the debates in fluid therapy for diabetic ketoacidosis (DKA) is whether to administer bicarbonate (bicarb).

Although it is generally accepted that bicarb should be given if pH is <7.1 (life-threatening), most authorities do not give bicarb otherwise. One reason bicarbonate is not administered relates to the oxygen-binding capacity of hemoglobin. In DKA, patients have low levels of 2,3-DPG. When acidosis is corrected rapidly, 2,3-DPG is much slower to recover, and may take 24 or more hours to return to normal.

Draw and label a graph of the normal oxygen-dissociation curve [● p. 582]. BRIEFLY explain and *draw in lines on the same graph* to show:
(a) what happens to oxygen release during DKA due to acidosis and low 2,3-DPG?
(b) what happens to oxygen release if the metabolic acidosis is rapidly corrected with bicarbonate?

LEVEL FOUR *Quantitative Problems*

30. One way to estimate obesity is to calculate a person's body mass index (BMI). A body mass index greater than 25 for men and 30 for women is considered a sign of obesity. To calculate your BMI, divide your body weight in kilograms by the square of your height in meters: kg/m^2. To convert weight from pounds, use 1 kg/2.2 lb. To convert height from inches, use 1 m/39.24 in.

31. What is the calorie content of the following food, and what percent of the calories come from fat?

Spaghetti and Meat Balls: Fat—6 g Carbohydrate—30 g Protein—8 g.

Explorations

Explorations are also available at *www.physiologyplace.com*

Exploration 1: Diabetes and the Discovery of Insulin

"Diabetes is a wonderful affection, not very frequent among men, being a melting down of the flesh and limbs into urine. Its cause is of a cold and humid nature as in dropsy. The course is the common one, namely the kidneys and bladder; for the patients never stop making water, but the flow is incessant, as if from the opening of aqueducts. The nature of the disease, then, is chronic, and it takes a long period to form; but the patient is short-lived, if the constitution of the disease be completely established; for the melting is rapid, the death speedy. Moreover, life is disgusting and painful; thirst unquenchable; excessive drinking, which, however, is disproportionate to the large quantity of urine, for more urine is passed; and one cannot stop them either from drinking or making water. Or if for a time they abstain from drinking, their mouths become parched and their bodies dry; the viscera seem as if scorched up; they are affected with nausea, restlessness, and a burning thirst; and at no distant time they expire."

This quote, from Aretaeus the Cappadocian (A.D. 81–138) describes two of the classic symptoms of type 1 diabetes: polyuria and polydispsia (excessive urination and drinking). Until the 1920s, anyone who developed type 1 diabetes died not long after. Fortunately, in one of the classic experiments of physiology, a young investigator named Oscar Minkowksi made the astonishing discovery that removal of the pancreas caused the symptoms of diabetes. The story often told in classrooms is that Minkowski noticed flies around the dog's cage and tasted its urine. To learn if this story is true, read for yourself the fascinating story of Minkowski's work in an era when most physicians still believed that the pancreas was not the cause of diabetes. The article was written in 1929 by Minkowski himself and provides a glimpse into the science and politics of late 19th century Germany.

"Historical development of the theory of pancreatic diabetes" can be accessed at *http://www.diabetesliving.com/basics/histdev.htm* (First published in *Diabetes*, Vol. 38, January 1989)

Exploration 2: Beta Cell Transplants

Endocrine cell transplants? Not too long ago, such operations sounded like medical science fiction. However, recent advances in beta cell (islet) transplants are promising, potentially offering complete freedom from insulin injections for type 1 diabetics. Besides the inconvenience and discomfort of multiple daily injections, exogenous insulin does not entirely mimic the release of endogenous insulin, leading to poor control of blood glucose, which is the basis of the serious, long-term complications of diabetes. Whole pancreas transplants have been performed successfully for years, but the necessity for immunosuppressant medication offsets the benefits for many. Furthermore, the shortage of donors is a limiting factor.

Transplantation of encapsulated islet cells appears to be a viable solution. Islet cells are enveloped in a semi-permeable membrane, and injected into the liver or peritoneal cavity. The membrane acts as a barrier from host antibodies and immune cells, but allows the diffusion of substances critical to cell survival as well as insulin release. For an interesting discussion of the pros and cons of various types of encapsulation, read "Considerations for successful transplantation of encapsulated pancreatic islets," *Diabetologia* 45: 159–173, 2002 Feb. To keep abreast of latest developments in islet cell technology, search for "islet cell transplants" on Google (*www.google.com*) or check *www.diabetes.org* (American Diabetes Association) for recent headlines.

ANSWERS

Page 697

1. The feeding center causes an animal to eat and the satiety center causes an animal to cease feeding. Both centers are located in the hypothalamus.

2. If the problem is not under- or over-secretion, then it must be abnormal tissue responsiveness. A target cell might not respond to leptin because it has no leptin receptors (or they are defective), or because there is a problem with the signal transduction/second messenger pathway.

Page 699

3. The six factors are age, gender, lean muscle mass, activity, diet, and hormones.

4. One gram of fat contains more than twice the energy of 1 gram of glycogen, making fat a more efficient form of energy storage.

5. $C_6H_{12}O_6 + 6O_2 \rightarrow 6CO_2 + 6H_2O$

6. RQ is 1.

Page 702

7. Glycogenesis is glycogen synthesis; gluconeogenesis is synthesis of glucose from amino acids or glycerol.

8. GLUT transporters are passive facilitated diffusion transporters.

Page 705

9. Amino acids used for energy become pyruvate or enter the citric acid cycle.

10. Plasma cholesterol is bound to a carrier protein and can't diffuse across the membrane.

Page 710

11. The primary target tissues are liver, muscle, and adipose tissue.

12. If glucose uptake depended on insulin, the intestine, kidney tubule, and neurons could not absorb glucose in the fasted state. Neurons use only glucose for metabolism and must always be able to take it up and metabolize it.

13. During fight-or-flight skeletal muscles need glucose for contraction. Inhibiting insulin causes the liver to put glucose into the blood and prevents adipose cells from taking it up. Exercising muscle, on the other hand, does not require insulin for glucose uptake and has access to blood glucose as needed.

Page 711

14. Glycogenolysis and gluconeogenesis both produce glucose but glycogenolysis does it from glycogen breakdown and gluconeogenesis uses amino acids or glycerol to make new glucose.

Page 715

15. Insulin is a protein and will be digested if administered as a pill.

16. Although dehydrated patients may have elevated K^+ concentrations in their blood, their total *amount* of K^+ is below normal. If fluid volume is restored to normal with no K^+ added, the result will be a below-normal K^+ concentration. Thus, these patients are given fluid with K^+ even though they have an initial hyperkalemia.

Page 717

17. Norepinephrine binds to α receptors to elicit vasoconstriction.

18. Researchers probably classified the neurons as sympathetic because of where along the spinal cord the neurons originate.

Page 720

19. Water has a high specific heat and will draw heat away from the body through conductive heat transfer. If this loss exceeds the person's heat production, they will feel cold.

20. A person exercising in a humid environment loses the benefit of evaporative cooling and is likely to overheat faster.

Page 701

Figure 22-3: 1. (a) next to up arrow leading to glycogen, (b) next to arrow from fatty acids to acetyl CoA, (c) next to arrow going from acetyl CoA to fatty acids, (d) next to down arrow from glycogen to G-6-P, (e) with electron transport system. 2. No, amino acids entering the citric acid cycle cannot be used to make glucose because the step from pyruvate to acetyl CoA is not reversible.

Page 724

30. Answers will vary. A 64-inch tall woman weighing 50 kg will have a BMI of approximately 19.

31. Fat—6 g × 9 kcal/g = 54 kcal. Carbohydrate—30 g × 4 kcal/g = 120 kcal. Protein—8 g × 4 kcal/g = 32 kcal. Total = 206 kcal. 54/206 = 26% of calories from fat.

Endocrine Control of Growth and Metabolism

23

Radioisotope bone scan of the ribs, spine, and pelvis, made using technetium-99m. Prostate cancer that has metastasized (spread) to the bone can be seen as pink or white spots in the thoracic spine near the top of the figure.

RUNNING PROBLEM

Hyperparathyroidism

"Broken bones, kidney stones, abdominal groans, and psychic moans." Medical students memorize this saying when they learn about a condition called *hyperparathyroidism,* a disease in which parathyroid glands [● Fig. 7-10, p. 221] work overtime and produce excess parathyroid hormone (PTH). Dr. Adiaha Spinks suddenly recalls this saying as she examines Professor Bob Magruder, who has arrived at the office in pain from a kidney stone lodged in his ureter. When questioned about his symptoms, Dr. Bob also mentions pain in his shin bones, muscle weakness, stomach upset, and a vague feeling of depression. "I thought it was all just the stress of getting my textbook published," he says. But to Dr. Spinks, his combination of symptoms sounds suspiciously like Dr. Bob might be suffering from hyperparathyroidism.

727 730 739 745 746

BACKGROUND BASICS

Matrix: **64** ı Intermediary metabolism: **100** ı Peptide hormones: **214** ı Steroid hormones: **216** ı Control pathways for pituitary hormones: **223** ı Permissiveness: **227** ı Primary and secondary endocrine disorders: **227** ı Adrenal medulla: **376** ı

Disorders of hormone action will be more common causes of endocrinopathy than states of hormone deficiency and excess combined.

—Jean D. Wilson, Endocrinology: Survival as a discipline in the 21st century? *Annual Reviews of Physiology* 62:947–950, 2000

In 1998 Mark McGwire made the news when he hit 70 home runs, surpassing the single season home run record of Roger Maris. McGwire also created a firestorm of controversy when he admitted to taking *androstenedione*, a performance-enhancing steroid prohormone that was banned by the International Olympic Committee and other groups but not by professional baseball. Some of McGwire's critics felt that his accomplishment should be omitted from the record books. Three years later, before the controversy had settled, Barry Bonds stepped up to the plate and broke McGwire's record, without a boost from androstenedione.

What is this prohormone and why was it so controversial? You will learn more about it in this chapter as we discuss the hormones that play a role in long-term regulation of metabolism and growth. In a normal person, these hormones can be difficult to study because their effects are subtle and their interactions with one another are complex. As a result, much of what we know about endocrinology comes from studying pathological conditions in which a hormone is over- or undersecreted. But in recent years advances in molecular biology and the use of transgenic animal models have allowed scientists to learn more about hormone action at the cellular level.

▶ REVIEW OF ENDOCRINE PRINCIPLES

Before we delve into the different hormones, here is a quick review of some basic principles and patterns of endocrinology that were introduced in Chapter 7. For additional information, please see the pages referenced.

1. **The hypothalamic-pituitary control system** [❧ p. 223] Several of the hormones described in this chapter are controlled by hypothalamic and anterior pituitary (*adenohypophyseal*) trophic hormones.

2. **Feedback patterns** [❧ p. 194]. The negative feedback signal for simple endocrine pathways is the systemic response to the hormone. For example, insulin secretion shuts off when blood glucose concentrations decrease. In complex pathways using the hypothalamic-pituitary control system, the feedback signal may be the hormone itself. In endocrine pathologies, feedback signals may not function correctly.

3. **Hormone receptors** [❧ p. 216]. Hormone receptors may be on the cell surface or inside the cell.

4. **Cellular responses** [❧ p. 214ff]. In general, target cells respond by altering existing proteins or making new proteins. The old distinctions between peptide and steroid hormone action are no longer valid. Some steroid hormones exert rapid, nongenomic effects, and some peptide hormones alter transcription and translation.

5. **Magnitude of target cell response** [❧ p. 186]. The magnitude of the target cell response will be determined by the amount of active hormone available to the cell and the number and activity of the target cell receptors for the hormone. Cells may upregulate or down-regulate their receptors to alter their response. Cells that do not have hormone receptors are nonresponsive.

6. **Endocrine pathologies** [❧ p. 227]. Endocrine pathologies result from excess hormone secretion, inadequate hormone secretion, and abnormal responsiveness of the target tissue to the hormone. It now appears that abnormal responsiveness and target cell resistance to hormone action are a major cause of endocrine pathologies.

In the following sections we first examine adrenal corticosteroids and thyroid hormones, two groups of hormones that influence long-term metabolism. We then consider the endocrine control of growth.

▶ ADRENAL GLUCOCORTICOIDS

The adrenal gland, like the pituitary gland, is two embryologically distinct tissues that merged during development. This complex organ secretes multiple hormones, both neurohormones and classic hormones (Fig. 23-1 ●). The *adrenal medulla* occupies a little over a quarter of the mass and is composed of modified sympathetic ganglia that secrete catecholamines (mostly epinephrine) to mediate rapid responses in "fight-or-flight" situations [❧ p. 370 and p. 377]. The adrenal cortex forms the outer three-quarters of the gland. It secretes a variety of steroid hormones including aldosterone, which was introduced in Chapter 20 [❧ p. 636].

The Adrenal Cortex Secretes Steroid Hormones

The **adrenal cortex** secretes three major types of steroid hormones: aldosterone (sometimes called a *mineralocorticoid* because of its effect on the minerals sodium and potassium), glucocorticoids, and sex hormones. Histologically, the adrenal cortex is divided into three layers, or zones (Fig. 23-1). The outer zone, or *zona glomerulosa*, secretes only aldosterone. The inner layer (*zona reticularis*) secretes mostly *androgens*, the sex hormones that are dominant in men. The middle section (*zona fasciculata*) secretes mostly **glucocorticoids**, named for their ability to increase plasma

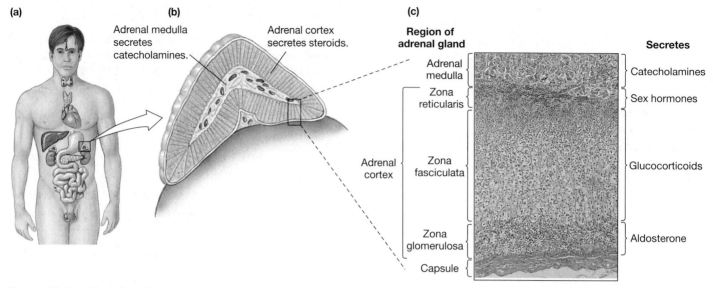

(a)

(b)

Adrenal medulla secretes catecholamines.

Adrenal cortex secretes steroids.

(c)

Region of adrenal gland

Secretes

Adrenal medulla — Catecholamines

Zona reticularis — Sex hormones

Adrenal cortex

Zona fasciculata — Glucocorticoids

Zona glomerulosa — Aldosterone

Capsule

● **Figure 23-1** *The adrenal cortex*

glucose concentrations. **Cortisol** is the main glucocorticoid secreted by the adrenal cortex (Table 23-1).

The generic synthesis pathway for steroid hormones is shown in Figure 23-2 ●. All steroid hormones begin with cholesterol that is modified by multiple enzymes to end up as aldosterone, glucocorticoids, or sex steroids (androgens, and estrogens and proges-

Table 23-1: Cortisol

Origin	Adrenal cortex
Chemical nature	Steroid
Biosynthesis	From cholesterol; made on demand, not stored
Transport in the circulation	On corticosteroid-binding globulin (made in liver)
Half-life	60–90 minutes
Factors affecting release	Circadian rhythm of tonic secretion; stress enhances
Control axis	CRH (hypothalamus) → ACTH (anterior pituitary) → cortisol (adrenal cortex)
Target cells or tissues	Most tissues
Target receptor	Intracellular
Whole body or tissue action	↑ Plasma [glucose]; ↓ immune activity; permissive for glucagon and catecholamines
Action at cellular level	↑ Gluconeogenesis and glycogenolysis; protein catabolism; blocks cytokine production by immune cells
Action at molecular level	Initiates transcription, translation, and new protein synthesis
Feedback regulation	Negative feedback to anterior pituitary and hypothalamus

terone, the sex hormones dominant in females). The pathways are the same in the adrenal cortex, gonads, and placenta, and what differs from tissue to tissue is the distribution of enzymes that catalyze the different reactions. For example, the enzyme that makes aldosterone is found in only one of the three adrenal cortex zones.

The close structural similarity of steroid hormones means that the binding sites on their receptors are also similar, and there can be *crossover* effects when a steroid binds to the receptor for a related molecule. For example, mineralocorticoid receptors (MR) for aldosterone are found in the distal nephron. MR receptors can also bind and respond to cortisol, which may be 100 times more concentrated in the blood than aldosterone. What is to keep cortisol from binding to the MR and influencing Na^+ and K^+ excretion? In this instance, the renal tubule cells with MR receptors also have an enzyme that converts cortisol to a less-active form with low specificity for the MR receptor, thus blocking crossover activation.

Crossover activity and the tightly linked synthetic pathways for the steroid hormones mean that in many endocrine pathologies, patients may exhibit symptoms related to more than one of the hormones.

CONCEPT CHECK   Answers: p. 750

1. Name the two parts of the adrenal gland and the major hormones secreted by each part.

2. For what hormones is androstenedione a prohormone? (See Fig. 23-2.) Why might this prohormone give an athlete an advantage?

Cortisol Secretion Is Controlled by ACTH

The control pathway for cortisol secretion follows the 3-part hypothalamic-pituitary-peripheral gland pattern to which you were introduced in Chapter 7. The pathway begins with hypothalamic **corticotropin-releasing hormone** (CRH) that is se-

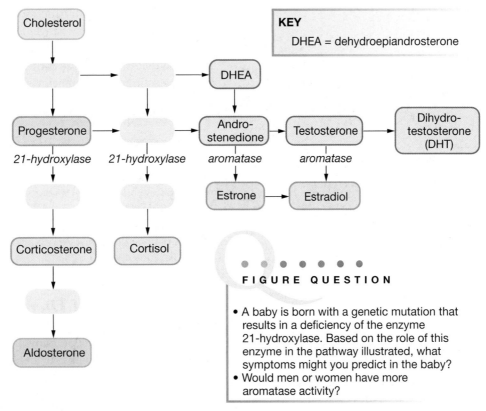

KEY

DHEA = dehydroepiandrosterone

FIGURE QUESTION

- A baby is born with a genetic mutation that results in a deficiency of the enzyme 21-hydroxylase. Based on the role of this enzyme in the pathway illustrated, what symptoms might you predict in the baby?
- Would men or women have more aromatase activity?

● **Figure 23-2** *Synthesis pathways of steroid hormones* Steroid hormones are all synthesized from cholesterol. The blank boxes represent intermediate compounds whose names have been omitted for simplicity. Each step is catalyzed by an enzyme but only two are shown in this figure.

creted into the *hypothalamic-hypophyseal portal system* and transported to the anterior pituitary (Fig. 23-3 ●). CRH stimulates release of **adrenocorticotropic hormone** (**ACTH** or *corticotropin*) from the anterior pituitary. ACTH in turn acts on the adrenal cortex to promote synthesis and release of cortisol. Cortisol acts as a negative feedback signal, inhibiting ACTH and CRH secretion.

Cortisol is continuously secreted with a strong diurnal rhythm (Fig. 23-4 ●). Secretion normally peaks in the morning and diminishes during the night. Cortisol secretion also increases with stress.

Cortisol follows the typical steroid hormone pattern for synthesis, release, transport, and action [● Fig. 7-7, p. 218]. Upon synthesis, it diffuses out of adrenal cells into the plasma, where most of it is transported by a carrier protein, *corticosteroid-binding globulin* (CBG or *transcortin*). The unbound hormone is free to diffuse into target cells.

All nucleated cells of the body have glucocorticoid receptors in their cytoplasm. The hormone-receptor complex enters the nucleus, binds to DNA with the aid of a *hormone-response element*, and alters gene expression, transcription, and translation. In general, the response to glucocorticoid hormones takes 60–90 minutes before it becomes evident in a tissue. However, cortisol's negative feedback effect on ACTH secretion is rapid and occurs within minutes.

● **Figure 23-3** *The control pathway for cortisol*

● **Figure 23-4** *Circadian rhythm of cortisol secretion*

Cortisol Is Essential for Life

The adrenal glucocorticoids are sometimes called the stress hormones of the body because of their role in the mediation of long-term stress. Adrenal catecholamines, particularly epinephrine, are responsible for more rapid metabolic responses in "fight-or-flight" situations.

Cortisol is essential for life. Animals whose adrenal glands have been removed will die if exposed to any significant environmental stress. The most important metabolic effect of cortisol is its protective effect against *hypoglycemia* [*hypo-*, low + *gly-*, glucose + *-emia*, blood]. In the last chapter you learned that pancreatic *glucagon* is secreted in response to decreased blood glucose. In the absence of cortisol, however, glucagon is unable to respond adequately to a hypoglycemic challenge. Cortisol must be present to ensure that glucagon is effective. Because cortisol is required for full activity of glucagon and catecholamines, it is said to have a *permissive effect* on those hormones [● p. 227].

The metabolic effects of cortisol are all aimed at preventing hypoglycemia. Overall, cortisol is considered to be catabolic (Fig. 23-3).

1. **Cortisol promotes gluconeogenesis** in the liver. The glucose produced is both released into the blood and stored as glycogen.
2. **Cortisol causes the breakdown of skeletal muscle proteins** to provide substrate for gluconeogenesis.
3. **Cortisol enhances lipolysis** (fat breakdown) so that fatty acids are available to peripheral tissues for energy use. The glycerol from fatty acids can be used for gluconeogenesis.

   Cortisol receptors are found in every tissue of the body, but for many targets we do not fully understand the physiological actions of cortisol. However, we can speculate on what they are based on tissue responses to high levels (*pharmacological doses*) of cortisol that are administered for therapeutic reasons or that occur with hypersecretion pathologies.

4. **Cortisol suppresses the immune system** through multiple pathways. This effect is discussed in more detail below.
5. **Cortisol causes negative calcium balance.** Cortisol decreases intestinal Ca^{2+} absorption and increases renal Ca^{2+} excretion, resulting in net Ca^{2+} loss from the body. In addition, cortisol is catabolic on bone tissue, with net breakdown of the calcified bone matrix. People who take therapeutic cortisol for extended periods therefore have a higher than normal incidence of broken bones.

6. **Cortisol influences brain function.** States of cortisol excess or deficiency cause mood changes and alteration in memory and learning. Some of these effects may be mediated by hormones in the cortisol release pathway, such as CRH. This will be discussed in more detail below.

CONCEPT CHECK   Answers: p. 750

3. What do ACTH, CRH, and CBG stand for? If there is an alternate name for the term, what is it?
4. You are mountain-biking in Canada and come across a bear, which chases you up a tree. Is your stress response mediated by cortisol? Explain.
5. The illegal use of anabolic steroids by bodybuilders and athletes periodically receives much attention. Do these illegal steroids include cortisol? Explain.

Cortisol Is a Useful Therapeutic Drug

Cortisol suppresses the immune system by preventing cytokine release and antibody production by white blood cells. It also inhibits the inflammatory response by decreasing leukocyte mobility and migration. These *immunosuppressant effects* of cortisol make it a useful drug for treating a variety of conditions including bee stings, poison ivy, and pollen allergies. Cortisol also helps prevent rejection of transplanted organs.

However, glucocorticoids also have potentially serious side effects because of their metabolic actions. Once *nonsteroidal anti-inflammatory drugs* (NSAIDs) such as ibuprofen were developed, the use of glucocorticoids for treating minor inflammatory problems was discontinued.

Exogenous administration of glucocorticoids will have a negative feedback effect on the anterior pituitary and shut down ACTH production (● Fig. 7-19, p. 228). Without ACTH stimulation, the adrenal cortisol-producing cells atrophy. For this reason it is essential that patients taking steroids taper their dose down gradually, giving the pituitary and adrenal gland a chance to recover, rather than stopping the drug abruptly.

RUNNING PROBLEM

High blood Ca^{2+} concentrations lead to high Ca^{2+} concentrations in the kidney filtrate. Calcium-based kidney stones occur when calcium phosphate or calcium oxalate crystals form and aggregate with organic material in the lumen of the kidney tubule. Once Dr. Bob's kidney stone passes into the urine, Dr. Spinks sends it for a chemical analysis.

Question 3: Only the free Ca^{2+} in the blood filters into Bowman's capsule at the nephron. A significant portion of plasma Ca^{2+} cannot be filtered. Use what you have learned about filtration at the glomerulus to speculate on why some Ca^{2+} cannot filter [● p. 607].

Cortisol Pathologies Result from Too Much or Too Little Hormone

Endocrine pathologies generally fall into three groups: too much hormone, too little hormone, or abnormal tissue response to hormone. Abnormal tissue responsiveness is an uncommon cause of adrenal steroid pathology.

Hypercortisolism Excess cortisol in the body is called **hypercortisolism**. It can arise from hormone-secreting tumors or from exogenous administration of the hormone. Cortisol therapy with high doses for more than a week has the potential to cause hypercortisolism, also known as *Cushing's syndrome*, after Dr. Harvey Cushing who first described the condition in 1932.

Most signs of hypercortisolism can be predicted from the normal actions of the hormone. Excess gluconeogenesis causes hyperglycemia, which mimics diabetes. Muscle protein breakdown and lipolysis cause tissue wasting. Paradoxically, excess cortisol deposits extra fat in the trunk and face. This may be due partly to increased appetite and food intake. The classic appearance of patients with hypercortisolism is thin arms and legs, obesity in the trunk, and a "moon face," with plump cheeks (Fig. 23-5a ●). CNS effects of too much cortisol include mood elevation initially, followed by depression, as well as difficulty with learning and memory.

Cushing's syndrome has three common causes:

1. **An adrenal tumor autonomously secretes cortisol.** These tumors are not under control of pituitary ACTH. This condition is *primary hypersecretion* [❧ p. 227].
2. **A pituitary tumor autonomously secretes ACTH.** This prompts the adrenal gland to oversecrete cortisol (*secondary hypercortisolism*). The tumor is not subject to the normal negative feedback pathway shown in Figure 23-3. This condition is also called *Cushing's disease* because it was the actual disease described by Dr. Cushing. (Hypercortisolism from any cause is called Cushing's *syndrome*.)
3. **Iatrogenic (physician-caused) hypercortisolism** occurs secondary to cortisol therapy for some other condition.

Hypocortisolism Hyposecretion pathologies are far less common than Cushing's syndrome. **Addison's disease** is hyposecretion of all adrenal steroid hormones, usually following autoimmune destruction of the adrenal cortex. Hereditary defects in enzymes for adrenal steroid production cause several related syndromes (see figure question in Fig. 23-2). These inherited disorders are often marked by excess androgen secretion because substrate that cannot be made into cortisol or aldosterone is converted to androgens. In newborn girls, excess androgens cause masculinization of the external genitalia, or *adrenogenital syndrome*.

CRH and ACTH Have Additional Physiological Functions

In recent years research interest has shifted away from glucocorticoids to the hormones of the cortisol control pathway, CRH and ACTH. Both peptides are now known to be part of larger families of related molecules, with multiple receptor types found in numerous tissues. Experiments with knockout mice lacking a particular receptor have revealed some of the physiological functions of the CRH and ACTH-related peptides.

One interesting finding of this research is that cytokines secreted by the immune system can stimulate the hormones of the hypothalamic-pituitary-adrenal (HPA) pathway, and that immune cells have receptors for ACTH and CRH. The association between stress and immune function appears to be mediated through the HPA pathway, and it provides one explanation for mind-body interactions in which mental state influences physiological function.

(a)   (b)

● **Figure 23-5** *Cushing's syndrome* (a) A "moon face" with red cheeks is typical in Cushing's. (b) Fat deposits in the trunk. The dark striations result from protein breakdown in the skin.

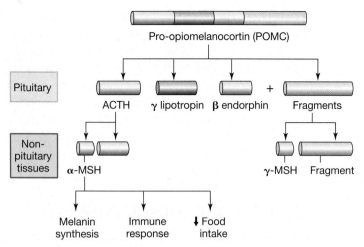

● **Figure 23-6** *POMC processes into a variety of active peptides.*

CRH Family   The CRH family includes CRH and a related brain neuropeptide that has been named *urocortin*. In addition to its involvement in inflammation and the immune response, CRH is known to decrease food intake (see Table 22-1, ❧ p. 696) and has been associated with signals that mark the onset of labor in pregnant women (discussed in Chapter 26). There is also evidence linking CRH to anxiety, depression, and other mood disorders.

ACTH and Melanocortins   CRH acting on endocrine cells of the anterior pituitary gland causes secretion of ACTH. ACTH is synthesized from a large glycoprotein called **pro-opiomelanocortin**, or **POMC** (Fig. 23-6 ●). POMC produces a variety of biologically active peptides in addition to ACTH. In the pituitary, POMC products include **β-endorphin**, an endogenous opiate that binds to receptors that block pain perception [❧ p. 334].

POMC is also processed in tissues outside the pituitary, where its products may be very different. Additional processing of POMC can create *melanocyte-stimulating hormone (MSH)*. **α-MSH** is produced in brain neurons, where it inhibits food intake, and in the skin, where it acts on melanocytes that influence skin color in humans and coat color in rodents.

The MSH hormones plus ACTH have been given the family name **melanocortins**. Five different melanocortin receptors (MC-R) have been identified. The MC2-R responds only to ACTH and is the adrenal cortex receptor. The MC1-R is found in skin melanocytes and responds equally to α-MSH and ACTH. When ACTH is elevated in Addison's disease, its action on this receptor leads to increased melanin production and the apparent "tan" or skin darkening that is characteristic of this pathology.

CONCEPT CHECK   Answers: p. 750

8. What might be the advantage for the body to co-secrete ACTH, and β-endorphin?

Melanocortins and the *Agouti* Mouse

Much of what we are learning about the functions of the hypothalamic-pituitary-adrenal (HPA) pathway comes from research done on mutant mice. One such animal is the *agouti* mouse, which results from a spontaneous mutation first described in 1905. Mice with one mutated gene overproduce agouti protein and have a characteristic yellow coat because agouti protein is an antagonist to the MC1-R that controls melanin synthesis in its hair. More of interest to physiologists, however, is the fact that these mice overeat and develop adult-onset obesity, hyperglycemia, and insulin resistance—in other words, the *agouti* mouse is a model for type 2 diabetes caused by obesity. Using molecular biology, researchers have learned that agouti protein blocks the MC4-R in the brain, the receptor that normally binds α-MSH and ACTH, and depresses feeding behavior. When the receptor is not functional, the animal overeats and develops obesity. This link between melanocortin receptors, eating behavior, and diabetes has opened a new area of research to scientists trying to find treatments to prevent type 2 diabetes. To learn more about the agouti mouse models, see the Jackson Laboratory website at *http://www.informatics.jax.org/searches/mlc.cgi?3*.

▶ THYROID HORMONES

The thyroid gland is a butterfly-shaped gland that lies across the trachea at the base of the throat, just below the larynx (Fig. 23-7a ●). It is one of the larger endocrine glands, weighing between 15 and 20 g. The thyroid gland has two distinct endocrine cell types: the *C* ("clear") *cells* that secrete a calcium-regulating hormone called *calcitonin*, and *follicle cells* that secrete thyroid hormones. Calcitonin is discussed later with calcium homeostasis.

Thyroid Hormones Contain Iodine

Thyroid hormones, like the glucocorticoids, have long-term effects on metabolism. Unlike glucocorticoids, however, they are not essential for life. They are essential for normal growth and development in children, and infants born with thyroid deficiency will be developmentally delayed unless treated promptly. Because of the importance of thyroid hormones in children, the United States and Canada test all babies at birth for thyroid deficiency.

Thyroid hormones are amines derived from the amino acid tyrosine and they are unusual because they contain the element iodine (Fig. 23-8 ●). Currently, thyroid hormones are the only known use for iodine in the body, although a few other tissues also concentrate the mineral.

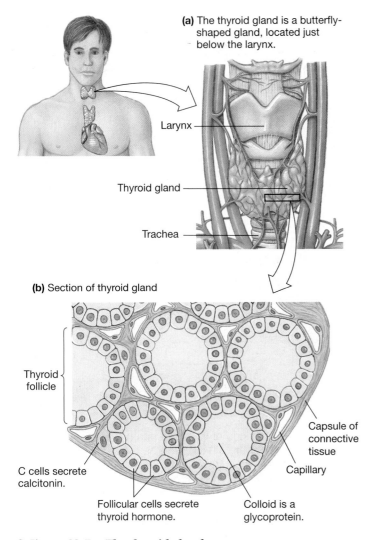

(a) The thyroid gland is a butterfly-shaped gland, located just below the larynx.

Larynx

Thyroid gland

Trachea

(b) Section of thyroid gland

Thyroid follicle

C cells secrete calcitonin.

Follicular cells secrete thyroid hormone.

Colloid is a glycoprotein.

Capsule of connective tissue

Capillary

● **Figure 23-7** *The thyroid gland*

Synthesis of thyroid hormones takes place in the thyroid **follicles** (also called *acini*), spherical structures whose walls are a single layer of epithelial cells (Fig. 23-7b). The hollow center of each follicle is filled with a sticky protein mixture called **colloid**. The colloid holds a 2–3 month supply of thyroid hormones at any one time.

Tyrosine → **Thyroxine (T$_4$)**

HO—C—C—C + I → HO—O—C—C—C

(2 tyrosine + 4 I)

Tyrosine → **Triiodothyronine (T$_3$)**

HO—C—C—C + I → HO—O—C—C—C

(2 tyrosine + 3 I)

● **Figure 23-8** *Thyroid hormones are made from tyrosine and iodine.*

Figure 23-9 ● is an overview of thyroid gland function.

① The follicular cells manufacture enzymes for thyroid hormone synthesis and a glycoprotein called **thyroglobulin**. These proteins are secreted into the center of the follicle.

② Follicular cells absorb dietary *iodide*, I$^-$, using a sodium-iodide cotransporter. I$^-$ movement into the colloid is apparently mediated by an anion transporter.

③ Enzymes combine I$^-$ with tyrosine on the thyroglobulin molecule to make the thyroid hormones T$_3$ and T$_4$. At this point, the hormones are still attached to thyroglobulin.

④ Thyroglobulin in the colloid moves back into the follicular cells by endocytosis.

⑤ Intracellular enzymes free T$_3$ and T$_4$ from the protein.

⑥ Both hormones are lipophilic, so they diffuse out of the follicular cells and into the plasma.

Figure 23-10 ● shows the details of thyroid hormone synthesis. The thyroid follicles actively transport iodide across the epithelium. As I$^-$ enters the colloid, the enzyme *thyroid peroxidase* removes an electron from the ion. Other enzymes then add iodine to tyrosine. The addition of one iodine to tyrosine creates **monoiodotyrosine** (MIT); addition of a second iodine creates **diiodotyrosine** (DIT). MIT and DIT then undergo *coupling reactions*. MIT + DIT creates **triiodothyronine**, or T$_3$. Two DIT couple to form **tetraiodothyronine**, also known as **thyroxine**, or T$_4$. Note the change from *tyrosine* to *thyronine* within the names.

Because T$_3$ and T$_4$ are lipophilic molecules, they have limited solubility in plasma. As a result, thyroid hormones bind to plasma proteins such as **thyroid-binding globulin** (TBG). Most plasma hormone is in the form of T$_4$. For years, it was thought that T$_4$ (thyroxine) was the active hormone. However, we now know that T$_3$ is three to five times more active biologically, and it is the active hormone in target cells.

About 85% of active T$_3$ is made in target cells by enzymatic removal of an iodine from T$_4$. Tissue activation of the hormone adds another layer of control because regulation of the tissue enzyme **deiodinase** allows individual target tissues to control their exposure to active thyroid hormone.

Thyroid receptors, with multiple isoforms, are in the nucleus of target cells. Hormone binding initiates transcription, translation, and synthesis of new proteins.

Thyroid Hormones Affect Quality of Life

The thyroid hormones are not essential for life, but they do affect the quality of life. Patients with thyroid excess or deficiency may exhibit decreased tolerance to heat or cold and mood disturbances in addition to other symptoms.

The overall effect of thyroid hormones in adults is to provide substrates for oxidative metabolism (Table 23-2). Thyroid hormones are *thermogenic* and increase oxygen consumption in most tissues. The exact mechanism is unclear, but is at least partly

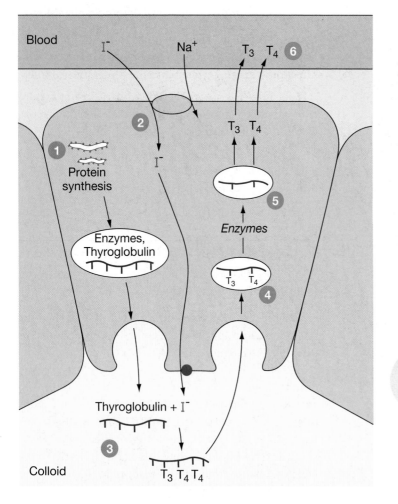

1 Follicular cell synthesizes enzymes and thyroglobulin for colloid.

2 I^- is co-transported into the cell with Na^+ and transported into colloid.

3 Enzymes add iodine to thyroglobulin to make T_3 and T_4.

4 Thyroglobulin is taken back into the cell.

5 Intracellular enzymes separate T_3 and T_4 from the protein.

6 Free T_3 and T_4 enter the circulation.

Q • • • • • • •

FIGURE QUESTIONS

• What kind of transport brings I^- into follicular cells?
• How does thyroglobulin get into the colloid?
• How does the cell take thyroglobulin back in?
• How do T_3 and T_4 leave the cell?

● **Figure 23-9** *Thyroid hormone synthesis*

KEY

MIT = monoiodotyrosine
DIT = diiodotyrosine
T_3 = triiodothyronine
T_4 = thyroxine

● **Figure 23-10** *Steps of thyroid hormone synthesis*

related to changes in ion transport across the cell and mitochondrial membranes. Also, thyroid hormones interact with other hormones to modulate protein, carbohydrate, and fat metabolism.

In children, thyroid hormones are necessary for full expression of growth hormone. They are essential for normal growth and development, especially in the nervous system. In the first few years after birth, myelin and synapse formation require T_3 and T_4. Cytological studies suggest that thyroid hormones regulate microtubule assembly, which is an essential part of neuronal growth.

The actions of thyroid hormones are most memorable in people who suffer from either hypersecretion or hyposecretion. Physiological effects that are subtle in normal people often become exaggerated in patients with endocrine pathologies.

Hyperthyroidism Hypersecretion of thyroid hormones causes changes in metabolism, the nervous system, and the heart.

1. Hyperthyroidism increases oxygen consumption and metabolic heat production. Because of the internal heat generated,

Table 23-2: Thyroid Hormones

Cell of origin	Thyroid follicle cells
Chemical nature	Iodinated amine
Biosynthesis	From iodine and tyrosine; formed and stored on parent protein thyroglobulin in follicle colloid
Transport in the circulation	Bound to thyroxine-binding globulin and albumins
Half-life	6–7 days for thyroxine (T_4); about 1 day for triiodothyronine (T_3)
Factors affecting release	Tonic release
Control axis	TRH (hypothalamus) $\rightarrow$ TSH (anterior pituitary) $\rightarrow T_3 + T_4$ (thyroid) $\rightarrow T_4$ deiodinates in tissues to form more T_3
Target cells or tissues	Most cells of the body
Target receptor	Nuclear receptor
Whole body or tissue action	$\uparrow$ Oxygen consumption (thermogenesis); protein catabolism in adults but anabolism in children; normal development of nervous system
Action at cellular level	Increases activity of metabolic enzymes and Na^+-K^+-ATPase
Action at molecular level	Production of new enzymes
Feedback regulation	T_3 has negative feedback effect on anterior pituitary and hypothalamus

● **Figure 23-11** *Hypothyroid myxedema* Mucopoly-saccharide deposits in myxedema cause bags under the eyes.

2. Hyposecretion of thyroid hormone decreases protein synthesis. In adults, this causes brittle nails, thinning hair, and dry, thin skin. Hypothyroidism also causes accumulation of *mucopolysaccharides* under the skin. These molecules attract water and cause the puffy appearance of *myxedema* (Fig. 23-11 ●). Hypothyroid children have slow bone and tissue growth and are shorter than normal for their age.

3. Nervous system changes in adults include slowed reflexes, slow speech and thought processes, and feelings of fatigue. Deficient thyroid hormone secretion in infancy causes **cretinism**, a condition marked by decreased mental capacity.

4. The primary cardiovascular change in hypothyroidism is slow heart rate, or *bradycardia*.

TSH Controls the Thyroid Gland

Problems with thyroid hormone secretion can arise in the thyroid gland itself or along the control pathway. The normal control path and feedback for thyroid hormones are illustrated in Figure 23-12 ●. **Thyrotropin-releasing hormone**, or **TRH**, from the hypothalamus controls secretion of the anterior pituitary hormone **thyrotropin**, also known as **thyroid-stimulating hormone**, or **TSH**. TSH in turns acts on the thyroid gland to promote thyroid hormone synthesis. The thyroid hormones normally act as a negative feedback signal to prevent oversecretion.

The trophic action of TSH on the thyroid gland causes enlargement, or *hypertrophy*, of the follicular cells. In pathological conditions with elevated TSH levels, the thyroid gland can enlarge to weights of hundreds of grams, forming a huge mass called a **goiter**. A large goiter can almost encircle the neck and will extend downward behind the sternum (Fig. 23-13 ●).

Goiters are the result of excess TSH stimulation of the thyroid gland. Simply knowing that someone has a goiter does not tell you what their pathology is, however. Let's see how both hypothyroidism and hyperthyroidism can be associated with goiter.

Our first example is *primary hypothyroidism* [● p. 227] caused by a lack of iodine in the diet. Without iodine, the thyroid gland

---

these patients have warm, sweaty skin and may complain of being intolerant of heat.

2. Excess thyroid hormone increases protein catabolism and may cause muscle weakness. Patients often report weight loss.

3. The effects of excess thyroid hormone on the nervous system include hyperexcitable reflexes and psychological disturbances ranging from irritability and insomnia to psychosis. The mechanism for psychological disturbances is unclear, but morphological changes in the hippocampus and effects on β-adrenergic receptors have been suggested.

4. Thyroid hormones are known to influence β-adrenergic receptors in the heart, and these effects are exaggerated with hypersecretion. A common sign of hyperthyroidism is rapid heartbeat and increased force of contraction due to up-regulation of β_1 receptors on the myocardium [● p. 467].

Hypothyroidism Hyposecretion of thyroid hormones affects the same systems altered by hyperthyroidism.

1. Decreased thyroid hormone secretion slows the metabolic rate and oxygen consumption. Patients become cold-intolerant because they are generating less internal heat.

● **Figure 23-12** *Thyroid hormone pathway*

● **Figure 23-13** *Goiter due to excessive TSH stimulation*

bug-eyed appearance caused by immune-mediated enlargement of muscles and tissue in the eye socket.

Therapy for thyroid disorders depends on the cause of the problem. Hypothyroidism is treated with oral thyroxine (T_4). Hyperthyroidism can be treated by surgical removal of all or part of the gland, by destruction of thyroid cells with radioactive iodine (^{131}I), or by drugs that block hormone synthesis (thiourea drugs) or peripheral conversion of T_4 to T_3 (propylthiouracil).

CONCEPT CHECK   Answers: p. 750

9. A patient who had her thyroid gland removed due to cancer was given pills containing only T_4. Why was this less-active form of the hormone an effective treatment for her hypothyroidism?

10. Why would excessive production of thyroid hormone, which uncouples mitochondrial ATP production and proton transport, cause a person to become intolerant of heat?

▶ **GROWTH HORMONE**

Growth in human beings is a continuous process that begins before birth. Growth rates in children are not steady. The first two years of life and the adolescent years are marked by spurts of rapid growth and development. Normal growth is a complex process that depends on a number of factors, including:

1. **Growth hormone and other hormones.** Without adequate amounts of growth hormone, children simply will not grow. Thyroid hormones, insulin, and the sex hormones at puberty also play both direct and permissive roles. A deficiency in any one of these hormones will lead to abnormal growth and development.

cannot make thyroid hormones (Fig. 23-14a ●). Low levels of T_3 and T_4 in the blood remove negative feedback on the hypothalamus and anterior pituitary. TSH secretion rises dramatically, and TSH stimulation enlarges the thyroid gland (goiter). Despite hypertrophy, the gland cannot obtain iodine to make hormone, and the patient remains hypothyroid. These patients will exhibit the signs of hypothyroidism previously described.

An enlarged thyroid gland may also be indicative of hypersecretion. In *Graves' disease*, the body produces antibodies called **thyroid-stimulating immunoglobulins**, or **TSI** (Fig. 23-14b). These antibodies combine with TSH receptors on the thyroid gland and activate them, enlarging the gland (goiter) and overstimulating thyroid hormone production.

In Graves' disease, negative feedback shuts down the body's TRH and TSH secretion but does nothing to block the TSH-like activity of TSI on the thyroid gland. The result is goiter, hypersecretion of thyroid hormone, and symptoms of hormone excess. Graves' disease is often accompanied by **exophthalmus** (Fig. 23-15 ●), a

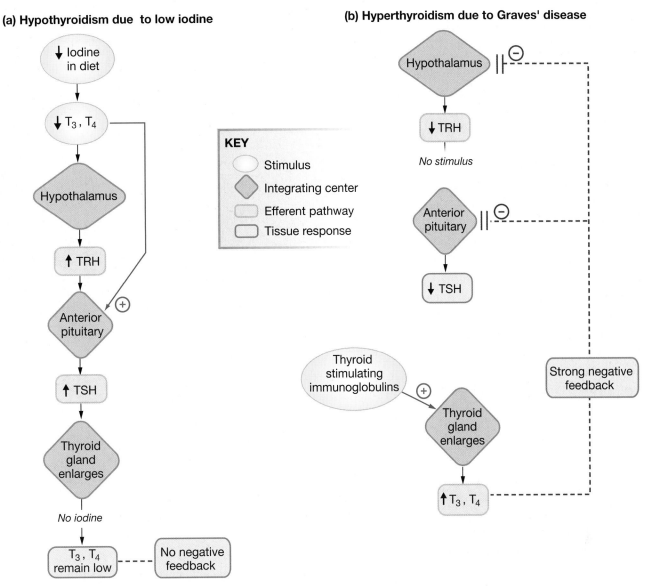

(a) Hypothyroidism due to low iodine

↓ Iodine in diet

↓ T_3, T_4

Hypothalamus

↑ TRH

Anterior pituitary (+)

↑ TSH

Thyroid gland enlarges

No iodine

T_3, T_4 remain low — — — No negative feedback

KEY
- ◯ Stimulus
- ◆ Integrating center
- ▢ Efferent pathway
- ▢ Tissue response

(b) Hyperthyroidism due to Graves' disease

Hypothalamus ⊖

↓ TRH

No stimulus

Anterior pituitary ⊖

↓ TSH

Thyroid stimulating immunoglobulins (+) Thyroid gland enlarges

Strong negative feedback

↑ T_3, T_4

● **Figure 23-14** *Goiter can occur with hyper- and hypothyroidism.*

● **Figure 23-15** *Exophthalmus is a sign of hyperthyroidism.*

2. **An adequate diet** that includes protein, sufficient energy (caloric intake), vitamins, and minerals. Many amino acids can be manufactured in the body from other precursors, but the essential amino acids [● p. 37] needed for protein synthesis must come from dietary sources. Among the minerals, calcium in particular is needed for proper bone formation.

3. **Absence of stress.** Cortisol from the adrenal cortex is released in times of stress and has significant catabolic, antigrowth effects. Children who are subjected to stressful environments may exhibit a condition known as *failure to thrive* that is marked by abnormally slow growth.

4. **Genetics.** Each human's potential adult size is genetically determined at conception.

Growth Hormone Is Anabolic

Growth hormone (GH or **somatotropin**) is released throughout life, although its biggest role is in children. Peak GH secretion

New Growth Charts

When you were growing up, did your parents mark your growth each year on your bedroom door frame or on a special wall chart? Monitoring growth is an important part of health care for children and adolescents, particularly now that we are seeing a growing problem with obesity in children in the United States. In 2000 the U.S. Centers for Disease Control and Prevention (CDC) issued new growth charts for the first time since 1977. The old charts were based on data gathered from 1929 to 1979 that had surveyed mostly bottle-fed, middle-class white children. We now know that breast-fed infants grow differently than bottle-fed babies: more rapidly in the first two months, then more slowly for the remainder of the first year. There are socioeconomic differences in growth, with babies in lower socioeconomic groups growing more slowly. The new charts take these differences into account. They also include body mass index (BMI) information up to age 20 (see question 30 on p. 724). To see the new charts and learn more about monitoring growth in infants and children, see the CDC website at *www.cdc.gov/growthcharts*.

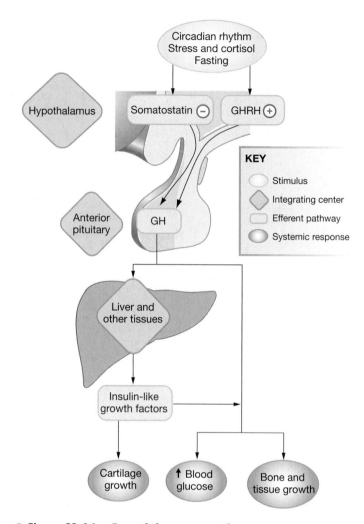

● **Figure 23-16** *Growth hormone pathway*

occurs during the teenage years. The stimuli for growth hormone release are complex and not well understood, but they include circulating nutrients, stress, and other hormones interacting with a daily rhythm of secretion (Fig. 23-16●).

The stimuli for GH secretion are integrated in the hypothalamus, which secretes two neuropeptides into the hypothalamic-hypophyseal portal system: **growth hormone–releasing hormone** (GHRH) and *growth hormone–inhibiting hormone*, better known as **somatostatin**. On a daily basis, pulses of GHRH from the hypothalamus stimulate GH release. In adults, the largest pulse of GH release occurs in the first two hours of sleep. It is speculated that GHRH has sleep-inducing properties, but the role of GH itself in sleep cycles is unclear.

GH is secreted by cells in the anterior pituitary. It is a typical peptide hormone in most respects except that nearly half the GH in blood is bound to a plasma **growth hormone binding protein**. The binding protein protects plasma GH from being filtered into the urine and extends its half-life from 7 to 18 minutes. In addition, the bound portion acts as a growth hormone reservoir in the blood. Researchers have hypothesized that genetic determination of binding protein concentration plays a role in determining adult height.

The target tissues for GH include both endocrine and nonendocrine cells. GH acts as a trophic hormone to stimulate secretion of **insulin-like growth factors** (IGFs; formerly called *somatomedins*) from the liver and other tissues. IGFs act with growth hormone to stimulate bone and soft tissue growth (Table 23-3).

Metabolically, growth hormone and IGFs are anabolic for proteins, directing energy and amino acids into protein synthesis, an essential part of tissue growth. Growth hormone also acts with the insulin-like growth factors to stimulate bone growth. IGFs are responsible for cartilage growth. GH increases fat breakdown and hepatic glucose output, raising plasma fatty acid and glucose concentrations.

Answers: p. 750

11. What pituitary hormone in addition to GH has two hypothalamic factors that regulate its release?

Growth Hormone Is Essential for Normal Growth in Children

The pathologies that reflect the actions of growth hormone are most obvious in children. Severe growth hormone deficiency in childhood leads to **dwarfism**. Dwarfism can result from a problem with growth hormone synthesis or defective tissue hormone receptors. Unfortunately, bovine or porcine growth hormone is not effective as replacement therapy, as only primate growth hormone is active in humans. Prior to 1985, when genetically engineered human growth

Table 23-3: Growth Hormone

Cell of origin	Anterior pituitary
Chemical nature	191–amino acid peptide; several closely related forms
Biosynthesis	Typical peptide
Transport in the circulation	Half is dissolved in plasma, half is bound to a binding protein whose structure is identical to that of the GH receptor
Half-life	18 minutes
Factors affecting release	Circadian rhythm of tonic secretion; influenced by circulating nutrients, stress, and other hormones in a complex fashion
Target cells or tissues	Trophic on liver for insulin-like growth factor production; also acts directly on many cells
Target receptor	Membrane receptor with tyrosine kinase activity
Whole body or tissue action (with IGFs)	Bone and cartilage growth; soft tissue growth; ↑ plasma [glucose]
Action at cellular level	Protein synthesis
Action at molecular level	Receptor linked to kinases that phosphorylate proteins to initiate transcription

hormone became available, donated human pituitaries harvested at autopsy were the only source for growth hormone. Fortunately, severe growth hormone deficiency is a relatively rare condition.

Oversecretion of growth hormone in children leads to **giantism**. Once bone growth stops in late adolescence, growth hormone cannot further increase height. GH and IGFs can continue to act on cartilage and soft tissues, however. Adults with excessive secretion of growth hormone develop a condition known as **acromegaly**, characterized by lengthening of the jaw, coarsening of facial features, and growth of hands and feet (Fig. 23-17 ●).

Genetically Engineered Human Growth Hormone Created Ethical Dilemmas

When genetically engineered human growth hormone became available in the mid-1980s, the medical profession was faced with a dilemma. Obviously the hormone should be used to treat children who would otherwise be dwarfs. But what about those with only partial GH deficiency or genetically short children with normal GH secretion? What about children whose parents want them to be taller for athletics? These questions are complicated

● **Figure 23-17** *Acromegaly* Excess growth hormone secretion in adults causes lengthening of the jaw, coarsening of the features, and growth in hands and feet.

by the difficulty of accurately determining exactly which children have partial growth hormone deficiency.

In 1997 the American Academy of Pediatrics issued a policy statement with recommendations and cautions about the use of human growth hormone in short children. This statement points out that long-term risks associated with GH treatment are unknown and that GH therapy itself has the potential to create psychological problems in children if the results are less than optimum.

▶ TISSUE AND BONE GROWTH

Growth can be divided into two general areas: soft tissue growth and bone growth. In children, bone growth is usually assessed by measuring height, and tissue growth by measuring weight. Multiple hormones have direct or permissive effects on growth. In addition, we are just beginning to understand how paracrine growth factors interact with classic hormones to influence the development and differentiation of tissues.

Tissue Growth Requires Hormones and Paracrines

Soft tissue growth requires adequate amounts of growth hormone, thyroid hormone, and insulin. Growth hormone and insulin-like growth factors are required for tissue protein synthesis and cell

---

RUNNING PROBLEM

Dr. Bob's blood work reveals that his Ca^{2+} level is 12.3 mg/dL plasma (normal 8.5–10.5 mg/dL). These results support the suspected diagnosis of hyperparathyroidism. "Do you take vitamin D or use a lot of antacids?" Dr. Spinks asks. "Those could raise your blood calcium." Dr. Bob denies the use of either substance. "Well, we need to do one more test before we can say conclusively that you have hyperparathyroidism," Dr. Spinks says.

Question 4: What one test could definitively prove that Dr. Bob has hyperparathyroidism?

division. Under the influence of these hormones, cells undergo both **hypertrophy** (increased cell size) and **hyperplasia** (increased cell number).

Thyroid hormones play a permissive role in growth and contribute directly to the development of the nervous system. At the target tissue level, thyroid hormone interacts synergistically with growth hormone for protein synthesis and development of the nervous system. Children with untreated hypothyroidism (cretinism) will not grow to a normal height despite having normal secretion of growth hormone.

Insulin supports tissue growth by stimulating protein synthesis and providing energy in the form of glucose. Because insulin is permissive for growth hormone, children who are insulin-deficient will fail to grow normally although they may have normal growth and thyroid hormone concentrations.

Bone Growth Requires Adequate Dietary Calcium

Bone growth, like soft tissue development, requires the proper hormones and adequate amounts of protein and calcium. Bone has calcified extracellular *matrix* formed when calcium phosphate crystals precipitate and attach to a collagenous lattice support. The most common form of calcium phosphate is **hydroxyapatite**, $Ca_{10}(PO_4)_6(OH)_2$.

Although the large amount of inorganic matrix in bone makes some people think of it as nonliving, bone is a dynamic tissue, constantly being formed and broken down, or *resorbed*. Spaces in the collagen and calcium matrix are occupied by living cells. They are well supplied with oxygen and nutrients by blood vessels that run through adjacent channels.

Bones generally have two layers: an outer layer of dense **compact bone** and an inner layer of spongy **trabecular bone** (Fig. 23-18 ●). In some bones, a central cavity is filled with *bone marrow* (see *Focus on Bone Marrow*, ● p. 531). Compact bone provides strength and is thickest where support is needed, such as in the long bones of the legs, or in places where muscles attach. Trabecular bone is less sturdy and has open cell-filled spaces between struts of calcified lattice.

Bone growth occurs when matrix is deposited faster than it is resorbed. Special bone-forming cells called **osteoblasts** produce enzymes and **osteoid**, a mixture of collagen and other proteins to which hydroxyapatite binds. Recent research has found two other proteins, *osteocalcin* and *osteonectin*, that appear to aid in deposition of the calcified matrix.

Bone diameter increases when matrix deposits on the outer surface of the bone. Linear growth of long bones occurs at special regions called the **epiphyseal plates**, found between the ends (*epiphyses*) and shaft (*diaphysis*) of the bone (Fig. 23-19 ●). On the epiphyseal side of the plate are continuously dividing columns of **chondrocytes**, cartilage-producing cells.

As the collagen layer thickens, the older collagen calcifies and older chondrocytes degenerate, leaving spaces that osteoblasts invade. The osteoblasts then lay down bone matrix on top of the cartilage base. As new bone is added at the ends, the shaft lengthens. Long bone growth continues as long as the epiphyseal plate is active. When osteoblasts complete their work, they revert to a less active form known as **osteocytes**.

Growth of long bone is under the influence of growth hormone and the insulin-like growth factors. In the absence of these hormones, normal bone growth will not occur. Long bone growth is also influenced by sex steroid hormones. The growth spurt of adolescent boys is attributed to increased androgen production. For girls, the picture is less clear-cut. In experiments, estrogens both stimulate and inhibit linear growth.

In all adolescents, the sex hormones eventually inactivate the epiphyseal plate so that long bones no longer grow. Because the epiphyseal plates of various bones close in a constant, ordered sequence, X-rays that show which plates are open and which have closed can be used to calculate a child's "bone age."

Linear bone growth ceases in adults, but bones are dynamic tissues that are constantly being remodeled throughout life under the control of hormones that regulate calcium metabolism in the body.

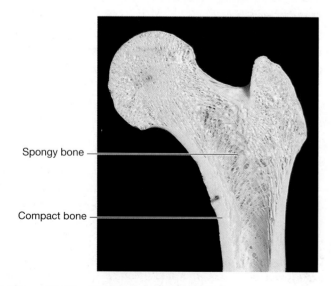

Spongy bone

Compact bone

● **Figure 23-18** *Compact and spongy bone*

CONCEPT CHECK   Answers: p. 750

12. Which hormones are essential for normal growth and development?

13. Why don't adults with hypersecretion of growth hormone grow taller?

▶ CALCIUM BALANCE

Most calcium in the body, 99% or nearly 2.5 pounds, is found in the bones. This pool is relatively stable. It is the body's small fraction of nonbone calcium that is most critical to physiological functioning (Table 23-4). You have encountered most of the physiological uses of Ca^{2+} in earlier chapters.

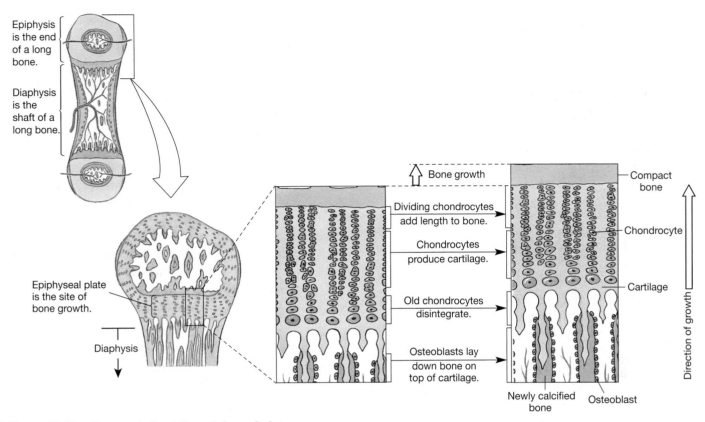

Epiphysis is the end of a long bone.

Diaphysis is the shaft of a long bone.

Epiphyseal plate is the site of bone growth.

Diaphysis

Bone growth

Dividing chondrocytes add length to bone.

Chondrocytes produce cartilage.

Old chondrocytes disintegrate.

Osteoblasts lay down bone on top of cartilage.

Compact bone

Chondrocyte

Cartilage

Newly calcified bone

Osteoblast

Direction of growth

● **Figure 23-19**   *Bone growth at the epiphyseal plate*

Table 23-4: Functions of Calcium in the Body

COMPARTMENT	PERCENTAGE OF ALL CALCIUM IN BODY	FUNCTION
Extracellular	99%	Calcified matrix of bone
Extracellular fluid	0.1%	"Cement" for tight junctions Myocardial and smooth muscle contraction Release of neurotransmitters at synapses Excitability of neurons due to effect on Na^+ permeability Cofactor for coagulation cascade
Intracellular	0.9%	Signal in second messenger pathways Muscle contraction

1. **Ca^{2+} is an important signal molecule.** The movement of Ca^{2+} between body compartments creates Ca^{2+} signals or "sparks." Calcium is most concentrated in the extracellular fluid and inside certain organelles such as endoplasmic and sarcoplasmic reticulum. If membrane channels for Ca^{2+} open, Ca^{2+} moves into the cytoplasm, creating a signal that initiates exocytosis of synaptic and secretory vesicles, contraction in muscle fibers, or altered activity of enzymes and transporters. Because of the concentration gradient, removal of Ca^{2+} from the cytoplasm requires active transport.

2. **Ca^{2+} is part of the intercellular cement that holds cells together at tight junctions.**

3. **Ca^{2+} is a cofactor in the coagulation cascade [✎ p. 539].** Although Ca^{2+} is essential for coagulation, body Ca^{2+} concentrations never decrease to the point that coagulation is inhibited. However, removal of Ca^{2+} from blood samples will prevent the specimen from clotting in the test tube.

4. **Plasma Ca^{2+} concentrations affect the excitability of neurons.** This function of Ca^{2+} has not been introduced before, but clinically, it is the function that is most obvious in Ca^{2+}-related pathologies.

   If plasma Ca^{2+} falls too low (**hypocalcemia**), neuronal permeability to Na^+ increases, neurons depolarize, and the nervous system becomes hyperexcitable. In its most extreme form, hypocalcemia causes sustained contraction (*tetany*) of the respiratory muscles, resulting in asphyxiation. **Hypercalcemia** has the opposite effect, causing depressed neuromuscular activity.

Calcium Concentrations in the Blood Are Closely Regulated

Because calcium is critical to so many physiological functions, the body's plasma Ca^{2+} concentration is very closely regulated. Calcium homeostasis follows the principle of mass balance:

$$\text{Total body calcium} = \text{intake} - \text{output}$$

1. **Intake** is dietary ingestion and uptake in the small intestine. Unlike organic nutrients whose intestinal uptake is unregulated, Ca^{2+} absorption is hormonally regulated. Most Americans do not eat enough Ca^{2+}-containing foods, however, and intake may not match output.

2. **Output** or Ca^{2+} loss from the body occurs primarily through the kidneys, with a small amount excreted in feces. Ionized Ca^{2+} is freely filtered at the glomerulus. Ca^{2+} is then reabsorbed along the length of the nephron, but hormonally regulated reabsorption takes place only in the distal nephron. In that part of the tubule reabsorption is accomplished by movement through an apical Ca^{2+} channel (*ECaC*) and basolateral Na^+-Ca^{2+} antiport and Ca^{2+}-ATPase.

3. **Total body Ca^{2+}**. Calcium in the body is divided between three compartments (Fig. 23-20 ●).

   (a) *Extracellular fluid/plasma*. Ionized Ca^{2+} is concentrated in the ECF, with an average value of 2.5 mM. In the plasma, about half of Ca^{2+} is carried bound to plasma proteins and other molecules; the remainder is unbound and free to diffuse across membranes through open Ca^{2+} channels.

   (b) *Intracellular Ca^{2+}*. The concentration of free Ca^{2+} in the cytosol is about 0.001 mM. In addition, Ca^{2+} is concentrated inside mitochondria and sarcoplasmic reticulum. These electrochemical gradients favor movement of Ca^{2+} into the cytosol when Ca^{2+} channels open.

   (c) *Extracellular matrix (bone)*. Bone is the largest reservoir of Ca^{2+} in the body, with most bone Ca^{2+} in the form of hydroxyapatite crystals. Bone Ca^{2+} forms a reservoir that can be tapped to maintain plasma Ca^{2+} homeostasis. Usually only a small fraction of bone Ca^{2+} is ionized and readily exchangeable, and this pool remains in equilibrium with Ca^{2+} of the interstitial fluid.

Osteoblasts control the deposition of Ca^{2+} into bone, as discussed above. The movement of Ca^{2+} out of hydroxyapatite and into the ionized Ca^{2+} pool is controlled by osteoclasts. **Osteoclasts** are large, mobile, multinucleate cells derived from hematopoietic stem cells [● p. 526]. They are responsible for dissolving bone, a process known as *resorption*.

Osteoclasts attach around their periphery to a section of calcified matrix, much like a suction cup (Fig. 23-21 ●). The central region of the cell next to the bone secretes acid (with the aid of H^+-ATPase) and *protease* enzymes that work at low pH. The com-

● **Figure 23-20** *Calcium balance in the body*

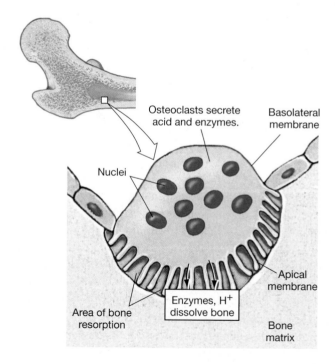

● **Figure 23-21** *Osteoclasts are responsible for bone resorption.*

bination of acid and enzymes dissolves the calcified matrix and its collagen support, freeing Ca^{2+}, which can enter the blood.

CONCEPT CHECK   Answers: p. 750

14. What does hypercalcemia do to neuronal membrane potential and why does that depress neuromuscular excitability?

15. Draw a picture of a distal tubule cell and label apical and basolateral membranes, lumen, and ECF. Use the description of distal tubule Ca^{2+} reabsorption in step 2 above to draw in the appropriate transporters.

16. Describe the transport of Ca^{2+} from the tubule cell to the ECF (active, passive, facilitated diffusion, etc.).

Three Hormones Control Calcium Balance

Three hormones regulate the movement of Ca^{2+} between bone, kidney, and intestine: parathyroid hormone, vitamin D_3, and calcitonin (Fig. 23-20). Of these, parathyroid hormone and vitamin D_3 are the most important in adult humans.

The **parathyroid glands** are the endocrine organs responsible for Ca^{2+} homeostasis. These four small structures monitor changes in plasma Ca^{2+} concentrations and secrete parathyroid hormone (Fig. 23-22 ●). The parathyroids are found on the dorsal surface of the thyroid gland, and their significance was first recognized in the 1890s by physiologists studying the role of the thyroid gland.

These scientists noticed that if they removed all of the thyroid gland, dogs and cats died in a few days. In contrast, rabbits died only if the little parathyroid "glandules" alongside the thyroid were removed. The scientists then looked for parathyroid glands in dogs and cats and found them tucked away behind the larger thyroid gland. If they left the parathyroids behind when

Table 23-5: Parathyroid Hormone (PTH)

Cell of origin	Parathyroid glands
Chemical nature	84–amino acid peptide
Biosynthesis	Continuous production, little stored
Transport in the circulation	Dissolved in plasma
Half-life	Less than 20 minutes
Factors affecting release	↓ Plasma Ca^{2+}
Target cells or tissues	Kidney, bone, intestine
Target receptor	Membrane receptor acts via cAMP
Whole body or tissue action	↑ Plasma Ca^{2+}
Action at cellular level	↑ Vitamin D synthesis; ↑ renal reabsorption of Ca^{2+}; ↑ bone resorption
Action at molecular level	Rapidly alters Ca^{2+} transport but also initiates protein synthesis in osteoclasts
Onset of action	2–3 hours for bone, with increased osteoclast activity requiring 1–2 hours; 1–2 days for intestinal absorption; within minutes for kidney transport
Feedback regulation	Negative feedback by ↑ plasma Ca^{2+}
Other information	Osteoclasts have no PTH receptors, so are affected by PTH-induced paracrines. PTH is essential for life; absence causes hypocalcemic tetany.

the thyroid was surgically removed, the animals lived. The scientists concluded that the parathyroids contained a substance that was essential for life, although the thyroid gland did not.

Parathyroid Hormone   We now know that parathyroid glands secrete **parathyroid hormone** (parathormone, or **PTH**), a peptide whose main effect is to increase plasma Ca^{2+} concentrations (Table 23-5). The stimulus for PTH release is a decrease in plasma Ca^{2+}, sensed by an extracellular Ca^{2+}-sensing membrane receptor. By acting on bone, kidney, and intestine, PTH raises plasma Ca^{2+} concentrations. The increase in plasma Ca^{2+} acts as negative feedback and shuts off PTH secretion.

Parathyroid hormone raises Ca^{2+} concentrations in three ways:

1. **PTH mobilizes calcium from bone.** Increased bone resorption by osteoclasts takes about 12 hours to become measurable. Curiously, although osteoclasts are responsible for dissolving the calcified matrix and would be logical targets for PTH, they do not have PTH receptors. PTH effects are mediated by a collection of paracrines including *osteoprotegerin* (OPG) and an osteoclast differentiation factor called *RANKL*.

Left superior thyroid artery

Pharynx

Right internal jugular vein

Thyroid

Right common carotid artery

Left parathyroid glands

Right parathyroid glands

Left inferior thyroid artery

Esophagus

Left subclavian artery

Trachea

The four parathyroid glands lie hidden behind the thyroid gland.

● **Figure 23-22**   *Parathyroid glands*

These paracrine factors are receiving intense scrutiny as potential pharmacological agents.

2. **PTH enhances renal reabsorption of calcium.** As mentioned above, regulated Ca^{2+} reabsorption takes place in the distal nephron. PTH simultaneously enhances renal excretion of phosphate by reducing its reabsorption.

The opposing effects of PTH on calcium and phosphate are needed to keep their combined concentrations below a critical level. If the concentrations exceed that level, calcium phosphate crystals form and precipitate out of solution. High concentrations of calcium phosphate in the urine are one cause of kidney stones. Additional aspects of phosphate homeostasis are discussed below.

3. **PTH indirectly increases intestinal absorption of calcium** by its influence on vitamin D_3, a process described below.

Calcitriol   Intestinal absorption of calcium is enhanced by the action of a hormone known as **1,25-dihydroxycholecalciferol** $\left(1,25(OH)^2D_3\right)$, also known as **calcitriol** or vitamin D_3.

Calcitriol is made from vitamin D obtained through diet or made in the body (Table 23-6). The body makes its own vitamin D_3 by the action of sunlight on a precursor made from acetyl CoA (Fig. 23-23 ●). Vitamin D_3 is then modified in two steps, first in the liver, then in the kidneys, to make calcitriol.

Calcitriol reinforces the plasma Ca^{2+}-increasing effect of parathyroid hormone by enhancing Ca^{2+} uptake across the small intestine. In addition, it facilitates renal reabsorption of Ca^{2+} and helps mobilize Ca^{2+} out of bone into the extracellular fluid.

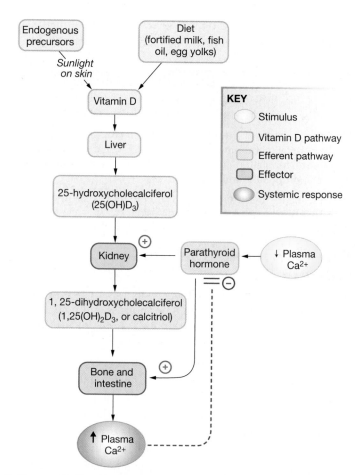

● **Figure 23-23**   *Endocrine control of calcium balance*

Table 23-6: Vitamin D (Calcitriol, 1,25-Dihydroxycholecalciferol)

Cell of origin	Complex biosynthesis; see below
Chemical nature	Steroid
Biosynthesis	Vitamin D formed by sunlight on precursor molecules or ingested in food; converted in 2 steps (liver and kidney) to $1,25(OH)_2D_3$
Transport in the circulation	Bound to plasma proteins
Stimulus for synthesis	$\downarrow$ Ca^{2+}; indirectly via PTH; prolactin also stimulates synthesis
Target cells or tissues	Intestine, bone, and kidney
Target receptor	Nuclear
Whole body or tissue action	$\uparrow$ Plasma Ca^{2+}
Action at molecular level	Stimulates production of calbindin, a Ca^{2+}-binding protein; associated with intestinal transport by unknown mechanism
Feedback regulation	Plasma Ca^{2+} shuts off PTH secretion

The production of calcitriol is regulated at the kidney by the action of PTH. Decreased plasma Ca^{2+} concentrations increase PTH secretion, which in turn stimulates calcitriol synthesis. Intestinal and renal absorption of Ca^{2+} raises blood calcium concentrations, turning off PTH and subsequently decreasing calcitriol synthesis.

Prolactin, the hormone responsible for milk production in breast-feeding women, also stimulates calcitriol synthesis. This action ensures maximal absorption of Ca^{2+} from the diet at a time when metabolic demands for calcium are high.

Calcitonin   The third hormone involved with calcium metabolism is **calcitonin**, a peptide produced by the **C cells** (parafollicular cells) of the thyroid gland (Table 23-7). Its actions are opposite to those of parathyroid hormone. Calcitonin is released when plasma Ca^{2+} goes up. Experimentally in animals, calcitonin decreases bone resorption and increases renal calcium excretion.

From what we can tell, calcitonin plays only a minor role in daily calcium balance in adult humans. Patients whose thyroid glands have been removed show no disturbance in calcium balance. People with thyroid tumors that secrete large amounts of calcitonin also show no ill effects.

Medically, calcitonin has been used to treat patients with *Paget's disease*, a genetically linked condition in which osteoclasts are overactive and bone is weakened by resorption. Calcitonin in these patients stabilizes the abnormal bone loss. This action of

Table 23-7: Calcitonin

Cell of origin	C cells of thyroid gland (parafollicular cells)
Chemical nature	32–amino acid peptide
Biosynthesis	Typical peptide
Transport in the circulation	Dissolved in plasma
Half-life	<10 minutes
Factors affecting release	↑ Plasma $[Ca^{2+}]$
Target cells or tissues	Bone and kidney
Target receptor	G protein-coupled membrane receptor
Whole body or tissue action	Prevents bone resorption; enhances kidney excretion
Action at molecular level	Signal transduction pathways appear to vary during cell cycle
Other information	Experimentally decreases plasma $[Ca^{2+}]$ but has little apparent physiological effect in adult humans; possible effect on skeletal development; possible protection of bone Ca^{2+} stores during pregnancy and lactation

calcitonin has led scientists to speculate that the hormone in humans is most important during childhood growth, when net bone deposition is needed, and during pregnancy and lactation (breast-feeding), when the mother's body must supply calcium for both herself and the child.

Calcium and Phosphate Homeostasis Are Linked

Phosphate homeostasis is closely linked to that of calcium. Phosphate is the second key ingredient in the hydroxyapatite of bone, $Ca_{10}(PO_4)_6(OH)_2$, and most phosphate in the body is found in bone. However, phosphates have other significant roles in physiology, including energy transfer and storage in high-energy phosphate bonds and activation or deactivation of enzymes, transporters, and ion channels. Phosphates also form part of the DNA and RNA backbone.

Phosphate homeostasis parallels that of Ca^{2+}. Phosphate is absorbed in the intestines, filtered and reabsorbed in the kidneys, and divided between bone, ECF, and intracellular compartments. Vitamin D enhances intestinal absorption of phosphate. Renal excretion is affected by both PTH, which promotes phosphate excretion, and vitamin D, which promotes phosphate reabsorption.

CONCEPT CHECK   Answers: p. 750

17. Name two compounds that store energy in high-energy phosphate bonds.

18. What is the difference between a kinase, a phosphatase, and a phosphorylase?

RUNNING PROBLEM

The results of Dr. Bob's last test confirm that he has hyperparathyroidism. He goes on a low-calcium diet, avoiding milk, cheese, and other dairy products, but several months later, he returns to the emergency room with another painful calcium phosphate kidney stone. Dr. Spinks sends him to an endocrine specialist, who recommends surgical removal of the overactive parathyroid glands. "We can't tell which of the parathyroid glands is most active," the specialist says, "and we'd like to leave you with some parathyroid hormone of your own. So I will take out all four glands, but we'll reimplant two of them in the muscle of your forearm. In many patients, the implanted glands secrete just enough PTH to maintain calcium homeostasis. And if they secrete too much PTH, it is much easier to take them out of your arm than do major surgery on your neck again."

Question 5: Why can't Dr. Bob simply take replacement PTH by mouth? (Hint: PTH is a peptide hormone.)

726 727 730 739 746

Osteoporosis Is a Disease of Bone Loss

One of the best-known pathologies of bone function is **osteoporosis**, a metabolic disorder characterized by bone resorption exceeding bone deposition. The result is fragile, weakened bones that are more easily fractured (Fig. 23-24 ●). Most bone resorption takes place in spongy trabecular bone, particularly in the vertebrae, hip, and wrist.

Osteoporosis is most common in women after menopause, when estrogen concentrations fall. However, older men also develop osteoporosis. Bone loss and small fractures and compression in the spinal column lead to *kyphosis* [hump-back], the stooped, hunchback appearance that is characteristic of advanced osteoporosis in the elderly.

Osteoporosis is a complex disease with genetic and environmental components. Risk factors include small, thin body type, postmenopausal age, smoking, and low dietary Ca^{2+} intake.

● **Figure 23-24** *Normal bone (left) and bone loss in osteoporosis (right)*

For many years estrogen or estrogen/progesterone *hormone replacement therapy* (HRT) was used to prevent osteoporosis. However, estrogen alone increases the risk of endometrial and possibly other cancers, and recent studies suggested that estrogen/progesterone HRT might increase risk of heart attacks and strokes. Newer drugs called *bisphosphonates* induce osteoclast apoptosis and effectively decrease bone loss, and they have become the treatment of choice for preventing osteoporosis.

To avoid osteoporosis in later years, younger women need to maintain adequate dietary calcium intake and participate in weight-bearing exercise that increases bone density. Loss of bone mass actually begins by age 30, long before most women start to think they are at risk, and many women suffer from low bone mass (*osteopenia*) long before they are aware of a problem. Bone mass testing can help with early diagnosis of osteopenia.

RUNNING PROBLEM CONCLUSION

Hyperparathyroidism

Dr. Bob had the surgery, and the implanted glands produced an adequate amount of PTH. He must have his plasma Ca^{2+} levels checked regularly for the rest of his life. To learn more about hyperparathyroidism, try a Google search (*www.google.com*). To find drawings, photographs, or micrographs, enter the search term *parathyroid* and click the IMAGES tab on the Google home page.

Check your understanding of this running problem by comparing your answers to those in the summary table.

	QUESTION	FACTS	INTEGRATION AND ANALYSIS
1	What role does Ca^{2+} play in the normal function of muscles and neurons?	Calcium triggers neurotransmitter release (● p. 263) and uncovers the myosin-binding sites on actin filaments of muscle (● p. 396).	Muscle weakness in hyperparathyroidism is the opposite effect of what you would predict from knowing the role of Ca^{2+} in muscles and neurons. However, calcium also affects the Na^+ permeability of neurons, and it is this effect that leads to muscle weakness and the CNS effects.
2	What is the technical term for "elevated levels of calcium in the blood"?	Prefix for elevated levels: *hyper-*. Suffix for "in the blood": *-emia*	Hypercalcemia is the technical term for elevated levels of calcium in the blood.
3	A significant portion of plasma Ca^{2+} cannot be filtered into Bowman's capsule. Speculate on why some plasma Ca^{2+} cannot filter.	Filtration at the glomerulus is a selective process that excludes blood cells and most plasma proteins (● p. 607).	A significant amount of Ca^{2+} is bound to plasma proteins and therefore does not filter at the glomerulus.
4	What one test could definitively prove that Dr. Bob has hyperparathyroidism?	Hyperparathyroidism is a condition in which excessive amounts of PTH are secreted.	A test for the amount of PTH in the blood would confirm the diagnosis of hyperparathyroidism.
5	Why can't Dr. Bob take replacement PTH by mouth?	PTH is a peptide hormone.	Ingested peptides will be digested by proteolytic enzymes. Thus PTH taken orally will not be absorbed intact into the body and will not be effective.

726 727 730 739 745

CHAPTER SUMMARY

Endocrinology is based on the physiological principles of *homeostasis* and *control systems*. Each hormone has stimuli that initiate its secretion and feedback signals that modulate its release. *Molecular interactions* and *communication across membranes* are also essential to hormone activity. In many instances, such as calcium and phosphate homeostasis, the principle of *mass balance* is the focus of homeostatic regulation.

Review of Endocrine Principles

1. Basic components of endocrine pathways include hormone receptors, feedback loops, and cellular responses. (p. 727)

Adrenal Glucocorticoids

2. The **adrenal cortex** secretes **glucocorticoids**, sex steroids, and aldosterone. (p. 727; Figs. 23-1, 23-2)

3. **Cortisol** secretion is controlled by hypothalamic **CRH** and **ACTH** from the pituitary. Cortisol is the feedback signal. (pp. 728–729; Fig. 23-3)

4. Cortisol is a typical steroid hormone in its synthesis, secretion, transport, and action. (p. 729)

5. Cortisol is catabolic and essential for life. It promotes gluconeogenesis, breakdown of skeletal muscle proteins and adipose tissue, Ca^{2+} excretion, and suppression of the immune system. (p. 730; Tbl. 23-1)

6. **Hypercortisolism** usually results from a tumor or therapeutic administration of the hormone. **Addison's disease** is hyposecretion of all adrenal steroids. (p. 731)

7. CRH and the **melanocortins** have physiological actions in addition to cortisol release. (p. 732; Fig. 23-6)

Thyroid Hormones

8. The thyroid **follicle** has a hollow center filled with **colloid** containing **thyroglobulin** and enzymes. (p. 733; Fig. 23-7)

9. Thyroid hormones are made from tyrosine and iodine. **Tetraiodothyronine** (thyroxine, or T_4) is converted in target tissues to the more active hormone **triiodothyronine** (T_3). (p. 733; Figs. 23-8, 23-9, 23-10)

10. Thyroid hormones are not essential for life, but they influence metabolic rate and protein, carbohydrate, and fat metabolism. (p. 733; Tbl. 23-2)

11. Thyroid hormone secretion is controlled by **thyrotropin** (thyroid-stimulating hormone, or **TSH**) and thyrotropin-releasing hormone, or **TRH**. (p. 735; Figs. 23-12, Fig. 23-14)

Growth Hormone

12. Normal growth requires growth hormone, thyroid hormones, insulin, and sex hormones at puberty. Growth also requires adequate diet and absence of stress. (pp. 736–737)

13. **Growth hormone** is secreted by the anterior pituitary and stimulates secretion of **insulin-like growth factors** (IGFs) from the liver and other tissues. These hormones promote bone and soft tissue growth. (pp. 737–738; Tbl. 23-3)

14. Secretion of growth hormone is controlled by growth hormone–releasing hormone (**GHRH**) and growth hormone–inhibiting hormone (**somatostatin**). (p. 738; Fig. 23-16)

Tissue and Bone Growth

15. **Bone** is composed of **hydroxyapatite** crystals attached to a collagenous support. Bone is a dynamic tissue with living cells. (p. 740; Fig. 23-18)

16. **Osteoblasts** synthesize bone. Long bone growth occurs at **epiphyseal plates**, where **chondrocytes** produce cartilage. (p. 740; Fig. 23-19)

Calcium Balance

 Fluids & Electrolytes: Electrolyte Homeostasis

17. Calcium acts as an intracellular signal for second messenger pathways, exocytosis, and muscle contraction. It also plays a role in cell junctions, coagulation, and neural function (p. 741; Tbl. 23-4)

18. Ca^{2+} homeostasis balances dietary intake, urinary output, and distribution of Ca^{2+} between bone, cells, and the ECF. (p. 742; Fig. 23-20)

19. Decreased plasma Ca^{2+} stimulates **parathyroid hormone** (**PTH**) secretion by **parathyroid glands**. (p. 743; Tbl. 23-5)

20. PTH mobilizes Ca^{2+} from bone, enhances renal Ca^{2+} absorption, and increases intestinal Ca^{2+} absorption through its effect on **calcitriol**. (p. 744; Figs. 23-20, 23-23; Tbl. 23-6)

21. Calcitonin from the thyroid gland plays only a minor role in daily calcium balance in adult humans. (p. 744; Tbl. 23-7)

QUESTIONS

LEVEL ONE *Reviewing Facts and Terms*

1. Name the zones of the adrenal cortex and the primary hormones secreted in each zone.

2. For each of the following hormones, draw the full control pathway and show feedback where appropriate. Do not use abbreviations. (a) cortisol (b) growth hormone (c) parathyroid hormone (d) T_3 and T_4

3. List four conditions that are necessary for people to achieve their full growth. Include five specific hormones known to exert an effect on growth.

4. Name the thyroid hormones. Which one has the highest activity? How and where is most of it produced?

5. Define the following terms and explain their physiological significance:

   (a) melanocortins   (e) trabecular bone

   (b) osteoporosis   (f) POMC

   (c) hydroxyapatite   (g) epiphyseal plates

   (d) mineralocorticoid

6. List seven functions of calcium in the body.

7. Make a table showing the effects of cortisol, thyroid hormones, growth hormone, insulin, and glucagon on protein, carbohydrate, and lipid metabolism.

LEVEL TWO *Reviewing Concepts*

8. Mapping exercise: Create a reflex map for each of the following situations.

   (a) Hypercortisolism due to an adrenal tumor

   (b) Hypercortisolism due to a pituitary tumor

   (c) Hyperthyroidism due to a hormone-secreting thyroid tumor

   (d) Hypothyroidism due to pituitary tumor that decreases TSH synthesis

9. Define, compare, and contrast or relate the terms within each set below:

   (a) cortisol, glucocorticoids, ACTH, CRH

   (b) thyroid, C cells, follicle, colloid

   (c) thyroglobulin, tyrosine, iodide, TBG, deiodinase, TSH, TRH

   (d) somatotropin, IGF, GHRH, somatostatin, growth hormone binding protein

   (e) giantism, acromegaly, dwarfism

   (f) hyperplasia, hypertrophy

   (g) osteoblast, osteoclast, chondrocyte, osteocyte

   (h) vitamin D, calcitriol, 1,25-dihydroxycholecalciferol, calcitonin, estrogen, PTH

10. Based on what you know about T_3's cellular mechanism of action, would you expect to see tissue response within a few minutes or in more than an hour?

11. If average plasma $[Ca^{2+}]$ is 2.5 mmol/L, what is the concentration in mEq/L?

12. Osteoclasts make acid (H^+) from CO_2 and H_2O. They secrete the acid at their apical membrane and put bicarbonate into the ECF. Draw an osteoclast and show this process, including the appropriate transporters on each membrane. How many different transporters can you think of that could be used to reabsorb bicarbonate?

LEVEL THREE *Problem Solving*

13. Diabetic patients who have surgery, become sick, or are otherwise under physiological stress are told to monitor their blood sugars carefully as they may need to increase their insulin dose temporarily. What is the physiological explanation behind this admonition?

14. One diagnostic test to determine the cause of hypercortisolism is a *dexamethasone suppression test*. Dexamethasone blocks secretion of ACTH by the pituitary. The following table shows the results from two patients given a dexamethasone suppression test.

	PLASMA CORTISOL BEFORE TEST	PLASMA CORTISOL AFTER TEST
Patient A	High	High
Patient B	High	Low

Can you tell from these results where the patients' pathologies originate? Explain for each patient.

15. When several lab results came back last week, someone in the office spilled a cup of coffee on them, smearing the names and some of the results. One lab slip shows elevated TSH levels but the thyroid levels are blurred. You have three charts waiting for thyroid hormone levels. Your tentative diagnosis, based on physical findings and symptoms, for those three patients is:

   Mr. A: primary hypothyroidism

   Ms. B: primary hyperthyroidism

   Mrs. C: secondary hyperthyroidism

   (a) Can you tell whose results are on the smeared lab slip, based on the TSH results and the tentative diagnosis?

   (b) Can you rule out any of the three people based on those same criteria? Explain.

16. The graph in the right column shows the results of a study done in Boston that compared blood vitamin D levels during summer and winter. Boston is located at 42-N latitude, and weak sunlight in winter there does not allow skin synthesis of vitamin D. (Data from "Vitamin D insufficiency among free-living healthy young adults," *American Journal of Medicine* 112:659–662, 2002 Jun 1)

    (a) Summarize the results of the study as shown in the graph. How many variables are shown in the graph that you must address in your summary?

    (b) Based on what you know, how could you explain the results of the study?

    (c) Would taking a multivitamin supplement affect the results?

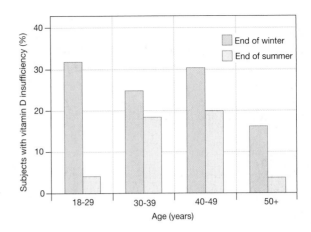

LEVEL FOUR *Quantitative Problems*

17. Plasma Ca^{2+} that can be filtered by the kidneys is about 5 mg/L. Assume a person has a GFR of 125 mL plasma filtered per minute.

    (a) How much Ca^{2+} will this person filter in a day?

    (b) Net dietary Ca^{2+} intake is 170 mg/day. To remain in Ca^{2+} balance, how much Ca^{2+} must he excrete?

    (c) What percentage of filtered Ca^{2+} is reabsorbed by the kidney tubule?

Explorations 🌐

Explorations are also available at *www.physiologyplace.com*

Exploration 1: Androstenedione and DHEA

The chapter opened with the story of baseball player Mark McGwire and his controversial use of the supplement *androstenedione*. Figure 23-2 shows that this prohormone is one step in the pathway to testosterone and DHT synthesis. One of its precursors, DHEA, is also used as a dietary supplement. In the United States, purchase of androstenedione and DHEA is not regulated, despite the conversion of these prohormones to testosterone, a controlled drug whose use is almost universally banned by sports associations.

To see what information is available to the general public from commercial sources, try a Google search (*www.google.com*) for *DHEA* or *androstenedione*. To read more about the McGwire controversy, search for *Mark McGwire androstenedione*. The Hormone Foundation, the public education affiliate of the Endocrine Society (*www.hormone.org*), has a page on hormone abuse that includes fact sheets and links to sites that discuss steroid abuse.

Exploration 2: A Link between Growth Hormone and Nephropathy?

Growth hormone seems an unlikely precursor to kidney disease. However, the two may be linked in the *nephropathy* (*nephros*, kidney + *pathos*, suffering) caused by diabetes mellitus. Uncontrolled diabetes is the most common cause for nephropathy and is responsible for a third of diagnosed end-stage renal diseases. One proposed model for the progression of the disease is that fluctuating glucose levels (particularly hypoglycemic episodes) stimulate overproduction of growth hormone (GH). GH has indirect effects via insulin-like growth factors (IGFs), but also has direct "anti-insulin" activity, greatly exacerbating the effects of diabetes.

The abnormal GH/IGF axis causes mesangial cells to overproduce growth factors, such as tumor necrosis factor. These growth factors initiate proliferation of epithelial cells and abnormal expansion of the basement membrane. As a result of these changes, the glomerulus begins to filter protein, while the amount of functional kidney tissue decreases. Renal disease progresses through several predictable stages, and if untreated, ends with renal failure.

To examine the growth hormone cascade more closely, read "Insulin-like growth factor-1 in diabetes mellitus: its physiology, metabolic effects, and potential clinical utility," *Diabetes Technology & Therapeutics* 2(1):69–80, 2000 Spring. *Review of Medical Physiology* by William F. Ganong (McGraw Hill 2001) has a good general discussion of the role of growth hormone in diabetes mellitus.

For more information on growth factors and nephropathy, read "Putative pathophysiological role of growth factors and cytokines in experimental diabetic kidney disease," *Diabetologica* 43(10):1205–1223, 2000 Oct. For a description of the stages of and treatments for diabetic nephropathy see the National Institutes of Health website at *www.niddk.nih.gov/health/diabetes/pubs/kdd/kdd.htm*.

(continues on next page)

Exploration 3: Learning more

The Endocrine Society (*www.endo-society.org*) has a number of well-written fact sheets about selected endocrine pathologies, including acromegaly, congenital adrenal hyperplasia, diabetes, osteoporosis, and thyroid disorders. From their home page, select *Patient Information* to see a list of topics.

For an interesting article on preventing and treating osteoporosis, go to the *Scientific American* website (*www.sciam.com*) and type *aging bones* in the SEARCH box. This will provide you with the link to an article titled "Restoring Aging Bones."

ANSWERS

Answers to Concept Check Questions

Page 728

1. Adrenal medulla secretes catecholamines (epinephrine, norepinephrine) and adrenal cortex secretes aldosterone, glucocorticoids, and sex hormones.

2. Androstenedione is a prohormone for testosterone. Testosterone is anabolic for skeletal muscle, which might give an athlete a strength advantage. Testosterone supplementation is banned by most athletic organizations.

Page 730

3. ACTH = adrenocorticotropic hormone or corticotropin. CRH = corticotropin-releasing hormone. CBG = corticosteroid-binding globulin or transcortin.

4. The immediate stress response to seeing the bear and being chased up the tree is too rapid to be mediated by cortisol. It is a fight-or-flight response mediated by the sympathetic nervous system and catecholamines. If you spend much time in the tree with the bear underneath, the cortisol-mediated stress response will have time to kick in.

5. Anabolic steroids, used to build muscle, do not include cortisol because cortisol is catabolic on muscle proteins.

Page 731

6. In primary hypercortisolism, ACTH is less than normal due to negative feedback. In secondary hypercortisolism, ACTH is greater than normal because of the ACTH-secreting tumor. In iatrogenic hypercortisolism, ACTH is less than normal due to negative feedback.

7. Someone with Addison's disease will have elevated ACTH due to reduced corticosteroid production and lack of negative feedback.

Page 732

8. ACTH is secreted during stress. If the stress is a physical one due to injury, the endogenous opioid can help decrease pain and allow the person to continue functioning.

Page 736

9. The T_4 given to the patient is converted to T_3 in the peripheral tissues. T_3 is the more active form of the hormone.

10. When mitochondria are uncoupled, the energy that is normally captured in the high-energy bond of ATP is released as heat instead. This will raise the body temperature of the person and cause heat intolerance.

Page 738

11. Prolactin also has two hypothalamic factors that regulate its release.

Page 740

12. Normal growth and development requires growth hormone, thyroid hormone, insulin, and insulin-like growth factors.

13. Adults who hypersecrete growth hormone do not grow taller because the epiphyseal plates of the long bones have closed.

Page 743

14. Hypercalcemia hyperpolarizes the membrane potential and makes it harder for the neuron to fire an action potential.

15. The figure should resemble Figure 20-12. Instead of a Na^+ channel on the apical membrane it will be a Ca^{2+} channel that allows Ca^{2+} into the cell. On the basolateral side, the ATPase will move Ca^{2+} from cytoplasm to interstitial fluid. On another transporter on the basolateral side, Na^+ enters the cell in exchange for Ca^{2+} leaving.

16. The Na^+-Ca^{2+} antiport is secondary active transport and the Ca^{2+}-ATPase is active transport.

Page 745

17. ATP and phosphocreatine store energy in high-energy phosphate bonds.

18. A kinase transfers a phosphate group between a substrate and ADP or ATP. A phosphatase removes a phosphate group and a phosphorylase adds one.

Answers to Figure Questions

Page 729

Figure 23-2: 1. A baby born with deficient 21-hydroxylase would have low aldosterone and cortisol and an excess of sex steroids, particularly androgens. Low cortisol would decrease the child's ability to respond to stress. Excess androgen would cause masculinization in female infants. 2. Women, who synthesize more estrogens, would have more aromatase.

Page 734

Figure 23-9: I^- comes into the cell by secondary active transport (co-transport with Na^+). Thyroglobulin moves between colloid and cytoplasm by exocytosis and endocytosis. Thyroid hormones leave the cell by simple diffusion.

Answers to Level Four Quantitative Problems

Page 750

17. (a) 5 mg Ca^{2+}/L plasma $\times$ 125 mL plasma filtered/min $\times$ 1440 min/day = 900 mg Ca^{2+} filtered/day (b) To remain in Ca^{2+} balance, he must excrete 170 mg/day. (c) 900 mg filtered -170 mg excreted = 730 mg reabsorbed. 730/900 = 81%

The Immune System

Although, at first sight, the immune system may appear to be autonomous, it is connected by innumerable structural and functional bridges with the nervous system and the endocrine system, so as to constitute a multisystem.

—Branislav D. Janković, *Neuroimmunomodulation: The State of the Art,* 1994

24

SEM of alveolar macrophage attacking E. coli *bacteria (turquoise). A red blood cell is at the upper right.*

RUNNING PROBLEM

Treatment for AIDS

In 1996, researchers developed a strategy that might offer long-term suppression of the HIV virus, the virus that causes acquired immuno-deficiency syndrome (AIDS). In that year, scientists confirmed that an AIDS "cocktail" consisting of several drugs, including a *protease inhibitor*, could be effective against the HIV virus that causes AIDS. But questions remained. How long would the new drug combination keep HIV at bay? Would the new drugs work for everyone with the infection? George, 28, pushed these questions to the back of his mind. After three years of relatively good health following his positive test for HIV infection, his T lymphocyte cell count had begun to slide. It would only be a matter of time before George became gravely ill.

753   768   771   774   776

BACKGROUND BASICS

Leukocytes: **524** | Pyrogens: **720** | Phagocytosis: **144** | Lymphatic system: **508** | Colony-stimulating factors: **528** | Skin: **78** | DNA, RNA: **41** | Transcription and translation: **112** | Coagulation: **539** | Cytokines: **174** | Mucus escalator of airways: **554** | Ligands: **129** | Types of epithelium: **67** |

Since the 1980s, it seems that human civilization has been visited by new plagues of infectious diseases. Some of them, such as tuberculosis and rabies, are old enemies that we thought we had under control. Others, such as SARS (severe acute respiratory syndrome), monkeypox, and the Hanta and the Ebola viruses, seem to be cropping up out of nowhere.

The conflict between humans and the invaders that cause disease is literally a fight for survival on both parts. The viruses and parasites need a host such as the human body to reproduce. But if they kill off all the hosts, they too will die out. We as hosts must fight off the invaders so that we can continue to survive as a species.

The ability of the body to protect itself from viruses, bacteria, and other disease-causing entities is known as **immunity**, from the Latin word *immunis*, meaning *exempt*. The human immune system consists of lymphoid tissues of the body, the immune cells, and chemicals, both intracellular and secreted, that coordinate and execute the immune functions.

▶ OVERVIEW OF IMMUNE SYSTEM FUNCTION

The key features of the immune system are *specificity* and *memory*. Together these processes enable the body to distinguish "self" from "non-self" and mount a targeted response to specific invaders when they are encountered a second time.

The immune system serves three major functions:

1. **It protects the body from disease-causing invaders** known as **pathogens**. Microscopic invaders, or **microbes**, include bacteria, viruses, fungi, and one-celled protozoans. Larger pathogens include multicellular **parasites** such as hookworms and tapeworms.

   In addition, virtually any molecule or cell not of the body has the potential to elicit an immune response. Pollens, chemicals, and foreign bodies are examples of substances to which the body may react. Substances that trigger an immune response from the body and that can react with products of that response are known as **antigens**.

2. **The immune system removes dead or damaged tissue and cells** or old red blood cells. Scavenger cells of the immune system patrol the extracellular compartment, gobbling up and digesting dead or dying cells.

3. **The immune system tries to recognize and remove abnormal cells** created when normal cell growth and development go wrong. For example, the diseases we call *cancer* result from abnormal cells that multiply repeatedly, crowding out normal cells and disrupting function. Scientists believe that cancer cells form on a regular basis, but are usually detected by the immune system and destroyed before they get out of control.

Sometimes the body's immune system fails to perform its normal functions. Pathologies of the immune system generally fall into one of three categories: incorrect response, overactive response, or lack of response.

1. *Incorrect responses.* If mechanisms for distinguishing self from non-self fail and the immune system attacks the body's normal cells, an *autoimmune disease* results. Type 1 diabetes mellitus, in which proteins made by immune cells destroy pancreatic beta cells, is an example of an autoimmune disease in humans.

2. *Overactive responses.* Allergies are conditions in which the immune system creates a response that is out of proportion to the threat posed by the antigen. In extreme cases, the systemic effects of allergic responses can be life-threatening.

3. *Lack of response.* **Immunodeficiency diseases** arise when some component of the immune system fails to work properly. *Primary immunodeficiency* is a family of genetically inherited disorders that range from severe to mild. *Acquired immunodeficiencies* may occur as a result of infection, such as *acquired immunodeficiency syndrome* (AIDS) caused by the HIV virus. They may also arise as a side effect of drug or radiation therapy such as that used to treat cancer.

▶ PATHOGENS OF THE HUMAN BODY

In the United States, our most prevalent infectious diseases are viral and bacterial infections. Worldwide, parasites are an additional significant public health concern. For example, malaria, a protozoan parasite whose life cycle alternates between human and mosquito hosts, is estimated to infect as many as 100 million people in the world.

Many parasitic organisms, such as the protozoan that causes malaria, are injected into the body by biting insects. Others enter the body through the digestive tract, brought in by contaminated food and water. Some, such as the fungi that cause valley fever and histoplasmosis, are inhaled. A few, such as the blood fluke *Schistosoma*, actually burrow through the host's skin until they reach a capillary. Once in the body, microbes and parasites may enter host cells in an effort to hide from the immune response, or they may remain in the extracellular compartment.

Bacteria and Viruses Require Different Defense Mechanisms

Bacteria and viruses differ from each other in several ways (Table 24-1):

1. **Structure.** Bacteria are cells, with a cell membrane that is usually surrounded by a cell wall. Some bacteria also produce an additional protective outer layer known as a *capsule*. Viruses are not cells. They consist of a nucleic acid (DNA or RNA) core enclosed in a *capsid* coat of viral proteins. Some viruses add an *envelope* of phospholipid and protein made from the host's cell membrane, and incorporate viral proteins into the envelope (Fig. 24-1a ●).

2. **Living conditions and reproduction.** Most bacteria can survive and reproduce outside a host if they have the required nutrients, temperature, pH, and so on. Viruses must use the intracellular machinery of a host cell to reproduce.

3. **Susceptibility to drugs.** Most bacteria can be killed by the drugs we call **antibiotics**. These drugs act directly on bacteria and destroy them or inhibit their growth. Viruses cannot be killed by antibiotics. Some viral infections can be treated with *antiviral drugs* that target viral reproduction.

Table 24-1: Bacteria and Viruses

	BACTERIA	VIRUSES
Structure	Cells; usually surrounded by cell wall	Not cells; nucleic acid core with protein coat
Living conditions	Most can survive and reproduce outside a host	Parasitic; must have a host cell to reproduce
Susceptibility to drugs	Most can be killed or inhibited by antibiotics	Cannot be killed with antibiotics; some can be inhibited with antiviral drugs

Viruses Must Reproduce Inside Host Cells

The life cycle of a virus begins when the virus invades the host cell (Fig. 24-1b). To cross the human host cell membrane, the virus binds to membrane receptors, triggering endocytosis of the entire virus particle. In an alternative scenario, the virus envelope fuses with the host cell membrane, allowing the core of the virus to enter the cytoplasm.

Once inside, the virus's nucleic acid takes over the cell. Viruses use their own nucleic acid and the host cell's resources to make new viral nucleic acid and viral proteins. These components assemble into additional virus particles that are released from the host cell to infect other cells.

Viruses can be released from host cells in one of two ways. (1) The virus causes the host cell to rupture, releasing virus particles to the ECF, or (2) virus particles surround themselves with a capsule of host membrane and bud from the host cell surface.

Viruses do other kinds of damage to host cells. In taking over the host cell, they may totally disrupt the host cell's metabolism so that the host cell dies. Some viruses (*Herpes simplex* type 1 and varicella-zoster virus, which cause cold sores and chicken pox, respectively) "hide out" in the host cell and reproduce only sporadically.

Other viruses incorporate their DNA into the host cell DNA. Viruses with this characteristic include *human immunodeficiency virus* (HIV), the virus that causes AIDS, and **oncogenic viruses** that cause cancer.

Retroviruses and AIDS

The RNA viruses known as **retroviruses** reproduce in an interesting manner. Because these viruses do not have their own DNA to act as a template for mRNA synthesis, they produce an enzyme called **reverse transcriptase**. This enzyme allows the viral RNA to make the complementary viral DNA. This is the reverse of the usual process—normally, DNA codes for the synthesis of RNA. In this case, the virus-controlled cell makes viral DNA. After a period of time (the *latent period*), the cell produces viral mRNA. The viral mRNA then interrupts the host cell's protein synthesis, using host ribosomes to make viral proteins. The disease we call AIDS (**acquired immunodeficiency syndrome**) is attributed to viral infection by the HIV retrovirus.

▶ THE IMMUNE RESPONSE

The defense system of the body has two components. Physical and chemical barriers, such as skin, mucus, and stomach acid, first try to keep pathogens out of the body's internal environment. If that line of defense fails, then the internal **immune response** takes over.

Whatever the nature of the foreign material that comes into the body, the internal immune response includes the same basic steps: (1) *detection* and *identification* of the foreign substance, (2) *communication* with other immune cells to rally an organized

(a) Influenza, an RNA virus

RNA core

Envelope and capsid

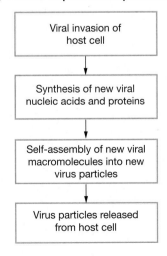

(b) General steps of viral reproduction

Viral invasion of host cell

↓

Synthesis of new viral nucleic acids and proteins

↓

Self-assembly of new viral macromolecules into new virus particles

↓

Virus particles released from host cell

● **Figure 24-1** *Viruses*

HIV is a retrovirus that uses an enzyme called *reverse transcriptase* to reproduce. Two drugs in the new AIDS "cocktail," AZT and ddC, are antiretroviral drugs that work by inhibiting reverse transcriptase. The other drug in the cocktail is a protease inhibitor. Protease is an enzyme needed to cleave key proteins during one step of HIV replication.

Question 1: Why can't retroviruses reproduce without reverse transcriptase?

response, (3) *recruitment* of assistance and *coordination* of the response among all participants, and (4) *destruction* or *suppression* of the invader.

Not all invaders can be destroyed by the body's immune system. In some cases, the best compromise the body can reach is to control the damage and keep the invader from spreading. Examples of pathogens that are suppressed by the immune system rather than destroyed include the bacterium that causes tuberculosis; the malaria parasite, which hides inside liver cells; and the herpes viruses that cause occasional outbreaks of cold sores or genital lesions.

The immune system is distinguished by its extensive use of chemical signaling. Detection, identification, communication, recruitment, coordination, and the attack on the invader depend on signal molecules such as antibodies and cytokines. **Antibodies**, proteins secreted by certain immune cells, bind antigens and make them more visible to the immune system. **Cytokines** are protein messengers released by one cell that affect the growth or activity of another cell [🔖 p. 174].

The human immune response is generally divided into two categories: nonspecific innate immunity and specific acquired immunity. **Innate immunity** is present from birth [*innatus*, inborn] and is the body's **nonspecific immune response** to invasion. Because the nonspecific innate response does not target a particular pathogen, it begins within minutes to hours. *Inflammation*, indicated on the skin by a red, warm, swollen area, is a hallmark reaction of innate immunity.

Acquired immunity, also called *adaptive* or **specific immunity**, is directed at specific invaders. One characteristic of acquired immunity is that a response to a first exposure to a pathogen may take days. With repeated exposures, the immune system remembers prior exposure to the pathogen and reacts more rapidly. Acquired immunity can be divided into *cell-mediated immunity* and *humoral immunity*, which refers to the presence of antibodies in the blood. (The term *humoral* comes from the Hippocratic school of medicine that divided the body's fluids into the four *humors* of blood, phlegm, black bile, and yellow bile.)

As we will see later in this chapter, the nonspecific and specific immune responses overlap. Although we have described them as if they are separate, in reality they are two interconnected parts of a single process. The innate response is the more rapid response but it does not target a specific invader. It is reinforced by the antigen-specific acquired response, which amplifies the efficacy of the innate response. Communication and coordination between the different pathways of immunity is vital for maximum protective effect.

Because of the huge numbers of molecules involved in human immunity, and because of the complex interactions between the different components of the immune system, the field of immunology is continuously expanding. In this chapter, you will learn the basics so that you can add the details with more advanced study.

▶ ANATOMY OF THE IMMUNE SYSTEM

The immune system is probably the least anatomically identifiable system of the body because most of it is integrated into the tissues of other organs such as the skin and gastrointestinal tract. Yet the mass of all the immune cells in the body equals the mass of the brain! The immune system has two anatomical components: the lymphoid tissues of the body and the cells that are responsible for the immune response.

Lymphoid Tissues Are Distributed Throughout the Body

The two primary lymphoid tissues are the **thymus gland** (see *Focus on the Thymus Gland*) and the **bone marrow**, both sites where cells involved in the immune response form and mature (Fig. 24-2a ●). Secondary lymphoid tissues are sites where mature immune cells interact with pathogens and initiate a response. Secondary tissues are divided into encapsulated tissues and unencapsulated diffuse lymphoid tissues.

The *encapsulated lymphoid tissues* are the spleen (Figs. 24-2 and 24-3 ●) and the lymph nodes (24-2b). Both have outer walls formed from fibrous collagenous capsules. The **spleen** contains immune cells positioned so that they monitor the blood for foreign invaders. Phagocytic cells in the spleen also trap and remove aging red blood cells.

The **lymph nodes** are associated with the lymphatic circulation. Recall from Chapter 15 that there is net flow of fluid out of capillaries and into the interstitial space [🔖 p. 506]. This fluid is picked up by the lymph capillaries. As the fluid called lymph moves through the lymph vessels, it passes through the encapsulated lymph nodes.

Inside the lymph nodes, clusters of immune cells intercept pathogens that have entered the interstitial fluid through breaks in the skin or mucous membranes. Once these bacteria, viruses, and other microbes have been swept into the lymph, the immune cells in the nodes help prevent their spread throughout the body. You have probably noticed that if you have a sinus infection or a sore throat, the lymph nodes in your neck become swollen. These sore, swollen nodes result from active immune cells that have collected within them to fight the infection.

Diffuse lymphoid tissues are aggregations of immune cells that appear throughout other organ systems of the body (see Fig. 24-2). They include the **tonsils** at the posterior nasopharynx, the **gut-associated lymphoid tissue** (GALT) that lies just under the epithelium of the esophagus and intestines, and clusters of lymphoid tissue associated with the skin and respiratory, urinary, and reproductive tracts.

In each case, these tissues contain immune cells that are in position to intercept pathogens before they get into the general circulation. Because of the large surface area of the digestive tract epithelium, some authorities consider the GALT to be the largest immune organ in the body. Anatomically, the cells of the immune system can be found in highest concentrations wherever they are most likely to encounter antigens that penetrate the epithelia.

Leukocytes Are the Primary Cells of the Immune System

The primary cells of the immune system are the white blood cells, or **leukocytes**, which were briefly introduced in Chapter 16 [p. 524]. Most leukocytes are much larger than red blood cells. They are also not nearly as numerous. A microliter (1 μL) of whole blood contains about 5 million red blood cells but only about 7000 leukocytes.

Although most leukocytes circulate in the blood, they usually leave the capillaries and carry out their job *extravascularly* (outside the vessels). Some types of leukocytes can live for several months out in the tissues, but others may live for only hours or days.

Leukocytes are divided into six basic groups: (1) *eosinophils*, (2) *basophils* in the blood and the related *mast cells* in the tissues, (3) *neutrophils*, (4) *monocytes* and their derivative *macrophages*, (5) *lymphocytes* and their derivatives such as *plasma cells*, and (6) *dendritic cells*. Dendritic cells are not usually found in the blood and therefore were not included with leukocytes in Chapter 16.

Leukocytes can be distinguished from one another in stained tissue samples by the shape and size of the nucleus, the staining characteristics of the cytoplasm and cytoplasmic inclusions, and the regularity of the cell border.

Immune Cell Names Reflect Cell Function or Appearance The terminology associated with immune cells can be very confusing. Some cell types have several variants, and others have been given multiple names (Fig. 24-4). In addition, immune cells can be grouped functionally or morphologically.

(a)

Tonsil is diffuse lymphoid tissue.

Thymus produces T lymphocytes.

Lymph nodes

Encapsulated lymphoid tissues

Spleen

Gut-associated lymphoid tissue (GALT) is a diffuse lymphoid tissue.

Lymph vessels

Bone marrow produces most blood cells.

(b)

Afferent lymph vessel

Capsule

Clusters of immune cells intercept pathogens that invade interstitial fluid.

Efferent lymph vessel

Lymph node artery and vein

● **Figure 24-2** *Anatomy of the immune system*

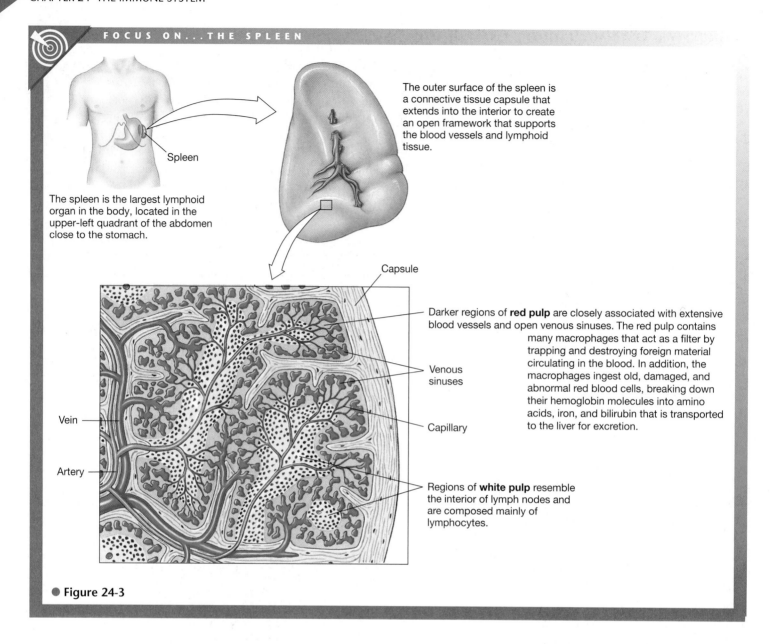

Spleen

The spleen is the largest lymphoid organ in the body, located in the upper-left quadrant of the abdomen close to the stomach.

The outer surface of the spleen is a connective tissue capsule that extends into the interior to create an open framework that supports the blood vessels and lymphoid tissue.

Capsule

Darker regions of **red pulp** are closely associated with extensive blood vessels and open venous sinuses. The red pulp contains many macrophages that act as a filter by trapping and destroying foreign material circulating in the blood. In addition, the macrophages ingest old, damaged, and abnormal red blood cells, breaking down their hemoglobin molecules into amino acids, iron, and bilirubin that is transported to the liver for excretion.

Venous sinuses

Vein

Capillary

Artery

Regions of **white pulp** resemble the interior of lymph nodes and are composed mainly of lymphocytes.

● Figure 24-3

One morphological group is the **granulocytes**, white blood cells whose cytoplasm contains prominent granules. The names of the different cell types come from the staining properties of the granules. Neutrophil granules do not stain darkly with standard blood stains and are therefore "neutral." Basophil granules stain dark blue with basic (alkaline) dye, and eosinophil granules stain dark pink with the acidic dye *eosin*. In all three types of granulocytes, the contents of the granules are released into the extracellular space in response to some type of signal, a process known as *degranulation*.

One functional group of leukocytes is the phagocytes. **Phagocytes** are white blood cells that engulf and ingest their targets by phagocytosis. This group includes the neutrophils, macrophages, monocytes (which are macrophage precursors), and eosinophils.

Cytotoxic cells kill the cells they are attacking. This group includes eosinophils and some types of lymphocytes. **Antigen-presenting cells** (APCs) display fragments of foreign proteins on their cell surface. This group includes certain lymphocytes, dendritic cells, macrophages, and monocytes.

The terminology associated with macrophages has changed along with the history of histology and immunology. For many years, tissue macrophages were known as the *reticuloendothelial system* and were not associated with white blood cells. To confuse matters, the cells were named when they were first described in different tissues, before they were all identified as macrophages. Thus, histiocytes in skin, Kupffer cells in the liver, osteoclasts in bone, microglia in brain, and reticuloendothelial cells in spleen are all names for specialized macrophages. The new name for the reticuloendothelial system is the **mononuclear phagocyte system**, a term that refers both to macrophages in the tissues and to their parent monocytes circulating in the blood.

In the sections that follow, we take a closer look at the six basic types of leukocytes. Figure 24-4, introduced on the following page, makes a good reference.

	Basophils and Mast Cells	Neutrophils	Eosinophils	Monocytes and Macrophages	Lymphocytes and Plasma Cells	Dendritic Cells
% of WBCs in blood	Rare	50–70%	1–3%	1–6%	20–35%	NA
Subtypes and nicknames		Called "polys" or "segs" Immature forms called "bands" or "stabs"		Called the mononuclear phagocyte system	B lymphocytes, Plasma cells T lymphocytes Cytotoxic T cells Helper T cells Natural killer cells Memory cells	Also called Langerhans cells, veiled cells
Primary function(s)	Release chemicals that mediate inflammation and allergic responses	Ingest and destroy invaders	Destroy invaders, particularly antibody-coated parasites	Ingest and destroy invaders Antigen presentation	Specific responses to invaders, including antibody production	Recognize pathogens and activate other immune cells by antigen presentation in lymph nodes
Classifications		Phagocytes				
		Granulocytes				
			Cytotoxic cells		Cytotoxic cells (some types)	
					Antigen-presenting cells	

● **Figure 24-4** *Cells of the immune system*

Eosinophils Fight Parasites and Contribute to Allergic Reactions

Eosinophils, easily recognized by the bright pink-staining granules in their cytoplasm, are associated with allergic reactions and parasitic diseases. Normally, few eosinophils are found in the peripheral circulation, where they account for only 1%–3% of all leukocytes. The life span of an eosinophil in the blood is estimated to be only 6–12 hours. Most functioning eosinophils are found in the digestive tract, lungs, epithelium of urinary and genital tracts, and connective tissue of the skin. These locations reflect their role in defense against parasitic invaders.

Eosinophils are known to attach to large antibody-coated parasites such as the blood fluke *Schistosoma* and release substances that damage or kill them. Because eosinophils kill pathogens, they are classified as *cytotoxic* [*cyto-*, cell] cells.

Eosinophils also participate in allergic reactions, where they contribute to inflammation and tissue damage by releasing toxic enzymes and oxidative substances. Although eosinophils have been shown to ingest particles such as pollen *in vitro*, there is no evidence that they do so *in vivo*.

Basophils Release Histamine and Other Chemicals

Basophils are rare in the circulation but are easily recognized in a stained blood smear by the large, dark blue granules in their cytoplasm. They are very similar to the **mast cells** of tissues. The granules of mast cells and basophils contain histamine, **heparin** (an *anticoagulant* that inhibits blood clotting), cytokines, and other chemicals involved in allergic and immune responses.

Mast cells, like eosinophils, are found concentrated in the connective tissue of skin, lungs, and the gastrointestinal tract. In these locations, they are ideally situated to intercept pathogens that are inhaled or ingested or that enter through breaks in the epidermis.

Neutrophils "Eat" Bacteria and Release Cytokines

Neutrophils are the most abundant white blood cell (50%–70% of the total) and the most easily identified, with a segmented nucleus of three to five lobes connected by thin strands of nuclear material. Because of the segmented nucleus, neutrophils are also called *polymorphonuclear leukocytes* ("polys") or "segs." Immature neutrophils, occasionally found in the circulation, can be identified by their horseshoe-shaped nucleus. These immature neutrophils go by the nicknames of "bands" and "stabs."

Neutrophils, like other blood cells, are formed in the bone marrow. Neutrophils are phagocytic cells that typically ingest and kill five to twenty bacteria during their programmed life span of one to two days. Most neutrophils spend their lives in the blood, but they can leave the circulation if attracted to an extravascular site of damage or infection. In addition to ingesting bacteria and foreign particles, neutrophils release a variety of cytokines, including fever-causing pyrogens [❧ p. 720] and chemical mediators of the inflammatory response.

Monocytes Mature into Macrophages   Monocytes are the precursor cells of tissue **macrophages**. Monocytes are not very common in the blood (1%–6% of all white blood cells). By some estimates, they spend only eight hours there in transit from the bone marrow to their permanent positions in the tissues.

Once out of the blood, rounded monocytes enlarge and differentiate into macrophages. Some tissue macrophages spend their lives patrolling the tissues, creeping along by amoeboid motion. Others find a location and remain fixed in place.

In either case, macrophages are the primary scavengers for the tissues. They are larger and more effective than neutrophils, ingesting up to 100 bacteria during their life span. Macrophages also remove larger particles such as old red blood cells and dead neutrophils.

Macrophages play an important role in the development of acquired immune responses. When they ingest molecular or cellular antigens and digest them, fragments of antigen are processed and inserted into the macrophage membrane as part of surface protein complexes (Fig. 24-5●). Immune cells that process antigen in this fashion are called *antigen-presenting cells* (APCs). Dendritic cells and the lymphocytes known as *B lymphocytes* also act as antigen-presenting cells.

Lymphocytes Mediate Acquired Immunity   Lymphocytes and their derivatives are the key cells that mediate the acquired immune response of the body. Only 5% of all lymphocytes

are found in the circulation, where they constitute 20%–35% of all white blood cells. Most lymphocytes are found in the lymphoid tissues where they are likely to encounter invaders, such as lymph nodes, spleen, and the gut-associated lymphoid tissue.

By one estimate, the adult body contains a trillion (10^{12}) lymphocytes at any one time. Although most of these cells look alike under the microscope, they have major differences in their function and specificity, as you will learn later in the chapter.

Dendritic Cells Activate Lymphocytes   Dendritic cells are antigen-presenting cells characterized by long, thin processes that resemble the dendrites of the neuron. Dendritic cells are found in the skin (where they are called Langerhans cells) and in various organs. When dendritic cells recognize and capture antigens, they migrate to the secondary lymphoid tissues. There they present antigen to lymphocytes to activate them.

▶ INNATE IMMUNITY: NONSPECIFIC RESPONSES

The body's first line of defense is to exclude pathogens by physical and chemical barriers. If those barriers fail to keep invaders out, then innate immunity provides the second line of defense. The innate immune response consists of patrolling and stationary leukocytes that attack and destroy invaders. These cells respond in the same fashion to any material they identify as foreign, which is why the innate response is considered nonspecific. Innate immunity either clears the infection or contains it until the acquired immune response is activated.

Physical and Chemical Barriers Are the Body's First Line of Defense

Physical barriers of the body include the skin [❧ p. 78], the protective mucous linings of the gastrointestinal and genitourinary tracts, and the ciliated epithelium of the respiratory tract. The digestive and respiratory systems are most vulnerable to microbial invasion because these regions have extensive areas of thin epithelium in di-

● **Figure 24-5**   *Antigen-presenting immune cells (APCs)*

rect contact with the external environment. In women, the reproductive tract is also vulnerable, but to a lesser degree. The opening to the uterus is normally sealed by a mucus plug that keeps bacteria from ascending from the vagina into the uterine cavity.

In the respiratory system, inhaled particulate matter is trapped by mucus lining the upper respiratory system. The mucus is then transported upward on the mucus escalator to be expelled or swallowed [❀ p. 554]. Swallowed pathogens may be disabled by the acidity of the stomach. In addition, respiratory tract secretions and tears contain **lysozyme**, an enzyme with antibacterial activity. Lysozyme attacks components of unencapsulated bacterial cell walls and breaks them down. It cannot digest the capsules of encapsulated bacteria, however.

Pathogens that get past the physical barriers of skin and mucus are dealt with by immune cells and the innate immune response. When patrolling and stationary phagocytes in the tissues detect invaders, their response is twofold: destroy or suppress the invader by ingesting it, and attract additional immune cells by secreting cytokines. Molecules that attract immune cells are called **chemotaxins**. They include bacterial toxins, products of tissue injury (fibrin and collagen fragments), and cytokines released by leukocytes.

Phagocytes Recognize Foreign Material and Ingest It

The primary phagocytic cells of the immune system are the tissue macrophages and the neutrophils. If the invader is in the tissue, phagocytes leave the circulation (*extravasation*) by squeezing through pores in the capillary endothelium Once the phagocytes reach the pathogen, they identify it by chemical cues, then ingest it.

Phagocytosis, the process in which cells engulf and digest particulate matter, is similar in macrophages and neutrophils [❀ p. 144]. It is a receptor-mediated event, which ensures that only unwanted particles are ingested.

In the simplest reactions, surface molecules on the pathogen act as ligands that bind directly to receptors on the phagocyte membrane (Fig. 24-6a ●). In a sequence similar to a zipper closing, the ligands and receptors combine sequentially, so that the phagocyte surrounds the unwanted foreign particle. The process is aided by actin filaments that push arms of the phagocytic cell around the invader.

Phagocytes recognize many different types of foreign particles, both organic and inorganic, using cell-surface receptors. Phagocytes ingest unencapsulated bacteria, cell fragments, carbon, and asbestos particles, among other materials. In the laboratory, scientists use the ingestion rate of tiny polystyrene beads as an indicator of phagocytic activity.

Phagocytes cannot instantly recognize all foreign material, however. For example, certain bacteria have evolved to hide their surface markers behind a polysaccharide capsule (Fig. 24-6b). Even so, these bacteria may not escape detection. If pathogens lack markers, they can be "tagged" so that the phagocytes will recognize them as something to be ingested. This tagging occurs when the pathogen becomes coated with proteins that occur in the blood of the host.

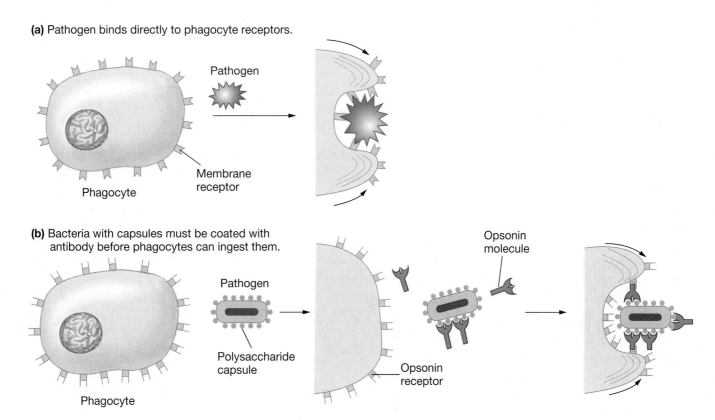

(a) Pathogen binds directly to phagocyte receptors.

Pathogen

Phagocyte

Membrane receptor

(b) Bacteria with capsules must be coated with antibody before phagocytes can ingest them.

Opsonin molecule

Pathogen

Phagocyte

Polysaccharide capsule

Opsonin receptor

● **Figure 24-6** *Phagocytosis*

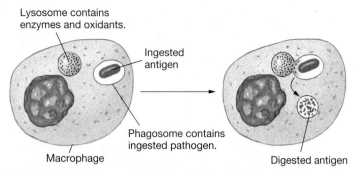

Lysosome contains enzymes and oxidants.

Ingested antigen

Phagosome contains ingested pathogen.

Macrophage

Digested antigen

● **Figure 24-7** *Enzymes digest material within phagosomes*

The tagging proteins are known collectively as **opsonins** (a term derived from the Greek word *opsonin*, meaning "to buy provisions.") In the body, opsonins convert unrecognizable particles into "food" for phagocytes. Opsonins act as a bridge between pathogens and phagocyte by binding to receptors on the phagocyte.

The ingested particle ends up in a cytoplasmic vesicle called a **phagosome** (Fig. 24-7 ●). Phagosomes fuse with intracellular lysosomes [● p. 61], which contain powerful chemicals that destroy and kill ingested pathogens. Lysosomal contents include enzymes and oxidizing agents such as hydrogen peroxide (H_2O_2), nitric oxide (NO), and the superoxide anion (O^-).

If an area of infection attracts a large number of phagocytes, the material known as **pus** may form. This thick whitish to greenish substance is a collection of living and dead neutrophils and macrophages, along with tissue fluid, cell debris, and other remnants of the immune process.

Natural Killer Lymphocytes Eliminate Virally Infected and Tumor Cells

One class of lymphocyte, the **natural killer cells**, or **NK cells**, participates in the innate response against viral infections. NK cells act more rapidly than other lymphocytes, with a response that can be measured within hours of a primary viral infection. NK cells induce virus-infected cells to commit suicide (*apoptosis*, ● p. 76) before the virus can reproduce. Complete elimination of the virus requires activation of a specific immune response. NK cells also attack some tumor cells.

NK cells and other lymphocytes secrete antiviral cytokines, including gamma-interferon. **Interferons** were named for their ability to interfere with viral reproduction in the body. **Alpha-interferon** (interferon-α) and **beta-interferon** (interferon-β) target uninfected cells and activate pathways to prevent viral replication. **Gamma-interferon** (interferon-γ) activates macrophages and other immune cells.

Chemical Mediators Create the Inflammatory Response

Inflammation, indicated on the skin by a red, warm, swollen area, is a hallmark reaction of the innate immune pathway. It has three important roles in fighting infection: (1) attracting immune cells and chemical mediators to the infection site, (2) producing a physical barrier to retard the spread of infection, and (3) promoting tissue repair once the infection is under control (a non-immunological function).

The inflammatory response is created when activated tissue macrophages release cytokines. These chemicals attract other immune cells, increase capillary permeability, and cause fever. Immune cells attracted to the site in turn release their own cytokines. Some representative examples of chemicals involved in the innate immune response are described below. They are also listed in Table 24-2, a convenient "scorecard" for following the players in these complex processes.

Acute Phase Proteins Are Released Early in the Immune Response In the time immediately following an injury or pathogen invasion (the acute phase), the body responds by increasing the concentration of various plasma proteins. Some of these proteins, produced mostly by the liver, are given the general name of **acute phase proteins**. They include molecules that act as opsonins, antiprotease molecules to help prevent tissue damage, and *C-reactive protein*.

Normally, the levels of these proteins fall back to normal as the immune response proceeds, but in *chronic inflammatory diseases* such as rheumatoid arthritis, elevated levels of acute phase proteins may persist. Increased levels of C-reactive protein are associated with increased risk of coronary heart disease [● p. 512].

Histamine Initiates Inflammation Histamine is an amine derived from the amino acid histidine [● p. 38]. Histamine is found primarily in the granules of mast cells and basophils, and

Table 24-2: Chemicals of the Immune Response

FUNCTIONAL CLASSES

Acute phase proteins: Liver proteins released during the acute phase that act as opsonins and enhance the inflammatory response

Chemotaxins: Molecules that attract phagocytes to a site of infection

Cytokines: Proteins released by one cell that affect the growth or activity of another cell

Opsonins: Proteins that coat pathogens so phagocytes recognize and ingest them: antibodies, acute phase proteins, and complement proteins

Pyrogens: Fever-producing substances

SPECIFIC CHEMICALS AND THEIR FUNCTIONS

Antibodies (immunoglobulins, gamma globulins): Proteins secreted by B lymphocytes that fight specific invaders

Bradykinin: Stimulates pain receptors; vasodilator

Complement: Plasma and cell membrane proteins that act as opsonins, cytolytic agents, and mediators of inflammation

C-reactive protein: Opsonin that activates complement cascade

Granzymes: Cytotoxic enzymes that initiate apoptosis in target cells

Heparin: An anticoagulant

Histamine: Vasodilator and bronchoconstrictor released by mast cells and basophils

Interferons: Proteins that inhibit viral reproduction and modulate the immune response

Interleukin-1: Macrophage cytokine that mediates inflammatory response and induces fever

Kinins: Inactive plasma proteins that activate in a cascade to form bradykinin

Lysozyme: An extracellular enzyme that attacks bacteria

Major histocompatibility complex (MHC): A family of membrane protein complexes used for cell recognition

Membrane attack complex: A membrane pore protein made in the complement cascade; allows ions and water to enter pathogens

Perforin: A membrane pore protein that allows granzymes to enter the cell; made by NK and cytolytic T cells

Superoxide anion (O^-): Powerful oxidant in phagocyte lysosomes

T-cell receptors: Membrane receptors on T lymphocytes that recognize and bind antigen presented by MHC receptors

it is the active molecule that helps initiate the inflammatory response when mast cells degranulate. Histamine's actions bring more leukocytes to the injury site to kill bacteria and remove cellular debris.

Histamine opens pores in capillaries, allowing plasma proteins to escape into the interstitial space. This process causes local edema, or swelling. In addition, histamine dilates blood vessels, increasing blood flow to the area. The result of histamine release is a hot, red, swollen area around a wound or infection site.

Mast cell degranulation is triggered by different cytokines in the immune response. Because mast cells are concentrated under mucous membranes that line the airways and digestive tract, the inhalation or ingestion of certain antigens can trigger histamine release. The resultant edema in the nasal passages leads to one annoying symptom of seasonal pollen allergies, the stuffy nose. Fortunately, pharmacologists have developed a variety of drugs called *antihistamines*. These drugs block the action of histamine at its receptor.

Scientists used to believe that histamine was the primary chemical responsible for the bronchoconstriction of asthma and for the severe systemic allergic reaction called *anaphylactic shock*. However, research has shown that mast cells release powerful lipid mediators in addition to histamine, including leukotrienes, platelet-activating factor, and prostaglandins. These mediators work with histamine to create the bronchoconstriction and hypotension that are characteristic of anaphylactic shock.

Interleukins Have Widespread Systemic Effects

Interleukins are cytokines that were initially thought to mediate communication between the different types of leukocytes in the body. As scientists learn more about them, however, they have discovered that many different tissues in the body respond to interleukins.

Interleukin-1 (IL-1) is secreted by activated macrophages and other immune cells. Its main role is to mediate the inflammatory response, but it also has widespread systemic effects on immune function and metabolism. Interleukin-1 modulates the immune response by:

1. *Altering blood vessel endothelium* to ease passage of white blood cells and proteins during the inflammatory response.
2. *Stimulating production of acute phase proteins* by the liver.
3. *Inducing fever* by acting on the hypothalamic thermostat [◉ p. 720]. IL-1 is a known pyrogen of the body.
4. *Stimulating cytokine and endocrine secretion* by a variety of other cells.

Bradykinin Stimulates Pain Receptors *Kinins* are a group of inactive plasma proteins that participate in a cascade similar to the coagulation cascade [◉ p. 539]. The end product of the kinin cascade is **bradykinin**, a molecule with the same vasodilator effects as histamine. Bradykinin also stimulates pain receptors, creating the tenderness associated with inflammation. Pain draws the brain's attention to the injury.

Complement Proteins Are Opsonins and Chemotaxins

Complement is a collective term for a group of more than 25 plasma proteins and cell membrane proteins. The complement pathway is a cascade of events similar to the blood coagulation cascade. Various intermediates of the complement cascade act as opsonins, chemical attractants for leukocytes, and agents to cause degranulation of mast cells.

Complement proteins insert themselves into the membrane of pathogens, creating a pore.

H₂O and ions

Water and ions enter cell.

Cell swells and lyses.

Pathogen

Pore of membrane attack complex

● **Figure 24-8** *Membrane attack complex*

The complement cascade terminates with the formation of **membrane attack complex**, a group of lipid-soluble proteins that insert themselves into the cell membranes of pathogens and form pores (Fig. 24-8 ●). These pores allow ions to enter the cell, and water follows osmotically. As a result, the cell swells and lyses.

CONCEPT CHECK   Answers: p. 781

1. Why does the action of histamines on capillary permeability result in swelling?

▶ ACQUIRED IMMUNITY: ANTIGEN-SPECIFIC RESPONSES

Acquired immune responses are *antigen-specific responses* in which the body recognizes a particular foreign substance and selectively reacts to it. Acquired immunity is mediated primarily by the family of immune cells called lymphocytes. There are three main types of lymphocytes: B lymphocytes, T lymphocytes, and natural killer cells. Activated *B lymphocytes* develop into *plasma cells* that secrete antibodies. Activated *T lymphocytes* develop either into cells that attack and destroy virus-infected cells or into cells that regulate other immune cells. The third group, the *natural killer* or *NK cells*, attacks and destroys virus-infected and tumor cells and was discussed earlier. All lymphocytes secrete cytokines that act on immune cells, on non-immune cells of the body, and, sometimes, on the pathogens themselves. The primary lymphocyte cytokines are the interleukins, discussed earlier.

The process of acquired immunity overlaps with the process of innate immunity. Cytokines released by the inflammatory response attract lymphocytes to the site of an immune reaction. The lymphocytes release additional cytokines that enhance the inflammatory response.

Acquired immunity can be subdivided into active immunity and passive immunity. **Active immunity** occurs when the body is exposed to a pathogen and creates its own antibodies. Active immunity can occur naturally, when the pathogen invades the body, or artificially, as when we are given vaccinations that contain dead or disabled pathogens. **Passive immunity** occurs when we

acquire antibodies made by another animal. The transfer of antibodies from mother to fetus across the placenta is one example. Immunoglobulin shots are another. Travelers going abroad may be injected with gamma globulin extracted from donated human plasma, but this passive immunity only lasts about three months.

Lymphocytes Are the Primary Cells Involved in the Acquired Immune Response

On a microscopic level, lymphocytes look alike. At the molecular level, however, the cells can be distinguished by their membrane receptors, which make each cell specific for a particular ligand (Fig. 24-9 ●). Each unique type of lymphocyte, along with the other cells exactly like it, forms a **clone** [*klon*, a twig].

If a specific type of lymphocyte fights each pathogen that enters the body, millions of different types of lymphocytes must be

A clone is a group of lymphocytes that are specific to one antigen.

Clone 1

Clone 2

Clone 3

FIGURE QUESTION

Which clone will this antigen activate?

● **Figure 24-9** *Lymphocyte clones* This figure shows three pairs of lymphocytes. Each pair is from a different clone.

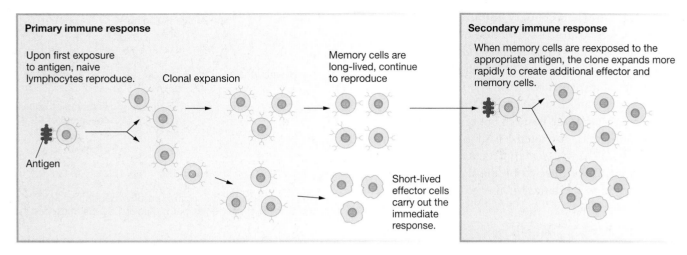

Primary immune response

Upon first exposure to antigen, naive lymphocytes reproduce.

Clonal expansion

Memory cells are long-lived, continue to reproduce

Antigen

Short-lived effector cells carry out the immediate response.

Secondary immune response

When memory cells are reexposed to the appropriate antigen, the clone expands more rapidly to create additional effector and memory cells.

● **Figure 24-10** *Clonal expansion and memory*

ready to combat millions of different pathogens. But where do all those lymphocytes come from?

At an individual's birth, each clone of lymphocytes is represented by only a few cells, called **naive lymphocytes** [*naif*, natural]. The small number of cells in each naive clone is not enough to fight off foreign invaders, so the first exposure to an antigen activates the appropriate clone and causes it to divide (Fig. 24-10●). This process, called **clonal expansion**, creates additional cells in the clone.

The newly formed lymphocytes in the clone differentiate into **effector cells**, which create the immune response, and into **memory cells**. Effector cells are short-lived and die within a few days. However, memory cells have longer lives and continue to reproduce themselves. With second and subsequent exposures to the antigen, the memory cells activate, creating a more rapid and stronger secondary response to the antigen.

B Cells Differentiate into Plasma Cells and Memory Cells

The first class of lymphocytes, **B lymphocytes** (also called **B cells**), develops in the bone marrow (a fact that provides a convenient way to remember their name.) Activated B lymphocytes differentiate primarily into specialized cells that secrete antibodies.

Antibodies, or immunoglobulins, are proteins that bind to pathogens and target them for destruction. The term *antibody*

describes what the molecules do: they work against the foreign body. The equivalent term *immunoglobulin* describes what the molecules are: globular proteins that participate in the humoral immune response.

Mature B lymphocytes insert antibody molecules into their cell membranes so that antibodies become surface receptors marking the members of each clone (see Fig. 24-9). When a clone of B cells activates in response to antigen exposure, some differentiate into plasma cells.

Plasma cells are the effector cells for the B lymphocytes. They do not have antibody proteins bound in their membranes. Instead, they synthesize and secrete additional antibody molecules at incredible rates, estimated to be as high as 2000 molecules of antibody per second! Plasma cell antibodies form *humoral immunity*, the soluble antibodies of the plasma.

After each invader has been successfully repulsed, a few *memory cells* of the clone remain behind, so that the next exposure to the same antigen is greeted by a more rapid response. This phenomenon can be demonstrated by measuring the concentration of antibodies in the circulation.

Figure 24-11● shows the antibody response after the first and second exposure to an antigen. Notice that the **primary immune response**, which occurs after the initial exposure, is slower and lower in magnitude because the body has not encountered

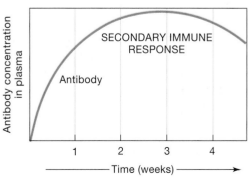

● **Figure 24-11** *Memory in the immune system* Antibody production in response to the first exposure to an antigen is both slower and weaker than the secondary response upon reexposure.

the antigen previously. The **secondary immune response**, which takes place after a second and subsequent exposure, is greater and quicker because it is enhanced by lymphocytes that carry a molecular memory of the first exposure to the antigen.

The existence of the enhanced secondary response is used in **vaccinations**. Patients are given a pathogen that has been altered so that it no longer harms the host cell but can be recognized as foreign by immune cells. The altered pathogen causes the creation of memory cells keyed to that particular pathogen. If the vaccinated person is later infected by the natural pathogen, a stronger and more rapid immune response results.

CONCEPT CHECK   Answers: p. 781

2. A child is stung by a bee for the first time. Why should the parent be particularly alert when the child is stung a second time?

Antibodies Are Proteins Secreted by Plasma Cells

Historically, antibodies were among the first aspects of the immune system to be discovered, and traditionally they were named by what they do—agglutinins, precipitins, hemolysins, and so on.

When plasma proteins are separated by a biochemical technique called *electrophoresis*, the antibodies move together in a group given the general name of **gamma globulins**. Today, antibodies are divided into five general classes of immunoglobulins (Ig): IgG, IgA, IgE, IgM, and IgD (pronounced *eye-gee-[letter]*).

IgG makes up 75% of the plasma antibody in adults because these immunoglobulins are produced in secondary immune responses. Maternal IgGs cross the placental membrane, so these antibodies give infants immunity in the first few months of life. IgG activates complement.

IgA antibodies are found in the external secretions of the body such as saliva, tears, intestinal and bronchial mucus, and breast milk. In these locations, they disable pathogens before they reach the internal environment.

IgE is associated with allergic responses. When mast cell receptors combine with IgE and antigen, the mast cells degranulate and release chemical mediators such as histamine.

IgM is associated with primary immune responses and with the antibodies that react to blood group antigens. IgM activates complement.

IgD appears on the surface of B lymphocytes along with IgM, but its physiological role is unclear.

CONCEPT CHECK   Answers: p. 781

3. Because antibodies are proteins, they are too large to cross cell membranes on transport proteins or through channels. How then do IgA and other antibodies become part of external secretions such as saliva, tears, and mucus?

An antibody molecule is composed of two identical light chains and two identical heavy chains, linked by disulfide bonds.

● **Figure 24-12** *Antibody structure* The arms contain binding sites for antigens.

Antibodies Are Y-Shaped Proteins The basic antibody molecule has four polypeptide chains linked into a Y-shape (Fig. 24-12 ●). Each side of the Y is identical, with one light chain attached to one heavy chain. The two arms, or **Fab regions**, contain antigen-binding sites that confer the antibody's specificity. The stem of the Y-shaped antibody molecule is known as the **Fc region**. The Fc region determines to which of the five Ig classes an antibody belongs. A *hinge region* between the arms and the stem allows flexible positioning of the arms.

In any given antibody molecule, both light chains are identical and both heavy chains are identical. However, the chains vary widely among different antibodies and depend upon the clone from which they come. In addition, the secreted forms of two immunoglobulin classes are polymers formed from multiple antibody molecules. IgM has five Y-shaped antibody molecules. IgA immunoglobulins have from one to four antibody molecules.

Antibodies Have Multiple Functions Most antibodies are found in the plasma, where they make up about 20% of the plasma proteins in a healthy individual. These antibodies are most effective against extracellular pathogens, such as bacteria, some parasites, antigenic macromolecules, and viruses before they have invaded their host cells.

Antibodies make antigens more visible to the immune system in the following three ways:

1. **Acting as opsonins.** Soluble antibodies coat antigens to facilitate recognition and phagocytosis by immune cells (Fig. 24-13 ●, ②). Antibodies are not toxic by themselves, but they label antigens so that other immune cells will attack them.
2. **Making antigens clump.** Antibodies cause antigens to clump for easier phagocytosis (Fig. 24-13, ③).
3. **Inactivating bacterial toxins.** Antibodies bind to and inactivate some toxins produced by bacteria. Macrophages can then ingest and destroy the inactivated toxins.

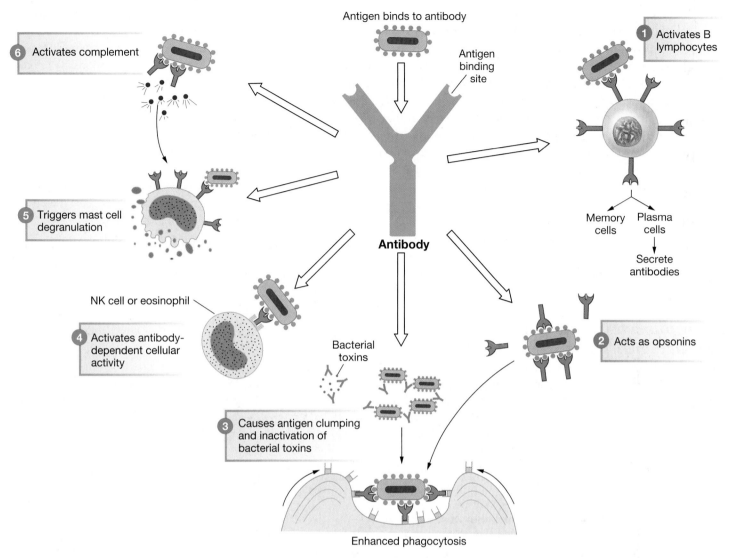

● **Figure 24-13** *Functions of antibodies* Adapted from J. Kuby, *Immunology 2e* (New York: W. H. Freeman, 1994).

Antibodies also enhance inflammation by acting in at least two other ways:

4. **Activating complement.** Antigen-bound antibodies use the Fc end of the antibody molecule to activate complement, proteins that play important roles in innate immunity (Fig. 24-13, ⑥).

5. **Activating mast cells.** Mast cells have IgE antibodies attached to their surface. When antigens or complement proteins bind to IgE, the mast cells degranulate and release chemicals that mediate the inflammatory response (Fig. 24-13, ⑤).

Finally, the last role of antibodies is to activate immune cells. In some instances, the antibody binds first to antigen, then to the immune cell. In the case of B lymphocytes, the antibodies are already an integral part of the lymphocyte. Antigen binding activates the cell.

6. **Antibody-bound antigen binds to and activates immune cells.** Phagocytic and some cytotoxic cells have membrane

receptors that bind to the Fc end of antigen-bound antibodies. This mechanism eliminates the need for these cells to have millions of different receptors to recognize different antigens. Instead, with a single Fc receptor, these cells can recognize any antibody-bound antigen. Fc binding may initiate phagocytosis (Fig. 24-13, ③), or it may cause the immune cell to kill the target cell by the release of toxic chemicals.

For example, natural killer cells can bind to the Fc stem of antibodies coating target cells. Activation causes the NK cell to kill the target by apoptosis. This nonspecific response of natural killer cells is called **antibody-dependent cell-mediated cytotoxicity**. Eosinophils also act in this fashion (Fig. 24-13, ④).

Antigen-Binding to Antibodies on B Cells Activates the B Cells
The surface of each B lymphocyte is covered with as many as 100,000 antibody molecules whose Fc ends are inserted into the lymphocyte membrane. This leaves the arms (Fab regions) free to bind to antigens in the extracellular fluid. Antigen binding

activates the B lymphocyte, causing it to differentiate into either a memory cell or an antibody-secreting plasma cell (Fig. 24-13, ①).

B cell antibodies bind to viruses or to toxins produced by bacteria, preventing these substances from affecting host cells. In addition to neutralizing the pathogen, some B cells then become memory cells to fight off a subsequent invasion.

For example, *Corynebacterium diphtheriae* is a bacterium that causes diphtheria, an upper respiratory infection. In this disease, the bacterial toxin kills host cells, leaving ulcers with a characteristic grayish membrane. Immunity to the disease occurs when the host develops antibodies that disable the toxin.

Researchers developed a vaccine for diphtheria by creating an inactivated toxin preparation that did not affect living cells. When administered to a person, the vaccine triggers antibody production without causing any symptoms of the disease. As a result, diphtheria has been almost eliminated in countries with good immunization programs.

T Lymphocytes Must Make Direct Contact with Their Target Cells

Antibodies are only effective with pathogens that are in the extracellular compartment. If a pathogen such as a virus gets inside a host cell, it is no longer "seen" by the humoral immune system. Defending the body against pathogens hidden inside cells is the role of the T lymphocytes that carry out *cell-mediated immunity*.

T lymphocytes (T cells) develop in the thymus gland from immature precursor cells that migrate there from the bone marrow (Fig. 24-14 ●). They differentiate into two major groups: cytotoxic T cells and helper T cells. T lymphocytes are responsible for **cell-mediated immunity**, a process in which the T cell binds to a target cell that displays foreign antigen fragments on its surface.

T lymphocytes bind to their target cells with the help of membrane **T-cell receptors** (Fig. 24-15 ●). T-cell receptors are not antibodies like B-cell receptors, although the peptide chains are closely related. When antigen binds to T-cell receptors, the T cell is activated. T cells cannot bind to free-floating antigens as B cells do. Instead, they only bind antigen that is part of a membrane protein complex called the *major histocompatibility complex.*

Major Histocompatibility Complexes Incorporate Antigen Fragments The **major histocompatibility complexes**, or **MHC**, are a family of membrane protein complexes encoded by a specific set of genes. These proteins were named when they were discovered to play a role in foreign tissue rejection following organ or tissue transplants. We know now that MHC is present in all nucleated cells of the body, where it is responsible for the extracellular presentation of processed antigen fragments.

All MHC proteins are related, but they vary from person to person because of the huge number of alleles that a person can inherit. There are so many alleles that it is unlikely that any two people other than identical twins will inherit exactly the same set. Major

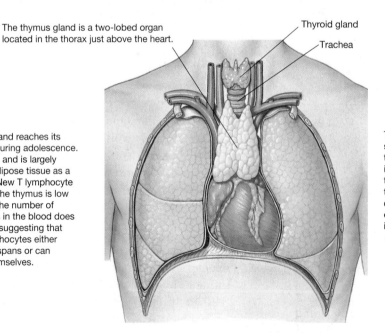

FOCUS ON... THE THYMUS GLAND

The thymus gland is a two-lobed organ located in the thorax just above the heart.

Thyroid gland

Trachea

The thymus gland reaches its greatest size during adolescence. Then it shrinks and is largely replaced by adipose tissue as a person ages. New T lymphocyte production in the thymus is low in adults, but the number of T lymphocytes in the blood does not decrease, suggesting that mature T lymphocytes either have long life spans or can reproduce themselves.

The epithelial cells of the thymus secrete peptide factors that influence the development of T lymphocytes, including interleukins, thymosins, thymopoietin, and thymulin. Although these peptides are sometimes classified as hormones, there is little evidence that they are released into the circulation to act on tissues outside the thymus gland.

During embryonic development, T lymphocytes synthesize their T-cell receptors and insert them into the membrane. Those cells that would be self-reactive are eliminated, while those that do not react with self tissues multiply to form clones.

● **Figure 24-14**

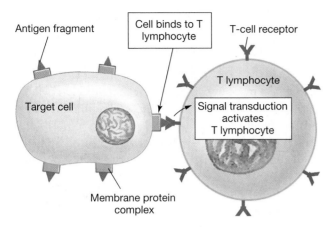

Labels: Antigen fragment; Cell binds to T lymphocyte; T-cell receptor; T lymphocyte; Target cell; Signal transduction activates T lymphocyte; Membrane protein complex

● **Figure 24-15** *Activation of T lymphocytes*

There are two types of MHC molecules. **MHC class I molecules** are found on all nucleated cells. When viruses and bacteria invade the cell, they are digested into peptide fragments and loaded onto MHC-I "platforms." If a **cytotoxic T cell** (T_C cell) encounters a target cell with foreign antigen fragment on its MHC-I, the T_C cell recognizes the target as virally infected or as a tumor cell and kills the cell to prevent it from reproducing (Fig. 24-16 ●).

MHC class II molecules are found primarily on macrophages, B lymphocytes, and dendritic cells. When the immune cell engulfs and digests an antigen, the fragments are returned to the immune cell membrane combined with MHC class II proteins. If a **helper T cell** (T_H cell) encounters a target cell with foreign antigen fragment on its MHC-II, the T_H cell responds by secreting cytokines that enhance the immune response.

histocompatibility complexes are one reason that tissues cannot be transplanted from one person to another without first establishing compatibility. More recently, it has been discovered that the inheritance of certain MHC alleles is linked to some disorders in which the body fails to recognize body components as "self."

How do antigen fragments get to the outside of antigen-presenting cells? Free antigen in the ECF cannot simply bind to unoccupied MHC receptors on the cell surface. A phagocyte must ingest the antigen and digest it. Fragments of digested antigen then combine with an MHC complex and are inserted into the phagocyte cell membrane by exocytosis (see Fig. 24-5). A similar process occurs when normal cells are invaded by viruses. In both instances, the antigen-containing extracellular complexes identify the cell as a target for immune cells.

Cytotoxic T Cells Kill Their Targets Cytotoxic T (T_C) cells attack and destroy cells that display MHC class I-antigen complexes. Although this may be an extreme response, it prevents the reproduction of intracellular invaders such as viruses, some parasites, and some bacteria.

How do cytotoxic T cells kill their targets? They do so in two ways. First they can release a cytotoxic pore-forming molecule called **perforin** along with **granzymes**, enzymes related to the digestive enzymes trypsin and chymotrypsin. When granzymes enter the target cell through perforin channels, they activate an enzyme cascade that induces the cell to commit suicide (**apoptosis**). Second, cytotoxic T cells can induce apoptosis by activating **Fas**, a "death receptor" protein on the target cell membrane that is linked to the enzyme cascade.

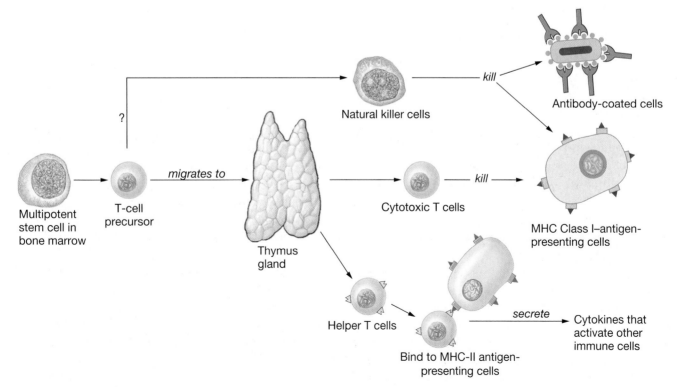

Labels: Multipotent stem cell in bone marrow; T-cell precursor; migrates to; Thymus gland; Natural killer cells; kill; Antibody-coated cells; Cytotoxic T cells; kill; MHC Class I–antigen-presenting cells; Helper T cells; Bind to MHC-II antigen-presenting cells; secrete; Cytokines that activate other immune cells

● **Figure 24-16** *T lymphocytes and NK cells* T lymphocytes and natural killer cells develop from the same precursor.

Early in the AIDS epidemic, researchers studied the effectiveness of a drug made with CD4. CD4 is a naturally occurring protein found on the surface of helper T cells, the primary target of the HIV virus. When the HIV virus binds to CD4, the virus is taken into the cell and begins replicating. Researchers found that adding high amounts of CD4 to test tubes containing HIV and helper T cells reduced the rate of HIV replication. But in human subjects, the results of clinical trials using the CD4 drug were disappointing. The drug did not slow the rate at which HIV infected helper T cells.

Question 2: What is the rationale for flooding the body with CD4 protein to prevent HIV replication?

Helper T (T$_H$) cells do not directly attack pathogens and infected cells, but they play an essential role in the immune response by secreting cytokines that influence other cells. The cytokines secreted by helper T cells include (1) *gamma-interferon*, which activates macrophages; (2) *interleukins* that activate antibody production and cytotoxic T lymphocytes; (3) *colony-stimulating factors* [p. 528] that enhance leukocyte production in the bone marrow; and (4) *interleukins* that promote growth of mast cells and eosinophils. Helper T cells also bind to B cells and promote their differentiation into plasma cells and memory cells.

The HIV virus that causes AIDS preferentially infects and destroys helper T cells, leaving the host unable to respond to pathogens that in other situations could be easily suppressed.

▶ IMMUNE RESPONSE PATHWAYS

How does the body respond to different kinds of immune challenges? Although the details depend on the particular challenge, the basic pattern is the same. The innate response starts first, reinforced by the more specific acquired response. The two pathways are interconnected, so cooperation and communication is essential. In the following sections, we examine the body's responses in four situations: an extracellular bacterial infection, a viral infection, an allergic response to pollen, and the transfusion of incompatible blood.

Inflammation Is the Typical Response to Bacterial Invasion

What happens when bacteria invade? As we've seen, if the passive barricades of the skin and mucous membranes fail, bacteria reach the extracellular fluid. There they usually cause an inflammatory response that represents the combined effects of many cells working to get rid of the invader. If bacteria enter the lymph, infection-fighting takes place in lymph nodes as well.

Inflammation is characterized by a red, swollen area that is tender or painful. In addition to the nonspecific inflammatory

response, lymphocytes attracted to the area produce antibodies keyed to the specific type of bacterium. The entry of bacteria sets off a series of reactions (Fig. 24-17 ●):

1. Components of the bacterial cell wall activate the complement system. Some complement proteins are chemical signals (*chemotaxins*) that attract leukocytes from the circulation to help fight the infection.

2. The complement cascade ends with formation of membrane attack complex molecules that insert themselves into the bacterial wall. The subsequent entry of ions and water lyses the bacteria, aided by the enzyme lysozyme. This is a purely chemical process that does not involve immune cells.

3. If the bacteria are not encapsulated, macrophages can begin to ingest the bacteria immediately.

4. If the bacteria are encapsulated, antibodies must coat the capsule before they can be ingested by phagocytes. Opsonins enhance the process for bacteria that are not encapsulated. Molecules that act as opsonins include complement, acute phase proteins, and antibodies.

5. Complement causes degranulation of mast cells and basophils. Cytokines secreted by mast cells act as additional chemotaxins, attracting more immune cells. Vasoactive chemicals such as histamine dilate blood vessels and increase capillary permeability. The enhanced blood supply to the site creates the red, warm appearance of inflammation. Plasma proteins that escape into the interstitial space pull water with them, leading to tissue edema.

6. Some elements of the acquired immune response are called into play with bacterial infections. If antibodies to the bacteria are already in the body, they will help the innate response by acting as opsonins and neutralizing bacterial toxins.

7. Memory B cells attracted to the infection site will be activated if they encounter an antigen they recognize. But this process is much slower than the innate responses described above, and it may be as long as several days before plasma cells begin secreting antibodies against the invader.

    If the infection is a novel one, some of the bacteria will activate naive B cells, and antigen presenting cells will present bacterial fragments to helper T cells to activate them. This triggers cytokine secretion from the T$_H$ cells, B cell clonal expansion, antibody production by plasma cells, and formation of memory B and T$_H$ cells.

8. If the initial wound damaged blood vessels underlying the skin, platelets and the proteins of the coagulation cascade will also be recruited to minimize the damage [p. 539]. Once the bacteria are removed by the immune response, the injured site is repaired under the control of growth factors and other cytokines.

Intracellular Defense Mechanisms Are Needed to Fight Viral Infections

What happens when viruses invade the body? First, they go through an extracellular phase of immune response similar to that described for bacteria. In the early stages of a viral infection,

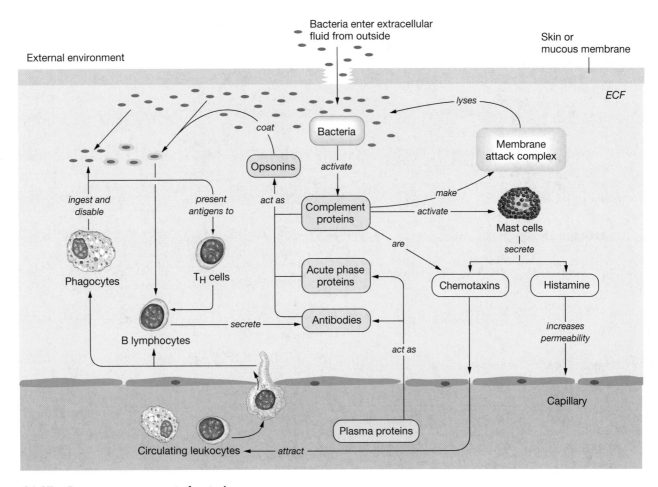

● **Figure 24-17** *Immune responses to bacteria*

innate immune responses and antibodies can help control the invasion of the virus.

Once the viruses enter the host's cells, humoral immunity in the form of antibodies is no longer effective. Cytotoxic T lymphocytes (and to a lesser extent, NK cells) are the main defense against intracellular viruses. They look for infected host cells, then destroy them.

For years, cell-mediated immunity of the T lymphocytes and humoral immunity controlled by B lymphocytes were considered independent processes. We now know that the two types of immunity are linked. Figure 24-18 ● shows the steps of a viral infection and shows how the two types of lymphocytes coordinate to destroy viruses and virus-infected cells. In this figure, we have assumed prior exposure to the virus and preexisting antibodies in the circulation. Antibodies can play an important defensive role in the early extracellular stages of a viral infection.

1. Antibodies act as opsonins, coating the viral particles to make them better targets for macrophages.
2. Antibody-bound viruses are prevented from entering their target cells. However, once the virus is inside the target host cell, antibodies are no longer useful.
3. Macrophages that ingest viruses insert fragments of viral antigen into MHC class II molecules on their membranes. Macrophages also secrete a variety of cytokines. Some of these cytokines initiate the inflammatory response. They produce alpha-interferon that causes host cells to make antiviral proteins. These keep invading viruses from replicating. Other macrophage cytokines stimulate NK cells and helper T cells.
4. Helper T cells bind to viral antigen on macrophage MHC class II molecules. Activated helper T cells are the primary mediators for additional antibody production by B lymphocytes.
5. Cytotoxic T cells use viral antigen-MHC class I complexes to recognize the infected host cells. When cytotoxic T cells and NK cells bind to the infected host cell, they secrete the contents of their granules onto the cell surface. Perforin molecules insert pores into the host cell membrane so that granzymes can enter the cell, inducing it to commit suicide and undergo apoptosis. Destruction of infected host cells is a key step in halting the reproduction of invading viruses.

Antibodies Against a Virus May Not Work in Subsequent Infections Although Figure 24-18 shows antibodies from a previous infection protecting the body, there is no guarantee that an antibody produced during one infection will be effective against the next invasion by the same virus. Why is this the case? Many viruses mutate constantly, and the protein coat forming the primary antigen may change significantly over time.

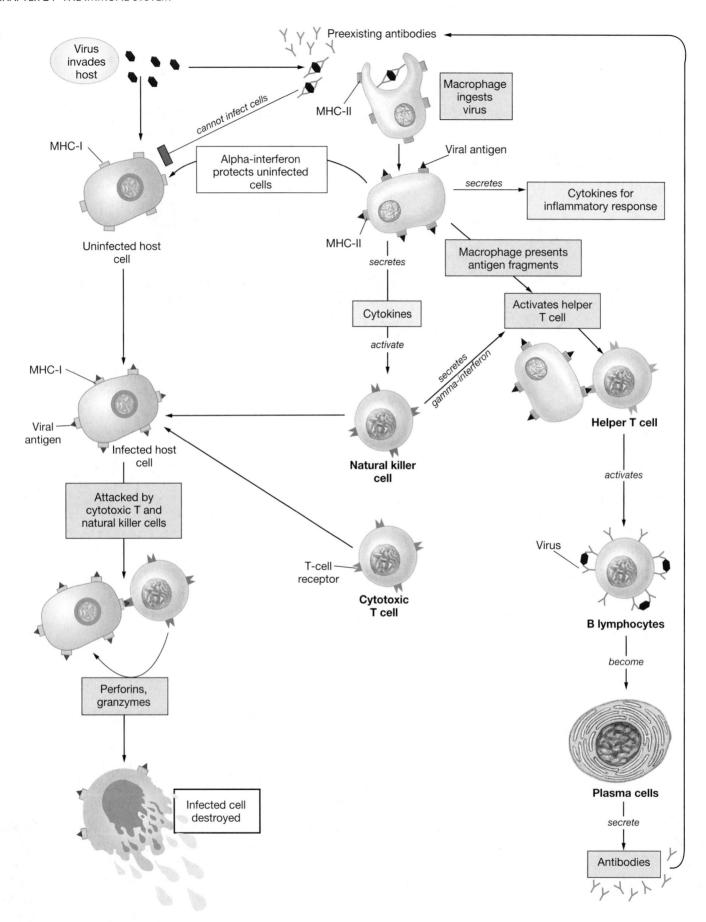

● **Figure 24-18** *Immune responses to viruses* This figure assumes prior exposure to the virus and preexisting antibodies.

The influenza virus is one virus that changes yearly. Consequently, the annual vaccines against influenza must be developed based on the virologists' predictions of what mutations have occurred. If the predictions do not match the mutations or the prevalent strains, the vaccine for that year will not keep someone from catching the flu.

The rapid mutation of viruses is also one reason that researchers have not yet developed an effective vaccine against HIV, the virus that leads to many cases of **acquired immunodeficiency syndrome** (AIDS). In addition, there is evidence that antibodies coating HIV virus particles can actually *enhance* virus uptake into target cells. HIV virus infects cells of the immune system, particularly T lymphocytes, monocytes, and macrophages. When HIV wipes out the helper T cells, cell-mediated immunity against the virus is lost. The general loss of immune response in AIDS leaves these patients susceptible to a variety of viral, bacterial, fungal, and parasitic infections.

Allergic Responses Are Inflammatory Responses Triggered by Specific Antigens

An **allergy** is an inflammatory immune response to a nonpathogenic antigen. Left alone, the antigen is not harmful to the body. But if an individual is sensitive to the antigen, the body creates an inflammatory response designed to get rid of it. Allergic inflammatory responses have symptoms ranging from mild tissue damage to fatal reactions.

The immune response in allergies is called **sensitivity** or **hypersensitivity** to the antigen. **Immediate hypersensitivity reactions** are mediated by antibodies and occur within minutes of exposure to antigen, which are called **allergens**. **Delayed hypersensitivity reactions** are mediated by helper T cells and macrophages and may take several days to develop.

Allergens can be practically any molecule not of the body: naturally occurring or synthetic, organic or inorganic. Certain foods, insect venoms, and pollen all trigger immediate hypersensitivity reactions. Allergens can be ingested, inhaled, or injected, or simply come in contact with the skin.

Delayed hypersensitivity reactions include contact allergies to copper and other base metals. These are common among people who wear costume jewelry. Poison ivy and poison oak also cause common contact allergies.

Allergies have a strong genetic component, so that if parents have a ragweed allergy, chances are good that their children will too. The development of allergies requires exposure to the allergen, a factor that will be affected by geographical, cultural, and social conditions.

What happens during an immediate hypersensitivity reaction to pollen? Let's follow the development of an allergy with Fig. 24-19 ●.

1. The first step, the sensitization phase ①, is equivalent to the primary immune response discussed previously (see Figs. 24-10, 24-11, and the right side of Fig. 24-18). The allergen is ingested and processed by an antigen-presenting cell such as a macrophage, which in turn activates a helper T cell.

    The helper T cell activates B lymphocytes, which have bound the allergen. This results in plasma cell production of antibodies to the allergen. Memory T and memory B cells store the record of the allergen exposure.

2. Upon reexposure ②, equivalent to the secondary immune response (see Figs. 24-10 and 24-11), the body reacts more strongly and more rapidly.

The type of allergic reaction depends on the antibody or immune cell involved. Allergen binding to IgE is the most common response to inhaled, ingested, or injected allergens. When allergens bind to IgE antibodies on mast cells, the cells degranulate, releasing histamine and other cytokines. The result is an inflammatory reaction.

The severity of the reaction varies, ranging from localized reactions near the site of allergen entry to systemic reactions such as total body rashes. The most severe IgE-mediated allergic reaction is called **anaphylaxis** [*ana-*, without + *phylax*, guard]. In an anaphylactic reaction, massive release of histamine and other cytokines causes widespread vasodilation, circulatory collapse, and severe bronchoconstriction. Unless treated promptly, anaphylaxis can result in death.

MHC Proteins Allow Recognition of Foreign Tissue

As surgical techniques have been developed that allow transplantation of organs between human beings or from animals to humans, physicians have had to wrestle with the problem of donor tissue rejection by the recipient's immune system. The ubiquitous major histocompatibility complex proteins (MHC) are the primary tissue antigens that determine whether tissue transferred between donor and recipient will be recognized as foreign by the recipient's immune system.

The MHC proteins are also known as **human leukocyte antigens (HLA)** and are classified in an international HLA system. A tissue graft or transplanted organ is more likely to be successful if the

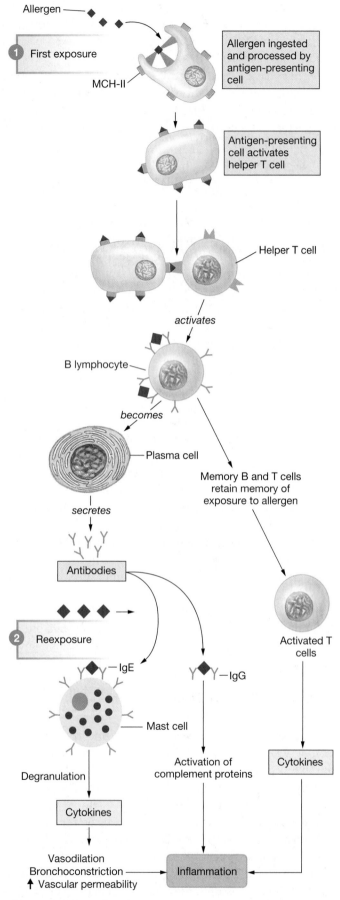

Allergen

① First exposure

MCH-II

Allergen ingested and processed by antigen-presenting cell

Antigen-presenting cell activates helper T cell

Helper T cell

activates

B lymphocyte

becomes

Plasma cell

Memory B and T cells retain memory of exposure to allergen

secretes

Antibodies

② Reexposure

Activated T cells

IgE

IgG

Mast cell

Degranulation

Activation of complement proteins

Cytokines

Cytokines

Vasodilation
Bronchoconstriction
↑ Vascular permeability

Inflammation

● **Figure 24-19** *Allergic responses*

donor and recipient share HLA antigens. Incompatible matches trigger antibody production and activate cytotoxic T cells and T_H cells. Generally it is the T cells that are responsible for acute rejection of solid tissue grafts.

One of the most widely used examples of donated tissue is blood transfusion. Human red blood cells contain antigenic membrane proteins and glycoproteins that are encoded by a number of different genes but lack the MHC proteins found on nucleated cells. In the absence of the usual cell marker for recognizing foreign tissue, the ABO blood group antigens and the Rh* antigens become the two surface proteins that are most likely to cause a rejection reaction after a blood transfusion. These reactions are similar to that described for hypersensitivities.

The ABO blood groups consist of four blood types created by combinations of two different antigens found on red blood cells. The antigens are membrane glycoproteins designated A and B (Fig. 24-20a ●). Each person inherits two alleles for ABO antigen, one from each parent. The three possible alleles are designated A, B, and O (neither A nor B will be expressed). Alleles A and B are dominant to allele O, leading to four possible physical expressions of antigen: A (AA, AO), B (BB, BO), AB, and O (OO) (Table 24-3).

Problems with blood transfusions arise because plasma normally contains antibodies to the ABO group antigens. These antibodies are thought to be stimulated early in life by common substances such as bacteria and food components. They can be measured in the blood of infants as early as age 3–6 months.

People express antibodies to the red blood cell antigen(s) that they do not possess. Thus, people with blood type A have anti-B antibodies in their plasma, people with no antigens on their red blood cells (blood type O) have both anti-A and anti-B antibodies, and people with both antigens on their red blood cells (blood type AB) have no antibodies to A or B antigens

How does the body respond to a transfusion of incompatible blood? If a person with blood type O is mistakenly given a transfusion of type A blood, for example, an immune reaction takes place. The anti-A antibodies of the type O recipient bind to the transfused type A red blood cells, causing them to clump (*agglutinate*). This reaction is easily seen in a blood sample and forms the basis for the blood-typing tests often performed in student laboratories.

Antibody binding also activates the complement system, resulting in production of membrane attack complex that causes the transfused cells to swell and leak hemoglobin. Free hemoglobin in the plasma can result in acute renal failure as the kidneys try to filter the large molecules from the blood. Cross-matching donor and recipient blood is critical prior to giving a blood transfusion.

CONCEPT CHECK   Answers: p. 781

4. A person with AB blood type is transfused with type O blood. What happens and why?

5. A person with O blood type is transfused with type A blood. What happens? Why?

*Rh stands for Rhesus (monkeys). This group is also called the D antigen.

Table 24-3: ABO Blood Groups

				Frequency in U.S. Population (%)			
BLOOD GROUP	POSSIBLE NATIVE GENOTYPES	RED BLOOD CELL ANTIGENS	SERUM ANTIBODY	WHITE	BLACK	ASIAN	AMERICAN
O	OO		anti-A, anti-B	45	49	40	79
A	AA, AO	A	anti-B	40	27	28	16
B	BB, BO	B	anti-A	11	20	27	4
AB	AB	A and B	no antibodies to A or B	4	4	5	<1

(a)

Blood type	Antigen on red blood cell	Antibodies in plasma
O	RBC No A or B antigens	"Anti-A" and "anti-B"
A	A antigens	"Anti-B"
B	B antigens	"Anti-A"
AB	A and B antigens	None to A or B

(b) A mixture of type O and type A blood

When red blood cells with group A antigens on their membranes are mixed with plasma containing antibodies to group A, the antibodies cause the blood cells to clump, or agglutinate.

● **Figure 24-20** *ABO blood groups* The antigens and antibodies characteristic of each blood type are shown.

Recognition of Self Is an Important Function of the Immune System

One amazing property of immunity is the ability of the body to distinguish its own cells from foreign cells. The lack of immune response by lymphocytes to cells of the body is known as **self-tolerance** and appears to be due to the elimination of self-reactive lymphocytes. During development of T lymphocytes in the thymus, T lymphocytes develop that can combine with MHC-self-antigen complexes. But by a mechanism that is not understood, these self-reactive T lymphocytes are eliminated, dying by apoptosis. If self-reactive lymphocytes arise after the thymus has become inactive, they are usually either destroyed or rendered ineffective.

When self-tolerance fails, the body makes antibodies against its own components through T cell-activated B lymphocytes. The body's attack on its own cells leads to **autoimmune diseases** [*auto-*, self]. The antibodies produced in autoimmune diseases are specific against a particular antigen and are usually restricted to a particular organ or tissue type. Table 24-4 lists some common autoimmune diseases of humans.

Table 24-4: Some Common Autoimmune Diseases in Humans

DISEASE	ANTIBODIES PRODUCED AGAINST
Graves' disease (hyperthyroidism)	Thyroid-stimulating hormone receptor on the thyroid cells
Insulin-dependent diabetes mellitus	Pancreatic beta cell antigens
Multiple sclerosis	Myelin of CNS neurons
Myasthenia gravis	Acetylcholine receptor of motor endplate
Rheumatoid arthritis	Collagen?
Systemic lupus erythematosus	Intracellular nucleic acid protein complexes (antinuclear antibodies)
Guillain-Barré syndrome (acute inflammatory demyelinating polyneuropathy)	Myelin of peripheral nerves

Why does self-tolerance suddenly fail? We still do not know. We have learned, however, that autoimmune diseases often begin in association with an infection. One potential trigger for autoimmune diseases is foreign antigens that are similar to human antigens. When the body makes antibodies to the foreign antigen, those antibodies have enough cross-reactivity with human tissues to do some damage.

One example of an autoimmune disease is type 1 diabetes mellitus, in which the body makes *islet cell antibodies* that destroy pancreatic beta cells but leave the other endocrine cells untouched. Another autoimmune condition, discussed in the last chapter, is the hyperthyroidism of Graves' disease. The body makes *thyroid-stimulating immunoglobulins* that mimic thyroid-stimulating hormone and cause the thyroid gland to oversecrete hormone.

Immune Surveillance Allows the Body to Remove Abnormal Cells

What is the role of the human immune system in protecting the body against cancer? This question holds considerable interest for scientists. One theory, known as **immune surveillance**, proposes that cancerous cells develop on a regular basis but are detected and destroyed before they can spread. Although some evidence supports that theory, it does not explain why so many cancerous tumors develop every year, or why people with depressed immune function do not develop cancerous tumors in all types of tissues. However, immune surveillance does appear to recognize and control virus-associated tumors, presumably because of the viral antigen presented to NK and cytotoxic T cells.

We do know that many types of endogenous tumors lack surface antigens that allow them to be recognized as abnormal by the immune system. One active area of cancer research is looking at ways to activate the immune system to fight cancerous tumor cells.

▶ NEURO-ENDOCRINE-IMMUNE INTERACTIONS

One of the most fascinating and rapidly evolving areas in physiology involves the link between the mind and the body. Although serious scientists scoffed at the topic for many years, the

RUNNING PROBLEM

George began taking the AIDS drug cocktail. On the day he filled the prescription, his T-cell count was 338. A normal T-cell count is 1000. Within four weeks, his T-cell count was 489. Within two months, it was inching toward 900. The amount of virus in George's blood (called the "viral load") had also decreased significantly. George was thrilled. "I don't know if this will last forever," he told his doctor. "But I'm going to enjoy this reprieve while I can."

Question 4: What correlation can be made between T-cell count and the viral load of HIV in George's blood?

interaction of emotions and somatic illnesses has been described for centuries. The Old Testament says, "A merry heart doeth good like a medicine; but a broken spirit drieth the bones" (Proverbs 17:22). Most societies have tales of people who lost the will to live and subsequently died without obvious illness, or of people given up for dead who made remarkable recoveries.

Today, the medical field is beginning to pay attention to the reality of the mind-body connection. Psychosomatic illnesses and the placebo effect have been accepted for many years. Why should we not investigate the possibility that cancer patients can enhance their immune function by visualizing their immune cells gobbling up the abnormal cancer cells?

The study of brain-immune interactions is now a recognized field known as **neuroimmunomodulation**. At this time, the field is still in its infancy, and the results of many studies raise more questions than they answer. For example, in one experiment, mice were repeatedly injected with a chemical that suppressed the activity of their lymphocytes. During each injection, they were also exposed to the odor of camphor, a chemical that does not affect the immune system. After a period of conditioning, the mice were exposed only to the camphor odor. When the researchers checked the mice's lymphocyte function, they found that the odor of camphor had suppressed the activity of the immune cells, just as the chemical suppressant had done previously. The pathways through which this conditioning occurred are still largely unexplained.

What do we know at this point about the relationship between the immune, nervous, and endocrine systems?

1. **The three systems share common signal molecules and common receptors for those molecules.** Chemicals that we once thought were exclusive to a single cell or tissue are being discovered all over the body. Immune cells secrete hormones, and leukocyte cytokines are produced by non-immune cells. For example, lymphocytes secrete thyrotropin (TSH), ACTH, growth hormone, and prolactin, as well as hypothalamic corticotropin-releasing hormone (CRH).

   Receptors for hormones, neurotransmitters, and cytokines are being discovered everywhere. Neurons in the brain have receptors for cytokines produced by immune cells. Natural killer cells have opiate receptors and beta adrenergic receptors. Everywhere we turn, the nervous, endocrine, and immune systems seem to share chemical signal molecules and their receptors.

2. **Hormones and neuropeptides can alter the function of immune cells.** It has been known for years that increased cortisol levels due to stress were associated with decreased antibody production, depressed lymphocyte proliferation, and diminished activity of natural killer cells. Substance P, a neuropeptide, has been shown to induce degranulation of mast cells in the mucosa of the intestines and the respiratory tract. And sympathetic innervation of the bone marrow increases antibody synthesis and production of cytotoxic T cells.

3. **Cytokines from the immune system can affect neuroendocrine function.** Stressors such as bacterial and viral infections or tumors can induce stress responses in the central nervous system through cytokines released by immune cells. Interleukin-1 is probably the best-studied cytokine for this response, but the induction of cortisol release by lymphocyte ACTH is receiving considerable attention.

Previously, it was believed that cortisol secretion depended on neural signals translated through the hypothalamic CRH-anterior pituitary ACTH pathway. Now it appears that pathogenic stressors can activate the cortisol pathway by causing immune cells to secrete ACTH.

The interaction of the three systems is summarized in the model shown in Fig. 24-21 ●. The nervous, endocrine, and immune systems are closely linked through bidirectional communication carried out using cytokines, hormones, and neuropeptides. The brain is linked to the immune system by autonomic neurons, CNS neuropeptides, and leukocyte cytokines. The nervous system controls endocrine glands by secretion of hypothalamic-releasing hormones, but immune cells secrete some of the same trophic hormones. Hormones from endocrine glands feed back to influence both the nervous and immune systems.

The result is a complex network of chemical signals subject to modulation by outside factors. These outside factors include physical and emotional stimuli integrated through the brain, pathogenic stressors integrated through the immune system, and a variety of miscellaneous factors including magnetic fields, chemical factors from brown adipose tissue, and melatonin from the pineal gland. It will probably take years for us to decipher these complex pathways.

Stress Alters Immune System Function

One area of interest is the link between inability to cope with stress and the development of illnesses. The modern study of stress is attributed to Hans Selye, beginning in 1936. He defined **stress** as nonspecific stimuli that disturb homeostasis and elicit an invariable stress response that he termed the *General Adaptation Syndrome*.

Selye's stress response consisted of stimulation of the adrenal glands, followed by suppression of the immune system due to high levels of circulating glucocorticoids [● p. 730]. For many years, Selye's experiment was the benchmark for defining stress. However, since the 1970s, the definition of stress and the stress response has broadened.

Stressors, the events or items that create stress, are highly variable and difficult to define in experimental settings. Acute stress is different from chronic stress. A person's reaction to stress is affected by loneliness and by whether the person feels in control of the stressful situation. Many stressors are sensed and interpreted by the brain, leading to modulation of the stressor by experience and expectations. A stressor to one person may not affect another.

Most physical and emotional stressors are integrated through the central nervous system. The two classic stress responses are (1) the rapid, neurally mediated **fight-or-flight reaction**, and (2) the elevation of adrenal cortisol levels with associated suppression of the immune response. Fight-or-flight is a rapid reaction to acute stress. The cortisol response is a better indicator of chronic or repetitive stress.

One difficulty in studying the body's response to stress is the complexity of integrating information from three levels of response (Fig. 24-22 ●). Scientists have gathered a good deal of information at each of three levels: cellular and molecular factors related to stress, systemic responses to stress, and clinical studies describing the relationships between stress and illness. In some cases, we have linked two different levels together, although in other cases, the experimental evidence is scanty or even contradictory. What we still do not understand is the big picture, the way that all three levels of response overlap.

Experimental progress is slow because many traditional study animals, such as rodents, are not good models for studying stress

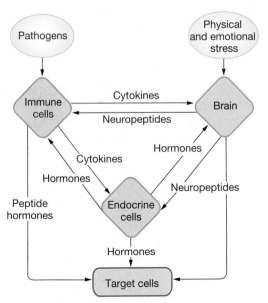

● **Figure 24-21** *Model for interaction between nervous, endocrine, and immune systems*

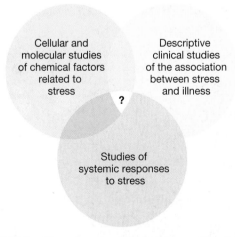

● **Figure 24-22** *Neuroimmunomodulation* The field of neuroimmunomodulation is still in its infancy, and we do not understand how the nervous, immune, and endocrine systems interact to affect health and well-being.

775

in humans. At the same time, controlled experiments that create stress in humans are difficult to design and execute for both practical and ethical reasons. Much of our research is observational.

For example, one study followed natural killer cell activity over an 18-month period in two groups of people: spouse caregivers of Alzheimer's disease patients, a group that reports high psychosocial stress, and a matched group of controls. For a subset of the caregivers, high perceived stress at the beginning of the study was associated with low NK cell activity at the end. However, there are many variables that must be taken into ac-

count before we can conclude that stress was the causative factor, including age, gender, exercise, alcohol use, and prescription drug use.

In spite of the difficulties, scientists are slowly beginning to piece together the intricate tapestry of mind-body interactions. When we can explain why students tend to get sick just before exam time, and why students tend to get sick just before exam time, and why relaxation therapy enhances the immune response, we will be well on the way to understanding the complex connections between the brain and the immune system.

RUNNING PROBLEM CONCLUSION

Treatment for AIDS

In this running problem, you learned about a treatment for AIDS that consists of two antiretroviral drugs and a protease inhibitor. The combination of these drugs seems to stem the tide of HIV replication more effectively than any of the drugs used alone. Additional drugs have been developed, and tests for an HIV vaccine are in progress. To learn more about current treatments, visit AIDS.org at *www.aids.org*. For statistics on HIV and AIDS in the United States, go to *http://www.cdc.gov/hiv/stats.htm*. For international statistics, go to the World Health Organization at *http://www.who.int/emc-hiv/*

Now check your understanding of this running problem by comparing your answers to those in the summary table.

	QUESTION	FACTS	INTEGRATION AND ANALYSIS
1	Why can't retroviruses reproduce without reverse transcriptase?	Retroviruses have no DNA to use as a template to make viral mRNA. Reverse transcriptase is an enzyme that allows retroviruses to make complementary DNA from their own viral RNA.	Without reverse transcriptase, the retroviruses cannot make viral DNA. Without viral DNA, the virus cannot make viral mRNA or viral proteins. Therefore, without reverse transcriptase, a retrovirus cannot replicate.
2	What is the rationale for flooding the body with CD4 protein to prevent HIV replication?	CD4 is one receptor on helper T cells to which HIV binds. Binding causes the virus to be taken into the T cell.	Introducing large amounts of CD4 protein into the body might "trick" HIV into binding to these free proteins instead of to CD4 on helper T cells. This would reduce the number of virus particles that could invade the T cells.
3	George asks, "If I'm making antibodies against HIV, why can't these antibodies kill the virus"? Answer George's question.	Viruses can only be affected by antibodies when they are outside the host cells. HIV bound to antibody is taken up into helper T cells.	HIV enters helper T cells and, over time, will destroy them.
4	What correlation can be made between T-cell count and the viral load?	George's T-cell count increases while the viral load decreases.	Apparently, the drug cocktail is effective in decreasing the rate of T-cell infection by HIV. This has two results. As fewer T cells are killed by the virus, the T-cell count rises. As the drugs suppress viral replication, the viral load decreases.

751 753 768 771 774

CHAPTER SUMMARY

The function of the immune system is based on *chemical communication* and *molecular interactions* between receptors, antibodies, and antigens. The *compartmentation* of the body affects how the

immune system fights pathogens: some pathogens hide inside cells while others remain in the extracellular compartment.

Overview of Immune System Function

1. **Immunity** is the ability of the body to protect itself from disease-causing entities. The human immune system consists of lymphoid tissues, immune cells, and cytokines. (p. 752)

2. **Antigens** are substances that trigger an immune response and react with products of that response. (p. 752)

Pathogens of the Human Body

3. Bacteria are cells that can survive and reproduce outside of a living host. Antibiotic drugs kill bacteria. (p. 752)

4. Viruses are DNA or RNA cores enclosed in protective proteins. They invade host cells and reproduce using the host's intracellular machinery. The immune system must find and destroy infected host cells to kill a virus. (p. 753; Fig. 24-1)

The Immune Response

5. The immune response includes: (1) *detection* and *identification*, (2) *communication* with other immune cells, (3) *recruitment* of assistance and *coordination* of the response, and (4) *destruction* or *suppression* of the invader. (pp. 753–754)

6. **Innate immunity** is the rapid, **nonspecific immune response. Acquired immunity** is slower and is an **antigen-specific response**. The two types of immunity overlap. (p. 754)

Anatomy of the Immune System

7. Immune cells form and mature in the **thymus gland** and **bone marrow**. The **spleen** and **lymph nodes** are encapsulated lymphoid tissues. Unencapsulated **diffuse lymphoid tissues** include **tonsils** and the **gut-associated lymphoid tissue (GALT)**. (p. 754; Figs. 24-2, 24-3, 24-14)

8. **Eosinophils** are **cytotoxic cells** that kill parasites. **Basophils** and **mast cells** release histamine, **heparin**, and other cytokines. (pp. 756–757; Fig. 24-4)

9. **Neutrophils** are phagocytic cells that release pyrogens and cytokines that mediate inflammation. (p. 757; Fig. 24-4)

10. **Monocytes** are the precursors of tissue **macrophages**. **Antigen-presenting cells (APCs)** display antigen fragments in proteins on their membrane. (pp. 756, 758; Figs. 24-4, 24-5)

11. **Lymphocytes** and **plasma cells** mediate antigen-specific immune responses. (pp. 758, 763)

12. **Dendritic cells** capture antigens and present them to lymphocytes. (p. 758)

Innate Immunity

13. Physical and chemical barriers, such as skin and mucus, are the first line of defense. (p. 758)

14. **Chemotaxins** attract immune cells to pathogens. (p. 759)

15. Phagocytosis requires that surface molecules on the pathogen bind to membrane receptors on the phagocyte. **Opsonins** are tagging proteins that coat foreign material. (p. 760; Figs. 24-6, 24-7)

16. **Natural killer cells** kill certain tumor cells and virus-infected cells. They bind to the Fc stem of antibodies that are bound to the target cell. This is called **antibody-dependent cell-mediated cytotoxicity**. (pp. 760, 765)

17. **Inflammation** results from cytokines released by activated tissue macrophages. (p. 760)

18. Inflammatory cytokines include **acute phase proteins**, histamine, **interleukins**, **bradykinin**, and **complement proteins**. (pp. 760–761)

19. The complement cascade forms **membrane attack complex**, pore-forming molecules that cause pathogens to rupture. (p. 762; Fig. 24-8)

Acquired Immunity

20. Lymphocytes mediate the acquired immune response. Lymphocytes that react to one type of antigen form a **clone**. (p. 762; Fig. 24-9)

21. First exposure to an antigen activates a **naive lymphocyte** clone and causes it to divide (**clonal expansion**). Newly formed lymphocytes differentiate into **effector cells** or **memory cells**. Additional exposure to the antigen creates faster and stronger **secondary immune response**. (pp. 763–764; Figs. 24-10, 24-11)

22. Lymphocyte cytokines include **interleukins** and **interferons**, which interfere with viral replication. **Gamma-interferon** activates macrophages and other immune cells. (pp. 760–761)

23. **B lymphocytes** secrete antibodies. **T lymphocytes** and natural killer cells attack and destroy infected cells. (pp. 763, 766)

24. **Antibodies** bind to pathogens. Mature B lymphocytes are covered with antibodies that act as surface receptors. Activated B lymphocytes differentiate into **plasma cells**, which secrete soluble antibodies. (p. 763)

25. The five classes of immunoglobulins (Ig), or antibodies, are IgG, IgA, IgE, IgM, and IgD, collectively known as **gamma globulins**. (p. 764)

26. The antibody molecule is Y-shaped and has four polypeptide chains. The arms, or **Fab regions**, contain antigen-binding sites. The stem is known as the **Fc region** and can bind to immune cells. Some immunoglobulins are polymers. (p. 764; Fig. 24-12)

27. Soluble antibodies act as opsonins, serve as a bridge between antigens and leukocytes, activate complement proteins and mast cells, and disable viruses and toxins. (p. 764; Fig. 24-13)

28. B cell antibodies bind antigens to activate B lymphocytes and initiate the production of additional antibodies. (p. 766; Fig. 24-13)

29. **T lymphocytes** are responsible for **cell-mediated immunity**, in which the lymphocytes bind to target cells using **T-cell receptors**. (p. 766; Fig. 24-15)

30. T lymphocytes react to antigen bound to **major histocompatibility complex** proteins on the target cell. **MHC class I molecules** displaying antigen cause cytotoxic T (T_C) cells to kill the target cell. **MHC class II molecules** with antigen activate helper T (T_H) cells to secrete cytokines. (pp. 766–767; Fig. 24-16)

31. **Cytotoxic T (T_C) cells** release **perforin**, a pore-forming molecule that allows **granzymes** to enter the target cell and trigger it into committing suicide (**apoptosis**). (p. 767)

32. **Helper T (T_H) cells** secrete cytokines that influence other cells. (p. 767; Fig. 24-16)

Immune Response Pathways

33. Bacteria usually cause a nonspecific inflammatory response. In addition, lymphocytes produce antibodies keyed to the specific type of bacterium. (p. 768; Fig. 24-17)

34. Innate immune responses and antibodies help control the early stages of a viral infection. Once the viruses enter the host's cells, cytotoxic T cells must destroy infected host cells. (p. 769; Fig. 24-18)

35. An **allergy** is an inflammatory immune response to a nonpathogenic **allergen**. The body gets rid of the allergen by using an inflammatory response. (p. 771; Fig. 24-19)

36. **Immediate hypersensitivity reactions** to allergens are mediated by antibodies and occur within minutes of exposure to antigen. **Delayed hypersensitivity reactions** are mediated by T lymphocytes and may take several days to develop. (p. 771)

37. The major histocompatibility complex (MHC) proteins found on all nucleated cells determine tissue compatibility. Lymphocytes will attack any MHC complexes that they do not recognize as self proteins. (p. 771)

38. Human red blood cells lack MHC proteins but contain other antigenic membrane proteins and glycoproteins such as the ABO and Rh antigens. (p. 772; Fig. 24-20)

39. **Self-tolerance** is the ability of the body to distinguish its own cells and not create an immune response. When self-tolerance fails, the body makes antibodies against its own components, creating an **autoimmune disease**. (p. 773)

40. The theory of **immune surveillance** proposes that cancerous cells develop on a regular basis but are detected and destroyed by the immune system before they can spread. (p. 774)

Neuro-Endocrine-Immune Interactions

41. The nervous, endocrine, and immune systems are linked together by signal molecules and receptors. The study of these interactions is a field known as **neuroimmunomodulation**. (p. 774; Fig. 24-21)

42. The link between inability to cope with stress and the development of illnesses is believed to result from neuroimmunomodulation. (p. 775)

QUESTIONS

LEVEL ONE *Reviewing Facts and Terms*

1. Define immunity. What is meant by memory and specificity in the immune system?

2. Name the anatomical components of the immune system.

3. List and briefly summarize the three main functions of the immune system.

4. Name two ways viruses are released from the host cell. Give three ways in which viruses damage host cells or disrupt their function.

5. If a pathogen has invaded the body, what four steps occur if the body is able to destroy the pathogen? What compromise may be made if the pathogen cannot be destroyed?

6. Define the following terms and explain their significance:
   (a) anaphylaxis
   (b) agglutinate
   (c) extravascular
   (d) degranulation
   (e) acute phase protein
   (f) clonal expansion
   (g) immune surveillance

7. How are histiocytes, Kupffer cells, osteoclasts, and microglia related?

8. What is the mononuclear phagocytic system and what role does it play in the immune system?

9. Match the cell type with its description:
   (a) lymphocyte
   (b) neutrophil
   (c) monocyte
   (d) dendritic cell
   (e) eosinophil
   (f) basophil

   1. most abundant leukocyte; phagocytic; lives 1–2 days
   2. cytotoxic; associated with allergic reactions and parasitic infestations
   3. precursor of macrophages; relatively rare in blood smears
   4. related to mast cells; releases cytokines such as histamine and heparin
   5. most of these cells reside in the lymphoid tissues
   6. these cells capture antigens, then migrate to the lymphoid tissues

10. Give examples of physical and chemical barriers to infection.

11. Name the three main types of lymphocytes and their subtypes. Explain the functions and interactions of each group.

12. Define self-tolerance. Which cells are responsible? Relate self-tolerance to autoimmune disease.

13. What is meant by the term *neuroimmunomodulation*?

14. Explain the terms *stress*, *stressor*, and *general adaptation syndrome*.

LEVEL TWO *Reviewing Concepts*

15. **Concept map:** Map the following terms. You may add additional terms.
    acquired immunity
    antibody
    antigen
    antigen-presenting cell
    B lymphocyte
    bacteria
    basophil
    chemotaxin
    cytokine
    cytotoxic cell
    cytotoxic T cell
    dendritic cell
    eosinophil
    Fab region
    Fc region
    helper T cell
    innate immunity
    memory cell
    MHC class I
    MHC class II
    monocyte
    natural killer cell
    neutrophil
    opsonin
    phagocyte
    phagosome
    plasma cell
    T lymphocyte
    viruses

16. Why do lymph nodes often swell and become tender or even painful when you are sick?

17. Summarize the effects of histamine, interleukin-1, acute phase proteins, bradykinin, complement, and gamma-interferon. Are these chemicals antagonistic or synergistic to one another?

18. Compare and contrast the following sets of terms:
    (a) pathogen, microbe, pyrogen, antigen, antibody, antibiotic
    (b) infection, inflammation, allergy, autoimmune disease
    (c) virus, bacteria, allergen, retrovirus
    (d) chemotaxin, opsonin, cytokine, interleukin, bradykinin, interferon
    (e) acquired immunity, innate immunity, humoral immunity, cell-mediated immunity, specific immunity, nonspecific immunity
    (f) immediate and delayed hypersensitivity
    (g) membrane attack complex, granzymes, perforin

19. Diagram and label the parts of an antibody molecule. What is the significance of each part?

20. Diagram the steps in the process of inflammation.

21. Diagram the steps in the process of fighting viral infections.

22. Diagram the steps in the process of an allergic reaction.

LEVEL THREE *Problem Solving*

23. One ABO blood type is called the "universal donor" and one is called the "universal recipient." Which types are these and why are they given those names?

24. Maxie Moshpit claims that Snidley Superstar is the father of her child. Maxie has blood type O. Snidley has blood type B. Baby Moshpit has blood type O. Can you rule out

Snidley as the father of the baby with only this information? Explain.

25. Every semester around exam time more students visit the Student Health Center with colds and viral infections. The physicians attribute the increased incidence of illness to stress. Can you explain the biological basis for that relationship? What are some other possible explanations?

26. Barbara is troubled with rheumatoid arthritis, characterized by painful, swollen joints from inflamed connective tissue. She notices that it flares up when she is tired and stressed.

This condition is regarded as an autoimmune disease. Outline the physiological reactions that would lead to this condition.

27. The differential white cell count [● Figure 16-2, p. 527] is an important diagnostic tool in medicine. If a patient's "diff" shows an increase in the number of immature neutrophils ("bands") in the circulation, do you think the cause is more likely to be a bacterial or a viral infection? If the count shows an increase in eosinophils (*eosinophilia*), is the cause more likely to be a viral or a parasitic infection? Explain your reasoning.

Explorations

Explorations are also available at *www.physiologyplace.com*

Introduction

In this chapter you learned how the body protects itself from pathogens and removes damaged tissue or abnormal cells using the cells and chemicals of the immune system. An immune response requires detection of the foreign substance, communication with other immune cells, and coordination of the response.

Exploration 1: Mind, Body, and Health

The role of mind-body interactions in traditional medicine is a hot topic. Although the placebo and nocebo effects are widely accepted, the role of complementary therapies such as laughter therapy is still controversial. Can watching a funny movie really help your immune system? Two recent studies suggest that it can. In both studies, researchers measured the activity of immune cells in subjects before and after watching a humorous movie or video. In both studies, immune cells became more active. But were these well-designed studies? How would you design a research study like this to ensure that it has the proper controls, and that any effect is due to laughter and not some other aspect of the experience? Compare your design to the actual studies.

"Modulation of neuroimmune parameters during the eustress of humor-associated mirthful laughter," *Alternative Therapies in Health and Medicine* 7(2): 62–72, 74–76, 2001 Mar. (*www.alternative-therapies.com*)

"The elevation of natural killer cell activity induced by laughter in a crossover designed study," *International Journal of Molecular Medicine* 8(6): 645–650, 2001 Dec.

You can learn more about *humor therapy* from the Association for Applied and Therapeutic Humor (*www.aath.org*)

Exploration 2: HIV Vaccine

The Holy Grail of HIV research is the development of an effective HIV vaccine. There are drugs to fight HIV, but these drugs are expensive and can be rendered ineffective by the fickle, highly mutable HIV particles. Therefore, tremendous efforts are being made to actually stop or at least slow the spread of the disease. A vaccine would be ideal, but a vaccine for HIV is tricky, and many researchers even believe it to be improbable.

In February of 2003, the initial results were reported for the first clinical trials of an HIV vaccine developed in part by an Australian company called VaxGen. The vaccine used a recombinant (denoted as "r") HIV envelope glycoprotein (gp), rgp120. The results were controversial, but overall disappointing. It appears that the vaccine was ineffective at preventing the spread of HIV.

However, there are many ways to create a vaccine—you might use a protein, nucleic acid, or another molecular piece from a virus to train an immune response—and current studies on vaccines for the closely related Simian Immunodeficiency Virus (SIV) are actually using dendritic cells that have been previously exposed to the virus, with some success, to immunize their simian patients. See "Therapeutic dendritic-cell vaccine for simian AIDS," *Nature Medicine* 9(1): 27–32, 2003.

What are some of the roadblocks that make the development of an HIV vaccine so difficult? Why have the early vaccine trials failed? What other research is currently under way for HIV vaccines? How much money is spent annually on HIV/AIDS research? Here are just a few sites you can visit for information. Each has its own list of links and resources helpful in answering questions about HIV.

- HIV Vaccine Trials Network: *www.hvtn.org*
- AIDScience: *www.aidscience.com*
- American Foundation for AIDS Research: *www.amfar.org*
- The Centers for Disease Control and Prevention: *http://www.cdc.gov/hiv/vaccine.htm*

ANSWERS

Page 762

1. When capillary permeability increases, proteins escape from the plasma to the interstitial fluid. This decreases the colloid osmotic force opposing capillary filtration, and additional fluid accumulates in the interstitial space (swelling or edema).

Page 764

2. The first exposure causes a slower and weaker primary immune response (see Fig. 24-11). The second exposure will cause a stronger response, which may be severe and cause anaphylaxis.

3. Protein antibodies can be moved across cells by transcytosis or released from cells by exocytosis.

Page 772

4. Type O blood has neither A nor B antigens, but even if it did, type AB blood does not produce anti-A or anti-B antibodies. Therefore no immune reaction will occur.

5. If type A blood is given to a type O recipient, the recipient will synthesize anti-A antibodies and agglutination will occur.

Page 762

Figure 24-9: The antigen will activate clone 1.

Integrative Physiology III:
Exercise

25

One fascinating aspect of the physiology of the human being at work is that it provides basic information about the nature and the range of the functional capacity of different organ systems.

— **Per-Olof Åstrand and Kaare Rodahl,**
Textbook of Work Physiology, 1977

RUNNING PROBLEM

Heat Stroke

"Dangerous heat continues for fifth straight day," reads the morning headline. Had Colleen Warren, 19, taken time to scan the story, she might have avoided the life-threatening health emergency she faced a few hours later. But Colleen overslept and was rushing to get to field hockey practice. The weather during this first week of the fall semester was unbearably hot and humid, just as it had been all summer. Today, the temperature was predicted to reach 105° F, with high humidity. In her haste to leave, Colleen forgot her water bottle. "Oh well," she thought as she hopped on her bike to peddle the two miles to the practice field.

785 787 789 790 792

BACKGROUND BASICS

Aerobic and anaerobic metabolism: **104** ǀ Up-regulation of receptors: **186** ǀ Feedforward reflexes: **195** ǀ Catecholamines: **218** ǀ Metabolism during muscle contraction: **401** ǀ Myoglobin: **404** ǀ Isometric and isotonic contraction: **409** ǀ Presynaptic inhibition: **272** ǀ Motor control of movement: **438** ǀ Control of heart rate and contractility: **467** ǀ Venous return: **481** ǀ Control of cardiac output: **482** ǀ Starling's law of the heart: **481** ǀ Control of blood pressure: **510** ǀ Local control of blood flow: **500** ǀ Hemoglobin: **532** ǀ Alveolar ventilation: **565** ǀ Blood gases: **577** ǀ Respiratory chemoreceptors: **589** ǀ Pulmonary stretch receptors: **592** ǀ Dehydration: **742** ǀ Insulin: **705** ǀ Diabetes mellitus: **711** ǀ Thermoregulation: **717** ǀ Glucagon: **710** ǀ Growth hormone: **736** ǀ Stress and the immune system: **775** ǀ

One of the most common challenges to the body's homeostasis is not a disease but an everyday activity, exercise. Exercise comes in many forms. Running is an example of a dynamic endurance exercise. Weight lifting and strength training are examples of dynamic resistive exercise.

In this chapter we examine dynamic exercise as a challenge to homeostasis that is met with an integrated response from multiple systems of the body. In many ways, exercise is the ideal teaching example of physiological integration. Everyone is familiar with it. It is an activity that does not require special environmental conditions, unlike high-altitude activities or deep-sea diving. Exercise is a normal physiological state rather than a pathological one, although it can be affected by disease and, if excessive, can result in injury.

Note that this chapter does not intend to be a complete description of exercise physiology. For further information, readers should consult an exercise physiology textbook.

▶ METABOLISM AND EXERCISE

Exercise begins with skeletal muscle contraction, an active process that requires ATP for energy. But where does the ATP for muscle contraction come from? A small amount is in the muscle fiber cytoplasm when contraction begins (Fig. 25-1●). As this

ATP is transformed into ADP, another phosphate compound, **phosphocreatine** (PCr), transfers energy from its high-energy phosphate bond to ADP. The transfer replenishes the muscle's supply of ATP [● p. 401].

Together, the combination of muscle ATP and phosphocreatine is adequate to support only about 15 seconds of intense exercise. Consequently, the muscle fiber must manufacture additional ATP from energy stored in the chemical bonds of nutrients. Some of these molecules are contained within the muscle fiber itself. Others must be mobilized from the liver and adipose tissue, then transported to muscles through the circulation.

The primary substrates for energy production are carbohydrates and fats. Recall from Chapter 4 that the most efficient production of ATP occurs through aerobic pathways such as the glycolysis-citric acid cycle pathway [● p. 103]. If the cell has adequate amounts of oxygen for oxidative phosphorylation, then both glucose and fatty acids can be metabolized to provide ATP (Fig. 25-1).

If the oxygen requirement of a muscle fiber exceeds its oxygen supply, energy production from fatty acids decreases dramatically and glucose metabolism shifts to anaerobic pathways. Anaerobic metabolism is also known as **glycolytic metabolism**, because in low oxygen conditions, anaerobic glycolysis is the primary pathway for ATP production. When a cell lacks oxygen for oxidative phosphorylation, the final product of glycolysis, pyruvate, is converted

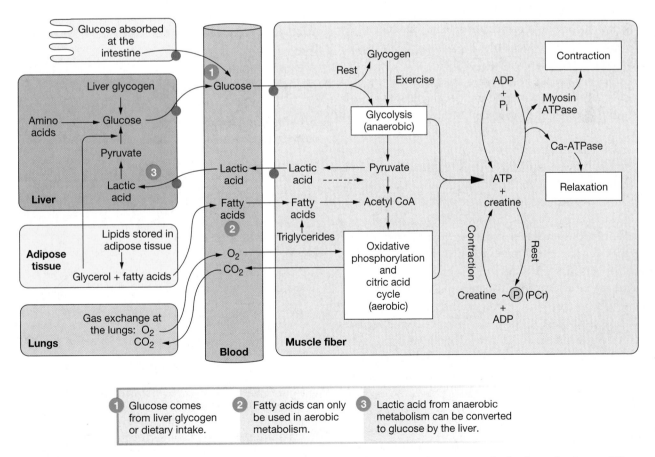

1. Glucose comes from liver glycogen or dietary intake.

2. Fatty acids can only be used in aerobic metabolism.

3. Lactic acid from anaerobic metabolism can be converted to glucose by the liver.

● **Figure 25-1** *Energy metabolism in skeletal muscle* Oxygen for aerobic metabolism enters the body at the lungs. When oxygen is not available, glucose is metabolized to lactic acid. Some energy in the muscle is stored in the form of phosphocreatine (PCr).

Phosphocreatine can only sustain 10 seconds of maximal exercise.

Anaerobic metabolism produces ATP 2.5 times faster than aerobic metabolism but can only support 1 minute of maximal exercise.

Aerobic metabolism supports exercise for hours.

● **Figure 25-2**   *Speed of ATP production compared with ability to sustain maximal muscle activity*

to lactic acid [● p. 104]. In general, exercise that depends on anaerobic metabolism cannot be sustained for an extended period.

Anaerobic metabolism has the advantage of speed, producing ATP 2.5 times as rapidly as aerobic pathways do (Fig. 25-2 ●). But this advantage comes with two distinct disadvantages: (1) anaerobic metabolism provides only 2 ATP per glucose, compared with the average of 30–32 ATP per glucose for oxidative metabolism, and (2) it contributes to a state of metabolic acidosis by producing lactic acid. (However, the CO_2 generated during metabolism is a more significant source of acid.)

Where does glucose for aerobic and anaerobic ATP production come from? The body has three sources: the plasma glucose pool, intracellular stores of glycogen in muscles and liver, and "new" glucose made in the liver from nonglucose precursors such as amino acids (*gluconeogenesis*; [● p. 110]). Muscle and liver glycogen stores provide enough energy substrate for about 2000 kcal (equivalent to about 20 miles of running in the average person), more than adequate for the exercise that most of us do. However, glucose alone cannot provide sufficient ATP for endurance athletes such as marathon runners. To meet their energy demands, they rely on the energy stored in fats, estimated at 70,000 kcal per person.

In reality, aerobic exercise of any duration uses both fatty acids and glucose as substrates for ATP production. About 30 minutes after aerobic exercise begins, the concentration of free fatty acids in the blood increases significantly, showing that fats are being mobilized from adipose tissue. However, the breakdown of fatty acids through the process of β-oxidation [● p. 109] is slower than glucose metabolism through glycolysis, so muscle fibers use a combination of fatty acids and glucose to meet their energy needs.

At lower exercise intensity, most of the energy for ATP production comes from fats (Fig. 25-3 ●). As the intensity of exercise increases and ATP is consumed more rapidly, the muscle fibers begin to use a larger proportion of glucose. When exercise reaches

about 70% of maximum, carbohydrates become the primary source of energy.

Aerobic training increases both fat and glycogen stores within the muscle fibers themselves. Training also enhances the ability of exercising muscle to use fatty acids as a source of ATP by increasing the activity of enzymes for β-oxidation.

Hormones Regulate Metabolism During Exercise

Several hormones that affect glucose and fat metabolism change their pattern of secretion with exercise. Plasma concentrations of glucagon, cortisol, the catecholamines (epinephrine and norepinephrine), and growth hormone all increase during exercise. Cortisol and the catecholamines, along with growth hormone, promote

● **Figure 25-3**   *Use of carbohydrates and fats with increasing exercise*   At low intensity exercise, muscles rely more on fats than glucose. Above exercise intensities exceeding 70% of maximum, glucose becomes the main energy source. Data from G. A. Brooks and J. Mercier, "Balance of carbohydrate and lipid utilization during exercise: The 'crossover' concept," *Journal of Applied Physiology* 76:2253–2261 (1994).

the conversion of triglycerides to glycerol and fatty acids. Glucagon, catecholamines, and cortisol also mobilize liver glycogen and raise plasma glucose levels. A hormonal environment that favors the conversion of glycogen into glucose is desirable, because glucose is a major energy substrate for exercising muscle.

Curiously, although plasma glucose concentrations rise with exercise, the secretion of insulin decreases. This response is contrary to what you would predict, because normally an increase in plasma glucose stimulates insulin release. During exercise, however, insulin secretion is suppressed, probably by sympathetic input onto the beta cells of the pancreas.

What could be the advantage of lower insulin levels during exercise? For one thing, less insulin means that cells other than muscle fibers reduce their glucose uptake, sparing blood glucose for the muscles to use. Actively contracting muscle cells, on the other hand, are not affected by low levels of insulin because they do not require insulin for glucose uptake. Experiments have shown that electrical stimulation of a muscle fiber in a glucose solution increases its glucose uptake in proportion to its contractile activity.

Oxygen Consumption Is Related to Exercise Intensity

The activities we call exercise range widely in intensity and duration, from the rapid burst of energy exerted by a sprinter or power lifter to the sustained effort of a marathoner. Physiologists traditionally quantify the intensity of a period of exercise by measuring oxygen consumption ($\dot{V}_{O_2}$). **Oxygen consumption** refers to the fact that oxygen is used up, or consumed, during oxidative phosphorylation, when it combines with hydrogen in the mitochondria to form water [❀ p. 106].

Oxygen consumption is a measure of cellular respiration and is usually measured in liters of oxygen consumed per minute. A person's *maximal rate of oxygen consumption* ($\dot{V}_{O_2max}$) is an indicator of a person's ability to perform endurance exercise. The greater the $\dot{V}_{O_2max}$, the greater the person's predicted ability to do work.

A metabolic hallmark of exercise is an increase in oxygen consumption that persists even after the activity ceases (Fig. 25-4 ●). When exercise commences, oxygen consumption increases rapidly, but it is not instantaneously matched by the oxygen supply to the muscles. During the lag time, ATP is provided by muscle ATP reserves, phosphocreatine, and aerobic metabolism supported by oxygen stored on muscle myoglobin and blood hemoglobin [❀ pp. 404, 582].

The use of these muscle stores creates an *oxygen deficit* because their replacement requires aerobic metabolism and oxygen uptake. Once exercise stops, oxygen consumption is slow to resume its resting level. The **excess postexercise oxygen consumption** (EPOC; formerly called the oxygen debt) represents oxygen being used to metabolize lactate, restore ATP and phosphocreatine levels, and replenish the oxygen bound to myoglobin. Other factors that play a role in elevating postexercise oxygen consumption include increased body temperature and circulating catecholamines.

● **Figure 25-4** *Changes in oxygen consumption during and after exercise*   The cells' use of energy begins as soon as exercise begins, but oxygen supply as measured by oxygen consumption lags behind, creating an oxygen deficit. The oxygen deficit is reflected by elevated oxygen consumption that persists after exercise stops.

Several Factors Limit Exercise

What are the factors that limit exercise capacity? To some extent, the answer depends on the type of exercise. Resistive exercise such as strength training is likely to be limited by the availability of oxygen, so the muscles rely heavily on anaerobic energy production. The picture is more complex with aerobic or endurance exercise. Is the limiting factor for aerobic exercise the ability of the exercising muscle to use oxygen efficiently? The ability of the cardiovascular system to deliver oxygen to the tissues? Or the ability of the respiratory system to provide oxygen to the blood?

One possible limiting factor in exercise is the ability of muscle fibers to use oxygen. If muscle mitochondria are limited in number, or if they have insufficient oxidative enzyme content, the muscle fibers are unable to produce ATP rapidly, even with adequate supply of oxygen and substrates. Although data suggest that muscle metabolism is not the limiting factor for maximum exercise capacity, it has been shown to influence submaximal exercise capacity—a finding that explains the increase in size and number of muscle mitochondria with endurance training.

RUNNING PROBLEM

By the time Colleen reached the practice field, she was already sweating and her face was flushed. At 9 A.M., the air temperature was 80° F, and the humidity was 54%. Colleen took a quick drink of water from the large water container and ran out to the field.

Question 1: "Humidity" is the percentage of water present in air. Why is the thermoregulatory mechanism of sweating less efficient in humid environments?

The question of whether the pulmonary system or the cardiovascular system limits maximal exercise was resolved when research showed that ventilation is only 65% of its maximum when cardiac output has reached 90% of its maximum. From that information, exercise physiologists concluded that the ability of the cardiovascular system to deliver oxygen and nutrients to the muscle at a rate that supports aerobic metabolism is a major factor in determining maximum oxygen consumption. Let's now examine the reflexes that integrate breathing and cardiovascular function during exercise.

▶ VENTILATORY RESPONSES TO EXERCISE

Think about what happens to your breathing when you exercise. Exercise is associated with both increased rate and increased depth of breathing, resulting in enhanced alveolar ventilation [❧ p. 565]. **Exercise hyperventilation**, or **hyperpnea**, results from a combination of feedforward signals from central command neurons in the motor cortex and sensory feedback from peripheral receptors.

When exercise begins, mechanoreceptors and proprioceptors [❧ p. 432] in muscles and joints send information about movement to the motor cortex. Descending pathways from the motor cortex to the respiratory control center of the medulla oblongata then immediately increase ventilation (Fig. 25-5 ●).

As muscle contraction continues, sensory information from a variety of sources feeds back to the control center to ensure that ventilation and oxygen use by the tissue remain closely matched. The sensory receptors involved in the secondary response probably include central and carotid body chemoreceptors that monitor P_{CO_2}, pH, and P_{O_2} [❧ p. 589], proprioceptors in the joints, and possibly receptors located within the exercising muscle itself. Pulmonary stretch receptors [❧ p. 592] were once thought to play a role, but recipients of heart-lung transplants display a normal ventilatory response to exercise although the neural connections between lung and brain are absent.

Exercise hyperventilation maintains nearly normal arterial P_{O_2} and P_{CO_2} by steadily increasing alveolar ventilation in proportion to the level of exercise. The compensation is so effective that when arterial P_{O_2}, P_{CO_2}, and pH are monitored during mild to moderate exercise, they show no significant change (Fig. 25-6 ●). This observation eliminates the traditional explanation for increased ventilation during exercise. Low arterial P_{O_2}, elevated arterial P_{CO_2}, and decreased plasma pH are the usual stimuli for the carotid and central chemoreceptors. The fact that these stimuli are not present in mild to moderate exercise means that the chemoreceptors or medullary respiratory control center, or both, must be responding to other exercise-induced signals.

Several factors have been postulated as the stimulus for exercise hyperventilation, including sympathetic input to the carotid body and changes in plasma K^+ concentration. During even mild exercise, plasma K^+ levels increase as repeated action potentials in the muscle fibers allow K^+ to move out of cells and into the extra-

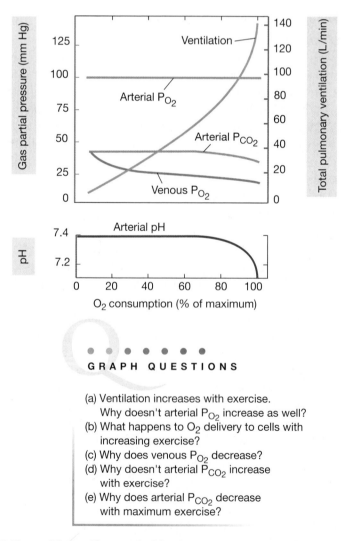

GRAPH QUESTIONS

(a) Ventilation increases with exercise. Why doesn't arterial P_{O_2} increase as well?

(b) What happens to O_2 delivery to cells with increasing exercise?

(c) Why does venous P_{O_2} decrease?

(d) Why doesn't arterial P_{CO_2} increase with exercise?

(e) Why does arterial P_{CO_2} decrease with maximum exercise?

● **Figure 25-6** *Changes in blood gas, partial pressures, and arterial pH with exercise*

● **Figure 25-5** *Changes in ventilation with submaximal exercise* Ventilation rate jumps as soon as exercise begins, despite the fact that neither arterial P_{CO_2} nor P_{O_2} has changed. Modified from P. Dejours, *Handbook of Physiology* (Washington, D.C.: American Physiological Society, 1964).

cellular fluid. Carotid chemoreceptors are known to respond to increased K^+ concentration by increasing ventilation. However, because K^+ concentration changes slowly, this mechanism does not explain the sharp initial rise in ventilation at the onset of activity.

It appears likely that the initial change in ventilation is caused by sensory input from muscle mechanoreceptors combined with parallel descending pathways from the motor cortex. Once exercise is under way, sensory input keeps ventilation matched to metabolic needs.

CONCEPT CHECK   Answers: p. 795

1. If venous P_{O_2} decreases as exercise intensity increases, what do you know about the P_{O_2} of the muscle cells as exercise intensity increases?

▶ CARDIOVASCULAR RESPONSES TO EXERCISE

When exercise begins, mechanosensory input from working limbs combines with descending pathways from the motor cortex to activate the cardiovascular control center in the medulla oblongata. The center responds with sympathetic discharge that increases cardiac output and causes vasoconstriction in many peripheral arterioles.

Cardiac Output Increases During Exercise

During strenuous exercise, cardiac output rises dramatically. In untrained individuals, cardiac output goes up fourfold, from 5 L/min to 20 L/min. In trained athletes, it may go up six to eight times, reaching as much as 35 L/min. As you learned in Chapter 14, cardiac output is influenced by several factors, including (1) heart rate, (2) force of contraction, and (3) venous return [●Fig. 14-31, p. 482]:

Cardiac output (CO) = heart rate × stroke volume

Or, if the factors that influence heart rate and stroke volume are considered,

CO = (SA node rate + influence of autonomic nervous system) × (venous return + force of contraction)

Let's examine which of those factors has the greatest effect on cardiac output during exercise in a healthy heart. During exercise, venous return is enhanced by skeletal muscle contraction and deep inspiratory movements [● p. 481]. It is therefore tempting to postulate that, owing to Starling's law of the heart, the cardiac muscle fibers simply stretch in response to increased venous return, thereby increasing contractility.

However, overfilling the ventricles is potentially dangerous, because overstretching may damage the fibers. One way the body offsets increased venous return is by speeding up heart rate. If the interval between contractions is less, the heart will not have as much time to fill and therefore is not likely to be damaged by excessive stretch.

The initial change in heart rate at the onset of exercise is due to decreased parasympathetic activity at the sinoatrial (SA) node [● p. 467]. As cholinergic inhibition lessens, heart rate rises from its resting rate to around 100 beats per minute, the intrinsic pacemaker rate of the SA node. At that point, sympathetic output from the cardiovascular control center escalates.

Sympathetic stimulation has two effects on the heart. First, it increases contractility so that the heart squeezes out more blood per stroke (increased stroke volume). Second, sympathetic innervation increases heart rate so that the heart does not have much time to relax, protecting it from overfilling. The combination of faster heart rate and greater stroke volume increases cardiac output during exercise.

Peripheral Blood Flow Redistributes to Muscle During Exercise

At rest, skeletal muscles receive less than a fourth of the cardiac output, or about 1.2 L/min. During exercise, because of local and reflex reactions, a significant shift in peripheral blood flow takes place. With strenuous exercise by highly trained athletes, the combination of increased cardiac output and vasodilation can increase blood flow through exercising muscle to more than 22 L/min! The relative distribution of blood flow to the tissues also shifts. About 88% of the cardiac output is diverted to the exercising muscle, up from 21% at rest (Fig. 25-7 ●).

The redistribution of blood flow during exercise results from a combination of vasodilation in skeletal muscle arterioles and vasoconstriction in other tissues. At the onset of exercise, sympathetic signals from the cardiovascular control center cause vasoconstriction in peripheral tissues. As muscles become active, changes in the microenvironment of the muscle tissue take place. Tissue O_2 concentrations decrease, while temperature, CO_2, and acid in the interstitial fluid around the muscle fibers increase. All of these factors act as paracrines to cause local vasodilation that overrides the sympathetic signal for vasoconstriction. The net

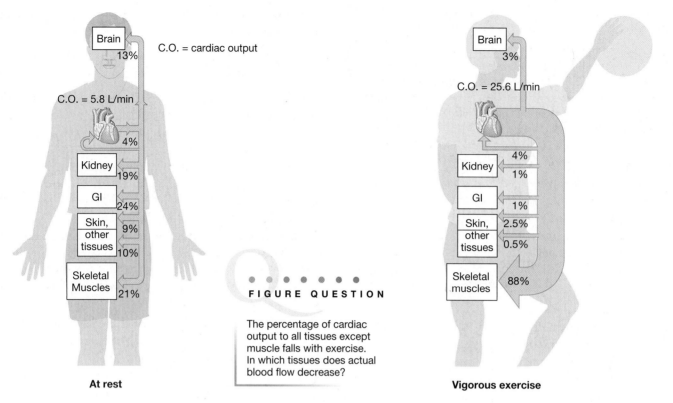

C.O. = cardiac output

FIGURE QUESTION

The percentage of cardiac output to all tissues except muscle falls with exercise. In which tissues does actual blood flow decrease?

At rest

Vigorous exercise

● **Figure 25-7** *Distribution of cardiac output at rest and during exercise*   The shunting of blood from nonexercising tissues to muscle is accomplished by sympathetically induced vasoconstriction in nonexercising tissues combined with vasodilation in exercising muscle.

result is shunting of blood flow from inactive tissues to the exercising muscles, where it is needed.

Blood Pressure Rises Slightly During Exercise

What happens to blood pressure during exercise? You learned in Chapter 15 that peripheral blood pressure is determined by a combination of cardiac output and peripheral resistance [● p. 510]. Cardiac output increases during exercise and therefore contributes to increased blood pressure. The changes resulting from peripheral resistance are harder to predict, however, because some peripheral arterioles are constricting while others are dilating:

Mean arterial blood pressure =

cardiac output × peripheral resistance

Skeletal muscle vasodilation decreases peripheral resistance to blood flow. At the same time, sympathetically induced vasoconstriction in nonexercising tissues offsets the vasodilation, but only partially. Consequently, total peripheral resistance to blood flow falls dramatically as exercise commences, reaching a minimum at about 75% of $\dot{V}_{O_2max}$ (Fig. 25-8a ●).

If no other compensation occurred, this decrease in peripheral resistance would dramatically lower arterial blood pressure. However, increased cardiac output cancels out decreased peripheral resistance. When blood pressure is monitored during exercise, mean arterial blood pressure actually increases slightly as

exercise intensity increases (Fig. 25-8b). The fact that it increases at all, however, suggests that the normal baroreceptor reflexes that control blood pressure are not functioning during exercise.

CONCEPT CHECK   Answers: p. 795

2. In Figure 25-8b, why does the line for mean pressure lie closer to the diastolic pressure line instead of being evenly centered between systolic and diastolic pressures? (Hint: what is the equation for calculating mean blood pressure?)

The Baroreceptor Reflex Adjusts to Exercise

Normally, homeostasis of blood pressure is regulated through peripheral baroreceptors in the carotid and aortic bodies: an increase in blood pressure initiates responses that return blood pressure to normal. But during exercise, blood pressure increases without activating homeostatic compensation. Why is the normal baroreceptor reflex absent during exercise?

There are several theories. According to one, signals from the motor cortex during exercise reset arterial baroreceptor threshold to a higher pressure. Blood pressure can then increase slightly during exercise without triggering the homeostatic counterregulatory responses.

Another theory suggests that signals in baroreceptor afferent neurons are blocked in the spinal cord by presynaptic inhibition

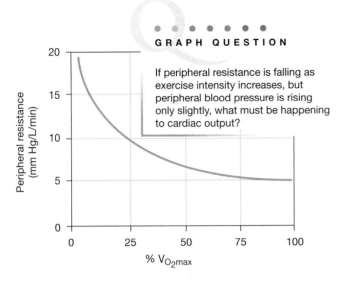

(a) Peripheral resistance decreases due to vasodilation in exercising muscle.

GRAPH QUESTION

If peripheral resistance is falling as exercise intensity increases, but peripheral blood pressure is rising only slightly, what must be happening to cardiac output?

(b) Mean arterial blood pressure rises slightly despite drop in resistance.

● **Figure 25-8** *Peripheral resistance and arterial blood pressure during exercise*

[◈ p. 272] at some point before the afferent neurons synapse with central nervous system neurons. This central inhibition inactivates the baroreceptor reflex during exercise.

A third theory is based on the postulated existence of muscle chemoreceptors that are sensitive to metabolites, probably H^+, produced during strenuous exercise. When stimulated, these chemoreceptors signal the CNS that tissue blood flow is not ade-

quate to remove muscle metabolites or keep the muscle in aerobic metabolism. The chemoreceptor input is reinforced by sensory input from mechanoreceptors in the working limbs. The CNS response to this sensory input is to override the baroreceptor reflex and raise blood pressure to enhance muscle perfusion. The same hypothetical muscle chemoreceptors may play a role in ventilatory responses to exercise.

▶ FEEDFORWARD RESPONSES TO EXERCISE

Interestingly, there is a significant *feedforward* element [◈ p. 195] in the physiological responses to exercise. It is easy to explain physiological changes that occur with exercise as reactions to the disruption of homeostasis. However, many of these changes occur in the absence of the normal stimuli or before the stimuli are present. For example, ventilation rates jump as soon as exercise begins, even though experiments have shown that arterial P_{O_2} and P_{CO_2} do not change (see Figs. 25-5 and 25-6).

How does the feedforward response work? As exercise begins, proprioceptors in the muscles and joints send information to the motor cortex of the brain. Descending signals from the motor cortex go not only to the exercising muscles but also along parallel pathways to the cardiovascular and respiratory control centers and to the limbic system of the brain.

Output from the limbic system and cardiovascular control center triggers generalized sympathetic discharge. As a result, an immediate slight increase in blood pressure marks the beginning of exercise. Sympathetic discharge causes widespread vasoconstriction, increasing blood pressure. Once exercise has begun, this increase in blood pressure compensates for decreases in blood pressure resulting from muscle vasodilation.

As exercise proceeds, reactive compensations become superimposed on the feedforward changes. For example, when exercise reaches 50% of aerobic capacity, muscle chemoreceptors sense the buildup of lactic acid and metabolites and send this information along to central command centers in the brain. The command centers then maintain changes in ventilation and circulation that were initiated in a feedforward manner. Thus, the integration of systems in exercise involves both common reflex pathways and some unique centrally mediated reflex pathways.

▶ TEMPERATURE REGULATION

As exercise continues, heat released through metabolism creates an additional challenge to homeostasis. Most of the energy released during metabolism is not converted into ATP but is released as heat. (Energy conversion from organic substrates to ATP is only 20%–25% efficient.) With continued exercise, heat production exceeds heat loss, and core body temperature rises. In endurance events, body temperature can reach 40°–42° C (104°–108° F), which we would normally call a fever.

When the paramedics arrived, they took Colleen's blood pressure. It was extremely low. Her body temeprature was 103° F. The paramedics quickly transported Colleen to the hospital.

Question 3: Why was Colleen's blood pressure so low?

 782 785 787 ® 790 792

This rise in body temperature with exercise triggers two thermoregulatory mechanisms: sweating and increased cutaneous blood flow [♠ p. 717]. Both mechanisms help regulate body temperature but have the potential to disrupt homeostasis in other ways.

Sweating lowers body temperature through evaporative cooling. At the same time, however, loss of fluid from the extracellular compartment has the potential to cause dehydration and significantly decrease circulating blood volume. Because sweat is a hypotonic fluid, the extra water loss increases body osmolarity. The combination of decreased ECF volume and increased osmolarity during extended exercise sets in motion the complex homeostatic pathway for dehydration described in Chapter 20, including thirst and renal conservation of water [♠ Fig. 20-17, p. 644].

The other thermoregulatory mechanism, increased blood flow to the skin, allows loss of body heat to the environment through convective heat loss [♠ p. 717]. However, increased sympathetic output during exercise tends to vasoconstrict cutaneous blood vessels and oppose the thermoregulatory response. The primary control of vasodilation in the skin during exercise appears to come from a sympathetic vasodilator system. Activation of these neurons as body core temperature rises results in dilation of cutaneous blood vessels without altering sympathetic vasoconstriction in other tissues of the body.

Although vasodilation of cutaneous vessels is essential for thermoregulation, it disrupts homeostasis by decreasing peripheral resistance and diverting blood flow from the muscles. In the face of these contradictory demands, the body initially gives preference to thermoregulation. However, if central venous pressure falls below a critical minimum, the body abandons thermoregulation in the interest of maintaining blood flow to the brain.

The degree to which the body can adjust to both demands depends on the type of exercise being performed and its intensity and duration. Strenuous exercise in hot, humid environments can severely impair normal thermoregulatory mechanisms and cause *heat stroke*, a potentially fatal condition. Unless prompt measures are taken to cool the body, core temperatures can go as high as 43° C (109° F).

It is possible for the body to adapt to repeated exercise in hot environments, however, through **acclimatization**. In this process, physiological mechanisms shift to fit a change in environmental conditions. As the body adjusts to exercise in the heat, sweating begins sooner and is more copious, doubling or tripling in volume, enhancing evaporative cooling. With acclimatization, sweat also becomes more dilute, as salt is reabsorbed from the sweat glands under the influence of increased aldosterone. Salt loss in an unacclimatized person exercising in the heat may reach 30 g NaCl per day, but that value decreases to as little as 3 g after a month of acclimatization.

▶ EXERCISE AND HEALTH

Physical activity has many positive effects on the human body. Our lifestyles have changed dramatically since humans were hunter-gatherers, but our bodies still seem to work best with a certain level of physical activity. Several common pathological conditions—including high blood pressure, strokes, and diabetes mellitus—can be improved by physical activity. Even so, developing regular exercise habits is one lifestyle change that many people find difficult to make. In this section, we look at the effect that exercise has on several common conditions.

Exercise Lowers the Risk of Cardiovascular Disease

As early as the 1950s, scientists showed that men who are physically active have a lower rate of heart attacks than do men who lead a sedentary lifestyle. These studies started many investigations into the exact relationship between cardiovascular disease and exercise. Subsequently, it has been demonstrated that exercise has positive benefits for both men and women. These include lowering blood pressure, decreasing plasma triglyceride levels, and raising plasma high-density lipoprotein (HDL) levels. High blood pressure (hypertension) is a major risk factor for strokes, while elevated triglycerides and LDL-cholesterol levels are associated with development of atherosclerosis and increased risk of heart attack.

Overall, exercise reduces the risk of death or illness from a variety of cardiovascular diseases. Even mild exercise, such as walking, has significant health benefits and could help the estimated 40 million American adults with sedentary lifestyles that put them at risk for developing cardiovascular disease, complications of obesity, or diabetes mellitus.

Type 2 Diabetes Mellitus May Improve with Exercise

Regular exercise is now widely accepted as effective in preventing and alleviating type 2 diabetes mellitus. With chronic exercise, skeletal muscle fibers up-regulate both the number of glucose transporters and the number of insulin receptors on their membrane. The addition of glucose transporters decreases the muscle's dependence on insulin for glucose uptake.

Increasing the number of insulin receptors makes the fibers more sensitive to insulin. A smaller amount of insulin then can

RUNNING PROBLEM

When Colleen arrived in the emergency room, she was quickly diagnosed with heat stroke, a life-threatening condition. She was immersed in a tub of cool water and given intravenous fluids. Heat stroke can occur in athletes who overexert in excessively hot weather. Heat stroke is also common in elderly individuals, whose thermoregulatory mechanisms are less efficient than those in younger people.

Question 4: What steps could Colleen have taken to avoid heat stroke? (Start at 9 A.M., when Colleen set out for the practice field.)

achieve a response that previously required more insulin. Because the cells are responding to lower insulin levels, the endocrine pancreas secretes less insulin. This lessens the stress on the pancreas, resulting in a lower incidence of type 2 diabetes mellitus.

Figure 25-9 ● shows the results of an experiment in which men with impaired glucose tolerance were started on an exercise

regimen. Their response to glucose ingestion was compared with that of normal subjects. In each experiment, the individual ingested 100 g of glucose, and researchers followed changes in his plasma glucose levels for 120 minutes (this procedure is known as a *glucose tolerance test*). Simultaneous measurements were made of plasma insulin.

After only seven days of exercise, both the glucose tolerance test and insulin secretion in exercising diabetic subjects had shifted to a pattern that was more like that of the normal subjects. Because of the beneficial effect of exercise on glucose transport and metabolism, patients with type 2 diabetes are encouraged to begin a regular exercise program.

Stress and the Immune System May Be Influenced by Exercise

Another area that is receiving much attention is the interaction of exercise with the immune system. Epidemiological evidence looking at large populations of people associates exercise with a reduced incidence of disease and with increased longevity. There is also widespread public belief that exercise boosts immunity, prevents cancer, and helps HIV-infected patients combat AIDS.

But so far, the results of rigidly controlled research studies have not supported those claims. Indeed, there is good evidence suggesting that strenuous exercise is a form of stress and will suppress the immune response. Some immune suppression may be due to corticosteroid release. Other studies suggest that it may be due to release of interferon-γ during strenuous exercise.

Researchers have proposed that the relationship between exercise and decreased immunity forms a J-shaped curve (Fig. 25-10 ●). People who exercise moderately have better immune systems

(a) Plasma glucose during glucose tolerance test

(b) Plasma insulin during glucose tolerance test

● **Figure 25-9** *The effect of exercise on glucose tolerance and insulin secretion* In this experiment, a group of men with abnormal glucose tolerance tests (red line, part a) exercised for seven days and were tested again. After seven days of exercise, their glucose tolerance test (green line, part a) had shifted in the direction of normal controls, as had their pattern of insulin secretion (green line, part b). Exercise is known to enhance glucose uptake and increase muscle sensitivity to insulin. Data from B. R. Seals et al., "Glucose Tolerance in Older and Younger Athletes and Sedentary Men," *Journal of Applied Physiology* 56(6):1521–1525 (1984); and M. A. Rogers et al., "Improvement in Glucose Tolerance After One Week of Exercise in Patients with Mild NIDDM," *Diabetes Care* 11:613–618 (1988).

● **Figure 25-10** *Immune function and exercise* The J-shaped curve of this graph shows one theory of the relationship between immune function and exercise.

than those who are sedentary, but people who exercise strenuously may see a decrease in immune function because of the stress of the exercise.

Another area of exercise physiology filled with interesting though contradictory results is the effect of exercise on stress, depression, and other psychological parameters. The literature shows an inverse relationship between exercise and depression: people who exercise regularly are significantly less likely to be clinically depressed than are people who do not exercise regularly. Although the association exists, assigning cause and effect to the two parameters is difficult. Are the exercisers less depressed because they exercise? Or do the depressed subjects exercise less because they are depressed?

Many published studies appear to show that regular exercise is effective in reducing depression. But a careful analysis of experimental design suggests that the conclusions of those studies may be overstated. The subjects in many of the experiments were being treated concurrently with drugs or psychotherapy, and it is difficult to attribute improvement in their condition solely to exercise.

In addition, participation in exercise studies gives subjects a period of social interaction, another factor that might play a role in the reduction of stress and depression. The assertion that exercise reduces depression would be better supported by a demonstration of the effects of exercise on biological parameters such as disturbances of monoamine neurotransmitters [🔍 p. 309]. To date, however, little work of that type has been done.

RUNNING PROBLEM CONCLUSION

Heat Stroke

Colleen was kept overnight in the hospital for observation. She was unable to practice with the team for the remainder of the season because victims of heat stroke are more sensitive to high temperatures for some time after the episode.

To learn more about the symptoms and treatment of heat stroke, try a Google search (www.google.com). Then check your understanding of this running problem by comparing your answers to those in the summary table.

	QUESTION	FACTS	INTEGRATION AND ANALYSIS
1	Why is the thermoregulatory mechanism of sweating less efficient in humid environments?	"Humidity" is the percentage of water present in air. Thermoregulation in warm environments includes sweating and evaporative cooling.	Evaporation is slower in humid air, therefore, evaporative cooling is less effective in high humidity.
2	Individuals with a heat emergency called heat exhaustion have cool, moist skin. Hot, dry skin indicates a more serious emergency called heat stroke. Why is heat exhaustion less serious than heat stroke?	Sweating helps the body regulate temperature through evaporative cooling. As evaporation occurs, the skin surface cools.	Cool, moist skin indicates that the sweating mechanism is still functioning. Hot, dry skin indicates that the sweating mechanism has failed. Subjects with hot, dry skin are likely to have higher internal temperatures.
3	Why was Colleen's blood pressure so low?	Blood volume decreases when the body loses large amounts of water through sweating.	Colleen had been sweating but not replacing the fluid she lost. This caused her blood volume to decrease, with a corresponding decrease in blood pressure.
4	What steps could Colleen have taken to avoid heat stroke? (Start at 9 A.M., when Colleen set out for the practice field.)	Heat stroke is caused by dehydration as a result of excessive sweating. This results in a drop in blood pressure. Peripheral blood vessels constrict in an effort to maintain pressure and blood flow to the brain. Constricted blood vessels in the skin cannot release excess heat. In addition, the sweating response is inhibited to prevent further fluid loss.	Colleen could have avoided heat stroke by taking the following measures: (1) consuming large amounts of fluid (water plus electrolytes) before and during practice; (2) avoiding excessive exercise such as bicycling to practice; (3) stopping exercising at the first signs of heat emergency (dizziness and nausea); (4) seeking shade and drinking large amounts of fluid to replenish lost fluids; (5) using cool wet towels or ice to bring body temperature lower.

782 785 787 789 790

CHAPTER SUMMARY

Exercise is an example of normal physiology at work. *Integration* and *coordination* between the body's *physiological control systems* allow the internal environment to remain relatively constant, despite the challenges to *homeostasis* that exercise presents.

Metabolism and Exercise

1. Exercising muscle requires a steady supply of ATP. A small amount of ATP is stored in the muscle fiber and additional ATP can be rapidly formed from phosphocreatine. (p. 783; Fig. 25-1)

2. Carbohydrates and fats are the primary substrates for energy production. Glucose can be metabolized through both oxidative and anaerobic pathways, but fatty acid metabolism requires adequate amounts of oxygen. (p. 783; Fig. 25-1)

3. Anaerobic **glycolytic metabolism** converts glucose to lactic acid. Glycolytic metabolism is 2.5 times more rapid than aerobic pathways but is not as efficient at ATP production. (p. 783; Fig. 25-2)

4. Glucagon, cortisol, catecholamines, and growth hormone influence glucose and fatty acid metabolism during exercise. These hormones favor the conversion of glycogen to glucose. (p. 784)

5. Although plasma glucose concentrations rise with exercise, the secretion of insulin decreases. This response reduces glucose uptake by most cells, making more glucose available for exercising muscle. (p. 785)

6. The intensity of exercise is indicated by **oxygen consumption** (V_{O_2}). A person's maximal rate of oxygen consumption (V_{O_2max}) is an indicator of that person's ability to perform endurance exercise. (p. 785)

7. Oxygen consumption increases rapidly at the onset of exercise. **Excess postexercise oxygen consumption** is due to ongoing metabolism, increased body temperature, and circulating catecholamines. (p. 785; Fig. 25-4)

8. Muscle mitochondria increase in size and number with endurance training. (p. 785)

9. In maximal exertion, the ability of the cardiovascular system to deliver oxygen and nutrients appears to be the primary factor limiting exertion. (p. 786)

Ventilatory Responses to Exercise

10. Exercise hyperventilation results from feedforward signals from the motor cortex and sensory feedback from peripheral sensory receptors. (pp. 786, 789; Fig. 25-5)

11. Arterial P_{O_2}, P_{CO_2}, and pH show no significant change during mild to moderate exercise. Exercise hyperventilation must be due to some other factor such as increased plasma K^+ levels. (p. 786; Fig. 25-6)

Cardiovascular Responses to Exercise

12. Cardiac output increases with exercise because of increased venous return and sympathetic stimulation of heart rate and contractility. (p. 787; Fig. 25-7)

13. Blood flow through exercising muscle increases dramatically when skeletal muscle arterioles dilate. Arterioles in other tissues constrict. (p. 787; Fig. 25-7)

14. Decreased tissue O_2 and glucose or increased muscle temperature, CO_2, and acid act as paracrines and cause local vasodilation. (p. 787)

15. Mean arterial blood pressure increases slightly as exercise intensity increases. The baroreceptor reflexes that control blood pressure change their set-points during exercise. (p. 788; Fig. 25-8)

Feedforward Responses to Exercise

16. When exercise begins, feedforward responses prevent significant disruption of homeostasis. (p. 789)

Temperature Regulation

17. Heat released during exercise is dissipated by sweating and increased cutaneous blood flow. (p. 790)

Exercise and Health

18. Physical activity has many positive effects and can help prevent or decrease the risk of developing high blood pressure, strokes, and type 2 diabetes mellitus. (p. 790)

QUESTIONS

LEVEL ONE *Reviewing Facts and Terms*

1. Name the two muscle compounds that store energy in the form of high-energy phosphate bonds.

2. The most efficient ATP production is through *aerobic/anaerobic* pathways. When these pathways are being used, then *glucose/fatty acids/both/neither* can be metabolized to provide ATP.

3. What are the differences between aerobic and anaerobic metabolism?

4. List three sources of glucose that can be metabolized to ATP, either directly or indirectly.

5. List four hormones that promote the conversion of triglycerides into fatty acids. What effects do these hormones have on plasma glucose levels?

6. What is meant by the term *oxygen deficit*? How is oxygen deficit related to excessive postexercise oxygen consumption?

7. What organ system is the limiting factor for maximal exertion?

8. In endurance events, body temperature can reach 40°–42° C. What is normal body temperature? What two thermoregulatory mechanisms are triggered by this change in temperature during exercise?

LEVEL TWO *Reviewing Concepts*

9. **Concept map:** Map the metabolic, cardiovascular, and respiratory changes that occur with exercise. Include the signals to and from the nervous system and show what specific areas signal and coordinate the exercise response.

10. What causes insulin secretion to decrease during exercise and why is this decrease adaptive?

11. State two advantages and two disadvantages of anaerobic glycolysis.

12. Compare and contrast the following terms, especially as they relate to exercise:
    (a) ATP, ADP, PCr
    (b) myoglobin, hemoglobin

13. Match the response to the brain area that controls it. Brain areas may be used once, more than once, or not at all. Some responses may have more than one answer.

    (a) pons
    (b) medulla oblongata
    (c) midbrain
    (d) motor cortex
    (e) hypothalamus
    (f) cerebellum
    (g) brain not involved; uses local control

    1. changes in cardiac output
    2. vasoconstriction
    3. exercise hyperventilation
    4. increased stroke volume
    5. increased heart rate
    6. coordination of skeletal muscle movement

14. Specify whether each parameter listed below stays the same, increases, or decreases when a person becomes better conditioned for athletic activities:
    (a) heart rate during exercise
    (b) resting heart rate
    (c) cardiac output during exercise
    (d) resting cardiac output
    (e) breathing rate during exercise
    (f) blood flow to muscles during exercise
    (g) blood pressure during exercise
    (h) total peripheral resistance during exercise

15. Why doesn't increased venous return during exercise overstretch the heart muscles?

16. Diagram the three theories that explain why the normal baroreceptor reflex is absent during exercise.

17. List and briefly discuss the benefits of a lifestyle that includes regular exercise.

18. Explain how exercise decreases blood glucose in type 2 diabetes mellitus.

LEVEL THREE *Problem Solving*

19. You have decided to manufacture a new sports drink that will help athletes, from football players to gymnasts. List at least five different ingredients you would put in your drink and tell why each one is important for the athlete.

LEVEL FOUR *Quantitative Problems*

20. You are a well-conditioned athlete. At rest, your heart rate is 60 beats per minute and your stroke volume is 70 mL/beat. What is your cardiac output? At one point during exercise, your heart rate goes up to 120 beats/min. Does your cardiac output increase proportionately? Explain.

Explorations

Explorations are also available at *www.physiologyplace.com*

In this chapter you learned about exercise and the physiological challenges that it presents. Exercise integrates responses from nearly every system of the body using neural control, feedback, and tissue-to-tissue communication.

Exploration 1: Exercise for Your Health

The evidence is growing that everyone should do some type of exercise on a regular basis, even if it is simply walking the dog around the neighborhood. Some recent articles that discuss the health benefits of exercise include:

"Effects of the amount and intensity of exercise on plasma lipoproteins," *New England Journal of Medicine* 347: 1483–1492, 2002 Nov 7.

"Additional steps for cardiovascular health," *New England Journal of Medicine* 347:755–756, 2002 Sep 5.

"Waging war on modern chronic diseases: primary prevention through exercise biology," *Journal of Applied Physiology* 88:774–787, 2000. (*http://jap.physiology.org*)

"Waging war on physical inactivity: using modern molecular ammunition against an ancient enemy," *Journal of Applied Physiology* 93:3–30, 2002.

The Health & Fitness Information page on the American College of Sports Medicine website (*www.acsm.org*) has information on calculating your exercise heart rate range.

Exploration 2: Lactate Threshold

One of the new trends in athletic training is monitoring the *lactate threshold*, the exercise intensity at which lactate begins to accumulate in the blood. There is debate among exercise physiologists about whether the lactate threshold marks the onset of anaerobic metabolism or whether it simply is the point at which lactate accumulates faster than it can be metabolized by other tissues. Whatever the cause, an increased lactate threshold is associated with increased exercise endurance, and training can increase the threshold. To learn more about this subject, do a Google search for *lactate threshold* (*www.google.com*).

ANSWERS

Answers to Concept Check Questions

Page 787

1. If venous P_{O_2} decreases, the cell's P_{O_2} is also decreasing.

Page 788

2. The line for mean blood pressure lies closer to the line for diastolic pressure because the heart spends more time in diastole than systole.

Answers to Figure Questions

Page 786

Figure 25-6: (a) Pulmonary ventilation increases steadily as exercise increases, but arterial P_{O_2} remains constant, indicating that ventilation is matched to blood flow through the lungs. (c) Venous P_{O_2} drops as exercise increases, showing that the cells are removing more oxygen from hemoglobin as oxygen consumption increases. (d) Arterial P_{CO_2} does not change until oxygen consumption reaches 80% of maximum. At that level, ventilation increases sharply. (e) As the person begins to hyperventilate, arterial (and alveolar) P_{CO_2} declines. Blood pH also remains fairly constant except at near-maximal exercise levels.

Page 788

Figure 25-7: To calculate blood flow to an organ, multiply cardiac output (L/min) times the percentage of flow to that organ. When rest and exercise values are compared, actual blood flow decreases only in the kidneys, GI tract, and other tissues.

Page 789

Figure 25-8: Mean arterial pressure is determined by cardiac output and resistance. If resistance is falling but MAP is increasing, cardiac output must be increasing.

Answers to Level Four Quantitative Problems

Page 794

20. 60 beats/minute × 70 mL/beat = 4200 mL/min cardiac output. If heart rate doubles to 120 beats/min, CO goes to 8400 mL/min, or doubles also.

Reproduction and Development

26

SEM of oocyte at fertilization showing zona pellucida (blue), sperm (pink), and partially cleared corona radiata cells (green).

RUNNING PROBLEM

Infertility

Peggy and Larry have just about everything to make them happy: successful careers, a loving marriage, a comfortable home. But one thing is missing. After five years of marriage, they have been unable to have a child. Today, Peggy and Larry have their first appointment with Dr. Charles Coddington, an infertility specialist. "Finding the cause of your infertility is going to require some painstaking detective work," he explains. Dr. Coddington will begin his workup of Peggy and Larry by asking detailed questions about their reproductive histories. Based on the answers to these questions, he will then order tests to pinpoint the problem.

806 815 818 821 826 829

BACKGROUND BASICS

Meiosis: **840** ǀ Prostaglandins: **37** ǀ Up- and down-regulation: **186** ǀ Positive and negative feedback: **194** ǀ Intestinal peptide absorption: **683** ǀ Oxytocin: **222** ǀ Hot flashes: **719** ǀ Flagella: **58** ǀ Mitosis: **840** ǀ Steroids: **216** ǀ Agonist/antagonist: **185** ǀ Spinal reflex: **295** ǀ Hypothalamic-pituitary axis: **223** ǀ Prolactin: **222** ǀ Osteoporosis: **745** ǀ

Birth, and copulation, and death. That's all the facts when you come to brass tacks.

—T. S. Eliot, *Sweeney Agonistes*

Imagine growing up as a girl, then at the age of 12 or so, finding that your voice is deepening and your genitals are developing into those of a man. This scenario actually happens to a small number of men with a condition known as *pseudohermaphroditism* [*pseudes*, false + *hermaphrodites*, the dual-sex offspring of Hermes and Aphrodite]. These men have the internal sex organs of a male but inherit a gene that causes a deficiency in one of the male hormones. Consequently, they are born with external genitalia that appear feminine, and they are raised as girls. At puberty, when they begin to secrete male hormones, they develop some, but not all, of the characteristics of men. As you can imagine, a conflict arises: change gender, or remain female? Most choose to continue life as men.

Reproduction is one area of physiology in which we humans like to think of ourselves as significantly advanced over other animals. We mate for pleasure as well as procreation, and the female human is always sexually receptive, not just during her fertile periods. But just how different are we?

Like other terrestrial animals, we have internal fertilization that allows motile flagellated sperm to remain in an aqueous environment. Because internal fertilization requires cooperation of the two parties, we have mating and courtship rituals as do other animals. Fetal development is also internal, protecting the growing embryo from dehydration by cushioning it with a layer of fluid.

Humans are *sexually dimorphic*; that is, the males and females are physically distinct. This distinction is sometimes blurred by dress and hairstyle, but these are cultural acquisitions. Although everyone agrees that male and female humans are physically dimorphic, we are still debating whether we are behaviorally and psychologically dimorphic as well.

Sex hormones play a significant role in the behavior of other mammals, acting on adults as well as influencing the brain of the developing embryo. The degree to which the results of nonprimate and primate animal studies can be extrapolated to humans is controversial. Human fetuses are exposed to sex hormones while in the uterus, but it is unclear how much gender imprinting occurs. Does the preference of little girls for dolls and of little boys for toy guns have a biological basis or a cultural basis? We have no answer yet, but evidence is growing that at least part of our brain structure is influenced by sex hormones before we ever leave the womb.

This chapter addresses the biology of human reproduction and development. We will begin with the fertilized egg, or **zygote**, and first examine sex determination and sexual differentiation.

▶ SEX DETERMINATION

The male and female sex organs consist of the gonads, the internal accessory ducts and glands, and the external reproductive structures known collectively as the **genitalia**. **Gonads** [*gonos*, seed] are the organs that produce **gametes** [*gamein*, to marry], the reproductive cells (eggs and sperm) that unite to form a new individual. The male gonads are the **testes** (singular *testis*), which produce **sperm** (spermatozoa). The female gonads are the **ovaries**, which produce

● **Figure 26-1** *Human chromosomes* Humans have 23 pairs of chromosomes, which in this figure have been arranged in homologous pairs. The presence of an X and a Y chromosome means that these chromosomes came from a male.

eggs, or **ova** (singular *ovum*). The undifferentiated gonadal cells destined to produce eggs and sperm are called **germ cells**.

The secret of sexual development is locked within the human genome [◉ Appendix B]. Each nucleated cell of the body except egg and sperm contains 46 chromosomes (the *diploid number*): 22 matched pairs of **autosomes** and one pair of **sex chromosomes**, each of which is designated either X or Y. The 22 pairs of autosomal chromosomes direct development of the human body form and of variable characteristics such as hair color and blood type.

The two sex chromosomes contain genes that direct development of internal and external sex organs. The X chromosome is larger than the Y chromosome and includes many genes that are missing from the Y chromosome. Figure 26-1 ● shows a set of human chromosomes laid out as pairs of *homologous* (corresponding) autosomes, with the sex chromosomes at the lower right.

Eggs and sperm each have a half set of 23 chromosomes (the *haploid number*) because of meiotic divisions during their creation. When egg and sperm unite, the resulting zygote contains a unique set of 46 chromosomes. Thus, humans inherit one chromosome of each pair from the mother and the other from the father.

The Sex Chromosomes Determine Genetic Sex

The sex chromosomes that a person inherits determine the genetic sex of the individual. Genetic females are XX, and genetic males are XY (Fig. 26-2 ●). Females inherit one X chromosome from each parent. Males inherit a Y chromosome from their father and an X from their mother. The Y chromosome is essential for development of the male reproductive organs.

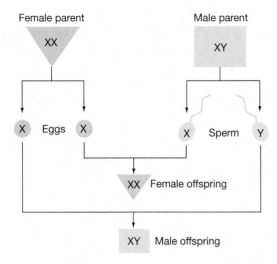

● **Figure 26-2** *Inheritance of X and Y chromosomes* Each egg produced by a female (XX) has an X chromosome. Each sperm produced by a male (XY) can have either an X or a Y chromosome.

If sex chromosomes are abnormally distributed during fertilization, the presence or absence of a Y chromosome determines whether the embryo will become male. The presence of a Y chromosome means that the embryo will be male, even if there are also multiple copies of the X chromosome present. For instance, an XXY zygote will become male. A zygote that inherits only a Y chromosome (YO) will die because the larger X chromosome contains unique genes that are essential for development.

In the absence of a Y chromosome, an embryo will develop into a female. Therefore, a zygote that gets only one X chromosome (XO; Turner's syndrome) will develop into a female. Two X chromosomes are needed for normal female reproductive function, however.

X-Linked Inherited Disorders

Normally, a person inherits two copies of the gene for a given trait, one copy from the mother and one from the father. But many traits that are found on the X chromosome have no matching gene on the much smaller Y chromosome (X-linked genes). Females, who inherit two X chromosomes, always get two copies of X-linked genes, so the expression of X-linked traits follows the usual pattern of gene dominance and recession. Males, on the other hand, receive only one copy of the gene on the X chromosome, so they *always* exhibit the characteristics of the X-linked gene. If the maternally inherited gene is defective, male offspring will exhibit the mutation. Among the X-linked diseases that have been identified are Duchenne muscular dystrophy [● p. 413], hemophilia [● p. 541], and color-blindness.

Once the ovaries develop, one X chromosome in each cell of the body inactivates and condenses into a clump of nuclear chromatin known as a *Barr body*. The selection of which X chromosomes become inactive is random, so some cells have an active maternal X chromosome and others have an active paternal X chromosome. Because inactivation occurs early in development, before cell division is complete, all cells of a given tissue will usually have the same active X chromosome, either maternal or paternal.

CONCEPT CHECK	Answers: p. 834

1. Name the male and female gonads and gametes.

Sexual Differentiation Occurs in the Second Month of Development

The gender of an early embryo is difficult to determine because reproductive structures do not begin to differentiate until the seventh week of development. Before differentiation, the embryonic tissues are considered *bipotential* because they cannot be morphologically identified as male or female (Fig. 26-3a ●). The bipotential external genitalia consist of a *genital tubercle, urethral folds, urethral groove,* and *labioscrotal swellings.* These structures differentiate into the male and female external genitalia as development progresses (Table 26-1).

In the early embryo, bipotential gonadal tissue is associated with two pairs of accessory ducts: **Wolffian ducts** from the embryonic kidney and **Müllerian ducts**. As development proceeds along either male or female lines, one pair of ducts develops while the other degenerates.

What is it that directs some single-cell zygotes to become males and others to become females? Gender determination depends on the presence or absence of the sex-determining region of the Y chromosome, or **SRY gene**. This gene produces a protein (**SRY protein** or *testis-determining factor*) that binds to DNA and activates additional genes. The products of these genes together direct the development of testes from the bipotential gonad (Fig. 26-4 ●).

The developing testes then secrete hormones to continue development of the male reproductive tract. **Anti-Müllerian hormone** (**AMH**; also called *Müllerian inhibiting substance*), a glycoprotein from testicular **Sertoli cells**, causes the Müllerian ducts to regress. **Testosterone**, a male sex hormone from testicular **Leydig cells**, and its derivative **dihydrotestosterone (DHT)** convert the Wolffian ducts into the *male accessory structures*: epididymis, vas deferens, and seminal vesicle (see Table 26-1).

Testosterone and DHT also cause the external genitalia to take on the male pattern. Later in fetal development, testosterone controls migration of the testes from the abdomen into the *scrotum*, or scrotal sac.

In female embryos, who have no SRY gene, the primitive ducts follow a different developmental course. In the absence of anti-Müllerian hormone, Müllerian ducts develop into the upper portion of the **vagina**, the **uterus**, and the **Fallopian tubes** (after

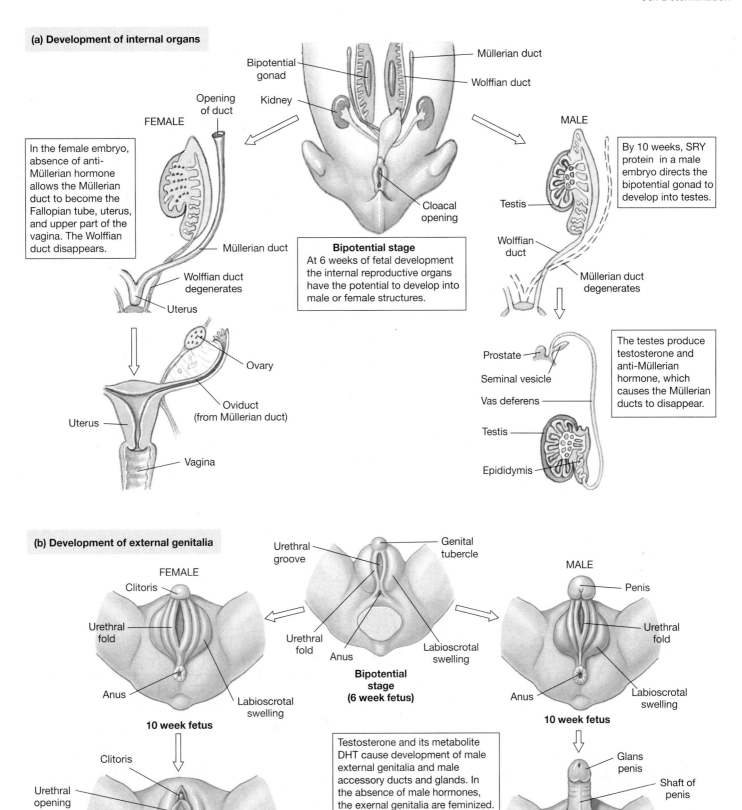

(a) Development of internal organs

Bipotential gonad

Müllerian duct

Wolffian duct

Opening of duct

Kidney

FEMALE

MALE

In the female embryo, absence of anti-Müllerian hormone allows the Müllerian duct to become the Fallopian tube, uterus, and upper part of the vagina. The Wolffian duct disappears.

Müllerian duct

Wolffian duct degenerates

Uterus

Cloacal opening

Testis

Wolffian duct

Müllerian duct degenerates

By 10 weeks, SRY protein in a male embryo directs the bipotential gonad to develop into testes.

Bipotential stage
At 6 weeks of fetal development the internal reproductive organs have the potential to develop into male or female structures.

Ovary

Oviduct (from Müllerian duct)

Uterus

Vagina

Prostate

Seminal vesicle

Vas deferens

Testis

Epididymis

The testes produce testosterone and anti-Müllerian hormone, which causes the Müllerian ducts to disappear.

(b) Development of external genitalia

Urethral groove

Genital tubercle

FEMALE

MALE

Clitoris

Urethral fold

Anus

Labioscrotal swelling

Urethral fold

Anus

Labioscrotal swelling

Penis

Urethral fold

Anus

Labioscrotal swelling

Bipotential stage (6 week fetus)

10 week fetus

10 week fetus

Clitoris

Urethral opening

Labia majora

Vaginal opening

Anus

Labia minora

At birth

Testosterone and its metabolite DHT cause development of male external genitalia and male accessory ducts and glands. In the absence of male hormones, the external genitalia are feminized.

By birth, the testes have descended from the abdominal cavity into the scrotum.

Glans penis

Shaft of penis

Scrotum

Anus

At birth

● **Figure 26-3   *Sexual development in the human embryo***

Table 26-1: Sexual Differentiation

MALE	BIPOTENTIAL STRUCTURE	FEMALE
Glans penis	← Genital tubercle →	Clitoris
Shaft of penis	← Urethral folds and groove →	Labia minora, opening of vagina and urethra
Shaft of penis and scrotum	← Labioscrotal swellings →	Labia majora
Regresses	← Gonad (cortex) →	Forms ovary
Forms testis	← Gonad (medulla) →	Regresses
Becomes epididymis, vas deferens, and seminal vesicle (testosterone present)	← Wolffian duct →	Regresses (testosterone absent)
Regresses (Müllerian inhibiting substance present)	← Müllerian duct →	Becomes Fallopian tube, uterus, cervix, and upper 1/3 of vagina (Müllerian inhibiting substance absent)

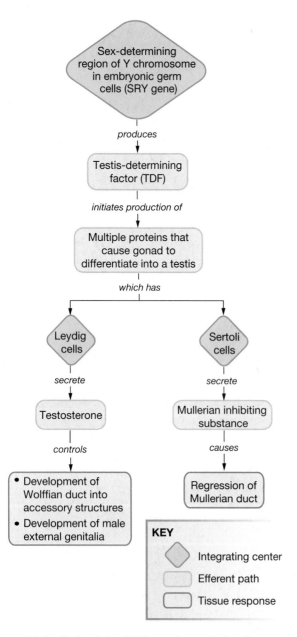

● **Figure 26-4** *Role of the SRY gene in male development*

the anatomist Fallopius who first described them; also known as **oviducts**). Without testosterone and DHT, the Wolffian ducts degenerate, and external genitalia take on the female pattern. Thus, absence of the SRY gene, anti-Müllerian hormone, testosterone, and DHT creates a female.

Exposure to testosterone during embryonic development is known to masculinize certain aspects of brain function, such as the brain's responsiveness to certain hormones. One controversial aspect of the masculinizing effects of testosterone is its influence on human sexual behavior and gender identity. It is well documented that in many nonhuman mammals, adult sexual behavior depends on the absence or presence of testosterone during critical periods of

CLINICAL FOCUS

Pseudohermaphrodites and DHT

The importance of dihydrotestosterone in male development came to the attention of scientists through studies of the male pseudohermaphrodites described in the opening of this chapter. These men inherit a defective gene for the enzyme *5-alpha-reductase*, which catalyzes the conversion of testosterone to dihydrotestosterone (DHT). The defect, like that of sickle cell anemia, results from the substitution of a single amino acid in the peptide sequence. Because of deficient DHT, the male external genitalia and prostate gland fail to develop fully during fetal development, despite normal testosterone secretion by the testes. At birth, the infants are usually considered female and are raised as such. However, at puberty, the testes again begin to secrete testosterone, causing masculinization of the external genitalia, pubic hair growth (although scanty facial and body hair), and deepening voice. By studying the defect in 5-alpha-reductase in these individuals, we have been able to separate the effects of testosterone from those of DHT.

brain development. However, a similar cause/effect relationship has never been proved in humans. In human behavior, it is very difficult to separate biological influences from environmental factors, and it will probably be years before this question is resolved.

2. Why was King Henry VIII of England wrong to blame his wives when they were unable to produce a male heir to the throne?

3. What sex will a zygote become if it inherits only one X chromosome (XO)?

4. If the testes are removed from an early male embryo, why does it develop a uterus and Fallopian tubes rather than the normal male accessory structures? Will the embryo have male or female external genitalia?

▶ BASIC PATTERNS OF REPRODUCTION

The male testis and female ovary both produce hormones and gametes, and they share other similarities, as might be expected of organs with the same origin. The gametes themselves are very different, however. Eggs are some of the largest cells in the body [● Fig. 3-2, p. 52]. They are nonmotile and must be moved through the reproductive tract on currents created by smooth muscle contraction or the beating of cilia. Sperm, on the other hand, are quite small. They are the only flagellated cells of the body and are highly motile so that they can swim up the female reproductive tract in their search for an egg to fertilize.

The timing of gamete production, or **gametogenesis**, is also very different in males and females. Women are born with all the eggs, or **oocytes**, they will ever have. During the reproductive years, the eggs mature. They are released from the ovaries roughly once a month for about 40 years. Then female reproductive cycles cease (**menopause**).

Men, on the other hand, manufacture sperm continuously from the time they reach reproductive maturity. Sperm and testosterone production diminish with age, but do not cease the way women's reproductive cycles do.

Gametogenesis Begins in Utero *and Resumes During Puberty*

Figure 26-5 ● compares the male and female patterns of gametogenesis. In both genders, germ cells of the embryonic gonads first undergo a series of mitotic divisions to increase their numbers. After that, the germ cells are ready to begin **meiosis**, the cell division process through which gametes are formed. Through meiotic divisions, the normal number of 46 chromosomes (written 46,XX for females or 46,XY for males) is reduced by half. All ova are 23,X. Sperm can be either 23,X or 23,Y.

When meiosis begins, the cell's DNA replicates to create the equivalent of 92 chromosomes. However, cell and chromosomal division do not take place. Each of the duplicated chromosomes is

in the form of two identical **sister chromatids** linked together at a region known as the **centromere**. The resultant cells, called *primary gametes*, still have 46 chromosomes, but each chromosome is duplicated and contains twice the normal amount of DNA.

In the **first meiotic division**, the chromosomes sort themselves into two *secondary gametes*, each of which has 23 chromosomes. Each chromosome is still composed of two sister chromatids. In the **second meiotic division**, the sister chromatids separate and go to different cells. The result of that division is two cells with 23 single chromatids.

The process of gametogenesis as described above is identical in males and females, but the timing of mitosis and meiosis is very different. The testes of a newborn boy contain immature spermatogonia that are capable of replicating through mitosis. However, the gonads become inactive after birth and do not resume mitotic activity until **puberty**, the period in the early teen years when the gonads mature.

Some of the germ cells, known as **spermatogonia** (singular *spermatogonium*), continually duplicate themselves throughout the male's reproductive life. Other spermatogonia are destined to develop into sperm. They undergo meiosis and become **primary spermatocytes**.

In the first meiotic division, one primary spermatocyte divides into two **secondary spermatocytes**. In the second meiotic division, each secondary spermatocyte divides into two **spermatids**. One spermatid has 23 single chromosomes, the haploid number characteristic of a gamete. The spermatids then mature into sperm. Thus, one primary spermatocyte creates four sperm.

In the embryonic ovary, germ cells are called **oögonia** (singular *oögonium*). Oögonia complete mitotic replication and the first stage of meiosis by the fifth month of embryonic development (Fig. 26-5). At this point, the embryonic ovary contains half a million primary gametes, or **primary oocytes**. Germ cell mitosis ceases before birth, and no further oocytes can be formed.

The oocyte's first meiotic division takes place following puberty. Each primary oocyte divides into two cells, a large **egg** (**secondary oocyte**) and a tiny **first polar body**. Despite the size difference, each cell contains 23 duplicated chromosomes. The first polar body disintegrates.

Meanwhile the egg begins the second meiotic division. After its sister chromatids separate, meiosis pauses again. The final step of meiosis, in which sister chromatids go to separate cells, does not take place unless the egg is fertilized.

The ovary releases the mature egg during a process known as **ovulation**. If the egg is not fertilized, meiosis never goes to completion, and the egg disintegrates or passes out of the body. If fertilization by sperm occurs, the final step of meiosis takes place. Half the sister chromatids remain in the fertilized egg (zygote), while the other half are released in a **second polar body**. The second polar body, like the first, degenerates. In the female, each primary oocyte thus gives rise to only one egg.

Gametogenesis in both males and females is under the control of hormones from the brain and from endocrine cells in the

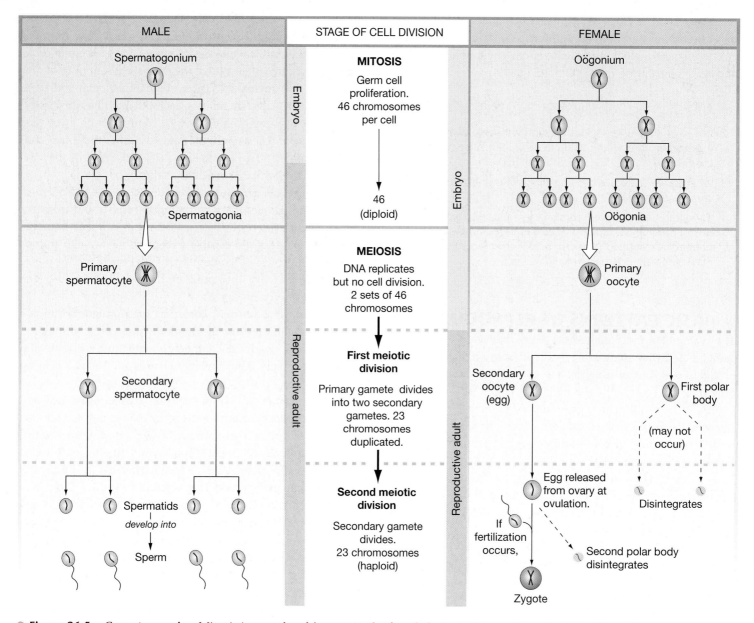

MALE	STAGE OF CELL DIVISION	FEMALE

MALE

Spermatogonium

Spermatogonia

Primary spermatocyte

Secondary spermatocyte

Spermatids
develop into

Sperm

STAGE OF CELL DIVISION

Embryo

MITOSIS

Germ cell proliferation. 46 chromosomes per cell

46 (diploid)

Reproductive adult

MEIOSIS

DNA replicates but no cell division. 2 sets of 46 chromosomes

First meiotic division

Primary gamete divides into two secondary gametes. 23 chromosomes duplicated.

Second meiotic division

Secondary gamete divides. 23 chromosomes (haploid)

FEMALE

Oögonium

Oögonia

Primary oocyte

Secondary oocyte (egg)

First polar body

(may not occur)

Disintegrates

Egg released from ovary at ovulation.

If fertilization occurs,

Second polar body disintegrates

Zygote

● **Figure 26-5** *Gametogenesis* Mitosis is completed *in utero* in the female but continues after puberty in the male.

gonads. Some of these hormones are identical in males and females, and others are different.

Answers: p. 834

5. At what stage of development is the gamete in a newborn male? In a newborn female?

6. Compare the amount of DNA in the first polar body with the amount of DNA in the second polar body.

7. How many gametes are formed from one primary oocyte? From one primary spermatocyte?

The Brain Directs Reproduction

The pathways that regulate reproduction begin with secretion of peptide, protein, and glycoprotein hormones by the hypothalamus and anterior pituitary. These trophic hormones con-

trol gonadal secretion of the steroid sex hormones, including **androgens**, **estrogens**, and **progesterone**. These hormones are closely related and arise from the same steroid precursors (Fig. 26-6 ●). Both sexes produce androgens and estrogens, but androgens predominate in males, and estrogens are dominant in females.

In men, most testosterone is secreted by the testes, but about 5% comes from the adrenal cortex. Testosterone is converted in peripheral tissues to its more potent derivative *dihydrotestosterone* (DHT). Some of the physiological effects attributed to testosterone are actually the result of DHT activity.

The testes also contain the enzyme **aromatase**, which converts testosterone to **estradiol**, the main estrogen in humans. A small amount of additional estrogen is made in peripheral tissues. Although males synthesize estrogens, their feminizing effects are usually not obvious.

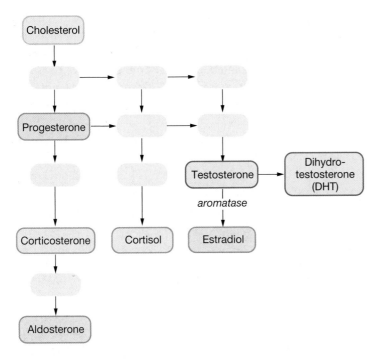

● **Figure 26-6** *Synthesis pathway for steroid hormones* The blank boxes represent intermediate compounds whose names have been omitted for simplicity.

In women, the ovary produces estrogens (estradiol and *estrone*) and progesterone, the other female hormone. The ovary and the adrenal cortex produce small amounts of androgens.

Control Pathways for Sex Steroids Are Similar in Males and Females
The hormonal control of reproduction in the two sexes follows the basic hypothalamus-anterior pituitary-peripheral gland pattern (Fig. 26-7 ●). **Gonadotropin releasing hormone (GnRH*)** from the hypothalamus controls secretion of two anterior pituitary **gonadotropins: follicle stimulating hormone (FSH)** and **luteinizing hormone (LH)**. FSH and LH in turn act trophically on the gonads.

FSH, along with steroid sex hormones, is required to initiate and maintain gametogenesis. LH acts primarily on endocrine cells, stimulating production of the steroid sex hormones.

Although primary control of gonadal function arises in the brain, the gonads also influence their own function. Both ovary and testis secrete peptide hormones that act directly on the pituitary. **Inhibins** are hormones that inhibit FSH secretion. A family of related peptides, called **activins**, stimulate secretion of FSH.

---

*GnRH is sometimes called *luteinizing hormone releasing hormone* (LHRH) because it was first thought to have its primary effect on LH.

● **Figure 26-7** *General pattern of hormonal control of reproduction*

Table 26-2: Effects of Sex Steroids on Gonadotropin Release

STEROID HORMONE	EFFECT	GONADOTROPIN LEVEL
Low estrogen or androgen	Absence of negative feedback	Increases
Moderate estrogen or androgen	Negative feedback	Decreases
High androgen	Negative feedback	Decreases
Sustained high estrogen	Positive feedback	Increases

Activins also promote spermatogenesis, oocyte maturation, and embryonic nervous system development. These gonadal peptides are produced in nongonadal tissues as well, and their other functions in the body are still being investigated.

Feedback Pathways Include Both Long-Loop and Short-Loop Feedback

The feedback pathways for trophic hormones from the hypothalamus and anterior pituitary follow the general pattern of long-loop and short-loop feedback described in Chapter 7 [p. 224]. Gonadal steroids suppress secretion of GnRH, FSH, and LH in a long-loop response. The pituitary gonadotropins inhibit GnRH release by the short-loop path (Fig. 26-7).

When circulating levels of gonadal steroids are low, the pituitary secretes FSH and LH (Table 26-2). Once steroid secretion reaches a certain level, negative feedback inhibits gonadotropin release. Androgens always maintain negative feedback on gonadotropin release: as androgen levels go up, FSH and LH secretion diminishes.

On the other hand, only lower concentrations of estrogen have a negative feedback effect. If estrogen secretion rises rapidly to above a threshold level for at least 36 hours, the feedback changes to positive feedback, and gonadotropin release (particularly LH) is *stimulated*. The paradoxical effects of estrogen on gonadotropin release play a significant role in the female reproductive cycle, as you will learn later in this chapter.

EMERGING CONCEPTS

Inhibins and Activins

The family of hormones known as *inhibins* and *activins* are part of a large family of related molecules known as the *transforming growth factor-β* family. Anti-Müllerian hormone is also a member of this family. Interestingly, although inhibin and activin have opposite effects on FSH secretion, their structures are similar. Inhibin is composed of two peptide subunits, α and β. Activin is composed of two β subunits. These two hormones are an interesting example of how a limited number of genes can create a variety of signal molecules.

We still do not fully understand the mechanism of the change from negative to positive feedback with estrogen. Some evidence suggests that high levels of estrogen increase the number of GnRH receptors, making the anterior pituitary more sensitive to GnRH (up-regulation of receptors; [p. 186]). Other evidence points to a direct effect of estrogen on hypothalamic release of GnRH.

The Hypothalamus Releases Gonadotropin in Small Pulses

Starting at puberty, tonic GnRH release from the hypothalamus occurs in small pulses every 1–3 hours in both males and females. The region of the hypothalamus that contains the GnRH neuron cell bodies has been called a **pulse generator** because it coordinates the periodic pulsatile secretion of GnRH from a group of neurons.

Scientists were puzzled for a time about why tonic GnRH release occurred in pulses rather than in a steady fashion, but several studies have shown the significance of the pulses. Children who suffer from a deficiency of GnRH will not mature sexually in the absence of gonadotropin stimulation of the gonads. If treated with steady infusions of GnRH through drug-delivery pumps, these children still fail to mature sexually.

But if the pumps are adjusted to deliver GnRH in pulses similar to those that occur naturally, the children will go through puberty. Apparently, steady high levels of GnRH cause down-regulation of the GnRH receptors on gonadotropin cells, making the pituitary unable to respond to GnRH.

This down-regulation is the basis for the therapeutic use of GnRH in treating certain disorders. For example, patients with prostate and breast cancers stimulated by androgens or estrogens may be given GnRH agonists to slow the growth of the cancer cells. It seems paradoxical to give these patients a drug that stimulates secretion of androgens and estrogens, but after a brief increase in FSH and LH, the pituitary becomes insensitive to GnRH. FSH and LH secretion then decreases, and gonadal output of steroid hormones also falls. In essence, the GnRH agonist creates chemical castration that reverses when the drug is no longer administered.

Some of the least-understood influences on reproductive hormones and gametogenesis are environmental effects. In men, factors that influence gametogenesis are difficult to monitor short of requesting periodic sperm counts. Disruption of the normal reproductive cycle in women is easier to study because physiological uterine bleeding in the menstrual cycle is easily monitored.

Some factors that affect reproductive function in women are stress, nutritional status, and changes in the day-night cycle such as those which occur with travel across time zones or with shift work. The hormone **melatonin** from the pineal gland mediates reproduction in seasonally breeding animals such as birds and deer, so researchers are investigating whether melatonin plays a role in seasonal and daily rhythms in humans.

Environmental estrogens are also receiving a lot of attention. These are naturally occurring compounds, such as the *phytoestrogens* of plants, or synthetic compounds that have been released into the environment. Some of these compounds bind to estrogen receptors and mimic estrogen's effects. Others are anti-estrogens that block estrogen receptors or interfere with second messenger pathways or protein synthesis.

Now that you have learned the basic patterns of hormone secretion and gamete development, let's look in detail at the male and female reproductive systems.

CONCEPT CHECK   Answers: p. 834

8. What does aromatase do?

9. What do the following abbreviations stand for? (Spelling counts!) FSH, DHT, SRY, LH, GnRH

10. Name the hypothalamic and anterior pituitary hormones that control reproduction.

▶ MALE REPRODUCTION

The male reproductive system consists of the testes, the accessory glands and ducts, and the external genitalia. The external genitalia consist of the **penis** and the **scrotum**, a saclike structure that contains the testes (Fig. 26-8 ●). The **urethra** serves as a common passageway for sperm and urine, although not simultaneously. It runs through the ventral aspect of the shaft of the penis, surrounded by a spongy column of tissue known as the **corpus spongiosum** [*corpus*, body; plural *corpora*]. The corpus spongio-

sum along with two columns of tissue called the **corpora cavernosa** constitute the erectile tissue of the penis.

The tip of the penis is enlarged into a region called the **glans** that at birth is covered by a layer of skin called the **foreskin**, or **prepuce**. In some cultures, the foreskin is removed surgically in a procedure called **circumcision**. In the United States, this practice goes through cycles of popularity. Proponents of the procedure claim that it is necessary for good hygiene and they cite evidence suggesting that the incidence of penile cancer, sexually transmitted diseases, and urinary tract infections is lower in circumcised men. Opponents claim that it is cruel to subject newborn boys to an unnecessary and possibly painful procedure.

The scrotum is an external sac into which the testes migrate during fetal development. This location outside the abdominal cavity is necessary because normal sperm development requires a temperature that is 2°–3° F lower than body core temperature. Men who have borderline or low sperm counts are advised to switch from jockey-style underwear, which keeps the scrotum close to the body, to boxer shorts, which allow the testes to stay cooler.

The failure of one or both testes to descend is known as **cryptorchidism** [*crypto*, hidden + *orchis*, testicle] and occurs in 1%–3% of all male births. If left alone, about 80% of cryptorchid testes spontaneously descend later. Those that remain in the abdomen through puberty become sterile and unable to produce sperm.

● **Figure 26-8**   *Reproductive structures in the male*

Although cryptorchid testes lose their gametogenic potential, they can produce androgens, indicating that hormone production is not as temperature-sensitive as sperm production. Because undescended testes are prone to become cancerous, authorities recommend that they be moved to the scrotum with testosterone treatment or surgery if necessary.

The male accessory glands and ducts include the **prostate gland**, the **seminal vesicles**, and the **bulbourethral (Cowper's) glands**. The bulbourethral glands and seminal vesicles empty their secretions into the urethra through ducts. The individual glands of the prostate open directly into the urethral lumen.

The prostate gland is the best known of the three accessory glands because of its medical significance. Cancer of the prostate is the second most common form of cancer in men (next to lung cancer), and *benign prostatic hypertrophy* (enlargement) creates problems for many men after age 50. Because the prostate gland completely encircles the urethra, its enlargement causes difficult urination by narrowing the passageway.

The Testes Produce Sperm and Testosterone

The human testes are paired ovoid structures about 5 cm by 2.5 cm (Fig. 26-9a ●). The word *testis* means "witness" in Latin, and its application to the male gonad comes from the fact that in ancient Rome, men taking an oath placed one hand on their genitals.

The testes have a tough outer fibrous capsule that encloses masses of coiled **seminiferous tubules** clustered into 250–300 compartments. Between the tubules is *interstitial tissue* consisting primarily of blood vessels and testosterone-producing *Leydig (interstitial) cells*. The seminiferous tubules constitute nearly 80% of the testicular mass in an adult. Each individual tubule is 0.3–1 meter long and, if stretched out and laid end to end, the entire mass would extend for about the length of two and a half football fields.

The seminiferous tubules leave the testis and join the **epididymis** [*epi-*, upon + *didymos*, twin], a single duct that forms a tightly coiled cord on the surface of the testicular capsule (Fig. 26-9b). The epididymis becomes the **vas deferens** [*vas*, vessel + *deferre*, to carry away from], also known as the **ductus deferens**. This duct passes into the abdomen, where it eventually empties into the urethra, the passageway from the urinary bladder to the external environment (see Fig. 26-8).

The Seminiferous Tubules Contain Developing Sperm

A seminiferous tubule is composed of two types of cells: spermatocytes in various stages of becoming sperm, and **Sertoli cells** (Fig. 26-9d, e). The developing spermatocytes stack in columns from the outer edge of the tubule to the inner lumen. Between each column of spermatocytes is a single Sertoli cell that extends from the outer edge of the tubule to the lumen.

Surrounding the outside of the tubule is a basement membrane (*basal lamina*). It acts as a barrier, preventing certain large molecules in the interstitial fluid from entering the tubule, yet al-

lowing testosterone to pass easily (Fig. 26-9c). Adjacent Sertoli cells in a tubule are linked to each other by tight junctions that form an additional barrier between the lumen of the tubule and the interstitial fluid outside the tubule.

The tight junctions are sometimes called the **blood-testis barrier** because functionally they behave much like the impermeable capillaries of the blood-brain barrier, preventing movement of material between two compartments. However, the testes have three functional compartments: (1) the interstitial fluid surrounding the tubules, (2) a basal compartment in the tubules between the basement membrane and basal ends of the Sertoli cells, and (3) an inner compartment that includes the lumen of the tubule. Because of the barriers between these three compartments, fluid in the tubule lumen has a composition different from that of interstitial fluid, with low concentrations of glucose and high concentrations of K^+ and steroid hormones.

Sperm Production Is Continuous *Spermatogonia*, the germ cells that undergo meiotic division to become sperm, are found clustered near the basal ends of the Sertoli cells, just under the basement membrane of the seminiferous tubules. In this basal compartment, they undergo mitotic replication to produce additional germ cells. Some of the spermatogonia remain near the outer edge of the tubule to produce future spermatogonia. Others enter meiosis and become primary spermatocytes.

As the spermatocytes differentiate into sperm, they move inward toward the tubule lumen, continuously surrounded by the Sertoli cells. The tight junctions of the blood-testis barrier break

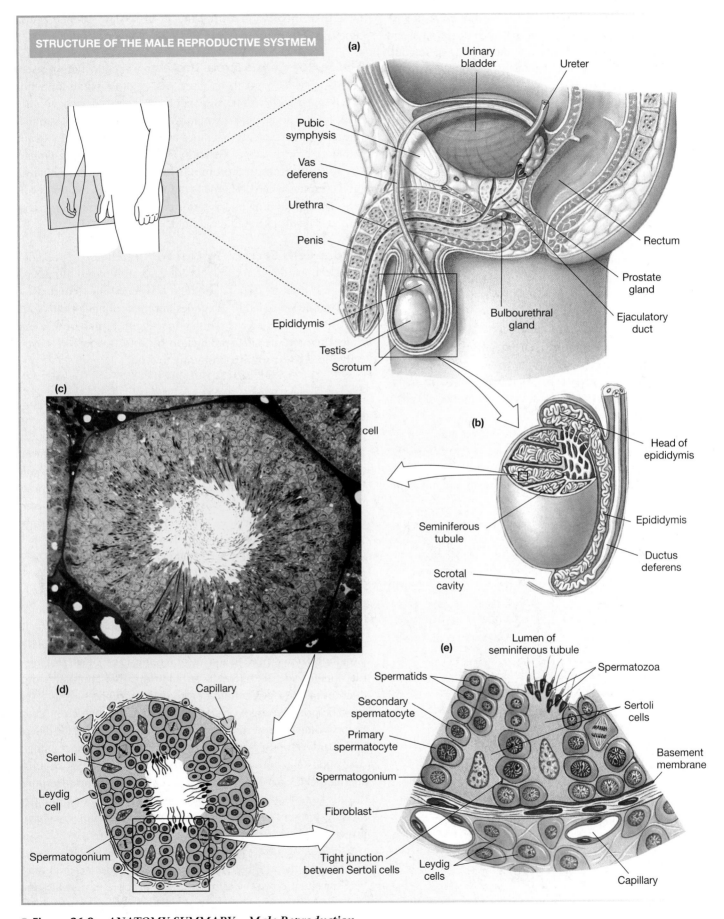

STRUCTURE OF THE MALE REPRODUCTIVE SYSTMEM

(a)

Urinary bladder

Ureter

Pubic symphysis

Vas deferens

Urethra

Penis

Epididymis

Testis

Scrotum

Rectum

Prostate gland

Bulbourethral gland

Ejaculatory duct

(b)

Head of epididymis

Seminiferous tubule

Epididymis

Ductus deferens

Scrotal cavity

(c)

cell

(d)

Capillary

Sertoli

Leydig cell

Spermatogonium

Tight junction between Sertoli cells

(e)

Lumen of seminiferous tubule

Spermatids

Spermatozoa

Secondary spermatocyte

Sertoli cells

Primary spermatocyte

Spermatogonium

Basement membrane

Fibroblast

Leydig cells

Capillary

● **Figure 26-9**   *ANATOMY SUMMARY*   *Male Reproduction*

and reform around the migrating cells, ensuring that the barrier remains intact. By the time the spermatocytes reach the luminal end of the Sertoli cell, they have divided twice and become spermatids.

The spermatids remain embedded in the apical membrane of the Sertoli cells while they complete the transformation into sperm. In this process, they lose most of their cytoplasm and develop a flagellated tail (Fig. 26-10●). The chromatin of the nucleus condenses into a dense structure, while a lysosome-like vesicle called an **acrosome** flattens out to cover most of the surface of the nucleus. The acrosome contains enzymes that are essential for fertilization.

Mitochondria to produce energy for sperm movement concentrate in the midpiece of the sperm body, along with microtubules that extend into the tail [● p. 58]. The result is a small, motile gamete that bears little resemblance to the parent spermatid. Sperm are released into the lumen of the tubule, along with secreted fluid. From there, they are free to move out of the testis.

The entire process takes about 64 days from spermatogonium division until sperm are released into the tubule. However, at any given time, different regions of the tubule contain spermatocytes in different stages of development. The staggering of developmental stages allows sperm production to remain nearly constant at a rate of *200 million* sperm per day. That may sound like an extraordinarily high number, but it is about the number of sperm released in a single ejaculation.

Sperm released from the Sertoli cells are not yet mature or capable of swimming. They are pushed through the lumen of the seminiferous tubule by other sperm and by bulk flow of fluid secreted by the Sertoli cells. Sperm entering the epididymis complete their maturation during the 12 or so days of their transit time, aided by protein secretions from epididymal cells.

The Sertoli Cells Secrete Proteins   The function of Sertoli cells is to regulate sperm development. Another name for Sertoli cells is *sustentacular cells* because they provide sustenance or nourishment for the developing spermatogonia. The Sertoli cells manufacture and secrete proteins that range from the hormones inhibin and activin to growth factors, enzymes, and androgen-binding protein.

Androgen-binding protein is secreted into the seminiferous tubule lumen where it binds to testosterone secreted by Leydig cells. Protein binding makes testosterone less lipophilic and concentrates it within the luminal fluid.

Leydig Cells Secrete Testosterone   The Leydig cells, located in the interstitial tissue outside the seminiferous tubules, secrete testosterone. They are first active in the fetus, when testosterone is needed to direct development of male characteristics. After birth, the cells become inactive until puberty. At that time, they resume their production of testosterone. The Leydig cells also convert some testosterone to estradiol.

Spermatogenesis Requires Gonadotropins and Testosterone

The hormonal control of spermatogenesis follows the general pattern described previously. Gonadotropin releasing hormone (GnRH) from the hypothalamus controls the release of LH and FSH from the anterior pituitary. Those hormones in turn stimulate the testes to produce testosterone. The gonadotropins were named originally from their effect on the female ovary, but the same names have been retained in the male.

GnRH release is pulsatile, peaking every 1.5 hours, and LH release follows the same pattern. FSH levels are not as obviously related to GnRH secretion because FSH secretion is also influenced by inhibin and activin.

The target of FSH is the Sertoli cells, because male germ cells do not have FSH receptors (Fig. 26-11●). FSH stimulates paracrine factors needed for mitotic replication of spermatogonia and for spermatogenesis. In addition, FSH stimulates production of androgen-binding protein and inhibin.

The primary target of LH is the Leydig cells, which produce testosterone. In turn, testosterone feeds back to inhibit LH release by the anterior pituitary. Testosterone is essential for spermatogenesis, but its actions appear to be mediated by the Sertoli cells. Spermatocytes have no androgen receptors and cannot respond directly to testosterone, but they do have receptors for androgen-binding protein. Sertoli cells, which produce androgen-binding protein, have androgen receptors.

Spermatogenesis is a very difficult process to study *in vivo* or *in vitro*, and the available animal models may not accurately reflect the situation in the human testis. Therefore, it may be some time before we can say with certainty how testosterone and FSH regulate spermatogenesis.

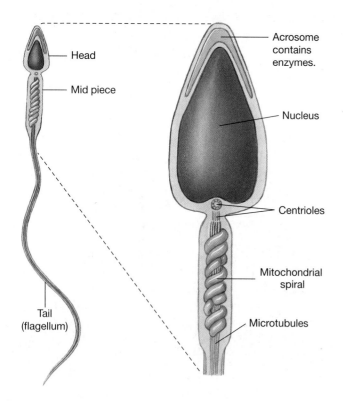

Head

Mid piece

Tail (flagellum)

Acrosome contains enzymes.

Nucleus

Centrioles

Mitochondrial spiral

Microtubules

● **Figure 26-10**   *Sperm structure*

KEY
◇ Integrating center
▢ Efferent path
▢ Effector
▢ Tissue response

● **Figure 26-11** *Hormonal control of spermatogenesis*

Male Accessory Glands Contribute Secretions to Semen

The male reproductive tract has several accessory glands whose primary function is to secrete various fluid mixtures. When sperm leave the vas deferens during ejaculation, they are joined by these secretions, resulting in a sperm-fluid mixture known as **semen**. About 99% of the volume of semen is fluid added from the accessory glands, the bulbourethral glands, seminal vesicles, and prostate.

The contributions of the accessory glands to the composition of semen are shown in Table 26-3. Semen provides a liquid medium for sperm delivery, including mucus for lubrication, buffers to neutralize the usually acidic environment of the vagina, and nutrients for sperm metabolism. The seminal vesicles contribute *prostaglandins* [● p. 37], lipid-related molecules that appear to influence sperm motility and transport within both the male and female reproductive tracts. Prostaglandins were originally believed to come from the prostate gland, and the name was well established by the time their true source was discovered.

In addition to providing a medium for sperm, the accessory gland secretions help protect the reproductive tract from pathogens that might ascend the urethra from the external environment. They physically flush out the urethra and supply immunoglobulins, lysozyme, and other compounds with antibacterial action. One interesting component of semen is zinc. Its role in reproduction is unclear, but concentrations of zinc below a certain level are associated with male infertility.

Androgens Influence Secondary Sex Characteristics

The androgens have a number of effects on the body in addition to gametogenesis. These effects are divided into primary and secondary sex characteristics. **Primary sex characteristics** are the internal sexual organs and external genitalia that distinguish males from females. As you have already learned, androgens are responsible for their differentiation during embryonic development and for their growth during the maturation period known as puberty.

DHT and the Prostate

Fetal development of the prostate gland, like that of the external genitalia, is under the control of dihydrotestosterone (DHT). The discovery of the role of DHT in prostate growth led to the development of finasteride, a 5-alpha-reductase inhibitor that blocks production of DHT. This drug provides the first nonsurgical treatment for benign prostatic hypertrophy. Studies are currently under way to determine if lowering (but not eliminating) DHT will decrease the incidence of cancer of the prostate gland.

The **secondary sex characteristics** are other features of the body, such as body shape, that distinguish males from females. The male body shape is sometimes described as an inverted triangle, with broad shoulders and narrow waist and hips. The female body is more pear-shaped, with broad hips and narrow shoulders. Androgens are responsible for typically male traits such as beard and body hair growth, muscular development, thickening of the vocal chords with subsequent lowering of the voice, and behavioral effects such as the sex drive (**libido**).

Androgens are anabolic hormones that promote protein synthesis, which gives them their street name of *anabolic steroids*. The illicit use of these drugs by athletes has been widespread despite possible adverse side effects such as liver tumors and infertility. One of the more interesting side effects researchers have noted is the apparent addictiveness of anabolic steroids.

Withdrawal from taking the drugs may be associated with behavioral changes that include excessive aggression (roid rage),

depression, or psychosis. These psychiatric disturbances suggest that human brain function can be modulated by sex steroids, just as the brain function of other animals can. Fortunately, many side effects of anabolic steroids are reversible once use is stopped.

13. Explain why the use of exogenous anabolic steroids by males might shrink the testes and make the man temporarily infertile.

▶ FEMALE REPRODUCTION

Female reproduction is a more complicated topic than male reproduction due to the cyclic nature of gamete production in the ovary.

The Female Reproductive Tract Includes Ovaries and Uterus

The female external genitalia are known collectively as the **vulva**, or **pudendum** [*vulva*, womb; *pudere*, to be ashamed]. They are shown in Figure 26-12a ●, the view seen by a health care worker who is about to do a pelvic exam or Pap smear [● p. 50].

Starting at the periphery are the **labia majora** [*labium*, lip], folds of skin that arise from the same embryonic tissue as the scrotum. Within the labia majora are the **labia minora**, derived from embryonic tissues that in the male give rise to the shaft of the penis (see Fig. 26-3). The **clitoris** is a small bud of erectile, sensory tissue at the anterior end of the vulva, enclosed by the labia minora and an additional fold of tissue equivalent to the foreskin of the penis.

In females, the urethra opens to the outside between the clitoris and the **vagina** [*vagina*, sheath], a cavity that acts as

Table 26-3: Composition of Semen

COMPONENT	FUNCTION	SOURCE
Sperm	Gamete	Seminiferous tubules
Mucus	Lubricant	Bulbourethral glands
Water	Provides liquid medium	All accessory glands
Buffers	Neutralize acidic environment of the vagina	Prostate, bulbourethral glands
Nutrients	Nourish sperm	
Fructose		Seminal vesicles
Citric acid		Prostate
Vitamin C		Seminal vesicles
Carnitine		Epididymis
Enzymes	Clot semen in vagina, then liquefy the clot	Seminal vesicles and prostate
Zinc	Unknown; possible association with fertility	Unknown
Prostaglandins	Smooth muscle contraction; may aid sperm transport	Seminal vesicles

the receptacle for the penis during copulation. At birth, the external opening of the vagina is partially closed by a thin ring of tissue called the **hymen**, or *maidenhead*. Contrary to what some young women believe, the hymen is external to the vagina, not found within it, so the normal use of tampons during menstruation will not rupture the hymen. However, it can be stretched by normal activities such as horseback riding and therefore is not an accurate indicator of a woman's virginity.

Now let's follow the path of sperm deposited in the vagina during intercourse. To continue into the female reproductive tract, sperm first must pass through the narrow opening of the **cervix**, the neck of the uterus that protrudes slightly into the upper end of the vagina (Fig. 26-12b, c). The cervical canal is lined with mucous glands whose secretions create a barrier between the vagina and uterus. Sperm that make it through the cervical canal find themselves in the lumen of the uterus, or *womb*, a hollow, muscular organ slightly smaller than a woman's clenched fist.

The uterus is the structure in which fertilized eggs implant and develop during pregnancy. It is composed of three tissue layers: a thin outer connective tissue covering, a thick middle layer of smooth muscle known as the **myometrium** (Fig. 26-12f), and an inner layer known as the **endometrium** [*metra*, womb]. The endometrial lining is an epithelium with glands that dip into a connective tissue layer below. The thickness and character of the endometrium vary during the menstrual cycle. Cells of the epithelial lining alternately proliferate and slough off, accompanied by a small amount of bleeding in the process known as **menstruation**.

Sperm swimming upward through the uterus leave its cavity through openings into the two Fallopian tubes. The Fallopian tubes are 20–25 cm long and about the diameter of a drinking straw. Their walls have two layers of smooth muscle, longitudinal and circular, similar to the walls of the intestine. A ciliated epithelium lines the inside of the tubes (Fig. 26-12c). Fluid movement created by the cilia and aided by muscular contractions transports eggs toward the uterus. Pathological conditions in which ciliary function is absent are associated with infertility or with pregnancies in which the embryo implants in the Fallopian tube rather than the uterus.

The dilated open end of the Fallopian tube divides into fingerlike projections called **fimbriae** that form a fringe around the open edge [*fimbriae*, fringe]. The fimbriae are held close to the adjacent ovary by connective tissue (Fig. 26-12c). The close association of Fallopian tube and ovary ensures that eggs released from the surface of the ovary will be swept into the tube rather than floating off into the abdominal cavity.

The Ovary Produces Eggs and Hormones

The ovary is an elliptical structure, about 2–4 cm long (Fig. 26-12d). It has an outer connective tissue layer and an inner connective tissue framework known as the **stroma** [*stroma*, mattress]. Most of the ovary is taken up by a thick *cortex* filled with ovarian follicles in various stages of development or decline. The small central *medulla* contains nerves and blood vessels.

The ovary, like the testis, produces both gametes and hormones. As discussed earlier, about 7 million oögonia in the embryonic ovary develop into half a million *primary oocytes*. Each oocyte is contained in a **primary follicle** with an outer layer of granulosa cells (Fig. 26-12e). As the primary follicle develops, it gains an additional outer layer of cells known as the **theca** [*theke*, case or cover]. The thecal cells are separated from granulosa cells by a basal lamina.

A Menstrual Cycle Lasts About One Month

Female humans produce gametes in monthly cycles (average 28 days; normal range 24–35 days). These cycles are commonly called **menstrual cycles** because they are marked by a 3–7 day period of bloody uterine discharge known as the **menses** [*menses*, months], or menstruation. The menstrual cycle is described according to changes that occur in follicles of the ovary (**ovarian cycle**) and in the endometrial lining of the uterus (**uterine cycle**). Figure 26-13 ● is a summary figure showing hormonal and morphological changes that take place during a typical menstrual cycle.

The ovarian cycle is divided into three phases:

1. **Follicular phase.** The first part of the ovarian cycle, known as the **follicular phase**, is a period of follicular growth in the ovary. This phase is the most variable in length and lasts from 10 days to 3 weeks.
2. **Ovulation.** Once one or more follicles has ripened, the ovary releases the egg(s) during **ovulation**.
3. **Luteal phase.** The phase of the ovarian cycle following ovulation is known as the postovulatory or **luteal phase**. The second name comes from the transformation of the follicular remnants into a **corpus luteum** [*corpus*, body + *luteus*, yellow], named from its yellow pigment and lipid deposits. The corpus luteum secretes hormones to continue the preparations for pregnancy. If a pregnancy does not occur, the corpus luteum ceases to function after about two weeks, and the ovarian cycle begins again.

The endometrial lining of the uterus goes through its own cycle regulated by ovarian hormones.

1. **Menses.** The beginning of the follicular phase in the ovary corresponds to menstrual bleeding from the uterus.
2. **Proliferative phase.** The later part of the ovary's follicular phase corresponds to the **proliferative phase** in the uterus, during which the endometrium adds a new layer of cells in anticipation of pregnancy.
3. **Secretory phase.** After ovulation, hormones from the corpus luteum convert the thickened endometrium into a secretory structure. Thus, the luteal phase of the ovary corresponds to the **secretory phase** of the uterus. If no pregnancy occurs, the superficial layers of the endometrium are lost during menstruation as the uterine cycle begins again.

STRUCTURE OF THE FEMALE REPRODUCTIVE SYSTEM

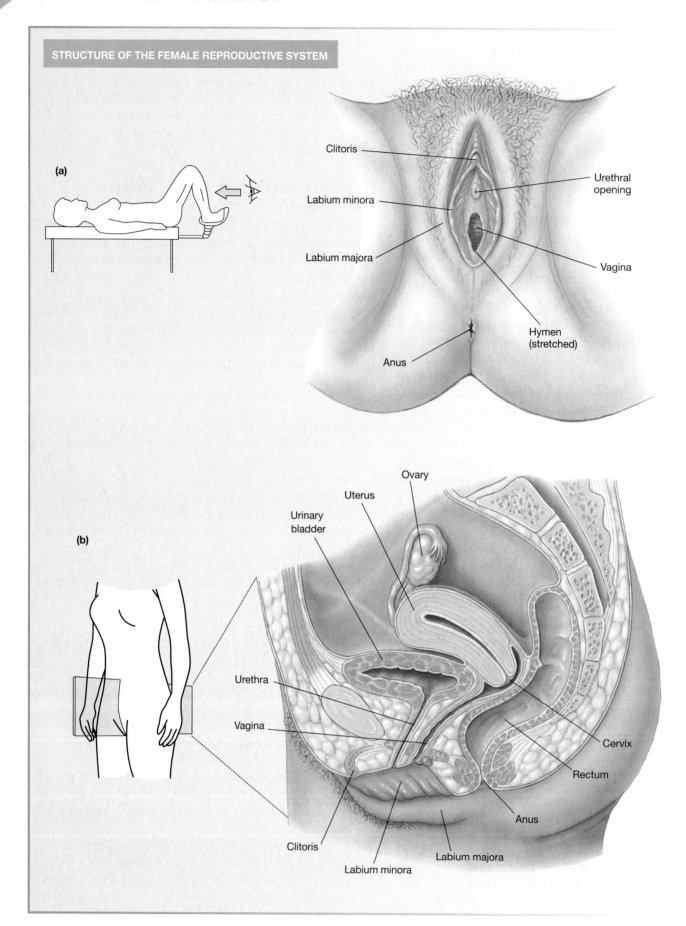

● **Figure 26-12** *ANATOMY SUMMARY* *Female Reproduction*

STRUCTURE OF THE UTERUS AND OVARY

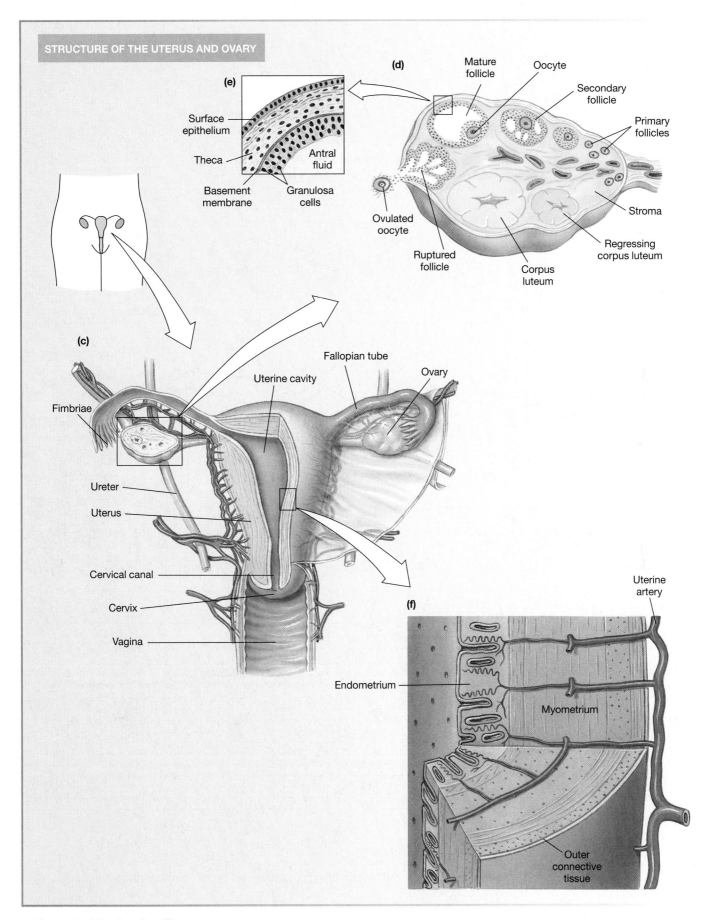

(e) Surface epithelium, Theca, Basement membrane, Granulosa cells, Antral fluid

(d) Mature follicle, Oocyte, Secondary follicle, Primary follicles, Stroma, Regressing corpus luteum, Corpus luteum, Ruptured follicle, Ovulated oocyte

(c) Fimbriae, Uterine cavity, Fallopian tube, Ovary, Ureter, Uterus, Cervical canal, Cervix, Vagina

(f) Endometrium, Myometrium, Uterine artery, Outer connective tissue

● **Figure 26-12** *(continued)*

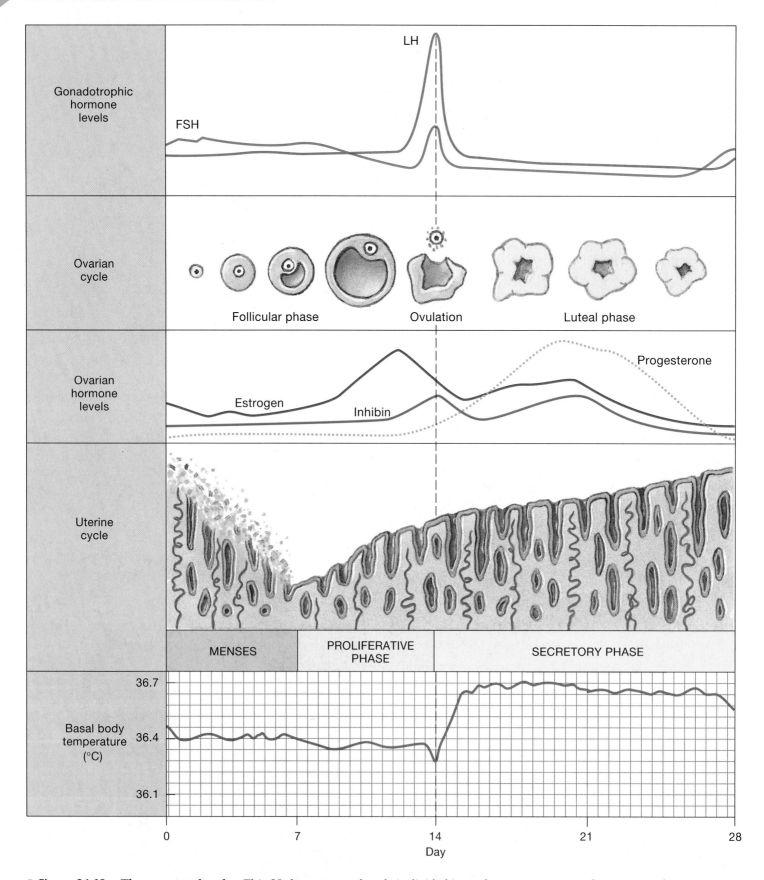

● **Figure 26-13**   *The menstrual cycle*   This 28-day menstrual cycle is divided into phases on events in the uterus and ovary.

Hormonal Control of the Menstrual Cycle Is Complex

The ovarian and uterine cycles are under the primary control of:

- Gonadotropin releasing hormone (GnRH) from the hypothalamus
- FSH and LH from the anterior pituitary
- Estrogen, progesterone, and inhibin from the ovary

During the follicular phase, the dominant steroid hormone is estrogen. In the luteal phase, progesterone is dominant, although estrogen is still present. Now let's go through an ovarian cycle in detail.

Early Follicular Phase   The first day of menstruation is day 1 of a cycle. This point was chosen to start the cycle because the bleeding of menstruation is an easily monitored physical sign. Just before the beginning of each cycle, gonadotropin secretion from the anterior pituitary increases. Under the influence of FSH, several follicles in the ovaries begin to mature (Table 26-4).

As the follicles grow, their **granulosa cells** under the influence of FSH and their thecal cells under the influence of LH start

RUNNING PROBLEM

The results of Larry's sperm analysis are normal. Dr. Coddington is therefore able to rule out sperm abnormalities as a cause of Peggy's and Larry's infertility. Peggy is instructed to track her body temperature daily and record the results on a chart. This temperature tracking is designed to determine if she is ovulating. Following ovulation, body temperature rises slightly and remains elevated through the remainder of the menstrual cycle.

Question 3: For what causes of female infertility is temperature tracking useful? For what causes is it not useful?

796 806 818 821 826 829

to produce steroid hormones. Thecal cells synthesize androgens that diffuse into the neighboring granulosa cells, where aromatase converts them to estrogens (Fig. 26-14a ●).

Gradually increasing estrogen levels in the circulation have several effects. Estrogen exerts a negative feedback effect on pituitary FSH and LH secretion, thus preventing the development of additional follicles in the same cycle. At the same time, estrogen stimulates estrogen production by the granulosa cells. This positive

Table 26-4: Development of Ovarian Follicles

The contents of the follicle are given starting at the center with the ovum and moving outward. The zona pellucida (see Fig. 26-16) is a glycoprotein coat that protects the ovum.

STAGE	OVUM	ZONA PELLUCIDA	GRANULOSA CELLS	ANTRUM	BASAL LAMINA	THECA
Primary follicle (before FSH stimulation)	Primary oocyte	Minimal	Single layer	None	Separates granulosa and theca	Single cell layer plus blood vessels
Secondary follicle (early follicular phase)	Primary oocyte	Increased in width	2–6 cell layer	None	Present	Single cell layer
Tertiary follicle (late follicular phase)	Primary oocyte, then secondary oocyte with division arrested	Present	3–4 cell layer	Develops within granulosa layer and fills with fluid; swells to 15–20 mm diameter	Present	*Inner layer:* secretory and small blood vessels *Outer layer:* connective tissue, smooth muscle cells, large blood vessels
Corpus luteum (luteal phase)	None	None	Converted to luteal cells	Fills with migrating cells	Disappears	Converted to luteal cells
Corpus albicans (post-luteal phase)	None	None	Cells degenerate	None	None	Cells degenerate

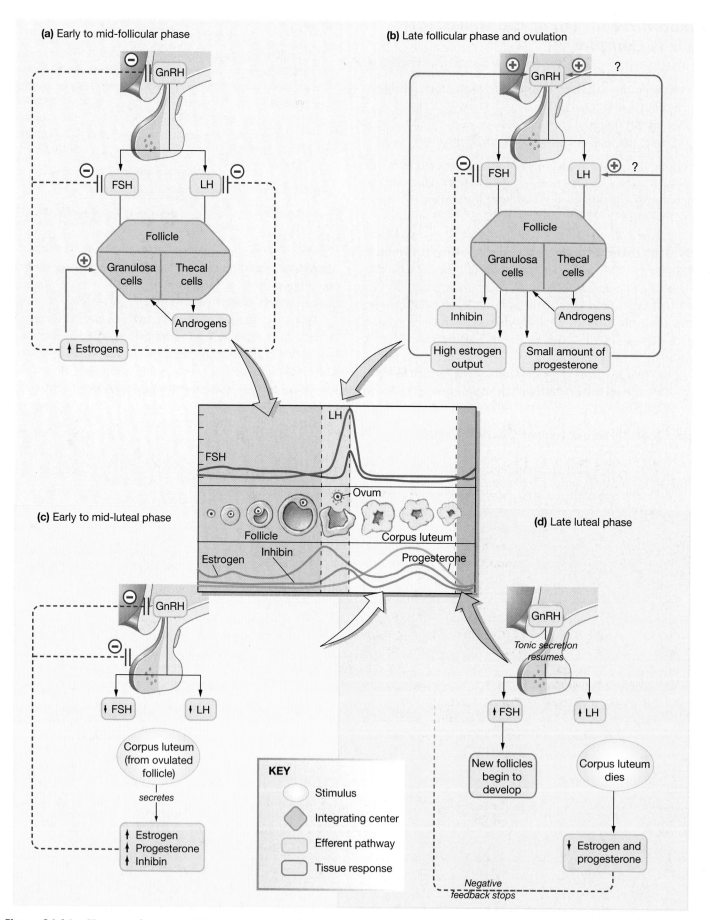

(a) Early to mid-follicular phase

(b) Late follicular phase and ovulation

(c) Early to mid-luteal phase

(d) Late luteal phase

LH

FSH

Ovum

Follicle

Corpus luteum

Estrogen

Inhibin

Progesterone

GnRH

FSH

LH

Follicle

Granulosa cells

Thecal cells

Androgens

Estrogens

GnRH

FSH

LH

Follicle

Granulosa cells

Thecal cells

Inhibin

Androgens

High estrogen output

Small amount of progesterone

?

?

GnRH

FSH

LH

Corpus luteum (from ovulated follicle)

secretes

Estrogen
Progesterone
Inhibin

GnRH

Tonic secretion resumes

FSH

LH

New follicles begin to develop

Corpus luteum dies

Estrogen and progesterone

Negative feedback stops

KEY

Stimulus

Integrating center

Efferent pathway

Tissue response

● **Figure 26-14** *Hormonal control of the menstrual cycle* feedback loop allows the follicles to continue estrogen production even though FSH and LH decrease.

As the follicles enlarge, granulosa cells secrete fluid that collects in a cavity known as the **antrum**. The antral fluid contains hormones and enzymes needed for ovulation. At each stage of development, some follicles undergo *atresia*, hormonally regulated cell death. Only a few follicles reach the final stage, and usually only one dominant follicle continues to ovulation.

In the uterus, menstruation ends during the early follicular phase. Under the influence of estrogens from the ovary, the endometrium begins to grow, or proliferate. This period is characterized by an increase in cell number and by enhanced blood supply to bring nutrients and oxygen to the thickening endometrium. Estrogen causes the mucous glands of the cervix to produce clear, watery mucus.

Late Follicular Phase   As the follicular phase nears its end, ovarian estrogen secretion peaks. By this point of the cycle, only one follicle is still developing. As the follicular phase ends, the granulosa cells of the dominant follicle begin to secrete inhibin and progesterone in addition to estrogen (Fig. 26-14b). Estrogen, which had exerted a negative feedback effect on GnRH and gonadotropins earlier in the follicular phase, changes to positive feedback.

Immediately before ovulation, the persistently high levels of estrogen, aided by rising levels of progesterone, enhance pituitary responsiveness to GnRH. As a result, LH secretion increases dramatically, a phenomenon known as the *LH surge*. FSH surges also, but to a lesser degree, presumably because it is being suppressed by inhibin and estrogen.

The LH surge is an essential part of ovulation. Without it, the final steps of oocyte maturation cannot take place. Meiosis, which had paused in the embryo, resumes in the developing follicle with the first meiotic division (see Fig. 26-5). The division converts a primary oocyte into a polar body, which is extruded, and an egg, or secondary oocyte. While this is taking place, antral fluid collects and the follicle grows to its greatest size, preparing to release the egg.

High levels of estrogen in the late follicular phase prepare the uterus for a possible pregnancy. The endometrium grows to a final thickness of 3–4 mm. Just before ovulation the cervical glands produce copious amounts of thin, stringy mucus to facilitate sperm entry. The stage is set for ovulation.

Ovulation   About 16–24 hours after LH peaks, ovulation occurs. The mature follicle secretes *collagenase*, dissolving collagen in the connective tissue that holds the follicular cells together. The breakdown products of collagen create an inflammatory reaction, attracting leukocytes that secrete prostaglandins into the follicle.

Our understanding of events from this point is unclear, but it is possible that the prostaglandins cause smooth muscle cells in the outer theca to contract, rupturing the follicle wall at its weakest point. The antral fluid spurts out along with the egg, which is sur-

rounded by two to three layers of granulosa cells. The egg is swept into the Fallopian tube and carried away to be fertilized or die.

In addition to promoting rupture of the follicle, the LH surge causes follicular thecal cells to migrate into the antral space, mingling with the former granulosa cells and filling the cavity. Both cell types then transform into *luteal cells* of the corpus luteum. This process, known as *luteinization*, involves biochemical and morphological changes. Newly formed luteal cells accumulate lipid droplets and glycogen granules in their cytoplasm and begin to secrete progesterone. Estrogen synthesis diminishes.

Early to Mid-Luteal Phase   After ovulation, the corpus luteum produces steadily increasing amounts of progesterone and estrogen. Progesterone is the dominant hormone of the luteal phase. Estrogen levels increase but never reach the peak seen before ovulation.

The combination of estrogen and progesterone exerts negative feedback on the hypothalamus and anterior pituitary. Gonadotropin secretion, further suppressed by luteal inhibin production, remains shut down throughout most of the luteal phase (Fig. 26-14c).

Under the influence of progesterone, the endometrium of the uterus continues its preparation for pregnancy. The endometrial glands coil, and additional blood vessels grow into the connective tissue layer. Endometrial cells deposit lipids and glycogen in their cytoplasm. These deposits will provide nourishment for a developing embryo while the placenta, the fetal-maternal connection, is developing.

Progesterone also causes cervical mucus to thicken. Thicker mucus creates a plug that blocks the cervical opening, preventing bacteria as well as sperm from entering the uterus.

One interesting effect of progesterone is its thermogenic ability. During the luteal phase of an ovulatory cycle, a woman's basal body temperature, taken immediately upon awakening and before getting out of bed, jumps 0.3°–0.5° F and remains elevated until menstruation. Because this change in the setpoint occurs after ovulation, it cannot be used effectively to predict ovulation. However, it is a simple way to assess if a woman is having ovulatory or *anovulatory* (not ovulating) cycles.

Late Luteal Phase and Menstruation   The corpus luteum has an intrinsic life span of approximately 12 days. If pregnancy does not occur, the corpus luteum spontaneously undergoes *apoptosis* to become an inactive structure called a **corpus albicans** [*albus*, white]. As the luteal cells degenerate, production of progesterone and estrogen decreases (Fig. 26-14d). This removes the negative feedback signal to the pituitary and hypothalamus, and secretion of FSH and LH increases.

Maintenance of the secretory endometrium depends on the presence of progesterone. When the corpus luteum degenerates and hormone production falls, blood vessels in the surface layer of the endometrium contract. Without oxygen and nutrients, the surface cells die. About two days after the corpus luteum ceases to function, or 14 days after ovulation, the endometrium begins to slough its surface layer, and menstruation begins.

The menstrual discharge from the uterus totals about 40 mL of blood and 35 mL of serous fluid and cellular debris. There are few clots of blood in the menstrual flow because of the presence of *plasmin*, a substance that breaks up clots. Menstruation continues for 3–7 days, well into the follicular phase of the next ovulatory cycle.

Estrogens and Androgens Influence Female Secondary Sex Characteristics

Estrogens control the development of primary sex characteristics in females, just as androgens control them in males. They also control the most prominent secondary sex traits: the female pattern of fat distribution (hips and upper thighs) and breast development. Other female secondary sex characteristics are actually governed by androgens produced in the adrenal cortex. Pubic and *axillary* (armpit) hair growth and libido (sex drive) are under the control of adrenal androgens.

Answers: p. 834

14. Name the phases of the ovarian cycle and the corresponding phases of the uterine cycle.

15. What side effects would you predict in women athletes who take anabolic steroids to build muscles?

16. Aromatase converts androgens to estrogen. What would happen to the ovarian cycle of a woman who is given an aromatase inhibitor?

▶ PROCREATION

Reproduction throughout the animal kingdom is marked by species-specific behaviors designed to ensure that egg and sperm meet. For aquatic animals that release gametes into the water, coordinated timing is everything. Interaction between males and females of these species may be limited to chemical communication by pheromones.

Results of the temperature tracking reveal after several months that Peggy is ovulating regularly. Dr. Coddington therefore believes that her ovaries are functioning normally. Other possible causes for this couple's infertility include abnormalities in Peggy's cervix, Fallopian tubes, or uterus. Dr. Coddington next decides to order a postcoital test. In this test, the couple is instructed to have intercourse 12 hours before the physician visit. The cervical mucus is then analyzed. This test will also analyze the interaction between sperm and mucus.

Question 4: What abnormalities in the cervix, Fallopian tubes, and uterus could cause infertility?

On the other hand, internal fertilization in terrestrial vertebrates requires interactive behaviors and special adaptations of the genitalia. The female must have an internal receptacle for sperm (the vagina) and the male must possess an organ (the penis) capable of placing sperm into the receptacle. The penis of human males is *flaccid* (soft and limp) in its resting state, not capable of penetrating the narrow opening of the female's vagina. The processes through which the penis stiffens and enlarges (**erection**), then releases sperm from the ducts of the reproductive tract (**ejaculation**), make up the main components of the male sex act. Without these events, fertilization cannot take place.

The Human Sexual Response Has Four Phases

The human sex act, also known as sexual intercourse, copulation, or **coitus** [*coitio*, a coming together] is highly variable in some ways and highly stereotyped in others.

Human sexual response in both sexes is divided into four phases: (1) excitement, (2) plateau, (3) orgasm, and (4) resolution. In the excitement phase, various erotic stimuli prepare the genitalia for the act of copulation. For the male, excitement involves erection of the penis. For the female, it includes erection of the clitoris and lubrication of the vagina by secretions from the vaginal walls. In both sexes, erection is a state of vasocongestion in which arterial blood flow into spongy erectile tissue exceeds venous outflow.

Erotic stimuli include sexually arousing tactile stimuli as well as psychological stimuli. The latter vary widely among individuals and among cultures, so what is erotic in one culture may be considered disgusting in another. Regions of the body that possess receptors for sexually arousing tactile stimuli are called **erogenous zones**. They include the genitalia as well as the lips, tongue, nipples, ear lobes, and other regions.

In the plateau phase, the changes that started during excitement intensify. They then peak in an **orgasm** (climax). In both sexes, orgasm is a series of muscular contractions accompanied by intense pleasurable sensations, increased blood pressure, heart rate, and respiration rate. In the female, the walls of the vagina and the uterus contract. In the male, the contractions usually result in the ejaculation of semen from the penis. Pregnancy cannot occur without ejaculation to place sperm in the female reproductive tract. However, female orgasm is not required for pregnancy.

The final phase of the sexual response is resolution, a period during which the physiological parameters that changed in the first three phases slowly return to normal.

The Male Sex Act Is Composed of Erection and Ejaculation

A key element to successful copulation is the ability of the male to achieve and sustain an erection. Sexual excitement from either tactile or psychological stimuli triggers the **erection reflex**, a spinal reflex that is subject to control from higher centers in the brain. The urination and defecation reflexes are similar types of reflexes [pp. 617, 685].

In its simplest form, the erection reflex begins with tactile stimuli sensed by mechanoreceptors in the glans penis or other erogenous zones (Fig. 26-15 ●). Sensory neurons signal the spinal integration center, which inhibits vasoconstrictive sympathetic input on penile arterioles. Simultaneously, increased parasympathetic input mediated primarily by nitric oxide actively vasodilates the penile arterioles.

As a result of vasodilation, arterial blood flows into the open spaces of the erectile tissue, passively compressing the veins and trapping blood. The erectile tissue becomes engorged, stiffening and lengthening the penis within 5–10 seconds.

The climax of the male sexual act coincides with emission and ejaculation. **Emission** is the movement of sperm out of the vas deferens and into the urethra, where they are joined by secretions from the accessory glands to make semen. The average semen volume is 3 mL (range 2–6 mL), of which less than 10% is sperm.

During **ejaculation**, semen in the urethra is expelled to the exterior by a series of rapid muscular contractions accompanied by sensations of intense pleasure, the orgasm. A sphincter at the base of the bladder contracts to prevent sperm from entering the bladder and urine from joining the semen.

Both erection and ejaculation can occur in the absence of mechanical stimulation. Sexually arousing thoughts, sights, sounds, emotions, and dreams can all initiate sexual arousal and even lead to orgasm in both men and women. In addition, nonsexual penile erection accompanies rapid eye movement (REM) sleep.

The inability to obtain or sustain an erection is known as *erectile dysfunction* or **impotence**. Impotence may result from a variety of physiological and psychological causes. For example, fear of interruption may result in loss of arousal. Alcohol also inhibits sexual performance, as noted by Shakespeare in *Macbeth* (III, iii, 26–30). When MacDuff asks, "What three things does

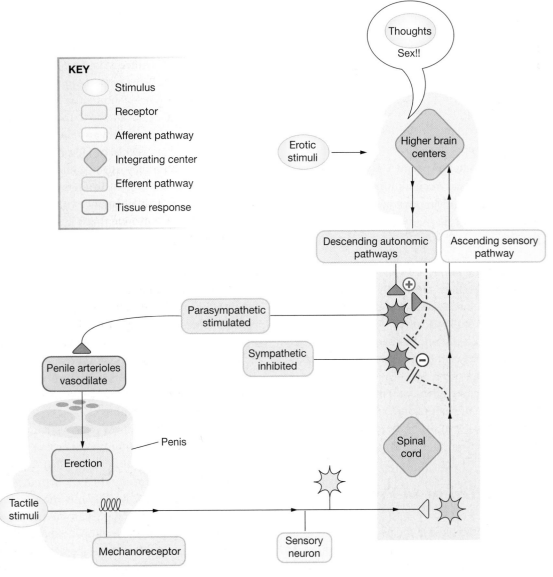

● **Figure 26-15** *The erection reflex* Erection can take place without input from higher brain centers. It can also be stimulated (and inhibited) by descending pathways from the cerebral cortex.

Viagra®

Erectile dysfunction is a matter of global concern, because inability to achieve and sustain an erection disrupts the sex act for both men and women. In 1998 the U.S. Food and Drug Administration approved sildenafil (Viagra®) for the treatment of erectile dysfunction. Erections occur when neurotransmitters released from pelvic nerves increase nitric oxide in vascular smooth muscle. Nitric oxide activates guanylyl cyclase, which increases cGMP and causes vasodilation. Sildenafil blocks the enzyme that degrades cGMP, thereby prolonging the effects of nitric oxide. The maximum effect occurs about one hour after taking the pill. Clinical trials showed that sildenafil corrected erectile dysfunction in 82% of the men who took it. However, it is not without side effects. Men who are taking nitrate-containing drugs should not take sildenafil because drug interactions could create very low blood pressure. The U.S. Federal Aviation Administration issued an order that pilots should not take sildenafil within six hours of flying because 3% of men report impaired color vision, a blue or greenish haze. This occurs because sildenafil also inhibits an enzyme in the retina. And finally, there is no evidence that sildenafil corrects orgasmic dysfunction in women. Although women do have clitoral erections, the female sexual response is more complicated. At least one drug company reportedly is testing a drug that affects brain receptors linked to sexual arousal.

drink especially provoke?" the porter answers, "Marry, sir, nose-painting, sleep, and urine. Lechery, sir, it provokes and unprovokes: it provokes the desire, but it takes away the performance."

Because of the influence of psychological factors on the erection reflex, clinicians often use the unconscious, nonsexual erection of REM sleep to determine whether impotence is physiological or psychological. In its simplest form, the REM erection test consists of gluing a band of postage stamps around the flaccid penis before the subject goes to sleep. If the perforations between stamps have torn by the time the subject awakens, he knows that he had normal erections during REM sleep.

Contraceptives Attempt to Prevent Pregnancy

One disadvantage of sexual intercourse for pleasure rather than reproduction is the possibility of an unplanned pregnancy. On average, 80% of young women who have intercourse without using any form of birth control will get pregnant within a year. But many women get pregnant after a single unprotected en-counter. Couples who hope to avoid unwanted pregnancies generally use some form of birth control, or **contraception**.

Contraceptive practices fall into several broad groups. **Abstinence**, the avoidance of intercourse, is the surest method to avoid pregnancy (and sexually transmitted diseases). Interventional methods include (1) barrier methods that prevent union of eggs and sperm, (2) methods that prevent implantation of the fertilized egg, and (3) hormonal treatment that decreases or stops gamete production.

Barrier methods of contraception are among the earliest recorded means of birth control. Once people made the association between pregnancy and semen, they concocted a variety of substances to block or kill the sperm. An ancient Egyptian papyrus with the earliest known references to birth control describes the use of vaginal plugs made of leaves, feathers, figs, and alum held together with crocodile and elephant dung. Sea sponges soaked in vinegar and disks of oiled silk have also been used at one time or another. Later, women used douches of garlic, turpentine, and rose petals to rinse the vagina after intercourse. As you can imagine, many of these items also caused vaginal or uterine infections.

The modern version of the female barrier is the **diaphragm**, introduced into the United States in 1916. These rubber domes and a smaller version called a *cervical cap* are usually filled with a spermicidal cream, then inserted into the top of the vagina so that they cover the cervix. One advantage to the diaphragm is that it is non-hormonal. In theory, diaphragms are highly effective (97%–99%) if used properly and regularly. However, in real life they are not always used because they must be inserted close to the time of intercourse, and about 20% of women who depend on diaphragms for contraception are pregnant within the first year. Another female barrier contraceptive that was recently re-introduced is the **contraceptive sponge**, which contains a *spermicidal chemical*.

The male barrier contraceptive is the **condom**, a closed sheath that fits closely over the penis to catch ejaculated semen. Males have used condoms made from animal bladders and intestines for centuries. Condoms lost popularity when oral contraceptives came into widespread use in the 1960s and 1970s, but in recent years they have regained favor because they combine pregnancy protection with protection from many sexually transmitted diseases. However, latex condoms may cause allergic reactions and there is evidence that the HIV virus can pass through pores in some contemporary condoms.

Sterilization is the most effective contraceptive method for sexually active people, but it is a surgical procedure and is not easily reversed. Female sterilization is called **tubal ligation** and consists of tying off and cutting the Fallopian tubes. A woman with a tubal ligation will still ovulate, but the eggs remain in the abdomen. The male form of sterilization is the **vasectomy**, in which the vas deferens is tied and clipped. Sperm are still made in the seminiferous tubules, but since they cannot leave the reproductive tract, they are reabsorbed.

Some contraceptive methods do not prevent fertilization but do keep a fertilized egg from establishing itself in the endometrium.

They include **intrauterine devices (IUDs)** as well as chemicals that change the properties of the endometrium. IUDs are plastic devices that are inserted into the uterine cavity, where they create an inflammatory reaction that prevents implantation. They have low failure rates (0.5% per 100 women per year) but a high incidence of side effects ranging from pain and bleeding to infertility from pelvic inflammatory disease and blockage of the Fallopian tubes.

Techniques for decreasing gamete production depend on altering the hormonal milieu of the body. These methods were tried historically when women would eat or drink various plant concoctions for contraception. Some actually worked because of estrogen-like compounds contained in the plants. Modern pharmacology has improved on this method, and now women can choose between oral contraceptive pills, injections lasting three months, or capsules that are inserted under the skin and release hormones.

The **oral contraceptives**, also known as birth control pills, were first made available in 1960. They rely on various combinations of estrogen and progesterone that inhibit gonadotropin secretion from the pituitary. Without adequate FSH and LH, ovulation is suppressed. In addition, progesterones in the contraceptive pills thicken the cervical mucus and help prevent sperm penetration. These hormonal methods of contraception are highly effective when taken correctly but also carry some risks, including an increased incidence of blood clots and strokes (especially in women who smoke).

Two new forms of hormonal contraception for women may become available soon: the **GnRH agonists** discussed earlier and *mifepristone*, also known as RU 486. The GnRH agonists appear to effectively and reversibly block ovulation, as do the estrogen-progesterone combination pills. Mifepristone, a controversial drug because of its use in inducing abortions in Europe, is an antiprogesterone drug that blocks progesterone receptors. In low doses, it blocks ovulation or alters the endometrium so that fertilized eggs do not implant.

Most attempts to devise a male hormonal contraceptive have failed so far because of undesirable side effects. To inhibit sperm production, the contraceptive must block testosterone secretion or action. In doing so, the contraceptive is also likely to decrease the male libido or even cause impotence. Both side effects are unacceptable to the men who would be most interested in using the contraceptive. The search for a male contraceptive pill is concentrating on something that will affect only sperm production in the seminiferous tubules and not testosterone production by the Leydig cells.

The newest methods of contraception being tested are based on development of antibodies to various components of the male and female reproductive systems. These contraceptives would be long-lasting and reversible, and they would be administered as vaccines. One contraceptive vaccine is currently in clinical trials in Sweden.

Infertility Is the Inability to Conceive

While some couples are trying to prevent pregnancy, other couples are spending thousands of dollars trying to get pregnant. Infertility is the inability of a couple to conceive a child after a year of unprotected intercourse. For years, infertile couples had no choice but adoption if they wanted to have a child, but incredible strides have been made in this field since the 1970s. As a result, many infertile couples today are able to have children.

Infertility can arise from a problem in the male, the female, or both. Male infertility usually results from a low sperm count or an abnormally high number of defective sperm. Female infertility can be mechanical (blocked Fallopian tubes or other structural problems), or hormonal, leading to decreased or absent ovulation. One problem involving both partners is production of antibodies to sperm by the female. In addition, not all pregnancies go to a successful conclusion. By some estimates, as many as a third of all pregnancies spontaneously terminate, many within the first weeks, before the woman is even aware that she is pregnant.

Some of the most dramatic advances have been made in the field of *in vitro* fertilization, in which a woman's ovaries are hormonally manipulated to ovulate multiple eggs at one time. The eggs are collected surgically and fertilized outside the body. The developing embryos are then placed in the woman's uterus, which has been primed for pregnancy by hormonal therapy. Because of the expense and complicated nature of the procedure, multiple embryos are usually placed in the uterus at one time, which may result in multiple births. *In vitro* fertilization has allowed some infertile couples to have children, although the success rate is less than 25%.

▶ PREGNANCY AND PARTURITION

Now let's return to a recently ovulated egg and some sperm deposited in the vagina and follow them through fertilization, pregnancy, and **parturition**, the birth process.

Fertilization Requires Capacitation

Once an egg is released from the ruptured follicle, it is swept into the Fallopian tube by the beating cilia. Meanwhile, sperm deposited in the vagina must go through their final maturation step, capacitation. **Capacitation** of sperm confers the ability to swim rapidly and fertilize an egg. The process is not well understood, but apparently it involves the reorganization of molecules in the outer membrane of the sperm head.

Normally, capacitation takes place in the female reproductive tract. This location presents a problem for sperm being used for *in vitro* fertilization. These sperm must be artificially capacitated by physiological salines supplemented with human serum before being exposed to the harvested eggs. Much of what we know

---

RUNNING PROBLEM

Analysis of the cervical mucus in Peggy's and Larry's postcoital test shows that sperm are present but not moving. Dr. Coddington explains that it is likely that Peggy's cervical mucus contains antibodies that destroy Larry's sperm.

Question 5: Speculate on how this kind of infertility problem might be treated.

(a)

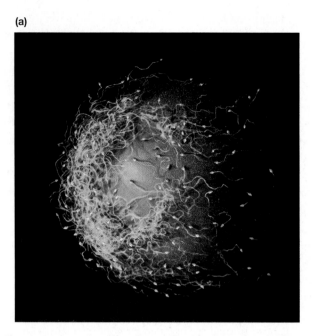

(b) Capacitated sperm release enzymes from their acrosomes in order to penetrate the cells and zona pellucida surrounding the egg.

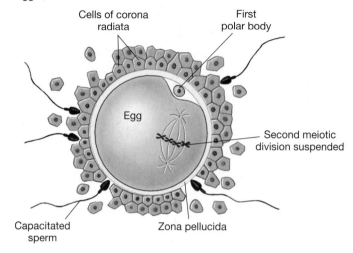

● **Figure 26-16** *Fertilization* The photograph (a) shows the tremendous difference in the size of human sperm and egg.

about human fertilization has come from infertility research aimed at improving the success rate of *in vitro* fertilization.

Fertilization of an egg by a sperm is the result of a chance encounter, possibly aided by chemical attractants produced by the egg. An egg can be fertilized for about 12–24 hours after ovulation. Sperm remain viable for four to six days.

Fertilization normally takes place in the distal part of the Fallopian tube. Of the millions of sperm in a single ejaculation, only about 100 or so reach this point (Fig. 26-16a ●). To fertilize the egg, a sperm must first penetrate an outer layer of loosely connected granulosa cells (the *corona radiata*), then the protective glycoprotein coat of the **zona pellucida**. To get past these barriers, capacitated sperm release powerful enzymes from the acrosome in the sperm head, a process known as the **acrosomal reaction**.

The enzymes dissolve cell junctions and the glycoprotein coat, allowing the sperm to wiggle their way toward the egg. The first sperm to reach the egg quickly finds sperm-binding receptors on the membrane and fuses its membrane to the oocyte membrane (Fig. 26-17 ●). The fused section of membrane opens, and the sperm nucleus sinks into the cytoplasm of the egg.

Fusion of the egg and sperm membranes signals the egg to resume meiosis and complete its second division. The final meiotic division creates a second polar body that is discarded. At this point, the 23 chromosomes of the sperm join the 23 chromosomes of the egg, creating a zygote nucleus with a full set of genetic material.

The fusion of sperm and oocyte membrane also prevents **polyspermy**, more than one sperm fertilizing an egg. A major part of this process is a chemical reaction called the **cortical reaction**. Membrane-bound **cortical granules** in the peripheral cytoplasm of the egg release their contents into the space just outside the oocyte membrane. These chemicals rapidly alter the membrane and surrounding zona pellucida so that additional sperm cannot penetrate or bind.

Once the egg is fertilized and becomes a zygote, it begins mitotic division as it slowly makes its way to the uterus, where it will settle for the remainder of the **gestation** period [*gestare*, to carry in the womb].

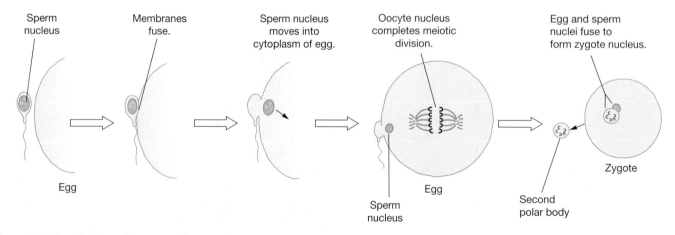

● **Figure 26-17** *Fusion of sperm and egg to form a zygote*

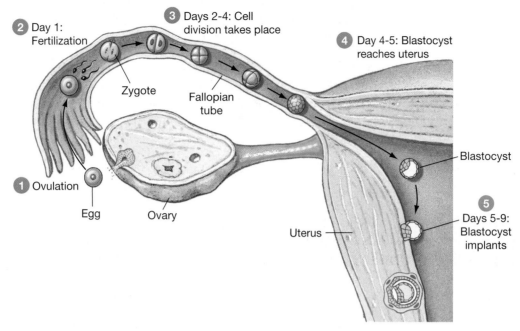

● **Figure 26-18**   *Ovulation, fertilization, and implantation of an ovum*

The Developing Zygote Implants in the Endometrium

The dividing zygote takes about four days to move from the distal end of the Fallopian tube into the uterine cavity (Fig. 26-18 ●). Under the influence of progesterone, the smooth muscle of the tube relaxes, and transport proceeds slowly. By the time the de-

veloping embryo reaches the uterus, it consists of a hollow ball of about 100 cells called a **blastocyst**.

Some of the outer layer of blastocyst cells will become the **chorion**, an *extraembryonic membrane* that encloses the entire embryo and forms the placenta (Fig. 26-19a ●). The inner cell mass of the blastocyst will develop into the embryo itself and other extraembryonic membranes. These membranes include

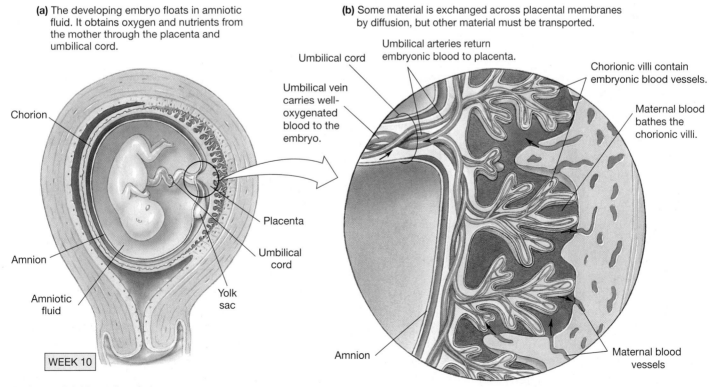

(a) The developing embryo floats in amniotic fluid. It obtains oxygen and nutrients from the mother through the placenta and umbilical cord.

(b) Some material is exchanged across placental membranes by diffusion, but other material must be transported.

● **Figure 26-19**   *The placenta*

the **amnion**, which secretes *amniotic fluid* in which the developing embryo floats; the **allantois**, which becomes part of the umbilical cord that links the embryo to the mother; and the **yolk sac**, which degenerates early in human development.

Implantation of the blastocyst into the uterine wall normally takes place about 7 days after fertilization. The outer cells of the blastocyst secrete enzymes that allow them to invade the endometrium, like a parasite burrowing into its host. As they do so, the cells of the endometrium grow out around the blastocyst until it is completely engulfed in endometrial cells.

As the embryo grows, cells that will become the placenta form fingerlike **chorionic villi** that penetrate into the vascularized endometrium. Enzymes from the villi break down the walls of maternal blood vessels until the villi are surrounded by pools of maternal blood (Fig. 26-19b). The blood of the embryo and that of the mother do not mix, but nutrients, gases, and wastes are exchanged across the membranes of the villi. Many of these substances move by simple diffusion, but some, such as maternal antibodies, are transported across the membrane.

The placenta continues to grow during pregnancy until, by delivery, it is about 20 cm in diameter (the size of a small dinner plate). The placenta receives as much as 10% of the total maternal cardiac output. The tremendous blood flow to the placenta is one reason that the sudden abnormal separation of the placenta from the uterine wall is a medical emergency.

The Placenta Secretes Hormones During Pregnancy

As the embryo implants in the uterine wall and the placenta begins to form, the corpus luteum is coming to the end of its preprogrammed 12-day life span. Unless the embryo sends a hormonal signal, the corpus luteum will disintegrate, progesterone and estrogen levels will drop, and the embryo will be washed from the body along with the surface layers of endometrium during menstruation. The hormones that prevent this from happening are secreted by the placenta. They include human chorionic gonadotropin, chorionic somatomammotropin, estrogen, and progesterone.

Human Chorionic Gonadotropin The corpus luteum remains active during early pregnancy because of the peptide hormone **human chorionic gonadotropin** (hCG), secreted by the chorionic villi and developing placenta. hCG is structurally related to LH, and it binds to LH receptors. Under the influence of hCG, the corpus luteum keeps producing progesterone to keep the endometrium intact.

By the seventh week of development, however, the placenta has taken over progesterone production and the corpus luteum is no longer needed. At that point, it finally degenerates. Human chorionic gonadotropin production by the placenta peaks at three months of development, then diminishes.

A second function of hCG is stimulation of testosterone production by the developing testes in male fetuses. As you learned in the opening sections of this chapter, fetal testosterone and its metabolite DHT are essential for expression of male characteristics and for descent of the testes into the scrotum before birth.

Human chorionic gonadotropin is the chemical detected by pregnancy tests. Because hCG can induce ovulation in rabbits, years ago the urine from a woman who suspected she was pregnant was injected into a rabbit. The rabbit's ovaries were then inspected for signs of ovulation. It took several days for the results of this test to get back to the patient. Today, with modern biochemical techniques, women can perform their own pregnancy tests in a few minutes in the privacy of their home.

Human Placental Lactogen (hPL) Another peptide hormone produced by the placenta is **human placental lactogen** (hPL), also known as *human chorionic somatomammotropin* (hCS). This hormone, structurally related to growth hormone and prolactin, was initially believed to be necessary for breast development during pregnancy and for milk production (**lactation**). Although hPL probably does contribute to lactation, women who do not make hPL during pregnancy because of a genetic defect still have adequate breast development and milk production.

The other postulated role for hPL is alteration of the mother's glucose and fatty acid metabolism to support fetal growth. Maternal glucose moves across the membranes of the placena by facilitated diffusion and enters the fetal circulation. During pregnancy, about 4% of women develop *gestational diabetes mellitus*, with elevated blood glucose levels due to insulin resistance, similar to type 2 diabetes. After delivery, glucose metabolism in most of these women returns to normal.

Estrogen and Progesterone Estrogen and progesterone are produced continuously during pregnancy, first by the corpus luteum under the influence of hCG and then by the placenta. With high circulating levels of the steroid hormones, feedback suppression of the pituitary continues throughout pregnancy, preventing another set of follicles from beginning development.

Estrogen during pregnancy contributes to the development of the milk-secreting ducts of the breasts. Progesterone is essential for maintaining the endometrium and in addition helps suppress uterine contractions. The placenta makes a variety of other hormones including inhibin, prolactin, and prorenin, but at present, the function of most of them is unclear.

Pregnancy Ends with Labor and Delivery

Parturition, the birth process, normally occurs in the 38th to 40th week of gestation. What triggers this process? For many years, researchers developed animal models of the signals that initiate parturition, only to discover in recent years that many of those theories do not apply to humans. Parturition begins with **labor**, the rhythmic contractions of the uterus designed to push the fetus out into the world. Signals that initiate these contractions could begin with either the mother or the fetus, or they could be a combination of signals from both.

In many nonhuman mammals, a decrease in estrogen and progesterone levels marks the beginning of parturition. A decrease in progesterone levels is logical, as progesterone inhibits

EMERGING CONCEPTS

Gestational Diabetes

Carbohydrate and fat metabolism change during pregnancy under the influence of maternal and placental hormones. In about 3% of all pregnancies, the mother develops elevated fasting blood glucose levels, similar to those observed in diabetes mellitus. This condition is called *gestational diabetes*. We still do not know what causes it, but we do know that mothers with gestational diabetes and their babies are at higher risk of developing type 2 diabetes later in life. The babies also have a higher rate of complications around birth. The elevated glucose in the mother's blood crosses the placenta and the babies secrete more insulin in response. As a result, they tend to weigh more at birth. They are also at risk of developing hypoglycemia around birth due to their elevated insulin levels.

uterine contractions. However, in humans, these hormones do not decrease until labor is well under way.

Another possible trigger that has been investigated is oxytocin, the peptide hormone that causes uterine muscle contraction. Could oxytocin be the signal that starts parturition? As a pregnancy nears full term, the number of uterine oxytocin receptors increases. However, studies have shown that secretion of the hormone does not increase until labor begins. Synthetic oxytocin is often used to induce labor in pregnant women, but it is not always effective. Apparently, the start of labor requires something more than adequate amounts of oxytocin.

Another possibility for the induction of labor is that the fetus somehow signals that it has completed development. One theory supported by clinical evidence is that corticotropin-releasing hormone (CRH) secreted by the placenta is the signal to begin labor. (CRH is also a hypothalamic releasing factor that controls release of ACTH from the anterior pituitary.) In the weeks prior to delivery, maternal blood CRH levels increase rapidly. In addition, women with elevated CRH levels as early as 15 weeks of gestation are more likely to go into premature labor.

Although we do not know for certain what initiates parturition, we do understand the sequence of events. Once the contractions of labor begin, a positive feedback loop consisting of mechanical and hormonal factors is set into motion. The fetus is normally oriented head down (Fig. 26-20a ●). At the beginning of labor, in late pregnancy, it repositions itself lower in the abdomen

● **Figure 26-20 *Parturition: the birth process***   (a) As labor begins, the fetus is normally head down in the uterus. (b) Rhythmic uterine contractions push the head against the softened cervix, stretching and dilating it. (c) Once the cervix is fully dilated and stretched, the uterine contractions push the fetus out through the vagina. (d) Shortly after the fetus is delivered, the placenta detaches from the uterine wall and is expelled.

(a) Fully developed fetus

Umbilical cord

Cervix

Vagina

Placenta

Cervical canal

(b) Cervical dilation

(c) Delivery

Uterus

(d) Expulsion of the placenta

Ejection of the placenta

("the baby has dropped") and begins to push on the softened cervix (Fig. 26-20b). The peptide hormone **relaxin**, secreted by the ovaries and placenta, apparently helps soften the cervix and also loosens ligaments holding the pelvic bones together.

Cervical stretch triggers uterine contractions that move in a wave from the top of the uterus down, pushing the fetus farther into the pelvis. The lower portion of the uterus stays relaxed and the cervix stretches and opens (dilates). Cervical stretch starts a positive feedback cycle of escalating contractions (Fig. 26-21 ●). The contractions are reinforced by secretion of oxytocin from the posterior pituitary [※ p. 222], with continued cervical stretch reinforcing oxytocin secretion.

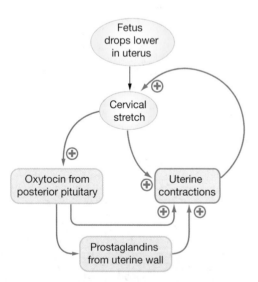

● **Figure 26-21** *The positive feedback loop of parturition*
The initiation of this loop depends on a uterus that is ready to respond to these stimuli. The positive feedback loop stops when the fetus is delivered and the cervix is no longer being stretched.

Prostaglandins, another chemical factor involved in labor and delivery, are produced in the uterus in response to CRH and oxytocin secretion. Prostaglandins are very effective at causing uterine muscle contractions at any time. They are the primary cause of menstrual cramps and have been used to induce abortion in early pregnancy. During labor and delivery, prostaglandins reinforce the uterine contractions induced by oxytocin.

As the contractions of labor intensify, the fetus moves down though the birth canal and out into the world, still attached to the placenta (Fig. 26-20c). The placenta then detaches from the uterine wall and is expelled a short time later (Fig. 26-20d). The contractions of the uterus clamp the maternal blood vessels and help prevent excessive bleeding, although typically about 240 mL of blood is lost.

The Mammary Glands Secrete Milk During Lactation

When an infant is born, it loses its source of maternal nourishment through the placenta and must rely on an external source of food instead. Primates, who normally have only one or two offspring at a time, have two functional mammary glands. A mammary gland is composed of about 20 milk-secreting lobules, each made of branched hollow ducts of secretory epithelium surrounded by *myoepithelial* contractile cells (Fig. 26-22b ●). Interestingly, the mammary gland epithelium is closely related to the secretory epithelium of sweat glands, so milk and sweat secretion share some common features.

The breasts first begin to develop under the influence of estrogen during puberty. The milk ducts grow and branch, while fat is deposited behind the glandular tissue. During pregnancy, the glands develop further under the direction of estrogen, aided by growth hormone and cortisol. The final development step also requires progesterone, which converts the duct epithelium into a secretory structure. This process is similar to progesterone's effect on the uterus, in which progesterone makes the endometrium into a secretory tissue during the luteal phase.

Although estrogen and progesterone stimulate mammary development, they inhibit the actual secretion of milk. Milk production is under the control of the hormone prolactin from the anterior pituitary [※ p. 222]. Prolactin is an unusual pituitary hormone in that it is primarily controlled by an inhibiting hormone from the hypothalamus. There is good evidence that the **prolactin inhibiting hormone** (PIH) is actually dopamine, an amine neurohormone related to epinephrine and norepinephrine.

During the later stages of pregnancy, PIH secretion falls, and prolactin levels reach 10 or more times those found in nonpregnant women. Prior to delivery, when estrogen and progesterone are also high, the mammary glands produce only small amounts of a thin, low-fat secretion called **colostrum**. After delivery, when estrogen and progesterone decrease, the glands produce greater amounts of milk that contains 4% fat and substantial amounts of calcium. Proteins in colostrum and milk include maternal

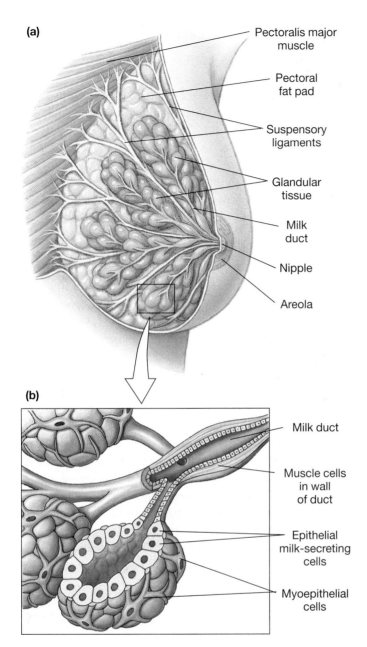

(a)

Pectoralis major muscle

Pectoral fat pad

Suspensory ligaments

Glandular tissue

Milk duct

Nipple

Areola

(b)

Milk duct

Muscle cells in wall of duct

Epithelial milk-secreting cells

Myoepithelial cells

● **Figure 26-22** *Mammary glands* Epithelial cells of the mammary glands secrete milk into the lumen of the gland.

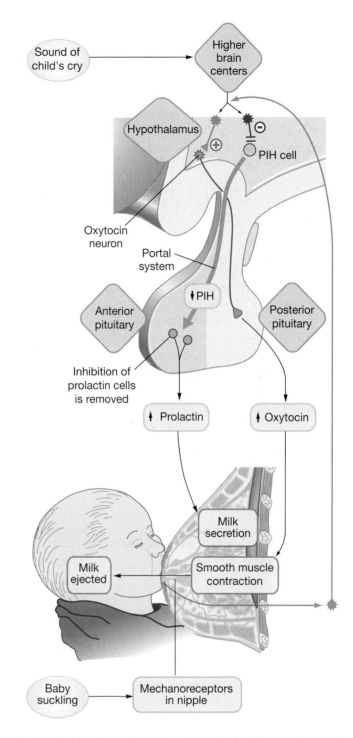

Sound of child's cry

Higher brain centers

Hypothalamus

PIH cell

Oxytocin neuron

Portal system

↓PIH

Anterior pituitary

Posterior pituitary

Inhibition of prolactin cells is removed

↑ Prolactin

↑ Oxytocin

Milk secretion

Milk ejected

Smooth muscle contraction

Baby suckling

Mechanoreceptors in nipple

● **Figure 26-23** *The hormonal control of milk secretion and release*

immunoglobulins, secreted into the duct and absorbed intact by the infant's intestinal epithelium [p. 683]. This transfers some of the mother's immunity to the infant during early weeks of life.

Suckling, the mechanical stimulus of the infant's nursing at the breast, inhibits release of PIH from the hypothalamus (Fig. 26-23 ●). In the absence of this inhibition, the pituitary increases spontaneous prolactin secretion, resulting in milk production. Pregnancy is not a requirement for lactation, and some women who have adopted babies have been successful in breast-feeding.

The ejection of milk from the glands, known as the **let-down reflex**, requires the presence of oxytocin from the posterior pituitary. Oxytocin initiates smooth muscle contraction in the uterus and

breasts. In the *postpartum* (after delivery) uterus, oxytocin-induced contractions help return the uterus to a prepregnancy size.

In the lactating breast, oxytocin causes contraction of myoepithelial cells surrounding the mammary glands. This contraction creates high pressure that sends the milk literally squirting into the infant's mouth. Although prolactin release requires the mechanical stimulus of suckling, oxytocin release can be stimulated by various cerebral stimuli, including the thought of the child. Many nursing mothers experience inappropriate milk release triggered by hearing someone else's child cry.

Prolactin Has Other Physiological Roles

Although we discussed prolactin in the context of nursing mothers, all nonnursing women and men have tonic prolactin secretion that shows a diurnal cycle, peaking during sleep. Prolactin is related to growth hormone, and plays a role in other reproductive and nonreproductive processes. For example, prolactin is synthesized in the uterine endometrium during normal menstrual cycles. Male knockout mice who lack prolactin or a prolactin receptor have decreased fertility.

Some of the most interesting research has established a role for prolactin in neuroimmunomodulation [p. 774]. Prolactin and growth hormone both appear to be necessary for normal differentiation of T lymphocytes in the thymus gland, an observation supported by impaired immune function in animals with *hypoprolactinemia*. In contrast, several autoimmune diseases, including multiple sclerosis, systemic lupus erythematosus, and autoimmune thyroiditis, have been linked to elevated levels of prolactin.

▶ GROWTH AND AGING

The reproductive years begin with the events surrounding puberty and end with decreasing gonadal hormone production.

Puberty Marks the Beginning of the Reproductive Years

Puberty is the period when a person makes the transition from being nonreproductive to being reproductive. In girls, the onset of puberty is marked by budding breasts and the first menstrual period, or **menarche**, a time of ritual significance in many cultures. In the United States, the average age at menarche is 12 years (normal range is considered 8 to 13 years).

In boys, the signs of puberty are more subtle. They include growth and maturation of the external genitalia, development of secondary sex characteristics such as pubic and facial hair, change in body shape, and growth in height. The age range for male puberty is 9 to 14 years.

Puberty requires the maturation of the hypothalamic-pituitary control pathway. Before puberty, the child has low levels of both steroid sex hormones and gonadotropins. Because low sex hormone levels normally enhance gonadotropin release, the combination of low steroids and low gonadotropins indicates that the pituitary is not yet sensitive to steroid levels in the blood.

What is the signal that initiates sexual development? The basis for the onset of puberty, like many other areas in reproductive physiology, is still uncertain. One theory says that it is the genetically programmed maturation of hypothalamic neurons. We know that puberty has a genetic base because inherited patterns of maturation are common. If a woman did not start her menstrual periods until she was 16, it is likely that her daughters will also have late menarche.

More recent theories have included the adipose tissue hormone *leptin* [p. 696] as a contributing factor to the onset of puberty. Undernourished women with low leptin levels often stop having menstrual periods (*amenorrhea*) and knockout mice without leptin are infertile. The improvement in nutrition over the last century presumably increased prepubertal leptin secretion, which could interact with other factors to initiate puberty.

Menopause and Andropause Are a Consequence of Aging

Several centuries ago in America, most people died of acute illnesses while still reproductively active. Now modern medicine has overcome most acute illnesses, and we are living well past the time that most of us are likely to have children. Women's reproductive cycles stop completely at the time known as **menopause**.

In men, testosterone production decreases with age. The existence of **andropause** in men is still controversial because the physical and psychological symptoms of aging men are not clearly linked to the decline in testosterone. Many men remain reproductively active as they age, and it is not uncommon for men in their fifties or sixties to have children with younger women. Postmenopausal women also remain sexually active, although not reproductively active, and some report a more fulfilling sex life once the fear of unwanted pregnancy has been removed.

The physiology of menopause has been well studied. After about 40 years of menstrual cycles, a woman's periods become irregular and finally cease. The failure of reproductive cycles is not due to the pituitary but to the ovaries, which can no longer respond to gonadotropins. In the absence of negative feedback, gonadotropin levels increase dramatically in an effort to stimulate the ovaries into maturing more follicles.

The absence of estrogen in postmenopausal women leads to symptoms of varying severity. These may include hot flashes [p. 719], atrophy of genitalia and breasts, and osteoporosis due to loss of calcium from the bones [p. 745]. Hormone replacement therapy (HRT) for women in menopause traditionally consists of estrogen or a combination of estrogen and progesterone. This treatment has become controversial, however, because of studies that suggest that the risks of HRT outweigh possible benefits.

The newest drug therapy for menopause is the *selective estrogen receptor modulators*, or SERMs [p. 186]. These drugs bind with different affinities to the two estrogen receptor subtypes. This allows the therapy to mimic the beneficial effects of estrogen on bone while avoiding the potentially detrimental effects on breast and uterus.

Infertility

In this running problem, you learned how the cause of infertility is diagnosed in a typical couple. To learn more about infertility, see the American Society for Reproductive Medicine at *www.asrm.org* or go to Medline Plus (*www.nlm.nih.gov/medlineplus*) and look under Health Topics. Test your understanding of the running problem by checking your answers against those in the summary table.

	QUESTION	FACTS	INTEGRATION AND ANALYSIS
1	Name the male reproductive structures that carry sperm from the testes to the external environment.	The male reproductive structures include the testes, accessory glandular organs, and a series of ducts.	Sperm leaving the seminiferous tubules pass into the epididymis, vas deferens, and finally the urethra.
2	For what causes of male infertility might retrival of sperm from the epididymis be necessary?	The epididymis is the first duct the sperm enter upon leaving the seminiferous tubules.	If the infertility problem is due to blockage or congenital defects in the vas deferens or urethra, removal of sperm from the epididymis might be useful. If the problem is due to low sperm count or abnormal sperm morphology, this technique would probably not be useful.
3	For what causes of female infertility is temperature tracking useful? For what causes is it not useful?	Basal body temperature rises slightly following ovulation.	Temperature tracking is a useful way to tell if someone is ovulating but it cannot be used to predict structural problems in the female reproductive tract.
4	What abnormalities in the cervix, Fallopian tubes, and uterus could cause infertility?	The cervix, Fallopian tubes, and uterus are hollow structures through which sperm must pass.	Any blockage of these organs due to disease or congenital defects would prevent normal movement of sperm and cause infertility. Hormonal problems might cause the endometrium to develop incompletely, preventing implantation of the embryo.
5	Speculate on how infertility due to cervical mucus antibodies to sperm might be treated.	Antibodies in the cervical mucus react with antigenic material in the semen or on the sperm, causing the sperm to become immobile.	If the antigenic material can be removed from the semen, this might help the problem of sperm immobilization. A semen sample can be washed to remove nonsperm components. If the antigens are part of the sperm, this method will not work.
6	Should ART or intrauterine insemination be recommended for Peggy and Larry? Why?	ART is used when ovulation is abnormal. Intrauterine insemination is used when ovulation is normal.	Intrauterine insemination can be used to overcome cervical factors, as this technique bypasses the cervix. Peggy can ovulate; therefore, intrauterine insemination should be recommended for Peggy and Larry.

796 806 815 818 821 826

CHAPTER SUMMARY

The reproductive system has some of the most complex *control systems* of the body, where multiple hormones interact in an ever-changing fashion. *Homeostasis* in the reproductive system is anything but steady-state, particularly during the female menstrual cycle, when the feedback effects of estrogen change from negative to positive and back again. Another example of *positive feedback* occurs with the stimulus for oxytocin secretion during labor and delivery. The testis provides a nice example of *compartmentation*, with the lumen of the seminiferous tubules, where sperm develop, isolated from the rest of the extracellular compartment.

Sex Determination

1. The sex organs consist of gonads, accessory ducts and glands, and **genitalia**. (p. 797)

2. The male **testes** produce **sperm**. The female **ovaries** produce eggs, or **ova**. Embryonic cells that will produce gametes are called **germ cells**. (p. 797)

3. Humans have 46 chromosomes. (p. 797; Fig. 26-1)

4. The genetic sex of an individual depends on the **sex chromosomes**: females are XX and males are XY. In the absence of a Y chromosome, an embryo will develop into a female. (p. 797; Fig. 26-2)

5. The **SRY gene** on the Y chromosomes produces **SRY protein**, a *testis determining factor*. (p. 798)

6. Testicular **Sertoli cells** secrete **anti-Müllerian hormone** that causes the **Müllerian ducts** to regress. **Leydig cells** secrete **testosterone** that converts **Wolffian ducts** to male accessory structures. (p. 798; Figs. 26-3, 26-4)

7. Absence of testosterone and anti-Müllerian hormone causes Müllerian ducts to develop into **Fallopian tubes (oviducts)**, **uterus**, and **vagina**. In females, the Wolffian ducts regress. (p. 798; Figs. 26-3, 26-4)

Basic Patterns of Reproduction

8. **Gametogenesis** begins with mitotic divisions of **spermatogonia** and **oögonia**. The first step of meiosis creates **primary spermatocytes** and **primary oocytes**. The first meiotic division creates two identical **secondary spermatocytes** or a large secondary oocyte (egg) and a tiny **first polar body**. (p. 801; Fig. 26-5)

9. The second meiotic division in males creates haploid **spermatids** that mature into sperm. In females, the second meiotic division of the oocyte does not take place unless the egg is fertilized. (p. 801; Fig. 26-5)

10. In both sexes, **gonadotropin releasing hormone** (GnRH) controls the secretion of **follicle stimulating hormone** (FSH) and **luteinizing hormone** (LH) from the anterior pituitary. FSH and steroid sex hormones regulate gametogenesis in gonadal gamete-producing cells. LH stimulates production of steroid sex hormones. (p. 803; Fig. 26-7)

11. The steroid sex hormones include **androgens, estrogens,** and **progesterone**. **Aromatase** converts androgens to estrogens. Peptide hormones **inhibin** and **activin** inhibit and stimulate secretion of FSH, respectively. (pp. 802–803; Fig. 26-6)

12. Gonadal steroids generally suppress secretion of GnRH, FSH, and LH in a long-loop response. However, if estrogen secretion rises rapidly above a threshold level for at least 36 hours, its feedback changes to positive and stimulates gonadotropin release. (p. 804)

13. After puberty, tonic GnRH release occurs in small pulses every 1–3 hours from a region of the hypothalamus called a **pulse generator**. (p. 804)

Male Reproduction

14. The **corpus spongiosum** and **corpora cavernosa** comprise the erectile tissue of the penis. The **glans** is covered by the **foreskin**. The urethra runs though the **penis**. (p. 805; Fig. 26-8)

15. The testes migrate into the scrotum during fetal development. Failure of one or both testes to descend is known as **cryptorchidism**. (p. 805)

16. The testes consist of **seminiferous tubules** and interstitial tissue with blood vessels and Leydig cells. The seminiferous tubules join the **epididymis**, which becomes the **vas deferens**. The vas deferens empties into the urethra. (p. 806; Fig. 26-9)

17. A seminiferous tubule has spermatocytes and Sertoli cells, which form a **blood-testis barrier**. (p. 806; Fig. 26-9)

18. Spermatogonia in the tubule enter meiosis, becoming primary spermatocytes, spermatids, and finally sperm in about 64 days. (p. 806; Fig. 26-9)

19. Sertoli cells regulate sperm development. They also produce inhibin, activin, growth factors, enzymes, and **androgen-binding protein**. (p. 808; Fig. 26-11)

20. The Leydig cells produce testosterone. About 5% of testosterone comes from the adrenal cortex. (p. 808)

21. FSH stimulates Sertoli cell production of androgen-binding protein, inhibin, and paracrines. The Leydig cells produce testosterone under the direction of LH. (p. 808; Fig. 26-11)

22. The **prostate gland, seminal vesicles,** and **bulbourethral glands** secrete the fluid component of **semen**. (pp. 806, 809)

23. **Primary sex characteristics** are internal sexual organs and external genitalia. **Secondary sex characteristics** are other features of the body, such as body shape. (pp. 809–810)

Female Reproduction

24. The female external genitalia are the **labia majora, labia minora**, and **clitoris**. The urethra opening is between the clitoris and the vagina. (p. 810; Fig. 26-12)

25. The uterine tissue layers are outer connective tissue, **myometrium**, and inner **endometrium**. (p. 813; Fig. 26-12)

26. The Fallopian tubes are lined with ciliated epithelium. The bulk of an ovary consists of ovarian follicles. (p. 813; Fig. 26-12)

27. Eggs are produced in monthly **menstrual cycles**. (p. 813; Fig. 26-13)

28. In the ovarian cycle, the **follicular phase** is a period of follicular growth. **Ovulation** is the release of an egg from its follicle. In the **luteal phase**, the ruptured follicle becomes a **corpus luteum**. (p. 813; Fig. 26-13)

29. The **menses** begin the uterine cycle. This is followed by a **proliferative phase**, with endometrial thickening. Following ovulation, the endometrium goes into a **secretory phase**. (p. 813; Fig. 26-13)

30. Follicular **granulosa cells** secrete estrogen. As the follicular phase ends, a surge in LH is necessary for oocyte maturation. (p. 815; Fig. 26-13)

31. The corpus luteum secretes progesterone and some estrogen, which exert negative feedback on the hypothalamus-anterior pituitary. (p. 817; Fig. 26-14)

32. Estrogens and androgen control primary and secondary sex characteristics in females. (p. 818)

Procreation

33. The human sex act is divided into four phases; (1) excitement, (2) plateau, (3) orgasm, and (4) resolution. (p. 818)

34. The male **erection reflex** is a spinal reflex that can be influenced by higher brain centers. Parasympathetic input mediated by nitric oxide actively vasodilates the penile arterioles. (p. 818; Fig. 26-15)

35. **Emission** is the movement of sperm out of the vas deferens and into the urethra. **Ejaculation** is the expulsion of semen to the exterior. (p. 819)

36. Contraceptive methods include **abstinence**, **barrier methods**, prevention of implantation of the fertilized egg, and hormonal treatment that decreases or stops gamete production. (p. 820)

37. Infertility can arise from a problem in the male, the female, or both. *In vitro* fertilization has allowed a number of infertile couples to have children, although the success rates are not high. (p. 821)

Pregnancy and Parturition

38. Sperm must go through **capacitation** before they can fertilize an egg. (p. 821)

39. Fertilization normally takes place in the Fallopian tube. Capacitated sperm release acrosomal enzymes (the **acrosomal reaction**) to dissolve cell junctions and the **zona pellucida**. The first sperm to reach the egg fertilizes it. (p. 822; Fig. 26-16)

40. Fusion of egg and sperm membranes initiates a **cortical reaction** that prevents **polyspermy**. (p. 822; Fig. 26-16)

41. The developing embryo is a hollow **blastocyst** when it reaches the uterus. Once the embryo implants, it develops **chorionic, amniotic**, and **allantoic membranes**. (p. 823; Figs. 26-18, 26-19a)

42. The **chorionic villi** of the placenta are surrounded by pools of maternal blood where nutrients, gases, and wastes are exchanged. (p. 824; Fig. 26-19b)

43. The corpus luteum remains active during early pregnancy due to **human chorionic gonadotropin** (hCG) from the developing embryo. (p. 824)

44. The placenta secretes hCG, estrogen, progesterone, and **human placental lactogen**. This hormone plays a role in maternal metabolism. (p. 824)

45. Estrogen during pregnancy contributes to the development of the milk-secreting ducts of the breasts. Progesterone is essential for maintaining the endometrium and, along with **relaxin**, helps suppress uterine contractions. (p. 826)

46. **Parturition** normally occurs in the 38th–40th week of gestation. It begins with **labor** and ends with delivery of the fetus and placenta. A positive feedback loop of oxytocin secretion causes uterine muscle contraction. (p. 824; Figs. 26-20, 26-21)

47. Following delivery, the mammary glands produce milk under the influence of prolactin. Milk is released during nursing by oxytocin. (p. 826; Figs. 26-22, 26-23)

48. Prolactin plays a role in immune function in both sexes. (p. 828)

Growth and Aging

49. **Puberty** is the period when a person makes the transition from being nonreproductive to being reproductive. (p. 828)

50. The cessation of reproductive cycles in women is known as the **menopause**. The existence of a male **andropause** is controversial. (p. 828)

QUESTIONS

1. Match each item below with all the terms that it applies to:

   (a) X or Y
   (b) inactivated X chromosome
   (c) XX
   (d) XY
   (e) XX or XY
   (f) autosomes

   1. chromosomes other than sex chromosomes
   2. fertilized egg
   3. sperm or ova
   4. sex chromosomes
   5. germ cells
   6. male chromosomes
   7. female chromosomes
   8. Barr body

2. The Y chromosome contains a region for male sex determination that is known as the _____ gene.

3. List the functions of the gonads. How do the products of gonadal function differ in males and females?

4. Trace the anatomical routes to the external environment followed by a newly formed sperm and by an ovulated egg. Name all structures that the gametes pass on their way.

5. Define the following terms and give their significance to reproductive physiology:

   (a) aromatase
   (b) blood-testis barrier
   (c) androgen binding protein
   (d) first polar body
   (e) acrosome

6. Decide if the statements below are true or false, and defend your answer.

   (a) All testosterone is produced in the testes.
   (b) Each sex hormone is produced only by members of one gender.
   (c) Anabolic steroid use appears to be addictive, and withdrawal symptoms include psychological disturbances.
   (d) High levels of estrogen in the late follicular phase help to prepare the uterus for menstruation.
   (e) Progesterone is the dominant hormone of the luteal phase of the female cycle.

7. What is semen? What are its main components and where are they produced?

8. List and give a specific example of the various methods of contraception. Which is/are most effective? Least effective?

9. **Concept maps:** Map the following groups of terms. You may add additional terms.

   List 1
   anti-Müllerian hormone
   DHT
   Leydig cells
   Müllerian ducts
   Sertoli cells
   sperm
   spermatids
   spermatocytes
   spermatogonia
   SRY
   testosterone
   Wolffian ducts

   List 2
   antrum
   corpus luteum
   endometrium
   follicle
   granulosa cells
   myometrium
   ovum
   thecal cells

10. Diagram the hormonal control of gametogenesis in males.

11. Diagram the menstrual cycle in the female, including the ovarian cycle and the uterine cycle. Include all hormones.

12. Why are X-linked traits exhibited more frequently by males than females?

13. Define and relate the terms in each of the following groups.

   (a) gamete, zygote, germ cell, embryo
   (b) coitus, erection, ejaculation, orgasm, emission, erogenous zones
   (c) capacitation, decapacitation factor, zona pellucida, acrosomal reaction, cortical reaction, cortical granules
   (d) puberty, menarche, menopause, andropause

14. Compare the actions of the hormones below in males and females.

   (a) FSH
   (b) inhibin
   (c) activin
   (d) GnRH
   (e) LH
   (f) DHT
   (g) estrogen
   (h) testosterone
   (i) progesterone

15. Compare and contrast the events of the four phases of sexual intercourse in each gender.

16. Discuss the roles of the following hormones in pregnancy, labor and delivery, and mammary gland development and lactation.

   (a) human chorionic gonadotropin
   (b) luteinizing hormone
   (c) human placental lactogen
   (d) estrogen
   (e) progesterone
   (f) relaxin
   (g) prolactin

17. Down syndrome is a chromosomal defect known as "trisomy" (three copies instead of two) of the twenty-first chromosome. The extra chromosome usually comes from the mother. Speculate what causes trisomy, using what you have learned about the events surrounding fertilization.

18. Sometimes eggs fail to leave the ovary at ovulation, even though they appear to have gone through all of the correct stages of development. This condition results in benign ovarian cysts, and the unruptured follicles can be palpated as bumps on the surface of the ovary. If the cysts persist, symptoms of this condition often mimic pregnancy, with missed menstrual periods and tender breasts. Can you explain or diagram how these symptoms occur?

19. The graph shows the results of an experiment in which normal men were given testosterone over a period of months (indicated by bar from A to E). Control values of hormones were measured prior to the start of the experiment. From time B to C, the men were also given FSH. From time D to E, LH was given in addition to testosterone. Based on the information given, answer the following questions.

(a) Why did testosterone increase at point A?

(b) Why did LH and FSH decrease at point A?

(c) Predict what happened to the men's sperm production in the A-B interval, the B-C interval, and the D-E interval.

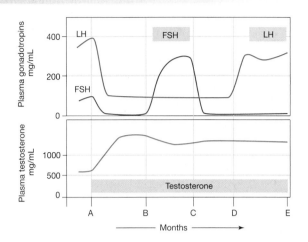

Explorations 🌐

Explorations are also available at *www.physiologyplace.com*

In this chapter you learned how the human race perpetuates itself through reproduction. Feedback mechanisms are crucial to coordinate and dictate development of zygotes into males and females. The discussion delved into the numerous signals that regulate growth and development during gestation and puberty.

Exploration 1: The Hunt for a Female Viagra®

When the U.S. Food and Drug Administration approved sildenafil (Viagra®) for male erectile dysfunction in 1998, it became the focus of conversation for many sexually active people. Women wondered if the drug, which increases nitric oxide levels and promotes the erection reflex, would improve their sexual response. By early 2003, the answer was still not clear. In one study, 51 women were given Viagra® to take before intercourse and were asked to self-report on their arousal and orgasms. The investigators of this study concluded that Viagra® was more effective than placebo in relieving female arousal disorder. Another study with 782 women did not find Viagra® any more effective than placebo.

Were these equivalent studies or are the results of one study more believable than those of the other? To evaluate the experimental designs for yourself, go to PubMed at *www.ncbi.nlm.nih.gov/entrez* and type *sildenafil women double-blind* into the SEARCH box. Look for the abstracts to these articles:

"Efficacy and safety of sildenafil citrate in women with sexual dysfunction associated with female sexual arousal disorder," *Journal of Women's Health & Gender-Based Medicine* 11(4):367–377, 2002 May.

"Premenopausal women affected by sexual arousal disorder treated with sildenafil: a double-blind, cross-over, placebo-controlled study," *British Journal of Obstetrics and Gynaecology* 108(6):623–628, 2001.

While the efficacy of Viagra® for women is still being investigated, a new over-the-counter herbal remedy, Avlimil®, is being touted as a "proven formula [that] works to improve libido, sexual feeling, and sexual response" (from the Avlimil® website *www.avlimil.com*). The manufacturers of Avlimil® claim that it was "proven effective through a randomized, double-blind clinical trial." The results of the trial are published on the website for you to evaluate. Once you have read them, go to PubMed and search for *salvia sexual* or *rubus sexual*. Have the Avlimil® trial results been subjected to peer review? Compare the Avlimil® trial to the Viagra® trials. Which is more believable?

Exploration 2: Determining Gender

The first question new parents typically ask about their child is, "Is it a boy or a girl?" Sometimes the answer is not obvious because in approximately 1 in 3000 births, the gender of the child cannot easily be determined. Make a list of the criteria that you might use to establish the gender of an individual. Consider genetic, chromosomal, gonadal, morphological, or even psychological characteristics. For example, *presence of the Y chromosome* (with the SRY gene) results in testes development and could be one criterion for "maleness." However, it is possible for an infant to have a Y chromosome and not appear to be male because of a pathology in the complicated cascade of steps that occur after testes development.

To explore the variety in sex determination and differentiation developmental patterns, do a Google search (*www.google.com*) for *sex determination* and *sex differentiation*. A good source for information is the online *Merck Manual of Diagnosis and Therapy* at *www.merck.com/pubs/mmanual*. This is a searchable source for standard medical information. Some examples of sex chromosome or reproductive system problems that you can look

(continues on next page)

for include Turner's syndrome (XO), Klinefelter's syndrome (XXY), Kallman's syndrome (no GnRH), congenital adrenal hyperplasia, testicular feminizing syndrome (no functional androgen receptor), and 5-alpha reductase deficiency. A list of sex chromosome abnormalities can be found in the Merck manual chapter on pediatric congenital anomalies (Section 19, Chapter 261: chromosomal abnormalities).

To test your understanding of sex determination and differentiation mechanisms, answer the following questions for each syndrome:

1. Are the gonads testes or ovaries?

2. Do Wolffian ducts or Müllerian ducts develop (or both or neither)?

3. Are the external genitalia masculinized or feminized (or ambiguous)?

4. Does the person have male-like or female-like behavior? How is this defined?

5. Will the person develop masculine or feminine secondary sex characteristics at puberty?

6. When people with this syndrome grow up, will they think of themselves as male or female?

For example, an XY individual with no functional androgen receptors will be genetically male and will have internal testes, but will have the external appearance of a female. This person also will have no Wolffian or Müllerian ducts.

Currently there is ongoing debate about how best to decide gender in cases where there is doubt. Recent discussions among patients, scientists, and health care workers have centered on question #6 above, what gender does the person consider himself or herself to be? This is called the person's *gender identity*. Traditionally gender determination has been based on the appearance of the external genitalia at birth, but the idea that individuals should be allowed to choose their own gender when they become old enough is gaining ground. You can read more about causes of ambiguous genitalia and the current criteria used to decide a child's gender in the American Academy of Pediatrics policy statement "Evaluation of the Newborn with Developmental Anomalies of the External Genitalia," *Pediatrics* 106(1):138–142, 2000 July (available online at *www.aap.org/policy/re9958.html*).

ANSWERS

Answers to Concept Check Questions

Page 798

1. The female gonad is the ovary and the gamete is the egg or ovum. The male gonad is the testis and the gamete is the sperm.

Page 801

2. The male donates the chromosome that determines the sex of the zygote; therefore, King Henry was at fault.

3. In the absence of a Y chromosome, the XO fetus will be a female. She will be infertile, however, as two X chromosomes are required for functioning ovaries.

4. Developing testes secrete testosterone, which (along with its derivative DHT) prompts the development of the male accessory structures. The external genitalia of the embryo will be female.

Page 802

5. (Use Figure 26-5 to answer questions 5–7.) The gamete in a newborn male is at the spermatogonia stage; it is at the primary oocyte stage in females.

6. The first polar body has twice as much DNA as the second polar body.

7. Each primary oocyte forms one egg; each primary spermatocyte forms four sperm.

Page 805

8. Aromatase converts testosterone to estradiol.

9. FSH = follicle-stimulating hormone, DHT = dihydrotestosterone, SRY = sex-determining region of the Y chromosome, LH = luteinizing hormone, GnRH = gonadotropin releasing hormone

10. GnRH from the hypothalamus and FSH and LH from the anterior pituitary control reproduction.

Page 809

11. Sertoli cells secrete inhibin, activin, androgen-binding protein, enzymes, and growth factors. Leydig cells secrete testosterone.

12. GnRH agonists cause reduced secretion of GnRH and therefore of FSH and LH. In the absence of the gonadotropins, the testes stop producing sperm. However, they also stop producing testosterone. This causes reduced sex drive, which is usually an undesirable side effect.

Page 810

13. Exogenous anabolic steroids are usually androgens, which have a negative feedback effect on the anterior pituitary gonadotropins. In the absence of FSH and LH, the testes will shrink and stop producing sperm.

Page 818

14. The ovarian cycle phases are follicular, ovulation, and luteal. The menses and proliferative phases of the uterus correspond to the follicular phase (and ovulation, which is very brief); the secretory uterine phase corresponds to the luteal phase.

15. Women who take anabolic steroids may experience growth of facial and body hair, deepening of the voice, increased libido, and irregular menstrual cycles.

16. A woman given an aromatase inhibitor would have decreased estrogen production during the follicular phase of the menstrual cycle.

Answers to Level Four Quantitative Problems

Page 833

19. (a) Testosterone increased at point A because it was being administered to the subjects.
(b) LH and FSH decreased at point A because of the negative feedback effect of testosterone.

(c) Sperm production decreased in the A-B interval because FSH and LH decreased. It increased toward the end of the B-C interval because FSH allowed sperm production to resume. Sperm production did not increase significantly during the D-E interval.

Appendix A
Physics and Math

Richard D. Hill and Daniel Biller, *University of Texas*

▶ INTRODUCTION

This appendix discusses selected aspects of **biophysics**, the study of physics as it applies to biological systems. Because living systems are in a continual exchange of force and energy, it is necessary to define these important concepts. According to the seventeenth-century scientist Sir Isaac Newton, a body at rest tends to stay at rest, and a body in motion tends to continue moving in a straight line unless the body is acted upon by some force (Newton's First Law). Newton further defined **force** as an influence, measurable in both intensity and direction, that operates on a body in such a manner as to produce an alteration of its state of rest or motion. Put another way, force gives **energy** to a quantity, or mass, thereby enabling it to do work. In general, a driving force multiplied by a quantity yields energy, or work. Some relevant examples of this principle include:

Mechanical force × distance =
    mechanical energy or mechanical work

Gas pressure × volume of gas =
    mechanical energy or mechanical work

Osmotic pressure × molar value =
    osmotic energy or osmotic work

Electrical potential × charge = electrical energy or electrical work

Temperature × entropy = heat energy or mechanical work

Chemical potential × concentration =
    chemical energy or chemical work

Energy exists in two general forms: kinetic energy and potential energy. **Kinetic energy** [*kinein*, to move] is the energy possessed by a mass in motion. **Potential energy** is energy possessed by a mass because of its position. Kinetic energy (*KE*) is equal to one-half the mass (*m*) of a body in motion multiplied by the square of the velocity (*v*) of the body:

$$KE = 1/2 \, mv^2$$

Potential energy (*PE*) is equal to the mass (*m*) of a body multiplied by acceleration due to gravity (*g*) times the height (*h*) of the body above the earth's surface:

$$PE = mgh \quad \text{where } g = 10 \, \text{m/s}^2$$

Both kinetic and potential energy are measured in joules.

▶ BASIC UNITS OF MEASUREMENT

For physical concepts to be useful in scientific endeavors, they must be measurable and should be expressed in standard units of measurement. Some fundamental units of measure include the following:

Length (*l*): Length is measured in meters (m).

Time (*t*): Time is measured in seconds (s).

Mass (*m*): Mass is measured in kilograms (kg), and is defined as the weight of a body in a gravitational field.

Temperature (*T*): Temperature is measured in degrees Kelvin (K),

    where K = degrees Celsius (°C) + 273.15

    and °C = (degrees Fahrenheit − 32)/1.8

Electric current (*I*): Electric current is measured in amperes (A).

Amount of substance (*n*): The amount of a substance is measured in moles (mol).

Using these fundamental units of measure, we can now establish standard units for physical concepts (Table A-1). Although these are the standard units for these concepts at this time, they are not the only units ever used to describe them. For instance, force can also be measured in dynes, energy can be measured in calories, pressure can be measured in torr or mm Hg, and power

Table A-1: Standard Units for Physical Concepts

MEASURED CONCEPT	STANDARD (SI*) UNIT	MATHEMATICAL DERIVATION/ DEFINITION
Force	Newton (N)	$1 \, \text{N} = 1 \, \text{kg m/s}^2$
Energy/Work/Heat	Joule (J)	$1 \, \text{J} = 1 \, \text{N} \cdot \text{m}$
Power	Watt (W)	$1 \, \text{W} = 1 \, \text{J/s}$
Electrical charge	Coulomb (C)	$1 \, \text{C} = 1 \, \text{A} \cdot \text{s}$
Potential	Volt (V)	$1 \, \text{V} = 1 \, \text{J/C}$
Resistance	Ohm (Ω)	$1 \, \Omega = 1 \, \text{V/A}$
Capacitance	Farad (F)	$1 \, \text{F} = 1 \, \text{C/V}$
Pressure	Pascal (Pa)	$1 \, \text{Pa} = 1 \, \text{N/m}^2$

* SI = Système International d'Unités

can be measured in horsepower. However, all of these units can be converted into a standard unit counterpart, and vice versa.

The remainder of this appendix will discuss some biologically relevant applications of physical concepts. This discussion will include topics such as bioelectrical principles, osmotic principles, and behaviors of gases and liquids relevant to living organisms.

▶ BIOELECTRICAL PRINCIPLES

Living systems are composed of different molecules, many of which exist in a charged state. Cells are filled with charged particles such as proteins and organic acids, and ions are in continual flux across the cell membrane. Therefore, electrical forces are important to life.

When molecules gain or lose electrons, they develop positive or negative charges. A basic principle of electricity is that opposite charges attract and like charges repel. A force must act on a charged particle (a mass) to bring about changes in its position. Therefore, there must be a force acting on charged particles to cause attraction or repulsion, and this electrical force can be measured. Electrical force increases as the strength (number) of charges increases, and it decreases as the distance between the charges increases. This observation has been called **Coulomb's law**, and can be written:

$$F = \frac{q_1 q_2}{\epsilon d^2}$$

where q_1 and q_2 are the electrical charges (coulombs), d is the distance between the charges (meters), ϵ is the dielectric constant, and F is the force of attraction or repulsion, depending on the type of charge on the particles.

When opposite charges are separated, a force acts over a distance to draw them together. As the charges move together, work is being done by the charged particles and energy is being released (Fig. A-1 ●). Conversely, to separate the united charges, energy must be added and work done. If charges are separated and kept apart, they have the potential to do work. This electrical potential is called **voltage**. Voltage is measured in **volts (V)**.

If electrical charges are separated and there is a potential difference between them, then the force between the charges will allow electrons to flow. Electron flow is called an electric **current**. The amount of current that flows depends on the nature of the material between the charges. If that material hinders the electron flow, then it is said to offer **resistance (R)**, measured in ohms. Current is inversely proportional to resistance, such that current decreases as resistance increases. If a material offers a high resistance, then that material is called an **insulator**. If resistance is low, and current flows relatively freely, then the material is called a **conductor**. Current, voltage, and resistance are related by **Ohm's law**, which states:

$$V = IR$$

where V = potential difference in volts

I = current in amperes

R = resistance in ohms

If you separate two opposite charges, there will be an electric force between them.

If you increase the number of charges that are separated, the force increases.

If you increase the distance between the charges, the force decreases.

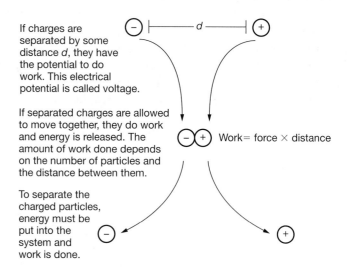

If charges are separated by some distance d, they have the potential to do work. This electrical potential is called voltage.

If separated charges are allowed to move together, they do work and energy is released. The amount of work done depends on the number of particles and the distance between them.

Work = force × distance

To separate the charged particles, energy must be put into the system and work is done.

● **Figure A-1** *Electrical force*

In biological systems, pure water is not a good conductor, but water containing dissolved NaCl is a fairly good conductor because ions provide charges to carry the current. In biological membranes, the lipids have few or no charged groups, so they offer high resistance to current flow across them. Thus, different cells can have different electrical properties depending on their membrane lipid composition and the permeability of their membranes to ions.

▶ OSMOTIC PRINCIPLES

Freezing point, vapor pressure, boiling point, and osmotic pressure are properties of solutions collectively called **colligative properties**. These properties depend on the number of solute particles present in a solution. **Osmotic pressure** is the force that drives the diffusion of water across a membrane. Because there are no solutes in pure water, it has no osmotic pressure. However, if one adds a solute like NaCl, the greater the concentration (c) of a solute dissolved in water, the greater the osmotic pressure. The osmotic pressure (π) varies directly with the concentration of solute (number of particles (n) per volume (V)):

$$\pi = (n/V)RT$$
$$\pi = cRT$$

where R is the ideal gas constant (8.3145 joules/K·mol) and T is the absolute temperature in Kelvin.

Osmotic pressure can be measured by determining the mechanical pressure that must be applied to a solution so that osmosis ceases.

Water balance in the body is under the control of osmotic pressure gradients (concentration gradients). Most cellular membranes allow water to pass freely because water is a small, uncharged molecule. Therefore, by making cellular membranes selectively permeable to solutes, the body can control the movement of water by controlling the movement of solutes.

▶ RELEVANT BEHAVIORS OF GASES AND LIQUIDS

The respiratory and circulatory systems of the human body have developed according to the physical laws that govern the behaviors of gases and liquids. This section will discuss some of the important laws that govern these behaviors and how our body systems utilize these laws.

Gases   The **ideal gas law** states:

$$PV = nRT$$

where

P = pressure of gases in the system
V = volume of the system
n = number of moles in gas
T = temperature
R = ideal gas constant ($8.3145 \, \text{J/K} \cdot \text{mol}$)

If n and T are kept constant for all pressures and volumes in a system, then any two pressures and volumes in that system are related by Boyle's Law,

$$P_1 V_1 = P_2 V_2$$

where P represents pressure and V represents volume.

This principle is relevant to the human lungs because the concentration of gas in the lungs is relatively equal to that in the atmosphere. In addition, body temperature is maintained at a constant temperature by homeostatic mechanisms. Therefore, if the volume of the lungs is changed, then the pressure in the lungs will change inversely. For example, an increase in pressure causes a decrease in volume, and vice versa.

Liquids   **Fluid pressure** (or hydrostatic pressure) is the pressure exerted by a fluid on a real or hypothetical body. In other words, the pressure exists whether or not there is a body submerged in the fluid. Fluid exerts a pressure (P) on an object submerged in it at a certain depth from the surface (h). **Pascal's law** allows us to find the fluid pressure at a specified depth for any given fluid. It states:

$$P = \rho g h$$

where:

P = fluid pressure (measured in pascals, Pa)
ρ = density of the fluid
g = acceleration due to gravity ($10 \, \text{m/s}^2$)
h = depth below the surface of the fluid

Fluid pressure is unrelated to the shape of the container in which the fluid is situated.

▶ REVIEW OF LOGARITHMS

Understanding logarithms ("logs") is important in biology because of the definition of pH:

$$\text{pH} = -\log_{10}[\text{H}^+]$$

This equation is read as "pH is equal to the negative log to the base 10 of the hydrogen ion concentration." But what is a logarithm?

A logarithm is the exponent to which you would have to raise the base (10) to get the number in which you are interested. For example, to get the number 100, you would have to square the base (10):

$$10^2 = 100$$

The base 10 was raised to the second power; therefore, the log of 100 is 2:

$$\log 100 = 2$$

Some other simple examples include:

$10^1 = 10$	The log of 10 is 1.
$10^0 = 1$	The log of 1 is 0.
$10^{-1} = 0.1$	The log of 0.1 is -1.

What about numbers that fall between the powers of 10? If the log of 10 is 1 and the log of 100 is 2, the log of 70 would be between 1 and 2. The actual value can be looked up on a log table or ascertained with most calculators.

To calculate pH, you need to know another rule of logs that says:

$$-\log x = \log(1/x)$$

and a rule of exponents that says:

$$1/10^x = 10^{-x}$$

Suppose you have a solution whose hydrogen ion concentration $[\text{H}^+]$ is 10^{-7} mEq/L. What is the pH of this solution?

$$\text{pH} = -\log[\text{H}^+]$$
$$\text{pH} = -\log(10^{-7})$$

Using the rule of logs, this can be rewritten as

$$\text{pH} = \log(1/10^{-7})$$

Using the rule of exponents, this can be rewritten as

$$\text{pH} = \log 10^7$$

The log of 10^7 is 7, so the solution has a pH of 7.

Natural logarithms (ln) are logs in the base e. The mathematical constant e is approximately equal to 2.7183.

Genetics

Richard D. Hill, *University of Texas*

▶ WHAT IS DNA?

Deoxyribonucleic acid (DNA) is the macromolecule that stores the information necessary to build structural and functional cellular components. It also provides the basis for inheritance when DNA is passed from parent to offspring. The union of these concepts about DNA allows us to devise a working definition of a gene. A **gene** (1) is a segment of DNA that codes for the synthesis of a protein, and (2) acts as a unit of inheritance that can be transmitted from generation to generation. The external appearance (**phenotype**) of an organism is determined to a large extent by the genes it inherits (**genotype**). Thus, one can begin to see how variation at the DNA level can cause variation at the level of the entire organism. These concepts form the basis of **genetics** and evolutionary theory.

▶ NUCLEOTIDES AND BASE-PAIRING

DNA belongs to a group of macromolecules called **nucleic acids. Ribonucleic acid (RNA)** is also a nucleic acid, but it has different functions in the cell that are not discussed here (see Chapter 4, p. 112). Nucleic acids are polymers made from monomers [*mono-*, one] called **nucleotides.** Each nucleotide consists of a *nucleoside* (a pentose, or 5-carbon, sugar covalently bound to a nitrogenous base) and a phosphoric acid with at least one phosphate group (Fig. B-1a ●). Nitrogenous bases in nucleic acids are classified as either **purines** or **pyrimidines**. The purine bases are **guanine (G)** and **adenine (A)**; the pyrimidine bases are **cytosine (C), thymine (T)**, found in DNA only, and **uracil (U)**, found in RNA only. To remember which DNA bases are pyrimidines, look at the first syllable. The word "pyrimidine" and names of the DNA pyrimidine bases all have a "y" in the first syllable.

When nucleotides link together to form polymers such as DNA and RNA, the phosphate group of one nucleotide bonds covalently to the sugar group of the adjacent nucleotide (Fig. B-1b, c). The end of the polymer that has an unbound sugar is called the 3' ("three prime") end. The end of the polymer with the unbound phosphate is called the 5' end.

▶ DNA STRUCTURE

In humans, many millions of nucleotides are joined together to form DNA. Eukaryotic DNA is commonly in the form of a double-stranded double helix (Fig. B-2 ●) that looks like a twisted ladder

or twisted zipper. The sugar-phosphate sides, or backbone, are the same for every DNA molecule, but the sequence of the nucleotides is unique for each individual organism.

The backbone of the double helix is formed by covalent **phosphodiester bonds** that link a deoxyribose sugar from one nucleotide to the phosphate group on the adjacent nucleotide. The "rungs" of the double helix are created when the nitrogenous bases on one DNA strand form hydrogen bonds with nitrogenous bases on the adjoining DNA strand. This phenomenon is called **base-pairing**. The base-pairing rules are as follows:

1. Purines pair only with pyrimidines.
2. Guanine (G) forms three hydrogen bonds with cytosine (C) in both DNA and RNA.
3. Adenine (A) forms with two hydrogen bonds with thymine (T) in DNA or with uracil (U) in RNA.

The number of hydrogen bonds is directly related to the amount of energy necessary to break the base pair. Thus, more energy is required to break G:::C bonds (each ":" represents a hydrogen bond) than A::T bonds. This fact can be useful experimentally to make general conjectures about the similarity of two DNA samples.

The two strands of DNA are bound in **antiparallel** orientation, so that the 3' end of one strand is bound to the 5' end of the second strand (see Fig. B-1c). This organization has important implications for DNA replication.

DNA Replication Is Semi-Conservative

In order to be transmitted from one generation to the next, DNA must be replicated. Furthermore, the process of replication must be accurate and fast enough for a living system. The base-pairing rules for nitrogenous bases provide a means for making an appropriate replication system.

In DNA replication, special proteins unzip the DNA double helix and build new DNA by pairing new nucleotide molecules to the two existing DNA strands. The result of this replication is two double-stranded DNA molecules, such that each DNA molecule contains one DNA strand from the template and one newly synthesized DNA strand. This form of replication is called **semi-conservative replication**.

Replication of DNA is bidirectional. A portion of DNA that is "unzipped" and has enzymes performing replication is called a **replication fork** (Fig. B-2). Replication begins at many points

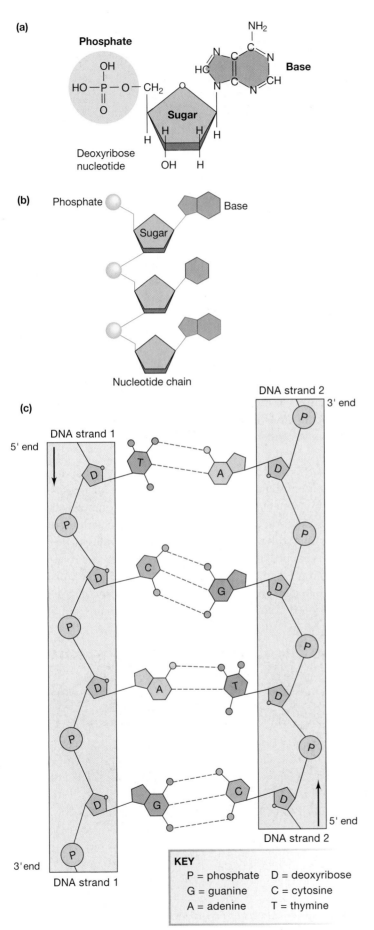

● **Figure B-1** *Nucleotides and DNA* (a) A nucleotide is composed of a pentose sugar, a nitrogenous base, and a phosphate group. (b) Nucleotides link sugar to phosphate to form nucleic acids. The end of the nucleic acid with an unbound sugar is designated the 3′ end; the end with the unbound phosphate is the 5′ end. (c) Two nucleotide strands link by hydrogen binding between complementary base pairs to form the double helix of DNA. Adenine always pairs to thymine, and guanine pairs to cytosine.

(**replicons**), and it continues along both parent strands simultaneously until all the replication forks join.

Nucleotides bond together to form new strands of DNA with the help of an enzyme called **DNA polymerase**. DNA polymerase can only add nucleotides to the 3′ end of a growing strand of DNA. For this reason, DNA is said to replicate in a 5′ to 3′ direction.

The antiparallel orientation of the DNA strands and the directionality of DNA polymerase force replication into two different modes: **leading strand replication** and **lagging strand replication**. The DNA polymerase can replicate continuously along only one parent strand of DNA: the parent strand in the 3′ to 5′ orientation. The DNA replicated continuously is called the **leading strand**.

The DNA replication along the other parent strand is discontinuous because of the strand's 5′ to 3′ orientation. DNA replication on this strand occurs in short fragments called **Okazaki fragments** that are synthesized in the direction away from the replication fork. Another enzyme known as **DNA ligase** later connects these fragments into a continuous strand. The DNA replicated in this way is called the **lagging strand**. Because the 5′ ends of the lagging strand of DNA cannot be replicated by DNA polymerase, a specialized enzyme called **telomerase** has arisen to replicate the 5′ ends.

Much of the accuracy of DNA replication comes from base pairing, but on occasion, mistakes in replication happen. However,

KEY

P = phosphate	D = deoxyribose
G = guanine	C = cytosine
A = adenine	T = thymine

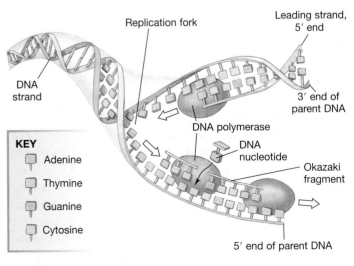

● **Figure B-2** *DNA replication*

several quality control mechanisms are in place to keep the error rate at 1 error/10^9 to 10^{12} base pairs. **Genome** (the entire amount of DNA in an organism) sizes in eukaryotes range from 10^9 to 10^{11} base pairs per genome, so this error rate is low enough to prevent many lethal mutations, yet still allows genetic variation to arise.

▶ FUNCTIONS OF DNA

One of the key functions of DNA is its ability to code for the synthesis of proteins that participate in structural or functional aspects of a cell. The processes by which information coded in DNA is converted into proteins are known as *transcription* and *translation*. Both are discussed in Chapter 4. The other key function of DNA is its ability to form genes that act as units of inheritance when transmitted across generations.

Before we begin our discussion of DNA as a unit of inheritance, a few terms need to be introduced. A **chromosome** is one complete molecule of DNA, and each chromosome contains many genes. An **allele** is a form of a gene; interactions between the cell products of alleles determine how that gene will be expressed in the phenotype of an individual. **Somatic** [*soma,* body] **cells** are those cells that comprise the majority of the body (e.g., a skin cell, a liver cell); they are not directly involved with passing on genetic information to future generations. Each somatic cell in a human contains two alleles of each gene, one allele inherited from each parent. For this reason, human somatic cells are called **diploid** ("two chromosome sets"), meaning that they have two complete sets of all their chromosomes. In contrast, **germ cells** pass genetic information directly to the next generation. In human males, the germ cells are the **spermatozoa** (sperm), and in human females, the germ cells are the **oocytes** (eggs). Human germ cells are called **haploid** ("half of the chromosome sets") because each germ cell only contains one chromosome set, which is equal to half of the number of chromosome sets in somatic cells. When a human male germ cell joins with a human female germ cell, the result is a fertilized egg (zygote) containing the diploid number of chromosomes. If this zygote eventually develops into a healthy adult, that adult will have diploid somatic cells and haploid germ cells.

Cells alternate between periods of cell growth and cell division. There are two types of cell division: mitosis and meiosis. **Mitosis** is cell division by somatic cells that results in two daughter cells, each with a diploid set of chromosomes. **Meiosis**, in contrast, is cell division that results in four daughter cells, each with a haploid set of chromosomes. The daughter cells develop into germ cells.

The period of cell growth is called **interphase**. It is divided into three stages: G_1, a period of cell growth, protein synthesis, and organelle production; S, the period during which DNA is replicated in preparation for cell division; and G_2, a period of protein synthesis and final preparations for cell division (Fig. B-3 ●). During interphase, the DNA in the nucleus is not visible under

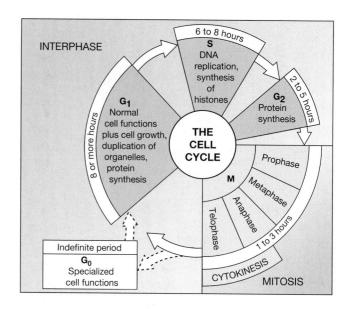

● **Figure B-3**   *The cell cycle*

the light microscope without dyes because it is uncoiled and diffuse. However, as a cell prepares for division, it must condense all its DNA to form more manageable packages. Each eukaryotic DNA molecule has millions of base pairs, which, if laid end-to-end, could stretch out to about 6 cm. If this DNA did not coil tightly and condense for cell division, imagine how difficult moving it around during cell division would be.

There is a hierarchy of DNA packaging in the cell (Fig. B-4 ●). Each chromosome begins with a linear molecule of DNA about 2 nm in diameter. Then proteins called **histones** associate with the DNA to form a fiber about 10 nm in diameter that looks like "beads on a string." The "beads" are **nucleosomes** made up of histones wrapped in DNA. The beaded string can twist into a **solenoid** [*solen,* pipe] coil about 30 nm in diameter, consisting of about 6 nucleosomes per turn. The 30-nm solenoid structure can form looped domains, and these looped domains can attach to

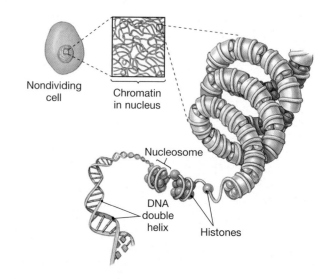

● **Figure B-4**   *Levels of organization of DNA*

nonhistone (structural) **proteins** to form a 250–300-nm fiber called **chromatid fiber**. The chromatid fiber then coils to form the **chromosome fiber** (about 700 nm in diameter) that is visible during cell division. In this state of packaging, the cell is ready for division.

Mitosis Creates Two Identical Daughter Cells

As stated earlier, mitosis is the cellular division of a somatic cell that results in two diploid daughter cells. The steps of mitosis are **prophase, metaphase, anaphase,** and **telophase** (Fig. B-5 ●). The entire somatic cell cycle can be remembered by the acronym, IPMAT, in which the "I" stands for interphase and the other letters stand for the steps of mitosis that follow.

Prophase   During prophase, chromatin becomes condensed and microscopically visible as duplicate chromosomes. The duplicated chromosomes form **sister chromatids**, which are joined to each other at the **centromere**. The centriole pair [● p. 58] duplicates and the centriole pairs move to opposing ends of the cell. The **mitotic spindle**, composed of microtubules, assembles between them. The nuclear membrane begins to break down and disappears by the end of prophase.

Metaphase   In metaphase, mitotic spindle fibers extending from the centrioles attach to the centromere of each chromosome. The forty-six chromosomes, each consisting of a pair of sister chromatids, line up at the "equator" of the cell.

Anaphase   During anaphase, the spindle fibers pull the sister chromatids apart, so that an identical copy of each chromosome moves toward each pole of the cell. By the end of anaphase, an identical set of forty-six chromosomes is present at each pole. At this point, the cell has a total of ninety-two chromosomes.

Telophase   The actual division of the parent cell into two daughter cells takes place during telophase. In **cytokinesis**, the cytoplasm divides when an actin contractile ring tightens at the midline of the cell. The result is two separate daughter cells, each with a full diploid set of chromosomes. The spindle fibers disintegrate, nuclear envelopes are formed around the chromosomes in each cell, and the chromatin returns to its loosely coiled state.

Mutations Change the Sequence of DNA

Over the course of a lifetime, there are countless opportunities for mistakes to arise in the replication of DNA. A change in a DNA sequence, such as the addition, substitution, or deletion of a base, is a **point mutation**. If a mutation is not corrected, it may cause a change in the gene product. These changes may be relatively minor, or they may result in dysfunctional gene products that could kill the cell or the organism. Only rarely does a mutation result in a beneficial change in a gene product. Fortunately, our cells contain enzymes that can detect and repair damage to DNA.

Some mutations are caused by **mutagens**, which are factors that increase the rate of mutation. Various chemicals, ionizing radiation such as X-rays and atomic radiation, ultraviolet light, and other factors can behave as mutagens. Mutagens either alter the base code of DNA or interfere with repair enzymes, thereby promoting mutation.

Mutations that occur in body cells are called **somatic mutations**. Somatic mutations are perpetuated in the somatic cells of an individual, but they are not passed on to subsequent generations. However, **germ-line mutations** can also occur. Because these mutations arise in the germ cells of an individual, they are passed on to future generations.

Oncogenes and Cancer

Proto-oncogenes are normal genes in the genome of an organism that primarily code for protein products that regulate cell growth, cell division, and cell adhesion. Mutations in these proto-oncogenes give rise to **oncogenes**, genes that induce uncontrolled cell proliferation and the condition known as **cancer**. The mutations in proto-oncogenes that give rise to cancer-causing oncogenes are often the result of viral activity.

(a) Interphase   (b) Early prophase   (c) Late prophase   (d) Metaphase   (e) Anaphase   (f) Telophase

● **Figure B-5** *Mitosis*

Anatomical Positions of the Body

ANTERIOR (situated in front of): in humans, toward the front of the body (see VENTRAL) C-1

POSTERIOR (situated behind): in humans, toward the back of the body (see DORSAL) C-2

MEDIAL (middle, as in *median strip*): located nearer to the midline of the body (the line that divides the body into mirror-image halves)

LATERAL (side, as in a *football lateral*): located toward the sides of the body

DISTAL (distant): farther away from the point of reference or from the center of the body

PROXIMAL (closer, as in *proximity*): closer to the center of the body

SUPERIOR (higher): located toward the head or the upper part of the body

INFERIOR (lower): located away from the head or from the upper part of the body

PRONE: lying on the stomach, face downward

SUPINE: lying on the back, face up

DORSAL: refers to the back of the body

VENTRAL: refers to the front of the body

IPSILATERAL: on the same side as

CONTRALATERAL: on the opposite side from

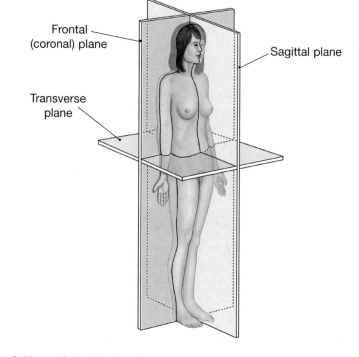

● Figure C-1   *Sectional planes*

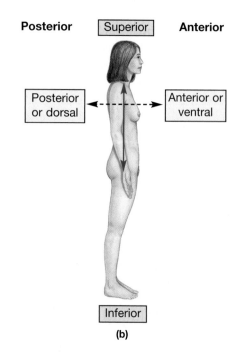

● **Figure C-2**

(a)

(b)

Glossary/Index

A band Band of striated muscle sarcomere whose length equals that of the thick filament (Ch 12), 394

abdominal muscles, 557

abducens nerve Cranial nerve that controls eye movement (Ch 9), 298t

abnormal hemoglobin (HbS), 534

abnormal tissue responsiveness, 228

ABO blood group Membrane glycoproteins found on red blood cells that are one determinant of blood type (Ch 24), 772, 773f, 773t

absorption Transfer of substances from the lumen of the kidney or gastrointestinal tract to the extracellular space (Ch 5, 15, 21), 147, 506, 668–669

absorptive cell Small intestinal epithelial cell whose primary function is absorption. Synonym: enterocyte (Ch 21), 663f, 680

absorptive state, 700, 702–704

abstinence, 820

accessory nerve (XI) Cranial nerve that controls muscles of oral cavity, neck, and shoulders, 298f

acclimation Physiological adaptation to an environmental change under laboratory conditions (Ch 6), 193

acclimatization Physiological adaptation to an environmental change (Ch 6, 25), 193, 790

accommodation The process by which the eye adjusts the shape of the lens to keep objects in focus (Ch 10), 354, 354f

ACE. *See* angiotensin converting enzyme.

acetoacetic acid Metabolic acid formed from oxidation of fatty acids (Ch 22), 705

acetylcholine (ACh) Neurotransmitter used by neurons of the central and peripheral nervous system (Ch 6, 8, 9, 11, 12, 14, 17), 185, 187, 265t, 266, 266f, 268, 305, 306t, 374, 376, 378, 379t, 400, 461, 562

acetylcholinesterase Enzyme that breaks down acetylcholine in the synapse (Ch 8, 11), 266, 382

acetyl CoA. *See* acetyl coenzyme A.

acetyl coenzyme A (acetyl CoA) metabolic intermediate that links glycolysis and β-oxidation to the citric acid cycle (Ch 4, 22), 99, 105, 109, 111, 704

acetylsalicylic acid. *See* aspirin.

ACh. *See* acetylcholine.

acid A molecule that ionizes and contributes an H^+ to a solution (Ch 2, 20), 33, 646

acid-base balance The homeostatic regulation of body pH (Ch 2, 20), 33–34, 645–650

acid-base disturbance, 651–653

acid lipase Fat-digesting enzyme of the mouth and stomach (Ch 21), 684

acidosis Extracellular pH less than 7.38 (Ch 18, 20), 584, 647, 648, 651–652

acid phosphatase, 93t

acini. *See* follicle, thyroid.

acquired immune deficiency syndrome. *See* AIDS.

acquired immunity Immune responses directed at specific invaders and mediated by antibodies (Ch 24), 754, 758, 762–768

acquired immunodeficiencies, 752

acromegaly Abnormal growth of cartilage and soft tissues due to excess growth hormone secretion in an adult (Ch 23), 739, 739f

acrosomal reaction Release of enzymes from the sperm head when it contacts an egg (Ch 26), 822

acrosome Lysosome-like vesicle of sperm that contains powerful enzymes essential for fertilization (Ch 26), 808

ACTH. *See* adrenocorticotrophic hormone.

actin A globular protein (G-actin) that polymerizes to form thin filaments (F-actin) (Ch 3, 12), 57, 394, 398, 399, 415, 416

action potential Rapid and uniform electrical signal conducted down an axon to the axon terminal or along the membrane of a muscle fiber (Ch 3, 8, 10, 13, 14, 15)

  absolute refractory period, 255–256

  from baroreceptors, 74, 240t, 249, 253f, 510

  coding for stimulus intensity, 256

  compared to graded potential, 250t

  conduction of, 261–262

  in contractile cell, 461

  distance traveled without losing strength, 252–253

  falling phase of, 254

  ion concentration gradients, 256

  large neurons and, 258–260

  in motor neuron, 436

  myelinated axons, 260–261

  in myocardial cells, 464–465

  in primary sensory neuron, 345

  rising phase of, 253–254

  from trigger zone to axon terminal, 257–258

  voltage-gated Na^+ channels, 254–255

activation energy Energy needed to initiate a chemical reaction (Ch 4), 88, 88f, 90–91, 90f

activation gate Sodium channel gate that opens to initiate an action potential (Ch 8), 254

active expiration Expiration that is assisted by contraction of abdominal and internal intercostal muscles (Ch 17), 557

active hyperemia An increase in blood flow that accompanies an increase in metabolism (Ch 15), 500

active immunity, 762

active site Region of an enzyme or transport protein to which the substrate binds. Synonym: binding site (Ch 4), 91

active transport Movement across a membrane that requires the input of energy from ATP (Ch 5, 19), 138, 140–144, 610

  secondary, 143

activin Peptide hormone from the gonads that stimulates FSH secretion (Ch 26), 803, 804

acuity Keenness of vision (Ch 10), 357

acupuncture, 335

acute motor axonal polyneuropathy (AMAN) (Ch 8), 274

acute mountain sickness, 579

acute phase protein Liver proteins that act as opsonins and enhance the inflammatory response (Ch 24), 671t, 760

adaptation of receptors Process in which sensory receptors decrease their response to a stimulus over time (Ch 10), 328, 330

adaptive immunity, 754

addictive behavior (Ch 9), 309

Addison's disease, 731, 732

addition reaction Chemical reaction that adds a functional group to a molecule (Ch 4), 99

adenine Nucleotide base found in ATP, DNA, RNA, and cAMP (Ch 2, 4, Appendix), 41, 112, 838

adenohypophyseal, 727

adenohypophysis. *See* anterior pituitary.

adenoma, 228

adenosine Nucleoside composed of adenine and ribose purine (Ch 8, 9, 15), 265t, 308, 501, 512

adenosine diphosphate (ADP) Precursor of ATP, composed of adenine, ribose, and two phosphate groups (Ch 2, 3, 12, 16), 41, 58, 101, 397, 398, 538f

adenosine triphosphate (ATP) An energy-storing compound composed of adenine, ribose, and three phosphate groups (Ch 2, 3, 4, 12, 25), 34, 41, 59, 784

  ATP/ADP ratio, 101

  electron transport system and, 106

  energy for active transport, 140–144

  energy release from, 103, 104

  in muscle contraction, 397, 398, 401–404, 783

  production of, 102–109

  transfer of energy between reactions, 101–102

adenylyl cyclase Membrane-bound enzyme that converts ATP to cyclic AMP (Ch 6), 179

adequate stimulus The form of energy to which a particular receptor is most responsive (Ch 10), 324

ADH. *See* antidiuretic hormone.

adherens junctions Bands that link action microfilaments in adjacent cells together with the help of cadherins (Ch 3), 66

adipocyte Fat cells (Ch 16, 22), 73, 526, 696

adipose tissue, 73, 74f

adjunct therapy, 430

ADP. *See* adenosine diphosphate.

adrenal cortex Outer portion of adrenal gland that produces steroid hormones (Ch 7, 11, 23), 216, 217, 376, 727–728, 728f

adrenal gland Endocrine and neuroendocrine gland that sits on top of the kidney (Ch 7, 11), 216, 376

adrenal glucocorticoids, 727–732

adrenaline. *See* epinephrine.

adrenal medulla Modified sympathetic ganglion, the inner portion of the adrenal gland that produces catecholamines (Ch 11), 376–377, 377f, 727

adrenergic Adjective pertaining to epinephrine (adrenaline) or norepinephrine (Ch 6, 11), 184, 378, 379t

adrenergic agonist, 562

adrenergic neuron Neuron that secretes norepinephrine (Ch 8), 266

adrenergic receptor Receptor that binds to norepinephrine or epinephrine (Ch 6, 8, 11), 184–185, 267, 372, 374, 377–378, 378f

adrenocorticotropic hormone (ACTH) Anterior pituitary hormone that regulates secretion of cortisol from the adrenal cortex (Ch 7, 9, 10, 23, 24, 26), 222, 225, 228, 229, 293, 307, 334, 729–732, 775, 825

adrenogenital syndrome, 731

aerobic Adjective pertaining to a process that requires oxygen (Ch 4), 102

aerobic exercise, 784

aerobic metabolism, 102, 104t, 105–106, 108, 547, 784

affective behavior Behaviors related to feeling and emotion (Ch 9), 285

afferent arteriole Renal arterioles that bring blood to the glomerulus (Ch 19), 602, 607, 608

afferent neuron. *See* sensory neuron.

afferent pathway The pathway that connects a receptor to an integrating center (Ch 6), 191, 192

agglutination Clumping of cells together (Ch 24), 772

aging, 828

agonist Molecules that combine with a receptor and mimic a response (Ch 6, 11), 185, 378, 379t

agouti mouse, 732

AIDS, 751, 753, 768, 771. *See also* HIV

AIDS treatment, 774, 776

air flow, 552–555

airway Anatomical structures from mouth to bronchioles that carry air to the alveoli (Ch 17), 548, 550f, 551f

airway resistance, 560–561, 561t

albumin Plasma protein made in the liver (Ch 7, 15), 217, 524, 525f

alcohol addiction, 335

alcohol consumption, 819–820

aldosterone A steroid hormone that stimulates Na^+ reabsorption and K^+ secretion in the kidney (Ch 20, 23), 635–638, 727

alkaline phosphatase, 93t

alkaline solution, 34

alkalosis Extracellular pH greater than 7.42 (Ch 20), 649, 650, 652–653, 652t

allantois Extraembryonic membrane that becomes part of the umbilical cord (Ch 26), 824

allele, 840

allergen Any substance capable of triggering an allergic reaction (Ch 24), 771

allergic response, 757, 771, 772f

allergy, 771

all-or-none phenomenon, 253

allosteric activation, 95

allosteric modulator Bind to an enzyme away from the binding site and change the shape of the active site (Ch 4), 95, 96f

alpha cell pancreatic cell that secretes glucagon. Synonym: A cell (Ch 22), 705

alpha-gamma coactivation Simultaneous activation of alpha and gamma motor neurons so that during muscle contraction, the intrafusal fibers continue to monitor tension in the muscle (Ch 13), 434

alpha helix Spiral configuration formed by some amino acid chains (Ch 2), 38

alpha interferon, 760

alpha motor neuron Neurons that innervate extrafusal muscle fibers and cause muscle contraction (Ch 13), 432

5-alpha-reductase, 800

alpha wave Low amplitude, high-frequency brain waves characteristic of the awake-resting state (Ch 9), 306f

alternative splicing The processing of mRNA to make different proteins from a single strand of DNA (Ch 4), 112–114

altitude hypoxia, 579

alveolar blood flow, 566–567

alveolar macrophage Immune cells that patrol the alveoli (Ch 17), 559

alveolar P_{O_2}, 577–578

alveolar pressure, 555–557

alveolar ventilation The volume of air that reaches the alveoli each minute (Ch 17, 18), 564–566, 578

alveoli The exchange surface of the lungs, where oxygen and carbon dioxide transfer between air and the blood (Ch 17), 548, 550f, 551–552, 567f, 577f, 579

Alzheimer's disease (Ch 9), 311

AMAN. *See* acute motor axonal polyneuropathy.

amenorrhea, 828

amination Addition of an amino group to a molecule (Ch 4), 99

amine hormone Derivatives of single amino acids, either tyrosine or tryptophan (Ch 7), 214t, 218

amine neurotransmitter Neurotransmitters made from amino acids, including the catecholamines, histamine, and serotonin (Ch 8), 266–267

amino acid Molecule with a central carbon atom linked to a hydrogen atom, an amino group, a carboxyl group, and a variable group of atoms designated "R." The building blocks of proteins (Ch 2, 4, 8, 21, 22), 37–38, 38t, 112, 265t, 669, 683, 700, 710

glucose made from, 110–111

mRNA translation producing, 114–115

from protein catabolism, 108, 109f

proteins made from, 702–703

structure of, 38–39, 38f

amino acid pool, 700–701

amino group Functional group whose composition is $-NH_2$ (Ch 2), 28t, 37, 38

aminopeptidase Digestive enzyme that removes amino acids from the NH_2 terminal end of a peptide (Ch 21), 668t

ammonia, 648

ammonium ion NH_4^+ (Ch 20), 649

amnion Extraembryonic membrane that secretes amniotic fluid (Ch 26), 824

amniotic fluid, 824

AMPA receptor Glutamate receptor-channel that allows net Na^+ influx (Ch 8), 268

amplification, 343

amplifier enzyme A membrane enzyme that creates two or more second messengers during signal transduction (Ch 6), 176, 177f

amplitude, 342

ampulla (vestibular apparatus) (Ch 10), 347

amu. *See* atomic mass unit.

amygdala Portion of the brain linked to emotion and memory (Ch 9), 300, 300f, 308

amylase Enzyme that digests starch to maltose (Ch 4, 21), 93, 668t

amylin, 714

amyloid protein (Ch 9), 311

amyotrophic lateral sclerosis (Ch 12), 426

anabolic metabolism, 702–704

anabolic pathway, 700

anabolism Metabolic pathways that require a net input of energy and that synthesize small molecules into larger ones (Ch 4, 22), 99, 707–710

anaerobic Adjective pertaining to a process that does not require oxygen (Ch 4), 102

anaerobic glycolysis, 402

anaerobic metabolism, 102, 104–106, 104t, 784

analgesic drugs Drugs that relieve pain (Ch 10), 334

anaphase, 841

anaphylactic shock Allergic reaction marked by bronchoconstriction and vasodilation with hypotension (Ch 15), 498

anaphylaxis, 771

anatomical positions of human body, 842

anatomic dead space The portions of the airways that do not exchange gases with the blood (Ch 17), 564f

anatomy The study of structure (Ch 1, 9), 2, 210f–211f

of central nervous system, 287–293, 289f

system to cell, 54–56

anaxonic neuron (Ch 8), 242

anchoring junction Form cell-cell or cell-matrix junctions (Ch 3), 64, 65t, 66

anchor protein (Ch 6), 181

Andersson, Bengt, 640

androgen Steroid hormone produced in the gonads and adrenal cortex; dominant hormone in males (Ch 23, 26), 727, 802, 804, 809–810, 818

androgen-binding protein Protein produced in Sertoli cells that binds testosterone to keep it in the lumen of the seminiferous tubule (Ch 26), 808

andropause, 828

androstenedione, 727, 729f, 749

anemia Pathological state with low hemoglobin (Ch 16, 21), 534, 676

anemone, 671

ANG. *See* angiotensin.

angina, 502

Na$^+$ and water excretion (Ch 15, 20), 501, 638–639, 639f

atrial natriuretic peptide. *See* atrial natriuretic factor.

atrial systole, 476

atriopeptin. *See* atrial natriuretic peptide.

atrioventricular (AV) valves Heart valves that separate the atria from the ventricles (Ch 14), 459

atrioventricular bundle. *See* bundle of His.

atrioventricular node (AV node) The electrical gateway to the ventricles, located near the floor of the right atrium (Ch 14), 468

atrioventricular node (AV node) delay Slowing of electrical conduction through the AV node that allows atria to complete contraction before the ventricles begin (Ch 14), 470

atrium Upper chamber of the heart that receives blood from the blood vessels (Ch 14), 450

atrophy, 62, 228, 413

atropine Muscarinic receptor antagonist (Ch 8), 265t

auditory cortex, 304, 313, 346

auscultation, 476, 563

autocrine A local chemical signal that acts on the cell that secreted it (Ch 6, 7), 172f, 173, 190, 212

autoimmune disease Diseases in which the immune system creates antibodies against the body's own tissues (Ch 22, 24), 711, 752, 773, 774

autoimmune thyroiditis, 828

autonomic division Efferent division of the nervous system that controls smooth muscle, cardiac muscle, glands, and some adipose tissue (Ch 8, 11), 240, 370–381

autonomic ganglion, 372

autonomic interneurons, 372

autonomic neuron Efferent neurons that control smooth muscle, cardiac muscle, many glands, and some adipose tissue (Ch 11, 14, 19), 370, 374, 375, 480, 608, 610

autonomic pathways, 372, 380f

autonomic reflex, 371, 429–431, 431f

autorhythmic cell Cardiac cells that spontaneously and rhythmically depolarize and fire action potentials (Ch 14), 460

autosome The 22 pairs of chromosomes that contain information for non-sex-related development (Ch 26), 797

AV node. *See* atrioventricular node.

AV valves. *See* atrioventricular valves.

axon An extension of a neuron that carries signals to the target cell (Ch 8), 240t, 242, 254–255, 259–261

axonal transport Movement of material between the axon terminal and the cell body (Ch 8), 240t, 244

axon diameter, 259

axon hillock Region of the axon where it joins the cell body. Often contains the trigger zone (Ch 8), 244

axon terminal The distal end of a neuron where neurotransmitter is released into a synapse (Ch 8, 11), 240t, 244, 257–258, 382

axoplasm Cytoplasm of the axon, 240t

axoplasmic flow Movement of cytoplasm in the axon. Used for slow axonal transport, (Ch 8), 244

AZT, 753

bacteria, 659, 687, 752, 753t

bacterial infection, 526, 768, 769f

bar graph, 12, 13

baroreceptor Stretch-sensitive mechanoreceptors that respond to changes in pressure (Ch 15), 510, 511

baroreceptor reflex The primary reflex pathway for homeostatic control of blood pressure (Ch 15, 25), 510–512, 511f, 513f, 788–789

Barr body The inactivated X chromosome in each female cell (Ch 26), 798

barrier method Contraception based on putting a physical or chemical barrier between egg and sperm (Ch 26), 820

basal body Cytoplasmic structure where microtubules of cilia and flagella terminate (Ch 3), 58

basal ganglia Nuclei surrounding the thalamus that help with planning movement (Ch 9), 299f, 300

basal lamina An acellular layer of extracellular matrix that lies beneath an epithelium, holding the epithelial cells to underlying cell layers. Synonym: basement membrane (Ch 15, 19, 26), 493, 606, 806, 811, 815t

basal metabolic rate (BMR) An individual's lowest metabolic rate (Ch 22), 698–699

basal nuclei, 440, 441

base (1) A molecule that decreases the H$^+$ concentration of a solution by combining with free H$^+$; (2) A carbon-nitrogen molecule with a ring structure that is an essential component of a nucleotide (Ch 2, 20), 33, 646

basement membrane. *See* basal lamina.

base pairing, 112, 838

bashful bladder, 618

basilar membrane Membrane that separates the cochlear duct from the tympanic duct. It supports the organ of Corti (Ch 10), 343, 344f, 345

basolateral membrane The sides of transporting epithelial cells that face the extracellular fluid. Synonym: serosal membrane (Ch 3, 5, 21), 69, 147, 667, 668

basolateral surface, 664

basophil Leukocyte that releases histamine, heparin, and other chemicals for the immune response (Ch 16, 24), 526, 755, 757, 757f

Bayliss, W. M., 672

B cell, 763–764, 766, 768

Beaumont, William, 660

behavior, 285

behavioral state system, 301, 304–305

benign positional vertigo (BPV), 349

benign prostatic hypertrophy, 806, 810

benzoate, 599

Bernard, Claude, 188

Berthold, A. A., 209, 210

beta-adrenergic receptor (β-adrenergic receptor) Membrane receptors that bind and transduce signals from norepinephrine and epinephrine (Ch 6), 185

beta-blocker Drugs that are beta-adrenergic receptor antagonists. Used to treat hypertension (Ch 11, 14), 378, 483

beta cells of the pancreas Endocrine cells that secrete insulin (Ch 5), 162, 705

β-hydroxybutyric acid Strong metabolic acid produced from beta-oxidation of fats, 705

beta-interferon, 760

beta-oxidation Process in which fatty acids are broken down by the sequential removal of 2-carbon units (Ch 4, 12), 109, 403

β-pleated sheet Sheet-like structure formed from some chains of amino acids (Ch 2), 38, 39f, 42f

bicarbonate buffer, 34, 646–647

bicarbonate-chloride antiporter Membrane transporter that exchanges one Cl$^-$ for one used for the chloride shift (Ch 18), 585f, 586

bicarbonate ion (HCO$_3^-$), 585–586, 646–648

bicarbonate secretion, 681

biceps, 390–391

bicuspid valve The left AV valve of the heart. Synonym: mitral valve (Ch 14), 459

bile A solution secreted by the liver and composed primarily of bile acids, bile pigments and cholesterol (Ch 16, 21), 533, 668, 669, 680, 681

bile acid Steroid detergents made from cholesterol by the liver (Ch 21), 681

bile canaliculi Liver channels through which bile is secreted (Ch 21), 682f

bile salt Bile acids conjugated with amino acids (Ch 21), 681, 683, 685

bilirubin Breakdown product of heme groups from hemoglobin (Ch 16, 21), 533, 685

binding site, 91, 581

binocular vision Three-dimensional vision from overlapping visual fields of two eyes (Ch 10), 361f

binocular zone, 361

bioartificial pancreas, 77

bioelectrical principles, 836

bioenergetics, 88

biological rhythm The cyclic variation of a biological process (Ch 5), 196–197

biomolecule Organic molecules associated with living organisms: carbohydrates, lipids, proteins and nucleotides (Ch 2), 34–35

biophysics, 835

biorhythm. *See* biological rhythm.

biosphere, 3

biotin, 94

bipolar neuron Neuron with a single axon and single dendrite (Ch 8, 10), 242, 355, 360

bipotential gonad Embryonic tissue that has the potential to develop into either testis or ovary (Ch 26), 798, 799f

birth control, 820–821

birth control pill, 821

bisphosphonates, 746

blackwater fever, 599

bladder. *See* urinary bladder.

blastocyst Early embryo, consisting of a hollow ball of cells (Ch 26), 823, 824

bleaching of visual pigment Process in which light activation causes opsin to release retinal (Ch 10), 359

bleb Weakened section of the lung tissue (Ch 3), 76

blind spot Region of the retina with no photoreceptors because the optic nerve and

coenzyme Organic cofactors that act as receptors and carriers for atoms or functional groups that are removed from substrates during the course of a reaction (Ch 4), 94

cofactor An inorganic or nonprotein organic molecule required for activation of an enzyme (Ch 4), 93, 94

cognitive behavior Behaviors that deal with thought processes rather than emotion (Ch 9), 285

language and, 312–313

cognitive system, 301

coitus Synonym: intercourse (Ch 26), 818

cold receptor, 332

colipase A protein cofactor that allows lipase to break through the bile salt coating of an emulsion (Ch 21), 685

collagen Flexible but inelastic protein fibers of connective tissue (Ch 3, 16, 26), 64, 72, 73f, 538f, 540t, 817

collagenase Enzyme that degrades collagen (Ch 26), 817

collateral Branch of an axon (Ch 8), 244

collecting duct Terminal region of the kidney tubule (Ch 19), 603

College of American Pathologists, 526

colligative property, 153–154, 836

colloid, 733

colloid osmotic pressure (π) Osmotic pressure that is due to the presence of plasma proteins that cannot cross the capillary endothelium. Synonym: oncotic pressure (Ch 15, 19), 506, 509, 606

colon Proximal portion of the large intestine. Divided into ascending, transverse, descending, and sigmoid regions (Ch 21), 661, 685–686

colonocyte Transporting epithelial cell of the large intestine (Ch 21), 686, 687

colony-stimulating factor (CSF) Cytokines made by endothelial cells and white blood cells that direct the production and development of white blood cells (Ch 16, 24), 528–530, 529t, 768

color-blindness, 357, 798

color vision, 820

colostrum Watery, low-fat secretions of the mammary gland prior to delivery (Ch 26), 826–827

column, of cerebral cortex, 293–294

columnar epithelia, 67

common bile duct, 681, 682f

common hepatic duct, 682f

compact bone, 740, 740f

comparative endocrinology, 231

compartmentation The internal division of the body or cell into compartments so that functions can be isolated from one another (Ch 1), 9

competition, 138–139

competitive inhibitor Molecules that bind to the active site of the enzyme, preventing substrate binding (Ch 4, 5), 95, 95f, 139

complement A group of plasma enzymes that are involved in immune function (Ch 24), 761–762, 761t, 765

complete (fused) tetanus Sustained maximal contraction of a muscle in response to repeated stimuli (Ch 12), 407f

complete heart block Pathological condition in which electrical signals cannot pass from the atria to the ventricles (Ch 14), 470

compliance The ability of the lung or other tissue to stretch (Ch 17), 559

compound molecule A molecule that contains more than one element (Ch 2), 26

computerized tomography (CT scan), 26

concave lens, 353, 353f

concentration The amount of solute per unit volume of solution (Ch 2), 32

concentration gradient A difference in the concentration of a substance between two places (Ch 4, 5, 8, 18), 86, 87, 135, 246, 575

concentric action, 409

concept map. *See* mapping for physiology.

condom, 820

conducting system The airways that lead from the environment to the exchange surface of the lungs (Ch 17), 548

conduction (1) The movement of an action potential through the axon at high speed; (2) Transfer of heat from one substance to another. (Ch 8), 257–258, 258f, 260–261

conductive hearing loss Hearing loss due to failure of sound transmission though outer or middle ear (Ch 10), 346

conductive heat gain, 715

conductor, 157, 836

cone A photoreceptor for high acuity vision and color vision during the daytime (Ch 10), 357

congestive heart failure Pathological condition in which the left ventricle fails to pump blood adequately, causing backup of fluid into the lungs (Ch 15), 516

conjugated protein Molecules of protein combined with either lipid or carbohydrate (Ch 2), 40

connective tissue, 71, 75t

structure of, 71–72, 73f, 75f

types of, 72–74, 72t

connexin Membrane-spanning proteins that form gap junctions; capable of opening and closing (Ch 6), 172, 173

connexon The protein channel of a gap junction, made of connexins (Ch 3, 6), 64, 172

consciousness, 305

consensual reflex Light shined in one pupil constricts both pupils (Ch 10), 352

consolidation Process that converts short-term memory to long-term memory (Ch 9), 311

constipation, 670

constitutive Any essential bodily function that is always taking place (Ch 5), 145

constitutive process (Ch 5), 147

contact-dependent signals Cell-cell signals that require surface molecule binding between two cells (Ch 6), 172, 172f, 173

continuous capillary Capillary whose endothelial cells are tightly joined (Ch 15), 504

contraception. *See* birth control.

contractile fibers, 432

contractility The intrinsic ability of a cardiac muscle fiber to contract at any given fiber length (Ch 14), 481–482

contraction, muscle. *See* muscle contraction.

contraction-relaxation cycle, 396

contralateral On the opposite side from (Ch 10), 346

control Part of an experiment designed to ensure that any observed changes are due to the experimental manipulation and not to an outside factor (Ch 1), 10

control pathways, 190–200, 199f, 224f, 803–804

control system, 197, 197f, 198t

coding for stimulus intensity, 200

duration of action, 198, 200

nature of signal, 198

specificity of, 197–198

convective heat loss Heat that is carried away by warm air that rises (Ch 22), 717

convergence A number of presynaptic neurons provide input to a smaller number of postsynaptic neurons (Ch 10, 13), 325, 429

convex lens, 353, 353f

COPD. *See* chronic obstructive pulmonary disease.

cornea The clear covering of the anterior surface of the eye (Ch 10), 351

coronary artery Artery supplying blood to the heart muscle (Ch 14), 451, 456

coronary heart disease, 51, 483, 494, 512, 760

coronary sinus, 451

coronary vein, 451, 456

corpora cavernosa Two columns of erectile tissue in the penis (Ch 26), 805

corpus albicans The remnants of a degenerated corpus luteum (Ch 26), 817

corpus callosum The central region where neurons pass from one hemisphere of the cerebrum to the other (Ch 9), 300

corpus luteum Ovarian structure that produces estrogen and progesterone after ovulation (Ch 26), 811, 817–818

corpus spongiosum A column of spongy erectile tissue in the penis (Ch 26), 805

corpus striatum. *See* basal ganglia.

cortex Literally, bark; the outer or surface portion of an organ (Ch 19), 602

cortical collecting duct, 636

cortical granules Cytoplasmic granules in the egg that contain chemicals to prevent polyspermy (Ch 26), 822

cortical reaction Chemical reaction that changes the zona pellucida after fertilization so that additional sperm cannot reach the egg (Ch 26), 822

corticospinal tract Neurons from motor cortex to spinal cord (Ch 9, 13), 297, 441, 442f

corticosteroid-binding globulin, 217, 729

corticosterone, 729f

corticotropin. *See* adrenocorticotropic hormone.

corticotropin-releasing hormone (CRH) Hypothalamic hormone that regulates secretion of ACTH from the anterior pituitary (Ch 7, 22, 23, 26), 228, 229, 696, 728–729, 731–732, 825

cortisol Steroid hormone from the adrenal cortex that regulates metabolism, particularly during stress (Ch 2, 7, 9, 23), 37f, 217, 224f, 225, 230f, 307, 728–731, 729f, 737

Corynebacterium diphtheriae, 766

co-secretion Secretion of more than one compound from a secretory vesicle (Ch 7), 215

cotransporter A protein that moves more than one kind of molecule at one time (Ch 5), 138

Coulomb's law, 836

coumarin anticoagulant, 541

countercurrent exchange system Anatomical arrangement of vessels so that flow in one vessel is in the opposite direction from flow in the adjacent vessel (Ch 20), 633

FAD (flavin adenine dinucleotide) Molecule that captures and transfers energy with high-energy electrons (Ch 2), 41
FADH$_2$, 106, 107
failure to thrive, 220
fainting, 512
falling phase of action potential, 254
Fallopian tube Tube that transport eggs from the ovary to the uterus (Ch 26), 798, 800, 811, 817, 821, 826
Fas A "death receptor" on cell membranes whose activation causes a cell to commit suicide by apoptosis (Ch 24), 767
fast axonal transport Rapid movement of particles along an axon using microtubules and kinesin foot proteins (Ch 8), 244
fasted state. *See* postabsorptive state.
fasting, 701
fast pain Sharp, rapidly transmitted pain (Ch 10), 333
fast synaptic potential Graded potential in postsynaptic cells that begins quickly and lasts only a few milliseconds (Ch 8), 268
fast-twitch glycolytic fiber Fast muscle fibers that rely on anaerobic metabolism and therefore fatigue rapidly (Ch 12), 404, 405f, 405t
fast-twitch muscle fiber Muscle fibers that develop tension rapidly (Ch 12), 404, 405t
fast-twitch oxidative fiber Fast muscle fibers that use a combination of aerobic and anaerobic metabolism and therefore do not fatigue as fast as glycolytic fibers (Ch 12), 404, 405t
fat, 668, 668t, 669, 680, 699, 703f. *See also* lipid.
fat absorption, 684–685
fat digestion, 669, 684–685
fatigue, muscle Inability of a muscle to continue to generate or sustain tension (Ch 12), 403–405
fat substitute, 676
fat synthesis, 710
fatty acid Long chain of carbon atoms bound to hydrogens and terminating with a carboxyl (Ch 2, 22), 35, 37f, 700, 704
fatty acid pool, 700–701
fatty acid synthetase Enzyme for fatty acid synthesis (Ch 4), 111
fatty streak, 514, 515f
F cell Pancreatic endocrine cell that secretes pancreatic polypeptide (Ch 22), 705
Fc region Stem of antibody molecule that binds to receptors on immune cells (Ch 24), 764, 765
feces, 661
fed state. *See* absorptive state.
feedback inhibition The end product of a metabolic pathway acts as an inhibitory modulator of the pathway. Synonym: end-product inhibition (Ch 4), 101
feedback loop Information about a homeostatic response that is sent back to the integrating center (Ch 6, 7), 194–195, 223–224
feedforward control Anticipatory responses that start a response loop in anticipation of a change that is about to occur (Ch 6), 196
feedforward reflex, 429, 440, 441, 442f, 670, 789
feeding. *See* eating.
feeding center Tonically active hypothalamic center that promotes food intake (Ch 22), 696

fenestrated capillary Capillary with large pores in the endothelium (Ch 15, 19), 504, 605
ferritin Protein that binds and stores iron in the body (Ch 16), 533
fertilization, 798, 818, 821–822, 823f
fetal development, 825
fetal hemoglobin (HbF), 532, 535, 584f
fever, 720, 758
F$_1$F$_0$ ATPase. *See* ATP synthase.
fibrillation, 478
fibrillin Straight protein fibers of connective tissue (Ch 3), 72
fibrinogen Plasma protein that becomes fibrin in blood clots (Ch 16), 524, 525f, 536, 540t
fibrinolysin. *See* plasmin.
fibrinolysis Dissolution of fibrin by plasmin (Ch 16), 539f, 540, 541t
fibrinolytic drug, 540
fibrin Plasma protein that forms polymer fibers that stabilize platelet plugs (Ch 16), 536, 539, 540, 540f
fibrin-stabilizing factor, 540t
fibroblast, 72, 559
fibronectin A protein fiber that helps connect cells to their extracellular matrix (Ch 3), 64, 66, 72
fibrotic lung disease, 559, 578
Fick's law of diffusion Diffusion through a membrane is directly proportional to the surface area and concentration gradient and inversely proportional to the thickness of the membrane and its resistance (Ch 5, 17), 137, 137f, 575
fight-or-flight response, 370, 381, 502, 503, 670, 727, 730, 775
filgrastim, 530
filtration Bulk flow of plasma-like fluid from the glomerular capillaries into Bowman's capsule (Ch 15, 17, 19), 506, 507f
of air, 554
function of, 605–610
in nephron, 603–604, 603f
filtration coefficient, 607
filtration fraction The percentage of total plasma volume that filters at the glomerulus (Ch 19), 606, 606f
filtration rate, 615
filtration slit Opening between podocyte foot processes through which renal filtration takes place (Ch 19), 606
fimbriae The fringed opening of the Fallopian tube (Ch 26), 811
first heart sound Sounds created by vibrations from closure of AV valves (Ch 14), 476
first law of thermodynamics Energy in the universe is constant (Ch 4), 87
first meiotic division, 801
first messenger Chemical signal molecules released by cells (Ch 6), 174
first polar body, 801
fistula, 660
5-alpha-reductase Enzyme that converts testosterone to DHT (Ch 26), 800
fixed cell, of connective tissue, 72
fixed ribosome, 53
flagellum A long hair-like extension of the cell whose microtubules create movement (Ch 3), 58, 59f
flatus Intestinal gas (Ch 21), 687
flavin adenine dinucleotide. *See* FAD.
Fleming, Alexander, 614

flexion reflex A polysynaptic reflex that causes an arm or leg to be pulled away from a painful stimulus (Ch 13), 436–438
flexor A muscle that brings connected bones closer together when it contracts (Ch 12), 390
flow Synonym: flow rate (Ch 8), 244
flow rate The volume of blood that passes one point in the system per unit time (Ch 14), 455
fluid balance, 512, 626–627, 638
fluid mosaic model Membrane composed of phospholipid bilayer with proteins inserted wholly or partially into the bilayer (Ch 5), 126, 127f, 129
fluid pressure Pressure created by the presence of fluid within an enclosed space (Appendix A), 837
flutamide An androgen receptor-blocking drug (Ch 6), 185
focal adhesion Junction between intracellular actin and matrix proteins (Ch 3), 66
focal distance The distance from the center of a lens to the focal point (Ch 10), 353
focal length, 353
focal point The point where parallel light waves passing through a lens converge (Ch 10), 353
folic acid, 94
follicle (1) Group of cells in the ovary that contains a developing egg; (2) Group of thyroid cells that produce thyroid hormones (Ch 23), 733
follicle-stimulating hormone (FSH) Anterior pituitary hormone that stimulates gamete production in the gonads (Ch 7, 26), 222, 225, 803, 808, 817
follicular phase Phase of the menstrual cycle during which ovarian follicles mature and prepare to release an egg (Ch 26), 811, 815, 817
food intake, 696–697
foot process Long cytoplasmic extension of a podocyte that wraps around glomerular capillaries (Ch 19), 606
force, 835
forebrain region, 286
foreskin, 805
fovea The region of most acute vision and the point on which light is focused when you look at an object (Ch 10), 351, 351f, 355, 356f
Framingham Heart Study, 513
Frank, Otto, 481
Frank-Starling law of the heart The principle that within physiological limits, the heart will pump all the blood that returns to it (Ch 14), 481
free energy The amount of energy stored in the chemical bonds of a molecule and available to perform work (Ch 4), 88
free nerve ending, 332t
free radical Unstable molecule with one more unpaired electrons (Ch 2), 28, 232
frequency coding The frequency of action potentials encodes the intensity of a stimulus (Ch 10), 327
friction rub, 563
frontal lobe, 302f, 304
fructose, 35, 36f, 810t
FSH. *See* follicle stimulating hormone.
fulcrum, 410

glycogenolysis The breakdown of glycogen (Ch 4, 22), 108, 701, 710

glycohemoglobin, 532

glycolipid Molecule that is a combination of carbohydrate and lipid (Ch 2, 5), 40, 41, 127f

glycolysis Metabolic pathway that converts glucose to pyruvate (aerobic) or lactic acid (anaerobic) (Ch 4, 16, 22), 102–104, 103f, 104t, 110, 531, 700, 710

glycolytic metabolism, 783–784

glycoprotein Molecule that is a combination of carbohydrate and protein (Ch 2, 3, 5), 40–41, 64, 127f, 132

glycosuria. *See* glucosuria.

GnRH. *See* gonadotropin releasing hormone.

GnRH agonist, 821

goblet cell Single exocrine cells that produce mucus (Ch 3, 17, 21), 70, 71f, 554, 667

goiter Enlarged thyroid gland (Ch 7, 23), 209, 735–736, 736f

Goldman-Hodgkin-Katz (GHK) equation Calculates resting membrane potential using membrane permeability and ion concentrations gradients (Ch 8), 248

G_{olf} G protein for olfactory transduction (Ch 10), 336

Golgi, Camillo, 61

Golgi apparatus Organelle that modifies and packages proteins into vesicles (Ch 3, 7), 61, 61f, 62, 215

Golgi tendon organ Receptors are found at the junction of the tendons and muscle fibers that respond to both stretch and contraction of the muscle (Ch 13), 435–436

Golgi tendon reflex, 435f

gonad The organs (ovaries and testes) that produce gametes (Ch 3, 7, 26), 70, 216, 800t

gonadotropin (FSH and LH) Peptide hormone from the anterior pituitary that acts on the gonads (Ch 7, 26), 225, 803, 804–805, 808, 828

gonadotropin releasing hormone (GnRH) Hypothalamic hormone that stimulates release of gonadotropins from the anterior pituitary (Ch 26), 803, 804, 808, 815

gout, 139, 598, 603, 612, 619

G protein Membrane proteins that couple membrane receptors to ion channels or membrane enzymes (Ch 6), 179, 187

G protein-coupled adenylyl cyclase-cAMP system The first signal transduction system discovered (Ch 6), 179, 180f

G protein-coupled receptor, 178–179, 181, 697

graded contraction Muscle contraction whose force varies with the amount of Ca^{2+} that enters the cell (Ch 14), 462

graded potential A change in membrane potential whose magnitude is proportional to the stimulus and that decreases with distance as it spreads through the cytoplasm (Ch 8), 249, 250, 250t, 251

gradient A pressure, concentration, or electrical difference (Ch 1), 9

gram molecular weight The weight of one mole of a substance, expressed in grams (Ch 2), 32

granulocyte White blood cells whose cytoplasmic inclusions give them a granular appearance: basophils, eosinophils, and neutrophils (Ch 16, 24), 526, 756

granulosa cell Cell of the ovarian follicle that secretes estrogen (Ch 26), 815t

granzyme Enzyme of cytotoxic T cells that triggers apoptosis in target cells (Ch 24), 761t

graphs, 12–13

Graves' disease Hyperthyroid disorder caused by TSH-like antibodies (Ch 7, 23, 24), 208, 227, 230, 231, 233, 736, 773t

gray matter Nerve cell bodies, dendrites, and axon terminals (Ch 9), 287, 300

ground, 158

ground substance A cellular portion of matrix consisting of glycoproteins and water (Ch 3), 71–72, 75f. *See also* matrix

growth, 736, 738

growth cone Tips of growing axons (Ch 8), 274

growth factor, 212

growth hormone Protein hormone from the anterior pituitary that controls tissue growth (Ch 7, 23, 26), 222, 225, 226f, 736–739, 738f, 739t, 824

growth hormone binding protein, 738

growth hormone-inhibiting hormone. *See* somatostatin.

growth hormone-releasing hormone (GHRH) Hypothalamic hormone that influences growth hormone secretion (Ch 23), 738

GTP. *See* guanosine triphosphate.

guanine, 41, 112, 838

guanosine diphosphate (GDP), 179

guanosine triphosphate (GTP), 101, 179

guanylyl cyclase Enzyme that controls formation of cyclic GMP (Ch 6, 26), 178, 820

guided imagery, 430

Guillain-Barré syndrome Rare autoimmune paralytic neural condition with loss of both sensory and motor function (Ch 8, 24), 239, 241, 244, 261, 276, 773t

gustation. *See* taste.

gustatory cortex, 304, 339

gut-associated lymphoid tissue (GALT) Immune cells and tissues of the GI tract (Ch 21, 24), 661, 664, 754

gyrus Convolution of cerebral surface (Ch 9), 299

habituation A decreased response to a stimulus that is repeated over and over (Ch 9), 310

hair cell Sensory cells for transduction of sound and equilibrium (Ch 10), 322, 323f, 343, 345f, 349

hair follicle, 78

half-life The amount of time required to reduce the concentration of hormone by one-half (Ch 6, 7), 183, 213, 216–217

Hanta virus, 752

haploid, 840

Harvey, Sir William, 449, 524

haustra Bulging pockets of the large intestine wall (Ch 21), 686

HbS Abnormal hemoglobin of sickle cell disease (Ch 16), 534

hCG. *See* human chorionic gonadotropin.

hCS. *See* human chorionic somatomammotropin.

HDL. *See* high-density lipoprotein.

head trauma, 314

health, exercise and, 790–792

hearing, 327, 341–346

hearing loss, 346

heart Muscular organ that serves as the pump for the circulatory system (Ch 14), 448–483
  anatomy of, 450–452, 457f, 458f
  cardiac muscle and, 421, 456, 459–465
  electrical conduction in, 468, 469f, 470, 473f
  embryological development of, 459f
  as pump, 468–482, 499f

heart attack. *See* myocardial infarction.

heartburn. *See* reflux esophagitis.

heart failure, 463, 509

heart rate, 189f, 466–468, 480

heart valve Connective tissue valves that prevent back flow of blood in the heart (Ch 14), 459, 460f

heat balance, 715, 716f

heat exhaustion, 720

heat index Synonym: humidex (Ch 22), 718

heat production, 715, 719

heat stroke, 720, 782, 790, 792

Hebeler, Peter, 592

Helicobacter pylori Bacterium associated with peptic ulcer disease (Ch 21), 659, 665, 667, 679, 689

helicotrema, 343, 344f

helper T cell Immune cells that secrete cytokines to help other immune cells (Ch 24), 757f, 768

hematocrit Percentage of the total blood volume that is packed red blood cells (Ch 16), 527f, 530, 563

hematological disorder, 524

hematology, 526

hematopoiesis Blood cell production in the bone marrow (Ch 16), 526–529, 528f

heme A carbon-hydrogen-nitrogen porphyrin ring with an iron atom in the center (Ch 16), 532, 533

hemidesmosome Strong junction that ties a cell to matrix (Ch 3), 66, 78

hemispherectomy, 312

hemoglobin Oxygen-carrying pigment of red blood cells (Ch 16, 18), 526, 527f, 532f, 533, 534
  binding of oxygen, 582–583, 583f
  carbon dioxide and, 586
  H^+ and, 586
  oxygen transport by, 580–582
  percent saturation of, 582
  synthesis of, 532–533

hemoglobin A_{1c}, 532

hemolytic anemia, 534

hemophilia, 541

hemorrhage Excessive blood loss (Ch 16, 20), 535, 642

hemostasis Process of keeping blood within the blood vessels by repairing breaks without compromising the fluidity of the blood (Ch 16), 535–536, 537f

heparin An anticoagulant molecule (Ch 14, 16, 24), 461, 541, 757, 761t

hepatic artery, 451

hepatic portal system Specialized region of the circulation that transports material absorbed at the intestine directly to cells of the liver (Ch 21), 680f

hepatocyte Liver cell (Ch 21, 22), 681, 709

hereditary spherocytosis, 534

Hering-Breuer inflation reflex Reflex to prevent overinflation of the lungs (Ch 18), 592

herpes virus, 753

hertz (Hz) Measure of sound wave frequency (Ch 10), 341
hexavalent ion, 33
hexokinase, 709
hexosaminidase Enzyme whose absence causes Tay-Sachs disease (Ch 4), 91, 98
hexose A six-carbon sugar (Ch 5), 138
high-altitude acclimatization, 574, 582, 592
high-altitude cerebral edema, 579
high-altitude pulmonary edema (HAPE), 579
high blood pressure. See hypertension
high-density lipoprotein (HDL) The "good" plasma carrier for cholesterol (Ch 15, 22), 514, 703
high-energy electron, 25, 106
high-energy phosphate bond, 101, 402
Hillary, Sir Edmund, 592
hinge region, 764
hippocampus Portion of the brain associated with learning and memory (Ch 9), 300, 300f, 310
Hippocrates, 2
histamine Paracrine secreted by mast cells and basophils; acts as a vasodilator and bronchoconstrictor (Ch 6, 7, 8, 15, 17, 21, 24), 173, 213, 265t, 501, 562, 689, 760–761, 761t
histamine receptor, 680
histamine receptor blocker, 680
histiocyte Old name for skin macrophages (Ch 24), 756
histogram Graph that quantifies the distribution of one variable over a range of values (Ch 1), 12, 13
histology The study of tissue structure and function (Ch 3), 63–64
histone, 840
histoplasmosis, 752
HIV, 751–753, 768, 776. See also AIDS.
HIV positive test, 771
HLA. See human leukocyte antigen.
homeostasis The ability of the body to maintain a relatively constant internal environment (Ch 1, 6, 9, 11, 15, 16, 19, 20), 7–8, 7f, 172, 190, 370f, 371, 509, 529
  control systems, 8, 187–189
  dehydration and, 644f
  diencephalon and, 298
  fluid and electrolyte, 626–627, 640–643
  local control, 190–191
  reflex control, 191–193
  salt, 635, 640–642
homeostatic regulation, 547, 599
homeothermic Mammals and birds that maintain relatively constant body temperature (Ch 22), 715
honey-urine disease, 599
Hooke, Robert, 51
hormone A chemical secreted by a cell or group of cells into the blood for transport to a distant target where it acts in very low concentrations to affect growth, development, homeostasis, or metabolism (Ch 3, 6, 7, 12, 15, 19, 22, 23, 24, 25), 70, 173, 173f, 209, 210f
  antagonistic, 227
  arteriolar resistance and, 499
  bound to target receptors, 213
  calcium balance and, 743–745
  cellular mechanism of action, 213
  classic, 211
  classification of, 214–218

concentration of, 213
control of release of, 198, 219–225
definition of, 211–212
endocrine pathologies, 227–229
enzyme activity and, 701–702
evolution of, 230–231
exercise and, 784–785
glomerular filtration rate and, 608, 610
growth, 736–739
historical background, 209–211
immune cell function and, 774
interactions of, 226–227
list of, 211
metabolic rate and, 699
muscle contraction and, 421
pathologies of, 227–230
permissiveness of, 227
production of, 599
replacement therapy, 746, 828
secreted by placenta, 824
secretion of, 212, 225
sex, 240, 727, 728, 797, 803, 811
terminating action of, 213–214
thyroid, 214t, 218, 699, 732–736
tissue growth and, 739–740
transport of, 212–213
hormone receptor, 727
hormone replacement therapy, 11, 746
hormone-response element, 729
horseradish peroxidase (HRP), 286
hot flash, 719, 828
HRT. See hormone replacement therapy
H₂ receptor. See histamine receptor.
5-HT. See serotonin.
human chorionic gonadotropin (hCG) Hormone secreted by the developing placenta (Ch 26), 824
human chorionic somatomammotropin See human placental lactogen (hPL)
Human Genome Project (HGP), 2, 85, 116
human immunodeficiency virus (HIV). See HIV
human leukocyte antigen (HLA) Name for classification of human MHC proteins (Ch 24), 771–772
human placental lactogen (hPL) Peptide placental hormone that influences maternal metabolism. Synonym human chorionic somatomammotropin (hCS)
Human Proteomics Initiative, 85
humidex. See heat index
humidity, 785, 792
humoral immunity Immunity conferred by antibodies (Ch 24), 754, 764
Huxley, Sir Andrew, 396
hydraulic pressure Pressure exerted by fluid in motion. Uses synonymously with hydrostatic pressure in the circulatory system (Ch 14), 452
hydrogen, 23–24, 24f, 27f
hydrogen bond Weak attractive forces between hydrogens and other atoms, especially oxygen and nitrogen (Ch 2), 26, 31
hydrogen ion, 33–34, 649, 650
hydrogen-potassium ATPase, 649–650
hydrolysis-dehydration reactions, 98–99, 99t
hydrolysis Reaction in which large molecules are broken into smaller ones by addition of water (Ch 4), 98, 99t
hydrophilic molecule Molecules that dissolve readily in water (Ch 2), 32
hydrophobic molecule Molecules that do not dissolve readily in water (Ch 2), 32

hydrostatic pressure The pressure exerted by a stationary column of fluid in a tube (Ch 14, 15, 16, 19), 452, 506, 509, 552, 606
hydroxyapatite Calcium phosphate crystals of bone (Ch 23), 740, 745
21-hydroxylase, 729f
5-hydroxytryptamine. See serotonin.
hydroxyurea, 535
hymen, 811
hyperbilirubinemia, 533
hypercalcemia, 187t
hypercapnia Elevated P_{CO_2} in the blood (Ch 18), 576, 584
hypercholesterolemia, 145, 704
hypercortisolism, 731
hyperemia, 500, 501f
hyperglycemia, 711, 712, 714
hyperinsulinemia, 229
hyperkalemia, 262, 639
hyperkalemic periodic paralysis, 403, 412, 422
hyperopia Far-sightedness (Ch 10), 354, 355f
hyperosmotic solution, 153
hyperparathyroidism, 333, 726, 727, 739, 745, 746
hyperplasia Increased cell number due to cell division (Ch 23), 740
hyperpnea Increase in ventilation rate to match an increase in metabolic rate. Synonym: exercise hyperventilation (Ch 17, 25), 565, 786
hyperpolarization A membrane potential that is more negative than the resting potential (Ch 8, 10), 248, 324
hyperpolarization-activated cyclic nucleotide-gated channel (HCN channel), 467
hypersecretion, 228
hypersensitivity, 771
hypertension Chronically elevated blood pressure (Ch 14, 15, 24), 452, 496, 501, 514–516, 760
hyperthermia, 720
hyperthyroidism, 734–735, 737f
hypertonic solution A solution that causes net movement of water out of a cell (Ch 5), 154
hypertrophy An increase in cell size without an increase in cell number (Ch 23), 735, 740
hyperuricemia, 603, 619
hyperventilation An increase in alveolar ventilation that is not associated with an increase in metabolic rate (Ch 17, 18), 565t, 566, 591
hypocalcemia, 741–742
hypochromic, 534
hypocortisolism, 231f
hypocretin, 697
hypodermis. See subcutis.
hypoglossal nerve, 298t
hypoglycemia, 293, 705, 730
hypokalemia, 262, 639, 642
hypokalemic periodic paralysis, 403, 421
hyponatremia, 642
hypoprolactinemia, 828
hyposecretion, 228
hyposmotic solution, 153
hypotension, 496, 512
hypothalamic-anterior pituitary pathway, 223f, 224f
hypothalamic-hypophyseal portal system Modified section of the circulation that takes neurohormones directly from the hypothalamus to the anterior pituitary (Ch 7, 9), 225, 225f, 299

hypothalamic trophic hormone, 727

hypothalamus Region of the brain that contains centers for behavioral drives and plays a key role in homeostasis (Ch 9, 15), 298–299, 304, 511, 804–805

hypothermia Low body temperature (Ch 22), 720

hypothesis, 10

hypothyroidism, 735

hypotonic solution A solution that causes a net influx of water into a cell (Ch 5), 154

hypoventilation A decrease in alveolar ventilation without a change in metabolic rate (Ch 17, 18), 565–566, 565t, 578

hypovolemic shock, 498

hypoxia Lack of oxygen in the cells (Ch 14, 15, 16, 18), 450, 501, 529, 534, 576, 577f, 579, 582, 591

hz. *See* hertz.

H zone Region of sarcomere with only thick filaments (Ch 12), 394

iatrogenic condition Physician-caused condition (Ch 7), 228

iatrogenic hypercortisolism, 731

I band Region of the sarcomere occupied only by thin filaments (Ch 12), 394

I cell. *See* intercalated cell.

IDDM. *See* type 1 diabetes mellitus.

ideal gas law, 837

I$_f$ channel Monovalent cation channels in cardiac autorhythmic cells that contribute to the pacemaker potential (Ch 14), 466

Ig. *See* immunoglobulin.

IGF. *See* insulin-like growth factor.

Ignarro, Louis, 502

IL. *See* interleukin.

ileocecal valve Muscular region whose contraction separates the large and small intestines (Ch 21), 685

ileum Distal portion of the small intestine (Ch 21), 661

immediate hypersensitivity reaction Allergic reaction that occurs within minutes (Ch 24), 771

Immune cells, 755–757, 757f, 765, 774

immune response, 753–754, 761t, 763–764

immune response pathways, 768–774

immune surveillance Theory that cancer cells develop regularly but are usually detected and destroyed by the immune system (Ch 24), 774

immune system The cells and tissues and their products that defend the body against invaders (Ch 1, 7, 9, 23, 24, 25), 4, 752–776

  acquired immunity and, 762–768

  anatomy of, 754–758, 755f

  cells of, 755–757, 757f, 765

  exercise and, 791–792

  functions of, 3, 775–776

  neuro-endocrine interactions with, 774–776

  response pathways and, 768–774

  sleep and, 307

  stress and, 220, 731, 775–776, 791–792

  suppression of, 730

immunity The ability of the body to protect itself from pathogens (Ch 24), 752, 762

immunization *See* vaccination

immunocyte General name given to any of the immune cells (Ch 16), 526

immunocytochemistry Laboratory technique in which antibodies to a molecule are produced, then tagged with a fluorescent or radioactive marker (Ch 7), 213

immunodeficiency disease, 752

immunoglobulin Synonym for antibody (Ch 16, 21, 24, 26), 524, 673, 674, 688, 763, 827

immunoglobulin A (IgA), 764

immunoglobulin D (IgD), 764

immunoglobulin E (IgE), 764

immunoglobulin G (IgG), 764

immunoglobulin M (IgM), 764

immunosuppressant effect, 730

impermeable membrane A membrane that does not allow substances to cross (Ch 5), 125

implicit memory. *See* reflexive memory.

impotence, 379, 819, 820

inactivation gate The slow gate of the Na$^+$ channel that closes to stop ion flow (Ch 8), 254

inclusion Particle of insoluble material in the cytoplasm such as glycogen granules and lipid droplets (Ch 3), 53

incontinence Inability to voluntarily control urination or defecation (Ch 11), 379

incus Middle of the three small bones of the middle ear (Ch 10), 340f, 341

independent variable The parameter manipulated by the investigator in an experiment; graphed on the x-axis (Ch 1), 10

indirect calorimetry Estimation of metabolic rate by measuring oxygen consumption (Ch 22), 698

induced-fit model of enzyme activity The active site changes shape to fit either substrate or product molecules (Ch 4), 91

infantile spasm, 293, 314

infarct An area of tissue that is damaged or dead due to inadequate blood supply explanation of(Ch 16), 540

inferior vena cava Great vein that returns blood from the lower body to the right atrium (Ch 14), 451

infertility Inability to conceive (Ch 25), 796, 806, 818, 821, 826, 829

inflammation A nonspecific reaction of the immune system to a foreign invader (Ch 15, 24), 501, 509, 754, 760, 768

inflammatory bowel diseases, 688

influenza, 753f

inherited disease, 541

inhibin Peptide hormone from the gonads that inhibits FSH secretion (Ch 26), 803, 804

inhibitory modulation, 326

inhibitory postsynaptic potential (IPSP) Hyperpolarizing graded potentials that make a neuron less likely to fire an action potential (Ch 8), 252

initial segment The axon hillock and first part of an axon; often the location of the neuron's trigger zone (Ch 8), 251

initiation factor Protein that binds to the promoter region of a gene to start transcription (Ch 4), 113

innate immunity The nonspecific responses of the body to invasion by foreign substances (Ch 24), 754, 758–762

inner ear Portion of the ear containing the cochlea and hair cells (Ch 10), 340f, 341, 343, 347, 348f

inner mitochondrial membrane, 126t

innervated Controlled by a neuron (Ch 8), 244

inositol triphosphate (IP$_3$) A second messenger made from a membrane phospholipid by phospholipase C; opens Ca^{2+} channels on the endoplasmic reticulum to create a calcium signal (Ch 6), 181

inotropic agent Any chemical that affects contractility (Ch 14), 482

inotropic receptor Neurotransmitter receptor that alters ion channel function (Ch 8), 267

insensible water loss Water loss across the skin and in exhaled air of which we are not normally aware (Ch 20), 628

insertion of the muscle The end of the muscle attached farthest from the trunk or to the more mobile bone (Ch 12), 390

insomnia Inability to sleep well (Ch 9), 307

inspiration The movement of air into the lungs (Ch 17), 548, 555, 557

inspiratory capacity, 563

inspiratory muscles The external intercostals, diaphragm, scalenes, and sternocleidomastoids (Ch 17), 556

inspiratory neuron Somatic motor neurons controlling the inspiratory muscles (Ch 18), 587–588

inspiratory reserve volume The volume of air that can be inhaled in addition to a normal inspiration (Ch 17), 563

insulator, 157, 836

insulin Pancreatic hormone that decreases plasma glucose concentration (Ch 4, 5, 6, 7, 22), 110, 162, 163f, 179, 209, 705–710

insulin-dependent diabetes mellitus. *See* type 1 diabetes mellitus

insulin-like growth factor, 738, 749

insulin-receptor substrates (IRS), 708

insulin-resistant diabetes. *See* type 2 diabetes mellitus

integral protein. *See* membrane-spanning protein.

integrated membrane processes, 162

integrating center The control center that evaluates incoming signal and decides on an appropriate response (Ch 1, 6), 8, 191, 192

integrin Membrane-spanning proteins that link the cytoskeleton to extracellular matrix proteins or ligands; involved in blood clotting, wound repair, immune cell adhesion and recognition, and cell movement during development (Ch 3, 6), 65f, 66, 176

integrin receptors, 181

integumentary system, 3–4

intensity of stimulus, 327–328, 329f, 342

intercalated cell (I cell) Cell of the collecting duct that transports H$^+$ and bicarbonate for acid-base balance (Ch 20), 649

intercalated disk Specialized cell junctions in cardiac muscle that contain gap junctions (Ch 12, 14), 421, 461

interconnecting neuron. *See* interneuron.

intercostal muscles Muscles associated with the rib cage; used for breathing (Ch 17), 548

interferon Cytokines secreted by lymphocytes that aid in the immune response (Ch 24), 761t

interleukin-1 (IL-1), 761, 761t

interleukin (IL) A group of cytokines whose amino acid sequence is known. Most interleukins are released by one type of white blood cell and act on another (Ch 16, 22), 528, 529t, 720, 768

intermediary metabolism, 100

intermediate, 100

intermediate filament Cytoplasmic protein fiber made of myosin, keratin, neurofilament, and other proteins (Ch 3), 57

intermembrane space Region between the two outer membranes of a mitochondrion (Ch 3), 59

internal environment The extracellular fluid that surrounds the cells of the body (Ch 1), 4, 7f

internal intercostal muscles, 557

internal sphincter, 618

interneuron A neuron that is completely contained within the central nervous system (Ch 8), 240t, 243

internodal pathway Conduction pathway from the SA node to the AV node (Ch 14), 468

interphase, 840

interstitial cells of Cajal (ICC) Modified smooth muscle cells of the digestive tract that appear to initiate slow waves (Ch 21), 665, 666

interstitial fluid Extracellular fluid that surrounds the cells and lies between the cells and the plasma (Ch 5, 14), 133, 493

interstitial protein, 509

intestinal epithelium, 680

intestinal phase, 679–687

intracellular fluid Fluid within the cells (Ch 1, 5, 23), 7, 126, 741

intrafusal fiber Modified muscle fibers of the muscle spindle that lack myofibrils in their central portions (Ch 13), 432

intraocular pressure, 350

intrapleural pressure Pressure within the pleural fluid (Ch 17), 555, 557–559

intrauterine device (IUD), 821

intravenous (IV) injection, 627

intravenous solutions, 156t

intrinsic factor Protein secreted by gastric parietal cells that is required for vitamin B_{12} absorption in the intestine (Ch 21), 676–677, 685

intrinsic pathway Coagulation reaction that begins with collagen exposure and uses proteins already present in plasma (Ch 16), 538

intrinsic protein. *See* membrane-spanning protein.

intron Non-coding segments of DNA within a gene (Ch 4), 113

inulin A polysaccharide isolated from plants; used to determine extracellular fluid volume and glomerular filtration rate (Ch 19), 615

inulin clearance, 615, 615f

in vitro fertilization Fertilization of an egg outside the body (Ch 26), 821, 822, 826

in vitro motility assay, 399

iodide, 733

iodine, 733, 736

ion An atom with a net positive or negative charge due to gain or loss of one or more electrons (Ch 2, 20, 21), 30, 30f, 626, 667–668, 685

ion balance, 599

ion channel, 130, 249

ion concentration, 248, 249

ionic bond A bond between ions attracted to each other by opposite charge (Ch 2), 26, 30, 30f

ion movement, 248–249, 255

ion permeability, 162, 249–250

IP₃ *See* inositol triphosphate.

IP₃-activated receptor-channel Ca^{2+} channels in smooth muscle sarcoplasmic reticulum that open in response to IP_3 (Ch 12), 415, 419

ipsilateral On the same side as (Ch 10), 346

IPSP. *See* inhibitory postsynaptic potential.

iron, 533, 534

iron-deficiency anemia, 534

iron metabolism, 533f

irreversible reaction A reaction can proceed in one direction but not in the reverse direction (Ch 4), 90, 101f

irritant receptors Stimulated by inhaled particles or noxious gases in the airway mucosa (Ch 18), 592

ischemia Lack of adequate blood flow and oxygen to a tissue (Ch 10, 14), 334, 452

islet cell antibody, 774

islets of Langerhans Clusters of endocrine tissue within the pancreas (Ch 22), 705

isoform Related forms of a molecule (Ch 6), 173, 185, 186

isometric contraction A contraction that creates force without movement (Ch 12), 409, 409f

isoproterenol, 379t

isosmotic solution, 153

isotonic contraction A contraction that creates force and moves a load (Ch 12), 409, 409f

isotonic solution A solution that results in no net water movement when a cell is placed in it (Ch 5, 14), 154, 483

isotope Atoms of the same element that have different numbers of neutrons (Ch 2), 23–24

isovolumic relaxation Phase of the cardiac cycle when the ventricles are relaxing but the volume of blood in them is not changing (Ch 14), 478

isovolumic ventricular contraction Phase of the cardiac cycle when the ventricles are contracting but all valves are closed and the volume of blood in them is not changing (Ch 14), 476

isozyme Related forms of a single enzyme (Ch 4, 14), 14, 92, 93, 483

itch (pruritis), 333

IV fluid therapy. *See* intravenous fluid therapy.

IV injection. *See* intravenous injection.

jaundice A yellow tint to the skin and sclera due to excessive levels of bilirubin (Ch 16), 533

jejunum The middle section of the small intestine (Ch 21), 661, 680

jellyfish, 671

jet lag, 232

JG cell. *See* juxtaglomerular cell.

joint, 390

joint receptor Sensory receptors that send information about the relative positioning of bones linked by flexible joints (Ch 13), 432

juvenile-onset diabetes, 712

juxtaglomerular (JG) apparatus Region where the distal tubule of the nephron passes between afferent and efferent arterioles (Ch 19), 603, 608, 609f

juxtaglomerular (JG) cell Specialized cells in the walls of renal arterioles that synthesize and release renin (Ch 19, 20), 608, 638

K_{ATP} channel. *See* ATP-gated K^+ channel.

kcal. *See* kilocalorie.

keratin Insoluble protein prevalent in hair and nails (Ch 3), 69

keratinocyte, 78

ketoacid, 646

ketoacidosis A state of acidosis that results from excessive ketone production (Ch 20, 22), 646, 705

ketogenic diets, 705

ketone body, 109, 705

ketosis, 705

key intermediate, 100

kidney cortex (cortical collecting duct), 636

kidneys, 602

  in blood pressure control, 498

  excretion and, 614–617

  filtration and, 605

  functions of, 599, 603–604

  nephron and, 602–603

  ph homeostasis and, 646, 648–650

  reabsorption and, 610–613

  secretion and, 613–614

  structure of, 600f

  urine concentration and, 629, 633

  water balance and, 628, 628f

kidney stone, 603, 616, 619, 726, 730, 745

killer T cell. *See* cytotoxic T cell.

kilocalorie (kcal or Calorie) Amount of energy needed to raise the temperature of 1 liter of water by 1° C (Ch 4, 22), 99, 698

kinase An enzyme that adds a phosphate group to the substrate (Ch 4), 99t

kinesin A motor protein (Ch 3), 58

kinetic energy The energy of motion (Ch 4), 86, 87

kinin, 501, 761

kininogen, 540t

kinocilium The longest cilium on a vestibular hair cell that imparts directionality (Ch 10), 347

kiss-and-run pathway Secretion in which the secretory vesicle fuses transiently with the membrane, then pulls away (Ch 8), 264

kit receptors, 666

knee jerk reflex, 304, 436, 437f

knock-out mouse A mouse whose DNA has been altered to remove ("knock out") a specific gene (Ch 8, 26), 261, 828

Korotkoff sound, 496

Krakauer, Jon, 575

Krebs, Sir Hans, 105

Krebs cycle, 105

kyphosis, 745

labeled line coding The 1:1 association of a sensory receptor with a sensation (Ch 10), 327

labia majora Outer lips of the vulva (Ch 26), 810

second law of thermodynamics Natural processes move from an ordered state to a state of disorder known as entropy (Ch 4), 87

second meiotic division, 801

second messenger Intracellular molecules that translate the signal from a first messenger into an intracellular response (Ch 6), 177, 178f

secretin Intestinal hormone that stimulates bicarbonate secretion and pepsin release; inhibits gastric acid (Ch 21), 672, 672t, 679, 681

secretion (1) The movement of selected molecules from the blood into the nephron; (2) The process by which a cell releases a substance into the extracellular space (Ch 3, 5, 7, 19, 21), 69, 147, 212, 613–614
 in gastrointestinal tract, 661, 666–668
 in nephron, 604
 penicillin, 614

secretory diarrhea, 687

secretory epithelia Epithelia that secrete hormones or exocrine secretions (Ch 3), 67, 68t, 69

secretory phase Postovulatory phase of the uterus when it develops into a secretory structure (Ch 26), 811

secretory vesicle, 61

segmental contraction, 666

selectins, 65t

selective estrogen receptor modulator (SERM) Drugs that act as agonists or antagonists on estrogen receptors (Ch 26), 186, 828

selectively permeable membrane A membrane that allows some molecules to cross freely but excludes others (Ch 9), 292

selective serotonin reuptake inhibitor (SSRI) Antidepressant drugs that extend the active life of serotonin in the synapse by slowing their reuptake into the axon terminal (Ch 6, 9), 187, 309

selenium, 23

self-tolerance The lack of immune response to cells of the body (Ch 24), 773, 774

Selye, Hans, 775

semen Sperm plus secretions from accessory glands (Ch 26), 809, 818, 819

semicircular canal Three fluid-filled canals of the vestibular apparatus that sense rotational movements (Ch 10), 340f, 341, 347, 348f

semi-conservative replication, 838

semilunar valve Heart valves between the ventricles and major arteries (Ch 14), 457f, 458f, 459

seminal vesicle Male accessory glands that contribute enzymes and other secretions to semen (Ch 26), 806

seminiferous tubule Region of the testes where sperm and hormones are produced (Ch 26), 806

sense organ, 322

sense strand The strand of DNA that serves as the template for mRNA (Ch 4), 113

sensitivity, 771

sensitization Exposure to a noxious or intense stimulus creates an enhanced response upon subsequent exposure (Ch 9), 310

sensor, 8, 191

sensorineural hearing loss, 346

sensory areas of cerebral cortex, 302

sensory coding, 345, 346

sensory deprivation tanks, 322

sensory neuron A neuron that transmits sensory information to the central nervous system (Ch 8, 10, 13), 240, 240t, 242–243, 324, 327–328, 330, 432

sensory pathway, 325, 326f, 330f, 331t

sensory receptor, 78, 191–192, 432

sensory system, 301
 properties of, 322–330

sensory transduction Conversion of a sensory stimulus to an action potential (Ch 10), 324

septic shock, 498

septum A dividing wall (Ch 14), 450

SER. *See* smooth endoplasmic reticulum.

series elastic element Elastic fibers in the muscle that stretch during isometric contraction (Ch 12), 410

SERM. *See* selective estrogen receptor modulator.

serosal surface. *See* basolateral surface.

serosa Outer surface of the digestive tract created by a continuation of the peritoneum (Ch 21), 664–665

serotonin A CNS neurotransmitter. Synonym: 5-hydroxytryptamine (5-HT) (Ch 8, 9, 15, 16), 265t, 267, 305, 306t, 501, 538, 538f

serous secretion Watery exocrine solution that often contains enzymes (Ch 3), 70

Sertoli cell Testicular cells that secrete anti-Müllerian hormone and support sperm production (Ch 26), 798, 806, 808

setpoint, 191, 193
 oscillations around, 194, 195f

sex chromosomes (Ch 26), 797–798

sex determination, 797–801

sex drive. *See* libido.

sex hormones, 727, 728, 740, 797, 803

sexual arousal, 819

sexual development, 828

sexual differentiation, 798, 800–801, 800t

sexual dimorphism Males and females have different physical characteristics (Ch 26), 797

sexual intercourse, 818–820

sexual response, 818, 820

shell, electron A series of energy levels around the nucleus of an atom that contain electrons (Ch 2), 25

shigella, 688

shivering thermogenesis, 719

shock Generalized, severe circulatory failure (Ch 15), 498

short-loop negative feedback, 224

short reflex, 670, 675f

short-term energy storage, 698

short-term memory, 310–311

sickle cell anemia, 534, 534f, 535

sight. *See* vision.

sigmoid colon, 685

signal amplification Process by which a single signal molecule can generate multiple intracellular effector molecules (Ch 6), 176

signal molecule, 174, 182–184, 212, 213, 741

signal sequence Initial segment of a newly-made protein that directs the protein to the proper organelle for processing, packaging, and delivery (Ch 4, 7), 115, 215

signal transduction The transmission of information from one side of a membrane to the other using membrane proteins (Ch 1, 6, 7, 10), 9, 176–178, 182f, 187, 216, 339, 345f

biological, 177

sildenafil, 820

simple diffusion Diffusion across the phospholipid bilayer of a cell (Ch 5), 136t

simple epithelia, 67

simple squamous epithelia, 67

Singer, S. J., 126

single-unit smooth muscle Smooth muscle fibers that are electrically coupled by numerous gap junctions (Ch 12), 414–415

sinoatrial node (SA node) A group of autorhythmic cells in the right atrium of the heart; the main pacemaker of the heart (Ch 14), 468

sinusoid, 505, 682f

sister chromatid, 801

skeletal muscle pump, 512

skeletal muscle reflex, 431–438

skeletal muscle Striated muscle usually attached to bones; responsible for positioning and movement of the skeleton (Ch 9, 12, 14, 15, 25), 304, 390–408, 421, 495
 action potential in, 461, 467f
 anatomy of, 391, 392f–393f
 contraction of, 396–403, 406–407
 disorders of, 412–413
 energy metabolism in, 783f
 mechanics of body movement, 408–412
 micturition reflex, 618f
 muscle fatigue, 403–404
 muscle fibers of, 391, 394–395, 394f, 404–407

skin, 77, 78

skull, 288, 350

sleep, 307

sleep apnea, 307

sleep disorder, 307

sleep-wake cycle, 232, 306–308

sleepwalking, 308

sliding filament theory of contraction The current model for muscle contraction in which muscle proteins slide past each other to generate force (Ch 12), 396–398

slow axonal transport, 244

slow pain, 333

slow synaptic potential Slower onset and longer lasting response of postsynaptic cells to certain neurotransmitters and neuromodulators (Ch 8), 268

slow-twitch fiber, 404, 405f, 405t, 407

slow wave potential Cyclic depolarization and repolarization of membrane potential in smooth muscle (Ch 12, 21), 419, 665, 665f

slow-wave sleep, 307

small intestine The segment of the gastrointestinal tract where most absorption and digestion take place (Ch 21), 661, 662f, 680, 685

smell. *See* olfaction.

smoking, 379, 382–384, 514, 546, 555, 579

smooth endoplasmic reticulum (SER) Organelle that is the primary site of fatty acid, steroid, and lipid synthesis (Ch 3), 60

smooth muscle Muscle found in internal organs that lacks organized sarcomeres (Ch 11, 12, 14), 372, 375, 390, 390f, 413–421
 actin and myosin in, 415
 activity regulation in, 420–421
 anatomy of, 416f
 contraction in, 416–419, 665, 666
 features of, 414–415
 membrane potentials of, 419–420

Photo Credits

ABOUT THE AUTHOR
Author: Dee Silverthorn.

OWNER'S MANUAL AND PREFACE
Dee Silverthorn.

CHAPTER 1
page 1: GJLP/CNRI/Phototake NYC.

CHAPTER 2
page 20: Paul Axelsen, M.D./Phototake NYC.

CHAPTER 3
page 50: Todd Derksen; 3-1a & c: Todd Derksen; 3-1b: David M. Phillips/Visuals Unlimited; 3-6: Dr. Don Fawcett/Photo Researchers, Inc.; 3-7: Fawcett/Hirokawa/Heuser/Science Researchers, Inc.; 3-8: Robert W. Riess; 3-9: CNRI/Science Source/Photo Researchers, Inc.; 3-10: Robert W. Riess; 3-11: Robert W. Riess; 3-12: Dr. Don Fawcett/Photo Researchers, Inc.; 3-13, left: Don Fawcett, M.D., Harvard Medical School; 3-13, middle: Biophoto Associates/ Photo Researchers, Inc.; 3-14a & b: Todd Derksen; 3-14c: Photo Researchers, Inc.; 3-16: Todd Derksen; 3-19: Custom Medical Stock Photo, Inc.; 3-20: Todd Derksen; 3-22: Ward's Natural Science Establishment, Inc.; 3-23: John D. Cunningham/Visuals Unlimited; 3-24: Frederic H. Martini; 3-26: Cabisco/Visuals Unlimited.

CHAPTER 4
page 84: Photo Researchers, Inc.

CHAPTER 5
page 124: Richard Feldmann/Phototake NYC.

CHAPTER 7
page 208: CNRI/Phototake NYC; 7-1: Dee Silverthorn.

CHAPTER 8
page 239: David Scott/Phototake NYC; 8-16a: Todd Derksen; 8-16b: David M. Phillips/Visuals Unlimited; 8-30: Timothy M. Gomez, University of Wisconsin–Madison.

CHAPTER 9
page 284: McGill University/CNRI/Phototake NYC; 9-12b: Robert Brons/BPS/Stone; 9-17: Marcus Raichle, Washington University School of Medicine.

CHAPTER 10
page 321: Prof. P. Motta, Department of Anatomy, University La Sapienza, Rome/Science Photo Library/Photo Researchers, Inc.; 10-15b: Todd Derksen; 10-28b: Custom Medical Stock Photo, Inc.

CHAPTER 11
page 369: Dr. Flora Love.

CHAPTER 12
page 389: University of Texas at Austin; 12-1a & c: Todd Dersken; 12-1b: Phototake NYC; 12-15, left: D. Comack, ed. *Ham's Histology 9th ed.* Philadelphia: J. B. Lippincott, 1987. By permission. 12-15, right: Frederic H. Martini; 12-26: Biophoto Associates/ Photo Researchers, Inc.

CHAPTER 14
page 448: Carole L. Moncman, Ph.D.

CHAPTER 15
page 490: Simon Fraser/Newcastle General Hospital/SPF/Photo Researchers, Inc.; 15-20: World Health Organization.

CHAPTER 16
page 523: Phototake NYC; 16-6a: Todd Derksen; 16-9: Photo Researchers, Inc.; 16-10a & b: Todd Derksen.

CHAPTER 17
page 546: Custom Medical Stock Photo, Inc.; 17-6: Frederic H. Martini.

CHAPTER 18
page 574: Phototake NYC.

CHAPTER 19
page 598: Daniel Casellas, Ph.D.; 19-1f: Todd Derksen; 19-4b: Todd Derksen.

CHAPTER 21
page 659: Mark L. Tamplin, Anne L. Gauzens, and Rita R. Colwell.

CHAPTER 22
page 695: McGill University/CNRI/Phototake NYC; 22-7c: Ward's Natural Science Establishment, Inc.

CHAPTER 23
page 726: CNRI/SPL/Photo Researchers, Inc.; 23-1c: Ward's Natural Science Establishment, Inc.; 23-5a: Biophoto Associates/ Science Source/Photo Researchers, Inc.; 23-5b: Biophoto Associates/Photo Researchers, Inc.; 23-11: Martin Rotker/Phototake NYC; 23-13: Alison Wright/CORBIS; 23-15: Ralph Eagle/Science Source/Photo Researchers, Inc.; 23-17: American Journal of Medicine, 20:133, 1956; 23-18: Ralph T. Hutchings; 23-24: Dr. Michael Klein/Peter Arnold, Inc.

CHAPTER 24
page 751: Dr. Dennis Kunkel/Phototake NYC.

CHAPTER 26
page 796: David Scharf/Peter Arnold, Inc.; 26-1: CNRI/Science Photo Library/Photo Researchers, Inc.; 26-9c: David M. Phillips/Visuals Unlimited; 26-16a: Francis Leroy/Biocosmos/ Science Photo Library/Custom Medical Stock Photo, Inc.

Measurements and Conversions

PREFIXES

deci-	(d)	1/10	0.1	1×10^{-1}
centi-	(c)	1/100	0.01	1×10^{-2}
milli-	(m)	1/1000	0.001	1×10^{-3}
micro-	(μ)	1/1,000,000	0.000001	1×10^{-6}
nano-	(n)	1/1,000,000,000	0.000000001	1×10^{-9}
pico-	(p)	1/1,000,000,000,000	0.000000000001	1×10^{-12}
kilo-	(k)		1000.	1×10^{3}

METRIC SYSTEM

1 meter (m)	=	100 centimeters (cm)	=	1000 millimeters (mm)
1 centimeter (cm)	=	10 millimeters (mm)	=	0.01 meters (m)
1 millimeter (mm)	=	1000 micrometers (μm; also called micron, μ)		
1 angstrom (Å)	=	1/10,000 micrometer	=	1×10^{-7} millimeters
1 liter (L)	=	1000 milliliters (mL)		
1 deciliter (dL)	=	100 milliliters (mL)	=	0.1 liters (L)
1 cubic centimeter (cc)	=	1 milliliter (mL)		
1 milliliter (mL)	=	1000 microliters (μL)		
1 kilogram (kg)	=	1000 grams (g)		
1 gram (g)	=	1000 milligrams (mg)		
1 milligram (mg)	=	1000 micrograms (μg)		

CONVERSIONS

1 yard (yd)	= 0.92 meters
1 inch (in)	= 2.54 centimeters
1 meter	= 1.09 yards
1 centimeter	= 0.39 inches
1 liquid quart (qt)	= 946 milliliters
1 fluid ounce (oz)	= 8 fluid drams = 29.57 milliliters (mL)
1 liter	= 1.05 liquid quarts
1 pound (lb)	= 453.6 grams
1 kilogram	= 2.2 pounds

TEMPERATURE

FREEZING

0 degrees Celsius (°C)	=	32 degrees Fahrenheit (°F)	=	273 degrees Kelvin (°K)

To convert degrees Celsius (°C) to degrees Fahrenheit (°F):
(°C × 9/5) + 32

To convert degrees Fahrenheit (°F) to degrees Celsius (°C):
(°F − 32) × 5/9

NORMAL VALUES OF BLOOD COMPONENTS

SUBSTANCE OR PARAMETER	NORMAL RANGE	MEASURED IN
Calcium (Ca^{2+})	4.3–5.3 meq/L	Serum
Chloride (Cl^{-})	100–108 meq/L	Serum
Potassium (K^{+})	3.5–5.0 meq/L	Serum
Sodium (Na^{+})	135–145 meq/L	Serum
pH	7.35–7.45	Whole blood
P_{O_2}	75–100 mm Hg	Arterial blood
P_{CO_2}	34–45 mm Hg	Arterial blood
Osmolality	280–296 mosmol/kg water	Serum
Glucose, fasting	70–110 mg/dL	Plasma
Creatinine	0.6–1.5 mg/dL	Serum
Protein, total	6.0–8.0 g/dL	Serum

Modified from W. F. Ganong, *Review of Medical Physiology* (Norwalk: Appleton & Lange), 1995.